EISLER'S ENCYCLOPEDIA OF

ENVIRONMENTALLY HAZARDOUS PRIORITY CHEMICALS

EISLER'S ENCYCLOPEDIA OF

ENVIRONMENTALLY HAZARDOUS PRIORITY CHEMICALS

By

RONALD EISLER

ELSEVIER

Amsterdam • Boston • Heidelberg • London • New York • Oxford
Paris • San Diego • San Francisco • Singapore • Sydney • Tokyo

Elsevier
Radarweg 29, PO Box 211, 1000 AE Amsterdam, The Netherlands
Linacre House, Jordan Hill, Oxford OX2 8DP, UK

First edition 2007

Library of Congress Cataloging-in-Publication Data
A catalog record for this book is available from the Library of Congress

British Library Cataloguing in Publication Data
A catalogue record for this book is available from the British Library

ISBN: 978-0-444-53105-6

For information on all Elsevier publications
visit our website at books.elsevier.com

Transferred to Digital Printing in 2010

Dedicated to my family:

Jeannette, Renée, David, Charles, Julie, and Eb

PREFACE

The chemicals featured in this volume are at the top of "List of Substances Discharged into the Environment as a Result of Human Activities and Considered Hazardous to Sensitive Species of Natural Resources." The List was prepared over the period 1985–2003 by environmental specialists of the U.S. Department of the Interior.

The metals, metalloids, organics, and radioactive substances chosen originated in wastes from agricultural, industrial, military, domestic, mining, and municipal sources. Some of these compounds were selected for inclusion because they had no known biological function and their presence in tissues is associated with adverse effects on growth, development, reproduction, and survival itself. Some have been incorporated into powerful biocides to control pestiferous organisms and, inadvertently, impact-desirable species of nontarget organisms. Others are highly prized by society, but the environmental consequences of extraction and refining them has adversely impacted habitats of plants and wildlife, sometimes for more than a hundred years. Several are essential to normal metabolism; however, insufficiency as well as excesses may be fatal. Most occur in a variety of chemical forms, some of which are comparatively benign and others extremely toxic. For each chemical or group of chemicals, basic information is presented on its sources, uses, properties, concentrations in living organisms, lethal and sublethal effects, identification of research opportunities, and proposed criteria to protect human health and natural resources.

It is emphasized that all proposed criteria listed were recommended by local, regional, national, and international regulatory agencies, as well as knowledgeable university and industrial researchers. In general, regulatory agencies are required to periodically update all criteria incorporating the most recent scientific findings. Unfortunately, criteria – unlike legislatively mandated standards – are not legally binding, although in certain extraordinary cases, such as massive discharge of a chemical to the biosphere, regulatory agencies are known to impose financial and other penalties. Ultimately, as chemical risk assessment predictions based on suitable databases become increasingly reliable, standards will be established for individual chemicals, together with adequate funds for enforcement, and stipulated penalties for violators.

This single volume compendium will provide a ready reference to professionals and students concerned with ecotoxicological aspects of numerous chemical wastes.

Ronald Eisler
2nd July 2007

ABOUT THE AUTHOR

Ronald Eisler received the B.A. degree from New York University in biology and chemistry, and the M.S. and Ph.D degrees from the University of Washington in aquatic sciences and radioecology, respectively. He retired as a senior research biologist in 2004 after a 45-year career with the U.S. federal government, mainly with the U.S. Environmental Protection Agency in Rhode Island, and the U.S. Department of the Interior in the Territory of Alaska, New Jersey, Washington, D.C., and Maryland. He has held a number of special assignments and teaching appointments, including senior science advisor to the American Fisheries Society, adjunct professor of zoology at the American University in Washington, D.C., adjunct professor at the Graduate School of Oceanography of the University of Rhode Island, and visiting professor of marine biology and resident director of the Marine Biology Laboratory of Hebrew University in Eilat, Israel. Eisler is the author of approximately 150 research publications on ecotoxicological aspects of contaminants discharged into the environment as a result of human activities. In retirement, he continues to write and consult.

BOOKS BY RONALD EISLER

Mercury Hazards to Living Organisms, 2006. CRC Press, Boca Raton, Florida, 312 pp.

Biogeochemical, Health, and Ecotoxicological Perspectives on Gold and Gold Mining, 2004. CRC Press, Boca Raton, Florida, 355 pp.

Handbook of Chemical Risk Assessment: Health Hazards to Humans, Plants, and Animals. Volume 1. Metals; Volume 2, Organics; Volume 3, Metalloids, Radiation, Cumulative Index to Chemicals and Species, 2000. Lewis Publishers, Boca Raton, Florida, 1903 pp.

Trace Metal Concentrations in Marine Organisms, 1981. Pergamon Press, Elmsford, New York, 687 pp.

LIST OF TABLES

LIST OF FIGURES

CONTENTS

ACROLEIN[a]

Chapter 1

1.1 Introduction

Acrolein ($CH_2{=}CHCHO$) is an aldehyde that was first isolated in 1843 from the dry distillation of fats and glycerol. It is now known that acrolein is ubiquitous in the environment; it is often present in trace amounts in foods and as a component of smog, fuel combustion products such as wood smoke, exhaust emissions from internal combustion engines, and cigarette smoke. Atmospheric concentrations of acrolein over urban areas are between 2.0 and 7.0 μg/L; cigarette smoke, however contains about 10,000 μg of acrolein/L. Acrolein is classified as a hazardous chemical because of its reactivity and flammability. At low sublethal concentrations, acrolein is widely known for its acrid pungent odor and strong irritating effects on mucous membranes of the eyes and upper respiratory tract, its toxicity to cilia in all organisms, and its interference with nucleic acid synthesis in bacteria. In bulk, acrolein during storage or transfer is potentially hazardous if it becomes overheated or contaminated with water. For example, in 1982, 17,000 residents from Toft, Louisiana, were evacuated when two large tanks of acrolein began to burn.

Acrolein enters the aquatic environment from its use as an aquatic herbicide, industrial discharges, and as a by-product of the chlorination of organic compounds in wastewater and drinking water treatment. Dilute solutions of acrolein kill undesirable plant life in irrigation streams and ditches and have been used routinely in about 4000 km of irrigation canals in southeastern Australia to control submerged weeds, including *Potamogeton tricarinatus*, *Elodea canadensis*, and *Vallisneria gigantia*. Acrolein has also been used for many years in channel maintenance in the United States (especially in western states), Canada, Egypt, Argentina, Mexico, and Turkey. Unlike most other aquatic herbicides, acrolein rapidly dissipates from water by volatilization and degradation without leaving phytotoxic residues. However, acrolein provides only temporary control of submerged weeds and also kills fish and other aquatic life at recommended treatment concentrations. In one Montana stream, acrolein killed all fish in a 4-km stretch after application to control submerged weeds; some fish deaths were recorded as far as 6.4 km downstream.

1.2 Sources and Uses

Acrolein enters the environment as a result of normal metabolic processes; incomplete combustion of coal, wood, plastics, tobacco, and oil fuels; and industrial emissions. Acrolein has been detected in smog, food, and water. It is used extensively in chemical manufacture, for control of fouling organisms, and as herbicide to control submerged weeds in irrigation canals.

1.2.1 Sources

Acrolein is ubiquitous in the environment as a result of natural and anthropogenic sources.

[a]All information in this chapter is referenced in the following sources:

Eisler, R. 1994. Acrolein hazards to fish, wildlife, and invertebrates: a synoptic review. *U.S. Natl. Biol. Surv. Biol. Rep. 23*, 29 pp.

Eisler, R. 2000. Acrolein. Pages 739–766 in *Handbook of Chemical Risk Assessment: Health Hazards to Humans, Plants, and Animals.* Volume 2, Organics. Lewis Publishers, Boca Raton, Florida.

Sources of atmospheric acrolein include smog; incomplete combustion of coal, wood, gasoline, plastics and fats; tobacco smoke; and industrial emissions. The total amount of acrolein released into the atmosphere is unknown. In 1978, production losses of acrolein by emission from the four main U.S. plant locations were estimated at 34,682 kg; however, the gaseous emission streams are now either burned on emergence from the exhaust stack or sent to a furnace to destroy residual material. Acrolein is found in photochemical smog and contributes to the smog's irritant capacity to the eye and respiratory pathways. Recorded maximum acrolein concentrations in smog ranged from 12.0 to 14.0 μg/L (0.025–0.032 mg/m^3) in Los Angeles, between 1961 and 1963, and were 13.0 μg/L in Hudson County, New Jersey. For humans, exposure to atmospheric acrolein is greatest in the vicinity of incompletely combusted organic materials such as coal, wood, and petrol; highest acrolein concentrations are reported near forest fires and urban area fires. The burning of southern pine (*Pinus* sp.), for example, generates 22.0–121.0 mg of acrolein/kg of wood burned. Acrolein is also in the smoke of burning plastic materials. Air samples from more than 200 fires in Boston, Massachusetts, contained greater than 3000.0 μg acrolein/L (greater than 6.8 mg/m^3) in more than 10% of all samples; greater than 3000.0 μg acrolein/L air is an immediately hazardous concentration for human life and health. Cigarette smoke in some enclosed areas may account for as much as 12,400.0 μg of acrolein/L air. In the case of an enclosed room of 30 m^3 capacity, smoking 5 cigarettes raises the air concentration to about 50.0 μg acrolein/L and 30 cigarettes to 380.0 μg/L.

Acrolein is also generated when animal or vegetable fats are subjected to high temperatures. Acrolein was detected aboard submarines in trace concentrations as a degradation product during the heating of lubrication oils and edible fats. Large amounts of acrolein are generated from exhausts of internal combustion engines. Acrolein concentrations of 10,000.0 μg/L (23.0 mg/m^3) have been measured in nondiesel automobile exhausts, 2900.0 μg/L in diesel engine emissions, and 2600.0–9600.0 μg/L in other internal combustion engines. Acrolein concentrations in air from several U.S. urban areas ranged from a maximum of 10.0 μg/L in 1960 to 1.8–3.4 μg/L in 1968; during this period, the air in Tokyo had an average acrolein concentration of 7.2 μg/L. Urban acrolein pollution is derived mainly from automobile exhaust and incomplete burning of refuse. Acrolein is formed during normal metabolic degradation of spermine and spermidine, glycerol, allyl formate, allyl alcohol, and cyclophosphamide. Acrolein was also in spores from the wheat stem fungus (*Puccinia graminis*) of infected wheat (*Triticum aestivum*); acrolein was the major chemical factor that normally induced infection processes in *Puccinia*.

Acrolein has been detected in effluent water streams from industrial and municipal sources. Municipal effluents from Dayton, Ohio, for example, contained between 20.0 and 200.0 μg acrolein/L in 6 of 11 analyzed samples. Acrolein is also a component of many foods, and processing may increase the acrolein content. Acrolein has been identified in raw turkey, potatoes, onions, coffee grounds, raw cocoa beans, alcoholic beverages, hops, white bread, sugarcane molasses, souring salted pork, and cooked bluefin tuna (*Thunnus thynnus*).

Occupational exposure to acrolein may occur during its production and isolation as a chemical intermediate or during the manufacture of acrylic acid, acrylic acid esters, and methionine. Other sources of acrolein in the workplace include emissions from rubber vulcanization plants, welding of metals treated with anticorrosion primers, and pitch-cooking plants; and skin contact during herbicidal applications for aquatic weed control, and from its use as a slimicide in paper and paperboard manufacture. Acute acrolein poisoning from occupational exposure is improbable. However, the risks of poisoning are significant in certain industries including welding of fat and oil cauldrons, smelting work and foundry operations, printing plants, linoleum and oil cloth factories, manufacture of insulators, tin plating of sheet iron, and processing of linseed oil.

1.2.2 Uses

Since its discovery in 1843, acrolein has been known to polymerize readily in the presence of many chemicals, and since 1947 it has been used safely in a wide variety of commercial applications. Acrolein is presently produced by the catalytic oxidation of propylene for the manufacture of methionine, glutaraldehyde, 1,2,6-hexane thiol, and other chemicals. The largest quantity of acrolein produced by this process is converted directly to acrylic acid and acrylic acid esters. In 1975, global production of acrolein was 59,000 metric tons; in 1980, this value was 102,000 tons – including 47,600 tons produced by the United States. In 1983, about 250,000 tons (about 550 million pounds) of acrolein were produced, and 92% were converted to acrylic acid, 5% to methionine, and 3% used as an aquatic herbicide. Acrolein copolymers are used in photography, in textile treatment, in the paper industry, as builders in laundry and dishwater detergents, and as coatings for aluminum and steel panels. Acrolein is used to scavenge sulfides from oil-field floodwater systems, to cross-link protein collagen in the leather tanning industry, and to fixate tissue of histological samples. The use of acrolein as a military poison gas has been advocated because of its lacrimatory and blistering properties; during World War I (1914–18) the French used acrolein – under the name of Papite – in hand grenades because of its irritating effect on the respiratory airways and the ocular mucosae.

Acrolein has been used since 1960 to control submerged aquatic weeds in irrigation systems in the United States, Australia, and other countries where open channels distribute water for crop production. Acrolein – as Magnacide H herbicide – is applied directly into agricultural irrigation systems at 1.0–15.0 mg/L. Water in treated canals is required by the Magnacide H label to be held for 6 days before discharge into receiving waters. Acrolein is preferable to sodium arsenite for herbicidal control of submerged weeds because arsenicals are persistent (up to 1 year), and the high arsenic concentrations that are attained in water may be hazardous to humans and livestock. Acrolein is extremely effective in killing submerged weeds that are difficult to control with other herbicides. Acrolein has also been used as a herbicide in ponds, drains, and other water bodies. In Australia, the concentration of acrolein in irrigation canals to control various species of *Elodea*, *Potamogeton*, and *Vallisneria* is usually less than 15,000.0 μg/L. In general, acrolein has a low order of toxicity to terrestrial plants. Most field and garden crops can tolerate water with as much as 15,000.0 μg acrolein/L without serious adverse effects. Acrolein, as discussed later, has comparatively low persistence and low accumulation in aquatic ecosystems. One disadvantage of its use as a herbicide is its pungent, irritating odor. Also, at recommended treatment concentrations, acrolein kills fish and other aquatic organisms; therefore, acrolein should be used only in aquatic systems where these resources are considered expendable.

Acrolein has been used to control bacteria, fungi, algae, and mollusks in cooling water systems: 1500.0 μg/L killed as much as 95% of the target species in a once-through treatment. Acrolein has been applied directly to the marine environment to control the growth and settlement of mussels (*Mytilus edulis*), and other fouling organisms in cooling water systems of coastal steam electric station power plants. Mussels in marine cooling water systems are controlled with 600.0 μg acrolein/L for 8 h daily for 3 days or with 700.0 μg/L for 3 h daily for 2 weeks. Acrolein prevents growth of microorganisms in liquid fuels such as jet fuels, in feed lines of subsurface wastewater injectors, and in water conduits of paper manufacturing plants.

1.3 Environmental Chemistry

Acrolein, the simplest member of the class of unsaturated aldehydes, has a pungent, irritating odor. It is volatile, flammable, and explosive, and requires elaborate and specific conditions for storage and use. The half-time persistence of acrolein in freshwater is usually less than 50 h; in seawater it is less than 20 h, and in the atmosphere less than 3 h. Biochemical and toxic effects of acrolein are caused by its rapid and essentially irreversible reaction with sulfhydryl compounds to form a stable thiol ether; however, many compounds can mitigate or block its toxicity. Acrolein is eventually metabolized to acrylic acid and glyceraldehyde; glycidaldehyde – an intermediate metabolite with mutagenic and carcinogenic properties – has been produced in vitro but not in vivo.

1.3.1 Chemical Properties

Acrolein is soluble in water and in many organic solvents including ethanol, acetone, and ether (Table 1.1). Acrolein is a highly reactive molecule with two reactive centers: one at the carbon–carbon double bond, and the other at the aldehydic group. Acrolein is extremely volatile, flammable, and explosive (Table 1.1), especially in sunlight or in the presence of alkali or strong acid. A potential hazard in handling acrolein is its rapid exothermic polymerization caused by the use of insufficient hydroquinone inhibitor or lack of strict control of pH. Commercial acrolein should be maintained at pH 6.0, contain less than 3% water, and 0.1–0.25% hydroquinone as a polymerization inhibitor. A typical commercial sample contains about 97% acrolein, 0.5% other carbonyls, and 2.5% water. The addition of hydroquinone (0.1–0.25%) prevents the vinyl polymerization of acrolein, and controlling the pH between 5 and 6 by acetic acid increases stability of commercial acrolein by preventing aldol condensation. Elaborate and specific conditions are now prescribed for the storage of acrolein and include vents and safety valves, construction materials, fire control, spills, and waste disposal. Commercial acrolein is stored and shipped under a blanket of oxygen-free inert gas.

Spectrophotometric determination with 4-hexyl-resorcinol and a fluorometric method with *m*-aminophenol are the most commonly

Table 1.1. Some properties of acrolein.

Variable	Datum
CHEMICAL NAME	2-Propenal
CAS NUMBER	107-02-8
STRUCTURAL FORMULA	$CH_2{=}CHCHO$
MOLECULAR WEIGHT	56.06
SPECIFIC GRAVITY	0.8427–0.8442
PHYSICAL STATE	Colorless or yellow liquid at 25°C
ODOR	Pungent, irritating
BOILING POINT	52.5–53.5°C
MELTING POINT	−86.95°C
SOLUBILITY	
Water	206.0–208.0 g/L
Organic solvents	Miscible
LOG K_{ow}	0.01
VAPOR PRESSURE	215–220 mm Hg at 20°C
EXPLOSIVE LIMITS OF VAPOR AND AIR	
Upper limit	31% acrolein
Lower limit	2.8% acrolein

used procedures for the determination of acrolein; however, gas chromatography and high-performance liquid chromatography procedures are also used. Acrolein concentrations in rainwater between 4.0 and 200.0 μg/L can be measured rapidly (less than 80 min) without interference from related compounds; the method involves acrolein bromination and analysis by gas chromatography with electron capture detection. Water samples from potential acrolein treatment systems require the use of water from that system, in preparing blanks, controls, and standards; further, acrolein measurements should be made at the anticipated use concentrations.

1.3.2 Persistence

Degradation and evaporation seem to be the major pathways for acrolein loss in water; smaller amounts are lost through absorption and uptake by aquatic organisms and sediments. The half-time persistence of acrolein in freshwater is 38 h at pH 8.6, and 50 h at pH 6.6; degradation is more rapid when initial acrolein concentrations are less than 3000.0 μg/L. Acrolein has a half-time persistence of 2.9–11.3 h at initial nominal concentrations of 20.0 μg/L, and 27.1–27.8 h at 101.0 μg/L. At pH 5, acrolein reacts by reversible hydrolysis to produce an equilibrium mixture with 92% beta-hydroxypropionaldehyde and 8% acrolein; in alkali, the primary reaction is consistent with a polycondensation reaction. Microbial degradation plays a major role in the transformation of acrolein in aquatic systems. In natural waters, acrolein degradation proceeds to carboxylic acid via a microbial pathway; beta-hydroxypropionaldehyde is readily biotransformed in about 17.4 days.

Acrolein is applied to irrigation canals to control submerged aquatic weeds at greatly different time–concentration treatments. Regardless of time–concentration regimens – which vary from 100.0 μg/L for 48 h in the United States to 15,000.0 μg/L for several hours in Australia – the daily decay-rate constants are remarkably similar, ranging from 0.14 to 0.21, and are probably affected by variations in weed density. In one case, acrolein applied to the Columbia River at an average initial concentration of 125.0 μg/L degraded to 25.0 μg/L after 48 h in samples greater than 65 km from the application point – a loss of 80%. High initial concentrations, 50,000.0–160,000.0 μg/L, of acrolein in natural waters degraded 57–80% in 192 h, suggesting that high concentrations can alter the rate of hydrolysis. In seawater, the half-time persistence of acrolein was less than 20 h. In photochemical smog, acrolein is comparatively unstable and not likely to persist; the dominant removal mechanism involves hydroxide attack on acrolein, and the atmospheric half-life persistence is 2–3 h under these conditions.

1.3.3 Metabolism

Biochemical and toxic effects of acrolein are probably caused by its reaction with critical protein and nonprotein sulfhydryl groups. The reaction of acrolein with sulfhydryl compounds is rapid and essentially irreversible, resulting in the formation of a stable thiol ether. Metabolism of acrolein is believed to result in the formation of acrylic acid and glyceraldehyde (Figure 1.1). The postulated metabolites of acrolein can be oxidized to carbon dioxide. Acrylic acid does not seem to represent a significant toxic hazard when compared to the parent acrolein because at low airborne concentrations of less than 1000.0 μg acrolein/L, the quantity of acrylic acid produced by metabolism is negligible. Thus, metabolism to acrylic acid after inhalation should be regarded as a detoxification pathway. Conjugation of acrylic acid with glutathione represents another elimination and detoxification pathway. In vitro studies of acrolein metabolism in mammals suggest that acrolein exposures may result in exposure to glycidaldehyde, an intermediate in acrolein metabolism (Figure 1.1). The major toxic effects of acrolein exposure – including irritation, ciliastasis, and hypersensitivity – are probably either due to the parent acrolein or to the reaction of glycidaldehyde with cell proteins. Glycidaldehyde is a potent mutagen and carcinogen; however, no evidence is

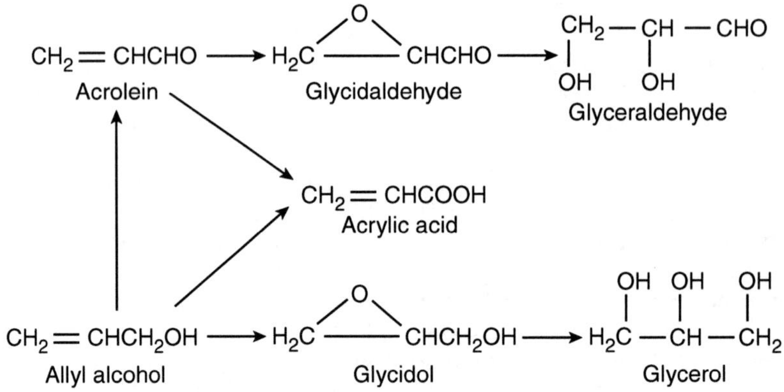

Figure 1.1.
Proposed scheme for in vitro mammalian metabolism of acrolein and allyl alcohol, a precursor of acrolein.

available showing that acrolein can produce glycidaldehyde in vivo. Acrolein is more toxic when inhaled than when taken orally. Inhalation of acrolein decreased the concentrations of protein and nonprotein sulfhydryl groups in nasal mucosal tissue. Acrolein is highly reactive towards thiol groups and rapidly conjugates with glutathione and cysteine. When glutathione is depleted, acrolein potentiates the nasal toxicity of formaldehyde to rats.

Acrolein is a metabolite of allyl alcohol and cyclophosphamide, and these compounds should be considered in acrolein metabolism schemes. Allyl alcohol in the presence of nicotinamide adenine dinucleotide phosphate (NADPH) and liver or lung microsomes degrades to acrolein, acrylic acid, and glycidol.

When added to water as an aquatic herbicide, acrolein undergoes rapid decomposition, especially in the sunlight. At the same time, it reacts rapidly with amines, alcohols, and mercaptans of aquatic plants, destroying cell structure and killing the plants. Mammals drinking acrolein-contaminated water rapidly convert acrolein to saturated alcohol compounds because of the low pH in the upper portion of their gastrointestinal tracts; the primary breakdown product is beta-propionaldehyde.

Many compounds including glutathione, 2-mercaptoethanol, beta-dimethylcysteamine, penicillamide, gamma-mercaptopropionylglycine, and *N*-acetylcysteine mitigate or block the toxic effects of acrolein. In frogs (*Rana japonica*), sulfhydryl compounds reduce the effects of acrolein on excitation–contraction uncoupling in skeletal muscle. In mice, cysteine reduced the cytotoxic effects of acrolein on tumor cells; in rabbits, cysteine mitigated acrolein-induced alveolar macrophage calcium-dependent ATP-ase, phagocytosis, and adhesiveness. In male rats, cysteine and ascorbic acid antagonized the acute lethal effects of orally administered acrolein, and 2-mercaptoethanol antagonized the inhibitory effect of acrolein on liver DNA-polymerase.

1.4 Lethal and Sublethal Effects

Acrolein degrades quickly in soils and in plant tissues regardless of mode of administration. Most terrestrial crop plants easily tolerate 25,000.0 μg of acrolein/L of irrigation water and some can tolerate 70,000.0–80,000.0 μg/L without adverse effects. Terrestrial plants were adversely affected at atmospheric concentrations of 500.0 μg acrolein/L air, but this concentration exceeds the recommended value of 110.0 μg/L (0.25 mg/m^3) air for protection of human health in occupational settings.

Adult fruit flies (*Drosophila* sp.) were comparatively resistant to acrolein and had lowered survival when reared in culture media with greater than 3,700,000.0 μg acrolein/L. At recommended concentrations for control of nuisance submerged aquatic weeds (frequently 100.0–1000.0 μg/L, often greater than 9600.0 μg/L), acrolein was lethal or harmful to almost all aquatic vertebrates and invertebrates tested in short-term exposures. The most sensitive groups of tested aquatic organisms in short-term assays were frog tadpoles (dead at 7.0 μg/L), representative species of fish (reduced survival at 14.0–62.0 μg/L), and crustaceans (death or immobilization at 34.0–80.0 μg/L). Adverse effects of acrolein on birds were observed at acute oral doses of 9100.0 μg/kg body weight (BW) (reduced survival), concentrations greater than 51.0 μg/kg egg for egg injection (abnormal development and reduced survival), and at greater than 50,000.0 μg/L air (respiratory tract histopathology). In mammals, acrolein is a strong cytotoxic and ciliostatic agent that is irritating to mucous membranes of dermal, ocular, gastrointestinal, and respiratory systems, and is systemically toxic by all routes of exposure. Adverse effects of acrolein are documented in sensitive species of mammals under the following regimens: 50.0 μg/L air for 1 min (increased blood pressure and heart rate); 300.0 μg/L air for 10 min (ocular and nasal irritation); 500.0–1000.0 μg/L air (repelled by odor); 660.0 μg/L air for 24 days (reduced survival); 8000.0–11,000.0 μg/L air for 4–6 h, or 875,000.0 μg/L air for 1 min (death); dietary concentrations equivalent to 500.0 μg/kg BW for 102 weeks (decreased survival); 850.0–6000.0 μg/kg BW by way of intravenous injection (liver necrosis, embryo resorption); and single oral doses between 4000.0 and 28,000.0 μg/kg BW (death).

Acrolein was mutagenic to certain microorganisms and to the fruit fly; mutagenicity may be due, in part, to glycidaldehyde, an acrolein metabolite. Injected into the amniotic fluid, acrolein is teratogenic to rats; teratogenicity may be due to acrylic acid, another acrolein metabolite. There is limited evidence that acrolein acts as a weak carcinogen and tumor promoter. Acrolein interacts with other chemicals, sometimes synergistically, additively, or antagonistically. Also, some chemicals normally contain acrolein as an impurity or yield acrolein as a metabolite.

1.4.1 Terrestrial Plants and Invertebrates

Most crop plants easily tolerate irrigation water with 25,000.0 μg of acrolein/L and many tolerate 70,000.0–80,000.0 μg/L without adverse effects – including corn (*Zea mays*), cotton (*Gossypium hirsutum*), milo (*Sorghum* spp.), squash (*Cucurbita* spp.), castor bean (*Ricinus communis*), tomato (*Lycopersicon esculentum*), alfalfa (*Medicago sativa*), and sugarcane (*Saccharum officinarum*). Acrolein degrades quickly in soils and plant tissues regardless of mode of administration. Atmospheric concentrations of 500.0 μg acrolein/L and higher were harmful to certain plants. Leaves of the pinto bean (*Phaseolus* spp.) and morning glory (*Ipomoea* spp.) developed brown foliar lesions after exposure to 500.0 μg/L air for 4–7 h; damage was more severe if the plants were moist during exposure. Leaves of the radish (*Raphanus* spp.) developed lesions after exposure to 1500.0 μg acrolein/L air for 6–7 h; however, leaves of the geranium (*Germanium* spp.) and the tomato showed no adverse effects after exposure to 1500.0 μg/L air for 7 h.

Acrolein inhibits DNA, RNA, and protein synthesis in the bacterium *Escherichia coli*, and this inhibition probably accounts for its cytotoxic and inhibitory effects on *E. coli* cell division. Acrolein is demonstrably mutagenic to microorganisms and to larvae of the fruit fly (*Drosophila melanogaster*). Acrolein-induced mutagenicity – including point mutations, sister chromatid exchanges, and chromosome breakages – has been observed in selected strains of bacteria (*E. coli*, *Salmonella typhimurium*), yeast (*Saccharomyces cerevisiae*), fruit fly larvae, and cultured Chinese hamster ovary cells. Acrolein's mutagenicity may be due to the metabolite glycidaldehyde: glycidaldehyde was mutagenic to bacteria and yeast under controlled conditions. Studies with

D. melanogaster show that acrolein is mutagenic in the sex-linked recessive lethal test when injected but not when fed. Acrolein caused 2.2% sex-linked mutations in *D. melanogaster* – the highest percentage recorded among several tested aldehydes. Early embryonic stages of fruit flies were most sensitive to the mutagenic properties of acrolein, and sensitivity decreased with increasing development to the point that adults showed negligible mutagenic responses. Adults of the fruit fly were generally resistant to acrolein; mortality was 25% when the culture medium contained 3,700,000.0 μg of acrolein/L, 50% at 8,600,000.0 μg/L, and 75% at 22,100,000.0 μg/L.

1.4.2 Aquatic Organisms

Adverse effects of acrolein on sensitive groups of aquatic organisms are documented at concentrations – in μg acrolein/L medium – as low as 7.0 for frog tadpoles (death), 14.0–62.0 for fish (death), 34.0–80.0 for crustaceans (death, immobilization), 50.0 for oysters (reduction in shell growth rate), 100.0–200.0 for freshwater algae (DNA and RNA reduction, photosynthesis inhibition), 151.0 for gastropods (death), >151.0 for insects (death), 500.0–2000.0 for macrophytes (leaf cell deterioration, death), 1250.0 for trematodes (death of miracidia in 20 min), and 62,000.0 for bacteria (growth reduction). Aquatic vertebrates were more sensitive than invertebrates, and younger fish were more sensitive than older fish. Aquatic insects do not avoid acrolein at concentrations that repel fish. Freshwater fishes and macroinvertebrates, when exposed under static conditions to sublethal concentrations of ^{14}C-labeled acrolein, metabolize acrolein so rapidly that neither acrolein nor its major oxidative and reductive metabolites (acrylic acid, allyl alcohol) were detected in edible tissues within 24 h after dosing.

As a herbicide, acrolein is most effective in controlling dense accumulations of submerged weeds in habitats where water flow is rapid and uniform, such as irrigation canals and rapidly flowing streams. Acrolein is lethal to various genera of submerged plants (*Hydrodictyon*, *Spirogyra*, *Potamogeton*, *Zannichellia*, *Cladophora*, *Ceratophyllum*, *Elodea*, *Chara*, *Najas*) at 1500.0–7500.0 μg/L. But some floating plants (*Pistia*, *Eichornia*, *Jussiaea*) are more resistant to acrolein than submerged plants and require concentrations that are at least double than those necessary for submerged forms. Also, acrolein has little effect on emergent aquatic macrophytes and should not be used to control emergents. Acrolein is the only herbicide now used in Australia for control of submerged aquatic weeds in larger irrigation canals; effective plant control was obtained at 9.6–28.8 mg/L for 3 h. In the United States, the U.S. Bureau of Reclamation controls aquatic algae and weeds at lower concentrations (0.1 mg/L) and longer exposures (48 h). In the Columbia River Basin in the state of Washington, acrolein is used to control submerged aquatic macrophytes at concentrations of 0.1 mg/L for 48 h or 1.0 mg/L for 4–8 h with applications every 3–5 weeks. Vegetation destruction by acrolein is maximal 1 week after application, and green filamentous algae are usually the first plants to return after 1 month. Biomass and species diversity were altered in acrolein-treated phytoplankton populations in Egyptian irrigation canals, 1 year after treatment. Although acrolein is a powerful cytotoxic agent, its inhibitory effects at sublethal concentrations on plant mitosis, nucleic acid synthesis, and protein synthesis are considered completely reversible.

Acrolein in concentrations sufficient to control nuisance submerged aquatic weeds may also kill snails, crayfish, shrimp, fish, and toads. In one case, acrolein was used to control *Potamogeton* and *Chara* in an Ohio farm pond during June. Acrolein was applied at 16,100.0 μg/L to a 0.1 ha portion of the 0.7-ha pond. Within 1 h of application, many dead amphibian tadpoles and small bluegills (*Lepomis macrochirus*) were recovered. In 24 h, *Chara* had turned white and *Potamogeton* brown; both plant species seemed dead; fish were swimming in the treated area. In 72–96 h, several large dead walleyes (*Stizostedium vitreum*) were found. One week posttreatment, all algae and weeds in the treated area were dead; weeds were present in the untreated areas. The treated section remained weed-free for 4–6 weeks; after 8 weeks, the treated area was heavily infested with *Chara*. It was concluded that tadpoles, walleyes, and small bluegills were more susceptible to acrolein toxicity than were larger bluegills and bass (*Micropterus* spp.) in the pond. Acrolein is also effective in controlling trematodes that cause schistosomiasis wherein snails are the intermediate host, especially in irrigation systems. For example, native species of snails (*Lymnae*, *Helisoma*), along with *Potamogeton* weeds, were destroyed within 12 km in the main irrigation canal of Kern County, California, after a single application of acrolein.

Acrolein was the most toxic of 15 herbicides tested for toxicity to fish. Responses by rainbow trout (*Oncorhynchus mykiss*) surviving 77.0 μg acrolein/L, a concentration that killed 50% in about 21 h, were characteristic of respiratory irritants. These responses included a steady increase in cough rate; decreases in ventilation rate, oxygen utilization, and heart rate; increases in hematocrit; and decreases in total arterial oxygen, carbon dioxide, and pH. No-observable-effect concentrations of acrolein for rainbow trout were 240.0 μg/L for exposures of 4.8 h and 48.0 μg/L for exposures of 48 h; these values are below the concentrations that control aquatic weeds. In the same study, rainbow trout that survived exposure to high sublethal concentrations for 48 h were unable to recover completely after acrolein treatments were ended. Trout and other teleosts are poorly adapted to detoxify acrolein and other xenobiotic aldehydes. The low metabolic capacity of fish liver aldehyde dehydrogenase for aldehydes, in general, suggests that these compounds may be hazardous to fish populations. Applications of acrolein to waters where fish may be taken for human consumption should be made with caution; rainbow trout surviving exposure to acrolein in reservoirs or connecting canals frequently presented odor and taste problems to human consumers.

In addition to weeds, acrolein is used to control fouling organisms in cooling water systems. Effective control was established in a once-through cooling system of a steel mill with continuous application of 200.0 μg acrolein/L. Acrolein, controlled bacteria in condenser pipes of a power-plant cooling system, but only at extremely high concentrations of 125,000.0 μg/L for 120 h or 500,000.0 μg/L for 2 h. Acrolein reduced settlement of young mussels (*Mytilus* sp.) in cooling seawater systems of power plants. In recirculating cooling water systems, algae and bacteria can be controlled at 500.0 μg/L for 5 months or at 5000.0 μg/L for one week.

1.4.3 Birds

Acrolein was lethal to birds at single oral doses of 9100.0 μg/kg BW. Observed signs of acrolein poisoning in subadult mallards (*Anas platyrhynchos*) after oral administration included regurgitation, a reluctance to leave the swimming area, slow responses, muscular incoordination, heavy-footed walking, phonation, wing tremors, running and falling, weakness, and withdrawal. Treatment concentrations as low as 3300.0 μg/kg BW have produced signs of acrolein poisoning. These signs appeared as soon as 10 min after administration and persisted for as long as 36 days. At lethal oral concentrations, deaths occurred as soon as 32 min posttreatment and continued for several days. Acrolein was lethal to developing avian embryos when whole eggs were injected with 51.0–182.0 μg/kg fresh weight (FW); in descending order, embryos were most sensitive when acrolein was administered by way of the yolk sac (51.0 μg/kg), by inner shell (82.0 μg/kg), and by air sac (182.0 μg/kg). Acrolein is 50 times more toxic to embryos of the domestic chicken (*Gallus* sp.) than acrylonitrile, and 100 times more toxic than acrylamide. Acrolein inhibits mucous transport in the trachea of the domestic chicken, probably through ciliostatic action. Adverse effects of acrolein were observed in chicken respiratory-tract physiology and pathology at greater than 50,000.0 μg/L air.

Malformations of the eye, coelom, neck, back, wings, and legs were observed in surviving acrolein-treated chicken embryos after whole eggs were injected with greater than 51.0 μg acrolein/kg FW. In other studies, acrolein showed no clear evidence of teratogenicity in chicken embryos, although there is a dose-dependent embryotoxic effect. Acrolein-treated chicken embryos had a higher frequency of abnormal limbs, abnormal neck, and everted viscera than the controls, but the frequency was not dose related. The overall incidence of abnormal embryos when treated at age 48 h was 24%, but only 4% in controls; in embryos, when treated with acrolein at age 72 h, these values were 26% and 12% in controls.

1.4.4 Mammals

Acrolein is a strong cytotoxic and ciliostatic agent; its irritating effects on mucous membranes and its acute inhalation toxicity in mammals are well documented. A characteristic of acrolein is its pungent, offensive, and acrid smell that is highly irritating to ocular and upper respiratory-tract mucosae. Acrolein is toxic by all routes of exposure, and many of its toxic and biochemical effects are produced by interfering with critical sulfhydryl groups. In isolated rat liver fractions, acrolein is a potent inhibitor of the high-affinity aldehyde dehydrogenase isozymes in mitochondrial and cytosolic fractions. Acrolein impairs DNA replication in vitro and inhibits certain mitochondrial functions. Studies with isolated rat liver-membrane proteins revealed that acrolein inhibits plasma membrane enzymes and alters the membrane protein profile; this may be due to acrolein-induced polymerization of plasma-membrane proteins.

Measurable adverse effects of acrolein have been documented in representative species of mammals, but the severity of the effects is contingent on the mode of administration, concentration, dose, and duration of exposure. Single oral doses of 4000.0 μg/kg BW were lethal to guinea pigs and 28,000.0 μg/kg BW to mice; diets containing the equivalent of 500.0 μg/kg BW and more decreased survival in rats after 102 weeks. Concentrations of 60,000.0 μg acrolein/L in drinking water had no measurable adverse effects on cows (*Bos* sp.) after 24 h; rats initially rejected drinking water containing 200,000.0 μg/L but eventually tolerated this concentration. Dermal toxicity seems low; rabbits that were immersed up to their necks in water containing 20,000.0 μg acrolein/L for 60 min showed no adverse effects. No dermal sensitization occurred in healthy female guinea pigs (*Cavia* spp.) after repeated skin exposures to acrolein. In undiluted liquid or pungent vapor form, however, acrolein produces intense irritation of the eye and mucous membranes of the respiratory tract, and direct contact with the liquid can produce skin or eye necrosis. A single intravenous injection of 850.0 μg acrolein/kg BW produced liver necrosis in rats; 6000.0 μg/kg BW caused increased embryo resorption in mice. Rats receiving near-lethal doses of acrolein by subcutaneous injection had liver and kidney damage and lung pathology. Although subcutaneous injections revealed LD50 values between 164,000.0 and 1,022,000.0 μg/kg BW in rabbits, these results are questionable because acrolein may be sequestered at the injection site and delay delivery to the systemic circulation. A single intraperitoneal injection of 1000.0 μg/kg BW caused peritonitis in rats and 7000.0 μg/kg BW was lethal to mice; daily injections of 1000.0 μg/kg BW were eventually lethal to rats. Sublethal intraperitoneal injections of acrolein induced ascites, increased hematocrit, and prolonged sleeping times. Acquired tolerance to acrolein in mice given repeated intraperitoneal injections suggests that an increased metabolism can partially explain the acquired tolerance.

The largest number of studies of the toxicity of acrolein in animals was conducted by way of inhalation, probably because acrolein has an appreciable vapor pressure under ambient conditions and inhalation is the principal exposure for humans. Because of their intolerance to sharp and offensive odor and to intense irritation of conjunctiva and upper respiratory tract, humans have not suffered serious intoxication from acrolein. The strong lacrimatory effect of acrolein usually is a warning to occupational workers. Physiological perception of acrolein by humans begins at about 500.0–1000.0 μg/L air with eye and nasal irritation; the irritating effects compel afflicted individuals to immediately leave the polluted area. Laboratory animals died from inhalation of 8000.0–11,000.0 μg/L after

4–6 h, mice from 875,000.0 μg/L after 1 min, and rats from 660.0 μg/L for 24 days. Animals dying from acute and subacute exposure to acrolein vapor had lung injury with hemorrhagic areas and edema. Repeated exposures of hamsters, rats, and rabbits to high sublethal concentrations of acrolein caused ocular and nasal irritation, growth depression, and respiratory tract histopathology in all species. However, repeated exposures to low, tolerated concentrations of acrolein did not produce toxicological effects, suggesting that acrolein effects are not cumulative and that minimal damage is quickly repaired.

Inhaled acrolein – in μg acrolein/L air – had sublethal effects at 10.0–50.0 for 1 min on rats (increased blood pressure and heart rate); at 10.0 for 5 weeks on mice (reduction in pulmonary compliance); at 140.0–150.0 for 2 min on humans (eye irritation in 30%); at 300.0–500.0 on humans (odor threshold); at 300.0 for 10 min on humans (acute irritation); at 400.0 for 13 weeks on rats (nasal histopathology); at 400.0–600.0 for 1–3 min on dogs (accumulations in upper respiratory tract); and at 1000.0 for 90 days on dogs, monkeys, and guinea pigs (ocular and nasal discharges). Sublethal effects of inhaled acrolein in representative small laboratory mammals were greatest on the upper respiratory tract and bronchial airways and included edema, ciliastasis, inflammation, degenerative loss of epithelia, altered ventilatory function, and bronchoconstriction. Typical signs of toxicity from inhaled acrolein in small mammals include ocular and nasal irritation; growth depression; shortness of breath; lesions in the urinary tract, respiratory tract, trachea, and nasal passages; laryngeal edema; reduced resistance to bacterial infection; enlarged liver and heart; elevated blood pressure and heart rate; altered enzyme activities; and protein synthesis inhibition. Signs of inhaled acrolein toxicity varied significantly with dose and species. For example, acrolein toxicity in rats at environmental concentrations was confined to local pathologic nasal changes, including metaplastic, hyperplastic, and dysplastic changes in the mucous, respiratory, and olfactory epithelium of the nasal cavity. Some inhaled toxicants, including acrolein, can prolong bacterial viability in the lung and thus enhance severity of the disease. Mice convalescing from viral pneumonia became severely deficient in antibacterial defenses when exposed to acrolein. But acrolein-treated mice subjected to 100.0 μg/L air (5 consecutive daily 3-h exposures) were not significantly sensitive to pulmonary bacteria *Klebsiella pneumoniae* or *Streptococcus zooepidemicus*.

Acrolein may be a carcinogen, co-carcinogen, or tumor initiator. As an aldehyde with strong affinity to sulfhydryl groups, acrolein is theoretically expected to remove free tissue thiols – compounds that protect bronchial epithelia against attack by carcinogens. But no carcinogenicity from inhalation of acrolein has been reported. Nor was acrolein an evident cofactor in studies of respiratory-tract carcinogenesis with hamsters (*Cricetus* spp.) exposed to benzo[a]pyrene or diethylnitrosamine. Moreover, long-term studies with rodents given acrolein by gavage did not increase incidences of neoplastic or non-neoplastic lesions. Other studies, however, suggest that acrolein is carcinogenic. Compounds closely related to acrolein are carcinogenic to rodents and humans and include acrylonitrile (vinyl cyanide) and vinyl acetate. Glycidaldehyde – an acrolein intermediate metabolite – is classified as an animal carcinogen by The International Agency for Research on Cancer; however, no convincing data are available on the carcinogenic potential of acrylic acid and other acrolein metabolites. Acrolein can account, at least partially, for the initiating activity of cyclophosphamide carcinogenesis. Cyclophosphamide and its analogs are a group of chemotherapeutic and immunosuppressive drugs; toxic side effects of this drug group are attributed to its metabolites, especially acrolein. Acrolein is a suspected carcinogen because of its 2,3-epoxy metabolite and its weak mutagenic activity in the *Salmonella* screen. Acrolein may be a weak carcinogen, as judged by the increased frequency of adrenal adenomas in female rats after exposure for 2 years to drinking water with 625,000 μg acrolein/L. Acrolein has cancer-initiating activity in the rat urinary bladder, but studies with *N*-[4-(5-nitro-2-furyl)-2 thiazoyl] formamide precluded evaluation of acrolein as promoting a complete carcinogenic activity from low rodent survival. Additional studies seem needed to evaluate the carcinogenic potential of acrolein.

After intraamniotic injection, acrolein is teratogenic to rats in vivo but not in vitro. When administered intraamniotically to the whole embryo culture system of the rat on day 13 of gestation, acrolein caused edema, hydrocephaly, open eyes, cleft palate, abnormal umbilical cord, and defects of the limbs and face, suggesting that acrolein-associated teratogenicity is caused by acrylic acid, an acrolein metabolite. Acrylic acid injected into amniotic fluid of rats on day 13 of gestation produced a dose-dependent increase in the percentage of fetuses with skeletal and other abnormalities.

Acrolein can react synergistically, additively, or antagonistically with other chemicals. Rat embryos were protected by glutathione against acrolein-induced mortality, growth retardation, and developmental abnormalities – provided that glutathione was concurrently present with acrolein. When rat embryos were cultured in the presence of acrolein for 2 h prior to glutathione exposure, there was no protection against acrolein-induced embryo lethality, teratogenicity, and growth retardation. Acrolein effects – including altered liver enzyme activity in rats – were reduced by pretreatment of animals with chemicals that inhibited protein synthesis. Exposure to acrolein is sometimes accompanied by exposure to formaldehyde and other short-chain saturated aliphatic aldehydes, which in combination cause allergic contact dermatitis. A 40-mL puff of cigarette smoke contains 8.2 μg of acrolein and 4.1 μg of formaldehyde; irritation, ciliastasis, and pathologic changes of the respiratory tract from both compounds have been widely studied. The toxicities of acrolein and formaldehyde seem similar; both exert their principal effects in the nasal passages. Acrolein in combination with formaldehyde was synergistic in reducing respiratory rate in mice; however, mixtures of sulfur dioxide and acrolein were antagonistic. Formaldehyde pretreatment (15,000.0 μg/L, 6 h daily for 9 days) of rats protects against respiratory-rate depression by acrolein. Rats pretreated with formaldehyde had a 50% respiratory-rate depression at 29,600.0 μg acrolein/L vs. 6000.0 μg/L from acrolein alone, suggesting cross-tolerance. Effects of interaction of acrolein with other toxicants are not comparable between rodents and humans. In rodents, the presence of irritant gases in smoke – such as acrolein – may delay the effects of other toxicants. In humans, however, the inhalation of acrolein and other irritant gases may cause a hypoxemic effect that can enhance the effects of hypoxia-producing gases.

Some chemicals normally contain acrolein as a metabolite or impurity. For example, allylamine toxicity to the rat cardiovascular system is believed to involve metabolism of allylamine to the highly reactive acrolein. Certain mercapturic acids can be used as biological markers of exposure for chemicals that are metabolized to acrolein and excreted as mercapturic acid in the urine. In one case, rats given 13,000.0 μg acrolein/kg BW by gavage excreted 79% of the acrolein and 3-hydroxypropylmercapturic acid (3-OHPrMCA) in urine within 24 h. These data suggest that 3-OHPrMCA can be used as a marker of exposure to allylic and other compounds that lead to the formation of acrolein. The common industrial chemical MDP (2-methoxy-3,4-dihydro-2PH-pyran) is frequently contaminated with acrolein during its synthesis; MDP causes severe irritancy and death of rats from accumulation of acrolein vapor. Sparging acrolein-contaminated MDP with nitrogen gas before atmospheric release significantly reduced or abolished lethal toxicity to rats.

1.5 Recommendations

Agricultural crops can usually tolerate as much as 15,000.0 μg of acrolein/L of irrigation water; however, aquatic invertebrates and fish die in acute exposures to 55.0–68.0 μg/L or in chronic exposures to greater than 21.0 μg/L (Table 1.2). Those who use acrolein to control submerged aquatic macrophytes are strongly advised that acrolein treatment at recommended application concentrations also eliminates nontarget fish and aquatic invertebrates.

Table 1.2. Proposed acrolein criteria for the protection of living resources and human health.

Resource, Criterion, and Other Variables	Concentration
AGRICULTURAL CROPS	
Irrigation water, tolerated level	<15,000.0 μg/L
AQUATIC LIFE	
Freshwater organisms	
Sensitive species, tolerated level	
Acute exposures	<68.0 μg/L
Chronic exposures	<21.0 μg/L
Rainbow trout, safe level	20.0 μg/L for <48 h or 200.0 μg/L for <4.8 h
Marine organisms; acute exposures, tolerated level	<55.0 μg/L
LABORATORY WHITE RAT	
Air	
Maximum daily average	<13.0 μg/L (<0.03 mg/m^3)
Maximum daily	<44.0 μg/L (<0.1 mg/m^3)
HUMAN HEALTH	
Air	
Maximum allowable emission concentration in populated areas of former Soviet Union	132.0 μg/L (0.3 mg/m^3)
No-observable-effect level	<22.0 μg/L (<0.05 mg/m^3)
90-day confined space (i.e., submarines) guideline	22.0 μg/L (0.05 mg/m^3)
Odor threshold	<44.0 μg/L (<0.1 mg/m^3)
Maximum acceptable concentration in room air of former Soviet Union	44.0 μg/L (0.1 mg/m^3)
Irritation threshold	44.0–88.0 μg/L (0.1–0.2 mg/m^3)
Occupational exposure standard (8-h daily, 40-h work week) in United States; not to exceed in most European countries, Australia, and Japan	100.0–110.0 μg/L (0.25 mg/m^3)
Occupational exposure standard in Hungary and former Soviet Union	308.0 μg/L (0.7 mg/m^3)
Maximum 15-min exposure limit in USA workplace	300.0–352.0 μg/L (0.8 mg/m^3)
Ceiling standard for occupational exposure in the former Czechoslovakia	440.0 μg/L (1.0 mg/m^3)
Acceptable ambient air concentrations New York, Florida, North Dakota, North Carolina	0.83 μg/m^3 for 1 year; 2.5 μg/m^3 for 8 h 8.0 μg/m^3 for 1 h 80.0 μg/m^3 for 15 min

Continued

Table 1.2. cont'd

Resource, Criterion, and Other Variables	Concentration
Diet	
Water plus consumption of contaminated aquatic organisms from that water body	<320.0 μg/L medium
Consumption of contaminated aquatic organisms alone	<780.0 μg/L medium
Food packaging materials; food starch	<0.6%
Total daily intake	<47.8 μg = <0.68 μg/kg BW daily for a 70-kg person

No acrolein criteria are now available or promulgated by regulatory agencies for the protection of avian and terrestrial wildlife; this seems to be a high-priority research need. Additional research is recommended in several areas: long-term effects of acrolein inhalation on carcinogenicity and respiratory histology with rodent models; biochemical mechanisms of acrolein toxicity; genotoxic potential with chromosome breakage and exchange systems; acute and chronic toxicity from interaction effects of acrolein with other gases; and fate of accumulated acrolein in animals.

The human threshold concentration of acrolein in the United States for an 8-h workday and 40-h workweek is 110.0 μg/L (0.25 mg/m^3) air; the short-term exposure limit is 350.0 μg/L (0.8 mg/m^3) air and is predicated on continuous exposure of workers for short intervals. Humans can tolerate a total daily intake of 47.8 μg of acrolein, equivalent to 0.68 μg/kg BW by a 70-kg individual (Table 1.2).

For handling acrolein, gloves, vapor-proof goggles or a full face mask, and other protective clothing are mandatory. Acrolein spills should be neutralized with 10% sodium bisulfite solutions. Air packs or fresh-air breathing masks, safety showers, and eye baths should be available wherever acrolein is handled. Purging confined areas with nitrogen is recommended prior to entering a suspected acrolein-contaminated enclosure. The eyes are particularly susceptible to liquid acrolein and, if exposed, should receive prompt treatment, although severe residual injury is probable regardless of treatment; dilute solutions of acrolein may also cause residual eye injury. Acrolein represents a serious fire hazard because of its high flammability and potential for vapors to form explosive mixtures with air. Flameproof electrical equipment and proper grounding are required to prevent acrolein ignition. Individuals exposed to acrolein by inhalation should be removed from the area and given oxygen; subsequent treatment by physicians of pulmonary inflammation with corticosteroids and hydroxocobalamin is recommended even if there are no symptoms because adverse effects from acrolein exposure may not become apparent until 4–24 h after exposure. Oxygen therapy should be continued and analgesics given for relief of other symptoms as necessary. There are many synthetic and natural sources of acrolein; however, special precautions are recommended when acrolein occurs as a contaminant in the synthesis of widely used chemicals, such as 2-methyoxy-3,4-dihydro-2H pyran.

1.6 Summary

Acrolein (CH_2=CHCHO) is the simplest member of the class of unsaturated aldehydes and enters the environment from incomplete combustion of fossil fuels, industrial discharges, herbicides, chemical control agents of fouling organisms, and normal metabolic processes of animals. Acrolein is volatile, flammable, and explosive. Biochemical and toxic effects of acrolein are caused by its reaction with sulfhydryl compounds to form a stable thiol ether. Acrolein metabolites under certain conditions are reportedly mutagenic, teratogenic, or carcinogenic. Acrolein degrades quickly in soils and in plant tissues; in water, the half-time persistence is usually less than 50 h, and in the atmosphere, less than 3 h. In treated irrigation canals, acrolein probably eliminates or seriously depletes all populations of aquatic fauna in treated areas. Recommended herbicidal concentrations of acrolein for the control of submerged aquatic weeds usually exceed 1000.0 μg/L; however, short-term tests with various species show that frog tadpoles die at 7.0 μg/L, representative fish are killed at 14.0–62.0 μg/L, and sensitive crustaceans are immobilized or die at 34.0–80.0 μg/L. Terrestrial plants and insects are comparatively resistant to acrolein: terrestrial plants tolerated 500.0 μg acrolein/L air and 25,000.0 μg/L in irrigation water, and adult fruit flies (*Drosophila melanogaster*), 3,700,000.0 μg acrolein/L culture medium. Birds are adversely affected by concentrations greater than 51.0 μg acrolein/kg whole egg by injection of eggs, greater than 9100.0 μg/kg BW by single oral doses, and greater than 50,000.0 μg/L (greater than 113.0 mg/m^3) by air concentrations. Mammals were affected by 50.0 μg acrolein/L air for 1 min, and 4000.0–28,000.0 μg/kg BW by single oral doses, or when fed diets equivalent to 500.0 μg/kg BW for 102 weeks. Proposed acrolein criteria for the protection of various resources include less than 15,000.0 μg/L in irrigation water of agricultural crops, less than 68.0 μg/L for aquatic fauna in acute exposures and less than 21.0 μg/L in chronic exposures, and less than 44.0 μg/L (less than 0.1 mg/m^3) in air for rats. No acrolein criteria are now available for the protection of avian and terrestrial wildlife. Acrolein criteria for the protection of human health include less than 320.0 μg/L in drinking water, less than 110.0 μg/L in air (less than 0.25 mg/m^3), and less than 0.68 μg/kg BW daily intake from all sources. More research is needed on acrolein and its metabolites.

ARSENIC[a]

Chapter 2

2.1 *Introduction*

Anxiety over arsenic (As) is understandable, and frequently justifiable. Arsenic compounds were the preferred homicidal and suicidal agents during the Middle Ages, and arsenicals have been regarded largely in terms of their poisonous characteristics in the nonscientific literature. Acute arsenic poisoning was reported in the 1800s in horses, cows, deer, and foxes as a result of emissions from metal smelters. In 1885, arsenic accounted for nearly one-third of the homicide poisonings in France. Data collected on animals, including humans, indicated that inorganic arsenic could cross the placenta and produce mutagenic, teratogenic, and carcinogenic effects in offspring. Correlations between elevated atmospheric arsenic levels and mortalities from cancer, bronchitis, and pneumonia were established in an epidemiological study in England and Wales, where deaths from respiratory cancer were increased at air concentrations >3.0 μg As/m^3. Chronic arsenical poisoning, including skin cancer and a gangrenous condition of the hands and feet called Blackfoot's disease, has occurred in people from several communities in Europe, South America, and Taiwan that were exposed to elevated concentrations of arsenic in drinking water. Inorganic arsenic in dietary staples such as yams and rice may have substantially contributed to exposure and adverse health effects observed in an endemic Taiwanese population, historically exposed to arsenic in drinking water. Severe health effects, including cancer and death, have been documented in Mongolia, Bangladesh, Chile, and India from ingestion of naturally elevated levels of arsenic in the drinking water; and in Thailand from the use of traditional arsenic-containing medicines and arsenical wastes from mining activities. In Japan, about 12,000 infants were poisoned (128 deaths) after consuming dry milk containing 15.0–24.0 mg inorganic As/kg, which originated from contaminated sodium phosphate used as a milk stabilizer. Fifteen years after exposure, the survivors sustained an elevated frequency of severe hearing loss and brain wave abnormalities.

Reviewers of ecotoxicological aspects of arsenic in the environment agree on six points: (1) arsenic is a relatively common element, and is present in air, water, soil, plants, and in all living tissues; (2) arsenicals have been used in medicine as chemotherapeutics since 400 BCE, and organoarsenicals were used extensively for this purpose until about 1945, with no serious effects when judiciously administered; (3) large quantities of arsenicals are released into the environment as a result of industrial and especially agricultural activities, and these may pose potent

---

[a]All information in this chapter is referenced in the following sources:

Eisler, R. 1981. *Trace Metal Concentrations in Marine Organisms*. Pergamon Press, New York, 687 pp.

Eisler, R. 1988. Arsenic hazards to fish, wildlife, and invertebrates: a synoptic review. *U.S. Fish. Wildl. Serv. Biol. Rep. 85(1.12)*, 1–92.

Eisler, R. 1994. A review of arsenic hazards to plants and animals with emphasis on fishery and wildlife resources. Pages 185–259 in J.O. Nriagu, ed. *Arsenic in the Environment. Part II: Human Health and Ecosystem Effects*. John Wiley, New York.

Eisler, R. 2000. Arsenic. Pages 1501–1566 in *Handbook of Chemical Risk Assessment: Health Hazards to Humans, Plants, and Animals. Volume 3. Metalloids, Radiation, Cumulative Index to Chemicals and Species*. Lewis Publishers, Boca Raton, Florida.

Eisler, R. 2004. Arsenic hazards to humans, plants, and animals from gold mining. *Rev. Environ. Contam. Toxicol.* 180, 133–165.

ecological dangers; (4) exposure of humans and wildlife to arsenic may occur through air (emissions from smelters, coal-fired power plants, herbicide sprays), water (mine tailings runoff, smelter wastes, natural mineralization), and food (especially seafoods); (5) chronic exposure to arsenicals by way of air, diet, and other routes have been associated with liver, kidney, and heart damage, hearing loss, brain-wave abnormalities, and impaired resistance to viral infections; and (6) exposure to arsenic has been associated with different types of human cancers such as respiratory cancers and epidermoid carcinomas of the skin, as well as precancerous dermal keratosis. The epidemiological evidence of human carcinogenicity is supported by carcinogenesis in experimental animals.

2.2 *Sources, Fate, and Uses*

Global production of arsenic is estimated to be 75,000–100,000 tons annually, of which the United States produces about 21,000 tons and uses about 44,000 tons; major quantities are imported from Sweden, the world's leading producer. Imports of arsenic trioxide (As_2O_3) to the United States have increased from about 16 million kg in 1985 to about 30 million kg in 1989; imports of elemental arsenic have increased from about 0.45 million kg during 1985–88 to 1.2 million kg in 1989. The United States exports about 0.3 million kg of arsenic compounds annually, mainly as arsenic acid, sodium arsenate, and lead arsenate. Almost all (97%) of the arsenic made worldwide enters end-product manufacture in the form of arsenic trioxide, and the rest is used as additives in producing special lead and copper alloys. About 74% of the total arsenic trioxide is in products used for wood preservation. Most of the rest (about 19%) is used in production of agricultural chemicals, such as insecticides, herbicides, fungicides, algicides, and growth stimulants for plants and animals. Smaller amounts are used in the production of glass, in the electronic industry, dyestuffs, and in veterinary and human medicines, including medicines for the eradication of tapeworm in sheep and cattle. The sole producer and refiner of As_2O_3 in the United States is a copper smelter in Tacoma, Washington.

Arsenic naturally occurs as sulfides and as complex sulfides of iron, nickel, and cobalt. In one form or another, arsenic is present in rocks, soils, water, and living organisms at concentrations of parts per billion to parts per million. Soil arsenic levels are normally elevated near arseniferous deposits, and in mineralized zones containing gold, silver, and sulfides of lead and zinc. Secondary iron oxides formed from the weathering of pyrite act as scavengers of arsenic. Pyrite is a known carrier of arsenic and may contain up to 5600 mg/kg; for example, total arsenic is 10 times above normal background levels in soils derived from pyritic shale. Natural weathering of rocks and soils adds about 40,000 tons of arsenic to the oceans yearly, accounting for <0.01 mg/L input to water on a global basis. Many species of marine plants and animals often contain naturally high concentrations of arsenic, but it is usually present in a harmless organic form. Anthropogenic input of arsenic to the environment is substantial, and exceeds that contributed by natural weathering processes by a factor of about 3.

The most important concept with respect to arsenic cycling in the environment is constant change. Arsenic is ubiquitous in living tissue and is constantly being oxidized, reduced, or otherwise metabolized. In soils, insoluble or slightly soluble arsenic compounds are constantly being resolubilized, and the arsenic is being presented for plant uptake or reduction by organisms and chemical processes. Humans reportedly modify the arsenic cycle only by causing localized high concentrations. The speciation of arsenic in the environment is affected partly by indiscriminate biological uptake, which consumes about 20% of the dissolved arsenate pool and results in measurable concentrations of reduced and methylated arsenic species. The overall arsenic cycle is similar to the phosphate cycle; however, regeneration time for arsenic is much slower – of the order of several months. The ubiquity of arsenic in the environment is evidence of the redistribution processes that have been operating since early geologic time. A pre-human steady-state solution to the global arsenic cycle indicates that major reservoirs of arsenic (in kilotons) are magma (50 billion), sediments (25 billion), oceanic deep waters (1.56 million), land (1.4 million), and ocean mixed layers (270,000); minor amounts occur in ocean particulates (100), and in continental (2.5) and marine troposphere (0.069). Arsenic is significantly mobilized from the land to the troposphere by both natural and anthropogenic processes. Industrial emissions account for about 30% of the present-day burden of arsenic in the troposphere. Agronomic ecosystems, for example, may receive arsenic from agricultural sources such as organic herbicides, irrigation waters, and fertilizers, and from such non-agricultural sources as fossil fuels and industrial and municipal wastes. Arsenic is mobile and nonaccumulative in air, plant, and water phases of agronomic ecosystems; arsenicals sometimes accumulate in soils, but redistribution mechanisms usually preclude hazardous accumulations.

Arsenic compounds have been used in medicine since the time of Hippocrates, ca. 400 BCE. Inorganic arsenicals have been used for centuries, and organoarsenicals for at least a century in the treatment of syphilis, yaws, amoebic dysentery, and trypanosomiasis. During the period 1200–1650, however, arsenic was used extensively in homicides. In 1815, the first accidental death was reported from arsine (AsH_3) poisoning, and in 1900–1903, accidental poisonings from consumption of arsenic-contaminated beer were widely reported. In 1938, it was established that arsenic could counteract selenium toxicity. The introduction of arsphenamine, an organoarsenical, to control venereal disease earlier this century gave rise to intensive research by organic chemists, which resulted in the synthesis of at least 32,000 arsenic compounds. But the advent of penicillin and other newer drugs nearly eliminated the use of organic arsenicals as human therapeutic agents. Arsenical drugs are still used in treating certain tropical diseases, such as African sleeping sickness and amoebic dysentery, and are used in veterinary medicine to treat parasitic diseases, including filariasis in dogs (*Canis familiaris*), blackhead in turkeys (*Meleagris gallopavo*), and chickens (*Gallus* spp.). Today, abnormal sources of arsenic that can enter the food chain from plants or animals include arsenical pesticides such as lead arsenate; arsenic acid, $HAsO_3$; sodium arsenite, $NaAsO_2$; sodium arsenate, Na_2AsO_4; and cacodylic acid, $(CH_3)_2As(OH)$.

The major uses of arsenic are in the production of herbicides, insecticides, desiccants, growth stimulants for plants and animals, and especially in wood preservatives. Much smaller amounts are used in the manufacture of glass (nearly all of which contains 0.2–1% arsenic as an additive – primarily as a decolorizing agent) and textiles, and in medical and veterinary applications. Arsenic is also an ingredient in lewisite, a blistering poison gas developed (but not used) during World War I, and in various police riot control agents. The availability of arsenic in certain local areas has been increased by various human activities: smelting and refining of gold, silver, copper, zinc, uranium, and lead ores; combustion of fossil fuels, such as coal and gasoline; burning of vegetation from cotton gins treated with arsenical pesticides; careless or extensive use of arsenical herbicides, pesticides, and defoliants; dumping of land wastes and sewage sludge (1.1 mg/L) in areas that allow leaching into groundwater; use of domestic detergents in wash water (2.5–1000.0 mg As/L); manufacture of glass; and by the sinking of drinking water wells into naturally arseniferous rock. There are several major anthropogenic sources of environmental arsenic contamination: industrial smelters – the effluent from a copper smelter in Tacoma, Washington, contained up to 70 tons of arsenic discharged yearly into nearby Puget Sound; coal-fired power plants, which collectively emit about 3000 tons of arsenic annually in the United States; and production and use of arsenical pesticides, coupled with careless disposal of used pesticide containers. Elevated levels of arsenic have been reported in soils near smelters, in acid mine spoils, and in orchards receiving heavy applications of lead arsenate. Air concentrations of arsenic are elevated near metal smelters, near sources of coal burning, and wherever arsenical pesticides are applied. Atmospheric deposition of arsenic has steadily increased for at least 30 years, as judged by sedimentary evidence from lakes in upstate New York. Arsenic is introduced into the aquatic environment through atmospheric deposition of combustion products and through runoff from flyash storage areas near power plants and nonferrous smelters. Elevated arsenic concentrations in water were recorded near mining operations, and from mineral springs and other natural waters – usually alkaline, and with high sodium and bicarbonate contents. In the United States, the most widespread and frequent increases in dissolved arsenic concentrations in river waters were in the northern Midwest; all evidence suggests that increased atmospheric deposition of fossil fuel combustion products was the predominant cause of the trend.

Agricultural applications provide the largest anthropogenic source of arsenic in the environment. Inorganic arsenicals (arsenic trioxide; arsenic acid; arsenates of calcium, copper, lead, and sodium; and arsenites of sodium and potassium) have been used widely for centuries as insecticides, herbicides, algicides, and desiccants. Paris green (cuprous arsenite) was successfully used in 1867 to control the Colorado potato beetle (*Leptinotarsa decemlineata*) in the eastern United States. Arsenic trioxide has been applied widely as a soil sterilant. Sodium arsenite has been used for aquatic weed control, as a defoliant to kill potato vines before tuber harvest, as a weed killer along roadsides and railroad rights-of-way, and for control of crabgrass (*Digitaria sanguinalis*). Calcium arsenates have been applied to cotton and tobacco fields to protect against the boll weevil (*Anthonomus grandis*) and other insects. Lead arsenate has been used to control insect pests of fruit trees, and for many years was the only insecticide that controlled the codling moth (*Carpocapsa pomonella*) in apple orchards and the horn worm larva (Sphyngidae) on tobacco. Much smaller quantities of lead arsenate are now used in orchards because fruit growers rely primarily on carbamate and organophosphorus compounds to control insect pests; however, lead arsenate is still being used by some growers to protect orchards from certain chewing insects. The use of inorganic arsenicals has decreased due to the banning of sodium arsenite and some other arsenicals for most purposes, although they continue to be used on golf greens and fairways in certain areas to control annual bluegrass (*Poa annua*). In recent decades, inorganic arsenicals have been replaced by organoarsenicals for herbicidal application, and by carbamate and organophosphorus compounds for insect control. By the mid-1950s, organoarsenicals were used extensively as desiccants, defoliants, and herbicides. Organoarsenicals marketed in agriculture today, which are used primarily for herbicidal application, include cacodylic acid (also known as dimethylarsinic acid (DMA)) and its salts – monosodium and disodium methanearsonate. Organoarsenicals are used as selective herbicides for weedy grasses in turf, and around cotton and noncrop areas for weed control; at least 1.8 million ha (4.4 million acres) have been treated with more than 8000 tons of organoarsenicals. In 1945, it was discovered that one organoarsenical (3-nitro-4 hydroxyphenyl arsonic acid) controlled coccidiosis and promoted growth in domestic chicken. Since that time, other substituted phenylarsonic acids have been shown to have both therapeutic and growth promoting properties as feed additives for poultry and swine (*Sus* spp.), and are used for this purpose today under existing regulations – although the use of arsenicals in poultry food was banned in France in 1959.

2.3 Chemical and Biochemical Properties

Elemental arsenic is a gray, crystalline material characterized by atomic number 33, atomic weight of 74.92, density of 5.727, melting point of 817°C, sublimation at 613°C, and chemical properties similar to those of phosphorus. Arsenic has four valence states: −3, 0, +3, and +5. Arsines and methylarsines, which are characteristic of arsenic in the −3 oxidation state, are generally unstable in air. Elemental arsenic, As^0 is formed by the reduction of arsenic oxides. Arsenic trioxide (As^{+3}) is a product of smelting operations and is the material used in synthesizing most arsenicals. It is oxidized catalytically or by bacteria to arsenic pentoxide (As^{+5}) or orthoarsenic acid (H_3AsO_4). Arsenic in nature is rarely in its free state. Usually, it is a component of sulfidic ores, occurring as arsenide; and arsenates, along with arsenic trioxide, which is a weathering product of arsenides. Most arsenicals degrade or weather to form arsenate, although arsenite may form under anaerobic conditions. Biotransformations may occur, resulting in volatile arsenicals that normally are returned to the land where soil adsorption, plant uptake, erosion, leaching, reduction to arsines, and other processes occur. This natural arsenic cycle reflects a constant shifting of arsenic between environmental compartments. Atomic absorption spectrometry is the most common procedure for measuring arsenic in biological materials, although other methods are used including, neutron activation. A variety of sensitive techniques have been used to obtain speciation data for the forms of arsenic at trace levels.

Arsenic species in flooded soils and water are subject to chemically and microbiologically mediated oxidation or reduction and methylation reactions. At high Eh values (i.e., high oxidation–reduction potential) typical of those encountered in oxygenated waters, pentavalent As^{+5} tends to exist as H_3AsO_4, H_2AsO_4, $HAsO_2$, and AsO_4^{-3}. At lower Eh, the corresponding trivalent arsenic species can be present, as well as AsS_2. In aerobic soils, the dominant arsenic species is As^{+5}, and small quantities of arsenite and monomethylarsonic acid (MMA) are present in mineralized areas; in anaerobic soils, As^{+3} is the major soluble species. Inorganic arsenic is more mobile than organic arsenic, and thus poses greater problems by leaching into surface waters and groundwater. The trivalent arsenic species are generally considered to be more toxic, more soluble, and more mobile than As^{+5} species. Soil microorganisms metabolize arsenic into volatile arsine derivatives. Depending on conditions, 17–60% of the total arsenic present in soil may be volatilized. Estimates of the half-life of arsenic in soil vary from 6.5 years for arsenic trioxide to 16 years for lead arsenate.

In water, arsenic occurs in both inorganic and organic forms, and in dissolved and gaseous states. The form of arsenic in water depends on Eh, pH, organic content, suspended solids, dissolved oxygen, and other variables. Arsenic in water exists primarily as a dissolved ionic species; particulates account for less than 1% of the total measurable arsenic. Arsenic is rarely found in water in the elemental state (0), and is found in the −3 state only at extremely low Eh values. Common forms of arsenic encountered in water are arsenate, arsenite, methanearsonic acid, and DMA. The formation of inorganic pentavalent arsenic, the most common species in water, is favored under conditions of high dissolved oxygen, basic pH, high Eh, and reduced content of organic material; reverse conditions usually favor the formation of arsenites and arsenic sulfides, although some arsenite is attributed to biological activity. Water temperature seems to affect arsenic species composition in the estuary of the River Beaulieu in the United Kingdom, where reduced and methylated species predominate during warmer months and inorganic As^{+5} predominates during the colder months; the appearance of methylated arsenicals during the warmer months is attributed both to bacterial and abiotic release from decaying plankton and to grazing by zooplankton. Also contributing to higher water or mobile levels are the natural levels of polyvalent anions, especially phosphate species. Phosphate, for example, displaces arsenic held by humic acids, and it sorbs strongly on the hydrous oxides of arsenates.

Marine algae transform arsenate into non-volatile methylated arsenic compounds such as methanearsonic acid and DMA. Freshwater algae and macrophytes, like marine algae, synthesize lipid-soluble arsenic compounds and do not produce volatile methylarsines. Terrestrial plants preferentially accumulate arsenate over arsenite by a factor of about 4. Phosphate inhibits arsenate uptake by plants, but not the reverse. The mode of toxicity of arsenate in plants is to partially block protein synthesis and interfere with protein phosphorylation – a process that is prevented by phosphate.

Physical processes play a key role in governing arsenic bioavailability in aquatic environments. For example, arsenates are readily sorbed by colloidal humic material under conditions of high organic content, low pH, low phosphate, and low mineral content. Arsenates also co-precipitate with, or adsorb on, hydrous iron oxides and form insoluble precipitates with calcium, sulfur, aluminum, and barium compounds. Removal of arsenic from seawater by iron hydroxide scavenging seems to be a predominant factor in certain estuaries. The process involves both As^{+3} and As^{+5} and results in a measurable increase in arsenic levels in particulate matter, especially at low salinities. Arsenic sulfides are comparatively insoluble under conditions prevalent in anaerobic, aqueous and sedimentary media containing hydrogen sulfide; accordingly, these compounds may accumulate as precipitates and thus remove arsenic from the aqueous environment. In the absence of hydrogen sulfide, these sulfides decompose within several days to form arsenic oxides, sulfur, and hydrogen sulfide.

In reduced environments, such as sediments, arsenate is reduced to arsenite and methylated to methylarsinic acid or dimethylarsinic acids: these compounds may be further methylated to trimethylarsine or reduced to dimethylarsine, and may volatilize to the atmosphere where oxidation reactions result in the formation of DMA. Arsenates are more strongly adsorbed to sediments than are other arsenic forms, the adsorption processes depending strongly on arsenic concentration, sediment characteristics, pH, and ionic concentration of other compounds. An important mechanism of arsenic adsorption onto lake sediments involves the interaction of anionic arsenates and hydrous iron oxides. Evidence suggests that arsenic is incorporated into sediments at the time of hydrous oxide formation, rather than by adsorption onto existing surfaces. Arsenic concentrations in lake sediments are also correlated with manganese; hydrous manganese oxides – positively charged for the adsorption of Mn^{+2} ions – play a significant role in arsenic adsorption onto the surface of lake sediments. The mobility of arsenic in lake sediments and its release to the overlying water is partly related to seasonal changes. In areas that become stratified in summer, arsenic released from sediments accumulates in the hypolimnion until turnover, when it is mixed with epilimnetic waters; this mixing may result in a 10–20% increase in arsenic concentration. Microorganisms (including four species of fungi) in lake sediments oxidized inorganic As^{+3} to As^{+5} and reduced inorganic As^{+5} to As^{+3} under aerobic conditions; under anaerobic conditions, only reduction was observed. Inorganic arsenic can be converted to organic alkyl arsenic acids and methylated arsines under anaerobic conditions by fungi, yeasts, and bacteria – although biomethylation may also occur under aerobic conditions.

Most arsenic investigators now agree on the following points: (1) arsenic may be absorbed by ingestion, inhalation, or through permeation of the skin or mucous membranes; (2) cells accumulate arsenic by using an active transport system normally used in phosphate transport; (3) arsenicals are readily absorbed after ingestion, most being rapidly excreted in the urine during the first few days, or at most a week (the effects seen after long-term exposure are probably a result of continuous daily exposure, rather than of accumulation); (4) the toxicity of arsenicals conforms to the following order, from greatest to least toxicity: arsines > inorganic arsenites > organic trivalent compounds (arsenoxides) > inorganic arsenates > organic pentavalent compounds > arsonium compounds > elemental arsenic; (5) solubility in water and body fluids appear to be directly related to toxicity (the low toxicity of elemental arsenic is attributed to its virtual insolubility in water and body fluids, whereas the highly toxic arsenic trioxide, for example, is soluble in water to 12.0 g/L at 0°C, 21.0 g/L at 25°C, and 56.0 g/L at 75°C); and (6) the mechanisms of arsenical toxicity differ considerably among arsenic species, although signs of poisoning appear similar for all arsenicals.

The primary toxicity mode of inorganic As^{+3} is through reaction with sulfhydryl (SH) groups of proteins and subsequent enzyme inhibition; inorganic pentavalent arsenate does not react as readily as As^{+3} with SH groups, but may uncouple oxidative phosphorylation. Inorganic As^{+3} interrupts oxidative metabolic pathways and sometimes causes morphological changes in liver mitochondria. Arsenite in vitro reacts with protein-SH groups to inactivate enzymes such as dihydrolipoyl dehydrogenase and thiolase, producing inhibited oxidation of pyruvate and beta-oxidation of fatty acids. Inorganic As^{+3} may also exert toxic effects by the reaction of arsenous acid (HAsO) with the SH groups of enzymes. In the first reaction, arsenous acid is reduced to arsonous acid ($AsOH_2$), which then condenses to either monothiols or dithiols to yield dithioesters of arsonous acid. Arsonous acid may then condense with enzyme SH groups to form a binary complex.

Methylation to methylarsonic acid [$(CH_3)_2AsO_3H_2$] and dimethylarsinic acid [$(CH_3)_2AsO_2H$] is usually the major detoxification mechanism for inorganic pentavalent arsenates and trivalent arsenites in mammals. Methylated arsenicals rapidly clear from all tissues except perhaps the thyroid. Methylated arsenicals are probably common in nature. Methylation of arsenic (unlike methylation of mercury) greatly reduces toxicity and is a true detoxification process. Before methylation (which occurs largely in the liver), As^{+5} is reduced to As^{+3} – the kidney being an important site for this transformation. Arsenate reduction and subsequent methylation is rapid: both arsenite and dimethylarsinate were present in hamster (*Cricetus* sp.) plasma only 12 min post-injection of inorganic As^{+5}. Demethylation of methylated arsenicals formed in vivo has not yet been reported. Although terrestrial biota usually contain much lower total concentrations of arsenic than do marine biota, metabolism of arsenic is similar in marine and terrestrial systems.

Toxic effects of organoarsenicals are exerted by initial metabolism to the trivalent arsenoxide form, and then by reaction with SH groups of tissue proteins and enzymes to form an arylbis (organylthio) arsine. This form, in turn, inhibits oxidative degradation of carbohydrates and decreases cellular ATP, the energy storage molecule of the cell. Among the organoarsenicals, those physiologically most injurious are methylarsonous acid [$CH_3As(OH)_2$] and dimethylarsinous acid [$(CH_3)_2AsOH$]. The enzyme inhibitory forms of organoarsenicals (arsonous acid) are formed from arsenous acid and the corresponding arsonic acids by a wide variety of enzymes and subcellular particles. Organoarsenicals used as growth promoters and drugs are converted to more easily excreted (and sometimes more toxic) substances, although most organoarsenicals are eliminated without being converted to inorganic arsenic or to dimethylarsinic acids.

2.4 Essentiality, Synergism, and Antagonism

Limited data are available on the beneficial, protective, and essential properties of arsenic, and on its interactions with other chemicals. Arsenic, apparently behaves more like an environmental contaminant than as a nutritionally essential mineral. Nevertheless, low doses ($<2.0\ \mu g$/daily) of arsenic stimulated growth and metamorphosis in tadpoles, and increased viability and cocoon yield in silkworm caterpillars. Arsenic deficiency has been observed in rats: signs include rough hair coat, low growth rate, decreased hematocrit, increased fragility of red cells, and enlarged spleen. Similar results have been documented in goats and pigs fed diets containing <0.05 mg As/kg. In these animals, reproductive performance was impaired, neonatal mortality was increased, birth weight was lower, and weight gains in second-generation animals were decreased; these effects were not evident in animals fed diets containing 0.35 mg As/kg.

The use of phenylarsonic feed additives to promote growth in poultry and swine and to treat specific diseases does not seem to constitute a hazard to the animal or to its consumers. Animal deaths and elevated tissue arsenic residues occur only when the arsenicals are fed at excessive dosages for long periods. Arsenic can be detected at low levels in tissues of animals fed organoarsenicals, but it is rapidly eliminated when the arsenicals are removed from the feed for the required 5-day period before marketing.

Selenium and arsenic are antagonists in several animal species. Dietary arsenic, as arsenate, alleviates the toxic effects of selenium, as seleno-DL-methionine, on mallard (*Anas platyrhynchos*) reproduction, duckling growth, and survival. Mallard ducklings fed arsenate in the diet at 200.0 mg As/kg ration were protected against the toxic effects of 60.0 mg Se/kg ration as selenomethionine, including selenium-induced mortality, impaired growth, hepatic lesions, and enzyme disruptions. In rats, dogs, swine, cattle, and poultry, arsenic protects against selenium poisoning if arsenic is administered in the drinking water and selenium through the diet. Inorganic arsenic compounds decrease the toxicity of inorganic selenium compounds by increasing biliary excretion. However, in contrast to antagonism shown by inorganic arsenic–inorganic selenium mixtures, the toxic effects of naturally methylated selenium compounds (trimethylselenomium chloride and dimethyl selenide) are markedly enhanced by inorganic arsenicals.

The toxic effects of arsenic can be counteracted with (1) saline purgatives, (2) various demulcents that coat irritated gastrointestinal mucous membranes, (3) sodium thiosulfate, and (4) mono- and dithiol-containing compounds and 2,3-dimercaptopropanol. Arsenic uptake in rabbit intestine is inhibited by phosphate, casein, and various metal-chelating agents. Mice and rabbits are significantly protected against sodium arsenite intoxication by *N*-(2,3-dimercaptopropyl)phthalamidic acid. Conversely, the toxic effects of arsenite are potentiated by excess dithiols, cadmium, and lead, as evidenced by reduced food efficiency and disrupted blood chemistry in rodents.

Arsenic effectively controls filariasis in cattle; new protective uses are under investigation. The control of parasitic nematodes (*Parafilaria bovicola*) in cattle was successful after 30 weekly treatments in plungement dips containing 1600 mg As_2O_3/L; however, the muscle of treated cattle contained up to 1.3 mg As/kg, or 12 times the amount in controls. Existing anionic organic arsenicals used to control tropical nematode infections in humans have sporadic and unacceptable lethal side effects. Cationic derivatives have been synthesized in an attempt to avoid the side effects and have been examined for effects on adult nematodes (*Brugia pahangi*) in gerbils (*Meriones unguiculatus*). All arsenicals were potent filaricides; the most effective compounds tested killed 95% of adult *B. pahangi* after five daily subcutaneous injections of 3.1 mg As/kg body weight (BW).

Animals previously exposed to sublethal levels of arsenic may develop tolerance to arsenic on re-exposure. Although the mechanism of this process is not fully understood, it probably includes the efficiency of in vivo methylation processes. For example, resistance to toxic doses of As^{+3} or As^{+5} increases in mouse fibroblast cells pretreated with a low As^{+3} concentration. Also, growth is better in arsenic-conditioned mouse cells in the presence of arsenic than in previously unexposed cells, and inorganic arsenic is more efficiently methylated. In vivo biotransformation and excretion of inorganic arsenic as MMA and DMA have been demonstrated in a number of mammalian species, including man. It seems that cells may adapt to arsenic by increasing the biotransformation rate of the element to methylated forms, such as MMA and DMA. Pretreatment of ovary cells of Chinese hamster (*Cricetus* spp.) ovary cells with sodium arsenite provided partial protection against adverse effects of methyl methanesulfonate (MMS), and may even benefit the MMS-treated cells; however, posttreatment dramatically increases the cytotoxic, clastogenic, and mitotic effects induced by MMS.

Although arsenic is not an essential plant nutrient, small yield increases have sometimes been observed at low soil arsenic levels, especially for tolerant crops such as potatoes, corn, rye, and wheat. Arsenic phytotoxicity of soils is reduced with increasing lime, organic matter, iron, zinc, and phosphates. In most soil systems, the chemistry of As becomes the chemistry of arsenate; the estimated half-time of arsenic in soils is about 6.5 years, although losses of 60% in 3 years and 67% in 7 years have been reported. Additional research is warranted on the role of arsenic in crop production, and in nutrition, with special reference to essentiality for aquatic and terrestrial wildlife.

2.5 Concentrations in Field Collections

In abundance, arsenic ranks twentieth among the elements in the earth's crust (1.5–2.0 mg/kg), fourteenth in seawater, and twelfth in the human body. It occurs in various forms, including inorganic and organic compounds and trivalent and pentavalent states. In aquatic environments, higher arsenic concentrations are reported in hot springs, in groundwater from areas of thermal activity or in areas containing rocks with high arsenic content, and in some waters with high dissolved salt content. Most of the other elevated values reported in lakes, rivers, and sediments are probably due to anthropogenic sources, which include smelting and mining operations; combustion of fossil fuel; arsenical grasshopper baits; synthetic detergent and sewage sludge wastes; and arsenical defoliants, herbicides, and pesticides. Most living organisms normally contain measurable concentrations of arsenic, but except for marine biota, these are usually less than 1.0 mg/kg fresh weight (FW). Marine organisms, especially crustaceans, may contain more than 100.0 mg As/kg dry weight (DW), usually as arsenobetaine, a water-soluble organoarsenical that poses little risk to the organism or its consumer. Plants and animals collected from naturally arseniferous areas or near anthropogenic sources may contain significantly elevated tissue residues of arsenic.

2.5.1 Abiotic Materials

Arsenic is a major constituent of at least 245 mineral species, of which arsenopyrite is the most common. In general, background concentrations of arsenic range from 0.2 to 15.0 mg/kg in the lithosphere, 0.005–0.1 μg/m^3 in air, $<$10.0 μg/L in water, and $<$15.0 mg/kg in soil. The commercial use and production of arsenic compounds have raised local concentrations in the environment far above the natural background concentrations.

Weathering of rocks and soils adds about 45,000 tons of arsenic to the oceans annually, accounting for less than 0.01 mg/L on a global basis. However, arsenic inputs to oceans increased during the past century both from natural sources and as a result of industrial use, agricultural and deforestation activities, emissions from coal and oil combustion, and loss during mining of metal ores. If present activities continue, arsenic concentrations in oceanic surface waters may increase overall by about 2% in the year 2000 and possibly beyond, with most of the increased burden in estuaries and coastal oceans – e.g., Puget Sound, Washington; the Tamar, England; and the Tejo, Portugal. Estimates of the residence times of arsenic are 60,000 years in the ocean and 45 years in a freshwater lake. In the hydrosphere, inorganic arsenic occurs predominantly as As^{+5} in surface water, and significantly as As^{+3} in groundwater containing high levels of total arsenic. The main organic species in freshwater are methylarsonic acid and DMA, and these are usually present in lower concentrations than inorganic arsenites and arsenates. Total arsenic concentrations in surface water and groundwater are usually $<$10.0 μg/L; in certain areas, however, levels above 1.0 mg/L have been recorded.

In air, most arsenic particulates consist of inorganic arsenic compounds, often as As^{+3}. Burning of coal and arsenic-treated wood, and smelting of metals are major sources of atmospheric arsenic contamination (i.e., $>$1.0 μg/m^3); in general, atmospheric arsenic levels are higher in winter, due to increased use of coal for heating.

The main carrier of arsenic in rocks and in most types of mineral deposits is iron pyrite (FeS_2) which may contain $>$2000.0 mg/kg of arsenic. In localized areas, soils are contaminated by arsenic oxide fallout from smelting ores (especially sulfide-containing ores) and combustion of arsenic-rich coal.

Arsenic in lacustrine sediment columns is subject to control by digenetic processes and adsorption mechanisms, as well as anthropogenic influences. For example, elevated levels of arsenic in surface or near surface sediments may have several causes, including natural processes (Loch Lomond, Scotland) and human activities such as smelting (Lake Washington, Washington; Kelly Lake, Ontario, Canada), manufacture of arsenical herbicides (Brown's Lake, Wisconsin), and mining operations (Northwest Territories, Canada; Clark Fork River, Montana). Elevated levels of arsenic in sediments of the Wailoa River, Hawaii, are the result of As_2O_3 applied as an anti-termite agent between 1932 and 1963.

These elevated levels are found mainly in anaerobic sediment regions where the chemical has been relatively undisturbed by activity. Low levels of arsenic in the biota of the Wailoa River estuary suggest that arsenic is trapped in the anaerobic sediment layers.

Arsenic geochemistry in Chesapeake Bay, Maryland, depends on anthropogenic inputs and phytoplankton species composition. Inputs of anthropogenic arsenic into Chesapeake Bay are estimated at 100 kg daily, or 39 tons/year – probably from sources such as unreported industrial discharges, use of arsenical herbicides, and from wood preservatives. The chemical form of the arsenic in solution varies seasonally and along the axis of the Bay. Arsenic is present only as arsenate in winter, but substantial quantities of reduced and methylated forms are present in summer in different areas. The forms and distribution patterns of arsenic during the summer suggest that separate formation processes exist. Arsenite, present in low salinity regions, may have been formed by chemical reduction in anoxic, subsurface waters and then mixed into the surface layer. Methylated arsenicals are highly correlated with landing crops of algae. One particular form, methylarsonate, is significantly correlated with the dominant alga *Chroomonas*. Since both arsenic reactivity and toxicity are altered by transformation of chemical form, the observed variations in arsenic speciation have considerable geochemical and ecological significance.

2.5.2 Biological Samples

Background arsenic concentrations in living organisms are usually <1.0 mg/kg FW in terrestrial flora and fauna, birds, and freshwater biota. These levels are higher, sometimes markedly so, in biota collected from mine waste sites, arsenic-treated areas, near smelters and mining areas, near areas with high geothermal activity, and near manufacturers of arsenical defoliants and pesticides. For example, bloaters (*Coregonus hoyi*) collected in Lake Michigan near a facility that produced arsenical herbicides consistently had the highest (1.5–2.9 mg As/kg FW whole body) arsenic concentrations measured in freshwater fishes in the United States between 1976 and 1984. Marine organisms normally contain arsenic residues of several to more than 100.0 mg/kg DW; however, as will be discussed later, these concentrations present little hazard to the organism or its consumers.

Arsenic concentrations in tissues of marine biota show a wide range of values, being highest in lipids, liver, and muscle tissues, and varying with the age of the organism, geographic locale, and proximity to anthropogenic activities. In general, tissues with high lipid content contained high levels of arsenic. Crustacean tissues sold for human consumption and collected in U.S. coastal waters usually contained 3.0–10.0 mg As/kg FW, or 1.0–100.0 mg/kg DW, and were somewhat higher than those reported for finfish and molluscan tissues. Marine finfish tissues usually contained 2.0–5.0 mg As/kg FW. However, postmortem reduction of As^{+5} to As^{+3} occurs rapidly in fish tissues, suggesting a need for additional research in this area. Maximum arsenic values recorded in elasmobranchs (mg/kg FW) were 30.0 in the muscle of a shark (*Mustelus antarcticus*) and 16.2 in the muscle of a ray (*Raja* sp.). The highest arsenic concentration recorded in a marine mammal, 2.8 mg As/kg FW lipid, was from a whale.

Arsenic appears to be elevated in marine biota because of their ability to accumulate arsenic from seawater or food sources, and not due to localized pollution. The great majority of the arsenic in marine organisms exists as water-soluble and lipid-soluble organoarsenicals that include arsenolipids, arsenosugars, arsenocholine, arsenobetaine $[(CH_3)_3AsCH_2COOH]$, monomethylarsonate $[CH_3AsO(OH)_2]$, demethylarsinate $[(CH_3)_2AsO(OH)]$, as well as other forms. There is no convincing hypothesis to account for the existence of all the various forms of organoarsenicals found in marine organisms. One suggested hypothesis is that each form involves a single anabolic–catabolic pathway concerned with the synthesis and turnover of phosphatidylcholine. Arsenosugars (arsenobetaine precursors) are the dominant arsenic species in brown kelp (*Ecklonia radiata*), giant clam (*Tridacna maxima*), shrimp (*Pandalus borealis*), and ivory shell (*Buccinum striatissimum*). For most marine species, however, there is general agreement that arsenic exists primarily as arsenobetaine, a water-soluble organoarsenical that has been identified in tissues of western rock lobster (*Panulirus cygnus*), American lobster (*Homarus americanus*), octopus (*Paroctopus* sp.), sea cucumber (*Stichopus japonicus*), blue shark (*Prionace glauca*), sole (*Limanda* sp.), squid (*Sepioteuthis australis*), prawn (*Penaeus latisulcatus*), scallop (*Pecten alba*), and many other species including teleosts, mollusks, tunicates, and crustaceans. Degradation of arsenobetaine in muscle and liver of the star spotted shark (*Mustelus manazo*) to inorganic arsenic occurs in a natural environment and suggests that arsenobetaine biotransformed from inorganic arsenic in seawater is degraded to original inorganic arsenic. About 13% of the arsenobetaine in shark muscle and 4% in liver were degraded to inorganic arsenic within 40 days. The potential risks associated with consumption of seafoods containing arsenobetaine seem to be minor. The chemical was not mutagenic in the bacterial *Salmonella typhimurium* assay (Ames test), had no effect on metabolic inhibition of Chinese hamster ovary cells at 10,000.0 mg/L, and showed no synergism or antagonism on the action of other contaminants. Arsenobetaine was not toxic to mice at oral doses of 10,000.0 mg/kg BW during a 7-day observation period, and it was rapidly absorbed from the gastrointestinal tract and rapidly excreted in urine without metabolism, owing to its high polar and hydrophilic characteristics.

Shorebirds (seven species) wintering in the Corpus Christi, Texas, area contained an average of 0.3 mg As/kg FW in livers (maximum of 1.5 mg/kg), despite the presence of smelters and the heavy use of arsenical herbicides and defoliants; these values probably reflect normal background concentrations. Similar arsenic levels were reported in livers of brown pelicans (*Pelecanus occidentalis*) collected from South Carolina. Bone arsenic concentrations in 23 species of birds collected in southwestern Russia during 1993–95 ranged from 0.1 to 1.7 mg As/kg DW; arsenic concentrations were similar for terrestrial and aquatic birds, and for urban and rural environments. The highest arsenic concentration recorded in seemingly unstressed coastal birds was 13.2 mg/kg FW lipids. This tends to corroborate the findings of others that, arsenic concentrates in lipid fractions of marine plants, invertebrates, and higher organisms. An abnormal concentration of 16.7 mg As/kg FW was recorded in the liver of an osprey (*Pandion haliaetus*) from the Chesapeake Bay region. This bird was alive but weak, with serious histopathology including the absence of subcutaneous fat, presence of serous fluid in the pericardial sac, and disorders of the lung and kidney. The bird died shortly after collection. Arsenic concentrations in the livers of other ospreys collected in the same area usually were <1.5 mg As/kg FW. Chicks of the avocet (*Recurvirostra americana*) reared from eggs taken near arsenic-contaminated ponds in California (127.0–1100.0 μg As/L) had reduced hatch, and impaired growth and immune function when compared to chicks from a reference site (29.0 μg As/L); other contaminants present included boron and selenium, although authors concluded that arsenic played a significant role in observed effects.

Arsenic concentrations in tissues of small mammals (field mouse, *Apodemus sylvaticus*; bank vole, *Clethrionomys glareolus*; field vole, *Microtus agrestis*; common shrew, *Sorex araneus*) from the vicinity of an English arsenic refinery were usually less than 1.0 mg/kg FW and did not reflect arsenic levels of the surrounding soil or vegetation. In general, mean arsenic concentrations were highest in spleen, followed in descending order by bone, heart, kidney, brain, muscle, and liver. For human adults, seafood contributes 74–96% of the total daily arsenic intake, and rice and rice cereals most of the remainder; for infants, 41% of the estimated total arsenic intakes arise from seafood and 34% from rice and rice cereals. Effective biomarkers of arsenic exposure in humans include elevated arsenic concentrations in hair, fingernails, and especially urine. In Taiwan, urinary levels of inorganic and organic arsenic metabolites are associated with previous exposure to high-arsenic artesian well water. Humans that had previously been exposed to high-arsenic drinking water and had switched to tap water containing <50.0 μg total As/L had – after 30 years – elevated levels of arsenic in urine, especially MMA and DMA.

2.6 Lethal and Sublethal Effects

As will be discussed later, most authorities agree on 10 points: (1) inorganic arsenicals are more toxic than organic arsenicals, and trivalent forms are more toxic than pentavalent forms; (2) episodes of arsenic poisoning are either acute or subacute; cases of chronic arsenosis are rarely encountered, except in humans; (3) sensitivity to arsenic is greatest during the early developmental stages; (4) arsenic can traverse placental barriers; as little as 1.7 mg As^{+5}/kg BW at critical stages of hamster embryogenesis, for example, can produce fetal death and malformation; (5) biomethylation is the preferred detoxification mechanism for inorganic arsenicals; (6) arsenic is bioconcentrated by organisms, but not biomagnified in the food chain; (7) in soils, depressed crop yields were recorded at 3.0–28.0 mg of water-soluble As/L, or about 25.0–85.0 mg total As/kg soil; adverse effects on vegetation were recorded at concentrations in air >3.9 μg As/m^3; (8) some aquatic species were adversely affected at water concentrations of 19.0–48.0 μg As/L, or 120.0 mg As/kg in the diet, or tissue residue of 1.3–5.0 mg As/kg FW; (9) sensitive species of birds died following single oral doses of 17.4–47.6 mg As/kg BW; and (10) adverse effects were noted in mammals at single oral doses of 2.5–33.0 mg As/kg BW, at chronic oral doses of 1.0–10.0 mg As/kg BW, and at feeding levels of 50.0 mg – and sometimes only 5.0 mg – As/kg in the diet.

The literature emphasizes that arsenic metabolism and toxicity vary greatly between species and that its effects are significantly altered by numerous physical, chemical, and biological modifiers. Adverse health effects, for example, may involve respiratory, gastrointestinal, cardiovascular, and hematopoietic systems, and may range from reversible effects to cancer and death, depending partly on the physical and chemical forms of arsenic tested, the route of administration, and the dose.

2.6.1 Carcinogenesis, Mutagenesis, and Teratogenesis

Arsenic is a known human carcinogen. Epidemiological studies show that an increased risk of cancers in the skin, lung, liver, lymph, and hematopoietic systems of humans is associated with exposure to inorganic arsenicals. These increased cancer risks are especially prevalent among smelter workers and in those engaged in the production and use of arsenical pesticides where atmospheric levels exceed 54.6 μg As/m^3. Skin tumors, mainly of low malignancy, have been reported after consumption of arsenic-rich drinking waters; a total dose of several grams, probably as As^{+3} is usually required for the development of skin tumors. High incidences of skin cancer and hyperpigmentation were noted among several population groups, especially Taiwanese and Chileans, consuming water containing more than 0.6 mg As/L; the frequency of cancer was highest among people over age 60 who demonstrated symptoms of chronic arsenic poisoning. In some areas of India, however, as many as 60% of the children between age 4 and 10 years had arsenical melanosis because of the exceptionally high levels of arsenic in the drinking water. Elimination of arsenic in drinking water decreased the mortality incidence of arsenic-related cancers of the liver, lung, kidney, and skin in communities where Blackfoot disease is endemic.

Arsenic reportedly inhibits cancer formation in species having a high incidence of spontaneous cancers. In fact, arsenic may be the only chemical for which there is sufficient evidence for carcinogenicity in humans but not in other animals. In general, animal carcinogenicity tests with inorganic and organic arsenicals have been negative, even when the chemicals were administered at or near the highest tolerated dosages for long periods. Most studies of arsenic carcinogenesis in animals were presumably of insufficient duration to simulate conditions in long-lived species such as humans. However, mice developed leukemia and lymphoma after 20 subcutaneous injections of 0.5 mg As^{+5}/kg BW: 46% of the experimental group developed these signs vs. none of the controls. And mice given 500 μg As/L, as sodium arsenate, in the drinking water for lifetime exposures developed tumors in the lung, liver, and GI tract. Pulmonary tumorigenicity has been demonstrated in hamsters administered calcium arsenate intratracheally. Inorganic arsenic interacts with benzo[a]pyrene in the induction of lung adenocarcinomas in hamsters. Cacodylic acid and other organoarsenicals are not carcinogenic, but may be mutagenic at very high doses. Monomethylarsonic acid (MMA) and dimethlarsinic acid (DMA) are primary metabolites of inorganic arsenic, a known human carcinogen. However, the rapid elimination (Tb1/2 of 2 h) and low retention (<2%) of MMA and DMA explain, in part, their low acute toxicity and cancer risk.

Several inorganic arsenic compounds are weak inducers of chromosomal aberrations, sister chromatid exchange, and in vitro transformation of mammalian and piscine cells; however, there is no conclusive evidence that arsenic causes point mutations in any cellular system. Studies with bacteria suggest that arsenite is a comutagen, or may inhibit DNA repair.

Arsenic is a known teratogen in several classes of vertebrates, and has been implicated as a cause of birth defects in humans. Specific developmental malformations have been produced experimentally in mammals using inorganic As^{+3} or As^{+5} either through a single dose or a continuous dose during embryogenesis. Teratogenic effects are initiated no later than 4 h post-administration of arsenic; fetal abnormalities are primarily neural tube defects, but may also include protruding eyes, incomplete development of the skull, abnormally small jaws, and other skeletal anomalies. Inorganic As^{+3} and As^{+5}, but not organoarsenicals, cross placental barriers in many species of mammals, and result in fetal deaths and malformations. Studies with hamsters, for example, showed that sodium arsenite can induce chromatid breaks and chromatid exchanges in Chinese hamster ovary cells in a dose-dependent manner. In an earlier study, As^{+3} was about 10 times more potent than As^{+5} in causing transformations. The birth defects were most pronounced in golden hamsters exposed to As^{+5} during the 24-h period of critical embryogenesis – day 8 of gestation – when 1.7 mg As^{+5}/kg BW induced neural tube defects in about 90% of the fetuses. Hamsters exposed to As^{+5} and heat stress (39°C for 50 min) on day 8 of gestation produced a greater percentage of malformed offspring (18–39%) than did hamsters exposed to As^{+5} alone (4–8%).

2.6.2 Terrestrial Plants and Invertebrates

In general, arsenic availability to plants is highest in coarse-textured soils having little colloidal material and little ion exchange capacity, and lowest in fine-textured soils high in clay, organic material, iron, calcium, and phosphate. To be absorbed by plants, arsenic compounds must be in a mobile form in the soil solution. Except for locations where arsenic content is high, e.g., around smelters, the accumulated arsenic is distributed throughout the plant body in nontoxic amounts. For most plants, a significant depression in crop yields was evident at soil arsenic concentrations of 3.0–28.0 mg/L of water-soluble arsenic and 25.0–85.0 mg/kg of total arsenic. Yields of peas (*Pisum sativum*), a sensitive species, were decreased at 1.0 mg/L of water-soluble arsenic or 25.0 mg/kg of total soil arsenic; rice (*Oryza sativa*) yields were decreased by 75% at 50.0 mg/L of disodium methylarsonate in silty loam; and soybeans (*Glycine max*) grew poorly when residues exceeded 1.0 mg As/kg. Forage plants grown in soils contaminated with up to 80.0 mg total As/kg from arsenical orchard sprays contained up to 5.8 mg As/kg DW; however, these plants were considered nonhazardous to grazing ruminants.

Attention was focused on inorganic arsenical pesticides after accumulations of arsenic in soils eventually became toxic to several agricultural crops, especially in former orchards and cotton fields. Once toxicity is observed, it persists for several years even if no additional arsenic treatment is made. Poor crop growth was associated with bioavailability of arsenic in soils. For example, alfalfa (*Medicago sativa*) and barley (*Hordeum vulgare*) grew poorly in soils containing only 3.4–9.5 mg As/kg, provided the soils were acidic, lightly textured, low in phosphorus and aluminum, high in iron and calcium, and contained excess moisture. Use of inorganic arsenical herbicides, such as calcium arsenate, to golf course turfs for control of fungal blight sometimes exacerbates the disease. The use of arsenicals on Kentucky bluegrass (*Poa pratensis*) is discouraged under conditions of high moisture and root stress induced by previous arsenical applications.

Methylated arsenicals, whether herbicides or defoliants, are sprayed on plant surfaces. They can reach the soil during application or can be washed from the plants. Additional arsenic enters soils by exchange from the roots or when dead plant materials decay. Cacodylic acid and sodium cacodylate are nonselective herbicides used in at least 82 products to eliminate weeds and grasses around trees and shrubs, and to eradicate vegetation from rights-of-way and other noncrop areas. Normal application rates of various organoarsenicals for crop and noncrop purposes rarely exceed 5 kg/ha. At recommended treatment levels, organoarsenical soil residues are not toxic to crops, and those tested (soybean, beet, wheat) were more resistant to organoarsenicals than to comparable levels of inorganic arsenicals.

Air concentrations up to 3.9 μg As/m^3 near gold mining operations were associated with adverse effects on vegetation; higher concentrations of 19.0–69.0 μg As/m^3, near a coal-fired power plant in Czechoslovakia, produced measurable contamination in soils and vegetation in a 6-km radius.

The phytotoxic actions of inorganic and organic arsenicals are different and physical processes significantly modify each. The primary mode of action of arsenite in plants is inhibition of light activation, probably through interference with the pentose phosphate pathway. Arsenites penetrate the plant cuticle to a greater degree than arsenates. One of the first indications of plant injury by sodium arsenite is wilting caused by loss of turgor, whereas stress due to sodium arsenate does not involve rapid loss of turgor. Organoarsenicals, such as cacodylic acid, enter plants mostly by absorption of sprays; uptake from the soil contributes only a minor fraction. The phytotoxicity of organoarsenical herbicides is characterized by chlorosis, cessation of growth, gradual browning, dehydration, and death. In general, plants cease to grow and develop after the roots have absorbed much arsenic. Plants can absorb arsenic through the roots and foliage, although translocation is species dependent. Concentrations of arsenic in plants correlate highly and consistently with water extractable soil arsenic, and usually poorly with total soil arsenic. For example, concentrations of arsenic in corn (*Zea mays*) grown in calcareous soils for 25 days were significantly correlated with the soil water extractable arsenic fraction, but not other fractions; extractable phosphorus was correlated positively to both arsenic in corn and to the water-soluble arsenic fraction. In the moss *Hylocomium splendens*, arsenate accumulation from solution was through living shoots, optimum uptake being between pH 3 and 5. Some plants, such as beets (*Beta vulgaris*) accumulated arsenic more readily at elevated temperatures, but the addition of phosphate fertilizers markedly depressed uptake.

Soils amended with arsenic-contaminated plant tissues were not measurably affected in CO_2 evolution and nitrification, suggesting that the effects of adding arsenic to soils do not influence the decomposition rate of plant tissues by soil microorganisms. The half-life of cacodylic acid is about 20 days in untreated soils and 31 days in arsenic-amended soils. Estimates of the half-time of inorganic arsenicals in soils are much longer, ranging from 6.5 years for arsenic trioxide to 16 years for lead arsenate.

Data on arsenic effects to soil biota and insects are limited. In general, soil microorganisms are capable of tolerating and metabolizing relatively high concentrations of arsenic. This adaptation seems usually to be due to decreased permeability of the microorganism to arsenic. Tolerant soil microbiota can withstand concentrations up to 1600.0 mg/kg; however, growth and metabolism were reduced in sensitive species at 375.0 mg As/kg and, at 150.0–160.0 mg As/kg, soils were devoid of earthworms and showed diminished quantities of bacteria and protozoans. Earthworms (*Lumbricus terrestris*) held in soils containing 40–100 mg As^{+5}/kg DW soil for 8–23 days had significantly reduced survival, especially among worms held in soils less than 70 mm in depth when compared to worms held at 500–700 mm; survivors had negligible arsenic residues; dead worms had higher concentrations and suggest that arsenic homeostasis breaks down after death. Honeybees (*Apis mellifera*) that were killed accidentally by sprayed As^{+3} contained 4.0–5.0 μg As per bee, equivalent to 21.0–31.0 mg/kg BW. Larvae of the western spruce budworm (*Choristoneura occidentalis*) continued to feed on As^{+3}-contaminated vegetation until a threshold level of about 2300–3300 mg As/kg DW whole larvae was reached; death then sometimes occurred. Larvae that had accumulated sufficient energy reserves completed the first stage of metamorphosis, but developed into pupae of subnormal weight; larvae containing <2600.0 mg As^{+3}/kg ultimately developed into adults of less than normal weight, and some containing >2600.0 mg/kg DW died as pupae.

2.6.3 Aquatic Biota

Adverse effects of arsenicals on aquatic organisms have been reported at concentrations of 19.0–48.0 μg/L in water, 33.0 mg/kg in diets, and 1.3–5 mg/kg FW in tissues. The most sensitive of the aquatic species tested that showed adverse effects were three species of marine algae, which showed reduced growth in the range of 19.0–22.0 μg As^{+3}/L; developing embryos of the narrow-mouthed toad (*Gastrophryne carolinensis*), of which 50% were dead or malformed in 7 days at 40.0 μg As^{+3}/L; and a freshwater alga (*Scenedesmus obliquis*), in which growth was inhibited 50% in 14 days at 48.0 μg As^{+5}/L. Chronic studies with mass cultures of natural phytoplankton communities exposed to low levels of arsenate (1.0–15.2 μg/L) showed that As^{+5} differentially inhibits certain plants, causing a marked change in species composition, succession, and predator–prey relations; the significance of these changes on carbon transfer between trophic levels is unknown. Adverse biological effects have also been documented at water concentrations of 75.0–100.0 μg As/L. At 75.0 μg As^{+5}/L, growth and biomass in freshwater and marine algae was reduced; at 85.0–88.0 μg/L of As^{+5} or various methylated arsenicals, mortality was 10–32% in amphipods (*Gammarus pseudolimnaeus*) in 28 days; at 95.0 μg As^{+3}/L, marine red alga failed to reproduce sexually; and at 100.0 μg As^{+5}/L, marine copepods died and goldfish behavior was impaired.

Rainbow trout (*Oncorhynchus mykiss*) fed diets containing up to 90.0 mg As^{+5}/kg were slightly affected, but those given diets containing >120.0 mg As/kg (as As^{+3} or As^{+5}) grew poorly, avoided food, and failed to metabolize food efficiently; no toxic effects were reported over 8 weeks of exposure to diets containing 1600.0 mg/kg, as methylated arsenicals. Dietary disodium heptahydrate (DSA) is more toxic to juvenile rainbow trout than dietary arsenic trioxide, DMA, or arsanilic acid. Diets containing 55.0–60.0 mg As^{+5}/kg ration as DSA were associated with changes in the hepatobiliary system of juvenile rainbow trout after 12 weeks of feeding. The most sensitive indicator of DSA insult in juvenile rainbow trout was chronic inflammation of the gallbladder wall and found in 71% of trout exposed to 33.0 mg As/kg ration for 24 weeks and 100% of those exposed to 65.0 mg As/kg ration for 24 weeks; there was no damage at 13.0 mg/kg ration and lower in the 24-week study. The whole body arsenic concentrations in moribund rainbow trout poisoned by arsenate compounds in 11-week exposures ranged between 4.0 and 6.0 mg As/kg FW (vs. 2.0 mg/kg FW in controls) – and dead whole trout had 8.0–12.0 mg As/kg FW – suggesting that a critical arsenic body concentration is reached before death. In bluegills (*Lepomis macrochirus*), tissue residues of 1.35 mg As/kg FW in juveniles and 5.0 mg/kg in adults are considered elevated and potentially hazardous.

Toxic and other effects of arsenicals to aquatic life are significantly modified by numerous biological and abiotic factors. The LC50 values, for example, are markedly affected by water temperature, pH, Eh, organic content, phosphate concentration, suspended solids, and presence of other substances and toxicants, as well as arsenic speciation, and duration of exposure. In general, inorganic arsenicals are more toxic than organoarsenicals to aquatic biota, and trivalent species are more toxic than pentavalent species. Early life stages are most sensitive, and large interspecies differences are recorded, even among species closely related taxonomically.

Arsenic is accumulated from the water by a variety of organisms; however, there is no evidence of magnification along the aquatic food chain. In a marine ecosystem based on the alga *Fucus vesiculosus*, arsenate (7.5 μg As^{+5}/L) was accumulated by all biota. After 3 months, arsenic was concentrated most efficiently by *Fucus* (120.0 mg/kg DW in apical fronds) and filamentous algal species (30.0 mg/kg DW); little or no bioaccumulation occurred in invertebrates, although arsenic seemed to be retained by gastropods and mussels. In a simplified estuarine food chain, there was no significant increase in arsenic content of grass shrimp (*Palaemonetes pugio*) exposed to arsenate-contaminated food or to elevated water concentrations. In a freshwater food chain composed of algae, daphnids, and fish, water concentrations of 0.1 mg cacodylic acid/L produced residues (mg As/kg DW), after 48 h of 4.5 in algae and 3.9 in daphnids, but only 0.09 in fish. Microcosms of a Delaware Cordgrass (*Spartina alterniflora*) salt marsh exposed to elevated levels of As^{+5} showed that virtually all arsenic was incorporated into plant tissue or strongly sorbed to cell surfaces. Studies with radioarsenic and mussels (*Mytilus galloprovincialis*) showed that accumulation varied with nominal arsenic concentrations, tissue, age of the mussel, and temperature and salinity of the medium. Arsenate uptake increased with increasing arsenic concentration in the medium, but the response was not linear, accumulation being suppressed at higher external arsenic concentrations. Smaller mussels took up more arsenic than larger ones. In both size groups, arsenic was concentrated in the byssus and digestive gland. In general, arsenic uptake and loss increased at increasing temperatures. Uptake was significantly higher at 1.9% salinity than at 3.8%, but loss rate was about the same at both salinities. Radioarsenic loss followed a biphasic pattern; biological half-life was 3 and 32 days for the fast and slow compartments, respectively; secretion of the byssal thread played a key role in elimination. Factors known to modify rates of arsenic accumulation and retention in a marine shrimp (*Lysmata seticaudata*) include water temperature and salinity, arsenic concentration, age, and especially frequency of molting.

Bioconcentration factors (BCFs) experimentally determined for arsenic in aquatic organisms are, except for algae, relatively low. The BCF values for inorganic As^{+3} in most aquatic invertebrates and fish exposed for 21–30 days did not exceed 17 times; the maxima were 6 times for As^{+5}, and 9 times for organoarsenicals. Significantly higher BCF values were recorded in other aquatic organisms, but they were based on mean arsenic concentrations in natural waters that seemed artificially high. A BCF of 350 times was reported for the American oyster (*Crassostrea virginica*) held in 5.0 μg As^{+5}/L for 112 days. There was no relation between oyster body burdens of arsenic and exposure concentrations; however, diet seemed to contribute more to arsenic uptake than did seawater concentrations. An arsenic-tolerant strain of freshwater alga (*Chlorella vulgaris*) from an arsenic-polluted environment showed increasing growth up to 2000.0 mg As^{+5}/L, and it could survive at 10,000.0 mg As^{+5}/L. Accumulations up to 50,000.0 mg As/kg DW were recorded, suggesting a need for additional research on the extent of this phenomenon and its implications on food-web dynamics.

Some investigators have suggested that arsenic in the form of arsenite is preferentially utilized by marine algae and bacteria. Arsenate reduction to arsenite in seawater depends on phosphorus in solution and available algal biomass. During algal growth, as phosphate is depleted and the P^{+5}:As^{+5} ratio drops, the rate of As^{+5} reduction increases. The resultant As^{+3}, after an initial peak, is rapidly oxidized to As^{+5}, indicating the possibility of biological catalysis of oxidation as well as mediation of As^{+5} reduction. Researchers generally agree that As^{+3} is more toxic than arsenates to higher organisms; however, As^{+5} had a more profound effect on growth and morphology of marine algae than does As^{+3}. Possibly marine algae erect a barrier against the absorption of As^{+3}, but not against As^{+5}. Within the cell, As^{+5} can then be reduced to the possibly more toxic As^{+3}. For example, the culture of two species of marine algae (*Tetraselmis chui, Hymenomonas carterae*) in media containing various concentrations of As^{+5} or As^{+3} showed that arsenic effects varied with oxidation state, concentration, and light intensity. Arsenate was incorporated and later partly released by both species. Differences between rates of uptake and release suggest that arsenic undergoes chemical changes after incorporation into algal cells. When bacterial cultures from the Sargasso Sea and from marine waters of Rhode Island were grown in As^{+3}-enriched media, the bacteria reduced all available As^{+5} and utilized As^{+3} during the exponential growth phase, presumably as an essential trace nutrient. The arsenate reduction rate per cell was estimated to be 75×10^{-11} mg As/min.

The ability of marine phytoplankton to accumulate high concentrations of inorganic arsenicals and transform them to methylated arsenicals that are later efficiently transferred in the food chain is well documented. Algae constitute an important source of organoarsenic compounds in marine food webs. In the food chain composed of the alga *Dunaliella marina*, the grazing shrimp *Artemia salina*, and the carnivorous shrimp *Lysmata seticaudata*, organic forms of arsenic were derived from in vivo synthesis by *Dunaliella* and efficiently transferred, without magnification, along the food chain. Laboratory studies with five species of euryhaline algae grown in freshwater, or seawater, showed that all species synthesized fat-soluble and water-soluble arseno-organic compounds from inorganic As^{+3} and As^{+5}. The BCF values in the five species examined ranged from 200 times to about 3000 times – accumulations being highest in lipid phases. In Charlotte Harbor, Florida, a region that has become phosphate enriched due to agricultural activity, virtually all the arsenic taken up by phytoplankton was biomethylated and returned to the estuary, usually as MMA and dimethylarsinic acids. The ability of marine phytoplankton to methylate arsenic and release the products to a surrounding environment varies between species and even within a particular species in relation to their possession of necessary methylating enzymes. The processes involved in detoxifying arsenate after its absorption by phytoplankton are not firmly established, but seem to be nearly identical in all plants, suggesting a similar evolutionary development. Like phosphates and sulfates, arsenate may be fixed with ADP, reduced to the arsonous level, and successfully methylated and adenosylated, ultimately producing the 5-dimethylarsenosoribosyl derivatives accumulating in algae.

Sodium arsenite has been used extensively as an herbicide for control of mixed submerged aquatic vegetation in freshwater ponds and lakes; concentrations of 1.5–3.8 mg As^{+3}/L have usually been effective and are considered safe for fish. However, As^{+3} concentrations considered effective for aquatic weed control may be harmful to several species of freshwater teleosts, including bluegills, flagfish (*Jordenella floridae*), fathead minnows (*Pimephales promelas*), and rainbow trout (*Oncorhynchus mykiss*). Fish exposed to 1.0–2.0 mg total As/L for 2–3 days may show one or more of several signs: hemorrhagic spheres on gills; fatty infiltration of liver; and necrosis of heart, liver, and ovarian tissues. In green sunfish (*Lepomis cyanellus*), hepatocyte changes parallel arsenic accumulations in the liver. Organoarsenicals are usually eliminated rapidly by fish and other aquatic fauna. Rainbow trout, for example, fed a marine diet containing 15.0 mg organic arsenic/kg had only negligible tissue residues 6–10 days later, although some enrichment was noted in the eyes, throat, gills, and pyloric caeca. Oral administration of sodium arsenate to estuary catfish (*Cnidoglanis macrocephalus*) and school whiting (*Sillago bassensis*) resulted in tissue accumulations of trimethylarsine oxide. Arsenobetaine levels, which occur naturally in these teleosts, were not affected by As^{+5} dosing. The toxicity of trimethylarsine oxide is unknown, but the ease with which it can be reduced to the highly toxic trimethylarsine is a cause for concern. Recent studies, however, suggest that humans are capable of metabolizing trimethylarsine to the comparatively innocuous arsenobetaine.

2.6.4 Birds

Signs of inorganic trivalent arsenite poisoning in birds (muscular incoordination, debility, slowness, jerkiness, falling hyperactivity, fluffed feathers, drooped eyelids, huddled position, unkempt appearance, loss of righting reflex, immobility, seizures) were similar to those induced by many other toxicants and did not seem to be specific for arsenosis. Signs occurred within 1 h and deaths within 1–6 days post-administration; remission took up to 1 month. Internal examination suggested that lethal effects of acute inorganic arsenic poisoning were due to the destruction of blood vessels lining the gut, which resulted in decreased blood pressure and subsequent shock. Coturnix (*Coturnix coturnix*), for example, exposed to acute oral doses of As^{+3} showed hepatocyte damage, i.e., swelling of granular endoplasmic reticulum; these effects were attributed to osmotic imbalance, possibly induced by direct inhibition of the sodium pump by arsenic.

Arsenic, as arsenate, in aquatic plants (up to 430.0 mg As/kg plant DW) from agricultural drainwater areas can impair normal development of mallard ducklings. Pen studies with ducklings showed that a diet of 30.0 mg As/kg ration adversely affected growth and physiology, and 300.0 mg As/kg diet altered brain biochemistry and nesting behavior. Decreased energy levels and altered behavior can further decrease duckling survival in a natural environment.

Western grasshoppers (*Melanophis* spp.) poisoned by arsenic trioxide were fed, with essentially no deleterious effects, to nestling common bobwhites (*Colinus virginianus*), mockingbirds (*Mimus polyglottos*), American robins (*Turdus migratorius*), and other songbirds. Up to 134 poisoned grasshoppers, containing a total of about 40.0 mg As, were fed to individual nestlings without any apparent toxic effect. Species tested that were most sensitive to various arsenicals were the brown-headed cowbird (*Molothrus ater*) with an LD50 (11 day) value of 99.8 mg of copper acetoarsenite/kg diet; California quail (*Callipepla californica*) with an LD50 single oral dose value of 47.6 mg of sodium arsenite/kg BW; and chicken with 33.0, and turkey with 17.4 mg/kg BW of 3-nitro-4-hydroxy phenylarsonic acid as a single oral dose.

Chickens rapidly excrete arsenicals; only 2% of dietary sodium arsenite remained after 60 h, and arsanilic acid was excreted largely unchanged. Excretion of arsanilic acid by chickens was affected by uptake route: excretion was more rapid if administration was by intramuscular injection than if it was oral. Studies with inorganic As^{+5} and chickens indicated that (1) arsenates rapidly penetrated mucosal and serosal surfaces of epithelial membranes, (2) As^{+5} intestinal absorption was essentially complete within 1 h at 370.0 mg As^{+5}/kg BW but only 50% complete at 3700.0 mg/kg BW, (3) Vitamin D_3 was effective in enhancing duodenal As^{+5} absorption in rachitic chicks, and (4) As^{+5} and phosphate did not appear to share a common transport pathway in the avian duodenum.

2.6.5 Mammals

Mammals are exposed to arsenic primarily through the ingestion of naturally contaminated vegetation and water, or through human activity. In addition, feed additives containing arsonic acid derivatives are often fed to domestic livestock to promote growth and retard disease. Some commercial pet foods contain up to 2.3 mg As/kg DW. Uptake may occur by ingestion (the most likely route), inhalation, and absorption through skin and mucous membranes. Soluble arsenicals are absorbed more rapidly and completely than are the sparingly soluble arsenicals, regardless of the route of administration. In humans, inorganic arsenic at high concentrations is associated with adverse reproductive outcomes including increased rates of spontaneous abortion, low birth weight, congenital malformations, and death; however, at environmentally relevant levels and routes of exposure, humans are not at risk for birth defects due to arsenic. In vitro tests with human erythrocytes demonstrate that inorganic As^{+5} as sodium arsenate was up to 1000 times more effective than inorganic As^{+3} as sodium arsenite after exposure for 5 h to 750.0 mg As/L in causing death, morphologic changes, and ATP depletion.

Acute episodes of poisoning in warm-blooded organisms by inorganic and organic arsenicals are usually characterized by high mortality and morbidity over a period of 2–3 days. General signs of arsenic toxicosis include intense abdominal pain, staggering gait, extreme weakness, trembling, salivation, vomiting, diarrhea, fast and feeble pulse, prostration, collapse, and death. Gross necropsy shows a reddening of gastric mucosa and intestinal mucosa, a soft yellow liver, and red edematous lungs. Histopathological findings show edema of gastrointestinal mucosa and submucosa, necrosis and sloughing of mucosal epithelium, renal tubular degeneration, hepatic fatty changes and necrosis, and capillary degeneration in the gastrointestinal tract, vascular beds, skin, and other organs. In subacute episodes, where animals live for several days, signs of arsenosis include depression, anorexia, increased urination, dehydration, thirst, partial paralysis of rear limbs, trembling, stupor, coldness of extremities, and subnormal body temperatures. In cases involving cutaneous exposure to arsenicals, a dry, cracked, leathery, and peeling skin may be a prominent feature. Nasal discharges and eye irritation were documented in rodents exposed to organoarsenicals in inhalation toxicity tests. Subacute effects in humans and laboratory animals include peripheral nervous disturbances, melanosis, anemia, leukopenia, cardiac abnormalities, and liver changes. Most adverse signs rapidly disappear after exposure ceases.

Arsenic poisoning in most animals is usually manifested by acute or subacute signs; chronic poisoning is infrequently seen. The probability of chronic arsenic poisoning from continuous ingestion of small doses is rare, because detoxification and excretion are rapid. Chronic toxicity of inorganic arsenicals is associated with weakness, paralysis, conjunctivitis, dermatitis, decreased growth, and liver damage. Arsenosis, produced as a result of chronic exposure to organic arsenicals, was associated with demyelination of the optic and sciatic nerves, depressed growth, and decreased resistance to infection.

Research results on arsenic poisoning in mammals show general agreement on eight points: (1) arsenic metabolism and effects are significantly influenced by the organism tested, the route of administration, the physical and chemical form of the arsenical, and the dose; (2) inorganic arsenic compounds are more toxic than organic arsenic compounds and trivalent species are more so than pentavalent; (3) inorganic arsenicals can cross the placenta in most species of mammals; (4) early developmental stages are the most sensitive, and humans appear to be one of the more susceptible species; (5) animal tissues usually contain low levels (<0.3 mg As/kg FW) of arsenic; after the administration of arsenicals these levels are elevated, especially in liver, kidney, spleen, and lung; several weeks later, arsenic is translocated to ectodermal tissues (hair, nails) because of the high concentration of sulfur-containing proteins in these tissues; (6) inorganic arsenicals are oxidized in vivo, biomethylated, and usually excreted rapidly in the urine, but organoarsenicals are usually not subject to similar transformations; (7) acute or subacute arsenic exposure can lead to elevated tissue residues, appetite loss, reduced growth, loss of hearing, dermatitis, blindness, degenerative changes in liver and kidney, cancer, chromosomal damage, birth defects, and death; (8) death or malformations have been documented at single oral doses of 2.5–33.0 mg As/kg BW, at chronic doses of 1.0–10.0 mg As/kg BW, and at dietary levels >5.0 and <50.0 mg As/kg diet.

Episodes of wildlife poisoning by arsenic are infrequent. White-tailed deer (*Odocoileus virginianus*) consumed, by licking, fatal amounts of sodium arsenite used to debark trees. The practice of debarking trees with arsenicals for commercial use has been almost completely replaced by mechanical debarking equipment. In another incident, white-tailed deer were found dead of arsenic poisoning in a northern New York forest and had 102.0 mg As/kg FW in liver and 56.0 mg As/kg FW in kidney; these tissue concentrations are 2–3 times higher than those in cattle that died of arsenic poisoning – estimated at 241.0–337.0 mg As/kg BW. It is speculated that these deer licked trees injected with Silvisar 550, which contains monosodium methanearsonate, probably because of its salty taste. Snowshoe hares (*Lepus* sp.) appear to be especially sensitive to methylated arsenicals; hares died after consuming plants heavily contaminated with monosodium methanearsonate as a result of careless silviculture practices.

Unlike wildlife, reports of arsenosis in domestic animals are common in bovines and felines, less common in ovines and equines, and rare in porcines and poultry. In practice, the most dangerous arsenic preparations are dips, herbicides, and defoliants in which the arsenical is in a highly soluble trivalent form, usually as trioxide or arsenite. Accidental poisoning of cattle with arsenicals, for example, is well documented. In one instance, more than 100 cattle died after accidental overdosing with arsenic trioxide applied topically to control lice. On necropsy, there were subcutaneous edematous swellings and petechial hemorrhages in the area of application, and histopathology of the intestine, mucosa, kidney, and epidermis. In Bangladesh, poisoned cattle showed depression, trembling, bloody diarrhea, restlessness, unsteady gait, stumbling, convulsions, groaning, shallow labored breathing, teeth grinding, and salivation. Cattle usually died 12–36 h after the onset of signs; necropsy showed extensive submucosal hemorrhages of the gastrointestinal tract, and tissue residues > 10.0 mg/kg FW in liver and kidney. It sometimes appears that animals, especially cattle, develop an increased preference for weeds sprayed with an arsenic weed killer, not because of a change in the palatability of the plant, but probably because arsenic compounds are salty, and thus attractive to animals.

When extrapolating animal data from one species to another, the species tested must be considered. For example, the metabolism of arsenic in the rat (*Rattus* sp.) is unique, and very different from that in humans and other animals. Rats store arsenic in blood hemoglobin, excreting it very slowly – unlike most mammals which rapidly excrete ingested inorganic arsenic in the urine as methylated derivatives. Blood arsenic, whether given as As^{+3} or As^{+5}, rapidly clears from humans, mice, rabbits, dogs, and primates; half-life is 6 h for the fast phase and about 60 h for the slow phase. In rat, however, blood arsenic is mostly retained in erythrocytes, and clears slowly; half-life is 60–90 days. In rats, the excretion of arsenic into bile is 40 times slower than in rabbits and up to 800 times slower than in dogs. Most researchers now agree that the rat is unsatisfactory for use in arsenic research.

Dimethylarsinic acid is the major metabolite of orally administered arsenic trioxide, and is excreted rapidly in the urine. The methylation process is true detoxification, since methanearsonates and cacodylates are about 200 times less toxic than sodium arsenite. The marmoset monkey (*Callithrix jacchus*), unlike all other animal species studied to date, was not able (for unknown reasons) to metabolize administered As^{+5} to dimethylarsinic acid; most was reduced to As^{+3}. Only 20% of the total dose was excreted in urine as unchanged As^{+5}, and another 20% as As^{+3}. The rest was bound to tissues, giving distribution patterns similar to arsenite. Accordingly, the marmoset, like the rat, may be unsuitable for research with arsenicals.

Arsenicals were ineffective in controlling certain bacterial and viral infections. Mice experimentally infected with bacteria (*Klebsiella pneumoniae*) or viruses (pseudorabies, encephalitis, encephalmyocarditis) showed a significant increase in mortality when treated with large doses of arsenicals compared to nonarsenic-treated groups.

It has been suggested, but not yet verified, that many small mammals avoid arsenic-supplemented feeds and consume other foods if given the choice, and that cacodylic acid, which has negligible effects on wildlife, reduces species diversity due to selective destruction of vegetation. Both topics merit more research.

2.7 *Recommendations*

Numerous criteria for arsenic have been proposed to protect natural resources and human health (Table 2.1). But many authorities

Table 2.1. Proposed arsenic criteria for the protection of human health and selected natural resources.

Resource, Criterion, and Other Variables	Criterion or Effective Arsenic Concentration
HUMAN HEALTH	
Diet	
Permissible levels	
Total diet	<0.5 mg As/kg dry weight (DW) diet; 0.0003–0.0008 mg/kg body weight (BW) daily
Total intake	No observable effect at <0.021 mg arsenic daily based on 0.0003 mg/kg BW daily for a 70-kg adult
Fruits, vegetables	The tolerance for arsenic residues as As_2O_3 resulting from pesticidal use of copper, magnesium, and sodium arsenates is 3.5 mg/kg
Muscle of poultry and swine, eggs, swine edible byproducts	<2.0 mg total As/kg (FW)
Edible by-products of chickens and turkeys, liver and kidney of swine	<2.0 mg total As/kg FW
Seafood products	In Hong Kong, limited to <6.0 mg As^{+3}/kg FW for edible tissues of finfish and <10.0 mg As^{+3}/kg for mollusks and crustaceans; in Yugoslavia, these values are 2.0 for fish and 4.0 for mollusks and crustaceans; in Australia, <1.0 mg inorganic As/kg FW and in New Zealand <2.0 mg inorganic As/kg FW. There is no limit on organoarsenicals. In the UK, seafood products should contain <1.0 mg As/kg FW contributed as a result of contamination
Shellfish diet, USA	
Crustaceans, edible tissues	<76.0 mg total As/kg FW tissue
Tolerable daily intake	<0.13 mg
Maximum allowable	<30.0 mg total As/kg FW diet
Adverse effects	
Consumption of aquatic organisms living in arsenic-contaminated waters: cancer risk	
10^{-5}	0.175 μg As/L
10^{-6}	0.0175 μg As/L[a]
10^{-7}	0.00175 μg As/L
Drinking water	
Allowable concentrations	
Total arsenic, USA	<10.0 μg/L
Total arsenic, Maine	<30.0 μg/L

Continued

Table 2.1. cont'd

Resource, Criterion, and Other Variables	Criterion or Effective Arsenic Concentration
Adverse effects	
Cancer risk of	
10^{-5}	0.022 μg As/L
10^{-6}	0.0022 μg As/L[a]
10^{-7}	0.00022 μg As/L
Symptoms of arsenic toxicity observed	9% incidence at 50.0 μg As/L, 16% at 50.0–100.0 μg/L, and 44% at >100.0 μg As/L
Harmful after prolonged exposure	>50.0−960.0 μg As/L
Cancer frequency	In Chile, cancer rate estimated at 0.01% at 82.0 μg As/L, 0.17% at 600.0 μg As/L
Skin cancer	0.26% frequency at 290.0 μg/L and 2.14% at 600.0 μg/L
Total intake	
No observable effect	
North America	0.007–0.06 mg As daily; 2.0 μg/kg BW daily
Japan	0.07–0.17 mg As daily
USA; 1960s vs. 1974	0.05–0.1 vs. 0.015 mg As daily
Canada	0.03 mg As daily
Netherlands	2.0 μg total inorganic As/kg BW daily (about 0.14 mg daily for a 70-kg adult); 0.094 mg daily through fishery products
Adverse effects (prolonged exposure)	
Subclinical symptoms	0.15–0.6 mg As daily
Intoxication	3.0–4.0 mg As daily
Blackfoot disease	Total dose of 20.0 g over several years increases prevalence of disease by 3%
Mild chronic poisoning	0.15 mg As daily or about 2.0 μg/kg BW daily
Chronic arsenic poisoning	Lifetime cumulative absorption of 1.0 g As, or intake of 0.7–2.6 g/year for several years (in medications) can produce symptoms after latent period of 4–24 years
Tissue residues	
No observed effect levels	<0.05 mg As/L urine; <0.5 mg/kg liver or kidney; 0.7 mg/L blood; <2.0 mg/kg hair[b]; <5.0 mg/kg fingernail
Arsenic-poisoned; liver or kidney	2.0–100.0 mg As/kg FW; confirmatory tests >10.0 mg As/kg FW; residues in survivors several days later were 2.0–4.0 mg/kg FW
Arsenic-poisoned; whole body; children vs. adults	Symptoms of chronic arsenicism evident at 1.0 mg As/kg BW, equivalent to intake of about 10.0 mg/month for 3 months vs. 80.0 mg/kg BW, equivalent to about 2.0 g/year for 3 years

Table 2.1. cont'd

Resource, Criterion, and Other Variables	Criterion or Effective Arsenic Concentration
Air	
Allowable concentrations	
Arsine	<200.0 μg/m^3 for USA industrial workers; proposed mean arsine limit of <4.0 μg/m^3 in 8-h period and <10.0 μg/m^3 maximum in 15 min
Inorganic arsenic; occupational vs. residential	<2.0 vs. <10.0 μg/m^3
Organic arsenic	<500.0 μg/m^3
Total arsenic	<3.0 μg/m^3 in former USSR and former Czechoslovakia; <500.0 μg/m^3 for USA industrial workers
Total arsenic (threshold limit value, time-weighted mean: 8 h/day, 40-h-work week)	Proposed limit of <50.0 μg/m^3, maximum of 2.0 μg/m^3 in 15 min, <10.0 μg airborne inorganic As/m^3
Arsenic trioxide	<0.3 μg/m^3 in former USSR, <0.1 μg/m^3 in USA
Adverse effects	
Increased mortality	Associated with daily time-weighted average arsenic exposure of >3.0 μg/m^3 for 1 year
Respiratory cancer (increased risk)	Lifetime occupational exposure >54.6 μg As/m^3; 50.0 μg As/m^3 for more than 25 years
Skin diseases	60.0–13,000.0 μg As/m^3
Dermatitis	300.0–81,500.0 μg As/m^3
AQUATIC LIFE	
Freshwater biota: medium	96-h mean water concentrations not to exceed 190.0 μg total recoverable inorganic As^{+3}/L more than once every 3 years; 1-h mean not to exceed 360.0 μg inorganic As^{+3}/L more than once every 3 years. Insufficient data for criteria formulation for inorganic As^{+5}, or for any organoarsenical
Freshwater biota: tissue residues	Diminished growth and survival in immature bluegills (*Lepomis macrochirus*), when total arsenic residues in muscle are >1.3 mg/kg FW or >5.0 mg/kg in adults
Saltwater biota: medium	96-h average water concentration not to exceed 36.0 μg As^{+3}/L more than once every 3 years; 1-h mean not to exceed 69.0 μg As^{+3}/L more than once every 3 years. Insufficient data for criteria formulation for inorganic As^{+5}, or for any organoarsenical
Saltwater biota: tissue residues	Depending on chemical form of arsenic, certain marine teleosts may be unaffected at muscle total arsenic residues of 40.0 mg/kg FW

Continued

Table 2.1. cont'd

Resource, Criterion, and Other Variables	Criterion or Effective Arsenic Concentration
BIRDS	
Single oral dose fatal to 50%, sensitive species	17.0–48 mg As/kg BW
Tissue residues	Residues, in mg total As/kg FW, liver or kidney in 2.0–10.0 range are considered elevated; residues >10.0 mg/kg are indicative of arsenic poisoning
Mallard, *Anas platyrhynchos*; sodium arsenate in diet	Reduced growth in ducklings fed >30.0 mg As/kg diet
Turkey, *Meleagris gallopavo*; arsanilic acid in diet	Maximum dietary concentration for turkeys less than 28 days old is 300.0–400.0 mg/kg feed
Phenylarsonic feed additives for disease control and improvement of weight gain in domestic poultry; safe dietary levels	Maximum levels in diets, in mg/kg feed, are 50.0–100.0 for arsanilic acid, 25.0–188.0 for 3-nitro-4-hydroxy-phenylarsonic acid (for chickens and turkeys, not recommended for ducks and geese), and 180.0–370.0 for others
DOMESTIC LIVESTOCK	
Prescribed limits for arsenic in feedstuffs	
Straight feedstuffs, except those listed below	<2.0 mg total As/kg FW
Meals from grass, dried lucerne, or dried clover	<4.0 mg total As/kg FW
Phosphate mealstuffs, fish meals	<10.0 mg total As/kg FW
Tissue residues	
Poisoned; liver and kidney	5.0–>10.0 mg total As/kg FW
Normal; muscle	<0.3 mg total As/kg FW
TERRESTRIAL VEGETATION	
No observable effects	<1.0 mg total water-soluble As/L, <25.0 mg total As/kg soil, <3.9 μg As/m^3 air
Adverse effects, crops and vegetation	3.0–28.0 mg water soluble As/L, equivalent to 25.0–85.0 mg total As/kg soil; air concentrations >3.9 μg As/m^3
Phytotoxic or growth inhibition of tolerant genotypes	>1000 mg/kg DW soil

[a]One excess cancer per million population (10^{-6}) is estimated during lifetime exposure to 0.0022 μg arsenic per liter of drinking water, or to lifetime consumption of aquatic organisms residing in waters containing 0.0175 μg As/L.

[b]Thai children, aged 6–9 years from the Ronpiboon district with >5.0 mg As/kg DW hair, had abnormally low IQs when compared to those with 2.01–5 mg As/kg DW. Both groups had significantly lower IQs than did controls (<1.0 mg As/kg DW hair), as measured by the Wechsler Intelligence Scale Test for Children. This study needs verification.

recognize that these criteria are not sufficient for adequate or (in some cases) reasonable protection, and that many additional data are required if meaningful standards are to be promulgated. Specifically, data are needed on the following subjects: (1) cancer incidence and other abnormalities in natural resources from areas with elevated arsenic levels, and the relation to potential carcinogenicity of arsenic compounds; (2) interaction effects of arsenic with other carcinogens, co-carcinogens, promoting agents, inhibitors, and common environmental contaminants; (3) controlled studies with aquatic and terrestrial indicator organisms on physiological and biochemical effects of long-term, low-dose exposures to inorganic and organic arsenicals, including effects on reproduction and genetic makeup; (4) methodologies for establishing maximum permissible tissue concentrations for arsenic; (5) effects of arsenic in combination with infectious agents; (6) mechanisms of arsenical growth-promoting agents; (7) role of arsenic in nutrition; (8) extent of animal adaptation to arsenicals and the mechanisms of action; (9) identification and quantification of mineral and chemical forms of arsenic in rocks, soils, and sediments that constitute the natural forms of arsenic entering water and the food chain; and (10) physicochemical processes influencing arsenic cycling. In addition, the following techniques should be developed and implemented: (1) more sophisticated measurements of the chemical forms of arsenic in plant and animal tissues; (2) correlation of biologically observable effects with particular chemical forms of arsenic; and (3) management of arsenical wastes that accommodates recycling, reuse, and long-term storage.

Some proposed arsenic criteria merit additional comment, such as those on aquatic life protection, levels in seafoods and drinking water, and use in food-producing animals as growth stimulants or for disease prevention and treatment.

For saltwater life protection, the current water quality criterion of 36.0 μg As^{+3}/L seems to offer a reasonable degree of safety; only a few species of algae show adverse effects at <36.0 μg/L (e.g., reduced growth at 19.0–22.0 μg/L). In 1980, this criterion was 508.0 μg/L, about 14 times higher than the current criterion. The downward modification seems to be indicative of the increasingly stringent arsenic criteria formulated by regulatory agencies. The current criterion for freshwater-life protection of 190.0 μg As^{+3}/L, however, which is down from 440.0 μg As^{+3}/L in 1980, is unsatisfactory. Many species of freshwater biota are adversely affected at <190.0 μg/L of As^{+3}, As^{+5}, or various organoarsenicals. These adverse effects include death and malformations of toad embryos at 40.0 μg/L, growth inhibition of algae at 48.0–75.0 μg/L, mortality of amphipods and gastropods at 85.0–88.0 μg/L, and behavioral impairment of goldfish (*Carassius auratus*) at 100.0 μg/L. A downward adjustment in the current freshwater aquatic-life protection criterion seems warranted.

In Hong Kong, permissible concentrations of arsenic in seafood destined for human consumption range from 6.0 to 10.0 mg/kg FW; however, these values are routinely exceeded in 22% of finfish, 20% of bivalve mollusks, 67% of gastropods, 29% of crabs, 21% of shrimp and prawns, and 100% of lobsters. The highest arsenic concentrations recorded in Hong Kong seafood products were in gastropods (*Hemifusus* spp.), in which the concentrations of 152.0–176.0 mg/kg FW were among the highest recorded in any species to date. A similar situation exists in Yugoslavia, where almost all seafoods exceed the upper limit prescribed by food quality regulations. Most of the arsenic in seafood products is usually arsenobetaine or some other comparatively harmless form. In effect, arsenic criteria for seafoods are neither enforced nor enforceable. Some toxicologists from the U.S. Food and Drug Administration believe that the average daily intake of arsenic in the different food commodities does not pose a hazard to the consumer. It is now clear that formulation of maximum permissible concentrations of arsenic in seafoods for health regulation purposes should recognize the chemical nature of arsenic.

For maximum protection of human health from the potential carcinogenic effects of exposure to arsenic through drinking water or contaminated aquatic organisms, the ambient water concentration should be zero, based on the nonthreshold assumption for arsenic. But zero level may not be attainable. Accordingly, the levels established are those that are estimated to increase cancer risk over a lifetime to only one additional case per 100,000 population. These values are estimated at 0.022 μg As/L for drinking water, and 0.175 μg As/L for water containing edible aquatic resources.

Various phenylarsonic acids – especially arsanilic acid, sodium arsanilate, and 3-nitro-4-hydroxyphenylarsonic acid – have been used as feed additives for disease control and for improvement of weight gain in swine and poultry for more than 50 years. The arsenic is present as As^{+5} and is rapidly excreted; present regulations require withdrawal of arsenical feed additives 5 days before slaughter for satisfactory depuration. Under these conditions, total arsenic residues in edible tissues do not exceed the maximum permissible limit of 2.0 mg/kg FW. Organoarsenicals probably will continue to be used as feed additives unless new evidence indicates otherwise.

2.8 Summary

Arsenic (As) is a relatively common element that occurs in air, water, soil, and all living tissues. It ranks twentieth in abundance in the earth's crust, fourteenth in seawater, and twelfth in the human body. Arsenic is a teratogen and carcinogen that can traverse placental barriers and produce fetal death and malformations in many species of mammals. Although it is carcinogenic in humans, evidence for arsenic-induced carcinogenicity in other mammals is scarce. Paradoxically, evidence is accumulating that arsenic is nutritionally essential or beneficial. Arsenic deficiency effects, such as poor growth, reduced survival, and inhibited reproduction, have been recorded in mammals fed diets containing <0.05 mg As/kg, but not in those fed diets with 0.35 mg As/kg. At comparatively low doses, arsenic stimulates growth and development in various species of plants and animals.

Most arsenic produced domestically is used in the manufacture of agricultural products such as insecticides, herbicides, fungicides, algicides, wood preservatives, and growth stimulants for plants and animals. Living resources are exposed to arsenic by way of atmospheric emissions from smelters, coal-fired power plants, and arsenical herbicide sprays; from water contaminated by mine tailings, smelter wastes, and natural mineralization; and from diet, especially from consumption of marine biota. Arsenic concentrations are usually low (<1.0 mg/kg FW) in most living organisms, but they are elevated in marine biota (in which arsenic occurs as arsenobetaine and poses little risk to organisms or their consumers) and in plants and animals from areas that are naturally arseniferous or are near industrial manufacturers and agricultural users of arsenicals. Arsenic is bioconcentrated by organisms but is not biomagnified in the food chain.

Arsenic exists in four oxidation states, as inorganic or organic forms. Its bioavailability and toxic properties are significantly modified by numerous biological and abiotic factors, including the physical and chemical forms of arsenic tested, the route of administration, the dose, and the species of animal. In general, inorganic arsenic compounds are more toxic than organic compounds, and trivalent species are more toxic than pentavalent species. Arsenic may be absorbed by ingestion, inhalation, or through permeation of the skin or mucous membranes; cells take up arsenic through an active transport system normally used in phosphate transport. The mechanisms of arsenic toxicity differ greatly among chemical species, although all appear to cause similar signs of poisoning. Biomethylation is the preferred detoxification mechanism for absorbed inorganic arsenicals; methylated arsenicals usually clear from tissues within a few days.

Episodes of arsenic poisoning are either acute or subacute; chronic cases of arsenosis are seldom encountered in any species except humans. Single oral doses of arsenicals fatal to 50% of sensitive species tested ranged from 17.0 to 48.0 mg/kg BW in birds and from 2.5 to 33.0 mg/kg BW in mammals. Susceptible species of mammals were adversely affected at chronic doses of 1.0–10.0 mg As/kg BW, or 50.0 mg As/kg diet. Sensitive aquatic species were damaged at water concentrations of 19.0–48.0 μg As/L (the U.S. Environmental Protection Agency drinking water criterion for human health protection is 50.0 μg/L), 120.0 mg As/kg diet, or (in the case of freshwater fish) tissue residues >1.3 mg/kg FW. Adverse effects to crops and vegetation were recorded at 3.0–28.0 mg of water-soluble As/L (equivalent to about 25.0–85.0 mg total As/kg soil), and at atmospheric concentrations >3.9 μg As/m^3.

Numerous arsenic criteria have been proposed for the protection of sensitive natural resources; however, the consensus is that many of these criteria are inadequate and that additional information is needed in at least five categories: (1) developing standardized procedures to permit correlation of biologically observable effects with suitable chemical forms of arsenic; (2) conducting studies under controlled conditions with appropriate aquatic and terrestrial indicator organisms to determine the effects of chronic exposure to low doses of inorganic and organic arsenicals on reproduction, genetic makeup, adaptation, disease resistance, growth, and other variables; (3) measuring interaction effects of arsenic with other common environmental contaminants, including carcinogens, co-carcinogens, and promoting agents; (4) monitoring the incidence of cancer and other abnormalities in the natural resources of areas with relatively high arsenic levels, and correlating these with the possible carcinogenicity of arsenic compounds; and (5) developing appropriate models of arsenic cycling and budgets in natural ecosystems.

ATRAZINE[a]

Chapter 3

3.1 Introduction

Atrazine (2-chloro-4-ethylamino-6-isopropylamino-1,3,5-triazine) is the most heavily used agricultural pesticide in North America, and is registered for use in controlling weeds in numerous crops, including corn (*Zea mays*), sorghum (*Sorghum vulgare*), sugarcane (*Saccharum officinarum*), soybeans (*Glycine max*), wheat (*Triticum aestivum*), pineapple (*Ananas comusus*), and various range grasses. Atrazine was first released for experiment station evaluations in 1957 and became commercially available in 1958. In 1976, 41 million kg (90 million pounds) were applied to 25 million ha (62 million acres) on farms in the United States, principally for weed control in corn, wheat, and sorghum crops; this volume represented 16% of all herbicides and 9% of all pesticides applied in the United States during that year. By 1980, domestic usage had increased to 50 million kg. In Canada, atrazine was the most widely used of 77 pesticides surveyed. Agricultural use of atrazine has also been reported in South Africa, Australia, New Zealand, Venezuela, and in most European countries. Global use of atrazine is estimated at 70–90 million kg annually, although Germany banned atrazine in 1991. Resistance to atrazine has developed in various strains of weeds typically present in crop fields – sometimes in less than two generations – suggesting that future agricultural use of atrazine may be limited.

Atrazine has been detected in lakes and streams at levels ranging from 0.1 to 30.3 μg/L; concentrations peak during spring, which coincides with the recommended time for agricultural application. In runoff waters directly adjacent to treated fields, atrazine concentrations of 27.0–69.0 μg/L have been reported, and may reach 1000.0 μg/L. Some of these concentrations are demonstrably phytotoxic to sensitive species of aquatic flora. Although atrazine runoff from Maryland cornfields was suggested as a possible factor in the decline of submerged aquatic vegetation in Chesapeake Bay, which provides food and habitat for large populations of waterfowl, striped bass (*Morone saxatilis*), American oysters (*Crassostrea virginica*), and blue crabs (*Callinectes sapidus*), it was probably not a major contributor to this decline.

3.2 Environmental Chemistry

Atrazine is a white crystalline substance that is sold under a variety of trade names for use primarily as a selective herbicide to control broadleaf and grassy weeds in corn and sorghum (Table 3.1; Figure 3.1). It is slightly soluble in water (33.0 mg/L at 27°C), but comparatively soluble (36.0–183,000.0 mg/L) in many organic solvents. Atrazine is usually applied in a water spray at concentrations of 2.2–4.5 kg/ha before weeds emerge. Stored atrazine is stable for several years, but degradation begins immediately after application (Table 3.1). The chemical is available as a technical material at 99.9% active ingredient and as a manufacturing-use product containing 80% atrazine for formulation of wettable powders, pellets, granules,

---

[a] All information in this chapter is referenced in the following sources:

Eisler, R. 1989. Atrazine hazards to fish, wildlife, and invertebrates: a synoptic review. *U.S. Fish Wildl. Serv. Biol. Rep. 85(1.18)*, 53 pp.

Eisler, R. 2000. Atrazine. Pages 767–797 in *Handbook of Chemical Risk Assessment: Health Hazards to Humans, Plants, and Animals. Volume 2, Organics.* Lewis Publishers, Boca Raton, Florida.

Table 3.1. Some properties of atrazine.

Variable	Datum
CHEMICAL NAME	2-chloro-4-ethylamino-6-isopropylamino-1,3,5-triazine
PRIMARY USES	Selective herbicide for control of most annual broadleaf and grassy weeds in corn, sugarcane, sorghum, macadamia orchards, rangeland, pineapple, and turf grass sod. Nonselective herbicide for weed control on railroads, storage yards, along highways, and industrial sites. Sometimes used as selective weedicide in conifer reforestation, Christmas tree plantations, and grass seed fields
APPLICATION METHODS	Usually as water spray or in liquid fertilizers applied pre-emergence, but also may be applied pre-plant or post-emergence. Rates of 2–4 pounds/acre (2.24–4.48 kg/ha) are effective for most situations; higher rates are used for nonselective weed control, and on high organic soils
COMPATIBILITY WITH OTHER PESTICIDES	Compatible with most other pesticides and fertilizers when used at recommended rates. Sold in formulation with Lasso®, Ramrod®, and Bicep®
STABILITY	Very stable over several years of shelf life, under normal illumination and extreme temperatures. Stable in neutral, slightly acid, or basic media. Sublimes at high temperatures and when heated, especially at high temperatures in acid, or basic media, hydrolyzes to hydroxyatrazine (2-hydroxy-4-ethylamino-6-isopropylamino-*s*-triazine), which has no herbicidal activity
EMPIRICAL FORMULA	$C_8H_{14}ClN_5$
MOLECULAR WEIGHT	215.7
MELTING POINT	173–175°C
VAPOR PRESSURE	5.7×10^{-8} at 10°C, 3.0×10^{-7} at 20°C, 1.4×10^{-6} at 30°C, and 2.3×10^{-5} mmHg at 50°C
HENRY'S LAW CONSTANT	6.13×10^{-8} to 2.45×10^{-7} atm-m^3/mol
PHYSICAL STATE	The technical material is a white, crystalline, noncombustible, noncorrosive substance
PURITY	No impurities or contaminants that resulted from the manufacturing process were detected
SOLUBILITY	
Water	22.0 mg/L at 0°C, 32.0 mg/L at 25°C, 320.0 mg/L at 85°C
***N*-pentane**	360.0 mg/L at 27°C
Petroleum ether	12,000.0 mg/L at 27°C
Methanol	18,000.0 mg/L at 27°C
Ethyl acetate	28,000.0 mg/L at 27°C
Chloroform	52,000.0 mg/L at 27°C
Dimethyl sulfoxide	183,000.0 mg/L at 27°C
Log K_{ow}	2.71

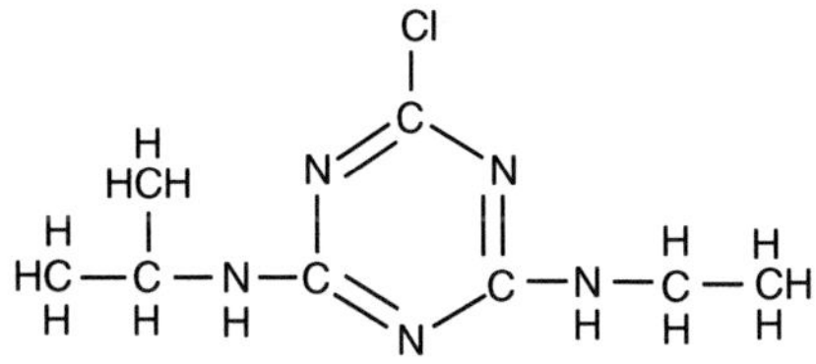

Figure 3.1.
Structural formula of atrazine.

flowable concentrates, emulsifiable concentrates, or tablets.

There are three major atrazine degradation pathways: hydrolysis at carbon atom 2, in which the chlorine is replaced with a hydroxy group; N-dealkylation at carbon atom 4 (loss of the ethylpropyl group) or 6 (loss of the isopropyl group); and splitting of the triazine ring. The dominant phase I metabolic reaction in plants is a cytochrome P450-mediated N-dealkylation, while the primary phase II reaction is the glutathione-*S*-transferase (GST)-catalyzed conjugation with glutathione. The presence of GST isoenzymes that metabolize atrazine has been demonstrated in at least 10 species: in liver of the rainbow trout (*Oncorhynchus mykiss*), starry flounder (*Pleuronectes stellatus*), English sole (*Pleuronectes vetulus*), rat (*Rattus norvegicus*), mouse (*Mus musculus*), leaves of common groundsel (*Seneco vulgaris*), and soft tissues of the cabbage moth (*Mamestra brassica*), and the Hebrew character moth (*Orthosia gothica*).

The major atrazine metabolite in both soil and aquatic systems is hydroxyatrazine. In soils, it accounts for 5–25% of the atrazine originally applied after several months compared to 2–10% for all dealkylated products combined, including deethylated atrazine and deisopropylated atrazine. Atrazine may be converted to nonphytotoxic hydroxyatrazine by chemical hydrolysis, which does not require a biological system. Bacterial degradation, however, proceeds primarily by N-dealkylation. In animals, N-dealkylation is a generally valid biochemical degradation mechanism. In rats, rabbits, and chickens, most atrazine is excreted within 72 h; 19 urinary metabolites – including hydroxylated, N-dealkylated, oxidized, and conjugated metabolites – were found. There is general agreement that atrazine degradation products are substantially less toxic than the parent compound and not normally present in the environment at levels inhibitory to algae, bacteria, plants, or animals.

Residues of atrazine rapidly disappeared from a simulated Northern Prairie freshwater wetland microcosm during the first 4 days, primarily by way of adsorption onto organic sediments. This is consistent with the findings of others who report 50% loss (Tb1/2) from wetlands in about 10 days and freshwater in 3.2 days, 82% loss in 5 days, and 88–95% loss in 55–56 days, although one report presents evidence of a 300-day half-life for atrazine, and another for months to years in the water column of certain Great Lakes. In estuarine waters and sediments, atrazine is inactivated by adsorption and metabolism: half-time persistence in waters has been estimated to range between 3 and 30 days, being shorter at elevated salinities; for sediments this range was 15–35 days. The comparatively rapid degradation of atrazine to hydroxyatrazine in estuarine sediments and water column indicates a low probability for atrazine accumulation in the estuary, and a relatively reduced rate of residual phytotoxicity in the estuary for the parent compound.

Atrazine is leached into the soil by rain or irrigation water. The extent of leaching is limited by the low water solubility of atrazine and by its adsorption onto certain soil constituents. Runoff loss in soils ranges from 1.2 to 18% of the total quantity of atrazine applied, but usually is less than 3%. Surface runoff of atrazine from adjacent conventional tillage and no-tillage corn watersheds in Maryland was measured after single annual application of 2.2 kg/ha for 4 years. Most of the atrazine in surface runoff was lost during the first rain after application. In 1979, the year of greatest precipitation, 1.6% of the atrazine applied moved from the conventional tillage compared to 1.1% from the no-tillage watershed, suggesting that no-tillage should be encouraged as an environmentally sound practice. Lateral and downward movement of atrazine was measured in cornfield soils to a depth of 30 cm when applied at 1.7 kg/ha to relatively moist soils; in lower elevation soils, atrazine accumulated by way of runoff and infiltration. Downward movement of atrazine through the top 30 cm of cornfield soils indicates that carryover of atrazine to the next growing season is possible: between 5 and 13% of atrazine was available one year after application. Atrazine is not usually found below the upper 30 cm of soil in detectable quantities, even after years of continuous use; accordingly, groundwater contamination by atrazine is not expected at recommended application rates.

Atrazine persistence in soils is extremely variable: reported Tb1/2 values ranged from 20 to 100 days in some soils to 330–385 days in others; intermediate values were reported by several scientists. Atrazine activity and persistence in soils are governed by many physical, chemical, and biological factors. In general, atrazine loss was more rapid under some conditions than others: it was more rapid from moist soils than from dry soils, during periods of high temperatures than during periods of low temperatures, from high organic and high clay content soils than from sandy mineral soils, during summer than in winter, from soils with high microbial densities than from those with low densities, from soils of acidic pH than from those of neutral or alkaline pH, during storm runoff events than during normal flows, at shallow soil depths than at deeper depths, and under conditions of increased ultraviolet irradiation. Microbial action, usually by way of N-dealkylation and hydrolysis to hydroxyatrazine, probably accounts for the major breakdown of atrazine in the soil, although non-biological degradation pathways of volatilization, hydroxylation, dealkylation, and photodecomposition are also important. The photolytic transformation rate of atrazine is enhanced at higher atrazine concentrations and the presence of dissolved organic carbon (DOC) and DOC mimics.

3.3 Concentrations in Field Collections

Although annual use of atrazine in the United States is about 35 million kg, atrazine concentrations in human foods are negligible. Monitoring of domestic and imported foods in the human diet by the U.S. Food and Drug Administration between 1978 and 1982 showed that only 3 of 4500 samples analyzed had detectable atrazine residues. Two samples in 1980 contained 0.01 and 0.08 mg atrazine/kg and one in 1978, following a known contamination incident, contained 47.0 mg/kg.

Atrazine was present in 100% of 490 samples analyzed in Lakes Michigan, Huron, Erie, and Ontario during 1990–92; concentrations were highest in Lake Erie at 0.11 μg/L. Atrazine concentrations in river waters of Ohio between 1995 and 1998 show strong seasonality, with the period of higher concentrations lasting 6–12 weeks, beginning with the first storm runoff following application, usually in May. The use of atrazine in the U.S. Great Lakes Basin is estimated at 2.7 million kg annually, and more than 600,000 kg of atrazine have entered the Great Lakes. Atrazine and its metabolites have been observed in freshwater streams contiguous to agricultural lands; 0.1–3% of the atrazine applied to the fields was lost to the aquatic environment. Atrazine concentrations as high as 691.0 μg/L were reported in agricultural streams during storm runoff events. In some cases, atrazine concentrations in runoff waters from treated cornfields may exceed 740.0 μg/L. Elevated levels were associated with high initial treatment rates, major storms shortly after application, conventional tillage practices (vs. no tillage), and increased flow rates, increased suspended solids, and increased dissolved nitrates and nitrites. Concentrations in runoff water usually declined rapidly within a few days. In 1991, maximum atrazine concentrations in the Des Plaines River, Illinois, after spring rains, briefly exceeded the Federal proposed drinking water criterion of 3.0 μg/L. Groundwater contamination by way of atrazine treatment of cornfields has been unexpectedly reported in parts of Colorado, Iowa, and Nebraska. Contamination was most pronounced in areas of highly permeable soils that overlie groundwater at shallow depths.

The total amount of atrazine reaching the Wye River, Maryland, estuary depended on the quantity applied in the watershed and the timing of runoff. In years of significant runoff,

2–3% of the atrazine moved to the estuary within 2 weeks after application and effectively ceased after 6 weeks. In Chesapeake Bay waters, a leakage rate of 1% of atrazine from agricultural soils resulted in aqueous concentrations averaging 17.0 μg/L – concentrations potentially harmful to a variety of estuarine plants. The maximum recorded atrazine concentration in runoff water entering Chesapeake Bay was 480.0 μg/L. However, these concentrations seldom persisted for significant intervals, and only rarely approached those producing long-term effects on submerged aquatic vegetation.

Atmospheric transport of atrazine-contaminated aerosol particulates, dusts, and soils may contribute significantly to atrazine burdens of terrestrial and aquatic ecosystems. The annual atmospheric input of atrazine in rainfall to the Rhode River, Maryland, as one example, was estimated at 1016.0 mg/surface ha in 1977, and 97.0 mg/ha in 1978. A similar situation exists with fog water. When fog forms, exposed plant surfaces become saturated with liquid for the duration of the fog.

3.4 Effects

In terrestrial ecosystems, atrazine effectively inhibits photosynthesis in target weeds, and may also affect certain sensitive crop plants. Atrazine metabolites are not as phytotoxic as the parent compound. Degradation is usually rapid, although atrazine may persist in soils for more than one growing season. Soil fauna may be adversely affected shortly after initial atrazine application at recommended levels, but long-term population effects on this group are considered negligible.

Sensitive species of aquatic flora experience temporary adverse effects at concentrations as low as 1.0–5.0 μg/L; however, most authorities agree that potentially harmful levels, i.e., >10.0 μg/L for long periods, have not been documented and are probably unrealistic under current application protocols and degradation rates. The observed declines in submerged aquatic vegetation in the Chesapeake Bay are not now directly attributable to atrazine use. Atrazine indirectly affects aquatic fauna at concentrations of 20.0 μg/L and higher by reducing the food supply of herbivores and, to some extent, their macrophyte habitat. Direct adverse effects on growth and survival of aquatic fauna were evident in the range of 94.0–500.0 μg/L. Bioaccumulation of atrazine is limited and food chain biomagnification is negligible in aquatic ecosystems.

Birds show a low probability for atrazine uptake and accumulation. Known acute oral LD50s and dietary LD50s are high: >2000.0 mg/kg BW, and 5000.0 mg/kg diet. Indirect ecosystem effects of atrazine on insect and seed-eating birds are not known, and seem to merit study. Data are lacking for mammalian wildlife, but tests with domestic livestock and small laboratory animals strongly indicate that this group is comparatively resistant to atrazine. Acute oral LD50s are >1750.0 mg/kg BW, and no adverse effects are evident at dietary levels of 25.0 mg/kg food (about 1.25 mg/kg BW) and sometimes 100.0 mg/kg food over extended periods.

3.4.1 Terrestrial Plants and Invertebrates

Atrazine enters plants primarily by way of the roots and secondarily by way of the foliage, passively translocated in the xylem with the transpiration stream, and accumulates in the apical meristems and leaves. The main phytotoxic effect is the inhibition of photosynthesis by blocking the electron transport during Hill reaction of photosystem II. This blockage leads to inhibitory effects on the synthesis of carbohydrate, a reduction in the carbon pool, and a buildup of carbon dioxide within the leaf, which subsequently causes closure of the stomata, thus inhibiting transpiration.

Atrazine is readily metabolized by tolerant plants to hydroxyatrazine and amino acid conjugates. The hydroxyatrazine can be further degraded by dealkylation of the side chains and by hydrolysis of resulting amino groups on the ring and some carbon dioxide production. Resistant plant species degrade atrazine before it interferes with photosynthesis. Corn, for example, has an enzyme (2,4-dihydroxy-7-methoxy-1, 4 [2H]-berzoxazin-3

[4H]-one) that degrades atrazine to nonphytotoxic hydroxyatrazine. In sensitive plants, such as oats, cucumber, and alfalfa, which are unable to detoxify atrazine, the compound accumulates, causing chlorosis and death. Corn and sorghum excrete about 50% of accumulated atrazine and metabolize the rest to insoluble residues that are indigestible to sheep (*Ovis aries*) and rats (*Rattus* spp.). These results strongly suggest that the final disposition of atrazine metabolites does not occur in either plants or animals, but ultimately through microbial breakdown.

Long-term applications of atrazine for weed control in corn result in degradation products, mainly hydroxylated analogs, that may persist in soil for at least 12 months after the final herbicide application, and may enter food crops planted in atrazine-treated soil in the years after cessation of long-term treatment. In one example, atrazine was applied to a cornfield for 20 consecutive years at rates of 1.4–2.2 kg/ha. Soils collected 12 months after the last application contained atrazine (55.0 μg/kg DW), hydroxyatrazine (296.0 μg/kg), and various mono dealkylated hydroxy analogs (deethylatrazine at 14.0 μg/kg, deethylhydroxyatrazine at 17.0 μg/kg, and deisopropylhydroxyatrazine at 23.0 μg/kg). Oat (*Avena sativa*) seedlings grown in this field contained hydroxyatrazine (64.0–73.0 μg/kg FW) and deisopropylhydroxyatrazine (84.0–116.0 μg/kg); similar results were obtained with timothy, *Phleum pratense*. In areas with a relatively long growing season, a double cropping of soybeans (*Glycine max*) – planted after corn is harvested for silage or grain – is gaining acceptance. Under conditions of warm weather, relatively high rainfall, and sandy soils, soybeans can be safely planted after corn (14–20 weeks after atrazine application) when rates of atrazine normally recommended for annual weed control (1.12–4.48 kg/ha) are used.

Seed germination of sensitive species of plants was reduced by 50% at soil atrazine concentrations between 0.02 and 0.11 mg/kg. Mustard (*Brassica juncea*) was especially sensitive, and died shortly after germination. Soil atrazine residues of this magnitude were typical of those remaining at the beginning of a new growing season, following corn in sandy loam under tropical conditions. Reduction in seed germination was also noted at soil atrazine concentrations of 0.25–0.46 mg/kg for the lentil (*Lens esculenta*), the pea (*Pisum sativum*), and the gram (*Cicer arietinum*). Many species of mature range grasses are tolerant of atrazine but are susceptible as seedlings; seedlings of the three most sensitive species of eight tested were adversely affected in soils containing 1.1 mg atrazine/kg.

Soil fungi and bacteria accumulated atrazine from their physicochemical environment by factors of 87–132, probably through passive adsorption mechanisms. Atrazine stimulated the growth of at least two common species of fungal saprophytes known to produce antibiotics: *Epicoccum nigrum* and *Trichoderma viride*. *Trichoderma*, for example, grew rapidly at all treatments tested (up to 80.0 mg/kg soil) and showed optimal growth 3–10 days post-inoculation. Atrazine suppressed the growth of various species of soil fungi, including *Rhizoctonia solani*, *Sclerotium rolfsii*, and *Fusarium* spp., and stimulated the growth of other species known to be antagonistic to *Fusarium*. This selectivity is likely to induce a shift in the fungal population of atrazine-treated soil that would be either harmful or beneficial to subsequent crops, depending on whether saprophytic or pathogenic fungi attained dominance.

At 2.5 mg atrazine/kg soil, equivalent to 2 kg/ha in the top 10 cm, field and laboratory studies demonstrated that mortality in arthropod collembolids (*Onchiurus apuanicus*) was 47% in 60 days; however, fecundity was not affected at dose levels up to 5.0 mg/kg soil. It was concluded that atrazine applications at recommended treatment levels had negligible long-term population effects on sensitive species of soil fauna. At 5.0 or 8.0 kg atrazine/ha, all species of soil fauna tested, except some species of nematodes, were adversely affected. One month post-application, population reductions of 65–91% were recorded in protozoans, mites, various insect groups, and collembolids at 5.0 kg/ha; after 4 months, populations were still depressed by 55–78%. At 8.0 kg atrazine/ha, soil faunal populations of beetles, collembolids, and earthworms remained depressed for at least 14 months after initial treatment. Final instar larvae of the cabbage moth (*Mamestra brassica*) fed synthetic diets for 48 h containing 500.0 or 5000.0 mg atrazine/kg ration, had significant changes in xenobiotic metabolizing activities of soft tissues and midgut, especially in aldrin epoxidase substrates; growth was retarded in the high-dose group.

3.4.2 Aquatic Plants

Since the mid-1960s, seagrasses and freshwater submerged vascular plants have declined in many aquatic systems, especially Chesapeake Bay. These plants provide food and habitat to diverse and abundant animal populations. In Chesapeake Bay, this decline has been associated with an overall decline in the abundance of fish and wildlife, and has been interpreted as an indication of serious disturbance in the ecological balance of the estuary. More than 10 native species of submerged aquatic plants in Chesapeake Bay have decreased in abundance. In the upper estuary, this decline was preceded by an invasion of Eurasian watermilfoil (*Myriophyllum spicatum*), which eventually also died back. Runoff of herbicides, including atrazine, from treated agricultural lands has been suggested as a possible factor involved in the disappearance of Chesapeake Bay submerged vegetation. During the past 20 years, the most widely used herbicide in the Chesapeake Bay watershed – and in the surrounding coastal plain – has been atrazine. Since its introduction into the region in the early 1960s, atrazine use has grown to about 200,000 kg annually in Maryland coastal communities alone. Potentially phytotoxic concentrations of atrazine would be expected in estuaries with the following characteristics (which seem to apply in most of upper Chesapeake Bay): immediately adjacent to cornfields in the watershed; rains occur shortly after atrazine application; clay soils in fields producing more rapid runoff; soils with circumneutral pH and relatively low organic content; and large estuarine areas of low salinity and poor mixing.

Most authorities agree that atrazine could induce some loss in aquatic vegetation but was not likely to have been involved in the overall decline of submerged plants in Chesapeake Bay, and that nutrient enrichment and increased turbidity probably played major roles. In the open waters of Chesapeake Bay, atrazine concentrations have rarely exceeded 1.0 μg/L; in major tributaries, such as the Choptank and Rappahanock Rivers, concentrations of 5.0 μg/L may occur after a major spring runoff. These runoffs sometimes generate transient, 2–6 h concentrations up to about 40.0 μg/L in secondary tributaries. In some small coves on the Chesapeake Bay, submerged plants may be exposed periodically to atrazine concentrations of 5.0–50.0 μg/L for brief periods during runoffs; however, dilution, adsorption, and degradation tend to reduce concentrations in the water phase to <5.0 μg/L within 6–24 h. Since atrazine degrades rapidly in estuarine conditions (half-time persistence (Tb1/2) of 1–6 weeks), concentrations of atrazine on suspended and deposited estuarine sediments were seldom >5.0 μg/kg, suggesting little potential for accumulation. The photosynthesis of redhead grass (*Potamogeton perfoliatus*) was significantly inhibited by atrazine concentrations of 10.0–50.0 μg/L; however, it returned to normal levels within 1 h after atrazine was removed. Recovery of redhead grass within several weeks has also been documented after exposure to 130.0 μg/L for 4 weeks. In Chesapeake Bay, potential long-term exposure of submersed aquatic plants to concentrations of atrazine in excess of 10.0 μg/L is doubtful; therefore, any observed reductions in photosynthesis by these plants under such conditions would be minor and reversible.

Some authorities, however, suggest that the effects of atrazine on aquatic plants may be substantial. For example, atrazine concentrations between 1.0 and 5.0 μg/L adversely affect phytoplankton growth and succession; this, in turn, can adversely affect higher levels of the food chain, beginning with the zooplankton. Also, exposure to environmentally realistic concentrations of 3.2–12.0 μg atrazine/L for about 7 weeks was demonstrably harmful to wild celery

(*Vallisneria americana*), a submersed vascular plant in Chesapeake Bay. At the highest concentrations of 13.0–1104.0 μg/L for 3–6 weeks, growth of representative submerged macrophytes in Chesapeake Bay was significantly depressed, and longer exposures were fatal to most species. Atrazine concentrations of 100.0 μg/L reportedly cause permanent changes in algal community structure after exposure for 14 days, including decreased density and diversity, altered species composition, and reduced growth. It seems that additional research is needed on the role of atrazine and on its interactions with other agricultural chemicals with regard to observed declines in submerged plants. It is emphasized that degradation products of atrazine did not play a role in the disappearance of the submerged vascular plants from the Chesapeake Bay. For example, 500.0 μg/L of deethylated atrazine was needed to produce 20–40% photosynthetic inhibition in four major species of submerged macrophytes in 2 h, but only 95.0 μg/L of the parent atrazine caused 50% inhibition in a similar period.

Many studies have been conducted on the effects of atrazine on various species of aquatic flora under controlled conditions. At concentrations of 1.0–5.0 μg/L, and exposure periods of 5 min to 7 weeks, documented adverse effects in sensitive species included inhibition in photosynthesis, growth, and oxygen evolution. Higher concentrations were associated with altered species composition, reduced carbon uptake, reduced reproduction, high accumulations of atrazine, decreased chlorophyll *a* production, ultrastructural changes on chloroplasts, and death. Phytotoxic effects were significantly increased at elevated levels of incident illumination, elevated water temperatures, decreased water pH, decreased dissolved oxygen concentrations, decreased nutrient content, and at increasing atrazine concentrations in the water column. Phytotoxicity was not significantly influenced by atrazine concentrations in the sediments or hydrosoils, or by the salinity of the medium. There are marked differences in sensitivity to atrazine among estuarine marsh plant species. Atrazine, at typical concentrations occurring in areas draining agricultural fields, should pose no significant adverse effects to *Spartina alterniflora*. In contrast, *Juncus roemerianus* at 250.0 μg atrazine/L or greater are likely to die or decline.

Atrazine was 4–10 times more effective than its degradation products in producing growth reduction, photosynthesis inhibition, and acetylene-reducing ability in two species of green algae, *Chlorella pyrenoidosa*, *Scenedesmus quadricauda*, and three species of cyanobacteria, *Anabaena* spp. Atrazine reduced growth by 50% at 0.03–5.0 mg/L and inhibited photosynthesis 50% at 0.1–0.5 mg/L. Comparable values for deethylated atrazine were 1.0–8.5 mg/L for growth reduction, and 0.7–4.8 mg/L for photosynthesis inhibition; for deisopropylated atrazine, these values were 2.5–> 10.0 mg/L for growth reduction and 3.6–9.3 mg/L for photosynthesis inhibition; hydroxyatrazine and diaminoatrazine were nontoxic to most cultures tested. Smooth cordgrass (*Spartina alterniflora*), the major emergent species in North American salt marshes, is only slightly affected by relatively high levels of atrazine, possibly due to its ability to metabolize this compound. Studies with radiolabeled atrazine and *Spartina* roots were conducted during 2-day exposures, followed by 28 days in atrazine-free solution. After 2 days 90% of the atrazine had translocated to the shoots. Atrazine was readily metabolized to chloroform-soluble substances, then to water-soluble substances, and finally to insoluble substances. Atrazine in the chloroform-soluble fraction decreased from 85 to 24% by day 28; the aqueous fraction contained 15% at the start and 60% at day 28. The basis of *Spartina* resistance is primarily due to its ability to convert atrazine to N-dealkylation products, such as 2-chloro-4-amino-6-isopropylamino-*s*-triazine. However, at least 14 water-soluble metabolites were isolated; about half contained the fully alkylated triazine rings, and most of the others had the 4-amino-6-isopropylamino derivative. Acid hydrolysates of the metabolites contained small amounts of amino acids, suggesting that a conjugation pathway, in addition to N-dealkylation, may be operative in *Spartina*.

Freshwater species of algae are among the most sensitive of all aquatic species tested.

The ability of freshwater algal cells to accumulate atrazine was significantly correlated with cell volume and surface area, and accumulations were higher in the more sensitive species. Uptake of radiolabeled atrazine by 4 species of freshwater green algae and 4 species of diatoms was rapid: about 90% of the total uptake occurred within the first hour of exposure during exposure for 24 h; maximum levels were reached 3–6 h after initial exposure; and accumulations were higher in algae than in diatoms. A green alga (*Chlorella kessleri*) showed numerous adverse effects when subjected to sublethal concentrations of atrazine over a 72-h period including dose-dependent growth inhibition, protein synthesis decrease, photosynthesis reduction, and stimulation of fatty acid synthesis.

Estuarine fungi contribute substantially to plant detritus due to their abundance and potential for degradation. Fungi are known to accumulate soluble atrazine from seawater through sorption, and release up to 2.2% as hydroxyatrazine and other atrazine metabolites; another 4.6% is more tightly associated and less available to the external environment. The combined processes result in atrazine accumulation, and may contribute to its transport and redistribution through the estuary.

3.4.3 Aquatic Animals

A marine copepod (*Acartia tonsa*) was the most sensitive aquatic animal tested against direct effects of atrazine, having a 96-h LC50 of 94.0 μg/L. Atrazine was most toxic to estuarine crustaceans at low salinities; however, it was most toxic to estuarine fishes at high salinities. Adverse effect levels to selected species of aquatic invertebrates and fishes ranged from 120.0 to 500.0 μg/L, based on lifetime exposure studies. The most sensitive criterion measured during long-term chronic exposure varied among species. Among freshwater invertebrates, for example, the most useful criterion was survival for *Gammarus*, the number of young produced for *Daphnia*, and developmental retardation for *Chironomus*.

Ambient concentrations as low as 20.0 μg atrazine/L have been associated with adverse effects on freshwater aquatic fauna, including benthic insects and teleosts, although effects were considered indirect. For example, the abundance of emerging chironomids (*Labrundinia pilosella*), and other aquatic insects declined at 20.0 μg atrazine/L. Richness of benthic insect species and total emergence declined significantly with atrazine addition. The effects were primarily indirect, presumably by way of reduction in food supply of non-predatory insects, and to some extent their macrophyte habitat. Dietary habits and reproductive success were negatively affected in three species of fish after exposure for 136 days in ponds containing 20.0 μg atrazine/L. About 70% of the original concentration applied was present in water at the end of the study. The reproduction of channel catfish (*Ictalurus punctatus*) and gizzard shad (*Dorosoma cepedianum*) failed, and that of bluegills, as measured by number of young per pond, was reduced by more than 95%. Also, the number of prey items in the stomachs of bluegills was significantly higher in control ponds (25.6) than in a treated pond (3.8), and number of taxa represented was significantly greater. Macrophyte communities in treated ponds were reduced by more than 60% in 2 months. It was concluded that the effects of atrazine on bluegills were probably indirect, and that the reduction of macrophytes that had provided habitat for food items led to impoverished diets and more cannibalism by adult bluegills.

Bioaccumulation of atrazine from freshwater is limited, and food chain biomagnification is negligible. Rainbow trout fed diets containing 100.0 mg atrazine/kg ration for 84 days had no significant accumulations in tissues, although some accumulation occurred (maximum of 0.9 mg/kg lipid weight in liver) at 1000.0 mg/kg ration. In a farm pond treated once with 300.0 μg atrazine/L, residues at 120 days posttreatment ranged between 204.0 and 286.0 μg/kg in mud and water, and from not detectable in bullfrog (*Rana catesbeiana*) tadpoles to 290.0 μg/kg (all FWs) in whole bluegills; values were intermediate in zooplankton and clams. No residues were detectable in biological components one year posttreatment, when residues were <21.0 μg/kg in water and mud. In a laboratory stream treated four times with 25.0 μg atrazine/L for 30 days, followed by depuration for 60 days, maximum accumulation factors ranged from about 4 in annelids to 480 in mayfly nymphs; however, residue concentrations declined to posttreatment levels within a few days after depuration began. Maximum atrazine concentrations recorded, in mg/kg whole organism FW, were 0.2 in the clam *Strophitis rugosus*, 0.4 in the snail *Physa* sp., 0.9 in crayfish *Orconectes* sp., 2.4 in the mottled sculpin *Cottus bairdi*, 3.0 in the amphipod *Gammarus pseudolimnaeus*, and 3.4 in mayflies *Baetis* sp. In studies with the freshwater snail *Ancylus fluviatilis* and fry of the whitefish *Coregonas fera*, atrazine was rapidly accumulated from the medium by both species and saturation was reached within 12–24 h; bioconcentration factors (BCFs) were 4–5 at ambient water concentrations of 50.0–250.0 μg atrazine/L. Elimination of atrazine was rapid: 8–62 min for *Coregonus*, and 18 min for *Ancylus*. No accumulation of atrazine was recorded in mollusks, leeches, cladocerans, or fish when contamination was by way of the diet. Atrazine accumulations in *Daphnia pulicaria* were significantly correlated with whole body protein content at low (8°C) water temperatures, and with fat content at elevated (20°C) water temperatures.

Atrazine is rapidly degraded in boxcrabs (*Sesarma cinereum*) feeding on smooth cordgrass (*Spartina alterniflora*) grown in radiolabeled atrazine solution. After 10 days, only 1.2% of the total radioactivity in the crab was unchanged atrazine, compared to 24% in the food source. The accumulation of water-soluble atrazine metabolites (86% of total radioactivity) in *Sesarma* suggested that glutathione conjugation, or a comparable pathway, was responsible for the almost complete degradation and detoxification of atrazine in crabs. Atrazine does not appear to be a serious threat to crabs in Chesapeake Bay, where water concentrations of 2.5 μg/L have been recorded, although it could have an indirect effect on crabs by decreasing the algal population, which composes a portion of their diet.

3.4.4 Birds

Atrazine is not acutely lethal to birds at realistic environmental levels, i.e., oral LD50 values were >2000.0 mg/kg BW and dietary LC50s were >5000.0 mg/kg ration. Also, the probability is low for chronic effects of atrazine on wetland aquatic organisms and for biomagnification of toxic residues through waterfowl food chains. However, indirect effects of atrazine on insect- and seed-eating birds have not been investigated, and this may be critical to the survival of certain species during nesting and brood-rearing. Studies are needed on the potential indirect ecosystem effects of atrazine, with special reference to seed-eating birds.

Domestic chickens (*Gallus* sp.) rapidly metabolized atrazine by way of partial N-dealkylation accompanied by hydrolysis; dealkylation occurred mainly at the ethylamino group, resulting in intermediate degradation products. In vitro studies with bird liver homogenates also demonstrated active transformation of atrazine and its metabolites. Chicken liver homogenates released nonextractable atrazine residues that had accumulated in corn plants, present mainly as 2-chloro mono N-dealkylated compounds, and subsequently metabolized them to 2-hydroxy analogs. Liver homogenates in the goose (*Anser* sp.) contained enzyme systems that metabolized atrazine by partial N-dealkylation and hydrolysis; hydrolysis predominated and resulted in the formation of hydroxyatrazine, which does not undergo further degradation by dealkylation. But partly N-dealkylated metabolites, such as deethylatrazine and deisopropylatrazine, were further hydrolyzed to the corresponding hydroxy analogs.

3.4.5 Mammals

Data are lacking for atrazine's effects on mammalian wildlife, although there is a growing body of evidence on domestic and small laboratory mammals. Available data demonstrate that mammals are comparatively resistant to atrazine, and that the compound is not carcinogenic, mutagenic, or teratogenic. However, there is a reported increase in the incidence of mammary gland tumors in rats given dietary equivalents of a lifetime dose of 70.0 mg atrazine/kg BW. There have been no established cases of skin irritation resulting from experimental or commercial applications of atrazine, and no documented cases of poisoning in humans. No observable ill effects were detected in cattle, dogs, horses, or rats fed diets that included 25.0 mg atrazine/kg food over extended periods. Most members of the triazine class of herbicides, including atrazine, have low acute oral toxicities – usually >1000.0 mg/kg BW. But at dosages bordering on lethality, rats showed muscular weakness, hypoactivity, drooped eyelids, labored breathing, prostration, altered liver morphology and renal function, and embryotoxicity. There seems to be a causal link between tumor formation and triazine-mediated hormonal balance, suggesting the existence of a threshold value below which contact with atrazine will have no effect on tumor formation.

Biomarkers of atrazine exposure is a developing field that merits additional research. For example, concentrations of atrazine in saliva of rats were significantly correlated with rat-free atrazine plasma concentrations. About 26% of the atrazine in rats is bound to plasma proteins (and is unavailable for transport from blood to saliva) and is independent of plasma levels of atrazine. Salivary concentrations of atrazine reflect total plasma-free atrazine concentration – in the 50.0–250.0 μg/L range – which may be of toxicological significance.

Animals feeding on atrazine-treated crops are at limited toxicological risk. Crop plants metabolize atrazine to hydroxyatrazine, dealkylated analogs, and cysteine- and glutathione-conjugates of atrazine; mature plants contain little unchanged atrazine. Bound atrazine residues in plants are of limited bioavailability to animals. Metabolic degradation of atrazine in mammals is usually rapid and extensive; unchanged atrazine was recovered only from the feces. Liver enzyme systems in pigs, rats, and sheep metabolize atrazine by partial N-dealkylation and hydrolysis. However, atrazine is reportedly converted in vivo to *N*-nitrosoatrazine in mice (*Mus* sp.). Since *N*-nitrosoatrazine is carcinogenic and mutagenic to laboratory animals, more research is recommended on the extent of nitrosation of atrazine in the environment.

3.5 Recommendations

Labels on products containing atrazine are required to contain information on acceptable uses and potential hazards to groundwater, and to fish and wildlife. At present, atrazine is approved for use as an herbicide to control broadleaf and grassy weeds on corn, sorghum, sugarcane, pineapple, macadamia nuts, rangeland, turf grass sod, conifer reforestation areas, Christmas tree plantations, grass seed fields, noncrop land, guava, grass in orchards, millet, perennial ryegrass, and wheat. Because atrazine is expected to leach into groundwater, it was recommended that labels of atrazine products bear the following statement: "*Atrazine leaches readily and accepted label rates have been found to result in contamination of water supplies by way of groundwater. Therefore users are advised to avoid use of atrazine in well drained soils, particularly in areas having high groundwater tables.*" A cautionary statement on potential hazards to living resources is another labeling requirement: "*This pesticide is toxic to aquatic invertebrates. Do not apply to water or wetlands. Runoff and drift from treated areas may be hazardous to aquatic organisms in neighboring areas. Do not contaminate water by cleaning of equipment or disposal of wastes. Do not discharge into lakes, streams, ponds, or public water supplies unless in accordance with an [approved EPA] permit.*"

Permissible tolerances for atrazine range from 0.02 mg/kg in meat, milk, and eggs, to 15.0 mg/kg in orchard grass forage, fodder, and hay. However, the 15.0 mg/kg tolerance in forage is considered high, and a new upper limit of 4.0 mg/kg is proposed. This limit would be expressed in terms of atrazine and three major metabolites: 2-amino-4-chloro-6-isopropylamino-1,3,5-triazine; 2-amino-4-chloro-6-ethylamino-1,3,5-triazine; and 2-chloro-4, 6-diamino-1,3,5-triazine.The maximum recommended safe level of atrazine to algal diatoms is 10.0 μg/L, although temporary inhibition of chlorophyll production in sensitive algal species has been reported in the range of 1.0–5.0 μg/L. Proposed atrazine concentrations for aquatic life protection range from about 1.0 to 11.0 μg/L; 1.0–2.0 μg/L for protection of estuarine productivity; 1.0–7.0 μg/L for no adverse effect levels to most species of submerged aquatic vegetation; less than 5.0 μg/L to prevent gill and kidney histopathology in rainbow trout and disrupted swimming behavior in zebrafish and goldfish; 5.0–10.0 μg/L for minor reductions in photosynthesis in sensitive species of aquatic macrophytes; 9.0 μg/L for sensitive aquatic invertebrates, as judged by an uncertainty factor of 10 applied to a 96-h LC50; and 11.0 μg/L for salt marsh algae, based on the least effect level of 110.0 μg/L, and an uncertainty factor of 10. Atrazine concentrations >11.0 μg/L sometimes occur during periods of runoff and non-flushing, but rarely persist at levels necessary to markedly inhibit photosynthesis in aquatic plants, i.e., 60.0–70.0 μg/L. At 80.0 μg/L, rainbow trout show kidney necrosis of endothelial cells after exposure for 28 days, and this suggests that atrazine criteria that protect sensitive plants will also protect aquatic vertebrates.

In laboratory animals, atrazine is only slightly toxic on an acute basis; no carcinogenic, mutagenic, or reproductive effects have been seen at low doses, and reduced food intake and body weight were the primary adverse effects seen at high doses in chronic studies with rats and dogs. However, data are lacking on indirect ecosystem effects of atrazine application on terrestrial wildlife, especially on insectivores and granivores; studies should be initiated in this subject area.

No allowable daily intake of atrazine in the human diet has been established, although 0.0375 mg/kg BW daily has been proposed – equivalent to 2.25 mg daily for a 60-kg adult, or 1.5 mg/kg diet, based on 1.5 kg food daily. In humans, the theoretical maximum residue contribution (TRMC) – a worst case estimate of dietary exposure – is 0.77 mg daily, assuming 1.5 kg of food eaten daily; this is equivalent to 0.51 mg/kg diet, or 0.013 mg/kg BW daily for a 60-kg person. Another TRMC calculation is based on 0.233 mg daily per 1.5 kg diet, equivalent to 0.156 mg/kg diet, or 0.0039 mg/kg BW daily for a 60-kg person. Both TRMC estimates are substantially below the proposed limit of 0.0375 mg/kg BW daily. Lifetime exposure to drinking water concentrations of 2.3 μg atrazine/L poses negligible risk to human health, as judged by the no adverse effect level of 7.5 μg/L when 1% of the allowable daily intake is obtained from this source. Higher allowable concentrations are proposed over short periods: 123.0 μg/L for adults and 35.0 μg/L for children, over a 10-day period. The proposed drinking water criterion to protect human health in Western Europe is <0.1 μg/L. In the United States, it should not exceed 3.0 μg atrazine/L drinking water, although some authorities recommend less than 3.6 μg atrazine/L.

Additional data are needed on toxicity, environmental fate, and chemistry of atrazine in order to maintain existing registrations or to permit new registrations. Specifically, data are needed on mobility and degradation rates of atrazine and its metabolites in soils; accumulation studies in rotational crops, fish, and aquatic invertebrates; and chronic testing with representative flora and fauna on survival, reproduction, carcinogenesis, teratogenesis, and mutagenesis. Animal metabolism studies are required if tolerances for residues in animal products are expressed in terms of atrazine and its metabolites. Finally, more research on aquatic species is merited on synergistic and additive effects of atrazine in combination with other agricultural chemicals at realistic environmental levels of 1.0–5.0 μg/L, and on the toxic effects of dealkylated atrazine metabolites.

3.6 Summary

The herbicide atrazine (2-chloro-4-ethylamino-6-isopropylamino-1,3,5-triazine) is the most heavily used agricultural pesticide in North America. In the United States alone, more than 50 million kg (110 million pounds) are applied annually to more than 25 million ha (62 million acres), primarily to control weeds in corn and sorghum. Residues have been detected at phytotoxic concentrations in groundwater, lakes, and streams as a result of runoff from treated fields. Atrazine degrades rapidly, usually by way of hydrolysis, nitrogen dealkylation, and splitting of the triazine ring to less toxic compounds not normally inhibitory to plants and animals. The half-time persistence of atrazine in soils is usually about 4 days, but may range up to 385 days in dry, sandy, alkaline soils, under conditions of low temperature and low microbial densities. Half-time persistence is about 3 days in freshwater, 30 days in marine waters, 35 days in marine sediments, and less than 72 h in vertebrate animals.

Sensitive species of aquatic plants experience temporary, but reversible, adverse effects at concentrations in the range of 1.0–5.0 μg atrazine/L. However, potentially harmful phytotoxic concentrations of atrazine, i.e., >10.0 μg/L for extended periods, have not been documented in the environment, and are probably unrealistic under current application and degradation rates. Aquatic fauna are indirectly affected at atrazine concentrations of 20.0 μg/L and higher, partly through reduction of the food supply of herbivores, and partly through loss of macrophyte habitat. Direct adverse effects to aquatic invertebrates and fishes were measured at 94 μg/L and higher. Bioaccumulation of atrazine is limited, and food chain biomagnification is negligible in aquatic ecosystems. Birds are comparatively resistant to atrazine, having a low probability for uptake and retention. Known acute oral LD50 values for birds are >2000.0 mg/kg BW, and dietary LD50s are >5000.0 mg/kg ration. However, indirect ecosystem effects of atrazine on seed and insect-eating birds are unknown, and should be investigated. Data are lacking for atrazine toxicity to mammalian wildlife, but tests with domestic livestock and small laboratory animals indicate that this group is also comparatively resistant. Acute oral LD50s for mammals are >1750.0 mg/kg BW; no adverse effects were measured at chronic dietary levels of 25.0 mg/kg (about 1.25 mg/kg bod BW) and, for some species, 100.0 mg/kg diet.

Proposed criteria for aquatic life protection include <5.0 μg atrazine/L for sensitive species of aquatic flora and fauna, and <11.0 μg/L for most species of aquatic plants and animals. No criteria have been promulgated for human or animal health protection, although it has been suggested that <3.0 μg/L in drinking water, and <0.0375 mg atrazine/kg BW (<2.25 mg daily for a 60-kg adult, <1.5 mg/kg diet, based on consumption of 1.5-kg food daily) would pose a negligible risk to human health. Additional data are needed on toxicity, environmental fate, and chemistry of atrazine and its metabolites in order to maintain existing registrations or to permit new registrations. In particular, more research is needed on possible synergistic or additive effects of atrazine with other agricultural chemicals in aquatic environments.

BORON[a]

Chapter 4

4.1 *Introduction*

Borax ($Na_2B_4O_7 \cdot 10H_2O$) was the first of the boron (B) minerals to be traded by the Babylonians more than 4000 years ago for use in the working and welding of gold. Borax has been known as a cleaning agent since the days of the ancient Greek and Roman empires and was used as a food preservative in Europe and America, although its use for the latter purpose has been discontinued. Boron and its compounds were used in Egyptian and Roman eras to prepare borosilicate glass. Borax glazes were known from the year 200 ca; by 1556, borax was widely used throughout Europe as a flux. Boric acid (H_3BO_3) was first synthesized in 1707. Boric acid and borates are the main boron compounds of ecological significance; other boron compounds usually degrade or are transformed to borates or boric acid.

Boron is an essential trace element for the growth and development of higher plants, although the range between insufficiency and excess is generally narrow, varying with the plant; boron is not required in fungi and animals. In the southwestern United States, naturally elevated boron concentrations in surface waters used for irrigation may be sufficiently high to cause toxicity to plants of commercial importance. Another major source of boron entering ground and surface waters results from the use of borax-containing laundry products coupled with ineffective removal of boron by conventional sewage processes. Agricultural drainwaters contaminated with boron are considered potentially hazardous to waterfowl and other wildlife populations throughout areas of the western United States.

Medical and household uses of boric acid solutions as antiseptics have led to numerous accidental poisonings by ingestion or absorption through abraded skin, particularly in infants. Poisonings have been reported in English children consuming milk containing 0.7 g boric acid/L, and in burn patients treated topically with saturated boric acid solutions. In the 1940s, topical preparations of boric acid became a popular remedy for diaper rash in England. By 1953, at least 60 fatal cases of boric acid poisoning had been reported in English infants. Inhalation of boranes, especially decaborane ($B_{10}H_{14}$) (which is used as a rocket propellent), diborane (B_2H_6), and pentaborane (B_5H_9), is toxic to exposed workers. Boron compounds, especially boric acid, can also accumulate in animal tissues and produce a reduction in fertility, an increase in developmental abnormalities – especially those involving the skeletal system – stillbirth, and death. There seems to be a reasonable margin between a toxic dose in man and other vertebrates and in boron levels that may occur as incidental residues from the use of borax and boric acid in agriculture and industry.

4.2 *Environmental Chemistry*

The United States and Turkey supply about 80% of the global boron demand. Of the total annual U.S. production of about 500,000 tons, 45% is used in the manufacture of glass and glassware, 15% in laundry products, 10% in

---

[a]All information in this chapter is referenced in the following sources:

Eisler, R. 1990. Boron hazards to fish, wildlife, and invertebrates: a synoptic review. *U.S. Fish Wildl. Serv. Biol. Rep. 85(1.20)*, 32 pp.

Eisler, R. 2000. Boron. Pages 1567–1612 in *Handbook of Chemical Risk Assessment: Health Hazards to Humans, Plants, and Animals. Volume 3, Metalloids, Radiation, Cumulative Index to Chemicals and Species*. Lewis Publishers, Boca Raton, Florida.

enamels and glazes, and 8% in agricultural chemicals. It is estimated that boron compounds enter the North American environment at a rate of 32,000 tons annually as a result of human activities, primarily from laundry products, irrigation drainwater, agricultural chemicals, coal combustion, and mining and processing. Boron compounds tend to accumulate in aquatic ecosystems because of the relatively high water solubility of these compounds.

The chemistry of boron is exceedingly complex and rivals that of carbon in its diversity. Most boron compounds, however, enter or degrade in the environment to borates (B–O compounds), such as borax and boric acid, and these are considered to be ecologically most significant.

Toxicosis in animals has resulted from ingestion of boric acid or borax solutions, from topical applications of boric acid solutions to damaged skin, and from inhalation of boranes; the exact mechanisms of action are not understood. Boron and its compounds are potent teratogens when applied directly to the embryo, but there is no evidence of mutagenicity or carcinogenicity. Boron's unique affinity for cancerous tissues has been exploited in neutron capture radiation therapy of malignant human brain tumors.

4.2.1 Sources and Uses

Boron is a dark brown element that is widespread in the environment but occurs naturally only in combined form, usually as borax, colemanite ($Ca_2B_6O_{11} \cdot 5H_2O$), boronatrocalcite ($CaB_4O_7NaBO_2 \cdot 8H_2O$), and boracite ($Mg_7Cl_2B_{16}O_{30}$). In the United States, boron deposits in the form of borax are concentrated in the desert areas of southern California, especially near Boron, California. Proven deposits of sodium tetraborates – from which borax is prepared and from which boron can be isolated – also exist in Nevada, Oregon, Turkey, Russia, and China. About 300,000 metric tons of boron are removed from mined ore each year. The United States supplies about 70% of the world boron demand, and Turkey supplies 18%; the most common commercial compounds are boric acid and borax.

In 1988, the United States produced 566,093 metric tons of boric oxide, imported an additional 60,000 metric tons of boron-containing minerals, and exported 589,680 metric tons of boric acid and borates. The majority of the boron produced annually at facilities in Oklahoma, New Jersey, Nevada, and Pennsylvania is in the form of sodium tetraborate compounds. Of the total production, about 42% occurs as anhydrous borax ($Na_2B_4O_7$), 29% as borax pentahydrate ($Na_2B_4O_7 \cdot 5H_2O$), 10% as borax decahydrate or borax, and 16% as boric acid or boric oxide (B_2O_3). Boron and its compounds are used in the manufacture of glassware (40–45%); soaps and cleansers (15%); enamels, frits, and glazes (10%); fertilizers (5%); and herbicides (2–3%); 22–28% goes for other uses including cosmetics, insecticides, antifreeze, as neutron absorbers in atomic reactors, and leather tanning. Borates have some toxicity to insects and, in relatively high concentrations, can control cockroaches, woodboring insects, gypsy moths, and larvae of flies in manure piles and in dog runs. Some organoboron compounds are used to sterilize fuel distribution and storage systems against fungi and bacteria. Radioboron-10 is widely used in radiation therapy against brain tumors, especially in Japan. In medicine, certain amine-carboxyborane derivatives show promise in reducing serum cholesterol and triglyceride concentrations, in alleviating some forms of chronic arthritis, and as antineoplastic agents. Other boron compounds are used widely as thermal protection materials in space probes, in fireproofing of fabrics and wood, in leather manufacture, in numerous pharmaceuticals and hygienic products, in steel hardening, in deoxidation of bronze, as a high energy fuel, as neutron-absorbing shielding near atomic reactors, and as water softeners, pH adjusters, emulsifiers, neutralizers, stabilizers, buffers, and viscosifiers.

The major global environmental reservoirs of boron (metric tons) include continental and oceanic crusts (10^{15}), oceans (10^{12}), groundwater (10^8), ice (10^8), coal deposits (10^7), commercial borate deposits (10^7), biomass (10^7), and surface waters (10^5). The largest flows of boron in the environment arise from the movement of boron into

Table 4.1. Environmental sources of domestic boron.

Source	Metric Tons Annually
Laundry products	14,000
Agricultural chemicals and fertilizers	7000
Coal combustion	4000
Mining and processing	3000
Glass and ceramics	1500
Miscellaneous	2500
Total	32,000

the atmosphere from the oceans at 1.3–4.5 × 10^6 tons annually. Drainage from soil systems into groundwaters and surface waters accounts for 1.3 × 10^5 to 1.3 × 10^6 tons yearly. And boron mining and volcanic eruptions account for 4 × 10^5, and 2 × 10^5 tons of boron per annum, respectively.

Boron enters the environment at about 32,000 metric tons annually in the United States (Table 4.1); most ends up in the aquatic environment because of the relatively high water solubility of all boron compounds, especially boron-containing laundry products and sewage. Conventional sewage treatment removes little or no boron. Studies of domestic wastewaters in California and Israel, using boron isotopic composition techniques, show that boron in sewage is derived from sodium borate components used in household detergents. Of the total boron in coal, as much as 71% may be lost to the atmosphere upon combustion; more than 50% of the boron found in coal ash is readily water soluble. The release of boron from coal flyash to leachate water is dependent upon the ash to water ratio: at 1.0 g ash/L, up to 90% of the boron is soluble; at 50 g/L, only 40% is released; at 100.0 g/L, less than 30% is released. Coating of coal ash with aluminum solution reduces boron solubility by about 90% due to the formation of an insoluble aluminum–borate complex.

Boron compounds listed in the "Commodity List of Explosives and Other Dangerous Articles" are boron trichloride (BCl_3), boron trifluoride (BF_3), decaborone ($B_{10}H_{14}$), and pentaborane (B_5H_9). Boron trichloride is a corrosive liquid; the maximum quantity allowed in containers by rail is one liter, and by air, only one container is permitted per aircraft. Boron trifluoride is a nonflammable gas restricted to 140.0 kg in one outside container by rail, and to 140.0 kg in cargo planes only. Decaborane is a flammable solid, and transport by rail or air is limited to 12 kg. Pentaborane is a flammable liquid and is prohibited for transport by air or rail. Diborane (B_2H_6) and higher boranes are unstable and are classified as dangerous articles in transport; no more than 0.1 kg can be shipped in a cylinder. Organic boron–oxygen compounds readily hydrolyze and should be stored and transferred in an inert atmosphere; usually, glass containers are used for shipping small quantities, and steel containers or tank cars are used for bulk items. Hazardous atmospheric conditions resulting from high concentrations of boron compounds are localized and are not considered a serious environmental problem.

4.2.2 Chemical Properties

The element boron has an atomic number of 5, a molecular weight of 10.811, an oxidation state of 3 for simple compounds (but other oxidation states for carboranes and other polyhedral cage boron compounds), a specific gravity of 2.34, a melting point of 2300°C, sublimation at 2550°C, and is almost insoluble in water. Boron exists as B-10 (19.78%) and B-11 (80.22%) isotopes, and it contributes about 0.001% to the earth's crust, although it does not occur free in nature. The chemistry of boron is exceedingly complex and rivals that of carbon in diversity.

Most boron compounds degrade in the environment to B–O (borate) compounds, and these are the boron compounds of ecological significance – especially borax and boric acid. Sodium tetraborate decahydrate (borax) has a melting point of 75°C, a boiling point of 320°C, and is soluble in water to 20.0 g/L at 0°C and to 1700.0 g/L at 100°C. Boric acid has a melting point of 169°C, a boiling point of 300°C and, like borax, is exceedingly soluble in water: 63.5 g/L at 30°C and 276.0 g/L at 100°C.

Boron exists in several forms in the soil, and in soil solution it exists largely as the undissociated weak monobasic acid that accepts hydroxyl groups. Most plant-available boron in soils is associated with soil organic matter, with the hot-water-soluble boron fraction, and with soil solution pH ranges of 5.5–8.5 and 10–11.5. It is assumed that boron adsorbs to soil particles and aluminum and iron oxide minerals. Boron mobility in soils is reduced under conditions of pH 7.5–9.0, and with high abundance of amorphous aluminum oxide, iron oxide, and organic content.

In water, boron readily hydrolyzes to form the electrically neutral, weak monobasic acid H_3BO_3 and the monovalent ion $B(OH)_4^-$. Waterborne boron may be adsorbed by soils and sediments. The predominant boron species in seawater is boric acid; concentrations are higher at higher salinities and in proximity to industrial waste discharges. In seawater, borate or boric acid occurs naturally at 4.5–5.5 mg/L. About 76% of the total inorganic boron in seawater occurs as undissociated boric acid $[B(OH)_3]$, and the remainder is identified as the borate ion $[B(OH)_4^-]$. Of the total borate ion, 44% appears to be complexed with sodium, magnesium, and calcium. Other evidence suggests additional complexation of borate with ferric ions and polyhydroxylated organic compounds.

Atmospheric boron is in the form of particulates or aerosols of borides, boron oxide, borates, boranes, organoboron compounds, trihalide boron compounds, or borazines. The half-time persistence of airborne boron particles is short, usually in the order of days.

Despite the development of sophisticated instrumentation and techniques, the accurate determination of boron in biological materials is difficult at concentrations less than 1.0 mg B/kg. Problems associated with analysis of boron from biological sources include contamination from teflon vessels during microwave digestion; losses due to freeze-drying; variations in boron isotope ratios, standards preparation, and reagent backgrounds; and instrumental interference. Inductively coupled plasma-mass spectrometry now allows quantitation of percutaneous absorption of ^{10}B in ^{10}B-enriched boric acid, borax, and disodium octaborate tetrahydrate in biological materials, although absorption through intact human skin is significantly less than the mean daily dietary intake.

4.2.3 Mode of Action

A proposed essential role for boron is as a regulator of enzymatic pathways closely involved with energy substrate metabolism, insulin release, and the immune system. Boron influences the activities of at least 26 enzymes – including reductases, transferases, hydrolases, and isomerases – examined in various biological systems by acting on the enzyme directly and binding to cofactors or substrates. The complexing ability of the boron atom is considered to be the key explanation of why it is essential to higher plants, although the exact mechanism of action is still unknown. In biological systems, boron probably is complexed with hydroxylated species and inhibition or stimulation of enzymes and coenzymes are pivotal in its mode of action. Boron interacts with substances of biological interest, including polysaccharides, pyridoxine, riboflavin, dehydroascorbic acid, and pyridine nucleotides. Boron's complexing ability is thought to beneficially influence transport of sugars and other organic compounds, production of plant growth regulators, biosynthesis of nucleic acids and phenolic acid, carbohydrate metabolism, respiration, and pollen germination.

Boron poisoning in animals is primarily an experimental phenomenon, although livestock in certain regions may be exposed to high concentrations in drinking water – up to 80.0 mg B/L – that have not shown to be toxic. Toxicosis in humans has resulted from ingestion of boric acid or borax solutions, topical applications of boric acid solutions to burn-damaged skin, and inhalation of boranes. In mammals, boron is thought to regulate parathyroid function through metabolism of phosphorus, magnesium, and especially calcium. Boron has a close relation with calcium metabolism, most likely at the cell membrane level. The toxicological effects of boric acid and borax are similar for different species. Other inorganic borates that dissociate to boric acid display similar toxicity, whereas those that do not dissociate to boric acid may display a different toxicological profile.

Dietary boron at nontoxic concentrations, as sodium borate or boric acid, is rapidly and almost completely absorbed from the gastrointestinal tract, does not seem to accumulate in healthy tissues, and is excreted largely in urine, usually within hours, but sometimes as long as 23 days; similar patterns are evident in humans, dogs, cows, rabbits, rodents, and guinea pigs. Urinary boron excretion changes rapidly with changes in boron intake, suggesting that the kidney is the site of homeostatic regulation. Boron does not seem to accumulate in soft tissues of animals, but does accumulate in bone; cessation of exposure to dietary boron resulted in a rapid drop in bone boron, usually within 24 h. Boric acid poisoning in animals, regardless of route of administration, is characterized by the following signs: generalized erythema (boiled lobster appearance) starting in the axillary, inguinal, and face regions, eventually covering the entire body with conjunctival redness, followed by massive desquamation 2–3 days later; acute gastroenteritis, including nausea and vomiting; diarrhea; anorexia; cardiac weakness; excessive urinary excretion of riboflavin; decreased oxygen uptake by the brain; hypoacidity; altered enzyme activity levels; impaired growth and reproduction; and death from circulatory collapse and shock, usually within 5 days.

Boron hydrides or boranes, such as B_2H_6, B_4H_{10}, and B_5H_9, from chemical processes produce acute central nervous system (CNS) pulmonary damage and lung disease through inhalation. Boranes produce toxic effects by creating embolisms of hydrogen gas as they react with tissue, and by depleting biogenic amines of the CNS and inhibiting aminotransferases and other pyridoxal-dependent enzymes. Boranes produce similar effects in humans and animals, and these are generally ascribable to CNS depression and excitation. Symptoms of borane intoxication include pulmonary irritation, headache, chills, fatigue, muscular weakness and pain, cramps, dizziness, chest tightness, and pneumonia. Boranes may adversely affect male reproductive capacity, but this requires verification. Decaborane ($B_{10}H_{14}$), as one example, is a highly lipid-soluble compound that can enter the body through inhalation, ingestion, or the skin. In water, decaborane is rapidly transformed into intermediate products that are eventually degraded to boric acid. The intermediate products, but not decaborane or boric acid, reduce phosphomolybdic acid and inhibit glutamic-oxaloacetic transaminase; treatment of intermediates with pyridoxal phosphase tends to reverse the inhibitory activity. Low decaborane doses cause behavioral effects such as depression, catatonia, and convulsions.

Inorganic borates are comparatively toxic, apparently complexing hydroxy compounds and interfering with protein synthesis. Organoborate compounds exert physiological effects on the CNS and peripheral nervous system, acting as spasmolytics, sedatives, and convulsants, depending on their structure. Boron trihalides, such as BBr_3, BCl_3, and BF_3, are corrosive to the eyes, skin, and mucous membranes, and will cause burns on the skin – apparently due to the hydrolysis of the trihalides to their halogen acids, and not to boron.

Boron is a potent teratogen when applied directly to the embryo. Boric acid injected into chicken and amphibian embryos produced abnormal development of the neural tube, notochord, tail, and limbs, perhaps through complexing polyhydroxy compounds and interfering with riboflavin metabolism. Boron and its compounds, however, are neither mutagenic nor carcinogenic. Nonmutagenicity is based on results of the *Salmonella typhimurium*–mammalian microsome mutagenicity assay; boron neither enhances nor inhibits the activity of benzo[a]pyrene, a known mutagen. There is no evidence that boron is a possible carcinogen, although long-term, selective uptake of boron by tumors has been reported.

Boron seems to have an affinity for cancerous tumors, and this property has been exploited in radiation therapy. Boron-10 has been used in neutron capture therapy to cure malignant sarcomas implanted in the hind legs of mice, as well as spontaneous malignant melanomas in pigs. The sulfhydryl borane monomer $(B_{12}H_{11}SH)^{2-}$ is used as a B-10 carrier in neutron therapy of malignant human brain tumors and seems to be most effective at 30 μg B-10/kg tissue. Polyhedral boranes attached to monoclonal antibodies that are tumor-specific may become useful in tumor therapy by neutron irradiation. It is possible, however, that uptake of boron may be a nonspecific attribute of tumors and of a variety of normal tissues that lack a blood–brain barrier. Thus, the potential usefulness of selected B-10 carriers for treating extracranial neoplasms seems questionable at this time.

4.3 Concentrations in Field Collections

Terrestrial plants are normally rich sources of boron. Levels in meat and fish are usually low. But these generalizations are based on limited data. Boron is ubiquitous in the environment as a result of natural weathering processes; however, human activities such as mining, coal burning, and use of borax laundry detergents have resulted in elevated boron loadings in air, water, and soils. Comparatively high levels of boron occur in fishes, aquatic plants, and insects at Kesterson National Wildlife Refuge (KNWR) that was contaminated by agricultural drainwater. The availability of inductively coupled plasma-mass spectrometry (ICP-MS) technology enables measurement of boron concentrations and isotope ratios in a large number of biological samples with minimal sample preparation at detection limits of 0.11 μg/L.

4.3.1 Nonbiological Materials

Boron is distributed widely in the environment. Naturally, elevated boron levels are usually associated with marine sediments, thermal springs, large deposits of boron minerals, seawater, and certain groundwaters. Human activities, however, have resulted in elevated boron concentrations near coal-fired plants, in mine drainage waters, in municipal wastes, and in agricultural drainage waters. In one case, agricultural drainwater practices in western California produced boron concentrations in local rivers, groundwaters, and surface waters that exceeded the established limits for the protection of crops and aquatic life. Contamination of pristine groundwaters (<0.05 mg B/L) by domestic wastewater and agriculture-return flows (0.5–1.0 mg B/L) is documented by the isotopically distinguished signature of borate compounds. For example, in areas where calcium borates are applied as fertilizers, the $B^{11}:B^{10}$ ratio of the soil water and leachates are expected to be low and can be used as diagnostic tools for tracing agriculture-return flows.

Coal-fired power plants are major sources of atmospheric boron contamination; at least 30% of boron in coal is lost in this manner. The apparently large amounts of boron lost to the environment through stack emissions may be directly related to the organic content of coal. Also, disposal of B-laden drainage waters from boron mines is a major problem in certain geographic areas. In Turkey, for example, which possesses about 60% of the world's boron reserves – localized in a rectangular area about 100 × 200 km near the Simav River – drainage waters discharged from the mines as a result of borate production have elevated boron concentrations in the Simav River to levels unsuitable for crop irrigation purposes. About 68,000 ha of agricultural land irrigated by the Simav River are now threatened by boron pollution. In the United States, laundry detergents originating from household use may contribute as much as 50% of the boron loadings in effluents discharged into aquatic environments; lesser amounts are contributed by soil minerals, rainfall, and industry and sewage effluents.

4.3.2 Plants and Animals

Boron accumulates in both aquatic and terrestrial plants but it does not seem to biomagnify in the food chain. Boron does not biomagnify in aquatic food chains and has low potential to accumulate in aquatic organisms, as judged by studies in the San Joaquin River, California, and its tributaries. Marine and freshwater plants, fishes, and invertebrates concentrated boron from the medium by factors of less than 100, suggesting that biota is not a significant removal mechanism of boron from water. Boron concentrations in livers of birds collected from Baja California, Mexico, in 1986 were highest in the seed-eating mourning dove *Zenaida macroura* (maximum 28.5 mg B/kg FW) and lowest (maximum 8.7 mg B/kg FW) in fish-eating and omnivorous species.

Boron occurred at high concentrations in plants, insects, and fishes at KNWR in California – the recipient of contaminated agricultural drainwater – when compared to a nearby control area. Both boric acid and borax produced mortality and teratogenic development when injected into eggs. The effects of boron on waterfowl growth, physiology, and reproduction are discussed later.

Terrestrial plants, especially nuts and some fruits and vegetables, are rich sources of boron. Honey is another good source of boron, and concentrations up to 7.2 mg B/kg dry weight (DW) are reported. Boron concentrations are also elevated in marine plants, zooplankton, and corals, but are low in fishes and certain marine invertebrates. No data were found on boron levels in terrestrial mammalian wildlife. The average daily intake of boron in humans ranges between 1.0 and 25.0 mg; however, populations residing in areas of the western United States with natural boron-rich deposits may be exposed to higher-than-average levels of boron. Boron intake from drinking water is highly variable and dependent on the geographic source, the quantity of water consumed, and the water sources used to bottle other beverages. Current estimates of boron in domestic diets for normal human adults is about 1.0 mg daily; however, toddlers aged 2 years consumed 3.7 times more boron than mature males when adjusted for body weight. There is great variability in the human diet, and people from different countries have different sources of dietary boron. In the United States, for example, major sources of dietary boron include coffee, milk, orange juice, peanut butter, wine, pinto beans, and other juices and fruits. By contrast, in Mexico and Kenya, major sources of dietary boron include, corn, kidney beans, maguey (an alcoholic drink from fermented bananas), cacti, and mangos. In rats, increasing concentrations of boron in drinking water were associated with increasing tissue boron concentrations, plasma testosterone, and vitamin D, and a decrease in HDL cholesterol. Data for humans and domestic animals indicate that boron levels are elevated in bony tissues, but are always less than 0.6 mg B/kg (FW) or 1.5 mg/kg DW in other tissues examined.

4.4 Effects

Boron is essential for the growth of higher plants and has been applied to boron-deficient soils for at least 50 years to improve yields of many crops. Phytotoxic levels of boron occur usually as a result of human activities, such as boron-contaminated irrigation waters and excess applications of boron-rich fertilizers, sewage sludges, and flyashes. Boron compounds at comparatively high concentrations are used to control pestiferous insects through direct biocidal action, through enhancement of disease sensitivity, or through use as a chemosterilant.

Representative species of aquatic plants, invertebrates, fishes, and amphibians can usually tolerate up to 10.0 mg B/L medium for extended periods without adverse effects, although it has been suggested that concentrations greater than 0.1 mg B/L may ultimately affect reproduction in rainbow trout (*Oncorhynchus mykiss*), and greater than 0.2 mg/L may impair survival of other fish species. In waterfowl, diets that contain 30.0 or 100.0 mg B/kg FW adversely affect growth rate. Elevated tissue residues were recorded in ducks fed diets containing between 100.0 and 300.0 mg B/kg, and reduced survival occurred at dietary levels of 1000.0 mg B/kg. Boron is a potent avian teratogen when injected directly into embryos during the first 96 h of development. In mammals, the lethal dose of boron, as boric acid, varies according to species, and usually ranges between 210.0 and 603.0 mg B/kg body weight (BW); early development stages are especially sensitive. Excessive boron consumption adversely affects growth and reproduction in sensitive species of mammals, i.e., >1000.0 mg B/kg diet, >15.0 mg B/kg BW daily, >1.0 mg B/L drinking water, or >3000.0 mg B/kg BW single dose on the first day of pregnancy. Boron is not considered essential for mammalian growth, but does protect against fluorosis and bone demineralization.

4.4.1 Terrestrial Plants

Boron was accepted as being an essential micronutrient for higher plants in 1923, with toxicity owing to excess boron much less common in the environment than is boron deficiency. The role of boron in nutrition and toxicity of terrestrial crops and forest trees has been reviewed extensively. It is generally agreed that boron is essential for the growth of higher plants and some species of fungi, bacteria, and algae, and that excess boron is phytotoxic. It is also agreed that plants vary greatly in their sensitivity to B toxicity. The exact mode of action of boron is unknown; however, its complexing ability facilitates the movement of sugars and other materials, and it is involved in cell wall bonding, conversion of glucose-1-phosphate to starch, and metabolism of nucleic acids. Some authorities suggest that boron toxicity to plants is attributed, in part, to interactions between borates and divalent cations like manganese, resulting in altered metabolic pathways of allantoate aminohydrolase in the case of manganese. Boron level in plants depends on the content and availability of soil boron, season, disease state, inherent species or variety differences, and interactions with other substances. Most of the plant-available boron comes from the decomposition of soil organic matter and from boron adsorbed and precipitated onto soil surface particles; however, soil solution boron is the most important form, and plants take it up directly from this source. Boron availability to plants is strongly associated with the hot-water-soluble fraction. This usually ranges from 0.4 to 4.7% of the total boron; the highest percentage occurs in fine-textured soils, and the lowest occurs in coarse-textured soils. Uptake of boron by plants is about 4 times higher at pH 4 than at pH 9, highest in the temperature range 10–30°C, and higher with increased light intensity.

For the past 70 years, boron has been applied to B-deficient soils to improve crop yields of grains, fruits, vegetables, legumes, pine trees, tobacco, cotton, sunflowers, and peanuts. Boron is unique among the essential micronutrients because it is the only element normally present in soil solution as a nonionized molecule over the pH range suitable for plant growth. Boron deficiency in plants is widespread and has been reported in one or more crops in at least 43 U.S. states, almost all Canadian Provinces, and many countries. Boron deficiency in crops is more widespread than that of any other micronutrient. It is more likely to occur in light-textured acid soils in humid regions because of boron's tendency to leach; however, deficiency may also occur in heavy-textured soils with high pH because boron is readily adsorbed under these conditions. Deficiency signs include browning and spotting of leaves, chlorosis, abnormal thickening of cell walls, increased production of indoleacetic acid, accumulation of polyphenolic compounds, changes in membrane permeability, necrosis, and finally death. Visible signs of deficiency in corn are accentuated by calcium deficiency, and are least evident when calcium is added in excess. Under conditions of boron and calcium deficiency combined, yields are low, and starch phosphorylase activity in corn leaves increases markedly, as does that of ribonuclease and polyphenol oxidase. Interaction effects were also measured between boron and potassium in alfalfa. Boron deficiency is usually corrected by application of 0.5–3.0 kg B/ha, depending on crop and formulation. Adding boron promotes translocation rate of photosynthetic products and increases CO_2 incorporation into free amino acids.

Boron toxicity has been reported in many species of grasses, fruits, vegetables, grains, trees, and other terrestrial plants. Toxic levels generally do not occur on agricultural lands unless boron compounds have been added in excessive quantities, such as with fertilizer materials, irrigation water, sewage sludge, or coal ash. Boron-contaminated irrigation water is one of the main causes of boron toxicity to plants. The continued use and concentration of boron in the soil due to evapotranspiration is the reason for eventual toxicity problems. Borates have also been used as herbicides for complete kill of vegetation at application rates of 2244 kg/ha (equivalent to 2000 pounds/acre). Borates are frequently applied at elevated concentrations (i.e., >2.0 g/kg soil) in combination with organic pesticides in order to produce bacteriostatic effects; the resultant B-produced reduction in microbial degradation of the pesticide effectively extends the pesticide's biocidal properties. In some cases, cooling tower drift from geothermal steam containing boron may cause foliar boron toxicity in the vicinity of generating units.

Boron poisoning in plants is characterized by stunted growth, leaf malformation, browning and yellowing, chlorosis, necrosis, increased sensitivity to mildew, wilting, and inhibition of pollen germination and pollen tube growth. In barley (*Hordeum vulgare*), for example, excess boron caused decreased growth and grain yield, elevated residues in leaves, and increased rate of leaf senescence. Barley grown on zinc-deficient soils tended to accumulate boron up to 2.5 times within 7 days; a similar pattern was evident for excess phosphorus. Thus, under conditions of marginally high boron in the rooting zone, low zinc, and high phosphorus, boron may accumulate to toxic levels in plants. Toxic effects in plants – including leaf injury – were observed in 26% of plants at or below substrate concentrations that resulted in greatest growth, indicating considerable overlap between injurious and beneficial effects of boron in plants. In general, deficiency effects in plants were evident when boron concentrations in soil solution were <2.0 mg/L; optimal growth occurred at 2.0–5.0 mg/L; and toxic effects were evident at 5.0–12.0 mg B/L. However, there is considerable variation in resistance to boron between species. Sensitive species are known to include citrus, stone fruits, and nut trees; semitolerant species include cotton, tubers, cereals, grains, and olives; tolerant species usually include most vegetables.

4.4.2 Terrestrial Invertebrates

Relatively high concentrations of boron compounds are used to control fruit flies, cockroaches, gypsy moth larvae, houseflies, and woodboring insects. Boric acid is an effective stomach poison for several insect species, including German cockroaches (*Blattella germanica*), that are unable to detect the presence of boric acid. Insect infestation of wood and other substrates can be prevented by pretreatment with boric acid or borax at doses of 0.25–0.55 kg/m^3 of wood. Boric acid and other boron compounds are effective chemosterilants of the cotton boll weevil (*Anthonomus grandis*) and houseflies.

4.4.3 Aquatic Organisms

Boron effects on aquatic plants are highly species-specific. Borate, like silicate, is an essential micronutrient for the growth of aquatic plants, such as diatoms, and it seems that a chemical combination of both nutrients in the form of silicoborate may be required by certain diatoms. In aquatic plants, boron affects nucleic acid metabolism, carbohydrate biosynthesis and transport, and membrane integrity, and it interacts with growth substances. Diatoms (*Cylindrotheca fusiformis*) cultured under boron-deficient conditions stop dividing and swell in size despite increased photosynthetic rates. Boron-deficient diatoms accumulate rubidium, phenolic compounds, nitrates, and phosphates, and they show increased activity of various enzymes, especially glucose-6-phosphate dehydrogenase; however, respiratory adjustment is negligible until nutrient stress becomes irreversible in about 48 h. Boron, under conditions of excess, alleviates nutrient deficiency in some phytoplankters and may cause temporal variations of phytoplankton composition in coastal waters. Phytoplankton can tolerate up to 10.0 mg inorganic B/L in the absence of stress from pH adversity and nutrient deficiency, although higher borate concentrations up to 100.0 mg/L are expected to cause species redistribution by favoring the growth of some species and suppressing that of others.

Available data for aquatic invertebrates and boron suggest that the no-observable-effect levels were 13.6 mg B/L for freshwater organisms and 37.0 mg B/L for marine biota. Juvenile Pacific oysters (*Crassostrea gigas*) accumulated boron in relation to availability, but showed no prolonged retention following cessation of exposure. At industrial discharge levels of about 1.0 mg B/L, no hazard is apparent to oysters and aquatic vertebrates.

Boron may be an essential nutrient in several species of aquatic vertebrates. Insufficient boron (<3.0 μg B/L; 62.0 μg B/kg ration) interfered with the normal development of the South African clawed frog, *Xenopus laevis*, during organogenesis, and substantially impaired normal reproductive function in adult frogs. Impaired growth of rainbow trout (*Oncorhynchus mykiss*) embryos was documented at <90.0 μg B/L, and death of zebrafish (*Brachydanio rerio*) embryos at <2.0 μg B/L.

The most sensitive aquatic vertebrates tested for which data are available were coho salmon (*Oncorhynchus kisutch*), with an LC50 (16-day) value of 12.0 mg B/L in seawater, and sockeye salmon (*O. nerka*), showing elevated tissue residues after exposure for 3 weeks in seawater containing 10.0 mg B/L. Boron concentrations between 0.001 and 0.1 mg/L had little effect on survival of rainbow trout embryos after exposure for 28 days. These low levels may represent a reduction in reproductive potential of rainbow trout, and concentrations more than 0.2 mg B/L may impair survival of other fish species; however, additional data are needed to verify these speculations. Scientists report that concentrations of 100.0–300.0 mg B/L killed all species of aquatic vertebrates tested, that embryonic mortality and teratogenesis were greater in hard water than in soft water, but larval mortality of fish and amphibians was higher in soft water than in hard water, and that boron compounds were more toxic to embryos and larvae than to adults. Moreover, no measurable effect on boron toxicity to aquatic vertebrates was found due to water temperature in the range of 13–29°C, dissolved oxygen between 6.4 and 10.3 mg/L, and pH between 7.5 and 8.5. However, elevated boron concentrations of 50.0–100.0 mg B/L adversely affect the development of amphibian embryos. In central Pennsylvania ponds, embryos from two species of salamanders (spotted salamander, *Ambystoma maculatum*; Jefferson salamander, *A. jeffersonianum*), the wood frog, *Rana sylvatica*, and the American toad, *Bufo americanus*, were exposed to wastewater effluents of 0.0, 50.0, or 100.0 mg B/L. At 50.0 and 100.0 mg B/L, there were significant increases in the frequency of deformed larvae and reduced hatching success.

4.4.4 Birds

Boron stimulated growth in vitamin D_3 deficient chicks. Supplemental dietary boron alleviated or corrected cholecalciferol deficiency-induced elevations in plasma glucose concentrations in chicks. There is no need to supplement the diets of laying hens with boron, provided that basal diets contained about 11.0 mg B/kg ration.

Boron is a potent teratogen to domestic chicken embryos when injected into eggs. Injection of boron into the yolk sac of chicken embryos during the first 96 h of development with 1.0–2.5 mg of boric acid – equivalent to 3.0–8.0 mg B/kg FW egg (55-g egg) – produced a wide range of developmental abnormalities. Several compounds are known to counteract boron-induced avian developmental abnormalities, or to reduce the frequency of malformations, although the mode of action is unclear. These compounds include sodium pyruvate, to counteract rumplessness; nicotinamide, to decrease frequency of facial defects and melanin formation; and riboflavin, which greatly reduced the teratogenic effects of boric acid. Other polyhydroxy compounds, such as D-ribose, pyridoxine hydrochloride, and D-sorbitol hydrate, also reduced or abolished boric acid-induced teratogenicity in chick embryos.

High concentrations of boron have been found in the San Joaquin Valley of California in irrigation drainwater and in aquatic plants consumed by waterfowl. Measured boron concentrations in that locale exceeded 20.0 mg/L in subsurface agricultural drainage waters,

400.0 mg/kg DW in widgeon grass (*Ruppia maritima*) and algae, 150.0 mg/kg DW in aquatic insects, 1860.0 mg/kg DW in some aquatic plants, and up to 3390.0 mg/kg DW in seeds consumed by waterfowl. At present, only selenium has been implicated as the cause of abnormal development among waterfowl in western areas impacted by irrigation drainwaters. However, studies with mallards demonstrate that dietary boron concentrations well below levels that can occur in the environment represent a toxicological hazard that has not been considered in the management of agricultural drainwater. For example, dietary concentrations of 300.0–400.0 mg B/kg of feed on a fresh weight basis – substantially lower than boron levels reported in the vicinity of some western wildlife refuges contaminated by agricultural drainwater – adversely affect mallard growth, behavior, brain biochemistry, and are often associated with elevated tissue boron levels. Dietary levels of 100.0 mg B/kg FW result in reduced growth of female mallard ducklings, and diets containing as little as 30.0 mg B/kg FW fed to mallard adults adversely affected growth rate of their ducklings. Resource managers must now consider boron, as well as selenium, and their possible interactions, as a toxic hazard to wildlife populations throughout areas of the western United States.

4.4.5 Mammals

No requirement for boron in mammals is proven, although evidence is accumulating suggesting that boron may be an essential nutrient. Boron is related to normal energy utilization, immune function, and metabolism of bone, minerals, and lipids. Boron deficiency (<0.04 mg B/kg ration of dams) impairs early embryonic development in rodents; these effects were not observed at 2.0 mg B/kg ration. Boron deprivation in animals and humans results in decreased brain electrical activity similar to that observed in nonspecific malnutrition, and reduced cognitive and psychomotor function. Learning performance (manual dexterity, eye–hand coordination, memory, attention, perception) in humans was significantly higher when daily boron ingestion rate was 3.0 mg vs. 0.23 mg. Boron dietary supplements to postmenopausal women aged 48–82 years induced changes consistent with the prevention of calcium loss and bone demineralization. In rats, adequate dietary boron protected against premature senescence, and alleviated the signs of vitamin D_3 deficiency through improved absorption and retention of calcium and phosphorus, and retention of femur magnesium. In cattle, increases in boron ingestion were associated with elevated boron levels in plasma and urine, increased boron excretion, decreased plasma phosphate concentrations, and increased renal and urinary clearance of phosphates. Boron accumulations in rat testes were associated with progressive germ cell depletion that persisted long after toxic exposure to boron had occurred.

Boron effectively counteracts symptoms of fluoride intoxication in humans and in rabbits poisoned experimentally. Humans suffering from skeletal fluorosis experienced 50–80% improvement after drinking solutions containing 300.0–1100.0 mg of borax per liter daily, 3 weeks a month for 3 months. Boron enhances sequestration of fluoride from bone and excretion through kidneys and possibly the intestinal tract.

Inorganic borates, including boric acid, and sodium-, ammonium-, potassium-, and zinc borates display low acute toxicity to mammals via oral, dermal, and inhalation routes of exposure. The critical effects in several species of mammals during chronic exposure to boron compounds are male reproductive toxicity and developmental abnormalities. For example, prenatal exposure to elevated levels of boric acid causes reduced incidences of supernumerary ribs and a shortening or absence of the thirteenth rib in several species of laboratory animals. The doses that cause these effects are far higher than any levels to which human populations could be exposed. Humans would need to consume 3.3 g of boric acid or 5.0 g of borax to ingest the same dose level at the lowest animal no-observed-adverse-effect-level (NOAEL). Boron has no measurable effect on human fertility or reproduction among workers exposed to borates or to populations exposed to high environmental borate levels. Adult Turkish females, for example, residing in boron-rich areas (29.0 mg B/L drinking water) or boron-poor areas (0.3–0.5 mg B/L drinking water) did not differ in rate of spontaneous abortions, stillbirths, or congenital malformations.

Long-term exposure of humans to airborne boron dust may cause irritation of the nose, throat and eyes, and large amounts of boron ingested over short periods of time can adversely affect the gastrointestinal tract, liver, kidney, and brain, and may lead to death. However, borax mean air exposures of 18.0 mg/m^3 measured for high-exposure workers together with dietary boron resulted in an estimated absorption of only 0.38 mg B/kg BW daily. At this level, there was no progressive accumulation across the workweek. Epidemics and sporadic cases of oral intoxication in people are often due to inadvertent addition of boric acid to infant formulas. Five of eleven human infants died within 3 days of exposure after ingesting formula prepared with a 2.5% aqueous solution of boric acid, equivalent to 4.5–14.0 g of boron ingested. Prior to death, these infants were lethargic and vomiting; postmortem degenerative changes were observed in liver, kidney, and brain. Some products containing boron compounds, such as pacifiers, have been sold in Northern Ireland despite a recommendation from the Pharmaceutical Society of Great Britain that they should not be sold because of hazards to infants. Fatal cases of boron poisoning have involved misuse of boron compounds in hospitals, either from accidental substitution of boric acid solution for water in infant formula or from accidental use of boric acid as a diapering powder. In an adult fatality, the victim died after inundation by borax solution. In one case, a 12-month-old girl developed violent vomiting, coughing, irritability, tremors, seizures, and a delirious reaction after accidentally swallowing a mixture containing 3 g of boric acid and 300.0 mg of cinchocaine chloride prescribed due to a painful dental protrusion. Her plasma boric acid level 6 h later was 26.0 mg/L; the half-time persistence ($Tb1/2$) for boric acid in plasma is about 7 h. The lethal dose of boric acid varies according to the species. In mammals it ranges from 210.0 to 603.0 mg B/kg BW, and death is due to CNS paralysis and gastrointestinal irritation. Human newborns are especially sensitive, and accidental deaths have been recorded at doses between 50.0 and 140.0 mg B/kg BW.

In mammals, excessive boron consumption results in a reduced growth rate and sometimes loss in body weight; these may not be entirely due to reduced feed and water consumption. Growth retardation has been reported in cattle given 150.0 mg B/L drinking water (about 15.0 mg B/kg BW daily), in dogs consuming diets containing 1750.0 mg B/kg, in rabbits eating rations equivalent to >140.0 mg B/kg BW daily, and in rats given 150.0 mg B/L drinking water or 1060.0 mg B/kg diet. In some cases, animals will avoid B-contaminated drinking water if given a choice. Rats, for example, will reject drinking water containing as little as 1.0 mg B/L, and cattle will avoid water containing >29.0 mg B/L.

Male workers engaged in boric acid production showed weakened sexual activity, decreased seminal volume, low sperm count and motility, and increased seminal fructose. Animal studies demonstrated that the testes atrophy or degenerate if large amounts of boron are eaten or drunk; these effects have not been reported in humans. Adverse effects on reproduction of laboratory animals have been reported in sensitive species fed diets 1000.0 mg B/kg, or given drinking water containing 1.0 mg B/L (equivalent to about 0.3 mg B/kg BW daily), or given a single oral dose of 3000.0 mg B/kg BW on the first day of pregnancy. Boric acid caused developmental toxicity – including fetal weight reduction, prenatal mortality and malformations, decreased survival – in rats, mice, and rabbits in the range of 16.0–80.0 mg B/kg BW daily given either throughout gestation or only during major organogenesis.

Volatile boron compounds, especially boranes, are usually more toxic than boric acid or soluble borates. However, there is little commercial production of synthetic boranes, except for sodium borohydride – one of the least toxic boranes. Boron trifluoride is a gas used as a catalyst in several industrial systems, but on exposure to moisture in air it reacts to form a stable dihydride.

For boric oxide dusts, occupational exposures to 4.1 mg/m^3 (range 1.2–8.5) are associated with eye irritation; dryness of mouth, nose and throat; sore throat; and cough.

4.5 Recommendations

Many boron criteria have been proposed for the protection of crops, aquatic life, waterfowl, livestock, and human health (Table 4.2). The risk to aquatic ecosystems from boron is low. Boron concentrations in contaminated industrial effluents seldom exceed 1.0 mg B/L, a level considered nonhazardous to aquatic life. In a few boron-rich areas, natural levels will be higher, although organisms may adapt to local conditions. However, future accumulations of boron in groundwaters through wider uses of boron-containing cleansing agents may adversely affect aquatic organisms and other species of plants and animals, as now occurs in areas where natural boron deposits exist. Long-term monitoring of groundwaters and surface waters for boron levels seems warranted.

Results of chronic feeding studies using mallards demonstrate that diets containing 13.0 mg B/kg FW produce no adverse effects, but those containing 30.0 or 100.0 mg B/kg FW are associated with elevated tissue boron residues

Table 4.2. Proposed boron criteria for the protection of natural resources and human health.

Resource	Criterion
CROPS	
Irrigation waters	
Sensitive crops	0.3–0.75 mg B/L
Semitolerant crops	0.67–2.5 mg B/L
Tolerant crops	1.0–4.0 mg B/L
Maximum safe concentration	4.0 mg B/L
Residues in crops	
Boron deficiency	<15.0 mg B/kg dry weight (DW) plant
Toxicosis	>200.0 mg B/kg DW plant
Soil concentrations	
Optimal growth of several species	>0.1 – <0.5 mg B/kg DW soil
Deficiency, Bangladesh	<0.2 mg B/kg DW surface layer
FOREST TREES	
Conifers, sensitive species	
Deficient	<4.0 mg B/kg DW foliage
Low	>4.0 – <8.0 mg B/kg DW foliage
Intermediate	13.0–20.0 mg B/kg DW foliage
Toxic	>75.0 mg B/kg DW foliage
Angiosperms, sensitive species	
Deficient	8.0–16.0 mg B/kg DW foliage
Toxic	>180.0 mg B/kg DW foliage
AQUATIC ORGANISMS	
Nonhazardous levels in water	
Fish, oysters	<1.0 – 5.0 mg B/L
Aquatic communities	1.0–2.0 mg B/L
Aquatic plants	4.0 mg B/L
Aquatic invertebrates	6.0–10.0 mg B/L

Continued

Table 4.2. cont'd

Resource	Criterion
Safe levels in water	
Largemouth bass	<30.0 mg B/L
Bluegill	<33.0 mg B/L
Rainbow trout, embryos and larvae	0.75–1.0 mg B/L
Adverse effects expected, sensitive species	10.0–12.0 mg B/L
WATERFOWL	
Diet	
No observed adverse effect	<13.0 mg B/kg fresh weight (FW) diet
Adverse effects observed	30.0–100.0 mg B/kg FW diet
Lethal	1000.0 mg B/kg FW diet
LABORATORY ANIMALS	
No observed adverse effect	
Rat	<15.6 mg/kg body weight (BW) daily during gestation
Rabbit	<25.0 mg B/kg BW daily
Mouse	<50.0 mg B/kg BW daily
Adverse effect level	
Rat	>15.6 mg B/kg BW daily during gestation
Rabbit	>50.0 mg B/kg BW daily
Mouse	90.0 mg B/kg BW daily during gestation
LIVESTOCK	
Diet	
Boron deficiency	<0.4 mg B/kg DW
Toxic signs probable	>100.0 mg B/kg DW
Maximum tolerable level, as borax	150.0 mg B/kg DW
Total dose, toxic	100.0–300.0 g of boron (equivalent to 200.0–600.0 mg B/kg BW)
Drinking water	
Maximum allowable	5.0 mg B/L
Maximum tolerated	40.0 mg B/L
Safe	40.0–150.0 mg B/L
Adverse effects observed	>150.0 mg B/L
PESTICIDE APPLICATIONS	
Boric acid, 99% powder	Effective for control of household cockroaches, ants, and fleas
Boric acid, 8% solution	Fungicide control agent for vegetables, fruits, and trees
HUMAN HEALTH	
Air	
Threshold Limit Value (8 h daily, 5 days weekly)	
Pentaborane	0.01 mg/L
Diborane	0.1 mg/L

Table 4.2. cont'd

Resource	Criterion
Decaborane	0.5 mg/L
Sodium borate	1.0–5.0 mg/m^3
Boron trifluoride	<3.0 mg/m^3
Calcium borate	4.0–6.0 mg/m^3
Boron tribromide	10.0 mg/m^3
Boron oxide	
Total dust	10.0 mg/m^3
Respirable fraction	<5.0 mg/m^3
Sodium salts	
Tetraborate	10.0 mg/m^3
Decahydrate	<5.0 mg/m^3
Anhydrous and pentahydrate	<1.0 mg/m^3
Borate dusts	
Safe	<1.0 mg B/m^3 daily
Infrequent effects	1.1 mg B/m^3 daily
Adverse effects	4.01–4.6 mg B/m^3 daily
Daily intake (all sources)	
Total tolerable	0.4 mg/kg BW[a]
Worldwide	Range 0.34–1.0 mg B; means usually 10.0–20.0 mg B[b]
Finland	1.7 mg B
England	2.8 mg B
USA	3.0 mg B
No effect level	4.0 g boric acid
Adverse effect level	
Chronic intoxication	4.0–5.0 g boric acid
Lethal to infants and small children	5.0–6.0 g boric acid
Lethal to adults	18.0–20.0 g boric acid, single dose
Dermal, ocular	
Sodium borate and boric acid	Safe as cosmetic ingredients at <5% concentrations; not recommended on infant skin or injured skin
Diet	
Citrus fruit	<8.0 mg B/kg FW
Cottonseed	<30.0 mg B/kg FW
Hop extracts	<310.0 mg B/kg FW
Minimal risk level	<3.2 mg B/kg FW diet
Adverse effects, including death	>4161.0 mg B/kg FW diet
Drinking water	
Recommended	<0.3 mg B/L
Former Soviet Union	<0.5 mg B/L
USA	<1.0 mg B/L
Safe	<20.0 mg B/L
No toxic effects observed	20.0–30.0 mg B/L

Continued

Table 4.2. cont'd

Resource	Criterion
Tissue residues	
Blood	
Normal, children and infants	<1.25 mg B/L FW
Adverse systemic effects	20.0–150.0 mg B/L FW
Fatal	200.0–1600.0 mg B/L FW
Serum, adults	
No significant toxicity	<2320.0 mg B/L FW
Urine, adults, normal	0.7–1.5 mg B/L FW

[a]Based on NOAEL of 9.6 mg B/kg BW daily for reproductive effects in rats and an uncertainty factor of 25.
[b]Global mean daily intake of B by humans estimated at 1.9 mg, mostly from food (65%) and drinking water (30%). For a 70-kg adult, this is equivalent to 0.027 mg B/kg BW daily.

and growth reduction, and diets containing 1000.0 mg B/kg are fatal. More research is needed on the fate and effects of boron on waterfowl and raptors, especially in those areas where high dietary boron loadings are encountered as a result of agricultural drainwater disposal practices.

Minimum concentrations of dietary boron needed to maintain animal health are not known with certainty. However, diets containing <0.4 mg B/kg FW may adversely affect metabolism of rats and chicks; accordingly, animal diets should contain >0.3 mg B/kg FW until necessary feeding data become available. Also, the defensible boron maximum for livestock drinking water may be considerably higher than 5.0 mg/L (Table 4.2) because several "safe" water sources in Nevada exceeded this upper maximum and approached 80.0 mg B/L. Data are unavailable on boron effects on terrestrial wildlife. Until these data become available, it seems reasonable to apply the same criteria proposed for livestock protection (Table 4.2) to mammalian wildlife, that is, diets should contain more than 0.4 mg B/kg DW but less than 100.0 mg/kg, and drinking water less than 5.0 mg/L.

Medicinal use of boric acid and borax for babies has resulted in anorexia, nausea, vomiting, diarrhea, marked cardiac weakness, a red eruption over the entire body, and (rarely) death. The medical community has since abandoned the use of boric acid solutions as irrigants and antiseptics, abandoned all medical uses in Denmark, and severely limited availability (prescription only) in Ireland. Increased use of boric acid as a household pesticide should be viewed with concern, especially in households where children have access to non-safety capped boric acid containers. However, the amine-carboxyborane derivatives show promise as therapeutic agents for a number of disease states. More research is needed on medical aspects of amine-carboxyborane compounds and their ability to reduce serum cholesterol and to relieve through their anti-inflammatory properties the effects of chronic arthritis. This group of compounds was effective antineoplastic agents with selective activity against single cell and solid tumors derived from human and rodent leukemias, lymphomas, sarcomas, and carcinomas. Health benefits of borates and boron compounds and their role in fertility and pregnancy merit additional investigation. Boron, for example, may be essential to normal bone growth and composition and protect against bone loss associated with aging.

The fact that boron is essential to plants is firmly established. However, when boron concentrations in irrigation waters exceed 2.0 mg/L, extensive plant toxicity should be expected. High concentrations of boron in some potential irrigation waters in parts of the western United States at levels capable of causing crop damage have prompted implementation of boron criteria for irrigation waters (Table 4.2), although no legally enforceable boron standards have been promulgated. More information is needed on crop plants in the following subjects: interaction of boron with other elements in the soil and its effects on boron availability to plants, the role of boron on pollination as it affects seed yield and sugar content of crops, and distinguishing signs of boron deficiency in plants from similar signs of molybdenum deficiency.

More research is needed on the accurate measurement of boron in biological materials when the concentrations are <1.0 mg B/kg. Standard biological reference materials with low boron levels need to be produced for use in interlaboratory comparisons. This becomes especially important in studies on boron-deficiency states and the ability of the organism to conserve boron at very low intakes. More research is needed on homeostatic regulation of boron and functional markers of boron metabolism. Scientists now recommend additional studies to establish the availability of boron from the diet and its distribution to the tissues; boron essentiality in higher organisms; beneficial effects of boron on health; and the role of borates in behavioral disorders and cognitive performance. New advances in boron nutrition research should include better characterization of the mechanisms through which boron modulates immune function and insulin release. Epidemiological studies should be initiated to identify health conditions associated with inadequate dietary boron. Finally, more research is recommended on uncertainty factors used in establishing tolerable daily intake values for the protection of human health, with emphasis on variations in interspecies and intraspecies differences in resistance to boron.

4.6 Summary

The United States is the major global producer of boron compounds and supplies about 70% of the annual demand. Although boron is ubiquitous in the environment, human activities such as mining, coal burning, drainwater disposal, and use of borax laundry detergents have resulted in elevated boron loadings in the atmosphere and in irrigation waters. The chemistry of boron is complex and rivals that of carbon in its diversity. However, most boron compounds enter or degrade in the environment to B–O compounds (borates) – such as borax and boric acid – and these are considered to be the most significant ecologically.

Boron is an essential trace element for the growth of terrestrial crop plants and for some species of fungi, bacteria, and algae, but excess boron is phytotoxic. Representative species of aquatic organisms, including plants, invertebrates, fishes, and amphibians, usually tolerated up to 10.0 mg B/L of medium for extended periods without harm. In waterfowl, growth was adversely affected at dietary levels of 30.0–100.0 mg B/kg FW, tissue boron concentrations were elevated at 100.0–300.0 mg B/kg diet, and survival was reduced at dietary levels of 1000.0 mg B/kg; all these dietary levels currently exist near agricultural drainwater disposal sites in the western United States. Boron is not considered essential in mammalian nutrition now, although low dietary levels protect against fluorosis and bone demineralization. Excessive consumption (i.e., >1000.0 mg B/kg diet, >15.0 mg B/kg BW daily, >1.0 mg B/L drinking water, or >210.0 mg B/kg BW in a single dose) adversely affects growth, survival, or reproduction in sensitive mammals. Boron and its compounds are potent teratogens when applied directly to the mammalian embryo, but there is no evidence of mutagenicity or carcinogenicity. Boron's unique affinity for cancerous tissues has been exploited in neutron capture radiation therapy of malignant human brain tumors.

Boron criteria recommended for the protection of sensitive species include <0.3 mg B/L in irrigation waters of crops, <1.0 mg B/L for aquatic life, <5.0 mg B/L in livestock drinking waters, <30.0 mg B/kg in diets of waterfowl, and <100.0 mg B/kg in diets of livestock.

CADMIUM[a]

Chapter 5

5.1 Introduction

There is no evidence that cadmium (Cd) is biologically essential or beneficial. On the contrary, it has been implicated as the cause of numerous human deaths and various deleterious effects in fish and wildlife. In sufficient concentration, it is toxic to all forms of life, including microorganisms, higher plants, animals, and humans. It is a relatively rare metal, usually present in small amounts in zinc ores, and is commercially obtained as an industrial by-product of the production of zinc, copper, and lead. Major uses of cadmium are in electroplating, in pigment production, and in the manufacture of plastic stabilizers and batteries. Anthropogenic sources of cadmium include smelter fumes and dusts, the products of incineration of cadmium-bearing materials and fossil fuels, fertilizers, and municipal wastewater and sludge discharges; concentrations are most likely highest in the localized regions of smelters or in urban industrialized areas. Industrial consumption of cadmium in the United States, estimated at 6000 metric tons in 1968, is increasing; recent use is about 14,000 tons, primarily for electroplating of motor parts and in the manufacture of batteries. The cadmium load in soils and terrestrial biota in other industrialized countries also appears to be increasing and is of great concern in Scandinavia, Germany, and the United Kingdom.

5.2 Environmental Chemistry

Cadmium, as cadmium oxide, is obtained mainly as a by-product during the processing of zinc-bearing ores and also from the refining of lead and copper from sulfide ores. In 1989, the United States produced 1.4 million kg of cadmium (usually 0.6–1.8 million kg) and imported an additional 2.7 million kg (usually 1.8–3.2 million kg). Cadmium is used mainly for the production of nickel–cadmium batteries (35%), in metal plating (30%), and for the manufacture of pigments (15%), plastics and synthetics (10%), and alloys and miscellaneous uses (10%).

Cadmium is a silver-white, blue-tinged, lustrous metal that melts at 321°C and boils at 767°C. This divalent element has an atomic weight of 112.4, an atomic number of 48, and a density of 8.642 g/cm^3. It is insoluble in water, although its chloride and sulfate salts are freely soluble. The availability of cadmium to living organisms from their immediate physical and chemical environments depends on numerous factors, including adsorption and desorption rates of cadmium from terrigenous materials, pH, Eh, chemical speciation, and many other modifiers. The few selected examples that follow demonstrate the complex behavior of cadmium in aquatic systems.

Microbial extracellular polymeric substances (EPSs) – ubiquitous features in aquatic environments – actively participate in binding dissolved overlying and pore-water metals in sediments. Organic sediment coatings in the form of bacterial EPS equivalent to about 0.5% organic matter can adsorb cadmium under estuarine conditions.

[a]All information in this chapter is referenced in the following sources:

Eisler, R. 1985. Cadmium hazards to fish, wildlife, and invertebrates: a synoptic review. *U.S. Fish Wildl. Serv. Biol. Rep. 85(1.2)*, 1–46.

Eisler, R. 2000. Cadmium. Pages 1–43 in *Handbook of Chemical Risk Assessment: Health Hazards to Humans, Plants, and Animals. Volume 1. Metals*. Lewis Publishers, Boca Raton, Florida.

EPS aggregates rapidly sorbed up to 90% of cadmium from solution. Changes in pH affected cadmium sorption, with the proportion of freed Cd to sorbed Cd changing from 90% at pH 5 to 5% at pH 9; desorption was enhanced with increasing salinity.

Adsorption and desorption processes are likely to be major factors in controlling the concentration of cadmium in natural waters and tend to counteract changes in the concentration of cadmium ions in solution. Adsorption and desorption rates of cadmium are rapid on mud solids and particles of clay, silica, humic material, and other naturally occurring solids. Concentration factors for river muds varied between 5000 and 500,000 and depended mainly on the type of solid, the particle size, the concentration of cadmium present, the duration of contact, and the concentration of complexing ligands; humic material appeared to be the main component of river mud responsible for adsorption. Changes in physicochemical conditions, especially pH and redox potential, that occur during dredging and disposal of cadmium-polluted sediments may increase chemical mobility and, hence, bioavailability of sediment-bound cadmium. For example, cadmium in Mississippi River sediments spiked with radiocadmium was transformed from potentially available organic forms to more mobile and readily available dissolved and exchangeable forms (i.e., increased bioavailability) under regimens of comparatively acidic pH and high oxidation. The role of dissolved oxygen and aquatic plants on cadmium cycling was studied in Palestine Lake, a 92-ha eutrophic lake in Kosciusko County, Indiana, a long-term recipient of cadmium and other waste metals from an electroplating plant. The maximum recorded concentration of dissolved cadmium in the water column was 17.3 μg/L; for suspended particulates, it was 30.3 μg/L. During anaerobic conditions in the lake's hypolimnion, a marked decrease in the dissolved fraction and a corresponding increase in the suspended fraction were noted. The dominant form of cadmium was free, readily bioavailable, cadmium ion, Cd^{2+}; however, organic complexes of cadmium, which are comparatively nonbioavailable, made up a significant portion of the total dissolved cadmium. Cadmium levels in sediments of Palestine Lake ranged from 1.5 mg/L in an uncontaminated area of the lake to 805 mg/L, near the outlet of a metal-bearing ditch that entered the lake. The dominant form of cadmium in sediments was a carbonate. Levels of cadmium in water varied over time and between sites, but usually ranged from 0.5 to 2.5 μg/L. It is possible that significant amounts of cadmium are transferred from the sediments into rooted aquatic macrophytes and later released into the water after macrophyte death (natural or herbicide-induced), particularly in heavily contaminated systems. In Palestine Lake, cadmium levels in pondweed (*Potomogeton crispus*), a rooted aquatic macrophyte, were about 90 mg/kg DW; a maximum burden of 1.5 kg was retained by the population of *P. crispus* in the lake. Release of the total amount could raise water concentrations by a maximum of 1.0 μg/L. This amount was considered negligible in terms of the overall lake cadmium budgets; however, it might have limited local effects.

5.3 Concentrations in Field Collections

Small amounts of cadmium enter the environment from the natural weathering of minerals, but most is released as a result of human activities such as mining, smelting, fuel combustion, disposal of metal-containing products, and application of phosphate fertilizers or sewage sludges. In 1988, an estimated 306,000 kg of cadmium entered the domestic environment as a result of human activities, mostly from industrial releases that were subsequently applied to soils. Where cadmium is comparatively bioavailable, these values are very near those that have been shown to produce harmful effects in sensitive biological species, as will be discussed later.

Loadings of cadmium in uncontaminated, nonbiological compartments extended over several orders of magnitude, with most of the cadmium in lithosphere and ocean reservoirs (Table 5.1). Concentrations (microgram

Table 5.1. Cadmium burdens and residence times in the principal global reservoirs.

Reservoir	Concentration	Total Cd in Reservoir, in Metric Tons[a]	Residence Time
LITHOSPHERE			
Surface	0.6 mg/kg	1,303,600	42 years
Subsurface (down to 45 km)	0.5 mg/kg	28,000,000,000,000	1 billion years
HYDROSPHERE			
Oceans			
Dissolved	0.06 μg/kg	84,000,000	21,000 years
Suspended particulates (total)	1.0 mg/kg	1400,000	nd[b]
Particulate organic matter	4.5 mg/kg	320,000	1.3 years
Freshwater			
Dissolved	0.05 μg/kg	16,000	nd[b]
Sediments	0.16 mg/kg	100,000	3.6 years
Groundwater	0.1 μg/kg	400	nd[b]
Glaciers	0.005 μg/kg	82,000	nd[b]
Combined fresh and marine waters			
Sediment pore waters	0.2 μg/kg	64,000,000	nd[b]
BIOSPHERE			
Marine plants	2.0 mg/kg	400	18 days
Marine animals	4.0 mg/kg	12,000	nd[b]
Land plants	0.3 mg/kg	720,000	20 days
Land animals	0.3 mg/kg	6000	nd[b]
Freshwater biota	3.5 mg/kg	7000	3.5 years
Human biomass	50.0 mg/person	200	1–40 years
ATMOSPHERE	0.00003 μg/m^3	1500	7 days

[a]Of the estimated 28 trillion metric tons of cadmium in the environment, almost all (99.9995%) is located in the subsurface lithosphere. The remaining 151.6 million metric tons reside in the hydrosphere (98.6%), surface lithosphere (0.9%), biosphere (0.5%), and atmosphere (0.001%).
[b]nd = no data.

per liter or microgram per kilogram) of cadmium reported in uncontaminated compartments ranged from 0.05 to 0.2 in freshwater, up to 0.05 in coastal seawater, from 0.01 to 0.1 in open ocean seawater, 0.1–14.0 in stormwater runoff, up to 5000.0 in riverine and lake sediments, 30.0–1000.0 in marine sediments, 10.0–1000.0 in soils of nonvolcanic origin, as much as 142.0 in human dietary items, up to 4500.0 in soils of volcanic origin, 1.0–600.0 in igneous rock, up to 100,000.0 in phosphatic rock. In air, cadmium values in uncontaminated ranged from 0.001 to 0.005 μg/m^3. Higher values are reported in abiotic materials and living organisms from cadmium-contaminated environments, being highest near cadmium-emitting industries such as smelters, municipal incinerators, and fossil fuel combustion facilities.

Cadmium, unlike synthetic compounds, is a naturally occurring element, and its presence has been detected in more than 1000 species of aquatic and terrestrial flora and fauna. At least six trends are evident from these data. First, marine biota generally contain significantly higher cadmium residues than their freshwater or terrestrial counterparts, probably because total cadmium levels are higher in seawater. Second, cadmium tends to concentrate in the viscera of vertebrates, especially the liver and kidneys. Third, concentrations of cadmium are higher in older organisms than in younger stages; this relationship is especially pronounced in carnivores and marine vertebrates. Fourth, higher concentrations reported for individuals of a single species collected at several locations are almost always associated with proximity to industrial and urbanized areas or to point source discharges of cadmium-containing wastes. Fifth, background levels of cadmium in crops and other plants are usually less than 1.0 mg/kg (ppm). Little is known about the cadmium concentrations required to reduce plant yields; however, plants growing in cadmium-contaminated soils contain abnormally high residues that may be detrimental to plant growth and to animal and human consumers. And finally, it is clear that species analyzed, season of collection, ambient cadmium levels, and sex of organism all modify concentrations of cadmium in organisms. In freshwater isopods, for example, cadmium was stored mainly in the hepatopancreas; cadmium concentrations were higher in juveniles than in adults; and seasonal fluctuations accounted for as much as 79% of the within population variability.

Cadmium tends to accumulate in avian tissues in the order of kidney, liver, brain, bone, and muscle, with highest concentrations in older birds and those found closest to a point source of cadmium. In Norwegian birds, cadmium in tissues is generally higher in adults than in juveniles, higher in winter than other seasons, positively correlated with tissue copper burdens, and positively correlated with selective consumption of seeds of the willow (*Salix* sp., known to have a high level of cadmium), and insects living on the willow. However, cadmium concentrations in kidneys of canvasback ducks (*Aythya valisineria*) foraging on the submerged plant *Vallisneria americana*, do not seem to reflect dietary cadmium intake. Cadmium and other metals also tend to concentrate in feathers, and molting frequency is important in the depuration process. Most birds molt their body feathers once a year, but some – such as Franklin's gulls (*Larus pipixcan*) – molt twice, thus they have a greater opportunity than other birds to rid the body of contaminants.

The relation between reported tissue cadmium concentrations of "unstressed" populations and hazard to the organism or its consumer is not well documented. For example, cadmium in eggs of successful nests of Cooper's hawks (*Accipiter cooperii*) collected in Arizona and New Mexico ranged from 0.015 to 0.24 mg/kg fresh weight (FW); concentrations were higher in eggs from unsuccessful nests. Cadmium concentrations in livers of breeding birds were higher in two declining colonies of puffins in St. Kilda and Clo Mor (12.9–22.3 mg/kg DW) than in colonies of puffins from other areas, or in livers of other seabirds examined; however, the link to cadmium requires elucidation. Among marine teleosts, whole body levels exceeding 5 mg/kg FW or 86 mg/kg ash weight (AW) in laboratory-stressed fish suggested that death would follow within 4 weeks. Marine bivalve mollusks occasionally contain more than 13.0 mg Cd/kg soft parts FW, a level considered acutely toxic to human consumers; however, many species of marine and terrestrial mammals frequently contained more than 20.0 mg Cd/kg FW in various tissues without apparent adverse effects to the organism. The significance of cadmium residues to organism health is further developed later.

In terrestrial mammals, cadmium tends to accumulate with increasing age in kidneys and livers of hares, moles and shrews, deer, caribou (*Rangifer tarandus*), and musk ox (*Ovibos moschatus*). Concentration of cadmium in kidney of deer increased from <0.002 mg/kg DW at age zero to 10.5 at age 7 years in males; in females, these values were 6.5 mg/kg DW at age 3, 12.5 at age 7, and 20.0–40.0 at age 8. Cadmium concentrations were positively correlated with age in kidneys of caribou and musk ox collected from the Canadian Yukon and Northwest Territories between 1985 and 1990. The highest cadmium concentration measured at 166.0 mg/kg DW was in renal tissue of a 15-year-old caribou; caribou diets rely heavily on lichens which accumulate cadmium to a greater extent than do sedges (musk oxen diet). The mean kidney concentration of 467.0 mg Cd/kg DW of beavers from a cadmium-contaminated estuary in Germany is the greatest reported in free-ranging herbivores.

5.4 *Lethal Effects*

The lethal effects of cadmium are thought to be caused by free cadmium ions, that is, cadmium not bound to metallothioneins or other proteins. Free cadmium ions may inactivate various metal-dependent enzymes; however, cadmium not bound to metallothionein may have the capacity to directly damage renal tubular membranes during uptake.

A substantial toxicological database for cadmium and freshwater biota demonstrates that ambient cadmium water concentrations exceeding 10.0 μg/L are associated with high mortality, reduced growth, inhibited reproduction, and other adverse effects. In as much as one recommended drinking water criterion for human health protection is 10.0 μg Cd/L, it is noteworthy that several species of freshwater aquatic insects, crustaceans, and teleosts exhibited significant mortality at cadmium concentrations of 0.8–9.9 μg/L during exposures of 4–33 days; mortality generally increased as exposure time increased, water hardness decreased, and organism age decreased. Daphnids (*Daphnia magna*), for example, were more resistant to the biocidal properties of cadmium under conditions of increasing concentrations of dissolved organic materials and water hardness, especially Ca^{2+}. Prior exposure to elevated sublethal concentrations of waterborne cadmium protects a minnow (*Phoxinus phoxinus*) against subsequent exposure to a lethal waterborne dose of cadmium; this phenomenon was associated with reduced uptake of ^{109}Cd into the gills of a minnow when this species was challenged with a lethal dose of waterborne cadmium. In the case of larval toads (*Bufo arenarum*) and northwestern salamanders (*Ambystoma gracile*), adverse effects on survival were documented at 250.0–468.0 μg Cd/L, with mortality greatest at elevated temperatures and early developmental stages.

Resistance to cadmium is higher in marine than in freshwater organisms and survival usually is higher at the lower temperatures and higher salinities for any given level of cadmium in the medium. Cadmium was fatal to 50% of nauplii of copepods (*Eurytemora affinis*) at 51.0 μg/L and 5 ppt salinity, and 83.0–213.0 μg/L at higher salinities; a similar pattern was evident in larvae of the sheepshead minnow (*Cyprinodon variegatus*); the free ion (Cd^{2+}) accounted for 20% of total cadmium at 5 ppt, 8% at 15 ppt, and 4.5% at 25 ppt salinity. Decapod crustaceans are also sensitive in short-term tests; LC50 (96 h) values ranged from 320.0 to 420.0 μg/L for the grass shrimp (*Palaemonetes vulgaris*), the hermit crab (*Pagurus longicarpus*), and the sand shrimp (*Crangon crangon*). Studies of longer duration demonstrated that survival of shrimp groups was low at >250.0 μg/L during 6 weeks of exposure and that hermit crab deaths were recorded at 60.0 μg/L after 6 weeks, although some survivors remained at 10 weeks when the studies ended. In another study, an LC50 range of 14.8–19.5 μg Cd/L was reported for two species of mysid shrimp subjected to lifetime (i.e., 23–27 days) exposure to cadmium salts. Studies with *Leptocheirus plumulosus*, an estuarine amphipod, showed that gravid females were more resistant than males or mature females to the biocidal properties of aqueous Cd^{2+}; juveniles were more sensitive than were adults; sensitivity was greatest among starved animals and immediately after molting; and field-collected animals were more sensitive to dissolved cadmium than were laboratory animals, regardless of season of collection.

Birds are comparatively resistant to the biocidal properties of cadmium. All the adult drake mallards (*Anas platyrhynchos*) fed up to 200.0 mg Cd/kg diet for 90 days survived with no loss of body weight. Laying hens fed 200.0 mg Cd/kg diet also survived; egg production was suppressed at that concentration but not at lower concentrations tested. Marine and terrestrial animals, including ducks, have been shown to be particularly abundant in a wildlife community associated with a marine sewer outfall; these animals were contaminated with high levels of cadmium, as well as zinc and copper, but were apparently protected from the deleterious effects of high metal body burdens by metallothioneins. Amounts of these metal-binding proteinaceous metallothioneins and heavy metal loading appear to depend primarily on the degree of pollution and secondarily on the species of animal and its position in the food web. Ducks contained the highest levels of metallothioneins of all groups examined.

Mammals, like birds, are comparatively resistant to cadmium. The lowest oral dose, in mg/kg body weight (BW) of cadmium (as fluroborate) producing death, was 250.0 in rats (*Rattus* sp.) and 150.0 (as cadmium fluoride) in guinea pigs (*Cavia* sp.). Exposure to high levels of cadmium by inhalation or orally is fatal in humans and other animals. Inhalation of a lethal dose of cadmium can occur without signs of acute distress during exposure, although high oral doses of cadmium sometimes induce vomiting in humans. Cause of death is pulmonary edema after inhalation exposure, and massive fluid imbalance and widespread gastrointestinal, liver, and other organ damage after oral exposure – mainly due to the destruction of lung cell membranes (inhalation) and gastrointestinal tract membranes (oral) at the point of entry. Suicidal doses in two humans are reported at 25.0 and at 1500.0 mg Cd/kg BW.

5.5 Sublethal Effects

Studies of 30–60 days duration with three comparatively sensitive species of freshwater fishes demonstrated that concentrations of >1.0 and <3.0 μg Cd/L in water of low alkalinity caused reductions in growth, survival, and fecundity of brook trout (*Salvelinus fontinalis*), the most sensitive species tested. Under conditions of increasing alkalinity, the maximum allowable cadmium concentration range for brook trout increased to >7.0 and <12.0 μg/L; a similar case was made for the walleye (*Stizostedion vitreum vitreum*). Among all species of freshwater biota examined, cadmium concentrations of 0.47–5.0 μg/L were associated with decreases in standing crop, decreases in growth, inhibition of reproduction, immobilization, and population alterations. Juvenile rainbow trout (*Oncorhynchus mykiss*) exposed to 1.0 or 5.0 μg Cd/L for 30 days showed reductions in liver size, glycogen content, growth rate, and plasma calcium with changes attributed to disrupted endocrine function, in part. Cadmium interactions with other metals and compounds are significant. For example, 4.0 μg Zn/L or 11.0 μg Se/L were effective in counteracting the decrease in oxygen uptake induced by ionic cadmium in *Channa punctatus*, freshwater teleost. There is an abundant technical literature documenting numerous sublethal effects at higher concentrations of cadmium salts; however, these were excluded if the effects were observed at >10.0 μg/L, a recommended criterion for drinking water. Delayed effects of cadmium intoxication, however, need to be considered. Studies with daphnids, amphipods, and fathead minnows indicate that exposure to cadmium concentrations as low as 370.0 μg/L for as little as 30 min caused increasing immobility for up to 172 h after exposure.

For marine organisms, ambient cadmium concentrations between 0.5 and 10.0 μg/L resulted in decreases in growth, respiratory disruption, molt inhibition, shortened life span of F1 generation crustaceans, altered enzyme levels, and abnormal muscular contractions. Effects, in general, were more pronounced at the lower salinities and higher temperatures tested. Marine algae accumulated more cadmium at lower salinities than at higher salinities; algae bound only free cadmium ions and these are inversely related to the concentration of the suspended particulate matter. Adaptation to environmental cadmium stress is reported for mussels (*Mytilopsis sallei*) exposed to high (>50.0 μg Cd/L) sublethal concentrations for at least 96 h. The decrease in oxygen consumption and the increased metabolism of glycogen and carbohydrates during exposure to cadmium suggest that *M. sallei* might shift to anaerobic metabolism to counter the environmental cadmium stress.

One of the more sensitive indicators of cadmium exposure is the inhibition of non-thioneine hepatic metal-binding proteins; inhibition was observed in juvenile bluegills

(*Lepomis macrochirus*) at concentrations as low as 0.8 μg Cd/L. Cadmium at elevated concentrations (100.0 μg/L) significantly inhibits the Na^+/K^+-ATPase activity in different tissues of rat, frog, rabbit, and trout, and sometimes as low as 16.0 μg/L in the case of gill basolateral membrane vesicle of European eels (*Anguilla anguilla*). Inhibition of ATPase activities may lead to severe perturbations of osmoregulation processes, causing disturbances in migration and perhaps the survival of feral eel populations. In the scorpionfish (*Scorpaena guttata*), cadmium disrupts cytosol balance and inhibits copper–zinc-superoxide dismutase activity in the intestine, with important implications for detoxification. Dose-dependent induction of liver metallothionein proteins within 24 h were observed in juvenile winter flounders (*Pleuronectes americanus*) after subcutaneous injection of high doses of cadmium (5.0–20.0 mg Cd/kg BW), with RNA disruption reported at the highest dose.

Sublethal effects in birds are similar to those in other species and include growth retardation, anemia, renal effects, and testicular damage. However, harmful damage effects were observed at higher concentrations when compared to aquatic biota. For example, Japanese quail (*Coturnix japonica*) fed 75.0 mg Cd/kg diet developed bone marrow hypoplasia, anemia, and hypertrophy of both heart ventricles at 6 weeks. In zinc-deficient diets, effects were especially pronounced and included all of the signs mentioned plus testicular hypoplasia; a similar pattern was evident in cadmium-stressed quail on an iron-deficient diet. In all tests, 1% ascorbic acid in the diet prevented cadmium-induced effects in Japanese quail. In studies with Japanese quail at environmentally relevant concentrations of 10.0 μg Cd/kg BW daily (for 4 days, administered orally), absorbed cadmium was transported in blood in a form that enhanced deposition in the kidney; less than 0.7% of the total administered dose was recovered from liver, kidneys, and duodenum.

In wood ducks (*Aix sponsa*) fed rations containing 10.0 mg Cd/kg, no renal effects were noted although kidneys contained 62.0 mg Cd/kg FW. Renal damage was noted when wood ducks were fed diets containing 100.0 mg Cd/kg and their kidneys contained 132.0 mg Cd/kg FW. Adult male white leghorn chickens (*Gallus* sp.) given 2.0 mg $CdSO_4$ daily by intraperitoneal injection for 15–22 days, or a total dose of 60.0 mg of cadmium per chicken, developed anemia, an enlarged heart, myocardial infarction, and other abnormalities. Testicular damage was observed in ring doves (*Streptopelia* sp.) 20 days after intramuscular injection of 6.6 mg Cd/kg BW; in domestic pigeons (*Columba livia*), however, testicular damage was observed after a single subcutaneous injection of only 0.5 mg/kg BW, and cardiovascular disease developed after exposure to 600.0 μg/L in drinking water. In mallard ducklings fed 20.0 mg Cd/kg ration for 12 weeks, blood chemistry was altered, and mild to severe kidney lesions developed. But mallard juveniles were unaffected when given diets containing 50.0 mg Cd/kg ration for 6 weeks. Mallard juveniles fed diets containing 150.0 mg Cd/kg ration and higher for 6 weeks had elevated liver (>135.0 mg Cd/kg DW) and kidney (>335.0 mg Cd/kg DW) burdens, disrupted plasma fatty acid concentrations, and increased adrenal and kidney weights. Altered avoidance behavior in the form of hyperresponsiveness was observed in young American black ducks (*Anas rubripes*), produced from parents fed 4.0 mg/kg dietary cadmium for about 4 months before egg laying; this behavioral effect was observed only at comparatively low dietary cadmium levels and is considered harmful to wild birds. Cadmium readily reacts with sulfhydryl groups and may compete, especially with zinc, for binding sites on proteins and, thus, may inhibit a variety of enzymatic reactions. The addition of zinc, iron, ascorbic acid, calcium, or selenium to diets ameliorated cadmium damage effects, whereas the addition of lead or mercury exacerbated them.

In male rats (*Rattus* sp.) and mice (*Mus* sp.), acute oral exposure to near-fatal doses (7.0–14.0 mg Cd/kg BW daily for 90–120 days) can cause testicular atrophy and necrosis, and decreased fertility. Oral intake of cadmium disrupts calcium metabolism of laboratory and freeliving rodents. Bank voles (*Clethrionomys glareolus*) on low calcium diets given diets equivalent to 1.5–1.7 mg Cd/kg BW daily had significantly poorer calcium net gut absorption efficiency than animals fed cadmium-free diets and were in negative calcium balance. Cadmium-mediated impairment of calcium assimilation may be important in calcium-poor habitats, suggesting more research in this area. Among small laboratory mammals it appears that physiologically bound cadmium is more effective than $CdCl_2$ in producing metabolic iron irregularities. For example, in young mice fed oysters containing 1.8 mg Cd/kg ration for 28 days, hematocrit and hemoglobin values were depressed and other blood chemistry factors were altered. Diets containing intrinsic oyster cadmium at 1.8 mg/kg were more effective in producing hematopoietic alterations than were diets containing $CdCl_2$ at 3.6 mg/kg. Adequate dietary iron supplementation markedly reduced cadmium retention and cadmium-induced anemia in rats. Iron supply and the increased iron demand during growth of rats can be disturbed within one week by a daily cadmium intake as low as 0.7–1.3 mg Cd/kg BW.

A study of metal contamination in wildlife from the vicinity of two zinc smelters in Palmerton, Pennsylvania, demonstrated the difficulties in interpretation of cadmium residues from biota in the presence of other potentially hazardous metal contaminants. The soil litter horizon at Palmerton was heavily contaminated with lead (2700.0 mg/kg), zinc (24,000.0 mg/kg), copper (440.0 mg/kg), and cadmium (710.0 mg/kg). Invertebrates that fed on soil litter or soil organic matter, such as earthworms, slugs, and millipedes, were rare or absent in the vicinity of the smelters, but not at more distant sampling sites. Concentrations of all metals tended to be higher in these invertebrates than in other invertebrate groups collected. Amphibians and reptiles were also rare or absent at the Palmerton site, but not at more distant stations. Mean cadmium concentrations, in mg/kg dry weight (DW), were highest in carrion insects (25.0), followed by fungi (9.8), leaves (8.1), shrews (7.3), moths (4.9), mice (2.6), songbirds (2.5), and berries (1.2). By contrast, average concentrations of lead, in mg/kg DW, were highest in shrews (110.0), followed by songbirds (56.0), leaves (21.0), mice (17.0), carrion insects (14.0), moths (4.3), berries (4.0), and fungi (3.7). Evidence for lead poisoning in shrews included high residues in kidney (280.0 mg/kg wet weight (WW)) and reduced blood enzyme levels. In addition, livers from two yellow-billed cuckoos (*Coccyzus americanus*) from Palmerton had lead concentrations of 18.0 and 25.0 mg/kg WW; however, the cuckoos and other songbirds appeared to be healthy. Concentrations of zinc and copper tended to be highest in the same organisms that contained the highest concentrations of cadmium, emphasizing the importance of documenting organism body burdens of all suspected contaminants before significance is attributed to any single component. A small portion of all metals measured in the soil became incorporated into plant foliage and suggested that most of the metal contamination detected in biota came from aerial deposition.

The kidney is the critical organ in mammalian cadmium toxicity and is the first organ in which damage is observed or adverse functional changes start to occur. Cadmium concentrations in excess of 200.0 mg/kg FW kidney cortex results in renal dysfunction in about 10% of the exposed human population; a similar pattern is evident in mice, rats, and rabbits. In male rats, given a single intravenous injection of 0.15 mg metallothionein-bound Cd/kg BW, DNA fragmentation was seen in kidney 12 h after injection; cycloheximide (3.0 mg/kg BW) inhibited Cd-induced DNA fragmentation, suggesting that protein synthesis is impaired. In human tissues, there was a significant increase in cadmium burdens in 1980 when compared to the period 1897–1939. Cadmium content in the renal cortex portion of the kidney increased by a factor of 47 during this interval, and whole body burden increased by a factor near 5. The significance of this increase is not fully clear; however, one study has suggested that cadmium and lead are associated with increased risk of heart-related death, even in the light of known conventional causes of such fatalities. Similar data for wildlife are lacking, and this clearly indicates an area for additional research.

5.6 Bioaccumulation

Cadmium biomagnifies in terrestrial food chains, and tends to accumulate in liver and kidneys of older apex organisms. This process was documented in the chain of soil to vegetation to invertebrates to upper trophic level consumers, including roe deer (*Capreolus capreolus*), barn owls (*Tyto alba*), weasels (*Mystela nivalis*), and kestrels (*Falco tinnunculus*). Radish (*Raphanus sativa*) accumulated cadmium from the soil in roots and shoots over a 75-day period; uptake was decreased markedly with liming or increased soil pH. However, centipedes (*Lithobius forficatus*) near a zinc smelter, with body burdens as high as 80.0 mg Cd/kg DW, on transfer to an uncontaminated site for 10 weeks lost all but 18.0 mg Cd/kg DW despite the very high cadmium diet provided.

Biological half-times of cadmium in humans are lengthy. Based on body burden and excretion data, cadmium may remain in the human body for 13–47 years. Although cadmium is excreted primarily in urine and feces, cadmium tends to increase in concentration with age of the organism and eventually acts as a cumulative poison. These phenomena have not been documented adequately in wildlife species.

In marine mammals, cadmium was present in all liver and kidney samples analyzed. Cadmium concentrations in livers of beluga whales (*Delphinapterus leucas*) were positively correlated with age, and in ringed seals (*Phoca hispida*) with length, increasing from <0.7 to 3.6 mg Cd/kg FW in belugas and <0.14 to 8.8 mg/kg FW in seals. A similar case is made for kidney and liver tissues of the Baikal seal (*Phoca sibirica*). Pilot whales (*Globicephala melas*) contained higher concentrations of cadmium in liver and kidney tissues than did other marine mammals, and this is attributed, in part, to the elevated cadmium (as much as 5.8 mg Cd/kg DW) content in squids (*Loligo forbesi*) – a major dietary item. Similarly, elevated levels of cadmium in Pacific walruses (*Odobenus rosmarus divergens*) are considered related to elevated cadmium burdens in clams (*Mya* sp.), a major food item. And uptake of cadmium from cadmium-contaminated prey by fish plays an important role in contaminated waters.

Red-eared turtles (*Trachemys scripta*) when treated with very high doses of cadmium (10.0 mg Cd/kg BW daily for 6 days) by intraperitoneal injection accumulated 42% of the cadmium body burden in liver, followed by kidney (20%), spleen (12%), heart (8%), gonads (8%), and shell (5%); lesser amounts were measured in lung, muscle, brain, and blood. Turtle metallothioneins, which are similar to those of other vertebrate metallothioneins, seemed to control the accumulation process. Frogs seem relatively resistant to cadmium. Adult female water frogs (*Rana ridibunda*) held in solutions containing 200.0 mg Cd/L, as $CdCl_2$, for 30 days. Cadmium accumulated in kidneys and livers in a time-dependent manner, and metallothionein and glutathione concentrations increased with increasing liver cadmium burdens. After 30 days, livers had 240.0 mg Cd/kg DW and kidneys 657.0 mg Cd/kg DW.

Freshwater and marine aquatic organisms accumulate cadmium from water containing cadmium concentrations not previously considered hazardous to public health or to many species of aquatic life. In American oysters (*Crassostrea virginica*), held for 40 weeks in flowing seawater containing 5.0 μg Cd/L, edible meats contained 13.6 mg Cd/kg FW, a level considered to be an emetic threshold for human consumers. These oysters retained virtually all accumulated cadmium (12.5 mg/kg) during a 16-week posttreatment immersion in clean seawater. Human emetic thresholds for cadmium in oysters were surpassed in 5 weeks at 25.0 μg/L and in only 2 weeks at 100.0 μg/L. The emetic threshold for juvenile northwestern salamanders (*Ambystoma gracile*) is higher than that of humans; regurgitation occurred between 2458.0 and 5701.0 mg Cd/kg diet with no regurgitation at 982.0 mg Cd/kg diet and lower.

Two species of freshwater aquatic mosses (*Fontinalis dalecarlica, Platyhypnidium riparoides*) exposed to concentrations between 0.5 and 6.5 μg Cd/L for 28 days had accumulation factors as high as 137,000 and 158,000, respectively. Accumulations increased with increasing cadmium concentration and decreasing water hardness. Cadmium tended to persist in these mosses. During a depuration period of 28 days following the 28-day exposure, only 37–48% of the accumulated cadmium was eliminated.

Cadmium uptake from the medium by aquatic organisms usually increased with increasing water temperature in the range of 5–25°C; in the case of midge (*Chironomus riparius*) larvae, about 60% of the increased uptake is due to increased respiration. However, mayfly (*Hexagenia rigida*) nymphs accumulated more cadmium at 15°C than at 25°C; in that study, more cadmium was bioavailable for uptake owing to the higher pH of 7.5 at 15°C vs. pH 5.0 at 25°C. In another study with *Hexagenia rigida* nymphs, cadmium accumulation from sediments containing 0.0–41.0 mg Cd/kg DW was higher with increasing exposure duration (0–60 days), increasing water temperature (12–24°C), and increasing sediment cadmium concentration. Cadmium was accumulated by fish hepatoma cells to a greater degree than were other metals tested; in declining order of accumulation: cadmium, nickel, copper, cobalt, zinc, and lead. There is considerable variation in the ability of teleost tissues to accumulate cadmium from the ambient medium. Cadmium accumulation rates in larvae of tilapia (*Oreochromis mossambica*) increased with increasing larval development and were inversely related to LC50 (96 h) values. Sequestering agents, such as EDTA, reduced by 34% the accumulation of cadmium from the medium (15.0 μg/L) by sac fry of the African tilapia (*Oreochromis niloticus*). Among rainbow trout exposed for two weeks to 9.0 μg Cd/L, bioconcentration factors (BCFs) were 260 for gill, 17 for liver, 26 for kidney, and zero for spleen and heart tissues. At slightly higher ambient dissolved cadmium levels of 10.0 μg/L and exposure for 3 months, BCF values were substantially higher: 1740 for gill, 4900 for liver, 740 for kidney, 160 for spleen, and 100 for heart tissues. The evidence for cadmium transfer through various trophic levels suggests that only the lower trophic levels exhibit biomagnification. In the freshwater food chain extending from the alga *Chlorella vulgaris*, to the cladoceran *Daphnia magna*, to the teleost *Leucaspius delineatus*, it was demonstrated that algae held 10 days in water containing 10.0 μg Cd/L had 30.0 mg Cd/kg DW, up from 4.5 mg/kg at the start. Cladocerans feeding on cadmium-loaded algae for 20 days contained 32.0 mg Cd/kg DW, up from 1.4 mg/kg at the start. However, fish fed cadmium-contaminated cladocerans for 4 days showed no change in body burdens.

Cadmium resistance has been documented in marine annelids. Oligochaetes (*Limnodrilus hoffmeisteri*) from Foundry cove, New York – a severely cadmium-contaminated site – were more tolerant of cadmium than conspecifics from a reference site, surviving twice as long in a 7-day acute toxicity bioassay (1.0 mg Cd/L) and with BCFs of 2020 vs. 577 for the controls (radiocadmium-109). The cadmium-resistant worms produced metal-rich granules and metallothioneins for cadmium storage and detoxification whereas nonresistant worms produced metallothioneins only. Grass shrimp (*Palaemonetes pugio*) fed cadmium-resistant worms absorbed 21% of the ingested cadmium vs. 75% for shrimp fed nonresistant worms.

In laboratory studies with chipping sparrows (*Spizella passerina*) fed radiocadmium-109 in their diets for 3 weeks, it was demonstrated that cadmium became localized in the liver and kidneys. During posttreatment on a radiocadmium-free diet, there was an initial rapid drop in radioactivity, and the remaining radiocadmium had an estimated biological half-life of 99 days. Marine killifish (*Fundulus heteroclitus*) containing radiocadmium-115m lost 90% of the accumulated radiocadmium during a 6-month posttreatment observation period; the liver usually contained 75–80% of the total body dose at any time. Mallards fed 200.0 mg Cd/kg diet for about 13 weeks, all survived but levels in liver and kidney were elevated at 110.0 and 134.0 mg/kg FW, respectively. Mallard ducklings fed only 20.0 mg Cd/kg in the diet for 12 weeks contained 42.0 mg Cd/kg liver.

The exact mechanism of acute cadmium poisoning is unknown, but, among teleosts, it depends, in part, on exposure period, concentration of dissolved and ionic cadmium in the medium, and water temperature and salinity. Under conditions of high cadmium concentration and short exposure, the gill seems to be the primary site of damage and accumulation; under conditions of prolonged exposure and low cadmium levels, the intestine, kidney, and possibly other tissues were measurably affected. Retention of cadmium by teleosts depends on tissue biomagnification potential, length of postexposure recovery period, and other factors. The significance of comparatively low concentrations of cadmium in tissues of fish, other aquatic organisms, and wildlife, and the implications for organism health, is not fully understood. Although numerous physical, chemical, and biological factors demonstrably modify uptake and retention of cadmium by fish and wildlife, the significance of relatively high cadmium residues to animal and plant health is difficult to interpret. There is some evidence, however, that life-threatening concentrations include 200.0 mg Cd/kg FW in the renal cortex portion of the mammalian kidney and 5.0 mg/kg FW whole body of estuarine teleosts.

5.7 Teratogenesis, Mutagenesis, and Carcinogenesis

Teratogenic effect on animals appears to be greater for cadmium than for other metals, including lead, mercury, copper, indium, and arsenic. Among amphibians, frog embryos reared in 5000.0–7500.0 μg Cd/L showed nonclosure of the neural tube. In embryos of fathead minnows from adults reared in water containing 37.0–57.0 μg Cd/L, and from eggs transferred directly to such media, percent hatching was reduced, deformities were increased, and various blood clots developed. Embryos of the bluegill held in water at 80.0 μg Cd/L and higher showed edema, microcephalia, and malformed caudal fins. Eggs of a marine killifish (*Fundulus heteroclitus*) were little affected at up to 10,000.0 μg Cd/L. Caudal and hind limb abnormalities were observed in chickens following injection of eggs with 0.1–1.0 mg/kg egg of cadmium chloride; excess zinc appeared to have a protective effect. Rats subjected to >6.0 mg Cd/kg BW daily during pregnancy produced fetuses with jaw defects, cleft palates, club feet, and pulmonary hyperplasia. Among hamsters (*Cricetus* spp.), cadmium administration was associated with embryonic tail defects; effects were synergized by salts of lead or mercury and antagonized by selenium. No conclusive evidence of cadmium teratogenesis in humans is available.

At high concentrations, cadmium is genotoxic to isolated cells. From a variety of studies in which mice and bacteria were used as models, it appears likely that cadmium has mutagenic effects. Mice injected with 3.0 or 6.0 mg $CdCl_2$/kg BW showed changes in chromosome number 12 h later; similar changes were observed in hamsters at 1.5–3.0 mg/kg. Very high dosages (>100.0 mg/kg) produced chromosomal abnormalities in plant seeds. Also, $CdCl_2$ had a mutagenic effect on indicator strains of *Salmonella* bacteria. However, the evidence for these effects is still diffuse and often contradictory.

Laboratory studies with mice and rats have conclusively demonstrated that the injection of cadmium metal or salts causes malignancies (sarcoma) at the site of injection and testicular tumors; however, the simultaneous administration of zinc is protective against sarcoma and interstitial cell tumor development. In rats, no dose-related increases in tumors were found at maximum oral daily doses of 4.4 mg Cd/kg BW. Among humans, the available epidemiological evidence is not sufficient to conclude that cadmium is definitely implicated as a carcinogen, although cadmium exposure is associated with lung cancer in humans.

5.8 Recommendations

Proposed limits for cadmium in water, diet, tissues, air, soils, and sewage sludge for the protection of human health, plants, and animals are shown in Table 5.2. It is noteworthy that the current upper limit of 10.0 μg Cd/L in drinking water for human health protection is not sufficient to protect many species of freshwater biota against the biocidal properties of cadmium or against sublethal effects, such as reduced growth and inhibited reproduction. Ambient water quality criteria formulated for

Table 5.2. Proposed cadmium criteria for the protection of human health and natural resources.

Resource, Criterion, and Other Variables	Effective Cadmium Concentration
HUMAN HEALTH	
Air[a]	
Ambient exposure	
Best case	0.001 μg/m^3
Average case	0.03 μg/m^3
Worst case	0.4 μg/m^3
Occupational exposure, worst case	
Cadmium dust	200.0 μg/m^3
Cadmium fumes	100.0 μg/m^3
Chronic inhalation; no adverse effect on kidney	<0.2 μg/m^3
No-observed-adverse-effect level (NOAEL); 8 h daily, 250 days/year, 70-year life span[b]	<1.6 μg/m^3
Threshold limit value	
Current	<50.0 μg/m^3
Proposed; total dust vs. respirable fraction	<10.0 vs. <2.0 μg/m^3
Diet	
Chronic oral, USA[c]	<0.7 μg/kg body weight (BW) daily
Best case, USA	<12.0 μg daily; <16.0 μg/kg diet daily
Average case, USA	<30.0 μg daily; <40.0 μg/kg diet daily
Worst case, USA	<75.0 μg daily; <100.0 μg/kg diet daily
Edible fish tissues, Spain	<3.0 mg/kg fresh weight (FW) fish
Weekly tolerable intake[d]	400–500 μg/adult or <7.0 μg/kg BW
Brown rice, Taiwan	<0.5 mg/kg rice
Tissue concentration	
Acceptable	
Blood	<10.0 μg/L
Kidney cortex	<200.0 mg/kg FW
Urine	<2.0 μg Cd/g creatinine
Indicative of substantial exposure; urine	>50.0 μg Cd/g creatinine
Drinking water	
Canada[e]	<10.0 μg/L
Europe, except Spain	<5.0 μg/L
Spain	<7.0 μg/L
International, goal	<5.0 μg/L
USA	
Best case	<0.5 μg/L
Average case	<1.3 μg/L
Worst case	<10.0 μg/L
Current	5.0–10.0 μg/L
Recommended	<0.5 μg/kg BW daily
Bottled water	<10.0 μg/L

Table 5.2. cont'd

Resource, Criterion, and Other Variables	Effective Cadmium Concentration
CROPS	
Soils	
Canada, except Alberta	1.0–6.0 mg/kg DW soil
Alberta	>1.0 mg/kg DW needs remediation
Europe, after application of sewage sludge	1.0–3.0 mg/kg DW soil
Japan	<9.0 mg/kg DW soil
The Netherlands	
Background	<1.0 mg/kg DW
Moderate contamination	5.0–20.0 mg/kg DW
Requires cleanup	>20.0 mg/kg DW
Taiwan	<10.0 mg/kg DW
USA, New Jersey	<3.0 mg/kg DW
Sewage sludge applied to agricultural soils	
Florida	
Moderately contaminated	>5.0 kg Cd/surface ha
Prohibited	>100.0 mg/kg DW sludge
Illinois	<11.0 kg Cd/surface ha
Massachusetts	<5.0 kg Cd/surface ha
Maryland; soils with low cation exchange capacity vs. high cation exchange	<5.0 vs. <10.0 kg Cd/surface ha
Minnesota, Missouri, Oregon; cation exchange capacity of soil, in meq/100 g	
Low (<5)	<5.0 kg Cd/surface ha
Medium (5–15)	<10.0 kg Cd/surface ha
High (>15)	<20.0 kg Cd/surface ha
New York	<3.4–<5.5 kg Cd/surface ha
Vermont; maximum amounts allowed on various soils	
Loamy sand or sandy loam	6.0 kg Cd/surface ha
Fine sandy loam, loam, or silt loam	11.0 kg Cd/surface ha
Clay loam, clay, silty clay	22.0 kg Cd/surface ha
Forest soil protection, New York	<11.0 kg Cd/surface ha
AQUATIC LIFE	
Freshwater	
Water hardness, in mg $CaCO_3$/L	
50	<1.5 μg/L
100	<3.0 μg/L
200	<6.3 μg/L

Continued

Table 5.2. cont'd

Resource, Criterion, and Other Variables	Effective Cadmium Concentration
Water; Chesapeake Bay; protection of 90% of species tested	
Acute	
Fish	<1.8 μg/L
All species	<5.1 μg/L
Benthos	<12.3 μg/L
Chronic	
All species[f]	<0.4 μg/L
Fish	<0.9 μg/L
Sediments disposed into Great Lakes	<1.0 mg/kg DW
Marine	
Water, 24-h average	<4.5 μg/L
Water, maximum allowable concentration (MAC)	
USA	43.0–59.0 μg/L
UK, estuaries and coastal waters	<5.0 μg/L
Chesapeake Bay; protection of 90% of species tested	
Acute	
All species	<31.7 μg/L
Benthos	<23.3 μg/L
Fish	<163.0 μg/L
Chronic, benthos[f]	<0.25 μg/L
Sediments, Spain	<35.0 mg/kg DW
BIRDS, NONMARINE SPECIES	
Diet	<2.0 mg/kg FW ration
Tissue concentrations	
Adverse effects expected	
Liver	>40.0 mg/kg FW
Kidney	>100.0 mg/kg FW
Indicative of increased environmental exposure	
Liver	>3.0 mg/kg DW
Kidney	>8.0 mg/kg DW
Significant renal tubular dysfunction	100.0–200.0 mg/kg FW renal cortex; 400.0–800.0 mg/kg DW renal cortex
Some kidney necrosis	100.0–200.0 mg/kg DW kidney
MAMMALS (LIVESTOCK AND WILDLIFE)	
Diet	<0.5 mg/kg DW ration
Whole body; adverse effects on reproduction or development	3.5–7.5 mg Cd/kg BW daily and higher

Table 5.2. cont'd

Resource, Criterion, and Other Variables	Effective Cadmium Concentration
Tissue concentrations	
Acceptable	
Kidney cortex	<150.0 mg/kg FW
Whole kidney	<100.0 mg/kg FW; <350.0 mg/kg DW
Adverse effects, kidney	
Cellular damage	105.0 mg Cd/kg DW and higher
Significant damage	>200.0 mg/kg FW renal cortex or >1000.0 mg/kg DW renal cortex

[a]Assumes consumption of 0.75 kg food per day by a 70-kg adult.
[b]Includes uncertainty factor of 10.
[c]Includes uncertainty factor of 3.
[d]Assumes an absorption rate of 5% and a daily excretion rate of 0.005%.
[e]However, surface water levels in numerous Canadian lakes subjected only to atmospheric deposition within mining areas are often >1.0 μg Cd/L and may not be sufficiently low to prevent adverse physiological effects on early life stages of sensitive aquatic species.
[f]Concentrations greater than 3.4 μg Cd/L are frequently encountered in the Chesapeake and Delaware Canal and infrequently greater than 1.4 μg Cd/L in other portions of the Bay.

the protection of freshwater aquatic life state that, for total recoverable cadmium, the criterion, in microgram per liter, is the numerical value given by $e^{(1.05(\ln(\text{hardness}))-8.53)}$ as a 24-h average and the concentration, in microgram per liter, should never exceed the numerical value given by $e^{(1.05(\ln(\text{hardness}))-3.73)}$. Thus, at water hardness of 50, 100, and 200.0 mg/L as $CaCO_3$, the criteria are 0.012, 0.025, and 0.051 μg Cd/L, respectively, and the concentration of total recoverable cadmium should never exceed 1.5, 3.0, and 6.3 μg/L, respectively. Unfortunately, data are accumulating that demonstrate that even these comparatively rigorous criteria are not sufficient to protect the most sensitive species of freshwater insects, plants, crustaceans, and teleosts. It now appears that levels in excess of 3.0 μg Cd/L in freshwater are potentially hazardous to aquatic biota and that levels near 1.0 μg/L are cause for concern in waters of low alkalinity. Not listed in Table 5.2, but still recognized as proposed criteria, are the comparatively high levels of 10.0 μg Cd/L allowed for agricultural use on all soils (except neutral and alkaline soils, which may be irrigated with water having levels as high as 50.0 μg Cd/L) and public water supplies for livestock purposes, which may not exceed 50.0 μg Cd/L.

The saltwater aquatic life protection criterion of 4.5 μg Cd/L seems adequate to prevent death, but will not prevent potentially deleterious physiological effects, including disrupted respiration in crustaceans and teleosts. Incidentally, at 5.0 μg Cd/L, the lowest concentration critically examined, oysters biomagnify ambient levels to concentrations hazardous to human consumers and possibly other animal consumers. The maximum allowable concentration (MAC) in saltwater during a 24-h period was recommended as 59.0 μg/L (Table 5.2). However, death of various species of marine crustaceans was reported at 60.0 μg Cd/L after exposure for 6 weeks and at 14.8–19.5 μg/L after 23–27 days. Furthermore, a MAC of 59.0 μg Cd/L may be met with daily discharges of 59.0 μg/L for 2 h and no discharge of cadmium for the rest of the day. The effects of exposure of marine life to 59.0 μg/L of cadmium salts for 2 h daily for protracted periods have not yet been investigated. Accordingly, seawater concentrations in excess of 4.5 μg/L of total cadmium at any time should be considered as potentially hazardous to marine life until additional data prove otherwise.

Food is recognized as the major source of cadmium in humans, except in comparatively rare cases of occupational air exposure. The recommended upper limit for cadmium in food is 75.0 μg/day (Table 5.2). On the basis of an absorption factor of 0.1, a total of 7.5 μg Cd will be retained daily. A 70-kg adult ingests an estimated 0.75 kg of food daily, which suggests that human diets should not exceed 100.0 μg/kg. Regular weekly consumption by humans of kidney tissue (range 23.0–166.0 mg Cd/kg DW) from Arctic caribou (*Rangifer tarandus*) and from musk oxen (*Ovibos moschatus*) older than 1 year (2.4–12.4 mg Cd/kg DW kidney) will probably cause the World Health Organization provisional weekly tolerable intake of cadmium (400.0 μg) to be exceeded. On cadmium-contaminated habitats, common shrews (*Sorex araneus*) may ingest more than 3.5–7.5 mg Cd/kg BW daily (associated with reproductive effects in laboratory animals) without apparent harm. Elevated cadmium concentrations in body organs of shrews may reflect their ability to store cadmium in a nontoxic metallothionein-bound state. Adverse effects were, however, observed in feral rodents that consumed more than 3.5 mg Cd/kg BW daily, and this suggests a need to establish dose–residue–effect relations for intakes and residues appropriate for feral organisms.

Ducks, geese, and other species of wildlife, unlike adult humans, may consume 6–7% of their total body weight daily and may graze extensively on crops directly affected by sewage and other wastes containing high cadmium residues. Feeding studies with mallards indicated that diets containing 200.0 mg Cd/kg produced no obvious deleterious effects after 13 weeks. At the end of that study, however, kidney cadmium levels under those conditions were about 134.0 mg/kg FW, a level near the 200.0 mg/kg FW designated a "critical threshold" (and presumably life-threatening) for the renal cortex portion of the human kidney. Field observations on ducks and laboratory studies are not strictly comparable. Under field conditions, birds and other wildlife may consume food containing high cadmium levels, but it is almost certain that these diets also contain other potentially harmful contaminants, as well as metals or compounds that may ameliorate cadmium toxicity. The significance of foods containing complex mixtures of contaminants and their resultant toxicological interactions are imperfectly understood. Until other data become available, wildlife dietary levels exceeding 100.0 μg Cd/kg diet FW on a sustained basis should be viewed with caution – as they are for humans.

Recommendations for cadmium in air and human health protection under the worst scenario (Table 5.2) assume that total daily air intake is 27.14 m^3 for an adult human who spends about 6.3 h in occupational exposure to air containing 100.0 μg Cd/m^3. Under these conditions, a 70-kg adult would retain about 361.0 μg Cd/day, based on an absorption factor of 0.5, and most of this cadmium would probably be translocated to the kidney; a critical threshold level of 200.0 mg Cd/kg in the kidney would be reached in about 1.52 years. It is not now known whether respiration rates of wildlife, particularly birds, are comparable to those of humans, or whether cadmium absorption energetics are similar, or whether wildlife species that frequent point sources of air contaminated by high cadmium levels for protracted periods are at greater risk than humans. Flora and fauna in the vicinity of industrial smelters were affected by cadmium and its associated heavy metals and this finding strongly suggests that proposed recommendations for cadmium levels under occupational air exposure should be revised downward for wildlife protection.

Additional research on cadmium is recommended in three areas: (1) effects on cancer, genotoxicity, and reproductive toxicity under conditions of acute, intermediate, and chronic durations of exposure, and administered by way of diet, inhalation, and dermal routes of exposure; (2) emphasis on studies with pregnant animals; and (3) methods for reducing toxic effects. Finally, the issue of the significance of cadmium residues in various body parts requires resolution. At this time, it appears that cadmium residues in the vertebrate kidney or liver that exceed 10.0 mg/kg FW or 2.0 mg/kg in whole body

FW should be viewed as an evidence of probable cadmium contamination. Elevated levels of 13.0–15.0 mg Cd/kg tissue FW probably represent a significant hazard to animals of the higher trophic levels, and residues of 200.0 mg Cd/kg FW kidney cortex or more than 5.0 mg Cd/kg whole animal FW should be considered life-threatening.

5.9 Summary

Cadmium contamination of the environment is especially severe in the vicinity of smelters and urban industrialized areas. There is no evidence that cadmium, a relatively rare heavy metal, is biologically essential or beneficial; on the contrary, cadmium is a known teratogen and carcinogen, a probable mutagen, and has been implicated as the cause of severe deleterious effects on fish and wildlife. The freshwater biota is the most sensitive group; concentrations of 0.8–9.9 μg Cd/L (ppb) in water were lethal to several species of aquatic insects, crustaceans, and teleosts, and concentrations of 0.7–570.0 μg/L were associated with sublethal effects such as decreased growth, inhibited reproduction, and population alterations. These effects were most pronounced in waters of comparatively low alkalinity. Marine organisms were more resistant than freshwater biota. Decapod crustaceans, the most sensitive saltwater group, died at concentrations of cadmium in seawater ranging from 14.8 to 420.0 μg/L. Sublethal effects to marine animals recorded at concentrations of 0.5–10.0 μg Cd/L included decreased growth, respiratory disruption, altered enzyme levels, and abnormal muscular contractions; effects were usually most obvious at relatively low salinities and high temperatures. Freshwater and marine aquatic organisms accumulated measurable amounts of cadmium from water containing concentrations not previously considered hazardous to public health or to many species of aquatic life; i.e., 0.02–10.0 μg Cd/L.

Mammals and birds are comparatively resistant to the biocidal properties of cadmium. The lowest single oral doses producing death in rats and guinea pigs ranged from 150.0 to 250.0 mg Cd/kg BW. Although mallards and chickens tolerated 200.0 mg Cd/kg diet for protracted periods, kidney cadmium exceeded 130.0 mg/kg FW under this regimen, a concentration considered life-threatening to some organisms. Sublethal effects of cadmium in birds, which were similar to those in other animals, included growth retardation, anemia, and testicular damage; however, these effects were observed at higher concentrations than in aquatic biota. Although the evidence is incomplete, wildlife populations, especially migratory birds that feed on crops growing on fields fertilized with municipal sewage sludges, may be exposed to considerable risk of harmful effects from cadmium.

It is now conservatively estimated that adverse effects on fish or wildlife are either pronounced or probable when cadmium concentrations exceed 3.0 μg/L in freshwater, 4.5 μg/L in saltwater, 100.0 μg/kg in the diet, or 100.0 μg Cd/m^3 in air. Cadmium residues in vertebrate kidney or liver that exceed 10.0 mg/kg FW or 2.0 mg/kg whole body FW should be viewed as an evidence of probable cadmium contamination; residues of 200.0 mg Cd/kg FW kidney, or more than 5.0 mg/kg whole animal FW, are probably life-threatening to the organism.

CARBOFURAN[a]

Chapter 6

6.1 Introduction

Carbofuran is a broad-spectrum systemic insecticide, acaricide, and nematicide that is widely used in forestry and in agricultural crop production of corn, alfalfa, peanuts, rice, sugar cane, tobacco, potatoes, strawberries, onions, mixed vegetables, mustard, carrots, sunflowers, turnips, and many other crops. Carbofuran, together with other carbamate compounds, organophosphorus insecticides, and pyrethroids, are the major substitutes for the more persistent pesticides such as DDT, chlordane, and heptachlor. In 1974, domestic carbofuran use was slightly over 3.2 million kg (7 million pounds) active ingredients (a.i.), most of which were applied to control corn pests. By 1989, annual use was about 4.5 million kg, mostly in the granular formulations. As a group, the carbamates, including carbofuran, have controlled insects effectively: their residual life in the environment is relatively short; excretion from the animal body is comparatively rapid and almost quantitative; and the terminal residues produced are polar and formed by chemical processes normally considered as steps in metabolic detoxication.

Flowable and granular formulations of carbofuran have histories of heavy wildlife losses associated with recommended rates of application as well as misuse. At recommended application rates, which ranged from 0.28 to 10.9 kg a.i./ha (0.25–9.7 pounds/acre), and in a variety of formulations, carbofuran was responsible for sporadic deaths of fish, wildlife, beneficial insects, and terrestrial and aquatic invertebrates. In California between 1984 and 1988, carbofuran residues up to 640.0 mg/kg fresh weight (FW) were measured in gizzard and crop content of dead birds found near rice fields treated with granular carbofuran; secondary intoxication was noted in raptors which fed on carbofuran-poisoned ducks. A decline of 80% in populations of striped bass (*Morone saxatilis*) in California is attributed, in part, to carbofuran and other contaminants in agricultural drainwater associated with rice culture. Carbofuran was implicated in the deaths of egrets and herons found dead in San Joaquin County, California in 1991, near an area treated a few days earlier to control grape phylloxera; brain cholinesterase activity in birds was inhibited and food items (crayfish) in crop contained 0.6 mg carbofuran/kg FW.

Aerial spraying of carbofuran to control grasshoppers killed California gulls (*Larus californicus*); in a carbofuran-sprayed field, gulls were found convulsing with their gullets packed with grasshoppers containing 4.0–7.0 mg carbofuran/kg. Among birds that only occasionally consume domestic crops, carbofuran applied to vegetables reportedly killed about 1400 ducks, largely green-winged teal (*Anas carolinensis*), pintail (*A. acuta*), and American widgeon (*A. americana*) in British Columbia between 1973 and 1975. Carbofuran applied to alfalfa killed 2450 American widgeons at one California location in 1974, 500 Canada geese (*Branta canadensis*) in southern Oklahoma in 1976, 1000 widgeons in Kansas in 1976, and more than

[a]All information in this chapter is referenced in the following sources:

Eisler, R. 1985. Carbofuran hazards to fish, wildlife, and invertebrates: a synoptic review. *U.S. Fish Wildl. Serv. Biol. Rep 85(1.13)*, 36 pp.

Eisler, R. 2000. Carbofuran. Pages 799–822 in *Handbook of Chemical Risk Assessment: Health Hazards to Humans, Plants, and Animals. Volume 2, Organics.* Lewis Publishers, Boca Raton, Florida.

1063 widgeons in California in 1976–77. Carbofuran-contaminated sand (7.0 g carbofuran and metabolites/kg DW sand) is associated with the death of Pacific Island seabirds, crabs, and insects which come into contact with the sand. Secondary poisoning of red-shouldered hawks (*Buteo lineatus*) was reported after the application of carbofuran to Maryland cornfields. Secondary poisoning was also documented in northern harriers (*Circus cyaneus*) feeding on a dead eastern cottontail rabbit (*Sylvilagus floridanus*). Aerial application to flooded rice fields in various portions of Texas between 1970 and 1975 at the rate of 0.56 kg/ha resulted in deaths of three species of sandpipers (*Erolia* spp.) and red-winged blackbirds (*Agelaius phoeniceus*), as well as frogs, crayfish, leeches, earthworms, and four species of fish; however, no carbofuran residues were detectable among survivors 2–11 days postexposure. Carbofuran was responsible for the deaths of American crows (*Corvus brachyrhyncos*), a red-tailed hawk (*Buteo jamaicensis*), and European starlings (*Sturnus vulgaris*) in a Pennsylvania cornfield in 1986; an intentional poisoning of starlings by a farmer is the most probable explanation of the high carbofuran residues (67.0–425.0 mg/kg) found in stomach contents. Application of granular carbofuran to Virginia cornfields in 1991 was accompanied by deaths from anticholinesterase poisoning of mammals, birds, and reptiles; carbofuran residues were found in the upper gastrointestinal tract of 81% of the birds examined. Many die-offs of adult waterfowl wintering in the southern United States have been attributed to carbofuran use. Granular formulations of carbofuran were especially toxic to birds and their sale and use was prohibited after September 1, 1994 except for some minor uses.

6.2 Chemical Properties and Persistence

Carbofuran (2,3-dihydro-2,2-dimethyl-1,7-benzofuranyl methylcarbamate; Figure 6.1) is also known as Furadan, Bay 70142, Brifur, Crisfuran, Cristofuran, CAS 1563-66-2, Curaterr, D-1221, ENT-27164, FMC 10242, Niagara NIA-10242, OMS 864, Pillarfuran, and Yaltox. Carbofuran has a molecular weight of 221.25 and a melting point of 150–152°C; it is comparatively stable under neutral or acidic conditions, but degrades rapidly in alkaline media. This white, crystalline, solid of empirical formula $C_{12}H_{15}NO_3$ is soluble at concentrations up to 700.0 mg/L in water, but at <30.0 mg/L in various organic solvents. It degrades at >130°C and supports combustion if ignited. The compound is available as a wettable powder, a granular formulation, and in solution as a flowable formulation.

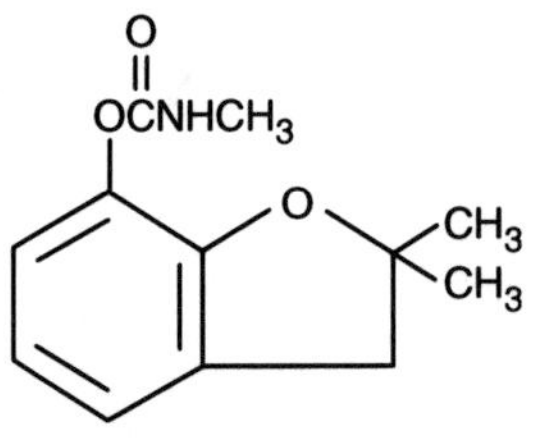

Figure 6.1.
Structural formula of carbofuran (2,3-dihydro-2,2-dimethyl-1,7-benzofuranyl methylcarbamate).

Pharmacologically, carbofuran inhibits cholinesterase, resulting in stimulation of the central, parasympathetic, and somatic motor systems. Sensitive biochemical tests have been developed to measure cholinesterase inhibition in avian and mammalian brain and plasma samples and are useful in the forensic assessment of carbamate exposure in human and wildlife pesticide incidents. Acute toxic clinical effects resulting from carbofuran exposure in animals and humans appear to be completely reversible and have been successfully treated with atropine sulfate. However, treatment should occur as soon as possible after exposure because acute carbofuran toxicosis can be fatal; younger age groups of various species are more susceptible than adults. Carbofuran labels indicate that application is forbidden to streams, lakes, or ponds. In addition, manufacturers have stated that carbofuran is poisonous if swallowed, inhaled, or absorbed through the skin; users are cautioned not to breathe carbofuran dust, fumes, or spray mist; and treated areas should be avoided for at least 2 days. Three points are emphasized at this juncture. First, some carbofuran degradation products have not been identified. Second, toxicologic, mutagenic, carcinogenic, and teratogenic properties of most carbofuran degradation products have not been satisfactorily evaluated. And third, numerous physical, chemical, and biological vectors modify carbofuran degradation processes, as well as biological uptake, retention, and translocation. Each of these points is developed later.

Carbofuran is metabolized by hydroxylation and hydrolysis in plants, insects, and mammals. The primary transformation product in most plants appears to be 3-hydroxycarbofuran. However, levels of 3-hydroxycarbofuran and other degradation products in plants are influenced by numerous factors, including plant age, soil type, pesticide formulation, application method and rate, and weather conditions. Oxidation of unconjugated 3-hydroxycarbofuran yields 3-ketocarbofuran, which is, in turn, rapidly hydrolyzed to the much less toxic 3-ketocarbofuran phenol. Accordingly, 3-ketocarbofuran is not likely to be detected as a terminal residue in plants above trace levels. Residue analyses indicated that carbofuran and 3-hydroxycarbofuran are the compounds that occur most often in plant tissues after treatment. In measurement of carbofuran and its degradation products in corn at 117 and 149 days after carbofuran application (Table 6.1), the decrease of 62% in the total carbamate residues detected between silage and harvest was attributed to cessation of root uptake, volatilization from drying plant surfaces, and further degradation to phenolic compounds. No losses of carbofuran or 3-hydroxycarbofuran were detected in fortified corn silage after storage at minus 18°C for 1 year.

Granular carbofuran is believed to persist for at least several months in the Fraser Delta of British Columbia under conditions of high humidity and low pH, and to kill waterfowl and cause secondary poisoning of raptors. During the winter of 1990 in British Columbia, bald eagles (*Haliaeetus leucocephalus*) and red-tailed hawks (*Buteo jamaicensis*) found dead or moribund had evidence of anticholinesterase exposure and crop contents that contained as much as 200.0 mg carbofuran/kg. Bald eagles, red-tailed hawks, and coyotes (*Canis latrans*) found dead in a field in Kansas in December 1992 were poisoned by flowable carbofuran

Table 6.1. Carbofuran and its degradation products, in mg/kg dry weight, in corn (*Zea mays*) at silage stage (117 days) and at harvest (149 days) following application of carbofuran (10%) granules at 5.41 kg/ha.

Plant Stage and Part	Carbofuran	3-Ketocarbofuran	3-Hydroxycarbofuran	Total Carbamates[a]
SILAGE				
Leaves	0.43	0.40	4.67	5.50
Stalks	0.24	0.00	0.04	0.28
Cobs	0.04	<0.02	<0.02	0.05
Kernels	0.00	<0.01	0.00	<0.01
HARVEST				
Leaves	0.21	0.34	1.51	2.06
Stalks	0.03	0.00	0.05	0.08
Cobs	0.06	0.00	0.00	0.06
Kernels	<0.01	<0.02	0.00	<0.01

[a] Sum of carbofuran, 3-ketocarbofuran, and 3-hydroxycarbofuran.

placed on sheep (*Ovis aires*) carcasses in October 1992 to kill coyotes. Flowable carbofuran can cause direct and secondary deaths of wildlife under some circumstances for at least 60 days. In this case, cold, dry weather and snow cover contributed to carbofuran preservation on the carcass.

Carbofuran residues of 10.8–13.3 mg/kg in vegetation usually declined by 50% in 24 h. Variation in content of carbofuran and its degradation products was evident among crop species. Strawberries (*Fragaria vesca*), for example, contained higher residues of phenol than of either carbamate or hydroxy products. Carbofuran can persist in Mugho pine needles for at least 2 years at insecticidally active concentrations. This unequal distribution of carbofuran in different parts of a plant has also been observed for tobacco (*Nicotiana tabacum*), in which more of the compound was in large leaves than in the tops of plants, suggesting that carbofuran moved in the plant fluids to the point of greatest transpiration in the leaves.

Carbofuran in animals may also be hydrolyzed to produce carbofuran-7-phenol; hydrolysis of the 3-hydroxyderivative leads to formation of 3-hydro-carbofuran-7-phenol. Other degradation products include *N*-hydroxymethyl carbofuran and, 3-hydroxy and 3-ketoderivatives. All these compounds may become conjugated and excreted by animals in urine and, presumably, bile. At least 10 metabolites of carbofuran are known at present.

Carbofuran accumulates in surface waters because of its relatively high water solubility and its relatively low adsorption on soils and sediments. It is stable in acid waters but is subject to increasing chemical hydrolysis as the water becomes more alkaline. In water, the carbofuran degradation rate is strongly influenced by pH. The time for 50% degradation of carbofuran in water is 3.2 years at pH 4.5, 13.3 years at pH 5.0 and 6.0, 1.9 months at pH 7.0, 1 week at pH 8.0, and only 5 h at pH 9.5. The rate of carbofuran loss is also influenced by sunlight, trace impurities, and temperature, but not as dramatically as by pH. Carbofuran is highly mobile and has the potential to leach into groundwater where it could persist under conditions of low temperature and low pH.

Persistence of carbofuran in soils is a function of many factors, including pesticide formulation, rate and method of application, soil type, pH, rainfall, temperature, moisture content, and microbial populations. Results of several studies indicate that loss from soil samples also takes place at low temperatures when air drying is used; this loss may present a problem to chemists who are unable to conduct analyses immediately after samples are collected. Soil pH is one of the more extensively documented variables affecting degradation; it may become increasingly important as acidic precipitation (acid rain) increases. Carbofuran decomposes rapidly at pH levels >7.0, but becomes increasingly stable as pH decreases. The hydrolysis half-life is about 16 years at a soil pH of 5.5; the half-lives are about 35, 6, and 0.25 days at pH levels of 7.0, 8.0, and 9.0, respectively. Similar results are shown in Table 6.2 for alumina and natural soils. Temperature and moisture content of soils were both positively related to degradation of carbofuran to 3-hydroxycarbofuran, 3-ketocarbofuran, carbofuran phenol, and 3-ketocarbofuran phenol. In general, an increase in temperature from 15 to 27°C had a greater influence on degradation than did an increase from 27 to 35°C, although 27–35°C was the range in which maximum degradation rates were observed.

The role of soil bacteria in carbofuran degradation is unclear. Most investigators agree that carbofuran is hydrolyzed to its phenol, which is immediately bound to soil constituents and then metabolized by microorganisms, either slowly or rapidly, especially when associated with a *Pseudomonas* sp. isolate. Others believe that carbofuran is degraded primarily by chemical hydrolysis, in which bacterial processes assume a negligible role. Evidence exists demonstrating that soil microbial populations increased by up to 3 times following application of carbofuran; that prior treatment with carbofuran produced rapid degradation attributed to acclimatized soil bacteria; that estuarine bacteria are comparatively resistant to carbofuran; that sterilized soils did not show evidence of carbofuran

Table 6.2. Effect of pH, soil type, and application rate on carbofuran degradation in soils.

Soil Type	pH	Initial Application Rate of Carbofuran (mg/kg soil)	Percent Carbofuran Remaining After 3 Weeks
ALUMINA SOILS			
Acid	5.4–6.1	1	76
Acid	5.4–6.1	20	82
Neutral	6.9–7.1	1	85
Neutral	6.9–7.1	20	79
Basic	8.3–8.5	1	55
Basic	8.3–8.5	20	72
NATURAL SOILS			
Mineral	8.0	1	95
Mineral	8.0	20	92
Mineral	6.8	1	100
Mineral	6.8	20	100
Organic	6.1	1	58
Organic	6.1	20	73
Organic	5.2	1	47
Organic	5.2	20	73
Sandy	6.6	1	8[a]
Sandy	8.0	1	28[a]

[a]Carbofuran remaining after 8 weeks (rather than 3 weeks as indicated in box heading).

degradation; and that degradation to carbofuran phenol was most rapid under anaerobic conditions. It appears that additional research is required on bacterial degradation of carbofuran, with special emphasis on acid-resistant strains.

6.3 Lethal Effects

In acute toxicity tests with aquatic organisms, LC50 (96 h) values, with only one exception, exceeded 130.0 μg/L. The exception was the larva of a marine crab with an LC50 (96 h) value of 2.5 μg/L. In tests of longer duration with fish, safe concentrations were estimated to range between 15.0 and 23.0 μg/L. Among the most sensitive species of birds tested, the acute oral LD50 was 238.0 μg/kg body weight (BW), the dietary carbofuran LD50 value was 190.0 mg/kg ration, dermal LD50s exceeded 100.0 mg/kg BW, and the LC100 value in drinking water was 2.0 mg/L. Mammals were comparatively resistant, having LD50 acute oral toxicities >2.0 mg/kg BW, a dietary LD38 of 100.0 mg/kg ration after 8 months, and dermal LD50s >120.0 mg/kg BW. However, only 2.0 μg/L as an aerosol killed 50% of rhesus monkeys in 6 h, and 40.0 μg/L killed all pheasants within 5 min. Bees and earthworms were relatively sensitive to carbofuran, but test conditions were sufficiently different to preclude a strict comparison with vertebrate species. Among photosynthetic species, concentrations of 200.0 mg/L carbofuran partly inhibited germination of rice seeds, but not other species tested, after exposure for 24 h. Effects of carbofuran on plants are considered negligible when contrasted to faunal damage effects.

6.3.1 Aquatic Animals

Among freshwater organisms, LC50 values for carbofuran ranged from 130.0 to 14,000.0 μg/L in tests of 72–96 h; fishes were the most sensitive and annelid worms the most resistant. A relatively narrow toxic range for carbofuran in the climbing perch (*Anabas testudineus*) was indicated by the LC-0 (120 h) value of 560.0 μg/L and the LC100 (24 h) value of 1560.0 μg/L. Carbofuran was not as toxic to aquatic biota as were various cyclodiene chlorinated hydrocarbon insecticides, almost all of which were subsequently withdrawn from commercial use and replaced by carbofuran and other carbamates, and organophosphorus and other compounds.

In flow-through toxicity tests with the marine sheepshead minnow (*Cyprinodon variegatus*), LC50 values had stabilized by day 60 of exposure with no significant mortality afterwards; however, the LC50 value was 386.0 μg/L at 96 h, or 7.8 times greater than that (49.0 μg/L) at 131 days. At concentrations up to 49.0 μg/L, carbofuran did not significantly affect the growth of parent fish or the number of eggs produced. But mortality of fry from fish exposed to 23.0 and 49.0 μg/L was measurably greater than that of controls. On the basis of these and other observations that indicate the growth of surviving fry in all concentrations was not affected and that carbofuran was degraded rapidly in seawater and in sheepshead minnows, it was concluded that the maximum allowable toxicant concentration (MATC) for carbofuran and sheepshead minnow lies between 15.0 and 23.0 μg/L. This observation is similar to that of others who demonstrated that adult dungeness crabs (*Cancer magister*) showed no deleterious effects on growth, survival, or reproduction during exposure to 25.0 μg/L of carbofuran for 69 days. Larvae of dungeness crabs were substantially more sensitive than adults in 96-h tests. Moreover, 1.5 μg/L of carbofuran inhibited swimming ability in zoeal stages of dungeness crabs and 1.0 μg/L inhibited molting, and prevented metamorphosis to more advanced larval stages. These observations require verification because mortality in control groups was high, a typical problem in bioassays with larvae of marine invertebrates.

6.3.2 Aquatic and Terrestrial Plants

Carbofuran was more toxic to blue-green alga (*Nostoc muscorum*) at pH 5–6 than at pH 7.5–10; toxicity was lessened under conditions of reduced illumination and low population density. All effects were observed at comparatively high carbofuran concentrations of 25.0–100.0 mg/kg. Seeds of okra (*Abelmoschus esculentus*) treated with carbofuran, at 1, 3, or 5% a.i. carbofuran by weight of seed, germinated normally after 90 days of storage. After 6 months of storage, however, germination was measurably reduced at all carbofuran treatments. Okra plants developed normally except for a reduction in plumule length, but this effect was also observed among okra seeds tested with a wide variety of agricultural chemicals.

The effect of carbofuran on the germination of seeds of cotton (*Gossypium hirsutum*), rice, and groundnut (*Arachis hypogea*) were investigated. Rice seeds exposed for 24 h to 100.0 or 200.0 mg/L carbofuran had decreased germination of 8 and 23%, respectively; seeds of the other two species were not affected at these exposure rates. Treated rice seedlings that germinated grew two to three times faster than controls, especially in the roots and leaves; no reasons were offered to account for these differences in rice plants. Carbofuran residues in seeds of the three test species exposed for 24 h to 100.0 mg/kg carbofuran ranged from 17.5 to 28.1 mg/kg; at 200.0 mg/kg carbofuran these values ranged from 24.0 to 30.4 mg/kg. At 72 h posttreatment, residues had declined markedly to 0.1–4.7 mg/kg in the groups treated with 100.0 mg/L carbofuran and 2.3–3.8 mg/kg in the groups treated with 200.0 mg/L. Observed growth promotion effects in certain plants by carbofuran and some of its metabolites may be due to effects on plant oxidase systems, rather than on insecticidal or nematocidal properties of the compound; however, the source of the effects has not been demonstrated conclusively.

6.3.3 Terrestrial Invertebrates

At recommended field application rates of granular carbofuran formulations, some losses of earthworms, springtails, and other soil-inhabiting organisms should be expected; spray and dust formulations adversely affect honeybees and other airborne crop pollinators. Bees are extremely susceptible to carbofuran. In one study with honeybees (*Apis* spp.), subjected to high levels of carbofuran, some young adults in the contaminated hive were unable to emerge from their cells, and those that did emerge remained weak and unfed. Eventually, the hive became vulnerable to invasion by the greater wax moth (*Galleria melonella*), an insect that subsequently destroyed the entire hive. The LD50 dose for honeybees was estimated at 0.16 μg/bee. If 1.12 kg carbofuran/ha were uniformly distributed at a height of 10 m, flying bees could encounter a lethal dose in only 2 s.

Studies with susceptible and selectively bred carbofuran-resistant carbofuran-resistant houseflies (*Musca domestica*) susceptible and resistant strains were 0.1 and 1.3 μg/insect, respectively. Resistant flies contained up to 34% more cholinesterase than susceptible strains and could excrete carbofuran almost twice as fast. Carbofuran resistance among pestiferous insects is not yet widely known or adequately documented.

Among earthworms, the characteristic symptoms of carbofuran poisoning were rigidity, immobility, lesions, and segmental swelling, as well as cholinesterase inhibition. Worms maintained in soils to which commercial applications of carbofuran had been applied developed two types of lesions within 72 h: multisegmental swelling that often ulcerated, causing death of the worm, and a discrete nodular mass protruding from the surface of the worm. The LC50 values were 0.5 mg/kg soil at 5 h for *Lumbricus herculeus*, but 2.4 and 13.0 mg/kg soil at 5 days for *Lumbricus terrestris* and *Eisenia foetida*, respectively; the differences in sensitivity were attributed to a greater excretion rate of carbofuran by *Eisenia*. However, mortality of *Eisenia* was 50% after 14 days in soils containing 3.1 mg carbofuran/kg DW. When application of carbofuran to soils was 9.1 kg/ha, 50% of the *Lumbricus* died in 72 h. At lower application rates of 2.0 kg/ha, populations of two species of Australian earthworms were reduced; juvenile stages were most severely affected. The loss of earthworms could result in reduced food for many wildlife species.

Field application of granular carbofuran can result in contamination of earthworms under rainy conditions. Earthworms contaminated with carbofuran may cause secondary poisoning of birds of prey, including buzzards (*Buteo buteo*), red kites (*Milvus milvus*), and black kites (*Milvus migrans*). In one case, buzzards found dead in fodder and sugar beet fields treated with granular carbofuran were poisoned by consuming dead and dying earthworms, mostly *Lumbricus terrestris*, from these fields. Remains of earthworms – which contained as much as 3.2 mg carbofuran/kg FW – were detected in all buzzard crop contents. A similar case is documented for American robins (*Turdus migratorius*). It has been suggested that the woodcock (*Philohela minor*), a species that consumes up to 50% of its body weight (about 125 g of food) daily from earthworms, may be at special risk from carbofuran poisoning. If each worm contained 1.3 mg/kg BW carbofuran, a woodcock would then ingest 0.16 mg of carbofuran or the equivalent of 0.65 mg/kg BW, an oral dose lethal to many bird species. To date, secondary poisoning of woodcocks has not been verified under controlled conditions.

6.3.4 Birds and Mammals

Acute oral toxicities of carbofuran to birds ranged from 238.0 μg/kg BW for fulvous whistling-ducks (*Dendrocygna bicolor*) to 38,900.0 μg/kg BW for domestic chickens. The fulvous whistling-duck has been listed as endangered since 1972 by the Texas Organization for Endangered Species. Concentrations of 1.0 mg/L of carbofuran in drinking water of the ducks caused symptoms of intoxication in 7 days, and 2.0 mg/L was lethal in the same period. Acute symptoms of carbofuran poisoning in birds, which may persist up to 7 days, include a loss in muscular coordination, wings crossed high over the back, head nodding, vocal sounds, salivation, tears, diarrhea, immobility with wings spread, labored breathing, eye pupil constriction, arching of back, and arching of neck over back; death may occur within 5 min after ingestion. Birds given a fatal oral dose of carbofuran showed a depression in brain cholinesterase activity of 83–91% within 8 h of dosing. Among mallards (*Anas platyrhynchos*), sensitivity to carbofuran was greater in ducklings than in older birds; this relation appears to hold true for other birds for which data are available.

Acute oral toxicities of carbofuran to various species of mammals ranged from 2000.0 μg/kg BW in mice to 34,500.0 μg/kg BW in rats. Mammals were generally more resistant than birds to acute biocidal properties of carbofuran.

The ingestion of carbofuran by mallard ducklings walking through carbofuran-sprayed vegetation appears to be the critical mode of intake, with dermal absorption being minimal. In one case, deaths were observed among mallard ducklings exposed to vegetation sprayed with 132.0 or 264.0 g carbofuran/ha; mortality was associated with depression of brain acetylcholinesterase (Ache) activity by more than 53%. Treated ducklings seemed to spend a larger proportion of time than controls hidden in emergent vegetation, but this was not statistically significant at the 0.05 level. Mallard ducklings force-fed carbofuran for 10 days at 0.85 mg/kg BW daily had survival of 31% and delayed fledging in survivors; this was not evident at 0.45 mg/kg BW daily for a similar period. Carbofuran administered to birds in the diet for 5 days, plus 3 days postexposure on an untreated diet, produced 50% kill values of 140.0–1459.0 mg carbofuran/kg ration; younger birds were more sensitive than older ones. Food consumption in groups of Japanese quail (*Coturnix japonica*) with high carbofuran-induced mortality was markedly depressed during the first 3 days of treatment. Red-winged blackbirds, the most sensitive bird species tested in food repellency tests, consumed a normal ration of food contaminated with carbofuran. As a result, carbofuran has a high potential for causing acute poisoning episodes in birds.

Secondary poisoning of avian raptors with carbofuran has been documented. Consider the case of a female red-shouldered hawk in adult plumage weighing 683 g, found in a cornfield near Beltsville, Maryland, in May of 1981. The field had been treated the previous day with Furadan 10 granules (10% carbofuran), applied at 1.12 kg/ha a.i. The bird was entirely paralyzed except for some head and neck movement, salivating a brown fluid, and respiring in rapid pants. These signs are consistent with those observed in birds dosed in the laboratory with carbofuran. Stomach contents contained remains of a northern short-tailed shrew (*Blarina brevicauda*) and a common grackle (*Quiscalus quiscula*). A total of 96.6 μg of carbofuran was extracted from the gastrointestinal tract and stomach contents and tissues. Judging by the body weight of the hawk, and an LD50 range of 0.26–5.6 mg/kg BW in various nondomesticated birds, this amount of carbofuran would constitute between 2.5 and 59% of the known LD50 values; however, carbofuran in birds is readily absorbed from the gut and widely transported in the body. Accordingly, the amount of toxicant extracted from the digestive tract was probably only a portion of that ingested by the hawk. In the same cornfield, at the same time, a smaller adult red-shouldered hawk (possibly the female's mate) was found that showed similar, but less severe, signs. Within 24 h, it appeared to have recovered completely and was released. As judged by carbofuran residues in small mammals and birds at this site, the residues present in the digestive tract of the female hawk, and the nature of the toxic symptoms observed, the two red-shouldered hawks were probably poisoned by carbofuran acquired from small vertebrate prey or scavenged from the treated areas.

Field application of carbofuran granules to corn, at planting, in Maryland during 1980 was presumed to be responsible for deaths of songbirds (order Passeriformes) and white-footed mice (*Peromyscus leucopus*); all organisms contained high levels of carbofuran in the gastrointestinal tract and liver, suggesting extensive feeding in treated fields. A similar situation occurred in Perry, Florida, after treatment of pine seed orchards. Laboratory studies with house sparrows (*Passer domesticus*) and red-winged blackbirds demonstrated that ingestion of a single carbofuran granule is fatal to either species. In groups of old-field mice (*Peromyscus polionotus*) fed diets containing 100.0 mg carbofuran/kg ration for 8 months, mortality was 38%; however, growth, development and behavior was normal among survivors from this group and their offspring. In a preliminary study with rats and old-field mice fed 100.0 mg carbofuran/kg ration, parents lost weight (but none died), and the survival of young was reduced.

Aerosol toxicity of carbofuran to warm-blooded animals ranged from about 2.0 μg/kg for rhesus monkeys to 110.0 μg/kg for rats. These values substantially exceed the established Threshold Limit Value (TLV) of 0.05 μg/kg (50.0 $\mu g/m^3$) for the protection of human health. The TLV is a time-weighted concentration for a 40-h workweek that nearly all workers can withstand without adverse effects, including eye and skin irritations, and other minor irritations. Inhalation doses to humans were estimated during and immediately after aerial spraying of Furadan 4-Flowable at the rate of 446 g a.i. carbofuran/ha, a concentration that generally controls most pests. During aerial sprayings of this level, the concentration of carbofuran in ambient air did not exceed 0.0033 μg/L at any location, suggesting that most birds and wildlife are afforded a high degree of protection during aerial spraying at recommended dosages. Studies with rats subjected to 1.2 μg/L of carbofuran aerosols for 50–70 min showed a substantial (55%) decrease in red blood cell cholinesterase 10 min posttreatment and a return to normal levels in 2 h. After 8 h, a maximum of 55% of the carbofuran was excreted by respiration (38%) or in the urine (12%), or feces (5%); the remainder was located primarily in the liver and gastrointestinal tract. Plasma half-lives in rats for carbofuran (36 min) and 3-hydroxycarbofuran (62 min) were similar to those previously determined after oral and intravenous exposures.

Carbofuran is not considered a chronic health hazard to humans because all existing evidence demonstrates that carbofuran is neither carcinogenic nor mutagenic, or teratogenic. The database is considered acceptable and complete. Dermal toxicity of carbofuran to birds and mammals is comparatively low. The LD50 dermal values ranged from about 1000.0 mg/kg BW in cattle down to 100.0 mg/kg BW in birds; i.e., house sparrows and queleas; rats and rabbits were intermediate in sensitivity at 120.0 and 885.0 mg/kg BW, respectively. Birds contaminated by carbofuran spray could possibly ingest significant amounts while preening, but such ingestion has not been demonstrated. For humans, the maximum potential dermal exposure based on exposed face, hands, forearms, back and front of the neck, and "V" of the chest is 3.1 mg or 0.04 mg/kg BW for a person weighing 70 kg. This relation suggests that human populations would be at greater risk than wildlife populations under recommended carbofuran spray application protocols.

6.4 Sublethal Effects

Most investigators agree that carbofuran degrades or is biotransformed rapidly, with negligible accumulations in biota. Numerous studies have demonstrated that carbofuran, at high sublethal concentrations, was capable of disrupting enzyme and lipid metabolism, but these effects were reversible with no observable permanent damage. Three major data gaps still appear to exist. First, latent biochemical and physiological effects that appear at substantial intervals posttreatment have not been explained. Second, interaction of carbofuran with other environmental compounds, especially other agricultural chemicals, is largely unknown, and the effects may cause more than additive damage. Third, and most important, data are scarce or lacking on chronic toxicity, teratogenicity, mutagenicity, and carcinogenicity of the degradation products of carbofuran, especially degradation products that may also form nitroso compounds; nitrosated carbofuran metabolites, for example, are demonstrably mutagenic.

6.4.1 Terrestrial Invertebrates

In decomposing the dead organic matter in a deciduous forest ecosystem, the detritus food chain may account for more than half the energy flowing through the ecosystem. Carbofuran can significantly disturb decomposition rates of litter communities, with profound consequences for nutrient recycling and incorporation of organic matter into the soils. For example, application of 0.29 kg/ha of carbofuran to a red maple (*Acer rubrum*) litter community near Ottawa, Canada, reduced daily decomposition rates by about 40%; all of the groups of macrodecomposers present, including Collembola, Acarina, Lepidoptera, Coleoptera, Diplopoda, and Annelida, have been shown to be susceptible to carbofuran and may have been affected by the treatment.

6.4.2 Aquatic Biota

Aquatic plants are comparatively resistant to carbofuran. Green alga (*Selenastrum capricornutum*) and several species of submergent macrophytes tolerated 1.0 mg carbofuran/L for 30 days without any measurable adverse effect. Elevated concentrations of 10.0 mg carbofuran/L did not affect growth or survival of duckweed (*Lemna minor*) or tubers of sago pondweed (*Potamogeton pectinatus*). Growth inhibition (50%) of algae (*Chlorella pyrenoidosa*) occurred in 96 h at 205.0 mg carbofuran/L, with no growth observed at 562.0 mg/L; a similar case is documented for *Chlorella emersonii*. However, first instar daphnids (*Daphnia magna*) and fourth instar midge (*Chironomus raparius*) were effectively immobilized within 48 h at only 48.0–64.0 μg carbofuran/L. And larvae of the Japanese medaka (*Oryzias latipes*) exposed to 88.0–110.0 μg/L for 4 days had impaired swimming performance immediately after exposure, which remained impaired after 10 days in uncontaminated water.

Carbofuran reportedly disrupts enzyme and lipid metabolism in fishes and may not degrade as rapidly under field conditions as suggested by laboratory studies. However, most investigators argue that carbofuran, under recommended application rates, does not accumulate to a significant extent in aquatic systems and rapidly degrades under field and model microcosm study conditions.

In studies with the African catfish (*Mystus vittatus*) exposed to 31.0 or 62.0 μg/L of carbofuran for 30 days, serum transaminases were significantly elevated. In comparison with catfish exposed to concentrations of 21.0 μg/L or less during the same period, there were also significant depressions in alkaline phosphatase activity in the liver; acid phosphatase activity in the liver, kidneys, and gills; and glucose-6-phosphatase in the liver and kidneys. In climbing perch, mean lipid levels in muscle and liver were elevated after exposure to an LC-0 (120 h) dose of 560.0 μg/L carbofuran for 120 h; a similar pattern was observed following exposure to an LC100 (24 h) concentration of 1560.0 μg/L for 6 h. Carbofuran-induced alterations have also been documented in serum chemistry of the African catfish during immersion in 21.0 μg/L for 30 days; in brain Ache activity of climbing perch and milkfish, (*Channa punctatus*) 30 days after exposure to high sublethal levels for 48 h; and in blood and tissue enzyme and ammonia levels in the air-breathing catfish (*Clarias batrachus*), 1 month after exposure for 30 days to 500.0 μg/L carbofuran. In field studies with *Trichogaster pectoralis*, a fish extensively cultured in flooded Malaysian rice paddies, degradation of carbofuran in the liver was slower than that reported for laboratory animals and suggested that caution be exercised in the extrapolation of rates of carbofuran oxidative hydroxylation activity from laboratory organisms to fishes cultured in rice fields. In fish, neurotoxic effects of carbofuran were localized to the brain regions which regulate motor activity and behavior. For example, adult snakeheads (*Channa punctatus*) exposed to 600.0 μg carbofuran/L for 15 days had a reduction in the level of neurotransmitters in the cerebral cortex of the brain. Histopathology of the liver and thyroid are reported in snakeheads after exposure for 6 months to extremely high sublethal (4.5 mg/L) concentrations of carbofuran.

Negligible accumulations of carbofuran were observed in egg masses of the caddisfly (*Triaenodus tardus*) during immersion for 120 h in water containing 8.0 μg/L of carbofuran; the low uptake was apparently related to the low partition coefficient of carbofuran. Rapid equilibrium and low accumulation was also reported for the sheepshead minnow (*Cyprinodon variegatus*); in a 28-day flow-through study, maximum tissue concentrations were measured between days 3–10 when upper concentration factors of 5–20 were recorded. Field applications of carbofuran in farm ponds in Arkansas and Kansas were associated with low mortality in fish or negligible effects on fish and plankton. Kansas farm ponds subjected to 25.0 μg carbofuran/L contained 10.6 μg/L in surface waters 1 day later, but non-detectable residues thereafter; residues were <0.4 μg/kg at 1 day in mud, zooplankton, and fish. Farm ponds treated with 50.0 μg/L of carbofuran after 3 days contained 15.0 μg/L carbofuran in surface water and 26.0–46.0 μg/kg in mud, but non-detectable residues in biota; no measurable residues were found in any sample after 25 days. When atrazine at 300.0 μg/L was applied in combination with 50.0 μg/L carbofuran, carbofuran was detectable in surface water at 23 days posttreatment at 1.5 μg/L, but not in the soil, biota, or any other compartment.

In an agromicrocosm study, influence of percolating water on soils containing 3.6 mg/L of radiolabeled (C-14) carbofuran was evaluated. After 3 weeks, 49% of the carbofuran had been removed with percolating water from soils, and 37% was later recovered from soils and corn. In nonpercolated soils, 80% of the carbofuran was still associated with soils and corn. The aquatic components, including water, lake mud, plants (*Elodea*), and fish (the guppy *Poecilia*), contained 25% of the soil-applied carbofuran, although 49% had been initially added to the aquariums by way of percolated water. This loss of 24% was attributed partly to the degradation of carbofuran to CO_2. About 75% of all the radiocarbon was in lake mud, most of it unextractable. Carbofuran was the major compound recovered from control and percolated soils, accounting for 39 and 15%, respectively; 3-ketocarbofuran and 3-hydroxycarbofuran were identified as the major metabolites. The addition of captafol, a fungicide, to carbofuran-treated soils resulted in a more rapid disappearance of the insecticide from terrestrial soils and reduced uptake by corn. The addition of EPTC, a herbicide, had no measurable effect on terrestrial components, but both EPTC and captafol caused increased recoveries of C-14 labeled carbofuran residues from lake bottom mud. In another study, radiolabeled carbofuran was applied at 1.12 kg/ha to a model ecosystem containing seedling sorghum plants (*Sorghum halopense*), saltmarsh caterpillar larvae (*Estigmene acrea*), the alga *Oedogonium cardiacum*, freshwater clams (*Corbicula manilensis*), crabs (*Uca minax*), a cladoceran (*Daphnia* sp.), mosquito larvae, unidentified species of frogs and snails, and the freshwater macrophyte, *Elodea canadensis*. Carbofuran was rapidly, but not completely, degraded in water to carbofuranphenol, 3-ketocarbofuran, 3-hydroxycarbofuranphenol, *N*-hydroxy-methyl carbofuran, 3-hydroxycarbofuran, and several unknown compounds. Carbofuran was highly toxic to crabs, clams, and *Daphnia* immediately after application to the model ecosystem, but all animals, except one crab, survived restocking 20 days later.

The freshwater bivalve mollusks, *Glebula rotundata* and *Rangia cuneata* absorbed waterborne carbofuran but did not appear to concentrate it. Both species of clams were very tolerant, even though symptoms of poisoning, such as shell gaping, foot extension, and incoordination, were evident when carbofuran exposures were high (75.0 mg/L). *Glebula* converted injected radiolabeled carbofuran to a variety of free metabolites, primarily hydrolysis products, and also polar carbofuran metabolites that were not degraded by conditions known to hydrolyze glycosidic conjugates. These polar metabolites may contain some type of amino acid moiety. The rate of carbofuran metabolism by *Glebula* was slower than that reported for most other animals, but was more rapid than that of plants and microorganisms. Bacterial metabolism of carbofuran was negligible in both in vivo and in vitro studies with bivalve mollusks.

6.4.3 Birds

Short-term studies demonstrate that juvenile ring-necked pheasants (*Phasianus colchicus*) given 3.6 mg carbofuran/kg grain diet (equivalent to 132 g/ha) for 5 days followed by 3 days on a clean diet had normal survival, growth, and brain Ache activity. Similarly, adult and nestling passerines were able to tolerate the dietary exposure resulting from ingestion of grasshoppers sprayed at the rate of 134 g/ha. Mallard ducklings led up to 300 m through vegetation plots sprayed with 132 or 164 g carbofuran/ha had normal growth and survival, but brain cholinesterase activity inhibition was directly related to spray rate and exposure distance. Growth and cholinesterase activity levels in mallard ducklings were reduced at 0.25 mg carbofuran/kg BW daily (but not 0.15 mg/kg BW daily) for 10 days.

Birds may encounter carbofuran through respiratory, dermal, and oral routes. Depending on the dietary requirements of particular species, ingestion of contaminated vegetables and poisoned invertebrates may be important exposure routes. Carbofuran may prove harmful alone or in combination with other substances. For example, male Japanese quail fed 0.5 mg carbofuran/kg ration for 18 weeks exhibited a 79% inhibition of plasma cholinesterase activity. The reduction was slightly greater (84%) when carbofuran was fed in combination with 0.05 mg morsodren/kg ration, a methyl mercury compound, although morsodren had no measurable effect on cholinesterase activity when fed alone at that dosage. Since many species of fish-eating birds frequently contain 0.05 mg/kg BW of mercury in various tissues, interaction effects of mercury with carbofuran and other cholinesterase-inhibiting compounds may produce synergistic, deleterious effects.

Low oral dosages or high dietary levels of carbofuran produced no permanent damage effects in northern bobwhites. A single oral dose of 2.0 mg carbofuran/kg BW did not affect brain cholinesterase levels at 48 h, or growth, metabolic efficiency, or metabolized energy at 8 days. The activity of bobwhites fed 131.0 mg carbofuran/kg ration for 14 days was reduced, but this effect was temporary and recovery was complete within 14 days on a carbofuran-free diet. The temporarily reduced activity was attributed to the rapid metabolic breakdown of carbofuran. Among laying white leghorn hens, 80% of a single oral dose of 2.7 mg carbofuran/kg BW was eliminated in feces within 10 days. All eggs contained detectable carbofuran; an egg with the highest concentration of 0.13 mg/kg developed on day 4. Residues in liver and kidney were about 2.6 mg/kg at 6 h but declined to 0.2 mg/kg in 24 h. Muscle and fat contained about 0.3 mg/kg at 6 h and <0.1 mg/kg at 24 h. Hydroxylation of carbofuran and hydrolysis of the carbamate ester were the predominant pathways in the metabolism of carbofuran by laying hens; similar results were obtained at single oral doses of 2.7 or 0.3 mg carbofuran/kg BW.

6.4.4 Mammals

Granular carbofuran applications at 7.9–11.2 kg/ha on conventionally tilled or no-till fields in Maryland and Pennsylvania in 1986 had no adverse effects on resident white-footed mice (*Peromyscus leucopus*). Exposed mice had normal blood chemistry, liver function, growth, migration, and populations. Among larger mammals, carbofuran is associated with a variety of stress symptoms, including increased salivation, muscle tremors, prostration, labored breathing, loss of appetite, and (in rare cases) death. These signs were observed in 1–2-week old calves given single doses of carbofuran at 0.25–5.0 mg/kg BW orally or 0.05–0.1% dermally, in cattle yearlings at 1.0–5.0 mg/kg orally or 0.1% dermally, and in sheep at 2.5–5.0 mg/kg orally. All survivors had completely recovered at 5 days posttreatment. Ewes given 0.3 mg carbofuran/kg BW orally three times weekly for 43 days had elevated serum thyroxine levels, suggesting additional research on the use of metabolic hormones as biomarkers of carbofuran exposure.

Lactating cows fed corn silage containing 1.4–3.9 mg carbofuran/kg ration for 8 weeks, or about 74.0 mg carbofuran daily, showed no decrease in blood cholinesterase; furthermore, no carbofuran residues were detected in the milk. Other studies with lactating cows dosed orally with carbofuran showed almost complete excretion in 10 days, mostly through urine (94%), feces (0.7%) and milk (0.2%). Carbofuran metabolites in urine, feces, and excreted milk included the 3-hydroxy-, 3-keto-, and 3-hydroxy-*N*-hydroxymethyl derivatives, both conjugated and free, and unknown constituents, perhaps carbon dioxide formed by carbofuran hydrolysis.

In investigations of the effects of carbofuran or its metabolites on mice and rats, pregnant mice receiving 0.01 or 0.5 mg dietary carbofuran/kg daily throughout gestation gave birth to viable, overtly normal offspring at term. Significant elevation of serum immunoglobins was measured in a 101-day-old male offspring of female parents receiving 0.5 mg/kg dietary carbofuran. This effect was not observed at day 400 or 800. In female offspring from the group receiving 0.01 mg/kg carbofuran, serum immunoglobins were significantly depressed at day 101, but not thereafter. Disturbances in immunoglobulin contents may decrease immunocompetence and, thus, indirectly contribute to morbidity and premature mortality. In rats fed comparatively high dietary levels of 30.0 mg carbofuran/kg ration for 90 days, with mean daily intake of 1.97 mg carbofuran/kg BW, growth was significantly reduced and ventral prostate gland metabolism of RNA, DNA, and protein was altered. Prenatal exposure of mice to 0.01 mg carbofuran/kg BW daily, administered orally during gestation, resulted in persistent postnatal endocrine dysfunction in adults; specifically, the impairment of hepatic metabolism and elevation of plasma corticosterone. Unexpectedly, however, at a higher dose of 0.05 mg/kg, there were no significant differences from controls, and the endocrine function of tested mice was normal. In female rats given a single dose of 0.05 mg carbofuran per kg BW orally on the eighteenth day of gestation, Ache activity decreased significantly in maternal and fetal blood, and in the maternal liver within 1 h. At higher dosages of 0.25 and 2.5 mg/kg, Ache was also depressed in the fetal liver and in the maternal and fetal brains; the effects were not measurable 24 h postadministration.

Carbamate pesticides can easily be converted to *N*-nitroso derivatives in the presence of sodium nitrite under acidic conditions. The *N*-nitroso form of carbofuran could possibly be formed in the human stomach. Since carbofuran is routinely used on a variety of crops and nitrite is a common component of the human diet and is present in human saliva, nitrosation of carbamates under conditions simulating those in the human stomach is possible. Nitrosocarbofuran and five other nitrosated carbamate pesticides were tested for carcinogenicity in rats. Nitrosocarbofuran, at 16.5 mg/kg BW administered orally once weekly for 23 weeks, was the most toxic compound tested and caused the death of several animals by liver damage early in the experiment. Among survivors, nitrosocarbofuran was the most carcinogenic, as judged by the numbers of carcinomas and tumors that developed. Nitrosation rates of carbofuran in the environment are not now adequately documented, but conceivably could represent an environmental risk to wildlife. Surprisingly, nitrosocarbofuran was among the least mutagenic compounds tested in rats; no obvious explanation is available for the differences in carcinogenic and mutagenic properties. It is noteworthy that data on chronic toxicity, teratogenicity, mutagenicity, and carcinogenicity of degradation products of carbofuran, especially carbofuran-7-phenol, and 3-hydroxycarbofuran-7-phenol are either scarce or lacking; a similar case is made for nitrosocarbofuran and other degradation products of carbofuran which may also form nitroso compounds. Nitrosated 3-hydroxycarbofuran and 3-ketocarbofuran produced mutagenic responses in bacterial strains of *Salmonella typhimurium* and chromosome aberrations in ovary cells of Chinese hamsters. Nitrosocarbofuran and 3-hydroxynitrosocarbofuran also induced large numbers of sister chromatid exchanges in the same cells. Further, nitroso derivatives of carbofuran were considerably more active than nitroso forms of other carbamate pesticides in producing mutagenicity in *Salmonella*. On the other hand, technical formulations of the parent carbofuran were neither genotoxic nor mutagenic to bacteria, yeast, or corn.

6.5 Recommendations

In Canada, for regulatory purposes, the tolerance level for carbofuran in animal tissues or food, feed, and fiber crops is based on the total carbamate content of the sample, as indicated by total carbofuran, 3-hydroxycarbofuran, 3-ketocarbofuran, and their conjugates, presumably carbofuran phenol, 3-ketocarbofuran phenol, and 3-hydroxycarbofuran phenol. In the United States, the tolerance level is based on carbofuran and four metabolites: 3-hydroxycarbofuran; carbofuran phenol; 3-hydroxycarbofuran phenol; and 3-ketocarbofuran phenol. Carbofuran levels considered safe range from 0.05 mg/kg (including 0.02 mg/kg carbofuran metabolites) in meat, fat, and meat by-products to 40.0 mg/kg (including 20.0 mg/kg carbofuran metabolites) in alfalfa hay; intermediate values are 0.1 mg/L in milk, 0.2 mg/kg in corn grain, and 25.0 mg/kg in corn fodder and forage. The acceptable daily intake for the protection of human health should not exceed 0.01 mg carbofuran/kg BW. No recommended carbofuran level is currently being promulgated by any regulatory agency for the protection of sensitive species of aquatic biota and wildlife.

I conservatively estimate that, in terms of total carbofuran in water, damage is possible to aquatic invertebrates at >2.5 μg/L and to teleosts at >15.0 μg/L. These levels could be attained during a heavy rainfall shortly after carbofuran treatment of adjacent fields. Among sensitive species of warm-blooded animals, dietary concentrations as low as 10.0 μg/kg ration have demonstrable effects, which were measurable only after extended periods postingestion; for comparison, this level is about one-fifth of that allowed in meat by-products for human consumption. Current maximum permissible aerosol levels of 0.05 μg/L (50.0 μg/m^3) appear sufficient to protect wildlife with the proviso that concentrations do not exceed 2.0 μg/L at any time.

Sporadic kills of migratory birds were associated with carbofuran formulations containing 3% a.i. For example, migratory sandpipers died after eating Furadan 3G granules (3% a.i.) applied to rice crops in Texas. The granules probably were ingested while the sandpipers were probing and skimming the surface of wet soil for insects and crustaceans.

Other species of migratory waterfowl may have mistaken the small size and density of Furadan granules for seed, particularly in areas where concentrations of granules were abundant after misuse and careless applications. It appears that granular carbofuran formulations need to be developed that contain less than 3% a.i. in order to protect waterfowl, yet maintain their effectiveness against target organisms. Nevertheless, the sale and use of granular formulations of carbofuran in the United States has been prohibited since September 1, 1994, except for five minor uses: bananas in Hawaii, spinach grown for seed, pine tree progeny tests, cucurbits (cucumbers, squash, pumpkins, cantaloupes, and watermelons), and dry-harvested cranberries. Total sales of granular carbofuran for these five uses is limited to 2500 pounds (1136 kg) annually. All other uses of granular carbofuran have been deleted from the label and remaining stocks stopped from use after August 31, 1995.

In rice field pest control, carbofuran should be applied before the fields are flooded and delayed to avoid peak bird migration. Research also appears warranted on the effects on fish and wildlife of the numerous carbofuran formulations used, especially liquid spray formulations (flowables), and on applications to crops other than rice, such as corn, alfalfa, and hay. Additional long-term research is urgently needed on potential impacts of degradation products of carbofuran on sensitive species of aquatic organisms and wildlife, with special attention to nitrosated carbofuran metabolites; these data are now scarce or lacking. Research is also needed on chemical and biological interactions of carbofuran with other agricultural chemicals applied to the same locations, which are imperfectly understood. Finally, researchers must elucidate the significance of metabolic upset recorded in various species of laboratory mammals at considerable periods after carbofuran insult.

The American Ornithologists' Union passed a resolution in 1990 calling for the cancellation and immediate suspension of all carbofuran granular products – which has since been implemented – and urging the United States to also ban the liquid formulations. Carbofuran flowable formulations are considered hazardous to the burrowing owl (*Athene cunicularia*) and nests have been abandoned following exposure. In Canada, all carbofuran formulations are prohibited within 250 m of owl burrows. In the United States, carbofuran prohibitions to protect endangered species of mammals and birds in fragile ecological areas are under consideration.

6.6 Summary

Carbofuran (2,3-dihydro-2,2-dimethyl-1,7-benzo-furanyl methyl carbamate) and other carbamate compounds, together with organophosphorus compounds, have virtually replaced the more persistent and hazardous organochlorine systemic pesticides used in agriculture. In general, carbofuran effectively controls insects through an anticholinesterase mode of action. Compared with chlorinated hydrocarbon insecticides, it has a relatively short residual life in the environment, degrades rapidly, and is almost completely excreted by nontarget organisms. Carbofuran degradation is complex and demonstrably modified by numerous biological and physicochemical factors; little is known of the biological properties of the degradation products, especially nitrosated metabolites, in relation to chronic toxicity, teratogenicity, mutagenicity, or carcinogenicity.

At currently recommended application rates and in present formulations, carbofuran has caused sporadic, and sometimes extensive, field kills of fish, wildlife, and invertebrates. In short-term laboratory tests, significant death rates were observed at concentrations of about 200.0 μg carbofuran/L (in water) for sensitive species of aquatic biota, 238.0 μg/kg BW (acute oral), and 190,000.0 μg/kg ration (dietary) for birds, and 2000.0 μg/kg BW (acute oral) and 100,000.0 μg/kg ration (dietary) for mammals. Among representative indicator species, harmful and sometimes life-threatening effects of carbofuran have been recorded for fish at nominal water concentrations of >15.0 μg/L and for aquatic invertebrates at >2.5 μg/L. For birds and mammals, harmful effects were observed at 10.0–50.0 μg/kg in the diet and 1000.0 μg/L in drinking water. For comparison, the "safe" level of carbofuran in meat products for human consumption is 50.0 μg/kg. Current maximum permissible aerosol levels of 0.05 μg/L (50.0 $\mu g/m^3$) carbofuran appear sufficient to protect wildlife; however, evidence suggests that aerosol concentrations should never exceed 2.0 μg/L. Plants are significantly more resistant to carbofuran than are invertebrates and higher organisms. Carbofuran hazards to migratory waterfowl have been reduced by effectively prohibiting granular formulations. In rice culture, carbofuran should be applied before the fields are flooded and after the peak of bird migration. More research is merited on the biotic effects of various formulations of carbofuran, especially flowable formulations, and on applications to crops other than rice, such as corn, alfalfa, and hay.

CHLORDANE[a]

Chapter 7

7.1 Introduction

Technical chlordane is a mixture of chlorinated hydrocarbons that has been used as an insecticide since its introduction in 1947. Chlordane was the first cyclodiene insecticide to be used in agriculture and was the second most important organochlorine insecticide in the United States in 1976–77, behind toxaphene, with an estimated annual production of 9 million kg. Chlordane is a leading insecticide in controlling termites, with about 1.2 million homes in the United States alone treated annually for this purpose.

Chlordane has been detected in human milk in Canada, Japan, Mexico, Spain, Hawaii, and Mississippi. Chlordane compounds have been detected in oysters from the South Atlantic Ocean and Gulf of Mexico, in fish from the Great Lakes and major river basins of the United States, in the blubber of cetaceans from the coastal waters of North America, and in the Antarctic atmosphere. In fact, all available evidence suggests that chlordane is ubiquitous in the environment. Air and water transport of technical chlordane has resulted in the detection of chlordane and its metabolites in rainwater, drinking water, air, surface waters, soils, sediments, plankton, earthworms, fish, shellfish, birds and their eggs, aquatic invertebrates, cats, dogs, livestock, and humans. Despite its widespread use, persistence, and tendency to accumulate in fat, there was no firm evidence of direct lethal or sublethal effects on terrestrial vertebrate wildlife until 1983 when several chlordane-related mortalities were recorded. A North Dakota marsh treated with chlordane had decreased reproductive success and some deaths of young of several bird species, but this was attributed to depletion of invertebrate prey and not to acute poisoning. Chlordane was implicated as the principal toxicant in 30 pesticide poisoning cases of hawks, owls, herons, and other birds in New York between 1982 and 1986. In New York, Maryland, and New Jersey between 1986 and 1990, chlordane was implicated in the poisoning of opossum (*Didelphis virginiana*), red fox (*Vulpes vulpes*), striped skunk (*Mephitis mephites*), big brown bat (*Eptesicius fuscus*), eleven species of songbirds, and nine species of raptors.

The U.S. Environmental Protection Agency (EPA) considers chlordane as a probable human carcinogen (defined as inadequate evidence from human studies and sufficient evidence from animal studies), as judged by chlordane-induced cancer of the liver in domestic mice. In 1978, EPA restricted chlordane use to subterranean termite control, nonfood plants, and root dip. Limited agricultural use was permitted until 1980. In 1987, EPA registered chlordane again, limiting its sale and use to licensed applicators for subterranean termite control. However, it seems that significant home and garden use exists, especially for control of termites and undesirable lawn insects.

7.2 Chemical and Biochemical Properties

Technical chlordane (64–67% chlorine) is produced by the condensation of cyclopentadiene

---

[a]All information in this chapter is referenced in the following sources:

Eisler, R. 1990. Chlordane hazards to fish, wildlife, and invertebrates: a synoptic review. *U.S. Fish Wildl. Serv. Biol. Rep. 85(1.21)*, 49 pp.

Eisler, R. 2000. Chlordane. Pages 823–881 in *Handbook of Chemical Risk Assessment: Health Hazards to Humans, Plants, and Animals. Volume 2, Organics.* Lewis Publishers, Boca Raton, Florida.

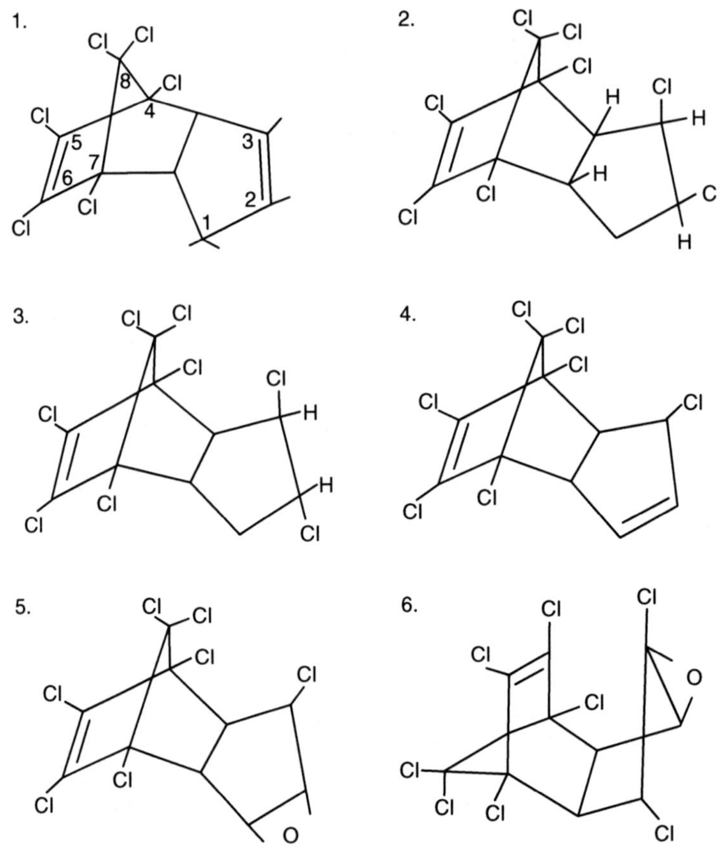

Figure 7.1.
Chemical structure of chlordane-related compounds: 1, chlordene (4,5,6,7,8,8-hexachloro-3a,4,7,7a-tetrahydro-4,7-methanoindene); 2, *cis*-chlordane, also known as *alpha*-chlordane (1-*exo*, 2-*exo*, 4,5,6,7,8,8-octachloro-2,3,3a,4,7,7a, hexahydro-4,7-methanoindene); 3, *trans*-chlordane, also known as *gamma*-chlordane (1-*exo*, 2-*endo*, 4,5,6,7,8,8-octachloro-2,3,3a,4,7,7a-hexahydro-4,7-methanoindene); 4, heptachlor (1,4,5,6,7,8,8-heptachloro-3a,4,7,7a-tetrahydro-4,7-methanoindene) – technical heptachlor contains about 15% *cis*-chlordane and 2.5% *trans*-chlordane; 5, heptachlor epoxide (1,4,5,6,7,8,8-heptachloro-2,3-epoxy-3a,4,7,7a-tetrahydro-4,7-methanoindene); and 6, oxychlordane, also known as octachlor epoxide (1-*exo*, 2-*endo*, 4,5,6,7,8,8-octachloro,2,3-*exo*-epoxy-2,3,3a,4,7,7a-hexahydro-4,7-methanoindene).

and hexachlorocyclopentadiene to yield chlordene (Figure 7.1). Addition of chlorine across the 2–3 olefinic bond of chlordene forms *cis*-chlordane and *trans*-chlordane; substitution of chlorine into position 1 of chlordene forms heptachlor, and further addition of chlorine across the 2–3 olefinic bond forms *cis*-nonachlor and *trans*-nonachlor. Technical chlordane includes about 45 components. Its approximate composition is 19% *cis*-chlordane ($C_{10}H_6Cl_8$), 24% *trans*-chlordane ($C_{10}H_6Cl_8$), 21.5–25% chlordene isomers ($C_{10}H_6Cl_6$), 7–10% heptachlor ($C_{10}H_5Cl_7$), 7–10% *cis*- and *trans*-nonachlor ($C_{10}H_5Cl_9$),

2% Diels-Alder adduct of cyclopentadiene and penta-chlorocyclopentadiene, 1% hexachlorocyclopentadiene, 1% octachlorocyclopentene, and 15.5% miscellaneous constituents. Oxychlordane and heptachlor epoxide are toxicologically significant degradation products (Figure 7.1).

Chlordane produced before 1951 contained a significant quantity of hexachlorocyclopentadiene – a toxic irritant to warm-blooded animals; chlordane produced after 1951 contains little or none of this compound. A high-purity chlordane formulation containing about 74% *cis*-chlordane and 24% *trans*-chlordane is also available.

Chemical analysis of technical chlordane is difficult because of frequent variations in both the number and relative composition of components in weathered chlordane. Other difficulties are encountered from analytical interferences from various organochlorine compounds; furthermore, the exact structure has not been determined for a number of compounds in technical chlordane, and the majority of compounds have not been isolated or synthesized for use as comparative standards.

Cis-chlordane (CAS number 5103-71-9) and *trans*-chlordane (CAS number 5103-74-2) are characterized by the following properties: molecular weight of 409.76; chemical formula of $C_{10}H_6Cl_8$; viscous, amber-colored liquid; boiling point between 104 and 105°C for *trans*-chlordane, and between 106 and 107°C for *cis*-chlordane; density of 1.59–1.63 at 25°C; soluble in most organic solvents, but only sparingly soluble in water, that is, 9.0 μg/L at 25°C; vapor pressure of 0.00001 mm mercury at 25°C; and a log K_{ow} (octanol–water partition coefficient) of 5.16. Pesticides containing chlordane or technical chlordane have been sold under a variety of names including 1068, Aspon, Belt, CD-68, Chlor-dan, Chlorindan, Chlor-kil, Chlorodane, Chlortox, Cortilan-neu, Corodane, Dichlorochlordene, Dichlorodene, Dowchlor, ENT 9932, HCS 3260, Kypclor, M 140, M 410, Niran, Niran 5% granular bait, Octachlor, Octa-klor, Octaterr, Ortho-klor, Synklor, Tat Chlor 4, Topiclor 20, Toxichlor, and Velsicol 1068.

Technical chlordane is stable under ultraviolet (UV) light, although some components, such as chlordene, heptachlor, *cis*-chlordane, and *trans*-chlordane will form photoisomers under high intensity UV in the presence of sensitizers, such as ketones. Several compounds were measured in alfalfa grown on soils treated with chlordane, including 1,2-dichlorochlordene, oxychlordane, and photo-*cis*-chlordane, as well as the parent chlordane compounds. The half-life (Tb1/2) of *cis*-chlordane in water is comparatively short, between 1.1 and 17.5 h. In soils, technical chlordane has a half-life ranging from 0.5 to 1.0 years for some samples, and 4 to 10 years for other samples. The lower Tb1/2 values refer to the initial rapid disappearance of chlordane from the soil; if studies continue over several years, the remaining chlordane is relatively persistent, with a Tb1/2 of 5–7 years. Measurable residues of chlordanes in soils were present more than 14 years after application. Chlordane persists in soils because of its low solubility in water, relatively low vapor pressure, and high tendency to adsorb to soil particles; accordingly, soil-bound chlordanes are not likely to become serious contaminants of the lower soil strata or deep-water sources. Transport into the hydrosphere from contaminated soils will occur through erosion of soil particles or sediments, not by desorption and dissolution.

Chlordane is a nerve stimulant; at low chronic doses, it produces hyperexcitability and lack of coordination in animals, and at high acute doses causes tremors and convulsions. Chlordane induces hepatic microsomal drug-metabolizing enzymes, resulting in enhanced biotransformation at low doses, although high doses may result in liver hypertrophy. The physiological target sites are in nerve and muscle membranes, presumably on proteins and phospholipids; the ultimate effect is axonic with membrane disruption, resulting in spasmic muscle twitching and death.

Chlordane is readily absorbed by warm-blooded animals through skin, diet, and inhalation. It is quickly distributed in the body and tends to concentrate in liver and fat. Up to 75% of a single oral dose of chlordane administered to rats and mice was absorbed in the gut, and up to 76% of an aerosol dose was absorbed in the respiratory tract; rabbits absorbed 33% in the gut following oral administration.

Chlordane residues in mammals were usually not measurable 4–8 weeks after cessation of exposure. Chlordane persistence in human serum and whole body was estimated at 88 and 21 days, respectively; this compares to a Tb1/2 of about 23 days in rats fed chlordane for 56 days.

Excretion kinetics of chlordane are complex, and different isomers exit through different pathways. In rats, chlordane elimination was almost complete 7 days after receiving single oral doses up to 1.0 mg/kg body weight (BW); 24 h after treatment, 70% of the *cis*-chlordane and 60% of the *trans*-chlordane had been excreted. In rodents, chlordane and its metabolites were usually excreted in feces, regardless of the administration route; the *cis*-isomer was excreted slightly faster than the *trans*-isomer, although identical metabolites seemed to be formed. In rabbits, however, up to 47% of the administered dose was voided in the urine, and *cis*- and *trans*-chlordane were excreted at the same rate.

Microorganisms such as *Nocardiopsis* sp., an actinomycete, can metabolize *cis*- and *trans*-chlordane to at least 8 solvent-soluble substances, including dichlorochlordene, oxychlordane, heptachlor, heptachlor endo-epoxide, chlordene chlorohydrin, and 3-hydroxy-*trans*-chlordane. Based on studies of chlordane metabolism in animals, four metabolic pathways are proposed: (1) hydroxylation to form 3-hydroxychlordane, which on dehydration forms 1,2-dichlorochlordene, with subsequent epoxidation to oxychlordane (*trans*-chlordane is converted to oxychlordane 7 times faster than *cis*-chlordane); (2) dehydrochlorination to form heptachlor, from which heptachlor epoxide and other hydroxylation products may be formed; (3) dechlorination to monochlorodihydrochlordene; and (4) the replacement of chlorine by hydroxyl groups resulting in the formation of hydroxy metabolites, which are excreted or further transformed by conjugation with glucuronic acid. Metabolism of chlordanes and nonachlors to oxychlordane, in orders of magnitude, greater in fish-eating and carnivorous birds than in marine mammals. The reasons for this are unclear and merit further research.

Trans-nonachlor, a major component of technical chlordane, was frequently found as the major chlordane residue in humans, whereas oxychlordane was the major component in rats fed technical chlordane. *Trans*-nonachlor is converted efficiently by rat liver microsomes to *trans*-chlordane, but this ability is lacking in humans, resulting in the accumulation of *trans*-nonachlor in humans.

Although technical chlordane is a mixture of compounds, two metabolites – heptachlor epoxide and oxychlordane – can kill birds when administered through the diet. These two metabolites originate from biological and physical breakdown of chlordanes in the environment, or from metabolism after ingestion. Heptachlor can result from breakdown of *cis*- and *trans*-chlordane, eventually oxidizing to heptachlor epoxide; oxychlordane can result from the breakdown of heptachlor, *cis*-chlordane, *trans*-chlordane, or *trans*-nonachlor. Heptachlor epoxide has been identified in soil, crops, and aquatic biota, but its presence is usually associated with the use of heptachlor, not technical chlordane – which also contains some heptachlor. Various components in technical chlordane may inhibit the formation of heptachlor epoxide or accelerate the decomposition of the epoxide, but the actual mechanisms are unclear.

In mammals, oxychlordane ($C_{10}H_4Cl_8O$) is a metabolite of *cis*- and *trans*-chlordanes and *trans*-nonachlor, and has proven much more toxic and persistent than the parent chemicals. Oxychlordane has been measured in the fat of rats, dogs, and pigs fed either isomer, and in milk and cheese from cows fed alfalfa treated with technical chlordane. Oxychlordane was isolated and identified from adipose tissues of pigs fed diets for 90 days containing 300.0 mg/kg of *cis*-chlordane or *trans*-chlordane. Sharply elevated oxychlordane levels were detected in milk from cows fed chlordane for 60 days; when chlordane was removed from their diet, oxychlordane residues in milk dropped rapidly during the week following termination, and stabilized after two weeks. The Tb1/2 for oxychlordane in beef cattle grazing in heptachlor-contaminated pastures for 4 weeks was about 92 days.

Rats and rabbits given chlordane orally or through the diet retained the highest levels in adipose tissue, followed by liver, kidney, brain, and muscle; oxychlordane was the most persistent residue after chlordane was removed from the diet.

7.3 Uses

Chlordane was first produced commercially in the United States in 1947 and became available in five basic formulations, including 5% granules, oil solutions containing 2.0–200.0 g/L, and emulsifiable concentrates containing chlordane at 400.0–800.0 g/L. Production of chlordane in the United States in 1971 was estimated at 11.3 million kg. By 1974, about 9.5 million kg of chlordane were used domestically to control commercial pests (35%); in homes, lawns, and gardens (30%); on corn (20%); turf (6%); potatoes (5%); tomatoes (2%); and other uses. On March 6, 1978, the EPA issued a cancellation proceeding on chlordane, allowing limited use on certain crops and pests until July 1, 1983, but no use thereafter except for underground termite control. A similar situation exists in Japan, where the only permitted use of chlordane is for control of termites and powder post beetles. Use in Japan is estimated at 500,000 kg per year.

In Canada, chlordane had been used in soils (usually at 0.45–4.5 kg/ha) against corn rootworms, strawberry root weevils, wireworms, white grubs, and subterranean cutworms infesting a wide range of crops. In the past, at least 75 different formulations containing chlordane as the active insecticidal ingredient had been registered for sale in Canada; the most widely sold formulation, accounting for about 60% of chlordane in soils, was the 25% granular type that was used extensively for corn rootworm control. Sales of chlordane in Canada increased about 10 times between 1969 and 1971 because of restrictions on DDT and other organochlorines; however, chlordane use was restricted in Canada in 1978.

7.4 Concentrations in Field Collections

Chlordanes and their metabolites are ubiquitous in the environment at low concentrations, but at a high occurrence in samples analyzed. Atmospheric transport is considered to be the major route of global dissemination. Some chlordane isomers persist in soils for 3–15 years, although there seems to be little accumulation of chlordanes by crop plants grown in these soils. Lengthy persistence of various chlordane isomers, especially *cis*-chlordane and *trans*-nonachlor, has been reported in certain organisms, but this has varied greatly between species and tissues.

In living organisms, chlordane concentrations are usually highest in samples collected near areas where chlordane was applied to control termites or other pests, in predatory species, and in tissues with high lipid content. Food chain biomagnification is usually low except in certain marine mammals. In some fishes, chlordane levels in muscles have been sufficient to endanger fish health (>100.0 μg/kg fresh weight (FW)) or human fish consumers (>300.0 μg/kg FW).

7.4.1 Abiotic Materials

Air and water transport of technical chlordane has resulted in the detection of chlordane and its metabolites in nonbiological samples worldwide. Chlordane enters the atmosphere mainly through aerial applications of dust and spray formulations, soil erosion by wind, and volatilization from soil and water. In aquatic systems, chlordane enters by way of surface runoff and rainfall; chlordane is rapidly adsorbed onto bottom sediments, where it persists. Atmospheric transport of chlordanes is considered the major route of global dissemination. Levels of chlordane compounds in the marine atmosphere of the southern hemisphere are nearly the same as those of DDT and its metabolites; this strongly suggests that chlordane compounds are globally distributed and dispersed. The yearly input of *cis*-chlordane to the Arctic Ocean from atmospheric sources is estimated at 3000 kg; if *cis*-chlordane constitutes 19% of technical chlordane, then more than 600,000 kg of technical chlordane has entered the Arctic Ocean since 1948. Chlordane is frequently measured in the air of buildings where the compound has been used for insect control. Chlordane has been found in household dust in the homes of farmers and pesticide formulators at exceedingly high mean levels: 5.8–23.1 mg/kg air-dried dust.

Chlordane has been detected in both groundwater and surface water at low levels of 0.001–0.01 μg/L. A high frequency of chlordane detection was noted in seawater samples collected from a Hawaiian marina: up to 90% of all samples contained *cis*-chlordane, and 68% contained *trans*-chlordane. Owing to chlordane's use as a soil-injected insecticide and its persistence, it has the potential to contaminate groundwater, particularly when it is applied near existing wells.

In soils, chlordane is comparatively immobile and persistent and has only a limited capacity for translocation into edible portions of food crops. Total chlordane content in cropland soils nationwide in 1971–72 averaged 0.05–0.06 mg/kg dry weight (DW), and ranged between 0.01 and 7.9 mg/kg DW; maximum values, in excess of 3.0 mg/kg, were recorded in soils from Illinois (7.0), Ohio (5.0), Indiana (4.1), and Iowa (3.4). The half-life of chlordane in soil when used at agricultural rates is about 1 year, but residues may be measurable much longer, depending on soil type. For example, 10 years after application of 8.5 kg technical chlordane/ha, up to 20% of the active ingredients were still measurable; in another study, 15% of the active ingredients remained in turf soils after 15 years. *Cis-* and *trans*-chlordanes were less persistent in mineral soils than in organic mucky soils. Chlordanes were usually detected in surface soils of basins receiving urban runoff water at a maximum concentration of 2.7 mg/kg; this decreased with soil depth to <0.03 mg/kg at depths below 24 cm. Chlordane levels in soils near Air Force bases in the United States in 1975–76 were similar to those found in nonmilitary urban environments.

Chlordanes in sediments usually were highest in those sediments with the highest organic content, especially downstream from the center of anthropogenic activities. Sediments from a lake in which the overlying water column initially was treated to contain 10.0 μg technical chlordane/L contained measurable residues 2.8 years after application: total chlordanes – consisting of *cis*-chlordane, *trans*-chlordane, and *trans*-nonachlor – averaged 20.0 μg/kg and ranged up to 46.0 μg/kg. The yearly flux of chlordanes from sediments to the overlying water column has been estimated at 0.02 $\mu g/m^2$, based on measurements made in the Sargasso Sea and deep North Atlantic Ocean between 1978 and 1980.

7.4.2 Terrestrial Crops

Maximum total chlordane concentrations in corn (*Zea mays*) and sorghum (*Sorghum halepense*) samples collected nationwide in 1971, in microgram per kilogram dry weight, were 480.0 in corn kernel, 1260.0 in cornstalk, and 420.0 in sorghum; these values were somewhat lower in 1972: 150.0 in kernels, 410.0 in stalks, and 150.0 in sorghum. Concentrations in various crops grown in soils treated with 15 kg technical chlordane/ha were always <260.0 μg/kg DW when clay content was 12%, and <150.0 μg/kg when clay content was 28%.

7.4.3 Aquatic Invertebrates

Extremely high levels of chlordanes (e.g., 1746.0–7643.0 μg/kg FW) were measured in several species of South Florida corals collected in 1985. It has been speculated that the elevated levels were due to the illegal disposal of chlordanes off Key Largo, Florida, in 1982.

Maximum concentrations of chlordanes in American oysters (*Crassostrea virginica*) taken in the Gulf of Mexico in 1976 were near 0.1 μg/kg DW. Chlordane concentrations were substantially lower than concentrations of other organochlorines measured in oysters, such as DDT (28.0 μg/kg) and polychlorinated biphenyls (90.0 μg/kg), suggesting a need for additional studies on interaction effects of chlordane residues with those of other environmental chemicals.

Marine clams and worms tended to underrepresent chlordane concentrations in the ambient sediments; concentration factors were less than 0.2 for clams and 0.6 for worms. Similarly, chlordane concentrations in clams from the Shatt al-Arab River in Iraq closely reflected chlordane concentrations in water particulates when compared to levels in water columns or in sediments.

7.4.4 Fishes

Health advisories have been issued near Lawrence, Kansas, based on chlordane levels in edible fish tissues. In fish from the Kansas River, Kansas, in 1986, chlordanes were detected more frequently and at higher levels than other contaminants measured. More than 80% of the sites sampled in Kansas had detectable chlordanes in fish; at more than 50% of these sites, levels exceeded 0.1 mg/kg FW – a guideline for the protection of predatory fish. At three urban sites in Kansas, concentrations of chlordanes in fish have approached or exceeded the Food and Drug Administration action level of 0.3 mg chlordane/kg FW for protection of human health. The most likely source of chlordane in fish from the Kansas River is urban and suburban use of chlordane as a termite control agent. Other health advisories based on chlordane contamination have been issued. In 1985, people were warned not to eat shovelnose sturgeon (*Scaphirhyncus platorynchus*) from the Missouri and Mississippi Rivers. In 1987, advisories warned against sturgeon consumption in the Missouri River between Kansas City and St. Louis, and against bullhead catfishes, suckers, carps, sturgeons, and sturgeon eggs in the Mississippi River near St. Louis. Chlordane residues in fish muscle decreased by 25–33% with baking, char broiling, and frying.

In general, chlordane residues in fishes were elevated in the vicinity of sewage outfalls, in older fishes, and near industrialized areas. Chlordane residues were detected in 36% of all fish samples collected in major domestic watersheds in 1976. In the Great Lakes region in 1979, chlordane residues in fish tissues exceeded 100.0 μg/kg on a FW basis in about 40% of the samples measured; residues were highest in samples collected near Alton, Illinois, and Fairborn, Ohio. In 1987, chlordane was ubiquitous in catfish tissues throughout a 1968-km stretch of the Mississippi River and its tributaries, being highest in catfish from the Illinois River, Missouri River, Ohio River, and the Mississippi River near Chester IL, Helena AR, Arkansas City AR, and at the confluence of the Old River LA, and at Belle Cahise LA. Elevated concentrations of chlordanes in sediments, stomach contents, liver, and bile of winter flounder (*Pleuronectes americanus*) from 22 sites in the northeastern United States were significant risk factors for the development of several lesion types, including proliferative and necrotic lesions.

The two most abundant components of technical chlordane found in fish tissues from Tokyo Bay, Japan, were *trans*-nonachlor and *cis*-chlordane. However, this may vary between locales. For example, *cis*-chlordane and *trans*-chlordane were the most abundant components in fish samples collected throughout Japan during the past 20 years, followed, in order, by *cis*-nonachlor, *trans*-nonachlor, and oxychlordane. Of the total chlordanes measured in muscle of northern pike (*Esox lucius*) from the Baltic Sea, 37% was *cis*-chlordane, 34% *trans*-chlordane, and 15% each *trans*-nonachlor and oxychlordane. In liver tissue of northern pike, 35% was oxychlordane, 28% *trans*-chlordane, 22% *cis*-chlordane, and 14% *trans*-nonachlor. In the United States, only chlordanes and nonachlors have been detected as significant residues in fish collected nationwide. The most abundant component was *cis*-chlordane, followed by *trans*-nonachlor, *trans*-chlordane, and *cis*-nonachlor. The two most abundant components were detected in about 93% of all fish samples collected in 1978 and 1979; residues were usually highest in Hawaii, the Great Lakes, and the Corn Belt. Fish from Manoa Stream in Hawaii had high residues because of heavy use of technical chlordane in pineapple culture and termite control. Nationwide monitoring of freshwater fishes showed that mean chlordane concentrations in whole fish did not change during 1980–84, following a period of decline; however, *trans*-nonachlor replaced *cis*-chlordane as the most abundant component, suggesting a lower influx of chlordane to the aquatic environment from terminated use of chlordane in agriculture in the mid-1970s. Residues of *cis*-chlordane and *trans*-nonachlor – the most abundant and persistent of the chlordane components measured – were present at 85 and 89%, respectively, of the stations sampled in 1984. Maximum chlordane levels in fish in 1984 occurred in the Great Lakes, Hawaii, watersheds of the Ohio, Missouri, and Mississippi rivers, and in the Delaware and Raritan rivers in the northeast.

Atmospheric transport may be the main source of chlordane in Finland – a country that prohibits chlordane use – because chlordanes are distributed evenly in the Finnish environment. No chlordane compounds were detected in rainbow trout (*Oncorhynchus mykiss*) taken from lakes in eastern Finland, although measurable residues were detected in other fish species. This phenomenon is attributed to the superior ability of rainbow trout to metabolize chlordanes to oxychlordane.

7.4.5 Amphibians and Reptiles

Chlordane residue data for amphibians and reptiles are extremely limited. Maximum concentrations of chlordane isomers did not exceed 70.0 μg/kg FW of oxychlordane in eggs of the American crocodile, *Crocodylus acutus*, or 250.0 μg/kg FW in carcass of the common garter snake, *Thamnophis sirtalis*. However, California newts, *Tarichia torosa*, taken near a lake treated with 10.0 μg/L technical chlordane had greatly elevated chlordane residues in liver and comparatively low concentrations in carcass, stomach, and stomach contents. After 14 days, livers contained about 34.0 mg/kg total chlordanes lipid weight (LW) – about 19% chlordanes, 9% nonachlors, and 6% chlordenes. After 2.8 years, 98% of the total chlordanes was lost. *Trans*-nonachlor was the most persistent component in newt liver, accounting for up to 55% of the total chlordanes in specimens collected 2.8 years after application.

7.4.6 Birds

Technical chlordane components and their metabolites – especially oxychlordane – are comparatively elevated in tissues with high lipid content, in older birds, and in raptors. The oxychlordane concentration of 5.2 mg/kg FW in the liver of one sharp-shinned hawk (*Accipiter striatus*) from the eastern United States was in the range (3.0–10.0 mg/kg FW liver) associated with acute toxicity of raptors. Other organochlorine compounds were frequently associated with chlordanes, sometimes at life-threatening concentrations. For example, eggs of peregrine falcons (*Falco peregrinus*) collected nationwide in 1986–89 had DDT levels that ranged between 8.8 and 11.0 mg/kg FW, and total PCB concentrations as high as 14.0 mg/kg FW.

Chlordane isomers occur frequently in birds collected nationwide. In 1976, for example, 41% of European starlings (*Sturnus vulgaris*) contained chlordane isomers; in 1979, 60% contained chlordane isomers. In 1982, oxychlordane was detected in 45% of all starlings analyzed, *trans*-nonachlor in 40%, *cis*-nonachlor in 9%, and *cis*- and *trans*-chlordanes in fewer than 2%. Chlordane isomers were detected at frequencies exceeding 50% in wings of American black ducks (*Anas rubripes*) and mallards (*Anas platyrhynchos*) from the Atlantic Flyway in 1976–77, in eggs of 19 species of Alaskan seabirds in 1973–76, and in carcasses of ospreys (*Pandion haliaetus*) found dead in the eastern United States between 1975 and 1982. Frequency of detection for chlordane isomers ranging between 14 and 40% has been reported in wings of black ducks and mallards from flyways other than the Atlantic Flyway, in 19 species of passeriformes from the western United States in 1980, and in 7 species of Texas shorebirds in 1976–77 – although residues in shorebirds were below levels known to adversely affect reproduction or survival.

Carcasses of bald eagles (*Haliaeetus leucocephalus*) collected between 1978 and 1981 usually contained oxychlordane at 45–56% frequency, *trans*-nonachlor at 62–74%, *cis*-chlordane at 38–45%, and *cis*-nonachlor at 38–47%. Frequency of occurrence in the brain was lower, ranging between 19 and 55% for individual isomers. However, a positive correlation was established in bald eagles between concentration of chlordanes in brain on a fresh weight basis and in carcass on a lipid-weight basis; this relation seems to extend to other avian species as well. Bald eagles also contained appreciable quantities of other organochlorine compounds, and a few – for example, dieldrin – were sometimes present at concentrations considered life-threatening. A similar situation exists in other species of raptors.

Some chlordane isomers tend to persist in avian tissues for lengthy periods. In northern gannets (*Sula bassanus*), the half-time persistence of *cis*-chlordane, *cis*-nonachlor, and oxychlordane was estimated at 11.2, 19.4, and 35.4 years. Oxychlordane residues in the thick-billed murre (*Uria lomvia*) tend to be high because of rapid excretion through uropygial gland secretions of *cis*- and *trans*-chlordanes and nonachlors, and to biotransformation of these and other chlordane components to oxychlordane. This observation is alarming because the metabolite oxychlordane has proven much more toxic and persistent than the parent chemicals. Secondary poisonings of raptors after consumption of poisoned bait or prey that had accumulated a large quantity of chlordane were documented for the red-shouldered hawk (*Buteo lineatus*) and the great horned owl (*Bubo virginianus*); concentrations of oxychlordane and heptachlor epoxide found in brain and carcass of both species were within the lethal range reported in experimental studies.

Chlordane-induced mortality of the long-billed curlew (*Numenius americanus*) has been documented at least four times since 1978, despite restriction of technical chlordane use since 1980 to subterranean applications for termite control. Death of these curlews was probably due to over-winter accumulations of oxychlordane of 1.5–5.0 mg/kg brain FW and of heptachlor epoxide at 3.4–8.3 mg/kg – joint lethal ranges for oxychlordane and heptachlor epoxide in experimental birds – compared to 6.0 mg/kg brain for oxychlordane alone, and 9.0 mg/kg for heptachlor epoxide alone. Additional research is needed on toxic interactions of chlordane components with each other and with other chemicals in the same environment.

7.4.7 Mammals

Chlordane levels in mammals were usually highest in lipids, in animals collected near areas of high chlordane use, and in aquatic mammals, especially marine species. In mammals, lipid-soluble and persistent organochlorines are transferred from maternal lipid stores to pups during lactation. The milk of marine mammals is especially rich in lipids and can range up to 60%; about 98% of the chlordane pesticide residues in pups of marine mammals is accumulated from maternal milk. In the ringed seal (*Phoca hispida*), lactational transfer of chlordanes is estimated at 30% of the whole body chlordane burden in the mature female. Biomagnification of total chlordanes through the food chain was strongly evident in marine mammals; chlordanes were concentrated gradually from zooplankton, through squid and fish, to porpoises and dolphins. Chlordane residues in marine mammals were positively related to lipid content and not to the age of the animal. Chlordanes and other organochlorine compounds in adipose tissues and milk of polar bears (*Ursus maritimus*) increased markedly on a lipid-weight basis with increased fasting, typical of hibernation. The possibility of chlordane-induced effects on the reproduction of polar bears exists and could be a factor in the slow decline of the western Hudson Bay, Canada, polar bears. Concentrations of chlordanes in adipose tissues of polar bears in their known range sampled between 1989 and 1993 ranged between 727.0 and 4632.0 μg/kg LW; adult males had 30% less chlordanes than females, and chlordane concentrations in both sexes were dominated by oxychlordane (46.8% of total chlordanes), nonachlors (19.9%), and heptachlor epoxide (7.7%).

A high death rate over a 2-year period was evident in the little brown myotis (*Myotis lucifugus*) following application of chlordane; young bats were most affected in the first year after application and adults in the second year. Residues were greatly elevated in the brain and carcass of another bat, the gray myotis (*Myotis grisescens*) – an endangered species – found dead near areas of high chlordane use. Animals reared in captivity for slaughter may contain chlordane and other pesticides in tissues obtained most probably from forage, and possibly from drinking water. Some animals, such as bison (*Bison bison*) habitually roll in the dust and residual pesticides – including oxychlordane – adsorbed on soil particles may contribute to the residues. Chlordane levels in raccoons (*Procyon lotor*) from Mississippi declined between 1978 and 1988 with the decline attributed to decreased pesticide use of chlordane compounds.

Chlordane levels in human blood were comparatively elevated among individuals living in residences treated with chlordane during the past 5 years, and in termite control operators; oxychlordane levels were usually significantly higher than *trans*-nonachlor except among those who consumed large quantities of fish.

7.5 Lethal and Sublethal Effects

Chlordane has been applied extensively to control pestiferous soil invertebrates, usually at rates between 0.6 and 2.24 kg/ha; within this range sensitive nontarget species, especially earthworms, were adversely affected. Nominal water concentrations between 0.2 and 3.0 μg/L were harmful to various species of fish and aquatic invertebrates. Effects included a reduction in survival, immobilization, impaired reproduction, histopathology, and elevated chlordane accumulations. *Cis*-chlordane, when compared to *trans*-chlordane, was more toxic, preferentially stored, and concentrated to a greater degree. In aquatic organisms, *cis*-chlordane photoisomers were frequently more toxic than the parent form. Oxychlordane was not a major metabolite in aquatic fauna. Sensitive bird species had reduced survival after consumption of diets as low as 1.5 mg chlordane/kg ration, or after a single oral dose as low as 14.1 mg/kg BW; accumulations were documented in tissues following consumption of diets containing 0.1–0.3 mg chlordane/kg feed. Oxychlordane was the most persistent metabolite in avian brain tissue.

Concern for the continued widespread use of chlordane centers on its ability to cause liver cancer in domestic mice. Other adverse effects in mammals, such as elevated tissue residues and growth inhibition, were frequently associated with diets containing between 0.76 and 5.0 mg chlordane/kg feed. Metabolism of technical chlordane by mammals results primarily in oxychlordane, a metabolite that is about 20 times more toxic than the parent compound and the most persistent metabolite stored in adipose tissues. Chlordane interactions with other agricultural chemicals produced significant biological effects in warm-blooded organisms, indicating a need for additional research on this subject.

7.5.1 Terrestrial Invertebrates

Chlordane has been used extensively to control grubs, ants, snails, and terrestrial invertebrates. Chlordane applied to wheat crops in India at 0.6 kg/ha and at higher concentrations controlled infestation by two species of termites (*Odontotermes obesus*, *Microtermes obesi*) and increased grain yield; chlordane applications of 0.4 kg/ha and at lower concentrations were ineffective. Chlordane has been used to control the imported fire ant (*Solenopsis invicta*), although registration for this purpose by EPA has now been withdrawn. Application of 4.5 g of chlordane per ant mound, applied as an emulsifiable concentrate, resulted in 83–94% control 4–5 weeks posttreatment. Cricket (*Acheta pennsylvanicus*) nymphs died within one minute of contact with technical chlordane; dead crickets showed cellular disruption of the caecal lining, the malpighian tubules, and the digestive tract.

Chlordane, at 1.12–2.24 kg/ha, was lethal to fly and beetle larvae and also caused reductions in populations of various species of soil invertebrates. Among nontarget soil species, earthworms were especially sensitive. Significant reductions in earthworm populations were recorded following application of 2.2 kg/ha; metabolism was adversely affected in 2 weeks at 13 kg/ha, and remained depressed for at least

5 years; at 80 kg/ha, 46% died in 4 days. In soil, chlordane effects decreased with increasing soil temperature and organic content; the heptachlor component in technical chlordane had the greatest biological activity to soil fauna.

7.5.2 Aquatic Biota

Signs of chlordane poisoning in fish included hyperexcitability, increased respiration rate, erratic swimming, loss of equilibrium, and convulsions; death frequently occurred within 12 h of exposure. Chlordane adversely affected sensitive species of fish and aquatic invertebrates at nominal water concentrations between 0.2 and 3.0 μg/L. Specifically, reduced survival was measured in shrimp and crabs at water concentrations of 0.2–2.0 μg/L, and in freshwater and marine fishes between 1.7 and 3.0 μg/L; immobilization, impaired reproduction, and histopathology were recorded in shrimp, fish, and planarians between 0.8 and 3.0 μg/L; and high accumulations were evident in fish, shrimp, and oysters between 0.2 and 4.2 μg/L. Growth stimulation and high residues were measured in resistant species of algae, such as *Scenedesmus quadricauda*, at media concentrations up to 100.0 μg chlordane/L; in sensitive algal species, however, growth was inhibited at water concentrations as low as 10.0 μg/L.

Large intra- and inter-specific differences in sensitivity to chlordane were evident. Some of this variability was attributed to variations in water temperature, salinity, and sediment loadings; some to the age, condition, and nutritional history of the test organism; and some to the chlordane formulation and isomer tested. In general, granular chlordane formulations were most toxic, organisms at a young developmental stage and organisms with reduced lipid content were most sensitive, and adverse effects were most pronounced under conditions of elevated water temperatures, reduced salinities, decreased sediment loadings, and increased duration of exposure. Reduced bioavailability and lessened toxicity of chlordane to daphnids was associated with increasing concentrations (up to 200.0 mg/L) of suspended solids and their associated carbon content. Sediment loadings of 5.8 mg chlordane/kg were fatal to 50% of sandworms (*Nereis virens*) in 12 days. Resistance or adaptation to chlordane has been reported in mosquitofish (*Gambusia affinis*) collected from ditches near treated cotton fields; these fish were up to 20 times more resistant than newly exposed fish.

Residues of *cis*-chlordane were preferentially stored and magnified over *trans*-chlordane by freshwater fish and invertebrates in ponds treated with technical chlordane at concentrations up to 1.14 μg/L; the *cis* isomer, with an estimated Tb1/2 of 46 days, persisted longer than the *trans* isomer. Tissue concentrations of 106,000.0 μg total chlordanes/kg, on a lipid-weight basis, were associated with reduced survival of estuarine invertebrates. Moribund amphipods (*Hyallela azteca*), for example, contained 137,000.0–2,180,000.0 μg/kg lipid of various chlordanes, heptachlors, and chlordenes. In fish, chlordane concentrations of 300,000.0–4,000,000.0 μg/kg LW in tissues were lethal.

Cis-chlordane was 8 times more toxic to bluegill (*Lepomis macrochirus*) than was *trans*-chlordane. *Cis*-chlordane was also more toxic to goldfish (*Carassius auratus*) than was *trans*-chlordane because of its comparatively rapid uptake from the medium and lengthy storage in body tissues, estimated at 99% after 25 days. The elimination rate of *cis*-chlordane from a cichlid (*Cichlasoma* sp.) was estimated at 2.9% weekly over a 20-week period, with a Tb1/2 of about 17 weeks; metabolites accounted for 12.5% (dichlorochlordene, oxychlordane, chlordene chlorohydrin, dihydroxyheptachlor, dihydroxydihydrochlordene, plus four unidentified compounds), and unchanged *cis*-chlordane for 87.5%. The assemblage of chlordane-related compounds present in lake trout (*Salvelinus namaycush*) from the Great Lakes is substantially different from technical chlordane, and is 3–5 times more toxic to mosquito larvae than the technical mixture. The increased toxicity is attributed to the presence of the stable metabolites oxychlordane and heptachlor epoxide.

Photoisomers seem to be more toxic than the parent form. For example, *cis*-photochlordane was about twice as lethal to bluegills and goldfish than was *cis*-chlordane. Bluegills exposed to 5 μg/L of radiolabeled *cis*-photochlordane or *cis*-chlordane for 48 h accumulated *cis*-chlordane from the medium by a factor of 78, and *cis*-photochlordane by a factor of 140. During the next 6 weeks, 20% of the *cis*-chlordane was eliminated in a linear pattern, and about 50% was eliminated in 46 days. Elimination of *cis*-photochlordane followed a biphasic pattern and was most rapid during the first 3 weeks; 40% was eliminated in the first 6 weeks, and 50% was eliminated in 15 weeks. Less than 7% of the radioactivity retained in *cis*-chlordane-treated bluegills was in the form of two conjugates, compared to 16% in the form of 14 metabolites for *cis*-photochlordane. No oxychlordane was found in bluegill tissues after treatment; this compound is one of the predominant metabolites found in chlordane-treated rodents and cockroaches. Thus, absence of epoxidation and presence of a mechanism of hydroxylation followed by conjugation seems to be the most active mode of chlordane metabolism in bluegill.

Cis-photochlordane was about 10 times less toxic to *Daphnia pulex* than was *cis*-chlordane. This is in sharp contrast to the pattern shown in bluegill and goldfish; further, *cis*-photochlordane and *cis*-chlordane toxicity to mice and houseflies was about the same, which demonstrates the difficulty in generalizing about the comparative toxicity of chlordane isomers.

7.5.3 Amphibians and Reptiles

Shortly after chlordane was applied to wooden huts in Australia for termite control, large numbers of dead skinks (*Morethia boulengeri*, *Lerista puctorittata*) and frogs (*Litoria caerulea*, *L. peronii*) were discovered, presumably killed by the chlordane. Toad (*Bufo arenarium*) embryos survived 0.5 mg technical chlordane/L for 8 days but died by day 20; all embryos held in 15.0 mg/L were dead by day 15. For tadpoles of the common toad (*Bufo bufo*) a 48-h LC50 of 2.0 mg/L was reported.

7.5.4 Birds

Signs of chlordane intoxication in birds include sluggishness, drooped eyelids, fluffed feathers, low crouching on perch, reduced food intake, and weight loss. Later, afflicted animals rested on their breast, wings spread, quivering and panting rapidly, back arched, neck arched over the back, and sometimes convulsing. Signs of intoxication appeared within 5 min, and death usually occurred in the first 8 days of exposure; remission took up to 4 weeks in some birds.

The most sensitive avian species tested against technical chlordane were California quail (*Callipepla californica*), with an acute oral LD50 of 14.1 mg/kg BW; the ring-necked pheasant (*Phasianus colchicus*), with an acute oral LD50 of 24.0–72.0 mg/kg BW; and European starlings (*Sturnus vulgaris*) fed diets containing 1.5 mg/kg ration for 57 days or 6.25 mg/kg for 24 days. Accumulations of various chlordane isomers and metabolites were evident in chickens (*Gallus* sp.) fed diets containing as little as 0.1 mg technical chlordane/kg feed for 6 weeks or 0.3 mg/kg for 4 weeks. Vapor toxicity of chlordane is persistent. In one instance, a room used for housing pigeons was sprayed with a chlordane solution; walls and floors were then scrubbed and the room left unoccupied for 2 months. When pigeons were returned to the room, enough chlordane remained to be lethal to all birds. Similar cases are reported for mice, presumably after use of very concentrated chlordane solutions.

Reproductive impairment was reported in several species of waterfowl from a marsh treated with 1.12 kg technical chlordane/ha. Studies with two species of ducks (*Anas* spp.) and the domestic chicken (*Gallus* sp.) demonstrated that various organochlorine compounds, including chlordane, interfered (in a dose-dependent manner) with reproduction by reducing the binding of progesterone to its cytoplasmic receptor in the shell gland mucosa of birds, especially ducks.

The lethal effect of technical chlordane in birds is attributed primarily to chlordane metabolites, especially oxychlordane, and to a lesser extent heptachlor epoxide. Oxychlordane was the most persistent chlordane component in avian brain tissues. The half-time persistence of oxychlordane in brain was 63 days, and 95% loss was estimated in 280 days; the Tb1/2 for heptachlor epoxide was 29 days, and for *trans*-nonachlor it was 19 days. Oxychlordane residues in brain tissue approaching 5.0 mg/kg FW were considered within the lethal hazard zone to birds.

Technical heptachlor contains about 15% *cis*-chlordane and 2.5% *trans*-chlordane. Diets containing 50.0 mg technical heptachlor/kg fed to brown-headed cowbirds (*Molothrus ater*), red-winged blackbirds (*Agelaius phoeniceus*), common grackles (*Quiscalus quiscula*), and European starlings produced 50% mortality in 9–24 days; birds that died contained 9.2–27.0 mg oxychlordane/kg FW brain, and survivors contained 2.7–7.8 mg/kg. Red-winged blackbirds fed diets containing 10.0 mg technical chlordane/kg for 84 days, 50.0 mg/kg for 42 days, or 100.0 mg/kg for 21 days, all contained about 17% of the total diet fed as *cis*-chlordane, with whole body residues in mg/kg FW of 1.8, 9.2, and 14.8, respectively; accumulations of *trans*-chlordane were negligible.

Chlordane interactions with other agricultural chemicals are significant and merit additional research. In one study, male Japanese quail (*Coturnix japonica*) pretreated for 8 weeks with 10.0 mg chlordane/kg diet had increased resistance to parathion, but not to paraoxon, as judged by cholinesterase activity. In another study, northern bobwhites (*Colinus virginianus*) treated with 10.0 mg chlordane/kg diet for 10 weeks, followed by endrin stress, had greater accumulations of chlordane in the brain than birds treated only with chlordane.

7.5.5 Mammals

Concern for the continued widespread use of chlordane is centered around its carcinogenicity in mice (*Mus* sp.). Chlordane produced liver cancer in both sexes of two different strains of domestic mice. A dose-dependent incidence of hepatocellular carcinoma was evident in mice fed chlordane in their diets; frequency of liver carcinomas were not significantly different from controls at dietary levels of 5.0 mg/kg and lower but were greatly elevated (i.e., >70% frequency) at dietary levels of 50.0 mg/kg and higher. In contrast to mice, chlordane was not a hepatic carcinogen in rats at dietary levels up to 64.0 mg/kg ration; however, a dose-related increase in follicular cell thyroid neoplasms and malignant fibrous histiocytomas were recorded in chlordane-exposed rats. In humans, no increased evidence of cancer was proven among employees in chlordane manufacturing facilities, although there is a statistically significant increase in death rate from cerebrovascular disease in that group.

Human toxicity data for chlordane usually is obtained after accidental exposure through spillage onto clothing or ingestion. In one case, a 15-month-old girl accidentally swallowed a mouthful of chlordane suspension and within 3 h displayed tremors and incoordination. Repeated seizures developed and she was treated with ethyl chloride, amobarbitol, and gastric lavage with magnesium sulfate; ataxia and excitability disappeared in about 3 weeks. At age 26 years, she was in excellent health and seemed not to have experienced latent effects from the childhood incident. Other cases of accidental chlordane poisoning in children are documented, and all appear to have recovered completely after treatment.

Symptoms of acute chlordane poisoning in humans include irritability, salivation, labored respiration, muscle tremors, brain wave abnormalities, incoordination, convulsions, deep depression, and sometimes death. Signs of acute chlordane intoxication in other mammalian species are similar to those in humans and may also include aplastic anemia and acute leukemia; cyanosis; pathology of gastrointestinal tract, liver, kidney, lung, and heart; pulmonary congestion; degenerative changes in the central nervous system; impaired uptake and utilization of glucose; interference with immunocompetence response; diarrhea; avoidance of food and water; enhanced estrone metabolism; increased production of hepatic mixed function oxidase enzymes; altered enzyme activity in brain and in kidney cortex; enlarged liver; hair loss; abdominal distension; hunched appearance; inhibited oxidative phosphorylation in liver mitochondria; and thyroid carcinoma.

Acute oral LD50 values for technical chlordane and sensitive mammals usually ranged between 25.0 and 50.0 mg/kg BW. Chlordane-related compounds (i.e., *cis*-chlordane, *trans*-chlordane, heptachlor, heptachlor epoxide) stimulate superoxide (O_2^-) generation in guinea pig leukocytes, alter membrane potential, and increase intracellular calcium concentration; toxicity of individual compounds seems to be related to superoxide generation. Metabolism of chlordane isomers results in oxychlordane, a metabolite that is about 20 times more toxic to rats than is the parent compound and is the most persistent metabolite stored in rat adipose tissue. Oxychlordane accounted for 53% in females and 63% in males of all chlordane isomers in fat of rats killed 24 h after a single oral dose of 1.0 mg/kg BW technical chlordane. Acute oral LD50 values in the rat, in milligram per kg body weight, were 19.1 for oxychlordane; 89.0–392.0 for *cis*-chlordane; 200.0–590.0 for technical chlordane; 327.0 for *trans*-chlordane; >4600.0 for chlordene, 3-chlordene, 1-hydroxychlordene, chlordene epoxide, 1-hydroxy, and 2,3-epoxy chlordene; and >10,000.0 for 2-chlorochlordene.

Chlordane adversely affects growth and fertility of laboratory animals. Neonatal exposure of mice to chlordane retards growth, as judged by lowered body weights during the first 12 weeks. No fetotoxic or teratogenic effects were observed in rats born to dams fed chlordane in their diets for 2 years at levels up to 300.0 mg/kg diet; however, pups nursed by dams consuming chlordane at 150.0 or 300.0 mg/kg diet developed signs of toxicity. In uterine mucosa of the rabbit, chlordane isomers (as well as isomers of DDE and polychlorinated biphenyls) reduced the binding of progesterone to its cytoplasmic receptor in a dose-dependent manner, which suggests a pathway to account for chlordane-induced reproductive impairment.

Chlordane tends to accumulate in adipose tissues and, to a lesser extent, in liver. In general, animals given a single oral dose of chlordane eliminated 80–90% of the dose within 7 days, usually via the feces; the *cis* isomer is eliminated more rapidly than the *trans* isomer and results in preferential accumulations of *trans*-chlordane. In rats, *trans*-chlordane is rapidly absorbed and distributed to liver and kidney at single oral dosages as low as 0.05 mg/kg BW. Rabbits fed *trans*-chlordane for 10 weeks excreted 70% of accumulated chlordane during the following 2 weeks on a chlordane-free diet. Treatment with *trans*-chlordane resulted in a greater percentage of oxychlordane in fat than did treatment with *cis*-chlordane. When chlordane was removed from the diet of treated animals, levels in fat declined 60% at a relatively steady rate over 4 weeks, but then only slightly thereafter; accumulations in liver, kidney, brain, and muscle were much lower than in fat, although excretion kinetics were similar.

Results of chronic feeding studies show that dietary concentrations of chlordane between 0.76 and 5.0 mg/kg ration did not affect survival but did produce adverse effects on various species of laboratory animals and livestock. Dietary concentrations of 0.76 mg/kg (equivalent to 0.09 mg/kg BW daily) were associated with enlarged livers in mice, 1.0 mg/kg produced elevated residues in cow's milk, 2.5 mg/kg resulted in liver pathology in rats, 3.0 mg/kg (equivalent to 0.075 mg/kg BW daily) produced high residues in fat of dogs, and 5.0 mg/kg caused liver pathology in mice.

Negative results for mutagenicity of *cis*-chlordane and *trans*-chlordane were reported in various strains of bacteria and in hepatocyte cultures of small mammals. But technical chlordane proved mutagenic to selected strains of *Salmonella typhimurium* and induced gene conversions in certain strains of the yeast, *Saccharomyces cervisiae*.

Chlordane interacts with other chemicals to produce additive or more-than-additive toxicity. For example, chlordane increased hepatotoxic effects of carbon tetrachloride in the rat, and in combination with dimethylnitrosamine acts more than additively in producing liver neoplasms in mice. Chlordane in combination with either endrin, methoxychlor, or aldrin is additive or more-than-additive in toxicity to mice. Protein deficiency doubles the acute toxicity of chlordane to rats. In contrast, chlordane exerts a protective effect against several organophosphorus and carbamate insecticides, protects mouse embryos against influenza virus infection, and mouse newborns against oxazolone delayed hypersensitivity response.

More research seems warranted on interactions of chlordane with other agricultural chemicals.

7.6 Recommendations

All use of chlordane was banned in Norway in 1967. In August 1975, EPA issued its intent to suspend registrations and prohibit production of all pesticides containing heptachlor or chlordane, based on evidence of carcinogenicity. On July 1, 1983, chlordane use was prohibited in the United States for any purpose except to control underground termites; a similar situation exists in Japan.

The continued use of chlordane, coupled with its general persistence in the environment, suggests that extreme caution be taken in all stages of its manufacture, transport, storage, and application. In particular, chlordane use near marine environments is not recommended because of chlordane's high toxicity to marine life. At elevated risk of chlordane toxicity in the human population are children, as a result of the milk they consume; fishermen and their families, because of high consumption of fish and shellfish; people living downwind from fields treated with chlordane; and individuals residing in houses treated with chlordane-containing pesticides.

The proposed criterion for marine life protection of 0.004 μg/L as a 24-h mean, not to exceed 0.09 μg/L at any time, seems to offer a reasonable degree of protection (Table 7.1). But the proposed freshwater criterion of 0.0043 μg/L, 24-h average, not to exceed 2.4 μg/L at any time, overlaps the range of 0.2–3.0 μg/L shown earlier to be harmful to sensitive species of fish and aquatic invertebrates; accordingly, the maximum permissible

Table 7.1. Proposed chlordane criteria for protection of natural resources and human health.

Resource	Criterion or Effective Chlordane Concentration
AQUATIC LIFE	
Water concentration, safe level	
Freshwater	<0.0043 μg/L, 24-h average; not to exceed 2.4 μg/L at any time
Marine	<0.004 μg/L, 24-h average; not to exceed 0.09 μg/L at any time
Tissue concentrations	
Fish	
Reduced survival	>300.0 mg/kg tissue, lipid weight (LW) basis
No observed adverse effect level (NOAEL)	<0.1 mg/kg fresh weight (FW) tissue
Estuarine invertebrates; lethal	>106.0 mg/kg tissue LW
BIRDS	
Concentration in brain	
Joint lethal range	1.1–5.5 mg/kg FW for oxychlordane and 3.4–8.3 mg/kg FW for heptachlor epoxide
Single lethal range	6.0 mg/kg FW for oxychlordane or 9.0 mg/kg FW for heptachlor epoxide
Diet, acceptable range, but producing slight elevation in tissue concentrations	0.1–0.3 mg/kg diet

Continued

Table 7.1. cont'd

Resource	Criterion or Effective Chlordane Concentration
MAMMALS	
Dog, *Canis familiaris*, NOAEL	<3.0 mg/kg diet, equivalent to <0.075 mg/kg body weight (BW) daily
Rat, *Rattus* sp., NOAEL	<5.0 mg/kg diet, equivalent to 0.25 mg/kg BW daily
Livestock water use, USA	<3.0 μg/L
HUMAN HEALTH	
Drinking water	
Worldwide	<0.3 μg/L
Canada, USA	<3.0 μg/L
Chronic, child, USA	<0.5 μg/L
Maximum 1-day exposure, adult, USA	63.0 μg/L
Increased lifetime risk of cancer, 70-kg adult, 2 L daily[a]	
10^{-4}	2.7 μg/L
10^{-5}	0.27 μg/L
10^{-6}	0.027 μg/L
Acceptable daily intake[b]; 70-kg adult	<70.0 μg, equivalent to <0.001 mg/kg BW
Diet	
U.S. Food and Drug Administration "action level"	0.3 mg/kg FW
Australia	<0.05 mg/kg FW in meats, including oxychlordane
Worldwide	Usually <0.3 mg/kg FW, but residue tolerances vary between 0.02 and 0.5 mg/kg FW, based on the sum of *cis*-chlordane, *trans*-chlordane, and oxychlordane
Air	
Former Soviet Union	Maximum allowable concentration of 0.01 mg/m^3
Romania	<0.3 mg/m^3, maximum allowed is 0.6 mg/m^3
Belgium, Finland, USA, Japan, The Netherlands	<0.5 mg/m^3
15-min exposure limit, USA	<2.0 mg/m^3

[a] One excess cancer per million (10^{-6}) is associated with lifetime exposure to chlordane in drinking water at concentrations as low as 0.027 μg/L, the most conservative estimate. A lifetime health advisory computation was not possible because chlordane is a probable human carcinogen.

[b] Consumed fish are considered to be the only source of chlordane; up to 98% of chlordane exposure results from aquatic organisms with high (up to 14,100*X*) bioconcentration potential. Urban residents should not consume more than 8 ounces (227.0 mg) of fish daily containing 0.03 mg total chlordane/kg FW, and nonurban residents up to 1135.0 mg of fish daily containing 0.03 mg/kg FW. The value of 0.001 mg/kg BW daily is based on the NOAEL of 5.0 mg/kg in the diet of the rat, equivalent to 0.25 mg/kg BW, and 3.0 mg/kg in the diet of the dog, equivalent to 0.075 mg/kg BW.

freshwater value should be adjusted downward. "Safe" residues in tissues of aquatic biota require clarification, and probably additional research effort. Criteria on chlordane for protection of mammalian wildlife are missing, and those formulated for birds are incomplete and require data on no-observable-effect levels from lifetime exposures. Until this information becomes available, it seems prudent to use criteria developed for human health protection as temporary guidelines for the protection of vertebrate wildlife. Specifically, daily intake should not exceed 0.001 mg total chlordane, including *cis*-chlordane, *trans*-chlordane, and oxychlordane/kg BW; and food items should not exceed 0.3 mg/kg FW (Table 7.1).

Additional research on chlordane is recommended in nine general areas: (1) monitoring of background concentrations of oxychlordane in wildlife, since this metabolite is more toxic and persistent than the parent chemical; (2) interpretation of the biological significance of residue levels found in wildlife; (3) adoption of improved uniform methods of quantitation so that residue levels can be compared, and so that a time estimate of their environmental significance can be made; (4) reexamination of aquatic toxicity data wherein concentrations tested exceeded the solubility of chlordane in water of 6.0–9.0 μg/L; (5) evaluation of interaction of chlordane with other agricultural chemicals, including heptachlor, to clearly delineate any additive, synergistic, or antagonistic effects; (6) reevaluation of the cancer risk of chlordane to experimental animals; (7) measurement of chronic exposures of fish and wildlife to realistic environmental levels; (8) measurement of effects of depleted soil fertility from chlordane-induced earthworm suppression on migratory birds and other wildlife; and (9) continuance of epidemiological studies on workers who have been exposed to chlordane.

7.7 Summary

Technical chlordane is an organochlorine compound first introduced in the United States in 1947 in a variety of formulations for use as a broad-spectrum pesticide. By 1974, about 9.5 million kg of chlordane were produced annually. Concern over the potential carcinogenicity of chlordane has led to sharply curtailed production. Since 1983, chlordane use in the United States has been prohibited, except for control of underground termites.

Technical chlordane consists of about 45 components, primarily *cis*-chlordane (19%), *trans*-chlordane (24%), heptachlor (10%), *cis*- and *trans*-nonachlor (7%), and various chlordane isomers (22%). Chemical analysis of technical chlordane is difficult because of analytical interferences from other organochlorine compounds, nonstandardization of analytical techniques, variations in the number and relative composition of components in weathered chlordane, and uncertainty of structural formulas and other properties of several compounds present.

Past chlordane use, coupled with atmospheric transport as the major route of dissemination, produced global contamination of fish and wildlife resources and human populations. The chemical and its metabolites were frequently detected in all species examined, but usually at low concentrations. Residues in fish muscle sometimes exceeded the U.S. Food and Drug Administration action level of 0.3 mg/kg FW, recommended for human health protection. In general, chlordane in animals is highest near areas where the chemical has been applied to control termites; concentrations are highest in fat and liver, especially in predatory species.

The half-life of chlordane in water is comparatively short; *cis*-chlordane, for example, usually persists less than 18 h in solution. In soils, however, some chlordane isomers persist for 3–14 years because of low solubility in water, high solubility in lipids, and relatively low vapor pressure. There seems to be little accumulation of chlordane in crops grown in contaminated soils.

Chlordane is readily absorbed by warm-blooded animals via skin, diet, and inhalation, and distributed throughout the body. In general, residues of chlordane and its metabolites are not measurable in tissues 4–8 weeks after exposure, although metabolism rates varied significantly between species. Food chain biomagnification is usually low, except in some marine mammals. In most mammals, the metabolite oxychlordane has proven much more toxic and persistent than the parent chemical.

Many species of aquatic organisms are adversely affected at concentrations in water between 0.2 and 3.0 μg/L technical chlordane. Sensitive bird species had reduced survival on diets containing 1.5 mg chlordane/kg ration, or after a single oral dose as low as 14.1 mg chlordane/kg BW. Chlordane has produced liver cancer in laboratory strains of domestic mice, but carcinogenicity has not been established in other mammals.

Chlordane criteria for protection of marine life (0.004 μg/L, 24-h mean; not to exceed 0.09 μg/L) appear satisfactory. Proposed criteria for freshwater life protection (0.0043 μg/L, 24-h mean; not to exceed 2.4 μg/L) however, overlap the range of 0.2–3.0 μg/L shown to adversely affect certain fish and aquatic invertebrates, suggesting that some downward modification in the maximum permissible level is needed. Chlordane criteria for protection of birds and mammals are inadequate because the database is incomplete. Until these data become available, a reasonable substitute is the criteria proposed for human health protection, namely, daily intake not to exceed 0.001 mg chlordane/kg BW, and diet not to exceed 0.3 mg chlordane/kg FW.

Most authorities agree that more studies are needed in several areas: monitoring of oxychlordane concentrations in wildlife; interpretation of the biological significance of residue levels found in wildlife; standardization of analytical extraction and other techniques for quantitation of chlordane and its metabolites; reexamination of aquatic toxicity data where test concentrations exceeded the solubility of chlordane in water (6.0–9.0 μg/L); interaction effects with other agricultural chemicals; reevaluation of the cancer risk of chlordane on representative organisms at realistic environmental levels; effects of depleted soil fertility from chlordane-induced earthworm suppression; and continuance of epidemiological studies on exposed workers.

CHLORPYRIFOS[a]

Chapter 8

8.1 Introduction

Chlorpyrifos (phosphorothioic acid *O*, *O*-diethyl *O*-(3,5,6-trichloro-2-pyridinyl) ester), also known commonly as Dursban and Lorsban, was first registered as a broad-spectrum insecticide in 1965, and subsequently was used widely to control a variety of pests such as fire ants, turf and ornamental plant insects, cockroaches, mosquitos, leatherjackets, termites, horn flies, lice, and fleas. In 1982, total agricultural use of chlorpyrifos was estimated at 2.2–3.2 million kg, and industrial uses ranged between 0.68 and 1.04 million kg. In 1984, about 0.15 million kg (0.33 million pounds) of chlorpyrifos was applied to about 600,000 ha (1.48 million acres) of wetlands in the United States for mosquito control. More than 1.1 million kg of chlorpyrifos was used in California in 1990. Atmospheric transport of chlorpyrifos from California's Central Valley to the Sierra Nevada mountains in California was estimated at 27 kg annually in 1995 and 1996. Treatment programs in which chlorpyrifos concentrations suitable for mosquito control and other insect pests were used have been shown to be detrimental to nontarget species, including aquatic organisms, waterfowl, and terrestrial organisms from surrounding ecosystems. Domestic use of chlorpyrifos has resulted in the death of an 11-day-old infant and the poisoning of office workers. Prophylactic use of chlorpyrifos on farm animals has caused reproductive impairment of livestock. Chlorpyrifos-resistant strains of insects have been detected; they include the German cockroach (*Blattella germanica*) in Florida and Nebraska and the saw-toothed grain beetle (*Oryzaephilus surinamensis*) in Australia.

8.2 Environmental Chemistry

Formulations of chlorpyrifos include emulsifiable concentrates, wettable powders, granules, pellets, microencapsulates, and impregnated materials. Suggested diluents for concentrates include water and petroleum distillates, such as kerosene and diesel oil. Carrier compounds include synthetic clays with alkyl/aryl sulfonates as wetting agents (Table 8.1). Little information is available to assess the influence of various use formulations on toxicity, dispersal, decomposition, and bioavailability. Chemical and other properties of chlorpyrifos are summarized in Table 8.2 and Figure 8.1.

The degradation half-life time (Tb1/2) of chlorpyrifos is 7.1 days in seawater, and 53 days in distilled water. Degradation is usually through hydrolysis to produce 3,5,6-trichloro-2-pyridinol, and phosphorothioic acid. Temperature, pH, radiation, and metal cations all significantly affect chlorpyrifos Tb1/2 in water: half-life is decreased with increasing water pH, temperature, sunlight, and metal cation concentrations.

In soil, Tb1/2 values for chlorpyrifos range from less than 1 week to more than 24 weeks, depending on soil moisture, microbial activity, clay and organic content, and temperature. In 411 soils studied, increasing temperature resulted in decreased Tb1/2 values. Degradation was more rapid in sandy loam than in

[a]All information in this chapter is referenced in the following sources:

Eisler, R. 2000. Chlorpyrifos. Pages 883–902 in *Handbook of Chemical Risk Assessment: Health Hazards to Humans, Plants, and Animals. Volume 2, Organics.* Lewis Publishers, Boca Raton, Florida.

Odenkirchen, E. W., and R. Eisler. 1988. Chlorpyrifos hazards to fish, wildlife, and invertebrates: a synoptic review. *U.S. Fish Wildl. Serv. Biol. Rep. 85(1.13)*, 34 pp.

Table 8.1. Selected chlorpyrifos formulations and carriers.

Compound[a]	Formulation	Carrier
Dursban 2E	Emulsifiable concentrate of 0.285 kg/L (2.4 pounds/gallon)	Solution in aromatic distillate with anionic/nonionic emulsifier blend and residual chlorinated solvent
Dursban M	Emulsifiable concentrate of 0.57 kg/L (4.8 pounds/gallon)	As above
Dursban 6	Solution of 0.855 kg/L (7.2 pounds/gallon)	Solution in an aromatic distillate
Dursban 2½G	Granular, 2.5%	Absorbed onto stabilized clay with release agents added
Lorsban 4C	Emulsifiable concentrate of 0.479 kg/L (4.0 pounds/gallon)	Solution in aromatic naphtha with emulsifiers
Lorsban 25W	Wettable powder, 25%	Dispersion on blended clays with alkyl/aryl sulfonates as wetting agents

[a] Dursban and Lorsban are registered trademarks of the Dow Chemical Company.

Table 8.2. Chemical and other properties of chlorpyrifos.

Variable	Datum
CHEMICAL NAME	Phosphorothioic acid *O*, *O*-diethyl *O*-(3,5,6-trichloro-2-pyridinyl) ester
ALTERNATE NAMES	CAS 2921-88-2; Dursban; Lorsban; 27311; Trichlorpyrphos; Brodan; Pyrinex; Chlorpyriphos-ethyl
PRIMARY USES	Insecticide, acaricide
PRODUCERS	Dow Chemical Company; India Medical Corp.; Makhteshim-Agan (Israel); Planters Products Inc.
EMPIRICAL FORMULA	$C_9H_{11}Cl_3NO_3PS$
MOLECULAR WEIGHT	350.57
PHYSICAL STATE AT 25°C	White granular crystalline solid
MELTING POINT	41.5–43.5°C
VAPOR PRESSURE	
25°C	1.87×10^{-5} mmHg
35°C	8.87×10^{-5} mmHg
HEAT OF SUBLIMATION	26,800 cal/mol
SOLUBILITY	
Water, 23–25°C	0.4–2.0 mg/L
Isooctane, 23°C	790.0 g/kg
Methanol, 23°C	450.0 g/kg
LOG *n*-OCTANOL–WATER PARTITION COEFFICIENT	5.2
SOIL ORGANIC CARBON–WATER PARTITION COEFFICIENT	13,600

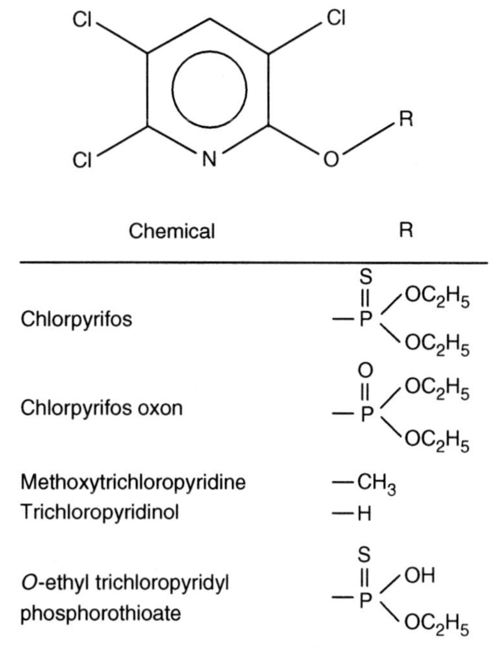

Figure 8.1.
Structures of chlorpyrifos and some of its metabolites.

organic muck soils, more rapid in moist than in dry soils, and more rapid in clay than in other soil types. The major routes of chlorpyrifos loss from soils are chemical hydrolysis in moist soils, clay-catalyzed hydrolysis in dry soils, and microbial degradation and volatilization. Chlorpyrifos is poorly metabolized by soil bacteria and frequent treatments of agricultural soils may result in accumulation with occasional adverse effects. For example, at 2–10 kg/ha, chlorpyrifos significantly reduces the growth and dinitrogen fixation of heterotrophic nitrogen-fixers. The negative effects of chlorpyrifos on nitrogen-fixing bacteria are important because root colonization by microorganisms such as *Azospirillum*, *Azotobacter*, and *Rhizobium* is positive for plant growth.

The half-life of chlorpyrifos in sediments is comparatively long; it was 24 days in a sediment-water slurry. In a pond treated with chlorpyrifos, total waterborne residues decreased by a factor of more than 10, while total sediment residues rose by about 3. Similar results were noted in an artificial lake treated with chlorpyrifos: lake water concentrations peaked 1 day after treatment at 0.9 μg/L and plateaued near 0.2 μg/L after 3 weeks.

Chlorpyrifos inhibited substrate-borne reception and emission of sex pheromone in *Trichogramma brassicae*, an entomophagus insect massively used as a biological control agent of corn borers, among survivors of an LC20 dose. Inhibition was probably due to nervous system effects and was not specific to pheromone communication.

Fish rapidly absorb, metabolize, and excrete chlorpyrifos from the diet. The mechanism of action of chlorpyrifos occurs via phosphorylation of the active site of acetylcholinesterase after initial formation of chlorpyrifos oxon by oxidative desulfuration. In studies with channel catfish (*Ictalurus punctatus*), the oral bioavailability of chlorpyrifos was 41%, substantially higher than in mammals. Catfish muscle contained less than 5% of the oral dose with an elimination half-life (Tb1/2) of 3.3 days. Chlorpyrifos residues in whole catfish were more than 95% chlorpyrifos, while bile and urine primarily contained metabolites. The dephosphorylated metabolite trichloropyridinol (TCP) was the major metabolite in the blood while the glucuronide conjugate of TCP was the major metabolite in urine and bile. The toxic metabolite, chlorpyrifos oxon, was not detected in blood, tissues, or excreta. Extensive metabolism resulted in a low potential for chlorpyrifos to accumulate in catfish from dietary exposure. In both fish and mammals, TCP is a major biotransformation product. Channel catfish rapidly distribute waterborne chlorpyrifos into the blood and more slowly to peripheral tissues, with concentrations highest in fat and lowest in muscle. As was true with dietary chlorpyrifos, TCP was the major metabolite in blood and the glucuronide conjugate of TCP was the major metabolite in urine and bile. Pharmacokinetics and metabolism of waterborne chlorpyrifos in channel catfish were similar to the disposition of chlorpyrifos in other vertebrates.

8.3 Laboratory Investigations

Lethal and sublethal effects of chlorpyrifos under controlled conditions are summarized for selected species of aquatic invertebrates and vertebrates, and for representative species of birds, and mammals.

8.3.1 Aquatic Organisms

During 96-h toxicity tests, several species of freshwater and marine invertebrates and fishes died at chlorpyrifos concentrations between 0.04 and 0.6 μg/L. LC50 (96-h) values, in microgram chlorpyrifos/liter, for sensitive species tested were 0.04 for mysid shrimp *Mysidopsis bahia*; 0.11 for amphipod, *Gammarus lacustris*; 0.38 for stonefly, *Pteronarcella badia*; and 0.6 for striped bass, *Morone saxatilis*. Toxicity was usually greater at elevated temperatures and at increasing pH levels. Aquatic invertebrates were usually more sensitive to chlorpyrifos than were vertebrates. Significant differences in genotype frequencies of the glucose phosphate isomerase (P_{gi}) locus were observed between chlorpyrifos-susceptible and chlorpyrifos-tolerant groups of mosquitofish, *Gambusia affinis*. In general, arthropods were the most sensitive group assayed and mollusks the most tolerant. Adult newts (*Triturus vulgaris*) had normal survival and activity patterns during exposure to 96.0 μg chlorpyrifos/L for 96 h. The bullfrog (*Rana catesbeiana*) also appears to be comparatively tolerant to chlorpyrifos, as judged by a single oral LD50 value of >400.0 mg/kg body weight (BW).

Sublethal effects of chlorpyrifos exposure have been documented for many species of freshwater and marine fauna; they include inhibition of cholinesterase (ChE) activity levels in brain and hematopoietic organs, reduction in blood glucose levels, sluggishness, motor incoordination, delayed maturation and growth, renal histopathology, reproductive impairment, and reduced feed intake. Reproductive impairment, for example, was observed in *Daphnia magna* at 0.08 μg chlorpyrifos/L. Reduction in settling rate was shown in oyster larvae after exposure to 0.1 μg/L for 8 days. Equilibrium loss was documented in 50% of brown shrimp (*Penaeus aztecus*) after exposure to 0.32 μg chlorpyrifos/L for 24 h.

In fish, growth of the California grunion (*Leuresthes tenuis*) was reduced by 20% in an early life stage during immersion in 0.5 μg chlorpyrifos/L for 35 days and by 26% in fry after exposure to 1.0 μg/L for 26 days. Nile tilapia (*Oreochromis niloticus*) exposed to 1.0 μg/L for 3 months had renal pathology and disrupted immune function. Larvae of fathead minnows exposed to 2.1 μg/L for 30 days had increased deformities at day 30, as did larvae exposed to 122.0 μg/L for 5 h. Guppies exposed to 3.0 μg/L for 2 weeks had 80–90% inhibition of acetylcholinesterase within 2 weeks; partial recovery occurred after 4 days in clean water, but ChE levels were still 56% inhibited after 14 days. In fathead minnows exposed to 120.0 μg chlorpyrifos/L for 200 days, ChE activity was significantly reduced, fecundity was reduced, maturation delayed and, in second-generation fish, growth and maturation were reduced. Chlorpyrifos was associated with deformities in fathead minnows.

Bioconcentration factors (BCFs) for tilapia (*Oreochromis aureus*) for individual tissues ranged from 85 in muscle to 939 in liver; gills (497) and bile (517) were intermediate. BCF of chlorpyrifos from the medium varied substantially among five species of fishes, but generally paralleled ambient levels of chlorpyrifos. Increases in BCFs in chlorpyrifos-exposed teleosts may be associated with three variables: increased metabolic rate, as indicated by hyperventilation, hyperactivity, and decreased growth; increased bioavailability of chlorpyrifos as a result of solvent-induced supersaturation or increased food availability; and decreased depuration rates due to possible physiological dysfunction. At high BCFs, adverse effects on growth and survival were observed in sheepshead minnow (*Cyprinodon variegatus*) and in Gulf toadfish (*Opsanus beta*). Chlorpyrifos is excreted rapidly from fish; the estimated Tb1/2 for many species is 8.7 h, and equilibration occurs with the surrounding medium in 24–72 h. Higher Tb1/2 values of 13.4–69.3 h were recorded for various species of freshwater fishes, with slow excretion rates associated with high BCF values and impaired metabolic breakdown. No detectable chlorpyrifos residues were found after 12 days in 10 species of estuarine invertebrates – including oligochaete annelids, mollusks, and crustaceans – after treatment with 0.046 kg chlorpyrifos/ha.

8.3.2 Birds and Mammals

Signs of chlorpyrifos intoxication, include excessive blinking, hypoactivity, hyperexcitability, excessive drinking, muscular incoordination, rapid breathing, muscular weakness, tremors, piloerection (mammals) or fluffed feathers (birds), salivation, lacrimation, diarrhea, excessive urination, prostration, loss of righting reflex, spasms, tetany, coma, and convulsions. Death usually occurs between 1 h and 9 days after exposure, Chlorpyrifos oxon (*O*, *O*-diethyl-*O*-(3,5,6-trichloro-2-pyridyl phosphate) is the active oxygen analog of chlorpyrifos and is probably responsible for most of the anticholinesterase mode of action of chlorpyrifos; the oxon is extensively and rapidly detoxified in mammalian liver via enzymatic hydrolysis by at least two microsomal esterases. Significant accumulations of chlorpyrifos were not detected in domestic turkeys (*Meleagris gallopavo*) and chickens. In birds kept in pens on soil treated with 4.5–9.0 kg active ingredients chlorpyrifos/ha, tissue residues were 0.16 mg/kg after 1 week; these decreased thereafter, although birds remained on the treated soil.

LD50 values, based on a single oral dose, ranged from 5.0 to 157.0 mg chlorpyrifos/kg BW in birds, and from 151.0 to 1000.0 in mammals; however, 7 of 14 avian species had reported LD50 values of <25.0 mg/kg BW. As little as 2.0 mg/kg BW to nestling red-winged blackbirds (*Agelaius phoeniceus*) was associated with reduced survival during the first 24-h postexposure interval; however, nestling European starlings had normal survival when given a single oral dose of 2.0 mg/kg BW. Many species of birds that survived chlorpyrifos poisoning showed gross pathological changes; furthermore, the slope of the acute dose–response curve was low. These findings suggest that decreasing dosage levels did not produce proportional decreases in response, and indicate a reduced safety margin for chlorpyrifos owing to mortalities that occur frequently at levels much lower than the calculated LD50 values.

Reduction in ChE activity levels of various tissues (blood, brain) is one of the earliest signs of chlorpyrifos intoxication. ChE reductions have been demonstrated in turkeys fed diets containing 50.0 mg chlorpyrifos/kg (estimated daily dose of 0.7 mg/kg BW) for 20 days, in chickens fed diets of 25.0 mg/kg (estimated daily dose of 0.94 mg/kg BW) for 20 days, in quail (*Coturnix coturnix*) given a single (sublethal) esophagal intubation of 13.0 mg/kg BW, and in mallard (*Anas platyrhynchos*) ducklings fed 75.0 mg chlorpyrifos/kg diet for 14 days. Brain ChE activity in quail was normal after 11 days despite 81% inhibition 8 h postadministration. Low temperatures (27.5°C vs. 35°C) potentiated dose-related ChE depression in juvenile northern bobwhite (*Colinus virginianus*), suggesting a need for more research on cold stress interactions between acute oral chlorpyrifos exposure. In adult female rats, brain ChE activity was inhibited by as much as 96% following subcutaneous injection of 280.0 mg/kg BW; all survived without extensive signs of toxicity, but weight loss was evident 2–7 days after treatment.

Dietary concentrations of 30.0–100.0 mg chlorpyrifos/kg feed produce some deaths in birds, and 136.0 to about 500.0 mg/kg feed usually kills at least 50%. In chickens fed diets of 100.0 mg chlorpyrifos/kg – equivalent to an estimated daily dose of 6.8 mg/kg BW – egg fertility was reduced by 15% and hatchability by 17%. Dietary levels lethal to mallard ducklings were 136.0–180.0 mg/kg feed, equivalent to 10.0 mg/kg BW fed daily for 5 days. In adult mallards, given diets containing 80.0 mg chlorpyrifos/kg for 60–84 days, body weight, food consumption, brain ChE activity levels, and egg production were all reduced; moreover, egg weight and eggshell thickness were reduced, the resultant ducklings weighed less than controls, and survival was comparatively poor at age 7 days. No effect on any variable was observed at diets of 8.0 mg/kg.

Dermal application routes are also toxic. Some deaths were recorded in turkeys from dermal treatments of 15.0–20.0 mg chlorpyrifos/kg BW. Higher levels applied to feathers killed turkeys within 8 h. Newborn piglets (*Sus* spp.) were especially more sensitive to cutaneous applications of chlorpyrifos than those 30–36-h-old; newborns showed clinical signs consistent with organophosphorus toxicosis after a 2.5% aerosol preparation (dosage unknown) was applied to the tail and umbilicus. Accidental poisoning of cattle (*Bos* spp.) by chlorpyrifos through dermal application to control ticks resulted in some deaths; among bulls that survived, sperm production was reduced by 43% in seriously affected animals and by 12% in those with no outward signs of poisoning.

Chlorpyrifos is not mutagenic, as judged by mitotic recombination assays, and did not increase sister chromatid exchange above background in tests with chick (*Gallus* spp.) embryos and Chinese hamster (*Cricetus* spp.) ovary cells. Chlorpyrifos altered serum cortisol and decreased thyroxine concentrations in sheep given oral doses of 12.5 mg chlorpyrifos/kg BW twice weekly for 43 days, indicting a need for more research on the role of chlorpyrifos in hormone metabolism.

Chlorpyrifos-impregnated ear tags are under development to control horn flies (*Haematobia irritans*) in U.S. cattle. Cattle fitted with ear tags (0.96 g chlorpyrifos per tag/365 kg animal, or about 2.6 mg chlorpyrifos/kg BW) had slightly elevated tissue residues (0.13 mg/kg fat) after 12 weeks, but residues were well within acceptable tolerance levels of 2.0 mg chlorpyrifos/kg fresh weight (FW) cattle fat, meat, or meat by-products. In dogs (*Canis familiaris*), chlorpyrifos-impregnated collars provided effective control of adult fleas (*Ctenocephalides* spp.) for up to 11 months, with no significant adverse reactions regardless of canine coat length, size, or age.

8.4 Field Investigations

There have been many accidental spills of chlorpyrifos, but little quantitative assessment of the environmental effects. One exception is a spill in April 1985 in England, in which a truck overturned, spilling 205 L of chlorpyrifos into an adjacent stream that drained into the Roding River. A resulting sharp decrease in the number and type of macroinvertebrate benthic organisms in affected parts of the river, compared to unaffected areas, lasted 6 months. In addition, certain chlorpyrifos-resistant benthic organisms were unusually abundant.

Chlorpyrifos controls mosquito larvae at applied dosages between 0.028 and 0.056 kg/ha, equivalent to 9.0–18.0 μg chlorpyrifos/L in 152 mm (6 inches) of water; in 1984 alone, chlorpyrifos was used for this purpose on about 600,000 ha. Surface waters of ponds on golf courses in North Carolina often contain 7.2–11.5 μg chlorpyrifos/L as a result of turf treatment with chlorpyrifos. At this time, no obvious deleterious effects of chlorpyrifos have been recorded in mammals, amphibians, or reptiles under field conditions of current use. For example, bullfrogs (*Rana catesbiana*) from an Iowa farm pond that received runoff from a 16-ha cornfield treated with the label rate of Lorsban 15G, when compared to a reference pond, had the same level of plasma activity and reactivation for total ChE, acetylcholinesterase, and butylcholinesterase. However, at recommended dosage application rates for control of mosquitoes and other pestiferous insects (usually 0.028–0.056 kg/ha), adverse effects have been documented on survival, reproduction, metabolism, and species diversity of a variety of fishes, terrestrial and aquatic invertebrates, freshwater flora, waterfowl, and horned larks (*Eremophila alpestris*), and on the marketability of various crops. For example, field populations of daphnids (*Daphnia pulex*) were reduced 50% by 0.38 μg chlorpyrifos/L after 2 days and by 0.25 μg/L in 7 days; similar values were found in laboratory populations of this species. It is emphasized that the effectiveness of chlorpyrifos under field conditions, like that of other organophosphorus pesticides, is significantly modified by numerous variables such as formulation, route of administration, pond substrate, dose, and water temperature.

8.5 Recommendations

Water quality criteria formulated for chlorpyrifos and aquatic life protection seem to afford a reasonable degree of safety, at least during short-term exposure. Specifically, the proposed criteria for freshwater are 0.041 μg/L (4-day average concentration) and 0.083 μg/L (1-h average concentration), neither of which should be exceeded more than once every 3 years; for saltwater, these criteria are 0.0056 μg/L and 0.011 μg/L, respectively. Levels of chlorpyrifos in the Sacramento–San Joaquin River system proposed by the state of California to protect aquatic life during chronic exposure are <0.02 μg/L in freshwater and <0.01 μg/L in saltwater. However, concentrations of chlorpyrifos in the San Joaquin River during 1991–93 ranged between 0.01 and 1.6 μg/L, and suggest that California's stringent water quality criteria proposed for chlorpyrifos should be reexamined. Proposed chlorpyrifos drinking water criteria by some agencies to protect human health are significantly higher than those recommended for aquatic life protection: 21.0 μg/L in Vermont, 30.0 μg/L to protect children in lifetime exposure in the United States, 90.0 μg/L in Canada, and 100.0 μg/L to protect adults in the United States.

The acceptable tolerance level of chlorpyrifos in meat and meat by-products destined for human consumption is 2.0 mg/kg FW, and for agriculture products it usually ranges between 0.05 and 15.0 mg/kg FW and up to 25.0 mg/kg for citrus oil. The significance of these concentrations to animal health, or to consumers other than humans, is unknown. More research is needed to establish maximum tolerable chlorpyrifos limits in tissues of sensitive fish and wildlife. Proposed air criteria for chlorpyrifos and human health include 200.0 $\mu g/m^3$ in the workplace, and much lower concentrations of 0.48–3.3 $\mu g/m^3$ in non-occupational settings. No air criteria are currently available or proposed for protection of wildlife.

Information is lacking on the effectiveness of chlorpyrifos in large-scale (>40 ha) coldwater ecosystems, typical of those found in Alaska or northern tier states; accordingly, initiation of long-term studies in these potential problem areas are recommended. As judged by the available literature, three courses of action now seem warranted:

(1) Restrict the use of chlorpyrifos for mosquito control in wetlands, estuaries, and waterfowl breeding areas because recommended treatment levels are demonstrably harmful to nontarget species, including mallard ducklings. The finding that certain mosquito populations in California are showing signs of chlorpyrifos resistance, and thus may require more aggressive future treatment programs further supports the unsuitability of chlorpyrifos for mosquito control.
(2) Curtail agricultural use of chlorpyrifos in watershed areas pending acquisition of additional data on its transport, fate, and effects, including data on chlorpyrifos flux rates from oils and sediments and its resultant bioavailability.
(3) Develop suitable replacements for chlorpyrifos in mosquito control programs. These replacement compounds should exhibit a relatively long half-life in aquatic environments while avoiding the broad-spectrum toxicity typical of chlorpyrifos to large numbers of nontarget organisms.

8.6 Summary

Chlorpyrifos (phosphorothioic acid *O*, *O*,-diethyl *O*-(3,5,6,-trichloro-2-pyridinyl) ester), an organophosphorus compound with an anticholinesterase mode of action, is used extensively in a variety of formulations to control a broad spectrum of agricultural and other pestiferous insects. Domestic use of chlorpyrifos in 1982 was about 3.6 million kg; the compound is not only used mostly in agriculture, but also to control mosquitos in wetlands (0.15 million kg applied to about 600,000 ha) and turf-destroying insects on golf courses (0.04 million kg). Accidental or careless applications of chlorpyrifos have resulted in the death of many species of nontarget organisms such as fish, aquatic invertebrates, birds, and humans. Applications at recommended rates of 0.028–0.056 kg/surface ha for mosquito control have produced mortality, bioaccumulation, and deleterious sublethal effects in aquatic plants, zooplankton, insects, rotifers, crustaceans, waterfowl, and fish; adverse effects were also noted in bordering invertebrate populations.

Degradation rate of chlorpyrifos in abiotic substrates varies, ranging from about 1 week in seawater (50% degradation) to more than 24 weeks in soils under conditions of dryness, low temperatures, reduced microbial activity, and low organic content; intermediate degradation rates reported have been 3.4 weeks for sediments and 7.6 weeks for distilled water. In biological samples, degradation time is comparatively short – usually less than 9 h in fishes, and probably the same in birds and invertebrates.

Chlorpyrifos is acutely toxic to some species of aquatic invertebrates and teleosts at nominal water concentrations ranging between 0.04 and 1.1 μg/L. Acute single-dose oral LD50 values of chlorpyrifos to susceptible avian species ranged from 5.0 to 13.0 mg/kg BW. Mammals were comparatively tolerant to chlorpyrifos: acute oral LD50s were reported at 151.0 mg/kg BW, and higher. Lethal dietary concentrations for sensitive species of birds ranged from 30.0 to 50.0 mg chlorpyrifos/kg food. Sublethal effects were recorded in all species of organisms examined at concentrations below those causing mortality. These effects included bioconcentration from the medium by teleosts (410–1000); Ache activity reduction in brain and hematopoietic tissues; reduced growth; impaired reproduction, including sterility and developmental abnormalities; motor incoordination; convulsions; and depressed population densities of aquatic invertebrates.

Three courses of action are recommended. (1) Restrict the use of chlorpyrifos for mosquito control in wetlands, estuaries, and waterfowl breeding areas because recommended treatment levels are demonstrably harmful to nontarget resident biota. (2) Curtail agricultural use in watershed areas pending acquisition of additional data on chlorpyrifos toxicokinetics. (3) Develop suitable replacements for chlorpyrifos in mosquito control programs; specifically, pesticides with more specificity to target organisms, and lower toxicity to nontarget biota.

CHROMIUM[a]

Chapter 9

9.1 *Introduction*

Environmental effects of chromium (Cr) have been extensively reviewed, with the majority of authorities agreeing that chromium is widely used in domestic and industrial products and that some chemical forms, notably hexavalent chromium (Cr^{+6}), are toxic, and others, notably trivalent chromium (Cr^{+3}), are essential nutrients. In North America, thousands of tons of chromium ore and concentrates are imported annually for the production of stainless steels, chrome-plated metals, pigments for inks and paints, and a wide variety of chemicals. Reports from Europe, Scandinavia, Asia, and North America all emphasize the high incidence of lung cancer and other respiratory diseases among workers involved in the manufacture of chromates. Others document that land dumping of wastes from chromate production and electroplating operations have been responsible for groundwater contamination; that discharge of chromium wastes into streams and lakes has caused damage to aquatic ecosystems and accidental poisoning of livestock; and that large amounts of Cr^{+3} and Cr^{+6} are reintroduced into the environment as sewage and solid wastes by the disposal of consumer products containing chromium.

[a]All information in this chapter is referenced in the following sources:

Eisler, R. 1981. *Trace Metal Concentrations in Marine Organisms*. Pergamon Press, Elmsford, New York, 687 pp.

Eisler, R. 1986. Chromium hazards to fish, wildlife and invertebrates: a synoptic review. *U.S. Fish Wildl. Serv. Biol. Rep. 85(1.6)*, 1–60.

Eisler, R. 2000. Chromium. Pages 45–92 in *Handbook of Chemical Risk Assessment: Health Hazards to Humans, Plants, and Animals. Volume 1, Metals*. Lewis Publishers, Boca Raton, Florida.

9.2 *Environmental Chemistry*

Chromium is the seventh most abundant element on earth with more than 2108 million tons of chromium metal, most of it residing in the core and mantle. The annual world production of chromium is estimated at 9 million metric tons; most of the ore, in the form of chromite ($FeOCr_2O_3$), is produced by the Former Soviet Union and the Republic of South Africa. In the United States, trivalent chromium compounds are used as pigments and in leather tanning, and hexavalent chromium compounds are used principally in the ferrochrome and chemical industries. The combined production of chromium ferroalloys and chromium metal in the United States in 1988 was 119,645 metric tons from producers in West Virginia, South Carolina, New Jersey, Ohio, Kentucky, New York, Indiana, Utah, Pennsylvania, California, Texas, and North Carolina. Export of chromium materials from the United States in 1988 was 39,887 tons, mostly to Mexico, Canada, the Netherlands, Japan, and Germany. In 1990, the United States produced 150,600 tons of sodium dichromate and 53,300 tons of chromic acid.

Although natural mobilization of chromium by weathering processes is estimated at 32,000 tons/year, the amounts of chromium added to the environment as a result of anthropogenic activities are far greater. New York City alone contributes about 440 tons of chromium annually to the environment. In the 1970s, major atmospheric emissions of chromium were from the chromium alloy and metal-producing industries, and lesser amounts came from coal combustion, municipal incinerators, cement production, cooling towers, use of chromium-containing phosphate fertilizers, and landfill dumping of chromium-contaminated sewage sludge and consumer products. In the United States in the 1990s, atmospheric chromium emissions from anthropogenic sources averaged 2800 tons/year, mostly from combustion of coal and oil (60%) and chromeplating (24%) sources. Atmospheric emissions contribute 4–6 times more chromium to aquatic ecosystems than liquid wastes. In aquatic environments, the major sources of chromium are the electroplating and metal-finishing industries and publicly owned treatment plants; relatively minor sources other than localized contamination are iron and steel foundries, inorganic chemical plants, tanneries, textile manufacturing, and runoff from urban and residential areas. Chromium in phosphates used as fertilizers may be an important source of chromium in soil, water, and some foods. In general, elevated levels of chromium in biological or other samples have been positively correlated with increased industrial and other uses of the element – especially uses associated with plating and foundry applications, chemical manufacturing, and corrosion inhibition.

Chromium in the crystalline form is a steel-gray, lustrous, hard metal characterized by an atomic weight of 51.996, an atomic number of 24, a density of 7.14 g/cm^3, a melting point of 1857°C, and a boiling point of 2672°C. Four chromium isotopes occur naturally: Cr-50 (4.3%), -52 (83.8%), -53 (9.6%), and -54 (2.4%), and seven are manufactured. Elemental chromium is very stable, but is not usually found pure in nature. Chromium can exist in oxidation states ranging from –2 to +6, but is most frequently found in the environment in the trivalent (+3) and hexavalent (+6) oxidation states. The +3 and +6 forms are the most important because the +2, +4, and +5 forms are unstable and are rapidly converted to +3 which, in turn, is oxidized to +6.

Most compounds prepared from chromite ore contain chromium in the more stable +3 and +6 states. The chromium in essentially all environmentally important chromium compounds is in one of these two oxidation states. Chromium in biological materials is usually in the +3 form, and is the form that functions as an essential element in mammals by maintaining efficient glucose, lipid, and protein metabolism. In general, the toxicity of trivalent chromium to mammals is low because its membrane permeability is poor and it is noncorrosive; further, there is little tendency for Cr^{+3} to biomagnify in food chains in the inorganic form. However, organo-trivalent chromium compounds may have significantly different accumulation tendencies although little is known about these compounds. Hexavalent chromium is more toxic than the +3 form because its oxidizing potential is high and it easily penetrates biological membranes. All toxic effects of Cr^{+6} seem to be related to the strong oxidizing action of chromates, and all biological interactions of chromates seem to result in reduction to the Cr^{+3} form and subsequent coordination to organic molecules. It is difficult to distinguish between the effects caused by Cr^{+6} and those caused by Cr^{+3} since Cr^{+6} is rapidly reduced to Cr^{+3} after penetration of biological membranes and in the gastric environment. However, Cr^{+6} can be readily transported into cells, whereas Cr^{+3} is unable to cross cell membranes. The reduction of Cr^{+6} to Cr^{+3} may be the most important mechanism for the toxicity of chromium.

Most of the Cr^{+6} found in nature is a result of domestic and industrial emissions. Interaction of +6 chromic oxide, dichromate, or chromate compounds with organic compounds can result in reduction to the comparatively less toxic trivalent form. Chromium compounds interact synergistically or antagonistically with many chemicals. For example, potassium dichromate administered by subcutaneous injection potentiated the effects of mercuric chloride, citrinin, and hexachloro-1,3-butadiene on rat kidneys. Chromium effects were lessened by ascorbic acid and vitamin E, and *N*-acetyl cysteine was effective in increasing urinary excretion of chromium in rats.

Little is known about the relation between concentrations of total chromium in a given environment and biological effects on the organisms living there. Depending on the physical and chemical state of chromium, the same element concentration has a wide variety of mobilities and reactivities and thus has different effects. Chromium toxicity to aquatic biota is significantly influenced by abiotic variables such as hardness, temperature, pH, and salinity of water; and biological factors such as species, life stage, and potential differences in sensitivities of local populations. In both freshwater and marine environments, hydrolysis and precipitation are the most important processes that determine the fate and effects of chromium, whereas adsorption and bioaccumulation are relatively minor. Both Cr^{+3} and Cr^{+6} can exist in water with little organic matter; Cr^{+6} is usually the major species in seawater. Under oxygenated conditions, Cr^{+6} is the dominant dissolved stable chromium species in aquatic systems. The hexavalent form exists as a component of a complex anion that varies with pH and Eh and may take the form of chromate (CrO_4^{-2}), hydrochromate ($HCrO_4^{-1}$), or dichromate ($Cr_2O_7^{-2}$), with dichromate predominating at acidic pHs. These ionic Cr^{+6} forms are highly soluble in water and thus mobile in the aquatic environment. All stable Cr^{+6} anionic compounds strongly oxidize organic matter on contact and yield oxidized organic matter and Cr^{+3}. Trivalent chromium tends to form stable complexes with negatively charged inorganic or organic compounds, and thus is unlikely to be found uncomplexed in aqueous solution if anionic or particulate compounds (such as decaying plant or animal tissues, or silt or clay particles) are present. Precipitated Cr^{+3} hydroxides remain in the sediments under aerobic conditions; under low pH and anoxic conditions, however, Cr^{+3} hydroxides may solubilize and remain as ionic Cr^{+3} unless oxidized to Cr^{+6} through mixing and aeration. Among estuarine sediments, chromium content tends to be highest in those of small grain size and high organic and iron content; concentrations in European estuaries ranged from 3.9 mg/kg in intertidal sands to 162.0 mg/kg in anaerobic muds. Adsorption of chromium by sediments is salinity dependent; adsorption is greatest at salinities of 0.1–1.0. Colloidal iron strongly scavenges Cr^{+3} from river water; flocculation of the colloids when they are mixed with seawater, coupled with lack of removal of the colloids to the sediments by gravitational settling or scavenging by suspended sediments, promotes the flux of Cr^{+3} through the estuary to the open ocean.

The solubility and potential bioavailability of waste chromium added to soils through sewage sludge, animal manures, and industrial wastewater are modified by soil pH and organic complexing substances. Although soil pH can affect oxidation rates of Cr^{+6} to Cr^{+3}, organic complexes appear to play a more significant role. For example, organically complexed Cr^{+3} added to soils may remain soluble for at least a year, whereas the free Cr^{+3} metal ion in the absence of soluble complexing ligands quickly becomes adsorbed, or hydrolyzed and precipitated. The biological effects of organochromium compounds, which are not well documented, appear to be high-priority subjects for further research.

In groundwater, hexavalent chromium tends to be mobile due to the lack of solubility constraints and the low adsorption of Cr^{+6} anion species by metal oxides in neutral to alkaline waters. Above pH 8.5, no Cr^{+6} adsorption occurs in groundwater; Cr^{+6} adsorption increases with decreasing pH. Trivalent chromium species tend to be relatively immobile in most groundwaters because of the precipitation of low-solubility Cr^{+3} compounds above pH 4 and high adsorption of the Cr^{+3} ion by soil clay below pH 4.

Data on the environmental cycling of chromium are lacking, and those on the biochemistry of chromium are incomplete and sparse; it is clear that these two subjects merit additional research. Furthermore, there is increasing concern about the uncertainties in the analysis of some types of biological and environmental samples. For example, collaborating laboratories have reported order-of-magnitude differences in persistence of chromium in standard bovine liver. Until more is learned about the reasons for these differences, caution should be exercised in interpreting past analytical results.

9.3 Concentrations in Field Collections

Chromium concentrations in selected nonbiological materials are elevated in the vicinity of industrial operations and municipal waste treatment facilities where chromium is a significant component of wastes discharged into the environment. It is generally agreed that suspended particulates are a major source of transport in aquatic systems, that most chromium in soil and sediment is unavailable to living organisms, that Cr^{+6} in air and water is hazardous to fish and wildlife, and that the grossly elevated levels of chromium (especially inorganic fractions) in sludge components may have serious implications to wildlife when the sludge is applied to croplands. Severe groundwater contamination, i.e., 40.0 mg Cr/L, has been reported in Nassau County, New York, from an aircraft plant that used chromium solutions for anodizing and plating metals. Chromium contamination of shallow aquifers has also been reported near Telluride, Colorado, from heavy metal mining and milling wastes discharged into a leaky holding pond, and from chromium wastes from automotive, electroplating, and wood treatment industries in Michigan. Hexavalent chromium added to or found in soils may be leached, reduced, adsorbed, precipitated, or taken up by a living organism. Chromium-contaminated sediments may act as a toxicant source by releasing chromium to interstitial sediment pore waters; the acute toxicity of chromium and other metals in sediments has been correlated with pore water concentrations.

Certain plants grown on serpentine soils containing 1000–50,000 mg Cr/kg dry weight (DW) may contain 10.0–100.0 mg Cr/kg DW – levels that may be toxic to wildlife, although no reports of this phenomenon are known. Most plant and invertebrate species die before accumulating amounts of chromium that are toxic to predators. Some species of terrestrial plants have been proposed for the removal of chromate from wastewaters. Terrestrial plants, such as wheat (*Triticum aestivum*), accumulate the greatest amount of chromium under conditions of sulfate deficiency or deprivation. Because sulfate is a strong inhibitor of chromate uptake in terrestrial plants, the presence of sulfates in the environment negates the usefulness of this approach.

Chromium concentrations in species of individual taxonomic groups tended to be elevated when collection localities were near electroplating plants, tanneries, oil drilling operations, sewage outfalls, drift cooling towers, dumpsites, or other sources of chromium-containing wastes that were being discharged into the environment. Among marine algae and invertebrates, for example, comparatively high concentrations of chromium were recorded in algae, clams, and annelids from the vicinity of electroplating plants; in crabs collected near an ocean dumpsite receiving large quantities of metals; and in algae and echinoderms near urbanized areas in Puerto Rico. Grossly elevated levels of chromium were also noted in selected plasma fractions of tunicate blood, in scales from a few species of teleosts, and in corals from chromium-rich areas containing high concentrations of scandium and titanium; however, these accumulations were not attributed to anthropogenic activities. Studies demonstrate that fish and sediments in Florida stormwater ponds do not have elevated concentrations of chromium, that chromium does not biomagnify in marine food chains in an Egyptian bay heavily contaminated with chromium, that chromium from sediments containing 717.0 mg Cr/kg DW in New Jersey wetlands is not biologically available to biota in the immediate vicinity, and that elevated chromium residues in biota from a Texas estuary are not reflective of residues of other heavy metals. Many factors are known to modify chromium levels. In marine mollusks, as one example, chromium concentrations tended to increase with the age of the organism, although uptake was significantly inhibited at high salinities. Accidental contamination of field samples by metal particles in the samples, rust from stainless hydrowire, or flaking paint from the hull of the collecting ship may also constitute significant sources of elevated chromium residues in aquatic environments.

Waterfowl from areas contaminated by mining wastes and consuming diets rich in chromium had elevated chromium concentrations in tissues, especially in gonads, gall bladder, and pancreas. Chromium burdens were highest in livers of seed-eating species of birds from Baja California. In feathers of passerine birds, chromium concentrations were lowest in species that ate mostly fruit and highest in older adults of long-lived species. Bones of 24 species of birds collected from southwestern Russia in 1993–95 contained 0.3–14.7 mg Cr/kg DW; concentrations were higher in conspecifics collected in urban areas than in rural areas, highest in sparrows and other seed-eating species and lowest in owls, and highest in bones of terrestrial species and lowest in those of aquatic species. Elevated concentrations of chromium were measured in brain, lung, kidney, and liver of rock doves (*Columba livia*) from Mexico City when compared to conspecifics collected from a rural area. In terrestrial ecosystems, elevated chromium levels were reported in cotton rats and plants collected near drift cooling towers, and in earthworms and plants from sludge-amended soils. However, the high levels of chromium reported in the hair of pronghorns and elk require verification.

A source of concern is the accuracy and precision of chromium analyses in biological samples. One interlaboratory calibration study, involving 87 laboratories, showed that an oyster homogenate averaged 1.1 mg/kg DW, with a standard deviation of 0.5 mg/kg. This means that about 67% of the laboratories were in the range of 0.6–1.6 mg/kg and about 33% of the laboratories reporting were outside this range. It seems clear that more rigorous and standardized sample preparation techniques and analyses for chromium are needed.

9.4 Beneficial and Protective Properties

Hexavalent chromium has no known essential function. However, trivalent chromium in the form of a dinicotinic acid–glutathione complex is an essential cofactor for insulin production and forms complexes with protein, amino acids, and other organic acids, and is the most biologically stable form of chromium. Dietary trivalent chromium deficiency results in an inability to clear glucose from the blood and pathology similar to diabetes. Although trivalent chromium is an essential nutrient, exposure to high levels via inhalation, ingestion, or dermal contact may cause adverse health effects. In humans, other mammals, and turkeys (*Meleagris gallopavo*), trivalent chromium is essential for the normal metabolism of carbohydrates, insulin, and glucose, and for regulating carbohydrate metabolism in mammals. Chromium deficiency has been described in rats, guinea pigs, and squirrel monkeys; signs include reduced growth, decreased life span, elevated serum cholesterol, increased formation of aortic plaques, and signs resembling those of diabetes mellitus. Subjecting chromium-deficient animals to stress can exacerbate the signs. In humans, chromium deficiency has been suggested as a possible factor in the incidence of diabetes and atherosclerosis. Autopsy data from 31 areas of the world suggested that many Americans, but few non-Americans, were deficient in chromium. One characteristic feature of chromium levels in human tissues is a decline with increasing age.

Chromium is beneficial but not essential for growth in higher plants. Residues in plants seldom exceed a few milligrams per kilogram, except in plants living on infertile serpentine soils containing high chromium concentrations, or grown on soils amended with sewage sludge. Plants with elevated chromium residues show no toxic effects, although concentrations in excess of 1.0 mg/kg in the aqueous medium may inhibit germination of the seed and growth of roots and shoots.

Chromium has proved effective in counteracting the deleterious effects of cadmium in rats and of vanadium in chickens. High mortality rates and testicular atrophy occurred in rats subjected to an intraperitoneal injection of cadmium salts; however, pretreatment with chromium ameliorated these effects. The Cr–Cd relation is not simple. In some cases, cadmium is known to suppress adverse effects induced in Chinese hamster (*Cricetus* spp.) ovary cells by Cr^{+6}. In southwestern Sweden, there has been an 80% decline in chromium burdens in liver of the moose (*Alces alces*) during 1982–92 from 0.21 to 0.07 mg Cr/kg fresh weight (FW). During this same period in this locale, moose have experienced an unknown disease caused by a secondary copper deficiency due to elevated molybdenum levels as well as chromium deficiency and trace element imbalance. In chickens (*Gallus* sp.), 10.0 mg/kg of dietary chromium counteracted adverse effects on albumin metabolism and eggshell quality induced by 10.0 mg/kg of vanadium salts. Additional research on the beneficial aspects of chromium in living resources appears warranted, especially where the organism is subjected to complex mixtures containing chromium and other potentially toxic heavy metals.

9.5 *Lethal Effects*

Biocidal properties of chromium salts to aquatic organisms are modified, sometimes by an order of magnitude or more, by a variety of biological and abiotic factors. These include the species, age, and developmental stage of the organism; the temperature, pH, salinity, and alkalinity of the medium; interaction effects of chromium with other contaminants; duration of exposure; and chemical form of chromium tested. For hexavalent chromium, LC50 (96 h) values for sensitive freshwater and marine species were between 445.0 and 2000.0 μg/L. For trivalent chromium, LC50 (96 h) concentrations were 2000.0–3200 μg/L for sensitive freshwater organisms and 3300.0–7500.0 μg/L for marine biota.

Among warm-blooded organisms, hexavalent chromium was fatal to dogs in 3 months at 100.0 mg/kg in their food and killed most mammalian experimental animals at injected doses of 1.0–5.0 mg Cr/kg body weight (BW), but had no measurable effect on chickens at dietary levels of 100.0 mg/kg over a 32-day period. Trivalent chromium compounds were generally less toxic than hexavalent chromium compounds, but significant differences may occur in uptake of anionic and cationic Cr^{+3} species, and this difference may affect survival.

9.5.1 Aquatic Organisms

Records of acute toxicities of hexavalent and trivalent chromium salts to representative species of aquatic life make it clear that Cr^{+6} is more toxic to freshwater biota in comparatively soft and acidic waters, that younger life stages are more sensitive than older organisms, and that 96 h is insufficient to attain stable mortality patterns. Euryhaline species of estuarine algae tested show increasing resistance to Cr^{+6} at increasing salinities. There are at least five ionic species of hexavalent chromium, of which two – the hydrochromate ion and the chromate ion – are the predominant species and probably the agents that are toxic to freshwater life. However, water pH dramatically affects the concentration of each: as pH decreased from 7.8 to 6.5 the hydrochromate ion increased by a factor of about 3, and the chromate ion decreased by a factor of about 6.8. To some species of freshwater fishes, Cr^{+6} was 50–200 times more toxic at pH 6.4–7.4 than at pH 7.8–8.0. Hexavalent chromium interacts with other metals in solution to produce additive or synergistic effects, as was the case with nickel salts in acute toxicity to guppies. More research is needed to fully elucidate chromium's mode of action in solution. The organisms most sensitive to Cr^{+6}, as judged by 96-h LC50 values, were freshwater crustaceans and rotifers, and marine crustaceans, for which LC50 values were 445.0–3100.0 μg/L; longer exposures of 28–84 days produced LC50 values of 200.0–500.0 μg/L. Other investigators had confirmed that Cr^{+6} is more toxic to freshwater daphnids and teleosts in water of comparatively low alkalinity, low pH, and low total hardness. In marine teleosts, the toxicity of Cr^{+6} increased at elevated temperatures; furthermore, chromium was additive in toxicity when present as a component in a complex mixture of cadmium, zinc, and Cr^{+6} salts.

For trivalent chromium and freshwater biota, toxicity was significantly increased in comparatively soft waters; this pattern was especially pronounced for daphnids. Among freshwater teleosts, survival was reduced at comparatively low pH. Also, organisms exposed previously to Cr^{+3} salts were not unusually sensitive or resistant when subjected to additional Cr^{+3}, suggesting that they were unable to acclimatize or to become sensitized to Cr^{+3}. As judged by 96-h LC50 values, Cr^{+3} was toxic to sensitive freshwater organisms at concentrations of 2000.0–3000.0 μg/L, or slightly less toxic than Cr^{+6}. Toxicity of Cr^{+3}, like that of Cr^{+6} increased with increasing exposure in rainbow trout (*Oncorhynchus mykiss*). However, in freshwater, Cr^{+3} was significantly less toxic than Cr^{+6} to salmon fingerlings, and was dramatically less toxic than Cr^{+6} to polychaetes and crustaceans (but not to mollusks or teleosts) in saltwater.

Maximum acceptable toxicant concentrations (MATCs) of chromium to aquatic life were derived from life cycle or partial life cycle exposures, and expressed as the highest concentration tested having no significant adverse effect on the characteristics measured usually survival, growth, and reproduction and the lowest concentration at which these effects were observed. For chromium and freshwater teleosts, MATC values ranged from as low as 51.0 to 105.0 μg/L in rainbow trout to as high as 1000.0–31,950.0 μg/L in fathead minnows (Table 9.1). The most sensitive saltwater organism tested was a polychaete worm with a MATC range of 17.0–38.0 μg/L (Table 9.1). For Cr^{+3}, the MATC range for freshwater organisms was 47.0–1400.0 μg/L, which was quite similar to that for Cr^{+6} for freshwater life. No MATC data were available for Cr^{+3} and marine biota.

9.5.2 Terrestrial Invertebrates

Data on toxicity of chromium to terrestrial invertebrates are sparse. Studies conducted in India showed that a concentration of 10.0–15.0 mg/L of Cr^{+6} in irrigation water, when applied to soils for agricultural purposes, was lethal to two species of earthworms in 58–60 days.

9.5.3 Mammals and Birds

Acute and chronic adverse effects of chromium to warm-blooded organisms are caused mainly by Cr^{+6} compounds; there is little conclusive evidence of toxic effects caused by Cr^{+2} or Cr^{+3} compounds. Most investigators agree that chromium in biological materials is probably always in the trivalent state, that greatest exposures of Cr^{+3} in the general human population are through the diet (but no adverse effects have been reported from such exposures), and that no organic trivalent chromium complexes of toxicological importance have been described. Studies with guinea pigs (*Cavia* spp.) fed Cr^{+3} for 21 weeks at concentrations up to 50.0 mg/kg dietary Cr^{+3} showed no adverse effects. Domestic cats were apparently unaffected after exposure to aerosol levels of 80.0–115.0 mg Cr^{+3}/m^3 for 1 h daily for 4 months, or after consuming diets with high amounts of chromic (Cr^{+3}) salts over a similar period. When chromium was administered by injection, trivalent salts were substantially less toxic than hexavalent salts in producing effects in embryos of golden hamsters. A similar pattern was evident in mice and in embryos of chickens. The LD50s for mice were 260.0 mg/kg BW for Cr^{+3}, but only 5.0 mg/kg BW for Cr^{+6}. In rats (*Rattus* sp.), route of exposure and compound tested were important. Rats given a single dermal dose of various hexavalent chromium compounds had LD50 values that ranged from 400.0 mg Cr/kg BW for sodium dichromate to 680.0 mg Cr/kg BW for ammonium dichromate; potassium dichromate and sodium chromate were intermediate in toxicity. When the same four hexavalent chromium compounds were administered via inhalation, LD50s after 4 h of exposure ranged between 33.0 and 82.0 mg Cr/m^3; females were more sensitive than males.

Studies with Cr^{+6} and dogs showed that 100.0 mg/kg in food for 3 months was fatal, that 11.2 mg/L in drinking water was not lethal over a 4-year period (although significant accumulation was observed), and that 6.0 mg/L in drinking water for 4 years had no measurable effects. In rats, 1000.0 mg/kg dietary Cr^{+6} represented the toxic threshold, but all animals survived 134.0 mg/L of Cr^{+6} in drinking water for 3 months. For most mammalian experimental animals, including mice (*Mus* sp.), dogs (*Canis familiaris*), rabbits

Table 9.1. Maximum acceptable toxicant concentration (MATC) values for hexavalent and trivalent chromium to aquatic life based on life cycle or partial life cycle exposures.

Chemical Species, Ecosystem, Organism	MATC[a] (μg/L, ppb)
HEXAVALENT CHROMIUM: FRESHWATER	
Rainbow trout, *Oncorhynchus mykiss*; water hardness, 34 mg $CaCO_3$/L vs. 45 mg $CaCO_3$/L	50.0–110.0 vs. 200.0–350.0
Fish, 7 species, embryo–larval stages in soft water	73.0–2167.0
Lake trout, *Salvelinus namaycush*	110.0–194.0
Channel catfish, *Ictalurus punctatus*	150.0–310.0
Brook trout, *Salvelinus fontinalis*	200.0–350.0
White sucker, *Catostomus commersoni*	290.0–538.0
Bluegill, *Lepomis macrochirus*	522.0–1122.0
Fathead minnow, *Pimephales promelas*	1000.0–3950.0
Northern pike, *Esox lucius*	538.0–963.0
Walleye, *Stizostedion vitreum*	>2161.0
HEXAVALENT CHROMIUM: SALTWATER	
Polychaete worm, *Neanthes arenaceodentata*	17.0–38.0
Mysid shrimp, *Mysidopsis bahia*	88.0–198.0
TRIVALENT CHROMIUM: FRESHWATER	
Rainbow trout	30.0–157.0
Cladoceran, *Daphnia magna*	47.0–93.0
Fathead minnow	750.0–1400.0

[a]Lower value in each pair represents the highest concentration tested with no-observed-adverse effect on survival, growth, poor reproduction; higher value is the lowest concentration tested producing a significant measurable adverse effect.

(*Oryctolagus* sp.), cats (*Felis domesticus*), and guinea pigs, the minimum injected fatal dose of Cr^{+6} ranged from 1.0 to 5.0 mg/kg BW, although doses of 0.2–0.5 mg/kg BW produced marked kidney damage. Repeated sublethal injections of Cr^{+6} did not promote tolerance in mice, but rather decreased the minimum lethal dose, suggesting that the animals were unable to develop tolerance to repeated chromium exposures. Investigators have not yet been able to identify a specific hexavalent chromium compound, or group of compounds, that could account for the most pronounced biological activity. A lethal oral dose of Cr^{+6} for a 14-year-old boy was estimated to be 10.0 mg/kg BW much lower than that tolerated by test animals on a repeated basis over a period of several months. A 44-year-old man who ingested 4.1 mg Cr^{+6}/kg BW as chromic acid died of severe gastrointestinal hemorrhage one month after ingestion.

Domestic chickens (*Gallus* sp.) appear to be more resistant than mammals. No adverse effects were observed in chickens exposed to 100.0 mg/kg dietary Cr^{+6} in a 32-day study, although embryolethal and teratogenic effects have been observed in the range of 0.2 mg/kg to 1.7–22.9 mg/kg, depending on the method of administration. For chicken embryos, the LD50 values were 22.9 mg/kg BW for Cr^{+3} and 1.7 mg/kg BW for Cr^{+6}.

9.6 Sublethal Effects

Under laboratory conditions, chromium is mutagenic, carcinogenic, and teratogenic to a wide variety of organisms, and Cr^{+6} has the greatest biological activity. However, information is lacking on the biological activities of water-soluble Cr^{+3} compounds, organochromium compounds, and their ionic states. Aquatic plants and marine polychaete worms appear to be the most sensitive groups tested. In exposures to Cr^{+6}, growth of algae was inhibited at 10.0 μg/L, and reproduction of worms at 12.5 μg/L. At higher concentrations, Cr^{+6} is associated with abnormal enzyme activities, altered blood chemistry, lowered resistance to pathogenic organisms, behavioral modifications, disrupted feeding, histopathology, osmoregulatory upset, alterations in population structure and species diversity indices, and inhibition of photosynthesis. Not all sublethal effects observed were permanent, but the potential for acclimatization of organisms to chromium is not well documented. The great variability among species and tissues in the accumulation or concentration of chromium is attributed partly to the route of administration, partly to the concentration of chromium and its chemical species, and partly to numerous biotic and physicochemical modifiers. High accumulations of chromium have been recorded among organisms from the lower trophic levels, but there is little evidence of biomagnification through food chains. Marine bivalve mollusks, for example, accumulated measurable concentrations at ambient water concentrations of 5.0 μg/L of Cr^{+6}, but the significance of chromium residues in mollusks and other organisms is not well understood. Depuration of accumulated chromium among organisms differs markedly, but usually follows a complex multicompartmental excretion pattern.

9.6.1 Aquatic Organisms: Freshwater

Sublethal effects of chromium salts are documented for selected species of freshwater microorganisms, plants, invertebrates, and fishes.

9.6.1.1 Bacteria

The role of sewage bacteria in chromium kinetics and cycling is unresolved and promises to be a fruitful field of research. Of 362 bacterial isolates from Cr^{+6} liquid sanitary sewage and chemical waste sludges, only 1–an isolate of *Arthrobacter* sp.—could tolerate 400.0 mg/L of Cr^{+6}; however, this isolate could not effectively accumulate chromium at comparatively low ambient levels of 5.0 mg/L of Cr^{+6}, whereas *Agrobacter* sp., another isolate, could. Hexavalent chromium in a wide array of forms showed dose-dependent responses for mutagenic activity in the bacterium *Salmonella typhimurium*; moreover, among 56 metal compounds tested, Cr^{+6} elicited the strongest mutagenic responses in *Bacillus subtilis*. In some tests, Cr^{+3} was genetically active, but only when present as a stable organic complex.

9.6.1.2 Algae and Macrophytes

Growth of freshwater algae was reduced at Cr^{+6} concentrations of 10.0 μg/L for *Chlamydomonas reinhardi*, 20.0 μg/L for *Chlorella pyrenoidosa*, and >45.0 μg/L for other species tested; effects were most pronounced in water of low alkalinity. Frond growth of the common duckweed, *Lemna minor*, the most sensitive aquatic plant tested, was reduced at 10.0 μg Cr^{+6}/L in 14 days. The green alga *Chlorella vulgaris* biomagnified Cr^{+6} from the medium about 1000 times in 28 days at ambient concentrations of 300.0 μg/L; growth was inhibited at 445.0 μg Cr^{+6}/L in 96 h, and adenosine triphosphate (ATP) production was reduced at 470.0 μg/L in 24 h. At 10.0 μg Cr^{+6}/L in the medium, bioconcentration factors (BCFs) for the chlorophytes *Hydrodictyon reticulatum* and *Oedogonium* sp. ranged from 200 to 600 in 14 days. Accumulation of chromium by living and dead plant tissue is extensive, uptake linearly approximating concentration on a logarithmic basis. Trivalent chromium is far less effective than Cr^{+6} in producing root weight inhibition in Eurasian water milfoil, *Myriophyllum spicatum*: 9900.0 μg Cr^{+3}/L vs. 1900.0 μg Cr^{+6}/L. A similar case is made for terrestrial plants, such as barley, wherein hexavalent chromium at 100.0 μg/L and trivalent chromium at 1000.0 μg/L produced comparable growth inhibition.

9.6.1.3 Invertebrates

Hexavalent chromium was associated with adverse effects in invertebrates of widely separated taxa: reduced survival and fecundity of the cladoceran, *Daphnia magna* at a concentration of 10.0 μg/L and exposure for 32 days; growth inhibition of the protozoan *Chilomonas paramecium* at 1100.0–3000.0 μg/L at temperatures of 10–30°C during exposures of 19–163 h; abnormal movement patterns of larvae of the midge *Chironomus tentans* at 100.0 μg/L in 48 h; and a temporary decrease in hemolymph glucose levels in the freshwater prawn *Macrobrachium lamarrei* surviving 1840.0 μg/L Cr^{+6} for 96 h.

Trivalent chromium was less effective than Cr^{+6} in reducing fecundity of *Daphnia magna*: 44.0 μg Cr^{+3}/L vs. 10.0 μg Cr^{+6}/L. Annelid worms (*Tubifex* sp.) accumulated about 1.0 mg total chromium/kg whole body during exposure for 2 weeks in sediments containing 175.0 μg Cr^{+3}/kg, suggesting that benthic invertebrates have only a limited ability to accumulate chromium from sediments or clays.

9.6.1.4 Fishes

Among sensitive species of freshwater teleosts, Cr^{+6} concentrations of 16.0–21.0 μg/L in the medium resulted in reduced growth of rainbow trout and Chinook salmon (*Oncorhynchus tshawytscha*) fingerlings during exposure of 14–16 weeks, and altered plasma cortisol metabolism in rainbow trout after 7 days. Rainbow trout avoid water containing 28.0 μg Cr^{+6}/L; however, avoidance thresholds increased linearly if pre-exposed to 800.0 μg Cr^{+6}/L for 7–20 weeks. Locomotor activity in bluegills (*Lepomis macrochirus*) increased after 2 weeks in 50.0 μg Cr^{+6}/L. The avoidance threshold for golden shiner (*Notemigonus crysoleucas*) is 73.0 μg Cr^{+6}/L. Exposure of cichlids (*Tilapia sparrmanii*) to 98.0 μg Cr^{+6}/L for 96 h led to clotting defects that caused internal bleeding with effects exacerbated by increasing water pH in the range of 4–9. Long-term exposure of rainbow trout for 180 days to high, but environmentally realistic, concentrations of 200.0 μg Cr^{+6}/L resulted in elevated levels of chromium in kidney (3.5 mg/kg FW), liver (2.0), and muscle (0.6); after 90 days in chromium-free media, chromium levels were 1.6, 1.3, and 0.5 mg/kg FW, respectively. Time required to reach median asymptotic uptake ranged from 36 to 55 days for various tissues; extrapolated values for almost complete equilibrium ranged from 237 to 365 days. In seaward migrating coho salmon (*Oncorhynchus kisutch*), salinity tolerance and serum osmolality were impaired during exposure to 230.0 μg Cr^{+6}/L for 4 weeks. In juvenile coho salmon, disease resistance and serum agglutinin production both decreased after 2 weeks in water containing 500.0 μg/L.

At high environmental concentrations of Cr^{+6} (i.e., 2.0 mg/L in water) and at alkaline pH, concentrations in rainbow trout tissues were greatest in gill, liver, kidney, and digestive tract; after transfer of the fish to chromium-free media, residues tended to remain high in kidney and liver; concentration in gill tissues tended to be greater at pH 7.8 than at pH 6.5. Studies with perfused gills showed that the transfer of chromium was directly coupled with the transfer of oxygen from the external solution to the internal perfusion medium and that this transfer was significantly more rapid at pH 6.5 than at alkaline pH. Uptake rate of Cr^{+6} was rapid, equilibrium usually being reached in 2–4 days of exposure for various tissues, except for gill, which continued to accumulate chromium with increasing exposure at acidic pH. In rainbow trout, the excretion pattern was biphasic. The biological half-life of the short-lived component (34% of the total chromium) was about 1 day, and that of the long-lived component about 26 days. Exposure of isolated intestine tissues of rainbow trout to chromium leads to an initial dose-dependent inhibition of alkaline phosphatase activity; this enzyme is sensitive to chromium ion, especially Cr^{+6}.

Various effects are reported in freshwater teleosts following exposure to comparatively high sublethal concentrations of hexavalent chromium. In the snakehead (*Channa punctatus*), enzyme activities were altered in a wide variety of organs and tissues after exposure for 30 days to 2.6 mg/L; the effects became life threatening after exposure for 120 days. Gill histopathology was documented in the air-breathing catfish (*Saccobranchus fossilis*) after immersion in 5.6 mg Cr^{+6}/L for 7 days. Growth rate of larvae of the fathead minnow (*Pimephales promelas*) was reduced at 6.0 mg/L during exposure for 7 days. The rudd (*Scardinus erythrophthalmus*), exposed to Cr^{+6} for 24 h, did not accumulate detectable levels of chromium in tissues during exposure to 16.0 mg/L, but did during exposures to 20.0 mg/L; the kidney contained the highest residues – 10.3 mg Cr/kg FW. Climbing perch (*Anabas scandens*) exposed for 30 days to 25.0 mg Cr^{+6}/L or 25.0 mg Cr^{+3}/L showed depletion of glycogen and glucose reserves of liver and kidney, and decreased activity of respiratory enzymes (various dehydrogenases) and ATPases; in all cases, Cr^{+6} produced the greater effect. In the mudskipper (*Boleophthalmus dussumieri*), chromosomal aberrations in the gill increased after injection of 1.0 mg/kg BW, or exposure to 24.0 mg/L in the medium for 24 h. Mudskippers exposed to 30.0–60.0 mg Cr^{+6}/L for 72 h had a dose-duration inhibition in specific activity of brain and muscle Na+, K+-ATPase and other ATPases.

Chromium is considered to be a cofactor for insulin activity and part of an organic glucose tolerance factor. In common carp (*Cyprinus carpio*) fed low protein diets, the addition of chromium chloride salts in the diet (equivalent to 60.0 μg Cr^{+3}/kg BW) or injected intraperitoneally (2.5 μg/kg BW), significantly decreased plasma glucose levels. Trivalent chromium supplementation in the diet improved glucose utilization in carp and tilapia when fed diets for 10 weeks containing 2.0 mg Cr/kg ration equivalent to 5% BW daily. Chromic oxide was more effective than chromium chloride, and both were more effective than sodium chromate tetrahydrate. Chromic oxide fed in the diet for 12 weeks at 0.5 and 2% to tilapia fingerlings improved glucose utilization and nutrient digestibility; the 0.5% diet was more efficient than the 2% diet.

Chromium uptakes and effects in teleosts were modified significantly by many biological and abiotic variables, including water temperature and pH, the presence of other contaminants or compounds, and sex and tissue specificity. In rainbow trout, only males showed significant changes in liver enzyme activity during exposure to 200.0 μg Cr^{+6}/L for 6 months; the effects were intensified by the presence of nickel and cadmium salts in solution. Rainbow trout are able to regulate chromium somewhat, either actively by reduced absorption or increased excretion, or passively by the limitation of binding sites for chromium in vivo. Tests with goldfish (*Carassius auratus*) and high Cr^{+6} concentrations indicated that lethal and sublethal effects were more pronounced at comparatively high water temperatures and reduced pH; further, chromium residue levels were abnormally high in dead or moribund fish, suggesting that residue values from dead or dying fish should be interpreted with extreme caution. In rainbow trout, acute chromium poisoning caused morphological changes in gills, kidney, and stomach tissues at pH 7.8, but only in the gills at pH 6.5. Chromium uptake in trout increased when 10.0 μg/L of ionic cadmium was present in solution – again demonstrating that uptake patterns are not necessarily predictable for single components in complex mixtures.

9.6.2 Aquatic Organisms: Marine

Sublethal effects of chromium are reported for representative species of marine organisms, including algae and higher plants, mollusks, nematodes, crustaceans, annelids, echinoderms, fishes, birds, and mammals.

9.6.2.1 Algae and Macrophytes

Algae and higher plants accumulated chromium from seawater by factors up to 8600, and from solutions containing 50 mg/L of chromium by a factor of 18 in 48 h. Algae also accumulated chromium from sewage sludge, showing increases in chromium burdens of 25.0–60.0 mg/kg DW. The unusually high chromium concentrations observed in some species of algae and macrophytes from Narragansett Bay, Rhode Island and from Puerto Rico almost certainly came from chromium wastes discharged from electroplaters (in Narragansett Bay) and from other anthropogenic sources (in Puerto Rico). A similar situation probably exists wherever grossly elevated chromium levels are observed.

Although chromium is abundant in primary producers, there is little evidence of biomagnification through marine food chains consisting of herbivores and carnivores. Successful transfer of assimilated and unassimilated radiochromium through an experimental food chain that included phytoplankton, brine shrimp, postlarval fish, and adult fish; however, concentrations declined after each transfer. Comparisons of the results from the food chain with laboratory studies on chromium uptake from seawater suggest that the food chain, despite the successive declines, was generally the more efficient pathway for uptake of chromium by all trophic levels.

Among sensitive species of marine algae, concentrations of 10.0 μg/L of Cr^{+6} partly inhibited growth of *Olisthodiscus lutens*. All cultures, including those in which growth was inhibited, contained viable, active (>75%) cells at the end of 10 days. Inhibitory effects were reversed by chelators such as EDTA, suggesting that naturally occurring ligands and sequestering agents in seawater may alleviate the toxicity of Cr^{+6}, and perhaps other metals. In the giant kelp (*Macrocystis pyrifera*), photosynthesis was inhibited by 20% in 5 days at 1000.0 μg/L of Cr^{+6}, and 50% in 4 days at 5000.0 μg/L; this kelp appears to be one of the more resistant aquatic plants.

9.6.2.2 Mollusks

Edible tissues of commercially important North American mollusks contained 0.1–0.6 mg Cr/kg FW. Although this concentration is in general agreement with molluscan data from other geographic areas, reported values (milligram of chromium per kilogram fresh weight) in edible portions were as high as 3.4 in oysters (*Crassostrea virginica*), 5.8 in hardshell clams (*Mercenaria mercenaria*), and 5.0 in softshell clams (*Mya arenaria*). The ability of marine mollusks to accumulate chromium far in excess of that in ambient seawater is well documented; chromium concentration in 5 species of bivalves from Greek waters exceeded that in seawater by 16,000 times (*Pinna nobilis*) to 260,000 times (*Astralium rogosum*). No deleterious health effects have been reported among consumers of mollusks that contained occasional high chromium residues. Anthropogenic and natural chromium gradients in sediments or the water column were reflected in the wide range of values reported for this element in field collections of clams and mussels.

Two factors known to modify chromium accumulations in mollusks are the weight of the organism and the salinity of the medium. Concentrations of chromium in clams were reported to decrease with increasing body weight and increasing salinity. Accumulation of chromium by oysters (*Crassostrea gigas*) was independent of sediment chromium levels and dependent on organism size suggesting some homeostatic regulation of this metal. In a 20-week laboratory study of chromium accumulation rates by oysters (*C. virginica*), the oysters were continuously subjected to seawater solutions containing 50.0 or 100.0 μg Cr^{+6}/L. After 5, 10, or 20 weeks in 50.0 μg/L, maximum whole body concentrations (milligram of chromium per kilogram fresh weight) were 2.4, 3.7, and 6.3, respectively (up from control values of <0.12); in 100.0 μg/L, the values were 4.4 (5 weeks), 6.4 (10 weeks), and 11.5 (20 weeks). It was concluded that *C. virginica*, under laboratory conditions, accumulated chromium more readily by direct absorption from the medium than from ingestion of radiochromium-labeled algae (*Chlamydomonas* spp.). In natural environments, however, chromium concentration is likely to be greater in the food supply than in the water. As a consequence, food might be the primary source of chromium to oysters, even though accumulation occurs more readily by direct absorption.

Clams, oysters, and mussels accumulate chromium from the medium or from contaminated sediments at comparatively low concentrations. For example, oysters subjected to 5.0 μg Cr^{+6}/L for 12 weeks contained 3.1 mg Cr/kg DW in soft parts and retained 52% of the accumulated chromium after they were transferred to chromium-free seawater for 28 weeks. Mussels (*Mytilus edulis*) subjected to the same dose–time regimen contained 4.8 mg/kg, but retained only 39% after 28 weeks of depuration. Both oysters and mussels contained higher residues after exposure to 10.0 μg Cr^{+6}/L for 12 weeks 5.6 and 9.4 mg Cr/kg DW in soft parts, respectively, and both contained substantial (30–58%) residues after 28 weeks in a chromium-free environment. In studies with mussels and softshell clams (*Mya arenaria*), it was demonstrated that chromium in New Hampshire sediments (contaminated with Cr^{+3} from tannery wastes) was bioavailable to clams by diffusion from seawater, and that both diffusion and particulate uptake were important pathways for mussels. Accumulation was observed at sediment chromium concentrations as low as 150.0 mg/kg. Kaolinite sediments containing up to 1200.0 mg Cr^{+3}/kg produced the most pronounced adverse effects on filtration rates and ciliary activity of bivalve mollusks, leading the authors to conclude that chromium that has accumulated in areas affected by industrial wastes might have serious consequences to filter feeding bivalves.

It is emphasized that Cr^{+3}, probably because of its very low solubility in seawater, appears to have a much lower bioavailability to most groups of marine animals than Cr^{+6}, which is more water soluble. The clam *Rangia cuneata* appears to be an exception: it accumulated up to 19 mg Cr/kg in soft parts, on a dry weight basis, during exposure for 16 days to chromium-contaminated muds, and retained most of it for an extended period; the estimated biological half-time was 11 days. In general, benthic invertebrates rarely accumulate chromium from contaminated sediments (82.0–188.0 mg Cr^{+3}/kg); only a few examples have been recorded.

9.6.2.3 Nematodes

Representatives of this phylum have been used extensively as indicators of stressed environments. Population structure and species diversity of free-living nematodes inhabiting sediments in the New York Bight were moderately influenced by the heavy metal content of sands. In medium-grained sands, species diversity was inversely correlated with increased concentrations of chromium and other metals. Sands containing 3.0–21.5 mg Cr/kg were also marked by high relative abundances of one or two nematode species; the tolerance of these species to chromium stress probably exceeded that of the normal nematode inhabitants of such sediments.

9.6.2.4 Crustaceans

In general, chromium seldom exceeds 0.3 mg/kg FW in edible crustacean tissues. The highest value (0.6 mg Cr/kg FW) reported in muscle of rock crab (*Cancer irroratus*) was from specimens collected near an ocean dumpsite receiving large quantities of metals. Digestive glands and gills from these crabs also contained the highest chromium residues for these tissues in crustaceans.

Uptake and loss of radiochromium by the crab *Podophthalmus vigil* was independent of sex and eyestalk hormone influences. Most of the radiochromium accumulated in gills. Equilibrium was reached in gill and muscle in 2–3 days, but in midgut and hemolymph in 4–5 days. Iron interfered with chromium uptake and retention. Chromium concentrates in setae, gills, and hepatopancreas of Dungeness crab (*Cancer magister*), suggesting that surface adsorption and physiological processes were both instrumental in chromium accumulation. Barnacles (*Balanus* sp.) incorporate Cr^{+6} in soft tissues up to 1000 times over ambient concentrations, reaching equilibrium in 7 days (biological half-life for some components was 120 days); however, Cr^{+3}, which precipitates in seawater, was quickly removed by filtering activity, was not concentrated in soft tissues, and was rapidly excreted by way of the digestive system.

Sediment chromium concentrations of 3200.0 mg/kg in the New Bedford (Massachusetts) Acushnet estuary, and 100.0 mg/kg in the New York Bight have been recorded. Massive cuticular lesions suggestive of shell disease characterized up to 30% of the lobsters, crabs, and shrimp collected from the New York Bight, and these lesions could also be induced in crustaceans exposed to New York Bight sediments in the laboratory. This shell disease syndrome has been induced in 41% of grass shrimp (*Palaemonetes pugio*) during exposure to 0.5 mg Cr^{+6}/L for 28 days. It is proposed that chromium interferes with the normal functions of subcuticular epithelium, particularly cuticle formation, and subsequently causes structural weaknesses or perforations to develop in the cuticle of newly molted shrimp. Because of these chromium-induced exoskeletal deficiencies, a viaduct for pathogenic bacteria and direct chromium influx is formed that perpetuates the development of the lesion.

Of the 65,000 tons of chromium compounds used annually in exploratory oil drilling, a significant portion enters the marine environment through the discharge of used drilling muds. More than 225 tons of drilling mud may be used in a single 3000-m well. One of the most frequently used muds in offshore drilling operations is a chrome lignosulfonate mud containing barium sulfonate, bentonite clay, and ferrochrome or chrome lignosulfonates. The bioavailability of chromium to grass shrimp from used chrome lignosulfonate drilling muds is most pronounced at the mud aqueous layers. At chromium concentrations of 248.0 μg/L in the mud aqueous fraction, grass shrimp accumulated 23.7 mg Cr/kg DW whole body after 7 days. Concentrations of drilling mud of 1% or greater in seawater were toxic to sensitive species of crustaceans; uptake of 4.0–5.0 mg/kg was reported in grass shrimp exposed to sediments containing 188.0 mg Cr/kg. The toxicity of chromium-contaminated drilling muds to grass shrimp may sometimes be attributable to large residuals of petroleum hydrocarbons in the sediments.

9.6.2.5 Annelids

Uptake and excretion studies of Cr^{+3} by *Hermione hystrix* show that Cr^{+3} was not readily accumulated from seawater, owing to the formation of particles and surface adsorption phenomena; furthermore, little accumulation was evident on contact with contaminated sediments. Hexavalent chromium in the medium was readily accumulated by *Hermione*; the process was slow and only small amounts were taken up in 19 days i.e., 0.03–0.10 mg/kg FW from media containing 3.0–10.0 μg Cr^{+6}/L. Higher body burdens of 0.5–1.8 mg Cr/kg fresh BW were reported at 100.0–500.0 μg Cr^{+6}/L, but some deaths were noted at these concentrations. Chromium accumulation by *Hermione* is a passive process and directly related to Cr^{+6} concentration in the medium. At least two rates of biological loss are involved, one of 8 days and another of 123 days. Most of the chromium accumulated by *Hermione* from long exposure is bound in a body component having a slow turnover rate and an estimated biological half-life of about 123 days.

Uptake of Cr^{+6} from seawater has been reported for *Neanthes arenaceodentata*. Whole *Neanthes* contained 30.0 mg Cr/kg DW after exposure for 150 days in 30.0 μg Cr^{+6}/L and 0.5–1.6 mg Cr/kg FW after exposure for 440 days; both of these observations were similar to those of other annelid species, after adjustment for wet and dry weights. Concentrations as low as 12.5 μg Cr^{+6}/L decreased brood size in *Neanthes*, although no significant body residues were evident. Uptake of Cr^{+6} by *Neanthes* was related to dose at low ambient chromium concentrations. Worms subjected to 2.6, 4.5, 9.8, or 16.6 μg Cr^{+6}/L for 309 days contained 0.5, 0.7, 2.2, and 2.5 mg Cr/kg whole fresh organism, respectively. There was no direct relationship between tissue concentration and brood size, suggesting that chromium in *Neanthes* attaches to proteins in the body wall, gut, and parapodial regions.

Neanthes arenaceodentata is the most sensitive marine organism yet tested. In worms exposed to sublethal concentrations of Cr^{+6}, feeding was disrupted after 14 days at 79.0 μg/L, reproduction ceased after 440 days (three generations) at 100.0 μg/L, brood size was reduced after 309–440 days at 12.5–16.0 μg/L, and abnormalities in larval development increased after 5 months at 25.0 μg/L. On the other hand, exposure for 293 days (two generations) in 50,400.0 μg Cr^{+3}/L caused no adverse effects on survival, maturation time required for spawning, or brood size. The polychaete *Capitella capitata* was more resistant than *Neanthes*; a decrease in brood size was noted only after exposure for 5 months to 50.0 and 100.0 μg Cr^{+6}/L.

9.6.2.6 Echinoderms

With the exception of two sea urchin samples collected from Puerto Rico, most chromium residues reported in echinoderms have been less than 1.0 mg/kg DW. The elevated levels of 24.0 and 43.0 mg/kg FW of whole organism in Puerto Rican sea urchins are exceptions, which were not reflected in sea cucumber muscle from the same vicinity, and thus should be viewed with caution. Echinoderms from the United Kingdom and environments were comparatively low in chromium; concentrations were less than 0.46 mg Cr/kg DW whole organism. Embryos of a sea urchin (*Anthocidaris* sp.) developed normally in solutions containing 3.2–4.2 mg Cr/L, but failed to develop at 8.4–10.0 mg Cr/L. Larvae of another species of sea urchin (*Hemicentrotus* sp.) were more sensitive, showing abnormal development or dying within 24 h at concentrations of <1.0 mg Cr/L. Hexavalent chromium at 6.0 mg/L was associated with abnormal development in embryos of *Anthocidaris crassispina*.

9.6.2.7 Fishes

Trivalent chromium is relatively innocuous to the gray mullet (*Chelon labrosus*). Mullet held for 60 days in aquaria with sediments containing 46.0 mg Cr^{+3}/kg DW and fed diets containing 4.4–13.8 mg Cr^{+3}/kg DW ration had normal growth, survival, macroscopic physiology, and behavior; however, when compared to controls (6.4 mg Cr/kg sediments, 0.3–0.9 mg/kg DW ration), liver concentrations were elevated: 34.0 mg/kg DW vs. 2.0 mg/kg DW. Trivalent chromium, as chromic oxide, in the diet has been used as an indigestible marker to measure nutrient digestibility in American lobsters and freshwater-reared Arctic charr (*Salvelinus alpinus*). The addition of 1% trivalent chromium to the diet of seawater-reared Arctic charr results in altered intestinal microflora and increased lipids in the feces, and suggests that Cr^{+3} use for this purpose be discontinued.

Individual tissues of most species of marine finfishes contained between 0.1 and 0.6 mg Cr/kg FW. For still unexplained reasons, chromium concentrated in the scales of some species collected in Greek waters with values ranging up to 97.0 mg Cr/kg DW. Chromium concentrations also vary significantly among different species of fish collected from the same geographic area. For example, muscle chromium concentration was 1430 times greater in a porgy (*Pachymetopan grande*) than in a goosefish (*Lophius piscatorius*) from the same collection.

Accumulation of chromium under controlled conditions has been documented for speckled sanddab, *Citharichthys stigmaeus*, and Atlantic croaker, *Micropogon undulatus*. Sanddabs held in seawater solutions containing 3.0–5.0 mg Cr^{+6}/L contained up to 100.0 mg Cr/kg intestine DW, 10.0 in liver, and 3.0 in muscle. Sanddabs accumulated significant concentrations of chromium in various tissues during long-term exposure in seawater concentrations as low as 16.0 μg Cr^{+6}/L. Retention of radiochromium-51 in croakers following a single intraperitoneal injection was expressed as two exponential rate functions: 70 days for the long-lived component and 20 days for the short-lived component.

9.6.3 Birds

Male domestic chickens fed diets containing up to 100.0 mg Cr^{+6}/kg ration for 32 days showed no adverse effects in survival, growth, or food utilization efficiency; however, teratogenic effects were documented in chicken embryos after eggs had been injected with Cr^{+6}. Deformities included short and twisted limbs, microphthalmia, exencephaly, everted viscera, growth stunting, and parrot beaks. The highest incidence of teratogenic effects was observed at Cr^{+6} concentrations that caused some deaths, and when the administration route was through the chorioallantoic membrane as opposed to the yolk; no teratogenic effects were observed with Cr^{+3} salts.

Young American black ducks (*Anas rubripes*) absorbed anionic chromium species more readily than cationic forms from the intestines, strongly indicating that ionic chromium state should be considered when avian dietary toxicity studies are being planned. In another study with black ducks, adults were fed diets containing 0.0, 20.0, or 100.0 mg/kg anionic Cr^{+3} and ducklings from these pairs were fed the same diets for 7 days; tests of avoidance responses of the ducklings to a fright stimulus showed that the chromium had no significant effect on their behavior.

9.6.4 Mammals

Chromium is causally associated with mutations and malignancy. Under appropriate conditions, chromium is a human and animal carcinogenic agent; its biological effects depend on chemical form, solubility, and valence. In general, nearly all Cr^{+6} compounds are potent mammalian mutagens and all forms of Cr^{+6} – both water-soluble and water-insoluble compounds – are respiratory carcinogens in humans; metallic chromium and Cr^{+3} are essentially nontoxic. However, exposure to water-solubilized Cr^{+3} has caused cancers and dermatitis in workers, and toxicity in rabbits. In the chromate-producing industry, workers who developed respiratory cancer had been exposed to 30.0–1100.0 μg/m^3 chromium in air for periods of 4–24 years, and workers producing chromate pigment who developed respiratory cancer had been subjected to an estimated Cr^{+6} exposure of 500.0–1500.0 μg/m^3 for 6–9 years. Carcinogens released in the chromate-manufacturing process have not yet been identified. Levels as low as 10.0 μg/m^3 of Cr^{+6} in air produced strong irritation in nasal membranes, even after short exposures. In some persons whose lower respiratory tissues became chromium sensitized, asthmatic attacks occurred at levels of Cr^{+6} as low as 2.5 μg/m^3. Cancer risks in occupational exposure have declined over the past 50 years due to improvements in the production process and industrial hygiene. There is no evidence of chromium sensitization in mammals other than humans. In the only animal study demonstrating a carcinogenic effect of an inhaled chromate, adenocarcinomas were reported in the bronchial tree of mice exposed throughout life to $CaCrO_4$ dust at 13.0 mg/m^3 (4330.0 μg Cr^{+6}/m^3) for 35 h weekly. Trivalent chromium compounds did not produce respiratory cancers. Interstitial fibrosis of the lungs was found in rats exposed via inhalation to 0.1 mg/m^3 of Cr^{+6} or Cr^{+3} for 22 h daily over 18 months. In rabbits, both Cr^{+3} and Cr^{+6}, given 1.7 mg/kg BW daily for 6 weeks, adversely affected blood and serum chemistry, and both produced significant morphological changes in liver; similar results were observed in rats. Although damage effects and residue accumulations were greater in rabbits treated with Cr^{+6}, water-soluble Cr^{+3} compounds may also have significant biological activity.

Hexavalent chromium compounds may cause skin ulceration, irritative dermatitis, ulcerations in mucous membranes, and perforations of the nasal septum. That inhalation of Cr^{+6} compounds may cause bronchial carcinomas has been well documented in humans. Inhalation exposure of mice to 1.8 mg Cr^{+6}/m^3 for 12 months, 2 days weekly, resulted in emphysema and nasal septum perforation. Skin lesions or ulcers were produced in guinea pigs when solutions containing 30,000.0 mg Cr^{+6}/L were applied to abraded skin or if the natural oils were removed from the skin beforehand; Cr^{+3} in concentrations as high as 100,000.0 mg/L had no ulcerogenic effects. Allergic guinea pigs developed dermatitis when exposed to solutions of either Cr^{+6} or Cr^{+3} at concentrations as low as 10.0 μg/L. In nonallergic animals, these effects were observed after repeated exposures to solutions containing 1000.0–3000.0 mg/L of Cr^{+3} or Cr^{+6} salts. Local sarcomas in muscle and local carcinomas of the skin have also been demonstrated in small laboratory animals exposed to Cr^{+6}. Kidney and liver lesions in rats were observed when the drinking water contained 134.0 mg Cr^{+6}/L for 2–3 months.

Hexavalent chromium has established its mutagenic activity in a wide array of screening tests, whereas insoluble Cr^{+3} forms appear to be inactive, in analogous evaluations, perhaps because Cr^{+3} absorption is poor in the systems analyzed. Studies with tissue cultures of ovary cells of the Chinese hamster showed that the addition of 52.0 μg Cr^{+6}/L not only induced sister chromatid exchanges but also inhibited cell proliferation; there was no measurable effect at 0.52 μg Cr^{+6}/L. Trivalent chromium at 520.0 μg/L did not measurably affect cell proliferation or chromatid exchanges. Genotoxic effects of Cr^{+6} are reversed by the addition of reducing agents or ascorbic acid. Chromosomal rearrangements and aberrations were recorded in rabbit cells after exposure to Cr^{+6}.

Chromium compounds, especially hexavalent chromium compounds, are associated with spermicidal, embryocidal, teratogenic, and other adverse effects on reproduction. Teratogenic effects induced by intravenous administration of 5.0 mg Cr^{+6}/kg BW to pregnant Syrian golden hamsters (*Mesocricetus auratus*) included cleft palates and defects in the ossification of the skeletal system. Pregnant hamsters that received an oral dose of 15.0 mg Cr/kg BW, as chromic trioxide, produced pups with a 64% malformation frequency. Pregnant mice given chromium-contaminated drinking water on gestation days 1–19 (equivalent to 57.0 mg Cr/kg BW daily) had increased fetal absorption and postimplantation loss, and males had decreased spermatogenesis after eating diets for 7 weeks that contained 3.5–4.6 mg Cr^{+6}/kg ration.

Chromium is the most common skin sensitizer known to human males. Up to 26% of all males tested and 10% of females were sensitive to potassium dichromate patch tests. The highest frequency of chromium-sensitive individuals was found in Brazil, Belgium, and North America, especially Detroit and New Orleans. Frequency of chromium dermatitis was highest in construction workers using cement. Other occupational exposures associated with chromium sensitivity include chromium plating, tanning of leather, application of anticorrosive agents, and printing. Oral ingestion of chromium compounds can sometimes lead to skin reactions in sensitive people. Hexavalent chromium compounds are more potent inducers and elicitors of skin sensitivity than trivalent chromium compounds, probably because Cr^{+6} compounds can penetrate the skin more readily than Cr^{+3} compounds.

Accumulations of chromium in tissues and organs depend heavily on its chemical form, route of entry, and amount administered. Trivalent chromium is the normal form of chromium in the mammalian diet. Trivalent chromium is localized primarily in blood, liver, spleen, kidney, and body soft tissues, with long-term storage in liver and spleen. Tissue accumulations were significant in dogs exposed to drinking water concentrations of 11.2 mg Cr/L; but were nil at 6.0 mg/L. Although both Cr^{+3} and Cr^{+6} accumulated in brain, kidney, and myocardium of rabbits, the accumulation of Cr^{+6} was highest in brain and that of Cr^{+3} in kidney; for both valence states there was no correlation between dose and concentration of stored chromium, or extent of tissue damage. Tissue residues in mice given 0.1 mg Cr^{+6}/kg in diet and water during lifetime exposure ranged from 0.1 mg Cr/kg FW in liver to 0.7 mg Cr/kg FW in heart; mice given 5.1 mg/kg diet for a similar period contained 0.5–1.8 mg Cr/kg FW in tissues, the residues being highest in the heart and spleen. Trivalent chromium was poorly absorbed from the intestinal tract of rats ($<1\%$ of an oral dose), whereas absorption of Cr^{+6} ranged from 3 to 6%. However, both Cr^{+3} and Cr^{+6} traverse placental barriers in mice when administered intravenously. All chemical forms of chromium, except chromates, cleared rapidly from the blood of rats. At dose levels of 60.0–250.0 μg/kg BW, Cr^{+6} tended to accumulate in the reticuloendothelial system, liver, spleen, and bone marrow; at the much lower doses of 10.0 and 1.0 μg/kg BW, major accumulation sites were bone marrow, spleen, testes, and epididymis. Female rats given a single intravenous injection of radiochromium-51 depurated the isotope primarily by urinary clearance, and secondarily by fecal and residual clearances over an 11-day period. Retained radiochromium-51 accumulated over time in bone, kidney, spleen, and liver. For multicompartmental excretion patterns recorded in rats, biological half-lives of the three components were estimated to be 0.5, 5.9, and 83.4 days; in mammals, chromium is excreted primarily in urine. At least three distinct Cr^{+6} excretion patterns exist in rats: blood has a single component, with a biological half-life of 13.9 days; testes, brain, kidney and lung have two components; and liver has three components with half-lives of 2.4 h, 52.8 h, and 15.7 days. Excretion patterns for Cr^{+3} in rats were unpredictable and difficult to calculate. The excretion patterns for fecal chromium among 40 grazing Angus cows given 20.0 g dietary Cr_2O_3 (13.6 g Cr^{+3}) daily for 72 days was diurnal; excretion was lowest at 8 p.m. and highest at 9 a.m.

9.7 Field Investigations

There is a wealth of data concerning the effects of chromium on living organisms under laboratory conditions simulating those encountered in the vicinity of high chromium discharges and accumulations typical of electroplating plants, tanneries, ocean dumping sites, and municipal waste outfalls. However, little research has been conducted under actual field conditions, except in three general fields: occupational exposures of humans in the chromate industry (discussed earlier), accidental poisoning of livestock resulting from oil-field activities, and chromium accumulations in ecosystems impacted by discharges associated with cooling waters or cooling towers.

All cases of accidental chromate poisoning in cattle have resulted from the exposure of animals to chromate compounds associated with oil-field activities. Chromates are used as a corrosion inhibitor between the pipe and casing and are often added to drilling fluids (in the form of chromelignosulfonate) to improve thermal stability. One recorded case involved 20 mature cows and their 8-month-old calves, grazing in a native pasture where an oil well had just been completed. One cow and calf died and another cow and calf became uncoordinated and thin, and the feces contained bloody mucus. The calf soon died. The cow aborted, but appeared to recover completely. Liver from the dead calf had 14.8 mg Cr/kg FW vs. 1.8 mg Cr/kg FW in controls; levels of arsenic and lead were not elevated. The cause of death was the consumption by the animals of concentrated sodium chromate found near the well site. In other cases, 2 of 80 heifers died after consuming concentrated zinc chromate, and 10 cows and one calf died after they had ingested ammonium chromate. In poisoned cows, chromium concentrations were 500.0 mg/kg in stomach contents, 15.8 mg/kg FW in kidney vs. 3.0 mg/kg in controls, and 1.1 mg/L in blood vs. 0.02 mg/L for controls.

Chromium is widely used as a corrosion inhibitor in cooling waters by the electric power industry. Its use in this capacity involves addition of a Cr^{+6} salt, typically sodium dichromate, which forms an oxide on metal surfaces. Chromates are subsequently released to surface waters in high concentrations, compared with background levels of chromium in most freshwaters. In White Oak Lake (eastern Tennessee), which received chronic inputs of chromates from cooling towers located on two tributary streams, typical Cr^{+6} concentrations of 3.0–10.0 mg/L in waste effluents produced 100.0–300.0 μg/L of Cr^{+6} in White Oak Lake vs. 5.0 μg/L in a control area. Concentrations of chromium in muscle of bluegills and largemouth bass from White Oak Lake did not differ significantly from those in fish from a control site suggesting that these species either effectively regulated chromium concentration or that the elevated chromium levels in White Oak Lake (where 20–73% of the total chromium was Cr^{+6}) were in a form that was unavailable for absorption into tissues. It is possible that chromium concentrations in fish are limited and that accumulation is independent of environmental concentration. This concept requires validation. Noteworthy is the observation that chromium concentrations were lower in muscle and body of older freshwater teleosts, an observation consistent with the trend of decreasing liver chromium with increasing age in marine teleosts. Cooling towers of uranium enrichment facilities and gaseous diffusion plants, similar to those of 1000-MW conventional steam electric stations, contain a chromate zinc-phosphate compound to inhibit corrosion and fouling within the cooling system. A small fraction of the cooling water, containing about 20.0 mg Cr^{+6}/L, becomes entrained within the exit air flow and is deposited as drift on the landscape, together with other salts found in the recirculating water system, such as sodium pentachlorophenate, chromated copper arsenate, and acid copper chromate. Effects of the chromium component on biological systems have been under investigation in Kentucky and Tennessee for many years. Analysis of vegetation along distance gradients from the cooling towers identified areas of significant drift deposition, accumulation, and magnitude of atmospheric transport over the landscape. At 13 m downwind from the point source, plant foliar concentrations of chromium were highest in winter at 1390.0 mg/kg DW, decreasing to 190.0 in spring, and to 173.0 in summer as the demand for cooling and hence the operation time of the facility decreased. Decreased accumulations on foliage probably reflected high mobility due to leaching, and the short life span of individual leaves. In contrast, chromium concentrations in plant litter at 13 m increased from 894.0 mg/kg DW during winter to 1890.0 mg/kg in summer and 2140.0 mg/kg in autumn. Accumulation of chromium in the litter was probably related to the higher surface-to-volume ratio in the litter biomass resulting from seasonal senescence of foliage. It is emphasized that no adverse biological effects were observed in native vegetation bearing high chromium residues. Concentrations in plants and litter decreased with increasing distance from the cooling towers: concentrations in foliage at 168, 530, and 923 m downwind were 157.0, 10.0, and 1.3 mg/kg DW, respectively; for litter, these values were 421.0, 24.0, and 5.8 mg/kg DW. Potted tobacco plants (*Nicotiana tabacum*) proved to be sensitive indicators of chromium contamination. Tobacco plants placed 15 m from the towers contained 30 times background levels after 1 week and up to 237.0 mg/kg DW in 5 weeks; in plants placed 200 m downwind, leaf growth was reduced by 75% after 7 weeks. Beetles and crickets collected near the towers contained 9.0–37.0 mg Cr/kg in gut contents (vs. 0.5–0.8 mg/kg for controls); however, assimilation rates were not measured. Cotton rats trapped in a fescue field adjacent to a large mechanical draft-cooling tower contained up to 10 times more chromium in hair, pelt, and bone than controls, but accumulations were negligible in viscera and other internal organs. Licking of the coat by rats appeared to be a primary route of chromium uptake – a likelihood confirmed experimentally. Feeding of radiochromium-51 to cotton rats demonstrated low assimilation (0.8%), and rapid initial loss of Cr^{+6} (99% in 1 day) suggesting that chromium is neither essential to cotton rats nor accumulated to any great extent through ingestion of drift-contaminated vegetation or inhalation of drift-contaminated air. Biological half-times of chromium assimilated by humans and cotton rats were similar: 616 and 693 days, respectively. The magnitude of the half-times suggests that chromium derived from a chromate has a high potential for biological interaction, but that fractional assimilation is very low thus reducing the likelihood of toxic effects.

Chromium does not appear to biomagnify in aquatic or terrestrial food chains. In virtually all studies, in both marine and freshwater environments involving birds and mammals, there was no biomagnification of chromium in the food web, but rather decreasing concentrations with increasing trophic level. Similar results were reported in freshwater and marine food webs involving invertebrates and fishes, and along the terrestrial food chain of soil–plant–animal.

Biomarkers that demonstrate chromium exposure under field conditions are under active investigation. Laboratory studies with Prussian carp (*Carassius auratus gibelio*) exposed for 3–9 days to 25.0–100.0 μg Cr^{+6}/L or 50.0–200.0 μg Cr^{+3}/L show a dose-dependent increase in the frequency of micronuclei in erythrocytes, and this increase is considered indicative of increasing DNA damage. Similar increases in micronuclei were observed in Prussian carp from the

River Ljubjanica near chromium-containing outfalls from leather waste products in the Republic of Slovenia.

9.8 Recommendations

Sensitive species of freshwater aquatic organisms showed reduced growth, inhibited reproduction, and increased bioaccumulation at about 10.0 μg/L and higher of Cr^{+6}, and other adverse effects at 30.0 μg/L and higher of Cr^{+3}. Among marine organisms, measurable accumulations were recorded in oysters and worms at 5.0 μg/L of Cr^{+6}, algal growth was reduced at 10.0 μg/L, and reproduction of polychaete annelid worms was inhibited at 12.5 μg/L; in all situations, Cr^{+3} was less damaging than Cr^{+6}. For mammals, 5.1 mg Cr^{+6}/kg dietary levels in food and water of mice were associated with elevated tissue residues. The significance of chromium residues is unclear, but available evidence suggests that organs and tissues of fish and wildlife that contain 4.0 mg total Cr/kg DW and higher should be viewed as presumptive evidence of chromium contamination. Aerosol concentrations in excess of 10.0 μg Cr^{+6}/m^3 are potentially harmful to human health; in the absence of supporting data, this value is recommended for protection of sensitive species of wildlife, especially migratory waterfowl. More research is recommended on carcinogenic and mutagenic properties of chromium on fish, and on algal and bacterial physiology.

Occupational exposure to hexavalent chromium compounds should not exceed 1.0 μg/m³ air for a 10-h workday and 40-h workweek because all hexavalent chromium compounds are potential carcinogens. Other recommendations include more research on: (1) the nature of speciation of chromium in soil and water, (2) bioavailability of chromium compounds from environmental media, (3) food chain biomagnification, (4) release data of chromium from anthropogenic sources to the biosphere, (5) reliable and more recent monitoring data of chromium in air, water, and food, with emphasis on chromium levels in tissues and body fluids of animals living near hazardous waste sites, and (6) methods for determining biomarkers of exposure and effect, for determining parent compounds and degradation products in environmental media.

Proposed criteria for the protection of various environmental compartments against chromium are numerous, disparate, and often contradictory (Table 9.2). Some of this confusion may be attributable to the general lack of confidence in analyses of chromium residues conducted some years ago, and some to the continued inability to quantify chemical species and ionic states of chromium. Uncertainties about the metabolic role of organochromium compounds, water-soluble Cr^{+3} species, and their interactions with other components in complex and potentially toxic mixtures, further confound the issue. The essentiality of chromium to some, but not all, species of mammals is recognized, but comparable data for other groups of organisms are missing. Finally, the wide range in sensitivities and accumulation rates documented between different taxonomic groups, and even among closely related species, to Cr^{+3} and Cr^{+6} salts merits elucidation.

9.9 Summary

Most authorities agree on eight points: (1) chromium levels are elevated in soil, air, water, and biota in the vicinity of electroplating and metal-finishing industries, publicly owned municipal treatment plants, tanneries, oil drilling operations, and cooling towers; (2) hexavalent chromium (Cr^{+6}) is the most biologically active chromium chemical species, although little is known about the properties of organochromium compounds, water-soluble species, or their interactions in complex mixtures; (3) chromium chemistry is imperfectly understood, and existing analytical methodologies are inadequate for quantification of chromium species and ionic states; (4) trivalent chromium (Cr^{+3}) is an essential trace element in humans and some species of laboratory animals, but the database is incomplete for other groups of organisms; (5) at high environmental concentrations, chromium is

Table 9.2. Proposed chromium criteria for the protection of human health and natural resources.

Resource and Criterion (Units in Parentheses)	Effective Chromium Concentration
HUMAN HEALTH	
Air, acceptable (μg total chromium/m^3)	
Arizona	11.0, 1-h average; 3.8, 24-h average
California, Maryland, Maine	0.00 at any time
Kansas, North Carolina	0.0000833, annual average
Rhode Island	0.00009, annual average
Montana	0.07, annual average
Texas	0.1, annual average
New York	0.167, annual average
Virginia	0.5–8.3, 24-h average
North Dakota	5.0, 8-h average
USA	<50.0 total chromium
Ceiling air, occupational exposure (mg total chromium/m^3)	
Chromic acid and hexavalent chromium compounds	0.05–<0.1
Water-soluble divalent and trivalent chromium compounds	<0.5
Metallic chromium and insoluble salts	0.5–<1.0
All hexavalent chromium compounds; 10-h daily, 40-h workweek; recommended	<0.001
Diet; normal dietary intake	50.0–200.0 μg Cr^{+3} daily, equivalent to 0.7–2.9 μg/kg BW daily for a 70-kg adult; 30.0–100.0 μg total chromium daily
Drinking water (μg/L)	
USA	<50.0 Cr^{+6}; <170,000.0 Cr^{+3}; <100.0 total chromium
Children	<240.0 total chromium for >10 days; <1400.0 for <10 days
Adults	<120.0 total chromium for lifetime exposure; <840.0 for >10 days
California, Florida, Brazil	< 50.0 total chromium
Colorado	<50.0 Cr^{+6}; <50.0 Cr^{+3}
Europe	<50.0 Cr^{+6}
Former Soviet Union	<600.0 total chromium
Tissue residues (μg total chromium/kg fresh weight (FW))	
Safe, soft tissues	<30.0
Normal	
Hair	<234.0
Lung	<204.0 (29.0–898.0)
Milk	<0.3 (0.06–1.6)

Continued

Table 9.2. cont'd

Resource and Criterion (Units in Parentheses)	Effective Chromium Concentration
Nail	<520.0
Serum	0.06 (0.01–0.17)
FRESHWATER AQUATIC LIFE PROTECTION; WATER (μg/L)	
USA	<0.29 Cr^{+6} as 24-h average; not to exceed 21.0 Cr^{+6} at any time
USA; water hardness in mg $CaCO_3$/L	
50	<2200.0 Cr^{+3} at any time
100	<4700.0 Cr^{+3} at any time
200	<9900.0 Cr^{+3} at any time
Colorado	<25.0 Cr^{+6}; <100.0 Cr^{+3}
Florida	
Effluent discharges	<500.0 Cr^{+6}; <1000.0 total chromium
Recovery waters	<50.0 total chromium
Indiana, most waters	Not to exceed 0.1 times the 96-h LC50 of aquatic species
Lake Michigan	<50.0 total chromium
Canada	<10.0 Cr^{+6}
MARINE AQUATIC LIFE PROTECTION; WATER (μg/L)	
USA	<18.0 Cr^{+6} as 24-h average; not to exceed 1260.0 Cr^{+6} at any time. Insufficient database for Cr^{+3} at this time, but presumably less stringent than Cr^{+6}
California	<2.0 total chromium, 6 month median; <8.0 total chromium, daily maximum; <20.0 total chromium, instantaneous mix
California; waste discharges into marine waters	<5.0 total chromium for 50% of measurements; <10.0 for 10% of measurements
GREAT LAKES BENTHOS; SEDIMENTS (mg TOTAL CHROMIUM/kg DW)	
Nonpolluted	<25.0
Moderately polluted	25.0–75.0
Heavily polluted	>75.0
AGRICULTURAL CROPS	
Sewage sludge (kg/ha)	
Missouri; maximum addition when soil cation exchange capacity (in meq/100 g) is:	
<5	560.0
5–15	1120.0
>15	2250.0

Table 9.2. cont'd

Resource and Criterion (Units in Parentheses)	Effective Chromium Concentration
New York, acceptable	336.0–500.0
Vermont	
Sandy loam	<140.0
Silt loam	<280.0
Clay loam	<560.0
Soils (mg total chromium/kg DW)	
Canada	
Cleanup indicated	>120.0
Nonagricultural use	<1000.0
Acidic soils (pH <6.5); Alberta, acceptable	<600.0
The Netherlands	
Acceptable	<100.0
Moderately contaminated	250.0–800.0
Requires cleanup	>800.0
USA, New Jersey	<100.0
Water (μg/L)	
Irrigation, Colorado	<100.0 of Cr^{+6}; <100.0 of Cr^{+3}
Groundwater, Florida	<50.0 total chromium
RUMINANT MAMMALS (mg TOTAL CHROMIUM/kg FW)	
Indicative of chromium intoxication	
Blood	>1.0
Kidney, liver	>15.0
Acutely toxic and perhaps fatal	
Blood	>4.0
Diet	>500.0
Liver	>30.0
WILDLIFE (mg TOTAL CHROMIUM/kg DW)	
Diet; potential adverse effects on health and reproduction	>10.0
Tissues	
Normal, depending on species and tissue	0.1–15.0 (up to 100 times higher in chromium-contaminated environments)
Probable exposure to chromium	>4.0

a mutagen, teratogen, and carcinogen; (6) no biomagnification of chromium has been observed in food chains, and concentrations are usually highest at the lowest trophic levels; (7) toxic and sublethal properties of chromium are modified by a variety of biological and abiotic factors; and (8) sensitivity to chromium varies widely, even among closely related species. Adverse effects of chromium to sensitive species occur at 10.0 μg/L (ppb) of

Cr^{+6} and 30.0 μg/L of Cr^{+3} in freshwater and 5.0 μg/L of Cr^{+6} in saltwater and, to wildlife, 10.0 mg of Cr^{+6}/kg of diet (ppm). Tissue levels in excess of 4.0 mg total Cr/kg DW should be viewed as presumptive evidence of chromium contamination, although the significance of tissue chromium residues is unclear. Some of these findings are in sharp contrast to chromium criteria proposed by regulatory agencies.

COPPER[a]

Chapter 10

10.1 Introduction

Copper (Cu) is plentiful in the environment and essential for normal growth and metabolism of all living organisms. Abnormal levels of copper intake may range from levels so low as to induce a nutritional deficiency to levels so high as to be acutely toxic. Copper is probably the first metal worked by humans about 70–80 centuries ago. The earliest known artifacts of hammered copper date from about 6000 BCE. After 4000 BCE, melting and casting of copper was common in the Near East. Smelting was developed about 3000 BCE and bronze around 2500 BCE. Brass, a copper alloy, was developed in Roman times. Copper derives from the Latin *cuprum*, a corruption of *cyprium*, Cyprus being the source of Egyptian and Roman copper. Copper was identified in terrestrial plants in 1817, in marine invertebrates in 1833, in vertebrates in 1838, and in hemocyanin – the blue respiratory pigment of mollusks and crustaceans – in 1880. But the metabolic importance of copper in plants and animals was not suspected until the 1920s when diseases due to copper deficiency began to be recognized. Copper deficiency in vertebrates, for example, is associated with anemia, gastrointestinal disturbances, aortic aneurisms, bone development abnormalities, and death.

Copper toxicosis in terrestrial higher plants is rare but occurs on mine spoils and where copper-rich manures or fungicides are used excessively. Copper is among the most toxic of the heavy metals in freshwater and marine biota, and often accumulates and causes irreversible harm to some species at concentrations just above levels required for growth and reproduction. Birds and mammals, when compared to lower forms, are relatively resistant to copper. But diets containing elevated concentrations of copper are sometimes fatal to ducklings and livestock when fed for extended periods. Domestic sheep (*Ovis aries*) are the most susceptible farm animals to chronic copper poisoning and effects include liver damage, impaired reproduction, reduced resistance to diseases, jaundice, and death.

10.2 Sources and Uses

The United States is the major world producer and consumer of copper and its compounds. Most of the copper produced is used to manufacture electrical equipment, pipe, and machinery. Copper releases to the global biosphere – which may approach 1.8 million metric tons per year – come mostly from anthropogenic activities such as mining and smelting, industrial emissions and effluents, and municipal wastes and sewage sludge. Copper compounds are widely used as biocides to control nuisance algae and macrophytes, freshwater snails that may harbor schistosomiasis and other diseases, ectoparasites of fish and mammals, marine fouling organisms, and mildew and other diseases of terrestrial crop plants. Copper compounds are also used in agricultural fertilizers, in veterinary and

---

[a]All information in this chapter is referenced in the following sources:

Eisler, R. 1979. Copper accumulations in coastal and marine biota. Pages 383–449 in J.O. Nriagu, ed., *Copper in the Environment. Part 1: Ecological Cycling*. John Wiley, New York.

Eisler, R. 1998. Copper hazards to fish, wildlife, and invertebrates: a synoptic review. *U.S. Geol. Surv. Biol. Sci. Rep. USGS/BRD/BSR–1997-0002*, 1–98.

Eisler, R. 2000. Copper. Pages 93–200 in *Handbook of Chemical Risk Assessment: Health Hazards to Humans, Plants, and Animals. Volume 1, Metals*. Lewis Publishers, Boca Raton, Florida.

medical products, in the food industry, and as a preservative of wood and other materials.

10.2.1 Sources

Global copper production during the past 60 centuries is estimated at 307 million metric tons, most of which (79%) occurred since 1900; annual global production of copper is now estimated at 13.6 million tons. Copper occurs naturally in many minerals and as uncombined metal. The three most important sources of copper are chalcocite (Cu_2S), chalcopyrite ($CuFeS_2$), and malachite ($CuCO_3 \cdot Cu(OH)_2$).

The United States is the largest global consumer and producer of copper. In 1986, domestic consumption of copper in the United States was 2.14 million metric tons and mine production was 1.14 million metric tons, mostly from mines in Arizona, New Mexico, and Michigan. The major copper deposits in the United States are of hydrothermal origin and uniformly distributed in fractures or veins. Copper is the major toxic component in streams impacted by active placer mines. About 60% of copper metal is eventually recycled; in 1986, smelting of scrap copper produced an additional 0.9 million metric tons of copper. Also in 1986, 1.1 million tons of copper were imported into the United States, mostly from Canada, Chile, Peru, and Mexico.

The amount of copper entering the global ecosystem annually is unknown, but estimates range from 211,000 metric tons to 1.8 million metric tons. About 80.7% of this copper is deposited in terrestrial compartments, 15.7% in the hydrosphere, and 3.6% in the atmosphere. The residence time for copper in the deep ocean is 1500 years; in soils it may be retained for as long as 1000 years; in air, copper persists for about 13 days. Copper in the atmosphere results mainly (73%) from human activities such as copper production and combustion of fossil fuels; the remainder is from natural sources that include sea salt sprays, windblown dusts, volcanogenic particles, and decaying vegetation.

Input of copper into aquatic ecosystems increased sharply during the past century and includes inputs from waste discharges into saline waters, industrial discharges into freshwater, and leaching of antifouling marine paints and wood preservatives. Present anthropogenic inputs of copper are two to five times higher than natural loadings; the atmosphere is a primary recipient of these inputs. In mining and industrial areas, precipitation of atmospheric fallout is a significant source of copper to the aquatic environment. More than 99.9% of oceanic copper fell as clay and manganese oxide particles in precipitation. In the lower Great Lakes, direct atmospheric inputs of copper – in metric tons per year – range from 55 to 2300 for Lake Michigan, 120 to 330 for Lake Erie, and 72 to 123 for Lake Ontario; regional disparities in atmospheric deposition of copper are related to the intensity of industrial activity and to the regional wind systems.

Copper in soils may come from a variety of anthropogenic sources: mining and smelting activities; other industrial emissions and effluents; traffic; flyash; dumped waste materials; contaminated dusts and rainfall; sewage and sludge; pig slurry; composted refuse; and agricultural fertilizers, pesticides, and fungicides. In the case of Florida citrus groves, copper-containing fertilizers applied during the early 1900s accounted for as much as 34.0 kg Cu/ha annually, and routine fungicidal sprays contributed another 10.0 kg Cu/ha annually. Surface soils (0–15 cm) from some mature citrus groves contained as much as 540.0 kg Cu/ha. Copper deposition rates in soils are higher in cities and near highways, railroads, power plants, and industrial activities. But a Kansas landfill near a freshwater stream had no significant effect on copper concentrations in water, sediments, crayfish, and sunfish.

10.2.2 Uses

Metallic copper end uses include electrical (70%), construction (15%), machinery (6%), transportation (4%), and ordinance (2%). The top domestic markets for copper and its alloys in 1986 were, in order of importance, plumbing, building wire, telecommunications, power utilities, in-plant equipment, air conditioning, automotive electrical, automotive nonelectrical, business electronics, and industrial valves and fittings. A small percentage of copper production is used to manufacture chemicals, mainly copper sulfate. Of the copper sulfate used domestically, 65% is used in agriculture for fungicides, algicides, nutritional supplements, insecticides, and repellents; 28% is used industrially in froth flotation production of chromated copper arsenate (CCA) wood preservatives, in electroplating, and in the manufacture of azo dyes; and 7% is used in water treatment to control nuisance algae.

Copper is widely used to control unwanted species of freshwater algae and macrophytes. Chelated copper products are claimed to be effective algicides in hard water; the chelation of copper by organic compounds, such as ethanolamines or ethanolamine complexes, protects copper from precipitation and complexation. Copper sulfate is approved by the U.S. Environmental Protection Agency as an algicide in waters used to raise fish for human consumption. In algae, copper inhibits photosynthesis, nitrogen fixation, and phosphorus uptake; it selectively eliminates cryptophytes but spares diatoms. Copper sulfate at low concentrations has been used to control freshwater algae in Wisconsin since 1918 without any conclusively proven effect on diversity or abundance of nontarget species. But reduced abundance of freshwater benthos was noted in Lake Monova, Wisconsin, which received 771 metric tons of copper to control algae over a 26-year period and had sediment levels as high as 1093.0 mg Cu/kg DW. Copper sulfate used to control algal blooms in Wisconsin lakes at 1.25 mg Cu/L killed nontarget fishes, crustaceans, snails, and amphibians in 14 days or less; however, 0.25 mg Cu/L was not fatal to these species in 20 days. Concentrations as low as 0.03 mg Cu/L inhibited growth in two of four species of nuisance aquatic weeds in Lake Mendota, Wisconsin, and 0.3 mg Cu/L was fatal to all four species. Copper sulfate controls algae in cranberry bogs at 0.4 mg Cu/L but this concentration also kills resident fishes. Copper was not measurable in the surface waters of cranberry bogs within 10 days of treatment, regardless of initial copper concentration; it is probable that copper was adsorbed onto bog soils. In England, copper sulfate was effective at 1.0 mg Cu/L in controlling algae and most species of aquatic weeds for 1.6 km downstream of treatment for 6 months during the summer; only 0.25 mg Cu/L was necessary in autumn and winter for effective aquatic weed control. During copper treatment to control plants, aquatic snails were greatly reduced or eliminated but fishes seemed unaffected. Sensitive aquatic weeds included *Myriophyllum spicatum*, *Elodea canadensis*, *Potamogeton* spp., and *Lemna minor*; sensitive genera of algae included *Oedogonium*, *Spirogyra*, *Enteromorpha*, and *Mougeotia*.

In California, 4.23 million kg of copper compounds are used each year in agriculture of which almost 1.4 million kg is applied as copper sulfate pentahydrate to flooded rice fields for control of algae and tadpole shrimp. Most of the copper precipitates as copper hydroxycarbonate within the first hour, with negligible loss to the atmosphere or from runoff. The remainder in rice field water is complexed with dissolved organic matter and is biologically unavailable to algae and shrimp. However, the major part accumulates in the soil, where it presents a threat to future plant agriculture at that site, although intentional mobilization may be possible.

In Iowa lakes, copper sulfate was used to control summer blooms of various species of toxic blue-green algae. Prior to treatment, blooms of *Anabaena flos-aquae*, *Aphanizomenon flos-aquae*, and *Microcystis aeruginosa* were associated with deaths of migratory waterfowl, game birds, songbirds, game, and domestic animals. Most of these species of algae were controlled within 24 h by 1.0 mg Cu/L as copper sulfate. Copper treatment had no adverse effects on bottom fauna but populations of crustaceans (daphnids, copepods, entomostracans) were reduced. One year after treatment no deaths of birds or mammals were recorded.

Copper salts are intentionally added to drinking water supplies of some municipalities to control growth of algae; concentrations as high as 59.0 μg Cu/L are maintained in New York City.

Copper compounds are used routinely and widely to control freshwater snails that serve as intermediate vectors of schistosomiasis and other diseases that afflict humans. These compounds include copper sulfate, copper pentachlorophenate, copper carbonate, copper-tartaric acid, Paris green (copper arsenite-acetate), copper oxide, copper chloride, copper acetyl acetonate, copper dimethyl dithiocarbamate, copper ricinoleate, and copper rosinate. Also, many species of oyster enemies are controlled by copper sulfate dips. All tested species of marine gastropods, tunicates, echinoderms, and crabs that had been dipped for 5 s in a saturated solution of copper sulfate died if held in air for as little as a few seconds to 8 h; mussels, however, were resistant.

Copper sulfate is used to control protozoan fish ectoparasites including *Ichthyopthirius*, *Trichodina*, and *Costia*; the effectiveness of the treatment diminishes with increasing total alkalinity and total hardness of the water. Copper compounds now used to control protozoan parasites of cultured red drum (*Sciaenops ocellatus*) include copper sulfate, copper sulfate plus citric acid, and chelated copper compounds (forms of copper bound by sequestering agents, such as ethanolamine); chelated copper compounds are considered less toxic to fish than copper sulfate and at least as effective in controlling parasites.

Copper is the active agent in many antifouling paints applied to watercraft. The growing use of copper-based paints subsequent to the prohibition in 1982 of tributyltin-based paints is associated with elevated copper concentrations in Pacific oysters (*Crassostrea gigas*) farmed in the Bay of Arcachon, France.

Copper compounds are used in agriculture to treat mildew and other plant diseases; in the food industry as preservatives, additives, or coloring agents; in preservatives of wood, leather, and fabrics; in coin manufacture; and in water treatment. The use of copper-containing pesticides is traditional along the Mediterranean Coast, especially the use of Bordeaux mixture, a copper sulfate-based fungicide that has been widely used for more than a century to prevent mildew on grape vines. However, at current application rates of about 0.8 mg Cu/cm^2, Bordeaux mixture significantly reduces the life span and breeding rate of the fruit fly (*Drosophila melanogaster*).

Copper is widely used in veterinary medical products. Copper sulfate is used by veterinarians to treat cattle and sheep for helminthiasis and infectious pododermatitis. Cuprol (a 1% solution of cupric oleinate) is used to control lice. Copper is routinely used as a growth supplement in the diets of swine (*Sus* sp.) in the United Kingdom and elsewhere; diets may contain as much as 250.0 mg Cu/kg ration. The intensity of pig farming within about 10 km from the coast may influence copper content in estuarine sediments. For example, intensive pig farming in coastal Brittany, France, increased soil copper concentrations by 0.6 kg/ha annually and increased coastal sediment copper concentrations to as high as 49.6 mg/kg DW.

In human medicine, metallic copper is used in some intrauterine devices, and various copper compounds are used as emetics and to treat rheumatoid arthritis. Some individuals wear copper bracelets as treatment for arthritis although its therapeutic value has little support.

10.3 Chemical and Biochemical Properties

This section demonstrates that (1) free ionic copper (Cu^{2+}) is the most toxic chemical species of copper and that copper bioavailability is modified by many biological and abiotic variables; (2) copper metabolism and sensitivity to copper of poikilotherms differs from that of mammals; and (3) copper interactions with inorganic and organic chemicals are substantial and must be considered when evaluating copper hazards to natural resources.

10.3.1 Chemical Properties

Copper is a soft heavy metal, atomic number 29, a density in elemental form at 20°C of 8.92 g/cc, and a melting point of 1083.4°C. Copper has two natural isotopes: Cu-63 (69.09%) and Cu-65 (30.91%).

Copper exists in four oxidation states: Cu^0, Cu^{+1}, Cu^{+2}, and Cu^{+3}. Elemental copper (Cu^0) is readily attacked by organic and mineral acids that contain an oxidizing agent and is slowly soluble in dilute ammonia; halogens attack elemental copper slowly at room temperature to yield the corresponding copper halide; elemental copper is not oxidized in water. Cuprous copper (Cu^{+1}) exists only in water solution when complexed, usually in a tetrahedral form, with affinity for sulfur and nitrogen ligands. Cuprous copper is unstable in aerated aqueous solution over the pH range 6–8, and will undergo auto-oxidation–reduction into elemental copper (Cu^0) and cupric ion (Cu^{+2}). The only cuprous ion (Cu^{+1}) compounds that are stable in water are extremely insoluble ones such as cuprous chloride, CuCl. Cuprous ion complexes may be formed in seawater by photochemical processes and persist for several hours. The cupric ion (Cu^{+2}) is the one generally encountered in water. Cupric ions are co-ordinated with six water molecules in solution. Cupric ion ordinarily forms planar, less stable chelates with nitrogen and oxygen ligands. In seawater and sediment interstitial waters, the free cupric ion (Cu^{+2}) is the most readily available and toxic inorganic species of copper; however, the free ion concentration is sensitive to complexation and is less available to aquatic biota in the presence of natural organic chelators and high salinities. Cupric ions account for about 1% of the total dissolved copper in seawater and less than 1% in freshwater. Trivalent copper (Cu^{+3}) probably does not occur naturally. Trivalent copper is strongly oxidizing and occurs only in a few compounds; none of these compounds is now considered industrially important or environmentally significant.

In freshwater, the solubility of copper salts is decreased under reducing conditions and is further modified by water pH, temperature, and hardness; size and density of suspended materials; rates of coagulation and sedimentation of particulates; and concentration of dissolved organics. The chemical form of copper in freshwater is important in controlling geochemical and biological processes. But the lack of knowledge on the adsorption characteristics of most cupric (Cu^{+2}) ion complexes contributes to uncertainties about the behavior of known copper species. Ionic copper (Cu^{+2}) and some copper hydroxyl species are correlated with high toxicity to aquatic life; however, carbonato species ($CuHCO_3^+$, $CuCO_3$, $Cu(CO_3)_2^{-2}$) are much less toxic than other copper complexes. The major chemical species of copper in freshwater are $Cu(CO_3)_2^{-2}$ and $CuCO_3$. Cupric ion is the dominant toxic copper species at pH less than 6; the aqueous copper carbonate complex is dominant from pH 6.0 to 9.3. This equilibrium is altered in the presence of humic acids, fulvic acids, amino acids, cyanide, certain polypeptides, and detergents. Most cupric salts dissolve readily in freshwater to produce the aquo ion, $Cu(H_2O)_6^{+2}$. Divalent copper chloride, nitrate, and sulfate are highly soluble in water, but copper carbonate, cupric hydroxide, cupric oxide, and cupric sulfide will precipitate from solution or form colloidal suspensions when excess cupric ions are present.

In seawater, the major chemical species of copper are $Cu(OH)Cl$ and $Cu(OH)_2$ and these account for about 65% of the total copper in seawater. The levels of copper hydroxide ($Cu(OH)_2$) increase from about 18% of the total copper at pH 7.0 to 90% at pH 8.6; copper carbonate ($CuCO_3$) dropped from 30% at pH 7.0 to less than 0.1% at pH 8.6. The dominant copper species in seawater over the entire ambient pH range are copper hydroxide, copper carbonate, and cupric ion. Bioavailability and toxicity of copper in marine ecosystems are promoted by oxine and other lipid-soluble synthetic organic chelators.

Copper concentrations in sediment interstitial pore waters correlate positively with concentrations of dissolved copper in the overlying water column and are now used to predict the toxicity of test sediments to freshwater amphipods. Sediment-bound copper is available to deposit-feeding clams, especially from relatively uncontaminated anoxic sediments of low pH. The bioavailability of copper from marine sediments, as judged by increased copper in sediment interstitial waters, is altered by increased acid volatile sulfide (AVS) content. But AVS is not an appropriate partitioning phase for predicting copper bioavailability of freshwater sediments.

10.3.2 Metabolism

Copper is part of several essential enzymes including tyrosinase (melanin production), dopamine beta-hydroxylase (catecholamine production), copper–zinc superoxide dismutase (free radical detoxification), and cytochrome oxidase and ceruloplasmin (iron conversion). All terrestrial animals contain copper as a constituent of cytochrome *c* oxidase, monophenol oxidase, plasma monoamine oxidase, and copper protein complexes. Excess copper causes a variety of toxic effects, including altered permeability of cellular membranes. The primary target for free cupric ions in the cellular membranes is thiol groups that reduce cupric (Cu^{+2}) to cuprous (Cu^{+1}) upon simultaneous oxidation to disulfides in the membrane. Cuprous ions are reoxidized to Cu^{+2} in the presence of molecular oxygen; molecular oxygen is thereby converted to the toxic superoxide radical O^{-2}, which induces lipoperoxidation.

10.3.2.1 Aquatic Organisms

Bioavailability and toxicity of copper to aquatic organisms depend on the total concentration of copper and its speciation. In hard, moderately polluted waters, 43–88% of the copper is associated with suspended solids and not available to biota. Copper toxicity to aquatic biota is primarily related to the dissolved cupric ion (Cu^{+2}) and possibly to some hydroxyl complexes. Soluble copper is largely complexed with carbonate, amino acids, or humic substances. Cupric copper – one of the most toxic forms – constitutes 0.1–0.2% of this soluble material. The toxicity of copper in its complexed, precipitated, or adsorbed form is less than that of the free ionic form. In aquatic invertebrates, copper causes gill damage at high concentrations, and in fishes it interferes with osmoregulation. Elevated concentrations of copper interfere with oxygen transport and energy metabolism; tissue hypoxia is the cause of death and is associated with reductions in the activities of regulatory enzymes of ATP-synthesizing pathways.

In freshwater algae, movement of copper into cells occurs mainly by physical transport; the plasmalemma is the initial site of copper binding. Copper on the plasmalemma increases its permeability, as shown by the leakage of potassium and other ions from copper-treated cells and entry of copper into intracellular sites.

Marine prosobranch gastropods, like several other groups of mollusks and arthropods, normally accumulate and store copper and use it in the synthesis of hemocyanin, a blood pigment. In gastropods, copper may elicit secretions of mucus by goblet cells; bind to hydrophilic regions of the external membranes of epithelial cells, altering their biochemical and biophysical properties; or disrupt the normal functioning of peroxidase and ferritin. Peroxidation products, such as hydroperoxides and malondialdehyde, are toxic to vital functions of membranes and cells; bivalve mollusks challenged with ionic copper show significant increases in these products. Exposure of gastropods to high sublethal concentrations of copper completely inhibits succinic dehydrogenase activity at whole body concentrations between 4.7 and 11.9 mg Cu/kg DW soft parts, causes a measurable decrease in heart beat rate, and adversely affects surface epithelia, especially those covering the head–foot and rectal ridge, disrupting osmoregulation and producing water accumulation in tissues. The primary lethal effect of copper in gastropod mollusks is caused by disruption of the transporting surface epithelium.

In crabs, the gills are a major target organ of copper toxicity; waterborne copper decreases hemocyanin–oxygen affinity. Exposure of shore crabs (*Carcinus maenas*) to lethal concentrations of copper is associated with reductions in activity of glycolytic enzymes but, unlike fishes, did not involve cellular energy deprivation. Copper-tolerant strains of aquatic mayflies (*Baetis thermicus*) have evolved in Japan. Tolerance is attributed to the ability to induce a metal-binding protein that preferentially sequesters copper over cadmium and zinc.

In fishes, the gill surface's low affinity for metal allows greater entry of the metal to the intracellular compartment. Once there, more complex binding sites are present. Binding to these ligands causes one or more of the following toxic mechanisms: (1) blocking of the essential biological functional groups of biomolecules; (2) displacing the essential metal ion in molecules; or (3) modifying the active conformation of biomolecules. These mechanisms may account for the specific inhibition of ion transport from ionic (Cu^{+2}) copper exposure. Studies with radiocopper-64 and rainbow trout (*Oncorhynchus mykiss*) show that the external gill epithelial surface has a relatively low affinity for copper, allowing copper to penetrate intracellular compartments. Copper disrupts gill function of rainbow trout by impairing transepithelial ion exchange, for example, impairing or upsetting electrolyte balance by inhibiting active uptake or stimulating passive loss. Copper toxicity to rainbow trout in hard water is related to the total concentration of soluble copper (copper carbonate, $CuCO_3$; cupric ion, Cu^{+2}) in the test medium rather than to either of these two forms alone. Long-term retention of copper accumulations in fish tissues is characterized by high half-time persistence after copper administration and binding of copper to proteins in a nonexchangeable or slowly exchangeable pool.

Copper detoxifying mechanisms in fishes include the induction of metallothioneins, allowing copper retention for weeks or months after absorption without producing toxic effects. Hepatic metallothionein contents of individual fishes usually reflect the accumulation of copper in that organ. This strongly supports the use of metallothionein as an indicator of copper stress.

In tench (*Tinca tinca*), hepatic alterations observed after exposure to lethal concentrations of copper (75 mg/L for 4–12 days) include accumulation of various pigments in Kupffer cells and hepatocytes. Death was attributed to deficient oxygen transport and consumption and to the lytic effect of copper on various cell membranes, eventually causing massive necrosis in large areas of the liver parenchyma. In rapidly growing juvenile flounders (*Paralichthys* spp.), copper blocks calcium transport, possibly through interference with gill chloride cells. Copper inhibition of calcium accumulation is alleviated by removing copper from the medium.

10.3.2.2 Mammals

Copper homeostasis plays an important role in the prevention of copper toxicity. After copper requirements are met, excess copper absorbed into gastrointestinal mucosal cells is bound to metallothionein and excreted when the cell is sloughed. Copper that eludes the intestinal barrier is stored in the liver or incorporated into bile and excreted in feces. The most likely pathway for the entry of toxic amounts of copper would be long-term inhalation or entry through the skin. Both of these pathways allow copper to pass unimpeded into the blood. The levels of copper in the mammalian body are held constant by alterations in the rate and amount of copper absorbed, its distribution, and rate and route of excretion. Many factors interfere with copper absorption including competition for binding sites, as with zinc; chelation, as with phytates; and interaction with ascorbic acid, which aggravates copper deficiency by decreasing copper absorption and – with excess copper – reduces the toxic effects.

Two inherited human diseases that represent abnormal copper metabolism are Menkes' syndrome and Wilson's disease. Menkes' syndrome, with symptoms similar to those of copper deficiency, is characterized by a progressive brain disease, abnormally low copper concentrations in liver and other tissues, and diminished ability to transfer copper across the absorptive cells of the intestinal mucosa. Wilson's disease (hepatolenticular degeneration) is the only significant example of copper toxicity in humans. Wilson's disease is an autosomal recessive disorder that affects normal copper homeostasis and is characterized by excessive retention of hepatic copper, decreased concentration of serum ceruloplasmin, impaired biliary copper excretion, and hypercupremia. Systemic manifestations of Wilson's disease are hepatic and renal lesions and hemolytic anemia. Certain strains of mutant rats with reduced excretion of biliary copper spontaneously develop hepatitis because of the extremely gross deposition of copper in the liver. Most humans afflicted with Wilson's disease usually die before puberty, although some survive to age 35. Postmortems of Wilson's patients show that livers had as much as 7.5% copper and kidney ash had up to 2.7% copper. There is no evidence, however, that persons with normal homeostatic mechanisms are subject to any chronic degenerative disorders resulting from modern exposures to copper. Unusually susceptible human populations to copper poisoning include those afflicted with Wilson's disease, infants and children aged under one year, those with liver damage or chronic renal disease, individuals undergoing dialysis (excess copper in the dialysate), and those with an inherited deficiency of the enzyme glucose-6-phosphate dehydrogenase (G-6-PD).

Ingested copper travels through the gastrointestinal (GI) tract, where some of it is absorbed into the blood and becomes associated with plasma albumin and amino acids or is used to maintain copper levels in erythrocytes. Albumin-bound copper is eventually transported to the liver; however, minor fractions are transported into the bone marrow, the erythrocytes, or other tissues. Most of the circulating copper is translocated within minutes. During the next few hours, blood copper concentrations increase and Cu becomes an integral part of the ceruloplasmin molecule. Gastrointestinal absorption is normally regulated by the copper status in the body. In general, up to 50% of small doses (i.e., less than 1.0 μg in rats) are absorbed, whereas large doses are absorbed to a lesser extent. In humans, about 40% of the dietary copper is absorbed. Absorbed copper is freely exchangeable with copper loosely bound to the alpha-globulin ceruloplasmin where it is exchanged in the cupric form. Copper is stored mainly in liver, brain, heart, kidney, and muscle; in tissues and blood cells, copper is bound to proteins, including many enzymes. Amino acids facilitate the entry of Cu into liver cells and a small proportion of copper in serum is bound to amino acids. About 80% of the absorbed copper is bound to metallothionein in the liver; the remainder is incorporated into compounds such as cytochrome *c* oxidase. Copper accumulations in animals are associated with increased number and increased size of copper-containing lysosomes in hepatocytes. In liver, copper is initially bound to a metallothionein-like, low-molecular-weight protein and later it appears in a high-molecular-weight protein, ceruloplasmin, which re-enters the circulation. Ceruloplasmin transports copper to tissues and also functions as an oxidase. The amount of copper absorbed is usually far in excess of metabolic requirements. Of the copper retained in the body, almost all play a particular physiological role in the function of at least 12 specific copper proteins, such as cytochrome *c* oxidase and tyrosinase. Thus, only extremely small concentrations of free copper ions are normally found in body fluids.

Retention of radiocopper injected into humans is high; only 10% is excreted within 72 h in urine and feces and 50% in four weeks. Most (72%) of the unabsorbed copper is excreted in the feces primarily by way of the biliary duct, the salivary glands, or the intestinal mucosa; a minor portion is excreted by way of sweat and menses. In mammals, copper is excreted mainly via the bile in association with glutathione or unidentified high-molecular-weight molecules; however, the transport mechanisms of copper from liver cells into bile are essentially unknown. In rats, biliary excretion of copper is increased by increased flow of bile, increased body temperature, or administration of adrenal steroids.

Mechanisms implicated in copper poisoning include free radical production, alteration in activities of several enzymes, and interference with metallothionein synthesis. At the cellular level, copper has several primary mechanisms of toxicity that alter protein configuration and biological activity. These include the catalysis of peroxidation reactions and subsequent generation of free radicals that damage lipids and proteins, interactions with R groups of proteins – particularly SH groups, and acting as a substituent for other metals in metalloproteins. Copper, in relative excess, is a cytotoxic metal with injury related to the process of lipid peroxidation. Isolated rat hepatocytes exposed to copper solutions for as long as 90 min show a concentration, and time-related decrease in cell viability as judged by loss of intracellular potassium and aspartate aminotransferase (AAT), an increase in lipid peroxidation, and a decrease in glutathione. Falling disease in cattle, dogs, and chickens is associated with a cardiovascular disorder caused by reduced activity of lysal oxidase, a Cu-requiring enzyme necessary for elastic tissue formation and maintenance. Metallothionein synthesis acts as a protective mechanism against buildup of excessive amounts of the essential, but potentially toxic, copper ions, possibly before the development of other control processes. In livers of newborn lambs, rabbits, mice, and hamsters, copper concentrations are usually directly related to the metallothionein content in cytoplasmic fractions. In sheep, elevated serum glutamic oxaloacetic transaminase (SGOT) levels were linked to elevated copper concentrations in blood at least one to six weeks before obvious external signs of copper poisoning. SGOT measurements in sheep serum seem to constitute an adequate early warning of the approach of the hemolytic crisis and eventual death of the animal from chronic copper poisoning.

10.3.3 Interactions

Copper interacts with numerous compounds normally found in natural waters. The amounts of the various copper compounds and complexes present in solution depend on water pH, temperature, and alkalinity and on the concentrations of bicarbonate, sulfide, and organic ligands. In animals, copper interacts with essential trace elements such as iron, zinc, molybdenum, manganese, nickel, and selenium and also with nonessential elements including silver, cadmium, mercury, and lead; interactions may be either beneficial or harmful to the organism. The patterns of copper accumulation, metabolism, and toxicity from these interactions frequently differ from those produced by copper alone. Acknowledgment of these interactions is essential for understanding copper toxicokinetics.

10.3.3.1 Aluminum

Mixtures of copper and aluminum were more than additive in toxicity to ova of brown trout, *Salmo trutta*.

10.3.3.2 Cadmium

Exposure of algae to low sublethal concentrations of copper (0.03 μg/L) increases their sensitivity towards additional copper challenge and to cadmium (Cd) salts. In freshwater clams (*Anodonta cygnea*) exposed for 46 days to a mixture of high concentrations of copper (139.0 μg/L) and cadmium (122.0 μg/L), cadmium accumulation is reduced by 90% and copper accumulation is reduced by 50%. Exposure of crayfish (*Cambarus bartoni*) to 12.5 μg Cd/L for 72 h results in significantly increased copper stores in the hepatopancreas; however, isopods similarly exposed had decreased copper stores in antennal glands. In the presence of copper, barnacles tend to accumulate cadmium. In fishes, copper–cadmium interactions occur in Mozambique tilapia (*Oreochromis mossambicus*) during single and combined exposures. Waterborne copper tends to increase whole body cadmium content of tilapia at all tested copper concentrations and exposure durations (as high as 400.0 μg Cu/L for 96 h); however, cadmium exposure tends to lower copper concentrations in tissues of tilapia.

In birds, copper concentrations in kidneys of the willow ptarmigan (*Lagopus lagopus*) are positively correlated with concentrations of cadmium.

In mammals, cadmium inhibits copper absorption across the intestinal mucosa. Intercorrelations of copper with cadmium and zinc in livers of polar bears (*Ursus maritimus*) are probably mediated by metallothioneins, which may contain all three metals. In rats, copper protects against nephrotoxicity induced by cadmium, provided that copper is administered 24 h prior to cadmium insult. Specifically, rats given 12.5 mg Cu/kg BW by way of subcutaneous injection 24 h before receiving 0.4 mg Cd/kg BW – when compared to a group receiving Cd alone – did not have excessive calcium in urine and renal cortex or excessive protein in urine; thus, 2.8 mg Cu/kg BW protects against 0.25 mg Cd/kg BW.

10.3.3.3 Iron

Iron-reducing bacteria from a copper-contaminated sediment were more tolerant of copper adsorbed to hydrous ferric oxide (HFO) than were pristine-sediment bacteria. Copper-tolerant bacteria were more efficient in reducing contaminated HFO, with greater potential for copper mobilization in aquatic sediments.

Mixtures of copper and iron salts were more than additive in toxicity to ova of brown trout.

In muscle of Weddell seals (*Leptonychotes weddelli*), copper is positively correlated with iron. In general, concentrations of copper in all tissues of all marine vertebrates examined are positively correlated with concentrations of iron.

The primary function of the mammalian red blood cell is to maintain aerobic metabolism while the iron atom of the heme molecule is in the ferrous (Fe^{+2}) oxidation state; however, copper is necessary for this process to occur. Excess copper within the cell oxidizes the ferrous iron to the ferric (Fe^{+3}) state. This molecule, known as methemoglobin, is unable to bind oxygen or carbon dioxide and is not dissociable. Simultaneous exposure of sheep to mixtures of cupric acetate, sodium chlorite, and sodium nitrite produced a dose-dependent increase in methemoglobin formation.

The addition of iron to diets of domestic pigs increases their resistance to copper poisoning, but this is an exception. High intake of iron, in general, adversely affects copper status in ruminants, guinea pigs (*Cavia* spp.), and rats (*Rattus* sp.); the mechanisms for this phenomenon are unknown. Genetically anemic and normal strains of rats fed high iron diets had reduced kidney copper concentrations in both groups; this was associated with decreased absorption and biliary excretion of copper.

10.3.3.4 Manganese

Copper in livers and muscles of Weddell seals was positively correlated with manganese. In general, manganese and copper are positively correlated in tissues of marine vertebrates. Uptake of copper from copper-contaminated freshwater sediments by annelid worms is related to the amount of reducible manganese oxide in the sediments.

10.3.3.5 Molybdenum

In terrestrial vegetation, molybdenum and sulfur interfere with copper-induced deficiencies. Copper poisoning in cattle (*Bos* sp.) and other ruminants is governed by dietary concentrations of molybdenum and sulfate. Molybdenum and sulfur in mammalian diets cause a decrease in the availability of copper because of the formation of the biologically unavailable copper-thiomolybdate complex. Cattle die when grazing for extended periods on pastures where the ratio of copper to molybdenum is less than 3 to 1, or if they are fed low copper diets containing molybdenum at 2–20 mg Mo/kg ration. Wilson's disease is induced in rabbits (*Oryctolagus* sp.) by feeding a diet high in molybdates and sulfates, suggesting that the disease is not solely the result of copper intoxication.

10.3.3.6 Zinc

Copper is positively correlated with zinc in gills of two species of fishes from the Mediterranean Sea. Mixtures of copper and zinc salts in marine or freshwater fishes are more-than-additive in toxicity, producing more deaths in 96 h than expected on the basis of individual components. Mixtures of copper and zinc are generally acknowledged to be more-than-additive in toxicity to a wide variety of aquatic organisms. But mixtures of copper (0.0–90.0 μg/L) and zinc (0.0–1200.0 μg/L) are only additive in action to a marine bacterium (*Photobacterium phosphoreum*), decreasing its luminescence after exposure for 30 min. And sometimes mixtures of copper and zinc salts are less-than-additive in action, as judged by DNA, RNA, and protein contents of newly hatched fathead minnows (*Pimephales promelas*) exposed for 4 days.

In birds, copper and zinc concentrations are positively correlated in kidneys of the willow ptarmigan (*Lagopus lagopus*) and in kidneys and livers of common murres (*Uria aalge*).

In mammals, copper absorption across the intestinal mucosa is inhibited by concomitant high oral intake of zinc. In livers from Weddell seals, copper is positively correlated with zinc. The addition of zinc to swine diets protects against copper toxicosis caused by eating diets containing 250.0 mg Cu/kg ration.

10.3.3.7 Other Inorganics

Copper interacts with lead, magnesium, silver, and other elements. Dose-dependent frequency of deformities was observed in chironomid larvae held in water containing 1.0–100.0 μg Cu/L for five generations; copper and lead mixtures – up to 500.0 μg Pb/L – interacted to produce more-than-additive deformity frequency. In mammals, supplemental copper promotes urinary excretion of lead from the body and loss of lead from tissues. In shore crabs (*Carcinus maenas*), ionic copper displaces ionic magnesium in gills, leading to inhibition of phosphoryl transfer. In embryos of the Pacific oyster (*Crassostrea gigas*), silver – at 0.5–15.5 μg Ag/L – enhances adverse effects when copper concentrations exceed 6.0 μg Cu/L. Silver positively correlates with copper in livers of Weddell seals, but in muscles the correlation is negative.

In fishes, additive or more-than-additive toxicity occurs with mixtures of salts of copper and mercury, copper–zinc–phenol, and copper–nickel–zinc. Accumulation of copper in gills of fathead minnows during exposure to 16.0 μg Cu/L is reduced by added ionic calcium, which competes with Cu for gill-binding sites.

10.3.3.8 Organic Compounds

Sequestering agents, increasing salinity, sediments, and other variables all reduce toxicity and accumulation of copper in tested species of aquatic plants and invertebrates. Chelating agents, such as nitrilotriacetic acid, reduce the toxicity of ionic copper to six species of estuarine phytoplankton. Sensitivity of freshwater zooplankton communities varies seasonally. Communities are most sensitive to copper stress (20.0 or 40.0 μg Cu/L) during exposure for 5 weeks in spring rather than in summer or autumn because, in part, of reduced dissolved organic carbon (DOC) concentrations in the spring. Adverse effects of copper on survival of marine copepods are reduced or eliminated by the presence of clay minerals, diatoms, ascorbic acid, sewage effluents, water extracts of humic acids, and certain soil types. Chelators, such as EDTA, and more alkaline pH increase the survival and larval developmental rates of copepods challenged with copper through increased complexation of cupric ions. Natural fulvic acids, which comprise 75% of dissolved humic substances, reduce the acute toxicity of copper to rotifers. A significant reduction in radiocopper-64 accumulation by clams (*Macoma balthica*) occurs at high concentrations of dissolved organic ligands; reduction is more pronounced at 3.0% salinity than 1.0% salinity. The presence of sediments in assay containers reduces the toxicity of copper to freshwater gastropods. Copper uptake by brine shrimp (*Artemia franciscana*) increases with decreasing pH and decreasing carbonate complexation. Studies with a freshwater shrimp (*Paratya australiensis*) and copper salts show that uncomplexed cupric ions are the most toxic chemical species in solutions containing nitrilotriacetic acid or glycine; however, the singly charged copper–glycine$^+$ complex also appears to be mildly toxic. Shrimp (*Paratya* sp.) are more resistant to copper in higher alkalinity waters and under conditions of increasing dissolved organic matter.

In freshwater fishes, mixtures of copper with anionic detergents or various organophosphorus insecticides cause more-than-additive toxicity. And in marine vertebrates, copper in tissues is positively correlated with metal-binding proteins. Accumulations of copper in gills of fathead minnows during exposure to 16.0 μg Cu/L are reduced by added EDTA, which reduces bioavailability of copper through complexation. Copper LC50 (96 h) values, i.e., concentrations of ionic copper in solution at start of the test estimated to kill 50% of the test species in 96 h, to larval fathead minnows range from a low of 2.0 μg/L at low pH and low DOC to 182.0 μg/L at pH 6.9, and DOC of 15.6 mg/L. Acidification and the removal of DOC increase the toxicity of copper to fathead minnows in natural waters of low alkalinity and explains 93% of the variability in field toxicity data for that species.

In mammals, phenobarbital and phenytoin increase serum ceruloplasmin concentrations. Chronic copper poisoning in sheep is exacerbated when diets contain heliotrope plants (*Heliotropium* sp., *Echium* spp., *Senecio* sp.). Aggravated effects of the heliotrope plants include reduced survival and a 2–3-fold increase in liver and kidney copper concentrations when compared to control animals fed copper without heliotropes. Rats given acutely toxic doses of 2,3,7,8-tetrachlorodibenzo-*para*-dioxin had elevated concentrations of copper in liver and kidney because of impaired biliary excretion of copper. Morphine increases copper concentrations in the central nervous system of rats and dithiocarbamates inhibit biliary excretion. In human patients, urinary excretion of copper is increased after treatment with D-penicillamine, calcium disodium EDTA, or calcium trisodium diethylenetriamine penta acetic acid.

10.4 Carcinogenicity, Mutagenicity, and Teratogenicity

No definitive evidence exists demonstrating that copper or copper compounds at environmentally realistic concentrations are the causative agents in the development of carcinogenicity, mutagenicity, or teratogenicity. However, under controlled conditions of grossly elevated exposures, some studies suggest that copper is a potential carcinogen in rodents; mutagen in rodents, sheep (*Ovis aires*), and grasshoppers (*Oxya velox*); and teratogen in fish, rodents, and other small laboratory animals.

10.4.1 Carcinogenicity

The carcinogenic classification of copper is Group 3 or D; that is, not classifiable as to its carcinogenicity in humans. No definitive evidence exists showing that copper or copper compounds cause cancer in mammals. Although hypercupremia is sometimes associated with neoplasms, some copper compounds seem to have an inhibitory effect on the development and growth of malignant tumor cells. Copper is not associated with an elevated incidence of cancer in humans or animals exposed by way of inhalation, oral, dermal, or intramuscular injection routes. A slightly increased incidence of reticulum cell sarcoma was noted in mice 18 months after a single subcutaneous injection of copper 8-hydroxyquinoline, but this needs to be verified.

Sensitivity of cancerous cells to copper may reflect cell DNA content. Two closely related rat hepatoma cell lines differed in sensitivity to copper toxicity by a factor of four; DNA content in each cell line decreased with increasing copper concentrations, but at different rates. Severity of toxicity was associated with increasing accumulations of copper in the cell nucleus and with decreasing DNA.

10.4.2 Mutagenicity

Grasshoppers injected intra-abdominally with relatively high concentrations of soluble copper showed a 1.6% frequency of chromosomal anomalies in meiotic cells of testes 24 h after injection; however, no control data were presented. Copper-induced DNA strand breaks in rats (*Rattus* sp.) and chromosomal aberrations and sperm abnormalities in mice (*Mus* sp.) suggest that copper is a potential human mutagen. Copper salts affect chromosomes in vitro in the presence of hydrogen peroxide and ascorbic acid and can also increase the frequency of noncomplementary nucleotides in the synthesized DNA double helix. Sheep, aged 1.5 years, given about 10.7 mg Cu/kg BW daily – in addition to other metals – until they died (65–84 days later) show a significant increase in sister chromatid exchanges in bone marrow; however, the specific role of copper on survival and mutagenicity is unclear and requires verification.

10.4.3 Teratogenicity

Grossly elevated concentrations of dissolved copper produce teratogenicity in fish embryos. A significant number of malformed fish larvae came from eggs treated with 500.0 μg Cu/L. In studies with laboratory animals and elevated concentrations of copper salts, copper penetrates the placental barrier into the fetus; intramuscular injection of 4.0 mg Cu/kg BW early in pregnancy adversely affects fetal central nervous system development. In humans, no definitive data are available on whether copper can cause birth defects; however, incubation of human spermatozoa with metallic copper results in loss of sperm motility.

10.5 Concentrations in Field Collections

Copper concentrations in air, soil, water, sediments, and other abiotic materials are elevated as a result of human activities, especially near copper smelters and mines, urban areas, municipal and industrial wastewater outfalls, marinas containing copper-based antifouling paints, and agricultural soils receiving prolonged applications of copper-based fungicides. Maximum copper concentrations in selected abiotic materials are 5.0 $\mu g/m^3$ in air, 5.0 μg/L in groundwater, 12.0 μg/L in rainwater, 300.0 mg/kg DW in black shales, 1200.0 mg/kg DW in poultry litter, 6500.0 mg/kg DW in marine sediments, 7000.0 mg/kg DW in soils, and 7700.0 mg/kg DW in sewage sludge. Copper concentrations in field collections of plants and animals are usually elevated in areas treated with copper-containing herbicides, near smelters, and from heavily urbanized and industrialized areas. The amount and distribution of copper in animal tissues varies with tissue, organism age, sex, and amount of copper in the diet.

In terrestrial vegetation, copper is usually less than 35.0 mg/kg DW except near smelters, where it may approach 700.0 mg/kg DW, and in copper-accumulator plants that may normally contain as much as 13,700.0 mg/kg DW. In aquatic vegetation, copper is elevated in metal-contaminated water bodies, reaching concentrations as high as 1350.0 mg/kg DW in eelgrass (*Zostera* spp.) from contaminated bays vs. 36.0 mg/kg DW in conspecifics from reference sites.

Copper concentrations in terrestrial invertebrates from industrialized areas range from 137.0 to 408.0 mg/kg DW. Soil invertebrates are not likely to accumulate copper but are important in recycling copper through terrestrial food webs. Aquatic invertebrates seldom contain as much as 95.0 mg Cu/kg DW, regardless of collection locale; exceptions include whole amphipods and lobster hepatopancreas (335–340 mg/kg DW) from copper-contaminated sites and many species of mollusks that normally contain 1100.0–6500.0 mg Cu/kg DW.

Maximum concentrations of copper in elasmobranchii and teleosts from all collection sites range from 7.0 to 15.0 mg/kg DW in eyeballs, intestines, muscles, scales, vertebrae, heart, and gonads and from 16.0 to 48.0 mg/kg DW in gills, kidneys, skin, and spleens and reach 53.0 mg/kg DW in whole animals, 155.0 mg/kg DW in stomach contents, 208.0 mg/kg DW in feces, and 245.0 mg/kg DW in livers.

Data on copper concentrations in field collections of amphibians and reptiles are scarce. Crocodile eggs contain as much as 60.0 mg Cu/kg DW; however, some toads (*Bufo* spp.) may contain as much as 2100.0 mg Cu/kg DW in livers without apparent adverse effects.

Birds from contaminated sites may contain as much as 9.0–28.0 mg Cu/kg DW in eggs, muscles, and stomach contents; 43.0–53.0 mg/kg DW in kidneys, feces, and feathers; and 367.0 mg/kg DW in livers.

Marine mammals usually contain less than 44.0 mg Cu/kg DW in all tissues except livers. Copper in livers seldom exceeds

116.0 mg/kg DW except in polar bears (146.0 mg/kg DW), and manatees (*Trichechus manatus*) (1200.0 mg/kg DW) from a copper-contaminated site. Maximum copper concentrations in terrestrial mammals from all collection sites are usually less than 29.0 mg/kg DW in all tissues except kidneys (108.0 mg/kg DW) and livers (1078.0 mg/kg DW).

10.5.1 Abiotic Materials

Copper concentrations in abiotic materials are comparatively elevated near copper smelters and urban areas. Copper concentrations are also elevated in drinking water from copper pipes, in poultry and livestock manures, mine tailings, fossil fuels, shales, sewage sludge, and in wastes from plating industries, foundries, and coking plants. Drinking water from certain locales contains elevated concentrations of copper added intentionally to control algal growth; drinking water may account for 10–20% of the daily intake of copper in humans.

Copper is found in the rocks and minerals of the earth's crust, occurring usually as sulfides and oxides, and sometimes as metallic copper. The mean concentration of copper in the upper lithosphere ranges from 70.0 to 100.0 mg/kg, ranking fourteenth of the trace elements in this compartment. Copper in the environmental crust averages 50.0 mg/kg, but is higher (140.0 mg/kg) in ferromagnesium minerals. Soil contamination by copper occurs around all known smelter locations; contamination may persist for decades, and plants and animals are often unable to survive the harsh chemical environments created. Italian soils have higher copper concentrations (51.0 mg/kg DW) than those of other European countries, probably as a result of the widespread and prolonged application of copper-based fungicides in Italian orchards and vineyards.

Copper concentrations in lake sediments within a radius of 80 km from a smelter in northern Sweden are positively correlated with proximity to the smelter. In some cases, lake sediments are sinks for copper, with little release to the overlying lake water. For example, copper-bearing mine tailings in Butte Lake, British Columbia, do not undergo oxidative diagenesis because of a rapid rate of accumulation and short exposure time to dissolved oxygen in bottom waters. In Michigan, lakes with elevated concentrations of copper (34.0 μg/L) have low densities of fish populations. In the Elizabeth River estuary of southern Chesapeake Bay, anthropogenic copper and other chelatable metals are present at concentrations sufficient to adversely affect growth and survival of the copepod *Acartia tonsa*. In Norway, freshwater fish are present only when copper is less than 60.0 μg/L and some humic acids are present. Successful reproduction of the spotted salamander (*Ambystoma maculatum*) occurs at low water concentrations of copper (<10.0 μg/L), lead, and aluminum, and high concentrations of silicon. Failed reproduction occurs at low water concentrations of silicon, and elevated concentrations of copper (>25.0 μg/L), lead, and aluminum.

In marine ecosystems, the high copper levels measured in heavily contaminated coastal areas sometimes approach the incipient lethal concentrations for some organisms. Elevated copper concentrations in marine and estuarine environments may result from atmospheric deposition, industrial and municipal wastes, urban runoff, rivers, and shoreline erosion. Chesapeake Bay, for example, receives more than 1800 kg of copper daily from these sources. Copper concentrations in abiotic marine materials are generally higher near shore than off shore. Copper is elevated in sediments of many marinas, probably from the copper antifouling bottom paints used on boats housed in these marinas. In New Zealand, copper concentrations in contaminated inshore sediments frequently exceed 100.0 mg Cu/kg DW vs. 14.0 mg Cu/kg DW at noncontaminated sites. The fine particle fraction of sediments collected near bulkheads made of CCA-treated wood contain elevated concentrations of copper, chromium, and arsenic; metal concentrations decreased with increasing distance from the bulkhead. Sediments, for example, decreased from 11.0 mg Cu/kg DW in the vicinity of treated bulkheads to <2.0 mg/kg DW at more distant sites.

10.5.2 Terrestrial Plants and Invertebrates

In general, copper concentrations in terrestrial vegetation seldom exceed 35.0 mg/kg DW, except near point sources of copper contamination and in certain copper-tolerant species. The highest copper concentration recorded in nonaccumulator plants is 726.0 mg/kg DW in hair grass (*Deschampia flexuosa*) near a smelter. Several species of terrestrial plants accumulate spectacular concentrations of copper. Mint plants (*Aeolanthus* spp., *Elsholtzia* spp.) growing in copper-rich soils contain unusually high concentrations and are used as economic indicators of copper deposits in the former Soviet Union and the People's Republic of China. The copper plant mint (*Aeolanthus biformifolius*), for example, normally contains as much as 13,700.0 mg Cu/kg DW whole plant. Copper-tolerant species of mosses, lichens, fungi, and higher plants occur in Greenland, Canada, the former Soviet Union, Africa, and elsewhere. In Zambia and Zimbabwe, the copper-tolerant *Becium homblei* is found only in soils containing more than 1000.0 mg Cu/kg and is believed to be responsible for the discovery of copper deposits in those nations. Some species of copper-indicator plants in Zambia tolerate as much as 70,000.0 mg Cu/kg in the soil and accumulate as much as 3000.0 mg Cu/kg in leaves.

Copper is not accumulated from soils by most crop plants, suggesting a soil–plant barrier for copper. Thus, corn (*Zea mays*) did not accumulate copper from soils treated with 365 kg Cu/surface ha (as copper-rich pig manure or copper sulfate) over a 13-year period; corn yield is not affected under these conditions.

Copper burdens in terrestrial invertebrates are highest in organisms collected near industrial locations and urban areas or from copper-contaminated soils. The highest copper concentration recorded among terrestrial invertebrates is 408.0 mg Cu/kg DW soft parts in gastropods from urban areas. Copper concentrations in pine moths (*Bupalus piniarius*) and pine noctuids (*Panolis flammea*) from industrialized areas range from 89.0 to 137.0 mg/kg DW, but are lower than dietary concentrations and suggest negligible accumulation. Accumulations of as much as 60.0 mg Cu/kg DW in 17-year-old cicadas (*Magicicada* spp.) pose no apparent dietary threat to insectivorous birds. Earthworms from soils heavily contaminated with copper (2740.0 mg/kg DW soil) can regulate copper more efficiently than cadmium and lead. However, copper is more toxic to earthworms than lead or zinc in the soil due to, in part, the inability of most soft tissues to synthesize copper-binding ligands when challenged with copper.

In woodland ecosystems, copper concentrations in the litter horizon are rarely exceeded by those in soil animals – which play a key role in copper cycling. Meiofaunal feces comprise an efficient distributing system through which copper and other nutrients are cycled through the food web of woodland ecosystems. During a 12-month cycle, the total copper bound in litter progresses through a cycle of chemical-binding states. It may be released from a strongly chelated organic complex as the litter is attacked by the digestive juices of animals or it may be discharged in soluble form with the feces and become complexed again by the activity of microorganisms in these feces. When feces are ingested by coprophagous animals, such as isopods, the copper may become trapped in proteins or membrane-bound vesicles.

10.5.3 Aquatic Organisms

Copper is essential for the successful growth and development of many species of aquatic organisms, but its rate and extent of accumulation and retention are modified by numerous biological and abiotic variables. Abiotic variables known to modify copper concentrations in tissues of aquatic biota include water temperature, pH, salinity, and depth; the presence of other inorganics, organics, and chelators; the chemical species of copper; and proximity to anthropogenic point sources of copper. Biological variables affecting copper accumulations in marine organisms include the organism's age, size, and developmental stage; physiological or genetic adaptation to high copper substrates; inherent species differences; and tissue specificity, such as the thorax of barnacles, gill and osphradium of gastropods, and livers of teleosts. Among marine organisms, the highest accumulations are generally found in molluscan tissues and soft parts, especially those of cephalopods and oysters. In order of decreasing copper accumulations, mollusks are followed by crustaceans, macrophytes, annelids, tunicates, algae, echinoderms, and coelenterates. Lowest concentrations of copper were consistently found among the vertebrates – elasmobranchii, fishes, and mammals – and strongly indicate a discrimination against copper among the highest marine trophic levels examined. Aquatic mollusks and arthropods that possess hemocyanin – a copper-containing respiratory pigment – have elevated tissue and plasma copper concentrations when compared to the ambient medium. Unlike many species of invertebrates, no vertebrate animal has a copper pigment as the main metallic constituent of blood. Marine organisms without hemocyanin have lower tissue concentrations of copper than those possessing this respiratory pigment.

Diet is the most important route of copper accumulation in aquatic animals, and food choice influences body loadings of copper. For example, whole body copper concentrations in aquatic insects from copper-contaminated rivers are highest in detritovores (as high as 102.0 mg/kg DW), followed by predators (54.0 mg/kg DW), and omnivores (43.0 mg/kg DW). Little or no biomagnification of copper is evident in freshwater food chains.

Copper concentrations in freshwater macrophytes near mining areas are elevated (as much as 256.0 mg/kg DW) compared to conspecifics collected from more remote sites. Bioconcentration factors (BCFs) (ratio of milligrams of copper per kilogram fresh weight organism to milligrams of copper per liter of ambient water) for copper by various species of freshwater algae range from 770 to 83,000. In general, copper accumulations in algae are higher at pH 8 than at pH 5; higher under conditions of low oxygen and reduced illumination; higher at low ambient concentrations of calcium, cobalt, zinc, magnesium, manganese, and organic chelators; and higher at elevated ambient concentrations of fluoride. Benthic communities in the vicinity of bulkheads made of CCA-treated wood had elevated concentrations of these elements, reduced species richness and diversity, and reduced numbers of total organisms when compared to reference sites. American oysters (*Crassostrea virginica*) from a canal lined with CCA-treated wood had 150.0 mg Cu/kg FW soft parts vs. 20.0 mg Cu/kg FW in oysters from a more distant site.

Copper concentrations in cephalopod mollusks are, in general, higher than those in bivalve mollusks; in cephalopods, 50–80% of the copper is localized in the digestive gland. Copper concentrations in tissues of clams (*Macoma balthica*) in San Francisco Bay are associated with seasonal variations in tissue weight, concentrations of copper in the sediments, and anthropogenic inputs from nearby sources. In cockles (*Cerastoderma edule*), copper concentrations in tissues decrease with increasing age, decrease in summer when compared to other seasons, and increase with increasing sediment copper concentrations. In the cockle (*Anadara trapezium*), tissue copper concentrations are positively related to dissolved copper concentrations in the water column and independent of sediment copper concentrations. Small freshwater clams (*Anodonta grandis*) have higher copper concentrations in soft tissues than large clams because small clams take up copper at a greater rate and excrete it more slowly than large clams; a similar case is made for oysters and other bivalve mollusks. Zebra mussels (*Dreissena polymorpha*) regulate body copper concentrations at water copper levels of 13.0 μg Cu/L and lower. Proximity to point sources, such as sewage discharge plants, is associated with elevated copper burdens in common mussels, *Mytilus edulis*. Copper concentrations increased in mussels (*Mytilus* spp.) analyzed in the coastal mussel watch program between the late 1970s and the late 1980s. This may be due to increased availability of copper from anthropogenic sources; however, concentrations of other metals (silver, nickel, cadmium, lead, and zinc) in mussels analyzed during this period also showed a decrease.

Pacific oysters near a copper recycling facility in Taiwan have grossly elevated concentrations of copper (as high as 4400.0 mg/kg DW soft parts), a characteristic green color, and low survival after exposure to waste effluents for 3 months. Diet is the major pathway by which greenish-colored Pacific oysters accumulate copper; initial daily accumulation rates are as high as 214.0 mg Cu/kg DW soft parts. Elimination of 50% of the copper from green Pacific oysters with elevated copper loadings takes only 11.6 days vs. 25.1 days in reference oysters. Elevated concentrations of copper in Pacific oysters (135.0 mg/kg DW soft parts) near a marina in Arcachon Bay, France, are attributed to the ban on tributyltin antifouling paints in 1982 and the subsequent growing use of copper-based antifouling paints. Copper concentrations in soft tissues of the American oyster are higher in oysters from low salinity waters than those from more saline waters; accumulations are not related to sediment copper concentrations in the immediate environment. In Maryland, copper concentrations in tissues of the American oyster are seasonally highest in July and lowest in October, and higher in low salinity waters than in high salinity waters. In Australia, copper concentrations in oyster soft parts from the Georges River, New South Wales, rose from 20.0 to 46.0 mg/kg FW in the 1970s to as high as 93.0 mg/kg FW in 1987, possibly as a result of urban and industrial discharges; this concentration exceeds the recommended limit of 70.0 mg Cu/kg FW in shellfish edible tissues for the protection of human health in Australia.

In amphipod (*Orchestia gammarellus*) crustaceans, copper concentrations, vary seasonally due to variable copper loadings, are higher in organisms from contaminated sites than reference sites, and higher in females with juveniles in the brood pouch than females without juveniles. The existence of copper-rich granules is common to all invertebrate phyla; these granules are usually found in the digestive gland or its evolutionary equivalent, and their formation is related to high concentrations of copper in the immediate environment. The tolerance of talitrid amphipods to high concentrations of ambient copper is attributable, in part, to the formation of intracellular granules within the cells of the ventral caeca. In the crab (*Carcinus mediterraneus*), tissue copper concentrations are lower in winter than in summer and correlate positively with total protein and hemolymph copper contents. Elevated copper burdens in hemolymph of crabs probably reflect the incorporation of copper atoms in the structure of hemocyanin, the major hemolymph protein. Marine decapod crustaceans regulate tissue copper concentrations within the range of 25.0–35.0 mg/kg DW.

In *Limnodrilus* sp., an oligochaete worm, copper bioavailability from surficial freshwater sediments is associated with the amount of copper present in the manganese oxide fraction of the sediment. The redox potential and pH in the gut of *Limnodrilus* allows the dissolution of the manganese oxide coating, making copper and other metals available for uptake.

Copper concentrations in freshwater fishes collected nationwide in the United States have not changed significantly since 1978. In 1984, samples with the highest copper concentrations were Mozambique tilapia (*Tilapia mossambica*) from Hawaii and white perch (*Morone americana*) from the Susquehanna River in Maryland. These locations have historically yielded fish with relatively high concentrations of copper; in Hawaii, this may develop from copper-containing pesticides. Copper concentrations in fishes are usually higher in liver than other tissues, higher in fish from copper-contaminated lakes than reference lakes, and higher in small fish than large fish of the same species. Residue data on copper in fish that are dead on collection are probably worthless for purposes of risk assessment owing to copper accumulation after death. Among sharks collected in British waters, copper concentrations in all tissues were highest from inshore demersal species and lowest from offshore pelagic species, with males having higher copper concentrations in liver than females.

Copper concentrations in tissues of marine vertebrates tend to decrease with increasing age of the organism. Concentrations of copper in marine and coastal vertebrates – including elasmobranchii, teleosts, and pinniped mammals – are related to the age of the animal. Regardless of species or tissues, except brain, concentrations decrease with increasing age of the organism; brain copper concentrations in marine mammals increase with organism age. Decreases in tissue copper content are also associated with spawning migrations of salmonids when entering freshwater from the sea and with reproductive cycles of cod and other gadoids. In the copper-contaminated Miramichi River, Canada, populations of Atlantic salmon (*Salmo salar*) are reduced in numbers due to poor survival and reproduction. Copper-containing mine wastes entering the Northwest Miramichi River cause many adult Atlantic salmon on their normal upstream spawning migration to return prematurely downstream; about 62% do not reascend. Downstream returns of salmon rose from 1 to 3% before pollution to 10–22% during four years of pollution. During some periods, dissolved copper and zinc concentrations exceed the lethal levels for immature salmon and the avoidance concentrations for subadults.

In polar bears, concentrations of copper in liver are 3–5 times higher than their seal diet. Copper concentrations in liver and kidney of polar bears are lower in juveniles than adults, which is contrary to a reverse trend noted in most species of vertebrates. Neonatal marine mammals, for example, have higher concentrations of copper in liver than those found in the mother. The use of copper herbicides in Florida to control aquatic plants may be hazardous to the endangered manatee. Copper concentrations in livers of these aquatic herbivores from areas of high copper herbicide use are as high as 1200.0 mg/kg DW. The maximum copper concentrations in livers of copper-challenged manatees are higher than any copper concentration measured in any species of free-ranging mammalian wildlife and are comparable to copper concentrations in livers of some species of domestic animals poisoned by copper.

10.5.4 Amphibians and Reptiles

Eggs of the Jefferson salamander (*Ambystoma jeffersonianum*) from a series of ponds that contain 1.0–25.0 μg Cu/L have – at the higher copper concentrations – a reduction in hatching success and an increase in embryonic mortality.

For reasons unknown, livers of some adult giant toads (*Bufo marinus*) normally contain grossly elevated concentrations of copper (>2000.0 mg/kg DW). The toads' livers are undamaged by this level of copper, and this lack of effect is in sharp contrast to human patients with Wilson's disease (2000.0 mfg Cu/kg DW liver) wherein hepatocyte degeneration, necrosis, and ultimately cirrhosis result. In toad livers, the copper is sequestered in lysosomes, which seems to protect the cell from the toxic effects of copper. In contrast, copper in liver of humans with Wilson's disease is diffusely distributed in the cytoplasm of hepatocytes and is associated with severe and often fatal pathological changes.

10.5.5 Birds

Season of collection and organism age affect copper concentrations in avian tissues. In livers of surf scoters (*Melanitta perspicillata*) from San Francisco Bay, copper concentrations are higher in March than in January; in livers of canvasbacks (*Aythya valisineria*) from Louisiana, concentrations are lower in November than later months; and in primary flight feathers of mallards (*Anas platyrhynchos*) and black ducks (*Anas rubripes*) from the vicinity of a smelter in Sudbury, Ontario, copper concentrations are highest in autumn. Copper concentrations in tissues of coastal seabirds tend to decrease with increasing age. In New Zealand, younger marine birds have higher concentrations of copper in livers than adults. But juveniles and adults of common murres (*Uria aalge*) from Scotland have similar concentrations of copper in kidneys, livers, and muscle.

In general, birds retain a very small portion of copper and other metals ingested. It is therefore noteworthy that livers of some canvasbacks collected in Louisiana, and livers of some mute swans (*Cygnus olor*) from England both contain more than 2000.0 mg Cu/kg DW. In the case of mute swans, several thousands of milligrams of copper per kilogram dry weight occur in the blackened livers; blackening is attributed to ingestion of flakes of copper-based antifouling paints. Tree swallows (*Tachycineta bicolor*) nesting near acidified aquatic ecosystems accumulate sufficient copper from the diet to induce elevated hepatic metallothionein concentrations. Copper concentrations in stomach contents of willets (*Catoptrophorus semipalmatus*) from San Diego Bay tend to reflect sediment copper concentrations. However, there is no evidence of copper biomagnification in the sediment food chain of sediment–pondweed–red-knobbed coot (*Fulica cristata*).

10.5.6 Mammals

Impalas (*Aepyceros melampus*) found dead in Kruger National Park, South Africa, had elevated concentrations of copper in livers (maximum 444.0 mg/kg FW) and kidneys (maximum 141.0 mg/kg FW); copper poisoning seems to be the most likely cause of death, but this needs verification. Copper concentrations in bones, kidneys, and livers of white-tailed deer (*Odocoileus virginianus*) near a copper smelter and from distant sites are about the same; however, deer near the smelter have significantly elevated concentrations of cadmium in kidneys and livers, lead in bone, and zinc in kidneys.

Only a small portion (0.037%) of copper mining wastes discharged into riparian wetlands is bioavailable to resident rodents, as judged by measurements of copper in carcasses of mice and voles. Populations of brown-backed voles (*Clethrionomys rufocanus*) and other microtine rodents (*Microtus* spp., *Lemmus*) are low or absent in the vicinity of Russian copper–nickel smelters. The reasons for this decline are unknown but may be due to a decrease in the abundance of important food plants (lichens, mosses, seed plants), and – as shown in preference studies – to an avoidance of plants from the contaminated area. Bank voles (*Clethrionomys glareolus*) from areas of Poland subjected to various degrees of industrial contamination have copper concentrations in tissues comparable to those in animals from polluted sites in North America and the United Kingdom. Compared to animals from a reference site, muskrats (*Ondatra zibethicus*) from a site contaminated by copper and other chemicals have higher concentrations of copper in kidneys, smaller spleens, larger adrenals, less fat, and lower body weight.

In Poland near copper foundries, livers from cattle (*Bos* sp.) have higher copper concentrations (35.0–140.0 mg/kg FW) than cattle from agricultural regions (7.0–32.0 mg/kg FW); however, kidney copper concentrations are comparable for both regions. Cattle found dead in South Africa near a copper smelter had elevated levels of copper in liver (600.0 mg/kg FW; 1078.0 mg/kg DW); airborne copper from the smelter is considered the most likely cause of death. Sheep held for 150 days in a paddock near a heavily traveled highway have significantly elevated copper concentrations in wool; these differences are not as pronounced in hair from horses (*Equus caballus*) and alpacas (*Lama pacos*) held under similar conditions.

Interspecies differences in copper contents are considerable. Serum from domestic dogs (*Canis familiaris*) lacks the strong copper-binding site available on the serum albumin molecule of humans and rats. Accordingly, copper concentrations in livers from dogs (82.0 mg/kg FW; 336.0 mg/kg DW) are normally about 12 times higher than those of human livers and 19 times higher than those of rat livers.

Human foods that are particularly rich in copper (20.0–400.0 mg Cu/kg) include oysters, crustaceans, beef and lamb livers, nuts, dried legumes, dried vine and stone fruits, and cocoa. In humans, copper is present in every tissue analyzed. A 70-kg human male usually contains 70.0–120.0 mg of copper. The brain cortex usually contains 18% of the total copper, liver 15%, muscle 33%, and the remainder in other tissues – especially the iris and choroid of the eye. Brain gray matter (cortex) has significantly more copper than white matter (cerebellum); copper tends to increase with increasing age in both cortex and cerebellum. In newborns, liver and spleen contain about 50% of the total body burden of copper. Liver copper concentrations were usually elevated in people from areas with soft water. Elevated copper concentrations in human livers are also associated with hepatic disease, tuberculosis, hypertension, pneumonia, senile dementia, rheumatic heart disease, and certain types of cancer.

10.6 Copper Deficiency Effects

Adverse effects of copper deficiency are documentable in terrestrial plants and invertebrates, poultry, small laboratory animals, livestock – especially ruminants – and humans. Data are scarce or missing on copper deficiency effects in aquatic plants and animals and in avian and mammalian wildlife. Copper deficiency in sheep – the most sensitive ruminant mammal – is associated with depressed growth, bone disorders, depigmentation of hair or wool, abnormal wool growth, fetal death and resorption, depressed estrous, heart failure, cardiovascular defects, gastrointestinal disturbances, swayback, pathologic lesions, and degeneration of the motor tracts of the spinal cord.

10.6.1 Terrestrial Plants and Invertebrates

Copper is an essential micronutrient of all higher plants studied, being a cofactor for the enzymes polyphenol oxidase, monophenol oxidase, laccase, and ascorbic acid oxidase. In copper-deficient soils, copper is strongly held on inorganic and organic exchange sites and in complexes with organic matter, causing reduced availability of copper to vegetation in these soils. Copper deficiency in terrestrial plants is usually associated with reduced growth, abnormally dark coloration in rootlets, and chlorotic leaves. In agricultural crops, copper deficiency occurs at <1.6 mg dissolved Cu/kg DW soil, and in sensitive plants at <2.0–<5.0 mg total Cu/kg DW leaves. In fruit trees, copper deficiency is characterized by death of apical buds, formation of multiple buds, and yellowing (chlorosis) of the leaf margins. Copper deficiency in alfalfa (*Medicago sativa*) and clover (*Trifolium* spp.) is associated with a faded green leaf color, growth inhibition, and withering. In grasses, copper deficiency is characterized by chlorosis, stunting, and necrosis and in cereals by pale color, reduced growth, and a reduction in the number of pollen grains.

Increased yields of various crops occur when copper salts are added to fertilizers at 300.0–800.0 mg Cu/m^3. In corn (*Zea mays*) and other vegetables, younger plants are more sensitive to copper deficiency than mature plants; in all cases, copper-deficient vegetables show chlorosis, reduced growth and reproduction, and low survival.

No evidence of copper deficiency exists in terrestrial species of invertebrates examined; however, relatively low concentrations of copper stimulated growth and reproduction. Reproduction in mites (*Platynothrus peltifer*) increases when fed diets containing 28.0 mg Cu/kg DW (vs. 13.0 mg/kg in controls) for 3 months. And juveniles of earthworms (*Eisenia andrei*) show increased growth at 18.0 mg Cu/kg DW soil after 12 weeks.

10.6.2 Aquatic Organisms

No documented report of fatal copper deficiency is available for any species of aquatic organism. And no correlation is evident in aquatic biota for the presumed nutritional copper requirements of a species and its sensitivity to dissolved copper. Extremely low copper concentrations (5.5 and 6.7 mg/kg DW) in whole bodies of 2 of 17 species of crustaceans from the Antarctic Ocean support the hypothesis that certain Antarctic species may show copper deficiencies or reduced metal requirements.

10.6.3 Birds and Mammals

Copper deficiency is not a major public health concern in the United States. Copper deficiency is rarely observed in humans except in cases of severely malnourished children or those with Menkes' disease – an X-linked recessively inherited disorder. This disease is a severe congenital copper deficiency marked by slow growth, progressive cerebral degeneration, convulsions, temperature instability, bone alterations, and peculiar steel-like hair. Treatment of Menkes' disease is now restricted to parenteral administration of copper salts, although complete prevention of neurodegradation is difficult to obtain. Copper deficiency is sometimes reported in humans after intestinal resection surgery (reduced absorptive surface), in people who consume high levels of zinc (zinc induces intestinal metallothionein that blocks copper transport), in infants who consume a diet based on cow milk (cow milk is a poor source of copper), and in genetic cases. Moderate copper deficiency also exists in burn and trauma patients, two groups at high risk for sepsis.

An inherited abnormal copper metabolism has been established in certain strains of mice, rats, and dogs. Feeding a copper-deficient diet to these animals may prevent acute hepatitis. In rats with abnormal copper metabolism and hereditary hepatitis, the feeding of a copper-deficient diet (0.5 mg Cu/kg ration for 35 days) prevents copper accumulation (94.0–139.0 mg Cu/kg DW liver) and dysfunction. But feeding a normal diet of 30.0 mg Cu/kg DW ration to these rats produces liver Cu concentrations of 375.0 mg/kg DW. Administered copper protects copper-deficient strains of mice against neurodegradation, and protects ponies against selenium poisoning when pretreated with 20.0 or 40.0 mg Cu/kg BW.

Chickens (*Gallus domesticus*) given diets deficient in copper (<2.7 mg Cu/kg ration) have anemia, poor growth, low survival, and a high frequency of cardiovascular and skeletal lesions. It is emphasized, however, that copper deficiency does not usually arise from eating a copper-poor diet because copper is found ubiquitously in foods. Chickens, turkeys (*Meleagris gallopavo*), cattle, and pigs deficient in copper are prone to die suddenly. Sudden death in some copper-deficient species is sometimes associated with rupture of a major blood vessel or rupture of the heart. Male weanling rats given a copper-deficient diet of 0.13 mg Cu/kg ration (vs. copper-normal diet of 5.7 mg Cu/kg ration) for 7 weeks show high mortality (24%) from cardiac rupture; ruptured hearts had elevated concentrations of sodium, potassium, and calcium, and depressed magnesium. Copper deficiency in weanling rats is confirmed by low activities of ceruloplasmin in serum and superoxide dismutase in liver and serum. Skeletal deformities and leg fractures occur in copper-deficient chickens, dogs, pigs, sheep, cattle, and children because of decreased tensile strength of bones. In lambs from copper-deficient ewes, locomotor disturbances of gait or posture occur because of lesions of excessive myelination of the central nervous system.

Copper deficiency in humans and other mammals is characterized by slow growth, hair loss, anemia, weight loss, emaciation, edema, altered ratios of dietary copper to molybdenum and other metals, impaired immune response, decreased cytochrome oxidase activity, central nervous system histopathology, decreased phospholipid synthesis, fetal absorption, and eventually death.

In laboratory white rats, signs of copper deficiency include reductions in tissue copper concentrations; reduced activities of cytochrome oxidase, superoxide dismutase, succinoxidase, and ceruloplasmin; increased activity of 7-ethoxyresorufin-*O*-deethylase (EROD) in small intestines; anemia associated with low hematocrit and hemoglobin; increased acute inflammatory response; increased sensitivity to endotoxins; central nervous system lesions; and reduced phospholipid synthesis. Copper-deficient rats also have prolonged sleeping times and significant reductions in activities of aniline hydroxylase and hexobarbital oxidase in liver. Earliest signs of copper deficiency in rats include low concentrations of copper in livers (1.4–<3.0 mg/kg DW vs. 12.6–15.0 mg/kg DW in controls); profound reductions in activities of cytochrome oxidase and succinoxidase; and reductions in hematocrit, hemoglobin, ceruloplasmin, and phospholipid synthesis. Severe copper deficiency in rats results in anemia characteristic of defective hemoglobin synthesis resulting from abnormal use of iron by mitochondria in heme synthesis. Copper-deficient rats are extraordinarily sensitive to endotoxins and die after receiving normally sublethal doses of various endotoxins. Copper deficiency-induced lesions in the central nervous system are produced experimentally in rats and guinea pigs and are a characteristic feature of Menkes' disease.

Sway disease of Bactrian camels (*Camelus bactrianus*) – characterized by anemia, emaciation, falling, fractures, and death – is caused by copper deficiency associated with high molybdenum content in soils and forage; deficiency effects are aggravated during reproduction. Sheep fed copper-deficient diets of less than 2.5 mg Cu/kg DW ration (vs. a normal diet of 11.0 mg Cu/kg DW) produce a high frequency of swaybacked lambs. Swaybacks have lower concentrations of copper in liver than nonswaybacked lambs from copper-deficient ewes; both groups have lower concentrations of copper in livers than normal lambs.

Copper deficiency effects are reported in mink (*Mustela vison*) and domestic swine. Copper deficiency in mink, as judged by reduced survival, occurs by feeding rations containing the equivalent of 3.5 mg Cu/kg BW daily for a period of 50 weeks. Swine, which seem to have higher copper requirements than mink, given low copper diets equivalent to 15.0–36.0 mg Cu/kg BW daily for 7 days have decreased hemoglobin, hematocrit, and growth rate.

Dietary copper deficiency increases the acute inflammatory response in rats and other small laboratory animals. The release of inflammatory mediators, such as histamine and serotonin, from mast cells increases the vascular permeability of postcapillary venules and results in edema. In copper-deficient rats, release of histamine from mast cells positively correlates with frequency of the acute inflammatory response. Copper-deficient rats (0.6 mg Cu/kg DW ration for 4 weeks) have more mast cells in muscle than copper-adequate controls given diets containing 6.3 mg Cu/kg DW ration; however, histamine content of mast cells is not affected. An early clinical sign of copper deficiency is a reduction in the number of circulating neutrophils; the mechanism for copper-deficient neutropenia (leukopenia in which the decrease in white blood cells is chiefly neutrophils) is unknown. Proposed mechanisms to account for neutropenia from copper deficiency include (1) early destruction of bone marrow progenitor cells; (2) impaired synthesis of neutrophils from progenitor cells; (3) a decrease in the rate of cellular maturation in the bone marrow; (4) impaired secretion of neutrophils from the bone marrow; and (5) rapid clearance of circulating copper-deficient neutrophils.

10.7 Lethal and Sublethal Effects

Copper is toxic to sensitive species of terrestrial vegetation at >40.0 μg/L nutrient solution (seedlings of pines, *Pinus* spp.), at >10.0 mg/kg DW leaves (cucumber, *Cucumis sativus*), and >60.0 mg extractable Cu/kg DW soil (sweet orange, *Citrus sinensis*). Among sensitive species of terrestrial invertebrates, adverse effects on survival, growth, or reproduction occur at 2.0 μg Cu/cm^2 on paper disks (earthworms), >50.0 mg Cu/kg diet (larvae of gypsy moth, *Lymantria dispar*), and 53.0–70.0 mg Cu/kg DW soil (earthworms and soil nematodes).

Sensitive species of representative freshwater plants and animals die within 96 h at waterborne copper concentrations of 5.0–9.8 μg/L. The most sensitive freshwater species have LC50 (96 h) values between 0.23 and 0.91 μg Cu/L and include daphnids (*Daphnia* spp.), amphipods (*Gammarus pseudolimnaeus*), snails (*Physa* spp.), and Chinook salmon (*Oncorhynchus tshawytscha*). In general, mortality of tested aquatic species is greatest under conditions of low water hardness (as measured by $CaCO_3$), starvation, elevated water temperatures, and among early developmental stages. Toxicity testing of copper-contaminated sediments to amphipods (*Hyalella azteca*) and daphnids (*Daphnia magna*) using techniques of enzyme inhibition and growth rate show that these variables are more sensitive in accurately predicting copper sensitivity than LC50 (48 h) values and should be considered when assessing risk of contaminated sediments to freshwater systems. The most sensitive saltwater species to copper have LC50 (96 h) values from 28.0 to 39.0 μg/L and include summer flounders (*Paralichthys dentatus*), copepods (*Acartia tonsa*), and softshell clams (*Mya arenaria*). Adverse sublethal effects of copper on representative species of estuarine algae, mollusks, and arthropods frequently occur at 1.0–10.0 μg/L.

No data are available on copper toxicity to avian wildlife. Experiments with domestic poultry show that copper accumulates in livers of mallard ducklings at dietary concentrations as low as 15.0 mg/kg DW ration; that gizzard histopathology and a reduction in weight gain of chicks (*Gallus* sp.) occur at 250.0–350.0 mg Cu/kg DW ration; and that growth of turkey poults is improved at 60.0 mg Cu/kg DW ration and inhibited at 120.0 mg/kg DW ration, with signs of gizzard histopathology at 500.0 mg/kg DW ration.

Copper salts are lethal to mammals through a variety of routes. Single oral doses of 6.0–637.0 mg Cu/kg BW are fatal to humans. A single oral dose of 200.0 mg/kg BW is usually fatal to cattle. Dietary copper is lethal when eaten for extended periods at >80.0 mg Cu/kg ration in sheep (equivalent to 5.1–10.7 mg Cu/kg BW daily), >238.0 mg/kg ration in pigs, and >4000.0 mg/kg ration in rats (equivalent to >133.0 mg Cu/kg BW daily). Adverse sublethal effects of copper to sensitive mammals occur in human infants at drinking water concentrations >3.0 mg Cu/L; in cattle at dietary levels >20.0 mg Cu/kg BW by way of intraperitoneal injection and >4.2 mg Cu/kg BW via drinking water; in sheep given daily oral doses of 7.5–15.0 mg Cu/kg BW or fed diets containing >37.3 mg Cu/kg ration; in rats at >100.0 mg Cu/kg ration (equivalent to >7.9 mg Cu/kg BW daily), >400.0 mg Cu/L drinking water, or >2.0–2.5 mg Cu/kg BW daily via injection; and in pigs at >14.5 mg Cu/kg BW daily via diet. Elevated copper concentrations (328.0 mg Cu/kg DW) occur in livers of surviving cattle fed diets containing 8.2 mg Cu/kg ration; of sheep (1109.0 mg/kg DW liver) fed diets containing 37.3 mg Cu/kg ration; and of rats (710.0 mg/kg FW liver) given intraperitoneal injections of 3.75 mg Cu/kg BW daily for 18 weeks.

10.7.1 Terrestrial Plants and Invertebrates

Copper is toxic to sensitive plants when plant nutrient solutions contain >40.0–200.0 μg Cu/L, when leaves have >10.0–12.0 mg Cu/kg DW, and when extractable copper in soils is >60.0 mg/kg DW soil. Excess copper inhibits root elongation and branching and reduces the ability of the plant to explore the soil for water and nutrients. Root damage occurs in pine seedlings (*Pinus* spp.) after exposure for 10 days to nutrient solutions that contain 40.0 μg Cu/L. A lower concentration of 4.0 μg Cu/L has no adverse effects on root growth and morphology, while a higher concentration of 400.0 μg Cu/L completely inhibits root growth within 3 days.

Severely reduced crop growth was measured in soils treated with 750 kg Cu/ha annually for 13 years wherein the soil pH declined from 6.1 to 4.7 during this period. Laboratory studies showed that planting of copper-tolerant grass (*Agrostis capillaris*) on these soils resulted within 10 weeks in faster bacterial growth, more protozoans and more nematodes in comparison to fallow controls. It seems metal-tolerant plant species can re-establish the necessary food base to support organism growth and can reverse the loss of soil function due to high copper levels under acidic conditions.

Poultry litter is a useful agricultural by-product with high nitrogen and phosphorus content and is frequently added to agricultural soils. Poultry litter from northern Georgia containing 1196.0 mg Cu/kg DW and applied at a final rate of 5.0–15.0 mg Cu/kg soil to fields of Sudex (*Sorghum bicolor* × *S. sudanense*) did not affect copper levels of treated Sudex or produce any evidence of toxicity. But most terrestrial vegetation in the United States, Sweden, Wales, and other locales is usually adversely affected by emissions from copper mines, brass foundries, and copper smelters. Damage to vegetation persists for at least 50 years after closure of a copper smelter because of copper, arsenic, and lead in the soil. Particularly sensitive to copper in the soils are white pine (*Pinus strobus*) and red maple (*Acer rubrum*); less sensitive are Douglas fir (*Pseudotsuga menziesii*) and lodgepole pine (*Pinus contorta*).

Earthworms (*Eisenia fetida*) held in soils containing 53.0 mg Cu/kg DW show a 50% reduction in cocoon production in 56 days; 32.0 mg Cu/kg soil had no effect on cocoon production. The LC50 (56 days) value for earthworms is 555.0 mg Cu/kg DW soil; no deaths occur at 210.0 mg/kg soil during this period. Copper is more toxic to *Eisenia fetida* than are salts of cadmium, zinc, or lead. Copper adversely affects the earthworm *Lumbricus rubellus*. Concentrations of 150.0 mg Cu/kg surface soil from an accidental spill of copper sulfate in grasslands reduced earthworm populations by about 50%; surface soil concentrations of 260.0 mg Cu/kg kill almost 100% of the *Lumbricus*. Copper is most toxic to *Lumbricus* at low soil pH (4.8–7.1) and at low temperatures. Earthworm cocoons, however, were relatively unaffected by copper at concentrations <12.0 mg Cu/L, and at 20°C unless stressed by desiccation or frost. Cocoons of *Aporrectodea calignosa* were unaffected at 20.0 mg Cu/L except under conditions of low relative humidity (95% relative humidity vs. 97.5 or 100% relative humidity), and temperatures of −3 and −6°C, but not at 0°C.

Tests show that the presence of soil reduces the toxicity of copper to the soil-dwelling nematode *Caenorhabditis elegans*; copper toxicity to nematodes increases with increasing densities of bacteria and increasing concentrations of sodium chloride or potassium chloride. Terrestrial isopods efficiently assimilate and store copper as detoxified granules in the hepatopancreas; this activity is in contrast to many species of marine crustaceans that are unable to assimilate, detoxify, or otherwise regulate copper.

10.7.2 Aquatic Organisms

Lethal and sublethal effects of copper compounds are documented for selected species of aquatic biota, including algae, macrophytes, cnidarians, mollusks, arthropods, annelids, and fishes.

10.7.2.1 Plants

Photosynthesis and growth in sensitive species of freshwater algae are inhibited by copper concentrations of 1.0–6.0 μg/L. For sensitive species of estuarine phytoplankton, copper is lethal at 50.0 μg/L and most toxic under conditions of decreasing salinity, pH, and concentrations of chelators. Sensitivity to copper varies widely among species of estuarine algae; some species, for example, grow normally at concentrations as high as 10.0 mg Cu/L during exposure for 9 days. In mesocosm studies, 50.0 μg Cu/L caused a reduction of about 80% in total zooplankton and total algal biovolumes; the algal assemblage that persisted was dominated by diatoms. Copper-resistant strains of *Euglena gracilis* challenged with high sublethal concentrations of copper for 5 days had an altered cysteine metabolism.

Some species of aquatic plants absorb or adsorb dissolved copper at extremely high rates. Bioconcentration factors for copper and freshwater alga (*Chlorella* sp.) range from 203 to 2000 after exposure for 14–30 h. Seagrass (*Heterozostera tasmanica*) in seawater containing 42.0 μg Cu/L for several weeks contain 2700.0 mg Cu/kg DW; seagrasses in media containing 0.3 μg Cu/L contain 2.5 mg Cu/kg DW; and intermediate values are reported for 10.0 μg Cu/L (306–564 mg/kg DW) and 20.0 μg/L (1280.0 mg/kg DW). Some freshwater aquatic macrophytes accumulate as much as 54,500.0 mg Cu/kg DW, as was the case for *Lemna* sp. during exposure to 1000.0 μg Cu/L; a lower dose regimen of 35.0 μg Cu/L results in 256.0 mg Cu/kg DW *Lemna*.

10.7.2.2 Cnidarians

Sea anemones (*Anemonia viridis*) in seawater solutions containing 50.0 or 200.0 μg Cu/L regulate copper by expelling zooxanthellae, which are shown to accumulate copper.

10.7.2.3 Mollusks

Initial effects of copper on mussels (*Mytilus* spp.) include valve closure, a reduction in filtration rates, and cardiac inhibition; all these responses serve to slow the uptake of copper through a reduction in mussel contact with the ambient environment and a reduction in blood flow within the organism. Copper impairs the structure and function of cellular membranes in mussels by stimulating the peroxidation of membrane lipids; end products of lipid peroxidation contribute to the formation of lipofuscins. Copper-induced lysosomal lipofuscin accumulations, together with metallothioneins, control copper residues at the cellular levels and are responsible for the short half-time persistence (6–8 days) of copper in the digestive gland of mussels. Concentrations of heat shock protein (hsp60) in mantle tissues of mussels exposed to copper increased in a dose-dependent manner; hsp60 may have potential as a biomarker of copper insult. Copper-stressed common mussels (*Mytilus edulis*) die more quickly under conditions of anoxia, high temperatures, and low salinities. Concentrations of copper that cause a decrease in yields of normal larvae in populations of *Mytilus edulis* from unpolluted or mildly contaminated sites did not affect embryonic development of mussels from polluted sites; cross breeding of mussels from these sites suggest that copper tolerance in mussels is mostly maternally determined. Embryos of common mussels are more sensitive to copper than veliger larvae or postlarval spat stages. A copper-induced decrease in glochidial viability is a possible explanation for the disappearance of freshwater unionid mussels from acid- and metal-contaminated waters. Some investigators aver that mussels of all ages are equally susceptible to copper and that their capacity to recover declines with increasing age; however, this phenomenon needs verification.

Bioconcentration factors for marine bivalves (ratio of milligrams of copper per kilogram fresh weight soft parts to milligrams of copper per liter of medium) vary from 85 to 28,200. Bioconcentration factors for copper are highest for American oysters after exposure for 140 days (20,700–28,200), and lowest for bay scallops (*Argopecten irradians*) after exposure for 112 days (3310) and for softshell clams after exposure for 35 days (3300). Copper is more toxic to embryos of the tropical giant clam (*Tridacna derasa*) than to embryos of bivalves from temperate regions, possibly because many tropical species of shellfish live near their upper lethal thermal limits and are unable to withstand additional environmental stressors. Juveniles of Asiatic clams (*Corbicula fluminea*) are more sensitive than adults to ionic copper.

On exposure to lethal concentrations of copper the channeled whelk (*Busycon canaliculatum*), a marine gastropod, accumulates the metal in gill and osphradium. These tissues show progressive histopathology including swelling of the gill filaments, amoebocytic infiltration of the connective tissue, and necrosis and sloughing of the mucosa. Copper-resistant strains of freshwater gastropods are found in media containing elevated concentrations of 35.0 μg Cu/L, suggesting physiological or genetic adaptation. Fine suspensions of copper and kaolinite mixtures are more toxic to freshwater gastropods than copper alone; toxicity is greater at pH 8 than at pH 7, suggesting that copper is strongly adsorbed by kaolinite in alkaline media and that the acidic pH of the snail gut enhances release of ionic copper. In freshwater gastropods, ionic copper causes hypersynthesis of lysosomal enzymes and acid and alkaline phosphatases; immature gastropods are more sensitive than adults.

10.7.2.4 Arthropods

Life-cycle exposures of four species of *Daphnia* to graded concentrations of copper show reductions in survival at more than 40.0 μg/L and reductions in growth and reproduction at 40.0–60.0 μg/L; heavier and larger species are the most resistant to copper. Starvation increases the sensitivity of most species of freshwater cladocerans to copper; however, there is no difference in LC50 (48 h) values between fed and starved *Daphnia magna*. Bioavailability and toxicity of copper to *D. magna* and other tested arthropods are usually higher under conditions of increasing acidification, ionic copper, alkalinity, and temperature, or of decreasing DOC. Mixtures of copper and other metals produce additive or more-than-additive effects in *D. magna* than would be expected on the basis of individual components. The concept that chronic exposures to pulses of the LC50 concentrations of copper or cadmium causes no damage to freshwater organisms – provided that the average daily concentration never exceeds the no-observable-effect concentration – was tested in daphnids. The concept was true for cadmium but not copper, and the use of pulsed exposures for establishing water quality criteria to protect aquatic life needs to be re-examined.

Copper uptake by aquatic arthropods occurs usually by way of the gut after eating or from the gills and other permeable surfaces in contact with the ambient medium. Copper accumulations by crustaceans are greatest at elevated (summer) temperatures and during molting. A relatively high BCF of 2000 is documented for copper and freshwater stoneflies (*Pteronarcys californica*), but the reasons for this phenomenon are unknown. The high tolerance to copper and other metals of mayfly larvae (*Baetis thermcius*), and high copper accumulations, is attributed, in part, to the selective induction of metal-binding proteins in the gut. Marine amphipods readily accumulate dissolved copper from seawater in a dose-dependent manner. But some species of talitrad amphipods are unable to meet their copper requirements from seawater alone and depend on dietary sources of copper. Mesocosm studies with freshwater zooplankton assemblages show that increasing copper concentrations in the range 0.0–50.0 μg/L causes a reduction in total zooplankton and changes in diversity; within 4 days, copepods became dominant at the expense of cladocerans.

Soldier crabs (*Mictyris longicarpus*) accumulate copper mostly from sediments rather than the water column. The fine particles of sediment trapped as food contain bioavailable fractions of copper and other metals, and these significantly correlate with metal concentrations in the body of the crab. However, copper accumulations from sediments by soldier crabs occur only at an artificially high concentration (1900.0 mg Cu/kg DW sediment), which also had toxic effects. Soldier crabs seem unable to regulate copper within their body.

In shore crabs, several days of exposure to sublethal concentrations of waterborne copper cause extensive damage to gill epithelium; at lethal concentrations, tissue hypoxia is probably the major effect of copper. Starved shore crabs show a reduction in carapace copper concentrations and heavier midgut glands; starvation in combination with copper exposure (500.0 μg/L) results in an increase in copper in the carapace and a decrease in carapace calcium. Shore crabs in seawater with high (10.0 mg/L) levels of waterborne copper show reductions in hemolymph sodium, gill sodium–potassium–ATPase activity, activities of various midgut gland enzymes (hexokinase, phosphofructokinase, pyruvate kinase), and hemolymph electrolytes.

In the rusty crayfish (*Orconectes rusticus*), toxicity of copper at high concentrations is due to the coagulatory action on cellular proteins and the interference with respiratory processes; at low concentrations, copper causes degenerative changes in certain tissues and interferes with glutathione equilibrium. Larvae of the red crayfish (*Procambarus clarkii*) exposed to copper as embryos are less sensitive than those exposed after hatching, suggesting acclimatization.

10.7.2.5 Annelids

Aquatic oligochaetes (*Lumbriculus variegatus*) do not accumulate significant amounts of copper when compared to controls after exposure for 30 days in sediments containing as much as 90.1 mg Cu/kg DW or in water containing as much as 2.3 μg Cu/L. Adult lugworms (*Arenicola marina*) living in sediments containing 182.0–204.0 mg Cu/kg DW sediment had inhibited digestive processes, suggesting that the digestive system of lugworms, and perhaps other deposit feeders, are vulnerable to copper-contaminated sediments. Larvae of the sandworm (*Nereis diversicolor*) are more resistant to copper with increasing organism age and with increasing temperature and salinity of the medium. In adult sandworms, whole body loadings of copper usually increase with increasing temperature in the range of 12–22°C and with decreasing salinity in the range 0.7–3.1%; however, copper–temperature–salinity interactions are significant and complex in this species.

10.7.2.6 Fishes

Adverse sublethal effects of copper on behavior, growth, migration, and metabolism occur in representative species of fishes at nominal water concentrations between 4.0 and 10.0 μg/L. In sensitive species of teleosts, copper adversely affects reproduction and survival from 10.0 to 20.0 μg Cu/L. Copper exerts a wide range of physiological effects in fishes, including increased metallothionein synthesis in hepatocytes, altered blood chemistry, and histopathology of gills and skin. At environmentally realistic concentrations, free copper adversely affects resistance of fishes to bacterial diseases; disrupts migration (that is, fishes avoid Cu-contaminated spawning grounds); alters locomotion through hyperactivity; impairs respiration; disrupts osmoregulation through inhibition of gill Na^{+}–K^{+}-activated ATPase; is associated with tissue structure and pathology of kidneys, liver, gills, and other hematopoietic tissues; impacts mechanoreceptors of lateral line canals; impairs functions of olfactory organs and brain; and is associated with changes in blood chemistry, enzyme activities, and corticosteroid metabolism. Copper-induced cellular changes or lesions occur in kidneys, lateral line, and livers of several species of marine fishes.

Copper-induced mortality in teleosts is reduced in waters with high concentrations of organic sequestering agents and in genetically resistant species. At pH values less than 4.9 (that is, at pH values associated with increased aluminum solubility and toxicity), copper may contribute to the demise of acid-sensitive fishes. Copper affects plasma Na^{+} and gill phospholipid activity; these effects are modified by water temperature and hardness. In red drum, copper toxicity is higher at comparatively elevated temperatures and reduced salinities. Copper is acutely toxic to freshwater teleosts in soft water at concentrations between 10.0 and 20.0 μg/L. In rainbow trout, copper toxicity is markedly lower at high salinities. Comparatively elevated temperatures and copper loadings in the medium cause locomotor disorientation of tested species. Copper may affect reproductive success of fish through disruption of hatch co-ordination with food availability or through adverse effects on larval fishes. Chronic exposure of representative species of teleosts to low concentrations (5.0–40.0 μg/L) of copper in water containing low concentrations of organic materials adversely affects survival, growth, and spawning; this range is 66.0–120.0 μg Cu/L when test waters contain enriched loadings of organic materials.

Larval and early juvenile stages of eight species of freshwater fishes are more sensitive to copper than embryos or adults. But larvae of topsmelt (*Atherinops affinis*) are increasingly sensitive to copper with increasing age. Topsmelt sensitivity is associated with increasing respiratory surface area and increasing cutaneous and branchial uptake of copper.

Sublethal exposure of fishes to copper suppresses resistance to viral and bacterial pathogens and, in the case of the air-breathing catfish (*Saccobranchus fossilis*), affects humoral and cell-mediated immunity, the skin, and respiratory surfaces. Rainbow trout exposed to 50.0 μg Cu/L for 24 h – a sublethal concentration – show degeneration of olfactory receptors that may cause difficulties in olfactory-mediated behaviors such as migration. The primary site of sublethal copper toxicity in rainbow trout is the ion transport system of the gills. In European sea bass (*Dicentrarchus labrax*), copper compromises the defense system of red blood cells against active forms of oxygen, leading to increased membrane lipid peroxidation.

Dietary copper is more important than waterborne copper in reducing survival and growth of larvae of rainbow trout. Simultaneous exposure of rainbow trout to dietary and waterborne copper results in significant copper assimilation. Diet is the main source of tissue copper; however, the contribution of waterborne copper to tissue burdens increases as water concentrations rise.

Rate and extent of copper accumulations in fish tissues are extremely variable between species and are further modified by abiotic and biological variables. Copper accumulations in fish gills increase with increasing concentrations of free copper in solution, increasing DOC, and decreasing pH and alkalinity. Starved Mozambique tilapia accumulate significantly more copper from the medium in 96 h than tilapia fed a diet containing 5.9 mg Cu/kg DW ration. The BCF for whole larvae of the fathead minnow was 290 after exposure for 30 h, but only 0.1 in muscle of bluegills after 660 h. Prior exposure of brown bullheads (*Ictalurus nebulosus*) to sublethal copper concentrations for 20 days before exposure to lethal copper concentrations produces higher copper concentrations in tissues of dead bullheads than in those not previously exposed; however, the use of tissue residues is not an acceptable autopsy procedure for copper. Rising copper concentrations in blood plasma of catfish (*Heteropneustes fossilis*) seem to reflect copper stress, although the catfish appear outwardly normal. Plasma copper concentrations of catfish increase from 290.0 μg Cu/L in controls at start to 380.0 μg Cu/L in survivors at 72 h (50% dead); a plasma copper concentration of 1060.0 μg Cu/L at 6 h is associated with 50% mortality. In rainbow trout, copper is rapidly eliminated from plasma; the half-time persistence is 7 min for the short-lived component and 196 min for the long-lived component.

Attraction to waters containing low (11.0–17.0 μg/L) concentrations of copper occurs in several species of freshwater teleosts, including goldfish (*Carassius auratus*) and green sunfish (*Lepomis cyanellus*); however, other species, including white suckers (*Catostomus commersonii*), avoid these waters. In avoidance–attraction tests, juvenile rainbow trout avoided waters containing 70.0 μg Cu/L but were significantly attracted to water containing 4560.0 μg Cu/L; a similar pattern was observed in tadpoles of the American toad, *Bufo americanus*. Copper concentrations in the range of 18.0–28.0 μg/L interfere with bluegill growth and prey choice. Copper interferes with the ability of fish to respond positively to L-alanine, an important constituent of prey odors; concentrations as low as 1.0 μg Cu/L inhibit this attraction response in some species.

Increased tolerance to copper was observed in fathead minnows after prolonged exposure to sublethal concentrations, but tolerance was not sustained on removal to clean water. Copper tolerance in fathead minnows is attributed to increased production of metallothioneins. Copper tolerance in rainbow trout seems dependent on changes in sodium transport and permeability.

10.7.2.7 Integrated Studies

Bioconcentration and biomagnification of copper occur in the food chain of diatom (*Skeletonema costatum*) to clam (*Donax cuneatus*) to prawn (*Penaeus indicus*). All species accumulate copper from the medium, and clams and shrimp from the diet. Maximum concentrations after exposure to 200.0 μg Cu/L and diets for 10 days, in milligrams of copper per kilogram fresh weight, are 2.8 in whole diatoms, 13.6 in clam soft parts, and 33.9 in whole shrimp. In the marine food chain of phytoplankton to clam (*Tellina tenuis*) to juvenile plaice (*Pleuronectes platessa*), copper accumulates in a concentration-dependent manner in viscera of plaice. All organisms held in 10.0, 30.0, or 100.0 μg Cu/L solutions for 100 days had reduced growth. Copper concentrations, in milligrams of copper per kilogram dry weight at day 100, in soft parts of clams *T. tenuis* were 270.0 in the 10.0 μg/L group, 470.0 in the 30.0 μg/L group, and 1100.0 in the 100.0 μg/L group vs. less than 50.0 in the controls; for plaice viscera, these values were 30.0 in controls, 71.0 in the 10.0 μg/L group, 147.0 in the 30.0 μg/L group, and 467.0 mg/kg DW in the 100.0 μg/L group. Accumulations in Pacific oysters held in copper-loaded sediments are similar to those of oysters contaminated through ingestion of diatoms (*Haslea ostrearia*). However, accumulations are highest in Pacific oysters when exposed through the medium; in that study, a concentration of 30.0 μg Cu/L medium for 21 days results in copper concentrations of 137.0 mg/kg DW in diatoms and 1320.0 mg/kg DW in oyster soft parts. Oysters fed contaminated diatoms in the study had 419.0 mg Cu/kg DW soft parts. Oysters held in sediments containing 108.0 mg Cu/kg DW – a level reached after exposure for 21 days to 300.0 μg Cu/L – had 401.0 mg Cu/kg DW. Copper-induced changes in population density and community metabolism occur in an aquatic mesocosm of algae, protozoans, rotifers, oligochaetes, and bacteria; death of rotifers, algae, and oligochaetes occur at concentrations as low as 700.0 μg Cu/L. Adverse effects occur at 300.0–700.0 μg Cu/L but are negated by increasing concentrations of dissolved organic matter.

Transfer of copper from wood treated with CCA occurs in estuarine algae (*Ulva*, *Enteromorpha*), American oysters, mud snails (*Nassarius obsoletus*), and fiddler crabs (*Uca* spp.). Algae, barnacles, and mussels from CCA-treated lumber show elevated concentrations of copper when compared to reference sites. The epibiotic estuarine community that forms on CCA-treated wood has lower species richness, diversity, and biomass when compared to untreated lumber. Copper is trophically transferred from CCA-exposed American oysters to predatory gastropods (*Thais* sp.), resulting in reduced gastropod feeding and growth.

10.7.3 Birds

No data are available on the toxicity of copper to avian wildlife. All studies with birds and copper use domestic chickens, ducks, or turkeys. Copper, however, may indirectly affect avian wildlife by curtailing certain prey species. For example, apple snails (*Pomacea paludosa*) are not only extremely susceptible to copper (LC50 of 24.0–57.0 μg/L in 96 h; immatures, most sensitive), but are the primary food of the snail kite (*Rostrhamus sociabilis*), an endangered species. The decline of the apple snail in southern Florida coincided with the use of copper-diquat to control hydrilla aquatic weeds (*Hydrilla verticillata*), with serious implications for snail kite survival.

In the domestic chicken, adverse effects of copper occur in chicks fed diets containing 350.0 mg Cu/kg ration for 25 days (reduced weight gain) and in adults given a dietary equivalent of more than 28.0 mg Cu/kg BW. Chicks fed diets of 500.0 mg Cu/kg ration show damage to the gizzard lining; damage effects are attributed to the shedding of gizzard glandular cells into the keratin-like koilin layer, disrupting koilin production. Copper-induced gizzard histopathology in growing chicks is not reversed by zinc or vitamins B_{12} or E. Supplementing chick diets with copper did not prove markedly advantageous, provided that normal rations had about 4.0 mg Cu/kg and adequate iron. Unlike mammals, chicks fed copper-supplemented diets do not have elevated copper concentrations in liver or signs of liver damage. Broiler hens housed on slats made of lumber pressure-treated with CCA show severe foot-pad dermatitis and excessive mortality after 17 weeks; however, arsenic and cresylic acid – not copper – may be the responsible agents.

Ducklings (*Anas* spp.), unlike chicks, accumulate copper in livers when fed diets supplemented with high loadings of copper. Domesticated mallards show a dose–time-dependent increase in copper liver concentrations, with a maximum concentration of 254.0 mg Cu/kg DW liver. Mallards seem to prefer drinking water containing 100.0 mg Cu/L over distilled water; however, these birds were molting and this may have influenced their response because trace mineral requirements rise during molting.

In turkeys, natural diets with as much as 800.0 mg Cu/kg ration have no adverse effects on growth or survival. But purified diets are toxic to turkeys in three weeks, and purified diets that contain as little as 50.0 mg Cu/kg ration produce adverse effects. Turkeys fed purified diets with supplemented copper show a dose-dependent increase in mortality and decrease in growth; these effects are attributed to a copper-accelerated dietary deterioration. Turkey growth and survival are acceptable when fed purified diets supplemented with as much as 800.0 mg Cu/kg ration, provided that effective levels of added antioxidant (0.02% ethoxyquin) and stabilized sources of vitamins A and D are present.

10.7.4 Mammals

Wilson's disease is the only naturally occurring neuropathological condition in humans and other mammals in which copper poisoning is implicated. People with Wilson's disease have severe pathological changes in the brain, especially in the basal ganglia, and in the liver; pathology is associated with excess copper in tissues. Copper concentrations in tissues from children that die from Wilson's disease are as much as 2217.0 mg/kg DW in liver and 1245.0 mg/kg DW in kidney. Long-term exposure of humans to copper dust irritates the nose, eyes, and mouth and causes headaches, dizziness, nausea, and diarrhea. Drinking water that contains higher than normal concentrations of copper may cause vomiting, diarrhea, stomach cramps, nausea, and greenish or bluish stools and saliva. Intentionally high intakes of copper may result in liver and kidney damage, and sometimes death, especially in children. The seriousness of the effects of copper is expected to increase with increasing dose and duration of exposure. Human tissues exposed directly to copper or copper salts will suffer adverse effects because of copper absorption. This is the case for copper bracelets on sweaty skin, for certain intrauterine devices, and for copper dental fillings. In monkeys, copper used as dental fillings in deciduous teeth causes more severe pulp damage than did other materials studied.

Mammals and birds are 100–1000 times more resistant to copper than other animals. But excessive dietary intakes of copper by 20–50-fold over normal levels may have serious effects in mammals. Depending on the species, growth and food intake may be reduced, anemia may develop, and liver, kidney, brain, and muscle may degenerate, often resulting in death. Copper poisoning in mammals may result from consumption of plants treated with copper-containing pesticides, from the veterinary use of copper sulfate to control helminthiasis and infectious pododermatitis in cattle and sheep, and from the ingestion of contaminated soils and vegetation near copper mining and refining operations. Emissions from copper mines and smelters are often associated with deaths of horses, cows, and sheep; pasture lands, in some cases, are fit for grazing only after heavy rains.

Ruminant mammals are significantly more sensitive to copper than nonruminant mammals and poultry. Signs of copper poisoning in ruminants include vomiting, excessive salivation, abdominal pain, diarrhea with greenish-tinted feces, pathology of internal organs, elevated copper concentrations in livers, altered enzyme activities in liver and serum, and collapse and death within 24–48 h. Young calves may develop copper toxicosis at relatively low copper intakes, especially when receiving milk-based diets; goats, however, seem resistant to copper toxicosis. Among ruminants, domestic sheep are particularly susceptible to copper insult from grazing on pastures treated with copper-containing fungicides and molluscicides or from inadvertently consuming diets specially formulated for pigs that contain large amounts of copper as a swine growth stimulant.

Chronic copper poisoning in domestic sheep is first characterized by a period of passive accumulation of copper in the tissues. This period varies from a few weeks to more than a year. During this time the animal appears outwardly normal although the liver may contain more than 1000.0 mg Cu/kg DW and plasma activities of aspartate transaminase, sorbitol dehydrogenase, lactic dehydrogenase, and arginase increase, indicating that liver damage has occurred. During the last few weeks of the passive phase, and prior to the so-called toxic phase, liver histopathology of parenchymal cells and copper-containing Kupffer cells occurs. The toxic phase, which is an acute illness and referred to as the hemolytic crisis, usually result in death 2–4 days later. During this phase sheep refuse to eat but have an excessive thirst. The eyes are usually sunken. The venous blood is chocolate colored. The liver is jaundiced. The kidneys are completely gorged with hemoglobin breakdown products and the medulla and cortex are black. The spleen is enlarged, with the parenchyma a deep brown to black color. The onset of these signs in sheep is associated with liberation of copper from the liver and a massive increase in blood copper concentrations. The increased blood copper concentrations lead to an increase in blood methemoglobin and a sudden fall in the erythrocyte glutathione level immediately followed by massive hemolysis and kidney damage, leading to uremia and death. At the time of crisis, elevated serum creatine phosphokinase activity suggests that muscle cell membranes are affected, and elevated SGOT and lactic dehydrogenase activities indicate progressive liver necrosis. It is emphasized that (1) blood copper status and liver function in sheep experimentally poisoned with copper sulfate are linked to elevated SGOT activities 1–6 weeks in advance of obvious external signs; (2) copper chloride is 2–4 times more toxic than copper sulfate to sheep; and (3) the use of copper-enriched feeding stuffs increases the risk of chronic copper poisoning in sheep fed purified rations. Also, sheep that accumulate higher than normal amounts of copper in the liver (i.e., 1900.0 mg Cu/kg DW) are more severely affected by lupinosis (acute liver atrophy due to poisoning by ingestion of plants of *Lupinus* spp.) than sheep with normal (40.0 mg/kg DW) concentrations of copper in the liver.

Copper toxicosis in lambs of domestic sheep occurs at dietary concentrations between 8.0 and 60.0 mg Cu/kg ration. The wide range of dietary concentrations is a function of copper availability. Availability, in turn, is influenced by dietary composition, genetic influence, age, breed, sex, physiological state, and interactions with other dietary constituents including iron, zinc, and molybdenum. Chronic copper poisoning in lambs occurs at dietary levels as low as 27.0 mg Cu/kg DW ration. During the passive phase, lambs – like adults – have normal plasma copper concentrations and seem outwardly unaffected. Unlike adults, copper accumulates in livers of lambs during a shorter period (several weeks to months vs. months to years). Signs of hemolytic crisis and death within a few days are similar for both adults and lambs. Elevated plasma AAT activity in lambs – up to 10 times higher than controls – occurs 4–8 weeks before the hemolytic crisis and strongly indicates a need for more research on the usefulness of AAT and other enzymes as early indicators of copper stress. A recommended treatment for lambs diagnosed with chronic copper poisoning is 20 mL of a mixture containing 100.0 mg of ammonium molybdate and 1.0 g of sodium sulfate administered orally 5 days weekly.

In domestic pigs, copper toxicosis results from eating diets containing 250.0 mg Cu/kg ration and is characterized by anemia, jaundice, elevated levels of Cu in serum and liver, and elevated serum AAT activity. Shortly before death, copper-poisoned pigs had white noses, poor balance, stomach histopathology, orange cirrhotic livers, anorexia, and anemia.

In rodents, copper administered by single intraperitoneal or subcutaneous injection is lethal at 3.0–7.0 mg Cu/kg BW. Mice died when their drinking water had 640.0 mg Cu/L. In rats, copper accumulation in kidneys and lungs are similar regardless of route of administration. Concentrations of copper in serum of rats (*Rattus* sp.) reflect dietary copper; concentrations in liver and kidney are directly related to serum Cu and ceruloplasmin. As serum Cu concentrations rise in rats, levels fall for serum cholesterol, triglycerides, and phospholipids.

10.8 Proposed Criteria and Recommendations

Proposed copper criteria for the protection of agricultural crops, aquatic life, terrestrial invertebrates, poultry, laboratory white rats, livestock, and human health are summarized in Table 10.1.

Copper is essential to normal plant growth, and copper deficiency is known in various agricultural crops such as vegetables and grains. Crops seem to be protected against copper deficiency when growing soils contain >10.0 mg Cu/kg DW and leaves >6.0 mg Cu/kg DW. With some exceptions, agricultural crops are protected against copper toxicosis when irrigation waters contain <1.0 mg Cu/L and soils <170.0 mg Cu/kg DW. But adverse effects occur on root development of seedling pines at irrigation water concentrations as low as 200.0 μg Cu/L and on growth of citrus trees when extractable copper in the soil exceeds 60.0 mg/kg DW. States allow application of sewage sludge to agricultural soils if total copper in the sludge does not exceed 1000.0 mg/kg DW (100.0 mg/kg DW in Florida), or if the application rate for sludge does not exceed 280 kg sewage sludge

Table 10.1. Proposed copper criteria for the protection of natural resources and human health.

Resource, Criteria, and Other Variables	Effective Copper Concentration
AGRICULTURAL CROPS	
Irrigation water	<1.0 mg/L
Leaves	
Severe deficiency	<4.0 mg/kg dry weight (DW)
Deficient	<5.0 mg/kg DW
Mild to moderate deficiency	4.0–5.0 mg/kg DW
Deficiency rare	>6.0 mg/kg DW
Sewage sludge	
Europe, acidic soils	50.0–140.0 kg/ha
USA	
All agricultural lands	<1000.0 mg/kg DW
Florida	<100.0 mg/kg DW
Illinois	<280.0 kg/ha
Maryland, Massachusetts	140.0–280.0 kg/ha[a]
Minnesota, Missouri	140.0–560.0 kg/ha[b]
New York	
Agricultural soils	<125.0 kg/ha
Forests	<280.0 kg/ha
Wisconsin, acidic soils	50.0–140.0 kg/ha
Soils	
Deficient	<10.0 mg/kg DW
Safe	<280.0 kg/ha[c]
M-3 extractable soil copper	<60.0 g/kg DW
Canada	
Agricultural lands	<100.0 mg/kg DW
Acidic soils, Alberta	<200.0 mg/kg DW
Industrial and other lands	<300.0 mg/kg DW
Former Soviet Union, maximum allowable concentration	3.0 mg/kg DW when extracted with ammonium acetate buffer
The Netherlands	
Normal	50.0 mg/kg DW
Moderately contaminated	100.0 mg/kg DW
Requires remediation	>500.0 mg/kg DW
USA, New Jersey	<170.0 mg/kg DW
AQUATIC LIFE, FRESHWATER	
Sediments	
Great Lakes	
Nonpolluted	<25.0 mg/kg DW
Moderately polluted	25.0–50.0 mg/kg DW
Heavily polluted	>50.0 mg/kg DW

Table 10.1. cont'd

Resource, Criteria, and Other Variables	Effective Copper Concentration
Reduced abundance of benthos	480.0–1093.0 mg/kg DW
Toxic to benthos	>9000.0 mg/kg DW
Tissue concentrations; rainbow trout, *Oncorhynchus mykiss*; ratio of zinc to copper in gill or opercle	
Normal	Ratio >1.5
Probably copper-poisoned	Ratio 0.5–1.5
Acute copper poisoning	Ratio <0.5
Water	
Safe. No adverse effects on rainbow trout exposed from fertilization through 4 days after hatching	
In soft or medium water	2.0–5.0 μg/L
In hard water	5.0–8.0 μg/L
Death or teratogenicity in eggs of sensitive species of fishes and amphibians	5.0–10.0 μg/L
USA	
Safe; total recoverable copper; 24-h average	<5.6 μg/L
Maximum allowable concentration at 50 mg $CaCO_3$/L	12.0 μg/L
Maximum allowable concentration at 100 mg $CaCO_3$/L	22.0 μg/L
Maximum allowable concentration at 200 mg $CaCO_3$/L	43.0 μg/L
Inhibits fish growth and ability of fish to discriminate prey	18.0–28.0 μg/L
Chesapeake Bay; proposed for the protection of 90% of species tested	
Acute exposures	
ALL SPECIES	<6.3 μg/L[e]
BENTHOS	<6.9 μg/L
FISHES	<10.8 μg/L
Chronic exposures	
ALL SPECIES	<3.8 μg/L[e]
FISHES	<3.9 μg/L
The Netherlands; total recoverable copper; maximum allowable concentration	<50.0 μg/L
AQUATIC LIFE, MARINE	
Seawater	
Safe. Total recoverable copper, 24-h average	<4.0 μg/L; not to exceed 23.0 μg/L at any time
Safe. Maximum concentration	<5.0 μg/L

Continued

Table 10.1. cont'd

Resource, Criteria, and Other Variables	Effective Copper Concentration
Chesapeake Bay; recommended for the protection of 90% of species tested	
Acute exposures	
All species	<6.3 μg/L[e]
Benthos	<4.1 μg/L
Fishes	<16.1 μg/L
Chronic exposures	
All species	<6.4 μg/L[e]
Sediments	
Avoidance by clams	>5.0 mg/kg DW
Clam burrowing ability inhibited (water concentrations of 113.0–120.0 μg Cu/L)	>15.0 mg/kg DW
Not polluted	<40.0 mg/kg DW
Moderately polluted	40.0–60.0 mg/kg DW
Very polluted	>60.0 mg/kg DW
Reduced species diversity; sensitive species absent	>200.0 mg/kg DW
Toxic to juvenile bivalve mollusks	>2000.0 mg/kg DW
Terrestrial Invertebrates	
Earthworms, whole; disrupted lysozyme activity in coelomic fluid and coelomocytes	>28.5 mg/kg DW
Isopod, *Porcellio scaber*; whole	
Deficiency	Unknown
Uncontaminated	<250.0 mg/kg DW
Low contamination	250.0–400.0 mg/kg DW
Medium contamination	400.0–600.0 mg/kg DW
High contamination	600.0–1000.0 mg/kg DW
Very high contamination	>1000.0 mg/kg DW
POULTRY, DIETS	
Deficient	<8.7 mg/kg DW ration; some deaths at 0.7–1.5 mg/kg DW ration; high frequency of vascular rupture at 2.7 mg/kg DW ration
Safe	<200.0 mg/kg DW feed
Recommended for growing chickens	>4.0 mg/kg DW diet plus adequate iron
LABORATORY WHITE RAT, *Rattus* sp.	
Minimal	3.0–6.0 mg/kg FW diet; 0.15–0.3 mg/kg body weight (BW) daily
Adequate	10.0 mg/kg DW diet

Table 10.1. cont'd

Resource, Criteria, and Other Variables	Effective Copper Concentration
LIVESTOCK	
All species except sheep; diet	
Deficient	<5.0 mg/kg DW
Minimal	>5.0–<15.0 mg/kg DW
Adequate	20.0–30.0 mg/kg DW
Cattle, *Bos* sp.; liver, copper-poisoned	>150.0 mg/kg fresh weight (FW); >450.0 mg/kg DW
Sheep, *Ovis aries*	
Toxic	20.0–30.0 mg/kg DW diet
Pig, *Sus* sp.	
Diet	
Safe	3.0–5.0 mg/kg DW
United Kingdom, maximum	200.0 mg/kg DW[d]
Tissue concentrations	
Fatal anemia with jaundice and stomach ulcerations; kidney vs. liver	95.0–800.0 mg/kg DW vs. 1300.0–2600.0 mg/kg DW
HUMAN HEALTH	
Air	
Montana	<0.26 μg/m^3 for 8 h; <1.57 μg/m^3 for 24 h
Massachusetts	<0.54 μg/m^3 for 24 h
Connecticut, North Dakota	<2.0 μg/m^3 for 8 h
Florida	<4.0 μg/m^3 for 8 h
Nevada	<5.0 μg/m^3 for 8 h
Virginia	<16.0 μg/m^3 for 24 h
New York	<20.0 μg/m^3 for 1 year
USA; workplace; 8 h daily	
Fumes	<0.1–<0.2 mg/m^3
Dusts and mists	<1.0 mg/m^3
Total	<1.0 mg/m^3
Daily intake, all sources	
Deficiency in children	<0.1 μg/kg BW
Infants, normal	14.0–80.0 μg/kg BW; 0.5–1.0 mg
Children, normal	40.0–100.0 μg/kg BW; 1.0–2.0 mg
Teenagers and adults, normal	28.0–40.0 μg/kg BW; 2.0–4.0 mg
Adults, safe and adequate	2.0–3.0 mg
Adults, maximum	40.0 μg/kg BW daily equivalent to 2.8 mg daily for a 70-kg adult
Adults, toxic	15.0 mg in single dose

Continued

Table 10.1. cont'd

Resource, Criteria, and Other Variables	Effective Copper Concentration
Diet	
Australia	
Seafood	<30.0 mg/kg FW
Shellfish; soft parts	<70.0 mg/kg FW; <266.0 mg/kg DW
Fish muscle	<15.0 mg/kg FW
Malaysia; bivalve mollusks; soft parts	<30.0 mg/kg FW
Spain, total diet	<20.0 mg/kg DW
Drinking water	
USA, safe	<1.0 mg/L (exceeded by about 1% of all samples)
Kansas, Rhode Island	<1.0 mg/L
Minnesota	<1.3 mg/L
Proposed, USA	<1.3 mg/L
Satisfactory smell and taste	<1.0–1.3 mg/L
Associated with diarrhea, abdominal cramps, and nausea	>1.3 mg/L
Health advisory for children and adults	Not to exceed 1.3 mg/L for more than 1 day
Adverse taste	>1.5 mg/L
Fish and shellfish collection locales; marine water	<4.0 μg/L
Tissues; human	
Blood; deficient vs. adequate	<0.8 vs. 1.03 mg/L
Serum; normal vs. toxic	1.64 vs. 2.86 mg/L

[a] Soil cation exchange capacity <5.0 meq/100.0 g for 140.0 kg/ha and >5.0 meq/100.0 g for 280.0 kg/ha.
[b] Soil cation exchange capacity ranges from <5.0 to >15.0 meq/100.0 g.
[c] Higher levels of 365.0 kg Cu/ha had no effect on corn yield or copper content in corn.
[d] Diet should also contain 150.0 mg Zn/kg and 200.0 mg Fe/kg to further reduce the chances of copper toxicity to pigs.
[e] Concentrations >50.0 μg Cu/L are routinely measured in the Chesapeake and Delaware Canal and infrequently in other portions of Chesapeake Bay.

per surface acre (50 kg/ha in Wisconsin). The practice by some localities of applying raw sewage sludge to crop soils on the basis of kilogram sludge: surface acre ratio should be discouraged unless the sludge is periodically analyzed for copper and other contaminants.

Proposed criteria to protect most species of freshwater aquatic life from copper toxicity or deficiency include maximum water concentrations over a 24-h period of 12.0 μg Cu/L in soft water and 43.0 μg/L in hard water, sediment concentrations <480.0 mg Cu/kg DW, and, in rainbow trout, a zinc: copper ratio in gill or opercle >1.5. However, the proposed maximum water concentration range of 12.0–43.0 μg Cu/L exceeds the 5.0–10.0 μg/L range that is lethal or teratogenic to sensitive species of fishes and amphibians and overlaps the 18.0–28.0 μg/L range that inhibits growth and ability to discriminate prey for other species. Some scientists state that laboratory studies tend to overestimate the adverse effects of copper on freshwater abundance and diversity and suggest more research on field mesocosms receiving water directly from the system under investigation. In marine ecosystems, copper concentrations should not exceed 23.0 μg Cu/L at any time, and sediments should contain less than 200.0 mg Cu/kg DW. But adverse sublethal effects of copper to representative species of estuarine algae, mollusks, and arthropods frequently occur at less than 10.0 μg Cu/L. Also, extrapolation of laboratory data on copper and marine benthos to actual field conditions is difficult because of changing environmental conditions such as thermosaline regimes and the nature of the sediment substrate.

Among sensitive species of terrestrial invertebrates, earthworms show disrupted enzyme activities at whole body concentrations as low as 28.5 mg Cu/kg DW. Soil copper concentrations between 53.0 and 100.0 mg/kg DW kill soil nematodes and soil faunal communities and cause a reduction in cocoon production of earthworms. Diets that contain between 50.0 and 63.0 mg Cu/kg ration inhibit development and reproduction in gypsy moths and oribatid mites. The wood louse (*Porcellio scaber*), an isopod, is proposed as a bioindicator of copper contamination in terrestrial ecosystems because whole body concentrations seem to reflect copper loadings in the isopod's immediate environment. More research is recommended on isopods and other sentinel organisms.

Quantitative data are missing on copper effects on avian and mammalian wildlife, and this represents a high priority research need. Some data are available for copper and poultry and livestock, but extrapolation of these results to wildlife species is contraindicated in view of the wide range in sensitivities to copper between species. Domestic chickens show good growth and survival when their diets contain adequate iron and more than 4.0 and less than 200.0 mg Cu/kg ration. In sheep – and some other mammals – prior knowledge of copper stress would allow adequate time for treatment (i.e., prophylactic dosing with ammonium molybdate plus sodium sulfate or intravenous injection of chelating agents) to prevent sudden death during copper-induced hemolytic crisis. In sheep, for example, elevated SGOT activity is an early indicator of copper poisoning and is measurable 1–6 weeks before the hemolytic crisis stage. Providing prophylactic licks containing zinc sulfate and sulfur to African cattle, buffaloes, and impalas seem to be successful in protecting against the lethal effects of excess airborne copper in the grazing area.

The proposed domestic drinking water criterion of <1.0 mg Cu/L for the protection of human health is not based on copper toxicosis but on the unpleasant taste which develops with higher levels of copper in drinking water. Increased copper levels (>1.3 mg Cu/L) in household water supplies caused by corrosion of copper plumbing materials may adversely affect infants and young children among residents of newly constructed or renovated homes. Human groups at greatest risk to copper toxicosis now include young children subjected to unusually high concentrations of copper in soft or treated water held in copper pipes or vessels, medical patients with Wilson's disease, medical patients treated with copper-contaminated fluids in dialysis or parenteral administration, people with a G-6-PD deficiency (about 13% of the Afro-American male population has a G-6-PD deficiency) that drink water containing >1.0 mg Cu/L, and occupationally exposed workers.

Other copper research areas that seem to merit additional effort include (1) establishment of specific biomarkers for copper toxicity; (2) development of a national system to verify incidents of deficiency and excess of copper and interrelated trace elements in species of concern; (3) clarification of copper interactions with molybdenum, sulfate, iron, and zinc in plant and animal metabolisms; (4) the role of copper in carcinogenesis, mutagenesis, and teratogenesis because preliminary evidence suggests that exposures to grossly elevated concentrations of copper produces teratogenicity in fish and mammals, carcinogenicity in rodents, and mutagenicity in rodents, sheep, and grasshoppers; (5) mechanisms by which copper deficiency results in neutropenia, with emphasis on the process of cellular differentiation and the viability of neutrophils in blood and marrow; (6) copper status effects on resistance to endotoxin-induced injuries because burn and trauma patients show moderate copper deficiency and high risk to sepsis, and copper-deficient rats are sensitive to endotoxins causing sepsis; (7) the role of aquatic organisms in copper cycling in aquatic ecosystems; (8) mechanisms of copper tolerance or acclimatization to high doses of copper; (9) the relation between copper toxicosis, copper absorption rates, and copper retention; (10) effects on reproduction, neurotoxicity, and immune response; (11) biochemistry and physiology of copper proteins; (12) measurement of flux rates of ionic copper from metallic copper; and (13) determination of safe levels of copper in livestock and poultry feeds, and in diets of avian and mammalian wildlife.

10.9 Summary

Copper discharges to the global biosphere are primarily due to human activities, especially from the mining, smelting, and refining of copper and from the treatment and recycling of municipal and industrial wastes. Some copper compounds, especially copper sulfate, also contribute to environmental copper burdens because they are widely and intensively used in confined geographic areas to control nuisance species of aquatic plants and invertebrates, diseases of terrestrial crop plants, and ectoparasites of fish and livestock.

Copper concentrations in field collections of abiotic materials and living organisms are usually elevated in the vicinity of human activities and intensive copper use. Maximum copper concentrations recorded in selected abiotic materials are 5.0 μg/m^3 in air, 5.0 μg/L in groundwater, 12.0 μg/L in rainwater, 1200.0 mg/kg DW in poultry litter, 7000.0 mg/kg DW in soils, and 7700.0 mg/kg DW in sewage sludge. In terrestrial vegetation, copper is usually less than 35.0 mg/kg DW except near smelters where it may approach 700.0 mg/kg DW and in certain copper-accumulator plants that may normally contain as much as 13,700.0 mg/kg DW. Aquatic vegetation from copper-contaminated sites contains as much as 1350.0 mg Cu/kg DW vs. 36.0 mg Cu/kg DW in conspecifics from reference sites. Terrestrial invertebrates from industrialized areas may contain from 137.0 to 408.0 mg Cu/kg DW whole organism. Aquatic invertebrates seldom contain as much as 95.0 mg Cu/kg DW, regardless of collection locale; exceptions include whole amphipods and lobster hepatopancreas (335.0–340.0 mg/kg DW) from copper-contaminated sites and many species of mollusks that normally contain 1100.0–6500.0 mg Cu/kg DW. Data are scarce on copper concentrations in field populations of amphibians and reptiles: crocodile eggs may contain as much as 60.0 mg Cu/kg DW and livers of some toads may contain as much as 2100.0 mg Cu/kg DW without apparent adverse effects. Maximum copper concentrations in tissues of fishes, elasmobranchii, birds, and marine mammals from all collection sites are low when compared to more primitive organisms and never exceed 53.0 mg Cu/kg DW except liver (146.0–367.0 mg/kg DW); an exception is liver from endangered manatees (1200.0 mg/kg DW) collected at a site treated with a copper-containing herbicide. Maximum copper concentrations in all tissues of terrestrial mammals, regardless of collection locale, are low and seldom exceed 29.0 mg/kg DW except kidneys (108.0 mg/kg DW) and livers (1078.0 mg/kg DW) from animals near a copper refinery.

Copper deficiency is not a major public health concern in the United States, although skeletal deformities and leg fractures may occur in some copper-deficient children. Copper deficiency effects occur, however, in various species of terrestrial plants (reduced growth, necrosis, reduction in number of pollen grains, death), chickens (poor growth, high frequency of cardiovascular and skeletal lesions, low survival), turkeys (sudden death), rats (defective hemoglobin synthesis, lesions of the central nervous system, low survival, altered blood and liver enzyme activities), guinea pigs (lesions of the central nervous system), dogs (leg fractures), sheep and other ruminant mammals (sudden death, skeletal deformities), pigs (poor growth, decreased hemoglobin and erythrocytes, skeletal deformities), mink (reduced survival), and camels (anemia, emaciation, falling, fractures, death). Data are scarce or missing on copper deficiency effects in aquatic flora and fauna and in avian and terrestrial mammalian wildlife; additional studies of copper deficiency in these groups are merited. In sensitive terrestrial agricultural crops, copper deficiency occurs at <1.6 mg dissolved Cu/kg DW soil and <5.0 mg total Cu/kg DW leaves. For domestic chickens, copper deficiency occurs when diets containing <2.7 mg Cu/kg ration are fed. Male weanling rats show deficiency effects when fed diets containing 0.13 mg Cu/kg ration vs. a copper-normal diet of 5.7 mg Cu/kg ration; earliest signs of copper deficiency in rats include low concentrations of copper in livers (<3.0 mg/kg DW vs. 12.6–15.0 mg/kg DW in controls), reductions in activities of cytochrome oxidase and succinoxidase, and prolonged sleeping times. Ewes of Bactrian camels fed copper-deficient diets of <2.5 mg Cu/kg DW ration (vs. normal diet of about 11.0 mg Cu/kg DW) produce a high frequency of swaybacked lambs. Copper deficiency in mink is produced at daily intake rates equivalent to 3.5 mg Cu/kg BW for 50 weeks. Swine require high intakes of copper to avoid deficiency; daily intakes of <36.0 mg Cu/kg BW are associated with reductions in growth rate, hemoglobin, and hematocrit.

Copper and its compounds are not carcinogenic, mutagenic, or teratogenic at environmentally realistic concentrations. But under controlled conditions of grossly elevated exposures some studies suggest that copper is a potential carcinogen in rodents; mutagen in rodents, sheep, and grasshoppers; and teratogen in fish and small laboratory animals. More research is needed in this area.

Bioavailability and toxicity of copper to aquatic organisms depend on the total concentration of copper and its speciation. Both availability and toxicity are significantly reduced by increased loadings of suspended solids and natural organic chelators and increased water hardness. Toxicity to aquatic life is primarily related to the dissolved cupric ion (Cu^{+2}) and possibly to some hydroxyl complexes. Cupric copper (Cu^{+2}) is the most readily available and toxic inorganic species of copper in freshwater, seawater, and sediment interstitial waters. Cupric ion accounts for about 1% of the total dissolved copper in seawater and less than 1% in freshwater. In freshwater, cupric copper and some copper hydroxyl species are correlated with high toxicity to aquatic life, although carbonato species are much less toxic than other copper complexes. More research seems needed on the adsorption characteristics of most cupric ion complexes. In solution, copper interacts with numerous inorganic and organic compounds resulting in altered bioavailability and toxicity. Acknowledgment of these interactions is essential to the understanding of copper toxicokinetics. In aquatic invertebrates, copper disrupts gill epithelium at high concentrations and in fishes it interferes with osmoregulation; death is caused by tissue hypoxia associated with disrupted ATP synthesis. Copper detoxifying mechanisms in fishes include the induction of metallothioneins, allowing copper retention for weeks or months after absorption without toxicity. In higher vertebrates, excess copper is cytotoxic and alters protein configuration and lipid peroxidation rates. Mechanisms implicated in copper poisoning of mammals include free radical production, alteration in activities of several enzymes, and inhibited metallothionein synthesis. In mammals, copper is normally excreted via the bile in association with glutathione or unidentified high-molecular-weight proteins.

Excess copper is toxic to representative species of plants and animals. Significant adverse effects in terrestrial plants occur at concentrations as low as 40.0 μg Cu/L of nutrient solution, >10.0 mg Cu/kg DW in leaves, and >60.0 mg extractable Cu/kg DW of soil. Sensitive species of terrestrial invertebrates show a reduction in growth, survival or reproduction at >50.0 mg Cu/kg diet or 53.0–70.0 mg Cu/kg DW of soil. Many species of freshwater plants and animals die within 96 h at waterborne copper concentrations of 5.0–9.8 μg Cu/L, and sensitive species of freshwater mollusks, crustaceans, and fishes die at 0.23–0.91 μg Cu/L within 96 h. The most sensitive tested species of marine mollusks, crustaceans, and fishes have an LC50 (96 h) range from 28.0 to 39.0 μg Cu/L; significant sublethal effects to representative species of estuarine algae, mollusks, and arthropods frequently occur at 1.0–10.0 μg Cu/L. Mammals and birds are at least 100 times more resistant to copper than other organisms, but ruminant mammals are significantly more sensitive to copper than nonruminant animals and poultry. Excessive dietary intakes of copper by 20–50-fold over normal levels may, however, have serious adverse effects on birds and mammals. No data are available on copper toxicity to avian wildlife. Studies with poultry demonstrate that copper accumulates in livers at dietary concentrations as low as 15.0 mg Cu/kg DW ration, inhibits growth at 120.0 mg Cu/kg DW ration, and causes gizzard histopathology at 250.0 mg Cu/kg DW ration. Copper is lethal to representative species of mammals through a variety of routes: single oral doses of 6.0–637.0 mg Cu/kg BW in humans and 200.0 mg/kg BW in cattle or diets with >80.0 mg Cu/kg ration (about 5.1–10.7 mg Cu/kg BW daily) fed to sheep or >238.0 mg/kg ration (more than 133.0 mg/kg BW daily) fed to rats. Adverse sublethal effects of copper to sensitive mammals occur in human infants at drinking water concentrations >3.0 mg/L, in cattle at >4.2 mg/kg BW by way of drinking water or >20.0 mg/kg BW via diet, in sheep given daily oral doses of 7.5–15.0 mg/kg BW or fed diets containing >37.3 mg/kg ration, in rats given >7.9 mg/kg BW daily by way of diet (equivalent to >100.0 mg Cu/kg DW ration), and in pigs at >14.5 mg/kg BW daily via diet.

Numerous and disparate copper criteria are proposed for protecting the health of agricultural crops, aquatic life, terrestrial invertebrates, poultry, laboratory white rats, and humans; however, no copper criteria are now available for the protection of avian and mammalian wildlife, and this needs to be rectified. Several of the proposed criteria do not adequately protect sensitive species of plants and animals and need to be re-examined. Other research areas that merit additional effort include biomarkers of early copper stress; copper interactions with interrelated trace elements in cases of deficiency and excess; copper status effects on disease resistance, cancer, mutagenicity, and birth defects; mechanisms of copper tolerance or acclimatization; and chemical speciation of copper, including measurement of flux rates of ionic copper from metallic copper.

CYANIDE[a]

Chapter 11

11.1 Introduction

The origin of terrestrial life probably depended on the presence and reactivity of hydrogen cyanide (HCN) and its derivatives; paradoxically, hydrogen cyanide is toxic to the majority of living matter. Cyanide is a general respiratory poison – although uptake can also occur through ingestion or dermal absorption – producing reactions within seconds, and death within minutes. The toxic mechanism of cyanide primarily involves the inhibition of cytochrome oxidase, the terminal oxidative enzyme of the mitochondrial electron transport chain, producing blockage of aerobic ATP synthesis. Because of their highly effective lethal potency, cyanides were used for genocidal programs in Germany in World War II, in mass suicides by members of the People's Temple religious sect in Guyana, and in the substitution of medication in Tyleno® capsules in drugstores in various cities in the United States. In fact, cyanides are responsible for more human deaths than any other chemicals known, owing to their deliberate use in suicide, murder, chemical warfare, genocide, and judicial execution. High sublethal doses of cyanide are rapidly detoxified, and accidental acute cyanide poisonings in humans are uncommon.

Cyanide compounds are useful to society in terms of their key role in synthetic and industrial processes, for certain fumigation and agricultural uses, and for some therapeutic applications. Cyanides are present in effluents from iron and steel processing plants, petroleum refineries, and metal-plating plants, and constitute a hazard to aquatic ecosystems in certain waste-receiving waters, and to livestock. Cyanide serves no useful purpose in the human body, yet it is present in our food, air, and water.

Natural sources of cyanide include various species of bacteria, algae, fungi, and higher plants that form and excrete cyanide. The most widely distributed major food crop with a high content of cyanogenic glycosides is cassava (*Manihot esculenta*), also known as manioc. Cassava is a staple food in human diets in over 80 countries, and it is sometimes added to animal feeds as a substitute for more expensive cereal grains. In humans, chronic cyanide intoxication caused by consumption of cassava is the main etiological factor in the debilitating tropical ataxic neuropathy. Other plants having comparatively elevated cyanide content include fruit pits, sweet potatoes (*Ipomoea batatas*), corn (*Zea mays*), bamboo shoots (*Bambusa* spp.), linseed (*Linum* sp.), lima beans (*Phaseolus lunatus*), and millet (*Panicum millaceum*). In higher plants that contain cyanogenic glycosides, at least 20 of these compounds have been identified. Amygdalin – one of the more intensively studied cyanogenic glycosides – is found

[a]All information in this chapter is referenced in the following sources:

Eisler, R. 1991. Cyanide hazards to fish, wildlife, and invertebrates: a synoptic review. *U.S. Fish Wildl. Serv. Biol. Rep. 85(1.23)*, 55 pp.

Eisler, R. 2000. Cyanide. Pages 903–959 in *Handbook of Chemical Risk Assessment: Health Hazards to Humans, Plants and Animals. Volume 2, Organics*. Lewis Publishers, Boca Raton, Florida.

Eisler, R., D.R. Clark, Jr., S.N. Wiemeyer, and C.J. Henny. 1999. Sodium cyanide hazards to fish and other wildlife from gold mining operations. Pages 55–67 in J.M. Azcue, ed. *Environmental Impacts of Mining Activities: Emphasis on Mitigation and Remedial Measures*. Springer-Verlag, Berlin.

Eisler, R. and S.N. Wiemeyer. 2004. Cyanide hazards to plants and animals from gold mining and related water issues. *Rev. Environ. Contam. Toxicol.*, 182, 21–54.

in seeds of the cherry (*Prunus* spp.), plum (*Prunus* spp.), peach (*Prunus persica*), apricot (*Prunus armeniaca*), apple (*Malus* spp.), pear (*Pyrus communis*), and many parts of the cherry laurel (*Prunus laurocerasus*). Apricot seeds and peach kernels are food delicacies in Turkey, and have caused at least nine poisonings (two fatal) in children from that country. Acute cyanide poisoning has occurred in the United States from the ingestion of almond-flavored milkshakes prepared from apricot kernels. Amygdalin is also the chief ingredient in laetrile, a medication prescribed by some physicians to control tumors. Both laetrile and amygdalin-containing fruit pits have been implicated as the causes of acute cyanide poisoning in humans. Another naturally occurring group of organic cyanides (nitriles) is the highly toxic pseudocyanogenic glycosides, especially cycasin, and these have been implicated in a variety of tropical diseases of the nervous system, and partial or total blindness. Other nitriles found in plants include the lathyrogenic compounds, glucosinolates, and the cyanopyridine alkaloids.

The recognition that certain plants, such as bitter almonds (*Amygdalus cummunis*), cherry laurel leaves, and cassava, are poisonous if consumed in sufficient quantities has been known for at least 2000 years. But it was not until the 1700s that cyanide was recognized as the basis for their lethal toxicity. The first account of an experimental administration of extract of bitter almonds and other poisons to dogs (*Canis familiaris*) dates from 1679. In 1731, two fatal cases of human poisoning in Ireland were caused by drinking cherry laurel water, in this instance used as a flavoring agent in cooking and to dilute brandy. In that same year, it was shown that cherry laurel water administered to dogs by various routes proved rapidly fatal. By 1781, it was well established that mammals, birds, reptiles, amphibians, fish, and insects could all be killed with small doses of laurel water, and that death was more rapid than that produced by other poisons tested. It was also at this time that cyanide was first implicated as a homicidal agent in England. In 1782, hydrocyanic acid was isolated from Prussian blue (a dye) by the Swedish chemist Scheele. In 1786, Scheele accidentally broke a vial of the material and died from vapor poisoning. In 1787, it was determined that hydrocyanic acid contained hydrogen, carbon, and nitrogen, but did not contain oxygen, formerly believed to be an essential component of all acids. Between 1802 and 1815, hydrocyanic acid was found to be lethal in small quantities to birds and dogs, and to act rapidly when given orally, intravenously, or applied to the eye surface. By 1803, it was known that cyanide occurred naturally and could be extracted from apricots or almonds. In 1815, hydrocyanic acid was prepared in a semi-pure form. Between 1817 and 1948, cyanide, in appropriate doses, was used therapeutically in England for the treatment of pulmonary diseases, tuberculosis, and as a sedative. By 1830, cyanogenic glycosides containing HCN were isolated from cassava; today, more than 800 species of cyanogenic plants have been identified. In 1876, it was first demonstrated that cyanide inhibited tissue oxidation. In 1894, cobalt compounds were suggested as antidotes due to their marked cyanide-binding capacity. Studies on cyanide detoxification conducted between 1877 and 1894 showed that thiosulfate administration caused the formation of thiocyanate – a relatively harmless metabolite. By the late 1800s, cyanide was regarded as a common plant metabolite rather than as an unusual poison. In 1929, it was conclusively demonstrated that cyanide combines with the trivalent iron atom in cytochrome oxidase, a respiratory enzyme that links the tricarboxylic acid cycle and formation of metabolic water.

Cyanide hazards to fish, wildlife, and livestock are well documented. Massive kills of freshwater fish by accidental discharges of cyanide wastes are fairly common. In one case, cyanide-containing mine effluents from a Canadian tailings pond released into a nearby creek killed more than 20,000 steelhead (*Oncorhynchus mykiss*). Many species of birds were found dead near burrows of the black-tailed prairie dog (*Cynomus ludovicianus*) after the burrows had been treated with calcium cyanide to control prairie dog populations; dead birds included the burrowing owl (*Athene cunicularia*), the bald eagle (*Haliaeetus leucocephalus*), and the golden eagle (*Aquila chrysaetos*). An endangered

California condor (*Gymnogyps californianus*) found dead in Kern County, California, in November 1983 had particles of a yellow fluorescent tracer in its mouth; these particles were similar to those mixed with sodium cyanide (NaCN) in M-44 spring-loaded ejector mechanism devices used in a U.S. Fish and Wildlife Service Animal Damage Control Program in that vicinity, suggesting that cyanide was a possible cause of death. M-44 devices are known to have caused the death of magpies (*Pica* sp.), ravens and crows (*Corvus* spp.), wild turkeys (*Meleagris gallopavo*), and various unidentified species of hawks and vultures. Between 1980 and 1989, 519 mammals – mostly rodents (35%) and bats (34%) – were found dead at cyanide-extraction, gold-mine leach ponds in California, Nevada, and Arizona; the list included coyote (*Canis latrans*), foxes, skunks, badger (*Taxidea taxus*), weasels, rabbits, deer, and beavers. Also found dead at these same leach ponds were 38 reptiles, 55 amphibians, and 6997 birds including many species of waterfowl and songbirds.

The major threat of cyanide poisoning to livestock and terrestrial mammalian wildlife is through ingestion of plants containing high levels of cyanogenic glycosides. Plants implicated in cyanide poisoning of animals include the sorghums (Johnson grass, *Sorghum halepense*; Sudan grass, *Sorghum almum*), arrowgrass (*Triglochin* spp.), elderberry (*Sambucus* spp.), wild cherry (*Prunus* spp.), and the pits of several common fruits, such as apple, peach, and apricot; these plants and fruit pits have the potential of releasing cyanide upon ingestion. Domestic goats (*Capra* spp.) died of cyanide poisoning after eating leaves and fruits of the crab apple (*Malus sylvestris*); the crab apple contains cyanogenic glycosides in its leaves and fruits. Cyanide poisoning of cattle (*Bos* spp.) by forage sorghums and various hybrid cultivars has been reported in India and elsewhere. Cattle appear to be more vulnerable to cyanide poisoning than are sheep (*Ovis aries*), horses (*Equus cabalus*), and pigs (*Sus* spp). *Equine sorghum cystitis ataxia* is a condition observed in horses grazing *Sorghum* or hybrid Sudan grass pastures; it is characterized by urinary incontinence, posterior incoordination, and degenerative central nervous system lesions. Grazing cyanogenic plants can induce sulfur deficiency in sheep, presumably because sulfur detoxifies the released cyanide. The increasing use of cassava and other cyanogenic plants in animal feeding portends a greater exposure to dietary cyanides.

11.2 Chemical Properties

The chemical speciation of cyanides varies according to their source. Specific terms used to describe cyanide include free cyanide, cyanide ion, simple cyanides, complex cyanides, nitriles, cyanogens, and total cyanide. The most common forms of cyanide in the environment are free cyanide, metallocyanide complexes, and synthetic nitriles. A brief description of each cyanide species follows.

Free cyanide is the primary toxic agent in the aquatic environment. Free cyanide refers to the sum of molecular HCN and the cyanide anion (CN^-), regardless of origin. In aqueous solution with pH 9.2 and lower, the majority of the free cyanide is in the form of molecular HCN. The chemical names for HCN include hydrogen cyanide, hydrocyanic acid, cyanohydric acid, and prussic acid. Hydrogen cyanide (Table 11.1) is a colorless, flammable liquid or gas that boils at 25.7°C and freezes at –13.2°C. The gas rarely occurs in nature, is lighter than air, and diffuses rapidly; it is usually prepared commercially from ammonia and methane at elevated temperatures with a platinum catalyst. It is miscible with water and alcohol, but is only slightly soluble in ether. In water, HCN is a weak acid with the ratio of HCN to CN^- about 100 at pH 7.2, 10 at pH 8.2, and 1 at pH 9.2. HCN can dissociate into H^+ and CN^-. Cyanide ion, or free cyanide ion, refers to the anion CN^- derived from hydrocyanic acid in solution, in equilibrium with simple or complexed cyanide molecules. Cyanide ions resemble halide ions in several ways and are sometimes referred to as "pseudohaline" ions. For example, silver cyanide is almost insoluble in water, as are silver halides. Cyanide ions also form stable complexes with many metals.

Simple cyanides typically refer to alkali water-soluble salts, such as NaCN, KCN, $Ca(CN)_2$, and $Hg(CN)_2$, but also include

Table 11.1. Some properties of potassium cyanide, hydrogen cyanide, and sodium cyanide.

Property	Potassium Cyanide	Hydrogen Cyanide	Sodium Cyanide
CAS Number	151-50-8	74-9-8	143-33-9
Chemical Formula	KCN	HCN	NaCN
Molecular Weight	65.12	27.03	49.01
Physical State	Solid	Gas or liquid	Solid
Boiling Point (°C)	–	25.7	1496.0
Melting Point (°C)	634.5	−13.21	563.7
Specific Gravity	1.5	0.7 (liquid)	1.6
Solubility in Water (g/L)	716.0 at 20°C	Miscible	480.0 at 10°C

several cyanide salts of alkali, alkaline earth, or heavy metals, that is, $Zn(CN)_2$, $Cd(CN)_2$, $Ni(CN_2)$, and AgCN, of varying degrees of solubility. In water, NaCN and KCN will completely dissociate to give free cyanide. All simple cyanides ionize in water to release cyanide ion which, depending on pH, will form hydrocyanic acid. For sodium cyanide, the reaction proceeds as follows:

$$(1)\quad NaCN \leftrightarrows Na^{+} + CN^{-}$$

$$(2)\quad CN^{-} + HOH \leftrightarrows HCN + OH^{-}$$

Increased pH will maintain a larger fraction of the cyanide as CN^-, and acidification will cause the reverse. At pH 7, about 99% of the free cyanide is in the form of HCN, whereas at pH 9.3 HCN composes 50%. Since HCN is extremely water soluble and is also one of the most toxic cyanide species, it is noteworthy that the toxicity of simple cyanides will not be affected measurably below pH 8.3.

Complex cyanides are compounds in which the cyanide anion is incorporated into a complex or complexes; these compounds are different in chemical and toxicologic properties from simple cyanides. In solution, the stability of the cyanide complex varies with the type of cation and the complex that it forms. Some of these are dissociable in weak acids to give free cyanide and a cation, while other complexes require much stronger acidic conditions for dissociation. The least stable complex metallocyanides include $Zn(CN)_4^{2-}$, $Cd(CN)_3^{-}$, and $Cd(CN)_4^{2-}$; moderately stable complexes include, $Cu(CN)_2^{-}$, $Cu(CN)_3^{2-}$, $Ni(CN)_4^{2-}$, and $Ag(CN)_2^{-}$; and the most stable complexes include, $Fe(CN)_6^{4-}$, and $Co(CN)_6^{4-}$. The toxicity of complex cyanides is usually related to their ability to release cyanide ions in solution, which then enter into an equilibrium with HCN; relatively small fluctuations in pH significantly affect their biocidal properties.

Cyanogen $[(CN)_2]$ is the simplest compound containing the cyanide group. Cyanogen is an extremely toxic, flammable gas that reacts slowly with water to form HCN, cyanic acid, and other compounds; it is rapidly degraded in the environment. Cyanogen and its halide derivations are comparable in toxicity to HCN.

Nitriles are defined as organic compounds (RCN) containing the cyanide group. Cyanide bound to carbon as nitriles (other than as cyanogenic glycosides) are comparatively innocuous in the environment, and are low in chemical reactivity and are biodegradable. For simple mononitriles there is a clear progression, with more cyanide being released as chain length increases. A similar pattern exists in dinitriles, but corresponding compounds require a longer carbon chain than mononitriles before free cyanide is produced. Based on studies with chicken liver homogenates, mononitriles were more toxic than dinitriles, and within each group the order of toxicity was $CH_3 > C_2H_5 > C_3H_7 > C_4H_9 > C_5H_{11} > C_7H_{15}$. Cyanohydrins $[R_2C(OH)CN]$ and cyanogenic glycosides $[R_1R_2C(OR)_3CN]$ are special classes of nitriles, in that under appropriate conditions they will decompose to HCN and cyanide ions. Cyanogens (not to be confused with cyanogen), such as acrylonitrile, propionitrile, and succinonitrile, are nitrile-containing materials of varying complexity and lability, and can liberate free and toxicologically available amounts of cyanide. But the non-nitrile portion of the cyanogen molecule may exert an independent or interactive toxicity, causing a complex response.

Cyanates contain the OCN group. Inorganic cyanates that are formed industrially by the oxidation of cyanide salts hydrolyze in water to form ammonia and bicarbonate ion. Alkyl cyanates are insoluble in water and form cyanurates. Alkyl isocyanates contain the OCN radical, are formed from cyanates, and, like cyanates, are readily hydrolyzed. Thiocyanates (SCN group) are formed from cyanides and sulfur-containing materials and are relatively stable.

Total cyanides refers to all cyanide-containing compounds, including simple and complex cyanides, cyanoglycosides, and free cyanide. Total cyanides is a chemical measurement of free cyanide present in solution or released by acidification or digestion. Only free cyanide is considered to be a biologically meaningful expression of cyanide toxicity. Under most circumstances, the concentration of total cyanide will exceed that of HCN. In some waters, however, the total cyanide concentration may consist almost entirely of free cyanide, or it may contain cyanides that readily photodecompose or dissociate to yield HCN. The relation between total cyanide and free cyanide in natural waters varies with receiving-water conditions, type of cyanide compounds present, degree of exposure to daylight, and presence of other chemical compounds.

Hydrogen cyanide has frequently been associated with the odor of bitter almonds. The threshold odor for olfactory detection of atmospheric HCN is 1.0 mg/L, but the odor may not be detected for various reasons, including the presence of other odors, and the fact that only 20–40% of those tested could detect a cyanide odor.

Analytical methods for determining free and bound cyanide and cyanogenic compounds in biological materials are under revision. Procedures include chromatography, enzymic post-column cleavage, electrochemical detection, and ultraviolet, infrared, proton, and carbon-13 nuclear magnetic resonance spectroscopies. Proposed newer analytical methodologies include chemiluminescence, deproteinization techniques, thin film dissociation coupled with preferential ultraviolet irradiation, differential pulse polarography, and modified spectrophotometric, colorometric, and ion chromatographic determinations. Analysis of cyanide and cytochrome oxidase is usually conducted with samples of whole blood, serum, plasma, brain, or ventricular myocardium tissues. Samples should be obtained as soon as possible after cyanide exposure and analyzed immediately otherwise erroneous analytical values will result. Brain and liver are recommended for cyanide analysis if removed and analyzed within a week. Cyanide measurements are further confounded by the presence of various antidotal agents; by various tissue preservatives, such as formaldoxime and sodium fluoride; and by the spontaneous postmortem production of cyanide in various tissues (e.g., sterile blood, brain, liver, kidney, uterus, and intestines) over time in cases of noncyanide death.

11.3 Mode of Action

Cyanide is a potent and rapid-acting asphyxiant. At lethal doses, inhalation or ingestion of cyanide produces reactions within seconds and death within minutes. Cyanide's toxic effect is due to its affinity for the ferric heme form of cytochrome a_3, also known as cytochrome *c* oxidase, the terminal oxidase of the mitochondrial respiratory chain. Inhibition of the enzyme cytochrome *c* oxidase is thought to involve a two-step reaction, initial penetration of cyanide into a protein crevice followed by binding to heme iron. Formation of a stable cytochrome *c* oxidase–CN complex in the mitochondria produces a blockage of electron transfer from cytochrome oxidase to molecular oxygen and cessation of cellular respiration, causing cytotoxic hypoxia in the presence of normal hemoglobin oxygenation. Tissue anoxia induced by the activation of cytochrome oxidase causes a shift from aerobic to anaerobic metabolism, resulting in the depletion of energy-rich compounds such as glycogen, phosphocreatine, and adenosine triphosphate (ATP), and the accumulation of lactate with decreased blood pH. The combination of cytotoxic hypoxia with lactate acidosis depresses the central nervous system – the most sensitive site of anoxia – resulting in respiratory arrest and death. If the absorption rate is significantly greater than the detoxification rate there will be a rapid accumulation of free cyanide in tissues and body fluids, resulting in the prompt onset of signs of acute cyanide poisoning. Acute cyanide poisoning is frequently encountered as a relatively massive overdose, where the amount of cyanide greatly exceeds the minimal concentration necessary to inhibit cytochrome *c* oxidase. In such cases, many enzymes and biological systems are disrupted, including various metalloenzymes, nitrate reductase, nitrite reductase, myoglobin, various peroxidases, catalase, and ribulose diphosphate carboxylase, resulting in severe signs of toxicity and rapid death.

The great majority of the absorbed cyanide reacts with thiosulfate in the presence of enzymes to produce thiocyanate, which is excreted in the urine over a period of several days. Owing to this rapid detoxification, animals can ingest high sublethal doses of cyanide over extended periods without harm. Authorities are also in general agreement on several points: thiosulfate is usually low in the body, and higher levels can significantly protect against cyanide toxicity; species vary considerably in both the extent to which thiocyanate is formed and the rate at which it is eliminated from the body; thiocyanate metabolites resulting from the transulfuration process are about 120 times less toxic than the parent cyanide compound; thiocyanate may accumulate in tissues and has been associated with developmental abnormalities and other adverse effects; the two enzyme systems responsible for the transulfuration process are thiosulfate-cyanide sulfurtransferase – also known as rhodanese – and beta-mercaptopyruvate cyanide sulfurtransferase. Rhodanese is widely distributed in the body, but activity levels in mammals are highest in the mitochondrial fraction of liver. Rhodanese activity levels in catalyzing the transformation of thiosulfate to thiocyanate are limited by the availability of sulfur. Minor detoxification pathways for cyanide include exhalation in breath as HCN, and as CO_2 from oxidative metabolism of formic acid; conjugation with cystine to form 2-iminothiazolidene-4-carboxylic acid or 2-aminothiazoline-4-carboxylic acid; combining with hydroxocobalamin (B_{12}) to form cyanocobalamin, which is excreted in urine and bile; and binding by methemoglobin in the blood.

Absorption of HCN liquid or gas readily occurs through inhalation, ingestion, or skin contact. Inhalation and skin absorption are the primary hazardous routes in cyanide toxicity in occupational exposure. Skin absorption is most rapid when the skin is cut, abraded, or moist. Inhalation of cyanide salts is also potentially hazardous because the cyanide dissolves on contact with moist mucous membranes. Regardless of route of exposure, cyanide is readily absorbed into the blood stream and distributed throughout the body. Cyanide concentrates in erythrocytes through binding to methemoglobin, and free cyanide concentrations in plasma are now considered one of the better indicators of cytotoxicity. Because of the affinity of cyanide for the mammalian erythrocyte, the spleen may contain elevated cyanide concentrations when compared to blood; accordingly, spleen should always be taken for analysis in cases of suspected cyanide poisoning. Cyanide also accumulates in various body cells through binding to metalloproteins or enzymes such as catalase and cytochrome *c* oxidase. Brain is probably the major target organ of cytotoxic hypoxia, and brain cytochrome oxidase may be the most active site of lethal cyanide action, as judged by distribution of cyanide, thiosulfate, and rhodanese. Significant positive correlations exist between cyanide concentrations in plasma, cerebrospinal fluid, and brain; these correlations need further exploration.

Hydrogen cyanide formation may contribute to the toxicity of snake venom, owing to the high levels of L-amino acid oxidase in some snake venoms. This enzyme is harmless on injection, but the tissue destruction caused by other venom components probably provides the required substrate and cofactor for HCN production. Cyanide inhibits ion transport mechanisms in amphibian skin, gall bladder, and proximal renal tubules. Measurable changes in cell membrane potentials of isolated gall bladder epithelium cells, for example, were induced by NaCN in a salamander (*Necturus maculosus*). Cyanide-induced hyperpolarization was caused primarily by an increase in permeability of the cell membrane to potassium, which, in turn, was mediated by an elevation of intracellular calcium ion activity, attributable to release from mitochondrial sources. The binding rate of CN to hemeproteins, specifically hemoglobin components III and IV, is 370–2300 times slower in a marine polychaete annelid (*Glycera dibranchiata*), when compared to guinea pig (*Cavia* spp.), soybean (*Glycine max*), and sperm whale (*Physeter macrocephalus*); the significance of this observation is unclear but warrants further exploration.

11.4 Clinical Features

Accidental exposure to cyanides or cyanogens through inhalation, skin exposure, and swallowing occurs in agricultural fumigation, laboratories, industrial operations, domestic abuse, and products of combustion. Intentional exposure is reported from homicides, suicides (usually uncommon), judicial executions, chemical warfare, and covert activities. Diagnosis of lethal cyanide poisoning is difficult because of the absence of gross pathology or histology, nonspecific congestion of viscera, and cerebral or pulmonary edema. Sometimes the blood is bright red, and sometimes the odor of bitter almonds is detected, but neither is sufficiently consistent for diagnostic purposes. At low lethal doses of cyanide, the effects are principally on cytochrome oxidase in the central nervous system. At higher doses, cardiovascular signs and changes in electrical activity of the brain are among the most consistent changes measured. Acute and subacute toxic effects of poisoning with cyanide can vary from convulsions, screaming, vomiting, and bloody frothing to less dramatic events, such as a slow, quiet onset to coma and subsequent death. In the first stage of cyanide poisoning, victims exhibit headache, vertigo, weak and rapid pulse, nausea, and vomiting. In the second stage, there are convulsions, falling, dilated pupils, clammy skin, and a weaker and more rapid pulse. In the final stage, heartbeat becomes irregular and slow, body temperature falls, there is cyanosis of lips, face, and extremities, coma, frothy bloody saliva flow from mouth, and death. If acute exposure is to a sublethal dose of cyanide, this may lead to signs of toxicity, but as detoxification proceeds these signs will become less obvious and eventually vanish, and cyanide will be excreted as thiocyanate without accumulating.

Chronic cyanide poisoning may develop in individuals who ingest significant quantities of cyanide or cyanide precursors in their diets; effects are exacerbated by dietary deficiencies in vitamin B_{12}, iodine, and sulfur amino acids, as well as by low levels and insufficient distribution of detoxifying enzymes such as rhodanese. Cyanide toxicity of dietary origin has been implicated in acute animal deaths and as a major etiologic factor in toxic ataxic neuropathy in man, and as a cause of blindness in humans suffering from tobacco amblyopia and Leber's hereditary optic atrophy. An increase in blood plasma cyanide is observed in healthy individuals who smoke cigarettes. An increase in blood plasma thiocyanate is also seen in smokers and in hemodialysis patients just before dialysis. Continuous intake of cyanide causes high levels of plasma thiocyanate and goiters in mammals; the antithyroid action (goiters) results from cyanide interference with iodine transport and thyroxine synthesis. Signs of chronic cyanide poisoning include demyelination, lesions of the optic nerve, decrease in sulfur-containing amino acids, increase in thiocyanate, goiter, ataxia, hypertonia, and depressed thyroid function. These effects are common in areas that depend on cyanogenic plants – such as cassava – as a major dietary component.

Biochemically, cyanide affects the citric acid cycle, strongly inhibits catalases and proteinases, induces glycolysis in protozoans, fish, and mammals, produces vitamin B_{12} deficiency, and modifies the phosphorylation mechanism of respiratory mitochondrial enzymes, causing arrested respiration due to inability to use oxygen. Cyanide biomagnification or cycling has not been reported, probably because of cyanide's high chemical reactivity and rapid biotransformation. There is no evidence that chronic exposure to cyanide results in teratogenic, mutagenic, or carcinogenic effects. Cyanide possibly has antineoplastic activity, as judged by a low therapeutic success against rat sarcomas, but this requires additional documentation.

Confirmatory evidence of cyanide poisoning includes elevated blood thiocyanate levels – except, perhaps, when death was rapid – and reduced cytochrome oxidase activity in brain and myocardium, provided that all tissues were taken within a day or so of death, frozen quickly, and analyzed shortly thereafter. Evaluation of cyanide poisoning and metabolism includes signs of toxicity, LD50 values, measurement of cyanide and thiocyanate concentrations, cytochrome *c* oxidase activity, metabolic modification of in vivo cyanogenesis, rate of cyanide liberation in vitro, and influence of modifying factors such as the animal species, dose, rate and frequency of administration, route of exposure, differential distribution of cyanide, detoxification rates, circadian rhythm interactions, age of the organism, and presence of antidotes. For example, the concentration of cyanide measured in body fluids and tissues in man and other animals following lethal administration of cyanide depends on several factors: route of exposure, with oral route yielding highest residues and inhalation route the lowest; amount and duration of exposure; nature of the material, with HCN and CN^- being most toxic; time to death; antidotes used; time to autopsy, with marked loss documented from simple evaporation, thiocyanate formation, hydrolysis, and polymerization; and time from autopsy to sample analysis, wherein cyanide concentrations may increase due to microbial action.

11.5 Antidotes

The antagonism of cyanide intoxication has been under investigation for at least 150 years. In 1840, cyanide lethality was reported to be antagonized by artificial respiration. In 1888, amyl nitrite was reported effective in antagonizing lethal effects of cyanide in dogs. In 1894, cobalt was shown to form a stable metal complex with cyanide and was used to antagonize cyanide. In 1933, the use of sodium thiosulfate as the sulfur donor was described. Many compounds are used today as cyanide antidotes including cobalt salts, rhodanese, sulfur donors, methemoglobin producers, carbohydrates, drugs used to treat acidosis, oxygen, methylene blue, 4-dimethylaminophenol, various aromatic amino- and nitro-compounds (such as aniline, *p*-aminopropiophenone, nitrobenzene), carbonyl compounds, and sodium pyruvate. Different antidotes are preferred in different countries: in the United States, a mixture of sodium nitrite and sodium thiosulfate; in France and the United Kingdom, cobaltedetate, also known as Kelocyanor; and in Germany, a mixture of 4-dimethylaminophenol and sodium thiosulfate.

The classic nitrite–thiosulfate treatment of cyanide poisoning, developed almost 70 years ago, is one of the antidotal combinations still employed. Excess oxygen improves this antidotal combination by potentiating the effectiveness of the nitrite–thiosulfate combination, as confirmed by studies in sheep and rats, even though, theoretically, oxygen should serve no useful purpose. This therapeutic regimen protected rats against 20 LD50 doses of cyanide. Nitrite converts hemoglobin to methemoglobin, which has a high affinity for cyanide. The methemoglobin–HCN complex then slowly releases cyanide, which is converted to thiocyanate by way of rhodanese. Sodium nitrite, administered intravenously, is now considered one of the more rapid therapeutic methods. The injection of sodium thiosulfate provides sulfur for the enzyme rhodanese to mediate the biotransformation of cyanide to the much less toxic thiocyanate. Multiple injections of sodium thiosulfate protected mice against death by organic cyanides and were more effective than sodium nitrite. The nitrite–thiosulfate antidotal combination is one of the most effective treatments of cyanide poisoning, even though the specific mechanism of action of these two compounds is now being questioned, and concerns have been raised because of the toxicity of nitrite. One accepted therapy is an intravenous combination of sodium nitrite (1 mL of 20% solution) and sodium thiosulfate (3 mL of 20% solution), giving 4 mL of this mixture per 45 kg of body weight (BW). For maximal effectiveness in treating cyanide intoxication in sheep, large doses of sodium thiosulfate (660.0 mg/kg BW) are given in combination with conventional doses of sodium nitrite (6.6 mg/kg BW). Livestock treatment in cases of suspected cyanide intoxication consists of intravenous administration of 10.0–20.0 mg sodium nitrite/kg BW followed by 30.0–40.0 mg sodium thiosulfate/kg BW; however, a sodium thiosulfate dose of 500.0 mg/kg BW, or more, may be more efficacious. Once clinical signs have abated, one gram of activated charcoal per kilogram body weight may be administered as a drench via a stomach tube. A 30-kg female goat (*Capra* sp.) was successfully treated after eating the leaves and fruits of the crab apple (*Malus sylvestris*), a plant that contains high levels of cyanogenic glycosides in leaves and fruits. Treatment consisted of four hourly treatments of 100.0 g of animal charcoal and bismuth subnitrate in water as a drench, followed by 300.0 mg sodium nitrite as a 1% aqueous solution, then 25.0 g of sodium thiosulfate. Another goat died despite identical treatment.

Cobalt compounds, such as hydroxocobalamin and its derivatives (i.e., cobalt histidine, cobalt chloride, dicobalt ethylenediamine tetracetic acid) have been used to treat cyanide poisoning for more than 100 years. Their efficacy was confirmed in pigeons (*Columba* sp.) and rabbits (*Oryctolagus* sp.), but cobalt compounds did not receive wide support as cyanide antagonists because of the inherent toxicity of cobalt ion. Nevertheless, proponents of the use of cobalt compounds (i.e., the United Kingdom, Scandinavia, and much of Europe) stress the rapidity of action in forming a stable metal complex with cyanide, thereby preventing its toxic effect. One of the more frequently used cobalt compounds in cyanide treatment is hydroxocobalamin, which reverses cyanide toxicity by combining with cyanide to form cyanocobalamin (vitamin B_{12}). Hydroxocobalamin has been used in guinea pigs and baboons (*Papio anubis*) to lower blood cyanide levels, and in humans after inhalation or ingestion of cyanide compounds.

Dimethylaminophenol (DMAP) forms methemoglobin by setting up a catalytic cycle inside the erythrocyte, in which oxygen oxidizes the DMAP to *N*-*N*-dimethylquinoneimine, the latter oxidizing the hemoglobin to methemoglobin. Dogs poisoned with KCN and given DMAP intravenously had restored respiration and decreased plasma cyanide levels. The 4-dimethylaminophenol induced ferrihemoglobin production, which combined with the cyanide in the red cells to form ferrihemoglobin cyanide.

No usable cyanide prophylactic therapy now exists for humans, although sodium thiosulfate, hydroxocobalamin, and other compounds have been used to protect against cyanide toxicity in laboratory animals. For example, pyridoxal 5-phosphate, the active form of vitamin B_6, readily forms complexes with cyanides, and was effective in providing significant protection to rats. Fructose fed prior to insult lessens cyanide-induced hepatotoxicity in rats. L-ascorbic acid and dehydroascorbic acid probably act as protectants against cyanide toxicity by way of nontoxic cyanohydrin formation. Carbon tetrachloride pretreatment was effective in protecting mice against death from most nitriles, and pretreatment with *p*-aminopropiophenone serves to protect against cyanide toxicity.

11.6 Sources and Uses

Production of cyanides in the United States increased from about 136 million kg in 1963 to 318 million kg in 1976. Cyanide consumption in North America was 64 million kg in 1988 and 98 million kg in 1989; about 80% of these amounts was used in gold mining. In Canada, more than 90% of the mined gold is extracted from ores with the cyanidation process. This process consists of leaching gold from the ore as a gold–cyanide complex, and gold being recovered by precipitation. Heap leaching occurs when crushed ore is stacked on an impermeable plastic pad on the ground surface, with spraying or dripping of a NaCN solution on the flattened top. Large leach heaps may include 272,000 metric tons or ore and tower 100 m or more. In the milling of gold ores, a NaCN solution is percolated through the crushed ores to dissolve the gold particles. In both leaching and milling processes, after the gold is chemically precipitated, the solution is adjusted for pH and cyanide concentration, and recycled to precipitate more gold. Eventually, the remaining solution must be treated to recycle the cyanide or to destroy it to prevent escape into the environment. Milling and heap leaching require cycling of millions of liters of alkaline water containing high concentrations of potentially toxic NaCN, free cyanide, and metal cyanide complexes that are frequently accessible to wildlife. Some milling operations result in tailings ponds of 150 ha and larger. Heap-leach operations that spray or drip cyanide solutions onto the top of the ore heap require solution processing ponds of about 1 ha in surface area. Although not intentional or desired, puddles of various sizes may occur on the top of heaps where the highest concentrations of NaCN are found. Exposed solution recovery channels are usually constructed at the base of leach heaps. All of these cyanide-containing water bodies are hazardous to wildlife if not properly managed.

About 84% of domestic HCN production is used to produce organic cyanides, also known as nitriles, including acrylonitriles, methyl methacrylate, and adiponitrile. Nitriles tend to polymerize, which is the basis for their use in the manufacture of synthetic fibers, resins, plastics, dyestuffs, vitamins, solvents, elastomers, agricultural insecticides, and high-pressure lubricants. The widespread usefulness of HCN is related to its strong tendency and that of its inorganic salts to form complexes with metals. For example, sodium cyanide is used in metallurgy for the extraction of gold and silver from ores and in electroplating baths because it forms stable soluble complexes. Similar behavior makes alkali cyanide solutions excellent for cleaning silverware and other precious metals and is responsible for their general use in industry as metal cleaners. In Canada, more than 90% of the gold mined is extracted from ores with the cyanidation process. This process consists of leaching gold from the ore as a gold–cyanide complex, and gold being precipitated with the addition of zinc dust. A variety of cyanide compounds are produced during gold cyanidation. In addition to their primary use in the metals and electroplating industries, and in the manufacture of synthetic fibers and plastics, various cyanide compounds have been used directly or as an intermediate to produce synthetic rubber, fumigants, rodenticides, insecticides, predator control agents, rocket fuels, paints and paint finishes, paper, nylon, pharmaceuticals, photographic chemicals, mirrors, cement, perfumes, bleaches, soaps and detergents, riot control agents, fertilizers, and weedicides.

Hydrogen cyanide vapor, because of its high and rapid acute lethal toxicity and ready diffusion, has been used widely to fumigate buildings, ships, and warehouses; to exterminate rabbits, rodents, and large predators; and in horticultural practice, to control insect pests that have developed resistance to other pesticides. Typically, fumigation powders containing either calcium cyanide, $Ca(CN)_2$, or sodium cyanide, NaCN, are blown into burrows or scattered over the floor in greenhouses. On coming into contact with water, such powders liberate HCN vapor. Hydrogen cyanide released from $Ca(CN)_2$ is registered for use on almonds, dried beans, citrus, cocoa beans, grains, nuts, and spices. Cyanide-containing compounds are used for a variety of agricultural and pesticidal agents. These compounds include cyanogen (NCCN), as an intermediate in the production of some commercial fertilizers; cyanogen chloride (CNCl), in the manufacture of triazine herbicides; cyanogen bromide (CNBr), as a pesticidal fumigant; hydrogen cyanide (HCN), in the synthesis of methionine for animal feeds; ammonium thiocyanate (NH_4SCN), as a cotton defoliant; sodium thiocyanate (NaSCN), as a weedkiller; and calcium cyanamide (CaNCN), as a plant fertilizer, herbicide, pesticide, and defoliant of cotton and tomatoes. Cyanide compounds have also been used as preservatives for raw vegetables.

Sodium cyanide has been used for about 50 years by the U.S. Fish and Wildlife Service for about 50 years against coyote in attempts to protect livestock, especially sheep. The Service has made extensive use of two NaCN ejector devices: "the coyote getter" from the late 1930s to 1970; and the M-44, from about 1968 to the present, except for the period 1972–74, when all uses of NaCN for predator control were canceled. Although both ejectors dispense toxicant when pulled, they differ in the way ejection is achieved. In the coyote getter, the toxicant is in a 0.38-caliber cartridge case and is expelled by the explosive force of the primer plus a small powder charge. The M-44 uses a spring-driven plunger to push out its toxic contents. M-44 capsules weigh about 0.94 g, and consist of about 89% NaCN, 6% Celatom MP-78 (mostly diatomaceous silica), 5% potassium chloride, and 0.25% FP Tracerite yellow – used as a fluorescent marker. Coyote getters and M-44s are set into the ground with only their tops protruding. Fetid scent or lure stimulates a coyote to bite and pull, whereupon a lethal dose of NaCN is ejected into its mouth; coma and death follow in 30–60 s. Although coyote getters were about 99% effective against coyotes, compared to 73% for M-44s, the Service decided that spring-driven plungers were less hazardous to operators than were explosive-driven plungers. The coyote getter was generally much more selective than the trap for the capture of coyotes. It was less destructive than traps to small mammals, birds of prey, ground-nesting birds, deer, antelope, and domestic sheep, but more destructive to dogs, bears, and cattle. In a 1-year test period (1940–41) in Colorado, Wyoming, and New Mexico, the following numbers of animals were killed by the coyote getter: 1107 coyotes, 2 bobcats (*Lynx rufus*), 24 dogs, 14 black-billed magpies (*Pica pica*), 7 foxes (*Vulpes* sp.), 8 unidentified skunks, 2 badgers, 2 unidentified eagles, 2 bears (*Ursus* sp.), and 1 each of hawk (unidentified), rockchuck (*Ochotona* sp.), and cow.

Cyanide compounds have been used to collect various species of freshwater fish. In England and Scotland, cyanides are used legally to control rabbits, and illegally to obtain Atlantic salmon (*Salmo salar*) and brown trout (*Salmo trutta*) from rivers, leaving no visible evidence of damage to the fish. Sodium cyanide has been applied to streams in Wyoming and Utah to collect fish through anesthesia; mountain whitefish (*Prosopium williamsoni*) were sensitive to cyanide and died at concentrations that were tolerable to salmon and trout. Sodium cyanide was also used as a fish control agent in Illinois, Nebraska, South Dakota, Missouri, and in the lower Mississippi River valley, but was never registered for this use because of human safety concerns. The widespread use of NaCN to collect exotic marine fishes is associated with high mortality in aquarium fish stocks in the Philippine islands and elsewhere. Cyanide fishing is banned in many Asia Pacific countries; however, widespread illegal fishing continues with significant adverse effects on coral reef ecosystems.

Cyanide compounds have been prescribed by physicians for treatment of hypertension and cancer. Sodium nitroprusside ($Na_2Fe(CN)_5NO \cdot 2H_2O$) was widely used for more than 30 years to treat severe hypertension and to minimize bleeding during surgery. Laetrile, an extract of ground apricot kernels, has been used for cancer chemotherapy and, in deliberate high intakes, as an attempted suicide vehicle.

Road salt in some areas may contribute to elevated cyanide levels in adjacent surface waters. In climates with significant snowfall, road salt is applied as a deicing agent. Road salts are commonly treated with anticaking agents to ensure uniform spreading. One anticaking agent, sodium hexacyanoferrate, decomposes in sunlight to yield the highly toxic free cyanide that contaminates surface waters by runoff. Another anticaking agent, yellow prussiate of soda (sodium ferrocyanide), has been implicated in fish kills when inadvertently used by fish culturists.

Napoleon III first realized the military uses of HCN, but it was not until World War I (WW I, 1914–18) that this application received widespread consideration. About 3.6 million kg of hydrogen cyanide were manufactured by France as a chemical weapon and used in WW I in various mixtures called Manganite and Bincennite, although its use was not highly successful because of limitations in projectile size and other factors. During WW II (1939–45), the Japanese were armed with 50 kg HCN bombs, and the United States had 500 kg bombs. More than 500,000 kg of HCN chemical weapons were produced during WW II by Japan, the United States, and the Soviet Union, but it is not known to what extent these weapons were used in that conflict.

Cyanides are widely distributed among common plants in the form of cyanogenic glycosides. Their toxicity following ingestion is primarily related to the hydrolytic release of HCN. Ingestion of cyanogenic plants probably has accounted for most instances of cyanide exposure and toxicosis in humans and range animals. Of chief agricultural importance among plants that accumulate large quantities of cyanogenic glycosides are the sorghums, Johnson grass, Sudan grass, corn, lima beans, flax, pits of stone fruits (cherry, apricot, and peach), vetch, linseed, sweet potatoes, bamboo shoots, southern mock orange, millet, almonds, and cassava. Factors favoring cyanide buildup in cyanogenic plants include high nitrogen and low phosphorus in soils; the potential for high glycoside levels is greatest in immature and rapidly growing plants. At present, more than 28 different cyanoglycosides have been measured in about 1000 species of higher plants. In cassava, for example, more than 90% of the cyanide is present as linamurin, a cyanogenic glycoside, and the remainder occurs as free (nonglycoside) cyanide. Laetrile, a preparation made from apricot kernels, contains high levels of amygdalin, a cyanogenic glycoside that can be degraded in the gut to cyanide and benzaldehyde. Several cases of cyanide poisoning in humans have been reported from intake of laetrile, either orally or anally. Cyanide formation in higher plants and microorganisms can also occur with compounds other than cyanogenic glycosides, such as glycine, glyoxylate plus hydroxylamine, or histidine. In some cases, plants may contain cyanide residues resulting from fumigation with HCN.

Many species of plants, including some fungi, bacteria, algae, and higher plants, produce cyanide as a metabolic product. Some species of soil bacteria suppress plant diseases caused by soilborne pathogens by producing metabolites with antibiotic activity. Certain strains of *Pseudomonas fluorescens*, a soil bacterium, suppress black root rot of tobacco caused by the fungus *Thielaviopsis basicola* by excreting several metabolites, including HCN. A wide variety of bacteria and fungi can degrade cyanide compounds, and may be useful in the treatment of cyanide wastes. For example, several species of fungi known to be pathogens of cyanogenic plants can degrade cyanide by hydration to formamide; dried mycelia of these species are now sold commercially to detoxify cyanide in industrial wastes.

Anthropogenic sources of cyanide in the environment include industrial processes, laboratories, fumigation operations, cyanogenic drugs, fires, cigarette smoking, and chemical warfare operations. Cyanides are present in many industrial wastewaters, especially those of electroplaters, manufacturers of paint, aluminum, and plastics, metal finishers, metallurgists, coal gasification processes, certain mine operations, and petroleum refiners. Electroplaters are a major source. In the United States alone electroplaters discharge about 9.7 million kg of cyanide wastes annually into the environment from 2600 electroplating plants. Paint residues annually contribute an additional 141,300 kg of cyanide wastes into the environment, and paint sludges 20,400 kg. Cyanide can also originate from natural processes, such as cyanide production by bacteria, algae, and fungi, and from many terrestrial plants that release free HCN when their cellular structure is disrupted. Hospital wastewaters usually contain no detectable cyanide, but concentrations up to 64.0 μg CN^-/L have been measured after alkali chlorination treatment. It seems that various compounds common in hospital wastewaters will produce 15.0–25.0 μg CN^-/L after alkali chlorination; these compounds include hydantoin (an antiepilepsy agent) and related nitrogenous compounds, such as hydantonic acid, 5,5-diphenyl hydantoin, imidazole, and 2-imidazolidinone.

Free hydrogen cyanide occurs only rarely in nature because of its high reactivity. The gas is sometimes found in the atmosphere, however, as a result of emissions from the petrochemical industry, malfunctioning catalytic converters on automobiles, fumigation of ships and warehouses, incomplete combustion of nitrogen-containing materials, and from tobacco smoke. Hydrogen cyanide is known to be produced in fires involving nitrogen-containing polymers and is probably the most important narcotic fire product other than carbon monoxide. Cyanide-related fire deaths and injuries, as judged by elevated blood cyanide and thiocyanate concentrations, have been documented in airplanes, jails, and high-rises. In a study of fire victims in Scotland, elevated blood cyanide levels were found in 78% of fatalities, and 31% had blood levels considered to be toxic. Major factors that influence HCN release include the chemical nature of the material, temperature, oxygen availability, and burning time. Substantial quantities of free HCN and organic cyanides are known to be produced in fire settings involving horsehair, tobacco, wool, silk, and many synthetic polymers, such as polyurethane and polyacrylonitriles. Polyacrylonitrile, for example, is used in fabrics, upholstery covers, paddings, and clothing; about 50% of the mass of the polymer is theoretically available as HCN under thermal decomposition.

11.7 Concentrations in Field Collections

The reactivity of HCN, and its ability to condense with itself and other compounds, was probably responsible for the prebiotic formation of the majority of biochemical compounds required for life. Cyanide is now known to be present in a number of foodstuffs and forage plants, as a metabolite of certain drugs, in various industrial pollutants, and may be formed by the combustion of cyanide-releasing substances, such as plastics in airplane fires, and tobacco in smoking. Hydrogen cyanide production may occur in hepatopancreas of mussels, *Mytilus edulis*, in rat liver, and in green and blue-green algae during nitrate metabolism. Except for certain naturally occurring organic cyanide compounds in plants, it is uncommon to find cyanide in foods consumed in the United States.

The cyanide anion is found in a variety of naturally occurring plant compounds as cyanogenic glycosides, glycosides, lathyrogenic compounds, indoleacetonitrile, and cyanopyridine alkaloids. Plants that contain cyanogenic glycosides are potentially poisonous because bruising or incomplete cooking can result in glycoside hydrolysis and release of HCN. Cyanide concentrations in cyanogenic plants are usually highest in leaves of young plants; levels drop rapidly after pollination. There are about 20 major cyanogenic glycosides, of which usually only one or two occur in any plant. They are synthesized from amino acids and sugars and are found in many economically important plants, such as sorghum, flax, lima bean, cassava, and many of the stone fruits. Cassava contains linamurin and lotaustralin, whereas the main cyanogenic glycoside in cereals is dhurrin; consumption of foods containing toxic cyanogens (primarily cassava) has been associated with death or morbidity – on an acute basis – or goiter, and tropical ataxic neuropathy on a chronic consumption basis. Cassava is a perennial shrub, native to the neotropics, grown for its tuberous starchy roots, and a traditional dietary staple of many indigenous populations in Amazonia, especially the Tukanoan Indians in northwestern Amazonia. Cassava is one of the few food plants in which the cyanide content may create toxic problems. All varieties of cassava contain cyanogenic glycosides capable of liberating HCN, but amounts vary greatly depending on variety and environmental conditions. Bitter cultivars of cassava provide over 70% of the Tukanoan's food energy, appearing in the diet as bread, meal, a starch drink, and boiled cassava juice. The greatly elevated total cyanide content in bitter varieties may contain 5.1–13.4% of the total as the toxic free cyanide.

The production of HCN by animals is almost exclusively restricted to various arthropods: 7 species of about 3000 species of centipedes; 46 of 2500 species of polydesmid millipedes; and 10 of 750,000 species of insects, including 3 species of beetles, 4 moths, and 3 butterflies. Millipedes – which are eaten frequently by toads and starlings – secrete cyanide for defensive purposes in repelling predators; in zygaenid moths, cyanide seems to be localized in eggs.

Cyanide concentrations in fish from streams that were deliberately poisoned with cyanide ranged between 10.0 and 100.0 μg total cyanide/kg whole body fresh weight (FW). Total cyanide concentrations in gill tissues of salmonids under widely varying conditions of temperature, nominal water concentrations, and duration of exposure ranged from about 30.0 to >7000.0 μg/kg FW. Unpoisoned fish usually contained <1.0 μg/kg FW in gills, although values up to 50.0 μg/kg occurred occasionally. Lowest cyanide concentrations in gill occurred at elevated (summer) water temperatures; at lower temperatures, survival was greater and residues were higher. Fish retrieved from cyanide-poisoned environments, dead or alive, can probably be consumed by humans because muscle cyanide residues were considered to be low (i.e., <1000.0 mg/kg FW).

Cyanide pollution is likely to occur in many places, ranging from industrialized urban areas to gold mines in the western United States and Northwest Territories of Canada. Cyanides are ubiquitous in industrial effluents and their increasing generation from power plants, and from the combustion of solid wastes, all these are expected to result in elevated cyanide levels in air and water. However, data are scarce on background concentrations of cyanides in various nonbiological materials. In soils, for example, high concentrations are unusual and are nearly always the result of improper waste disposal. Cyanides in soils are not absorbed or retained; under aerobic conditions, microbial metabolism rapidly degrades cyanides to carbon dioxide and ammonia; under anaerobic conditions, cyanides are converted by bacteria to gaseous nitrogen compounds that escape to the atmosphere. Heat treatment wastes from metal processing operations may contain up to 200.0 g CN/kg, mostly as NaCN, and are frequently hauled to landfills for disposal. The presence of cyanide in landfill waste is potentially hazardous because of the possibility that cyanide may leach to soil and groundwater, release HCN, and disturb natural microbiological degradation of organic materials. Measurements at landfills in England and the Netherlands showed total cyanide levels up to 560.0 g/kg in soil and 12.0 μg/L in groundwater. However, 7-month-long experimental studies of cyanide in heat treatment wastes in landfills showed that between 72 and 82% of the cyanide was converted, mostly to ammonium and organic nitrogen compounds; between 4 and 22% of the cyanide leached as free or complex cyanide; and up to 11% remained in the landfill.

Hydrogen cyanide is a common industrial pollutant and frequently occurs in water at concentrations between 0.1 and several milligrams per liter of free HCN. Total cyanides is the most often cited measurement in aqueous solutions, owing to limitations in analytical methodologies. Cyanides have been identified in freshwaters of rural and wilderness areas in Canada and Germany. Concentrations ranging between 30.0 and 60.0 μg total cyanides/L seem related to runoff, with cyanide peaks more frequent in fall and winter during periods of minimal runoff. In larger rivers, cyanide was low in winter owing to dilution by high runoff, but peaked in summer because of cyanide production by plants. Cyanides do not seem to persist in aquatic environments. In small, cold oligotrophic lakes treated with 1.0 mg NaCN/L, acute toxicity was negligible within 40 days. In warm shallow ponds, toxicity disappeared within 4 days after application of 1.0 mg NaCN/L. In rivers and streams, toxicity rapidly disappeared on dilution. Cyanide was not detectable in water and sediments of Yellowknife Bay, Canada, between 1974 and 1976, although the bay receives liquid effluents containing cyanides from an operating gold mine. Nondetection was attributed to rapid oxidation. Several factors contribute to the rapid disappearance of cyanide from water. Bacteria and protozoans may degrade cyanide by converting it to carbon dioxide and ammonia. Chlorination of water supplies can result in conversion to cyanate. An alkaline pH favors oxidation by chlorine, and an acidic pH favors volatilization of HCN into the atmosphere.

11.8 Persistence in Water, Soil, and Air

In water, cyanides occur as free hydrocyanic acid, simple cyanides, easily degradable complex cyanides such as $Zn(CN)_2$, and sparingly decomposable complex cyanides of iron and cobalt; complex nickel and copper cyanides are intermediate between the easily decomposable and sparingly degradable compounds. Cyanide has relatively low persistence in surface waters under normal conditions but may persist for extended periods in groundwater. Volatilization is the dominant mechanism for removal of free cyanide from concentrated solutions and is most effective under conditions of high temperatures, high dissolved oxygen levels, and at increased concentrations of atmospheric carbon dioxide. Loss of simple cyanides from the water column is primarily through sedimentation, microbial degradation, and volatilization. Water-soluble strong complexes, such as ferricyanides and ferrocyanides, do not release free cyanide unless exposed to ultraviolet light. Thus, sunlight may lead to cyanide formation in wastes containing iron-cyanide complexes.

Alkaline chlorination of wastewaters is one of the most widely used methods of treating cyanide wastes. In this process, cyanogen chloride, CNCl, is formed, which at alkaline pH is hydrolyzed to the cyanate ion, CNO^-. If free chlorine is present, CNO^- can be further oxidized. The use of sulfur dioxide in a high dissolved oxygen environment with a copper catalyst reportedly reduces total cyanide in high cyanide rinsewaters from metal plating shops to less than 1.0 mg/L; this process may have application in cyanide detoxification of tailings ponds. Other methods used in cyanide waste management include lagooning for natural degradation, evaporation, exposure to ultraviolet radiation, aldehyde treatment, ozonization, acidification–volatilization–reneutralization, ion exchange, activated carbon absorption, electrolytic decomposition, catalytic oxidation, and biological treatment with cyanide-metabolizing bacteria. Additional cyanide detoxification treatments include the use of $FeSO_4$, $FeSO_4$ plus CO_2, H_2O_2, $Ca(OCl)_2$, dilution with water, $FeSO_4$ plus H_2O_2, and $(NH_4)HSO_3$. In Canadian gold-mining operations, the primary treatment for cyanide removal is to retain gold mill wastewaters in impoundments for several days to months; removal occurs through volatilization, photodegradation, chemical oxidation, and, to a lesser extent, microbial oxidation. Microbial oxidation of cyanide is not significant in mine tailings ponds, which typically have pH >10, a low number of microorganisms, low nutrient levels, large quiescent zones, and cyanide concentrations >10.0 mg/L.

Cyanide seldom remains biologically available in soils because it is either complexed by trace metals, metabolized by various microorganisms, or lost through volatilization. Cyanide ions are not strongly adsorbed or retained in soils, and leaching into the surrounding groundwater will probably occur. Under aerobic conditions, cyanide salts in the soil are microbially degraded to nitrites or form complexes with trace metals. Under anaerobic conditions, cyanides denitrify to gaseous nitrogen compounds that enter the atmosphere.

Volatile cyanides occur only occasionally in the atmosphere, largely due to emissions from plating plants, fumigation, and other special operations. Under normal conditions, cyanide has relatively low persistence in air, usually between 30 days and one year, although some atmospheric HCN may persist for up to 11 years. Data are lacking on the distribution and transformation of cyanide in the atmosphere and should be acquired.

11.9 Lethal and Sublethal Effects

Cyanide effects on terrestrial plants and invertebrates, aquatic biota, birds, and mammals are numerous and disparate.

11.9.1 Terrestrial Flora and Invertebrates

Bacteria exposed to cyanide may exhibit decreased growth, altered cell morphology, decreased motility, mutagenicity, and altered respiration. Mixed microbial populations capable of metabolizing cyanide and not previously exposed to cyanide were adversely affected at 0.3 mg HCN/kg; however, these populations can become acclimatized to cyanide and can then degrade wastes with higher cyanide concentrations. Acclimatized populations in activated sewage sludge can often completely convert nitriles to ammonia, sometimes at concentrations as high as 60.0 mg total cyanides/kg. Cyanide can be degraded by various pathways to yield a variety of products, including carbon dioxide, ammonia, beta-cyanoalinine, and formamide. Several species of fungi can accumulate and metabolize cyanide, but the products of cyanide metabolism vary. For example, carbon dioxide and ammonia are formed as endproducts by *Fusarium solani*, whereas alpha-amino butyronitrile is a major cyanide metabolite of *Rhizoctonia solani*. Significant amounts of cyanide are formed as secondary metabolites by many species of fungi and some bacteria by decarboxylation of glycine. Rhizobacteria may suppress plant growth in soil through cyanide production. In one case volatile metabolites – including cyanide – from fluorescent pseudomonad soil bacteria prevented root growth in seedlings of lettuce, *Lactuca sativa*. Not all cyanogenic isolates inhibited plant growth. Some strains promoted growth in lettuce and beans by 41–64% in 4 weeks vs. 49–53% growth reduction by inhibitory strains.

In higher plants, elevated cyanide concentrations inhibited respiration (through iron complexation in cytochrome oxidase) and ATP production and other processes dependent on ATP, such as ion uptake and phloem translocation, eventually leading to death. Cyanide produces chromosomal aberrations in some plants, but the mode of action is unknown. At lower concentrations, effects include inhibition of germination and growth, but cyanide sometimes enhances seed germination by stimulating the pentose phosphate pathway and inhibiting catalase. The detoxification mechanism of cyanide is mediated by rhodanese. This enzyme is widely distributed in plants. The rate of production and release of cyanide by plants to the environment through death and decomposition is unknown.

Free cyanide is not found in intact plant cells. Many species of plants, such as cassava, sorghum, flax, cherries, almonds, and beans, contain cyanogenic glycosides that release HCN when hydrolyzed. Cyanide poisoning of livestock by forage sorghums, such as Sudan grass and various hybrid cultivars, is well known and has led to the development of several variations of sorghums that have a reduced capability of producing cyanide poisoning. Cyanogenesis has an important role in plant defense against predatory herbivores. This herbivore–plant interaction is not simple; the degree of selectivity by herbivores varies among individuals and by differences in hunger and previous diet.

Cyanide metabolism in higher plants involves amino acids, *N*-hydroxyamino acids, aldoximes, nitriles, and cyanohydrins. Cyanide is a coproduct of ethylene synthesis in higher plants. The increase in ethylene production that occurs during the senescence of certain flowers and the ripening of fruits is accompanied by a rise in beta-cyanoalanine activity; activity of this enzyme correlates closely with that of ACC (1-aminocyclopropane-1-carboxylic acid) oxidase, the last enzyme in the ethylene pathway. It has been suggested that ACC oxidase reacts with various amino acids to liberate cyanide. Cyanide added to isolated castor-bean (*Ricinus communis*) mitochondria significantly enhanced the rate and amount of protein synthesis. Cyanide stimulated mitochondrial protein synthesis in a dose-dependent manner, with an optimal stimulation of over 2-fold at 26.0 μg/L, but at this concentration mitochondrial respiration was inhibited by 90%. Cyanide is a weak competitive inhibitor of green bean (*Phaseolus vulgaris*) lipoxygenase, an enzyme that catalyzes the formation of hydroperoxides from polyunsaturated fatty acids. Because degradation of hydroperoxides causes unacceptable changes in bean flavor and color, compounds that inhibit lipoxygenase may enjoy wide application in the frozen vegetable industry. Corn seedlings from cold-resistant cultivars were more resistant to 65.0 mg KCN/L at low temperatures (13°C) than were seedlings from cold-susceptible cultivars (25°C), as judged by respiratory activity of mitochondria. Results suggest that cyanide-resistant respiration may play a role in cold resistance in maize seedlings, although more evidence is needed to demonstrate that cold-resistant plants actually use their greater potential for alternative respiration at low temperatures.

The cyanogenic system comprising cyanogenic glycosides, cyanohydrins, beta-glucosidases, and nitrile lyases is widespread in plants, but also occurs in several species of arthropods, including the tiger beetle (*Megacephala virginica*), leaf beetle (*Paropsis atomaria*), zygaenid moths, and certain butterflies. In *Zygaena trifolii*, cyanide compounds seem to function as protection against predators. Defensive secretions of cyanide have also been reported in polydesmid millipedes, and these organisms seem to be more tolerant than other species when placed in killing jars containing HCN. In a millipede (*Apheloria* sp.), cyanide is generated in a two-compartment organ by hydrolysis of mandelonitrile; cyanide generation occurs outside the gland when the components of the two compartments are mixed during ejection.

Highly toxic substances, such as cyanides, are sometimes feeding cues and stimulants for specialized insects. For example, instar larvae of the southern armyworm (*Spodoptera eridania*) strongly prefer cyanogenic foods, such as foliage of the lima bean, a plant with comparatively elevated cyanide content – up to 31.0 mg/kg in some varieties – in the form of linamurin. Feeding was stimulated in southern armyworms at dietary levels up to 508.0 mg KCN/kg (208.0 mg HCN/kg) for first to fourth instar larval stages, and between 1000.0 and 10,000.0 mg KCN/kg diet for fifth and sixth instar larvae. Sixth instar larvae pre-exposed to diets containing 5000.0 mg KCN/kg showed no adverse affects at dietary levels of 10,000.0 mg KCN/kg; however, previously unexposed larvae showed reversible signs of poisoning at 10,000.0 mg/kg diet, including complete inhibition of oviposition and 83% reduction in adult emergence. Experimental studies with southern armyworm larvae and thiocyanate – one of the in vivo cyanide metabolites – showed that 5000 mg thiocyanate/kg diet reduced pupation by 77%, completely inhibited oviposition, and reduced adult emergence by 80%, strongly suggesting that thiocyanate poisoning is the primary effect of high dietary cyanide levels in southern armyworms.

Resistant species, such as southern armyworms, require injected doses up to 800.0 mg KCN/kg BW (332.0 mg HCN/kg BW) or diets of 3600.0 mg KCN/kg for 50% mortality, but data are scarce for other terrestrial invertebrates. Exposure to 8.0 mg HCN/L air inhibits respiration in the granary weevil (*Sitophilus granarius*) within 15 min and kills 50% in 4 h; some weevils recover after cessation of 4-h exposure.

11.9.2 Aquatic Organisms

Numerous accidental spills of sodium cyanide or potassium cyanide into rivers and streams have resulted in massive kills of fishes, amphibians, aquatic insects, and aquatic vegetation; sources of poisonings were storage reservoirs of concentrated solutions, overturned rail tank cars, or discharge of substances generating free HCN in the water from hydrolysis or decomposition. Data on the recovery of poisoned ecosystems are scarce. In one case, a large amount of cyanide-containing slag entered a stream from the reservoir of a Japanese gold mine as a result of an earthquake. The slag covered the streambed for about 10 km from the point of rupture, killing all stream biota; cyanide was detected in the water column for only 3 days after the spill. Within 1 month flora was established on the silt covering the above-water stones, but there was little underwater growth. After 6–7 months, populations of fish, algae, and invertebrates had recovered, although species composition of algae was altered. Short-term exposure of 12 days to 10.0 μg HCN/L – a level frequently encountered in freshwater aquatic ecosystems – had produced cyanide-induced hormonal and physiological balance in sexually mature rainbow trout. Female trout had reduced plasma estradiol levels and reduced oocyte growth and yolk deposition within the ovary; males had decreased numbers of spermatocytes and selective loss of Type I basophils in the pituitary gland.

Fish were the most sensitive aquatic organisms tested. Significant adverse nonlethal effects, including reduced swimming performance and inhibited reproduction, were observed in the range of 5.0–7.2 μg free cyanide/L; deaths were recorded for most species between 20.0 and 76.0 μg free cyanide/L. Among invertebrates, adverse nonlethal effects were documented between 18.0 and 43.0 μg/L, and lethal effects between 30.0 and 100.0 μg/L – although some deaths were recorded in the range 3.0–7.0 μg/L for the amphipod *Gammarus pulex*. Algae and macrophytes were comparatively tolerant; adverse effects were reported at >160.0 μg free cyanide/L. The high tolerance of mudskippers (*Boleophthalmus boddaerti*), and perhaps other species of fishes, to HCN (LC50 of 290.0 μg/L in 96 h) is a result of a surplus of cytochrome oxidase and inducible cyanide-detoxifying mechanisms and not to a reduction in metabolic rate or an enhanced anaerobic metabolism.

Adverse effects of cyanide on aquatic plants are unlikely at concentrations that cause acute effects to most species of freshwater and marine fishes and invertebrates. Water hyacinth (*Eichhornia crassipes*) can survive for at least 72 h in nutrient solution containing up to 300.0 mg CN/L and can accumulate up to 6.7 g/kg DW plant material. On this basis, 1 ha of water hyacinths has the potential to absorb 56.8 kg of cyanide in 72 h, and this property may be useful in reducing the level of cyanide in untreated wastewater from various electroplating factories, where concentrations generally exceed 200.0 mg CN/L. Large-scale use of water hyacinths for this purpose has not yet been implemented, possibly due to disagreement over appropriate disposal mechanisms. Cyanide may also affect plant community structure. Some algae, for example, metabolized cyanide at water concentrations <1.0 mg CN/L, but at concentrations of 1.0–10.0 mg/L, algal activity was inhibited, leaving a biota dominated by *Actinomycetes* – a filamentous bacterium.

Cyanide adversely affects fish reproduction by reducing the number of eggs spawned, and the viability of the eggs by delaying the process of secondary yolk deposition in the ovary. Vitellogenin, a glycolipophosphoprotein present in plasma of fish during the process of yolk formation, is synthesized in liver under stimulation of estrogen and subsequently sequestered in the ovary; it is essential for normal egg development. Exposure of naturally reproducing female rainbow trout to as little as 10.0 μg HCN/L for 12 days during the onset of the reproductive cycle caused a reduction in plasma vitellogenin levels and a reduction in ovary weight. The loss of vitellogenin in the plasma would remove a major source of yolk. Reproductive impairment in adult bluegills (*Lepomis macrochirus*) has been reported following exposure to 5.2 μg CN/L for 289 days. Fertilized fish eggs are usually resistant to cyanide prior to blastula formation, but delayed effects occur at 60.0–100.0 μg HCN/L, including birth defects and reduced survival of embryos and newly hatched larvae. Concentrations as low as 10.0 μg HCN/L caused developmental abnormalities in embryos of Atlantic salmon after extended exposure. These abnormalities, which were absent in controls, included yolk sac dropsy and malformations of eyes, mouth, and vertebral column.

Other adverse effects of cyanide on fish include delayed mortality, pathology, impaired swimming ability and relative performance, susceptibility to predation, disrupted respiration, osmoregulatory disturbances, and altered growth patterns. Free cyanide concentrations between 50.0 and 200.0 μg/L were fatal to the more sensitive fish species over time, and concentrations >200.0 μg/L were rapidly lethal to most species of fish. Cyanide-induced pathology in fish includes subcutaneous hemorrhaging, liver necrosis, and hepatic damage. Exposure of fish for 9 days to 10.0 μg HCN/L was sufficient to induce extensive necrosis in the liver, although gill tissue showed no damage. Intensification of liver histopathology was evident at dosages of 20.0 and 30.0 μg HCN/L and exposure periods up to 18 days. Cyanide has a strong, immediate, and long-lasting inhibitory effect on the swimming ability of fish. Free cyanide concentrations as low as 10.0 μg/L can rapidly and irreversibly impair the swimming ability of salmonids in well-aerated water. Osmoregulatory disturbances recorded at 10.0 μg HCN/L may affect migratory patterns, feeding, and predator avoidance. In general, fish experience a significant reduction in relative performance (based on osmoregulation, growth, swimming, and spermatogenesis) at 10.0 μg HCN/L (Figure 11.1),

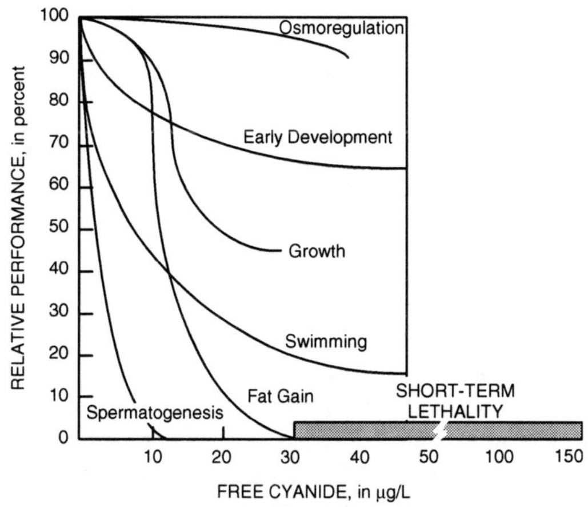

Figure 11.1.
Summary of lethal and sublethal effects of free cyanide on freshwater fish.

and although fish can survive indefinitely at 30.0 µg HCN/L in the laboratory, the different physiological requirements necessary to survive in nature could not be met. Increased predation by green sunfish (*Lepomis cyanellus*) on fathead minnows (*Pimephales promelas*) was noted at sublethal concentrations of HCN, but it was uncertain if fatheads became easier prey or if green sunfish had greater appetites.

Sodium cyanide has stimulatory effects on oxygen-sensitive receptors in lungfish, amphibians, reptiles, birds, and mammals. Facultative and aquatic air-breathers appear to rely on air breathing when external chemoreceptors are stimulated, whereas obligate air-breathing fish are more responsive to internal stimuli. Gill ventilation frequency of longnose gar (*Lepisosteus osseus*), for example, was little affected by external cyanide application, but responded strongly when cyanide was administered internally by injection. Cyanide, like many other chemicals, can stimulate growth of fish during exposure to low sublethal levels. This phenomenon, referred to as hormesis, is little understood and warrants additional research.

The observed toxicity to aquatic life of simple and complex cyanides was attributed almost entirely to molecular (undissociated) hydrocyanic acid (HCN) derived from ionization, dissociation, and photodecomposition of cyanide-containing compounds. The toxicity of the cyanide ion, CN^-, which is a minor component of free cyanide ($HCN + CN^-$) in waters that are not exceptionally alkaline is of little importance. The acute toxicity of stable silver cyanide and cuprocyanide complex anions is much less than that of molecular HCN, but is nevertheless important; these ions can be the principal toxicants, even in some very dilute solutions. The much lower toxicities of the ferrocyanide and ferricyanide complexes – which are of high stability but subject to extensive and rapid photolysis, yielding free cyanide on direct exposure to sunlight – and the nickelocyanide ion complex are not likely to be of practical importance. Toxicity to aquatic organisms of organic cyanide compounds, such as lactonitrile, is similar to that of inorganic cyanides because they usually undergo rapid hydrolysis in water to free cyanide. There is general agreement that total cyanide concentrations in water in most cases will overestimate the actual cyanide toxicity to aquatic organisms, and that the analytically determined HCN concentration in cyanide-polluted waters is considered to be the most reliable index of toxicity.

Cyanide acts rapidly in aquatic environments, does not persist for extended periods, and is highly species selective; organisms usually recover quickly on removal to clean water. The critical sites for cyanide toxicity in freshwater organisms include the gills, egg capsules, and other sites where gaseous exchange and osmoregulatory processes occur. On passing through a semipermeable membrane, the HCN molecules are usually distributed by way of the circulatory system to various receptor sites where toxic action or detoxification occurs. Once in the general circulation, cyanide rapidly inhibits the electron transport chain of vital organs. Signs of distress include increased ventilation, gulping for air at the surface, erratic swimming movements, muscular incoordination, convulsions, tremors, sinking to the bottom, and death with widely extended gill covers. The acute mode of action of HCN is limited to binding those porphyrins that contain Fe^{+3}, such as cytochrome oxidase, hydroperoxidases, and methemoglobin. At lethal levels, cyanide is primarily a respiratory poison and one of the most rapidly effective toxicants known. The detoxification mechanism of cyanide is mediated by thiosulfate sulfur transferase, also known as rhodanese. This enzyme is widely distributed in animals, including fish liver, gills, and kidney. Rhodanese plays a key role in sulfur metabolism, and catalyzes the transfer of a sulfane-sulfur group to a thiophilic group. Thiosulfate administered in the water with cyanide reduced the toxicity of cyanide to fish, presumably by increasing the detoxification rate of cyanide to thiocyanate.

Additive or more-than-additive toxicity of free cyanide to aquatic fauna has been reported in combination with ammonia or arsenic. However, conflicting reports on the toxicity of mixtures of HCN with zinc or chromium require clarification. Formation of the nickelocyanide complex markedly reduces the toxicity of both cyanide and nickel at high concentrations in alkaline pH. At lower concentrations and acidic pH, solutions increase in toxicity by more than 1000-fold, owing to dissociation of the metallocyanide complex to form HCN. Mixtures of cyanide and ammonia may interfere with seaward migration of Atlantic salmon smolts under conditions of low dissolved oxygen. The 96-h toxicity of mixtures of sodium cyanide and nickel sulfate to fathead minnows is influenced by water alkalinity and pH. Toxicity decreased with increasing alkalinity and pH from 0.42 mg CN/L at 5.0 mg $CaCO_3$/L and pH 6.5, to 1.4 mg CN/L at 70.0 mg $CaCO_3$/L and pH 7.5, to 730.0 mg CN/L at 192.0 mg $CaCO_3$/L and pH 8.0.

Numerous biological and abiotic factors are known to modify the biocidal properties of free cyanide, including water pH, temperature, and oxygen content; life stage, condition, and species assayed; previous exposure to cyanide compounds; presence of other chemicals; and initial dose tested. There is general agreement that cyanide is more toxic to freshwater fish under conditions of low dissolved oxygen; that pH levels within the range 6.8–8.3 had little effect on cyanide toxicity but enhanced toxicity at acidic pH; that juveniles and adults were the most sensitive life stages tested and embryos and sac fry the most resistant; and that substantial interspecies variability exists in sensitivity to free cyanide. Both initial dose and water temperature modify the biocidal properties of HCN to freshwater teleosts. At slow lethal concentrations (i.e., $<10.0\,\mu$g HCN/L), cyanide was more toxic at lower temperatures; at high, rapidly lethal HCN concentrations, cyanide was more toxic at elevated temperatures. By contrast, aquatic invertebrates were most sensitive to HCN at elevated water temperatures, regardless of dose. Season and exercise modify the lethality of HCN to juvenile rainbow trout; higher resistance to cyanide correlated with higher activity induced by exercise and higher temperatures, suggesting a faster detoxification rate or higher oxidative and anaerobic metabolisms. Low levels of cyanide that were harmful when applied constantly may be harmless under seasonal or other variations that allow the organism to recover and detoxify. Acclimatization by fish to low sublethal levels of cyanide through continuous exposure might enhance their resistance to potentially lethal concentrations, but studies with Atlantic salmon and rainbow trout indicate otherwise. Prior acclimatization of Atlantic salmon smolts to cyanide increased their resistance only slightly to lethal concentrations. Juvenile rainbow trout previously exposed to low sublethal concentrations showed a marked reduction in fat synthesis and swimming performance when challenged with higher cyanide doses; effects were most pronounced at low water temperatures. Experimental evidence is lacking on exposure to lethal concentrations after prior exposure to high sublethal concentrations; some investigators predict decreased resistance, and others increased survival.

11.9.3 Birds

Free cyanide levels associated with high avian death rates include 0.12 mg/L in air, 2.1–4.6 mg/kg BW via acute oral exposure, and 1.3 mg/kg BW administered intravenously. In cyanide-tolerant species, such as the domestic chicken (*Gallus domesticus*), dietary levels of 135.0 mg total cyanide/kg ration resulted in growth reduction of chicks, but 103.0 mg total cyanide/kg ration had no measurable effect on these chicks. First signs of cyanide toxicosis in sensitive birds appeared between 0.5 and 5 min after exposure, and included panting, eye blinking, salivation, and lethargy. In domestic chickens, signs of cyanide toxicosis began 10 min after exposure. At higher doses, breathing in all species tested became increasingly deep and labored, followed by gasping and shallow intermittent breathing. Death usually followed in 15–30 min, although birds alive at 60 min frequently recovered. Elevated cyanide concentrations were found in blood of chickens that died of cyanide poisoning; however, these concentrations overlapped those in survivors. Despite this variability, blood is considered more reliable than liver as an indicator of cyanide residues in exposed birds. No gross pathological changes in chickens related to cyanide dosing were observed at necropsy. The rapid recovery of some birds exposed to cyanide may be due to the rapid metabolism of cyanide to thiocyanate and its subsequent excretion. Species sensitivity to cyanide was not related to body size but seemed to be associated with diet. Birds that feed predominantly on flesh, such as vultures, kestrels, and owls, were more sensitive to cyanide than were species that feed mainly on plant material – with the possible exception of mallard (*Anas platyrhynchos*) – as judged by acute oral LD50 values. Mallards given single oral doses of KCN (1.0 mg KCN/kg BW) at cyanide concentrations and amounts similar to those found at gold-mining tailings ponds (40.0 mg CN^-/L) had elevated concentrations of creatinine kinase in serum, suggesting tissue damage. At 0.5 mg KCN/kg BW, mitochondrial function (an indicator of oxygen consumption) and ATP concentrations were significantly depressed in heart, liver, and brain. Rhodanese and 3-mercaptopyruvate sulfurtransferase – two enzymes associated with cyanide detoxification – were induced in brain but not in heart of KCN-dosed mallards. Although cyanide concentrations as high as 2.0 mg KCN/kg BW (at 80.0 mg CN^-/L) were not acutely toxic to mallards, the long-term effects of such exposures were not determined and may have serious consequences for migratory birds exposed sublethally to cyanide at gold-mine tailings ponds.

Many species of migratory birds – including waterfowl, shorebirds, passerines, and raptors – were found dead in the immediate vicinity of gold-mine heap-leach extraction facilities and tailings ponds, presumably as a result of drinking the cyanide-contaminated (>200.0 mg total cyanide/L) waters. About 7000 dead birds – mostly waterfowl and songbirds – were recovered from cyanide extraction, gold-mine leach ponds in the western United States between 1980 and 1989; no gross pathological changes related to cyanide were observed in these birds at necropsy. No gross pathology was evident in cyanide-dosed captive birds, and this is similar to the findings of laboratory studies with cyanide and other animal orders that were tested and examined. Migratory bird mortality from cyanide toxicosis may be eliminated at these facilities by screening birds from toxic solutions or lowering the cyanide concentrations with hydrogen peroxide to <50.0 mg total cyanide/L, although the latter procedure requires verification. Chemical bird repellents used at cyanide ponds with some success against European starlings (*Sturnus vulgaris*) include *O*-aminoacetophone and 4-ketobenztriazine.

Some birds may not die immediately after drinking lethal cyanide solutions. Sodium cyanide rapidly forms free cyanide in the avian digestive tract (pH 1.3–6.5), whereas formation of free cyanide from metal cyanide complexes is comparatively slow. A high rate of cyanide absorption is critical to acute toxicity, and absorption may be retarded by the lower dissociation rates of metal–cyanide complexes. In Arizona, a red-breasted merganser (*Mergus serrator*) was found dead 20 km from the nearest known source of cyanide, and its pectoral muscle tissue tested positive for cyanide. A proposed mechanism to account for this phenomenon involves weak-acid dissociable (WAD) cyanide compounds. Cyanide bound to certain metals, usually copper, is dissociable in weak acids such as stomach acids. It has been suggested that drinking of lethal cyanide solutions by animals may not result in immediate death if the cyanide level is sufficiently low; these animals may die later when additional cyanide is liberated by stomach acid. More research is needed on WAD cyanide compounds.

Cyanide–nutrient interactions are reported for alanine, which appears to exacerbate cyanide toxicity, and for cystine, which seems to alleviate toxicity. Dietary cyanide – at levels that do not cause growth depression – alleviates selenium toxicity in chickens, but not the reverse. For example, dietary selenium, as selenite, at 10.0 mg/kg for 24 days, reduced growth, food intake, and food utilization efficiency, and produced increased liver size and elevated selenium residues; the addition of 45.0 mg CN/kg diet (100.0 mg sodium nitroprusside/kg) eliminated all effects except elevated selenium residues in liver. The mechanism of alleviation is unknown and may involve a reduction of tissue selenium through selenocyanate formation, or increased elimination of excess selenium by increasing the amount of dimethyl selenide exhaled. At dietary levels of 135.0 mg CN/kg plus 10.0 mg selenium/kg chick growth was significantly decreased. This interaction can be lost if there is a deficiency of certain micronutrients or an excess of vitamin K.

11.9.4 Mammals

Microgold and silver mining are probably the most widespread sources of anthropogenic cyanides in critical wildlife habitat, such as deserts in the western United States. Many species of mammals, mostly rodents and bats, were found dead at cyanide extraction gold-mine tailings and heap leach ponds in California, Nevada, and Arizona, including 10 endangered, threatened, or otherwise protected species of mammals. A similar situation was documented for a vat-leach gold mine in South Carolina with a large tailings pond.

Much of the toxicological interest in cyanide relating to mammals has focused on its rapid lethal action; however, its most widely distributed toxicologic problems are due to its toxicity from dietary, industrial, and environmental factors. Chronic exposure to cyanide is correlated with specific human diseases: Nigerian nutritional neuropathy, Leber's optical atrophy, retrobulbar neuritis, pernicious anemia, tobacco amblyopia, cretinism, and ataxic tropical neuropathy. The effects of chronic cyanide intoxication are confounded by various nutritional factors, such as dietary deficiencies of sulfur-containing amino acids, proteins, and water-soluble vitamins.

Most authorities now agree on five points: (1) cyanide has low persistence in the environment and is not accumulated or stored in any mammal studied; (2) cyanide biomagnification in food webs has not been reported, possibly due to rapid detoxification of sublethal doses by most species, and death at higher doses; (3) cyanide has an unusually low chronic toxicity, but chronic intoxication exists and, in some cases, can be incapacitating; (4) despite the high lethality of large single doses or acute respiratory exposures to high vapor concentrations of cyanide, repeated sublethal doses seldom result in cumulative adverse effects; and (5) cyanide, in substantial but sublethal intermittent doses can be tolerated by many species for long periods, perhaps indefinitely.

The toxicity of cyanogenic plants is a problem for both domestic and wild ungulates. Poisoning of herbivorous ungulates is more prevalent under drought conditions, when these mammals become less selective in their choice of forage; dry growing conditions also enhance cyanogenic glycoside accumulations in certain plants. Animals that eat rapidly are at greatest risk, and intakes of 4.0 mg HCN/kg BW can be lethal if consumed quickly. In general, cattle are most vulnerable to cyanogenic plants; sheep, horses, and pigs – in that order – are more resistant than cattle. Deer (*Odocoileus* sp.) and elk (*Cervus* sp.) have been observed to graze on forages that contain a high content of cyanogenic glycosides; however, cyanide poisoning has not been reported in these species. Ruminant and nonruminant ungulate mammals that consume forage with high cyanogenic glycoside content, such as sorghums, Sudan grasses, and corn, may experience toxic signs due to microbes in the gut that hydrolyze the glycosides, releasing free HCN. Signs of acute cyanide poisoning in livestock usually occur within 10 min and include initial excitability with muscle tremors, salivation, lacrimation, defecation, urination, and labored breathing, followed by muscular incoordination, gasping, and convulsions; death can occur quickly, depending on the dose administered. Thyroid dysfunction has been reported in sheep grazing on star grass (*Cynodon plectostachyus*), a plant with high cyanogenic glycoside and low iodine content. Sheep developed enlarged thyroids and gave birth to lambs that were stillborn or died shortly after birth. Cyanogenic foods can exacerbate selenium deficiency, as judged by the increased incidence of nutritional myopathy in lambs on low-selenium diets. A secondary effect from ingesting cyanogenic glycosides from forage is sulfur deficiency as a result of sulfur mobilization to detoxify the cyanide to thiocyanate.

Cyanide poisonings of livestock by forage sorghums and other cyanogenic plants are well documented. Horses in the southwestern United States grazing on Sudan grass and sorghums developed posterior muscle incoordination, urinary incontinence, and spinal cord histopathology; offspring of mares that had eaten Sudan grass during early pregnancy developed musculoskeletal deformities. Salt licks containing sulfur (8.5%) have been used to treat sheep after they failed to gain weight when grazing on sorghum with high HCN content. Sugar gum (*Eucalyptus cladocalyx*) and manna gum (*Eucalyptus viminalis*) contain high levels of cyanogenic glycosides, and both have been implicated as the source of fatal HCN poisoning in domestic sheep and goats that had eaten leaves from branches felled for drought feeding, or after grazing sucker shoots on lopped stumps. In one case, 10 goats died and 10 others were in distress within 2 h after eating leaves from a felled sugar gum. Dead goats had bright red blood that failed to clot and subepicardial petechial hemorrhages. Rumens of dead goats contained leaves of *Eucalyptus* spp. and smelled of bitter almonds. The 10 survivors were treated intravenously with 3.0 mL of a 1.0-L solution made to contain 20.0 g of sodium nitrite and 50.0 g of sodium thiosulfate; four recovered and six died. Of 50 afflicted goats, 24 died within 24 h and the remainder recovered. In rare instances, HCN poisoning occurs when animals are exposed to chemicals used for fumigation or as a fertilizer, but there is general agreement that ingestion of plants containing high levels of cyanogenic glycosides is the most frequent cause of cyanide poisoning in livestock.

Cassava, also known as manioc, tapioca, yuca, or guacamate, is one of the very few – and, by far, the most important – food crops in which the cyanide content creates toxic problems. Cassava is a major energy source for people and livestock in many parts of the world; it accounts for an average of 40% of the human caloric intake in Africa, to more than 70% in some African diets. In comparison to other tropical crops it produces the highest yield per hectare. Cassava is native to tropical America from southern Mexico to northern Argentina and probably has been under cultivation there for 4000–5000 years. It has been introduced to East Africa, Indian Ocean islands, southern India, and the Far East. The global production of cassava roots was estimated at 50 million tons in 1950, and 100 million tons in 1980; about 44.2 million tons are grown annually in Africa, 32.7 million tons in tropical America, and 32.9 million tons in Asia. Linamurin is the principal cyanogenic glycoside in cassava; its toxicity is due to hydrolysis by intestinal microflora releasing free cyanide. Rabbits (*Oryctalagus cuniculus*) fed 1.43 mg linamurin/kg BW daily (10.0 mg/kg BW weekly) for 24 weeks showed effects similar to those of rabbits fed 0.3 mg KCN/kg BW weekly. Specific effects produced by linamurin and KCN included elevated lactic acid in heart, brain, and liver; reduced glycogen in liver and brain; and marked depletion in brain phospholipids.

The use of cassava in animal feed presents two major problems: the presence of cyanogenic glycosides in the tuber, and the remarkably low protein levels in fresh and dried cassava. Pigs fed low-protein cassava diets for 8 weeks had reduced food consumption and lowered liver weight; addition of protein supplement to the diet reversed these trends. Removal of cyanogenic glycosides from cassava tubers, mash, peels, and root meal is accomplished with several techniques. Usually, the cassava root is dried in the sun for several weeks, and this process removes most of the cyanogenic glycosides; however, under conditions of famine or food shortage, this process cannot be done properly. Long fermentation periods, especially under conditions of high moisture content, may be effective in substantial detoxification of cassava mash. Cassava peels containing as much as 1061.0 mg HCN/kg FW can be rendered suitable for feeding to livestock (4.0–625.0 mg/kg) by boiling for 7 min, roasting for 30 min, soaking for 15 h, or drying in the sun for 7.6 days. Cassava root meal (up to 40% of cassava meal) is satisfactory as a diet supplement for domestic pigs, provided that cyanide content is <100.0 mg/kg ration.

Neuropathies associated with cassava ingestion (i.e., cyanide intoxication) can develop into a syndrome in humans and domestic animals, characterized by nerve deafness, optic atrophy, and an involvement of the sensory spinal nerve that produces ataxia. Other symptoms include stomatitis, glossitis, and scrotal dermatitis. Potentially more serious are long-term effects such as ataxic neuropathy, goiter, and cretinism, which have been attributed to high cassava content in diets. Thiocyanate – one of the detoxification products – inhibits iodine absorption and promotes goiter, a common ailment in tropical countries. At high dietary cyanide intakes there is an association with diabetes and cancer, but this requires verification. The first case of cassava toxicity occurred almost 400 years ago. The toxic principle was later identified as a cyanogenic glycoside, shown to be identical with flax linamurin (2-(beta-D-glucopyranosyloxy)-isobutyronitrile). All parts of the plant, except possibly the seeds, contain the glycoside together with the enzyme linamarase. This enzyme effects hydrolysis of the nitrile to free HCN when the tissue cellular structure is damaged. Mantakassa disease is related to chronic cyanide intoxication associated with a diet consisting almost exclusively of cassava; in times of famine and sulfur-poor diets, Mantakassa effects were more pronounced. Symptoms of Mantakassa disease include the sudden onset of difficulty in walking, increased knee and ankle reflexes, elevated serum thiocyanate levels, fever, pain, headache, slurred speech, dizziness, and vomiting. Women of reproductive age and children were the most seriously affected. Symptoms persisted for up to 4 months after treatment with hydroxycobalamin, vitamin supplements, and a high protein, energy-rich diet. Mantakassa was reported in 1102 victims in Mozambique in 1981 from a drought-stricken cassava staple area; from Zaire in 1928, 1932, 1937, and again in 1978–81; in Nigeria; and in the United Republic of Tanzania. The mean serum thiocyanate level in patients with Mantakassa is 2.6 times higher than in non-Mantakassa patients in Nigeria, and 3.5 times higher than in Tanzanian patients. Pesticides, infections, viruses, and consumption of food other than cassava were eliminated as possible causative agents in Mantakassa disease. It is still unresolved whether the disease is triggered when a threshold level of thiocyanate is reached, or when a critical combination of cyanide intoxication plus nutritional deficiency occurs.

Routes of administration other than dietary ingestion should not be discounted. Livestock found dead near a cyanide disposal site had been drinking surface water runoff from the area that contained up to 365.0 mg HCN/L. The use of cyanide fumigant powder formulations may be hazardous by contact of the powder with moist or abraded skin, contact with the eye, swallowing, and inhalation of evolved HCN. In rabbits, lethal systemic toxicity was produced by contamination of the eye, moist skin, or abraded skin (but not dry skin) with cyanide powder formulations (40% NaCN plus 60% kaolin) administered at 1.0–5.0 g powder/m^3. Hydrogen cyanide in the liquid state can readily penetrate the skin, and skin ulceration has been reported from splash contact with cyanides among workers in the electroplating and gold extraction industries – although effects in those instances were more likely due to the alkalinity of the aqueous solutions. In one case, liquid HCN ran over the bare hand of a worker wearing a fresh air respirator; he collapsed into unconsciousness in 5 min, but ultimately recovered.

Use of poisons in livestock collars is both specific and selective for animals causing depredations, as is the case for cyanide collars to protect sheep against coyotes. These collars contain a 33% NaCN solution and are usually effective against coyotes. However, field results indicate that some coyotes kill by means other than neck attack, and some exhibit great wariness in attacking collared sheep.

Calcium cyanide in flake form was used in the 1920s to kill black-tailed prairie dogs and pocket gophers (*Geomys bursarius*) in Kansas, and various other species of rodents in Nova Scotia. For prairie dog control, the usual practice was to place 43.0–56.0 g of calcium cyanide 0.3–0.7 m below the rim of the burrow and to close the entrances. The moisture in the air liberated HCN, gas which remained in the burrow for several hours, producing 100% kill. A lower dose of 28.0 g per burrow was only about 90% effective. Control of prairie dogs with cyanide sometimes resulted in the death of burrowing owls that lived in the prairie dog burrows. Some animals can develop an aversion to food associated with sodium cyanide-induced illness. Thus, sodium cyanide-containing baits used in New Zealand against the brushtail possum (*Trichosurus vulpecula*) produced bait shyness through conditioned food aversion induced by sublethal (4.0–5.0 mg NaCN/kg BW) cyanide ingestion, effectively reducing NaCN poisoning operations in that country.

Clinical signs of acute cyanide poisoning in mammals last only a few minutes after ingestion and include rapid and labored breathing, ataxia, cardiac irregularities, dilated pupils, convulsions, coma, and respiratory failure; death may occur quickly depending on the dose administered. Despite the high lethality of large single exposures, repeated sublethal doses – especially in diets – are tolerated by many species for extended periods, perhaps indefinitely. Cyanide poisoning causes cardiovascular changes as well as its better known effects on cellular respiration. Cyanide increases cerebral blood flow in rabbits and cats, and disrupts systemic arterial pressure in dogs. Cyanide affects mammalian behavior, mostly motor functions, although these effects have not been quantified. Cyanide-induced motor alterations observed in rats and guinea pigs include muscular incoordination, increased whole-body locomotion, disrupted swimming performance, and altered conditioned avoidance responses. As a consequence of the cytotoxic hypoxia in acute cyanide poisoning, there is a shift from aerobic to anaerobic metabolism, and the development of lactate acidosis. A combination of rapid breathing, convulsions, and lactate acidosis is strongly suggestive of acute cyanide poisoning. As with other chemical asphyxiants, the critical organs that are most sensitive to oxygen depletion are the brain and heart. The only consistent postmortem changes found in animals poisoned by cyanide are those relating to oxygenation of the blood. Because oxygen cannot be utilized, venous blood has a bright red color and is slow to clot. Bright-red venous blood is not a reliable indicator of cause of death, however, because it is also associated with chemicals other than cyanide.

Cyanide poisoning is associated with changes in various physiological and biochemical parameters. The earliest effect of cyanide intoxication in mice seems to be inhibition of hepatic rhodanese activity, due to either blockage by excess binding to the active site or to depletion of the sulfane-sulfur pool. These changes do not seem to occur in blood, where rhodanese functions at its maximal rate, thus preventing cyanide from reaching the target tissues and causing death. Cyanide causes dose- and species-dependent responses on vascular smooth muscle; studies with isolated aortic strips indicate that rabbits are 80 times more sensitive than dogs or ferrets (*M. putorius furo*). Rabbits killed with HCN had higher concentrations of cyanide in blood and other tissues and lower tissue cytochrome oxidase activities than did those killed with KCN. Cyanide promotes dose- and calcium-dependent release of dopamine from tissues in the domestic cat, and reductions in ATP content of the carotid body. Cyanide-induced hypoxia is believed to produce decreases in the ATP content of Type I glomus cells. The decrease in the phosphate transfer potential is a crucial step in the overall transduction process, that is, the activation of the transmitter release from Type I cells, with resultant release and activation of sensory nerve endings. Studies with isolated heart of the domestic ferret demonstrate that cyanide affects intracellular ionic exchange of H^+, Na^+, and calcium;

inhibits cardiac action potential; and inhibits oxidative phosphorylation accompanied by an intracellular acidosis, a decrease in phosphocreatinine, and a rise in inorganic phosphate. When oxidative phosphorylation is inhibited in cardiac muscle, there is a rapid decrease of developed force or pressure; most of the decrease of developed pressure produced by cyanide in ferret heart is not produced by intracellular acidosis, and may result from increased inorganic phosphate. Observed changes in rat cerebral oxidative responses to cyanide may be due to redistribution of intracellular oxygen supply to mitochondria respiring in an oxygen-dependent manner or by branching effects within brain mitochondria. Hyperammonemia and the increase of neutral and aromatic amino acids may also be important in the loss of consciousness induced by cyanide. Rats exposed for 30 days to 100.0 or 500.0 mg KCN/L drinking water had mitochondrial dysfunction, depressed ATP concentrations in liver and heart, and a depressed growth rate; little effect was observed at 50.0 mg KCN/L. The adverse effect on growth is consistent with the biochemical indicators of energy depletion. However, the concentrations should be viewed with caution as CN may have volatilized from the water solutions prior to ingestion by the rats, due to presumed neutral pH.

Organic cyanide compounds, or nitriles, have been implicated in numerous human fatalities and signs of poisoning – especially acetonitrile, acrylonitrile, acetone cyanohydrin, malonitrile, and succinonitrile. Nitriles hydrolyze to carboxylic acid and ammonia in either basic or acidic solutions. Mice (*Mus* sp.) given lethal doses of various nitriles had elevated cyanide concentrations in liver and brain; the major acute toxicity of nitriles is CN release by liver processes. In general, alkylnitriles release CN much less readily than aryl alkylnitriles, and this may account for their comparatively low toxicity.

No human cases of illness or death due to cyanide in water supplies are known. Accidental acute cyanide poisonings in man are uncommon; however, a man accidentally splashed with molten sodium cyanide died about 10 h later. Human cyanide deaths usually involve suicides, where relatively large amounts of sodium cyanide or potassium cyanide are ingested and the victims die rapidly in obvious circumstances. Recovery after oral ingestion is rare. In one case, a spouse emptied capsules containing medicine and repacked them with 40% solid NaCN. The victim took one capsule and ingested about 0.05 g, but vomited and recovered completely. Human deaths are increasing from gas or smoke inhalation from urban fires, possibly owing to the increased toxicity of fire atmospheres caused by the use of organocyanide plastics in modern construction and furnishings. Hydrogen cyanide may be important in some fires in producing rapid incapacitation, causing the victims to remain in the fire and die from carbon monoxide or other factors, although HCN concentrations of 60.0 mg/L air and lower had minimal effects. Exposure to the mixture of HCN and carbon monoxide, with accompanying changes in cerebral blood flow during attempts to escape from fires, may be a cause of collapse and subsequent death. For example, cynomolgus monkeys (*Macaca* spp.) exposed to pyrolysis products of polyacrylonitrile and to low-level HCN gas had similar physiological effects in both atmospheres, specifically: hyperventilation, followed by loss of consciousness after 1–5 min; and bradycardia, with arrhythmias and T-wave abnormalities. Recovery was rapid following cessation of exposure. Because HCN is the major toxic product formed by the pyrolysis of PAN, HCN may produce rapid incapacitation at low blood levels of cyanide in fires, while death may occur later due to carbon monoxide poisoning or other factors.

Finally, cyanide does not appear to be mutagenic, teratogenic, or carcinogenic in mammals. In fact, there has been a long-standing hypothesis for an anti-cancer effect of the cyanogenic glycoside amygdalin (also called laetrile). The hypothesis is based on amygdalin's selective hydrolysis by a beta glucosidase, liberating cyanide and benzaldehyde at the neoplastic site. The cyanide then selectively attacks the cancer cell, which is presumed to be low in rhodanese, whereas normal cells are assumed to possess sufficient rhodanese and sulfur to detoxify the cyanide.

However, many tumors are neither selectively enriched in beta glucosidase nor low in rhodanese.

11.10 Recommendations

Proposed free cyanide criteria suggest that sensitive species of aquatic organisms are protected at <3.0 μg/L, birds and livestock at <100.0 mg/kg diet, and human health at concentrations of <10.0 μg/L drinking water, <50.0 mg/kg diet, and <5.0 mg/m^3 air (Table 11.2). In aquatic systems, research is needed in several areas: (1) long-term effects of cyanide on life cycles, growth, survival, metabolism, and behavior of a variety of aquatic organisms and microorganisms in addition to fish; (2) effects of seasonal pulses of cyanide on aquatic organisms in rural and

Table 11.2. Proposed free cyanide criteria for the protection of living resources and human health.

Resource, Criterion, and Other Variables	Criterion
FRESHWATER ORGANISMS (μg/L MEDIUM)	
Minimal impairment, most species of fish	3.0–5.0
Reduced survival, amphipods	>3.03–4.0
Safe, most fish species	3.5 (24-h average, not to exceed 52.0 at any time)
Significant impairment, most species of fish tested	8.0–16.0, exposure for at least 20 days
Hazardous	
Most fish species	>11.0
Microorganisms	>300.0
Reduced survival, chronic exposure	
Bivalve mollusks, larvae	>14.0
Fish, many species	30.0–150.0
Impaired reproduction, sensitive fish species	>25.0
Impaired swimming ability, growth, development, and behavior	>100.0
Lethal to rapidly lethal, acute exposure	300.0–1000.0
Great Lakes	
Acute exposure, safe	<22.0
Chronic exposure, safe	<5.2
MARINE ORGANISMS (μg/L SEAWATER)	
Acceptable	
Acute	<1.0
Chronic	<1.0
Adverse effects, chronic exposure	>2.0
Minimal risk	<5.0
Hazardous	>10.0
Lethal	>30.0

Continued

Table 11.2. cont'd

Resource, Criterion, and Other Variables	Criterion
SEDIMENTS, GREAT LAKES (mg TOTAL CYANIDE/kg DRY WEIGHT (DW))	
Nonpolluted	<0.10
Moderately polluted	0.1–0.25
Heavily polluted	>0.25
BIRDS	
Domestic chickens, diet, safe level (mg total cyanide/kg ration fresh weight (FW))	90–>100.0
Waterfowl, drinking water, safe level (mg/L)	<50.0
LIVESTOCK (mg/kg FW)	
Diet, safe level	
Free cyanide	<100.0
Total cyanide	<625.0
Forage, hazardous level	>200.0
Drinking water (μg/L)	<200.0
LABORATORY WHITE RAT	
Diet, safe level (mg/kg ration FW)	<1000.0
Blood (mg/L)	
Normal	0.25–0.45
Minimum lethal concentration (mg/kg FW)	2.6–2.9
Liver, minimum lethal concentration (mg/kg FW)	0.5–6.1
HUMAN HEALTH	
Drinking water (μg/L)	
Recommended	< 5.0 – <10.0
USA nationwide survey	Maximum 8.0
Safe	<10.0
Goal, USA	<10.0
Maximum allowable limit	
USA	10.0
Canada, goal	<20.0
Lifetime health advisory	<154.0–200.0
World health Organization, acceptable	<100.0
USA and Canada	
Acceptable	<200.0
Mandatory limit	200.0
Rejected	>200.0
10-day health advisory	
Child	<220.0
Adult	<770.0

Table 11.2. cont'd

Resource, Criterion, and Other Variables	Criterion
Diet	
Acceptable daily intake	
Water	1.5 mg, equivalent to 0.02 mg/kg body weight (BW) daily for a 70-kg adult
Food (mg/kg BW)	8.4
Food (mg/kg FW ration)	<50.0
Food (mg total cyanide/kg FW ration)	<415.0
Cassava (*Manihot esculenta*), roots, total cyanide (mg/kg FW)	
Safe	<50.0
Moderately toxic	50.0–100.0
Very poisonous	>100.0
Food items (mg/kg)	
Cocoa	<20.0 DW
Beans, nuts	<25.0 DW
Cereals, grains	<25.0 DW
Citrus fruits	<50.0 FW
Uncooked pork	<50.0 FW
Grains	<75.0 FW
Cereal flours	<125.0 DW
Spices	<250.0 FW
Frozen meat	<950 FW
Bakery products, yeast	<1500.0 DW
Egg white solids	<1000.0 DW
Tissue residues	
Blood and spleen (μg/L or μg/kg FW)	
Normal	<77.0
Suspected poisoning, whole blood (μg/L)	>1000.0
Usually fatal	1000.0–2000.0
Whole body (mg/kg BW), fatal	4.0, if taken rapidly
Air (mg/m^3)	
Recommended safe levels	
Soviet Union, Romania, Hungary, Bulgaria, Czechoslovakia	<0.3
USA	<5.0
Most nations	<11.0
Occupational exposure, proposed safe level, USA	<3.0
Safe ceiling concentration	<5.0
Hazardous levels	4.2–12.4

Continued

Table 11.2. cont'd

Resource, Criterion, and Other Variables	Criterion
Agricultural soils (mg/kg DW)	
Free cyanide	
Background	<1.0
Moderate contamination	10.0
Requires cleanup	>100.0
Complex cyanide	
Background	<5.0
Moderate contamination	50.0
Requires cleanup	>100.0

wilderness areas; (3) influence of various environmental parameters (e.g., oxygen, pH, temperature), if any, on adaptive resistance to cyanide; and (4) usefulness of various biochemical indicators of cyanide poisoning, such as cytochrome oxidase inhibition, estradiol and thyroxine levels in fish plasma, and pituitary gland histology.

To protect vertebrate wildlife from mine water poisoning with certainty, it is necessary to exclude them from cyanide solutions or to reduce cyanide concentrations to nontoxic levels. More research is needed on chemical repellents, and physical screening methods. Mortality of avian and terrestrial wildlife from cyanide toxicosis may be curtailed at small ponds associated with leach heaps by screening wildlife from toxic solutions or covering small solution ponds with polypropylene netting, provided that the fencing and netting are properly maintained. Some mines in Nevada are now covering surfaces of small ponds with 4-inch (10.2-cm) diameter high-density polyethylene balls; birds are no longer attracted to these ponds as water sources. Gold mine operators in southern California and Nevada used plastic sheeting to cover the cyanide leach pond, resulting in a cessation of wildlife mortality. The comparatively high cost of this process was soon recouped through reduced evaporation of water and cyanide. To reduce the potential for puddling on ore heaps, more research is recommended on physical exclusionary devices, chemical repellents, and monitoring of solution application rates.

Analytical methodologies need to be developed that differentiate between free cyanide (HCN and CN^-) and other forms of cyanide, and that are simple, sensitive (i.e., in the microgram per liter range), and accurate. Procedures need to be standardized that ensure prompt refrigeration and analysis of all samples for cyanide determination because some stored samples generate cyanide while others show decreases. Periodic monitoring of cyanide in waterways is unsatisfactory for assessing potential hazards because of cyanide's rapid action, high toxicity, and low environmental persistence. A similar case is made for cyanide in the atmosphere. Development of a continuous monitoring system of cyanides in waterways and air is recommended, with emphasis on point source dischargers, such as industrial and municipal facilities. Information is needed on the fate of cyanide compounds in natural waters, relative contributions of natural and anthropogenic sources, and critical exposure routes for aquatic organisms. Additional research is needed on the origin of cyanide in wilderness and rural watershed areas, specifically the roles of organic wastes and their associated bacterial flora, aquatic vegetation induced by nutrient enrichment, and terrestrial plant cover in the watershed.

The use of M-44 sodium cyanide capsules for predator control was suspended and canceled by the U.S. Environmental Protection

Agency on March 9, 1972. M-44, use was again permitted by the U.S. Environmental Protection Agency beginning on February 4, 1976, provided that "each authorized or licensed applicator shall carry an antidote kit on his person when placing or inspecting M-44 devices. The kit shall contain at least 6 pearls of amylnitrite and instructions on their use. Each authorized or licensed applicator shall also carry on his person instructions for obtaining medical assistance in the event of accidental exposure to sodium cyanide." The use of cyanide-containing paste baits to control pestiferous mammals is practiced widely in New Zealand; however, some species are repelled by these baits and more research is needed to improve their effectiveness.

Farmers need to be aware of factors that influence the cyanogenic potential of forage crops and to conduct regular inspections of grazing fields for cyanogenic plants. Moreover, hay and silage should be properly cured in order to minimize cyanide content before feeding to livestock. Selective breeding of plants with low cyanide content will help reduce livestock poisoning, but the most advisable prevention method at present is to prohibit grazing on fields where cyanogenic plants are present. More research seems needed on: (1) effects of drought and other factors that may increase the concentration of cyanogenic glycosides in livestock forage plants, (2) mechanisms of cyanide liberation by plants, and (3) effects of cyanide on wildlife and range animals that graze foliage with high cyanogenic glycoside content.

Research is needed on low-level, long-term cyanide intoxication in mammals by oral and inhalation routes in the vicinities of high cyanide concentrations, especially on the incidence of skin dermatitis, nasal lesions, and thyroid dysfunction, and on urinary thiocyanate concentrations. These types of studies may provide a more valid rationale in establishing standards and threshold limit values for HCN and inorganic cyanide. Data are scarce on the carcinogenic, teratogenic, and mutagenic properties of cyanide, and on the distribution and transformation of cyanides in air, land, or water. Additional analysis of available information and more research in these areas is recommended. Finally, more research is needed on cyanide toxicokinetics because cyanide is a very reactive nucleophile that distributes widely through the body, is permeable to cell membranes, and may accumulate in the fetus.

11.11 Summary

Cyanides are used widely and extensively in the manufacture of synthetic fabrics and plastics, in electroplating baths and metal-mining operations, as pesticidal agents and intermediates in agricultural chemical production, and in predator control devices. Elevated cyanide levels are normally encountered in more than 1000 species of food plants and forage crops, and this probably represents the greatest source of cyanide exposure and toxicosis to man and to range animals. Anthropogenic sources of cyanide in the environment include certain industrial processes, laboratories, fumigation operations, cyanogenic drugs, fires, cigarette smoking, and chemical warfare. Although cyanide is ubiquitous in the environment, levels tend to be elevated in the vicinity of metal-processing operations, electroplaters, gold-mining facilities, oil refineries, power plants, and solid waste combustion.

Many chemical forms of cyanide are present in the environment, including free cyanide, metallocyanide complexes, and synthetic organocyanides, also known as nitriles. But only free cyanide (i.e., the sum of molecular hydrogen cyanide, HCN, and the cyanide anion, CN^-) is the primary toxic agent, regardless of origin. Cyanides are readily absorbed through inhalation, ingestion, or skin contact and are readily distributed throughout the body via blood. Cyanide is a potent and rapid-acting asphyxiant; it induces tissue anoxia through inactivation of cytochrome oxidase, causing cytotoxic hypoxia in the presence of normal hemoglobin oxygenation. Diagnosis of acute lethal cyanide poisoning is difficult because signs and symptoms are nonspecific, and numerous factors modify its biocidal properties, such as dietary deficiencies in vitamin B_{12}, iodine, and sulfur amino acids. Among the more consistent changes measured in acute cyanide poisoning are inhibition of brain cytochrome oxidase activity, and changes in electrical activity in heart and brain. At sublethal doses, cyanide reacts with thiosulfate in the presence of rhodanese to produce the comparatively nontoxic thiocyanate, most of which is excreted in the urine. Rapid detoxification enables animals to ingest high sublethal doses of cyanide over extended periods without harm. Antidotes in current use to counteract cyanide poisoning include a combination of sodium nitrite and sodium thiosulfate (United States), cobalt edetate (United Kingdom, Scandinavia, and France), or a mixture of 4-dimethylaminophenol and sodium thiosulfate (Germany).

All available evidence suggests that cyanides are neither mutagenic nor teratogenic, or carcinogenic. Moreover, there are no reports of cyanide biomagnification or cycling in living organisms, probably owing to its rapid detoxification. Cyanide seldom persists in surface waters and soils owing to complexation or sedimentation, microbial metabolism, and loss from volatilization. More data are needed on cyanide distribution and transformation in the atmosphere. Analytical methods for the determination of free and bound cyanides and cyanogenic compounds in biological materials are under constant revision. Further, unless tissue samples are obtained promptly after cyanide exposure and analyzed immediately, erroneous analytical values will result.

Higher plants are adversely affected by cyanide through cytochrome oxidase inhibition; the rate of production and release of cyanide by plants to the environment through death and decomposition is unknown. Nonacclimatized soil bacteria are adversely affected at 0.3 mg HCN/kg; acclimatized populations, however, can degrade wastes containing up to 60.0 mg total cyanide/kg. In some cases, soil bacteria and fungi produce cyanides as secondary metabolites, with adverse effects on certain plants. Several species of arthropods normally contain elevated whole-body cyanide concentrations, and these confer protection against predators and allow consumption of cyanogenic plants.

Fish were the most sensitive aquatic organisms tested. Adverse effects on swimming and reproduction were observed between 5.0 and 7.2 µg free cyanide/L; lethal effects usually occurred between 20.0 and 76.0 µg/L. Biocidal properties of cyanide in aquatic environments were significantly modified by water pH, temperature, and oxygen content; life stage, condition, and species assayed; previous exposure to cyanides; presence of other chemicals; and initial dose tested. Birds that feed predominantly on flesh were more sensitive to cyanide than were herbivores. Free cyanide levels associated with high avian death rates include 0.12 mg/L in air, 2.1–4.6 mg/kg BW via acute oral exposure, and 1.3 mg/kg BW administered intravenously. Dietary levels of 135.0 mg total cyanide/kg ration resulted in growth reduction of chicks, but 103.0 mg total cyanide/kg ration had no measurable effect on domestic chickens. Cyanogenic plants represent a problem for various range animals and wildlife, primarily among species that eat rapidly. Intakes of 4.0 mg HCN/kg BW are lethal to these species if it is consumed quickly. Cassava (*Manihot esculenta*) is a cyanogenic plant that accounts for up to 70% of human caloric intake in some areas, and this is associated with serious, long-term toxic effects including ataxia, optic nerve lesions, altered thyroid function, demyelination, and increases in tissue thiocyanate levels. Acute oral LD50 values for representative species of mammals ranged between 2.0 and 3.6 mg HCN/kg BW. Despite the high lethality of large single exposures, repeated sublethal doses – especially in diets – can be tolerated by many species for extended periods, perhaps indefinitely. Mammalian deaths were also recorded at air concentrations of 140.0 mg HCN/m^3 (exposure for 60 min) and 4400.0 mg HCN/m^3 (exposure for 1 min), and at dermal applications between 2.3 mg HCN/kg BW for abraded skin and 100.0 mg HCN/kg BW for intact skin. Adverse nonlethal effects were noted at drinking water concentrations >150.0 mg HCN/L and at dietary concentrations >720.0 mg HCN/kg ration.

Free cyanide criteria currently proposed for natural resource protection include <3.0 µg/L medium for aquatic life, and <100.0 mg/kg diet for birds and livestock. For human health protection, free cyanide values are <10.0 µg/L drinking water, <50.0 mg/kg diet, and <5.0 mg/m^3 air.

DIAZINON[a]

Chapter 12

12.1 Introduction

Diazinon, an organophosphorus compound with an anticholinesterase mode of action, was released for experimental evaluation in the early 1950s. Diazinon is now used extensively by commercial and home applicators in a variety of formulations to control flies, cockroaches, lice on sheep, insect pests of ornamental plants and food crops (especially corn, rice, onions, and sweet potatoes), forage crops such as alfalfa, and nematodes and soil insects in turf, lawns, and croplands. Diazinon is the most widely used organophosphorus pesticide in Pakistan to control cabbage root fly and carrot fly. In 1992, more than 612,000 kg of diazinon were used in California on alfalfa, nuts, stone fruits, vegetables, and other crops.

Avian and terrestrial wildlife may acquire diazinon by drinking contaminated water, by absorbing it through legs and feet, by consuming treated grass or grain, or by ingestion of pesticide-impregnated carrier particles. Diazinon was detected at low concentrations (<0.2 mg/kg) in tissues of 29% of loggerhead shrikes (*Lanius ludovicianus*) collected in Virginia between 1985 and 1988. Diazinon poisonings of birds – involving 54 incidents in 17 States – have been recorded for at least 23 species, especially among waterfowl feeding on recently treated turfgrass; incidents involving agricultural applications may be less conspicuous, and thus not as well documented. Killings of Canada geese (*Branta canadensis*), brant (*Branta bernicla*), mallard (*Anas platyrhynchos*), American black duck (*Anas rubripes*), American wigeon (*Anas americana*), other species of waterfowl, and songbirds have all been associated with consumption of grass or grain shortly after diazinon application. Fatal diazinon poisonings have also been recorded in humans, domestic chickens (*Gallus gallus*), domestic ducklings (*Anas* spp.) and goslings (*Anser* spp.), in laboratory monkey colonies of the tamarin (*Saguinus fuscicollis*) and the common marmoset (*Callithrix jacchus*), and the honeybee (*Apis mellifera*). Mammals seem to be less sensitive than birds to diazinon poisoning. The lack of reported mammalian mortalities (only one suspected case of a pocket gopher, *Thomomys* sp., found dead in a park at Yakima, Washington, following aerial spraying of diazinon on shade trees) is consistent with the general findings for organophosphorus insecticides. Sublethal effects such as reduced food consumption and egg production in the ring-necked pheasant (*Phasianus colchicus*), and behavioral modifications, reduced food intake, alterations in liver enzyme activities, reductions in vitamin concentrations, reduced body temperature, and lowered resistance to cold stress in white-footed mice (*Peromyscus leucopus*) have been noted at diazinon concentrations markedly lower than those causing acute mortality. It has been suggested, but not proven, that wildlife partially disabled in the field as a result of diazinon poisoning would be more likely to die of exposure, predation, starvation, or dehydration, or face behavioral abnormalities, learning impairments, and reproductive

---

[a]All information in this chapter is referenced in the following sources:

Eisler, R. 1986. Diazinon hazards to fish, wildlife, and invertebrates: a synoptic review. *U.S. Fish Wildl. Serv. Biol. Rep. 85(1.9)*, 37 pp.

Eisler, R. 2000. Diazinon. Pages 961–982 in *Handbook of Chemical Risk Assessment: Health Hazards to Humans, Plants, and Animals. Volume* 2, Organics. Lewis Publishers, Boca Raton, Florida.

declines than would similarly treated domestic or laboratory animals. Sublethal effects of diazinon on fish populations include vertebral malformations, altered blood chemistry, inhibition of acetylcholinesterase (Ache) activity, reduced larval and adult growth, impaired swimming, abnormal pigmentation, histopathology of muscle and gills, and reduction of liver RNA, DNA, and protein content.

12.2 Environmental Chemistry

Diazinon is a broad-spectrum insecticide that is effective against a variety of orchard, vegetable, and soil pests, ectoparasites, flies, lice, and fleas. It exists as a technical grade product, a wettable powder, an emulsifiable concentrate, as granules, and in a variety of other formulations. The active ingredient in diazinon is phosphorothioic acid *O,O*-diethyl *O*-(6-methyl-2-1(methylethyl)-4-pyrimidinyl) ester (Figure 12.1). Its molecular formula and molecular weight are $C_{12}H_{21}N_2O_3PS$, and 304.35. The technical grade is light amber to dark brown, and boils at 83–84°C. Diazinon is soluble in water to 60.0 mg/L, and dissolves readily in aliphatic and aromatic solvents, alcohols, and ketones. Diazinon may be stored on the shelf for at least 3 years with negligible degradation. Diazinon is also known as G-24480, Sarolex, Spectracide, AG-500, Alfa-tox, Basudin, Dazzel, Diazajet, Diazide, Diazol, ENT 19507, Gardentox, Neocidol, Nucidol, CAS 333-41-5, Diagran, Dianon,

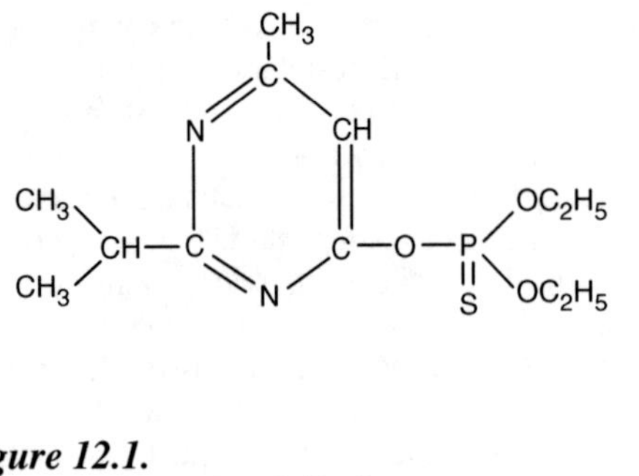

Figure 12.1.
Structural formula of diazinon.

DiaterrFos, Diazatol, Dizinon, Dyzol, D.z.n., Fezudin, Kayazinon, Kayazol, Knox Out, and Nipsan.

Some diazinon formulations contain 0.2–0.7% (2000.0–7000.0 mg/kg) of Sulfotep (tetraethyldithiopyrophosphate) as a manufacturing impurity; Sulfotep is reportedly at least 100 times more toxic than diazinon to some organisms. It seems that additional research is warranted on diazinon–Sulfotep interactions.

Diazinon degrades rapidly in plants, with half-time persistence usually less than 14 days; however, persistence increases as temperatures decrease, and is longer in crops with a high oil content. In water, diazinon breaks down to comparatively nontoxic compounds with little known hazard potential to aquatic species, although the degradation rate is highly dependent on pH. The half-time persistence of diazinon on sandy loam soil exposed to sunlight is 2.5–10 days. In most soils, diazinon seldom penetrates below the top 1.3 cm. But diazinon may remain biologically available in soils for 6 months or longer at low temperature, low moisture, high alkalinity, and lack of suitable microbial degraders. Bacterial enzymes, derived from *Pseudomonas* sp., can be used to hydrolyze diazinon in soil, although costs are prohibitive except in treating emergency situations involving spills of concentrated diazinon solutions. In one case, diazinon was enzymatically hydrolyzed within 24 h in an agricultural sandy soil at concentrations as high as 10,000.0 mg/kg.

In almost every instance of diazinon poisoning, there has been a general reduction in cholinesterase activity levels, especially in brain and blood. Diazinon exerts its toxicity by binding to the neuronal enzyme Ache for a considerable time postexposure. It is emphasized that all organophosphorus pesticide compounds, in sufficient dose, inhibit Ache in vivo, and all share a common mechanism of acute toxic action. Ache inhibition results in the accumulation of endogenous acetylcholine in nerve tissues and effector organs, resulting in signs that mimic the muscarinic, nicotinic, and central nervous system (CNS) actions of acetylcholine. The immediate cause of death in fatal organophosphorus compound poisonings, including diazinon, is asphyxia resulting from respiratory failure. Contributing factors are the muscarinic actions of bronchoconstriction and increased bronchial secretions, nicotinic actions leading to paralysis of the respiratory muscles, and the CNS action of depression and paralysis of the respiratory center.

Diazinon is not a potent inhibitor of cholinesterase, and must be converted to its oxygen analogs (oxons), especially diazoxon (diethyl 2-isopropyl-6-methylpyrimidin-4-yl phosphate) in vivo before poisoning can occur. Diazoxon is about 10,000 times more effective in reducing cholinesterase activity levels than diazinon. At least eight diazinon metabolites have been identified in vertebrates, of which four are oxons. It is generally agreed that diazinon is metabolized to diazoxon through the action of liver mixed-function oxidases and nicotinic adenine dinucleotide (NAD) phosphate. Diazinon toxicity will depend to some extent upon the relation between the rates of activation of diazinon to diazoxon, and of decomposition of the latter to harmless products. Birds are more sensitive to diazinon than mammals, probably because mammalian blood enzymes hydrolyze diazoxon rapidly, whereas bird blood has virtually no hydrolytic activity. It seems that diazoxon stability in blood is a major factor affecting susceptibility of birds and mammals to diazinon poisoning.

Diazinon poisoning effects in animals can be delayed or prevented by treatment with a variety of compounds. For example, Ache in diazinon-stressed birds can be reactivated by pralidoxime. Furthermore, pretreatment of large white butterfly (*Pieris brassicae*) larvae with methylene dioxyphenyl compounds will inhibit the diazinon to diazoxon activation. Added tryptophan and its metabolites may prevent teratogenic defects by maintaining NAD levels in diazinon-treated chicken embryos; diazinon reportedly acts to decrease the availability of tryptophan to bird embryos, subsequently interfering with NAD metabolism and causing birth defects. NAD metabolism in diazinon-stressed birds may also be maintained with nicotinamide. In contrast to many other organophosphorus insecticides, organisms that survive diazinon-inhibited cholinesterase levels can undergo considerable spontaneous reactivation (dephosphorylation), indicating that its dephosphorylation occurs more readily than that of cholinesterase inhibited by other organophosphorus compounds.

12.3 Lethal Effects

Diazinon toxicity varies widely within and among species, and is modified by organism age, sex, and body size, climatic conditions, pesticide formulation, chemistry of the environment, and other factors. Nevertheless, several trends are apparent as judged by available data. Among aquatic organisms, for example, freshwater cladocerans and marine shrimps were the most sensitive species tested, with LC50 (96 h) values of less than 5.0 μg/L; freshwater teleosts were more resistant, with the lowest LC50 (96 h) value recorded being 90.0 μg/L. Diazinon has considerable potential for causing acute avian poisoning episodes; sensitive species of birds, including ducks, turkey (*Meleagris gallopavo*), and red-winged blackbird (*Agelaius phoeniceus*), died at single oral doses of 2.0 mg of diazinon/kg body weight (BW). Mammals are more resistant than birds to diazinon; the lowest LD50 (acute oral) value recorded is 224.0 mg/kg BW for female rats (*Rattus*). Chronic oral toxicity tests with mammals suggest that daily intake exceeding 5.0 or 10.0 mg diazinon/kg BW is probably fatal over time to swine (*Sus scrofa*) and dogs (*Canis familiaris*), respectively. Finally, 9.0 mg/kg of dietary diazinon fed during gestation to pregnant mice (*Mus musculus*) was associated with significant mortality of pups prior to weaning.

12.3.1 Aquatic Organisms

Freshwater cladocerans and marine crustaceans were the most sensitive groups tested, with LC50 (96 h) values of less than 2.0 μg/L for the more sensitive species. European eels (*Anguilla anguilla*), rainbow trout (*Oncorhynchus mykiss*), and bluegills (*Lepomis macrochirus*) seemed to be most sensitive freshwater teleosts tested, with LC50 (96 h) values between 80.0 and 120.0 μg/L; however, the postlarval and juvenile stages of the striped knifejaw (*Oplegnathus fasciatus*) – a marine fish cultured intensively in Japan – were unusually sensitive. In general, technical grade formulations of diazinon seem to be more toxic than emulsifiable concentrates, dusts, and oil solutions. Also, large variations in acute toxicity values were evident, even among closely related species.

Outward signs of diazinon poisoning in fish included lethargy, forward extension of pectoral fins, darkened areas on posterior part of body, hyperexcitability when startled, sudden rapid swimming in circles, and severe muscular contractions. Internally, physiological mechanisms in teleosts preceding death involved the following sequence: cholinesterase inhibition, acetylcholine accumulation, disruption of nerve functions, respiratory failure, and asphyxia. Closely related species of fishes differ markedly in their sensitivity to diazinon. Guppies (*Poecilia reticulata*) are 5 times more sensitive to diazinon than are zebrafish (*Brachydanio rerio*), as judged by LC50 (96 h) values. Differences of resistance and accumulation between guppies and zebrafish are related to the rate of oxidative metabolism. Preexposure of guppies to a high sublethal concentration of diazinon increases resistance to diazinon by a factor of 5 when compared to nonpretreated guppies; zebrafish similarly pretreated were not more resistant. Pretreatment of guppies resulted in a strong inhibition of diazoxon formation and pyrimidinol during incubations of diazinon with the hepatic postmitochondrial supernatant. It was concluded that toxicity of diazinon in the guppy is due to its metabolism to a highly toxic metabolite, likely diazoxon; and in zebrafish or pretreated guppies having low rates of diazinon metabolism toxicity is due to the accumulation of the parent compound. Limited data indicated that the yellowtail (*Seriola quinqueradiata*), a marine teleost, was 84 times more sensitive to diazinon than were 4 species of freshwater fishes, as judged by LC50 (48 h) values, and by its inability to biotransform diazinon to nontoxic metabolites within 1 h. Diazinon has not been detected in marine waters, but the potential exists for contamination of estuarine areas from agricultural and urban runoff.

12.3.2 Birds

Diazinon adversely affects survival of developing mallard embryos when the eggshell surface is subjected for 30 s to concentrations 25–34 times higher than recommended field application rates; mortality patterns were similar for solutions applied in water or in oil. This laboratory finding suggests that eggs of mallards, and probably other birds, are protected when diazinon is applied according to label directions. Chickens dipped in solutions containing 1000.0 mg of diazinon/L, an accidentally high formulation, experienced 60% mortality within 3 days; no other deaths occurred during the next 4 months. Results of 5-day feeding trials with 2-week-old Japanese quail (*Coturnix japonica*), followed by 3 days on untreated feed, showed an LD50 of 167.0 mg diazinon/kg diet – a concentration considered "very toxic." No deaths were observed at dietary levels of 85.0 mg diazinon/kg, but 53% died at 170.0 mg/kg, and 87% at 240.0 mg/kg.

Diazinon has a potential for causing acute avian poisoning episodes. Ingestion of 5 granules of Diazinon 14G (14.3% diazinon) killed 80% of house sparrows (*Passer domesticus*), and all red-winged blackbirds to which they were administered. Ingestion of fewer than 5 granules of Diazinon 14G, each containing about 215.0 μg of diazinon, could be lethal to sparrow-sized birds (i.e., 15–35 g BW), especially juveniles of seed-eaters. Acute oral LD50s indicate that 15.0 mg of diazinon/kg BW is fatal to virtually all species tested, and that 2.0–5.0 mg/kg is lethal to the more sensitive species. Signs of diazinon poisoning in birds included muscular incoordination, wing spasms, wing drop, hunched back, labored breathing, spasmodic contractions of the anal sphincter, diarrhea, salivation, lacrimation (tear production), eyelid drooping, prostration, and arching of the neck over the back. Most of these signs have been observed in birds poisoned by other than diazinon; these compounds also act via an anticholinesterase mode of action.

12.3.3 Mammals

Signs of diazinon poisoning in mammals included a reduction in blood and brain cholinesterase activity, diarrhea, sweating, vomiting, salivation, cyanosis, muscle twitches, convulsions, loss of reflexes, loss of sphincter control, and coma. Other compounds that produce their toxic effects by inhibiting Ache, such as organophosphorus pesticides and many carbamates, show similar effects. Two species of marmoset monkeys accidentally poisoned by diazinon exhibited, prior to death, high-pitched voices, trembling, frog-like jumping, a stiff gait, and pale oral mucous membranes; internally, bone marrow necrosis and hemorrhages in several organs were evident. Internal damage was also observed in swine and dogs that died following controlled administration of diazinon. Swine showed histopathology of liver and intestinal tract, and duodenal ulcers; dogs showed occasional rupture of the intestinal wall, and testicular atrophy. Results of acute oral toxicity tests indicated that the rat was the most sensitive mammalian species tested, with an acute oral LD50 of 224.0 mg diazinon/kg BW. It is clear that mammals are significantly more resistant to acute oral poisoning by diazinon than birds. Diazinon was also toxic to mammals when administered dermally, through inhalation, and in the diet. The lowest dermal LD50 recorded was 600.0 mg diazinon/kg BW for rabbits (*Lepus* sp.) using an emulsifiable (4E) formulation. The single datum for inhalation toxicity indicated that 27.2 mg of diazinon/L of air killed 50% of test rabbits after exposure for 4 h. All pregnant mice fed diets containing 9.0 mg of diazinon/kg during gestation survived, but some pups died prior to weaning. Results of chronic oral toxicity tests of diazinon indicated that death was probable if daily doses exceeded 5.0 mg/kg BW for swine, or 10.0 mg/kg for dogs.

12.3.4 Terrestrial Invertebrates

Accidental spraying of beehives in Connecticut with diazinon resulted in a complete kill of resident honeybees. Dead bees contained up to 3.0 mg/kg of diazinon. Diazinon is an effective insecticide. LD50 values for diazinon and adult houseflies (*Musca domestica*), applied topically, were 0.4 μg/insect, or 4.6 mg/kg BW. LD50 values for larvae of the large white butterfly, applied topically, were 8.8 mg/kg BW for diazinon, and 11.0 mg/kg BW for diazoxon. Pretreatment of larvae with methylene dioxyphenyl compounds antagonized the action of diazinon by a factor of about 2, but synergized the action of diazoxon by an order of magnitude.

12.4 Sublethal Effects

Among sensitive species of aquatic organisms, diazinon was associated with reduced growth and reproduction in marine and freshwater invertebrates and teleosts, spinal deformities in fish, reduced emergence in stream insects, measurable accumulations in tissues, increased numbers of stream macroinvertebrates carried downstream by currents (drift), possible mutagenicity in fish, and interference with algal–invertebrate interactions. In birds, diazinon is a known teratogen; it is also associated with reduced egg production, decreased food intake, and loss in body weight. Diazinon fed to pregnant mice resulted in offspring with brain pathology, delayed sexual maturity, and adverse behavioral modifications that became apparent late in life. For all groups tested, diazinon directly or indirectly inhibited cholinesterase activity.

12.4.1 Aquatic Organisms

Atlantic salmon (*Salmo salar*) exposed to 0.3–45.0 μg diazinon/L for 120 h had reduced levels of reproductive steroids in blood plasma at all concentrations; exposure to 2.0 μg/L for only 30 min produced a significant reduction in olfactory response to prostaglandin F_{2a}.

Carp and other species of freshwater teleosts that survived high sublethal concentrations of diazinon had impaired swimming and abnormal pigmentation. Spinal deformities, mostly lordosis and scoliosis, were among the more insidious effects documented for diazinon. Malformations were observed in fathead minnows (*Pimephales promelas*) after 19 weeks in water containing 3.2 μg diazinon/L, in yearling brook trout (*Salvelinus fontinalis*) within a few weeks at 4.8 μg/L, and in various species of freshwater teleosts after exposure for 7 days to 50.0 μg diazinon/L. Exposure of bluegills (*Lepomis macrochirus*) to 15.0 μg diazinon/L for only 24 h resulted in mild hyperplasia of the gills that increased in severity with increasing concentration (30.0–75.0 μg/L) and may lead to death.

Diazinon is a noncarcinogen and reportedly has no significant mutagenic activity in microbial systems, yeast, and mammals including humans. However, a significant increase in the frequency of sister chromatid exchange (SCE) was measured in central mud minnows (*Umbra limi*) that were exposed in vivo for 11 days to solutions containing 0.16–1.6 μg of diazinon/L. This finding requires verification.

In general, diazinon does not bioconcentrate to a significant degree and is rapidly excreted after exposure. Diazinon in water is bioconcentrated by brook trout at levels as low as 0.55 μg/L, but tissue residues for all aquatic organisms seldom exceeded 213 times that of ambient water, even after months of continuous exposure. Common carp (*Cyprinus carpio*) exposed to 1.5–2.4 μg/L for 168 h had bioconcentration factors (BCFs) of 12 in muscle, 12 in gallbladder, 50 in kidney, and 51 in liver; almost all was excreted in 72 h on transfer to clean water, except for kidney which is the major organ for excretion. High BCFs of 800 in liver, 1600 in muscle, 2300 in gill, and 2730 in blood are reported for juvenile European eels (*Anguilla anguilla*) after exposure to 42.0–56.0 μg/L for 96 h; however, diazinon residues in tissues were usually not detected in tissues after 24 h in clean water. The half-time persistence of diazinon in tissues of European eels was estimated at 17–31 h in liver, 32–33 h in muscle, and 27–38 h in gill. Whole guppies exposed to high sublethal concentrations of diazinon show BCFs of 59 after 48 h and 188 after 144 h; the half-time persistence of diazinon was 10 h after 48-h exposure and 23 h after 144-h exposure. Diazinon and its metabolites are excreted rapidly posttreatment; the loss rate is approximately linear. The enzyme system responsible for diazinon metabolism in fish liver microsomes required NADPH and oxygen for the oxidative desulfuration of diazinon to diazoxon. Fish with high fat content contained greater residues of diazinon in fatty tissues than did fish with comparatively low lipid content, and this could account, in part, for inter- and intraspecies variations in uptake and depuration. Some organisms, such as the sheepshead minnow (*Cyprinodon variegatus*), have measurable diazinon residues during initial exposure to 6.5 μg/L, but no detectable residues after lengthy exposure, suggesting that physiological adaptation resulting in rapid detoxification is possible.

Freshwater and marine algae were unaffected at water diazinon concentrations that were fatal (i.e., 1000.0 μg/L) to aquatic invertebrates. However, diazinon at 1.0 μg/L induced extensive clumping of a freshwater alga (*Chlorella pyrenoidosa*) onto the antennae of *Daphnia magna* within 24 h. The affected daphnids were immobilized and settled to the bottom of the test containers. The causes of particulate matter adhesion are open to speculation, and additional research is merited.

Freshwater macroinvertebrates were comparatively sensitive to diazinon. Results of large-scale experimental stream studies showed that dose levels of 0.3 μg diazinon/L caused a 5–8-fold reduction in emergence of mayflies and caddisflies within 3 weeks; after 12 weeks, mayflies, damselflies, caddisflies, and amphipods were absent from benthic samples. Elevated (and catastrophic) drift of stream invertebrates was also documented in diazinon-treated streams, especially for amphipods, leeches, and snails. Short-term tests of 5-h duration with rotifers (*Brachionus calyciflorus*) show a 50% reduction in feeding rate on alga (*Nannochloris oculata*) at 14.2 mg/L, with long-term implications to population stability.

Freshwater fish populations can be directly damaged by prolonged exposure to diazinon at concentrations up to several hundred times lower than those causing acute mortality. Impaired reproduction and Ache inhibition occur concurrently in teleosts during long-term exposure to diazinon, but reproduction can be impaired for at least 3 weeks after fish are placed in uncontaminated water, even though Ache is normal and they contained no detectable diazinon residues. Furthermore, diazinon exposure during spawning caused complete, but temporary, inhibition of reproduction at concentrations which did not produce this effect in fish exposed since hatch. This could severely impact aquatic species with a short reproductive period.

12.4.2 Birds

Diazinon produces visible Type I and II teratisms when injected into chicken embryos. Type I teratisms (related to tissue NAD depression) included abnormal beaks, abnormal feathering, and shortened limbs. Type II teratisms, which included short and wryneck, leg musculature hypoplasia, and rumplessness were associated with disruptions in the nicotinic cholinergic system. The severity of effects depended on embryo age and was dose related. Chick embryos (age 48 h) receiving 25.0 μg or more of diazinon/embryo had cervical notochord and neural tube malformations at 96 h, and short neck at 19 days. Wryneck occurred at doses ranging from 6.2 to 100.0 μg/embryo, but was more frequent at higher doses. Type II teratisms were attributed to disruption of notochord sheath formation. Coinjection of 2-pyridinealdoxime methochloride (2-PAM) along with 200 μg of diazinon/embryo markedly reduced notochord and neural tube deformations. Similarly, the copresence of tryptophan – or its metabolites L-kynurenine, 3-hydroxyanthronilic acid, quinolinic acid – maintained NAD levels of diazinon-treated embryos close to, or above, normal, and significantly alleviated the symptoms of Type I teratisms.

Reduced egg production, depressed food consumption, and loss in body weight have been observed in ring-necked pheasants at daily diazinon intakes greater than 1.05 mg/bird; a dose-related delay in recovery of egg laying was noted after termination of diazinon treatment. Threshold levels in ring-necked pheasants of 1.05 and 2.1 mg of diazinon daily corresponded to 1/16 and 1/8 of daily ration (70 g) treated at commercial application rates. Food consumption of ring-necked pheasants was reduced significantly when only food treated with diazinon was available; pheasants avoided diazinon-treated food if suitable alternatives existed. Dietary levels above 50.0 mg/kg were associated with reduced food consumption, weight loss, and reduction in egg production in northern bobwhites. If food reduction is important, then diets containing more than 17.5 mg diazinon/kg (based on, empirical calculations) were potentially harmful to bobwhites. The mechanisms accounting for reduction in egg deposition are not clear, but are probably related primarily to decreased food intake. They may also be associated with diazinon-induced pituitary hypofunction at the level of the hypothalamus, resulting in reduced synthesis and secretion of gonadotrophic, thyrotrophic, and adrenocorticotrophic hormones.

12.4.3 Mammals

Diazinon exerts its toxic effects by binding to the neuronal enzyme Ache for long periods after exposure. Diazinon, in turn, is converted to diazoxon, which has a higher affinity for Ache (and thus greater toxicity) than the parent compound. There is a latent period in white-footed mice in reduction of cholinesterase activities, sometimes up to 6 h, until diazinon is converted to diazoxon. Effects of multiple doses of diazinon to mammals are not clear, e.g., rats exposed to a high dose of diazinon did not respond fully to a second dose until one month later. It is difficult to ascertain when complete recovery of diazinon-poisoned animals has occurred. It is speculated, but not verified, that wildlife recovering from diazinon poisoning may face increased predation, aberrant behavior, learning disabilities, hypothermia, and reproductive impairments. Data are now lacking on recovery aspects of diazinon-poisoned native mammal populations.

Diazinon is rapidly biotransformed and excreted in mammals. Estimated half-times of diazinon persistence were 6–12 h in rats and dogs. Most of the diazinon metabolites were excreted in the urine as diethyl phosphoric acid and diethyl phosphorothioic acid in dogs, and as hydroxy diazinon and dehydrodiazinon in sheep.

Determination of Ache activity in selected tissues following diazinon exposure provided an estimate of potential toxicity, but tissue sensitivity varied widely between and among taxa. In sheep, brain cholinesterase inhibition was pronounced after diazinon insult, and metabolism of diazinon in, or close to, the brain was the most likely source of toxicologically effective diazoxon. In rats, diazinon effectively reduced blood cholinesterase levels, with inhibition significantly more evident in erythrocytes than in plasma. All mammalian bloods hydrolyze diazoxon rapidly, whereas birds have virtually no hydrolytic activity in their blood, and, as a result, were more susceptible than mammals. The stability of diazoxon in the blood appears to be a primary factor in susceptibility to diazinon poisoning. In species lacking blood oxonases, the liver was probably the most important site of diazinon metabolism. Diazinon that accumulated in rat liver was biotransformed, usually within 24 h, by microsomal mixed-function oxidases and glutathione-*S*-transferases; however, diazinon residues in rat kidney were almost 500 times those in liver (and 11 times those in brain), and were measurable in kidney but not in liver. It now seems that diazinon residues in kidney, and cholinesterase inhibition in erythrocytes are the most useful indicators of acute diazinon poisoning in mammals.

Sublethal effects of diazinon have been recorded in rodents, the most sensitive mammal group tested. Effects were measured at 0.5 mg diazinon/kg in diets of rats for 5 weeks, at 0.18 mg/kg BW administered daily to pregnant mice, and at single doses of 1.8 mg/kg BW for rat and 2.3 mg/kg BW for white-footed mice. Many variables modify diazinon-induced responses, including the organism's sex. For example, female rats and dogs were more sensitive to diazinon than males, but male swine were more sensitive than females.

Behavioral deficits observed in offspring of mice exposed to diazinon during gestation indicated that prenatal exposure might produce subtle dysfunctions not apparent until later in life. Pregnant mice given a daily dose of 0.18 or 9.0 mg diazinon per kg BW throughout gestation gave birth to viable, overtly normal, offspring. But pups born to mothers of the 9.0 mg/kg groups grew more slowly than controls and were significantly smaller at 1 month than controls. Offspring of mothers receiving 0.18 mg/kg BW exhibited significant delays in the appearance of the contact-placing reflex, and in descent of testes or vaginal opening. Mature offspring of mothers exposed to either dose level, displayed impaired endurance and coordination on rod cling and inclined plane tests of neuromuscular function. In addition, offspring of the 9.0 mg/kg dose had slower running speeds and less endurance in a swimming test than controls. At 101 days, forebrain neuropathology was evident in the 9.0 mg/kg dose, but not in the 0.18 mg/kg group. The mechanisms responsible for these effects are unknown.

Diazinon is nonmutagenic to mammals, as judged by its inability to induce SCEs in Chinese hamster ovary cells (CHOCs) at 80.0 mg/kg culture medium; most organophosphorus insecticides tested induced SCE in CHOC at this concentration. Diazoxon, an oxygen analog of diazinon, did produce SCE at 304.0 mg/kg, but was 3–10 times less effective than oxygen analogs of other organophosphorus compounds screened.

12.4.4 Terrestrial Invertebrates

Tobacco hornworms (*Manduca sexta*) from a field sprayed with 840.0 mg diazinon/ha contained no detectable residues of diazoxon. Only one sample, collected about 4 h after spraying, exceeded 1.0 mg diazinon/kg BW. No diazinon residues in these insects were detectable after 18 days. It was concluded that the potential hazard to birds eating hornworms was minimal. In contrast, diazinon residues in molluscan slugs (*Agriolimax reticulatus*), collected from plats of spring wheat sprayed with 8000.0 mg diazinon/ha, increased linearly to about 200.0 mg/kg at 6 weeks postapplication, then declined to background levels after 16 weeks. During this same period, soil residues decreased from about 4.0 mg/kg immediately after application, to about 1.0 mg/kg at 6 weeks, and were not detectable after 12 weeks. The high residues observed in slugs may be , in part, due to physical adsorption of diazinon to slug mucous. It was concluded that slugs heavily contaminated by diazinon constituted a serious danger to birds and mammals feeding on them.

Depuration rates of diazinon differed significantly for two species of nematodes, *Panagrellus redivivus* and *Bursaphelenchus xylophilus*. Both species showed maximum uptake of radiolabeled diazinon between 6 and 12 h, and both metabolized diazinon to diazoxon and pyrimidinol. By 96 h, 95% of the diazinon in *P. redivivus* had been metabolized, but only 26% was transformed in *B. xylophilus*, again demonstrating variability in diazinon metabolism between related species.

12.5 Recommendations

Aquatic organisms were impacted by diazinon water concentrations between 0.3 and 1.2 μg/L; effects included lowered emergence and elevated drift of stream insects (0.3 μg/L), reduced fecundity of marine minnows (0.47 μg/L), accumulations in freshwater teleosts (0.55 μg/L), and daphnid immobilization (1.0 μg/L) and death (1.2 μg/L). These comparatively low levels are of concern because transient peak water concentrations of 4.0–200.0 μg diazinon/L have been recorded near diazinon sheep-dipping sites in England, and 36.8 μg/L in the Sacramento–San Joaquin River, California. For protection of sensitive aquatic organisms, a water diazinon level should not exceed 0.08 μg/L. This value represents a safety factor of about 4 over the lowest recorded adverse effect level of 0.3 μg/L. For protection of freshwater aquatic life, an average 4-day concentration of 0.04 μg diazinon/L is recommended, provided that this value is not exceeded more than once every three years and the maximum 1-h concentration does not exceed 0.08 μg/L more than once every three years. Safety factors may require adjustment as additional data become available. Establishment of safe levels is complicated by the fact that diazinon can persist for many months in neutral or basic waters, including seawater, but hydrolyzes rapidly in acidic waters. Data on chronic effects of fluctuating and intermittent exposures of fishes and invertebrates to diazinon are also needed, and these will aid in the establishment of safe concentrations for this organophosphorus pesticide.

Granular formulations were especially hazardous to seed-eating birds; ingestion of fewer than 5 granules of a Diazinon 14G formulation could be lethal. A reduction in diazinon content of existing granular formulations may become necessary in application areas frequented by high densities of seed-eating birds. Diazinon should not be used in areas where waterfowl feed, especially turfgrass. Suggested alternatives to diazinon for turfgrass use include Dursban (*O,O*-diethyl *O*-(3,5,6-trichloro-2-pyridyl)-phosphorothioate), Dylox (dimethyl (2,2,2-trichloro-1-hydroxyethyl) phosphonate), Carbaryl (1-naphthyl *N*-methylcarbamate), and Lannate (*S*-methyl-*N*-((methylcarbamoyl) oxy)-thioacetimidate). Diazinon should be used with caution in large-scale spray applications, such as grasshopper control, as judged by some deaths of horned larks (*Eremophila alpestris*), lark buntings (*Calamospiza melanocorys*), western meadowlarks (*Sturnella neglecta*), and chestnut-collared longspurs (*Calcarius ornatus*) when used for this purpose in Wyoming. Diazinon applications to agricultural crops comprised a relatively small percentage of the reported mortality incidents, but it is likely that this category is underreported since such incidents were probably less conspicuous than those noted on lawns and golf courses. Also, diazinon interactions with other agricultural chemicals, such as Captan (*cis*-*N*-((trichloromethyl)thio)-4-cyclohexene-1,2-dicarboximide), may produce more-than-additive (but reversible) adverse effects on food consumption and egg production of ring-necked pheasants. More research is needed on complex mixtures of agricultural pesticides that contain diazinon.

In female rats, the no-observable-effect level (NOEL) is 0.1 mg/kg of dietary diazinon; at 0.5 mg/kg there was a marked lowering of plasma cholinesterase activity in 5 weeks. But studies with male rats indicate that the NOEL is 2.0 mg/kg of dietary diazinon, or about 20 times higher than female rats. Accordingly, future studies should consider sex as a variable in toxicity evaluation of diazinon. It is generally agreed that mammals are more resistant than birds to diazinon owing, in part, to their ability to rapidly metabolize diazoxon. However, data are missing on the effects of diazinon to native mammals under field conditions, and this should constitute a priority research area. No diazinon criteria to protect human health have been proposed by the U.S. Food and Drug Administration or the state of California.

12.6 Summary

Diazinon (phosphorothioic acid *O*,*O*-diethyl *O*-(6-methyl-2-(1-methylethyl)-4-pyrimidinyl) ester) is an organophosphorus compound with an anticholinesterase mode of action. It is used extensively to control flies, lice, insect pests of ornamental plants and food crops, as well as nematodes and soil insects in lawns and croplands. Diazinon degrades rapidly in the environment, with half-time persistence usually less than 14 days. But under conditions of low temperature, low moisture, high alkalinity, and lack of suitable microbial degraders, diazinon may remain biologically active in soils for 6 months or longer.

At recommended treatment levels, diazinon-related kills have been noted for songbirds, honeybees, and especially waterfowl that consume diazinon-treated grass; moreover, incidents involving agricultural applications may be underreported. Accidental deaths through misapplication of diazinon have also been recorded in domestic poultry, monkeys, and humans. It has been suggested, but not yet verified, that wildlife partially disabled in the field as a result of diazinon poisoning would be more likely to die of exposure, predation, starvation, or dehydration, or face behavioral modifications, learning impairments, and reproductive declines than would similarly treated domestic or laboratory animals.

Among sensitive aquatic organisms, LC50 (96 h) values of 1.2–2.0 μg/L were derived for freshwater cladocerans, and 4.1–5.9 μg/L for marine shrimps; freshwater teleosts were comparatively resistant, with all LC50 (96 h) values >80.0 μg/L. Sublethal effects were recorded at 0.3–3.2 μg diazinon/L and included reduced emergence of stream insects (0.3 μg/L), reduced fecundity of a marine fish (0.47 μg/L), significant accumulations in freshwater teleosts (0.55 μg/L), daphnid immobilization (1.0 μg/L), potential mutagenicity in a freshwater fish (1.6 μg/L), and spinal deformities in teleosts (3.2 μg/L). Exposure to diazinon during spawning caused temporary, but complete, inhibition of reproduction at concentrations which did not produce this effect in fish exposed continuously since hatch.

Acute oral LD50s of about 2500.0–3500.0 μg diazinon/kg BW were determined for goslings (*Anser* spp.), ducks (*Anas* spp.), domestic turkey (*Meleagris gallopavo*), and the red-winged blackbird (*Agelaius phoeniceus*), the most sensitive birds tested. A dietary LD50 of 167,000.0 μg diazinon/kg was determined for Japanese quail (*Corturnix japonica*). Diazinon produced marked teratogenic effects in embryos of the domestic chicken (*Gallus gallus*) at 6.2–25.0 μg/embryo, reduced egg deposition in the ring-necked pheasant (*Phasianus colchicus*) at more than 1050.0 μg/bird, and (empirically) decreased food consumption and increased weight loss in the northern bobwhite (*Colinus virginianus*) at >17,500.0 μg diazinon/kg diet.

The rat (*Rattus rattus*) was the most sensitive mammal tested in acute oral toxicity screenings, with an LD50 of 224,000.0 μg diazinon/kg BW. Chronic oral toxicity tests with swine (*Sus scrofa*) indicated that death was probable if daily intakes were greater than 5000.0 μg diazinon/kg BW. Measurable adverse effects of diazinon have been recorded in rodents, the most sensitive mammalian group tested, at: 500.0 μg/kg in diets fed to rats for 5 weeks, causing blood cholinesterase inhibition; 180.0 μg/kg BW administered daily to pregnant mice (*Mus musculus*) during gestation, inducing behavioral modifications and delayed sexual maturity of progeny; and single oral doses of 1800.0 and 2300.0 μg/kg BW in rats and white-footed mice (*Peromyscus leucopus*), respectively, which produced altered blood chemistry and brain cholinesterase inhibition.

For protection of sensitive aquatic organisms, diazinon concentrations in water should not exceed 0.08 μg/L; however, more data are needed on the effects of fluctuating and intermittent chronic exposures of diazinon on reproduction of fish and aquatic invertebrates. Granular formulations of diazinon seem to be especially hazardous to seed-eating birds, suggesting a need to control or eliminate granular applications when these species are present. For additional protection of birds, diazinon should be used with extreme caution in areas where waterfowl feed, and in large-scale spray applications such as grasshopper control. Diazinon in combination with some agricultural chemicals produced more-than-additive adverse effects on bird growth and fecundity; accordingly, more research is needed on effects of complex mixtures of pesticides that contain diazinon. Most investigators agreed that mammals were less susceptible to diazinon than were birds, at least under controlled environmental regimens. Data are lacking on diazinon impacts to mammals under field conditions; acquisition of these data should constitute a priority research area.

DIFLUBENZURON[a]

Chapter 13

13.1 Introduction

Compounds collectively known as insect growth regulators are recognized as important insecticides. These compounds include juvenile hormone mimics, antijuvenile hormone analogs, and chitin synthesis inhibitors. The most widely studied chitin synthesis inhibitor, and the only one registered for use against selected insect pests in the United States, is diflubenzuron (1-(4-chlorophenyl)-3-(2,6-difluorobenzoyl)urea), also known as dimilin. Chitin is a major component of the tough outer covering, or cuticle, of insects. As insects develop from immature larvae to adults, they undergo several molts, during which new cuticles are formed and old ones shed. Diflubenzuron prevents successful development by inhibiting chitin synthetase, the final enzyme in the pathway by which chitin is synthesized from glucose.

Diflubenzuron is highly effective against larval stages of many species of nuisance insects. It has been used extensively to control mosquitos, midges, gnats, weevils (including the cotton boll weevil, *Anthonomus grandis*), various beetles, caterpillars of moths and butterflies (especially the gypsy moth, *Lymantria dispar*), flies, and rust mites. In Maryland, for example, more than 30,000 ha are sprayed annually to control gypsy moths. In general, less than 140 g/ha (2 ounces/acre) of diflubenzuron is sufficient to control susceptible species, although affected larvae do not die until they molt.

Most authorities agree that diflubenzuron has low mammalian toxicity, is not highly concentrated through vertebrate food chains or by absorption from water, remains stable on foliage, and seldom persists for extended periods in soil and water. Chitin synthesis inhibitors, however, are not specific to insect pests. Beneficial insects also produce chitin, as do all arthropods, including spiders, crabs, crayfish, lobsters, shrimp, daphnids, mayflies, stoneflies, barnacles, copepods, and horseshoe crabs. All of these groups are adversely affected by diflubenzuron, including effects on survival, reproduction, development, limb regeneration, and population growth.

13.2 Environmental Chemistry

Diflubenzuron breakdown by hydrolysis, soil degradation, or plant and animal metabolism initially yields 2,6-difluorobenzoic acid and 4-chlorophenylurea. Ultimately, the end products are either conjugated into mostly water-soluble products or are biologically acylated and methylated. At extremely low doses, diflubenzuron selectively inhibits the ability of arthropods to synthesize chitin at the time of molting, producing death of the organism from rupture of the cuticle or starvation. Other organisms that contain chitin (i.e., some species of fungi and marine diatoms), or polysaccharides similar to chitin (i.e., birds and mammals), seem unaffected. Mobility and leachability of diflubenzuron in soils is low, and residues are usually not detectable after 7 days. Degradation is most rapid when small-particle (2–5 μm) formulations are applied and soil bacteria are abundant. In water,

---

[a]All information in this chapter is referenced in the following sources:

Eisler, R. 1992. Diflubenzuron hazards to fish, wildlife, and invertebrates: a synoptic review. *U.S. Fish Wildl. Serv. Biol. Rep. 4*, 36 pp.

Eisler, R. 2000. Diflubenzuron. Pages 983–1019 in *Handbook of Chemical Risk Assessment: Health Hazards to Humans, Plants, and Animals. Volume 2, Organics.* Lewis Publishers, Boca Raton, Florida.

diflubenzuron usually persists for only a few days; degradation is most rapid under conditions of high organic and sediment loadings, and elevated water pH and temperature.

13.2.1 Chemical and Biochemical Properties

Selected properties of diflubenzuron are listed in Table 13.1. Diflubenzuron degradative pathways are almost entirely through cleavage between the carbonyl and amide groups of the urea bridge. Ultimately, the end products are either conjugated into predominantly water-soluble products or are acylated and ethylated biologically. Hydrolysis, soil degradation, and plant and animal metabolism of diflubenzuron yield the same initial products: 2,6-difluorobenzoic acid and 4-chlorophenylurea. Soil degradation and plant and animal metabolism involve further conversion of these compounds to 2,6-difluorobenzamide and 4-chloroaniline (Figure 13.1). Interspecies variations in ability to metabolize diflubenzuron are common, as judged by metabolic patterns in rat (*Rattus* spp.), cow (*Bos bovis*), and sheep (*Ovis aries*). In all three species, hydroxylation of either aromatic ring and scission of the ureido bridge constituted the main metabolic pathways. In cow and rat, the prevailing route was ring hydroxylation; in sheep, it was the scission reaction. In cow and sheep, about

Table 13.1. Selected properties of diflubenzuron.

Variable	Datum
CHEMICAL NAMES	1-(4-chlorophenyl)-3-(2,6-difluorobenzoyl)urea; *N*-[[(4-chlorophenyl)amino]carbonyl]-2,6-difluorobenzamide; 1-(2,6-difluorobenzoyl)-3-(4-chlorophenyl)urea
ALTERNATE NAMES	Deflubenzon, Diflubenuron, Dimilin, DU, DU 112307, Duphar BV, ENT-29054, Largon, Micromite, OMS 1804, PDD 6040-I, PH 60-40, TH 6040, Vigilante
ACTION	Insecticide, larvicide, ovicide; insect growth regulator acting by interference with deposition of insect chitin
CAS NUMBER	35367-38-5
EMPIRICAL FORMULA	$C_{14}H_9ClF_2N_2O_2$
MOLECULAR WEIGHT	310.68
FORMULATIONS	Granular, oil-dispersible concentrate; wettable powder
MANUFACTURING PROCESS AND IMPURITIES	Produced by reaction of 2,6-difluorobenzamide with 4-chlorophenyl isocyanate. The technical product is 95% pure. Impurities are of low toxicological concern in terminal residues
STABILITY	Stable under sunlight and in neutral or mildly acidic solutions; unstable in strong basic solutions
PHYSICAL STATE	White crystalline solid
MELTING POINT	210–230°C (technical); 230–232°C (pure)
SOLUBILITY	
Water	0.1–0.2 mg/L at 20°C; 1.0 mg/L at 25°C
Polar organic solvents	Moderate to good
OCTANOL–WATER PARTITION COEFFICIENT	3500

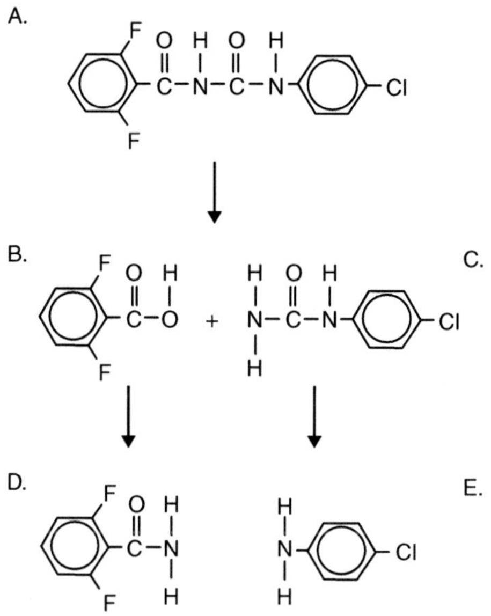

Figure 13.1.
Generalized degradation pattern for diflubenzuron. Diflubenzuron (A) degrades initially to 2,6-difluorobenzoic acid (B) and 4-chlorophenylurea (C). 2,6-difluorobenzoic acid (B) degrades to 2,6-difluorobenzamide (D) and 4-chlorophenylurea (C) degrades to 4-chloroaniline (E).

half the 2,6-difluorobenzoyl moiety excreted in urine was conjugated to glycine, but in rat the acid was excreted largely unchanged. In sheep, where cleavage splitting of the diflubenzuron molecule was the primary metabolic route, there was no evidence of 4-chlorophenylurea or 4-chloroaniline in urine.

The benzoylphenyl ureas – including diflubenzuron – control target insect populations at extremely low doses by selectively inhibiting their ability to synthesize chitin-bearing parts. Ingested diflubenzuron has no apparent adverse effects until the molting process is underway. Diflubenzuron caused increases in cuticle chitinase and cuticle phenoloxidase activity, producing a softened endocuticle through reduction of its chitin content and a hardened exocuticle as a result of increased phenoloxidase activity. Diflubenzuron inhibits serine protease, thus blocking the conversion of chitin synthetase zymogen into an active enzyme. Insect larvae treated with diflubenzuron develop cuticles that are unable to withstand the increased turgor occurring during ecdysis and that fail to provide sufficient muscular support during molting. These larvae are unable to cast their exuviae, resulting in death from starvation or rupture of the new, delicate, malformed cuticle. In addition to terrestrial insects, diflubenzuron is toxic to a wide variety of aquatic insects and crustaceans, but it does not seem to affect other organisms that contain chitin, including fungi and marine diatoms.

Chitin is a polymer of *N*-acetylglucosamine (AGA), and it rivals cellulose as the most abundant biopolymer in nature. Measured chitin concentrations in marine waters range between 4 and 21 μg/L and planktonic crustaceans are the most significant source of chitin in the sea. Insect chitin is synthesized during phosphorylation by uridine diphospho *N*-acetylglucosamine (UDPAGA) – the immediate precursor of chitin. Diflubenzuron inhibits the incorporation of chitin precursors into chitin, with a resultant accumulation of UDPAGA. Chitin is not found in vertebrates, although several important polysaccharides similar to chitin are found, including hyaluronic acid (HA). Hyaluronic acid is found in skin, synovial fluid, connective tissue, vitreous humor, and the covering of the ovum. Hyaluronic acid is a polysaccharide compound of alternating groups of glucuronic acid and AGA; the immediate precursor for glucuronic acid is uridine diphospho-glucuronic acid and that for AGA is UDPAGA. Because UDPAGA is used in the synthesis of chitin by insects and of HA by vertebrates, and because diflubenzuron interferes with the incorporation of UDPAGA into chitin by insects, diflubenzuron may interfere with the formation of HA in birds; this will be discussed later.

13.2.2 Persistence in Soil and Water

Mobility and leachability of diflubenzuron in soils is low, and residues are usually not detectable after 7 days; in water half-time persistence (Tb1/2) is usually less than 8 days and lowest at elevated temperatures, alkaline pH, and high sediment loadings. Increased concentrations of diflubenzuron in soils and waters are associated with increased application frequency, flooding of treated supratidal areas, wind drift, and excessive rainfall.

Diflubenzuron is persistent in postharvest soils during winter and spring months, especially if associated with plant litter; concentrations decline rapidly with the onset of high summer temperatures to <0.3 mg/kg DW soil in summer. Diflubenzuron particle size and soil flora may be important in the soil degradation process. Diflubenzuron adsorbed to smaller particles of 2 μm diameter had a short Tb1/2 of 3–7 days; diflubenzuron adsorbed to larger particles (10 μm diameter) persisted for 8–16 weeks. Diflubenzuron adsorbed to particles of 2 μm diameter had a low rate of degradation in sterile soils (<6% in 4 weeks), but in nonsterile soils 98% degraded in the same period, suggesting that soil bacteria are important in the degradation process. In Canada, data on mobility of a pesticidal chemical in forest soil must be collected before it can be registered for use under the Canadian Pest Control Products Act in order to assess its potential for groundwater contamination. Diflubenzuron used properly in forest management is unlikely to be leached into groundwater from a site of application.

Water concentrations of diflubenzuron in treated ponds are significantly higher in surface and middle samples than in bottom samples during the first 5 h after treatment; however, after 24 h, distribution is about the same for all depths. Diflubenzuron persists for only a few days in pasture waters at 22–45 g/ha applied to control pasture mosquitos (*Aedes nigromaculis*, *A. melanimon*); hydrolysis and adsorption onto organic matter limit persistence in water. Aerial spraying of 70 g/ha in a forest ecosystem resulted in pond water concentrations of 5.9–13.8 μg/L, which declined to <0.05 μg/L within 16–20 days. Water temperature and pH significantly affect persistence of diflubenzuron. Degradation is most rapid at elevated temperatures and alkaline pH values. Half-time persistence of diflubenzuron at pH 7.7 and various thermal regimes is 8 days at 38°C, 35 days at 24°C, and 29 days at 10°C; at pH 10, Tb1/2 values are 2 days at 38°C, 14 days at 24°C, and 32 days at 10°C; degradation is negligible at pH 4, and at low temperatures regardless of pH. In water, as in soil, small-particle (2–5 μm diameter) diflubenzuron formulations, such as WP-25%, degrade rapidly, usually in 2–8 days. Larger particle sand-granule formulations, developed for use in mosquito control programs wherein the compound needs to penetrate thick vegetation to reach the water, reduce drift during application, and also provide slower release of diflubenzuron into aquatic habitats.

The presence of sediments in diflubenzuron marine microcosms results in rapid removal from seawater and ultimately a reduction in mortality of larval crustaceans. But marine sediments that exceed 200.0 μg diflubenzuron/kg – levels normally encountered at application rates for control of salt marsh mosquitos – could be detrimental to juvenile and adult crustaceans that consume detritus and organic matter on the surface of the marsh or at the water–sediment interface.

13.3 Uses

Diflubenzuron effectively inhibits molting in many species of insect pests, especially among the lepidoptera, coleoptera, and diptera. In the United States, diflubenzuron was approved for use by the U.S. Environmental Protection Agency (EPA) against the gypsy moth in 1976, the cotton boll weevil in 1979, and foliar feeders on soybeans in 1982. By 1989 diflubenzuron was also registered for domestic use against mosquitos, forest lepidoptera, mushroom flies, and certain leaf-eating insect pests of citrus, woody ornamentals, vegetables, and fruits. In 1990, over 269,000 ha of forest were treated with diflubenzuron to suppress gypsy moth and tent caterpillar (*Malacosoma disstria*) populations.

In Europe and elsewhere, diflubenzuron is used in a variety of ways not permitted in the United States. For example, diflubenzuron and other insect growth regulators are fed as admixtures to rations of chickens, cattle, and swine in order to control fly larvae breeding in their manures, and also as a spray directly on manures prior to disposal. Diflubenzuron has been administered orally as a bolus to beef cattle for control of face flies (*Musca autumnalis*) and horn flies (*Haematobia irritans*), two serious pests of cattle in North America; immature insects develop in fresh manure on open pasture. A single bolus released diflubenzuron into feces that killed horn and face fly larvae for 8 weeks and remained partially effective for 16 weeks.

Three diflubenzuron formulations are now in general use: an oil-dispersible concentrate, a wettable powder (WP), and granules. Granular formulations are produced by applying diflubenzuron to sand granules. Since technical diflubenzuron (99.5% pure) is a crystalline material that is almost insoluble in water (i.e., 0.1 mg/L at 20°C), it is usually dispersed in an organic solvent carrier. Wettable powders (25% active ingredients), however, are dispersed in water for use in many commercial applications; diflubenzuron particle size in WP-25 formulations usually ranges between 2 and 5 μm.

13.4 Lethal and Sublethal Effects

Diflubenzuron applied to foliage of terrestrial plants tends to remain adsorbed for several weeks with little or no absorption or translocation from plant surfaces; loss is mainly by wind abrasion, rain washing, or shedding of senescent leaves. Among insect species, there is great variability in sensitivity to diflubenzuron. In general, diflubenzuron is toxic to early life stages of insects at concentrations as low as 0.1 mg/kg diet and at topical applications between 0.003 and 0.034 μg/larvae. Among aquatic organisms, early developmental stages of crustaceans and insects are the most sensitive groups tested; adverse effects on growth, survival, reproduction, and behavior occur between 0.062 and 2.0 μg/L. Groups highly resistant to diflubenzuron include the algae, gastropods, fishes, and amphibians. Birds are comparatively resistant: acute oral LD50 values exceed 2000.0 mg diflubenzuron/kg body weight (BW), and dietary levels of 4640.0 mg/kg ration are tolerated for 8 days. Also, forest birds seem unharmed by recommended diflubenzuron application procedures to control pestiferous insects. No data are available on mammalian wildlife. However, studies with small laboratory animals and domestic livestock suggest a high degree of resistance. No observable adverse effects occur in cows given 0.25 mg/kg BW daily for 4 months, in rabbits given 4.0 mg/kg BW daily on days 6–18 of gestation, in dogs fed diets containing 40.0 mg/kg for 13 weeks (equivalent to 1.6 mg/kg BW daily), in rats fed diets containing 160.0 mg/kg for 2 years, and in rabbits and rodents given single oral or dermal doses <2000.0 g/kg BW. These points are discussed later.

13.4.1 Terrestrial Plants

There is little to no absorption and translocation of diflubenzuron residues from plant surfaces. Due to its stability and low volatility, diflubenzuron residues adhering to plant surfaces are removed primarily through physical effects such as wind abrasion, rain washing, or the loss of dead leaves. A greenhouse study with corn (*Zea mays*), soybeans (*Glycine max*), cabbage (*Brassica oleracea capitata*), and apples (*Malus* sp.) showed no significant degradation of diflubenzuron residues in leaves for up to 16 weeks after treatment. In a study with radiolabeled diflubenzuron, a single dose applied to a cotton (*Gossypium hirsutum*) leaf showed <5% photodegradation in 4 weeks, <7% absorption in 7 weeks, <50% loss to weathering or volatilization in 4 weeks in samples not exposed to rain, and 77% loss in 3 weeks after a heavy rainfall. Edible portions of rotational crops treated repeatedly with diflubenzuron at recommended application levels had low, but detectable, residues. Maximum concentrations, in milligrams per kilogram dry weight, were always <0.01 in wheat (*Triticum* spp.), <0.02 in cotton, <0.09 in collards (*Brassica* spp.), and <0.16 in radish (*Raphanus* spp.).

Diflubenzuron applied aerially to forest leaves in the spring growing season did not persist when the leaves were placed in stream water. Residues on oak leaves decreased 36% in July, and 23% in August within the first

48 h of stream incubation, reaching >90% loss within 3 weeks. However, in December, after 54 days in the stream, there was no significant loss from leaves of red maple, oak, or poplar. The persistence of diflubenzuron to forest leaves under winter conditions is attributed to the increased stability of diflubenzuron at cold temperatures, the greater retention to leaves at low temperatures, and the reduction in microbial degradation. In view of the persistence of diflubenzuron on hardwood leaves at low stream temperatures, nontarget aquatic organisms that consume these leaves may be exposed for extended periods with possible adverse effects.

Foliage of cotton that initially contained 100.0 mg/kg DW contained about 60.0 mg/kg after 7 weeks; leaf residues consisted entirely of the parent diflubenzuron. Diflubenzuron applied topically to lima bean (*Phaseolus lunatus*) foliage was not absorbed by the plant, as expected. Injected diflubenzuron, however, was metabolized, and certain metabolites were similar to those isolated from mites.

Diflubenzuron mixed into compost layers of the cultivated mushroom (*Agaricus bisporus*) at 30.0 mg/kg compost to control dipteran pests of mushroom resulted in increased yield and size; however, at higher concentrations of 180.0 and 1080.0 mg/kg, mushroom yield and number were reduced, and this became more severe over time. Frequent applications of diflubenzuron to agricultural soils are not detrimental to nitrogen-fixing bacteria (i.e., *Azotobacter vinelandii*), and high concentrations could stimulate nitrogenase activity in soils. This conclusion is based on a study – nonsterile agricultural soils and sterilized soils inoculated with *A. vinelandii*. At diflubenzuron loadings between 100.0 and 500.0 mg/kg all concentrations tested had a stimulatory effect on nitrogen fixation in both soils.

13.4.2 Terrestrial Invertebrates

Diflubenzuron is most toxic to early life stages of some insects at 0.1 mg/kg diet, 0.034 μg/larvae (about 3.1 mg/kg BW), or in combination with various chemicals. Some beneficial insects, such as the honeybee (*Apis mellifera*), are adversely affected at dietary concentrations of 1.0 mg/kg for 12 weeks, 10.0 mg/kg for 10 weeks, or 59.0 mg/kg for 10 days. At 28.0–56.0 g/ha (0.025–0.05 pounds/acre), diflubenzuron effectively controls mosquitos for 8–15 days, especially organophosphorus insecticide-resistant strains of salt marsh mosquitos in California. Diflubenzuron was also effective in controlling strains of housefly (*Musca domestica*) that were resistant to organochlorine, organophosphorus, carbamate, and pyrethroid insecticides on a United Kingdom pig farm; 416.0 mg/m^2 to slurry pots of pig weaning rooms gave effective control 2–4 weeks after application.

Chemical control of larvae of gypsy moth and other forest-insect defoliators may cause indiscriminate reduction of nontarget arthropods, which, in turn, may affect food resources of forest birds and small mammals. This problem is of special concern in West Virginia, where two species of endangered bats (Indiana bat, *Myotis sodalis*; eastern big-eared bat, *Plecotus phyllotis*) occur in areas threatened by gypsy moth defoliation. Diflubenzuron applications, usually at 70.0 g/ha on two consecutive days, controlled gypsy moth larvae and also significantly reduced populations of canopy macrolepidoptera and nonlepidopteran mandibulate herbivores. Sucking herbivorous insects, microlepidoptera, and predaceous arthropods, however, were relatively unaffected, which suggests that although diflubenzuron can potentially affect food supply of forest birds and small mammals, these effects are probably minimal.

Researchers generally agree that diflubenzuron causes incomplete ecdysis by interfering with chitin synthesis. Diflubenzuron at lethal concentrations, however, causes an effect in chironomid larvae (*Chironomus decorus*, *Tanypus grodhausi*) other than inhibition of chitin synthetase, as judged by histopathology of the alimentary canal, especially the ventriculus. Dysfunction of the ventriculus, an organ that normally lacks chitin, results in a general breakdown of the digestive apparatus of exposed chironomid larvae. Exposure of nematodes and adults of several insect species, including boll weevil, housefly, and stable fly (*Stomoxys calcitrans*), to diflubenzuron results in deposition of eggs that appear normal but fail to hatch. This effect seems to be due to an ovicidal action and not to sterility of the treated adults, since the larvae appear to undergo normal development within the egg. Secretion of unmetabolized diflubenzuron into the eggs apparently accounts for observed ovicidal effects. Treated female boll weevils began to lay viable eggs 12 days after treatment and became as productive as controls in 24 days; additional treatment is required to maintain a significant suppression of egg hatch.

Diflubenzuron is the most investigated benzoylphenyl urea and has shown excellent potency for controlling mosquitos and certain lepidopterous and coleopterous pests. Some insect species, however, cannot be controlled efficiently by diflubenzuron. For example, the cotton leafworm (*Spodoptera littoralis*) is comparatively resistant because of reduced penetration through the exoskeleton, rapid elimination of unchanged diflubenzuron, and rapid metabolism, which occurs mainly through hydrolysis. To combat *Spodoptera* and other resistant pests, new benzoylphenyl urea compounds have been developed, including chlorfluzuaron, teflubenzuron, and hexafluron.

Beneficial insects associated with fruit orchards show different responses to diflubenzuron treatment. Lacewings (*Chrysopa oculata*) in contact with leaves containing 300.0 mg/kg DW had reduced survival and inhibited molting of first instar larvae, but the assassin bug (*Acholla multispinosa*) was not affected by contact with treated leaves. Lacewings and other beneficial predator insects fed diflubenzuron-treated two-spotted spider mites (*Tetranychus urticae*) for 3 days showed no adverse effects after 14 days. Spraying of diflubenzuron at 28.0 g/ha to control gypsy moth did not affect *Cotesia melanoscela*, a hymenopterid predator of the gypsy moth; however, another natural enemy, a virus, was adversely affected. Certain arthropod predators were unaffected by diflubenzuron at 70.0 mg/ha applied four times in 3 weeks to control the boll weevil; these include the convergent lady bug beetle (*Hippodamia convergens*), the big-eyed bug (*Geocoris punctipes*), and various species of *Coleomegilla*, *Orius*, *Nabis*, and *Chrysopa*.

Diflubenzuron can be either hydrolyzed at the urea bridge or oxidized by ring hydroxylation followed by conjugation. Hydrolytic cleavage seems to be a major route for diflubenzuron metabolism in many insect species. Two-spotted spider mites showed <10% absorption in 96 h of topically applied diflubenzuron. Of the amount absorbed, about 27% was metabolized in 96 h to 4-chlorophenyl urea, 2,6-difluorobenzoic acid, 4-chloroformanilide, 2,6-difluorobenzamide, and other metabolites. Effects of diflubenzuron were synergized by profenofos in cotton leafworm fourth instar larvae, and they were antagonized by 20-hydroxyecdysone in beetle (*Tenebrio molitor*) pupae. More information is needed on interaction effects of diflubenzuron with other chemicals.

13.4.3 Aquatic Organisms: Laboratory Studies

Studies with diflubenzuron and representative aquatic organisms under controlled conditions show several trends: (1) crustaceans are the most sensitive group of nontarget organisms tested. Adverse effects on growth, survival, reproduction, and behavior of copepods, shrimp, daphnids, amphipods, and crabs occur between 0.062 and 2.0 μg/L medium, and early developmental stages were the most vulnerable; (2) next in sensitivity are aquatic insects, including mayflies, chironomids, caddis flies, and midges. Diflubenzuron concentrations between 0.1 and 1.9 μg/L medium produce low emergence and survival; (3) other groups tested are comparatively resistant (i.e., adverse effects occur at >45.0 μg/L). In fish, for example, death occurred at >33,000.0 μg/L; and (4) elevated accumulations occur in aquatic plants during exposure to 100.0 μg/L and in fish during exposure between 1.0 and 13.0 μg/L. All species in these groups, however, seemed unaffected by elevated body burdens, as judged by normal growth and metabolism.

The major degradation products of diflubenzuron in water are 4-chlorophenylurea and 2,6-difluorobenzoic acid; these compounds are less toxic to aquatic organisms than the parent chemical. A minor metabolite, 4-chloroaniline, which is classified as a mutagen by The National Cancer Institute, and the Cancer Assessment group of the U.S. EPA, is significantly more toxic to fish and *Euglena gracilis* than is diflubenzuron. For example, LC50 (96 h) values for 4-chloroaniline and four species of freshwater teleosts are 16–56 times lower than comparable data for diflubenzuron, but 4-chloroaniline is 76 times less toxic to *Chironomus* midge larvae in 48 h than diflubenzuron. There is a dose-dependent effect of 4-chloroaniline on *Euglena* growth inhibition and glycine metabolism in the range of 1.0–200.0 mg/L during exposure for 30 h. The most sensitive organism to 4-chloroaniline was bluegill (*Lepomis macrochirus*) with an LC50 (96 h) value of 2.3 mg/L. It is highly unlikely, however, that this concentration will be encountered under recommended diflubenzuron application practices.

Diflubenzuron inhibits several enzyme systems in crab and insect larvae, resulting in disrupted glucose metabolism, reduced AGA incorporation into cuticle, and ultrastructural deformities of chitinous components of the cuticle. Specifically, diflubenzuron inhibits chitin synthetase, a magnesium-requiring enzyme that catalyzes the transfer of *N*-acetyl-D-glucosamine to chitin; the final result is relatively large accumulations of AGAs.

Diflubenzuron acts specifically on insects and crustaceans as a larvicide by interfering with chitin deposition into cuticles during juvenile development through ecdysis. The biosynthesis of chitin in arthropods is under hormonal control. Arthropods increase in size by resorbing a portion of the shell and initiating the secretion of a new exoskeleton under the old cuticle. At this time, chitin synthesis is maximal. After completion of about half the new shell, molting occurs, the old shell is discarded, and the new shell is synthesized. Diflubenzuron exposure produces disturbances in the cuticular structure, weakening the cuticle so that it fails mechanically during ecdysis of insects and crustaceans. In general, treated larvae appear healthy during the entire intermolt period until molting commences, at which time many larvae are unable to cast their molts completely and die within a few hours. Several genera of diatoms, including *Thalassiosira* and *Skeletonema*, produce up to 33% of their biomass as chitin. These diatoms synthesize chitin strands that extend outside their frustules to increase buoyancy. Chitin-producing diatoms, as well as nonchitanaceous diatoms, are seemingly unaffected at elevated concentrations of 1.0 mg/L for periods up to 14 days. Some species of algae, especially *Plectonema boryanum*, are reported to degrade diflubenzuron efficiently, but this requires verification.

Studies with laboratory stream communities dosed for 5 months confirm that insects and crustaceans are the most severely affected groups; adverse effects occur in the range of 1.0–1.1 μg diflubenzuron/L. Fish and mollusks, however, show no adverse effects at 45.0 μg/L. Freshwater clams (*Anodonta cygnea*) exposed to high concentrations of diflubenzuron for lengthy periods may experience blocked polycondensation reactions to chitin chains in the outer mantle epithelium secretory cells, producing unstabilized chitin and increasing shell fragility. On this basis, the comparatively resistant burrowing bivalve mollusks may be at risk if exposed over several calcification periods. Fish accumulated diflubenzuron from water up to 160 times water levels, but tissue concentrations during exposure declined steadily over time.

Exposure of *Aedes albopictus*, a mosquito vector of dengue and encephalitis in Taiwan, for 24 h to 0.00025–25.0 μg/L diflubenzuron resulted in dose-dependent aberrations in larvae, pupae, and adults. In general, most treated second and third instar *Aedes* larvae died during molting, while most fourth instar larvae developed abnormally. Unfortunately, levels of diflubenzuron used to control saltwater mosquitos and other insects are also toxic to zoeal stages of crustaceans and adversely affect growth and reproduction of adults. Treated larvae of estuarine crustaceans are characterized by the following: histological alterations in the cuticular layers of the exoskeleton at concentrations as low as 1.0 μg/L, higher mortality associated with molting and gross morphological deformities at concentrations as low as 0.5 μg/L, and behavioral modifications at concentrations as low as 0.1 μg/L. Behavioral effects in fiddler crabs (*Uca pugilator*) were the most sensitive indicator of diflubenzuron stress, and these effects potentially may influence the ability of juvenile crabs to avoid predation, construct burrows, or feed adequately in nature. Behavioral effects on cladocerans that may result in latent mortality include reduced filter feeding rates, reduced body movements, and inability to exhibit positive phototaxis, a characteristic of untreated individuals. Shrimp larvae exposed to >2.5 μg/L will not undergo daily vertical migration, and those exposed to 1.0 μg/L undergo only limited migration, which could affect horizontal transport and dispersal of populations and reduce recruitment to benthopelagic adult populations. In addition to its inhibitory effect on cuticle synthesis, diflubenzuron affects hormone balance by delaying or arresting the molt cycle, and it inhibits limb regeneration by inhibiting mitosis and differentiation. Regenerated limbs of diflubenzuron-stressed crabs that survived ecdysis had lesions in the form of black areas in which the cuticle was improperly developed. Also, diflubenzuron caused a reduction in metabolism of beta-ecdysone in larval insects, leading to an excess of this molting hormone in the tissues. Treatment of decapod crustaceans with ecdysones frequently causes high mortality and molt acceleration.

Toxicity and persistence of diflubenzuron in aquatic environments depend on formulation, frequency of application, quantity of organic matter, sediment type, and water pH and temperature. Biological variables are more important than physical variables in assessing diflubenzuron toxicity, especially the age of the test organism, and frequency and synchrony of molting during the exposure period. Crustaceans and other organisms that molt do not demonstrate a typical survival dose–response curve against diflubenzuron because death occurs only when molting is blocked. In general, the most sensitive species had comparatively short larval or nymphal periods, and the organism molted frequently. Susceptible species include mayflies (*Leptophlebia* sp., *Baetis pygmaeus*), while more resistant species includes a stonefly (*Paragnetina media*) and caddis fly (*Hydropsyche bettani*). Amphipods were especially sensitive at 25°C, but not at 10, 15, or 20°C.

Mortality patterns of megalops larvae of blue crab (*Callinectes sapidus*) were elevated at higher temperatures but were seemingly unaffected by water salinity. In studies on larvae of black fly (*Simulium vittatum*), diflubenzuron was more effective (1) against earlier larval instar stages than later ones, (2) against rapidly growing larvae than starved, slow-growing larvae, and (3) at 25°C than at 20°C. Among diflubenzuron-stressed barnacles (*Balanus eburneus*), mortality was higher in fed groups than in starved groups, perhaps due to an increased uptake from contaminated food or to an increased molting rate due to feeding. Increased fragility of cast exuviae from diflubenzuron-treated barnacles suggests mechanical weakening of the cuticle due to a decrease in chitin content.

13.4.4 Aquatic Organisms: Field Studies

Field use of diflubenzuron in aquatic habitats for control of pestiferous insects also affects other species. Diflubenzuron applications in marshes, ponds, streams, lakes, and rice fields routinely cause population reductions – sometimes irreversible – in many species of nontarget organisms, especially crustaceans and aquatic insects. Taxonomic groups that seem comparatively tolerant to diflubenzuron include algae, turbellarians, rotifers, aquatic beetles, mollusks, annelid worms, ostracods, and fish. Following multiple applications to lake and pond ecosystems, diflubenzuron was not measurable in water, sediment, and aquatic vegetation after several days. Algae (*Plectonema boryanum*) reportedly degrade 80% of absorbed diflubenzuron in 1 h, primarily to 4-chlorophenylurea and 4-chloroaniline.

Most authorities agree on four points: (1) rates as low as 28.0–56.0 g diflubenzuron/surface ha (0.025–0.05 pounds/surface acre), or 2.5–16.0 μg/L, are highly effective against pestiferous dipterans, including many species of chaoborids, chironomids, and culicids; (2) these same dosages suppress nontarget populations of cladocerans, copepods, mayfly nymphs, corixids, and springtails; (3) moderately resistant to diflubenzuron are larvae of diving beetles, dragonfly adults and naiads, ostracods (*Cypericercus*, *Cyprinotus*), backswimmers, and water boatmen; highly resistant species include mosquito fish (*Gambusia affinis*), frogs and toads, snails, and algae; and (4) all populations of survivors begin to recover within days or weeks, and recovery is usually complete within 80 days after the last treatment.

Unlike laboratory studies, diflubenzuron does not bioaccumulate markedly in fish or biomagnify through food chains, although altered feeding habits may occur. Under field conditions, marsh or pond sediments usually contain <50.0 μg/kg fresh weight (FW). This concentration presents negligible risk to channel catfish (*Ictalurus punctatus*) over a 28-day period, suggesting little hazard to catfish during multiple mosquito control applications of diflubenzuron. Bioaccumulation of diflubenzuron from marsh applications are minimal, as judged by results of uptake studies using marsh sediments containing 550.0 μg/kg; maximum residues in fish tissues after 3 days were 4.0 μg/kg FW in muscle and 10.0 μg/kg DW in viscera. Diflubenzuron residues are moderately persistent in algae, snails, salt marsh caterpillars (*Estigmene* spp.), and mosquito larvae but are not biomagnified in food chains ending in fish. Maximum diflubenzuron concentrations range from 50.0 to 720.0 μg/kg FW in whole body of three species of freshwater teleosts exposed to water treated up to 8 times with 135.0 g/ha. Feeding habits of freshwater fishes change in ponds showing marked reductions (94–99%) in copepod and cladoceran populations after diflubenzuron treatment, perhaps due to availability of various food items. In one study, black crappie (*Pomoxis nigromaculatus*) and brown bullhead (*Ictalurus nebulosus*) altered their diets for one month after treatment, eating about 3 times more insects and ostracods, and almost no cladocerans and copepods (usually, major items), than before treatment.

Although diflubenzuron is not sprayed directly on freshwaters in gypsy moth control, aerial spraying of large forest tracts may result in exposure of streams by way of leaf litter. Residual diflubenzuron was present for at least 4 months on leaves submerged in flowing water, and it was toxic to various invertebrates. Thus, treated leaves of the tulip poplar (*Liriodendron tulipifera*) that contain 10.0 mg diflubenzuron/m^2 after 4 months of submersion produce adverse effects on survival and growth when fed to craneflies (*Tipula abdominalis*, *Platycentropus radiatus*). The effects of diflubenzuron on leaf-litter processing rates in streams are unresolved and merit additional research.

13.4.5 Birds

Birds are comparatively resistant to diflubenzuron, as judged by the ability of the mallard (*Anas platyrhynchos*) to tolerate single oral doses up to 2000.0 mg/kg BW or dietary loadings up to 4640.0 mg/kg ration for 8 days. Poisoning of insectivorous birds by diflubenzuron, after spraying in orchards as recommended, is highly improbable. This conclusion is based on the maximum possible daily intake of insects by wild nestlings (15.0 mg/kg BW in Great tit, *Parus major*; 10.0 mg/kg BW in tree sparrow, *Passer montanus*), on a maximum whole body loading of 0.5 mg diflubenzuron/kg FW in insect prey, and on observations of normal growth and subsequent breeding of nestlings in orchards sprayed with diflubenzuron.

Despite the apparent absence of direct effects in forest birds, the widespread use of diflubenzuron in the suppression of forest insect defoliators may lead to potentially harmful effects by reducing populations of immature lepidoptera and other mandibulate herbivorous insects upon which they feed. All field evidence collected to date however, is either inconclusive or negative. In one study, 70.75 g diflubenzuron/ha was applied to an oak forest (*Quercus rubra*, *Q. velutina*, *Q. prinus*) in West Virginia to control first and second instars of gypsy moths. The maximum diflubenzuron residue recorded in a wide variety of canopy forager birds (great crested flycatcher, *Myiarchus crinitus*; eastern wood peewee, *Contopus virens*; black-capped chickadee, *Parus atricapillus*; tufted titmouse, *Parus bicolor*; blue-gray gnatcatcher, *Polioptila caerulea*; red-eye vireo, *Vireo olivaceous*; warblers, *Dendroica* spp.; scarlet tanager, *Piranger olivacea*) was 0.21 mg/kg whole body FW. A similar value, 0.20 mg/kg whole body FW, was recorded in ground or low foragers, including wood thrush (*Hylocichla mustelina*), ovenbird (*Seiurus aurocapillus*), rufous-sided towhee (*Pipilo erythrophthalmus*), chipping sparrow (*Spizellza passerina*), song sparrow (*Melospiza melodia*), and indigo bunting (*Passerina cyanea*). Neotropical migrants breeding in a diflubenzuron-treated forest had significantly lower fat reserves on a dry weight basis than did conspecifics from reference sites, possibly due to a reduction in food availability and quality. In another study, up to 280.0 mg diflubenzuron/ha applied to control the Douglas-fir tussock moth (*Orygia pseudotsugata*), an important defoliator of true firs (*Abies* spp.) and Douglas-fir (*Pseudotsuga menziesii*) in western North America, had no adverse effects on forest birds, as judged by population censuses, nesting studies, and bird behavior.

Domestic chickens (*Gallus* sp.) metabolize diflubenzuron to a greater extent than insects, but less than rodents and ruminants. The main pathway of diflubenzuron degradation in chickens is through cleavage of the urea bridge, whereas rats and cows tend to hydroxylate and conjugate the parent molecule. Metabolism studies in chickens showed that major residues in tissues and eggs were unchanged diflubenzuron and 4-chlorophenylurea; also present were 2,6-difluorobenzoic acid and 4-chloroaniline. Metabolites in chicken excreta included 4-chlorophenylurea, 4-chloroaniline, 2,6-difluorobenzamide, 2,6-difluorobenzoic acid, and several unidentified compounds. At high dietary loadings of 50.0–500.0 mg/kg ration, diflubenzuron accumulates in fat, egg, and muscle tissues of chickens; however, excretion is rapid and residues are usually negligible after 5 weeks on a clean diet. Diflubenzuron fed at levels up to 250.0 mg/kg ration to male broiler chickens for 98 days had no effect on HA concentration in the combs and wattles. Both chitin and HA are polysaccharides and have a common biochemical precursor, UDPAGA, which is used in the synthesis of chitin by insects and in the production of HA by vertebrates. Since diflubenzuron interferes with the incorporation of UDPAGA into chitin by insects but not with HA production, it would seem that diflubenzuron is relatively harmless to birds; however, more research is needed for verification.

Intraspecies differences in diflubenzuron metabolism are reported for domestic chickens. The White Leghorn breed, for example, produced eggs with significantly higher residues than other breeds tested after 3 weeks on a diet containing 50.0 mg diflubenzuron/kg, and it had elevated concentrations in fat tissues after 15 weeks on a 10.0 mg/kg diet. In chickens, diflubenzuron is usually eliminated more rapidly in feces than in eggs, but in the White Leghorn breed the major route of elimination is via egg production. The White Leghorn breed also differed significantly from the Rhode Island Red/Barred Plymouth Rock (RIR/BPR) breed in ability to metabolize diflubenzuron administered orally or intravenously. White Leghorn chickens accumulated diflubenzuron to a greater extent than RIR/BPR chickens, and they retained residues for longer periods. Also, White Leghorn chickens produced a higher percentage and greater number of diflubenzuron metabolites in their excreta than other breeds tested. Differences in ability to metabolize diflubenzuron between different strains of domestic chickens may be due to differences in lipid metabolism associated with egg production. No comparable database exists for avian wildlife, and one should be developed through research.

13.4.6 Mammals

No data are available on effects of diflubenzuron on mammalian wildlife. However, results of studies on small laboratory animals and domestic livestock are available, and these indicate several trends. Adverse effect levels occurred in dogs fed diets containing 160.0 mg/kg (6.2 mg/kg BW daily) for 13 weeks (abnormal blood chemistry), in mice given 125.0 mg/kg BW daily for 30 days (hepatocellular changes), in rabbits fed diets containing 640.0 mg/kg for 3 weeks (abnormal hemoglobin), and in rats given 5000.0 mg/kg BW daily for 13 weeks (abnormal hemoglobin). Accumulations of diflubenzuron occurred in several species. Elevated tissue residues – but no other measurable effects – occurred in cows given 0.05–0.5 mg/kg ration for 28 days or 1.0–16.0 mg/kg BW for 4 months, in pigs given a single oral dose of 5.0 mg/kg BW, and in sheep given a single oral dose of 10.0 mg/kg BW. No-observable-adverse-effect levels (NOAELs) occurred in cows given 0.25 mg/kg BW daily for 4 months, in rabbits given 4.0 mg/kg BW daily on days 6–18 of gestation, in dogs fed diets containing 40.0 mg/kg for 13 weeks (equivalent to 1.6 mg/kg BW daily), in rats fed diets containing 160.0 mg/kg for 2 years, and in rabbits and rodents given single oral or dermal doses <2000.0 mg/kg BW.

All available data indicate that diflubenzuron is not a mutagen, teratogen, or carcinogen. Diflubenzuron is not mutagenic, as judged by the results of (1) the mouse lymphoma forward mutation test at the thymidine kinase locus (detects mutations to a nonfunctional thymidine kinase in a line of culture mouse lymphoma cells), (2) the Ames *Salmonella typhimurium* microsome reverse mutation test (ability to produce point gene mutations of a base pair), (3) the mouse micronucleus test (which detects chromosome breakage or chromosome loss from mitotic abnormalities in bone marrow erythrocytes), and (4) a DNA damage study with yeast, *Saccharomyces cerevisiae*. No teratogenicity or reproductive effects were associated with elevated doses of diflubenzuron in all species of mammals tested. Diflubenzuron suppresses melanogenesis and uptake of nucleosides in mouse melanoma cells, and it inhibits growth of experimental tumors in mice, either alone or in combination with $CoCl_2$. Mixed-function oxidase, induced by 3-methylcholanthrene, enhances the antitumor properties of diflubenzuron, suggesting that aromatic hydroxylation may be required for tumor growth regulation. The most likely diflubenzuron metabolite that affects tumor growth regulation is the form oxidized at the 2 carbon of the phenyl ring; other metabolites tested (i.e., 4-chlorophenylurea, 3-OH-diflubenzuron) are only marginally effective. Diflubenzuron did not produce tumors in fetal cells of hamsters (*Cricetus* spp.) at whole body doses of 500.0 mg/kg, and this also suggests a relatively low oncogenic potential. Diflubenzuron is not cytotoxic and does not inhibit the synthesis of complex carbohydrates in animal cells, as judged by results of studies with cultured rat glial cells, wherein diflubenzuron was not metabolized to any measurable extent, and more than 98% could be recovered from particulate fractions of whole cells.

Intestinal absorption of diflubenzuron in laboratory rats, measured as the sum of urinary and biliary excretion, decreases with increasing dose: from 50% at a single oral dose of 4.0 mg/kg BW to 4% at 900.0 mg/kg BW. Excretion is almost complete after 75 h; at that time up to 4% of the administered dose is recovered from skinned carcasses. About 80% of diflubenzuron metabolites excreted by rats seem to have the basic diflubenzuron structure intact. Three metabolites are largely excreted as conjugates in the bile. One metabolite, 2,6-difluorobenzoic acid, is excreted largely in urine. Its counterpart, 4-chlorophenylurea, was not present in urine or bile in appreciable quantity, nor was 4-chloroaniline detected. Lifetime feeding studies of 4-chloroaniline, a relatively common diflubenzuron metabolite, showed no compound-related effects in laboratory mice and rats.

Oral treatment of sheep and cattle (*Bos* spp.) with diflubenzuron is followed by absorption of the compound through the gastrointestinal tract, metabolism, and elimination of residues through the urine, feces, and, to a very limited extent, milk. Intact diflubenzuron is eliminated in the feces of orally dosed cattle and sheep. Major metabolites of diflubenzuron excreted by cattle and sheep result from hydroxylation on the difluorobenzoyl and chlorophenyl rings, and by cleavage between the carbonyl and amide groups to produce metabolites that are excreted free or as conjugates. Cattle dosed repeatedly with diflubenzuron had detectable residues only in liver and milk. The parent compound, 4-chlorophenylurea, 2,6-difluorobenzoic acid, and 4-chloroaniline compose only 15% of the total residue in liver; the bulk of the residue is not extractable. Dietary levels of 5.0 mg/kg ration produce low (13.0 μg/L), but detectable, diflubenzuron concentrations in milk of cattle. The major hydroxylated diflubenzuron metabolite in cow milk (*N*-[[(4-chlorophenyl) amino] carbonyl]-2,6-difluoro-3-hydroxybenzamide) when fed to white rats is rapidly excreted with little biotransformation.

Metabolism of diflubenzuron by mammals and birds probably occurs by way of hydroxylation, conjugation, and cleavage of the urea moiety; however, interspecies differences are considerable. In cows, for example, the major identified metabolic transformation is hydroxylation at the 3 position of the 2,6-difluorobenzoyl ring. In sheep, however, major metabolites arise through cleavage of the amide bond at the benzoyl carbon to produce 2,6-diflurobenzoic acid, which is excreted in the urine either free or conjugated with glycine. The major diflubenzuron metabolite in cow urine is 2,6-difluoro-3-hydroxydiflubenzuron, accounting for 45%, and in feces 18%; unchanged diflubenzuron accounts for 43% of the administered dose in cow feces. In sheep urine, 2,6-difluorobenzoic acid and 2,6-difluorohippuric acid account for 57%; in sheep feces, unchanged diflubenzuron is 97%. In swine, the great majority of the administered dose is eliminated in feces unchanged; the urine contains mostly metabolites, indicating that most of the absorbed diflubenzuron is metabolized.

13.5 Recommendations

Since diflubenzuron toxicity seems to be similar in both insects and crustaceans, extreme care must be taken when this compound and other chitin synthesis inhibitors are used for insect control in areas where aquatic crustaceans occur. Otherwise, ecological instability may result, with consequences for feeding, metabolism, growth, reproduction, and survival of numerous nontarget organisms. Specifically, diflubenzuron use in salt marsh mosquito breeding areas or on agricultural lands less than 5 km from coastal areas is not recommended because of concerns that runoff may reach the adjacent estuaries, which are the primary hatcheries for many economically important species of crustaceans. Also, diflubenzuron concentrations in seawater should not exceed 0.1 μg/L, the minimum concentration known to produce measurable behavioral changes in estuarine crustacean larvae. Concentrations of 2.5 μg diflubenzuron/L and higher in freshwater are known to reduce arthropod populations by 67–71%, resulting in reduced growth of young of year bluegills; growth inhibition, in turn, may result in greater starvation, increased predation, reduction in over-winter survival, and diminished to poor recruitment. If diflubenzuron and other insect growth regulators continue to be used near productive aquatic habitats, then food chain transfer studies are recommended. High accumulations of diflubenzuron by aquatic algae – up to 4.5 mg/kg DW in some cases – strongly implicate food chain transfer as a potential mechanism of contaminant transfer in aquatic invertebrate food webs. To protect certain fishes, diflubenzuron use to control copepod vectors of human disease – including various species of *Cyclops* – is not recommended in areas where these fishes breed or feed on *Cyclops*.

For control of cotton pests, including the boll weevil, a maximum recommended treatment schedule is 421.0 g diflubenzuron/ha, applied 6 times, usually weekly, during the growing season. Honeybees (*Apis mellifera*) in heavily sprayed areas, however, may experience adverse effects if their diets exceed 1.0 mg diflubenzuron/kg FW. Diflubenzuron inhibits housefly development in poultry manure. A recommended cost-effective fly control program in poultry houses involves the feed-through method (5.0 mg diflubenzuron/kg FW poultry diet) during hot, wet summers for 3–4 months, coupled with good sanitation and good manure management.

For protection of domestic cattle, feeds should contain <0.05 mg diflubenzuron/kg FW; cottonseed may be added to cattle diets provided that diflubenzuron concentrations in the seed do not exceed 0.2 mg/kg FW and that cottonseed composes <17% of the total diet bulk. Diflubenzuron causes biochemical upset, as judged by lowered testosterone levels in chickens and rats, altered glutathione-*S*-transferase activity in mouse liver (which adversely affects the ability to detoxify foreign substances by way of conjugation), and disrupted hydroxylamine activity in human infants. Additional research seems needed on biochemical alterations induced by diflubenzuron. No diflubenzuron criteria are currently recommended for protection of avian and mammalian wildlife. All data available suggest that wildlife species are about as tolerant to diflubenzuron as are domestic poultry and livestock; however, the wildlife database seems inadequate for practicable criteria formulation.

Anti-cancer properties of diflubenzuron require elucidation. The indication that one or more hydroxylated forms of diflubenzuron can regulate growth of mouse tumor cells provides a basis for further studies to identify and isolate the most active analog of this compound, and it suggests that other benzoylphenyl ureas may have similar properties.

Diflubenzuron has a Surveillance Index Classification of Class IV, indicating a sufficiently low-hazard potential to human health from toxicological and exposure standpoints to justify only minimal monitoring efforts. Human cancer risk of lifetime dietary exposure to diflubenzuron in a worst case scenario is considered slight. Diflubenzuron has little potential for human dietary exposure because of its limited use on cotton and the low residues measured on cottonseed, meat, milk, poultry, and eggs. For protection of human health, tolerances of <0.05 mg/kg FW have been set for fat, meat, meat by-products, poultry, milk, dairy products, and eggs, and <0.2 mg/kg FW for cottonseed. These foods compose about 45% of the average human diet. If all these foods bore residues at the tolerance level, they would contribute 0.035 mg daily on the basis of 1.5 kg food eaten daily. For a 60-kg adult, the theoretical maximum residue concentration would be 0.6 μg/kg BW daily. Tolerances would be approached only when maximum quantities of cottonseed fraction (i.e., hulls, meal, and soap stock), all bearing tolerance level residues, are incorporated into lives tock diets. At present, however, no acceptable daily intake level in humans has been established.

13.6 Summary

Diflubenzuron (1-(4-chlorophenyl)-3-(2,6-difluorobenzoyl)urea), also known as dimilin, is a potent broad-spectrum insect growth regulator that interferes with chitin synthesis at time of molting and is effective in controlling immature stages of insects. Diflubenzuron was approved for domestic use in 1976 to control gypsy moth (*Lymantria dispar*), and in 1979 against the cotton boll weevil (*Anthonomus grandis*). By 1989, this compound was also registered for domestic use against mosquitos, forest lepidoptera, mushroom flies, and leaf-eating insect pests of citrus, woody ornamentals, vegetables, and fruits.

Diflubenzuron seldom persists for more than a few days in soil and water. When used properly in forest management, it is unlikely to be leached into groundwater from the application site. Degradation in water and soil is most rapid when small particle formulations are applied, microorganisms are abundant, and at elevated pH, temperature, and organic loading. Chemical and biological processes initially yield 2,6-difluorobenzoic acid and 4-chlorophenylurea. Soil degradation processes and plant and animal metabolism involve further conversion of these compounds to 2,6-difluorobenzamide and 4-chloroanaline. Ultimately, the end products are either conjugated into mostly water-soluble products or are biologically methylated.

Diflubenzuron applied to foliage of terrestrial plants tends to remain adsorbed for several weeks with little or no absorption or translocation from plant surfaces; loss occurs mainly from wind abrasion, rain washing, or shedding of senescent leaves. Among terrestrial insects, there is great variability in sensitivity to diflubenzuron. Sensitive pestiferous species of insects die at topical applications of 0.003-0.034 μg/larvae or after consuming diets containing 0.1 mg/kg. Some beneficial insects, such as the honeybee (*Apis mellifera*) are adversely affected at 1.0 mg/kg FW of diet.

Diflubenzuron application rates between 28.0 and 56.0 g/ha (0.025–0.05 pounds/acre) or 2.5–16.0 μg/L are highly effective against pestiferous aquatic dipterans, including representative chaoborids, chironomids, and culicids.

These same dosages temporarily suppress nontarget populations of cladocerans, copepods, mayfly nymphs, corixids, and springtails; population recovery is usually complete within 80 days. In general, crustaceans were the most sensitive nontarget aquatic organisms tested. Adverse effects on crustacean growth, survival, reproduction, and behavior occur between 0.062 and 2.0 μg/L. Next in sensitivity are mayflies, chironomids, caddis flies, and midges; concentrations between 0.1 and 1.9 μg/L produce low emergence and survival. Moderately resistant to diflubenzuron are larvae of diving beetles, dragonfly adults and naiads, ostracods, spiders, backswimmers, and water boatmen. Relatively tolerant of diflubenzuron (i.e., NOAELs at <45.0 μg/L) are the algae, mollusks, fishes, and amphibians. High accumulations occur on some aquatic plants during exposure to 100.0 μg/L and in fish during exposure to 1.0–13.0 μg/L, but all species in these groups seem unaffected by elevated body burdens and grow and metabolize normally.

Birds seem comparatively resistant to diflubenzuron: acute oral LD50 doses exceed 2000.0 mg/kg BW; dietary concentrations <4640.0 mg/kg FW are tolerated for at least 8 days; and forest birds seem unharmed by recommended diflubenzuron application procedures to control pestiferous insects, except for a possible loss in fat reserves. Intraspecies differences in ability to metabolize diflubenzuron are probably large; different strains of domestic chickens show significant differences in ability to accumulate and retain this compound.

No data were found on diflubenzuron effects on mammalian wildlife. However, studies on small laboratory animals and domestic livestock indicate no observable effects in cows (*Bos bovis*) given 0.25 mg/kg BW daily for 4 months, in rabbits (*Oryctolagus cuniculus*) given 4.0 mg/kg BW daily on days 6–18 of gestation, in dogs (*Canis familiaris*) fed diets containing 40.0 mg/kg for 13 weeks (equivalent to 1.6 mg/kg BW daily), in rats (*Rattus* spp.) fed diets containing 160.0 mg/kg for 2 years, and in rabbits and rodents given single oral or dermal doses <2000.0 mg/kg BW. All experimental studies conducted with laboratory animals indicate that diflubenzuron is nonmutagenic, nonteratogenic, and noncarcinogenic. Adverse effects occur in dogs fed diets containing 160.0 mg/kg (6.2 mg/kg BW daily) for 13 weeks (abnormal blood chemistry), in mice (*Mus* spp.) given 125.0 mg/kg BW daily for 30 days (hepatocellular changes), in rabbits fed diets of 640.0 mg/kg for 3 weeks (abnormal hemoglobin), and in rats given 5000.0 mg/kg BW daily for 13 weeks (abnormal hemoglobin). Elevated tissue residues – but no other measurable effects – occur in cows given 0.05–0.5 mg/kg ration for 28 days or 1.0–16.0 mg/kg BW for 4 months, in pigs (*Sus* spp.) given a single oral dose of 5.0 mg/kg BW, and in sheep (*Ovis aries*) given a single oral dose of 10.0 mg/kg BW.

Criteria now recommended for protection of various species include the following: dietary loadings, in milligrams per kilogram fresh weight ration, of <0.05 for human health, <0.05 for livestock, <1.0 for honeybees, and <5.0 for poultry; seawater concentrations <0.1 μg/L for estuarine crustacean larvae; and, for all aquatic life, restricted or prohibited use of diflubenzuron in salt marsh mosquito breeding areas and on agricultural lands less than 5 km from coastal areas. No criteria are available or proposed for protection of avian and mammalian wildlife against diflubenzuron, probably because of an incomplete toxicological database.

DIOXINS[a]

Chapter 14

14.1 Introduction

Accidental contamination of the environment by polychlorinated dibenzo-*para*-dioxins (PCDDs), especially 2,3,7,8-tetrachlorodibenzo-*para*-dioxin (2,3,7,8-TCDD), is the primary cause of poor reproduction of herring gulls (*Larus argentatus*) in Lake Ontario, and of other fish-eating birds in the Great Lakes and on Canada's west coast. Humans exposed to herbicides contaminated with 2,3,7,8-TCDD and higher chlorinated dioxins may experience an increase in overall cancer and specific tumor risk. The sale of striped bass (*Morone saxatilis*) and blue crabs (*Callinectes sapidus*) from Newark Bay, New Jersey, was prohibited because the levels of 2,3,7,8-TCDD exceeded 50.0 ng/kg (parts per trillion; ppt), a level of concern established by the U.S. Food and Drug Administration. PCDDs are associated with the closure of selected rivers in Missouri and Arkansas to anglers because of high residues in fish, with the destruction of fish and wildlife in Vietnam during military defoliation operations using phenoxy herbicides, and with the death of livestock and wildlife in Missouri and Italy. For example, in 1976, massive killing of small animals (predominantly rabbits and poultry) occurred within the first few weeks after a chemical plant explosion in Seveso, Italy, in which 2,3,7,8-TCDD was released; many humans were hospitalized. Levels of 2,3,7,8-TCDD in milk from dairy cows and tissues of pigs, chicken, cattle, goats, and sheep from Seveso were sufficiently elevated to pose a risk to human health. Accordingly, all domestic livestock in the most seriously afflicted areas were destroyed. In eastern Missouri during 1971, waste oil contaminated with 2,3,7,8-TCDD was applied to control road dust. Later, hundreds of horses kept in the riding arenas became sick, and 75 died; deaths were also observed among dogs, rodents, chicken, cats, and birds near the treated areas. The soil at the Times Beach, Missouri was so heavily contaminated with 2,3,7,8-TCDD that it was permanently evacuated in December 1982. The U.S. Environmental Protection Agency (EPA) had earlier announced that they would buy the dioxin-contaminated city of Times Beach; the purchase has been completed, and the city no longer exists officially. Approximately 22 kg (48.4 pounds) of 2,3,7,8-TCDD was involved in the Times Beach incident.

PCDDs are present as trace impurities in some commercial herbicides and chlorophenols. They can be formed as a result of photochemical and thermal reactions in flyash and other incineration products. Their presence in manufactured chemicals and industrial wastes is neither intentional nor desired. The chemical and environmental stability of PCDDs coupled with their potential to accumulate in fat has resulted in their detection throughout the global ecosystem. The number of chlorine atoms in PCDDs can vary between one and eight to produce up to 75 positional isomers. Some of these isomers are extremely toxic, while others are believed to be relatively innocuous. The most toxic and extensively studied PCDD isomer is 2,3,7,8-TCDD. In fact, it is the most toxic synthetic compound ever tested under laboratory conditions. This isomer is produced during

[a]All information in this chapter is referenced in the following sources:

Eisler, R. 1986. Dioxin hazards to fish, wildlife, and invertebrates: a synoptic review. *U.S. Fish. Wildl. Serv. Biol. Rep. 85(1.8)*, 37 pp.

Eisler, R. 2000. Dioxins. Pages 1021-1066 in *Handbook of Chemical Risk Assessment: Health Hazards to Humans, Plants, and Animals. Volume 2, Organics*. Lewis Publishers, Boca Raton, Florida.

the synthesis of 2,4,5-trichlorophenol, which is used in the manufacture of the herbicide 2,4,5-trichlorophenoxyacetic acid (2,4,5-T), and other trichlorophenoxy acids, and the germicide hexachlorophene. There is a general agreement that 2,3,7,8-TCDD is exceedingly stable, readily incorporated into aquatic and terrestrial ecosystems, extraordinarily persistent, and virtually impossible to destroy. PCDD-contaminated phenoxy herbicides are not the only sources of 2,3,7,8-TCDD, others include polychlorinated biphenyls and pentachlorophenols. The other 74 isomers enter the biosphere from a variety of sources. The fate and effects of PCDDs – with special reference to 2,3,7,8-TCDD and its role in the poisoning of humans, aquatic organisms, wildlife, livestock, poultry, and its contamination of vegetation, soils, and sediments – have been extensively reviewed.

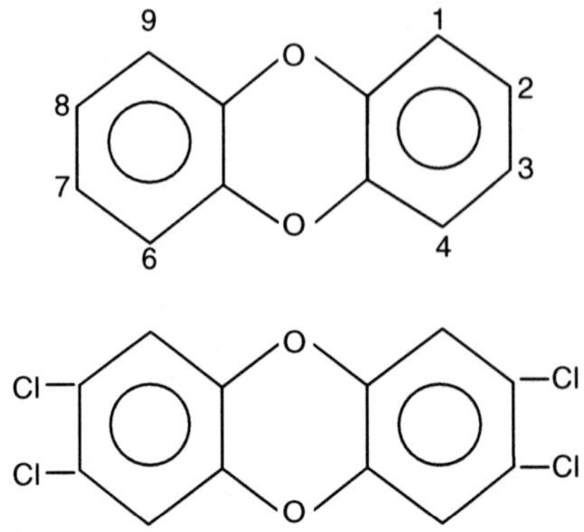

Figure 14.1.
Upper: Numbering system used for identification of individual PCDD isomers. Lower: The isomer 2,3,7,8-tetrachloro-dibenzo-*para*-dioxin (2,3,7,8-TCDD).

14.2 Environmental Chemistry

PCDDs consist of 75 isomers that differ in the number and position of attached chlorine atoms; each isomer has its own unique identity and toxicological properties. The most toxic of the chlorinated dioxin isomers is 2,3,7,8-TCDD (Figure 14.1). It is one of the 22 possible congeners of tetrachlorodibenzo-*p*-dioxin. There is a general agreement that PCDDS, including 2,3,7,8-TCDD, are (or were, until recently) found in chlorophenols, especially trichlorophenol and pentachlorophenol, in certain phenoxy pesticides (2,4,5-T; 2,4-D; Fenoprop; Silvex; Ronnel; Erbon; Agent Orange), in hexachlorophene, and in polychlorinated biphenyls (used in electrical transformers and capacitors, and contaminated with trichlorobenzenes). PCDDs enter the environment naturally through forest fires and volcanoes, and through human activities such as accidental release during chlorophenol production, aerial application of some phenoxy herbicides, and through improper disposal of waste into terrestrial and aquatic ecosystems from municipal incinerators and pulp and paper mills that use chlorine for the bleaching process. In Japan, major PCDD sources include waste incineration and the herbicide chlornitofen with metabolites of chlornitofen that include 1,3,6,8-TCDD and 1,3,7,9-TCDD. The PCDD content of technical products varies between manufacturers, between lots and grades, and between various formulations of pesticidal chemicals. PCDDs have been identified in effluents from combustion products of municipal and industrial incinerators, including flyash and flue gas. These PCDDs may be associated with small particles that have long residence time in the atmosphere and can become distributed over large areas. For example, in the Great Lakes, atmospheric transport of combustion products is the major source of PCDDs (mostly octa-, hepta-, and hexa-CDDs) in sediments. Increased retention capacity of the higher chlorinated PCDDs may account for the pattern of increasing dioxin concentrations in sediments, with increasing chlorine substitution observed in the Great Lakes and other aquatic environments. High-temperature combustion of bituminous coal in an oxidized and chlorinated atmosphere (produced experimentally) yields chlorodioxins, mostly octa-, hepta-, and hexa-CDDs and measurable quantities of tetra-CDDs. Other potential sources of PCDDs include microbial by-products in activated sludge basins fed with chlorophenol-containing wastes, fossil fuel power plants, internal combustion engines, home fireplaces, and cigarette smoke.

In general, PCDDs exhibit a relative inertness to acids, bases, oxidation, reduction, and heat. With increasing halogen content, they become more environmentally and chemically stable. PCDDs are usually destroyed at temperatures greater than 1000°C. They are resistant to biological breakdown, concentrated in fat, not readily excreted, and extremely toxic to some animals, and the cumulative effects of small doses to both animals and humans are a source of increasing concern. Most PCDDs are relatively insoluble in water, sparingly soluble in organic solvents, and will decompose on exposure to UV light, including sunlight, or to hydroxyl compounds. The isomer 2,3,7,8-TCDD is a colorless crystalline solid at room temperature and decomposes when heated at greater than 700°C (Table 14.1).

Table 14.1. Chemical and physical properties of 2,3,7,8-TCDD, also known as CAS Registry No. 1746-01-6.

Criterion	Property
EMPIRICAL FORMULA	$C_{12}H_4Cl_4O_2$
MOLECULAR WEIGHT	322
VAPOR PRESSURE, mm Hg AT 25°C	1.7×10^{-6}
MELTING POINT	305°C
DECOMPOSITION TEMPERATURE	>700°C
SOLUBILITY, IN g/L	
***O*-Dichlorobenzene**	1.4
Chlorobenzene	0.72
Benzene	0.57
Chloroform	0.37
***N*-Octanol**	0.05
Methanol	0.01
Acetone	0.11
Water	0.0000002

Data on the bioavailability of PCDDs are scarce. It is known that the PCDDs incorporated into wood as a result of chlorophenol (preservative) treatment are bioavailable. Swine and poultry using chlorophenol-treated wooden pens or litter have been found to be contaminated with PCDDs. Toxicities of individual PCDD isomers can vary by a factor of 1000 to 10,000 for isomers as closely related as 2,3,7,8-TCDD and 1,2,3,8-TCDD or 1,2,3,7,8-penta CDD and 1,2,4,7,8-penta CDD. Isomers with the highest biological activity and acute toxicity have 4 to 6 chlorine atoms, and all lateral (i.e., 2,3,7, and 8) positions substituted with chlorine. On this basis, the most toxic PCDD isomers are 2,3,7,8-TCDD, 1,2,3,7,8-penta CDD, 1,2,3,6,7,8-hexa CDD, 1,2,3,7,8,9-hexa CDD, and 1,2,3,4,7,8-hexa CDD. Toxic equivalencies for various PCDDs have been assigned with 2,3,7,8-TCDD given a value of 1 (highest biological activity), followed by a value of 0.5 for 1,2,3,7,8-penta CDD, a value of 0.1 for three PCDD isomers (1,2,3,4,7,8-hexa CDD, 1,2,3,4,7,8-hexa CDD, 1,2,3,7,8,9-hexa CDD), a value of 0.01 for 1,2,3,4,6,7,8-hepta CDD, and a value of 0.001 for 1,2,3,4,6,7,8,9-octa CDD.

Although PCDDs are highly persistent, volatilization and photolysis are major removal processes. In soils, 2,3,7,8-TCDD undergoes photolysis rapidly on the surface in a few hours, but more deeply buried 2,3,7,8-TCDD could have a chemical half-time greater than 10 years. When more than one degradation pathway was possible, microbial dechlorination of dioxins occurred at the most positively charged atom in the ring, which was usually a lateral carbon atom. Microbial degradation of 2,3,7,8-TCDD in soils is slow, with biological half-times estimated at 1.0 to 1.5 years. However, 2,3,7,8-TCDD was detected in northwestern Florida from samples of soils, rodents, birds, lizards, fishes, and insects, 12 years after application. This half-time in soil was estimated at 2.9 years. Uptake of 2,3,7,8-TCDD from soils by vegetation is considered negligible. The half-time persistence of 1,3,6,8-TCDD and OCDD in freshwater is short, of the order of 2.6–4 days; the rapid partitioning to dissolved and particulate organic matter in the water column and sediments limited their bioavailability. Techniques are now available to measure 2,3,7,8-TCDD at the part per quadrillion (0.1 parts per trillion) level in foods with high fat content, human blood, and in fish tissues to 1.0 ppt. Analyses at such low levels are complicated by interference from a multitude of other compounds, as well as by the large number of PCDD isomers and their differences in chemical properties. Although 2,3,7,8-TCDD is the most extensively studied PCDD isomer, data on its fate and persistence are generally poor, interpretations are frequently absent, and extrapolations from case to case usually impossible. The general result is a qualitative concept of this compound's behavior in environmental situations. This issue is further confounded by the presence in biological and abiotic samples of chemicals of similar structure and toxicological properties to that of 2,3,7,8-TCDD. These isosteric compounds include: 2,3,6,7-tetrachlorobiphenylene; 2,3,7,8-chlorine-substituted dibenzofurans; and 3,3′,4,4′-tetra-, 3,3′,4,4′,5-penta-, and 3,3′,4,4′,5,5′-hexa-chlorobiphenyl. For example, analytical results of fish, birds, and sediments indicated that every sample that was positive for 2,3,7,8-TCDD also contained 2,3,7,8-chlorine-substituted dibenzofurans.

In mammals, biochemical and pathological responses of 2,3,7,8-TCDD are mediated through binding to a cytoplasmic protein, the Ah receptor. Many of the toxic responses to 2,3,7,8-TCDD are associated with alterations of cell proliferation in affected tissues. Common examples of 2,3,7,8-TCDD-mediated differentiation and proliferation in both fishes and mammals include dermal hyperkeratinization (fin necrosis in fish), teratogenicity, lymphoid involution, immunotoxicity, and carcinogenesis. Lipid peroxidation in liver, kidney, thymus, and testes is induced in rats by 2,3,7,8-TCDD in a dose- and time-dependent manner. The enhanced lipid peroxidation by microsomes from 2,3,7,8-TCDD-treated rats may be associated with an increase in hydrogen peroxide production in conjunction with a decrease in glutathione peroxidase activity and an increase in free iron.

14.3 Concentrations in Field Collections

PCDDs are ubiquitous in the environment and have been measured in air, water, soil, sediments, and food. In Canada, all animals have been and continue to be exposed to these substances. Plasma dioxin levels of Canadian villagers are about eight times higher than the levels in urban residents, and this is attributed to the consumption of contaminated wildlife. Many PCDDS, in addition to 2,3,7,8-TCDD, are present in biological and abiotic samples. In general, wherever high levels of dioxins have been detected in the environment, a local application of 2,3,7,8-TCDD-contaminated herbicide, hazardous waste site, or industrial discharge has usually been implicated as the source. Apart from direct deposition, observed increases in 2,3,7,8-TCDD concentrations may result from microbial dechlorination as modified by the action of humic constituents such as resorcinol, 3,4-dihydroxybenzoic acid, and catechol.

At Eglin Air Force Base (EAFB), located in northwestern Florida, contamination of a 208-ha section with 2.8 kg of 2,3,7,8-TCDD (equivalent to 13.0 mg/ha) occurred between 1962 and 1970 as a result of repeated, massive herbicide applications. The 2,3,7,8-TCDD isomer was present as an impurity in 76,740 kg of 2,4-D and 73,010 kg of 2,4,5-T applied to this section of EAFB during the 9-year span. Ecological surveys conducted between 1970 and 1975 showed an apparently healthy and diverse wildlife fauna, although soil levels of 520.0 ppt (ng/kg) of 2,3,7,8-TCDD were frequently encountered, and 2,3,7,8-TCDD residues were elevated in some species examined. The highest residues recorded in various trophic levels were 283.0 ppt in whole beetle grubs, up to 1,360.0 ppt in whole southern toads (*Bufo terrestris*), 360.0 ppt in viscera and 430.0 ppt in carcass of a lizard, the six-lined racerunner (*Cnemidophorus sexlineatus*), 18.0 ppt in gonad and 85.0 ppt in gut contents of the spotted sunfish (*Lepomis punctatus*), 100.0–1200.0 ppt in stomach contents of the southern meadowlark (*Sturnella magna argutula*), and 300.0–2900.0 ppt in liver and

130.0–200.0 ppt in pelt of a beachmouse (*Peromyscus polionotus*). The significance of these elevated residues will be discussed later.

In some cases, 2,3,7,8-TCDD has constituted up to 95% of the total body PCDD burden, as was true in lake trout, (*Salvelinus namaycush*) and rainbow trout (*Oncorhynchus mykiss*) from Lake Ontario. Concentrations of 2,3,7,8-TCDD in whole carp (*Cyprinus carpio*) varied from 24% of total body PCDDs in Saginaw Bay, Michigan, to 45–56% in the Niagara River. In herring gulls from Saginaw Bay, 2,3,7,8-TCDD comprised 40–60% of the whole body PCDD content, and 72–78% in gulls from Lakes Huron and Ontario. In 1983, Forster's tern (*Sterna forsteri*) from Green Bay, Wisconsin, contained 114.0 ppt of PCDDs in egg, of which 41% was 2,3,7,8-TCDD; double-crested cormorants (*Phalacrocorax auritus*) from the same area contained 25.0 to 214.0 ppt of PCDDs in whole body, of which only 10–31% was 2,3,7,8-TCDD. The causes of the observed variations are not known, but may be associated with localized inputs from municipal sewage treatment plants and with atmospheric transport of incinerated domestic and industrial chemical wastes. For example, the PCDD composition of sewage sludge from Milwaukee, Wisconsin, was relatively constant, as judged by the analysis of samples from 1933, 1981, and 1982. Total PCDD content in sewage sludge samples ranged between 60,950.0 and 70,191.0 ppt, of which the great majority was in the form of octa-CDDs (82–86%), hepta-CDDs (11.0–15.4%), and hexa-CDDs (1.3–2.1%). However, the tetra-CDDs increased from 34.0 ppt in 1933, to 138.0 in 1981, and to 222.0 in 1981; corresponding values for the 2,3,7,8-TCDD isomer in 1933, 1981, and 1982 were 2.2, 11.0, and 16.0 ppt, respectively. Other TCDD isomers also showed increases from 6.0 ppt in 1933 to 22.0 ppt in 1982 (1,3,7,8-TCDD), and during that same period from 2.2 ppt to 140.0 ppt (1,2,3,7-, and 1,2,3,8-TCDD). It seems that chlorinated dibenzodioxins have been present in the dried sludge from this plant for at least 50 years. Their presence in this material suggests that they may have been formed by the condensation of chlorophenols resulting from the chlorination of naturally occurring phenolic compounds. PCDDs were also found in sediments from Siskiwit Lake on Isle Royale in Lake Superior, a location that can receive only atmospheric inputs. The source of these compounds is the atmospheric transport of dioxins formed by combustion of domestic and chemical wastes. For example, particulates from a chemical waste incinerator in Midland, Michigan, had 260,000,000 ppt of octa-CDDs and 170,000,000 ppt of hepta-CDDs; lower, but still elevated levels of 440,000 ppt of octa-CDDs and 310,000 ppt of hepta-CDDs were measured in municipal trash incinerator particulates.

Mussels and fish can accumulate 2,3,7,8-TCDD and other PCDDs from the medium and may be used as sentinel organisms. Mussels (*Elliptio complanata*) from an uncontaminated site exposed for 21 days in the contaminated Rainy River, Ontario, were sensitive monitors of PCDD sources, such as kraft pulp and paper mills. Fish in Maine rivers contaminated by the effluent of bleached kraft paper mills accumulated 2,3,7,8-TCDD from the medium by factors as high as 24,600 in the muscle of smallmouth bass (*Micropterus dolomieui*), 28,300 in the muscle of brown trout (*Salmo trutta*), 7500 in white perch (*Morone americana*) fillets, and 106,000 in whole white suckers (*Catostomus commersoni*). High levels of PCDDs were found in edible tissues of mountain whitefish (*Prosopium williamsoni*) exposed to effluent from bleached kraft pulp mills in Canada; uptake was attributed to the food prey selection of filter-feeding invertebrates that ingest suspended sediments. The primary route of exposure of 2,3,7,8-TCDD for pelagic fish is the diet, which accounts for 75% or more of the total uptake in lake trout. Food chain biomagnification of PCDDs is reported for fish to birds, fish to turtle, and in the Great Lakes food chains of phytoplankton to zooplankton to forage fish to lake trout. But food chain biomagnification has not been observed, using sophisticated stable nitrogen isotope techniques, in complex Baltic Sea food chains, and in fish to seal, and fish to bird (guillemot, *Uria aalge*) food chains. To confound matters, PCDD patterns in different fish species collected at the same site are highly variable, indicating species differences in accumulation. PCDD sources from pulp mill effluents, combustion, and magnesium production can now be recognized in aquatic fauna by the analysis of profiles of 2,3,7,8-substituted chlorinated dibenzodioxins, using statistical principal component analysis techniques.

Eggs of the lake trout (*Salvelinus namaycush*) from Lake Ontario had higher 2,3,7,8-TCDD toxic equivalents (10.7 ppt) than eggs from Lake Superior or a hatchery (0.3 ppt), but this was below the levels considered fatal to lake trout eggs (LD50 of 42.0–72.0 ppt) or harmful (NOAEL of 30.0–45.0 ppt). Concentrations of 2,3,7,8-TCDD in whole fish collected nationwide in the United States in 1983 from 395 sites usually contained 0.5–2.0 ppt fresh weight (FW), with a maximum of 85.0 ppt; concentrations were highest in predatory fish collected near pulp and paper manufacture discharge sites and lowest in estuarine areas. In the Great Lakes area, fish from the Tittabawasee and Saginaw Rivers, two tributaries of Lake Huron's Saginaw Bay, contained up to 695.0 ppt of 2,3,7,8-TCDD. High 2,3,7,8-TCDD levels (87.0 to 162.0 ppt) were also recorded in fish from the Niagara River, New York, and from parts of Lake Ontario; lower concentrations (2.0–28.0 ppt) were noted in fish from Lakes Erie, Huron, Michigan, and Superior. Muscle from larger specimens of commercial fish collected from Lake Ontario in 1980 had higher levels of 2,3,7,8-TCDD than smaller fish, suggesting that accumulation increases with age. The larger fish also contained high concentrations (1.2–4.9 mg/kg, FW) of polychlorinated biphenyls, demonstrating the need to elucidate TCDD interaction kinetics with other contaminants. Bottom-feeding fish, such as carp and catfish, from rivers in Michigan during 1978, contained higher 2,3,7,8-TCDD residues than surface feeders, indicating an association with contaminated sediments. Marine amphipods (*Ampelisca abdita*) held on sediments containing as much as 25.0 μg 2,3,7,8-TCDD/kg dry weight (DW) for 10 days had normal survival and growth, although authors did not measure 2,3,7,8-TCDD residues in amphipods. Sediments from the Spring River, Missouri, contained 12.0 ppt of 2,3,7,8-TCDD immediately downstream of a now-defunct hexachlorophene facility; concentrations in fish were measurable 111 km downstream from this disposal site. Fish from the Spring River (and also the Meremac River, Missouri) contained inordinately high levels of 18.0–78.0 ppt of 2,3,7,8-TCDD, prompting the U.S. Food and Drug Administration to issue a health advisory in 1982 against fish consumption from these areas. In Massachusetts, six ponds were surveyed for 2,3,7,8-TCDD in 1983 after prior treatment with phenoxy herbicides between 1958 and 1978. Only one fish, a brown bullhead (*Ictalurus nebulosus*), aged 3+ years, contained measurable (25.0 ppt) dioxin levels. Residues were not detectable in other species of fish sampled, including several species of ictalurid catfish, yellow perch (*Perca flavescens*), and chain pickerel (*Esox niger*). Negative results (less than 10.0 ppt 2,3,7,8-TCDD) were also documented in freshwater fish from Arkansas and Texas following spraying of the herbicide 2,4,5-T.

In general, high PCDD concentrations in avian tissues were correlated with poor growth, survival, and reproduction. Nest success, hatching success, and duckling production of wood ducks (*Aix sponsa*) were suppressed at nesting sites within 58 km of a PCDD point source (a former chemical plant that manufactured the herbicide 2,4,5-T). Egg toxicity equivalent factors were inversely correlated with productivity in nests, and teratogenic effects occurred in ducklings at the most contaminated sites. High PCDD levels in yolk sacs of cormorant (*Phalacrocorax carbo*) hatchlings may have been responsible for the observed low breeding success of Dutch colonies in 1989. The endangered Forster's tern (*Sterna forsteri*) population on Green Bay, Wisconsin, showed signs of embryotoxicity, congenital deformities, and poor hatching success. Eggs from this population contained 215.0 (90.0–245.0) ppt FW of 2,3,7,8-TCDD equivalents vs. 23.0 (14.0–34.0) ppt from a reference population. Great blue herons (*Ardea herodias*) located near a pulp mill in British Columbia failed to fledge young in 1987, with a concurrent sharp increase in PCDD levels in their eggs. In 1988, levels in heron eggs were – in ppt FW – 211.0 for 2,3,7,8-TCDD, 263.0 for 1,2,3,7,8-penta CDD, and 430.0 for 1,2,3,6,7,8-hexa CDD. These values were significantly higher than eggs from other sites, although other factors may have contributed to the decline of the heron population. Eggs of the osprey (*Pandion haliaetus*), however, collected near a bleached kraft mill facility in Wisconsin between 1992 and 1996 hatched and fledged normally despite 2,3,7,8-TCDD concentrations as high as 162.0 ppt on a FW basis.

Proposed indicators of 2,3,7,8-TCDD contamination in bird eggs include changes in yolk retinoid concentrations and ethoxyresorufin *O*-deethylase (EROD) enzyme activity levels. Changes in yolk retinoids, as measured by Vitamin A concentrations, of herring gulls (*Larus argentatus*) from the Great Lakes during 1986–87 were associated with increasing concentrations of 2,3,7,8-TCDD, and this may be an early indicator of survival and teratogenesis in birds from the Great Lakes. Eggs of the great blue heron (*Ardea herodias*) from British Columbia in 1990–92 had decreased concentrations of 2,3,7,8-TCDD when compared to 1988. Decreases in 2,3,7,8-TCDD content in eggs was associated with decreased EROD activity in eggs and resultant chicks, decreased incidence of chick edema, and increased growth rate of chicks. The use of avian hepatic microsomal EROD activity is a useful index of cytochrome P4501A1 induction by polychlorinated aromatic hydrocarbons, including PCDDs. A significant regression of hepatic microsomal EROD activity on TCDD-toxic equivalent was measured in chicks of the double-crested cormorant (*Phalacrocorax auritus*).

Fish-eating birds usually have the highest concentrations of PCDDs of all birds measured. Diet is the major route of PCDD accumulation in livers of fish-eating birds. Fish-eating birds and their food from the Netherlands are contaminated with 2,3,7,8-substituted PCDDs and polychlorinated dibenzofurans (PCDFs); pentachlorophenol and polychlorinated biphenyls are the two major contaminating sources of PCDDs and PCDFs. A linear relation exists between the retained PCDDs and PCDFs in the livers of cormorants (*Phalacrocorax carbo*), probably caused by continuous exposure to a relatively stable mixture of these compounds in their fish diet. Biomagnification of 2,3,7,8-TCDD and 1,2,3,6,7,8-hexa CDD probably occurs from the European eel (*Anguilla anguilla*) to the cormorant. In birds, the levels of 2,3,7,8-TCDD have been decreasing, according to an analysis of herring gull eggs from Lake Ontario. During the decade 1970–1980, there was a reduction of about 50% in 2,3,7,8-TCDD levels every two years. The reasons for the decline are unknown, and the relevance to higher-chlorinated PCDDs has not yet been determined. Until these questions are resolved and more substantive data are acquired on dioxin residues in birds, the predictive trends on decline rates should be interpreted with caution.

Some information on 2,3,7,8-TCDD levels in wildlife and domestic livestock are from the vicinity of Seveso, Italy. On July 10, 1976, a thick cloud of chemicals – including at least 34 kg of 2,3,7,8-TCDD, and perhaps as much as 250 kg – was released here into the atmosphere when a runaway reaction accidentally occurred in a trichlorophenol-producing facility. Its wind-driven settling contaminated large inhabited areas. It contaminated the food (hay, grass, cut-up corn) of dairy cows. Grossly elevated levels (7900.0 ppt) were measured in milk from these herds at concentrations considered hazardous to human health, i.e., more than 7000.0 ppt. Wildlife from the most heavily-contaminated area appeared to accumulate 2,3,7,8-TCDD. Field mice (*Microtus arvalis*), for example, contained very high whole-body concentrations of 2,3,7,8-TCDD (up to 49,000.0 ppt) almost 2 years after the critical contamination. The mechanisms for this phenomenon included ingestion of contaminated soil and licking of their dioxin-contaminated pelt. In 1996, humans from the most heavily contaminated area of Seveso (based on soil 2,3,7,8-TCDD concentrations) had 53.0 ppt of 2,3,7,8-TCDD in plasma on a fresh weight basis compared to 4.9 ppt in a reference population; women had significantly higher plasma 2,3,7,8-TCDD concentrations than men, especially women closest to the accident site and who consumed meat regularly. In another study, no 2,3,7,8-TCDD was detected in livers of mountain beavers (*Aplodontia rufa*) that fed for 45–60 days in Oregon forests that had been sprayed with 2.2 kg of 2,4,5-T/ha. Although it was presumed that the herbicide was heavily contaminated with dioxins, no chemical analysis of the 2,4,5-T was performed.

In Germany, humans ingest an average of 85 pg daily of 2,3,7,8-TCDD equivalents per person or 1.2 pg/kg body weight (BW) daily, mostly from fish and fish products (32%), beef (20%), and cow's milk (16%). In the United States, there is a potential for 2,3,7,8-TCDD to migrate from paperboard-based food packaging on contact with food. The migration rate of 2,3,7,8-TCDD from bleached paperboard cartons into whole milk was linear with the square root of exposure time. After 12 days of storage, 6.7% of the 2,3,7,8-TCDD in the paperboard carton migrated into the milk. The amount of 2,3,7,8-TCDD formed in the U.S. bleached kraft industry in 1988 was estimated at 0.64 kg annually. In Vietnam, 2,3,7,8-TCDD concentrations in food and wildlife in 1985–87 were higher in South Vietnam than in North Vietnam, with elevated values attributed to the increased industrialization of the South.

14.4 Lethal and Sublethal Effects

Information is lacking or scarce on the biological properties of PCDD isomers, except 2,3,7,8-TCDD. The latter has been associated with lethal, carcinogenic, teratogenic, reproductive, mutagenic, histopathologic, and immunotoxic effects. There are substantial inter- and intraspecies differences in sensitivity and toxic responses to 2,3,7,8-TCDD. Typically, animals poisoned by 2,3,7,8-TCDD exhibit weight loss, atrophy of the thymus gland, and eventually death. The toxicological mechanisms are imperfectly understood.

14.4.1 Terrestrial Plants and Invertebrates

Higher plants, such as corn (*Zea mays*) and beans (*Phaseolus vulgaris*) grown on soil contaminated with 2,3,7,8-TCDD (12.0–2750.0 ng 2,3,7,8-TCDD/kg FW) accumulated the toxin in the aerial parts progressively over time with increasing soil burdens of 2,3,7,8-TCDD; in hydroponic tests, TCDD uptake occurred only in light. Aerial parts of beans grown in soil containing 12.0 ng 2,3,7,8-TCDD/kg had 0.09 ng 2,3,7,8-TCDD/kg FW after 7 days; beans grown in soils with 2750.0 ng 2,3,7,8-TCDD/kg had 7.2 ng/kg FW after 7 days. After 57 days, beans contained a maximum of 5.0 ng/kg FW. Aerial parts of corn had 0.74–10.0 ng 2,3,7,8-TCDD/kg FW after 17 days; after 57 days this range was 1.2–3.2 ng/kg FW.

Two species of earthworms (*Allolobophora caliginosa, Lumbricus rubellus*) showed no adverse effects when held for 85 days in soils containing grossly elevated levels of 5.0 mg/kg of 2,3,7,8-TCDD, but both species died at 10.0 mg/kg. In soils containing lower concentrations of 50.0 μg/kg of 2,3,7,8-TCDD, earthworms accumulated five times the soil levels in 7 days. There was no avoidance of soils contaminated with 2,3,7,8-TCDD, suggesting indifference. No surface penetration of dioxins into the body of earthworms was noted, and there was no biological breakdown of 2,3,7,8-TCDD during digestion as judged by the absence of mono-, di-, and tri-CDDs in excrement. Worm-worked soils had 2,3,7,8-TCDD retention times of 80–400 days, suggesting that earthworms may significantly alter half-time patterns of 2,3,7,8-TCDD in soils.

Mutagenic responses were produced in *Escherichia coli* and certain strains of *Salmonella typhimurium* bacteria by 2,3,7,8-TCDD, but not by octa-CDD. Further, chromosomal aberrations were induced in at least one species of higher plant and mammal. It must be concluded at this time that 2,3,7,8-TCDD is mutagenic or has mutagenic potential.

14.4.2 Aquatic Organisms

Limited data were available on lethal and sublethal effects of any PCDD isomer to aquatic organisms, except for 2,3,7,8-TCDD and freshwater biota; 2,3,7,8-TCDD and liver microsomal enzyme activities in two marine species: winter flounder, *Pleuronectes americanus*, and the little skate, *Raja erinacea*; and 1,3,6,8-TCDD uptake and elimination by fathead minnows (*Pimephales promelas*) and rainbow trout (*Oncorhynchus mykiss*). Lymphomyeloid and epithelial tissues are the primary target organs for TCDD-induced pathologic lesions in rainbow trout. Cardiovascular systems seem to be the initial tissue affected in both the TCDD toxicity syndrome and in blue-sac disease of developing lake trout. Lesions in other organs, including brain, retina, and liver develop as a result of circulatory derangements, anemia and hypoxia. The mode of action of 2,3,7,8-TCDD in rainbow trout and mammals on the epidermal growth factor is mediated, in part, through protein kinase C activity.

Tissue residues of 55.0 ng 2,3,7,8-TCDD/kg FW in sac fry of the lake trout (*Salvelinus namaycush*) were lethal, and water concentrations of 0.038 ng 2,3,7,8-TCDD/L caused a reduction in feeding activity and general activity levels of rainbow trout. Less sensitive species of teleosts exhibited reduced growth and fin necrosis at concentrations as low as 0.1 ppt of 2,3,7,8-TCDD after exposure for 24–96 h. Concentrations of 1.0 ppt and higher were eventually fatal, and exposure to lower concentrations of 0.01 ppt for 24 h had no measurable effect. A typical 2,3,7,8-TCDD poisoning sequence in guppies (*Poecilia reticulatus*) and coho salmon (*Oncorhynchus kisutch*) during a postexposure observation period included: declining interest in feeding (5–8 days postexposure); skin discoloration and fin necrosis (30 days), with caudal fin most severely affected; reduced resistance to fungal infestations; reduced swimming; and, finally, death several weeks to months after exposure. In general, older and larger fish die last, and smaller or younger specimens succumb first.

In brook trout (*Salvelinus fontinalis*), feces was the most important excretion route, and spawning was not an important source of chemical elimination of 2,3,7,8-TCDD. However, others disagree and showed, using radioisotope techniques, 39% maternal transfer of 2,3,7,8-TCDD to eggs in brook trout, suggesting the need for additional research in this area.

Histopathologic and teratogenic effects were noted in fry of rainbow trout (*Oncorhynchus mykiss*) exposed to 10.0 ppt (ng/L) of 2,3,7,8-TCDD for 96 h as eggs, or as yolk sac fry. Some fry showed extensive degeneration and necrosis of the liver and subsequently developed edema prior to death. The remaining fry showed a high incidence of teratogenic changes, including opercular defects, and foreshortened maxillas. Invertebrates, plants, and amphibians were comparatively resistant to 2,3,7,8-TCDD. For example, there were no adverse effects on growth, reproduction, or food consumption of algae, daphnids, and snails during immersion for 32 days in solutions containing 2.4–4.2 ppt of 2,3,7,8-TCDD. Populations of the mummichog (*Fundulus heteroclitus*) from New Jersey coastal waters contaminated with 2,3,7,8-TCDD are unusually resistant to dioxins when compared to reference populations. Mummichogs from a TCDD-contaminated site were resistant to the toxicity of 2,3,7,8-TCDD and the ability of 2,3,7,8-TCDD to induce P4501A activity, which may imply an alteration in the Ah receptor complex of TCDD-contaminated mummichogs.

Bioconcentration factors of dioxins in fishes are relatively low compared to other chlorinated aromatic compounds because of the low metabolic conversion of dioxins, their low available concentrations in test systems, and their highly variable uptake rates. In general, bioconcentration factors for persistent superlipophilic chemicals, such as OCDD, derived for freshwater fishes from supersaturated solutions may seriously underestimate the true BCF. BCF values for OCDD in various fish species are about 4.3×10^6 on a fresh weight basis and 8.5×10^7 on a lipid weight basis; however, these values are by several orders of magnitude higher than those reported for guppies, rainbow trout, and fathead minnows because the lower values were derived from OCDD concentrations that exceeded the water solubility of OCDD. Although uptake from food predominates over direct uptake of dioxins from water, the accumulation of 2,3,7,8-TCDD from the aquatic environment occurred in all species examined. The isomer 1,3,6,8-TCDD was also accumulated from the environment by freshwater teleosts, but accumulations were much lower than predicted when compared to 2,3,7,8-TCDD, and elimination was 10–15 times more rapid than 2,3,7,8-TCDD. Of seven chlorinated dibenzodioxins tested, 2,3,7,8-TCDD was concentrated by rainbow trout and fathead minnows to the greatest extent. In outdoor pond studies, a major portion of the added 2,3,7,8-TCDD concentrated in aquatic plants and at the sediment–water interface; however, most (85–99%) of the 2,3,7,8-TCDD originally added to the ecosystem remained in the sediments at the end of the study, with significant reduction in bioavailability of PCDDs to fishes. Among teleosts, body burdens of 2,3,7,8-TCDD increased with increasing concentration in the water column and with increasing duration of exposure; on removal to uncontaminated water, less than 50% was lost in 109 days. Accumulation of PCDDs in biota shifted from direct equilibrium partitioning during the first few days when concentrations in the water column were relatively high, to a detrital food chain transfer, as the freely available PCDDs in the water column declined.

14.4.3 Birds

LD50 values computed 37 days after a single oral dose of 2,3,7,8-TCDD varied from 15.0 μg/kg body weight in Northern bobwhite (*Colinus virginianus*), with 95% confidence limits of 9.2 and 24.5 μg/kg, to more than 810.0 μg/kg body weight for the ringed turtledove (*Streptopelia risoria*). Mallards (*Anas platyrhynchos*) were intermediate in sensitivity with an acute oral LD50 value of more than 108.0 μg/kg BW. For all three species, death occurred 13–37 days after treatment; remission in survivors had apparently occurred by day 30 posttreatment. Gross necropsy of ringed turtledoves that survived treatment showed enlarged livers, about twice the normal size. Bobwhites showed severe emaciation, high accumulations of uric acid salts in connective tissues, and fluid accumulations in the pericardial and abdominal cavities. Some birds regurgitated within a few minutes after treatment. Signs of intoxication that began 7 days after treatment included excessive drinking, loss of appetite, hypoactivity, emaciation, weakness, debility, muscular incoordination, increased reaction to stimuli, fluffed feathers, huddled position, unkempt appearance, falling, tremors, spasms, convulsions, and immobility.

In ovo exposure to dioxins is associated with development of grossly asymmetric avian brains, especially the forebrain and tectum. Brain asymmetry was observed in herons, cormorants, eagles, and chickens exposed to 2,3,7,8-TCDD under controlled conditions. Asymmetry appears with increasing frequency and severity in embryos and hatchlings exposed to increasing doses of 2,3,7,8-TCDD beginning early in development. Injection of 1.3–11.7 μg 2,3,7,8-TCDD/kg FW egg into yolks of the double-crested cormorant (*Phalacrocorax auritus*) prior to incubation produced an LD50 at hatch of 4.0 μg/kg; hepatic EROD activities were elevated in all treatment groups when compared to controls, but development was normal.

Studies with 2,3,7,8-TCDD and various life stages of the ring-necked pheasant (*Phasianus colchicus*) showed several trends. (1) Hens given a single intraperitoneal (ip) injection of 25.0 μg/kg BW and higher had reduced growth and survival during an 11-week observation period. (2) Hens given weekly ip injections of 1.0 μg/kg BW for 10 weeks had reduced egg production and hatchability, and a wasting disease that proved fatal to 100% by 13 weeks after the last injection. (3) Hens given weekly ip injections for 10 weeks prior to egg laying of 0.1 μg/kg BW had no adverse effects on growth, reproduction, or survival; translocation of 2,3,7,8-TCDD to egg yolks indicates that egg laying is an important route of elimination. (4) The half-time persistence of 2,3,7,8-TCDD in hatchlings exposed as embryos is about 13 days; for adult hens not producing eggs, the half-time persistence of radiolabeled 2,3,7,8-TCDD is 378 days. (5) The most sensitive effect of in ovo exposures was induction of hepatic ethoxyresorufin *O*-deethylase (EROD) activity in 1-day-old chicks, with an

ED-50 dose of 0.312 μg/kg FW egg. (6) LD50 values for embryos were 1.35 μg/kg egg when injected into the egg albumin and 2.18 μg/kg when injected into the egg yolk. (7) No adverse effects on 1-day-old hatchlings and 28-day-old chicks on growth, survival, development, and metabolism were observed at egg 2,3,7,8-TCDD doses up to and including 1.0 μg/kg. (8) Embryo mortality is the most sensitive sign of 2,3,7,8-TCDD toxicity in the ring-necked pheasant; pheasant embryos were more sensitive to 2,3,7,8-TCDD than were embryos of the eastern bluebird (*Sialia sialia*), but less sensitive than embryos of the domestic chicken (*Gallus* sp.).

Domestic chickens were relatively sensitive to PCDDS, especially 2,3,7,8-TCDD, with an estimated 2,3,7,8-TCDD oral LD50 range of 25.0 to 50.0 μg/kg BW. EROD induction by 2,3,7,8-TCDD in avian hepatocytes of five species followed a concentration-response relation in livers of hatchlings; chickens were the most sensitive species tested, followed by the double-crested cormorant, ring-billed gull (*Larus delawarensis*), herring gull, and Forster's tern. Chickens fed 1.0 or 10.0 μg of 2,3,7,8-TCDD, 1,2,3,7,8,9-hexa CDD, or hepta-CDDs per kg BW daily for 21 days showed signs of chick edema disease, i.e., pericardial, subcutaneous, and peritoneal edema; liver enlargement and necrosis with fatty degeneration; and frequently resulted in death. Autopsies of poultry killed by 2,3,7,8-TCDD in Seveso, Italy, in 1976 showed signs characteristic of chick edema disease. Pathological signs of chick edema disease were also seen in herring gull chicks on the lower Great Lakes in the early 1970s. Concentrations of 2,3,7,8-TCDD in eggs of the herring gull declined from about 1000.0 ppt in 1971 to less than 80.0 ppt in 1981. This was accompanied by a decrease in the frequency of chick edema disease. Decreases in levels of other contaminants, notably mirex, were probably more important to the survival of gulls in these colonies than 2,3,7,8-TCDD; however, little data exist on the interaction of PCDDs, including 2,3,7,8-TCDD, with other contaminants appearing concomitantly in bird tissues or their diets. Although there presently is no evidence of biomagnification of PCDDs in birds, it is speculated that piscivorous birds have a greater potential to accumulate PCDDs than the fish that they eat.

14.4.4 Mammals

Many industrial accidents involving malfunctioning reaction vessels used to manufacture chlorinated phenols or phenoxy herbicides have exposed more than 1300 workers to short-term, high-level doses of the dioxins that occur as contaminants of these substances. Exposures have frequently been associated with acne-like skin lesions, dermatitis, altered liver enzyme concentrations, pulmonary deficiency, numbness, nausea, headache, hearing loss, sleep disturbance, tiredness, sexual dysfunction, depression, and appetite loss. Populations exposed to dioxin-contaminated materials through non-occupational sources – including dioxin-contaminated soils in Missouri, a trichlorophenol reactor explosion in Italy, dioxin-containing herbicide in Vietnam, and assorted laboratory accidents – have all experienced similar effects.

The International Agency for Research on Cancer concludes that there is inadequate human data but sufficient animal data for 2,3,7,8-TCDD to be a possible human carcinogen. In rats, carcinomas in liver, pharynx, skin, lung, and thyroid were documented at daily dosages of 0.01–0.1 μg of 2,3,7,8-TCDD/kg BW; comparable values for mice were 0.03–0.07 μg/kg BW. No response occurred at continuous daily dose levels of 0.001 to 0.0014 μg/kg BW in rats and 0.001–0.03 in mice. Carcinogenic or co-carcinogenic effects were also induced in rodents by 1,2,3,6,7,8-hexa CDD and 1,2,3,7,8,9-hexa CDD, but only at high dose levels. In rodents, 2,3,7,8-TCDD was a potent tumor promoter and inducer of the cytochrome P4501A family, especially in liver, lungs, and kidneys. In Seveso, Italy, a follow-up study of the human population between 1976 and 1986 showed an increased cancer incidence of soft tissue sarcoma, and hepatobiliary and hematologic neoplasms. Workers exposed to 2,3,7,8-TCDD or higher chlorinated dioxins had an increased risk for all neoplasms, especially soft tissue sarcomas, compared with workers from the same cohort exposed to phenoxy herbicides and chlorophenols but with minimal or no exposure to 2,3,7,8-TCDD and higher chlorinated dioxins.

The greater toxic potential of certain PCDD isomers involves two properties: halogen atoms occupying at least three of the four lateral ring positions (2,3,7,8 positions) and at least one of the adjacent ring positions being nonhalogenated. Comparative toxicity data for selected PCDD isomers to the guinea pig (*Cavia* sp.) and the mouse (*Mus* sp.) confirmed this generalization and demonstrated significant interspecies differences in sensitivity. Other PCDD isomers tested (2,8-di CDD, octa-CDD) were relatively nontoxic to mice and guinea pigs. In marmoset monkeys (*Callithrix jacchus*) given a subcutaneous injection of a mixture of PCDDs and PCDFs, only the 2,3,7,8-substituted congeners were detected at high concentrations in liver and fat, equivalent to 25% of the administered dose for 2,3,7,8-TCDD to 74% for 2,3,4,6,7,8-hexa CDD. The half-time persistence of 2,3,7,8-TCDD was about 8 weeks in hepatic tissues and 11 weeks in adipose tissue. Half-time persistence increased with increasing chlorination and, in the case of OCDD, there was no loss of accumulated OCDD after 28 weeks.

Acute toxicity studies with 2,3,7,8-TCDD have shown marked differences – up to 8400 times – between the single oral LD50 dose for the guinea pig and the hamster (*Cricetus* sp.). The acute oral LD50 value of 0.6 µg/kg BW for guinea pigs suggests that 2,3,7,8-TCDD may be the most toxic compound ever tested on small laboratory animals. The unusual resistance of hamsters may be associated with its enhanced rate of metabolism and excretion of 2,3,7,8-TCDD relative to other PCDD isomers examined. Poisoning in mammals by 2,3,7,8-TCDD is typically characterized by loss of body weight, shrinkage of the thymus, and delayed lethality; large interspecies differences exist in lethal dosages and toxic effects. Thus, 2,3,7,8-TCDD produces prominent chloracne-type skin lesions in man and monkeys, edema formation in birds, and severe liver damage in rats, mice, and rabbits.

Studies with mink (*Mustela vison*) indicate that route of administration, age, and sex of the animal influences 2,3,7,8-TCDD toxicity. Adult male minks given a single oral dose showed a dose-dependent decrease in food consumption and body weight and a calculated LD50 after 28 days of 4.2 µg/kg BW. Livers, spleens, and kidneys were discolored after a single oral dose of 2,3,7,8-TCDD and at high sublethal doses the brain, kidneys, heart, thyroid, and adrenals were enlarged. Newborn mink kits given 0.1 or 1.0 µg/kg BW daily by intraperitoneal injection for 12 consecutive days had mortality in excess of 50% in the high-dose group and growth rate reduction at 0.1 µg/kg BW daily. Adult minks fed diets containing 1.0–80.8 ng 2,3,7,8-TCDD equivalents/kg ration for 85 days (equivalent to total ingestion of 23.0–1019.0 ng 2,3,7,8-TCDD) prior to and throughout the reproductive period had impaired reproduction with reduced body weights and survival in a dose-dependent manner. Females in the highest-dose group whelped the fewest number of kits, all of which were stillborn or died within 24 h. For adult females, a value of 3.6 ng 2,3,7,8-TCDD equivalents/kg BW daily was determined for the lowest observable adverse effect level. Adult females fed diets containing 0.001, 0.01, 0.1, 1.0, 10.0, or 100.0 µg 2,3,7,8-TCDD/kg ration for up to 125 days showed a dose-dependent decrease in food consumption and body weight. At day 125, 63% of the 1.0 µg/kg group had died and 100% of the 10.0 and 100.0 µg/kg groups. Calculated LD50 values were 4.8 µg 2,3,7,8-TCDD/kg diet at day 28 and 0.85 µg/kg diet at day 125, equivalent to 0.264 µg 2,3,7,8-TCDD/kg BW daily for the 28-day exposure and 0.047 µg/kg BW daily for the 125-day exposure.

Intraspecies differences in sensitivity to 2,3,7,8-TCDD – up to 14 fold – are reported among three strains of mice; no reasons were given to account for these differences. Oral LD50 (30 day) values varied from 182.0 µg 2,3,7,8-TCDD per kg BW in strain C57, the most sensitive strain tested, and 296.0 for strain BD6, to 2570.0 for strain DBA. All three strains of mice evidenced a 25–34% weight loss prior to death; however, there was no measurable decline in food consumption.

Atrophy of the thymus is a consistent finding in mammals poisoned by 2,3,7,8-TCDD, and suppression of thymus-dependent cellular immunity, particularly in young animals, may contribute to their death. Although the mechanisms of 2,3,7,8-TCDD toxicity are unclear, research areas include the role of thyroid hormones, interference with plasma membrane functions, alterations in ligand receptors, the causes of hypophagia (reduced desire for food) and subsequent attempts to alter or reverse the pattern of weight loss, and excretion kinetics of biotransformed metabolites.

Teratogenic and fetotoxic effects of 2,3,7,8-TCDD are well-documented in several species of animals. Cleft palate in young mice was associated with daily dosages of 1.0 μg 2,3,7,8-TCDD per kg BW in pregnant females (no-effect level at 0.1 μg/kg), and intestinal hemorrhage was found in sensitive strains of rats given daily dosages of 0.125 μg/kg BW (no-effect level at 0.03 μg/kg). Developing mammalian fetuses are especially sensitive to 2,3,7,8-TCDD, and maternal exposure results in increased frequencies of stillbirths. Among live births, exposure to 2,3,7,8-TCDD produces teratogenic effects such as cystic kidney, cleft palate, and spinal column deformities. Effects of 2,3,7,8-TCDD on reproduction are reported for rats and monkeys. In a three-generation study with rats, daily dose levels of 0.01 μg of 2,3,7,8-TCDD/kg BW (equivalent to 120.0–290.0 ppt or ng/kg in the diet), produced decreased litter size at birth, increased number of stillborns, and reduced survival and growth of young in both the F_1 and F_2 generations. In rats, no adverse effect occurred at daily dosages of 0.001 μg/kg BW – equivalent to 12.0–30.0 ng 2,3,7,8-TCDD ng/diet – on growth, survival, reproduction, metabolism, or cancer incidence during exposure for three generations. Abortion and weight loss were reported in rhesus monkeys (*Macaca mulatta*) at dietary levels as low as 50.0 ppt 2,3,7,8-TCDD (about 0.0017 μg/kg BW daily) after 7 to 29 months. However, comparatively high dosages (200.0 ppt in diets equivalent to 0.0095 μg/kg BW daily) could be tolerated by monkeys for short periods (three times weekly for 3 weeks) with no adverse effects on reproduction. Higher dose levels for extended periods (i.e., 500.0 ppt in diets equivalent to about 0.011 μg/kg BW daily for 9 months) caused death (63%) or, among survivors, abortion, chloracne, nail loss, scaly and dry skin, and progressive weakness. Most treated monkeys remained fairly alert to external stimuli until just prior to death. On removal from the 500.0 ppt 2,3,7,8-TCDD diet and transfer to an uncontaminated diet, a severely affected monkey became pregnant and gave birth to a well-developed infant after an uneventful gestation. This suggests that some 2,3,7,8-TCDD damage effects are not permanent.

Androgenic deficiency in male rats given a single oral dose of 15.0 μg 2,3,7,8-TCDD/kg BW was evident as early as 2 days posttreatment, with persistence up to 12 days. These deficiencies may account for male reproductive pathology and dysfunction in rats treated with overtly toxic doses of TCDD. Findings included depression in plasma testosterone concentrations, as well as decreased weight of seminal vesicles (by 68%), ventral prostate gland (by 48%), testes, and epididymis.

Accumulation of 2,3,7,8-TCDD is reported in the liver of rats during lifetime exposure to diets containing 0.022 μg 2,3,7,8-TCDD/kg, or when administered orally at 0.01 μg/kg BW once a week for 45 weeks. Liver residues of rats fed 2,3,7,8-TCDD were 0.54 μg/kg, or about 25 times dietary levels; livers of rats dosed orally contained 1.05 μg/kg, or about 2.3 times the total dose received on a unit weight basis. Unlike toxicity, elimination rates of accumulated 2,3,7,8-TCDD were within a relatively narrow range. The estimated retention times of 2,3,7,8-TCDD in small laboratory mammals (rats, mice, guinea pigs, and hamsters) extended from 10.8 to 30.2 days for 50% elimination and seemed to be little influenced by species, concentration administered, duration of dose, or route of administration. Half-time persistence in primates is usually lengthy: about 1 year in most species of monkeys and 5.8 years in humans, although the half-time persistence of 2,3,7,8-TCDD in marmoset monkeys (*Callithrix jacchus*) is only 6–8 weeks.

Histopathological effects have been reported in rabbits and horses poisoned by 2,3,7,8-TCDD. Rabbits surviving exposure to an industrial accident in Seveso, Italy, in which 2,3,7,8-TCDD was released, had edema, hemorrhagic tracheitis, pleural hemorrhage, and dystrophic lesions of hepatic tissue. Horses from Missouri that died after waste oil contaminated with 2,3,7,8-TCDD was applied as a dust control agent in riding arenas had liver lesions, skin hyperkeratosis, gastric ulcers, and lung and kidney lesions. Since 2,3,7,8-TCDD is an extremely potent porphyrogenic agent, it is probable that these animals also exhibited porphyria, a condition characterized by fragility of the skin, photosensitivity, and accumulation of porphyrins in the liver.

Thermoregulatory function and vitamin A metabolism in rodents are altered by 2,3,7,8-TCDD. Perinatal exposure to 2,3,7,8-TCDD alters thermoregulatory function in adult rats and hamsters, as judged by reduced body temperature during sleep; however, normal behavioral regulation suggests that hypothalamic thermoregulatory centers are not permanently altered. For example, pregnant rats exposed on gestational day 15 to 1.0 μg 2,3,7,8-TCDD/kg BW by gavage produced male offspring with reduced body temperature during nocturnal phases; TCDD-treated rats given endotoxins had higher fevers than controls. Rats given a single oral dose of 10.0 μg 2,3,7,8-TCDD/kg BW had altered vitamin A turnover and inhibited hepatic accumulation of dietary vitamin A, and this is attributed to a combination of inhibited retinol esterification in hepatic cells, increased release of endogenous vitamin A, and increased hepatic catabolism of retinoids.

Interaction effects of PCDDs with other polychlorinated compounds or mixtures are not extensively documented. For example, certain polychlorinated hexachlorobiphenyls (PCBs) have a low toxic potency to induce cleft palate deformities in mice. However, mixtures of 2,3,7,8-TCDD and 2,3,4,5,3′,4′ hexachlorobiphenyl resulted in a tenfold increase in incidence of cleft palate in mice. Thus, the toxicity of compounds such as 2,3,7,8-TCDD may be enhanced by compounds of relatively low acute toxicity such as selected PCBs. The widespread environmental occurrence of such combinations suggests the need for further evaluation of the mechanism of this interaction.

14.5 Recommendations

No criteria or standards have been promulgated for any of the 75 PCDD isomers by any regulatory agency for the protection of sensitive species of wildlife and aquatic organisms except 2,3,7,8-TCDD (Table 14.2). Data are scarce or missing on the distribution and upper limits of background levels of PCDDs in natural resources (except 2,3,7,8-TCDD), on the identification of fish and wildlife resources potentially at risk, on the relative importance of PCDD sources, and on the comparative toxicities of various PCDDs to fish and wildlife, especially reproductive and immunosuppressive toxicities. Multigeneration reproductive studies are recommended for the most widespread PCDDs, including 1,2,3,7,8-penta CDD and OCDD. Improved analytical accuracy and standardized protocols are needed in PCDD sampling techniques to account for the organic content of sediments, dissolved organic content of water, and lipid content of samples. The stability and mobility of organic-associated PCDDs and their partitioning between substrates needs clarification.

Lethality of 2,3,7,8-TCDD to freshwater teleosts is documented at concentrations as low as 0.038 ng/L in water and 55.0 ng/kg FW in sac fry (Table 14.2). It is noteworthy that water column concentrations of 2,3,7,8-TCDD immediately downstream of 89% of 104 chlorine-bleaching pulp and paper mills exceeded 0.038 ng/L. More research is needed on dietary routes of PCDD exposure to aquatic organisms. A simple apparatus for loading 2,3,7,8-TCDD and other chemicals onto commercially available pelletized fish food is now available. Recommended safe levels for aquatic life protection are 0.01 ng/L in water and 34.0 ng/kg FW tissue, and these confer a degree of protection based on lethal action – of 1.6–3.8 times (Table 14.2). Values for 2,3,7,8-TCDD protection of freshwater aquatic life are in sharp contrast to those recommended for birds and mammals, which were usually

Table 14.2. Proposed 2,3,7,8-TCDD criteria for the protection of natural resources and human health.

Resource, Criterion, and Other Variables	Effective 2,3,7,8-TCDD Concentration
AQUATIC ORGANISMS, FRESHWATER	
Water	
Acceptable	<0.01 ng/L (parts per trillion)[a]
Safe	0.01–<0.038 ng/L[b]
No observed adverse effects on rainbow trout, *Oncorhynchus mykiss*; juveniles	<0.038 ng/L
Unacceptable	
Adverse effects on growth, survival and swimming of rainbow trout juveniles	0.038 ng/L
Adverse effects expected in sensitive species	0.1–1.0 ng/L and higher
Death observed in sensitive species several weeks after exposure for 6 days	10.0 ng/L
Death expected in sensitive species shortly after exposure	>1000.0 ng/L
Tissue residues	
Lake trout, *Salvelinus namaycush*	
Egg and sac fry	
No effect on survival	34.0 ng/kg FW
Some deaths	55.0 ng/kg FW
50% dead by swim-up	65.0 ng/kg FW
Whole body	
Reproduction inhibited	78.0 ng/kg FW
Rainbow trout; adverse effects expected; egg	More than 400.0 ng/kg FW
BIRDS	
Diet	
Safe	10.0–12.0 ng/kg FW ration
Adverse effects	
Domestic chicken, Gallus sp.	Equivalent to about 1.0 ng/kg BW daily
Woodcock, Philohela minor; adults; Maine	Daily intake of >6.0 ng/kg BW, and half-time persistence of 7.2 days
Soils, Maine	
No adverse effects expected in woodcocks after consuming earthworms and insects from 2,3,7,8-TCDD contaminated soils	<50.0 ng/kg FW soil
Adverse effects observed in woodcocks	>27.0–250.0 ng/kg FW soil
Tissue residues	
Wood duck, *Aix sponsa*; eggs; adverse effects on survival	>20.0–50.0 ng/kg FW
MAMMALS	
Diet	
No effect on carcinogenicity or reproduction in laboratory white rat, *Rattus* sp.	Equivalent to <1.0 ng/kg BW daily

Continued

Table 14.2. cont'd

Resource, Criterion, and Other Variables	Effective 2,3,7,8-TCDD Concentration
Safe; laboratory white rat; exposure for three generations	12.0–30.0 ng/kg FW ration, equivalent to about 1.0 ng/kg BW daily
Safe; mammalian wildlife	10.0–12.0 ng/kg FW ration
Acceptable; daily intake; lifetime exposure	<0.01 ng/kg BW[c]
Unacceptable, monkeys	>50.0 ng/kg FW ration equivalent to >1.7 ng/kg BW daily
Subcutaneous injection; marmoset monkey, *Callithrix jacchus*; adverse sublethal effects	Persistent altered blood chemistry after single injection of 10.0 ng/kg BW
HUMAN HEALTH	
Diet, acceptable	
Fish muscle	
New York	<10.0 ng/kg FW fillet
Canada	<20.0 ng/kg FW fillet
Most U.S. states	<25.0 ng/kg FW fillet
Diet, limited	
Fish muscle; may be eaten once weekly by occasional consumers of fish and up to twice monthly for those who eat contaminated fish year round	25.0–50.0 ng/kg FW fillet
Diet, unacceptable	>50.0 ng/kg FW ration
Soils, acceptable	
Residential areas	<1000.0 ng/kg FW
Of grazing dairy cattle	<6.0 ng/kg FW
Cancer risk	
Lifetime cancer risk increases from drinking water and eating fish from 2,3,7,8-TCDD contaminated waters	
Cancer risk of 1 in 1,000,000	0.000013 ng/L water
Cancer risk of 1 in 100,000	1.0 ng/L water
Drinking water, acceptable	<0.015 ng/L
Ambient water, acceptable	<0.01 ng/L
Ambient air, acceptable	<0.03 ng/m^3
Total daily intake from all sources; maximum allowed	<160.0 ng/kg BW
Total daily intake from all sources; recommended	<61.0–<100.0 ng/kg BW
Total daily intake from all sources; lifetime exposure; acceptable	<0.01 ng/kg BW[c]

[a]0.01 ng/L was the highest 2,3,7,8-TCDD concentration tested that had no measurable adverse effect on freshwater teleosts.
[b]Based on bioconcentration factor of 5000.
[c]Based on no-observable-adverse effect concentration of 1.0 ng/kg BW daily in three-generation rat study and an uncertainty factor of 100.

derived from the highest concentration tested that produced no observed adverse effect, and on uncertainty factors of 100 or more (Table 14.2). Proposed 2,3,7,8-TCDD criteria for birds include <12.0 ng/kg FW ration, whole-body concentrations of <1.0 ng/kg BW in domestic chickens and <6.0 ng/kg BW in woodcocks, and <20.0 ng/kg FW in wood duck eggs (Table 14.2). Concentrations of 2,3,7,8-TCDD as high as 50.0 ng/kg in soil were not likely to produce adverse effects in woodcocks, nor does this soil concentration present a significant risk to humans who may consume game birds inhabiting these sites. Proposed criteria for nonhuman mammals include <0.01 ng/kg BW daily from all sources and 10.0–30.0 ng/kg ration (Table 14.2). To protect human health, edible fish tissues should contain <25.0 ng/kg FW in most states and <10.0 ng/kg FW in New York; dairy cattle need to be grazed on soils containing <6.0 ng/kg FW; drinking water should contain <0.015 ng/L and ambient air <0.03 ng/m^3. Food items containing more than 50.0 ng/kg FW are considered unsafe for human consumption, but fish fillets containing between 25.0 and 50.0 ng/kg of 2,3,7,8-TCDD may be eaten once weekly by occasional consumers of fish, and twice monthly for those who eat contaminated fish year round (Table 14.2). It is not known at this time whether residues of 10.0 to 50.0 ng/kg (or higher) of 2,3,7,8-TCDD in fish flesh represents an unacceptable risk to the growth, survival, reproduction, metabolism, or behavior of the teleost, or to its predators; clearly, this is a high-priority research topic. Also, mechanisms of PCDD toxicity need to be established in mammals, birds, and poikilotherms to support extrapolation to humans and other species.

More research is needed on factors that alter bioconcentration, toxicokinetics, and metabolism of PCDDs. For example, field bioconcentration factors (BCF) for 2,3,7,8-TCDD in fish from contaminated rivers in Maine ranged from 3000 to 106,000, and most exceeded the value of 5000 recommended by the U.S. EPA. BCF values between 15,000 and 25,000 are now considered a reasonable estimate of the BCF for use in regulatory purposes in the State of Maine. Mixtures of 2,3,7,8-TCDD and PCB 77 were slighter greater than additive in toxicity to rainbow trout, indicating the need for more research on chemical interactions. Daily intakes of PCDDs vary substantially between and within species, and this needs to be incorporated into future models of risk assessment. In humans, for example, daily intake of 2,3,7,8-TCDD – on a ng/kg BW basis – is 78 times higher in neonates than adults, and 2.3 times higher in children and 5.2 times higher in infants than adults.

In the past, the major source of 2,3,7,8-TCDD in the environment was as a contaminant in phenoxy herbicides (such as 2,4,5-T; Silvex; 2,4-D; and Agent Orange), in hexachlorophene, and in other chlorophenol-type compounds. Concentrations of 2,3,7,8-TCDD in some of these products exceeded 60,000.0 μg/kg. This situation has been largely corrected by new manufacturing processes and by increasingly stringent federal regulations. For example, 2,3,7,8-TCDD level in 2,4,5-T has decreased from 60,000.0 μg/kg in 1957 to 2000.0 μg/kg in 1965 as a result of new manufacturing processes, and it was limited to 500.0 μg/kg in 1970 by the Canadian federal government. In 1970, the U.S. Department of Defense halted the spraying of Agent Orange. In 1972, the U.S. Food and Drug Administration banned the use of hexachlorophene in nonprescription soaps and deodorants. In 1978, 7 of 14 major producers of 2,4,5-T no longer manufactured this product, and the remainder claimed that their products contained less than 100.0 μg/kg of 2,3,7,8-TCDD. In 1979, production of 2,4,5-T and Silvex ceased in the United States, although stockpiles of both are still being distributed and permitted for use on rice fields, sugarcane fields, orchards, fence rows, vacant lots, and lumber yards. In 1982, the EPA required some industries to certify that chlorophenol-type compounds were no longer used as slime-control agents. On October 18, 1983, EPA published its intent to cancel the registration of pesticide products containing 2,4,5-T and Silvex, and to prohibit the transfer, distribution, sale, or importation of any unregistered product containing 2,4,5-T, Silvex, or their derivatives. Continued monitoring is recommended of food, air, and sediments, including time trends and determinations of isomer patterns.

Burning or heating of commercial and purified chlorophenates, and pyrolysis of polychlorinated biphenyls contaminated with trichlorobenzenes can result in the production of 2,3,7,8-TCDD and other PCDDs. These sources, together with discharges from various municipal and industrial incinerators of chlorinated compounds, probably constitute the largest source of PCDDs in the environment today. In 1983, the U.S. EPA proposed to monitor 2,3,7,8-TCDD in the environment. Specific goals of the monitoring program include: determination of 2,3,7,8-TCDD concentrations in soils and biota, with emphasis on geographic areas where PCDDs may have been manufactured, used, or stored, and where concentrations may be in excess of 1000.0 ng/kg; monitoring of industrial and municipal incinerators for 2,3,7,8-TCDD emissions; and establishment of background levels for PCDDs in areas where these compounds are not expected to occur in high levels. Information is also needed on the toxicological interactions of groups of polychlorinated chemicals (such as certain biphenyls, biphenylenes, and dibenzofurans) known to be isosteric with 2,3,7,8-TCDD and which frequently coexist with 2,3,7,8-TCDD in environmental samples. Acquisition of these data should provide the basis of a risk assessment analysis for dioxin and fishery and wildlife resources.

14.6 Summary

Polychlorinated dibenzo- *para*-dioxins (PCDDs) are present as trace impurities in some manufactured chemicals and industrial wastes. The chemical and environmental stability of PCDDs and their tendency to accumulate in fat have resulted in their detection within many ecosystems. In general, wherever high levels of PCDDs have been detected, the source has been a hazardous waste dump, an industrial discharge, or an application of PCDD-contaminated herbicide. There are 75 PCDD isomers; some are extremely toxic, while others are believed to be relatively innocuous. The most toxic and most extensively studied PCDD isomer is 2,3,7,8-tetrachlorodibenzo-*para*-dioxin (2,3,7,8-TCDD). In the United States and elsewhere, accidental contamination of the environment by 2,3,7,8-TCDD has resulted in deaths of many species of wildlife and domestic animals. High residues of 2,3,7,8-TCDD in fish, i.e., more than 50.0 ppt wet weight, have resulted in closing rivers to fishing. In the most seriously affected areas, hospitalization and permanent evacuation of humans has been necessary. Laboratory studies with birds, mammals, aquatic organisms, and other species have demonstrated that exposure to 2,3,7,8-TCDD can result in acute and delayed mortality as well as carcinogenic, teratogenic, mutagenic, histopathologic, immunotoxic, and reproductive effects. These effects varied greatly among species.

No regulations governing PCDD contamination exist at present to protect sensitive species of wildlife and aquatic organisms. Data available suggest that 2,3,7,8-TCDD concentrations in water should not exceed 0.01 ppt to protect aquatic life, or 10.0 to 12.0 ppt in food items of birds and other wildlife. Additional data are needed in several areas: concentrations of PCDDs in natural systems; identification of fish and wildlife populations at risk; relative importance of PCDD sources; toxicological effects of various PCDDs to aquatic biota and wildlife, especially reproductive and immunosuppressive effects; and toxic and other interaction effects of PCDDs with other groups of polychlorinated chemicals having similar structure and properties, such as biphenyls, dibenzofurans, and biphenylenes.

FAMPHUR[a]

Chapter 15

15.1 Introduction

Famphur (phosphorothioic acid, *O*-[4-[(dimethylamino) sulfonyl], phenyl] *O*,*O*-dimethyl ester) also known as Warbex, is a systemic organophosphorus insecticide found effective against lice, grubs, flies, and gastrointestinal nematodes of ruminants. Introduced commercially in 1961, the compound is especially effective against cattle grubs (*Hypoderma* spp.) when fed in the diet, injected subcutaneously or intramuscularly, or applied as a pour-on and oral drench treatment.

Many dead birds, including robins, hawks, and magpies, were found after cattle were treated with pour-on applications of famphur. The black-billed magpie (*Pica pica*) was especially sensitive; ranchers reported observations of magpies dying after famphur use on cattle as early as 1973. Dead magpies usually had famphur in the gizzard contents and severely depressed brain cholinesterase activity – a characteristic of organophosphorus poisoning. Populations of the black-billed magpie in western states declined between 1968 and 1979, which coincides with widespread use of famphur in that region; however, factors other than famphur may have caused the decline.

[a]All information in this chapter is referenced in the following sources:

Eisler, R. 1994. Famphur hazards to fish, wildlife, and invertebrates: a synoptic review. *U.S. Natl. Biol. Surv. Biol. Rep. 20*. 23.

Eisler, R. 2000. Famphur. Pages 1067–1088 in *Handbook of Chemical Risk Assessment: Health Hazards to Humans, Plants and Animals. Volume 2, Organics*. Lewis Publishers, Boca Raton, Florida.

15.2 Uses

The cholinesterase-inhibiting and intoxicating properties of organophosphorus compounds have been known since these products were first synthesized about 80 years ago. During World War II, the more toxic organophosphorus compounds – such as soman, tabun, and sarin – were stockpiled for use as potential chemical warfare agents. More than 50,000 organophosphorus compounds have been synthesized and screened for possible insecticidal and antihelminthic activity and several dozen, including famphur, are available commercially. During the 1970s, most organochlorine insecticides were removed from common use in North America, Europe, and most developed countries, and the removal increased reliance on carbamate and organophosphorus compounds, – the two major classes of cholinesterase-inhibiting pesticides. The relative lack of target specificity of these compounds and their high acute toxicity to many nontarget organisms were ignored in favor of their short-term environmental persistence and lack of accumulation in organisms. The anticholinesterase insecticides now account for the majority of globally registered insecticides.

Famphur is now not applied to forests or crops but is used almost exclusively as a veterinary chemical. A single treatment controls cattle grubs and reduces cattle-lice infestations. Famphur is especially effective against maggots of the botfly and warble fly (*Hypoderma* spp.) and of other oestrid dipterous flies. Eggs from this group of insects are laid on the feet and legs of cattle and other mammals and licked off by the host and hatch in the mouth or esophagus. The resultant larvae burrow through the tissues to the skin of the animal's back where they live until ready to pupate and cause warbles or swellings. When applied carelessly, famphur and other systemic insecticides are highly toxic and frequently produce acute poisoning in ruminants. Famphur is not now registered or regulated by the U.S. EPA.

Cholinesterase-inhibiting agents such as famphur vary widely in their effectiveness of controlling target pests and depend on the route of administration, dose rate, formulation, and timing and frequency of applications. Famphur can be administered to livestock by intramuscular injection, orally in the diet, as a dermal pour-on, or as a bolus. Intramuscular injections of a 35% famphur concentrate are usually given in the gluteal muscle. When fed in the diet, famphur is formulated as a 33.3% liquid feed premix. The topical use of famphur as a systemic insecticide was recommended in 1970. As a pour-on over part of the backline of cattle at dosages between 15.0 and 35.0 mg/kg body weight (BW), 12.5–13.2% w/v famphur applied in the autumn controls warbles before they develop into grubs the following year and controls various species of lice. When used as a pour-on for cattle-tick control, famphur may be transferred from treated cattle to untreated animals, presumably through body contact. The solvent used in preparing dermal formulations of famphur significantly affects absorption hazards. In the case of the laboratory white rat (*Rattus* spp.), corn oil proved to be the least hazardous solvent and acetone, the most hazardous; benzene was intermediate. Famphur can also be administered in a rumen bolus as a systemic insecticide against ticks in cattle. Boluses have been designed to release 200.0 mg famphur/bolus daily over a 65–75 day post-ingestion period; actual release rates range from 207.0 to 308.0 mg daily.

Technical information from the manufacturer lists five precautions and warnings for the use of famphur:

1. Famphur is "Toxic to fish, birds, and other wildlife. Keep out of lakes, ponds, and streams. Do not apply to areas where run-off occurs. Do not contaminate water by cleaning of equipment or disposal of wastes."
2. After use, all containers should be drained and rinsed several times with a solution of water, detergent, and lye ("bury rinse solution deeply in an isolated location with 18 inches of cover"); the empty container should be punctured and crushed to prevent reuse.
3. Famphur should not be used in combination with any compound having cholinesterase-inhibiting activity either simultaneously or within a few days before or after treatment.
4. Famphur use on livestock is contraindicated for calves less than 3 months old; animals stressed from castration, dehorning, or overexcitement; and sick or convalescent animals. Brahman and Brahman crossbreeds are less tolerant of cholinesterase-inhibiting insecticides than other breeds, and Brahman bulls are especially sensitive and should not be treated with famphur. Cattle should not be slaughtered for at least 35 days after treatment with famphur.
5. For humans, famphur is considered harmful or fatal if swallowed or absorbed through the skin, especially by children. If poisoning should occur, physicians are advised that atropine is antidotal and that pralidoxime chloride may be effective as an adjunct to atropine. Pour-on formulations are flammable, and users should keep them away from heat, sparks, and open flames including hot branding irons and cautery dehorning devices.

15.3 Chemistry and Metabolism

Some physical and chemical properties of famphur are listed in (Table 15.1). Gas chromatography is used to measure famphur and its oxygen analog famoxon in bovine milk, blood, and edible tissues; detection limits are 0.005 mg/L in milk and <0.01 mg/kg in tissues. The main degradation routes of famphur in mammals occur through hydrolysis of the *P–O* phenyl, *P–O* methyl, and *N*-methyl bonds; oxidative desulfuration and *N*-demethylation take place to a small extent (Figure 15.1). In the metabolic scheme for famphur in mammals (Figure 15.1), only famphur and its oxygen analog, famoxon, were

Table 15.1. Chemical and other properties of famphur.

Variable	Datum
CHEMICAL NAMES	Phosphorothioic acid *O*-[4-[(dimethylamino) sulfonyl], phenyl] *O*,*O*-dimethyl ester; Phosphorothioic acid, *O*,*O*-dimethyl-, *O*-ester with *p*-hydroxy-*N*, *N*-dimethylbenzene sulfonamide; Phosphorothioic acid, *O*,*O*-dimethyl *O*-*p*-(dimethylsulfamoyl) phenyl ester; *O*-Dimethyl hydrogen phosphorothioate, *O*-ester with *p*-hydroxy *N*, *N*-dimethylbenzenesulfonamide; *O*-[4-1-(Dimethylamino) sulfonyl] phenyl phosphorothioic acid *O*,*O*-dimethyl ester; *O*,*O*-Dimethyl *O*,*p*-(*N*, *N*-dimethylsulfamoyl) phenyl phosphorothioate; *O*,*p*-(Dimethylsulfamoyl) phenyl *O*,*O*-dimethyl phosphorothioate; *p*-(Dimethylsulfamoyl) phenyl dimethyl phosphorothioate; *O*,*O*-dimethyl *O*-[*p*-(dimethylsulfamoyl)-phenyl] phosphorothioate; Dimethyl *p*-(dimethylsulfamoyl) phenyl phosphorothionate; *O*,*O*-dimethyl-*O*,*p*-(dimethylsulfamoyl) phenyl phosphorothionate
ALTERNATE NAMES	AC 38023, American Cyanamid 38023, Bo-Ana, CL 38023, Cyflee, Dovip, ENT 25644, Famaphos, Famfos, Famophos, Famphos, Fanfos, Warbex, 38023
PRIMARY USE	Systemic livestock insecticide
CAS NUMBER	52-85-7
EMPIRICAL FORMULA	$C_{10}H_{16}NO_5PS_2$
MOLECULAR WEIGHT	325.36
MELTING POINT, CRYSTALS VS. POWDER	52.5–53.5°C vs. 55°C
SOLUBILITY	
Chlorinated hydrocarbons	Highly soluble
Water	−100 mg/L
Polar solvents	Slightly soluble
Aliphatic hydrocarbons	Insoluble

of toxicological significance, as judged by acute oral toxicity in mice. In studies with mice, acute oral LD50 values in mg/kg BW were 27.0 for famphur; 18.0 for famoxon; 2270.0 for *O*-desmethylfamphur; 860.0 for *O*, *N*-bisdesmethylfamphur; 2290.0 for *p*-(*N*, *N*-dimethylsulfamoyl)phenol; 2500.0 for *p*-(*N*-methylsulfamoyl)phenol; 6400.0 for *p*-hydroxybenzenesulfonic acid; and >5000.0 for *p*-(*N*,*N*-dimethylsulfamoyl)phenyl glucuronide.

Famphur residues of 1.0–3.0 mg/kg fresh weight (FW) are common in cattle tissues after normal pour-on applications of the chemical. The half-time persistence of famphur in subcutaneous fat of cattle after a single pour-on application was 0.9 days and was independent of dose within the range of 25.0–150.0 mg/kg BW or initial tissue residues between 1.8 and 12.3 mg/kg FW; fat residues were <0.08 mg/kg FW 5 days after treatment and <0.01 mg/kg FW after 11 days.

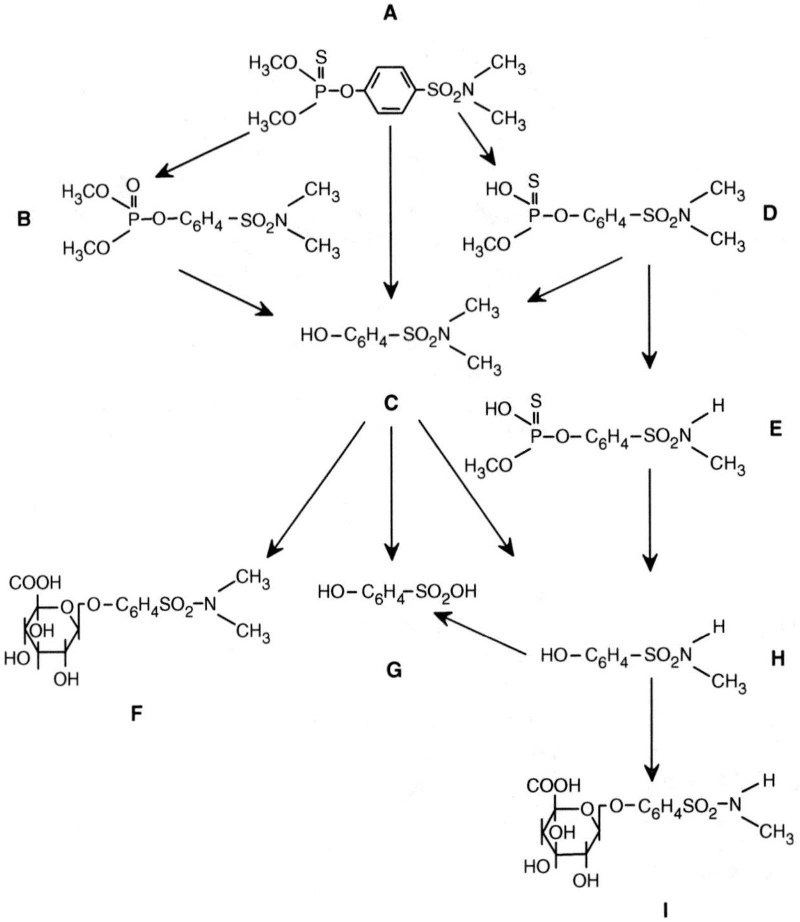

Figure 15.1.
Metabolic scheme for famphur in mammals. Major metabolic routes are indicated by an asterisk (*). A, famphur; B, famoxon; C, *p*-(*N*,*N*-dimethylsulfamoyl)phenol; D, *O*-desmethylfamphur; E, *O*,*N*-bisdesmethylfamphur; F, *p*-(*N*,*N*-dimethylsulfamoyl) phenyl glucuronide; G, *p*-hydroxybenzene sulfonic acid; H, *p*-(*N*-methylsulfamoyl) phenol; and I, *p*-(*N*-methylsulfamoyl) phenyl glucuronide. According to this scheme, famphur (A) initially undergoes oxidation at the P=S bond to yield famoxon (B), hydrolysis at the P-*O*-phenyl bond to yield the transitory *p*-(*N*,*N*-dimethylsulfamoyl) phenol (C)*, or hydrolysis at one of the P-*O*-methyl bonds to yield *O*-desmethylfamphur (D)*. *p*,*N*,*N*-dimethylsulfamoylphenol (C) may also arise by hydrolysis of famoxon (B) or *O*-desmethylfamphur (D)*. *P*-(*N*,*N*-dimethylsulfamoyl) phenol (C) is immediately conjugated to form *p*-(*N*,*N*-dimethylsulfamoyl) phenyl glucuronide (F)* or transformed to *p*-hydroxybenzene sulfonic acid (G) or *p*-(*N*-methylsulfamoyl) phenol (H). *O*-desmethylfamphur (D) can also give rise to *O*,*N*-bisdesmethylfamphur (E)* by removal of one of the methyl groups of the sulfonamide moiety. *O*,*N*-bisdesmethylfamphur (E) is hydrolyzed to the corresponding transitory *p*-(*N*-methylsulfamoylphenol) (H)*, which is immediately conjugated to yield the corresponding glucuronide, *p*-(*N*-methylsulfamoyl) phenyl glucuronide (I)*.

These observations suggest that famphur tissue residues are near or below detection levels within 1 week after treatment, even with gross misuse of the chemical. However, because famphur persists on cattle hair for >90 days at concentrations of >1000.0 mg/kg, this has serious implications on the local populations of birds.

Famphur and other organophosphorus compounds are metabolized and excreted with greater efficiency by mammals than the target pests before these compounds can bind to and ultimately inhibit the cholinesterase enzyme. Mice, for example, degrade famphur rapidly. Less than 1 h after an intraperitoneal famphur injection of 1.0 mg/kg BW, only 8.34% of the original administered dose remained in the mouse: 8.11% as the parent famphur, 0.22% as famoxon, and 0.01% as desmethylfamphur. Famphur's biocidal properties are associated with its ability to inhibit cholinesterase activity, blocking synapses at the neuromuscular junction. Brain cholinesterase inhibition is often used to diagnose death of wildlife after exposure to famphur and other organophosphorus insecticides. It is emphasized that the type and number of cholinesterase compounds and cholinesterase activities vary widely between species and tissues, and activities are further modified by metabolic factors, age, genotype, circadian rhythms, sex, reproductive status, nutritional status, ambient temperature, and disease.

15.4 Lethal and Sublethal Effects

Famphur controls many species of pestiferous insects that afflict poultry and livestock. The LD50 values for target insects ranged from 2.4 to 4.1 mg/kg BW from a dermal route and 8.0–11.8 mg/kg BW from abdominal injection. Toxicity of famphur is often associated with differential degradation and cholinesterase sensitivity among various species of target pests. Famoxon is more effective than famphur in producing cholinesterase inhibition and death, and this confirms the generalization that the corresponding oxons are the more potent anticholinesterase agents.

No published data are available on famphur toxicity to aquatic life. Other data, however, suggest that acute famphur toxicity to fishes may be comparable to that of other phosphorothioate insecticides. Among birds, sensitive species had reduced survival after single oral famphur doses of 1.8–3.0 mg/kg BW or when fed diets containing 35.0–49.0 mg famphur/kg ration. Daily oral famphur doses as low as0.3 mg/kg BW caused depressed cholinesterase activity in the brain and in plasma. Secondary poisoning of eagles and hawks foraging on famphur-killed vertebrates and tertiary poisoning of a great horned owl (*Bubo virginianus*) feeding on a famphur-poisoned hawk are documented. Famphur has also been used illegally to kill birds – including migratory waterfowl and other federally protected species – thought to be depredating crops. Famphur-induced mortality in mammals was documented at concentrations as low as 11.6 mg/kg BW in intraperitoneal injection (mouse), 27.0 mg/kg BW in oral exposure (mouse), >33.3 mg/kg BW in intramuscular injection (Brahman cattle), and 400.0 mg/kg BW in dermal application (rat). In reindeer, altered blood chemistry was evident one year after famphur exposure. Famphur is metabolized rapidly by mammals; residues in animal tissues and milk–regardless of mode of administration, length of exposure, or dose–were usually not detectable within 4 days of final exposure.

15.4.1 Terrestrial Invertebrates

Famphur controls many species of pestiferous insects that afflict poultry and livestock, especially warble flies (*Hypoderma* spp.). Famphur is one of the most toxic compounds for the control of adults and late instars of the lesser mealworm (*Alphitobius diaperinus*), the most abundant beetle inhabiting poultry litter and manure. *Alphitobius* can transmit several diseases to poultry, including avian leukosis – one of the most costly diseases for the poultry industry. By tunneling, *Alphitobius* can destroy polyurethane and polystyrene panels adjacent to manure. Famphur also controls the lesser mealworm in nests of birds and in bat roosts. The northern fowl mite (*Ornithonyssus sylvilarum*) is the most important ectoparasite of commercial breeders and laying hens in the United States. However, attempts to control northern fowl mite with famphur were ineffective regardless of tested mode of administration.

Cattle lice (*Haematopinus* spp.) were controlled when the equivalent of 2.5 mg famphur/kg BW in diets was fed to cattle for at least 30 days or 40.5 mg/kg BW were applied as a topical pour-on. Famphur was used in 1971 to control cattle lice at pour-on applications equivalent to 15.0–35.0 mg/kg BW. Pour-on treatments of Australian yearling heifers were especially effective in controlling the long-nosed cattle louse (*Linognathus vituli*) and the short-nosed cattle louse (*Haematopinus eurysternus*); untreated heifers grew more slowly than famphur-treated heifers. Larvae of the hornfly (*Haematobia irritans*) were controlled in manure of cattle fed famphur at 2.5–5.0 mg/kg BW daily. Manure of treated cows contained low concentrations of famphur (as much as 0.14 mg/kg FW) 1 day after diet cessation, but residues were undetectable thereafter.

In Alaska, reindeer (*Rangifer tarandus*) infested with reindeer warble fly (*Oedemagena tarandi*) produced hides of little value and low-quality meat. Reindeer warble flies were not controlled by pour-on applications of famphur because the product was unable to penetrate the hair coat of reindeer; however, intramuscular injections were effective. In Norway, Sweden, and Finland, famphur was the most promising control agent against reindeer warble fly and reindeer nostril fly (*Cephenomyis trompe*) – two parasites that together caused a 15–20% annual loss of total yield in reindeer husbandry.

Famphur was not very effective in the control of ticks. The tropical horse tick (*Anocentor nitens*) is a species of serious concern to horse breeders and raisers in Florida mainly because it transmits *Babesia caballi*, the causative agent of equine piroplasmosis. A secondary concern is that heavy tick infestations may cause injury to the ears of the horse. Data are unavailable on famphur control of ticks in horses; however, famphur was 99.9–100% effective in controlling *A. nitens* in Hereford steers and heifers when fed in the diet at 5.0 mg/kg BW for 14–21 days. Famphur at 2.5 mg/kg BW in cattle diets for 7 days was only partially effective (39–87.5%) in controlling horse ticks. Famphur – despite multiple treatments – was not effective in controlling cattle ticks (*Haemaphysalis longicornis*) when used as a pour-on at recommended application rates in weaned Hereford calves.

Results of selected studies of famphur and insects indicate several trends: males are more sensitive than females; the oxygen analog, famoxon, is more toxic than the parent chemical; dermal LD50 values range from 2.4–4.1 mg/kg BW; abdominal injection LD50's range from 8.0 to 11.8 mg/kg BW; and metabolic degradation rates vary widely between species. Famoxon is about 100 times more effective than famphur in controlling house flies (*Musca* spp.), which confirms the generalization that the corresponding oxons are the most effective anticholinesterase agents and are, in fact, the actual toxicants.

Differences in the toxicity of organophosphorus compounds among species are often associated with differential degradation rates, pathways, and metabolites. Although injections of famphur were equally toxic to mice (*Mus* sp.), the American cockroach (*Periplaneta americana*), and the milkweed bug (*Oncopeltus fasciatus*), famphur was rapidly degraded by mice (91.7% degraded within 1 h after injection) and cockroaches (81.5% in 1 h); however, milkweed bug degraded only 15.4% during a similar period. The variations in degradation rate among mice and cockroaches were relatively small, about 1.9-fold. Despite the great similarity in famphur toxicity to mice and cockroaches, net famoxon production – like famphur persistence – was very low in the mouse but ten times higher in the milkweed bug. The cholinesterase activity in the milkweed bug was 32 times more resistant to inhibition by famoxon than either mouse or cockroach cholinesterase, and this could account for the comparatively slow breakdown of famphur by the milkweed bug.

There is a correlation among cholinesterase-activity depression in rabbit blood, depression of cholinesterase activity in ectoparasites feeding on the blood of the host, and mortality of ectoparasites. In one case, rabbits (*Oryctolagus* sp.) parasitized by the yellow fever mosquito (*Aedes aegypti*) and Rocky mountain wood tick (*Dermacentor andersoni*) were treated with 5.0–50.0 mg of famphur/kg BW administered orally, subcutaneously, or intravenously. Regardless of dose or route of administration, tick and mosquito mortality was related to cholinesterase-activity levels in rabbit plasma and erythrocytes. Some ectoparasite deaths were noted when cholinesterase levels in rabbits were depressed 32%; ectoparasite mortality increased to 90% at 33% depression and to 100% at 68% cholinesterase inhibition. In general, wood ticks and mosquitoes reflected cholinesterase activity levels of the host rabbit. Surviving female ticks that fed on dosed hosts laid no eggs during a 32-day post-removal observation period. Mosquitoes that had fed on famphur-dosed hosts were more susceptible to cold than those that fed on control hosts.

15.4.2 Aquatic Organisms

An extensive literature search revealed no published data on famphur toxicity to aquatic animals. Unpublished studies of acute lethality were, however, conducted with the bluegill (*Lepomis macrochirus*) and rainbow trout (*Oncorhynchus mykiss*). In those studies, the range in LC50 values at 96 h was 18–21 mg/L in bluegills and 4.9–5.3 mg/L in rainbow trout; the no-observable-effect concentration at 96 h ranged from 14 to 18 mg/L in bluegills and was 2.1 mg/L in rainbow trout.

Although no data were available on the effects of famphur in aquatic ecosystems, there is a substantial data base on other organophosphorus insecticides. For example, methyl parathion (*O*,*O*-dimethyl *O*-(*p*-nitrophenyl) phosphorothioate), another phosphorothioate organophosphorus insecticide, had LC50 (96 h) values for bluegills and rainbow trout that were similar to those of famphur: 5.7 mg/L and 2.7 mg/L, respectively. But exposure for 96 h is not sufficient to satisfactorily evaluate the aquatic toxicity of organophosphorus insecticides. The mortality of adult northern puffers (*Sphaeroides maculatus*) continuously exposed to 20.2 mg/L of methyl parathion was <5% in 96 h but 100% in 40 days. Puffers refused to eat during exposure, and survivors between days 10 and 40 showed complete inhibition of serum esterase activity, zinc-depleted liver and gills, and altered blood chemistry. In another study, male guppies (*Poecilia reticulata*) held in sublethal concentrations (0.01–1.0 mg/L) of methyl parathion for 40 days or longer showed a dose-dependent decrease in spermatogenesis. Pesticide-induced mortality patterns of representative organophosphorus compounds are also modified by water temperature, pH, and salinity. The mummichog (*Fundulus heteroclitus*), an estuarine cyprinodontiform teleost, was most sensitive to organophosphorus compounds at elevated temperatures, reduced salinities, and low pH. Duration of exposure to organophosphorus compounds also affects mummichog survival: fish exposed to high (LC75, 24 h) concentrations of representative insecticides for more than 30 min died by day 21, postexposure; some insecticides were as much as 8.3 times more toxic after exposure for 240 h than 96 h, as judged by LC50 values. In general, crustaceans were more sensitive than teleosts – sometimes by several orders of magnitude – to organophosphorus insecticides in 96-h tests; grass shrimp (*Palaemonetes vulgaris*) and fishes were most sensitive to organophosphorus insecticides at high salinities in the 1.2–3.6% test range and high temperatures in the 10–30°C test range. Marine clams and gastropods were comparatively resistant to organophosphorus insecticides; none died in 96-h exposure to 25.0 mg/L of five organophosphorus insecticides, including methyl parathion. But during a post-exposure observation of 133 days, some bivalves and gastropods died, and these deaths are similar to the delayed mortality for some species of mammals and invertebrates after exposure to certain organophosphorus insecticides.

The expected continued use of famphur in the environment and its vehicular transport along roads that border navigable waters suggests a need for aquatic toxicity data. Famphur data – like those on other organophosphorus insecticides – should reflect the influence of dose, exposure duration, formulation, and other biological and abiotic variables on growth, survival, and metabolism of representative species of aquatic organisms.

15.4.3 Birds

The avian acute oral LD50 of famphur is usually between 1.0 and 9.5 mg/kg BW. Laboratory studies with sensitive species of birds show reduced survival after a 5-day consumption of diets containing 35–49 mg famphur/kg ration. Depressed cholinesterase activity in the brain and in plasma of European starling (*Sturnus vulgaris*) nestlings occurred after 15 daily oral exposures of concentrations as low as 0.3 mg famphur/kg BW. Signs suggesting famphur poisoning in mallards (*Anas platyrhynchos*) included regurgitation, goose-stepping, ataxia, wing drop, tremors, and tonic seizures.

Famphur is considered a Class-II-toxic compound to the Japanese quail (*Coturnix japonica*). Class-II compounds (very toxic) kill 50% of the test organisms on diets containing 40.0–200.0 mg chemical/kg ration for 5 days followed by a 3-day observation. In comparison, the 50% killing in other classes (in mg/kg diet) is <40.0 in Class I (highly toxic), >200.0–1000.0 in Class III (moderately toxic), >1000.0–<5000.0 in Class IV (slightly toxic), and >5000.0 in Class V (practically nontoxic). Some investigators have rated famphur as a Class-I-toxic compound, based on results from dietary tests with mallards.

Birds killed by organophosphorus compounds in the wild consistently show 80–95% depression of brain-cholinesterase activity. Depression of brain-cholinesterase activity by >20% in birds has been used as a conservative criterion to indicate significant exposure to organophosphorus chemicals. Depression of brain-cholinesterase activity by >50% and confirmation of suspected organophosphorus chemical residues in tissues or ingesta are criteria for cause-effect diagnosis of death in birds exposed to cholinesterase-inhibiting chemicals. Death occurs in many avian species when brain-cholinesterase inhibition is 60–90%; however, no barn owls (*Tyto alba*) died or showed signs of intoxication after consuming famphur-poisoned Japanese quail, although 70% of the owls had brain-cholinesterase inhibition within these lethal bounds. Barn owls fed famphur-poisoned quail, the digestive tracts of which had been removed, showed significant but lesser brain-cholinesterase activity inhibition than owls fed intact poisoned quail, indicating that famphur- or cholinesterase-inhibiting metabolites were most heavily concentrated in digestive tracts.

The black-billed magpie seems unusually sensitive to famphur. Dead famphur-poisoned magpies contained as much famphur as 290.0 mg/kg liver FW, 4770.0 mg/kg gizzard FW, and <0.2 mg/kg muscle or fat. There is a growing body of literature on adverse effects on magpies from pour-on (13.2% famphur) applications along the backline of cattle to control cattle warbles at the recommended rate of 0.326 mL/kg BW, not to exceed 118 mL/animal – equivalent to 43.0 mg/kg BW, not to exceed 15.6 g/animal. One investigator documented three occasions when dead birds were found after pour-on-famphur treatment of cattle against warble flies: (1) 12 black-billed magpies in a nearby field 2–3 days after cattle were treated; (2) 6 magpies during a 14-day period (although other species of corvids were present, only magpies were affected); and (3) 8 robins (*Erithacus rubecula*) and a single dunnock (*Prunella modularis*) near a cattle crush a few days after famphur treatment. The dead birds had no measurable brain-cholinesterase activity, and famphur was detected in the gizzards of birds in all three incidents. Partially-paralyzed magpies containing as much as 3500.0 mg famphur/kg gizzard contents were found in the vicinity of cattle recently treated with a pour-on formulation of famphur to control an infestation by warble-flies; another 20–30 dead magpies were found in the immediate area. Magpies and one red-tailed hawk (*Buteo jamaicensis*) were the only birds found dead where cattle had been topically treated with famphur, although several other species including killdeers (*Charadrius vociferus*) and European starlings were common in these pastures. Famphur residues were detected in all dead magpies and hawks, and brain-cholinesterase-activity depression ranged from 70 to 92%. Based on famphur residue concentrations in the gizzards, dead magpies contained 5.2–6.1 mg/kg whole body; these values were above the acute oral LD50 values for several species of birds. The most probable explanation for the sensitivity of magpies to famphur is associated with the contents of the poisoned magpies which consisted of as much as 12% cattle hair. Although most organophosphorus compounds degrade rapidly, famphur persists for >90 days on hair of Hereford bulls and steers and Angus yearlings. Famphur concentration in hair of a Hereford bull averaged 38,000.0 mg/kg FW one week after a single pour-on treatment and a maximum of 12,000.0 mg/kg FW 60 days posttreatment. High concentrations of famphur in the gizzards of magpies indicated that the material was ingested and not from dermal contact or inhalation. Tissue residues in mg famphur/kg FW in famphur-poisoned magpies were as much as 550.0 in the upper GI tract, 4.3 in the lower GI tract, and 3.0 in the whole body. Cow hair from gizzards of dead magpies averaged 4600.0 mg famphur and famoxon/kg FW; other animal matter in the gizzard contained 620.0 mg famphur and famoxon/kg FW and plant matter 340.0 mg famphur and famoxon/kg FW. A potentially lethal dose to magpies would be 8.0–19.0 mg of treated hair at day 7 and 26.0–60.0 mg of treated hair after 60 days. Coincidentally, magpie mortality persisted for more than 3 months; most deaths occurred 5–13 days after cattle were treated. The manure-insect-bird pathway of famphur translocation is untenable because of extremely low (<0.14 mg/kg FW) concentrations of famphur in cow manure.

Secondary poisoning of flesh-eating birds foraging on famphur-killed vertebrates is well-documented; the degree of hazard to the predator is related to the amount and type of consumed tissues and famphur concentrations in the prey tissues. Secondary poisoning of raptors killed by famphur that was topically applied to livestock include the bald eagle (*Haliaeetus leucocephalus*) – after eating cattle that died within 100 days of famphur treatment or famphur-poisoned brown-headed cowbirds (*Molothrus ater*) and European starlings – and a red-tailed hawk after eating famphur-poisoned black-billed magpies or European starlings. In one case, an adult-female bald eagle that was unable to fly near Lewes, Delaware, was brought to a national wildlife refuge where it died after a few days. Stomach contents included one lead shot and remains of brown-headed cowbirds and European starlings. A necropsy showed no signs of lead poisoning. Clinical signs, physical examination, and presence of a full crop suggested acute poisoning. Crop and stomach contents were analyzed for a variety of pesticides, metals, and herbicides, but only famphur was elevated at 1.9 mg/kg FW. As judged by famphur residues in the GI tract and by brain-cholinesterase-activity inhibition of 85%, the authors concluded that famphur was the probable cause of death. There is also a case of tertiary poisoning in which a great horned owl (*Bubo virginianus*) died after consuming a dead famphur-poisoned red-tailed hawk. In all of these cases brain-cholinesterase activity of poisoned birds was depressed >50% and undigested remains contained famphur.

Famphur has also been used to intentionally kill birds, including migratory waterfowl and other protected species, and should be added to the list of other toxic organophosphorus insecticides such as monocrotophos, dicrotophos, and parathion that have been used for this purpose. In 1988, for example, famphur was used illegally by farmers in Georgia and West Virginia to kill birds thought to be depredating crops. Corn and grain at the mortality sites contained between 4240.0 and 8500.0 mg/kg famphur. Dead birds at these locations included Canada geese (*Branta canadensis*), mallards, American black ducks (*Anas rubripes*), American crows (*Corvus brachyrhynchos*), common grackles (*Quiscalus quiscula*), red-winged blackbirds (*Agelaius phoniceus*), sandhill cranes (*Grus canadensis*), and a single red-tailed hawk. Most of the poisoned waterfowl, cranes, raptors, corvids, and songbirds from the five sites had severely depressed brain-cholinesterase activity (i.e., >50%), poisoned bait in the gizzards, and famphur concentrations in the gastrointestinal tracts ranging from 5.0 mg/kg

FW in the red-tailed hawk to 1480.0 mg/kg FW in Canada geese. It was concluded that all birds died from direct ingestion of the poisoned bait, except the red-tailed hawk that had eaten one or more famphur-poisoned crows.

15.4.4 Mammals

Famphur is a group-D compound that is not classifiable as a human carcinogen. However, a study of leukemia risk among males in Iowa and Minnesota indicated a slight but significant elevation in risk – especially chronic lymphocyte leukemia – for farmers but not for nonfarmers. Moreover, a significantly elevated leukemia risk was seen from exposure to specific animal insecticides including famphur. It is clear that more research is needed on the potential carcinogenicity of famphur.

Signs of famphur toxicosis in cattle include ataxia, muscular fasciculations, general weakness, lacrimation, salivation, and diarrhea. In comparison with European breeds of cattle (*Bos taurus*), the Brahman (*Bos indicus*) and European X Brahman hybrids are more sensitive to famphur, and Brahman bulls are more sensitive than cows. At a comparatively low famphur dose of 16.6 mg/kg BW, both *B. taurus* and *B. indicus* are tolerant of intramuscular injectable famphur; however, *B. indicus* is more sensitive and bulls sometimes died when treatment levels exceeded 33.3 mg/kg BW. In addition to cattle, famphur-induced mortality in other species of mammals was documented. Single exposures of famphur in mg/kg BW killed rabbits (*Oryctolagus* sp.) at 2730.0 in dermal exposure; mice (*Mus* sp.) at 27.0 in oral dose or 11.6 by intraperitoneal injection; domestic sheep (*Ovis aires*) at 400.0 in oral dose; and laboratory white rats (*Rattus* sp.) at 400.0 dermal exposure or >28.0 in oral dose. Mice receiving fatal or near-fatal intraperitoneal injections of famphur or famoxon began to convulse 10–20 min postinjection; death came within 45 min post-injection, usually from respiratory failure. Mice remaining alive at 60 min post-injection usually recovered.

Latent effects of famphur exposure in reindeer hinds strongly indicated a need for additional studies in this subject area. Intramuscular injections of reindeer hinds and their 4-week-old calves controlled warble-fly infection in treated animals. Treated calves did not differ significantly from controls during the following year in body weight, body temperature, or blood chemistry. Treated hinds, however, had significantly lower erythrocyte sedimentation rates and serum-gamma-globulin concentrations and significantly higher hemoglobin, serum calcium, serum inorganic phosphorus, and serum magnesium than untreated hinds 1 year after treatment.

Reduced brain-cholinesterase activity in avian and mammalian wildlife is associated with adverse effects on metabolism, reproduction, sensory behavior, motor activity, food and water intake, learning, and memory. Cholinesterase activity in mammals regenerates rapidly after a cessation from treatment with famphur. In humans, typical symptoms of organophosphorus-induced cholinesterase inhibition include headache, giddiness, nervousness, blurred vision, weakness, nausea, cramps, diarrhea, chest discomfort, sweating, salivation, vomiting, and tremors. In severe cases, victims show muscular weakness, convulsions, coma, loss of reflexes, loss of sphincter control, and eventually death. Effects of cholinesterase-inhibiting agents in humans are usually counteracted with repeated intravenous injections of atropine sulfate (2.0–4.0 mg), intravenous injections of pralidoxime chloride (1.0 g), and oxygen. Rats had depressed plasma-cholinesterase activity when fed diets containing as little as 1.0 mg famphur/kg for as many as 90 days, although growth and appetite seemed normal. Brahman bulls had maximum erythrocyte-cholinesterase inhibition 14 days after intramuscular injection of famphur; cholinesterase-activity levels recovered towards normal during the next 14 days, and recovery correlated with the formation of new erythrocytes. Except for cholinesterase-activity inhibition, there were no signs of organophosphate intoxication in Brahman heifers and steers given single dermal famphur doses of 20.0–61.0 mg/kg BW. Cholinesterase activity was inhibited for as many as 14 days posttreatment at the lower (20.0–41.0 mg/kg BW) doses and for at least 7 weeks at 61.0 mg/kg BW.

Famphur is metabolized rapidly in mammals. In cattle, famphur controlled target-insect pests when administered as a bolus, in the diet, as an oral paste, by intramuscular injection, or by pour-on. Regardless of mode of administration, length of exposure, or dose, famphur residues in tissues and milk were usually undetectable within 4 days of final exposure. A similar pattern was evident in other species of mammals. Rats and sheep metabolize famphur differently. During the first 24-h post-dosing period, urine of rats contained as much as two times more of the unchanged *O*–desmethyl compound than urine of sheep, about the same amount of dimethylsulfamoylphenyl glucuronide, about 0.3 times as much *O*,*N*-bisdesmethylfamphur, and about 0.5 times less methylsulfamoylphenyl glucuronide. With the exception of the oxon, metabolites of famphur were considerably less toxic to mammals than the parent chemical. In general, famoxon was 100 times more effective than famphur in depressing erythrocyte cholinesterase activity.

Famphur in pour-on applications penetrates skin at different rates depending on the solvent. In rat skin, penetration was most rapid when the solvent was acetone and least rapid in corn oil and benzene; the percent of remaining famphur in rat skin 3 h after a single dermal application was 38% from the acetone mixture, and 67% from both benzene and corn oil solvents. The penetrability of famphur pour-on formulations used in lice control on Angora goats was enhanced when applied in combination with a liquid-detergent wetting agent. Laboratory screening tests in which small mammals are treated with chemicals and parasitized by insects are now used to predict the effectiveness of systemic insecticides. Tests with mice and rodent botfly (*Cuterebra* sp.) were useful in predicting the effectiveness of famphur against larvae of the common cattle grub (*Hypoderma lineatum*) in cattle, and show promise for screening additional chemicals.

15.5 Recommendations

The four primary areas of concern about famphur use are:

(1) mortality of birds associated with topical applications to cattle
(2) latent effects on domestic livestock
(3) the absence of aquatic toxicity data
(4) potential carcinogenicity

Because of its high toxicity to birds and field and experimental evidence of primary and secondary poisoning of birds, famphur is considered hazardous to avian wildlife – especially magpies – where cattle are topically treated with this insecticide. The pour-on application for cattle is now preferred to systematic dipping or intramuscular injection; dipping is reportedly labor intensive and costly. Intramuscular injection is more labor intensive, causes greater tissue damage and higher famphur absorption at the injection site, and produces a greater depression in blood cholinesterase levels and a lower rate of weight gain in cattle than pour-on application. Nevertheless, famphur-induced mortality of magpies and other birds can be significantly reduced or eliminated by changing the insecticide application from the present pour-on method to other, now available modes of administration such as by diets, bolus, and intramuscular injection. Furthermore, a warning should be added to famphur labels; livestock dying within 3 months of famphur treatment should be removed from the range or farmland; this would offer partial protection to carrion-feeding raptors such as eagles and vultures.

Reindeer are considered safe for human consumption 6–7 weeks after famphur treatment by intramuscular injection (dermal applications of famphur seldom penetrate the thick hair coat of reindeer). Treated reindeer had no detectable residues in liver, kidney, and muscle after 3 weeks and none in fat and other tissues after 6–7 weeks. However, treated hinds during the following year had a significantly altered blood-chemistry profile when compared to untreated hinds, suggesting a need for additional research on latent effects of famphur exposure. A safe dosage for cattle (*Bos* spp.) is 7.0–25.0 mg/kg BW by intramuscular injection or 40.0–55.0 mg/kg BW by pour-on. The maximum concentration of famphur and famoxon allowed in cattle meat, fat, and meat by-products in the United States is 0.1 mg/kg. In Australia, the maximum value is 0.05 mg/kg FW. The recommended minimum time between famphur treatment and slaughter of Australian cattle is 14 days. The half-time persistence of famphur in cattle tissues is 0.9 days, implying that even with gross misuse of the chemical, residues fall to low levels within a week. At present, no published studies are available on latent effects of famphur to cattle. Evidence of latent effects of famphur in reindeer strongly suggests initiation of research into this subject area with cattle and other treated livestock.

No published data are available on effects and fate of famphur in aquatic ecosystems. This seems to be a high-priority research need in view of the increasing and illegal use of famphur to kill migratory waterfowl. In the absence of such data, it is recommended that concentrations of famphur and famoxon in water and in tissues of aquatic organisms not exceed current analytical detection limits of 0.005 mg/L in water or 0.01 mg/kg FW tissue.

The carcinogenicity of famphur has not been satisfactorily resolved. Studies indicate a significantly elevated risk for leukemia among farmers handling famphur, but this needs verification.

15.6 Summary

Famphur (phosphorothioic acid, *O*-[4-(dimethylamino) sulfonyl], phenyl] *O*,*O*-dimethyl ester), also known as Warbex, is a systemic organophosphorus insecticide used almost exclusively as a veterinary chemical to control parasites in livestock. Famphur has proven effective in controlling maggots of the botfly and warble fly (*Hypoderma* spp.), lice (*Haematopinus* spp., *Linognathus* spp.), hornfly (*Haematobia* spp.), reindeer warble fly (*Oedemogena* spp.), reindeer nostril fly (*Cephanomyia* spp.), and mealworms (*Alphitobius* spp.). The LD50 values for target insects ranged from 2.4 to 4.1 mg/kg body weight (BW) from a dermal route and 8.0–11.8 mg/kg BW from abdominal injection.

Only famphur and its oxygen analog, famoxon, were of toxicological significance; other famphur metabolites were 31–237 times less toxic, as judged by acute oral toxicity tests in the mouse (*Mus* sp.). Famoxon was more effective than famphur in producing cholinesterase inhibition and death and confirms the generalization that the corresponding oxons are the more effective anticholinesterase agents. Differential toxicity of famphur to target pests and nontarget species was often associated with differential degradation and cholinesterase sensitivity.

Famphur is administered to livestock by intramuscular or subcutaneous injection, through the diet, as a dermal pour-on, or as an oral bolus. In mammals, famphur induced mortality at concentrations as low as 11.6 mg/kg BW in intraperitoneal injection (mouse), 27.0 mg/kg BW in a single oral exposure (mouse), >33.3 mg/kg BW in an intramuscular injection (Brahman cattle, *Bos indicus*), and 400.0 mg/kg BW in a dermal application (rat, *Rattus* sp.). Latent effects of famphur exposure were reported in reindeer (*Rangifer tarandus*) hinds one year posttreatment (altered blood chemistry). Famphur is rapidly metabolized by mammals. The half-time persistence of famphur and famoxon in subcutaneous fat of cattle after a single pour-on application is 0.9 days and is independent of dose between 25.0 and 150.0 mg/kg BW or initial tissue residues between 1.8 and 2.3 mg/kg BW.

Famphur has been used illegally by U.S. farmers to kill wild birds – including migratory waterfowl – thought to be depredating crops. Pour-on applications of famphur to cattle at recommended doses are sometimes associated with bird die-offs, especially the black-billed magpie (*Pica pica*). Magpie mortality – which persisted for 3 months – was probably associated with the lengthy persistence (>90 days) of famphur on cattle hair, and the ingestion of cattle hair by magpies. Cattle hair composed as much as 12% of gizzard contents of dead magpies, and hair in the gizzards of dead magpies averaged 4600.0 mg famphur and famoxon/kg. Secondary poisoning of eagles and hawks foraging on famphur-killed vertebrates and tertiary poisoning of a great horned owl (*Bubo virginianus*) feeding on a famphur-poisoned hawk are documented. In the laboratory, sensitive species of birds died after single oral doses of 1.8–3.0 mg famphur/kg BW or when fed diets containing 35.0–49.0 mg famphur/kg ration. Depressed cholinesterase activity in the brain and in plasma occurred in nestlings at daily oral concentrations as low as 0.3 mg famphur/kg BW.

No published data are available on the fate or effects of famphur in aquatic ecosystems. In the absence of aquatic toxicity data on famphur, it is recommended that famphur and famoxon concentrations do not exceed the analytical detection limits of these compounds in water (0.005 mg/L) or in tissues of aquatic organisms (<0.01 mg/kg fresh weight).

Current recommendations include the discontinuance of topical applications of famphur to cattle because of its association with primary and secondary poisoning of birds – especially magpies, hawks, and eagles – and more research on famphur in three areas: latent effects on treated livestock; fate and effects in aquatic ecosystems; and carcinogenicity evaluation.

FENVALERATE[a]

Chapter 16

16.1 Introduction

Synthetic pyrethroids, including fenvalerate [(*RS*) α-cyano-3-phenoxybenzyl (*RS*) 2-(4-chlorophenyl)-3-methylbutyrate][b], are now broadly recognized as a major class of synthetic organic insecticides. Introduced commercially about 45 years ago, synthetic pyrethroids account for more than 30% of insecticide use worldwide, in household, agricultural, and veterinary applications. More than 1000 pyrethroids have been synthesized since 1973; they include compounds containing nitrogen, sulfur, fluorine, chlorine, and bromine, in addition to carbon, hydrogen, and oxygen. The most potent synthetic pyrethroid insecticides are the cyanophenoxybenzyl pyrethroids, and fenvalerate is the most widely used compound in this group.

Pyrethroid insecticides are synthetic analogs of natural pyrethrins. Natural pyrethrins were widely used in Europe during the 19th century, when few effective insecticides were available. Natural pyrethrins, which contain six insecticidally active components extracted from the dried heads of the pyrethrum flower (*Chrysanthemum cinariaefolium*), have high insecticidal properties and low mammalian toxicity; however, they are expensive to produce and have low photostability and high biodegradability. Modern synthetic pyrethroids have been designed to provide enhanced residual activity through greater photostability and greater resistance to chemical and biological degradation, greater insecticidal activity, diminished mammalian toxicity, and greater cost-effectiveness. The first synthetic pyrethroids, allethrin and cyclethrin, were produced around 1950, but lacked adequate photostability and were not as effective insecticidally as the natural pyrethrins. Tetramethrin was introduced in 1964, but it had inferior insecticidal activity. The first synthetic pyrethroids with greater insecticidal activity than natural pyrethrins were resmethrin and cismethrin, produced in 1968. Photostable pyrethroids were produced in the mid-1970s and included deltamethrin, cypremethrin, fenpropathrion, and fenvalerate.

Pyrethroid insecticides are generally recognized as potent neurotoxicants that interfere with nerve membrane function by interaction with the sodium channel. Synthetic pyrethroids are more toxic against insect pests, up to ten times more potent in some cases, than the other insecticides now in general use. However, the stereochemical structure of pyrethroid insecticides greatly influences their toxicity to insects and mammals, and this phenomenon is especially pronounced in fenvalerate.

As broad-spectrum insecticides, the synthetic pyrethroids are necessarily toxic to a wide range of arthropods. Most insect orders are extremely susceptible, including many types of beneficial predator and parasite species. Synthetic pyrethroids are also toxic to fish and nontarget aquatic insects and crustaceans. Fenvalerate, for example, enters freshwater aquatic environments in runoff from food crop use, in drift from forest-spray

[a]All information in this chapter is referenced in the following sources:

Eisler, R. 1992. Fenvalerate hazards to fish, wildlife, and invertebrates: a synoptic review. *U.S. Fish Wildl. Serv. Biol. Rep.* 2. 43 pp.

Eisler, R. 2000. Fenvalerate. Pages 1089–1131 in Handbook of Chemical Risk Assessment: Health Hazards to Humans, Plants, and Animals. Volume 2, Organics. Lewis Publishers, Boca Raton, Florida.

[b]The technical fenvalerate formulation is no longer manufactured, although existing stocks may be used until exhausted. The new fenvalerate formulation is sold as Asana or Esfenvalerate, and contains only the 2*S*, α*S* isomer.

procedures, and by direct spraying of water bodies. Estuarine organisms may be exposed to fenvalerate and other pyrethroids after applications to corn, cotton, rice, and vegetables in coastal areas or by discharges from pyrethroid manufacturers or formulating and distribution centers. Fenvalerate has been implicated in kills of coastal organisms in South Carolina, primarily from agricultural runoff into estuarine tidal creeks.

16.2 Environmental Chemistry

Synthetic pyrethroids now account for at least 30% of the world insecticide market and are rapidly replacing other agricultural chemicals for control of insect pests. Fenvalerate is one of the more widely used synthetic pyrethroid insecticides. It is derived from a combination of α-cyano-3-phenoxybenzyl alcohol and α-isopropyl phenylacetate ester. Technical fenvalerate is a mixture of four optical isomers, each occurring in equal amounts but with different efficacies against insect pests. Fenvalerate does not usually persist in the environment for >10 weeks, and it does not accumulate readily in the biosphere. Time for 50% loss (Tb1/2) in fenvalerate-exposed amphibians, birds, and mammals was 6–14 h; for reptiles, terrestrial insects, aquatic snails, and fish it was >14h–<2 days; and for various species of crop plants it was 2–28 days. Fenvalerate degradation in water is due primarily to photoactivity, and in soils to microbial activity. Half-time persistence in nonbiological materials is variable, but may range up to 6 days in freshwater, 34 days in seawater, 6 weeks in estuarine sediments, and 9 weeks in soils.

16.2.1 Chemical Properties

Synthetic pyrethroid insecticides are photostable analogs of the natural pyrethrins of botanical origin; they consist of a series of related esters derived from alcohols and acids that maintain critical isosteric relations with the natural product prototype. Small changes in substituents and stereochemistry are sufficient to produce compounds differing in their insecticidal potency, spectrum of activity, and mammalian toxicology. These halogenated, lipophilic, photostable compounds are exceptionally active against many species of insects. Although these compounds are relatively safe for birds and mammals, they are usually extremely toxic to certain freshwater and marine groups, including fish.

The first significant success in creating a photostabilized pyrethroid with high insecticidal activity was achieved through the use of the 3-phenoxybenzyl alcohol moiety. A further step was the finding that 2-aryl-3-methylbutyric acid esters of pyrethroid alcohols were both photostable and insecticidal. Fenvalerate is one of the more recently developed and widely used synthetic pyrethroid insecticides, and it is a highly active phenyl acetate ester of known pyrethroid alcohols, specifically, a combination of α-isopropyl phenyl acetate ester and α-cyano-3-phenoxybenzyl alcohol. The phenoxybenzyl group and the halogenated phenyl ring increase the photostability of the molecule. The cyano group, substituted on the benzylic carbon, stabilizes the ester bond against hydrolysis.

Fenvalerate, like most other synthetic pyrethroids, is a halogenated, lipophilic, stable compound with low solubility in water and high solubility in organic solvents (Table 16.1). Technical fenvalerate is a racemic mixture of four isomers, composed of equal amounts of dextrorotatory and levorotary forms; however, the four optical isomers (Figure 16.1) have very different efficacies against pest species. In general, fenvalerate stereoisomers with *S* configurations in both the acid and alcohol moieties are more active pharmacologically and toxicologically than those with *R* configurations. Fenvalerate can be isolated and concentrated from pond water and other solutions using solid-phase extraction, and analyzed to 1.0 μg/L on a capillary gas chromatograph equipped with an electron-capture or flame photometric detector.

16.2.2 Uses

Pyrethroids are used primarily for the control of household and agricultural insect pests, and

Table 16.1. Chemical and other properties of fenvalerate.

Variable	Datum
CHEMICAL NAME	(*RS*)-α-cyano-3-phenoxybenzyl(*RS*)-2-(4-chlorophenyl)-3-methylbutyrate; cyano(3-phenoxyphenyl)methyl4-chloro-α-(1-methylethyl)benzeneacetate; α-cyano-3-phenoxybenzyl 2-(4-chlorophenyl)-3-methylbutyrate; 4-chloro-α-(1-methylethyl)benzeneacetic acid cyano(3-phenoxyphenyl) methyl ester; α-cyano-3-phenoxybenzyl α-(4-chlorophenyl) isovalerate
ALTERNATE NAMES	Agmatrin, Belmark, Ectrin, Fenkill, Phenvalerate, Pydrin, S-5602, Sanmarton, SD 43775, Sumicidin, Sumifly, Sumipower, Sumitox, WL 43775
CAS NUMBER	51630-58-1
CHEMICAL FORMULA	$C_{25}H_{22}ClNO_3$
MOLECULAR WEIGHT	419.92
PHYSICAL STATE	Clear yellow, viscous, liquid at 23°C
PURITY	Technical grade compound is about 92% pure; nature and extent of impurities unknown
VAPOR PRESSURE AT 25°C	1.1×10^{-8} mm mercury
DENSITY AT 23°C	1.17 g/Ml
STABILITY	Stable in most solvents except alcohols at ambient temperature. Unstable in alkaline media. No significant breakdown after 100 h at 75°C; gradual degradation occurred in range 150–300°C
DEGRADATION	Cleavage of the ester linkage is the primary route
FORMULATIONS	Emulsifiable concentrate, dust, granules, wettable powder
LOG OCTANOL–WATER PARTITION COEFFICIENT	6.2
SOLUBILITY AT 20°C	
Acetone	>450.0 g/L
Chloroform	>450.0 g/L
Methanol	>450.0 g/L
Hexane	77.0 g/L
Freshwater	2.0–85.0 μg/L
Seawater	24.0 μg/L

secondarily in industrial, stored-product, and veterinary applications. They are especially advantageous for use in northern climates because their toxicity is enhanced at low temperatures. Synthetic pyrethroid insecticides, including fenvalerate, are used as alternatives to organochlorine, organophosphorus, carbamate, and natural pyrethrum insecticides, because they are highly toxic to insect pests, low to intermediate in persistence, and low in toxicity to warm-blooded organisms, although they are extremely toxic to many aquatic organisms. By 1982, more than 30% of the world market for insecticides consisted of synthetic pyrethroids, and this percentage is increasing.

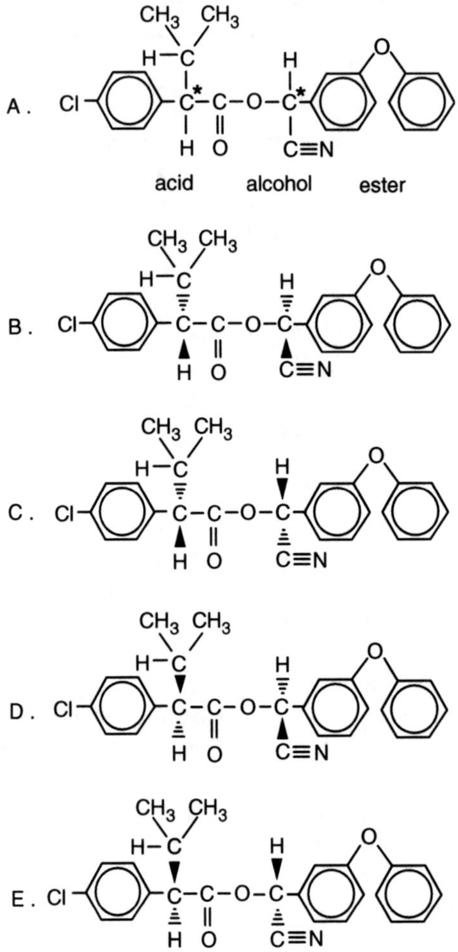

Figure 16.1.
Fenvalerate and its isomers.
A. Chemical structure of fenvalerate denoting two asymmetric carbon atoms (*): the 2C position of the acid moiety, and the αC position of the α-cyano-3-phenoxybenzyl alcohol moiety. These two chiral centers, at the 2C and αC positions, yield a mixture of four stereoisomers, in approximately equal amounts, but with greatly different biological properties.
B. (2*S*)-α-cyano-3-phenoxybenzyl (α*S*)-2-(4-chlorophenyl)-3-methylbutyrate. The 2*S*, α*S* isomer is extremely toxic to insects, and is the most active form of fenvalerate.
C. (2*S*)-α-cyano-3-phenoxybenzyl (α*R*)-2-4-chlorophenyl-3-methylbutyrate. The 2*S*, α*R* isomer has markedly reduced insecticidal activity when compared with the 2*S*, α*S* isomer, but is greatly elevated in this respect when compared with fenvalerate stereoisomers with an *R* configuration in the acid moiety, that is, the 2*R*, α*S*, and the 2*R*, α*R* isomers.
D. The 2*R*, α*S* isomer is the only fenvalerate isomer that caused granulatomous changes in liver, spleen, and mesenteric lymph node in rodents.
E. The 2*R*, α*R* isomer has greatly reduced biological and toxicological properties when compared with other fenvalerate isomers. Isomers with an *R* configuration in the acid moiety degraded slightly faster than the insecticidally-active 2*S*, α*S*, and 2*S*, α*R* isomers.

Outside of the United States, fenvalerate is used on cotton (*Gossypium hirsutum*) in Australia, Greece, and South Africa and on apples (*Malus* sp.), pears (*Pyrus* sp.), and potatoes (*Solanum* sp., *Ipomea* sp.) in Canada; uses in other countries, including Mexico, are anticipated, as is the increased use against agricultural, poultry, dairy, and household pests. In agricultural use, recommended application rates of fenvalerate range between 0.055 and 0.224 kg/ha for control of a broad spectrum of pestiferous insects.

Domestically, about 6500 kg of fenvalerate were used in 1979, all of which was imported. In 1980, in addition to registered use, the U.S. Environmental Protection Agency (EPA) allowed an additional 80,000 kg for crisis and experimental use. By 1981, fenvalerate had been registered for domestic use on apples, cotton, peanuts (*Arachis* sp.), pears, and potatoes. Additional uses were allowed under various experimental or crisis exemptions on beans (*Phaseolus* spp.), peppers (*Piper* spp.), broccoli (*Brassica* spp.), cabbage (*Brassica* spp.), cauliflower (*Brassica* spp.), celery (*Apium* sp.), corn (*Zea mays*), cucumbers (*Cucumis* sp.), eggplant (*Solanum melongena*), grapes (*Vitis* spp.), lettuce (*Lactuca* sp.), peas (*Pisum* sp.), squash (*Cucurbita* spp.), and tomatoes (*Lycopersicon esculentum*). By 1989, this list was expanded to include tobacco (*Nicotiana tabacum*); soybeans (*Glycine max*); sugarcane (*Saccharum* sp.); a wide variety of nuts, fruits, and vegetables; pine seed orchards and forest tree nurseries; and mosquitoes, biting insects, insect vectors of disease, mite control in poultry, and fly and tick control in cattle.

The delivery vehicle of fenvalerate-containing insecticides may account for wide variations in toxic action. For example, fenvalerate microcapsules used to control caterpillar pests (*Plutella xylostella*, *Spodoptera litura*) were most effective with thin-walled capsules and small particles; however, significant protection of nontarget organisms, such as fish, occurred with thicker-walled capsules and larger particles. The popularity of commercial synthetic pyrethroids and their widespread replacement of older, more-toxic compounds in various settings mandate a thorough understanding of the formulation used and of the active and inert components.

16.2.3 Persistence

In nonbiological samples, half-time persistence of fenvalerate is variable, but frequently ranges between 2 and 6 days in freshwater, 27 and 34 days in seawater, 3 and 9 weeks in soils, and up to 6 weeks in estuarine sediments. Persistence is longer at higher initial application rates and under conditions of reduced light, low microbial activity, and high organic content. Fenvalerate is not readily transported from upland field application sites into the aquatic environment. Fenvalerate that directly enters the aquatic environment by way of runoff has limited bioavailability to aquatic organisms owing to rapid adsorption onto soil particles, organic matter, or plants, and to chemical hydrolysis and photodecomposition. Under acidic conditions, fenvalerate in water is stable to hydrolysis for 100 h at 75°C; Tb1/2 at elevated recommended application rates is about 21 days, primarily as a result of photodegradation.

Fenvalerate is one of the more persistent synthetic pyrethroids in soils. In agricultural soils, fenvalerate is tightly adsorbed to soil particles, does not easily move laterally or to lower soil layers with ground water, and almost always localizes in the application site because of its extremely low solubility in water. Fenvalerate degradation rates from mineral soil surfaces are dependent on soil type, moisture, temperature, and microbial activity. Half-time persistence in soils usually range between 2 and 18 days, but 3 months has also been recorded. Although fenvalerate is susceptible to chemical degradation by hydrolysis and oxidation, most authorities agree that degradation in soils is due primarily to microbial activity, that microbial degradation is most rapid under aerobic conditions, and that transformed products do not persist longer than the parent compound.

In biological samples, fenvalerate neither persists for lengthy periods nor is readily accumulated. In general, fenvalerate is rapidly (i.e., Tb1/2 of 6–14 h) excreted by amphibians, birds, and mammals; has low persistence in various reptiles, terrestrial insects, aquatic snails, and fish; and has moderate (i.e., Tb1/2 of 2–28 days) persistence in various species of target plants. Animals collected after 5 days from a cotton field sprayed with 0.112 kg/ha, or from the immediate vicinity, had very low fenvalerate residues. In that study fenvalerate was detected in one of nine bird species sampled, in one of four mammals, in the western ribbon snake (*Thamnophis* sp.), in one of four amphibian species, and in fish and insects. The bird was a male dickcissel (*Spiza americana*) that had established a breeding territory within the sprayed cotton field. Carnivorous ground beetles, found moribund on the ground, contained the highest mean fenvalerate residue of 0.55 mg/kg fresh weight (FW) whole body; large numbers of dead insects were found in the fields during collection. The highest residues (0.32–0.55 mg/kg) in fish and invertebrates were in those collected from a small pool in a drainage ditch, which compares with 0.92 mg/kg found in common carp (*Cyprinus carpio*) after exposure in the laboratory for 7 days to 0.8 μg fenvalerate/L.

Fenvalerate is not significantly absorbed or translocated in plants. Cotton, apples, and lettuce treated with fenvalerate contained surface residues of parent fenvalerate 8 weeks after treatment. In addition to the parent compound, which accounted for 80% of all residues, identified metabolites included 3-phenoxybenzaldehyde, 3-phenoxybenzyl methylbutyric acid, and conjugates of these compounds. Half-time persistence of fenvalerate on plant surfaces is between 2 and 4 weeks, and degradation is primarily a result of weathering.

Various plants sprayed with 0.25 kg fenvalerate/ha all had measurable residues 7 days after application, and non-detectable residues 15–30 days after treatment. Washing plants in cold water to remove the pesticide was effective only on the initial day of application, removing 30–50%. Afterwards, only 3–13% could be removed by washing. Cooking removed 71–88% of the fenvalerate residues on the initial day of treatment, but in later samplings removal was 68–70% in spinach (*Spinacea oleracea*) and tomatoes, and 38–40% in okra (*Abelmoschus esculentus*) and cauliflower (*Brassica oleracea botrytis*).

Adsorption and persistence in plants can be modified by other chemicals or by selected carriers, although mechanisms to account for these phenomena are unclear. The application mixture influences adsorption and persistence of fenvalerate. For example, interception and persistence in sugarcane were increased when fenvalerate was applied in a 25% water – 75% soybean oil mixture vs. water or soybean oil alone. Also, biocidal properties of fenvalerate residues on cotton foliage were increased up to 100% due to enhanced persistence of fenvalerate in the presence of toxaphene.

Fenvalerate photoproducts merit consideration, as some may be comparatively toxic. Decarboxyfenvalerate is a major degradation product of fenvalerate that is formed by photochemical reactions in water and on plant foliage. This photoproduct composes up to 10% of the total residues in forage crops that have been exposed to prolonged sunlight and drying. Decarboxyfenvalerate did not persist in tissues of hens, rats, and cows when consumed with feed for extended periods; its residue levels in ova, milk, and meat were negligible.

Photolysis of fenvalerate in various solvents by sunlight yields products resulting from ester cleavage, primarily decarboxyfenvalerate, but also 15 other products. All sunlight photoproducts were relatively harmless to mice; LD50 values were >500.0 mg/kg body weight. When photolysis was by way of ultraviolet light, however, two of the photoproducts formed (3-phenoxybenzoyl cyanide, 3-phenoxybenzyl cyanide) were considerably more toxic than fenvalerate; LD50 values for intraperitoneal injection in mice were >500.0 mg/kg BW for fenvalerate, 2.0 mg/kg BW for 3-phenoxybenzoyl cyanide, and 105.0 mg/kg BW for 3-phenoxybenzyl cyanide. This finding strongly suggests a need for additional research on fenvalerate photoproduct persistence and toxicity.

16.3 Mode of Action

Two types of synthetic pyrethroids have been identified, as judged by different behavioral, neurophysiological, chemical, and biochemical profiles: Type I, those pyrethroids lacking the α-cyano group, and Type II, those possessing the α-cyano group (i.e., fenvalerate). Induction of repetitive activity in the nervous system is the principal effect of pyrethroids. Repetitive activity originates from a prolongation of the transient increase in sodium permeability of the nerve membrane associated with excitation. All pyrethroids affect sodium channel gating in a similar manner, although Type II pyrethroids are significantly more neurotoxic than Type I pyrethroids.

Metabolism of fenvalerate proceeds by way of oxidation and hydrolysis to produce metabolites considered pharmacologically inactive or inferior to the parent compound. Insects and fish are extremely susceptible to fenvalerate when compared to mammals and birds; interspecies differences are associated with rates of metabolism, excretion, absorption, esterase activity, and neurosensitivity.

Fenvalerate is neither mutagenic nor teratogenic. Tumor-like growths in rodent tissues, however, were associated with the 2*R*, α*S* isomer – heretofore believed innocuous – more specifically, with its cholesterol conjugate.

16.3.1 Types of Pyrethroids

Two distinct types of synthetic pyrethroids have been identified, as judged by different behavioral, neurophysiological, chemical, and biochemical profiles in rodents: Type I, also known as Class 1, or T – for tremor; and Type II, also known as Class 2, or CS – for choreoathetosis/salivation. In general, authorities agree that pyrethroids containing both a halogenated acid esterified with the α-cyano-3-phenoxybenzyl alcohol – such as fenvalerate, deltamethrin, and cypermethrin – produce the Type II poisoning syndrome, and that pyrethroids lacking either or both of these moieties (i.e., permethrin, resmethrin, cismethrin, allethrin, bromphenothrin, phenothrin, kadethrin, tetramethrin) tend to produce the Type I syndrome. Type I is characterized by sparring, aggressive behavior (in rats, but not mice), rapid onset of tremor in the extremities, increased body temperatures, and whole-body tremors. As toxicity progresses, mice show hyperactivity, whereas rats become prostrate and die with immediate onset of rigor mortis; in mice, death is often associated with spasmodic seizures. The Type I syndrome is very similar to that produced by *p,p′*-DDT. Type II is characterized by pawing and burrowing behavior, profuse salivation, a decrease in body temperature of rats (partially due to evaporation of saliva), tremors progressing to choreoathetosis (i.e., a sinuous, writhing movement), muscular contractions and seizures, and death. With repeated high doses sufficient to kill some rats, degenerative changes in sciatic and posteriad tibial nerves were observed. The same two types of pyrethroid actions are also evident among insects.

Regardless of route of administration, signs of fenvalerate poisoning in rodents were similar. Doses administered by intercerebroventricular injection of comparatively low concentrations were more toxic than higher doses given orally, or by intravenous or intraperitoneal injection, suggesting greater central nervous system involvement in Type II than in Type I poisoning. In fact, pyrethroids that produce the Type II syndrome – including fenvalerate – are 5–10 times more potent neurotoxicants than Type I pyrethroids, which suggests different sites of action in the central nervous system.

16.3.2 Sodium Gating Kinetics

Pyrethroids have an action at or near the sodium channel in the nerve, resulting in greatly altered ionic currents and disrupted nerve function through membrane depolarization. Based on studies with insects, crustaceans, frogs, and small mammals, there is general agreement that the sodium channel in the nerve membrane is the major target site for all synthetic pyrethroid insecticides and (many other neurotoxicants); that synthetic pyrethroids prolong the transient increase in sodium permeability of the nerve membrane during excitation, resulting in spontaneous depolarization and repetitive discharges; that persistent repetitive discharges lead to muscular fasciculations, acetylcholine depletion, and muscular weakness; that effects are enhanced at lower temperatures; and that α-cyano (Type II) pyrethroids are more potent neurotoxicants than noncyano (Type I) pyrethroids, differences in neurotoxic effects being attributed solely to the α-cyano substituent. Most authorities agree that fenvalerate was the most effective pyrethroid tested for inducing pronounced repetitive activity in nerve fibers and that the $2S$, αS isomer was up to 15 times more potent than other fenvalerate isomers. Pyrethroids induce the sodium channels to close more slowly than normal, resulting in a gradually decaying inward sodium current (called a tail current) after termination of membrane depolarization. Type I pyrethroids induce tail currents with time constants of decay in milliseconds, but Type II pyrethroids result in time constants of decay that are orders of magnitude longer and contain thousands of impulses, inducing a quickly reversible, frequency-dependent suppression of the action potential. Depolarization of axons by synthetic pyrethroids was most effective at low temperatures; the negative temperature dependence of the steady-state current seems to be due to the stabilizing effect of low temperature on the open-modified channel.

Myelinated nerves of vertebrates are thought to sequester the pyrethroid molecules, known to be soluble in the myelin sheath, thereby preventing a portion of their chemical effect on the nerve axon. Fenvalerate, unlike other α-cyano pyrethroids, had little effect on the electrophysiological function of single myelinated nerve fibers in the frog (*Rana esculenta*), suggesting that additional research is needed on mechanisms other than membrane sodium transport.

The role of calcium in pyrethroid interaction with nerve tissue is under active investigation. Fenvalerate affects calcium-ATPase enzyme and calmodulin-activated enzyme activities, such as phosphodiesterase. Fenvalerate inhibits calcium uptake by nerve cord of crayfish (*Procambarus clarki*) and axon of spiny lobster (*Panulirus japonicus*), an action that seems to be related to its lipophilic properties. Fenvalerate enhances the calcium-dependent potassium-stimulated release of norepinephrine from rat brain and could lead to an overall depletion of brain stores of this neurotransmitter, producing a convulsive state typical of Type II pyrethroid poisoning. Fenvalerate evoked a calcium-dependent release of dopamine and acetylcholine from rabbit brain that was concentration related and specific for the $2S$, αS isomer; release of dopamine and acetylcholine was antagonized completely by tetrodotoxin, a sodium channel blocker. The relatively low potency of fenvalerate and other Type II pyrethroids on potassium-stimulated calcium uptake in rat brain and other responses suggests that neither the sodium–calcium exchanger nor the voltage-dependent calcium channels are primary targets for pyrethroid toxicity.

Toxic isomers of Type II pyrethroids usually antagonize γ-aminobutyric acid (GABA) by interacting with the *t*-butyl bicyclophosphorothionate/picrotoxin binding site in the brain; antagonism of GABA leads to a reduction in inhibition. Fenvalerate seems to increase inhibition, however, and this may be explained by a differential effect on sodium channel kinetics. Fenvalerate also inhibits perhydrohistrionicotonin binding with electric organ membrane of the electric ray (*Torpedo* sp.) and interacts with binding sites for dihydropicrotoxinin and kainic acid in the brain, but the significance of these observations is unclear.

16.3.3 Metabolism

The most important metabolic degradation pathways for synthetic pyrethroids are oxidation on the phenoxy ring, hydrolysis of the ester linkage, and conjugation of metabolites; rates and pathways differ among taxonomic animal groupings, resulting in large differences in sensitivity.

All metabolic degradation products of fenvalerate are pharmacologically inactive or inferior to the parent compound, implying that metabolic modifications lead to detoxication.

Fenvalerate and other α-cyano pyrethroids, however, are consistently more resistant to oxidative attack than their noncyano analogs. Liver is the predominant site of fenvalerate metabolism via hydrolysis by one or more hepatic microsomal esterases; inhibition of these enzymes results in enhanced toxicity. Hydrolysis has also been demonstrated in plasma, kidney, stomach, and brain tissues. Except for brain, however, these tissues were relatively unimportant in the detoxification process.

Metabolism of the 2*S* isomers proceeds sequentially: hydroxylation at the phenoxy group, hydrolysis of the cyano group, and cleavage of the ester linkage. Fenvalerate and the 2*S* isomers yield two ester metabolites in feces from hydroxylation at the 4′ and 2′ phenoxy positions. Other significant metabolites were 3-phenoxybenzoic acid and its hydroxy derivatives from the alcohol moiety, 3-(4-chlorophenyl) isovaleric acid and its hydroxy derivatives from the acid moiety, and thiocyanate and carbon dioxide from the cyano moiety. A slow elimination rate characterizes fenvalerate and other α-cyano pyrethroids when compared with noncyano pyrethroids; it seems to be due to the release of the cyano group during ester cleavage, which is then incorporated into the body thiocyanate pool and retained in the skin and stomach. Decarboxyfenvalerate, a photolysis product of fenvalerate, is present in water and on plant surfaces, but it is extensively hydroxylated in mammals, and excreted rapidly and completely into feces, with no apparent toxic effects.

Signs of fenvalerate intoxication are similar in birds, fish, mammals, and insects, but insects and fish are extremely sensitive when compared with warm-blooded organisms, frequently by one to three orders of magnitude. Increased resistance to fenvalerate and other synthetic pyrethroid insecticides in mammals and birds, when compared with aquatic organisms and terrestrial insects, is attributed to their higher metabolism, more rapid excretion, lower absorption from diet or the surrounding envirosphere, higher esterase activity, higher fat content, and lower neurosensitivity. For example, rainbow trout (*Oncorhynchus mykiss*) – one of the more sensitive aquatic species – have significantly lower rates of metabolism and elimination of fenvalerate than those reported for birds and mammals; show little or no esterase activity towards pyrethroids and substantially lower oxidative activity than warm-blooded animals; efficiently accumulate fenvalerate from the medium, and show greater intrinsic sensitivity of the central nervous system when compared with birds and mammals.

Fenvalerate effects are antagonized or synergized by various compounds or chemicals. Dermal exposure to fenvalerate in mammals may produce a skin sensory response, most frequently on the face, characterized by itching and tingling. Administration of vitamin E up to 29 h before fenvalerate exposure partially reduced the fenvalerate-mediated skin sensation in guinea pigs (*Cavia* sp.). The effectiveness of vitamin E may be associated with its membrane stabilizing property, although the exact mode of action is unknown. Fenvalerate skin sensations were also reduced by piperonyl butoxide when applied directly to the skin or in conjunction with fenvalerate. Delayed toxic effects in rodents and insects were produced with various muscle relaxants, including propranolol and diazepam, perhaps through depolarization of nerve terminals. Mice given profenofos, an esterase inhibitor, were up to 27 times more susceptible than were nontreated animals.

16.3.4 Mutagenicity, Teratogenicity, and Carcinogenicity

Fenvalerate and other synthetic pyrethroids caused no oncogenic, reproductive, mutagenic, or teratogenic effects, as judged by results of 2-year feeding studies with rodents at 250.0–300.0 mg/kg diet, 3-generation rodent reproduction studies at 250.0 mg/kg diet, various mutagenicity assays, bone marrow cytogenicity up to 150.0 mg/kg BW, the dominant lethal bioassay at 100.0 mg/kg, and a host-mediated bioassay in mice at 50.0 mg/kg BW. Some chromosomal aberrations and alterations in the mitotic index were noted, however, in bone marrow and testis cells of rats given

100.0 mg fenvalerate/kg BW orally, a dose that killed 71% of the rats. A similar pattern was noted in mice, indicating that additional research is needed to establish mutagenicity of fenvalerate.

The carcinogenic potential of fenvalerate is based on negative or inconclusive evidence and centers on its ability to produce microgranulomas in various tissues, especially liver, in dogs and rodents. Beagle dogs exposed to 250.0, 500.0, or 1000.0 mg fenvalerate/kg diet for 6 months showed treatment-related microscopic effects, including histiocytic cell infiltrates in mesenteric lymph nodes and multifocal microgranulomas in liver. Female rats fed a diet containing 1000.0 mg fenvalerate/kg ration for 2 years showed a statistically significant increase in the incidence of mammary tumors; however, this was judged by the authors to be of unlikely biological significance. Their unusual conclusion was based on four points: (1) none of the mammary tumor incidences exceeded those expected or reported on aged female rats of this strain, (2) time and appearance of tumors in control and treated groups were unchanged by treatment, (3) the benign/malignant ratio of mammary tumors was the same in control and treated groups, and (4) the tumors were common in this strain of rat and did not seem to be related to treatment.

Fenvalerate inhibits intercellular communication between fibroblast cells and enhances the development of hepatocyte foci in rat liver at nonheptatotoxic dose levels. Chemicals that possess these properties are likely to be tumor promoters. Fenvalerate alone induced no hepatotoxic effects in rat liver, as judged by transaminase activities and histology. However, some rats that were partially hepatectomized and insulted with nitrosodiethylamine – a carcinogen and tumor initiator – had significantly elevated numbers of liver foci after administrations of fenvalerate. This response suggests that fenvalerate is a potential tumor promoter.

Linkage of the tumor-like formations in rodents with a specific fenvalerate stereoisomer was an important breakthrough. Granulatomous cells in spleen, lymph node, and liver of fenvalerate-stressed rats and mice tended to fuse, forming large multinucleated cells called giant cells. Researchers convincingly demonstrated that the $2R$, αS isomer, heretofore believed innocuous, was solely responsible for the observed microgranulomas. The residual metabolite in this instance is the cholesterol conjugate [cholesterol ($2R$)-2-(4-chlorophenol) isovalerate] known as CPIA-cholesterol ester. This lipophilic conjugate forms rapidly, usually peaking within 60 min, and tends to persist in tissues, especially in adrenal, spleen, liver, and mesenteric lymph node. Of the four fenvalerate isomers, only the $2R$, αS isomer yielded CPIA-cholesterol ester in tissue homogenates of mice, rats, dogs, and monkeys. Mouse tissues showed relatively higher activities than those of other animals. Kidney, brain, and spleen of mice showed relatively higher capacities to form CPIA-cholesterol ester when compared with other mouse tissues; in all cases, enzyme activity localized mainly in microsomal fractions. Researchers concluded that stereoselective formation of the CPIA-cholesterol ester resulted from the stereoselective formation of the CPIA-carboxyesterase complex only from the $2R$, αS isomer, which subsequently undergoes cleavage by cholesterol to yield the CPIA-cholesterol ester that produced giant cells in mice. These findings strongly support the need for more research on carcinogenic potential of fenvalerate stereoisomers.

16.4 Effects

Fenvalerate is extremely toxic to representative nontarget aquatic organisms and to some beneficial terrestrial arthropods at concentrations substantially lower than those recommended to control pestiferous insects. Toxic effects are associated primarily with the $2S$, αS isomer and are exacerbated at low temperatures. Birds, mammals, and terrestrial plants are normally tolerant.

Target insect species are usually killed at fenvalerate concentrations of 0.015 μg whole body, 0.11 kg/ha by way of aerial application, 5.4 mg/kg in the soil, or 50.0 mg/kg in the diet. Adverse effects on survival of sensitive aquatic organisms occur at 0.003–0.03 μg/L for crustaceans and 0.09–1.1 μg/L for fish and amphibians. Younger stages of sensitive birds had reduced survival at acute oral doses >500.0 mg/kg BW, and reduced growth at diets containing >750.0 mg/kg ration; poultry diets containing <50.0 mg fenvalerate/kg feed produced no appreciable residues in eggs and meat of exposed birds. Among sensitive mammals, adverse effects on survival were noted at acute oral doses of 50.0–450.0 mg/kg BW, dietary concentrations of 50.0–1000.0 mg/kg, and dermal applications of 1800.0 mg/kg BW.

16.4.1 Terrestrial Plants and Invertebrates

Terrestrial plants are relatively unaffected by fenvalerate at recommended application rates, as judged by negligible uptake of fenvalerate from treated soils, formation of numerous fenvalerate conjugates that are pharmacologically inactive, and metabolism of the liberated cyano group into amino acids and eventually carbohydrate and protein.

Adverse effects of fenvalerate on survival of terrestrial arthropods were observed at 0.002–0.015 μg whole body topical application, 0.11 kg/ha aerial application, 5.4 mg/kg in the soil, 50.0 mg/kg in the diet, and for ants 1.4 g/ant mound. Synthetic pyrethroids are more effective in biological systems at low temperatures. The relative sensitivity of insects when compared with mammals is attributed in part to this negative temperature coefficient; thus, warm-blooded animals are less affected than insects and other poikilotherms. Fenvalerate, for example, showed a negative correlation between temperature and toxicity to crickets (*Acheta pennsylvanicus*), being up to 1.9 times more toxic at 15E than at 32EC; a similar case is made for honey bees (*Apis mellifera*) and for many species of aquatic invertebrates and fish.

Signs of lethal pyrethroid poisoning in insects and other arthropods generally include hyperexcitation, tremors, and convulsions, culminating in paralysis and death. At sublethal doses, equivalent to about 10% of a lethal dose, signs of poisoning in sensitive insects include cessation of feeding, wandering, hyperactivity, restlessness, and flushing out of hiding. The American cockroach (*Periplaneta americana*) exposed to topical lethal concentrations of fenvalerate had uncoordinated rapid movements followed by inactivity, appearance of water drops under wings and abdomen, and blackening of the abdomen. Signs appeared in <1 h at lethal concentrations and <3 h at sublethal concentrations. Roaches exposed to sublethal doses began recovery 6 h after exposure, attaining full recovery at 24 h.

Field application rates of 0.05–0.2 kg fenvalerate/ha are recommended for insect control on many food crops. Under these conditions, fenvalerate remained completely effective for 5 days against adults and nymphs of aphids (*Lipaphis erysimi*), jassids (*Amrasca biguttula*), and white fly (*Bemisia tabaci*). Fenvalerate, applied as a drench to mounds, shows promise as an effective control agent of the fire ant, (*Solenopsis invicta*). Foliar applications of fenvalerate sprays at 135.0 mg/L effectively controlled various pests in pear orchards of northern California, including pear psylla (*Psylla pyricola*), codling moth (*Laspeyresia pomonella*), and pear rust mite (*Epitrimerus pyri*); populations of spider mites increased, especially the two-spotted spider mite, (*Tetranychus urticae*).

A concentration of 2.0 mg fenvalerate/L is frequently applied to soils to control insect pests. However, several species of soil protozoans (*Blepharisma undulans*, *Colpoda cucullus*, *Oikomonas termo*) have LC10 (9 h) values in the range of 0.1–0.18 mg/L, suggesting that some damage occurs to this group under recommended application protocols. In fact, all fenvalerate treatments applied to control insect pests of crops also reduced populations of beneficial nontarget organisms, including spiders, ground beetles, and crickets. For example, spiders (*Chiracanthium mildei*) exposed for 48 h to grapefruit leaves that had been dipped 1 h previously for 5 sec in aqueous emulsions of fenvalerate at field-recommended application rates all died within 2 days postexposure.

Fenvalerate-tolerant strains of arthropods include insect vectors of disease, flies and cockroaches, arthropods of veterinary importance, and agricultural pests. But serious control problems are restricted to only a few areas, such as Central America and Thailand, where insecticidal usage is often excessive. The exact mechanisms of resistance are unknown, although tolerance to fenvalerate in the diamondback moth (*Plutella xylostella*), a worldwide pest of cabbage-type crops, is about 20% genetic, involving several genes and multiple loci. Estimates of heritability in tolerance of insects to all biocides ranges between 14% and 47%. Tolerant insect species, such as larvae of the common green lacewing, and resistant strains of houseflies and lepidopterous larvae may hydrolyze fenvalerate faster than sensitive species or susceptible strains. Thus, fenvalerate-resistant strains of domestic houseflies (*Musca domestica*) when compared with susceptible strains absorbed up to three times less fenvalerate, had a metabolic rate up to eight times faster, began excretion of metabolites five times faster, and were twice as resistant to piperonyl butoxide, a synergist applied with fenvalerate.

The alfalfa leaf cutter bee (*Megachile rotundata*) is the most important insect pollinator of alfalfa grown for seed production in France. Alfalfa is parasitized by many insects, including the flower midge (*Contarina medicaginis*). Fenvalerate, at 0.05 kg/ha, controls the flower midge without harm to alfalfa leaf cutter bees. In general, fenvalerate-treated plants were usually nontoxic to bees after 24 h. Fenvalerate does not poison bees when they are in contact with contaminated (100.0 mg/kg) wax in combs. Fenvalerate does not pose a serious threat to honey bees except when dietary levels exceed 50.0 mg fenvalerate/kg. Field application of 0.22 kg fenvalerate/ha on blooming alfalfa, pollen-shedding corn, and blooming red raspberry resulted in reduced honey bee visitation and low to moderate adult bee mortality. Caged honey bees exposed to an equivalent dose of 0.11 kg fenvalerate/ha experienced >50% mortality within 24 h. However, field studies showed that 0.11 kg/ha caused no observable adverse effects to bee colonies located adjacent to a treated alfalfa field; researchers concluded that fenvalerate temporarily repelled bees, as judged by a 70% reduction in bee visits to the alfalfa field in the afternoon after application when compared with periods 24 h before and after application. Impaired response to scent stimuli, in addition to repellency, may account for a reduction in bee visits. Studies suggest that bees surviving LD50 doses of fenvalerate were unable to distinguish odor-mediated learned responses for up to 6 days after treatment. This finding indicates that more research is needed on fenvalerate-associated olfactory inhibition.

16.4.2 Aquatic Organisms

"Supertoxic" compounds are those with LC50 (96 h) values <10.0 µg/L. Fenvalerate is considered supertoxic, as judged by LC50 (96 h) values of <1.0 µg/L for sensitive aquatic organisms, and <10.0 µg/L for representative aquatic species.

Signs of fenvalerate poisoning in fish include loss of schooling behavior, swimming near the water surface, hyperactivity, erratic swimming, seizures, loss of buoyancy, elevated cough rate, increased gill mucous secretions, flaring of the gill arches, head shaking, and listlessness prior to death. Fenvalerate mainly affects the nervous system in fishes, as discussed earlier. It also produces osmoregulatory imbalance, as judged by altered calcium uptake, abnormal sodium and potassium excretion rates, and elevated urine osmolality. Histological damage to gill surfaces by fenvalerate is attributed to high accumulations in gills, irritation due to elevated mucous secretion, increased ventilation volume, and decreased gill-oxygen uptake efficiency. In fish, as in mammals, fenvalerate toxicity is primarily dependent on the $2S$, αS component of the technical mixture. Studies with individual isomers and various freshwater fishes indicate that the $2S$, αS isomer is 96 times more toxic than the $2S$, αR isomer, and at least 1766 times more toxic than the $2R$, αS or $2R$, αR isomers.

Laboratory studies with fenvalerate and aquatic organisms indicate marked differences in sensitivity (among taxonomic groups). Crustaceans were the most sensitive group: reduced survival was evident between 0.0032 and 0.03 μg/L and impaired feeding and reproduction between 0.0016 and 0.01 μg/L. Fish and amphibians were more tolerant to fenvalerate than were crustaceans: increased mortality was evident between 0.088 and 1.1 μg/L and no adverse effects were demonstrated in several species between 0.062 and 0.083 μg/L, although certain salmonids showed high uptake at concentrations as low as 0.0003 μg/L. Algae, molluscs, and chordates were comparatively resistant to fenvalerate. Survival patterns of fenvalerate-stressed aquatic organisms are significantly altered, sometimes by an order of magnitude or greater, by selected biological, chemical, and physical variables. In general, increased mortality was associated with the following: reduced metabolism and excretion, depleted glycogen stores due to starvation, larval and juvenile stages of development, low concentrations of humic acid and other dissolved materials, low particulate loadings, increased water hardness, increased exposure time and bioavailability, emulsifiable formulations, low temperatures, and the 2*S*, α*S* component. Fenvalerate intoxication effects may be reversible. Tadpoles of the northern leopard frog (*Rana pipiens*) that survived LC50 concentrations of the *S*, *S* isomer for 96 h appeared normal 7 days after being placed in clean water. Fenvalerate-protective agents include diazepam and endosulfan. Diazepam provides up to 14-fold protection to frogs against toxic doses of fenvalerate; endosulfan provides limited protection to estuarine fish and shrimp.

Bioaccumulation factors for fenvalerate by representative freshwater and estuarine organisms during exposure for 28–30 days to various sublethal doses ranged from 40 to 570 for fish, 356 to 4700 for invertebrates, and 477 to 933 for algae. Because of its unusually high lipophilicity, fenvalerate is accumulated at only 30% efficiency by aquatic fauna, and uptake is not dose dependent. Contamination of algal food of daphnids with fenvalerate does not seem to contribute to an increase in whole body burdens, although reduced filtration rates due to toxicity could also account for a reduced intake of fenvalerate adsorbed to algae.

Fenvalerate applications of 0.055–0.220 kg/ha are recommended for control of pestiferous crop insects, but these levels are rapidly fatal to nontarget organisms if introduced accidentally into aquatic environments. In one study, large earthen ponds containing red crawfish (*Procambarus clarki*) were treated with fenvalerate at concentrations equivalent to 28.0, 56.0, 112.0, or 224.0 g/ha. All crawfish died within 24 h at all concentrations tested. After 3 days, ponds dosed with 112.0 g/ha and lower were not lethal to crawfish exposed for 24 h. The 224.0 g/ha pond remained toxic to crawfish after 72 h (71% dead) and 120 h (32% dead); mortality was negligible (<10%) after 168 h. Fenvalerate applications of 28.0–112.0 g/ha 0.025–0.1 pounds/acre usually control 90–100% of floodwater mosquitoes and stagnant water mosquitoes. But at 2.0–11.0 g/ha equivalent, the following effects are reported: mayfly naiads are eliminated; populations of diving beetles, cladocerans, and dragonfly naiads are suppressed for up to 3 weeks; zooplankton filtration rates are reduced; colonization processes are altered; and algal and rotifer populations increase due to lack of cladoceran grazing and competition.

Several large-scale mesocosm studies were conducted with esfenvalerate, the most active form of fenvalerate. These investigators demonstrated that (1) the half-time persistence of fenvalerate in the water column is about 10 h; (2) initial concentrations of 1.0 or 5.0 μg/L were not detectable in the water column after 2 and 4 days, respectively; (3) in the first 2 days after application the water column contained the majority of the fenvalerate; (4) by day 4 postapplication, the sediments and macrophytes were the major reservoirs of fenvalerate; (5) fish usually contained less than 1% of the esfenvalerate at any time; (6) sensitive species of some copepod and insect genera showed declines in abundance at 0.08–0.2 μg/L and were unable to recover; (7) most species of zooplankton and benthic invertebrates decreased in abundance at 0.25 μg/L and higher; (8) and finally, at 1.0–5.0 μg/L, there were drastic reductions or elimination of most species of crustaceans and chironomids, juvenile bluegills, and larval cyprinid fishes.

Sediment/water interactions are important to the understanding of fenvalerate toxicokinetics. Addition of soil to fenvalerate-treated waters reduced toxicity to channel catfish (*Ictalurus punctatus*) through adsorption of fenvalerate to clay and organic components of soil; however, crayfish (*Procambarus* spp.) were not protected. Chironomid larvae held in water on sand initially spiked with 50.0 μg fenvalerate/kg accumulated up to 15 times more fenvalerate than did larvae held in water above spiked silt or clay; a similar pattern was evident at an initial concentration of 5.0 μg/kg. This phenomenon is attributed to the greater bioavailability of fenvalerate in sand, as judged by elevated sediment interstitial water concentrations in sand when compared with those of silt or clay. Mortality was observed in systems where fenvalerate concentrations in sediments were sufficient to establish lethal concentrations in the overlying water through sediment/water partitioning; lethal effects at nominal sediment concentrations of 0.1 mg fenvalerate/kg were observed for mysids (*Mysidopsis bahia*) and grass shrimp (*Palaemonetes pugio*), and at 10.0 mg/kg for pink shrimp (*Penaeus duorarum*). Because fenvalerate readily sorbs and binds to organic and inorganic particulate matter, it is difficult to predict its toxic effects on aquatic biota after runoff from agricultural areas or from discharges into particulate-laden habitats.

16.4.3 Birds

Birds that died of fenvalerate poisoning contained residues of 0.1 to 1.26 mg/kg brain fresh weight (FW) and 0.74 mg/kg liver FW, based on acute oral doses of 500.0 to 4000.0 mg/kg BW; juveniles were more sensitive than adults. When compared to other synthetic pyrethroids tested in laying hens, fenvalerate provided higher, more persistent residues in tissues. Birds given single oral doses as low as 250.0 mg fenvalerate/kg BW experienced significant weight loss (adults) or a reduction in the rate of weight gain (immatures); similar signs were noted at dietary levels of 15,000.0 mg/kg ration, but not at 7500.0 mg/kg feed. Poultry diets that contain <50.0 mg fenvalerate/kg feed do not produce an appreciable concentration of residues in eggs or meat of exposed birds.

Adult Japanese quail (*Coturnix japonica*) given a single oral dose of 4000.0 mg/kg BW started feeding normally, but in about 90 min they became hyperactive. Hyperactivity increased until 2 h postdosing, at which time feeding ceased. At 4 h, they had convulsions, irregular movements, jerking, and twitching; they became progressively ataxic and uncoordinated. One quail died at 4 h, another at 8 h. By 24 h, most of the survivors had resumed feeding, but they had an odd standing posture: head held high above the body, legs extended as far straight as possible, and wings held in an upright position close to the body. By 48 h the survivors seemed to be feeding and drinking regularly; growth was normal 14 days after exposure. Signs of intoxication in fenvalerate-poisoned bobwhites (*Colinus virginianus*) usually appeared within 2 h and included hyperactivity, irregular locomotion, ataxia, and spastic muscle contractions. Bobwhites use croplands for feeding, and insects are an important dietary item of chicks and adults in summer months. Little potential exists for adverse effects of fenvalerate on bobwhite and other gallinaceous bird populations from dietary exposures because insects from sprayed fields had maximum whole body residues of only 0.5 mg/kg, a level far below that associated with adverse effects.

Birds rapidly and efficiently metabolize fenvalerate by hydrolytic cleavage of the ester bond followed by extensive hydroxylation of the acid moiety at the carbon adjacent to the carboxyl group, the methyl group, or both. Major metabolites identified in liver preparations were 2-(4-chlorophenyl)-3-methylbutyric acid, 4-hydroxyfenvalerate, 3-phenoxybenzaldehyde, and 3-phenoxybenzoic acid. Liver microsomal drug-metabolizing enzymes usually play an important role in pesticide metabolism; however, fenvalerate and other synthetic pyrethroids are very weak inducers of avian microsomal enzymes. Birds are more resistant to fenvalerate than are mammals, as judged by studies with Japanese quail and rats (*Rattus* sp.). Quail excreted fenvalerate more rapidly, had lower absorption, and faster metabolism; the oral LD50 for quail was >4000.0 mg/kg BW vs. 450.0 mg/kg BW for rat, almost an order of magnitude higher.

16.4.4 Mammals

In general, fenvalerate administered to mammals was rapidly eliminated and had little tendency to accumulate in tissues. Fenvalerate killed sensitive species of mammals at a brain injection concentration of 1.0 mg/kg FW brain (equivalent to 14.0 μg/kg body weight = BW), an intraperitoneal injection concentration of 3.9 mg/kg BW, acute oral doses of 50.0 to 450.0 mg/kg BW, dietary levels of 50.0 to 1000.0 mg/kg feed, and an acute dermal concentration of 1800.0 mg/kg BW; in all cases the $2S$, αS isomer was the most toxic. Measurable residues of fenvalerate were detected in tissues of sensitive mammals at 0.15 mg/kg diet, 0.15 mg/kg BW applied dermally six times over a 3-week period, and at 2.5 mg/kg BW given orally; in all cases the $2R$, αS isomer was taken up 9–16 times over other isomers. Behavioral alterations (e.g., change in drinking water preference) occurred in mice after a single oral dose of 0.3 mg/kg BW. No significant adverse effects were observed in dogs on diets equivalent to 12.5 mg/kg BW daily for 90 days, or in rats on diets containing 250.0 mg fenvalerate/kg (equivalent to 12.5 mg/kg BW) for 2 years.

At sublethal doses in rodents (i.e., 100.0 mg/kg BW), fenvalerate produces neurological toxicity but no histological damage; at higher doses, pathological alterations in peripheral nerves occur. Rats given acutely toxic doses of fenvalerate showed histopathological changes such as axonal swelling and degeneration, and myelin fragmentation of the sciatic nerve; the significance of these findings is unclear.

Route of administration may account for wide variations in the toxic action of fenvalerate. Most authorities agree that fenvalerate is most toxic to rodents when administered by intercerebroventricular injection relative to other routes – indicating the importance of the brain in the Type II poisoning syndrome; fenvalerate was decreasingly toxic when administered intravenously, intraperitoneally, orally, and dermally.

Differences in fenvalerate metabolism occur, even among closely related species such as rats and mice. In both species, regardless of sex, dose level, or chiral isomer, fenvalerate is metabolized primarily by oxidation at the 2′-4′-phenoxy positions of the alcohol moiety and at the C-2 and C-3 positions of the acid moiety, by cleavage of the ester linkage, by conversion of the CN group to SCN^- and CO_2, and by conjugation of the resultant carboxylic acids and phenols with glucuronic acid, sulfuric acid, and amino acids. However, the taurine conjugate of 3-phenoxybenzoic acid was found in mice but not in rats; 4′-hydroxylation of the alcohol moiety and the sulfate conjugate of 3-(4′-hydroxyphenoxy) benzoic acid occurred to a greater extent in rats than in mice; and more thiocyanate was excreted in mice than in rats. Dogs (*Canis familiaris*) are remarkably different from rodents in their ability to metabolize fenvalerate. Four major differences have been observed: 1) rats and mice show hydroxylation of the 2′ position of the alcohol moiety, whereas dogs do not; 2) dogs produce 3-phenoxybenzyl alcohol and 3-4′-(hydroxyphenoxy) benzyl alcohol, whereas rodents do not; 3) the predominant conjugate of the alcohol moiety in dogs is 3-phenoxybenzoyl glycine, but this is only a minor conjugate in rodents; and 4) dogs produced more glucuronides of acid moiety and their hydroxy derivatives than did rats and mice. The proposed fenvalerate metabolic pathways in dogs suggest that species differences and pathways are important and require more research.

Cattle (*Bos* spp.) protected against various insect pests by fenvalerate-impregnated ear tags grow better than unprotected cattle. Beef cattle were protected against hornflies (*Haematobia irritans*) and other bloodsucking dipterans by fenvalerate-impregnated ear tags; during a 115-day grazing period, protected cattle had greater weight gain than unprotected cattle. This technique may have application in protecting fly-infested threatened or endangered species of mammals. Dairy cows tagged with 8% fenvalerate ear tags showed a 99.9% reduction in hornflies over a 16-week trial. But other species of flies, house fly, (*Musca domestica*); stable fly (*Stomoxys calcitrans*); face fly, (*Musca autumnalis*) were not controlled to the same extent, and they increased as horn fly populations decreased. Protected cows produced 117 kg more milk in 16 weeks than did unprotected cows; fat and protein percentages in milk were the same for both groups. The higher milk production in the fenvalerate-tagged group was attributed to more uninterrupted forage time, greater forage consumption, and more efficient energy utilization because less energy was expended on avoiding or removing flies. Similar results were reported in dairy cows in a 12-week study. Fenvalerate was adequately distributed over the entire body and persisted for at least 80 days on the hair of cattle with one ear tag containing 10.5 g active ingredients. Residues in hair were highest at 14 days (18.4 mg/kg FW) and lowest at 80 days (1.3–3.0 mg/kg FW). All four stereoisomers were present on cattle hair, and no stereoselective degradation occurred. Hair contained 14.8 mg/kg (FW) fenvalerate after 30 days with two ear tags.

Cows fed fenvalerate in grain at 10.0 mg/kg diet for 4 days excreted most of the fenvalerate in urine, essentially unchanged. A secondary excretion route is feces, accounting for about 25% of the ingested dose; milk accounted for 0.44–0.64% of the total excreted. Half-time persistence of fenvalerate in milk of treated cows is about 6.4 days. Effects of low concentrations (1.14–6.8 μg/L) of fenvalerate in milk of treated cows on newborn suckling calves are unknown and merit additional research.

Fenvalerate toxicity is antagonized by atropine sulfate or methocarbamol, which may be effective in treating severe cases of poisoning. Conversely, some compounds exacerbate the toxicity of fenvalerate and interfere with a desired use. Domestic cats (*Felis domesticus*) treated with Fendeet (an aerosol mixture of fenvalerate and N-N-diethyl-*m*-toluamide) to control fleas and ticks sometimes show signs of toxicosis, such as tremors, hypersalivation, ataxia, vomiting, depression, and seizures. Signs usually appeared within hours of topical application, and females and juveniles seem to be the most sensitive group. The demonstrated ability of N-N-diethyl-*m*-toluamide to enhance the dermal absorption of fenvalerate is the probable cause of toxicosis.

In occupational settings, fenvalerate produces temporary irritation and itching. Among human fenvalerate applicators, sensitive individuals complain of a burning and tingling skin sensation after using the insecticide, and sometimes they substitute a more toxic insecticide to nontarget species in order to avoid this uncomfortable sensation. This practice, if widespread, may compromise existing or proposed natural resource management practices.

16.5 Recommendations

Fenvalerate is listed under the Class IV Surveillance Index Classification, indicating a low hazard potential to humans from toxicological and exposure standpoints. This classification requires only nominal monitoring. Monitoring efforts of regulatory agencies to the present time, however, have been limited and of marginal worth in evaluating background concentrations of fenvalerate. Additional monitoring is recommended to measure fenvalerate residues in tissues of birds and mammals of concern to the U.S. Fish and Wildlife Service and other agencies charged with protecting natural resources.

Products that contain fenvalerate and are registered for use on corn, wheat, soybeans, sorghum, oats, barley, rye, or cotton are subject to the provisions of the Endangered Species Act. The Endangered Species Act requires that actions of Federal agencies not jeopardize threatened or endangered species or their habitats. Specifically, the U.S. Environmental Protection Agency, in consultation with the U.S. Fish and Wildlife Service, determines whether use of fenvalerate poses a threat to listed species of animals and plants in various locations. Clearly, fenvalerate and other synthetic pyrethroid insecticides should be used with extreme caution in habitats of endangered species.

No regulations exist for protection of sensitive natural resources against fenvalerate, although current application rates to control pestiferous crop insects are lethal to many species of nontarget organisms, including bees, fish, and crustaceans. Proposed fenvalerate guidelines for protection of livestock, poultry, and human health are as follows: <5.0 mg/kg in diets of livestock, <50.0 mg/kg in diets of poultry, <3.0 mg/kg in human diets (<1.0 mg/kg for vegetables, <0.5 mg/kg for meat, <0.25 mg/kg for milkfat), and <0.125 mg/kg BW for acceptable daily intake in man (Table 16.2).

Despite the high toxicity of fenvalerate and other pyrethroids to aquatic organisms, few environmental problems have been documented, presumably due to the very low application levels needed to control insects, adsorption onto soil and organic matter, and comparatively rapid degradation. Nevertheless, fenvalerate is extremely toxic to aquatic organisms (Table 16.2), has high bioaccumulation, and is persistent in sediments; these patterns are most pronounced in estuaries and other wetland environments. Fenvalerate use in areas adjacent to estuarine systems poses unacceptable risks to those ecosystems at concentrations not detectable by analytical methods. It seems reasonable to prohibit all uses of fenvalerate directly into aquatic environments, and to severely restrict usage in areas adjacent to drainage systems.

Additional research is needed on sublethal effects of fenvalerate in the following areas: (1) impaired response to scent stimuli as demonstrated in bees, (2) genotoxic potency as shown in positive genotoxic effects on mice bone marrow, (3) photoproduct formation – especially those formed through ultraviolet irradiation – wherein at least two photoproducts were more toxic than the parent chemical, (4) enhanced tumor formation in rodent liver, (5) development of analytical procedures to detect minute and short-lived reactive metabolites, (6) development of simplified and reliable laboratory test systems more representative of total natural ecosystems, and (7) interaction

Table 16.2. Proposed fenvalerate criteria for the protection of natural resources and human health.

Resource and Other Variables	Criterion
BEES (*Apis* spp., *Megachile* sp.)	
Adverse Effects	
Whole body	>0.1 μg/bee
Diet	>10.0 mg/kg fresh weight (FW)
Aerial application	0.05–>0.11 kg/ha
AQUATIC ORGANISMS	
Crustaceans, decreased survival	
Water column	0.003–0.022 μg/L
Sediments	97.0–190.0 μg/kg FW
Fish	
Water column	
Persistent tissue residues	>0.00028 μg/L
No adverse effects on growth survival, or reproduction	0.062–0.083 μg/L
Lethal	0.088–0.31 μg/L
Brain residues at death	>0.16 mg/kg FW

Continued

Table 16.2. cont'd

Resource and Other Variables	Criterion
BIRDS	
Acute oral exposure, single dose	
No deaths	<250.0 mg/kg body weight (BW)
Some deaths	>500.0 mg/kg BW
Persistent residues	>5.0 mg/kg BW
Tissue residues at death	
Brain	0.1–1.26 mg/kg FW
Liver	0.74 mg/kg FW
Dietary exposure	
Sublethal	
No residues in eggs or meat	<50.0 mg/kg diet
Biochemical upset	>2000.0 mg/kg diet
Lethal	>15,000.0 mg/kg diet
MAMMALS	
Dietary exposure	
Sublethal	
Persistent residues	0.15–15.0 mg/kg feed
Temporary tolerance level, livestock, dried apple pomace	5.0 mg/kg feed
No significant effects	12.5–15 mg/kg BW daily, 100.0–250.0 mg/kg diet
Significant adverse effects	250.0–1000.0 mg/kg diet
Lethal	>50.0 mg/kg BW, >1250.0 mg/kg diet
Single oral exposure	
Sublethal	
Behavioral changes	0.3–30.0 mg/kg BW
Persistent residues	>2.5 mg/kg BW
Lethal	50.0–450.0 mg/kg BW
Dermal exposure	
Sublethal	
Persistent residues	0.15–1.0 mg/kg BW
No deaths	310.0 mg/kg BW daily for 2 weeks
Lethal	1800.0 mg/kg BW, single application
HUMAN HEALTH	
Permanent tolerance level; meat and milk fat	<0.02 mg/kg FW
Temporary tolerance level	
Milk fat	<0.25 mg/kg FW
Meat	<0.5 mg/kg FW
Total diet	<3.0 mg/kg FW
Vegetables; "safe" level	<1.0 mg/kg FW
Acceptable daily intake (ADI), 60-kg person, 1.5 kg food daily	0.125 mg/kg BW
Theoretical daily exposure from diet	
Minimum	0.015 mg, 0.00025 mg/kg BW, 0.2% of ADI
Maximum	0.334 mg, 0.0056 mg/kg BW, 4.5% of ADI

effects of fenvalerate degradation products with other chemicals. More research is also needed on indirect effects on wildlife due to reductions in nontarget insects, and on bioavailability of fenvalerate to aquatic organisms from sediments and the sediment-water interface.

16.6 Summary

Synthetic pyrethroids are the newest major class of broad-spectrum organic insecticides used in agricultural, domestic, and veterinary applications, and now account for more than 30% of global insecticide use. Fenvalerate [(*RS*) α-cyano-3-phenoxybenzyl (*RS*) 2-(4-chlorophenyl)-3-methylbutyrate] is one of the newer synthetic pyrethroid insecticides, and the one most widely used. Technical fenvalerate is a mixture of four optical isomers, each occurring in equal amounts, but with different efficacies against insect pests. Insecticidal properties are largely associated with the 2*S*, α*S* isomer, and to a minor extent with the 2*S*, α*R* isomer. Isomers with a 2*R* configuration have negligible biocidal properties; however, tumor-like growths in rodent liver are associated with the comparatively innocuous 2*R*, α*S* isomer. Pyrethroid insecticides are potent neurotoxicants that interfere with nerve membrane function by interaction with the sodium channel. Fenvalerate is among the most effective pyrethroid neurotoxicants tested, and the 2*S*, α*S* isomer is up to 15 times more potent than other fenvalerate isomers.

Fenvalerate persists for <10 weeks in the environment and does not accumulate readily in the biosphere. Time for 50% loss (Tb1/2) in fenvalerate-exposed amphibians, birds, and mammals is 6–14 h; for reptiles, terrestrial insects, aquatic snails, and fish it is usually >14 h–<2 days; and for crop plants it is 2–28 days. In nonbiological compartments, Tb1/2 is up to 6 days in freshwater, 34 days in seawater, 6 weeks in estuarine sediments, and 9 weeks in soils.

At recommended application rates to control pestiferous crop insects, fenvalerate and other synthetic pyrethroids are relatively harmless to birds, mammals, and terrestrial plants; however, certain nontarget species, including bees, crustaceans, and fish, are at considerable risk, especially at low temperatures. Target insect species are usually killed at fenvalerate concentrations of 0.015 μg/insect, 0.11 kg/ha by way of aerial application, 5.4 mg/kg in soil, or 50.0 mg/kg in diet. Fenvalerate is especially toxic to aquatic organisms e.g., crustaceans died at 0.003–0.03 μg/L and fish and amphibians at 0.09–1.1 μg/L, and its use in or near aquatic environments now seems contraindicated. Birds and mammals are significantly more resistant than fish and invertebrates. Adverse effects on birds occur at acute oral doses >500.0 mg/kg body weight (BW), and 750.0 mg/kg ration; <50.0 mg fenvalerate/kg feed produced no appreciable residues in eggs and meat of exposed birds. Among sensitive mammals, adverse effects on survival occur at acute oral doses of 50–450 mg/kg BW, dietary loadings of 50.0–1000.0 mg/kg feed, and dermal applications of 1800.0 mg/kg BW.

Criteria have not yet been formulated by regulatory agencies for protection of sensitive fish and wildlife resources against fenvalerate. Proposed guidelines for protection of poultry, livestock, and human health include <50.0 mg/kg in poultry diets, <5.0 mg/kg in livestock diets, <3.0 mg/kg in human diets, and <0.125 mg/kg BW daily in humans.

GOLD[a]

Chapter 17

17.1 Introduction

Since antiquity, gold has been valued for its scarcity, beauty, and resistance to corrosion. Gold is the best known of all native elements and the most likely to be found in a metallic state. It is the universal standard of value and the common medium of exchange in world commerce. Gold is almost everywhere considered to be the symbol of everything precious and of enduring value because of the effort required to extract it from nature and its scarcity, relative to other metals. Gold was known and highly valued by the earliest civilizations, Egyptian, Minoan, Assyrian, and Etruscan, and from all these periods ornaments of great beauty and workmanship have survived several thousands of years, many of them being as perfect as when they were first made. The earliest mining work of which traces remain was on gold ores in Egypt; gold washing is depicted on Egyptian monuments of the fourth dynasty, about 2000 BCE. The legend of the Golden Fleece may actually describe an expedition around 1200 BCE to seize gold washed out from the river sands by Armenians using sheepskins. Investors regard gold as an excellent hedge against inflation. In recent years, the net buyers of gold have been central banks and government monetary agencies. In Asia, investment demand for gold has risen sharply. In 1988, about 75% of new gold output went to Taiwan, Japan, South Korea, Hong Kong, and Singapore, as a store of wealth and as adornment.

Gold, however, has produced deep political and social changes through the ages on whole continents. Examples include the Spanish conquistadors who destroyed indigenous cultures in Central and South America in their search for "El Dorado"; the discovery of rich gold mines in Bohemia, Silesia, the northern Carpathians, and northern Romania in the 12th and 13th centuries, which produced a large immigration of miners; the settling of the American West, started mainly by the California gold rush in 1849; the Australian gold rush of 1851, resulting in a doubling of that population in seven years to one million; and the Alaskan gold rush that started in 1897. Unfortunately, the human costs of mining, extraction, refining, marketing, and accumulation of gold include war, slavery, conscripted and convict labor, unhealthy and shortened life span, and degraded living conditions. Gold as an investment is difficult to justify. The intangibility of investor anxieties, the volatility of gold's international market price, and the infectious nature of the desire to hoard, all strongly indicate that complex forces are involved in gold as an investment. It is difficult to justify the production of gold in view of society's minimal need for it and the shortfalls in, for example, food and timber production. However, the appeal of gold has not diminished over the past century, nor the damage caused by its extraction.

Long believed to be relatively benign, it is now known that gold is, in fact, a relatively common allergen that induces dermatitis about the face and eyelids and at sites of direct skin contact. In patch tests worldwide the prevalence of gold allergy might be as high as 13%, with 9.5% the most recent estimate in North America. The main exposure sources of gold contact dermatitis are personal jewelry and dental alloys. Occupations most frequently

[a]All information in this chapter is referenced in the following source:

Eisler, R. 2004. *Biogeochemical, Health and Ecotoxicological Perspectives on Gold and Gold Mining*. CRC Press, Boca Raton, Florida. 355 pp.

causative of contact-dermatitis due to gold include photography, chinaware or glass decorating, manufacturing of dental alloys, and jewelry. Contact allergy to gold in humans may be lifelong or at least extend for years, unlike some strains of rodents that recover quickly when the gold stimulus is removed. The growing evidence of human health concerns related to gold, coupled with known adverse ecotoxicological aspects of various gold extraction techniques, has prompted a critical review of gold in the environment.

17.2 Geology, Sources, and Production

This section briefly summarizes the geological characteristics of gold-bearing deposits, and sources and production of gold. It is emphasized that reliable data on these subjects were scarce and difficult to obtain and that interpretations should be treated with caution.

17.2.1 Geology

Gold originates at considerable depth, and is carried upwards by hot fluids and magma that force their way into rock fractures. Crystallization, most often in quartz veins, occurs as the fluids cool and pressures diminish. As lode deposits break up by weathering, such materials are carried downslope and accumulate as gold-bearing sand or gravel in streams, along beaches, in front of melting glaciers, and in sand dunes. Placer gold in sand and gravels are the most readily available sources of gold for the recreational prospector. Geological events such as uplift and subsidence may cause prolonged and repeated cycles of erosion and concentration, and where these processes occur, deposits may be enriched. The greater specific gravity of gold (19.3) compared with that of residual rock (about 2.7) leads to gold settling out while lighter rocks are washed away. Alluvial gold, once discovered, is often easy to extract as nuggets or grains by simple gravity concentration, and this fact gave impetus to the gold rushes of California in 1848, Australia in 1850, and the Yukon in 1896. Much of the gold in the Ural Mountains of the Former Soviet Union is alluvial.

Gold is wide-spread in the environment. It occurs in minute quantities in almost all rocks, especially igneous and metamorphic rocks. Gold is usually obtained from quartz lodes or veins, or from deposits derived from them by denudation into river gravel. Gold occurs in about 40 minerals, but only native gold (Au) and electrum (Au-Ag) are common. Gold generally occurs in native form, and also in combination with tellurium as the ore calaverite ($AuTe_2$) and with silver and tellurium as the ore sylvanite [$(Au,Ag)Te_2$]. The mineral most commonly found with gold in lodes is iron pyrites, a yellow sulfide of iron. Others are copper pyrites, arsenical pyrites, and other metal sulfides. No mineral is an infallible guide to gold. Limonite, a yellow oxide of iron, is considered a reliable indicator of lode gold on ground surfaces. Magnetite (black iron sand) is useful as an indicator of placer deposits.

Gold veins fill bedrock cracks or fissures. They can range from about one cm in thickness to hundreds of meters thick or long. Lodes are ore deposits of veins that are close to each other. The "Mother Lode" on the western slopes of the Sierra Nevada Mountains in California is where the 1848–49 Gold Rush originated. Gold occurs as thin veins in quartz rocks in the gold fields of South Africa. The gold, known as vein or reef gold, is present as microscopic particles so that extraction is comparatively difficult. In general, reef gold is found in quartz or albite rocks, often together with iron pyrites. Deposits with silver and other metals often occur in volcanic regions controlled by major fault zones where concentration of gold takes place by thermal metamorphism from basic rocks and deposition in sedimentary rocks.

Certain locations in Canadian stream beds and drainage basins favor preferential accumulation of gold and other heavy minerals, and many models have been formulated to locate placer formations. One model was based on the process of erosion and redeposition of the bed during annual flood events that affected changes in equilibrium transport rates. The model postulates that increases in bed roughness (diameter of materials in the stream) results in preferential accumulation of heavy minerals, with enrichment increasing with increasing density, as is the case for gold. Decreasing channel slope also results in enrichment of heavy minerals. Distribution of gold predicted by the model agreed with field data from a 5 km stretch along a gravel bed stream in British Columbia.

In northeastern Nevada, high-grade ores containing as much as 24.7 g of gold per metric ton with an estimated total gold endowment of 2200 tons, was first discovered in 1962 in an areal extent of 8.5 km by 2 km. This deposit – the Carlin gold deposit – is one of the largest hydrothermal disseminated replacement deposits discovered in North America, with four zones of gold mineralization, mostly in the upper 250 m. It was formed in the late Tertiary, about 70 million years ago, over a period of at least 100,000 years, as a result of high-angle faulting, igneous and hydrothermal activity, and other processes. Ores contained, in mg/kg, about 8.0 Au, 21.0 Hg, 222.0–409.0 As, and 52.0–106.0 Sb. Gold, in unoxidized form was found mainly within arsenian pyrite and associated with mercury, antimony, and thallium. Characteristics of the host rocks that may enhance their favorability to gold deposition include the presence of reactive carbonate, porosity, permeability, and the presence of iron, which can be sulfidized to form auriferouspyrite. The mining districts of north-central Nevada are localized by major structural features, including the Roberts Mountains thrust fault on which clastic and volcanic rocks of early and middle Paleozoic age have ridden eastward over correlative carbonate rocks. Genesis of gold deposits in northern Nevada is not fully understood and subject to conflicting models. A general consensus of these models is that regional structures control the spatial distribution of deposits. Measurement of the earth's magnetic and electrical fields through magnetotelluric soundings demonstrate resistivity structure and fracture zones associated with subsurface gold deposits, and shows promise in locating future subsurface gold deposits.

Distribution of gold deposits in Far East Russia was mapped using information about gold discoveries, as well as geological, gravity, and magnetic data. Factors responsible for gold localization and the erosion of ore bodies were evaluated using mineral composition, geologic formation boundaries, lithosphere structure, and others. It is now established that the main source of gold is the mantle and its derivatives, and that the mobilization and transport mechanisms are granitic magmas and fluids occurring at various depths. Gold is mobilized in the fluid phase in the form of chlorides, hydrosulfides, and complex compounds, and deposited in sediments with high carbon, sulfide, and iron content. Gold can be mobilized by hydrothermal solutions and granitic intrusion. Deep crustal faults were important in the accumulation and transport of ore-bearing solutions.

The Circle Mining District in Alaska is a major gold repository. The area contained granite, quartzite, quartzite schist, and mafic schist overlain by colluvium gravel, fan deposits, silt, organic material, and several ages of gold-bearing gravel. Mafic schist seems to be the bedrock source of the gold. The Tintina fault zone in the northeast section of the Circle District is a dominant structure in the area. The zone contains at least three ages of superimposed fan gravel: late Tertiary, late Pleistocene, and Holocene, the latter being the most gold-enriched and with 16% silver; all samples contained antimony, which distinguishes it from other gold-bearing areas of Alaska.

Most aspects and characteristics of porphyry and epithermal systems containing at least 200 tons of gold are similar to those that typify their smaller and lower grade counterparts. Nevertheless, hypothetical mechanisms operating in the mantle, in high-level magma chambers, during exsolution of magmatic fluids and sites of gold deposition are considered to be particularly favorable for either the liberation, concentration, transport, or precipitation of gold, and hence for the formation of large deposits. When considering the 25 largest gold deposits over 200 tons in the circum-Pacific region, a number of criteria are identified as favorable indicators for large gold accumulations. Both gold-rich porphyry copper and various epithermal gold deposits seem to be more common in association with igneous rocks containing high concentrations of potassium. The porphyry deposits are characterized by high hydrothermal magnetite contents and impermeable host rocks, especially limestones. Epithermal gold deposits, in contrast, are controlled by marked lithologic differences and proximity to volcanic settings.

17.2.2 Sources and Production

Total world production of gold is estimated at 3.4 billion troy ounces of which more than 67% was mined in the past 50 years and with 45% of the total world production coming from the Witwatersrand district of South Africa. World gold production in 1979 was about 39 million ounces, with the Republic of South Africa (RSA) producing 58% of the global output. The Soviet Union was ranked second and Canada third. About 87% of the gold production came from primary sources, such as lodes and placers, and the remainder as a by-product from the refining of copper and base metal ores. The United States was fourth in 1979 with an output of 964,000 ounces, or 2.5% of the world production; however 57% of the output was primary and 43% by-product. In addition to the RSA, the Soviet Union, Canada, and the United States, significant gold production (>300,000 ounces) in 1979 was also documented in the Dominican Republic, Brazil, Ghana, Zimbabwe, Papua New Guinea, the Philippines, and Australia. During 1979, the three largest gold-producing states, with 75% of the U.S. total, were Utah, followed by South Dakota, and Nevada. The Utah production was almost entirely a by-product of copper mining. Gold was also produced in quantity in Alaska and eleven western states. Total world production in 1994 was estimated at 220 tons or 60 million ounces, but this needs verification.

In developing countries, gold mining activities increased substantially in the late 1960s following the end of the 1944 Breton Woods agreement which limited the price of gold to US $35 per troy ounce. The price rose gradually during the 1970s, leading to reworking of ores previously considered too low grade for gold extraction. In May 1985, representatives from 20 countries at a World Bank meeting concluded that legal title to gold-mined lands was the major concern, superseding lack of technology, environmental impact, and financial support. As of 1998, about 3 million people are directly involved in gold mining throughout the developing world and mercury pollution problems as a result of gold mining are particularly severe in Brazil, Peru, Ecuador, Columbia, Bolivia, Venezuela, Indonesia, and the Philippines.

In 1986, the largest gold producers were the RSA with 38.5% of the world's production, the Soviet Union (18.8%), the United States (7.0%), Canada (6.6%), Australia (5.5%), China (4.1%), and Brazil (4.1%). By 1988, the surge in Amazonian gold mining placed Brazil second in worldwide production of gold behind RSA (621 tons), and ahead of the United States (205 tons), Australia (152 tons), and Canada (116 tons). In 1990, gold production in the United States exceeded that of the Soviet Union for the first time in 50 years. Between 1980 and 1990, annual production of new gold increased from 962 to 1734 metric tons, with production significantly increased in the United States, Canada, and Australia. These three nations controlled about 33% of the new gold's market share in 1990, up from 12% in 1980. New gold output from less industrialized nations, such as Brazil, the Philippines, Papua New Guinea, Ghana, Zimbabwe, and Ecuador rose to 26% in 1990, with production increases being greatest in Brazil and Papua New Guinea. In industrialized countries, it was corporate mining interests that propelled this growth. However, in Latin America, Asia, and Africa (excluding RSA), most of the new gold was produced by small scale or informal sector mining operations, employing several million people worldwide and accounting for about 25% of global gold output. The market price of gold per troy ounce has moved from about $20 in 1873 to $35 in 1934, $100 in 1972, $200 in July 1978, $600 in late 1978, a record $850 in January 1980, $474 in March 1980, $750 in September 1980, $526 in December 1980, $275 in December 2001, and $425 in September 2005. The incentive to produce gold is high; at $500 an ounce, a cube of gold 30.5 cm on a side (one foot) weighs about 550 kg and is valued at approximately nine million dollars. Fluctuations in the price of gold are considered irrational and based on fear, greed, and anxiety, rather than on industrial production and associated costs.

17.2.2.1 Asia and Environs

In 1995, the People's Republic of China produced 105 tons of gold, about one-third from small-scale mines using mercury amalgamation extraction techniques. The Bo Hai Sea area of northeast China is adjacent to onshore occurrence of gold lodes that have made the region the largest gold-producing area in China. The best provenances for the placer gold materials that occur in the Bo Hai Sea are the Liadong-Koren and Shangdong peninsulas, which are rich in epithermal gold deposits. The depositional sandy coast is considered the most favorable area for the formation of placer gold due to a large and consistent source of detrital minerals and a suitable concentrating environment. Submarine morphologic features, such as sand bars and shoals, seafloor platforms, and submerged rivers are the most favorable features for formation of placer gold deposits within the Bo Hai Sea.

Since ancient times gold has been produced by territories in the Urals, Caucasus, Kazakhstan, and Central Asia, with 60% deriving from placer deposits. By the 1970s, the Soviet Union was considered to be the second largest producer of gold in the world, although the precise amounts mined were not published.

In recent years, India has produced about 2 tons of gold annually, using the cyanidation process.

17.2.2.2 Canada

Gold has been discovered and mined in Alberta, British Columbia, Manitoba, Nova Scotia, Quebec, and the Yukon Territory. The earliest recorded notice of gold in Canada was of placer findings in Quebec in 1823, but little effort was expended during the next 25 years. Lode gold was first mined in Nova Scotia in the 1850s. The gold rush to California in 1849 and to Australia in 1851 stimulated the gold prospecting fever in Canadians. In 1858, placer gold was discovered in northern British Columbia. In 1862, lode deposits were opened in Nova Scotia. In 1866, lode gold was discovered in southeastern Ontario. In 1896, the discovery of rich placer gravel along River Klondike in the Yukon Territory stimulated the great gold rush of 1898 that served as the springboard for intensive prospecting in Alaska. By 1913, Yukon gold production was about 11 tons. In 1909, important gold discoveries were made in Ontario and Manitoba.

Between 1858 and 1972, Canada produced 195.7 million troy ounces of gold, with peak production in 1941. Ontario was the leading producer, followed by Quebec, with most of the gold coming from quartz lode mines or as a by-product of base-metal refining. About 90% of the Canadian gold production is from the northern and eastern half of Canada, all of it from lode mines and placers. In 1975, Canada ranked second among the world's gold producing nations, first being the RSA. In 1992, Canada ranked fifth in world production of gold.

For more than 100 years, exploration and extraction of gold have taken place in the coastal regions of the western Canadian margin. There is a potential for significant marine placer deposits (up to 4.0 g Au/ton) in the nearshore and shelf regions off western Canada over an extensive area (60 km), with thicknesses up to one meter, especially in the Queen Charlotte Islands of British Columbia. The main source of the gold and minable minerals, mainly titanium, is the early Holocene beach deposits that form the coastal bluffs of the area. Deposits in any particular area are ephemeral, the result of wave action in a macrotidal setting along an eroding and unconsolidated coast.

17.2.2.3 Europe

Native gold and secondary minerals were identified in the quartz veins from Bastogne,

Belgium, and extend their range in this area. Native gold occurs as metallic flakes up to 5 mm in diameter and is associated with pyrrhotite and chalcopyrite. Gold mineralization was found at the site of abandoned kaolin quarries in southern Sardinia, Italy. Tailings from this quarry were of low pH and contained elevated concentrations of arsenic, zinc, copper and other potentially harmful metals. If open pit mining of gold is initiated using the cyanidation process, similar tailings problems are expected.

17.2.2.4 Republic of South Africa (RSA)

Gold production in all of Africa for the period 1976–1985 was about 230 million troy ounces, with RSA producing about 96% of the total; Zimbabwe, Ghana, and Zaire producing about 3.8%; and the rest by Ethiopia, Mali, Liberia, Gabon, Congo, Central Africa Republic, Rwanda, Tanzania, Sudan, Burundi, Cameroon, Kenya, and the Malagasy Republic. More than 95% of all gold produced in the RSA has its source in the Witwatersrand area, and almost all of it is lode gold. The largest producing gold mine in Witwatersrand is 1000–3000 m beneath the surface. By the 1970s, the gold fields of RSA had yielded more than 31 million kg of gold, or more than half the gold reserves of the world. The genesis and tectonic setting of the Witwatersrand gold deposits are somewhat similar in hydrothermal lode equivalents to other large deposits in eastern Russia, but differ in their sedimentary-hydrothermal metamorphic origin.

17.2.2.5 South America

The most recent gold rush in South America was triggered in 1980 by the discovery of a large gold mine in Serra Pelada, Amazon Region, Brazil. From the Amazon, the fever spread to neighboring countries including Venezuela, Guyana, French Guiana, and Suriname. Most of the gold produced in Brazil is from individuals operating in remote areas. Production figures are difficult to obtain; however, some authorities aver that Brazil produces more gold than any nation except RSA, with about 90% coming from informal mining in Amazonia using mercury amalgamation. This will be discussed later (see also Chapter 19).

Small-scale mining in Suriname started in 1876. At its height in 1907, the industry produced about 1200 kg of gold yearly. But after 1910, production declined, with annual production of about 200 kg between 1930 and 1970. Gold mining revived in the 1990s because of the rise in the price of gold, the deterioration of the economy in Suriname, the immigration of Brazilian miners, and the presence of foreign prospecting companies. About 35,000 workers are currently involved in uncontrolled small-scale gold mining in Suriname, or almost 9% of the total population of 400,000. The estimated production of gold increased from 10,000 kg in 1995 to 20,000 kg per year in 1998 and 1999. About 30% of the gold production is registered in official figures, the rest being sold illegally or smuggled out of the country.

17.2.2.6 United States

Except for small recoveries of gold by Indians and Spanish explorers, gold was produced in southern California as early as 1775, in the southern Appalachian region of the eastern United States in 1792, and in North Carolina in 1799. New discoveries were made in other Appalachian states in the 1820s and 1830s. Appalachian gold was mined by panning and dredging of stream gravels (placer mining). There were also a few underground mines in North Carolina, Virginia, and Georgia that used slave labor. At that time, gold was typically concentrated using the mercury amalgamation process. These states produced significant amounts of gold until the American Civil War (1861–1865). After the discovery of gold in California in 1848, California and other western states contributed the bulk of domestic gold production. In general, gold was derived from three types of ore: ore in which gold is the main metal of value, base-metal ore which yields gold as a by-product, and placers. In the early years, most of the gold was mined from placers, but after 1873 production came mainly from lode deposits. By 1905, gold deposits in Alaska and Nevada had been discovered and gold production in the United States exceeded 4 million troy ounces annually – a level maintained until 1917; during World War I (1914–1918) and for some years afterwards, annual domestic gold production declined to 2 million ounces. Since the 1930s, by-product gold became a measurable fraction of the annual domestic gold output. The largest single source of by-product gold in the United States is the porphyry deposit at Bingham Canyon, Utah, which has produced about 18 million troy ounces of gold since 1906.

In 1934, President Franklin Delano Roosevelt issued an order limiting the amount of gold per individual to a maximum of 200 ounces. In 1942, the War Production Board issued Order L-208 closing all gold mines for the duration of World War II (1939–1945). Only a few mines reopened after the war ended. On March 17, 1968, the U.S. Treasury discontinued buying and selling domestically mined gold and ceased to control its price. When U.S. Treasury buying ended, the official fixed price of pure gold was set at $35 a troy ounce, the price set in 1934. Since 1968, miners can sell gold to any willing purchaser at prices determined by supply and demand. The restriction on the 200-ounce limit of possession ended on December 31, 1974. With government controls lifted, the price of gold began rising until by 1975 it had risen from $35/ounces to almost $600/ounces. This sharp price increase resulted in increased gold prospecting and mining world-wide, although in 2002 the price of gold was less than $300/ounces. The industrial break-even cost of producing an ounce of gold in 1995 in the United States was estimated at $256. Non-cash costs and profitability added another $51, for a total of $307. This did not include production costs, such as exploration and development. One estimate for producing a sustainable supply of gold in the United States is $370–400/troy ounce.

From the end of World War II (1945) through 1983, domestic mine production of gold did not exceed 2 million ounces per year. Since 1985, annual production has risen about 1 million ounces every year, reaching about 9 million troy ounces in 1990, and – for the first time – exceeding domestic consumption, estimated at about 3.5 million ounces per annum.

The most abundant areas for gold in the conterminous United States are located in the Sierra Nevada and the Rocky Mountain ranges. Of the 50 states, 32 have reported the existence of gold in sufficient showings to interest the casual collector. At least 21 states produced more than 10,000 troy ounces of placer and lode gold through 1959; however, more than 77% of the 307.1 million ounces of gold mined between 1799 and 1965 came from only five states: Alaska, California, Colorado, Nevada, and South Dakota. There are more than 500 districts in the United States that have produced at least 10,000 ounces of gold, and 45 have produced more than 1 million ounces. Four districts have produced more than 10 million ounces: Lead SD, Cripple Creek CO, Grass Valley CA, and Bingham UT. Operating open pit mines in Nevada, which started production in 1965, now account for about 1.5 million troy ounces annually. Gold mining returned to the South Carolina Piedmont in the 1980s and 1990s, and renewed gold exploration is underway in parts of Appalachia.

In 1995, the United States became the second largest producer of gold in the world, behind the RSA. Although gold was mined from hundreds of lode mines and placer sites across the United States, almost 40% (3.9 million ounces) came from the two largest producers operating in north-central Nevada. In fact, the top four producers accounted for almost 55% of U.S production in 1995. About 90,000 people were directly involved in the production of precious metals in the United States, which includes gold and silver, with 51,457 jobs in Nevada alone. About 79% of the ore was processed by heap-leach methods to produce 32% of the gold, and 21% of the ore was processed using various milling techniques to produce 68% of the gold. There was also a shift of production from near surface, low grade oxide ores, to deeper, higher grade refractory ores. Nevada produced about 6.8 million ounces in 1995, or 67% of total

Table 17.1. U.S. gold production by state: 1995 vs. 2000.

State	Troy Ounces	Percentage of Total U.S. Production
ALASKA	141,800 vs. 546,000	1.4 vs. 4.8
CALIFORNIA	783,000 vs. 447,000	7.7 vs. 3.9
IDAHO	96,500 vs. 72,000	1.0 vs. 0.6
MONTANA	437,200 vs. 212,000	4.3 vs. 1.9
NEVADA	6,765,000 vs. 8,585,000	66.5 vs. 75.6
SOUTH DAKOTA	559,100 vs. 171,000	5.5 vs. 1.5
UTAH	729,700 vs. 700,000	7.2 vs. 6.1
OTHERS	660,400 vs. 616,100	4.4 vs. 5.4
TOTAL	10,172,700[a] vs. 11,349,100[b]	

[a]Worth $3.906 billion @ US$384/troy ounce.
[b]Worth $4.358 billion.

U.S. production, followed by California with 0.8 million ounces, and Utah with 0.7 million ounces (Table 17.1). By the year 2000, Nevada produced 8.58 million ounces of gold and accounted for 75.6% of all U.S. production (Table 17.1). In 2001, Nevada produced 8.125 million ounces of gold, the fourth year in a row above the 8 million ounce mark. In 2001, Nevada became the third largest producer of gold in the world – producing about 21% of all mined gold – ranking behind the RSA and Australia; underground operations contributed about 20% of the gold production in Nevada in 2001.

17.3 Uses

The most common uses of gold are in personal jewelry, coinage, and bullion. Because of its unique properties, gold is also used in electronics, human medicine, dentistry, physiology, and immunology. Radiogold isotopes have been used to tag various species of wildlife and to treat human carcinomas. Secondary uses of gold are expected to increase significantly.

17.3.1 Jewelry

For the past 7000 years gold jewelry has been sought by consumers for traditional, social, and stylistic reasons. As a result, jewelry fabrication always accounted for the largest share of global gold production. In 1991, for example, gold production including the use of scrap amounted to 2111.1 tons and more than 95% was used in jewelry manufacture. Gold alloys, especially those containing silver and copper, have been preferred by goldsmiths for the past 5000 years. Alloys used in jewelry today also contain zinc and nickel with the result that gold can appear yellow, white, green, or red depending on the mixture. But it is only since 1960 that the International Gold Council has decided to support research and investigations on metallurgy of gold alloys. This was clearly connected with the fact that jewelry accounts for the largest use of gold in every country of the world.

The gold concentration of alloys can be expressed either in terms of caratage, 24 carats representing pure gold, or in fineness which is the weight fraction expressed in 1000ths. In Europe, 8, 14, and 18 carat jewelry is most common (0.333, 0.585, and 0.750 fine, respectively). Low carat alloys are inferior to the widely used 14 and 18 carat alloys with respect to tarnish and superficial corrosion resistance. In general, white gold alloys show high nickel and silver and low copper and gold. Blue or purple gold, an aluminum alloy ($AuAl_2$), being 0.785 fine is rather brittle and not corrosion resistant. A recent addition to jewelry alloys is the so-called Spangold, an Au-Cu-Al alloy, about 18 carat, in yellow (76% Au, 19% Cu, 5% Al) and pink (76% Au, 10% Cu, 6% Al). The most recent alloy in jewelry is gold-titanium at 23.5 carat, which has unusual hardness and wear-resistance.

17.3.2 Coinage

Gold was the first metal used in currency, and has probably been used for this purpose for more than 5000 years. The first coins of pure gold were probably struck in 561 BCE by King Croesus. He was also the first to fix the relation of silver to gold in coinage at 13 to 1, a relation not far from that of the terrestrial abundance of the metals which is 0.08 mg/kg for silver and 0.004 mg Au/kg, or 20 to 1. Excluding the Orient, the total amount of gold produced during medieval times (a period of about 500 years) is estimated at 2000 tons. From 1837 until 1939, the United States minted 0.900 fine $10 and $20 coins. In 1932, the currency of 31 countries was based on gold 1.000 fine, nine countries on gold 0.91667 fine, and one country on gold 0.875 fine. In 1934, the United States fixed the price of gold at $35 per troy ounce (31.1035 g), which was accepted as the international standard. The price stayed constant and by the 1950s most of Alaska's gold mines were in decline. But in 1967, controls were lifted and the price of gold soared, with subsequent reopening of many mines. In 1980, the price of gold reached $850 per troy ounce, but fell to between $350 and $400 in the mid-1980s. With improved technology for working the deposits, miners found it cost effective to work poorer deposits and to reopen old mines.

Gold bars have become available in the past 100 years. There are three qualities of fine gold used in most bars. (1) For the international market, gold bars of about 12.5 kg (400 troy ounces) with a purity of at least 99.5% Au are prescribed. Each bar must be marked with the purity in 4 digits, the trademark of the refinery, and a serial number. The mark and the dimensions of the bar are registered by the London Bullion Market. The bars can then be traded worldwide without further control. Most of the official hoardings of national banks are in these bars. (2) Small bars of 1 g-1 kg of 99.99% purity are produced. The bars need to contain less than 100.0 mg Ag/kg, <20.0 mg Cu/kg, and <30.0 mg total other base metals/kg. These bars are also stamped with the purity and refiner's trademark and a serial number. These bars are mostly used for private investment purposes. (3) For some high-technical purposes, mainly in electronic systems, gold is required at 99.999% purity (0.99999 fine). These bars must contain, in mg/kg, less than 5.0 platinum, <5.0 palladium, <3.0 silver, <3.0 iron, <2.0 lead, <2.0 bismuth, <0.5 copper, and <0.5 nickel. Gold coins are no longer issued as official currency. The types minted today are but another form of gold bars.

17.3.3 Electronics

At ordinary temperatures, the electrical conductivity of gold is about 75% that of pure silver; however, the electrical resistance of gold, which is the reverse of conductivity, decreases with decreasing temperature, and at about 5° Absolute it has practically disappeared, at which point gold is a near perfect conductor of electricity. Gold is now used as soldering components in high technology electronics, electrical contacts, plating materials, wear-resistant contacts in rockets, submarines, computers, and signaling devices. Gold is the most resistant metal to corrosion, has low electrical resistivity, and excellent malleability, ductility, and softness. All these properties are useful in making very thin wires and other connectors for some applications, and in allowing joining of components by thermocompression binding in others. In 1994, 165,700 kg of gold was used in the field of electrical engineering and electronics, mostly in electromechanical devices such as connectors, switches, or relays in which gold is the electrical contact material.

17.3.4 Radiogold

Radiogold isotopes have been used to treat mucosal carcinomas of the mouth, and adenocarcinomas of the prostate; tag various species of wildlife, including mice, bats, and lizards; measure radiation exposures from nuclear accidents; and as a chemical label for water-soluble gold-compound pharmaceuticals.

17.3.5 Medicine

Up to the eighth century, metallic gold was considered a panacea for every known disease. By the 1400s, auric chloride (prepared from metallic gold and aqua regia, followed by neutralization with chalk) was used in the treatment of leprosy. In India, calcined gold preparations, colloidal gold, and gold bromide were alleged to exhibit antiepileptic properties. Gold drugs were used frequently during the 1700s and 1800s, with claims of great success. It was Koch's observation in 1890 that gold cyanide (AuCN) inhibited the growth of the bacteria that caused tuberculosis and represented the beginning of systematic gold molecular pharmacology. Koch's experiments showed that tubercle bacilli were killed at 0.5 mg AuCN in vitro, but met with limited success when introduced into the blood serum of infected animals. The "gold decade" between 1925 and 1935 marked extensive use of gold compounds in treatment of tuberculosis and syphilis, although toxic side effects remained a problem.

From about 1913 to 1927 marked a period of intense searching for Au^+ compounds of lower toxicity. Until the 1930s, Au^+ salt therapy was extended to the treatment of rheumatoid arthritis and lupus erythematosus because of the (mistaken) belief that these diseases were atypical forms of TB. Some Au^+ complexes have anti-microbial and anti-fungal activity.

Monovalent organogold salts have been used, with mixed results, in the treatment of pemphigus (skin and mucous membrane blisters), lupus erythematosus, ulcerative colitis, Crohn's disease, various types of arthritis, bronchial asthma, bullous skin conditions, discoid lupus, several forms of rheumatism, ankylosing spondylitis affecting peripheral joints, kala-azar (caused by the flagellate protozoan *Leishmania donovani* transmitted by the bite of sand flies), tuberculosis, and malaria. In malaria, for example, monovalent gold compounds show promise in treating a disease currently affecting about 400 million people and threatening another billion world-wide. Chloroquine is the preferred treatment against non-resistant strains of *Plasmodium falciparum*, the parasite responsible for malaria. Tests with chloroquine-resistant strains of *Plasmodium* and a monovalent gold-chloroquine complex show promise in controlling *Plasmodium*. The gold-chloroquine complex displayed high in vitro activity – markedly higher than other metal-chloroquine complexes tested – against the blood stage of two chloroquine-resistant strains of *P. falciparum*. It was also active in vitro and in vivo against *Plasmodium berghei* (rodent malaria), showing that incorporation of Au^+ produced a marked enhancement in chloroquine efficacy.

Gold compounds have been used to control various microbial pathogens. The concept of gold as an antibacterial therapy is traced back to the 8th century. In 1890, it was firmly established that gold cyanide was inhibitory to TB bacilli in vitro, but not in vivo. The introduction of gold compounds for the treatment of rheumatoid arthritis was based on the assumption that bacteria were responsible for this condition. In one study, in vivo tests with mouse skin demonstrated that a monovalent gold thiocyanate compound applied topically controlled antibiotic-resistant strains of bacteria and the yeast *Candida albicans*. In vitro tests with monovalent organogold treatments used in rheumatoid arthritis therapy showed antibacterial activity against *Helicobacter pylori*, a bacterium associated with gastritis and peptic ulcer. Antimicrobial activities of two isomeric Au^+-triphenylphosphine complexes with nitrogen-containing heterocycles were documented against two species of gram positive bacteria (*Bacillus subtilis*, *Staphylococcus aureus*) and *Candida albicans*, and were more effective against these organisms than were corresponding silver (Ag^+) complexes; however, they were not effective against tested species of gram-negative bacteria and molds.

Anti-tumor properties of monovalent and trivalent gold complexes are documented.

Intramuscular injections of sodium aurothiomalate (Au^{+}) were successful in controlling tumor-associated antigens in patients with tongue carcinoma or pulmonary cancer. Selected Au^{+3} complexes are reasonably stable under physiological conditions and show cytotoxic properties when tested in vitro on the human ovarian tumor cell line A2780. The cell-killing properties of stable Au^{+3} complexes were attributed to the Au^{+3} center with ligands of ethylenediamine, diethylenetriamine, or cyclam. Four monovalent gold compounds coordinated with different phosphines showed varying degrees of cytotoxicity when tested in vitro against three human ovarian cancer cell lines, and one – 1,2-(bisdiphenylphosphino)ethane bis[Au^{+} lupinylsulfide] dihydrochloride – was more effective than conventional antitumor platinum compounds. More research on the antitumor properties of gold compounds seems warranted.

Oral gold treatment with a monovalent organogold compound has been used successfully to treat psoriatic arthritis in a patient simultaneously infected with HIV. The use of gold compounds in HIV-associated inflammatory arthritis and the possible antiretroviral effects of gold merit additional study. And reports of anti-HIV activity for cyanide and thioglucose derivatives of gold compounds may be even more significant.

Gold needles are now used by acupuncturists, and gold prostheses by ophthalmologists and otologists. Japanese acupuncturists used gold needles in patients with rheumatoid arthritis to relieve severe pain and to slow the progression of joint destruction. Deficient eyelid closure is a major visual threat to patients with unresolved facial nerve palsy. Gold weight implants assisted closure in patients with incomplete paralysis of the orbicularis oculi, ameliorating complaints of dry eye, excessive tearing, and corneal epithelial breakdown. Gold has also been used successfully in synthetic middle ear prostheses; there were no postoperative infections, possibly due to gold's bacteriostatic action, or other post-operative complications.

Sporadic use of gold drugs in the treatment of tuberculosis led to the use of sodium gold thiomalate in the treatment of rheumatoid arthritis. Most gold drugs in current medical use are monovalent gold thiol complexes that are stable in air and water, and usually administered by injection. Some gold phosphine complexes, however, can be administered orally. The medical community views gold with mixed emotions: optimism, because gold drugs can cause remission of rheumatoid arthritis and other diseases; and concern, because of the high frequency of toxic side effects. In current medical practice, chrysotherapy – the treatment of rheumatoid arthritis (RA) with gold-based drugs – is well established. It derives its name from Chryseis, the golden-haired daughter of a priest of Apollo in Greek mythology. Five monovalent organogold complexes are now widely used throughout the world in these treatments. Thiomalatogold (myochrisin or gold sodium thiomalate), thioglucose gold (solganol), and thiopropanosulfonate gold (allochrysin) are oligomeric complexes that contain linear Au^{+} ions connected by bridging thiolate ligands. Bis(thiosulfate) gold (sanochrysin) contains gold bound to the terminal sulfur donor atoms of $S_2O_3^{2-}$. The newest drug, auranofin, licensed in the mid-1980s, contains coordinated triethylphosphine and 2,3,4,6-tetra-O-acetyl-B-1-D-thioglucose ligands. Auranofin has the advantage of being orally absorbed but is considered less effective than the injectable gold thiolates. After more than 75 years of use in the management of RA, chrysotherapy with injectable monovalent organogold salt formulations is now routinely used in the treatment of RA and other diseases having an autoimmune or inflammatory component to their pathogenesis, despite a significant incidence of adverse side effects. The Empire Rheumatism Council in 1961 confirmed the effectiveness of gold salt treatment for RA and today injectable gold salts are considered a major line of treatment of progressive RA. As recently as 1985, injectable monovalent organogold compounds were the initial drugs of choice for Canadian patients with RA; by 1992, gold and methotrexate were equally preferable in moderate RA, and methotrexate was preferable to gold in aggressive RA. It is noteworthy that the mechanism of action for chrysotherapy is not known with certainty despite almost 75 years of use and major advances in bioinorganic chemistry.

Gold thioglucose ($C_6H_{11}O_5SAu$), or GTG, was initially developed and marketed as a therapeutic agent for the treatment of arthritis and rheumatism. However a single intraperitoneal or subcutaneous injection in mice destroyed the appetite center located in the hypothalamus region of the brain producing hyperphagia (uncontrolled eating) and, within a few weeks, many of the characteristics of human obesity. The degree of obesity induced by GTG is dependent on the dose administered and the strain of the mouse.

17.3.6 Dentistry

Gold is now used in restorative dentistry mainly as the base element in precious metal alloys for gold inlays, crowns, and bridges. High gold dental alloys are distinguished by their superior corrosion resistance when compared to other metal alloys. Gold is usually the major component of dental alloys containing silver, copper, and small amounts of platinum and lead; these alloys can be heat-treated to develop strengths as great as 150,000 psi. Most dental gold alloys containing platinum or palladium show conspicuous age-hardening characteristics, due to the formation of a metastable gold–copper complex. The American Dental Association recommends that dental casting gold alloys should contain more than 75% gold and other platinum group metals. Amalgam (Au–Hg) fillings are a mainstay of restorative dental care for the population of many countries, but these are considered to be the main source of human mercury intake next to methylmercury contained in food. Mercury is discussed in detail in Chapter 19.

17.3.7 Others

Gold has been used as a delivery vehicle for various substances, for use in microscopy, and in food and liquor.

Minute metallic gold particles are used as delivery vehicles for introduction of exogenous proteins, antibodies, and for gene therapy. Proteins, such as serum albumin, prevent flocculation of colloidal gold sols; labeled antibodies adsorbed onto colloidal gold particles retained full activity, with major implications for immunization. DNA-coated gold beads were effective at introducing DNA into cells without serious cell injury. Immunization of ferrets, for example, with a plasmid DNA expressing influenza virus hemagglutinin, provided complete protection from influenza virus.

Colloidal gold was first made in the 1600s for use in microscopy. Colloidal gold particles are excellent markers in electron microscopy because they are electron-dense, spherical in shape, and can be prepared in sizes from 1 to 25 nm. Colloidal gold (10 nm) is a useful tracer to obtain information about the processes involved in the immunological response.

A minute portion of the world's inhabitants is exposed to powdered gold (Au) as a decoration for pastries, in chocolates, or in alcoholic beverages. A rare instance of severe lichen planus in response to metallic gold dust included in a liquor is documented.

17.4 Properties

Gold is a complex and surprisingly reactive element, with unique physical, chemical, and biochemical properties. Some of these properties are listed and discussed below.

17.4.1 Physical Properties

Gold is a comparatively rare native metallic element, ranking 50th in abundance in the earth's crust. The chemical symbol for gold is Au, from the Latin *aurum* for gold. Metallic gold is an exceptionally stable form of the element and most deposits occur in this form. The main elements with which gold is admixed in nature include silver, tellurium, copper, nickel, iron, bismuth, mercury, palladium, platinum, indium, osmium, iridium, ruthenium, and rhodium. The native gold-silver alloys have a color range from pale yellow to pure white, depending on the amount of silver present. Finely divided gold is black, like most other metallic powders, while colloidally suspended gold varies in color from deep ruby red to purple. Gold occurs as metallic gold (Au) and also as Au^+ and Au^{+3}, so that it occurs in combination with tellurium as calaverite ($AuTe_2$) and sylvanite ($AuAgTe_4$), and also with tellurium, lead, antimony, and sulfur as nagyagite, $Pb_5Au(TeSb)_4S_{5-8}$.

Gold is characterized by an atomic weight of 196.967, atomic number of 79, a melting point of 1063°C, and a boiling point of about 2700°C. In the massive form, gold is a soft yellow metal with the highest malleability and ductility of any element. A single troy ounce of gold can be drawn into a wire over 66 km in length without breaking, or beaten to a film covering approximately 100 m^2. Traces of other metals interfere with gold's malleability and ductility, especially lead, but also cadmium, tin, bismuth, antimony, arsenic, tellurium, and zinc. It is extremely dense, being 19.32 times heavier than water at 20°C. A cube of gold (30 cm; 12 in.) on a side weighs about 544 kg (1197 pounds). Gold has high thermal and electrical conductivity, properties that make it useful in electronics. It is extremely resistant to the effects of oxygen and will not corrode, tarnish, or rust. Pure (100%) gold is 1.000 fine, equivalent to 24 carats. Gold is usually measured in troy ounces, wherein 1 troy ounce equals 31.1 grams vs. 28.37 grams in an ounce avoirdupois.

Gold has 30 known isotopes, but only one, ^{197}Au, is stable. The nucleus of ^{197}Au contains 79 protons and 118 neutrons. Isotopes of mass numbers 177–183 are all α emitters and all have a physical half-life of <1 minute. Isotopes of mass numbers 185–196 decay by electron capture accompanied by γ radiation and in some cases by positron emission. The only long-lived isotope is ^{195}Au with a half-life of 183 days. The neutron-heavy isotopes of 198–204 all decay by β emission accompanied by γ radiation. The isotope ^{198}Au is widely used in radiotherapy, in medical diagnosis, and for tracer studies.

The color of gold alloys depends on the metal mixture. Red gold is comprised of 95.41% Au and 4.59% copper (Cu); yellow gold of 80% gold and 20% silver (Ag); and white gold of 50% Au and 50% Ag. The white gold commonly used in jewelry contains 75–85% Au, 8–10% nickel, and 2–9% zinc, while more expensive white alloys include palladium (90% Au–10% Pd) and platinum (60% Au, 40% Pt). Colloidally suspended gold varies in color from deep ruby red to purple, and is used in the manufacture of ruby glass.

Gold-silver-copper alloys are frequently used in coinage and gold wares. A purple alloy results with 80% Au and 20% aluminum, but this compound is too brittle to be made into jewelry. Gold forms alloys with many other metals, but most of these are also brittle. As little as 0.02% of tellurium, bismuth, or lead makes gold brittle.

Analytical methodologies to measure gold in biological samples and abiotic materials rely heavily on its physical properties. These methodologies include X-ray fluorescence, adsorptive stripping voltammetry, bacteria-modified carbon paste electrodes, inductively coupled plasma mass spectrometry [ICP-MS], atomic absorption spectrometry [AAS], fire assay, neutron activation [NA], and γ spectrometry. Analyses of gold based upon gravimetric, volumetric, and UV/Visible spectrophotometric techniques have been largely displaced by instrumental methods, such as NA, AAS, and more recently ICP-MS and ICP-AAS. In ICP-MS, for example, detection limits of gold after preconcentration of samples were as low as 0.04 ng/g ash in vegetation, 0.1–0.8 ng/L in water and urine, and 0.1 ng/g in soils and sediments.

17.4.2 Chemical Properties

The chemistry of gold is complex. Gold can exist in seven oxidation states: −1, 0, +1, +2, +3, +4, and +5. Apart from Au in the colloidal and elemental forms, only Au^+ and Au^{+3} are known to form compounds that are stable in aqueous media and important in medical applications. The remaining oxidation states of −1, +2, +4, and +5 are not presently known to play a role in biochemical processes related to therapeutic uses of gold. Neither Au^+ or Au^{+3} forms a stable aquated ion ($[Au(OH_2)^+_{2-4}]$ or $[Au(OH_2)^{3+}_4]$, respectively) analogous to those found for many transition metals and main group cations. Both are thermodynamically unstable with respect to elemental gold and can be readily reduced. The gold-based anti-arthritic agents are considered pro-drugs that undergo rapid metabolism to form new metabolites. In complexes containing a single gold atom, the oxidation states +1, +2, +3, and +5 are well established. Divalent gold (Au^{+2}) is rare, usually being formed as a transient intermediate in redox reactions between the stable oxidation states Au^+ and Au^{+3}. The first Au^{+5} complex containing the ion AuF_6^- was reported in 1972. The compound AuF_5 can also be prepared. Both are powerful oxidizing agents. Gold also forms many complexes with metal-metal bonds in which it is difficult to assign formal oxidation states.

Metallic gold (Au) is comparatively inert chemically. Gold is resistant to tarnishing and corrosion during lengthy underground storage or immersion in seawater. It does not oxidize or burn in air even when heated. However, gold reacts with tellurium at high temperatures to yield $AuTe_2$, and reacts with all the halogens. Bromine is the most reactive halogen and, at room temperature, reacts with gold powder to produce Au_2Br_6. At temperatures below 130°C, chlorine is adsorbed onto the gold forming surface compounds; at 130–200°C, further reactions occur but the rate is limited by the diffusion rate of chlorine through the surface layer of gold chlorides; at >200°C, a high reaction rate occurs as the gold chlorides sublime, continually exposing a gold surface. Atomic gold is considerably more reactive than the massive metal. Evaporation of gold at high temperatures under vacuum followed by condensation of the vapor with a suitable reagent onto an inert noble-gas matrix at liquid helium temperature produces $Au(O_2)$, $Au(C_2H_4)$, $Au(CO)$, and $Au(CO_2)$. Cocondensation of atomic gold with carbon monoxide and dioxygen gives the complex $Au(CO)_2O_2$; all these gold compounds decompose on warming the matrix. When auric oxide is treated with strong ammonia, a black powder is formed called fulminating gold ($AuN_2H_3 \cdot 3H_2O$). When dried, it is a powerful explosive, as it detonates either by friction or on heating to about 145°C. Caution is advised when handling this compound.

Halogen compounds of gold are well known, especially aurous chloride (AuCl) and auric chloride ($AuCl_3$). Aurous chloride is a yellowish-white solid that is insoluble in cold water, but it undergoes slow decomposition into Au and $AuCl_3$. Auric chloride takes the form of a reddish brown powder or ruby red crystals. The auric chloride of commerce is aurichloric or chloroauric acid, ($HAuCl_4 \cdot 3H_2O$), a brown deliquescent substance that is soluble in water or ether. Aurichloric acid forms a series of salts called aurichlorides or chloroaurates. Aurichlorides of Li, K, and Na are very soluble in water, and those of Rb and Cs much less soluble. The sodium salt, $NaAuCl_4 \cdot 2H_2O$, is sold as sodio-gold chloride and, unlike aurichloric acid, is not deliquescent. Two gold bromides are known, AuBr and $AuBr_3$, corresponding to their chlorine counterparts. Auric iodide (AuI_3) is unstable and decomposes into aurous iodide (AuI) and free iodine. Iodine in aqueous-alcoholic solutions combines with metallic gold to form aurous iodide, a white or lemon-yellow powder that is insoluble in water.

Gold is inert to strong alkalis and virtually all acids, except aqua regia – a mixture of concentrated nitric acid (1 part) and hydrochloric acid (3 parts). The nitric and hydrochloric acids interact forming nitrosylchloride (NOCl) together with free chlorine, which reacts with gold. In aqua regia, gold forms tetrachloroauric acid, $HAuCl_4$, which is the source of gold chloride. Gold is also soluble in hot selenic acid forming gold selenate, and in aqueous solutions of alkaline sulfides and thiosulfates. Gold will dissolve in hydrochloric acid in the presence of hypochlorite or ferric iron (Fe^{+3}) as oxidant. The dissolution of gold in cyanide solutions with air or hydrogen peroxide as oxidant is another example of this effect. The reaction with oxygen as oxidizing agent apparently takes place by adsorption of oxygen onto the gold surface, followed by reaction of this surface layer to yield AuCN, followed by the complex $Au(CN)_2^-$ which passes into solution. Gold is also soluble in liquid mercury and in dilute solutions of sodium or calcium cyanide.

The cyanide solvent was used in Australia in 1897 where it was used to remove finely disseminated gold from pulverized rock. The cyanide process is the only known method of profitably treating massive low-grade gold ores. Using the cyanide process, auriferous rocks containing as little as 1 part gold in 300,000 parts of worthless materials can be treated successfully.

Gold is readily dissolved by halide or sulfide ions in the presence of oxidizing agents to yield Au^{+3} or Au^{+} complexes. It is probably in this way that gold is dissolved when hot volcanic rock is buried or when a hot granite intrusion rises near the surface of the earth's crust. As the solution cools to 300–400°C, concentrations of oxygen and hydrochloric acid decrease sharply, and gold is redeposited. Hydrothermal transfer of gold as the complex ion $[Au(SH)_2]^-$ may occur in some cases. Dissolution and redeposition of gold in stream beds may also be responsible for the formation of large crystals of alluvial gold.

Solutions containing gold complexes, such as $AuCl_4^-$, are easily reduced to Au and under controlled conditions colloidal gold may be formed. Colloids of gold – first reported in the 18th century – may be red, blue, or violet depending on the mean particle size and shape. Various reducing agents can be used for preparing colloidal gold including tannin, phosphorous, formaldehyde, and hydrazine hydrate. The "purple of Cassius" is a mixed colloid of hydrated Sn^{+4} oxide and gold formed by reducing $AuCl_4^-$ with Sn^{+2} chloride, A purple or ruby-red precipitate is formed on heating the solution. A sensitive test for gold is based on this process, that is, a purple color is formed if a 10^{-8} M solution of $AuCl_4^-$ is added to a saturated solution of $SnCl_2$.

17.4.3 Biochemical Properties

Gold is not an essential element for living systems. Indeed, administration of gold drugs to patients resulted in effects more similar to that of toxic elements, such as mercury, than to that of biologically utilized transition elements such as copper and iron. Gold distributes widely in the body and the number of possible reactions and reaction sites is large. Most of the in vivo gold chemistry is concerned with the reaction of gold species with thiols. Within mammalian systems subjected to Au, Au^{+}, or Au^{+3}, gold metabolism resulted in both monomeric and polymeric species. Most gold complexes administered orally or parenterally were absorbed, but rate and extent of accumulation were highly variable between gold compounds. Gold circulated in blood mainly by way of the serum proteins, especially albumin. Gold was deposited in many tissues and was dependent on dose and compound administered. Likely storage forms included colloidal Au, insoluble Au^{+} deposits, and possibly Au^{+3} polymers. Accumulated gold containing sulfur was documented. There is no suitable animal model available for testing mechanisms of action of gold compounds used in human medicine.

Gold has a unique biochemical behavior. Biochemical behaviors of heavy metal ions show some similarities, particularly in their affinity for polarizable ligands. But they also show important differences. Gold, for example, has a comparatively low affinity for amino and carboxylate groups, a stable higher oxidation state in water, and proven anti-inflammatory activity of selected Au^{+} organic salts. The biochemistry of gold has developed mainly in response to prolonged use of gold compounds in treating rheumatoid arthritis and in response to efforts to develop complexes with anti-tumor and anti-HIV activity.

Chemical reactions of gold drugs exposed to body fluids and proteins are mainly ligand exchange reactions that preserve the Au^{+} oxidation state. Aurosomes (lysosomes that accumulate large amounts of gold and undergo morphological changes) taken from gold-treated rats contain mainly Au^{+}, even when Au^{+3} has been administered. However, the potential for oxidizing Au^{+} to Au^{+3} in vivo exists. Monovalent gold drugs can be activated in vivo to a Au^{+3} metabolite that is responsible for some of the immunological side effects observed in chrysotherapy. For example, treatment of rodents and humans with anti-arthritic, monovalent gold drugs generates

T-cells that react to Au^{+3} but not to the parent compound.

Although metallic gold (Au) is arguably the least corrosive and most biologically inert of all metals, it can be gradually dissolved by thiol-containing molecules such as cysteine, penicillamine, and glutathione to yield Au^{+} complexes. Metallic gold reacted with cysteine in aqueous or saline solution in the presence of oxygen to produce a Au^{+}-cysteine complex; Au^{+} and cysteine formed a 1:1 Au^{+}-cysteine complex; L-cysteine reduced most Au^{+3} compounds in solution to produce the Au^{+}-L-cysteine complex. With D-penicillamine, Au formed a Au-penicillamine complex; Au^{+} under a nitrogen environment formed a R_3PAu^{+}-penicillamine complex; and Au^{+3} formed a bis complex with penicillamine. With glutathione, Au^{+} formed a stable 1:1 complex in solution; Au^{+3} oxidized glutathione to sulfoxide, the gold being reduced to Au^{+} which was stabilized by complexing with unreacted glutathione. These processes were amplified at alkaline pH, significantly at pH 7.2, and perceptibly in acidic environments having pH values as low as 1.2. The rate of the reaction was controlled by the concentrations of thiol-containing molecules and by the pH; reactions might take place within cells and inside lysosomes. Under favorable conditions, reactions occurred at low rates on skin surfaces. Skin samples taken from beneath gold wedding bands of normal individuals averaged 0.8 mg/kg dry weight (DW) skin. In vitro studies designed to simulate conditions inside phagocytic lysosomes showed substantial dissolution of Au in the presence of hydrogen peroxide and amino acids such as histidine and glycine. There are reported instances of rheumatoid arthritis patients who, on initiation of gold drug treatment (chrysotherapy), have promptly produced rashes in the skin areas which have had regular contact with gold jewelry. Gold jewelry, if in close contact with skin, could be slowly dissolved by sweat. Thus, the thinning of gold rings over time, thought to be mainly due to abrasion, could also be due, in part, to dissolution. Colloidal gold is readily accumulated by macrophages. The gold particles are taken into small vesicles, which form by surface invagination, and into vesicles fusing to form vacuoles with subsequent transport to the centrosomic region. The part played by the surface of the Au particle may be due to Au^{+} ions on the surface, which promote uptake. A soluble gold-uptake stimulating factor of MW $<100{,}000$ is reportedly secreted by lymphocytes and acts upon the macrophages.

$Gold^{+}$ drugs were metabolized rapidly in vivo. The half-life for gold excretion in dogs was 20 days, but major metabolites had half-life times of 8–16 h. Within 20 min of administration, gold was protein-bound mainly in the serum. Injectable $gold^{+}$ drugs were not readily taken up by most cells, but bound to cell surface thiols where they affected cell metabolism. The high affinity of Au^{+} for sulfur and selenium ligands suggested that proteins, including enzymes and transport proteins, were critical in vivo targets. It was clear that extracellular gold in the blood was primarily protein bound, suggesting protein-mediated transport of gold during therapy. Metallothioneins play an important role in metal homeostasis and in protection against metal poisoning in animals. Metallothioneins are cysteine-rich (>20%), low-molecular weight proteins with a comparatively high affinity for gold, copper, silver, zinc, cadmium, and mercury. These heat-stable, metal-binding proteins were found in all vertebrate tissues and were readily induced by a variety of agents – including gold – to which they bind through thiolate linkages. The role of metallothioneins in maintaining low intracellular gold concentrations needs to be resolved.

Following a chrysotherapy type regimen with gold disodium thiomalate in mice, Au^{+3} generation was analyzed with a lymph node assay system using T lymphocytes sensitized to Au^{+3}. The findings were consistent with three separate anti-inflammatory mechanisms: (1), generation of Au^{+3} from Au^{+} scavenges reactive oxygen species, such as hypochloric acid; (2), Au^{+3} is a highly reactive chemical that irreversibly denatures proteins, including those lysosomal enzymes which non-specifically enhance inflammation when they are released from cells at an inflammatory focus; and (3), Au^{+3} may interfere with lysosomal enzymes involved in antigen processing or may directly alter molecules along the lysosomal-endosomal pathway, resulting in reduced production of arthritogenic peptides. If all these activities occurred within a redox system in phagocytic cells, then the anti-inflammatory actions of Au^{+}/Au^{+3} could be effective for protracted periods, and explain, in part, both the anti-inflammatory and the adverse effects of anti-rheumatic Au^{+} drugs. Deviation of proteins could also contribute to the rare instances of auto-immunity reported in association with chrysotherapy.

Knowledge of Au^{+} binding sites on large molecules, such as proteins, is limited to a few studies using $Au(CN)_2^-$. Although $Au(CN)_2^-$ is one of the most stable gold ions in solution, it is considered too toxic for clinical use. The simple Au^{+} cation does not appear to exist in solution and most Au^{+} compounds are insoluble or unstable in water. Mercaptides stabilize Au^{+} in water, and sodium gold thiomalate is now in widespread use as an anti-inflammatory drug. Ionic Au^{+} seems to enter many cells, but localize within the lysosomes of the phagocytic cells called macrophages. Here they may inhibit enzymes important in inflammation. Studies with sodium gold thiomalate suggest that anti-tumor mechanisms, like inflammation, are also macrophage-mediated.

$Gold^{+}$ drugs and their metabolites react in vivo with cyanide, forming dicyanoaurate^{+}, $Au^{+}(CN)_2)^{-}$; this ion has been identified as a common metabolite of Au^{+} drugs in blood and urine of chrysotherapy patients. Also, Au^{+} is the primary oxidation state found in vivo although there is increasing evidence for the generation of Au^{+3} metabolites. Biomimetic studies indicate that the oxidation of sodium $gold^{+}$ thiomalate and sodium $gold^{+}$ thioglucose by hypochlorite ion $(OCl)^{-}$, released when cells are induced to undergo the oxidative burst at inflamed sites, is rapid and thermodynamically feasible in the formation of Au^{+3} species. The OCl^{-} ion is involved in both the generation of $Au(CN)_2^-$ and the formation of Au^{+3} species in vivo.

The potential anti-tumor activity of gold complexes is driven by three rationales: (1), analogy to immunomodulatory properties underlying the benefit from Au^{+} complexes in treating rheumatoid arthritis; (2), the structural analogy of square-planar Au^{+3} to $platinum^{+2}$ complexes, which are potent anti-tumor agents; and (3), complexation of Au^{+} or Au^{+3} with other active anti-tumor agents in order to enhance the activity and alter the biological distribution of Au^{+3}. For example, the rate of hydrolysis of $AuCl_4^-$ in water is 375 times greater than that of $PtCl_4^-$. There is potential for developing new cytotoxic gold complexes that have anti-tumor properties, and this requires robust, new ligand structures that can move gold through cell membranes and into the cytoplasm, and perhaps into the cell nucleus. Trivalent gold (Au^{+3}) compounds are potential anticancer agents. These compounds are soluble in organic solvents, such as methanol or DMSO, but poorly soluble in water. In water, $AuCl_3$ undergoes hydrolysis of the bound chloride without loss of the heterocycle ligand. When Au^{+3} compounds react with proteins, like albumin or transferrin, Au^{+3} is easily reduced to Au^{+}. Cytotoxicity studies with tumorous cells showed marked anticancer activity of Au^{+3} complexes, probably mediated by a direct interaction with DNA. However, rapid hydrolysis of Au^{+3} to Au^{+} under physiological conditions may severely restrict their use. More studies are needed to understand the biological mechanisms of gold complexes, including extent of cell penetration and biodistribution.

Anti-HIV activity of monovalent gold compounds were associated with inhibition of reverse transcriptase (RT), an enzyme that converts RNA into DNA in the host cell. Other reports indicate that Au^{+} inhibits the infection of cells by HIV strains without inhibiting the RT activity, with the critical target site tentatively identified as a glycoprotein of the viral envelope. Other reports show that $Au(CN)_2^-$ at concentrations as low as 20 μg/L is incorporated into a T-cell line susceptible to HIV infection, and retards the proliferation of HIV in these cells. This concentration is well tolerated in patients with rheumatoid arthritis, suggesting that $Au(CN)_2^-$ may have promise for existing HIV patients.

17.5 Gold Concentrations in Field Collections

Maximum gold concentrations documented in abiotic materials were 0.001 μg/L in rainwater, 0.0015 μg/L in seawater near hydrothermal vents, 5.0 μg/kg DW in the Earth's crust, 19.0 μg/L in a freshwater stream near a gold mining site, 440.0 μg/kg DW in atmospheric dust near a high traffic road, 843.0 μg/kg DW in alluvial soil near a Nevada gold mine, 2.53 mg/kg DW in snow near a Russian smelter, 4.5 mg/kg DW in sewage sludge, 28.7 mg/kg DW in polymetallic sulfides from the ocean floor, and 256.0 mg/kg DW in freshwater sediments near a gold mine tailings pile. In plants, elevated concentrations of gold were reported in terrestrial vegetation near gold mining operations 19.0 μg/kg (DW), in aquatic bryophytes downstream from a gold mine 37.0 μg/kg (DW), in leaves of beans grown in soil containing 150.0 μg Au/kg 170.0 μg/kg (DW), in algal mats of rivers receiving gold mine wastes up to 1.06 mg/kg DW), and in selected gold accumulator plants 0.1–100.0 mg/kg (DW). Fish and aquatic invertebrates contained 0.1–38.0 μg Au/kg DW. In humans, gold concentrations of 1.1 μg/L in urine of dental technicians were documented vs. 0.002–0.85 μg/L in urine of reference populations, 2.1 μg/L in breast milk, 1.4 mg/kg DW in hair of goldsmiths vs. a normal range of 6.0–880.0 μg/kg DW, 2.39 mg/L in whole blood of rheumatoid arthritis patients receiving gold thiol drug therapy (chrysotherapy) vs. a normal range of 0.2–2.0 μg/L blood; and 60.0–233.0 mg/kg fresh weight (FW) in kidneys of rheumatoid arthritis patients undergoing active chrysotherapy vs. <42.0 mg/kg FW kidney in these same patients 140 months posttreatment.

17.5.1 Abiotic Materials

Most gold in ocean surface waters comes from fallout of atmospheric dust. Riverine sources of gold into seas and oceanic coastal waters are minor, as judged by studies of manganese transport. The presence of dissolved gold in seawater was discovered in 1872, and many unsuccessful attempts were made to recover the gold commercially from seawater. The most famous attempt was made by German scientists in the years 1920–27, with the intention of paying off the German war debt incurred in 1914–18. The method was based on reduction to metallic gold using sodium polysulfide. Unfortunately, the German calculations of 0.004 μg Au/L were 100–400 times higher than the recently calculated range for dissolved oceanic gold of 0.00001–0.00004 μg/L. At these low concentrations it was not possible to directly determine what gold species were present. However, based on redox potentials of gold compounds and seawater composition, it is probable that $AuCl_2^-$ predominates, with smaller amounts of $AuClBr^-$, as well as bromo-, iodo-, and hydroxy complexes of Au^+, in oxidation states of Au, Au^+, and Au^{+3}. Dissolved gold may be usable as a tracer of hydrothermal influence on bottom waters near vents. Concentration of gold in bottom water samples of the mid-Atlantic ridge in 1988 near hydrothermal vents was 0.0015 μg/L vs. 0.0007 μg/L at a reference site; hydrothermal vent samples also had elevated concentrations for manganese and turbidity.

Known gold-rich, sea-floor deposits in the southwest Pacific Ocean occur along the axis of a major gold belt extending from Japan through the Philippines, New Guinea, Fiji, Tonga, and New Zealand. Polymetallic sulfides recovered from the sea-floor hydrothermal systems of this region contain up to 28.7 mg Au/kg (about 1 ounce per ton) with an average of 3.1 mg Au/kg. These samples are among the most gold-rich hydrothermal precipitates reported from the sea floor. The gold is generally of high purity, containing less than 10% silver. In one hydrothermal vent field, gold concentrations averaged 30.0 mg/kg and visible gold was seen in the sulfide chimney. Gold concentrations decreased sharply to <0.02 mg/kg when the temperature dropped from about 280–300°C in the center of the chimney to about 200°C at its outer margin. Subsea-floor boiling and precipitation of sulfides is important in separating gold from base metals in the ascending hydrothermal fluids. Gold seemed to be precipitated largely from aqueous sulfur complexes [$Au(HS)_2$_] as a result of the combined effects of conductive cooling, mixing with seawater, and oxidation of H_2S. Sulfide deposits in this basin and elsewhere in the southwest Pacific Ocean are similar to some gold-rich massive sulfides on land. Gold enrichment in high-sulfide marine sediments is usually – but not always – associated with elevated concentrations of silver, arsenic, antimony, lead, zinc, and various sulfosalts, especially iron-poor sphalerite (zinc sulfide); in contrast, gold is typically depleted in samples with high levels of cobalt, selenium, or molybdenum. In one study, high gold concentrations in marine sediments were associated with elevated arsenic (1100.0–6600.0 mg/kg), antimony (85.0–280.0 mg/kg), and lead, but the correlations between these elements and gold were variable. There was no significant correlation between gold and other trace metals measured or with silicon, iron, and magnesium.

Abnormally high gold concentrations (>10.0 μg Au/kg) found in the sediments around Sado Island in the Sea of Japan were attributed to auriferous mineralization of the island and anthropogenic mining activities. Gold is probably supplied to marine sediments in dissolved form through rivers and seawater, and to a lesser extent as discrete minerals. Gold distribution in coastal sediments of the Sea of Japan is controlled by geologic characteristics of the catchment area of rivers, the grain size of the sediments, redox potential, water depths of the sampling locations, and dissolved oxygen. For example, gold is more abundant in the finer fraction sediments than in coarse ones. In cases where there is a clear negative correlation between gold content and redox potential of the sediments, the gold occurs mostly in the dissolved form; if the correlation is not significant, the gold occurs in metallic form. Dissolved gold is converted by reduction to Au in oxygen-depleted environments. The suspended gold particles are subsequently adsorbed on mineral surfaces or precipitated as hydroxide or sulfide.

Freshwater sediments in Murray Brook, New Brunswick, Canada received gold between 1989–92 from a vat leach cyanidation process used to separate gold from ores. The gossan (oxidized pyrites) tailings pile in Murray Brook leached gold into the adjacent freshwater stream sediments from complexation of gold to $Au(CN)_2^-$ by residual cyanide within the tailings. The elevated gold concentrations (up to 256.0 mg Au/kg) in stream sediments close to the headwaters of the creek near the tailings suggest that $Au(CN)_2^-$ is degraded and the gold is removed from solution via reduction of Au^+ by Fe^{2+}. Gold is converted from a complexed form to a colloidal form with increasing distance downstream, consistent with dissolved nitrate contents which decreased from 5.2 mg/L near the headwaters to 1.4 mg/L at the lower end of the stream.

Worldwide accumulation of gold in sewage is about 360 tons each year. Sewage is commonly dumped on land or at sea. Discharge of excessive sewage into coastal areas poses a threat to human health and coastal fisheries, diminishes the recreational use of the littoral zone, and may result in the formation of anthropogenic labile-metal deposits. Sewage solids from a southeastern Australian community with a gold mining history of more than 100 years contained 0.18–2.35 mg Au/kg DW. These concentrations are similar to those of ore deposits currently mined for gold. Gold in sewage sludge containing 0.35 mg Au/kg DW applied to agricultural surface soils migrates downwards; after 15 years, about 60% of the gold was found in subsurface soils.

Gold concentrations in different strata of snow/ice cores from the French-Italian Alps deposited over a period of 200 years were consistently low (0.07–0.35 μg/kg FW, detection limit of 0.03 μg/kg) and constant, except for minor increases resulting from atmospheric deposition from nearby smelters. In northwestern Russia, however, gold concentrations in the annual winter snow cover of 1995–96 were greatly elevated (>350.0 μg/kg DW). Dust and smokestack emissions from the local ore roasting and metal smelters were the sources. Concentrations of gold in snow increased with proximity to these industrial sources. The high concentrations of gold and other precious metals (rhodium, platinum, palladium) deposited on snow during a single winter season suggest that modernization of the industrial plants to recover these metals would result in substantial economic benefits.

17.5.2 Plants

Gold accumulator plants, such as *Artemisia persia*, *Prangos popularia*, and *Stripa* spp. grasses, routinely contain >0.1 mg Au/kg DW and may contain as much as 100.0 g of gold per metric ton or 100.0 mg Au/kg. Microorganisms in the plant roots may be responsible for solubilizing the gold, allowing ready uptake by these species. Some strains of *Bacillus megaterium*, for example, secrete amino acids, aspartic acid, histidine, serine, alanine, and glycine to aid in gold dissolution. Bioaccumulation of gold from metals-contaminated soils was documented in stems and needles of Corsican pine trees (*Pinus laricio*) from the Mount Olympus area of the island of Cyprus, and plants grown on soils containing 1.0–25.0 μg Au/kg DW soil had comparatively high concentrations of gold in seeds and pericarp and low concentrations in pods, leaves, and stems. Faba beans (*Vivia* sp.) seem to contain about the same amount of gold in leaves than did the soil on which they were grown (170.0 μg/kg DW vs. 150.0–180.0 μg/kg DW); however, leaves, sugar, and juice of sugarcane (*Saccharum officinarum*) grown in Egypt contained 17–130 times less gold than did the soil of their sugarcane fields.

Gold was detected in aquatic macrophytes from streams draining abandoned base-metal mines, suggesting use of these plants in biorecovery. Bryophytes collected downstream of a gold mine in Wales had slightly higher concentrations of gold than did upstream samples, with a maximum value of 37.0 μg Au/kg DW. In Poland and the Czech Republic, aquatic bryophytes reflected increased amounts of gold in a biotype, with high arsenic mineralization; highest values recorded were in *Fontinalis antypyretica* (18.8 μg Au/kg DW) and *Chiloscyphus pallescens* (20.2 μg Au/kg DW) from areas of former gold mining.

In the gold mining communities of Sri Lanka, peat and algal mats contain elevated concentrations of gold. In peat, gold is positively correlated with increasing depth as well as with increasing concentrations of iron, manganese, cobalt, zirconium, sodium, magnesium, and potassium. In euryhaline algal mats, gold concentrations increase in a seaward direction, suggesting a greater geochemical mobility of dissolved gold with increasing concentrations of chloride ions.

Gold exploration in tropical or subtropical countries has indirectly accelerated efforts to understand the behavior of gold within lateritic formations. Gold uptake by vegetation is a significant mechanism for mobilizing gold in tropical forests more than 100,000 years old. Pure gold dissolves only under organic conditions. The three primary gold complexes of mobilized gold are: $[Au(OH)_3 \cdot H_2O]^0$, $AuClOH^-$, and $Au(OH)_2FA^-$, where FA indicates fulvic acid from soil organic matter. These gold complexes are believed to be stable under surficial equatorial rain forest conditions and they could be leached from soils to rivers.

17.5.3 Animals

Gold concentrations in soft tissues of marine invertebrates ranged between 0.3 and 38 μg Au/kg DW; for fish muscle the mean concentrations were 0.12 μg/kg on a DW basis and 2.6 μg/kg on an ash weight basis. Insect galls induced by egg deposition of the chalcid wasp (*Hemadas nubilpennis*) on shoots of the low-bush blueberry (*Vaccinum angustifoloium*) had elevated levels of gold and other metals in epidermal tissues, especially near the stomata. Gold comprised up to 5.4% of the total weight of gall periderm and epiderm, but was not detectable in nutritive cells or other tissues. Emissions from the nearby Sudbury, Ontario, site of the largest nickel producer in the world, may have confounded the results of this study.

In humans, gold concentrations in breast milk ranged from 0.1 to 2.1 μg/L; it is speculated that the highest concentrations were due to gold dental fillings and jewelry of the mothers. In dental technicians, concentrations of gold in urine were significantly higher than urine from students or road construction workers. Dental technicians also had elevated urinary concentrations of platinum and palladium when compared to students and laborers. The comparatively high gold excretion rates of dental technicians were due to the greater number of noble-containing artificial dentures worn by that group.

Gold in scalp hair of Italian goldsmiths, when compared to controls, was significantly higher (1440.0 μg/kg DW vs. 670.0 μg/kg DW). Hair from goldsmiths also contained significantly higher concentrations, in μg/kg DW, of silver (1290.0 vs. 400.0), copper (13,300 vs. 11,100), and indium (0.0016 vs. 0.0008); there were no significant differences found for cadmium, cobalt, chromium, mercury, nickel, lead, platinum, or zinc. In Nigeria, gold concentrations in hair of normal adults were low (6.0–880.0 μg/kg DW) and there were significant positive correlations of gold with concentrations of arsenic, lanthanum and cobalt.

17.6 Gold Effects on Plants and Animals

Lethal and sublethal effects of Au, Au^{+}, and Au^{+3} are summarized for aquatic organisms and laboratory mammals. Gold accumulations from solution are documented for microorganisms and other living resources under various physicochemical conditions.

17.6.1 Aquatic Organisms

This section summarizes lethal and sublethal effects of Au^{+} and Au^{+3} for aquatic microorganisms, plants, fishes, and amphibians.

17.6.1.1 Monovalent Gold

Monovalent gold is toxic to aquatic biota at comparatively elevated concentrations of 7.9 mg Au/L and higher. Toxicity of gold to microorganisms is affected by concentration and oxidation state of gold, presence of competing metal ions in solution, pH, and composition of the growth medium. Exposure to gold may induce cell adaptation and cell resistance, as has been demonstrated for monovalent gold chloride, sodium aurothiomalate, and auranofin. Cellular adaptation is a potential mechanism for gold resistance. Antimicrobial activities of two isomeric Au^{+}-triphenylphosphine compounds were documented for two species of gram-positive bacteria (*Bacillus subtilis*, *Staphylococcus aureus*) and one species of yeast (*Candida albicans*) at concentrations as low as 7.9 mg Au^{+}/L for bacteria and 250.0 mg/L for yeast. Growth inhibition of *Tetrahymena pyriformis*, a ciliate protozoan, is reported after 24 h in 99.0–296.0 mg Au^{+}/L (as gold sodium aurothiomalate), and prolonged cell generation time at 390.0–2960.0 mg/L in 24 h. At 1576.0 mg Au^{+}/L, no cells died in 24 h; although endocytosis and cell proliferation were inhibited; after 2 days however, the cell density of the culture was sufficiently high to permit recovery. Exposure of *Tetrahymena* to 3050.0 mg Au^{+}/L (as gold sodium aurothiomalate) for 24 h, equivalent to eight normal cell generations, resulted in a growth reduction of 50% and visible amounts of gold accumulated. Gold remained detectable for at least 24 h. After dilution to a low cell density, gold turnover was slow except in rapidly proliferating cells. The protozoan recovers fully after heavy accumulation of Au^{+}, but only in low-density cultures. Proliferating *Tetrahymena* have a high metabolic rate associated with high lysosomal enzyme activity, which are presumed to be the prerequisite for a rapid turnover of accumulated gold.

Intact single fibers of skeletal muscle of bullfrogs (*Rana catesbeiana*) were subjected to varying concentrations of Au^{+} as gold sodium thiomalate. At 500 μM 98.5 mg/kg, Au^{+} decreased tension amplitude by 27% after 30 min, and resting membrane potential by 5.3% after 22 min. Results suggest that Au^{+}, as gold sodium thiomalate, could be used as an antirheumatic drug without severe side effects on skeletal muscle and that coexistent thiomalate probably contributes to the protection of muscle function from the side effects of Au^{+}.

17.6.1.2 Trivalent Gold

Trivalent gold is significantly more toxic to aquatic biota than monovalent gold. $Gold^{+3}$, as tetrachloroaurate ($AuCl_4^-$), depressed chlorophyll concentrations, photosynthetic rates, and thiol levels at concentrations greater than 98.5 μg Au^{+3}/L over a 21-day period in *Amphora coffeaeformis*, a marine diatom. Cells were able to recover at concentrations less than 985.0 μg Au^{+3}/L due to cellular and photoreduction of the $AuCl_4^-$. Adverse effects were exacerbated by Cu^{+2}. Uptake of Au^{+3} by *Amphora* is apparently not an energy dependent process. At 394.0–985.0 μg Au^{+3}/L, only 30% of the total gold uptake after 24 h was internal, although increased uptake by heat-killed cells and uptake by illuminated cells suggest otherwise. It was concluded that algal cells, alive or dead, rapidly accumulate Au^{+3} and begin to reduce it to Au and Au^+ within 2 days.

Growth inhibition of yeast (*Saccharomyces cerevisiae*) was observed in 40 h at the lowest concentration tested of 20.0 mg Au^{+3}/L, with no growth observed at 50.0 mg/L. Both calcium and magnesium enhanced the inhibitory effect of gold on the yeast cells.

Results of acute toxicity bioassays of 96-h duration with adults of *Fundulus heteroclitus*, an estuarine cyprinodontiform killifish, and salts of various metals and metalloids showed that gold, as auric chloride (Au^{+3}), was comparatively lethal, with 50% dead in 96 h at <0.8 mg/L. The relative order of lethality, with silver (Ag), most toxic and lithium (Li) least toxic was: Ag^+, Hg^{+2}, Au^{+3}, Cd, followed by As^{+3}, Be, Al, Cu, Zn, Y, Tl, Fe, La, Cr^{+6}, Ni, Co, Sb, and Li. Salts of 13 additional elements tested to *Fundulus* were less toxic than were salts of Li, including Rb, Si, Mo, Re, Ba, Mn, Ca, Sr, K, and Na, in that order.

When bullfrog skeletal muscle fibers previously pretreated with 98.5 mg/kg Au^+ (as gold sodium thiomalate) were subjected to 2.0 mg Au^{+3}/kg as $NaAuCl_4$, the fibers lost their ability to contract upon electrical stimulation, as was the case for 2.0 mg Au^{+3}/kg alone. However in the presence of thiomalic acid, Au^{+3} did not completely block tetanus tension, even at 10 mg Au^{+3}/kg. Thiomalic acid also inhibited Au^{+3}-induced membrane depolarization. In bullfrogs, skeletal muscle fibers spontaneously produced phasic and tonic contractures upon addition of 5–20 μM Ag^+ or more than 50 μM Au^{+3} 9.0 mg (Au^{+3}/L). Simultaneous application of 5 μM Ag^+ and 20 μM Au^{+3} inhibited contractures induced by Ag^+. Trivalent gold applied immediately after development of Ag^+-induced contractures shortened the duration of the phasic contracture and markedly decreased the tonic contracture through modification of the Ca^{+2} release channel. It was concluded that extracellular Au^{+3} at comparatively low concentrations inhibits the silver (Ag^+)-induced contractions in skeletal muscle and that intracellular Au^{+3} activates the sarcoplasmic reticulum Ca^{+2} release channel to partially contribute to the tonic contractions.

17.6.2 Laboratory Mammals

No satisfactory animal model studies exist that show the same responses to gold complexes as human rheumatoid arthritis patients. The models generally used included rats with adjuvant arthritis and resistant to penicillamine, rats with kaolin paw edema, and guinea pigs with erythema. In animal gold studies – as in human gold studies – gold was widely distributed in tissues, with major gold accumulations in kidney, liver, spleen, skin, lymph, and bone marrow. Significant gold accumulations were found in most other tissues examined, including brain. In rat liver cells, gold uptake from sodium gold thiomalate was via membrane binding to lysosomes, possibly to thiols; however, in blood plasma it was complexed to albumin. And in guinea pigs, different gold distributions occurred depending on oral or parenteral route of administration.

17.6.2.1 Metallic Gold

Submicroscopic gold particles (0.05–0.10 microns in diameter) in colloidal suspension when injected intravenously (iv) into rabbits *Oryctolagus* sp. at 2.0 mg/kg BW total dose of 6–8 mg (Au) produced significant elevation of rectal temperatures over a 7-h post-injection observation period. Similar observations were recorded with colloidal suspensions of glass, iron oxide, quartz, and thorium dioxide. The fine state of particle division, shown by all materials tested, was the factor which rendered them thermogenic.

Tissue injury in mice from intraperitoneal (ip) insertion of gold implants initiated an inflammatory response, involving the activation of the humoral and cellular defense systems, that terminated in healing or rejection. The early inflammatory reaction in vivo to gold was measured by the adherence and activation of inflammatory cells during ip implantation. After 1 h, gold implants inserted ip into mice had 18% of the surface covered with white blood cells. It was concluded that peritoneal leukocytes adhering to foreign materials produced a respiratory burst response via a phospholipase D-dependent and protein kinase C-independent pathway. Subcutaneous implantation of gold (1.000 fine) and gold alloys in rats caused only a mild tissue reaction when compared with other dental restorative materials, inducing relatively few inflammatory cells.

The distribution of colloidal gold coupled with albumin within lymph nodes of rats up to 10 h following intrapleural injection was studied using X-ray fluorescence analysis-synchrotron radiation beams (XFA-SR). Potentially, XFA-SR can detect very low concentrations of gold and other elements, and microscopical SR analysis can demonstrate differences in elemental concentrations within single cells. Gold appeared in the lysosomes of the follicular reticular cells 4 h post-injection; colloidal gold concentrations in the node periphery were maximal after 6–8 h. Dose enhancement in tumor therapy is reported at interfaces between high and low atomic number materials, which is significantly intense for low energy photon beams. Gold microspheres suspended in cell culture or distributed in tumorous tissues exposed to kilovoltage beams produced an increased biologically effective dose, with increasing tumor cell death related to increasing concentration of microspheres 1.5–3.0 μm in diameter; the mean effective dose increase in solutions that contained 1% gold particles was 42–43% for 200 kv X-rays.

17.6.2.2 Monovalent Gold

Gold thioglucose ($C_6H_{11}O_5SAu$) was initially developed and marketed as a therapeutic agent for the treatment of arthritis and rheumatism. However, a single subcutaneous or intraperitoneal injection of gold thioglucose (GTG), equivalent to 0.5–0.6 mg Au^+/kg body weight (BW), in juveniles of certain strains of mice produced irreversible hyperphagia and obesity 10–12 weeks later, with many of the characteristics of human obesity. In contrast to genetically obese mice, GTG-injected mice were relatively tolerant of gold. The effect of GTG on the brain of mice is specific. Other gold thiol compounds tested – including gold thiogalactose, gold thiosorbitol, gold thiomalate, gold thiocaproate, gold thioglycoanilide, and gold thiosulfate – do not induce the brain damage that results from the administration of GTG, and neither obesity nor increased appetite occur, although all were toxic.

In addition to hyperphagia, the GTG brain lesion also induces hyperglycemia, hyperinsulinemia, insulin resistance, triglyceride accumulation, and a range of tissue specific changes in regulatory enzyme activities of glucose and lipid metabolic pathways. Hypersecretion of insulin is evident from an early stage in the development of GTG-induced obesity. Hyperinsulinemia is an early abnormality in many animal models of obesity and non-insulin dependent diabetes mellitus, including the GTG-injected obese mouse. The hyperinsulinemia of GTG mice was accompanied by a developing insulin resistance in fat and skeletal muscle, and was evident in both young (age 8 weeks) and old (age 24 weeks) GTG-obese mice. Insulin resistance at the level of phosphatidylinositol 3-kinase (PIK) occurs very early both in muscle and adipose tissue at a time when alterations in glucose transport were moderate or absent. Removal of glucocorticoid hormones alters insulin release and glucose metabolism in both lean control and GTG-obese mice. GTG mice that were adrenalectomized and examined one week later for glucose tolerance and insulin secretion showed reductions in body weight, liver glycogen content, and plasma glucose. But adrenalectomy normalized plasma insulin concentrations.

Injection of GTG into C57BL/6J mice also damages glucose receptive neurons in the ventromedial hypothalamus, preventing metabolic regulation of circadian responses to light during shortage of glucose availability.

In addition to the obese mouse model, selected studies show that monovalent organogold compounds affect survival, carcinogenicity, teratogenicity, histopathology, metabolism, immune function, disease resistance, and gold accumulation dynamics. High survival of the BALB/C mouse strain – a nonresistant organomonovalent gold strain – was reported after administration of comparatively high doses of organomonovalent gold compounds used in chrysotherapy. No deaths were reported after 20 days in 4-week-old mice given either a single subcutaneous injection of 10.0 mg Au^+/kg BW, three subcutaneous injections at 48-h intervals of 2.0 mg Au^+/kg BW, or 30.0 mg Au^+/kg BW daily per os for 10 days. By comparison, monovalent organogold salts used to treat humans for rheumatoid arthritis were usually administered at the rate of 0.36 mg Au^+/kg BW every 7–14 days.

Survival was reduced in adult mice infected with various viruses – including Semliki Forest virus, various strains of yellow fever virus, and West Nile virus – after intraperitoneal injection of gold sodium thiomalate. Adverse effects of therapeutic monovalent gold compounds may be linked to their ability to induce membrane proliferation. For example, the virulent strain of Semliki Forest virus in adult mice is characterized by the development of numerous membrane vesicles in brain with mature virus budding from these structures. In contrast, infection of the avirulent strain of Semliki Forest virus results in the formation of very few membrane vesicles and no mature virus particles in adult mouse brain. Proliferation of smooth membrane vesicles from whole mouse brain was induced in mice treated with gold sodium thiomalate. In certain virus infections, smooth membranes are a prerequisite for virus RNA synthesis and maturation. The ability of a virus to stimulate the smooth membranes may be the limiting factor in determining both the extent of viral RNA synthesis and maturation. This mechanism could also be responsible for the striking variation in strains of Semliki Forest virus in adult mice. The concept of membrane proliferation might also be relevant to the increased virulence of encephalitogenic viruses in childhood when the brain is still developing and cellular membrane proliferation is more abundant.

Carcinogenicity and teratogenicity of Myochrisine ($C_4H_3AuNa_2O_4S$), the disodium salt of gold sodium thiomalate ($C_4H_4AuNaO_4S$), is reported at high doses in rats and rabbits. Renal adenomas are documented in rats at 2.0 mg Au/kg BW weekly for 45 weeks followed by 6.0 mg/kg BW daily for 47 weeks; this dosage is equivalent to twice that administered to humans at the low dose and 42 times at the high dose. Adenomas produced were similar to those produced in rats by chronic administration of other gold compounds, lead, and other heavy metals. There is no report of Myochrisine-associated renal adenomas in humans.

Teratogen effects were observed in rats and rabbits when given Myochrisine during the organogenetic period at doses of 140 times higher than that of humans for rats and 175 times higher for rabbits. Hydrocephaly and microphthalmia malformations were observed in rats given subcutaneous injection dose levels of 25.0 mg Au^+/kg BW daily from days 6 through 15 of gestation. In rabbits, limb malformations and gastroschisis were observed when subcutaneous injection doses were 20.0–45.0 mg Au/kg BW from gestation days 6 through 18.

Antitumor activity of gold sodium thiomalate was documented in mice when given subcutaneously or per os. Mice inoculated with Meth A tumor cells were given sc injections of gold sodium thiomalate every other day for 3 injections, or given in drinking water daily for 2 weeks. Survival was improved 50% by 30 mg/kg BW via injection, or 75 mg/kg BW daily per os. No significant toxicity of gold was observed in afflicted mice at doses up to 125.0 mg Au^+ kg/BW daily via drinking water; however, adverse effects were observed at 6.0 mg/kg BW daily and lethality at 125.0 mg/kg BW daily via injection. Malignant tumor cells from humans and from mice in vitro showed growth inhibition after 4 days when subjected to 2.0 mg Au^+/L as gold sodium thiomalate, with 50% inhibition recorded between 10.0 and 50.0 mg Au^+/L. Antitumor activity of the metal-bound aromatic cation Au^+ (1,2-(bisdiphenylphosphino)ethane was reported in mice. However preclinical development was abandoned after the identification of severe hepatotoxicity in dogs after administration of this cation, and was attributed to alterations in mitochondrial function. Antitumor activity of other gold phosphino complexes is a function of the drug's lipophilicity. Alteration of lipophilicity of aromatic cationic antitumor Au^+ drugs greatly affects cellular uptake and binding to plasma proteins. Changes in lipophilicity also affect host toxicity, and optimal lipophilicity may be a critical factor in the design of gold analogs with high antitumor activity.

Renal injury was documented in male Wistar rats weighing 130 grams receiving a single ip injection of 25.0 mg of gold sodium thiomalate. Nephrotoxic effects were observed mainly in the renal tubular region. Changes in the activity of urinary enzymes suggested that the renal tubules were selectively injured by gold salts. Similar results were observed in adjuvant arthritic rats produced by a single intradermal injection of *Mycobacterium butyricum* receiving 11 intramuscular injections of gold sodium thiomalate every other day for 21 days. Gold concentrations in these rats after the final injection were 316.0 mg/kg FW in kidney and 44.0 mg/kg FW in liver. Selected chelating agents reduced gold concentrations in kidney and liver by about 51%. Effects of various chelating agents on distribution, excretion, and renal toxicity of gold sodium thiomalate in rats was measured immediately after intravenous injection of 0.026 mmol gold sodium thiomalate/kg BW. Of all agents tested, tiopronin and captopril successfully ameliorated gold sodium thiomalate-induced renal toxicity. DMPS (2,3-dimercaptopropanesulfonate) was the most effective in removing gold from the kidney and in protecting against renal toxicity after gold sodium thiomalate injection of 0.2 mmol/kg BW.

Human patients with rheumatoid arthritis sometimes developed skin reactions, glomerulonephritis, and increased serum IgE concentrations when treated with Au^+ salts. [Note: elevated levels of immunoglobulin E indicate an increased probability of an IgE-mediated hypersensitivity, responsible for allergic reactions]. Brown Norway rats (*Rattus norvegicus*) injected with allochrysine (gold sodium thiopropanosulfonate) showed an increase in serum IgE concentration, produced anti-laminin antibodies, and developed glomerular liver immunoglobulin deposits; Lewis strain rats were resistant. Genetic studies of gold salt-induced immune disorders in brown Norway rats showed that susceptibility to allochrysine was linked mainly to histocompatibility complex genes. Like humans, brown Norway rats injected with allochrysine had an autoimmune glomerulonephritis and increased serum IgE concentration, with few complications. Homologous genes encoding for cytokines, genes involved in sulfoxidation or in the control of glutathione synthetase, and genes encoding for nuclear factors regulating the IgE response could be implicated in allochrysine manifestation in rats and in the regulation of IgE levels in humans. Brown Norway rats injected with allochrysine developed an autoimmune syndrome. Major histocompatibility complex (MHC) T cell lines were derived from gold-treated rats. On transfer into normal brown Norway rats, the T-cells produced an autoimmune syndrome similar to, or more severe than, that observed in the active gold model, including an increase in serum IgE concentration, and production of anti-DNA and anti-laminin antibodies. The T-helper cell lines may induce an autoantibody-mediated disease and may be responsible for cell-mediated immunity.

$Gold^+$ salts injected subcutaneously over a 10-day period induced an autoimmune syndrome similar to that of $HgCl_2$ in brown Norway rats. The auto-immune syndrome in both cases was characterized by a marked increase in IgE production, lymphoproliferation, T cell-dependent polyclonal B cell activation, hypergammaglobulinaemia, and tissue injury with necrotizing leucocytoclastic vasculitis in the gut at day 15 post-injection.

Gold also induced granulomata and neutrophil infiltrates in the lung at day 15 post-injection.

Gold toxicity as a consequence of chrysotherapy has been treated by various chelating agents to remove accumulated gold from the body. Gold in urine and bile of rats after gold sodium thiomalate administration was mainly bound to high molecular weight compounds. After administration of chelating agents, gold in urine was bound to the sequestering agents. In bile, the gold was excreted into the feces primarily as a gold-chelating agent compound and secondarily as gold-L-cysteine and high molecular weight compounds. Incidentally, gold accumulation rates in the rat kidney differed markedly between administration routes. Renal concentrations of gold in rats administered auranofin orally were 33 times lower than those given equivalent gold dosages from sodium gold thiomalate parenterally. These differences persisted for at least 1 year, and spotlight administration route as a factor in gold accumulation dynamics.

Distribution of radiogold-198 in rats 2 h postadministration was mainly in urine (40.1% of injected dose), small intestine (39.6%), liver (8.2%), and carcass (8.0%). Histological examination of thyroid, adrenal, suprarenal, and testicular glands of rats given a total of 540 mg of allochrysine via 9 subcutaneous injections over a 3-week period showed gold-sulfur compounds in aurosomes of all glands. In gold-containing tissues, gold was localized intracellularly and selectively concentrated in lysosomes as nonsoluble crystalline precipitates of a constant S/Au ratio.

Monovalent gold salts impacted metabolism of selenium, copper, and zinc. Intravenous Au^{+} may adversely affect the availability of selenium for synthesis of selenoenzymes. Rats given gold sodium thiomalate iv at 25.0 or 50.0 μmol/kg BW had significantly altered selenium deposition, as measured by radioselenium-75. These effects included the almost complete cessation of ^{75}Se exhalation as dimethyl sulfide and the accumulation of ^{75}Se in blood plasma. Direct chemical reaction with nucleophilic selenium metabolites in the body may underlie these alterations. Auranofin inhibited selenium-glutathione peroxidase in bovine erythrocytes; this enzyme protects the cell from initiation and propagation of free radical reactions. High doses of gold alter copper and zinc metabolism in liver and kidney of Sprague-Dawley rats. These, in turn were controlled by metallothioneins, low molecular weight proteins that serve as oxygen scavengers and as metal sequestering agents. At low doses of 0.5 μg Au^{+}/kg BW, given via intramuscular injection, the main changes occurred in the kidney where an increase of gold was found 0.5 h post-injection, followed by an increase in copper and metallothionein concentrations after 6 h. Zinc homeostasis did not change. Authors concluded that gold induces an increase of metallothionein-like peptides in the kidney cytosol, accompanied by an increase in copper bound to these peptides.

Gold compounds were among the few classes of antiarthritic drugs which retarded rheumatoid arthritis and are now widely used to treat this disease and other chronic immunologically mediated inflammatory conditions in humans and dogs. Healthy dogs given either 0.6–3.6 mg auranofin orally every 24 h, or gold sodium thiomalate via intramuscular injection of 0.5–2.0 mg/kg BW every 3 days, were examined for changes in 13 immune functions. None of the changes in these aspects of immune function, previously attributed to treatment with auranofin or gold sodium thiomalate, could be demonstrated in normal dogs after treatment with either drug after treatment for 6–7 years. It seems that longterm administration of auranofin or gold sodium thiomalate to normal dogs at doses up to 30 times therapeutic doses in humans does not result in many of the alterations in lymphocyte, monocyte, and neutrophil functions observed in previous studies using other animal models or in human patients with rheumatoid arthritis. Accordingly, the changes previously reported in these variables may not be due to the effects of monovalent gold compounds.

17.6.2.3 Trivalent Gold

Organometallic Au^{+3} complexes hold promise as possible antitumor agents, as judged by cytotoxic activity in vitro against cultured tumor cell lines and appreciable tumor inhibiting properties in vivo toward human tumors grown in xenografted rats, the probable target for newly developed cytotoxic Au^{+3} complexes. In vitro tests with five representative organogold complexes and calf thymus DNA showed that all complexes interacted with DNA and modified its behavior in solution. Another potential antitumor agent is chlorobischolylglycinatogold^{+3} $(C_{52}H_{84}N_2)_{12}AuCl$, a synthetic bile acid. This compound is soluble in water, methanol, ethanol, and dimethylsulfoxide, and can inhibit the growth of a variety of cell lines. The cytostatic effect was mild against human hepatoma, mouse hepatoma, rat hepatoma, and human colon adenocarcinoma cell lines, but stronger against mouse sarcoma S180-II and mouse leukemia L-1210 cells. The appearance of colloidal gold during the process of hydrolysis under physiological conditions may account for the limited cytostatic activity.

Studies with sodium tetrachloroaurate^{+3} showed that gold accumulates mainly in kidneys and in the reticuloendothelial cells of tissues, especially in the renal cortex. Male rats given daily intraperitoneal (ip) injections of 1.0 mg Au^{+3}/kg BW as sodium tetrachloroaurate dihydrate $[Na(AuCl_4)A2H_2O]$ and killed one day after the last injection had elevated concentrations of gold in kidneys (55.1 mg/kg FW) and liver (4.0), as was the case for copper and zinc. Rats given a single ip injection of sodium tetrachloroaurate dihydrate at 5.0, 10.0, or 20.0 mg Au^{+3}/kg BW and killed 14 h post-injection showed dose-dependent increases of gold and metallothioneins in kidneys. The increasing amount of gold was attributed to high molecular weight proteins and the metallothionein fractions. About 14% of the increased gold in renal cytosols of gold-injected rats was bound to metallothioneins and 79% to high molecular weight fractions. Trivalent gold had stronger binding affinity to metallothioneins than did zinc or cadmium. Zinc concentrations and metallothionein content in rat livers showed a dose-dependent increase in response to Au^{+3} injections of 5.0, 10.0, or 20.0 mg/kg/BW, but copper was unaffected. Increased zinc was due to high molecular weight proteins and the comparatively low molecular weight metallothioneins. About 68% of the increased zinc in the hepatic cytosol of Au^{+3}-injected rats was bound to metallothioneins, suggesting that the role of metallothioneins in zinc accumulation in the liver was similar for both Au^{+3}- and Zn^{+2}-injected rats.

Gold sensitivity in mice may be genetically determined. Chronic treatment of inbred mice strains with monovalent gold sodium thiomalate produced immune responses to Au^{+3} in 100% of A.SW mice, 70% of C57BL/6 mice, and zero in DBA/2 mice. The dose-dependent immune responses were directed towards the trivalent gold compounds $AuCl_3$ and $HAuCl_4$ and were T cell dependent and specific. Thus, in order to induce T cell sensitization, gold has to exist in the Au^{+3} state; however, mechanisms governing gold immunogenicity are imperfectly understood. Popliteal lymph node (PLN) assay reactions to Au^{+3} in mice were dose- and T-cell dependent. When Au^{+3} was reduced to Au^{+} by addition of disodium thiomalate or methionine before testing in the PNL assay its sensitizing capacity was significantly decreased, suggesting that Au^{+} of disodium thiomalate was oxidized to Au^{+3} before T cells were sensitized and adverse immunological reactions developed.

17.6.3 Accumulation

This section briefly reviews the potential of living and dead plants and animals to accumulate gold from solution, and some of the processes involved – including biooxidation, dissolution, bioreduction, bacterial leaching, and biosorption.

17.6.3.1 Microorganisms, Fungi, and Higher Plants

Biomining processes are used successfully on a commercial scale for the recovery of gold and other metals, and are based on the activity of obligate chemoautolithotrophic bacteria that use iron or sulfur as their energy source and grow in highly acidic media. Biooxidation of difficult to treat gold-bearing arsenopyrite ores occurs in aerated, stirred tanks and rapidly growing, arsenic-resistant bacterial strains of *Thiobacillus* and *Leptospirillium*. These bacterial species obtain their energy through the oxidation of ferrous to ferric iron or through the reduction of inorganic sulfur compounds to sulfate. Monetary costs of biooxidation are reported to be about 50% lower than roasting or pressure oxidation.

Several species of Fe^{+3}-reducing bacteria (*Bacteria* spp., *Archaea* spp.) can precipitate gold by reducing Au^{+3} to Au with hydrogen as the electron donor. Pretreatment of refractory gold concentrates with the bacterium *Thiobacillus ferrooxidans* ultimately results in sulfur and sulfide oxidation by ferric ions from bacterial oxidation of ferrous ions. The maximum concentration of attached *Thiobacillus* increases with increasing concentration of Fe^{+2} and decreases with increasing size of the refractory gold concentrate particles. In Chile, which produced 30,000 kg of gold in 1990, *Thiobacillus ferrooxidans* was used to recover gold from a complex ore under laboratory conditions. The ore contained 8.2% Fe, 0.78% Cu, 0.88% As, and 3.5 g Au/ton, with pyrite, hematite, arsenopyrite, and chalcopyrite as the main metal-bearing minerals. Initial gold recovery by conventional cyanidation on a crushed ore sample was 54%; concentration by flotation improved recovery to 56%. Concentrated samples (17.0 g Au/ton) were leached in reactors at pH 1.8. In the presence of bacteria, all dissolved iron was present as ferric ion; gold recovery by cyanidation increased from 13% for the initial concentrate to 97% after 10 days of bacterial leaching. To further increase gold recovery, flotation tailings were submitted to cyanidation. Some microorganisms isolated from gold-bearing deposits are capable of dissolving gold; dissolution was aided by the presence of aspartic acid, histidine, serine, alanine, glycine, and metal oxidants. Bacteriform gold is well known, for the uptake of Au^{+3} from chloride solutions documented for at least seven genera of freshwater cyanobacteria. Some bacteriform gold biogenic–the result of precipitation by bacteria – and may be useful indicators of gold deposits and of processes of gold accumulation.

Plectonema terebrans, a species of filamentous marine cyanobacteria, accumulates gold in its sheath from an aqueous solution of $AuCl_3$. Sheaths are among the few structures likely to be preserved in some form in microfossils of ancient bacteria. In marine media, it is expected that $AuCl_3$ (2.0 g Au/L) will form $AuCl_4^-$, AuO_2^-, and $AuCl_2^-$. Biosorption of Au^{+3}, as $AuCl_4^-$, by dried *Pseudomonas* strains of bacteria was inhibited by palladium, as Pd^{+2}, and possibly other metal ions.

Gold adsorption from cyanide solutions by dead biomass of bacteria (*Bacillus subtilis*), fungus (*Penicillium chrysogenum*), or seaweed (*Sargassum fluitans*) at pH 2.0 was 1.8 g Au/kg DW for bacteria, 1.4 g/kg DW for fungus, and 0.6 g Au/kg DW for seaweed. Anionic $AuCN_2^-$ adsorption was the major mechanism in gold biosorption from cyanide solutions, being most efficient at lower pH values. L-cysteine increased gold-cyanide biosorption of *Bacillus*, *Penicillium*, and *Sargassum*. At pH 2, the maximum gold uptakes were 4.0 g Au/kg DW for bacteria, 2.8 g/kg for fungus, and 0.9 g/kg for seaweed, or 150–250% greater than in the absence of cysteine. The anionic gold cyanide species were adsorbed by ionizable functional groups on cysteine-loaded biomass; deposited gold could be eluted from gold-loaded biomass at pH 5.0.

Gold-resistant strains of bacteria that also accumulate gold are documented, although the fundamental mechanism of resistance to gold in microorganisms is not known or understood. One strain of *Burkholderia* (*Pseudomonas*) *cepacia* contained millimolar concentrations of Au^+ thiolates. *Burkholderia* cells were large, accumulated polyhyroxybutyrate and gold, and excreted thiorin, a low molecular weight protein into the culture medium. This effect was not observed with the Au^{+3} complexes tested, which were reduced to metallic gold in the medium. Gold-resistant strains of fungi and heterotrophic bacteria are also known.

Metabolically active fungal cells of *Aspergillus fumigatus* and *A. niger* removed gold from cyanide leach liquor of a Brazilian gold extraction plant more efficiently than did dried fungal biomass or other species of *Aspergillus* tested. These two species of fungi removed 35–37% of gold from solutions containing 2.8 mg Au/L in 84 h. Gold removal from cyanide-containing solutions is documented for a strain of *Aspergillus niger*, a fungus isolated from the gold extraction plant at Nova Linda, Brazil. The leach liquor contained, in mg/L, 181.0 cyanide, 1.3 gold, 0.4 silver, 7.1 copper, 5.2 iron, and 4.5 zinc. After 60–72 h of incubation, *A. niger* removed from solution, probably by adsorption, 64% of the gold, 100% of the silver, 59% of the copper, 80% of the iron, and 74% of the zinc; all gold was removed after 120 h. Use of this fungus to develop a bioprocess to reduce metal and cyanide levels as well as recovery of valuable metals shows promise. Uptake patterns of gold from Au^{+3} solutions by dead fungal biomass followed mathematical uptake models of Langmuir and Freundlich; biomass was prepared from the fruiting body of a mushroom collected from the forests of Kerala, India. Dried fungus, *Cladosporium cladosporoides*, mixed with keratinous material of natural origin to form a bead, proved effective in absorbing gold from solution. The biosorpent beads adsorbed 100.0 g Au/kg beads from a solution containing 100.0 mg Au/L. Maximum biosorption of 80% occurred at acid pH (1–5) in less than 20 min. The biosorpent beads degraded in soil in about 140 days. The beads also removed 55% of the gold from electroplating solutions containing 46.0 mg Au/L, with observed gold loading capacity of 36.0 g/kg beads. Dried biosorpents encapsulated in polysulfone were prepared from microorganisms isolated from pristine or acid mine drainage environments. Biosorpent material rich in exopolysaccharides from the acid mine drainage site bound Au^{+3} three times more effectively than did other materials, and removed 100% of the Au^{+3} from solutions containing 1.0 mg Au/L within 16 h at 23°C and pH 3.0.

Algal cells, alive or dead, rapidly accumulate Au^{+3} and begin to reduce it to Au and Au^{+} within 2 days. Uptake of Au^{+3} by *Chlorella vulgaris*, a unicellular green alga, from solutions containing 10.0 or 20.0 mg Au^{+3}/L is documented. *Chlorella* accumulated up to 16.5 g Au/kg DW. Inactivating the algal cells by various treatments resulted in some enhancement in uptake capacity over the pristine cells. Inactivation by heat treatment yielded up to 18.8 g/kg DW; for alkali treatment, this was 20.2 g/kg DW; for formaldehyde treatment, 25.5 g/kg DW; and for acid treatment, 25.4 g/kg DW. Elemental gold (Au) was measured by X-ray photoelectron spectroscopy on the cell surface, indicating that a reduction had occurred. Studies with living *Chlorella vulgaris* suggest that accumulated Au^{+3} is rapidly reduced to Au^{+}, followed by a slow reduction to Au. With dead algae, Au initiates a seeding process which results in the formation of elemental gold.

Sequestering metal ions using living or dead plants is a proposed economical means of removing gold and other metals via intracellular accumulation or surface adsorption. However, in the case of live plants, this is frequently a relatively slow and time-consuming process. Nonliving plant material for surface adsorption offers several advantages over live plants, including reduced cost, greater availability, easier regeneration, and higher metal specificity. In South African mining effluents, gold usually ranges between 1.0 and 10.0 mg/L. In studies of 180-min duration, dried red water ferns, *Azolla filiculoides*, removed 86–100% of Au^{+3} from solutions containing 2.0 to 10.0 mg Au^{+3}/L; removal increased with increasing initial concentration of Au^{+3}. The biomass gave >95% removal efficiency at all biomass concentrations measured. Optimum (99.9%) removal of gold occurred within 20 min at pH 2, 42% removal at pH 3 and 4, 63% at pH 5, and 73% removal at pH 6; removal efficiency seemed independent of temperature. Similar results were observed with four species of ground dried seaweeds (*Sargassum* sp., *Gracilaria* sp., *Eisenia* sp., and *Ulva* sp.). Treated seaweeds removed 75–90% of the gold within 60 min at pH 2 from solutions containing 5.0 mg Au^{+3}/L. Gold (Au^{+3}) can be sequestered from acid solutions by dead biomass of a brown alga, *Sargassum natans*, and deposited in its elemental form, Au. The cell wall of *Sargassum* was the major locale for gold deposition, with carbonyl groups (C=O) playing a major role in binding, and N-containing groups a lesser role. Like activated carbon, the biomass of *Sargassum natans* is extremely porous, reportedly more than most biomaterials, and accounts, in part, for its ability to accumulate gold. Dried ground shoots of alfalfa, *Medicago sativa*, were effective in removing gold from solution. The accumulation process involved the reduction of Au^{+3} to colloidal Au, and was most efficient at elevated temperatures and acid pH. In solutions containing 60.0 mg Au^{+3}/L, about 90% of the Au^{+3} was bound to dried alfalfa shoots in about 2 h at pH 2 and 55°C. The mechanisms to account for this phenomenon are unknown, but may involve reduction of Au^{+3} to Au^{+}, the latter being unstable in water to form Au and Au^{+3}. Dried peat from a Brazilian bog accumulated up to 84.0 g Au/kg DW within 60 min from solutions containing 30.0 mg Au^{+3}/L.

17.6.3.2 Aquatic Macrofauna

Except for crab exoskeletons, gold recovery from the medium by various species of living molluscs, crustaceans, and fishes is negligible. The maximum concentration of stable gold measured in tissues of living marine organisms was 38.0 μg/kg FW.

Certain chitinous materials, such as exoskeletons of the swamp ghost crab, *Ucides cordatus*, can remove and concentrate gold from anionic gold cyanide solutions over a wide range of pH values. The maximum $AuCN_2^-$ uptake occurred at pH 3.7, corresponding to a final value of 4.9 g Au/kg DW; exoskeletons burnt in a non-oxidizing atmosphere removed 90% of the gold at pH 10. Phenolic groups created during the heat treatment seemed to be the main functional group responsible for $AuCN_2^-$ binding by burnt acid-washed crab shells.

17.6.3.3 Animal Fibrous Proteins

Gold recovery is proposed using animal fibrous proteins such as egg shell membrane, chicken feathers, wool, silk, elastin, and other stable water soluble fibers with high surface area. All animal fibrous proteins tested accumulated gold-cyanide ion from aqueous solution. Adsorption was highest at pH 2; accumulations were up to 9.8% of the DW for wool, 8.6% for eggshell membrane, 7.1% for chicken feathers, and <3.9% for other materials. In the case of eggshell membrane, adsorbed gold was desorbed with 0.1 M NaOH and the material can be used repeatedly. Eggshell membrane could remove gold-cyanide ion at concentrations near 1.0 μg/L.

17.7 Health Risks of Gold Miners

Health problems are documented for gold miners who worked mainly underground with little exposure to elemental mercury in Australia, North America, South America, Europe, and Africa. Major problems examined included life expectancy, cancer frequency, and pleural diseases. Health problems of miners who worked mainly on the surface and with extensive exposure to elemental mercury owing to its use in amalgamating and extracting gold are reported extensively in Chapter 19.

17.7.1 Australia

Australian gold miners are vulnerable to dengue fever (a mosquito-borne acute infectious viral disease characterized by headache, severe joint pain, and rash), silicosis (massive fibrosis of the lungs marked by shortness of breath and caused by inhalation of silica dusts, usually SiO_2), and phthisis (a historical term used to describe a wasting condition, possibly pulmonary tuberculosis).

Gold miners were the first recorded victims of dengue fever in 1885 in tropical northeastern Queensland. In the dengue epidemic of 1993, 2 percent of the population was infected despite source reduction of surface mosquito breeding grounds. In 1994, larvae and pupae of the dengue vector mosquito *Aedes aegypti* were found in flooded unused shafts of gold mines more than 45 m below ground. Copepods (*Mesocyclops aspericornis*) were also found in some flooded shafts and were found to be effective predators of mosquito larvae in the laboratory. Copepods (N = 50) were added to about half the mosquito-infested wells and the rest were untreated controls. After nine months, all copepod-inoculated shafts were free of mosquitos and all untreated wells contained *A. aegypti* larvae. The use of *M. aspericornis* is recommended as an effective control agent of *Aedes aegypti*, especially in comparatively inaccessible breeding sites, such as flooded gold mine shafts.

Gold miners from Bendigo suffered – in epidemic proportions for 100 years, from the 1860s to the 1960s – a wasting disease, possibly silicosis or pulmonary tuberculosis. Eventually, it was treated as an occupational sickness, with social, economic, and political implications that resulted in marked improvements in working conditions, better medical treatment, and improved productivity. In Western Australia, three retired gold miners were diagnosed with asbestos-related pleural disease after working in gold mines for 5–17 years. They had no other significant known asbestos exposure except for possible asbestos contamination of gold mine dust. Although air from these mines contained measurable concentrations of asbestos fibers, this is the first report of asbestos-related diseases among gold miners. In view of the large number of potentially exposed workers, additional assessment is recommended on the relation between dust exposure from gold mining and asbestos-related lung disease.

In Western Australia 2297 gold miners were examined in 1961, 1974, 1985, and 1993 for lung cancer and silicosis. The incidence of silicosis was clearly related to exposure to silica and the onset of silicosis conferred a significant increase in risk for subsequent lung cancer. But there was no evidence that exposure to silica caused lung cancer in the absence of silicosis. Silica has recently been reclassified as carcinogenic to humans based largely on the observed increase in rates of lung cancer in patients with silicosis. The International Agency for Research on Cancer has reclassified crystalline silica inhaled in the form of quartz or cristobalite from occupational sources as carcinogenic to humans (Class 1). Previously, silica was in Class 2A, that is, carcinogenic to animals and probably carcinogenic to humans.

17.7.2 North America

Canadian gold miners had an increased risk of cancer of the trachea, bronchus, lung, and stomach. In the United States, gold miners had significantly higher rates of lung cancer, silicosis, and tuberculosis when compared to the general population, and elevated risks for several debilitating diseases including diseases of the blood, skin, and musculoskeletal system.

A significant excess of mortality from carcinoma of the stomach was demonstrated in gold miners from Ontario, Canada, when compared to other miners. The increased frequency of stomach cancer appeared 5–19 years after they began gold mining in Ontario. Twenty or more years after the gold miners started work, stomach cancer cases were significantly greater in miners born outside North America when compared to a reference population, but not in the native-born. This late increase is similar to the excess of gastric carcinoma evident in residents of Ontario born in Europe. Possible explanations to account for the excess of stomach cancer in Canadian gold miners include exposures to arsenic, chromium, mineral fiber, diesel emissions, and aluminum powder. Diesel emissions and aluminum powder were rejected because gold miners and uranium miners were exposed to both agents but excess stomach cancer was noted only in gold miners. Exposure to dust was significant and the time-weighted duration of exposure to dust in gold mines was found in miners under age 60. A statistically significant time-weighted correlation for chromium – but not arsenic or mineral fiber – occurred, especially among gold miners under age 60. Exposure to chromium is associated with the development of the intestinal, rather than the diffuse, type of gastric cancer. Gold miners in Ontario with 5 or more years of gold mining experience before 1945 had a significantly increased risk of primary cancer of the trachea, bronchus, or lung. A minimum of 15 years' latency was recorded between first employment in a dusty gold mining occupation and diagnosis of primary lung cancer. For purposes of occupational exposure assessment in establishing work-relatedness, authors concluded that primary lung cancer in Ontario gold miners was related to exposure to silica, arsenic, and radon decay products and were consistent with miner's age at first exposure, length of exposure to dust, and latency.

The health of 3328 gold miners who worked underground in a South Dakota gold mine for at least one year (average was 9 years) between 1940 and 1965 was analyzed through 1990, with emphasis on exposures to silica and nonasbestiform minerals, by death certificates and radiographic surveys. Miners had been exposed to a median silicon level of $0.05\,mg/m^3$ after 1930 and $0.15\,mg/m^3$ for those hired before 1930. The risk of silicosis was less than 1% with a cumulative exposure under $0.5\,mg/m^3$-years, increasing to 68–84% for the highest cumulative exposure category of more than $4.0\,mg/m^3$-years. Cumulative exposure was the best predictor of silicosis, followed by duration of exposure and average exposure. After adjustment for competing causes of death, a 45-year exposure under the current U.S. Occupational Safety and Health Administration (OSHA) standard of 0.09 mg Si/m^3 would lead to a lifetime risk of silicosis of 35–47%, suggesting that the current OSHA silicon exposure level is unacceptably high. The lung cancer rate of these miners was 13% higher than the general U.S. population, 25% higher when the county was the referral group, and 27% higher 30 years post-exposure. Miners had significantly higher frequencies of tuberculosis and silicosis with clear exposure-response trends. Renal disease associated with silica exposure was elevated for those hired as young men, and also showed a positive correlation with length of exposure. This group also had significant excesses of arthritis, musculoskeletal diseases, skin diseases, diseases of autoimmune origin, and diseases of the blood and hematopoietic organs.

17.7.3 South America

Death from mining accidents in Columbia, increased prevalence of malaria in Brazil, and increased frequency of attacks by rabid vampire bats (*Desmodus rotundus*) in Venezuela are documented.

In Columbia, at least 28 gold miners were killed by landslides and dozens reported missing while digging at a condemned strip mine. The victims were said to be poor people who had ignored government warnings that erosion had made the mine unsafe. This incident was documented in a newspaper, and also, perhaps, in official mining records that were difficult to obtain. It is reasonable to conclude that gold mining fatalities are probably grossly underreported. The prevalence of malaria in Brazil has increased dramatically since the 1980s, particularly in Amazonian gold mining areas where increased colonization and deforestation is recorded. About 600,000 cases of malaria are reported annually in Brazil. The Amazon River Basin accounts for 99% of the cases in Brazil and for about 50% of all cases in the Americas. Infections by *Plasmodium vivax* protozoans represent about 58% of the cases, followed by *Plasmodium falciparum* (41%) and *Plasmodium malariae* (1.0%). Many of the infected miners have no obvious symptoms of malaria and often do not take prescribed antimalarial agents. Malarial control programs rely on early detection and treatment; however, special problems are associated with limited access to gold mining areas, the high mobility of the mining population, and the steady increase in drug-resistant *Plasmodium* species. These alluvial gold-mining sites are important reservoirs of drug-resistant *P. falciparum* and other parasites, and nonminers (Indians, farmers, loggers) who live there are at increased risk of malaria. An outbreak of attacks by rabid vampire bats (154 cases in 4 months in a population of about 1500) was documented for the gold mining village of Payapal in southeastern Venezuela. Cattle and horses were bitten by vampire bats in the two-month period preceding the human attacks. The outbreak may be due to loss of normal prey habitat of bats from mining, deforestation, and housing construction, and with human blood providing an alternative food source.

17.7.4 Europe

A high incidence of neoplasms of the respiratory system among gold extraction and refinery workers in Solsigne, France, was first reported in 1977, and again in 1985, and appears related to occupational exposure. Mine and smelter workers at this location were twice as likely to die from lung cancer than the general population. Soluble and insoluble forms of arsenic in combination with other risk factors, such as radon and silica in the mine, are likely determinants of the lung cancer excess.

17.7.5 Africa

Gold miners in Africa show increased prevalence of various bacterial and viral diseases (Gabon), noise-induced hearing loss (Ghana), lung cancer (Zimbabwe), carbon monoxide poisoning (Kenya) and, in the RSA – the largest producer of gold in the world – almost the entire spectrum of mining-related health problems, especially lung diseases and cancer.

Residents of five gold-panning villages in northeastern Gabon were analyzed for seroprevalence of leptospirosis and Ebola virus, both of which can cause lethal hemorrhagic fever. The villages surveyed were remote, isolated communities and their economies were entirely dependent on gold. The seroprevalence was 15.7% for leptospirosis (14.7% of gold miners, zero% of fishermen) and 10.2% for Ebola virus (11.3% of miners, 25.0% of fishermen), demonstrating the persistence of this infection among the endemic population and the need to consider it a potential cause of hemorrhagic fever in Gabon. In another survey, residents from these same villages had elevated (up to 8.5%) blood serum titers for spotted fever and typhus group *Rickettsia* bacteria. The influence of *Rickettsia* on public health in Africa remains unknown, but victims sometimes die as a result of infection by louse and flea vectors.

Noise pollution laws are usually not enforced in developing countries. This was the case at a large gold mining company in central Ghana where 20% of all workers experienced significant noise-induced hearing loss, with frequency rates of 34% for miners, 20% for machine operators, and zero percent for office workers. In general, hearing loss increased with increasing age and noise exposure. It was concluded that mining companies need to implement hearing conservation programs to protect workers exposed to hazardous noise levels. Lung cancers were reported in gold miners from Zimbabwe, with silica dust and arsenic considered relevant exposures.

The gold mining industry in the RSA began around 1886 when gold was discovered on the Witwatersrand. By 1920, about 200,000 migrant African laborers were employed in the RSA gold mines; in 1961, this number was 427,000, and in 1988 just over 500,000. Most worked underground at depths up to 3500 m. Until the mid-1970s, when recruiting patterns began to shift towards domestic sources of migrant labor, most workers were recruited from Mozambique and Malawi, with smaller numbers coming from Angola, Botswana, Zambia, and Zimbabwe. In the 1970s, critical studies appeared on the conditions of extreme social and physical deprivation governed by monetary interests and racist policies. These conditions reportedly rendered the labor force excessively prone to tuberculosis and pneumonia, parasitic infections, and traumatic injury or death as a result of poor safety procedures in the mines and the culture of violence from housing in ethnically segregated single sex hostels. During this period mining medicine improved to sustain productivity, although it was widely perceived by black miners as yet another means to repress the African persona.

Black miners in RSA comprise approximately 85% of all gold miners. Between 1975 and 1991, and based on 16,454 case histories, the prevalence of tuberculosis (TB) increased from 0.9% in 1975 to 3.9% in 1991; for silicosis, these values were 9.3% in 1975 and 12.8% in 1991. The frequency of both diseases increased with age and duration of service. Silicosis was the most significant predictor of TB. Lowering of dust levels in the mines was recommended to prevent an increased disease burden. In a seven-year study, it was shown that miners with chronic simple silicosis had a nearly three-fold greater risk of developing TB than did their fellow workers of similar age who did not show radiographic evidence of silicosis at the start of the study; about 25% of the miners with silicosis will have developed TB by age 60 years. Death rates of black RSA gold miners from pulmonary TB and silicosis were higher than those from their white counterparts, possibly because of greater severity of silicosis and a high rate of HIV infection. By 1996, the death rate from tuberculosis among black migrant miners had risen to 2476 per 100,000, accounting for the largest single cause of death among this group, apart from trauma in the workplace. Concomitantly, HIV prevalence in RSA miners with TB increased from 15% in 1993 to 45% in 1996; HIV is known to interfere with the accuracy of radiological TB screening programs. TB is likely to remain the most important health hazard in RSA mines during the new millennium, necessitating greater commitment to TB control and reduction of risk factors, such as silicosis and HIV infection. The role of HIV, a retrovirus that infects human T cells and causes acquired immune deficiency syndrome (AIDS) – a condition of deficiency of certain leukocytes resulting in infections and cancer – is discussed later.

During the periods 1980–89 and 1990–94, cancer deaths of black male gold miners in the early period were due primarily to liver cancer followed by esophagal and lung cancers. Primary liver cancer during this period was the fourth leading cause of death in the RSA, but first among black gold miners who worked underground. In the period 1990–94, esophagal cancer had overtaken liver cancer in numbers of deaths. New cases of esophagal cancer had doubled. New cases of respiratory cancer had also doubled. The reasons for these trends are uncertain but may be associated with repatriation of transient workers to their homelands outside RSA where health care was not as extensive. In another study, pulmonary dysfunction was measured in black South African gold miners with reactive airways. Reactive airways were found in 12% of 1197 older miners, and were not related to extent of exposure to the underground environment. However, those so afflicted were more susceptible to bronchial tree problems after correction for age, tobacco smoking, and presence of silicosis.

White South African miners who had spent at least 85% of their working life in gold mines and had worked underground at least 15% of their shifts, had a 30% chance of dying sooner than the general population due to higher frequencies of lung cancer (140%), heart disease (124%), pulmonary disease (189%), and cirrhosis of the liver (155%). However, very little of this increase could be attributed to gold mining and was instead associated with their unhealthy life style when compared to other South African white males, particularly in smoking and excessive alcohol consumption. There is, however, growing evidence that white RSA gold miners – like their black counterparts – were also vulnerable to silicosis, emphysema, lung cancer, asthma, and pulmonary tuberculosis.

RSA gold miners have among the highest rates of TB in the world. This is attributed, in part, to the high endemic rate of TB in rural regions from which miners are recruited, crowding, silica dust exposure, increasing age of the workforce, and HIV infection. Rates are rising, despite cure rates that meet WHO targets in patients with new TB. The incidence of pulmonary tuberculosis in RSA gold miners increased from 686 per 100,000 workers in 1989 to more than 1800 per 100,000 in 1995. Changes were associated with longer service and a rise in the average age of the work force. Miners with pulmonary mycobacterial disease were more likely to have nontuberculosis mycobacteria (NTM) than *Mycobacterium tuberculosis* (TB) if they worked longer underground, had silicosis, or were treated previously for TB. Attempts to reduce the incidence of all pulmonary mycobacterial disease among gold miners should include early diagnosis and treatment. Despite a control program that cures 86% of new cases, most TB in this mining community is due to ongoing transmission from persistently infectious individuals who have previously failed treatment and may be responsible for as many as one third of TB cases. There is a low incidence of NTM isolates and diseases in developed countries; however, this incidence is 27% in RSA miners, and is largely attributable to chronic chest disease from silicon dust inhalation and prior tuberculosis. Previous studies have shown that isolates of the most common NTM species, *M. kansasii* and *M. scrofulaceum*, occur with high incidence and are often associated with NTM risk factors such as silicosis and lung diseases than either patients with TB or control patients. NTM were isolated from 118 patients during the study period of whom 40 (34%) were HIV positive. HIV infection has recently become an additional risk factor for mycobacterial disease in miners and is likely to become increasingly important as the HIV epidemic progresses.

Silicosis reflects a failure in adequate control of occupational dust exposure. The risk of silicosis in a cohort of 2235 white RSA gold miners, with an average of 24 years of mining experience between 1940 and 1970, was followed up to 1991 for radiological signs of onset of silicosis. About 14% of the miners developed silicosis at an average age of 56 years, with radiological signs appearing, on average, 7.4 years after mining exposure ceased. The risk of silicosis was strongly dose-dependent, although the latency period was variable. Silicosis risk increased exponentially with the cumulative dust dose, the accelerated increase occurring after 7.0 mg/m^3-years. At the highest exposure level of 15.0 mg/m^3-years – equivalent to about 37 years of gold mining exposed to an average respirable dust concentration of 0.4 mg/m^3 – the cumulative risk for silicosis reached 77%. There is also a positive association between exposure to silica dust and risk of lung cancer; risks were higher among those exposed to higher dust exposures and also diagnosed with silicosis. Miners who had withdrawn from dusty occupations showed declines in lung function similar to those who continued to work underground for five years. RSA gold miners with chronic obstructive airway disease from working in a dusty atmosphere in scheduled mines or works were entitled to workmen's compensation, as judged by lung function tests for airflow obstruction.

The association between silicosis and pulmonary tuberculosis (PTB) is well established. Epidemiological and case studies show that workers exposed to silica dust have increased morbidity and mortality from PTB. In one study, a cohort of 2255 white RSA gold miners were evaluated for increased risk of PTB from 1968 to 1971 when they were 45–55 years of age to December 1995. During the followup, 1592 (71%) of this cohort died. Of these, 1296 (81%) were necropsied to determine the presence of silicosis and PTB. It was concluded that: exposure to silica dust is a risk factor for the development of PTB in the absence of silicosis, even after exposure to silica dust ends; the risk of PTB increases with the presence of silicosis, and in miners without silicosis with increasing exposure to dust; and severity of silicosis was associated with increasing risk of PTB. In addition to silicosis, TB, and obstructive airways disease, RSA gold miners show a high prevalence of previously undiagnosed and untreated pneumoconiosis, a lung disease caused by habitual inhalation of irritant mineral or metallic particles.

South Africa currently harbors one of the fastest-growing HIV epidemics in the world. The prevalence of HIV-1 in pregnant women has increased from 0.76% in 1990 to 14.1% in 1996, with more than 2.5 million South Africans infected. Migrant workers employed as RSA gold miners were found infected with HIV-1. HIV infection and silicosis are powerful risk factors for TB and are associated with an increased risk of death among RSA gold miners. The incidence of TB was almost 5 times greater in HIV-positive than HIV-negative miners. Among RSA gold miners with TB, the prevalence of HIV infection increased rapidly to about 50% of all cases between 1993 and 1997. NTM disease incidence, morbidity, and mortality are likely to increase further among miners as the HIV epidemic progresses.

RSA gold miners have a high prevalence of HIV infection. Most are migrants from rural areas within South Africa and others are from surrounding countries such as Lesotho, Botswana, and Mozambique. The vast majority of these workers are housed in single sex hostels close to their workplace. Despite extensive education from mine operators on the consequences of unprotected sex, this group perceives condom use as a diminishment of their masculinity and continue to practice risky behaviors with sex workers, and the incidence of sexually transmitted diseases in these men is extremely high. Many workers commented that the risk of HIV/AIDS appear minimal compared to the risks of death or injury underground and that this was the reason why many mine workers did not bother with condoms. It remains unclear how best to communicate risks of HIV and prevent transmission by altering risky behaviors in African populations.

17.8 Human Sensitivity to Gold

This section specifically reviews adverse reactions to gold exposure, including hypersensitivity, carcinogenicity, and teratogenicity.

17.8.1 Hypersensitivity

Proverbially stable and generally considered inert, gold was long overlooked as an allergen, and overt hypersensitivity to the metal was observed so rarely as to be virtually unknown. Gold is now gaining recognition as a major factor in the etiology of cellular and humoral immunity owing to increasing systemic exposure for therapeutic purposes and to new patterns of intimate cutaneous contact. Characteristic immunological responses to gold hypersensitivity include late reactions to challenge, extraordinary persistence of clinical effects, formation of intracutaneous nodules and immunogenic granulomas unresponsive to conventional steroid therapy, the occurrence of eczema at sites distant from the site of contact, and flareups of eczema upon systemic provocation with allergen characteristic of drug-induced therapy. Gold salts take one of the top positions among drugs causing cutaneous side effects, and gold dermatitis may have many presentations, including eczematous, lichenoidal, toxicodermal, and pityriasis rosea-like eruptions.

In the year 2001, gold was selected as the contact allergen of the year by the American Contact Dermatitis Society. In the United States, Europe, and Japan, gold is now ranked among the ten most frequent allergens; the greatest majority of those sensitized were women. The prevalence of gold allergy worldwide, as determined by patch tests with various gold salts, might be as high as 13%, with 9.5% the most recent estimate in North America. Positive reactions to gold salts may appear 7–10 days, or longer, after testing. Most patients with positive gold patch tests have dental gold. In Sweden, gold is now considered the second most common metal allergen after nickel, as based on sensitivity to gold sodium thiomalate in patch tests. In Sweden, hypersensitivity to gold sodium thiomalate was more frequent in patients with oral restorative materials containing gold and was associated with distal eczema. Beginning in the 1980s and continuing today, case reports have been increasing of gold causing dermatitis at sites of jewelry contact and eyelid dermatitis from gold allergy.

The clinical picture of allergic contact dermatitis to gold usually consists of a toxicoderma-like rash at the site of contact, and transient fever. Cell-mediated allergic responses to gold were accompanied by positive lymphocyte transformation and proliferation tests; gold was selectively accumulated in Langerhans cells of the epidermis. Intramuscular injection of gold sodium thiosulfate into patients allergic to gold are accompanied by immunological tissue reactions and release in blood of cytokines and acute phase reactants, including plasma tumor necrosis factor-alpha, soluble tumor necrosis factor receptor 1, interleukin-1 receptor antagonist, and neutrophil gelatinase associated lipocalin. Results of patch tests with gold sodium thiosulfate among Swedish dermatitis patients should take longer than 3 days – the usual post-observation period – in order to fully evaluate the findings. Only 46% of the positive patch test reactions appeared within 3 days, the rest appeared within 10 days. Reactions were still readable after two months in about a third of the tests; a supplemental reading of patch test results should be made at three weeks post-exposure.

The most common outcome of female patients who had a positive allergic response to gold sodium thiosulfate was eczema of the head and neck (62% frequency), limbs (46%), and anus and vulva (15%). The mean duration of eczema in this group was 15.8 months. Most

(54%) of the patients allergic to gold were also allergic to nickel. Contact allergy to gold sodium thiosulfate in humans (unlike certain strains of mice) is hypothesized to be either lifelong or at least to last for years, although evidence is incomplete. Experimental studies with gold sodium thiosulfate in humans indicate a 100% response that lasts for at least two months.

Gold dermatitis from occupational exposure is rare. Gold salts are usually the cause, rarely gold objects. The main exposure sources of gold contact dermatitis are personal jewelry and dental alloys. The occupations most frequently causative of contact dermatitis due to gold are photography, chinaware or glass decorating, jewelers, and manufacturers of dental alloys. Occupational allergic contact dermatitis due to gold is infrequent in automated industrial processes.

Apart from medical therapeutic purposes, the use of gold in jewelry brings the greatest risk of sensitization. The risk is greatest when the gold-containing alloys are introduced and left in permanent contact with live tissues, as occurs in piercing of ears and other body parts. Cases of contact dermatitis due to gold, especially in pierced earlobes, are increasing in number worldwide. Small fragments of gold may remain in the skin lesions of pierced earlobes for at least 4 months after the 24-carat gold studs have been removed, causing prolonged irritation and various cutaneous reactions. Insertion of gold earrings immediately following piercing may result – through gold solubilization and cellular response – in the formation of intracutaneous bodies in the earlobes at the site of piercing, with ultimate surgical removal of the nodules. The nodules were characterized by large macrophages, lymphoid cell infiltration and eosinophils, confirming the immunological nature of such nodules. However, metallic gold (Au) used both in jewelry and in prostheses is ordinarily alloyed with other metals that may contribute to acute contact dermatitis. High carat yellow gold contains minute quantities of copper and silver; low carat yellow gold contains these metals plus zinc and small amounts of nickel. White gold usually contains palladium and nickel. The nickel in white gold alloys is a strong sensitizer and contact dermatitis to nickel often coexists with rare instances of acute contact dermatitis following exposure to Au. Even the most highly purified forms of gold contain minute quantities of contaminating materials, mainly iron and sodium, which in total may represent about 0.1% or 1000 mg/kg. Defects in the gold coating on stems of some commercial ear-piercing studs, normally in contact with the pierced ears, allowed body fluids to contact the stem's substrate; the substrate contained nickel, cobalt, zinc, and copper, with cytotoxicity in at least one case attributed to copper.

In contact allergy to gold, a low rate of responsiveness and mild symptoms were typical, although some people developed strong and persistent reactions. Sensitivity to gold was based on responsiveness to patches applied to the skin containing either metallic gold (Au), gold chloride (Au^{+3}), or various organomonovalent gold compounds (Au^{+}). Gold sodium thiomalate (Au^{+}) was the best marker of gold contact allergy because Au often yielded false negative results due to the inadequate release of soluble gold, and Au^{+3} caused persistent allergic reactions more frequently than did other gold compounds. Patch tests in recent years using gold sodium thiomalate have indicated positive patch test frequencies as high as 8.6% in Asia, 10% in Europe, and 13% in North America. In patch tests, some studies suggested that gold sodium thiomalate produced few positive reactions in patients hypersensitive to gold sodium thiosulfate. But in tests of intracutaneous administration of equimolar concentrations, allergic reaction rates were similar for gold sodium thiomalate and gold sodium thiosulfate, suggesting that contact allergy rates were probably similar. The efficacy of gold salt patch tests needs to be critically reexamined.

Hypersensitivity to gold is variable. It is suggested that Au toxicity may be associated, in part, with the formation of the more reactive Au^{+} and Au^{+3} species; however, this has not yet been verified. Additional research is warranted at the molecular level of the unusual mechanisms of action induced by gold dermotoxicity.

17.8.2 Teratogenicity and Carcinogenicity

There are no adequate studies of teratogenicity for gold sodium thiomalate in pregnant humans; however, a potential risk to the fetus exists because gold was found in the serum and red blood cells of a nursing infant.

Trivalent gold complexes were potentially attractive as anticancer agents because of their cytotoxic effects on established human tumor cell lines. All tested Au^{+3} complexes substantially retained their antitumor potency against platinum-resistant tumor cell lines for leukemia and ovarian cancer. Cytotoxicity of these compounds in vitro is attributed to binding with DNA and modification and subsequent impairment of replication and transcription processes. The paucity of data on Au^{+3} complexes probably derives from their high redox potential and relatively poor stability, which makes their use problematical under physiological conditions.

17.8.3 Dental Aspects

Gold allergy was overrepresented in those having dental gold, and sensitization to gold seems to be more common than previously anticipated. Dental patients had a 12.4% positive patch test reaction to gold sodium thiomalate; of those who tested positive, 73% responded to gold compounds in vitro in the lymphocyte proliferation test. When sensitization does occur, it is usually from exposure to the salts of mercury or other metals in dental alloys and may manifest itself as oral lichenoid lesions; replacement of the gold filling with non-metallic restorations frequently leads to resolution of the inflammatory lesion. Many metals that are used today in dentistry may be hazardous to certain genetically predisposed individuals, and could limit future use of these metals. In one case, oral mucosal problems suspected to be associated with release of metal ions from dental restorations, coupled with chronic fatigue, was reported in a patient occupationally exposed to metals while working in a dental practice. Lymphocyte assays indicated that the patient was sensitive to gold, mercury, and palladium. Patients claiming various subjective symptoms related to dental restoration materials were tested for sensitivity to 19 metals by patch test; sensitivity was 23% to gold sodium thiomalate, 22% to nickel sulfate, and <8% for all other metals.

Dental restorations made of dissimilar metals may undergo a series of electrogalvanic reactions of corrosion when brought together, causing short-lived, but severe pain in a few patients. In opposing teeth, one with amalgam and another with gold restoration are in contact, the galvanic current generated by gold restoration is always smaller than that in the tooth restored with amalgam. If pain persists, treatment may consist of replacing the amalgam restoration with a composite restoration to break the interproximal dissimilar metal contact. Oral fluids slowly dissolve elemental gold used in dental restorative materials. Since gold is used in alloys with copper, silver, zinc, platinum, and palladium, solubilization can be accelerated by galvanic reactions with other adjoining restorative metals. The salts thus formed may provoke allergic response of the delayed type as they are absorbed through the mucous membrane. Oral lesions are seen as a consequence including erythema, mucosal erosions, lichen planus, and stomatitis.

The safety of amalgam, containing about 50% mercury, as restorative material in dentistry is controversial and its use has been restricted in several countries. Alternatives to dental amalgam include alloys in which gold is partly replaced by palladium (gold-reduced alloys) or the more expensive high-gold alloys. Insertion of high-gold dental alloys containing platinum and palladium did not contribute to increased gold or palladium in urine over a 3-month period in non-occupationally exposed volunteers. Platinum content of urine, however, was significantly elevated when compared to pre-insertion levels. In vitro release studies of gold from four different types of artificial alloys containing 4, 51, 70, or 74% gold into either artificial saliva or 1% lactic acid solutions showed that gold, as well as Pt and Pd are released from noble metal-containing dental alloys by corrosion. The possibility exists that the release of noble metals from dental alloys may cause local or systemic effects.

Contact allergy to gold sodium thiomalate was reported in 4.6% of females with suspected contact dermatitis. The most frequent site of eczema was the head and neck with 62% frequency. Up to 10% of patients tested positive to gold sodium thiomalate in patch-tested eczema patients and seems to reflect true contact allergy. In both studies, many tested patients were also allergic to nickel. Allergy to gold sodium thiomalate was overrepresented in those having dental gold. Gingivitis caused by gold in teeth, without eczema, is also reported. Contact allergy to dental gold can also lead to glossitis, oral lichen planus, and chrysiasis.

17.9 Gold Mine Wastes

Of the major metal mining industries, gold mining is the most waste intensive. Refined gold consists of but 0.00015% of all raw materials used in the gold-mining process. It is estimated that it takes 2.8 tons of gold ore to produce the gold in a single wedding band ring, the rest being waste. After waste rock is removed and the ore extracted, the ore is processed to separate the gold from the valueless portion. After the gold is processed and recovered, the remaining rock is known as tailings. Mine tailings, as is true for waste rock, contain heavy metals and acid-forming minerals. Tailings can also contain chemicals used in ore processing. Amounts of toxicants in tailings – including arsenic, lead, cyanide, and sulfuric acid – are deleterious to fish and other wildlife. Tailings are usually stored in piles on land or in containment ponds, but sometimes are pumped back into the underground space from which the ore was mined. Dumping of mine tailings directly into rivers or other water bodies is no longer allowed in the United States, but occurs with some frequency elsewhere, especially in developing countries.

This section touches on aspects of acid mine drainage, tailings, and especially cyanide. Other gold mining wastes include arsenic (Chapter 2) and mercury (Chapter 19).

17.9.1 Acid Mine Drainage

Gold mines in the United States and Canada – some more than 100 years old, some recently closed, and some still active – are leaking metal-rich acidic water into the environment, resulting in hundreds of millions of dollars in remediation costs annually. This acidic drainage, often referred to as acid mine drainage or AMD, is derived from sulfide-containing rock excavated from an underground mine or open pit. The sulfur reacts with water and oxygen to form sulfuric acid (H_2SO_4). Iron pyrite (FeS_2) is the most common rock type that reacts to form AMD, but marcasites and pyrrhotites also contribute significantly. On exposure to air and water, the acid will continue to leach from the source rock until the sulfides are leached out – a process that can last for centuries. The sulfur is released by weathering, oxidation, and erosion, with concurrent production of sulfuric acid. The rate of acid production from inorganic oxidation of iron sulfides is enhanced by various species of acidophilic bacteria, especially *Thiobacillus ferrooxidans*. The acidity of the water and its proximity to metal in the ore may generate waters of low pH that are high in copper, cadmium, iron, zinc, aluminum, arsenic, selenium, manganese, chromium, mercury, lead, and other elements released from the ores with increasing acidity. The resulting solution is sufficiently acidic to dissolve iron tools in underground mines and kill migratory waterfowl that shelter overnight in pit lakes. AMD seeps out of tailings, overburden, and rock piles being processed for gold removal. If left unchecked, it can contaminate groundwater. AMD is often transported from the mining site by rainwater or surface drainage into nearby watercourses where it severely degrades water quality, killing aquatic life, and making water virtually unusable.

Underground gold mines puncture ore bodies with adits, mine tunnels, and shafts that allow air and water to enter and react with sulfide materials that are exposed inside the mine. AMD can leach from underground mine openings into streams and aquifers. In open pit mines, sulfide minerals on the exposed sides of the pit excavation are moistened by precipitation or by groundwater seeps, generating intense AMD flows.

17.9.2 Tailings

In Canada, there are an estimated 6000 abandoned mine sites that pose potential tailings hazards to aquatic ecosystems. One such event occurred in October 1990 when 300,000 metric tons of water-saturated gold mine tailings spilled into the Montreal River in northern Ontario, Canada, via a small creek, as a result of collapse of a tailings dike. About 50 km of the river was contaminated and 5 km heavily contaminated. Bulk concentrations of copper in sediments (120.0 mg/kg DW) exceeded the severe effect level (110.0 mg/kg) listed in the Ontario Provincial Quality Guidelines; similar cases were made for cadmium (>0.6 mg Cd/kg DW sediment), chromium (>26.0 mg Cr/kg), manganese (>460.0 mg Mn/kg), nickel (>16.0 mg Ni/kg), lead (>31.0 mg Pb/kg), and zinc (>120.0 mg Zn/kg). Eggs from resident walleyes (*Stizostedion vitreum*) incubated on these sediments had low hatch, and this was attributed to lead and copper toxicity, and possibly hypoxia from the resuspension and settling of mine tailings. Similar observations were made near gold mining sites in Korea, Zimbabwe, Malaysia, and Columbia.

17.9.3 Arsenic

Gold-bearing ores worldwide contain variable quantities of sulfide and arsenic compounds that interfere with efficient gold extraction using current cyanidation technology. Some gold-containing ores in Columbia, South America, contain up to 32% of arsenic-bearing minerals, and surrounding sediments may hold as much as 6300.0 mg As/kg DW. In Fairbanks, Alaska, some groundwaters are contaminated with arsenic from gold mining activities 30 years earlier and considered unsafe for drinking. In environments such as acid mine drainage of abandoned gold mines, As^{+3} concentrations ranged from 2.0 to 13.0 mg/L. The As^{+3} can then be oxidized to arsenate (As^{+5}). Both these soluble forms of arsenic are toxic to living organisms, especially inorganic arsenite. Arsenic concentrations near gold mining operations are elevated in abiotic materials and biota: maximum total arsenic concentrations measured were 560.0 μg/L in surface waters, 5.16 mg/L in sediment pore waters, 5.6 mg/kg DW in bird liver, 27.0 mg/kg DW in terrestrial grasses, 50.0 mg/kg DW in soil, 79.0 mg/kg DW in aquatic plants, 103.0 mg/kg DW in bird diets, 225.0 mg/kg DW in soft parts of bivalve molluscs, 324.0 mg/L in mine drainage waters, 625.0 mg/kg DW in aquatic insects, 7700.0 mg/kg DW in sediments, and 21,000.0 mg/kg DW in tailings.

Gold miners had a number of arsenic-associated health problems including excess mortality from cancer of the lung, stomach, and respiratory tract. Miners and schoolchildren in the vicinity of gold mining activities had elevated urine arsenic of 25.7 μg/L (range 2.2–106.0 μg/L). Of the total population at this location, 20% showed elevated urine arsenic concentrations associated with future adverse health effects; arsenic-contaminated drinking water is the probable causative factor of elevated arsenic in urine.

The significance of these elevated values is discussed in detail in Chapter 2.

17.9.4 Mercury Hazards from Gold Mining

Gold mining currently accounts for about 10% of the global mercury emissions from human activities. Mercury contamination of the environment from historical and ongoing mining practices that rely on mercury amalgamation for gold extraction is widespread. Contamination was particularly severe in the immediate vicinity of gold extraction and refining operations; however, mercury, especially in the form of water-soluble methylmercury, may be transported to pristine areas by rainwater, water currents, deforestation, volatilization, and other vectors. Examples of gold mining-associated mercury pollution have been shown for Canada, the U.S., Africa, China, the Philippines, Siberia, and

South America. In parts of Brazil, for example, mercury concentrations in all abiotic materials, plants, and animals, including endangered species of mammals and reptiles, collected near ongoing mercury amalgamation gold mining sites were far in excess of allowable mercury levels promulgated by regulatory agencies for the protection of human health and natural resources. Although health authorities in Brazil are unable to detect conclusive evidence of human mercury intoxication, the potential exists in the absence of mitigation for epidemic mercury poisoning of the mining population and environs. In the U.S., environmental mercury contamination is mostly from historical gold mining practices, and portions of Nevada remain sufficiently mercury contaminated to pose a hazard to reproduction of carnivorous fishes and fish-eating birds. This subject, and other aspects of mercury pollution, are treated in detail in Chapter 19.

17.9.5 Cyanide Hazards to Plants and Animals from Gold Mining and Related Water Issues

Highly toxic sodium cyanide (NaCN) is used increasingly by the international mining community to extract gold and other precious metals through milling of high-grade ores and heap leaching of low-grade ores. The process to concentrate gold using cyanide was developed in Scotland in 1887 and used almost immediately in the Witwatersrand gold fields of the RSA. Heap leaching with cyanide was proposed by the U.S. Bureau of Mines in 1969 as a means of extracting gold from low-grade ores. The gold industry adopted the technique in the 1970s, soon making heap leaching the dominant technology in gold extraction. The heap leach and milling processes, which involve dewatering of gold-bearing ores, spraying of dilute cyanide solutions on extremely large heaps of ores containing low concentrations of gold, or the milling of ores with the use of cyanide and subsequent recovery of the gold-cyanide complex, have created a number of serious environmental problems affecting wildlife and water management. This section contains a review of the history of cyanide use in gold mining, with emphasis on heap leach gold mining, cyanide hazards to plants and animals, water management issues associated with heap leach gold mining, and proposed mitigation and research needs.

17.9.5.1 History of Cyanide Use in Gold Mining

About 100 million kg cyanide (CN) is consumed annually in North America, of which 80% is used in gold mining. In Canada, more than 90% of the mined gold is extracted from ores with the cyanidation process, which consists of leaching gold from the ore as a gold-cyanide complex and recovering the gold by precipitation. The process involves the dissolution of gold from the ore in a dilute cyanide solution and in the presence of lime and oxygen according to the following reactions:

$$(1)\ 2Au + 4NaCN + O_2 + 2H_2O = 2NaAu(CN)_2 + 2NaOH + H_2O_2$$

$$(2)\ 2Au + 4NaCN + H_2O_2 = 2NaAu(CN)_2 + 2NaOH$$

Depending on solution pH, free cyanide concentrations, and other factors, gold is recovered from the eluate of the cyanidation process using either activated carbon, zinc, or ion-exchange resins. Using zinc dust, for example, gold along with silver is precipitated according to the reaction:

$$(3)\ 2NaAu(CN)_2 + Zn = Na_2Zn(CN)_4 + 2Au$$

Milling and heap leaching require cycling of millions of liters of alkaline water containing high concentrations of NaCN, free cyanide, and metal cyanide complexes that are available to the biosphere. Some milling operations result in tailings ponds 150 ha in area and larger. Heap leach operations that spray or drip cyanide solution onto the flattened top of the ore heap require solution processing ponds of about 1 ha surface area. Puddles of various sizes may occur on the top of heaps where the highest concentrations of NaCN are found. Solution recovery channels are usually constructed at the base of leach heaps; sometimes, these are buried or covered with netting to restrict access of vertebrates.

All these cyanide-containing water bodies are hazardous to natural resources and human health if not properly managed. For example, cyanide-laced sludges from gold mining operations stored in diked lagoons have regularly escaped from these lagoons. Major spills have occurred in Guyana in 1995 and in Latvia and Kyrgyzstan in the 1990s. Failure of gold mine tailings ponds killed one child in Zimbabwe in 1978, 17 people in South Africa in 1994 after a heavy rainfall, and contaminated streams and rivers in New Zealand in 1995 and elsewhere.

In September 1980 the price of gold had increased to $750 per troy ounce [1 Troy ounce (oz) = 31.1035 g] from $35 a decade earlier. This economic incentive resulted in improved cyanide processing technologies to permit cost-effective extraction of small amounts of gold from low-grade ores. The state of Nevada is a major global gold-producing area, with at least 40 active operations. Increased gold mining activity is also reported in other Western states, Alaska, the Carolinas, and Northern Plains states. Where relatively high-grade ores (>0.09 troy ounces Au/t ore) are found, milling techniques are used, but heap leaching of low-grade ores (0.006–0.025 troy ounces Au/t) is the most commonly employed extraction technique. Heap leach facilities usually produce gold for less than $200 US/troy ounces.

The amount of gold produced in the U.S. by heap leaching rose 20 fold throughout the 1980s, accounting for 6% of the supply at the beginning of the decade and more than 33% at the end. In 1980, there were approximately 24 heap leach facilities in the U.S.; by 1991, there were 265, of which 151 were active. The rise in domestic gold production in this period from 31 t in 1980 to 295 t in 1990 is attributable mainly to cyanide heap leaching. Although more tons of gold ore are heap leached than vat leached in the U.S. today, a greater quantity of gold is actually produced by vat leaching because that method is used on higher-grade ores and has a higher gold recovery rate. In 1989, cyanide heap leaching produced 3.7 million troy ounces from 129.8 million t ore and cyanide vat leaching produced 4.3 million troy ounces gold from 40.6 million t ore.

Heap leaching occurs when ore is stacked on an impermeable liner on the ground surface, with spraying or dripping of a dilute (usually about 0.05%) NaCN solution on the flattened top over a period of several months. Large leach heaps may include 1–25 million t ore, tower 100 m or more, and occupy several hundred hectares. As the solution percolates through the heap, gold is complexed and dissolved. For best results, heap-leached ores need to be porous, contain fine-grained clean gold particles, have low clay content, and have surfaces accessible to leach solutions. After the gold-containing solution is collected in a drainage pond, it is chemically precipitated, and the remaining solution is adjusted for pH and cyanide concentration and recycled to precipitate more gold. Eventually the remaining solution is treated to recycle the cyanide or to destroy it to prevent escape into the environment.

Cyanide and other contaminants may be released through tears and punctures in pad liners; leaks in liners carrying the cyanide solution; open ponds, piles, and solution ponds that can overflow; nitrogen compounds released during cyanide degradation; and release of lead, cadmium, copper, arsenic, and mercury, present in ore, that can be mobilized during crushing or leaching. The amount of hydrogen cyanide that escapes into the atmosphere from gold mining operations is estimated at 20,000 t annually, where it is quite stable; the half-time persistence of HCN in the atmosphere is about 267 d.

Individual mines often cover thousands of hectares, and mining companies sometimes lease additional thousands of hectares for possible mining. Ultimately, mining converts the site into large flat-topped hills of crushed ores, waste rock, or extracted tailings and large open pits. This alteration may result in permanent damage to wildlife habitat, although most areas with the general exception of open pits are reclaimed through revegetation. Between 1986 and 1991, cyanide in heap leach solutions and mill tailings ponds at gold mines in Nevada alone killed at least 9500 birds, mammals, reptiles, and amphibians. Dead birds representing 91 species, especially species of migratory waterfowl, shorebirds, and gulls, comprised about 90% of the total number of animals found dead, mammals 7% (28 species), and amphibians and reptiles together 3% (six species). In more recent years, the Nevada Division of Wildlife, through its toxic pond permit program (Nevada Administrative Code 502.460 through 502.495) and cooperative work with mining companies, significantly reduced the number of cyanide-related deaths of vertebrate wildlife.

Under certain alkaline conditions, cyanide may persist for at least a century in groundwater, mine tailings, and abandoned leach heaps. Cyanide destruction by natural reaction with the ore, soil, clay, and microorganisms has been advanced as the major mechanism for returning a site to an environmentally safe condition. To legally shut down the operation, concentrations <0.2 mg/L of weak acid dissociable cyanide metal-bound cyanide dissociable in weak acids, (WAD) are required. The use of cyanide to extract gold was banned in Turkey by the Turkish Supreme Court in 1999 because of accidental releases into the environment of untreated cyanide wastes stored in open ponds and resultant harm to human and ecosystem health. In Turkey, where more than 250,000 t of crushed rocks with mean gold content of 3 g/t were subjected to 125,000 t sodium cyanide in 365,000 m^3 water every year, more than 2,000,000 m^3 untreated cyanide/heavy metals solution had accumulated in waste ponds. Other countries that are considering prohibition of the cyanide leaching gold recovery process include the Czech Republic, Greece, and Romania.

Alkaline chlorination of wastewaters is one of the more widely used methods of treating cyanide wastes. In this process, cyanogen chloride (CNCl) is formed, which is hydrolyzed to the cyanate (CNO^-) at alkaline pH. If free chlorine is present, CNO^- can be further oxidized. The use of sulfur dioxide in a high dissolved oxygen environment with a copper catalyst reportedly reduces total cyanide in high-cyanide rinsewaters from metal plating shops to less than 1.0 mg/L; this process may have application in cyanide detoxification of tailings ponds. In general, because chemical treatments do not degrade all cyanide complexes, biological treatments are used. Biological treatments include (1) oxidation of cyanide compounds and thiocyanate by *Pseudomonas paucimobilis* with 95–98% reduction of cyanides in daily discharges of 15,000,000 L; (2) metabolism of cyanides by strains of *Pseudomonas*, *Acinetobacter*, *Bacillus*, and *Alcaligenes* involving oxygenase enzymes; and (3) bacterial cyanide degraders involving cyanide oxygenase, cyanide nitrilase, and cyanide hydratase.

In soil, cyanide seldom remains biologically available because it is either complexed by trace metals, microbially metabolized, or lost through volatilization. Cyanide ions are not strongly adsorbed or retained in the soil, and leaching into the surrounding groundwater will probably occur. Under aerobic conditions, cyanide salts in the soil are microbially degraded to nitrites or form complexes with trace metals. Under anaerobic conditions, cyanides denitrify to gaseous nitrogen compounds that enter the atmosphere. Mixed microbial communities that can metabolize cyanide and were not previously exposed to cyanide are adversely affected at 0.3 mg HCN/kg; however, these communities can become acclimatized to cyanide and then degrade wastes with higher cyanide concentrations. Acclimatized microbes in activated sewage sludge can often convert nitriles to ammonia at concentrations as high as 60.0 mg total CN/kg.

With regard to cyanide use and toxicity on the recovery of gold and other precious metals, most authorities currently agree on nine points:

(1) Metal mining operations consume most of the current cyanide production.
(2) The greatest source of cyanide exposure to humans and range animals is cyanogenic food plants and forage crops, not mining operations.
(3) Cyanide is ubiquitous in the environment, with gold mining facilities in only one of many sources of elevated concentrations.

(4) Many chemical forms of cyanide are present in the environment, including free cyanide, metallocyanide complexes, and synthetic organocyanides, but only free cyanide the sum of molecular hydrogen cyanide [HCN], and the cyanide anion [CN^-]) is the primary toxic agent, regardless of origin.
(5) Cyanides are readily absorbed through inhalation, ingestion, or skin contact, and are readily distributed throughout the body via blood. Cyanide is a potent and rapid-acting asphyxiant; it induces tissue anoxia through inactivation of cytochrome oxidase, causing cytotoxic hypoxia in the presence of normal hemoglobin oxygenation.
(6) At sublethal doses, cyanide reacts with thiosulfate in the presence of rhodanese to produce the comparatively harmless thiocyanate, most of which is excreted in the urine. Rapid detoxification enables animals to ingest high sublethal doses of cyanide over extended periods without adverse effects.
(7) Cyanides are not mutagenic or carcinogenic.
(8) Cyanide does not biomagnify in food webs or cycle extensively in ecosystems, probably because of its rapid breakdown.
(9) Cyanide seldom persists in surface waters owing to complexation or sedimentation, microbial metabolism, and loss from volatilization.

17.9.5.2 Cyanide Hazards: Aquatic Ecosystems

Fish killings from accidental discharges of cyanide-containing gold mining wastes are common. In one case, mine effluents containing cyanide from a Canadian tailings pond released into a nearby creek killed more than 20,000 steelheads (*Oncorhynchus mykiss*). In Colorado, overflows of 760,000 L NaCN-contaminated water from storage ponds into natural waterways killed all aquatic life along 28 km of the Alamosa River. In 1990, 40,000,000 L cyanide wastes from a gold mine spilled into the Lynches River in South Carolina, from a breached containment pond after heavy rains, killing an estimated 11,000 fish. In 1995, 160,000 L cyanide solution from a gold mine tailings pond near Jefferson City, Montana, were released into a nearby creek with loss of all fish and greatly reduced populations of aquatic insects. In August 1995, in Guyana, South America, a dam failed with the release of more than 3.3 billion L cyanide-containing gold mine wastes into the Essequibo River, the nations' primary waterway, killing fish for about 80 km, and contaminating drinking and irrigation water.

On January 30, 2000, a dike holding millions of liters of cyanide-laced wastewater gave way at a gold extraction operation (owned jointly by Australian and Rumanian firms) in northwestern Romania, sending a waterborne plume into a stream that flows into the Somes, a Tisza tributary that crosses into Hungary. At least 200 t of fish were killed, and endangered European otters (*Lutra lutra*) and white-tailed sea eagles (*Haliaeetus albicilla*) that ate the tainted fish were threatened. After devastating the upper Tisza, the 50-km-long pulse of cyanide and heavy metals spilled into River Danube in northern Yugoslavia, killing more fish before the now-dilute plume filtered into the Danube delta at the Black Sea, more than 1000 km and 3 weeks after the spill. This entire ecosystem was previously heavily contaminated by heavy metals from mining activities. Villages close to the accident were provided with alternate water sources. Hungarian officials were most concerned that heavy metals in the Tisza River may enter flooded agricultural areas, with subsequent accumulation by crops and entry into the human food chain.

Data on the recovery of cyanide-poisoned ecosystems are scarce. In one case, a large amount of cyanide-containing slag entered a stream from the reservoir of a Japanese gold mine as a result of an earthquake. The slag covered the streambed for about 10 km from the point of rupture, killing all stream biota; cyanide was detected in the water column for only 3 d after the spill. Within 1 month, flora was established on the silt covering the above-water stones, but there was little underwater growth. After 6–7 months, populations of fish, algae, and invertebrates had recovered, although the species composition of algae was altered.

Fish are the most cyanide-sensitive group of aquatic organisms tested. Under conditions of continuous exposure, adverse effects on swimming and reproduction usually occurred between 5.0 and 7.2 μg free CN/L and on survival between 20.0 and 76.0 μg/L. Reproductive impairment in adult bluegills (*Lepomis macrochirus*) occurred following exposure to 5.2 μg CN/L for 289 d. Concentrations of 10.0 μg HCN/L caused developmental abnormalities in embryos of Atlantic salmon (*Salmo salar*) after extended exposure. These abnormalities, which were absent in controls, included yolk sac dropsy and malformations of eyes, mouth, and vertebral column. Exposure of naturally reproducing female rainbow trout (*Oncorhynchus mykiss*) to 10.0 μg HCN/L for 12 d during the onset of the reproductive cycle produced a reduction in plasma vitellogenin levels and a reduction in ovary weight; vitellogenin is a major source of yolk. Oocyte growth was reduced in female rainbow trout and spermatocyte numbers decreased in males following exposure to 10.0 μg HCN for 12 d. Free cyanide concentrations as low as 10.0 μg/L can rapidly and irreversibly impair the swimming ability of salmonids in well-aerated water. Exposure of fish to 10.0 μg HCN/L for 9 d was sufficient to induce extensive necrosis in the liver, although gill tissue showed no damage. Intensification of liver histopathology was evident at dosages of 20.0 and 30.0 μg HCN/L and exposure periods up to 18 d. Other adverse effects on fish of chronic cyanide exposure included susceptibility to predation, disrupted respiration, osmoregulatory disturbances, and altered growth patterns. Free cyanide concentrations between 50.0 and 200.0 μ/L were fatal to sensitive fish species over time, and concentrations >200.0 μg/L were rapidly lethal to most species of fish. The high tolerance of mudskippers (*Boleophthalmus boddaerti*; 96-hr LC50 of 290.0 μg/L), and perhaps other species of teleosts are attributed to a surplus of cytochrome oxidase and inducible cyanide-detoxifying mechanisms and not to a reduction in metabolic rate or an enhanced anaerobic metabolism.

Fish retrieved from cyanide-poisoned environments, dead or alive, can probably be consumed by humans because muscle cyanide residues were considered to be lower than the currently recommended value of 50.0 mg/kg diet for human health protection. Cyanide concentrations in fish from streams poisoned with cyanide ranged between 10.0 and 100.0 μg total CN/kg whole-body FW. Gill tissues of cyanide-exposed salmonids contain from 30.0 to >7000.0 μg/kg FW under widely varying conditions of temperature, nominal water concentrations of free cyanide, and duration of exposure. Unpoisoned fish usually contained <1.0 μg total CN/kg FW in gills, although values up to 50.0 μg/kg FW occurred occasionally. Lowest cyanide concentrations in gills occurred at elevated (summer) water temperatures; at lower temperatures, survival was greater and residues were higher.

Among aquatic invertebrates, adverse nonlethal effects occurred between 18.0 and 43.0 μg/L, and lethal effects between 30.0 and 100.0 μg/L, although some deaths occurred between 3.0 and 7.0 μg/L for the amphipod *Gammarus pulex*. Aquatic plants are comparatively tolerant to cyanide; adverse effects occurred at >160.0 μg free CN/L. Adverse effects of cyanide on aquatic plants are unlikely at concentrations that cause acute effects to most species of freshwater and marine fishes and invertebrates.

Biocidal properties of cyanide in aquatic environments are modified by water pH, temperature, and oxygen content; life stage, condition, and species assayed; previous exposure to cyanides; presence of other chemicals; and initial dose tested. There is general agreement that cyanide is more toxic to freshwater fishes under conditions of low dissolved oxygen; that pH levels within the range 6.8–8.3 have little effect on cyanide toxicity but enhance toxicity at more acidic pH; that juveniles and adults are the most sensitive life stages and embryos and sac fry the most resistant; and that substantial interspecies variability exists in sensitivity to free cyanide. Initial dose and water temperature both modify the biocidal properties of HCN to freshwater teleosts. At low lethal concentrations near 10.0 μg HCN/L, cyanide is more toxic at lower temperatures; at high, rapidly lethal HCN concentrations, cyanide is more toxic at elevated temperatures. By contrast, aquatic invertebrates are most sensitive to HCN at elevated water temperatures, regardless of dose.

Cyanides seldom persist in aquatic environments. In small, cold oligotrophic lakes treated with NaCN (1.0 mg/L), acute toxicity to aquatic organisms was negligible within 40 d. In warm shallow ponds, no toxicity was evident to aquatic organisms after application of 1.0 mg NaCN/L. In rivers and streams, cyanide toxicity fell rapidly on dilution. Cyanide was not detectable in water and sediments of Yellowknife Bay, Canada, between 1974 and 1976, despite the continuous input of cyanide-containing effluents from an operating gold mine. Nondetection was attributed to rapid oxidation. Several factors contribute to the rapid disappearance of cyanide from water: bacteria and protozoans may degrade cyanide by converting it to carbon dioxide and ammonia; chlorination of water supplies can result in conversion to cyanate; an alkaline pH favors oxidation by chlorine; and an acidic pH favors volatilization of HCN into the atmosphere.

Cyanide interacts with other chemicals, and knowledge of these interactions is important in evaluating risk to living resources. Additive or more than additive toxicity of free cyanide to aquatic fauna may occur in combination with ammonia or arsenic. Formation of the nickel-cyanide complex markedly reduced the toxicity of both cyanide and nickel at high concentrations in alkaline pH; at lower concentrations and acidic pH, nickel-cyanide solutions increased in toxicity by more than 1000 times, owing to dissociation of the metallocyanide complex to form hydrogen cyanide. In 96-hr bioassays with fathead minnows, (*Pimephales promelas*), lethality of mixtures of sodium cyanide and nickel sulfate were influenced by water alkalinity and pH. LC50 values decreased with increasing alkalinity and increasing pH, being 0.42 mg CN/L at 5 mg $CaCO_3$/L and pH 6.5, to 730.0 mg CN/L at 192 mg $CaCO_3$/L and pH 8.0.

17.9.5.3 Cyanide Hazards: Birds

Cyanide waste solutions following gold extraction are released into the environment to form ponds, sometimes measuring hundreds of hectares in surface area. In the U.S., these ponds are often located in arid regions of Western states, and attract wildlife including migratory birds. Between 1983 and 1992, at least 1018 birds representing 47 species were killed when they drank cyanide-poisoned water from heap leach solution ponds at a gold mine in the Black Hills of South Dakota; in 1995, heap leach ponds from this site overflowed after heavy rains, spilling into a nearby creek with fatal results to all resident fishes. Many species of migratory birds, including waterfowl, shorebirds, passerines, and raptors, were found dead in the immediate vicinity of gold mine heap leach extraction facilities and tailings ponds, presumably as a result of drinking the cyanide-contaminated waters. About 7000 dead birds, mostly waterfowl and songbirds, were recovered from cyanide-extraction gold mine leach ponds in the western U.S. between 1980 and 1989; no gross pathological changes related to cyanide were observed in these birds at necropsy. No gross pathology was evident in cyanide-dosed birds, which is consistent with laboratory studies with cyanide and other animal groups tested and examined. In one case, waterfowl deaths were recorded in cyanide-containing ponds of an operating gold mine located in western Arizona shortly after the mine began operations in 1987. Deaths ranged from single birds to flocks of more than 70. At least 33 species of birds, including waterfowl, wading birds, gulls, raptors, and songbirds, and three species of mammals (bats, fox) were found dead in these ponds. Most of the waterfowl deaths were located in desert areas where the nearest water was 8–80 km distant.

To protect wildlife, various techniques were used including cyanide recovery, cyanide destruction, physical barriers, hazing, and establishment of decoy ponds. Techniques that were 92% successful (i.e., 8% mortality) cost mine owners about $8.58 per dead bird. This 92% survival was considered unsatisfactory by the U.S. Bureau of Land Management, and mine owners were forced to spend $295 for each dead bird found to reach 99% protection. Under existing legislation, however, zero mortality (100% survival) is the only acceptable solution. It is probable that 100% protection may not be possible using the best available technology. Songbird deaths were associated with hard rock gold mining in the Black Hills, South Dakota. This operation used the cyanide heap leaching process. Exposed collection ditches resembled small streams and were particularly attractive to songbirds, mostly red crossbills (*Loxia curvirostrata*) and pine siskins (*Carduelis pinus*), with fatal results. These ditches are now covered to prevent wildlife contact. Ponds containing cyanide solution were found to attract migrant waterfowl, and flagging devices were installed to dissuade waterfowl from landing, with partial success.

Free cyanide levels associated with high avian death rates have included 0.12 mg/L in air, 2.1–4.6 mg/kg body weight (BW) via acute oral exposure, and 1.3 mg/kg BW administered intravenously. In cyanide-tolerant species, such as the domestic chicken (*Gallus domesticus*), dietary levels of 135.0 mg total CN/kg ration resulted in growth reduction of chicks, but 103.0 mg total CN/kg ration had no measurable effect on these chicks. First signs of cyanide toxicosis in sensitive birds appeared between 0.5 and 5 min post-exposure, and included panting, eye blinking, salivation, and lethargy. In more tolerant species, signs of toxicosis began 10 min post-exposure. At higher doses, breathing in all species tested became increasingly deep and labored, followed by gasping and shallow intermittent breathing. Death usually followed in 15–30 min, although birds alive at 60 min frequently recovered. The rapid recovery of some cyanide-exposed birds may be due to the rapid metabolism of cyanide to thiocyanate and its subsequent excretion. Species sensitivity to cyanide seems to be associated with diet, with birds that feed predominantly on flesh being more sensitive to NaCN than species that feed mainly on plant materials, with the possible exception of mallards, as judged by acute oral LD50 values Table (17.2).

Some birds may not die immediately after drinking lethal cyanide solutions. Sodium cyanide rapidly forms free cyanide in the avian digestive tract (pH 1.3–6.5), whereas formation of free cyanide from metal cyanide complexes is comparatively slow. A high rate of

Table 17.2. Single oral dose toxicity of sodium cyanide (Mg NaCN/Kg body weight) fatal to 50% of selected birds and mammals (listed from most sensitive to most tolerant).

Species	**Oral LD50 (95% Confidence Limits)**
Mallard, *Anas platyrhynchos*	2.7 (2.2–3.2)
Human, *Homo sapiens*	3.0 estimated
American kestrel, *Falco sparverius*	4.0 (3.0–5.3)
Coyote, *Canis latrans*	4.1 (2.1–8.3)
Black vulture, *Coragyps atratus*	4.8 (4.4–5.3)
Laboratory white rat, *Rattus norvegicus*	5.1–6.4
Little brown bat, *Myotis lucifugus*	8.4 (5.9–11.9)
Eastern screech-owl, *Otus asio*	8.6 (7.2–10.2)
House mouse, *Mus musculus*	8.7 (8.2–9.3)
Japanese quail, *Coturnix japonica*	9.4 (7.7–11.4)
European starling, *Sturnus vulgaris*	17.0 (14.0–22.0)
Domestic chicken, *Gallus domesticus*	21.0 (12.0–36.0)
White-footed mouse, *Peromyscus leucopus*	28.0 (18.0–43.0)

cyanide absorption is critical to acute toxicity, and absorption may be retarded by the lower dissociation rates of metal-cyanide complexes. In Arizona, a red-breasted merganser (*Mergus serrator*) was found dead 20 km from the nearest known source of cyanide, yet its pectoral muscle tissue tested positive for cyanide. A proposed mechanism to account for this phenomenon involves weak acid dissociable (WAD) cyanide compounds. Cyanide bound to certain metals, usually copper, is dissociable in weak acids such as stomach acids. Drinking of lethal cyanide solutions by animals may not result in immediate death if the cyanide level is sufficiently low; these animals may die later when additional cyanide is liberated by stomach acid. In Canada, regulations typically require measurement of total cyanide and WAD cyanide in mine effluents. More research seems needed on WAD cyanide compounds and delayed mortality.

Cyanide is a respiratory poison because of its affinity for the cytochrome oxidase complex of the mitochondrial respiratory chain. High dosages of cyanide are lethal through inhibition of cytochrome oxidase via cessation of mitochondrial respiration and depletion of ATP. Mallards given single oral doses of KCN (1.0 mg KCN/kg BW) at cyanide concentrations and amounts similar to those found at gold mine tailings ponds (40.0 mg CN/L), although it is NaCN that is used almost exclusively in mining, had elevated concentrations of creatine kinase in serum, suggesting tissue damage. At 0.5 mg KCN/kg BW, mitochondrial function, an indicator of oxygen consumption, and ATP concentrations were significantly depressed in heart, liver, and brain. Rhodanese and 3-mercaptopyruvate sulfurtransferase, two enzymes associated with cyanide detoxification, were induced in brain but not in liver or heart of KCN-dosed mallards. Although cyanide concentrations as high as 2.0 mg KCN/kg BW (at 80.0 mg CN/L) were not acutely toxic to mallards, the long-term effects of such exposures were not determined and may have serious consequences for migratory birds exposed sublethally to cyanide at gold mine tailings ponds.

Under the Migratory Bird Treaty Act, cyanide-containing ponds must be maintained at a level that does not result in deaths of migratory birds. At present, there is negligible mortality of most avian species at ponds maintained at 50.0 mg CN^-/L. However some deaths of migratory birds have been recorded at <50.0 mg CN^-/L, and sublethal effects have been demonstrated in mallards in water containing 20.0 mg CN^-/L. These effects include significant decreases in excised liver and brain tissue ATP levels, and significant decreases in mitochondrial respiration rates in heart, liver, and brain tissues. It is clear that water containing <50.0 mg CN^-/L can cause generalized tissue damage in birds, and this needs to be addressed in future regulatory actions.

17.9.5.4 Cyanide Hazards: Mammals

Gold and silver mining are probably the most widespread source of anthropogenic cyanides in critical wildlife habitat, such as deserts in the western U.S. Between 1980 and 1989, 519 mammals, mostly rodents (35%) and bats (34%), were found dead at cyanide extraction gold mine mill tailings and heap leach ponds in California, Nevada, and Arizona. The list also included coyote (*Canis latrans*), badger (*Taxidea taxus*), beaver (*Castor canadensis*), mule deer (*Odocoileus hemionus*), blacktail jackrabbit (*Lepus californicus*), and kit fox (*Vulpes macrotis*), as well as skunks, chipmunks, squirrels, and domestic dogs, cats, and cattle. Also found dead at these same ponds were 38 reptiles, 55 amphibians, and 6997 birds. At the time of this study (1980–1989) there were approximately 160 cyanide extraction gold mines operating in California, Arizona, and Nevada and these mines were operating within the geographic ranges of ten endangered, threatened, or otherwise protected species of mammals. Bats comprised six of the ten listed species. Because bats were not identified to species, members of these six protected species could have been among the 174 reported dead bats. A population of Townsend's big-eared bats (*Plecotus townsendii*), one of the 10 protected species, may have been extirpated by cyanide at a nearby mine in California. Badgers were another of the 10 protected species; 6 were counted among the 519 mammals found dead.

Bighorn sheep (*Ovis canadensis*) were found dead in August 1983 on a cyanide heap leach pile in Montana; in 1991 gulls died after landing on an unnetted cyanide pond, and deer died after consuming cyanide solution that had trickled beneath a fence. In Nevada, the state with the most heap leach sites, cyanide spills occurred weekly during the 1980s. In South Dakota, a company's state-of-the-art leach pond was leaking cyanide solution at the rate of 19,000 L daily. Also, some companies allegedly punched holes in the heap leach liner when mining ended to allow drainage for more than 1 billion L of cyanide solution.

Signs of acute cyanide poisoning in livestock usually occur within 10 min and include initial excitability with muscle tremors, salivation, lacrimation, defecation, urination, and labored breathing, followed by muscular incoordination, gasping, and convulsions; death may occur quickly, depending on the dose administered. Acute oral LD50 values for representative species of mammals ranged between 4.1 and 28.0 mg HCN/kg BW and overlapped those of birds (Table 17.2). Despite the high lethality of large single exposures, repeated sublethal doses, especially in diets, are tolerated by many species for extended periods, perhaps indefinitely. Livestock found dead near a cyanide disposal site had been drinking surface water runoff that contained up to 365.0 mg HCN/L. Rats exposed for 30 d to 100.0 or 500.0 mg KCN/L drinking water had mitochondrial dysfunction, depressed ATP concentrations in liver and heart, and a depressed growth rate; little effect was observed at 50.0 mg KCN/L. The adverse effect on growth is consistent with the biochemical indicators of energy depletion. However, the concentrations should be viewed with caution as CN may have volatilized from the water solutions before ingestion by the rats, due to presumed neutral pH.

17.9.5.5 Cyanide Hazards: Terrestrial Flora

Mixed microbial populations capable of metabolizing cyanide and not previously exposed to cyanide were adversely affected at 0.3 mg HCN/kg substrate; however, these populations can become acclimatized to cyanide and can then degrade wastes containing cyanide concentrations as high as 60.0 mg/kg.

Cyanide metabolism in higher plants involves amino acids, *N*-hydroxyamino acids, aldoximes, nitriles, and cyanohydrins. Cyanide is a weak competitive inhibitor of green bean (*Phaseolus vulgaris*) lipooxygenase, an enzyme that catalyzes the formation of hydroperoxides from polyunsaturated acids. In higher plants, elevated cyanide concentrations inhibited respiration, through iron complexation in cytochrome oxidase, and ATP production and other processes dependent on ATP. At lower concentrations, effects include inhibition of germination and growth, although sometimes cyanide enhances seed germination by stimulating the pentose phosphate pathway and inhibiting catalase. The detoxification mechanism of cyanide is mediated by rhodanese, an enzyme widely distributed in plants. The rate of production and release of cyanide by plants to the environment through death and decomposition is unknown.

17.9.5.6 Cyanide Mitigation and Research Needs

Exclusion from cyanide solutions or reductions of cyanide concentrations to nontoxic levels are the only certain methods of protecting avian and mammalian wildlife from cyanide poisoning. Mortality of migratory birds from cyanide toxicosis may be curtailed at small ponds associated with leach heaps by screening birds from toxic solutions. Fencing and covering of small solution ponds with polypropylene netting have proved effective for excluding most birds, bats, and larger mammals, provided that the fencing and netting are properly maintained. Fences installed around cyanide-containing ponds at a heap leach gold mine in South Dakota successfully prevented deer and elk from entering; this, and other practices, reduced overall wildlife mortality at this site by about 95%.

Reclamation of leach heaps to establish suitable wildlife areas is ongoing by mining corporations, and involves mechanical creation of new wildlife habitats of slopes, ledges, and crevices, and revegetation of these habitats, usually with native plants, but sometimes with introduced species.

A few mines in Nevada are now covering surfaces of small ponds with 4-in. (10.2-cm) diameter high-density polyethylene balls; birds are no longer attracted to these ponds as water sources. Although initial costs of the balls are higher than installation of netting, there are no maintenance expenses for the balls, whereas netting needs continual maintenance. Gold mine operators in southern California and Nevada used plastic sheeting to cover the cyanide leach pond, resulting in a cessation of wildlife mortality. The comparatively high cost of this process was soon recouped through reduced evaporation of water and cyanide.

To reduce the potential for puddling on ore heaps, ores should be less compacted; this can be accomplished by reducing the clay content of the ores and stacking ores using conveyer belts rather than trucks. Puddling can also be reduced by careful monitoring of solution application rates and maintenance of solution distribution systems. Wildlife have been excluded from leaching solution on the heaps by substituting drip lines for sprinklers and covering the drip lines with a layer of gravel. Some mines use small net panels over areas of puddling to exclude birds.

Water hyacinth (*Eichornia crassipes*) has been proposed as the basis of a cyanide removal technology, although large-scale use has not been implemented. Hyacinths can survive for at least 72 hr in a nutrient solution containing up to 300.0 mg CN/L and can accumulate up to 6700.0 mg CN/kg DW plant material. On this basis, 1 ha of hyacinths has the potential to absorb 56.8 kg cyanide in 72 hr, and this property may be useful in reducing the level of CN in untreated wastewaters where concentrations generally exceed 200.0 mg CN/L. Large-scale use of water hyacinths for this purpose has not yet been implemented, possibly due to disagreement over appropriate disposal mechanisms.

17.9.5.7 Proposed Cyanide Criteria for the Protection of Natural Resources and Human Health

Free cyanide criteria currently proposed for the protection of natural resources include $<3.0\ \mu g/L$ medium for aquatic life, and <100.0 mg/kg diet for birds and livestock. For human health protection, free cyanide values are $<10.0\ \mu g/L$ drinking water, <50.0 mg/kg diet, and $<5.0\ mg/m^3$ air. Additional research is needed to establish legally enforceable standards and threshold limit values for potentially toxic cyanides in various forms, including HCN and inorganic cyanide. More research is merited on low-level, long-term cyanide intoxication in birds and mammals by oral and inhalation routes in the vicinities of high cyanide concentrations, especially on the incidence of nasal lesions, thyroid dysfunction, and urinary thiocyanate concentrations. Research is also needed on threshold limits in water where birds and mammals may be exposed, including the role of CN-metal complexes, and on sublethal effects of free cyanide on vertebrate wildlife. In aquatic systems, research is needed on (1) long-term effects of low concentrations of cyanide on growth, survival, metabolism, and behavior of a variety of aquatic organisms; (2) adaptive resistance to cyanide and the influence, if any, of oxygen, pH, temperature and other environmental variables; and (3) usefulness of various biochemical indicators of cyanide poisoning, such as cytochrome oxidase inhibition and vitellogenin levels in fish plasma.

17.9.5.8 Water Management Issues

To provide the quantities of ore needed for heap leach facilities, large pits are dug. One prospective open pit mine is expected to measure 0.9 km deep, 1.44 km wide, 2.4 km long, and involves more than 1 billion t of rock. Many mining pits intrude below the water table and must be continually pumped dry. After the mine closes, the pumping ceases and the pit fills to become a small lake. Pit lakes have the potential to become acidic and may eventually contain elevated concentrations of various elements. As the level of potentially toxic water rises, it can begin to infiltrate into groundwater.

Gold mine operators in Nevada and elsewhere are digging large open pits to reach extensive deep deposits of low-grade gold ore. To prevent flooding in the mine pits and to permit efficient earth moving of surface soils, it is necessary to withdraw groundwater and use it for irrigation, discharge it to rapid infiltrations basins, or in some cases discharge it into a nearby watercourse. Surface waters are diverted around surface mining operations. After cessation of mining operations and pumping is stopped, lakes will form in the open pit mines dug to levels below that of the surrounding groundwater. At least six of Nevada's open pit mines have filled with water that does not meet federal criteria for the protection of human drinking water and aquatic life for heavy metals and acidity.

Miners usually pump water by using a combination of in-pit wells and the perimeter wells. In-pit wells pump out water that has entered the mine site, and perimeter wells intercept groundwater before it can seep into the pit. The lowering of the water table decreases groundwater elevation kilometers away from the mining site. In the Humboldt River basin, Nevada, gold miners divert pumped water to irrigate fields, to create wetlands in what is naturally a desert, or to supply water to another user. Only a small fraction of the pumped water is restored to the original aquifer. Under Nevada law, miners must replace the lost water from dried-out springs or wells, deepen the well that has been dried out, or reduce pumping so that prior water levels are restored.

The potential effects of water management actions on resources in gold mining communities in northern Nevada beginning in 1992 are considerable. Specific resources impacted include geological structures, groundwater and surface water resources, riparian areas and wetlands, terrestrial wildlife, aquatic habitat and fisheries, special status species, livestock grazing, socio-economics, and Native American religious concerns. Geological impacts were demonstrated by the formation of at least three sinkholes within 5 km of groundwater drawdown and discharge into the Humboldt River. As a result of mine dewatering operations, groundwater levels at the end of 1998 had decreased 110–466 m in the vicinity of mines examined. Several springs near the mines have dried up or shown a reduction in flow as a result of mine dewatering and drought. The flow and vegetation in a nearby creek were significantly reduced. Reductions in the baseflow of perennial springs and streams could affect surface water rights within the drawdown area, directly impacting rights for irrigation, stock watering, domestic, mining and milling, municipal, and other uses. Contaminated soils and sediments from mine sites affect bed, bank, and floodplain sediments, as well as riparian areas and wetlands downgradient from the mine. Terrestrial vegetation diversity may be altered by construction of roads or use of offroad vehicles related to development of mines.

Reduction in surface or subsurface flows could result in a reduction or loss of vegetation cover for wildlife. This, in turn, could reduce breeding, foraging, and cover habitats for wildlife; increase animal displacement and loss; reduce prey availability; reduce overall biological diversity; produce possible genetic isolation in Lahontan cutthroat trout (*Oncorhynchus clarki henshawi*); reduce carrying capacity for terrestrial wildlife; and may result in population declines. Incremental habitat loss would affect a variety of big game species, upland gamebirds, waterfowl, shorebirds, raptors, songbirds, nongame mammals, and area reptiles and amphibians. The eventual reduction in flows within artificially created wetlands in the study area would result in a transition back into an upland plant community with reduction in use by waterfowl.

Reduction in stream flows as a result of mine dewatering could result in decreased biodiversity and biomass in these communities. If stream segments become dry as a result of reduced flows, aquatic habitat and associated biota would disappear. Increased flows to the Humboldt River would increase the habitat for fish and their food organisms. However, the possible elimination of shallow pools and channels could decrease nursery ground habitat. Increased sediment levels associated with increased flows may affect aquatic biota near outfalls.

The potential long-term loss of water sources may cause long-term loss of permitted active grazing use or affect livestock distribution and forage use within grazing allotments. In addition, perennial creeks located within these allotments could be affected by groundwater drawdown, which could affect grazing management operations. Slightly increased water levels within the Humboldt River during the mine dewatering period would probably increase the areal extent of wetlands adjacent to the river channel producing increased forage and increased water availability for livestock; this would reverse after mine dewatering ceases.

Reductions in groundwater levels may impose economic hardship on those using water for domestic, industrial, commercial, agricultural, and husbandry purposes. Excess mine water discharged into the Humboldt River would be a positive effect to water right holders in the basin, although adverse effects from increased flow may include flooding of irrigated fields during periods of high flow. Native Americans in the vicinity, primarily members of the Western Shoshone, believe that disruption of water resources would impact life and spirit forces associated with these waters, plants, and animals to the extent that Shoshone cultural traditions could not be maintained.

17.9.5.9 Pit Lakes

In Nevada, very low concentrations of gold (0.7 g/metric t) can be mined economically from bulk mineral deposits and large open pit mines have been and are currently being developed for this purpose. Many of these mining operations in Nevada are extracting ore from below the water table and are withdrawing large volumes of water to maintain dry operating conditions. A single gold mine in Eureka County, Nevada, allegedly has pumped as much groundwater from its open mining pit in 1 year as the entire annual consumption of a city with a population of 500,000. In 1994, gold mining operations in the Humboldt River basin withdrew enough water to supply all domestic water users in the greater Seattle area population (1.1 million) over that same period. After mining is completed and dewatering activities has stopped, the pits will begin to fill with water with the ultimate lake surface approaching the elevation of the premining groundwater level. Between 1992 and 2002, about 35 mines in Nevada have, or will have, a lake in their open pit mines after dewatering and cessation of mining. Of the existing pit lakes at eight different gold mines, most had near-neutral pH but exceeded drinking water standards for arsenic, sulfate, or total dissolved solids for at least one sampling period. Pit lakes have lower evaporation rates than natural lakes because they are at higher elevations and the surface-to-depth ratio of pit lakes is more than 1000 times smaller than that of natural lakes. It seems that the ultimate water quality and limnology of the pit lakes and their potential impact on wildlife have not been adequately evaluated.

When pit lakes exceed local or federal surface water or groundwater quality standards, three main options for mitigation are recommended: neutralize the pit lake in place through treatment; prevent the formation of a pit lake by pumping groundwater; and regulate the pit lake level at a certain height to prevent commingling with other aquifers. Treatment options include physical treatment processes that consist mainly of screening and filtration techniques to remove particulate matter; chemical treatment, the most common, to raise pH through lime precipitation of sulfates, and other methods to remove contaminants, such as arsenic; and biological processes, including sulfate reduction processes to treat acid drainage and remove sulfates as metals. These treatment options are preferable to the use of clean water for dilution or relying on faulty land application disposal systems.

Goldmining pit lakes in Nevada, when filled, will contain more water than all reservoirs combined within the borders of this arid state. Pit lakes are important to Nevada, and their water quality will determine their future use, as well as their effects on the aquifer, wildlife, and ecosystems. State positions on pit water quality issues vary; however, the death of migratory birds at a pit lake is of concern to the U.S. Fish and Wildlife Service, the lead agency in enforcing violations of the Migratory Bird Treaty Act. Waters of the Berkeley pit in Butte, Montana, were lethal to lesser snow geese (*Chen caerulescens caerulescens*) that used the lake in 1995. Nevada regulates pit lake water quality standards on a case-by-case basis, enforcing a state regulation that loosely states that pit lake water cannot degrade surrounding groundwater quality or adversely affect the health of human or terrestrial life. Other states may require that pit lake water conform to aquifer water quality standards (Arizona); suitability with human drinking, cooking, and food processing purposes after conventional treatment (Montana); and meet livestock, agriculture, or domestic water quality standards, depending on geographic location (Wyoming). Most states now require a bond from mine operators, with amounts ranging from nominal to millions of dollars, depending on the state. Aquatic communities may also form in pit lakes; these organisms have the potential to biomagnify various compounds and trace elements from pit lake waters. If pit lakes become attractive to migratory birds or other species, these species will be exposed to the contaminants that may be present. The introduction of fish to pit lakes may, in some cases, present unacceptable risks of contaminant exposure to fish-eating birds.

17.9.5.10 Water Quality and Management Research Needs

The long-term environmental impacts of pit lakes are poorly known at this time and long-term predictions are currently made on the basis of short-term data. Accordingly, the relation between predicted and actual outcomes needs to be evaluated. Research is needed on the chemistry, hydrology, and biology of pit lakes and their surroundings to minimize the environmental influence of future pit lakes. Pit lakes now filling need to be monitored over time to evaluate lake chemistry changes. Studies are also needed on the potential development of biological communities in pit lakes and their influence on both aquatic biota and avian and terrestrial wildlife.

Food chain effects in pit lakes should be determined before stable biological habitats are artificially established. Additional investigations are recommended to establish water quality criteria on selected metals for protection of aquatic life and for establishment of acceptable metal burdens in aquatic prey of migratory birds; this is of particular interest in pit lakes located in arid areas with increasing metals concentrations attributed to high evaporation rates.

More research is recommended on mine area dewatering and discharge of surplus water, especially to surface waters. Water balance models for different hydrogeological settings need to be developed to address local and regional interrelations among surface flow, pit lake hydrology, and hydraulic head of shallow and deep aquifers. These models may permit long-term predictions of the consequences of alteration of surface waters and the interruption, use, and withdrawal of groundwater. Results of studies on mine water discharges into nearby watercourses, for example, mine water discharges into the Humboldt River in Nevada, will aid in providing long-term predictions of the response of riverine ecosystems to hydrologic and biological changes. Finally, a better understanding is needed of the risks associated with increased loads of various contaminants from mine watering discharges to surface waters that flow to important wetlands in terminal systems where evapoconcentration is a factor.

17.10 Recommendations

Despite all evidence to the contrary, and for reasons that seem neither logical nor rational, it is probable that society will continue to value gold as a commodity in the foreseeable future and to provide continued employment to about 30 million individuals world-wide who derive a significant portion of their income from the mining, refining, and sale of raw and finished gold. Jewelry, coins, and bullion will doubtless continue to account for the great majority of all new gold mined – about 2000 tons annually – although new gold and gold products are constantly being developed for use in electronics, medicine, and other disciplines. As is true for other commodities, it is expected that the price of gold will continue to fluctuate over time based on demand and supply. As demand for gold decreases, the price for raw gold will decrease and marginally productive mines will be closed or abandoned, often with little or no regard for the environmental consequences. Gold miners and refiners will continue to suffer from increased illnesses and death rates compared to the general population, although this is expected to lessen for miners through additional training and education, implementation of existing safety regulations, and installation of adequate mine safety equipment. Environmental degradation associated with the extraction of gold will continue or increase in developing nations, and at a reduced rate among nations with strong and enforceable environmental laws. New mining technologies now under development to produce gold from very low grade ores are expected to be both cost-effective and environmentally friendly. Additional research efforts and information needs on various aspects of gold and gold mining are merited. Some of the more pressing needs are listed below.

(1) Reliable production data.

Major producers of gold include the RSA, the United States, the former Soviet Union, Canada, Australia, the People's Republic of China, and Brazil. However, official gold production data from many localities and nations are unreliable, and are usually underreported. Accurate reporting of gold production is needed in order to establish a realistic price for this commodity.

(2) Geologic characterization of gold deposits.

Geologic characteristics of commercial gold-bearing strata are not known with certainty, although many deposits are related by proximity to volcanic settings, granitic magmas and fluids, pyrites of iron and other metals, and potassium-containing igneous rocks. Cost-effective technologies need to be developed to locate gold and other valuable mineral ore deposits.

(3) Gold properties database.

Gold is a complex and reactive element, with unique chemical, physical, and biochemical properties. Additional research is strongly recommended on expansion of knowledge concerning properties of gold and its salts, and to centralize all findings in a single accessible database.

(4) Significance of gold concentrations in abiotic materials and living organisms.

Gold concentrations are unusually high in certain abiotic materials, such as sewage sludge from a gold mining community (4.5 mg Au/kg DW), polymetallic sulfides from the ocean floor (28.7 mg/kg DW), and freshwater sediments near a gold mine tailings pile (256.0 mg/kg DW), suggesting that gold recovery is commercially feasible from these materials. A similar situation exists for gold accumulator plants, which may contain as much as 100.0 mg Au/kg DW. However, the significance of gold concentrations in various organisms and materials, and mechanisms governing its uptake and retention are relatively unknown when compared to most metals. For a more complete evaluation of the role of gold in the environment, systematic monitoring of gold concentrations is recommended in abiotic materials and organisms comprising complex food chains; samples should also be analyzed for selected metals, metalloids, and other compounds known to modify gold uptake and retention.

(5) Expanded use of gold and gold salts.

The unique properties of gold have led to increasing use of the metal and its salts in the disciplines of electronics, dentistry, physiology, immunology, electron microscopy, and human medicine. Additional applications need to be explored.

(6) Gold effects on biota.

Recommended areas of research emphasis include: sublethal effects of Au^{+3} species to aquatic organisms because concentrations as low as 98.0 μg Au^{+3}/L adversely affect algal growth; mechanisms of accumulation of Au, Au^{+}, and Au^{+3} by microorganisms and algae; and effects of injected colloidal gold Au and monovalent gold drugs Au^{+}) in mammals on temperature regulation, brain chemistry, carcinogenicity, teratogenicity, and nephrotoxicity, with potential application to human medicine.

(7) Biorecovery of gold.

Gold recovery technologies based on bacteria, fungi, and other organisms are now commercially available; additional research is recommended on biorecovery of gold under different physicochemical conditions.

(8) Health protection of gold miners.

To protect the health of underground workers, continued intensive monitoring of atmospheric dust levels is recommended for conformance with safe occupational levels, implementation of regular and frequent medical examinations with emphasis on early detection and treatment of disease states, and continuation of educational programs on hazards of risky behaviors outside the mine environment. Miners who use elemental mercury to extract gold need to control mercury emissions in confined environments and limit consumption of larger carnivorous fishes. Intensive monitoring by physicians and toxicologists of populations at high risk for mercury poisoning is recommended in order to provide adherence to existing mercury criteria, as is critical examination of current mercury criteria to protect human health.

(9) Gold sensitivity to humans.

There is increasing documentation of allergic contact dermatitis and other effects to metallic gold from jewelry, dental restorations, and occupational exposure; these effects are most frequent in females wearing body-piercing gold objects. Research is needed to determine the extent of sensitivity (one estimate lists gold allergy at 13% worldwide), the mechanisms of action, and whether nursing infants are at risk. Similar studies are recommended for gold salts.

(10) Gold drugs in medicine.

Gold and gold drugs have been used in human medicine for centuries, and certain gold drugs have been used routinely for more than 75 years to treat rheumatoid arthritis. However, anti-inflammatory properties of gold metabolites and other mechanisms of action of gold drugs are not known with certainty and merit additional research, as does development of gold drugs with minimal side effects, alternate routes of gold drug administration for maximum efficacy, evaluation of gold drugs in combination with other drugs to enhance relief, and development of more sensitive and uniform indicators for evaluation of gold drug therapy.

(11) Gold mine wastes.

Acidic metal-rich water and tailings wastes from active gold mines devastate receiving aquatic ecosystems. More research is needed on prediction of extent of acid mine drainage and its prevention through physical, chemical, and biological remediation technologies, and on appropriate storage of tailings wastes.

(12) Amalgamation contaminant problems.

The use of mercury to recover gold has resulted in wholesale and persistent contamination of the biosphere, with direct – and frequently fatal – consequences to all members of the immediate biosphere, including humans. The use of mercury for this purpose must be abandoned, and improved remediation methodologies developed for mercury-contaminated environments using physical, chemical, and biological technologies.

(13) Cyanide extraction and water management issues.

The use of cyanide to extract gold from low grade ores by major mine operators has caused a variety of lethal and significant sublethal effects on wildlife, especially in desert-like arid areas. But this situation has been steadily improving as wildlife are shielded from cyanide-containing solutions and new habitat has been created from waste rock, overburden, and tailings piles. Pit lakes, however, resulting from surface mining technologies, including cyanidation, may create a variety of problems, as yet poorly documented, to wildlife and human health. Additional research is merited on ecotoxicology of pit lakes.

(14) Arsenic wastes.

Since arsenic interferes with gold extraction, and since most gold-containing ores contain significant quantities of arsenic, these ores are roasted to remove the arsenic, with resultant arsenic contamination of air, water, and soil ecosystems. Cost-effective methods to remove the arsenic without release to the surrounding environment need to be refined, especially those involving microorganisms.

17.11 Summary

Despite the esteemed nature of gold in society, evidence of adverse ecotoxicological effects and risk to human health in various mining and extraction techniques has generated increasing interest in the biological and environmental implications of gold. The effect of gold production and use on human health and the environmental impact of gold mining and extraction is examined through topics that include gold geology; sources; production; physical, chemical, and metabolic properties; concentrations in field collections of abiotic materials and biota; lethal and sublethal effects of gold and gold salts on plants and animals; health risks of gold miners; and human sensitivity to gold jewelry, dental implants, and medicinal use. The cyanidation gold extraction technology is examined with attendant risk to wildlife and water resources.

Gold is ubiquitous in the environment, usually occurring in minute amounts. Geological characteristics of commercial gold-bearing strata are related, in part, to proximity to volcanic settings, mobilization and transport rates of granitic magmas and fluids, pyrites of iron and other metals, and igneous rocks containing potassium. Accurate production figures for new gold are difficult to obtain, but probably exceed 39 million troy ounces annually (1209 metric tons). The major commercial producer of gold is the RSA; others include the Former Soviet Union, the United States, Canada, Australia, China, and Brazil. In 1995, the United States ranked second in global gold production, with most of its gold produced in Nevada.

The most common use for privately owned gold is jewelry and personal ornaments. Coinage and bullion ranks second, especially gold coins, minted by many nations, which are attractive to small investors. Secondary – and increasing – uses of gold include the disciplines of electronics, dentistry, physiology, immunology, electron microscopy, and human medicine.

Elemental gold is a soft yellow metal with the highest malleability and ductility of any known element. It is dense, being 19.32 times heavier than water at 20°C; a cube of gold 30 cm on a side weighed about 544 kg. Metallic gold is inert to strong alkalis and virtually all acids; however, solubility is documented for aqua regia, hot selenic acid, aqueous solutions of alkaline sulfides and thiosulfates, cyanide solutions, and liquid mercury. Sensitive analytical methodologies developed to measure gold in biological samples and abiotic materials relied heavily on its physical properties. Gold has 30 known isotopes and exists in seven oxidation states. Apart from Au° in the colloidal and elemental forms, only Au^{+} and Au^{+3} are known to form compounds that are stable in aqueous media and important in medical applications. The remaining oxidation states of -1, $+2$, $+4$, and $+5$ are not presently known to play a role in biochemical processes related to therapeutic uses of gold. The biochemistry of gold has developed mainly in response to prolonged use of gold compounds in treating rheumatoid arthritis and in response to efforts to develop complexes with anti-tumor and anti-HIV activity. Most of the in vivo gold chemistry is concerned with the reaction of gold species with thiols, especially Au^{+}. Gold is not considered essential to life, although it distributes widely in the body and the number of possible reactions and reaction sites is large. Monovalent organogold drugs were metabolized rapidly in vivo, usually within 20 min of administration; however, half-time excretion rates ranged between 8 h and 20 days, depending on the metabolite.

The significance of gold concentrations in various environmental compartments, gold's mode of action, and mechanisms governing its uptake, retention and translocation are not known with certainty. In order to facilitate a more detailed evaluation of the role of gold in the biosphere, systematic measurements of gold levels is recommended in abiotic materials and organisms comprising diverse multitrophic food chains using sensitive analytical methodologies. Samples should also be analyzed for various metals, metalloids, and compounds known to modify ecological and toxicological properties of gold.

Monovalent gold compounds have negligible impact on aquatic organisms at medium concentrations less than 98.0 mg Au^{+}/L; however, trivalent Au^{+3} compounds are significantly more toxic than Au^{+} compounds and inhibit growth of diatoms at 98.0 μg Au^{+3}/L, kill marine teleosts at less than 800.0 μg Au^{+3}/L, inhibit contraction ability of frog muscle at 2.0 mg Au^{+3}/L, and reduce growth of yeasts at 20.0 mg Au^{+3}/L. Accumulation of ionic and metallic gold from a wide variety of solutions by selected species of bacteria, yeasts, fungi, algae, and higher plants is documented. Gold accumulations are up to 7.0 g/kg DW in various species of bacteria, 25.0 g/kg DW in freshwater algae, 84.0 g/kg DW in peat, and 100.0 g/kg DW in dried fungus mixed with keratinous material. Crab exoskeletons accumulate up to 4.9 g Au/kg DW; however, gold accumulations in various tissues of living teleosts, decapod crustaceans, and bivalve molluscs are negligible. Uptake patterns are significantly modified by the physicochemical milieu. The mechanisms of accumulation – which include oxidation, reduction, dissolution, leaching and sorption – are not known with certainty and merit additional research effort.

In tests with small laboratory mammals, injected doses of colloidal gold are associated with increased body temperatures, gold accumulations in reticular cells, and dose enhancement in tumor therapy. In certain strains of gold-resistant mice, a single injection of sodium gold thioglucose at 0.5–0.6 mg Au^{+}/kg body weight destroyed the appetite center in the hypothalamus portion of the brain, resulting in obese mice with many of the characteristics of human obesity. Other strains of mice and other species of rodents tested either died or produced variable responses to gold thioglucose. High doses of monovalent organogold compounds administered parenterally are usually tolerated by mice, but survival is reduced in adult mice infected with various viruses. Carcinogenicity, teratogenicity, and other effects are documented in small laboratory mammals given extremely high injected doses of various monovalent organogold compounds; effects include renal adenomas, limb malformations, nephrotoxicity, elevated gold accumulations in kidney (316.0 mg/kg FW) and liver (44.0 mg/kg FW), increased serum immunoglobulin E, and altered metabolism of selenium, copper, and zinc. At comparatively low doses, monovalent organogold compounds may have antitumor properties. Exposure of rodents to trivalent inorganic gold compounds results in high accumulations in the renal cortex and other tissues, increased metallothionein concentrations, and variable immune responses dependent on rodent strain tested.

Health problems of gold miners who work underground include decreased life expectancy; increased frequency of cancer of the trachea, bronchus, lung, stomach, and liver; increased frequency of pulmonary tuberculosis, silicosis, and pleural diseases; increased frequency of insect-borne diseases, such as malaria and dengue fever; noise-induced hearing loss; increased prevalence of certain bacterial and viral diseases; and diseases of the blood, skin, and musculoskeletal system. These problems are briefly documented in gold miners from Australia, North America, South America, Europe, and Africa. In general, HIV infection or excessive alcohol and tobacco consumption tended to exacerbate existing health problems. Miners who used elemental mercury to amalgamate and extract gold were heavily contaminated with mercury. Among individuals exposed occupationally, concentrations of mercury in their air, fish diet, hair, urine, blood, and other tissues significantly exceeded all criteria proposed by various national and international regulatory agencies for protection of human health. However, large-scale epidemiological evidence of severe mercury-associated health problems in this cohort was not demonstrable (see Chapter 19).

In humans – especially among females wearing body-piercing gold objects – there is increasing documentation of allergic contact dermatitis and other effects to gold from jewelry, dental restorations, and occupational exposure, as judged by patch tests with monovalent organogold salts; one estimate of the prevalence of gold allergy world-wide is 13%. Eczema of the head and neck was the most common response of individuals hypersensitive to gold, and sensitivity may last for at least several years. The toxic action of gold° may be attributed, in part, to the formation of the more reactive Au^{+} and Au^{+3} species, although this has not been verified. Gold salts may also be lethal or teratogenic.

Cyanide extraction of gold through milling of high-grade ores and heap leaching of low-grade ores requires cycling of millions of liters of alkaline water containing high concentrations of potentially toxic sodium cyanide (NaCN), free cyanide, and metal-cyanide complexes. Some milling operations result in tailings ponds of 150 ha and larger. Heap leach operations that spray or drip cyanide onto the flattened top of the ore heap require solution processing ponds of about 1 ha in surface area. Puddles of various sizes occur on the top of heaps, where the highest concentrations of NaCN are found. Exposed solution recovery channels are usually constructed at the base of leach heaps. All these cyanide-containing water bodies are hazardous to wildlife, especially migratory waterfowl and bats, if not properly managed. Accidental spills of cyanide solutions into rivers and streams have produced massive killings of fish and other aquatic biota. Freshwater fish are the most cyanide-sensitive group of aquatic organisms tested, with high mortality documented at free cyanide concentrations >20.0 μg/L and adverse effects on swimming and reproduction at >5.0 μg/L.

LEAD[a]

Chapter 18

18.1 Introduction

Lead (Pb) has been known for centuries to be a cumulative metabolic poison; however, acute exposure is lessening. Of greater concern is the possibility that continuous exposure to low concentrations of the metal as a consequence of widespread environmental contamination may result in adverse health effects. Environmental pollution from lead is now so high that body burdens in the general human population are closer than the burdens of any other toxic chemical to those that produce clinical poisoning. Further, lead is a mutagen and teratogen when absorbed in excessive amounts, has carcinogenic or cocarcinogenic properties, impairs reproduction and liver and thyroid functions, and interferes with resistance to infectious diseases. In the U.S., about 9% of the children aged between 1 and 5 years have blood lead levels >100.0 μg/L, high enough to produce adverse health effects. Among non-Hispanic black children in this age group, about 21% had blood lead levels >100.0 μg/L and 4.2% had >200.0 μg/L. In some undeveloped Asiatic and eastern European countries blood lead concentrations >1000.0 μg/L occur in children living near lead smelters.

Ecological and toxicological aspects of lead and its compounds in the environment have been extensively reviewed. There is agreement by all authorities on five points. First, lead is ubiquitous and is a characteristic trace constituent in rocks, soils, water, plants, animals, and air. Second, more than 4 million metric tons of lead is produced worldwide each year, mostly for the manufacture of storage batteries, gasoline additives, pigments, alloys, and ammunition. The widespread broadcasting of lead through anthropogenic activities, especially during the past 50 years, has resulted in an increase in lead residues throughout the environment – an increase that has dislocated the equilibrium of the biogeochemical cycle of lead. Third, lead is neither essential nor beneficial to living organisms; all existing data show that its metabolic effects are adverse. Fourth, lead is toxic in most of its chemical forms and can be incorporated into the body by inhalation, ingestion, dermal absorption, and placental transfer to the fetus. Fifth, lead is an accumulative metabolic poison that affects behavior, as well as the hematopoietic, vascular, nervous, renal, and reproductive systems. In humans, lead causes stillbirths, miscarriages, inhibited development of fetuses, decreased male fertility, and abnormal sperm. Severe damage to the central nervous system from exposure to large amounts of lead may result in stupor, convulsions, coma, and death. Children who survive lead poisoning are often permanently retarded or have permanent neurological handicaps. At subclinical injury levels, lead causes slight, but irreversible, damage to the brain development of growing children.

Natural resources are also affected by environmental lead contamination, and some wildlife species numbers may be reduced, as a result. For example, waterfowl deaths resulting from the ingestion of spent lead shot pellets from shotgun shells were discovered more

---

[a] All information in this chapter is referenced in the following sources:

Eisler, R. 1988. Lead hazards to fish, wildlife, and invertebrates: a synoptic review. *U.S. Fish Wildl. Serv. Biol. Rep. 85(1.14)*, 1–134.

Eisler, R. 2000. Lead. Pages 201–311 in *Handbook of Chemical Risk Assessment: Health Hazards to Humans, Plants, and Animals. Vol. 1, Metals*. Lewis Publishers, Boca Raton, FL.

than 100 years ago in Italy and in the U.S. Since then, lead poisoning of waterfowl has occurred in 20 countries. In North America alone, approximately 3000 tons of lead shot are expended annually into lakes, marshes, and estuaries by several million waterfowl hunters. Spent pellets are eaten by waterfowl and other birds, either mistaken for seeds or as pieces of grit. These pellets may be retained in the gizzard for weeks, where they are reduced chemically and mechanically, form soluble toxic salts, and cause characteristic signs of lead intoxication, especially lethargy and emaciation. At least 2% of all North American waterfowl – or about 2 million ducks and geese – die each year as a direct result of ingestion of lead shot. These deaths contribute to the decline of some species, such as the canvasback, pintail, northern pintail, black duck, common pochard, spectacled eider, graylag goose, and mute swan. About seven times more waterfowl died from lead toxicosis as a result of ingesting spent pellets than from wounding by hunters. In addition, lead-poisoned waterfowl show delayed mortality from lead-induced starvation, are readily captured by predators, are susceptible to disease, and reproduce poorly. Susceptibility is markedly influenced by species, by the number and size of shot ingested, and by the types of food eaten. Swans are among the more vulnerable waterfowl. In England, lead poisoning through the ingestion of discarded lead fishing sinkers was the major cause of death in the mute swan; for all species of swans in England, about half died as a direct result of lead poisoning. In Washington State, 30% of the endangered trumpeter swans found dead had died of lead poisoning from ingestion of lead shot. Fatal lead poisoning of swans through ingestion of shotgun pellets, fishing sinkers, and lead-contaminated sediments is reported in several geographic locations. Lead toxicosis from ingested fishing sinkers was the most common cause of death in adult breeding- loons in New England, and may be an important factor limiting loons.

Lead poisoning mortality has occurred in over 30 species of birds other than waterfowl. Lead toxicosis caused by ingestion of spent shot and other lead objects was reported for the whooping crane, sandhill crane, mourning dove, and wild turkey. Secondary poisoning has been documented in at least six species of raptors that ate food containing lead shot (especially hunter-wounded animals): golden eagle, Andean condor, bald eagle, honey buzzard, king vulture, and California condor.

The availability of lead-based paints, discarded oil filters, used crankcase oil, lead storage batteries, or pastures contaminated by industrial lead operations make lead one of the most common causes of accidental poisoning in domestic animals. Cattle and horses in the vicinity of a lead smelter in California developed signs of lead poisoning, and many died between 1880, when the smelter opened, and 1971, when the smelter closed. Of the mules used in the early mining of lead, all died during their first year of service. Lead toxicosis has been reported in buffalos and cattle in India after they ate green fodder near a factory that recycled lead from old batteries. Total milk yield declined sharply, and stillbirths and abortions increased significantly in cattle that ingested lead-contaminated hay; the field from which the hay had been cut had a history of use for clay pigeon shoots and contained an estimated 3.6 tons of lead shot pellets. In sheep grazing in areas near lead mines, the frequency of abortions was high, and the learning behavior of the lambs was impaired. Many species of zoo animals, including monkeys, fruit-eating bats, and parrots, have been fatally poisoned from ingestion of flaking lead-based paint on the walls and bars of their cages. Ingestion of lead-based paint chips was one cause of epizootic mortality of fledgling Laysan albatross at Midway Atoll in 1983. At present, there is no known dietary requirement for lead in domestic animals, nor has it been shown unequivocally that lead plays any beneficial role. On the contrary, lead demonstrably and adversely affects weight, survival, behavior, litter size, and skeletal development, and induces teratogenic and carcinogenic responses in some species of experimental animals.

Lead is not essential for plants, and excessive amounts can cause growth inhibition, as well as reduced photosynthesis, mitosis, and water absorption. The decline of some European spruce forests has been attributed to excessive concentrations of lead in the atmosphere. Lead is toxic to all phyla of aquatic biota, though effects are modified significantly by various biological and abiotic variables. Wastes from lead mining activities have severely reduced or eliminated populations of fish and aquatic invertebrates, either directly through lethal toxicity or indirectly through toxicity to prey species. Health advisories warning anglers against eating lead-contaminated fish have been posted in Missouri. The significant increases in lead concentration shown by marine corals between 1954 and 1980 were representative of the increases noted in other biota as a direct result of increased global lead availability during that period.

18.2 Sources and Uses

Lead is a comparatively rare metal, with an average abundance in the earth's crust of 16.0 mg/kg; it is also a major constituent of more than 200 identified minerals, of which only 3 are sufficiently abundant to form mineral deposits: galena (PbS), anglesite ($PbSO_4$), and cerusite ($PbCO_3$). Galena, the primary form of lead in the natural state, is often associated with sphalerite (ZnS), pyrite (FeS_2), chalcopyrite ($CuFeS_2$), and other sulfur salts. Most (88%) of the domestic primary lead production originates from stratabound deposits in southeastern Missouri, another 8% from Idaho's Couer D'Alene district, and the rest from deposits in Colorado and Utah. Primary lead is smelted and refined at plants in Texas, Montana, Nebraska, Missouri, and Idaho. Scrap or secondary lead accounted for about half the domestic consumption in 1978; by 1980, more lead was produced from secondary sources than from domestic ores. In 1989, nine mines in southeastern Missouri produced 89% of the total domestic output of lead ores and concentrates; other mines in Idaho, Alaska, and Montana contributed 10.4% of the domestic output; the remainder was a byproduct from the mining of other commodities in Arizona, California, Colorado, Idaho, Illinois, Nevada, New Mexico, Tennessee, and Utah.

About 4 million tons of lead is refined annually worldwide. Domestic lead consumption is 1.3 million tons annually, of which about half is used in storage battery manufacture and, until recently, about 20% in the manufacture of gasoline antiknock additives such as tetramethyl lead (TML) and tetraethyl lead (TEL). Domestic use in 1990 was 1.28 million tons, most of which was used in the manufacture of lead-acid storage batteries (80%), as well as ammunition and metal alloys (12.4%), and the rest in ceramics, ballast, oxides, and gasoline additives. New uses of lead that seem environmentally innocuous include protection shielding against radiation exposure in computers, televisions, and certain medical instruments; in lead alloy solders; in super conductors; to manufacture certain ceramics and precision glass products; and in energy generation. Lead enters the atmosphere mainly through smelter emissions, primarily as $PbSO_4$ and $PbO\text{-}PbSO_4$, and through vehicle emissions, which include unburned lead, TEL, TML, and various lead halides, sulfates, phosphates, and oxides. The primary method of lead disposal is recycling, estimated at 50,000 tons annually in the U.S. Domestically, 70–75% of the consumed lead is considered to be recyclable, mostly from lead-acid storage batteries. Municipal and hazardous waste landfills receive about 25,000 tons of lead wastes annually, usually from ammunition and ordnance.

Lead and its compounds have been known to humans for about 7000 years, and lead poisoning has occurred for at least 2500 years. In Egypt, between 5000 and 7000 BCE, lead was used for glazing pottery, solder, ornaments, net sinker, anchors, caulking, coins, weights, aqueducts, piping, and cooking utensils. The biocidal properties of lead were familiar to the ancient Egyptians, and lead salts were sometimes used by them for homicidal purposes. Lead encephalopathy (inflammation of the brain) has been recognized since 400 BCE among workers in the lead trades; initial symptoms are dullness, irritability, ataxia, headaches, memory loss, and restlessness. These symptoms often progressed to delirium, mania, coma, convulsions, and sometimes death. The same general effects were described in young children and infants, among whom mortality rate was sometimes 40%. Extensive use of lead by the Romans, in pipes for water transport, in cosmetics, and as a wine sweetener, is estimated to have increased environmental lead levels to about five times the existing background levels. The decline of the Roman Empire may have been hastened by endemic lead poisoning – a theory supported by residue data showing high lead concentrations in bones and remains of Roman aristocrats – perhaps through ingestion of excessive amounts of wine laced with lead. After the fall of the Romans, the use of lead declined sharply. In the 14th century, gunpowder was introduced into Europe and was the impetus for the development of a weapon that fired a malleable metal pellet: a lead shot. Otherwise, the metal's resistance to corrosion led to its use as lead sheets applied as roofing for cathedrals and as protective encasement of underground pillars.

In 1721, the first lead mine was established in the New World by English settlers at Falling Creek, Virginia, primarily to supply bullets and shot. By 1750, European and British lead smelting operations were flourishing. In 1763, lead deposits in southeastern Missouri were permanently opened. The 18th century's Industrial Revolution produced an estimated tenfold increase in existing lead background levels. In the late 1700s, symptoms of acute lead poisoning recorded among industrial workers were called "Mill Reek" or "Devonshire Colic." Lead poisoning in Mexico was documented in 1878 among users of lead salts for glazing ceramics, a practice that persists even today. Lead poisoning was frequently recorded among U.S. lead miners in 1870–1900, especially in Utah, Colorado, and New Mexico. By 1880, the U.S. had surpassed Germany and Spain in the mining and refining of lead, and has continued as the leader in the output of refined lead. Air pollution from combustion of leaded gasoline containing TEL rose in the 1920s. In the mid-1940s, atmospheric lead concentrations increased sharply due to massive increases in lead emissions from automobiles; since then, increased lead emissions to the atmosphere have matched trends in gasoline lead content and consumption. In 1957, the U.S. was overtaken by Australia and the USSR in domestic mine production of lead; however, in 1967, the opening of the "New Lead Belt" in Missouri revived mining in the U.S., and subsequently lead was produced at the annual rate of 450,000–550,000 metric tons. In 1975, the U.S. was again the leading lead producer from mine sources, accounting for 16% of the world total; at that time, about 70% of the world lead production came from the U.S., the USSR, Australia, Canada, Peru, Mexico, China, Yugoslavia, and Bulgaria. In 1986, world mine production of lead was 2,352,000 tons of which the U.S.'s mine production was 353,000 tons, or 15% of the world total, and production in Missouri was 308,000 tons, or 87% of the U.S. total. In 1987, the leading lead producers of the world produced 2,110,000 metric tons: the USSR produced 24% of the total, Australia 23%, Canada 20%, the U.S. 15%, Mexico 9%, and Peru 9%.

Human exposure to lead sources is highly variable. In Mexico, for example, lead exposure is comparatively elevated in workers who manufacture or use lead-glazed pottery, urban populations exposed to high air lead concentrations due to the continued use of lead fuel additives, workers in industries that manufacture ballast and pigments, consumers who routinely eat canned foods such as hot peppers (2.4 mg Pb/kg) and fruit products (2.1 mg Pb/kg), and for the general population living in the vicinity of smelters, refineries, and other industries that emit lead. In the U.S., consumption of leaded gasoline in automobiles was completely phased out in 1980; however, leaded gasoline is still available at airport pumps for propelled aircraft and this as well as leaky underground fuel storage tanks comprise potential sources of human exposure.

18.3 Chemical Properties

Elemental lead is a bluish-gray, soft metal of atomic weight 207.19 and atomic number 82; it melts at 327.5°C, boils at 1749°C, and has a density of 11.34 g/cm^3 at 25°C. Metallic lead is sparingly soluble in hard, basic waters to 30.0 μg/L, and up to 500.0 μg/L in soft, acidic waters. Lead has four stable isotopes: Pb-204

(1.5%), Pb-206 (23.6%), Pb-207 (22.6%), and Pb-208 (52.3%). Of its 24 radioactive isotopes, two (Pb-210, Tb1/2 of 22 years; Pb-212, Tb 2 of 10 h) have been used in tracer experiments. Lead occurs in four valence states: elemental (Pb°), monovalent (Pb^{+}), divalent (Pb^{2+}), and tetravalent (Pb^{4+}); all forms are environmentally important, except possibly Pb^{+}. In nature, lead occurs mainly as Pb^{2+}; it is oxidized to Pb^{4+} only under strong oxidizing conditions, and few simple compounds of Pb^{4+}, other than PbO_2, are stable. Some lead salts are comparatively soluble in water (lead acetate, 443.0 g/L; lead nitrate, 565.0 g/L; lead chloride, 9.9 g/L), whereas others are only sparingly soluble lead sulfate, (42.5 mg/L; lead oxide, 17.0 mg/L; lead sulfide, 0.86 mg/L); solubility is greatest at elevated temperatures in the range 0.0–40.0°C. Of the organoleads, tetraethyl lead (TEL) and tetramethyl lead (TML) are the most stable and the most important because of their widespread use as anti-knock fuel additives. Both are clear, colorless, volatile liquids, highly soluble in many organic solvents; however, solubility in water is only 0.18 mg/L for TEL, and 18.0 mg/L for TML. Boiling points are 199.0°C for TEL and 110.0 C for TML; both undergo photochemical degradation in the atmosphere to elemental lead and free organic radicals, although the fate of automotive organoleads has yet to be fully evaluated.

Lead chemistry is complex. In water, for example, lead is most soluble and bioavailable under conditions of low pH, low organic content, low concentrations of suspended sediments, and low concentrations of the salts of calcium, iron, manganese, zinc, and cadmium. Accordingly, solubility of lead is low in water, except in areas of local point source discharges. Lead and its compounds tend to concentrate in the water surface microlayer (i.e., the upper 0.3 mm), especially when surface organic materials are present in thin films. Organolead compounds are generally of anthropogenic origin and are found mostly in the aquatic environment as contaminants; however, some organolead complexes form naturally, and their rate of formation may be affected by man-made organoleads. In surface waters, lead exists in three forms: dissolved labile (e.g., Pb^{2+}, $PbOH^{+}$, $PbCO_3$), dissolved bound (e.g., colloids or strong complexes), or as a particulate. The labile forms represent a significant part of the lead input from washout of atmospheric deposits, whereas particulate and bound forms are common in urban runoff and ore-mining effluents. The solubility of lead compounds in water is pH dependent, and ranges from about 10.0 g Pb/L at pH 5.5–<1.0 μg Pb/L at pH 9.0; little detectable lead remains in solution at pH >8.0. At pH 6.5 and water alkalinity of 25.0 mg $CaCO_3$/L, elemental Pb^{2+} is soluble to 330.0 μg/L; however, Pb^{2+} under the same conditions is soluble to 1000.0 μg/L. In acidic waters, the common forms of dissolved lead are salts of $PbSO_4$ and $PbCl_4$, ionic lead, cationic forms of lead hydroxide, and (to a lesser extent) the ordinary hydroxide $(PbOH)_2$. In alkaline waters, common species include the anionic forms of lead carbonate and hydroxide, and the hydroxide species present in acidic waters. Unfortunately, the little direct information available about the speciation of lead in natural aqueous solutions has seriously limited our understanding of lead transport and removal mechanisms.

Most lead entering natural waters is precipitated to the sediment bed as carbonates or hydroxides. Lead is readily precipitated by many common anions; desorption and replacement by other cations is extremely slow. In some acidic lakes, the deposition of particulate lead was strongly correlated with the deposition of aluminum and carbon, especially during periods of increasing pH. Precipitation of sparingly soluble lead compounds is not a primary factor controlling the concentration of dissolved lead in stream waters. Migration and speciation of lead was strongly affected by water flow rate, increasing flow rate resulting in increased concentrations of particulate and labile lead and a decrease in bound forms. At low stream flow, lead was rapidly removed from the water column by sedimentation.

In the sediments, lead is mobilized and released when the pH decreases suddenly or ionic composition changes. However, there was no significant release of lead from dredge spoils suspended in estuarine waters of different salinities for 4 weeks. Some Pb^{2+} in sediments may be transformed to tetraalkyl lead compounds, including TML, through chemical and microbial processes. There is also the possibility of methylation of ionic lead in vivo by fish and other aquatic biota, but the mechanisms are unclear. Methylation of lead in sediments was positively related to increasing temperatures, reduced pH, and microbial activity, but seemed to be independent of lead concentration. In general, the concentration of tetraalkyl leads in sediments is low, representing less than 10% of total lead.

Degradation of tetraethyl lead in soils not contaminated with gasoline is complete in 14 days. In subsoils contaminated with lead gasoline, however, 4–17% of the applied TEL was still present after 77 days. The retardation of TEL degradation in leaded gasoline-contaminated soils is attributed to the presence of gasoline hydrocarbons. As long as gasoline hydrocarbons remain in the soil, TEL may also remain, most likely in the gasoline hydrocarbon phase.

18.4 Mode of Action

Lead modifies the function and structure of kidney, bone, the central nervous system, and the hematopoietic system and produces adverse biochemical, histopathological, neuropsychological, fetotoxic, teratogenic, and reproductive effects. The mechanism underlying lead-induced growth suppression is thought to involve disruption of pituitary growth hormone during puberty. Inorganic lead absorbed into the mammalian body enters the bloodstream initially and attaches to the red blood cell. There is a further rapid distribution of the lead between blood extracellular fluid and other storage sites that is so rapid that only about half the freshly absorbed lead remains in the blood after a few minutes. The storage sites for lead are uncertain, although they are probably in soft tissues as well as bone; the half-time residence life (Tb1/2) of inorganic lead is estimated to be 20 days in blood, 28 days in whole body, and 600–3000 days in bone. Lead levels in bone exert an influence on plasma lead levels. There is concern that previously accumulated lead stores in bone may constitute an internal source of exposure, particularly during periods of increased bone mineral loss, such as pregnancy or lactation. Inorganic lead in the environment can be biologically methylated to produce alkyl lead compounds. Bile is an important route of excretion; ingested lead probably proceeds sequentially from gut, to blood, to bone and soft tissue, and by way of the bile to small intestine and fecal excretion.

Tetraalkyl lead mode of action differs from that of inorganic lead. Although initial entry is still into the bloodstream, the lead is evenly distributed between blood plasma and the red blood cells. Tetraalkyl leads are lost rapidly from the bloodstream, although some reappear in 5–10 h associated exclusively with the red blood cells, probably as trialkyl lead, though a fraction may be converted to inorganic lead. The organoleads concentrate in liver, and it is there that tetraalkyl lead is probably converted to trialkyl lead. Otherwise, the lead is widely dispersed throughout the body with Tb1/2 values of 200–350 days. Tetraethyl lead, by virtue of its high solubility in lipids, is rapidly accumulated in non-bony tissues, particularly the brain, where the onset of signs of poisoning is rapid. Short-term repeated exposures of rats to TEL results in a neurotoxic syndrome consisting of altered reactivity to noxious stimulation through disruption of forebrain-area function. In marine systems, unstable TEL is rapidly accumulated by marine organisms, probably through controlled diffusion processes. Once assimilated, this compound can be dealkylated into an ionized chemical species and react at the molecular level within the cells. Several fish species metabolize tetraalkyl leads to trialkyl lead compounds by way of their mixed function oxidase system. The trialkyl lead derivatives are considered responsible for the toxicity of the parent compound. Trialkyl leads and dialkyl leads rapidly traverse biological membranes in bird eggs and accumulate in the yolk and developing embryo. At present, the organolead mode of action is poorly understood, but organolead compounds are known to inhibit amino acid transport, uncouple oxidative phosphorylation, and inhibit cerebral glucose metabolism.

Biochemically, lead exerts deleterious effects on hematopoiesis through derangement of hemoglobin synthesis, resulting in a shortened life span of circulating erythrocytes, often resulting in anemia. Two essential enzymes in heme formation that are extremely sensitive to lead are delta aminolevulinic acid dehydratase (ALAD), which catalyzes the dehydration of delta amino levulinic acid (ALA) to form porphobilinogen (PBG), and ferrochelatase (= heme synthetase), which catalyzes the insertion of Fe^{2+} into protoporphyrin IX (PP). This second reaction requires the presence of glutathione and ascorbic acid. Some of the intermediates in heme follow sequentially: ALA, PBG, uroporphyrinogen III, coproporphyrinogen III, protoporphyrinogen IX, and PP. It is now well established that ALAD and ferrochelatase are the most sensitive biochemical indicators of lead exposure, the net result being lowered ALAD activity and elevated PP activity. An increase in glutathione content observed in lead-exposed fish may be due to an increase in glutathione synthesis rather than to a decrease in the use of the tripeptide or to its sequestration by lead.

Inhibition of blood ALAD activity after exposure to lead has been documented in many species of freshwater and marine teleosts, in freshwater crustaceans, in ducks, quail, doves, swallows, raptors, and songbirds, and in humans, sheep, mice, rats, rabbits, and calves. Lead-induced ALAD inhibition has been recorded not only in blood, but also in brain, spleen, liver, kidney, and bone marrow. Time for ALAD recovery to normal levels is dose dependent, organ specific, and usually directly correlated with blood lead concentrations. ALAD activity levels in lead-stressed teleosts were normal 3–11.7 weeks post-administration; this range was 2–14 weeks in birds. The physiological significance of depressed blood ALAD activity levels, except perhaps as an early indicator of lead exposure, is debatable. Aside from a few instances of moderate anemia in workers at lead smelters, other abnormalities noted were not regarded as serious. Lead-induced depression in ALAD activity in mallard ducklings and ring-necked pheasant chicks was not associated with signs of overt toxicity; a similar case is made for lead-stressed domestic chickens showing 98% reduction in ALAD activity, and for American kestrel showing an 80% reduction. Birds may be more sensitive than mammals to lead-induced depressions in blood ALAD activity. In ducks, for example, inhibition of ALAD would be more harmful than a comparable depression in mammals, for three reasons: first, metabolic activity is greater in nucleated duck erythrocytes than in human erythrocytes; second, ducks require porphyrin synthesis not only for hemoglobin production (as in humans), but also for production of respiratory heme-containing enzymes; finally, the half-life of erythrocytes is shorter in ducks than in humans: 40 days vs. 120 days.

Elevated blood protoporphyrin IX activity resulting from lead-inhibition of heme synthetase has been documented for humans and small mammals and for many species of birds; recovery to normal levels occurs in a lead-free environment in 2–7 weeks. The blood protoporphyrin IX technique is preferable to that of ALAD inhibition as a means of measuring lead stress because of its comparative simplicity and lower cost.

Other chemical changes that have been observed as a result of lead exposure include increased serum creatinine and serum alanine aminotransferase in birds, suggestive of kidney and liver alterations; changes in potassium, chloride, glucose metabolism and calcium binding proteins in rainbow trout; and a decrease in brain acetylcholinesterase activity in rats. In kidney, lead tends to accumulate in the proximal convoluted tubule cells of the renal cortex, producing morphological changes such as interstitial fibrosis, edema, and acid-fast intranuclear inclusion bodies, as well as biochemical changes. Renal intranuclear inclusion bodies occurred in 83% of mallards experimentally poisoned by dietary lead acetate or lead shot; similar results have been reported in other species of birds and in primates, cattle, and bats.

In a freshwater daphnid, about 90% of the total body lead burden is adsorbed to the exoskeleton. In animals with a vertebral column, total amounts of lead tend to increase with age; by far the most lead is bound to the skeleton, especially in areas of active bone formation. The retention of lead stored in bone pools poses a number of difficulties for the usual multicompartmental loss-rate models.

Some lead in bones of high medullary content, such as the femur and sternum, have relatively long retention times – i.e., Tb1/2 of >20 years in humans – whereas lead stored in bones of low medullary content have Tb1/2 values of 20–200 days, similar to the values for lead in soft tissues and blood. In birds, medullary bone undergoes sequences of bone formation and destruction associated with the storage and liberation of calcium during eggshell formation, indicating that sex and physiological condition primarily influence lead kinetics in avian bone. The use of diffusion models based on the exchange of lead between blood in canaliculi and the crystalline bone of the osteon may account for retention and bioavailability. More research is needed on the role of bone in lead kinetics.

Lead damages nerve cells and ganglia, and alters cell structure and enzyme function. Axonal degenerative changes, especially in neuronal cell bodies, were recorded in lead-poisoned freshwater snails, leading to altered protein synthesis. Mallards dosed orally with lead shot developed demyelinating lesions in vagal, branchial, and sciatic nerves, and showed vascular damage in the cerebellum; lesions were similar to those in lead-intoxicated guinea pigs, rats, and guinea hens. Crop stasis in birds, which is characterized by paralysis of the alimentary tract, impaction of food in the gizzard and proventriculus, and regurgitation of crop fluid, has been produced by lead shot or lead acetate solutions. Lead induces crop dysfunction by acting either directly on the smooth muscle or on associated nerve plexuses of crop tissue, depending on the route of administration. Mammals, including humans, undergo similar alimentary distress following intakes of lead.

Effects of lead on the nervous system are both structural and functional, involving the cerebellum, spinal cord, and motor and sensory nerves; the result may be deterioration of intellectual, sensory, neuromuscular, and psychological functions. The pathogenesis of lead-induced injury to the nervous system is poorly understood, but may be mediated through vascular damage, the direct action of lead on neurons, or alterations in porphyrin metabolism. Retarded brain growth in prenatal guinea pigs has been recorded at subclinical levels of lead (i.e., at concentrations producing no elevation in blood lead and no change in body weight), and this effect is potentiated at temperatures of 42.0°C. Lead may cause a transient disturbance in the blood-brain barrier during early postnatal growth of rats. This effect is associated with the presence of hemorrhagic lesions, suggesting focal damage to the vessels as an important event in the pathogenesis of lead encephalopathy to suckling rats. Brain histopathology has been recorded in lead-poisoned chickens and cattle. Brain lead concentrations are usually among the lowest in body organs, but the brain is one of the main sites of action. During chronic lead poisoning, distribution of lead in the brain is positively related to both dose and duration of exposure; preferential accumulation is in the hippocampus area of the brain. Significant amounts of lead persisted in rat brain tissue up to 4 weeks after the withdrawal of lead treatment. The role of organolead compounds in hippocampal function is largely unknown.

Absorption and retention of lead from the gastrointestinal tract, the major pathway of intake, varies widely because of the age, sex, and diet of the organism. Diet is the major modifier of lead absorption and of toxic effects in many species of domestic and laboratory animals, waterfowl, and aquatic organisms. In fact, the lack of certain major minerals in the diet often affected toxicity and storage of lead in tissue, more than did doubling the dosages of lead in the diet. Dietary deficiencies in calcium, zinc, iron, vitamin E, copper, thiamin, phosphorus, magnesium, fat, protein, minerals, and ascorbic acid increased lead absorption and its toxic effects. Toxic effects of lead-stressed fauna also were exacerbated when animals were fed diets containing excess cadmium, lactose, ethylenediaminetetraacetic acid, zinc, fat, protein, sodium citrate, ascorbate, amino acids, vitamin D, copper, mercury, fiber content, and nitrilotriacetic acid. Protection against various toxic effects of ingested lead was provided by measured dietary supplements of calcium, iron, zinc, ascorbic acid, vitamin E, and thiamin. Many other conditions affect lead absorption, including size of lead particle, type of lead compound ingested, presence of other compounds that act synergistically or antagonistically, and dosage. For example, smaller lead particles, <180.0 μm in diameter, were absorbed from the intestinal tract up to seven times more rapidly than larger particles of 180.0–250.0 μm. However, when large pieces of lead are ingested, such as lead shot, these may lodge in the gastrointestinal tract, dissolve slowly, and cause lead poisoning. Also, lead phthalates were absorbed more rapidly than carbonates, acetates, sulfides, and naphthanates, in that sequence. It is evident that all of these variables, as well as diet, need to be considered in risk assessment of lead.

18.5 Concentrations in Field Collections

Lead concentrations were usually highest in ecosystems nearest to lead mining, smelting, and refining activities; lead storage battery recycling plants; areas of high vehicular traffic; urban and industrialized areas; sewage and spoil disposal areas; dredging sites; and areas of heavy hunting pressure. In general, lead does not biomagnify in food chains. Older organisms usually contain the greatest body burdens, and lead accumulations are greatest in bony tissues. It seems that resources that are now at high risk (i.e., increased mortality, reduced growth, or impaired reproduction) from lead include the following: migratory waterfowl that congregate at heavily-hunted areas; raptors that eat hunter-wounded game; domestic livestock near smelters, refineries, and recycling plants; wildlife that forage extensively near heavily traveled roads; aquatic life in proximity to mining activities, lead arsenate pesticides, metal finishing industries, lead alkyl production, and lead aerosol fallout; and crops and invertebrates growing or living in lead-contaminated soils.

18.5.1 Abiotic Materials

Average lead concentrations in nonbiological materials worldwide were much higher in sediments (47,000.0 μg/kg), soils (16,000.0 μg/kg), and sediment interstitial waters (36.0 μg/L) than in atmospheric and other hydrospheric compartments (Table 18.1). Most of the lead discharged into surface waters is rapidly incorporated into suspended and bottom sediments, and most will ultimately be found in marine sediments. Sediments now constitute the largest global reservoir of lead; sediment interstitial waters and soils constitute secondary reservoirs.

Lead concentrations were elevated in certain nonbiological materials as a result of nonhunting human activities and natural processes. In sediments, lead concentrations ranged from 3.0 mg/kg in carbonate marls off the Florida coast to more than 11,000.0 mg/kg at Sorfjord, Norway, the site of massive discharges of lead-containing industrial and domestic wastes. Lead contaminates sediments from sources as diverse as steelworks, shipyards, crude oil refineries, cement and ceramic factories, lead storage battery recycling plants, and heavy traffic. Mining activities are also important. High concentrations of lead were measured in sediments (up to 2200.0 mg/kg) and detritus (up to 7000.0 mg/kg) of the Big River in southeastern Missouri. The Big River drains what was once the largest lead-mining district in the world; commercial mining was extensive between the early 1700s and 1972. During this period more than 200 metric tons of tailings accumulated within the Big River watershed as a result of seepage from tailings ponds, from erosion of tailings piles on the banks, and through accidental discharges.

In soils, lead concentrates in organic-rich surface horizons. In one instance, only 17.0 mg of soluble Pb/kg was found in soils days after the addition of 2784.0 mg of lead (as lead nitrate)/kg. The estimated residence time of lead in soils is about 20 years; complete turnover in topsoil is expected every few decades. In forest litter, however, the mean residence time of lead is lengthy; estimates range from 220 years to >500 years. Lead may leach from loamy soils of clay target shooting sites, where soils contain about 50,000.0 mg Pb/kg (about 40% Pb as particulate lead shot in the most contaminated areas); leachates at 100 mm depth contained up to 3.4 mg Pb/L vs. little or no lead in leachates collected

Table 18.1. Estimated amounts of lead in global reservoirs.

Reservoir	Concentration, in μg/kg	Total Lead in Pool, in Millions of Metric Tons
ATMOSPHERE	0.0035	0.018
LITHOSPHERE		
Soils	16,000.0	4800.0
Sediments	47,000.0	48,000,000.0
HYDROSPHERE		
Oceans	0.02	27.4
Sediment interstitial waters	36.0	12,000.0
Lakes and rivers	2.0	0.061
Glaciers	0.003	0.061
Groundwater	20.0	0.082
BIOSPHERE		
Land biota		
Living	100.0	0.083
Dead	3000.0	2.1
Marine biota		
Living	500.0	0.0008
Dead	2500.0	2.5
All freshwater biota	2500.0	0.825

from soil containing background concentrations of lead.

Lead deposited on roadways is removed in drainage water, and later accumulated in roadside soils. Amounts of lead in roadside soils are increased as a direct result of the combustion of gasoline containing organolead additives. In general, the amounts of lead were greatest along roads with the highest density of vehicular traffic, and amounts decreased rapidly with increasing distance from the roadway. Elevated levels of lead in soils also were recorded from the vicinity of storage battery reclamation plants, smelting activities, and mining and milling operations. Fly ash from coal burned in homes or privately hauled from power plants, which contains 100.0–450.0 mg Pb/kg, and is frequently used to reclaim land for the growth of forage and pasture crops, and as an alkaline amendment in the reclamation of strip-mined areas, is considered another source of soil lead. Two additional sources of lead in soils are municipal sewage sludge and lead-arsenate pesticides. Sewage sludge, which contains up to 100.0 mg Pb/kg and is applied as a fertilizer and soil conditioner at the rate of 50 million tons annually, may increase top soil levels by as much as 25.0 mg Pb/kg. Lead arsenate, a pesticide used to reduce bird hazards near airport runways by controlling earthworm abundance, and also to control pests in fruit orchards, represents another local source of lead contamination to soils.

Lead reaches the aquatic environment through industrial and municipal discharges, in atmospheric deposition, from weathering processes in areas of natural lead mineralization, and in highway runoff. Industrial lead input to aquatic environments is estimated at ten times that introduced by natural weathering processes; sewage and aerosols are the major sources. Snowmobile exhausts are considered a major source of lead in some locations; concentrations up to 135.0 μg Pb/L have been recorded in surface waters at the time of ice breakup. On the other hand, lead content in water (and sediments) of a fly ash settling pond of a coal-fired power plant did not increase as a result of plant operations. Isotopic lead ratios ($^{207}Pb/^{206}Pb$) and Pb/Ca atomic ratios can demonstrate anthropogenic perturbations of the lead cycle in present day food webs. As judged by these ratios, lead in teeth of contemporary California sea otters from Alaska have increased by 2- to 15-fold over their pre-industrial counterparts that died about 2000 years ago; lead in pre-industrial otters derived from natural deposits while lead in contemporary animals was primarily from Asia and western Canada and dominated by lead aerosols or industrial waste lead deposits.

Anthropogenic activities leading to increased air lead levels include primary and secondary lead smelting, the burning of gasoline containing lead antiknock agents, coal combustion, storage battery manufacture, and pigment production. It is generally agreed that combustion of leaded gasoline is the primary source of atmospheric lead. Atmospheric lead is usually attached to aerosols $<0.2\,\mu m$ in diameter, is efficiently scavenged by precipitation, has a short atmospheric residence time that is usually measured in days but may range up to 14 weeks depending on meteorological conditions, and may be transported long distances (i.e., hundreds or thousands of kilometers) from emitting sources. Along roadways, more than 90% of lead emissions are dispersed by the atmosphere away from the immediate vicinity of the road; air lead levels stabilize at low levels about 30 m from the road as a result of rapid settling of particles $>5.0\,\mu m$ in diameter, and from the downwind traverse of particles entrained in the turbulent atmosphere. Since 1970, the lead content in gasoline has decreased; profiles of lead in dated sediment cores and lead in atmospheric aerosols suggest that the environment is responding to decreasing use of leaded gasoline, and that atmospheric lead concentrations and fluxes will continue to decrease substantially if use of lead in gasoline is further decreased. Atmospheric deposition of lead to the Great Lakes in 1983 was conservatively estimated at 891 tons, with an upper limit of 1252 tons; these values are considerably lower than those from the early 1970s and may be due, in part, to the decrease in the lead content of gasoline.

18.5.2 Fungi, Mosses, and Lichens

Concentrations of lead were highest in specimens collected near metal smelters, lead mines, industrial areas, and urban locations. Lead concentrations were 9–13 times greater in a lichen collected in Washington, DC, in 1970 than in the same species collected 34 years earlier.

18.5.3 Terrestrial Plants

Elevated lead contents were recorded in various species of plants from the vicinity of metal smelters, roadsides, soils heavily contaminated with lead, natural ore deposits, and lead recycling factories. Bioavailability of lead in soils to plants is limited, but is enhanced by reduced soil pH, reduced content of organic matter and inorganic colloids, reduced iron oxide and phosphorus content, and increased amounts of lead in soils. Lead, when available, becomes associated with plants by way of active transport through roots and by absorption of lead that adheres to foliage. Lead concentrations were always higher in the older parts of plants than in shoots or flowers.

Damage to plants with elevated lead contents is usually negligible, but varies widely among species. Atmospheric lead may have contributed to the decline of European spruce forests. The mean lead content of needles and litter was significantly higher where tree decline was most pronounced than in areas where forests were unaffected. Lead can have deleterious effects on plant growth processes at reported lead levels in urban areas and may similarly affect plants in rural areas in the future. A reduction in yield of corn or soybeans is expected in low-binding capacity soils with lead levels greater than 200.0 mg/kg. Hay grown near roadsides may be toxic to horses and cattle. In extreme cases, reforestation has been initiated in areas where forage is so heavily contaminated with lead that it has become necessary to slaughter domestic livestock because the amounts of lead in their livers and kidneys became unacceptably high. Typical area reforestation includes removal of contaminated forage by cutting, bailing, and burying native grasses; burning of stubble and litter; and adding of agricultural lime at the rate of 2244 kg/ha (2000 pounds/acre) to all soils within 1525 m (5000 feet) of sites where lead levels exceed 175.0 mg/kg.

18.5.4 Terrestrial Invertebrates

In earthworms, lead levels were highest in those closest to highways and in areas with high volumes of traffic. Various species of insects and soil invertebrates from roadsides, from areas receiving sewage sludge, and from metal smelter environs also contain high amounts of lead. Amounts of lead in whole body were higher in earthworms, millipedes, and woodlice collected from soil and plant litter near highways than away from highways; soil and litter seem to be major reservoirs of lead in roadside communities. In contrast, lead concentrations in the eastern tent caterpillar were lower than those reported for roadside soil and litter invertebrates, and were about 76% of that in leaves of its host, the black cherry.

The use of terrestrial invertebrates as sentinel organisms has been suggested for monitoring lead. Some species of spiders, for example, contain lead body burdens that correlated with that in lichen used to monitor atmospheric lead. Similarly, the woodlouse seems to reflect lead concentrations in adjacent soil or leaf litter.

18.5.5 Aquatic Biota

Nationwide, there has been a significant decline in lead concentrations of whole freshwater fish from 1976–77 to 1984, continuing a decrease that became evident in 1978–79. Nationwide monitoring of freshwater fishes conducted periodically by the U.S. Fish and Wildlife Service showed that whole body lead burdens were highest for Atlantic coast streams, the Great Lakes drainage, the Mississippi River system, the Columbia River system, and in certain Hawaiian streams. Major sources of lead in Atlantic coast streams included wastes from metal finishing industries, brass manufacturing, lead alkyl production, primary and secondary lead smelting, coal combustion, and manufacture of lead oxide. For the Great Lakes, especially for the Lake St. Clair collection site, industrial sources and urban lead aerosol fallout from the Detroit area were major sources. In the Mississippi River system, naturally occurring deposits of lead ores, and effluents from zinc producers and industrial dischargers were prevalent. The Columbia River system was characterized by lead inputs from natural geologic deposits, industrial effluents, and the mining and smelting of lead. Hawaiian streams received most of their lead from urban runoff, vehicle sources, and agricultural and residential use of lead arsenate. Fish collected in 1979–1981 in the Big River, Missouri, near a ruptured tailings pond dam where lead concentrations in tailings approached 4000.0 mg/kg, contained greatly elevated whole body lead burdens of 9.0–18.0 mg/kg fresh weight (FW). Increased blood lead concentrations in longear sunfish from this area were associated with decreased blood ALAD activity and altered mechanical and biochemical properties of bone. By comparison, the highest lead concentration recorded to date in U.S. nationwide monitoring is 6.7 mg/kg FW in whole Mozambique tilapia from Honolulu in 1979. Suckers from contaminated portions of the Big River contained elevated blood lead levels, depressed blood ALAD activity levels, and lead concentrations in edible tissues exceeding 0.3 mg/kg FW–a level considered hazardous to human health. The Missouri Department of Health later issued an advisory against eating catostomids caught in a 65-km section of the Big River. Whitefish from lead-contaminated Swedish lakes showed depressed blood ALAD and blood chemistry derangement when compared to fish from a reference lake – suggesting that lead affects natural populations of fish in a manner similar to that observed in laboratory studies.

In general, freshwater algae, invertebrates, and fish had comparatively elevated lead concentrations when collected near industrialized areas, ponds with high numbers of lead shot, urban areas, lead mines, tailings ponds, areas of historic mining activities, and near highways. For marine biota, lead residues were generally highest where lead concentrations were high in the water: near bridges, near industrial disposal areas, near sewage and disposal areas, near dredging sites, and at mining sites. Lead concentrations in soft parts of American oysters were higher in smaller oysters than in larger oysters, much lower than the lead burden of the surrounding sediments, and strongly correlated with iron content of its geochemical environment. However, soft parts of larger bivalve mollusks from Greenland had higher lead burdens than did smaller bivalves. Lead concentrations in tissues of Arctic char from Austria were positively correlated with increasing water temperature, increasing age, and decreasing water alkalinity; enhanced lead accumulation during the summer was a consequence of increasing metabolic rate. The ability of parasitic acanthocephalans to accumulate high concentrations of lead may prove to be a useful indicator of lead contamination in the aquatic environment; however, the underlying mechanism of accumulation requires additional research.

Among aquatic biota, lead concentrations were usually highest in algae and benthic organisms, and lowest in upper trophic level predators. No significant biomagnification of lead occurs in aquatic food chains. Lead concentrations in cartilaginous and bony fishes – and also birds and mammals – were usually highest in areas of high human and vehicular density, and near lead mines and ore concentration plants. Lead concentrations in (aquatic and terrestrial) vertebrates tend to increase with increasing age of the organism, and to localize in hard tissues such as bone and teeth. In stream sediments, lead was highest in urban streams and lowest in the rural streams, reflecting lead inputs from storm runoff; species diversity was greater in the rural streams, due partly to lowered contaminant loadings, including lead.

The significance of organolead residues in aquatic life is unknown, and merits additional research. In Ontario, Canada, about 16% of all fish sampled contained tetraalkyl lead compounds, although none were recorded in water, vegetation, or sediments from the collection sites. Tetramethyl lead reportedly was produced from biological and chemical methylation of several inorganic and organic lead compounds in the aquatic environment, and has been detected at low concentrations in marine mussels, lobsters, and bony fishes.

18.5.6 Amphibians and Reptiles

Tadpoles of bullfrogs and green frogs from drainages along highways with different daily average traffic volumes (4272–108,800 vehicles per day) contained elevated amounts of lead (up to 270.0 mg/kg dry weight [DW]), which were positively correlated with lead in sediments and with average daily traffic volume. Lead in tadpoles living near highways may contribute to the lead levels reported in wildlife that eat tadpoles. Diets with amounts of lead similar to those in tadpoles collected near heavily traveled highways have caused adverse physiological and reproductive effects in some species of birds and mammals. Elevated lead concentrations also were recorded in various species of amphibians and reptiles collected near lead smelters and mines.

18.5.7 Birds

In general, lead concentrations were highest in birds from urban locations (perhaps reflecting greater exposure to automotive and industrial contamination) and in birds near lead mining and smelting facilities. Lead residues also are greatest in older birds (especially in bone, because of accumulation over time), in sexually mature females, and in waterfowl that have ingested lead shot pellets and other lead objects. In eastern Canada, elevated lead concentrations in wing bones of ducks are attributed to waterfowl hunting, nonferrous metal mining and smelting, and urban and industrial development. Ingestion of spent lead shot pellets from waterfowl hunting is the primary source of elevated lead exposure for wild ducks in eastern Canada; however, proximity to metal mining sites, especially silver, gold, copper, and zinc mines was also important.

Continued deposition of lead shot by hunters into wetlands habitats exposes birds to lead. Lead shot is a substantial localized source of contamination, especially in prime waterfowl habitat. Several million hunters are estimated to deposit more than 6000 metric tons of lead shot annually into lakes, marshes, and estuaries; this represents about 6440 pellets per bird bagged. Shot densities as great as 330,000–860,000 pellets/ha (up to 2,124,000/acre) have been estimated in some locations, and more than 2.6 million shot/ha in the Ebro Delta, Spain in 1992–93, although concentrations of 34,000–140,000/ha are more common. For example, lead shot in bottom sediments from Merrymeeting Bay, Maine, a prime waterfowl staging area, averaged 99,932 shot/ha (274,000/acre), and ranged from 59,541 to 140,324/ha; shot were significantly more numerous in silt than in sand sediments. In general, shot sink more rapidly in soft than in firm substrates, and there is only slight carryover of shot from one season to the next in areas with silt or peat bottoms. In agricultural soils, normal tillage practices may reduce lead shot availability by as much as 73%. In some locations, however, lead exposure and lead poisoning in waterfowl will continue to occur despite the conversion to steel shot for waterfowl hunting.

Waterfowl and other birds ingest spent shot during feeding and retain them as grit in the gizzard; the pellets are eroded and soluble lead is absorbed from the digestive tract. In many species, the ingestion of a single pellet is often fatal. Most deaths, however, go unnoticed and unrecorded. Species such as the mallard and pintail that feed in shallow water by sifting through bottom mud are more likely to encounter shot than species that feed on submerged vegetation or at the surface. Several techniques are used to determine the presence of ingested shot in gizzards of birds; these included fluoroscopy of intact gizzards, X-ray of gizzard contents (the most reliable), and microscopic visual examination. Shot ingestion in waterfowl is higher where shot densities in sediments are high and grit is absent, and higher during spring than in autumn for diving ducks. Incidence of ingested shot in 5 species of waterfowl harvested in Yucatan, Mexico in 1986–88 was 10.3%. Ingested lead shot frequency was 8.1% in American black ducks sampled in Maine during the hunting seasons of 1976–80. Gizzards of black ducks from Prince Edward Island, Canada, examined between 1988 and 1991 had a 7.1% frequency of ingested shot, and those from Wallace Bay, Nova Scotia in 1987 had a 13% frequency of ingested shot. In Dartmouth, Nova Scotia, in 1988, 96% of overwintering black ducks had elevated blood lead concentrations (>0.1 mg Pb/L) and 76% had detrimental concentrations in excess of 0.2 mg/L. In dry seasons, species that probe for food deep in the sediment are especially susceptible. In England, ingested pellets occurred in 3.2% of the total waterfowl in 16 species examined. Incidences of shot were relatively high (7.1–11.8%) in four species: greylag goose, gadwall, pochard, and tufted duck. A very high incidence (>22–100%) of shot ingestion was recorded in the gizzards of various species of hunter-shot waterfowl in Japan, the Netherlands, Greece, France, Spain, and Italy. At least 8000 mallards in Britain die each winter of lead toxicosis from ingestion of spent shot. It is estimated that about 2.4 million ducks die worldwide of lead shot poisoning each year – and this estimate does not include population losses resulting from the sublethal effects of lead. Among larger species of waterfowl, outbreaks of lead poisoning have been documented in Canada geese, whistling swans, trumpeter swans, and mute swans. Lead-poisoned waterfowl tend to seek seclusion and often die in areas of heavy cover; these carcasses are rapidly removed by predators and scavengers, and may result in secondary lead poisoning, especially among raptors such as the bald eagle. Of 293 bald eagles found dead nationwide between 1978 and 1981, (17 5.8%) probably died of lead poisoning after ingesting hunter-killed or hunter-crippled waterfowl containing lead pellets. Unlike the emaciated state of birds with chronic lead poisoning, bald eagles acutely poisoned by lead may be in good body condition if high concentrations of lead are rapidly absorbed.

Raptors are susceptible to lead poisoning wherever lead shot is used for hunting. As raptor populations are often small and reproduction rates low, their populations are particularly vulnerable to adult mortality. Lead exposure, especially poisoning, was a major factor affecting the California condor population during 1982–86. The probable source of lead was bullet fragments in carrion on which condors were feeding. Of 162 golden eagles sampled in California in 1985–86, 5.6% had elevated blood lead concentrations (>0.6 mg Pb/kg), and 2.5% had >1.0 mg Pb/kg blood; lead was probably from feeding on deer carcasses and offal from hunter-killed deer.

Elevated tissue lead concentrations in field collections of birds were associated with a variety of adverse effects. Lead concentrations in livers of the Canada goose and goldeneye found dead near mining and smelting activities in northern Idaho ranged from 8.0–38.0 mg Pb/kg FW; these levels exceeded the lower lethal limit of 5.0 mg Pb/kg FW liver in experimental birds. At the Ebro Delta in Spain in 1991–92, as many as 27% of the mallards had tissue lead concentrations sufficiently elevated to qualify as clinically lead-poisoned: >1.5 mg Pb/kg FW liver, >3.0 mg Pb/kg FW kidney. In Poland, livers of dead and sick nestlings of sparrows had >2.0 mg Pb/kg DW, and healthy birds <2.0 mg Pb/kg DW. Toxic lead concentrations (up to 70.0 mg/kg FW liver) in chicks of the Laysan albatross in Hawaii were associated with proximity to buildings and the presence of lead paint chips in the proventriculus. Elevated blood lead concentrations were associated with ingested shot in gizzards of diving ducks, black ducks, and Canada geese, and with reduced blood ALAD activity in mallards. Blood lead concentrations in Canada geese from the eastern prairie population in 1986–88 were below levels of recent lead exposure in areas where lead shot had never been used. In areas where lead shot was still in use, as many as 10% of the geese had blood lead levels >0.18 mg/L. The use of steel shot is substantially reducing the effect of lead poisoning in this population and other populations of geese and ducks.

Lead poisoning through the ingestion of anglers' lead weights was the major cause of death in the mute swan in England. Between 1983 and 1985, lead poisoning was most common in areas where the largest number of lead weights were found and accounted for at least 76% and perhaps up to 94% of the local swan deaths. In Ireland, about 60% of all mute swans found dead in 1984–86 died from lead poisoning; lead sources included spent gunshot from clay pigeon shooting sites in Northern Ireland and lost or discarded anglers' weights elsewhere. Ingestion of lead artifacts (shotgun pellets, fishing sinkers) accounted for about 20% of the known mortality of trumpeter swans in Idaho, Wyoming, and Montana, where the population has been declining for several decades. In Washington, the incidence of lead-induced mortality was higher and accounted for nearly 50% of the known mortalities. In Minnesota, lead poisoning of trumpeter swans from ingestion of shot was responsible for 23% of the documented deaths in 1987, and 54% in 1988–89. The increased mortality was attributed to drought conditions that lowered water levels and allowed swans to reach previously inaccessible lake bottoms containing spent lead shot. The treatment protocol for lead-poisoned trumpeter swans (blood Pb > 0.4 mg/L) included supportive therapy rehydration fluids, force feeding, injection of vitamins, iron salts, and 5-fluorocystine, chelation therapy calcium (EDTA), and lead removal from the gizzard using gastric lavage; treatment was successful in about 53% of the cases. Ingestion of spent shot was associated with mortality of black swans in South Australia, and up to 47% of all whooper swans in Scotland in 1980–86. Sediment ingestion by swans and other waterfowl is sometimes the main route of exposure to lead. Sediments – not shot or diet – was the primary source of lead ingested by tundra swans from the Coeur d'Alene River Basin in Idaho, as judged by fecal analysis. Sediment ingestion by tundra swans is associated with ingestion of wild rice. Mortality of tundra swans found dead near a lead mining complex in Idaho was attributed to ingestion of sediments that contained up to 8700.0 mg Pb/kg, and plants that contained up to 400.0 mg Pb/kg. Chelation therapy using sodium calcium edetate was successful in treating the clinical signs of lead poisoning (i.e., those with blood lead concentrations >0.4 mg/L) in mute swans in 49% of the cases. Subcutaneous injections of 0.25 mL/kg BW of a 25% w/v solution of sodium calcium edetate was given 1–3 times daily for at least 7 days. About 22% of treated swans survived for at least 2 years. However, results indicate that lead-poisoned swans, despite treatment, have a 59% reduction in survival when compared with untreated swans living in flocks.

Lead concentrations in feathers reflected population increases or declines of common terns from Long Island, New York. Lead in feathers of adult common terns decreased significantly from 5.6 mg/kg DW in 1978 to 1.0 mg/kg DW in 1985, were stable until 1988, and then increased (to 3.0 mg/kg DW) through 1992. Lead declines in feathers coincided with environmental decreases in lead from the phasing out of leaded gasoline in vehicles. Source of the increased lead from 1989 to 1994 is unclear, but may result from increased dredging in the New York area in the early 1990s or from increased amounts of lead paint removed from bridges during repainting operations in the late 1980s and early 1990s. There was a positive correlation between lead concentrations in maternal feathers of common terns and their eggs. Lead concentrations in feathers of pigeons from the Czech Republic were unaffected by season of collection, sex, area of collection, or diet; however, concentrations were lowest in nestlings and increased with increasing age.

The relation between embedded shot and lead toxicosis is unclear. The incidence of embedded shot in various species of waterfowl ranged from 11 to 43% in adults, and 2 to 11% in immature ones. Many birds that were struck by shotgun pellets but survived may have died prematurely or been eaten by predators. In one study, the bodies of 23% of adult Atlantic brant that died from starvation in New Jersey in 1977 contained embedded lead shot. The effects on survival and fecundity of receiving and carrying relatively high frequencies of embedded shot might be significant, and during years of low adult numbers might have substantial population consequences.

Lead in seeds and invertebrates within rights-of-way of major highways probably are not a hazard to adult ground-foraging songbirds, as judged from experiments with the European starling. However, the effects of lead on survival of fledglings are unknown, although lead causes reductions in blood hemoglobin, hematocrit, ALAD activity, and brain weight. In another study, lead concentrations in feather, carcass, and stomach contents of adult and nestling barn swallows were greater near a major U.S. highway than in a rural area; however, the number of eggs and nestlings, the body weight of nestlings at 17 days of age, and body weights of adults were similar in the two colonies, suggesting that contamination of roadsides with lead from automobile emissions is not a major hazard to birds that feed on flying invertebrates.

Signs of lead poisoning, i.e., depressed blood ALAD activity or elevated blood lead levels, were reported for birds near a metal smelter, in 17% of canvasbacks from Chesapeake Bay in 1974, and in three species of waders from the Dutch Wadden Sea living in an urban postnuptial molting area. In some species and locales, such as mallards from California national wildlife refuges, males had significantly higher blood lead concentrations than did females. Reduced body mass of 10% was documented in overwintering canvasbacks that had lead shot in their gizzards when compared to birds with no lead shot in gizzards. The decline in submerged aquatic vegetation in Chesapeake Bay and the later shift in diet of some waterfowl species of Chesapeake Bay from the vegetation lead content (2.2–18.9 mg/kg DW), to the softshell clam (1.3–7.6 mg Pb/kg DW), or to other bivalve mollusks (0.8–20.4 mg Pb/kg DW), probably did not increase dietary lead burdens in these species.

The significance of trace amounts of organolead residues in birds is unknown. Trialkyl lead seems to concentrate in avian kidney, but contributes less than 5% of the total amount of lead in kidneys.

18.5.8 Mammals

The highest body burdens of lead reported in mammals were near urban areas of dense vehicular traffic, near metal mines and smelters, adjacent to petrochemical refinery industries, or near plants that reclaimed storage batteries; concentrations were higher in older organisms, especially in bone and hematopoietic tissues. A similar pattern of lead occurrence and distribution was evident for human populations. In Italy, tissue lead concentrations in mice correlated positively with vehicular traffic densities; mice from the highest traffic flow areas also had genetic damage as judged by an increased frequency of micronucleated erythrocytes and of abnormal sperm cells.

Diet provides the major pathway for lead exposure, and amounts in bone are indicative of estimated lead exposure and metabolism. Amounts of whole body lead and feeding habits of roadside rodents were correlated: body burdens were highest in insectivores such as shrews; intermediate in herbivores, and lowest in granivores. Food chain biomagnification of lead, although uncommon in terrestrial communities, may be important for carnivorous marine mammals, such as the California sea lion; accumulations were highest in hard tissues, such as bone and teeth, and lowest in soft tissues, such as fat and muscle. A similar pattern was observed in the harbor seal.

The most sensitive index of lead intoxication in populations of deer mice was the formation of acid-fast-staining intranuclear inclusion bodies within proximal convoluted tubule cells of kidney; secondary indicators included decreased body weight, renal edema, reticulocytosis, increased urinary ALA excretion, and decreased hematocrit. It was concluded that lead pollution from automobile exhausts has had little impact on deer mice, and that severe lead poisoning is unlikely at traffic densities below 200,000 vehicles per day. Others, however, believe that lead emissions from automotive exhausts may pose unnecessary risks to various species of bats, rodents, and mule deer. Estimated doses of lead ingested by the little brown bat and highway populations of shrews and voles equaled or exceeded dosages that have caused death or reproductive impairment in domestic animals; further, mean lead concentrations in bats and shrews near highways exceed those reported for small rodents with lead-induced renal abnormalities collected from abandoned lead-mining sites. Mule deer from the Rocky Mountain National Park, Colorado, that graze on (heavily contaminated) roadside forage must consume 1.4% of their daily intake from roadsides before harmful amounts of lead (3.0 mg Pb/day) are obtained; however, this value needs to be verified.

Dairy cows adjacent to a lead battery reclamation plant showed signs of lead toxicosis, including muscle tremors, blindness, dribbling urine, and drooling. Mice trapped within 400 m of the plant had acid-fast-staining intranuclear inclusions in renal tubular epithelial cells – a useful diagnostic indicator of lead poisoning. A faulty air pollution control system at the plant caused deposition of particulate lead on the cornfield used for cattle forage, and was the probable source of the lead toxicosis in the animals. Industrial airborne lead pollution is responsible for contamination of cattle and horses within 1000 m of the source, resulting in elevated blood lead levels in both species, stillbirths and abortions in cattle, and some deaths in horses.

Proximity to the smokestacks of metal smelters is positively associated with increased levels of lead in the hair (manes) of horses and in tissues of small mammals, and is consistent with the results of soil and vegetation analyses. Lead concentrations were comparatively high in the hair of older or chronically impaired horses. However, tissues of white-tailed deer collected near a zinc smelter did not contain elevated levels of lead. Among small mammals near a metal smelter, blood ALAD activity was reduced in the white-footed mouse but was normal in others, e.g., the short-tailed shrew. The interaction effects of lead components in smelter emissions with other components, such as zinc, cadmium, and arsenic, are unresolved, and warrant additional research.

18.6 Lethal and Sublethal Effects

Lead adversely affects survival, growth, reproduction, development, and metabolism of most species under controlled conditions, but its effects are substantially modified by numerous physical, chemical, and biological variables. In general, organolead compounds are more toxic than inorganic lead compounds, food chain biomagnification of lead is negligible, and the younger, immature organisms are most susceptible. Uptake of lead by terrestrial plants is limited by the low bioavailability of lead from soils; adverse effects seem to occur only at total concentrations of several hundred mg Pb/kg soil. In aquatic environments, waterborne lead was the most toxic form. Adverse effects were noted on daphnid reproduction at 1.0 μg Pb^{2+}/L, on rainbow trout survival at 3.5 μg tetraethyl Pb/L, and on growth of marine algae at 5.1 μg Pb^{2+}/L. High bioconcentration factors were recorded for filter-feeding bivalve mollusks and freshwater algae at 5.0 μg Pb^{2+}/L.

Ingestion of spent lead shot by migratory waterfowl and other birds is a significant cause of mortality in these species, and also in raptors that eat the waterfowl killed or wounded by hunters. Forms of lead other than shot are unlikely to cause clinical signs of lead poisoning in birds, except for certain alkyl lead compounds that bioconcentrate in aquatic food items. Among sensitive species of birds, survival was reduced at doses of 75.0–150.0 mg Pb^{2+}/kg BW or 28.0 mg alkyl lead/kg BW, reproduction was impaired at dietary levels of 50.0 mg Pb^{2+}/kg, and signs of poisoning were evident at doses as low as 2.8 mg alkyl lead/kg BW.

The veterinary medical literature on lead toxicosis is abundant for domestic livestock and small laboratory animals, but notably lacking for feral mammals. Among sensitive species of mammals, survival was reduced at acute oral doses as low as 5.0 mg/kg BW in the rat, at chronic oral doses of 0.3 mg/kg BW in the dog, and at dietary levels of 1.7 mg Pb/kg BW in the horse. Sublethal effects were documented in monkeys given doses as low as 0.1 mg Pb/kg BW daily (impaired learning 2 years post-administration), or fed diets containing 0.5 mg Pb/kg (abnormal social behavior). Reduction in ALAD activity was recorded in blood of rabbits given 0.005 mg Pb/kg BW, and in mice given 0.05 mg Pb/kg BW. Tissue residues increased in mice given 0.03 mg Pb/kg BW, and in sheep given 0.05 mg Pb/kg BW.

18.6.1 Terrestrial Plants and Invertebrates

Fruits and vegetables acquire lead by surface deposition from rainfall, dust, and soil, and by biological uptake through the root system. Foliar absorption of lead and transport to the root could account for a significant portion of the lead in root tissues; however, this transport process varies widely among species. For example, this pathway accounted for 35% of the root lead content in the radish, but less than 3% in carrots and beans. Corn contained 30.0 mg Pb/kg DW when grown in soils containing lead concentrations of 924.0 mg/kg, but only 17.0 mg/kg when grown in soils containing 786.0 mg Pb/kg. It was concluded that contamination of soils with up to 800.0 mg Pb/kg probably does not elevate concentrations of lead in corn plants. Within any plant species, however, there are lead-resistant and lead-sensitive breeds; some genetically fixed resistant species grow in soils containing up to 10,000.0 mg Pb/kg.

Plants readily accumulate lead from soils of low pH or low organic content; however, uptake is significantly reduced after the application of lime or phosphate, which converts lead to hydroxides, carbonates, or phosphates of relatively low solubility. Lead persists for lengthy periods in forest litter; the estimated Tb1/2 is 220 years. High levels of lead persisted for at least 6 years in litter, soil, amphibians and mammals after zinc smelting was discontinued in Palmerton, Pennsylvania. Lead seems to be tightly bound by most soils, and substantial amounts must accumulate before it affects the growth of higher plants. Although lead is preferentially bound in soils by organics and oxides, interaction kinetics of lead with other metals are complex and largely unknown. For example, uptake of lead from soils by oat seeds was inhibited by cadmium salts, and reduced in loamy or organic soils; further, lead in soils interfered with manganese uptake, and also increased the availability of cadmium and other heavy metals.

Lead inhibits plant growth, reduces photosynthesis, and reduces mitosis and water absorption. Inhibition of photosynthesis is attributed to the blocking of protein sulfhydryl groups and to changes in phosphate levels in living cells. For two species of roadside weeds, pollen germination was reduced by 90% and seed germination by 87% at lead levels of about 500.0 mg/kg (DW) in soil and about 300.0 mg/kg (DW) in foliage. Normal germination rates were recorded at lead levels of 46.0 mg/kg in soil and 22.0 mg/kg (DW) in foliage; however, some adverse effects were evident at lead levels of 12.0–312.0 mg/kg in soil, and 55.0–97.0 mg/kg (DW) in foliage. Tetraethyl lead from automobile exhaust fumes is known to react in the light to produce the highly phytotoxic triethyl lead cation, which can freely permeate the plasma membranes of plant cells. Growth of cultures of soybean cells exposed to 207.0 μg Pb/L (as triethyl lead salts) was inhibited before the cells died. There is no evidence for biomagnification of lead in the food chain of vegetation, to cattle, to dung, to the dung beetle, nor is there convincing evidence that any terrestrial vegetation is important in food chain biomagnification of lead.

Concentrations of lead in soil litter ranged from 3200.0 mg/kg in locations near a zinc smelter in Palmerton, Pennsylvania, to 150.0 mg/kg at sites 105 km distant; relative concentrations of cadmium, zinc, and copper were similar. In woodlice fed litter from these locales for 8 weeks, survival decreased as metal content in the litter increased, but the major cause of death was zinc poisoning and not lead poisoning. Woodlouse hepatopancreas that were collected 3 km downwind of a metal smelter contained large amounts of zinc, copper, cadmium and lead. Centipedes that ate woodlice hepatopancreas did not assimilate lead even though the food contained concentrations that were many times greater than normally encountered. However, survival and reproduction were reduced in woodlice fed soil litter treated with 12,800.0 mg Pb/kg, as lead oxide, for 64 weeks, or two generations. This amount of lead is similar to the amounts reportedly associated with reductions in natural populations of decomposers, such as fungi, earthworms, and arthropods. The poisoning of decomposers may disrupt nutrient cycling, reduce the number of invertebrates available to other wildlife for food, and contribute to food chain contamination. The effect of lead on soil microbial populations is unknown.

Herbivorous land snails are important in lead cycling through contaminated ecosystems. Snails fed lettuce enriched with lead (about 1000.0 mg Pb/kg DW lettuce) for 32 days contained 1301.0 mg Pb/kg DW in the mid-gut gland (vs. 52.0 in controls), and much lower amounts (<30.0 mg/kg) in other tissues. After the snail had fed on uncontaminated lettuce for 50 days, lead remained elevated at 1203.0 mg/kg in the mid-gut gland, which contained more than 90% of the total body burden.

18.6.2 Aquatic Biota

In general, the responses of aquatic species to lead insult differed markedly. Among sensitive species, however, several trends were evident: (1) dissolved waterborne lead was the most toxic form; (2) organic lead compounds were more toxic than inorganic forms; (3) adverse effects on daphnid reproduction were evident at 1.0 μg Pb^{2+}/L; (4) high bioconcentrations were measured in oysters at 1.0 μg Pb/L and in freshwater algae at 5.0 μg Pb^{2+}/L; (5) tetramethyl lead was acutely toxic to rainbow trout at 3.5 μg/L; (6) growth inhibition of a marine alga was reported at 5.1 μg Pb^{2+}/L; and (7) for all species, effects were most pronounced at elevated water temperatures and reduced pH, in comparatively soft waters, in younger life stages, and after long exposures.

Lead is toxic to all phyla of aquatic biota, but its toxic action is modified by species and physiological state, and by physical and chemical variables. Only soluble waterborne lead is toxic to aquatic biota, and free cationic forms are more toxic than complexed forms. The biocidal properties of soluble lead are also modified significantly by water hardness: as hardness increased, lead becomes less bioavailable because of precipitation increases. In salmonids, for example, the toxicity and fate of lead are influenced by the calcium status of the organism, and this relation may account for the reduced effects of lead in hard or estuarine waters. In coho salmon, an increase in waterborne or dietary calcium reduced uptake and retention of lead in skin and skeleton.

Organolead compounds are, in general, more toxic than inorganic lead compounds to aquatic organisms. Ethyl derivatives were more toxic than methyl derivatives, and toxicity increased with increasing degree of alkylation, tetralkyl lead being the most toxic. Tetraethyl lead was about ten times more effective than tetramethyl lead in reducing oxygen consumption by coastal marine bacteria, and was 1.5–4 times more toxic than tetramethyl lead to marine teleosts. Tetramethyl lead chloride was 20 times as toxic as $Pb(NO_3)_2$ to freshwater algae, and twice as toxic as trimethyl lead acetate. In seawater, the release of tetraalkyl lead compounds is more likely than accumulation to result in acutely toxic effects; however, alkyl lead compounds degrade rapidly to trialkyl lead chlorides, which are only 0.1–0.01 as toxic as TEL compounds. Alkyl lead compounds are accumulated more readily by freshwater teleosts than are inorganic lead compounds. The BCF values for tetramethyl lead and rainbow trout, for example, ranged from 124 in lipids after exposure for 1 day, to 934 after 7 days. Depuration of tetramethyl lead is rapid; the estimated Tb1/2 values range from 30 h for intestinal lipids to 45 h for skin and cephalic fat deposits. Some microorganisms in lake sediments transform certain inorganic and organic lead compounds into the more toxic tetramethyl lead, but the pathways are not well understood.

Lethal solutions of lead (as well as of many other heavy metals) cause increased mucus formation in fishes. The excess coagulates over the entire body and is particularly prominent over the gills, interfering with respiratory function and resulting in death by anoxia. Increasing waterborne concentrations of lead over 10.0 μg/L is expected to provide increasingly severe long-term effects on fish and fisheries. Fish that are continuously exposed to toxic concentrations of waterborne lead show various signs of lead poisoning: spinal curvature, usually as lordosis; anemia; darkening of the dorsal tail region, producing a black-tail effect due to selective destruction of chromatophores but not of melanophores; degeneration of the caudal fin; destruction of spinal neurons; ALAD inhibition in erythrocytes, spleen, liver, and renal tissues; reduced ability to swim against a current; destruction of the respiratory epithelium; basophilic stippling of erythrocytes; elevated lead concentrations in blood, bone, gill, liver, and kidney; muscular atrophy; paralysis; renal pathology; growth inhibition; retardation of sexual maturity; altered blood and lipid chemistry; testicular and ovarian histopathology; and death. The prevalence of signs is closely correlated with duration of exposure to lead and to its uptake. Toxic effects of lead uptake in fishes are increased under conditions favoring their rapid growth. The rate of intoxication by lead – as judged by uptake rates into tissues and incidence and prevalence of black tail – did not increase with fish size, but rather with growth rate.

Rooted aquatic plants, such as wild rice, can accumulate up to 67 mg Pb/kg (DW) when cultured in tanks contaminated with high concentrations of powdered lead (equivalent to 7400 kg Pb/ha); however, this level is not considered hazardous to waterfowl feeding on wild rice. Lead content in plants collected from heavily hunted areas near refuges did not differ from those collected in the protected areas, which suggests that lead bioavailability to rooted aquatics is substantially lower from shot than from powdered lead. Rooted macrophytes, rapidly accumulated lead from solutions containing 1.0 mg Pb^{2+}/L, i.e., 70.0 mg Pb/kg DW per minute; the process was overwhelmingly passive. Depuration was rapid; 90% of the lead sorbed during the first hour by macrophyte shoots was released within 14 days after transfer to clean water, though 10% seemed to be irreversibly bound. Certain emergent aquatic plants, removed 97–98% of all soluble lead from solution in the vicinity of a lead battery site in Tampa, Florida, suggesting that phytoremediation may be feasible as the basis of a lead removal technology. High accumulations of lead from ambient seawater by marine plants is well documented; concentration factors vary from 13,000 to 82,000 for algae from Raritan Bay, New Jersey, and from 1200 to 26,000 for algae from Sorfjorden, Norway. Studies on the kinetics of lead uptake and retention in two species of marine algae showed that both species accumulated lead from the medium at ambient concentrations of 20.0 μg/L and higher. In the first phase, usually completed within minutes after addition of lead, one species became saturated when the lead reached a remarkable 11,640.0 mg/kg (DW). In the second phase, the lead content continued to rise slowly, but in the other species it declined after 2 or 3 days. In both species the content of bound lead increased with increasing exposure time, suggesting that during prolonged exposure lead is initially adsorbed to the cell surface, then translocated into the cell wall, the plasma membrane, and, eventually, the cytoplasm.

Sediments are not only sinks for lead but may act as a source of lead to aquatic biota after contamination from the original source has subsided. The uptake of lead from artificially contaminated pond sediments was recorded in roots and foliage of submersed aquatic macrophytes and in the exoskeleton of crayfish. Accumulation of lead in crayfish primarily was through adsorption; most was lost through molting, though some internal uptake and elimination occurred without molting. Crustacean molts represent 15% of the lead body burden and are probably more significant than fecal pellets in lead cycling processes. Uptake of lead from sediments containing 7.0 mg Pb/kg DW is reported for the stone loach, but this needs verification.

Median bioconcentration factor BCF values in aquatic biota exposed to various concentrations of Pb^{2+} for 14–140 days varied from about 42 in fish to 2570 in mussels; intermediate values were 536 for oysters, 500 for insects, 725 for algae, and 1700 for snails. There are several notable exceptions to this array: significantly higher values have been reported in crustacean hepatopancreas, in various species of freshwater invertebrates, in fish bone and liver, and in whole oysters. In oysters, for example, BCF values varied from 3450 to 6600 after exposure to solutions containing 1.0–3.3 μg Pb^{2+}/L for 140 days, but oysters and their progeny were apparently unaffected at whole body burdens (less shell) up to 11.4 mg Pb/kg DW. Many species of aquatic biota contain lead in amounts >1000.0 mg/kg FW (>10,000.0 mg/kg DW) including some marine seaweeds, freshwater macrophytes and algae, annelids, crustaceans, echinoderms, mollusks, and teleosts; presumably, the lead was sorbed passively and little, if any, was incorporated biologically. Variations in lead concentrations in aquatic biota probably reflect the ability of individual species to adsorb waterborne lead, and may be a direct function of the ratio of surface to body weight. The residence time of lead in aquatic biota seems to be related to the route of administration: Tb1/2 values were 9 days by waterborne routes and 40 days by diet.

Although lead is concentrated by biota from water, there is no convincing evidence that it is transferred through food chains. Lead concentrations tended to decrease markedly with increasing trophic level in both detritus-based and grazing aquatic food chains. In the marine food chain of seawater (<0.08 μg Pb/L), to a brown alga, to the red abalone (a marine gastropod), lead concentrations in the alga and abalone were both <0.04 mg Pb/kg FW after 6 months, indicating negligible biomagnification. When seawater contained 1000.0 μg Pb/L, young abalones that fed on algae for 6 months contained up to 21.0 mg Pb/kg FW, but neither growth nor activity was affected; lead selectively accumulated in the digestive gland (38.0 mg/kg), and was lowest in muscle(<l.0 mg/kg) – the part normally consumed by humans. In the freshwater food chain of an alga, to a daphnid, to the guppy, lead accumulation progressively decreased from the alga to the guppy. Thus, in organisms held for 28 days in solutions containing 5.0 μg Pb/L, lead content was 460.0 mg/kg DW in the alga, 23.0 mg/kg in the grazing daphnids, and 4.0–16.0 mg/kg in the guppies that fed on the daphnids. Concentrations of lead in a freshwater snail collected near an abandoned lead mine were positively correlated with the lead content in its diet; the digestive glands contained

up to 5600.0 mg/kg DW. The gut contents of eels grazing on contaminated snails contained up to 4350.0 mg Pb/kg, but the lead was rapidly released; feces from both snails and eels return the lead to the ecosystem as particulates and detritus.

As discussed earlier, lead clearly inhibits the formation of heme at several points, adversely affects blood chemistry, and accumulates in hematopoietic organs of aquatic organisms. In addition, lead interferes with chlorophyll formation in plants by inhibiting the conversion of coproporphyrinogen to proporphyrinogen by competing with iron, inhibits allantois formation in annelids, inhibits alpha-glycerophosphate dehydrogenase activity in trout, increases glutamic oxalacetate transaminase activity in daphnids, affects neural and hormonal systems that control activity and metabolic rates in fish, interacts with polar sites of glycoproteins in epidermal mucus of fish, and may inhibit vitamin C and tryptophan metabolism. Fish may be more resistant than mammals to lead. For example, isolated liver hepatocytes of channel catfish were about 40 times more resistant to lead than rat liver hepatocytes as judged by ALAD inhibition.

Some populations of freshwater isopods are tolerant to lead. Inasmuch as intolerant isopods from an unpolluted site can be made tolerant by exposure to low levels, it is suggested that naturally occurring tolerance may be achieved by acclimatization. Research is needed on lead transformation mechanisms, on toxic forms of lead and interaction effects with other compounds, and on effects of lead-contaminated sediments on benthos.

18.6.3 Amphibians and Reptiles

Lead poisoning in adult leopard frogs is indicated by a series of signs: sloughing of integument; sluggishness; decreased muscle tone; decreases in red blood cells, white blood cells, neutrophils, and monocytes; erosion of the gastric mucosa; and before death, excitement, salivation, and muscular twitching. The 30-day LC50 value for adult leopard frogs was 105.0 mg Pb/L, but some deaths and elevated liver residues were noted at water concentrations as low as 25.0 mg/L. Toad eggs are comparatively sensitive to ionic lead, with LC50 (48 h) values of 0.47–0.9 mg/L, and a high incidence of malformations in survivors; zinc seems to confer a degree of protection against lead toxicosis in toad eggs. In soft water (99 mg $CaCO_3$/L) some marbled salamanders exposed to 1.4 mg Pb/L died in 8 days. At about 1.0 mg/L, lead blocked synaptic transmission by competitive inhibition of calcium in the bullfrog. At 0.5 mg Pb/L, frog tadpoles required additional time to metamorphose; and at 1.5 mg Pb/L, thyroid histopathology was recorded and the delay in metamorphosis was more pronounced. Tadpoles of the bullfrog and the green frog do not avoid lead concentrations shown to produce behavioral deficiencies, i.e., >0.5 mg Pb/L. Lead interferes with the normal development of newly-fertilized eggs of toads at concentrations of 250.0 μg Pb/L and higher.

No data were available on toxic or sublethal effects of lead to reptiles under controlled conditions.

18.6.4 Birds

Lead poisoning resulting from the ingestion of lead shotgun pellets has been recognized as a cause of waterfowl deaths since the late 1800s. More than a million ducks – especially mallards – and geese die annually from lead shot poisoning. The principal cause is the ingestion of spent shot by migrating birds feeding in heavily hunted areas. The pellets are retrieved from the marshy bottoms of shallow and deep water by waterfowl in search of feed and grit. Shot retained in the gizzard is solubilized by a combination of the powerful muscular grinding action and the low pH (2.0–3.5) of gizzard contents. The released lead is available for absorption, producing weakened birds whose reproductive abilities are reduced and that may starve or fall prey to predators. There does not appear to be a no-effect level for lead in waterfowl, and the activities of certain enzymes are inhibited at blood lead concentrations of <50.0 μg/L. Ionic lead was 10–100 times more effective in reducing avian blood ALAD activity than were ionic copper, cadmium, and inorganic or methyl mercury. Absorbed lead causes a variety of effects leading to death, including damage to the nervous system, muscular paralysis, inhibition of heme synthesis, immunotoxic effects, and damage to kidneys and liver. Lead poisoning in waterfowl is a debilitating disease in which death follows exposure by an average of 2–3 weeks. During this time, affected birds lose mobility, tend to avoid other birds, and become increasingly susceptible to predation and other causes of mortality. Accordingly, acute large-scale die-offs of lead-poisoned waterfowl are uncommon.

The relation between incidence of lead shot in waterfowl gizzards and biological effects varies widely, and is probably a function of shot availability caused by differences in shooting intensity, size of pellets, availability of grit, firmness of soil and sediments, and depth of surface water. Also, lead accumulations and the frequency of avian lead toxicosis following ingestion of lead shot are modified by the age and sex of the bird, geographic location, habitat, and time of year. There is a growing body of evidence linking waterfowl poisoning with ingestion of lead-contaminated sediments, especially in certain areas of Idaho impacted by mining wastes. In the common bobwhite, tissue lead accumulation can occur from ingestion of lead-contaminated sediments, although there were no signs of overt lead toxicity.

The effect of diet on vulnerability to lead makes interpretation of published information on experimental lead poisoning in waterfowl extremely difficult. For example, many mallards on a diet of corn die within 10–14 days after ingesting a single lead shot, whereas similar birds on a balanced commercial duck ration appear outwardly normal after ingesting as many as 32 pellets of the same size. Also, multiple nutritional deficiencies may have additional effects in potentiating the toxicity of lead in mallards. Under conditions of reduced dietary calcium availability, such as can occur in acid-impacted environments, birds risk increased uptake of lead (and other metals) and may accumulate toxic concentrations more rapidly. Enhanced accumulation of lead was accompanied by an increased synthesis of metallothioneins and a greater inhibition of ALAD activity.

Birds of prey may ingest lead in the form of shot from dead or crippled game animals, or as biologically incorporated lead from lead-poisoned waterfowl, small roadside mammals, and invertebrates. Lead poisoning in carnivorous birds has been reported in various species of eagles, condors, hawks, harriers, owls, vultures, and falcons, and most – if not all – cases seem to result from ingestion of lead shot in food items. Some raptors ingest many shot in a short time. Thus, the stomach of a bald eagle suspected of dying from lead poisoning contained 75 shot. Results of experimental lead shot poisoning of bald eagles confirmed results of nationwide monitoring showing that 5.4% of all dead eagles found in 1974–1975 died of lead poisoning, as evidenced by liver lead levels of 23.0–38.0 mg/kg FW. Ingestion of food containing biologically incorporated lead, although contributing to the lead burden of carnivorous birds, is unlikely in itself to cause clinical lead poisoning. A similar case is made for lead residues in soil and biota following field applications of lead arsenate, powdered lead, and forms of lead other than shot; the strong indication is that the form in which lead is ingested is crucial.

Signs of lead poisoning in birds have been extensively documented. Outwardly, lead-poisoned birds show the following signs: loss of appetite, lethargy, weakness, emaciation, tremors, drooped wings, green liquid feces, and impaired locomotion, balance, and depth perception. Internally, lead-poisoned birds show microscopic lesions of the proventricular epithelium, pectoral muscles, brain, proximal tubular epithelium of the kidney, and bone medullary osteocytes; an enlarged bile-filled gall bladder; anemia; elevated protoporphyrin IX levels in blood; decreased ALAD activity levels in blood, brain, and liver; reduced brain weight; abnormal skeletal development; cephalic edema; and esophageal impaction. Postmortem examination of lead-poisoned birds may show edematous lungs; serous fluid in the pleural cavity; bile regurgitation; abnormal gizzard lining; an usually pale, emaciated, and dehydrated carcass; and elevated lead levels in liver (>2.0 mg/kg FW, >10.0 mg/kg DW), kidney (>6.0 mg/kg DW), and blood (>0.2 mg/L). The most reliable indicators of lead poisoning in waterfowl include impactions of the upper alimentary tract, submandibular edema, myocardial necrosis, biliary discoloration of the liver, and hepatic lead concentrations of at least 38.0 mg/kg DW or 10.0 mg/kg FW.

Toxic and sublethal effects of lead and its compounds on birds held under controlled conditions vary widely with species, with age and sex, and with form and dose of administered lead. Several generalizations are possible: decreased blood ALAD and increased protoporphyrin IX activity levels are useful early indicators of lead exposure; lead shot and certain organolead compounds are the most toxic forms of lead; nestlings are more sensitive than older stages; and tissue lead concentrations and pathology both increase in birds given multiple doses over extended periods. Blood lead concentrations of lead-poisoned waterfowl may be reduced with initiation of disodium calcium ethylenediamine tetraacetate therapy; however, ALAD levels remained depressed.

Trialkyl lead salts are 10–100 times more toxic to birds than are inorganic lead salts; they tend to accumulate in lipophilic soft tissues in the yolk and developing embryo, and have high potential as neurotoxicants; accordingly more research is needed on alkyl lead toxicokinetics. Some alkyl lead compounds have been implicated in bird killings. In autumn 1979, about 2400 birds of many species were found dead or disabled on the Mersey estuary, England, an important waterfowl and marsh bird wintering area; smaller killings were observed in 1980 and 1981. Affected birds contained elevated lead concentrations in liver (>7.5 mg/kg FW), mostly as organolead. It has been suggested that trialkyl lead compounds were discharged from a petrochemical factory producing alkyl leads, into the estuary where they were accumulated (up to 1.0 mg/kg FW) by clams and other invertebrates on which the birds could feed. Birds dosed experimentally with trialkyl lead compounds died with the same behavioral and internal signs found in Mersey casualties; tissue levels of trialkyl lead were similar in the two groups of birds. Sublethal effects that might influence survival in the wild were found in both sublethally dosed and apparently healthy wild birds when tissue levels of trialkyl lead compounds were matched in the two groups of birds. It was concluded that trialkyl lead compounds were the main cause of the observed mortalities and that many apparently healthy birds were still at risk.

Nestlings of altricial species (those confined to the nest for some time after hatch) may be considerably more sensitive to lead exposure than adults, and also more sensitive than hatchlings of many precocial species. Hatchlings of precocial species, including chickens, Japanese quail, mallards, and pheasants, are relatively tolerant to moderate lead exposure, i.e., there was no effect on growth at dietary levels of 500.0 mg Pb/kg, or survival at 2000.0 mg Pb/kg.

Some species of domestic birds are comparatively resistant to lead toxicosis. For example, blood lead levels of 3.2–3.8 mg/L in lead-stressed cockerels were much higher than residues considered diagnostic for lead poisoning in most domestic mammals, except swine – which tolerated up to 143.0 mg Pb/L blood. Captive wild ducks dosed with No. 4 lead shot in summer or winter were more sensitive than their domesticated counterparts, as judged by lower survival, and increased weight loss following lead shot administration and may be related to increased stress and unnatural diet. Fatal lead poisoning (50% mortality) is documented for a captive colony of Gentoo penguins in a zoo in Omaha, Nebraska. The source of lead was ankle weights worn by divers that cleaned the pool; weights contained 1–2 mm lead pellets that were expelled through faulty seams and subsequently ingested by the penguins.

18.6.5 Mammals

Three stages of recognizable lead poisoning, or plumbism, have been reported in humans: (1) mild or severe dysfunction of the alimentary tract as shown by loss of appetite, constipation, abdominal cramps, headaches, general weakness, and fatigue; (2) atrophy of forearm extensor muscles., or paralysis of these muscles and more striking atrophy; and (3) lead encephalopathy, which occurs frequently in lead-poisoned infants and young children, but only rarely in industrial workers. In general, people with hepatitis, anemia, and nervous disorders were more susceptible to lead poisoning. The transfer of lead across the human placenta and its potential threat to the fetus have been recognized for more than 100 years; women occupationally exposed to lead showed a comparatively high abortion rate. Sensitivity of the brain to the toxic effects of lead is considerably greater in the fetus than in the infant or young child. Lead is not considered carcinogenic to humans. However, reports of chromosomal aberrations in human blood lymphocytes suggest that lead is a probable mutagen.

Signs of plumbism in domestic and laboratory animals, which are similar to those in humans, have been well documented. There is general agreement on several details: significant differences occur between species in response to lead insult; effects of lead are more pronounced with organolead than inorganic lead compounds; younger developmental stages are the most sensitive; and the effects are exacerbated by elevated temperatures, and by diets deficient in minerals, fats, and proteins. Tetramethyl lead, for example, is about seven times more toxic than tetraethyl lead to animals, and both compounds showed toxic effects earlier than did inorganic lead compounds. In severe cases, death is usually preceded by impairment of normal functions of the central nervous system, the gastrointestinal tract, and the muscular and hematopoietic systems. Signs include vomiting, lassitude, loss of appetite, uncoordinated body movements, convulsions, stupor, and coma. In nonfatal cases, signs may include depression, anorexia, colic, disturbed sleep patterns, diarrhea, anemia, visual impairment, blindness, susceptibility to bacterial infections, excessive salivation, eye blinking, renal malfunction, peripheral nerve diseases affecting the motor nerves of the extremities, reduced growth, reduced life span, abnormal social behavior, and learning impairment. Lead crosses the placenta and is passed in milk, producing early intoxication of the fetus during pregnancy and the newborn during lactation. High lead doses in mammals induce abortion, reduce or terminate pregnancy, or can result in stillbirths or an increase in skeletal malformations. These signs, together with lead levels in blood and tissues and histopathological examination, are used to diagnose lead poisoning. The best overall prediction of the risk of clinical lead effects in mammals are the blood lead concentrations and the erythrocyte protoporphyrin values as they correlate with most toxicity endpoints of either neurological, renal, hematopoietic, or musculoskeletal nature.

Lead adversely affected the survival of sensitive mammals tested at different concentrations: 5.0–108.0 mg Pb/kg BW in rats (acute oral), 0.32 mg Pb/kg BW daily in dogs (chronic oral), and 1.7 mg Pb/kg diet in horses (chronic dietary). Adverse sublethal effects were noted in monkeys given 0.1 mg Pb/kg BW daily (impaired learning 2 years post-administration) or fed diets containing 0.5 mg Pb/kg (abnormal social behavior); in rabbits given 0.005 mg Pb/kg BW (reduced blood ALAD activity) or 0.03 mg Pb/kg BW (elevated blood lead levels); in mice at 0.05 mg Pb/kg BW (reduced ALAD activity); or in sheep at 0.05 mg Pb/kg BW (tissue accumulations).

Although lead is undeniably toxic at high levels of exposure, the implications of lower levels of exposure are poorly defined. Behavioral effects such as hyperactivity, distractibility, and decreased learning ability, as well as certain peripheral neuropathies, have been ascribed to subclinical lead exposure. Impaired learning ability of lead-stressed animals showing no obvious signs of lead intoxication have been documented for rats, sheep, and primates, although variability was great in all studies. Some learning deficits may be reversible and may not persist beyond a period of rehabilitation, and some may be induced only at relatively high exposure levels. Abnormal social behavior (usually aggression) has been reported in baboons and monkeys, although mice showed inhibited development of isolation-induced aggression. Altered parent-child relationships were suggested when suckling rats were used as surrogates. In that study, pregnant rats fed diets containing powdered lead nursed for longer periods than normal, and the resultant offspring were slower to explore their environment. Lead-exposed pups, with blood lead levels as low as 200.0 μg/L (considered elevated but within the "normal" range) at weaning, showed an altered dam-pup interaction that resulted in the dam spending longer periods in the nest than usual. Retarded development of lead-treated pups may account for the longer bouts of nesting by lead-stressed dams, and the delay in age at which pups explore and learn. It was concluded that maternal behavior was related to delays in pup development, and that the functional isolation of pups from their environment may be the antecedent to altered behavior later in maturity. No data are currently available on effects of lead-induced altered parent-offspring relationships, impaired learning ability, or abnormal social behavior for any population of free-ranging wildlife.

Ingestion of lead-containing paint from bars or walls is a significant cause of death among captive wild animals – including many species of apes, monkeys, bears, ferrets, pinnipeds, foxes, panthers, bats, raccoons, and armadillos – and is probably underreported. A similar situation exists for domestic animals – including dogs, cats, goats, horses, swine, cattle, and sheep. Passage of laws regulating the amount of lead in paint has decreased the frequency of lead poisoning, but many animals are still at risk from this source. Lead also occurs in used motor oils, gasoline, batteries, shot, putty, golf balls, linoleum, and printer's ink – all of which are considered sources of lead poisoning to domestic animals. Although the use of lead arsenate as an insecticide in orchards is diminishing, residues of lead still remain in the upper soil surface of treated orchards and will continue to remain bioavailable almost indefinitely.

Naturally occurring radiolead-210, which has a half-life of 22 years, is a significant contributor to the natural radiation dose in humans; comparatively high levels have been reported in certain grasses and lichens, and their consumers, such as reindeer, caribou, and ptarmigan, as well as in lanternfishes. The implications of this finding on wildlife health are unknown.

Treatment of lead-poisoned animals usually involves the removal of ingested lead objects and application of antibiotics. For example, a captive bottlenose dolphin that had 40 lead-containing air pellets in its second stomach, as determined by radiography, was treated with 250.0 mg penicillamine/kg BW given orally three times daily for 5 days after the pellets had been removed from the stomach using an endoscope. Anemia in chimpanzees is sometimes associated with lead toxicity. In one case, a 19-year-old female chimpanzee with a history of excessive menstrual bleeding had a blood serum level of 1.03 mg Pb/L. The animal was successfully treated using oral chelation therapy: 2,3-dimercaptosuccinic acid at 10.0 mg/kg BW per os for 5 days, then 10.0 mg/kg BW for 2 weeks.

18.7 Recommendations

Proposed lead criteria for the protection of natural resources and human health are numerous and disparate (Table 18.2). Some of these criteria do not provide adequate protection. Criteria for aquatic life protection, for example, range from 1.3 to 7.7 μg total waterborne Pb/L; however, within this range, high accumulations and adverse effects on growth and reproduction were recorded among sensitive species. Moreover, certain organolead compounds were lethal to some species of aquatic biota within this range, but no criteria have been formulated yet for this highly toxic group of chemicals. Nor have any criteria been proposed for lead in tissues of aquatic biota connoting elevated or hazardous levels to the organism. It is noteworthy that health effects to humans through ingestion of lead-contaminated seafood (and probably other fishery products) are considered negligible. Total lead concentrations observed in highly polluted areas in the 1970s were usually about one-tenth of those showing effects on marine organisms.

Organolead compounds are more toxic than ionic forms. Since methylation of ionic lead in vivo or in stored tissues is possible, and since some liver enzyme systems are capable of converting tetraethyl lead to the more toxic triethyl

Table 18.2. Proposed lead criteria for the protection of natural resources and human health.

Resource, Units, and Other Variables	Criterion
CROPS	
Irrigation water (mg/L)	
USA	
Neutral and alkaline soils	<101.0
Acidic soils	<51.0
Chronic use	<52.0
Short-term use	<202.0
Canada	
Continuous use	<51.0
Intermittent use	<101.0
Australia	<5.0
AQUATIC LIFE	
Freshwater (μg total Pb/L)	
USA	
Water hardness, in mg $CaCO_3$/L	
50	1.3[a], 34.0[b]
100	3.2[a], 82.0[b]
200	7.7[a], 200.0[b]
Great Lakes	
Superior	<100.0
Huron	<200.0
Others	<250.0
England	<100.0
The Netherlands	
Maximum in surface waters	<10.0
Adverse effects on fish embryo survival and development	>25.0 at acidic pH
Seawater (μg total Pb/L)	
USA	5.6[a], 140.0[b]
California	
6-month median	<2.0
Daily maximum	<8.0
Instantaneous maximum	<20.0
Water (μg/L)	
Tetraalkyl lead	<1.0
Trialkyl lead	<100.0
Sewage effluent limits (μg/L)	
California	<4000.0
Industrial discharge limits to surface waters (μg/L)	
Illinois	<100.0
USA	<500.0
Canada	<2000.0
Switzerland	<5000.0

Continued

Table 18.2. cont'd

Resource, Units, and Other Variables	Criterion
BIRDS	
Most species	
Diet (mg/kg dry weight [DW] ration)	<5.0
Blood, normal (mg/L)	<0.15
Mallard, *Anas platyrhynchos*; liver (mg/kg DW)	
No adverse effects	<5.0
Adverse effects on body condition and body weight	>5.0
Canvasback, *Aythya valisneria*; elevated	
Wingbones, immatures (mg/kg DW)	>20.0
Blood (mg/L)	>0.2
Lesser snow geese, *Chen caerulescens caerulescens*; adverse effects on height and fat (mg/kg DW)	
Kidney	>30.0
Liver	>30.0
Trumpeter swan, *Cygnus buccinator*; blood (mg/kg fresh weight [FW])	
No clinical signs	<0.5
Subclinical signs: mild depression, weight loss, anemia, and green diarrhea. Good chance of recovery with treatment	0.5–0.99
More pronounced clinical signs: as above plus weakness and neurological abnormalities. Fair to good chance of recovery with treatment	1.0–1.99
Progressively worse clinical signs; poor to fair chance of recovery with treatment	2.0–3.99
Similar to above; prognosis very poor with treatment	>4.0
Mute swan, *Cygnus olor*; diagnostic of lead poisoning (mg/kg FW)	
Blood	>3.0
Kidney	>31.0
Liver	>12.0
American kestrel, *Falco sparverius*; nestlings (mg/kg FW)	
Elevated	
Liver	>2.0
Kidney	>6.0
Poisoned	
Liver	>5.0
Kidney	>15.0
Bald eagle, *Haliaeetus leucocephalus*	
Uncontaminated; liver (mg/kg FW)	<2.0
Elevated (mg/kg DW)	
Kidney	>2.0–20.0
Liver	>10.0

Table 18.2. cont'd

Resource, Units, and Other Variables	Criterion
Sublethally poisoned (mg/kg FW)	
Blood	>0.2
Liver	2.0–10.0
Adverse effects; liver (mg/kg FW)	>8.0
Poisoned (mg/kg FW)	
Blood	>0.6–>0.8
Kidney	>5.0
Liver	>10.0
Poisoned; kidney (mg/kg DW)	>20.0
Herring gull, *Larus argentatus*; feather	
Adverse effects in fledglings (delayed parental recognition; impaired thermoregulation, locomotion and depth perception; lower chick survival)	>4.0
Raptors	
Subclinical effects (mg/kg FW)	
Blood	>0.2
Kidney, liver	>2.0
Toxic (mg/kg FW)	
Blood	>1.0
Kidney, liver	>3.0
Lethal (mg/kg FW)	
Blood, liver, kidney	>5.0
Waterfowl	
Safe (mg/kg FW)	
Blood	<0.1–<0.2
Liver	<2.0–<2.3
Elevated (mg/kg FW)	
Blood	>0.2–>0.5
Liver	>2.0
Elevated (mg/kg DW)	
Bone	>10.0–>20.0
Liver	>6.0
Poisoned (mg/kg FW)	
Liver	
Total lead	>2.0–>8.0
Trimethyl lead	>0.5
Blood	>0.2–>0.6
Bone	>20.0
Poisoned (mg/kg DW)	
Kidney	>30.0
Liver	10.0–>30.0
Lethal; liver (mg/kg FW)	>10.1

Continued

Table 18.2. cont'd

Resource, Units, and Other Variables	Criterion
Gizzards containing ingesting lead shot (percent frequency)	
Further study indicated	>5.0%
Lead shot ban in area	>10.0%
MAMMALS	
Cattle, *Bos* spp.; poisoned (mg/kg FW)	
Blood	>1.0
Liver	>20.0
Kidney	>40.0
Feces	>35.0
Domestic livestock	
Drinking water (μg/L)	
USA	<100.0
Australia	<250.0
Canada	
Horse, *Equus caballus*	<500.0
Others	<1000.0
Forage (mg/kg FW)	
Horse	<80.0
Cattle	<200.0
Tissue residues; unstressed (mg/kg FW)	
Blood	<0.2
Liver	<1.1
Kidney	<1.2
Mammals; selected species; adverse effects expected	
Daily lead intake (mg/kg body weight)	>20.0
Kidney (mg/kg DW)	>25.0
Liver (mg/kg FW)	>20.0
Liver (mg/kg DW)	>10.0
Mouse, *Mus* sp.; elevated (mg/kg body weight daily)	
Total intake	>0.05
Mule deer, *Odocoileus hemionus*; excessive	
Total intake (mg/day)	>3.0
Raccoon, *Procyon lotor*; elevated	
Liver (mg/kg FW)	>10.0
Wildlife	
Blood (mg/kg FW)	
Normal	0.02–0.08
Safe	<0.2
Poisoned	0.35, usually >0.6
Liver (mg/kg DW)	
Elevated	>10.0
Poisoned	>30.0

Table 18.2. cont'd

Resource, Units, and Other Variables	Criterion
Kidney (mg/kg DW)	
Elevated	>25.0
Poisoned	>90.0
HUMAN HEALTH	
Air (μg Pb/m^3)	
Safe	
USA	<1.5 (3-month arithmetic mean)
Connecticut	<3.0[e]
Kansas	0.357 for 1 year
Massachusetts	
Annual	<0.07
24-h	<0.14
Philadelphia, PA	
Annual	<1.5
24-h	<2.5
Occupational, USA	
Current	<50.0[e]
Goal	<30.0[e]
Lead chromate	<50.0
Lead arsenate	<150.0
Metallic lead	<100.0
Proposed, worldwide	<2.0
Hazardous	>2220.0
Blood (μg Pb/L)	
Acceptable (ALAD inhibition, protoporphyrin elevation)	100.0–300.0
Target	150.0, maximum 300.0
Level of concern	
USA	
Children	>100.0
Adults	>400.0
International	>200.0
Anemia, neurobehavioral effects, some poisoning in children	>400.0
Evidence of exposure	>500.0
Central nervous system deficits, peripheral neuropathy, intellectual deficits	500.0–700.0
Brain structure alterations, encephalopathy	>800.0
Life-threatening	>1000.0

Continued

Table 18.2. cont'd

Resource, Units, and Other Variables	Criterion
Drinking water (μg/L)	
USA	
1975	<500.0
1977	<250.0
1980–93	<50.0
Most states	<20.0
Goal	<15.0
South Africa	<500.0
Canada, Australia	<50.0
Russia, Japan	<100.0
India	10.0–<100.0
International	<50.0
World Health Organization	<100.0
Intake, all sources (mg)	
Daily	
Unacceptable	>2.3
Average	
Adult	<0.3
Child	<0.21
Weekly	
Adults	<3.0
Adult, 70 kg; tolerable	1.75
Children	<3.0
Food	
Citrus (mg/kg FW)	<1.0
Dried fish (mg/kg DW)	
England	<5.0
World Health Organization	<8.0
Raw fruits and vegetables (mg/kg FW)	<7.0
Fishery products (mg/kg FW)	
Canada	<10.0
Denmark	
Fish	<0.3
Shellfish, soft parts	<1.0
USA	<0.3
United Kingdom	
Fish	<2.0 (<14.0 DW)
Shellfish	<5.0 (35.0 DW)
Meat products (mg/kg FW)	
Muscle	<0.3
Kidney	Max. 1.0
Liver	<0.8–1.0
Total diet (mg/kg FW)	<0.3

Table 18.2. cont'd

Resource, Units, and Other Variables	Criterion
Gasoline (mg/L)	
USA	
Recent	473.0–658.0[c]
Proposed	131.0[d]
UK	
1972	840.0
1978	450.0
1981	400.0
Proposed	150.0
Germany	150.0
Groundwater (μg/L)	
Wisconsin	
Preventive action limit	5.0
Enforcement standard	50.0
House paints	
Product (mg/L)	<600.0
Application (mg/m^3)	<1.0
Urinary lead levels (μg/L)	
Normal	<80.0
Acceptable	80.0–120.0
Excessive	120.0–220.0
Dangerous	>200.0

[a] 4-day average, not to be exceeded more than once every 3 years.
[b] 1-h average, not to be exceeded more than once every 3 years.
[c] Equals 1.8–2.5 g/gallon.
[d] Equals 0.5 g/gallon.
[e] Average 8-h period.
[f] Blood lead levels, usually expressed as μg/deciliter, have been converted to μg/L, for uniformity, in the present work.

lead species, it would appear that the proposed Canadian permissible concentration limit of 10.0 mg Pb/kg FW in fishery products should be re-evaluated downwards. Downward evaluation has also been recommended for the standard of 2.0 mg/kg in the UK, where new guidelines have been recommended for total lead and for tetralkyl lead compounds in fishery products. Increasing use of organolead compounds as wood preservatives, as biocides, and as catalysts in the manufacture of plastics, polyurethanes, and polyvinyl chlorides may adversely affect survival, sensory responsiveness, and behavioral reactivity in aquatic organisms, and avian wildlife. It seems that additional research is needed on organolead toxicokinetics, with special reference to fishery and wildlife resources.

The evidence implicating ingestion of spent lead shot as a major cause of mortality in waterfowl and other birds is overwhelming. Moreover, forms of inorganic lead – besides lead shot or other ingestible-sized lead objects – are rarely known to produce subclinical signs of lead toxicosis in avian populations. Accordingly, in the 1986 advent of the lead shot phaseout, steel shot nontoxic zones were established for the protection of bald eagles and waterfowl in 44 States. Possession of shot shells containing lead shot by hunters of waterfowl in a steel shot zone is a violation of Federal regulations. By 1991–1992, all uses of lead shot for hunting waterfowl and coots were eliminated nationwide, including Alaska. The conversion to a nontoxic shot zone was deferred in some cases until – but not beyond – the 1991–1992 hunting season in States that demonstrate, through monitoring, compliance with the following criteria: minimum of 100 birds sampled; less than 5% of birds examined having one or more lead shot in the gizzard; and less than 5% of the birds collected having >2.0 mg Pb/kg FW in liver, or with >0.2 mg Pb/L in blood, or with blood protoporphyrin concentrations >0.4 mg/L. In addition, the occurrence of three or more individual specimens confirmed as lead-poisoned during the monitoring year disqualified the area for deferral. States may elect to forego monitoring and convert to nontoxic shot zones on a countywide or statewide scheduled or accelerated basis. Steel shot is now approved by the U.S. Fish and Wildlife Service as nontoxic. Alternatives to steel shot are under investigation and include bismuth-tin, tungsten-iron, and tungsten-polymer shot. Despite the ban on lead shot in the state of Washington since 1986, mortality of swans from lead poisoning continues to occur as a result of ingestion of previously deposited lead pellets. More aggressive enforcement of steel shot regulations in Washington may help reduce or eliminate the illegal use of lead shot. This, and continued settling of lead shot deeper into bottom sediments should eventually result in reduced lead-poisoning mortality of swans from ingestion of lead shot.

In addition to the U.S., the use of lead shot is now banned – sometimes voluntarily, sometimes by legislation – locally or nationally in hunting waterfowl in Australia, Canada, Italy, Mexico, Denmark, Finland, the Netherlands, Norway, Sweden, Switzerland, Germany, Japan, and the United Kingdom. In 1995, the British Department of the Environment recommended that lead-free shot be used when shooting over or within 300 m of a wetland if there is a possibility that the shot would be deposited in it. Beginning in 1997, all migratory game bird hunting in Canada required the use of nontoxic shot. Legislation has been introduced – and signed by 54 countries – to protect African-Eurasian migratory birds by the phase out of the use of lead shot for hunting in wetlands by the year 2000. To fully protect raptors, upland game birds, and waterfowl, many researchers now recommend a global ban on the use of lead shot in hunting, and in trap and skeet shooting. Continued monitoring of lead residues for waterfowl, in addition to gizzard examination for shot, should also include blood and soft tissues as indicators of short-term exposure, and bone as an indicator of longer exposure; monitoring is recommended until lead shot is no longer widespread in the wetland environment.

In Louisiana, deep tillage is a management option for reducing the availability of lead shot to foraging waterfowl. Agricultural deep tillage with a vegetable plow can result in the redistribution of artificially seeded lead shot from the top 10 cm of soil to lower strata. In control sites, 92% of the shot was recovered above 10 cm vs. tillage wherein 97% of the shot was recovered below 10 cm and 75% below 20 cm. In wetlands areas lacking naturally available grit and with high shot densities in soils and sediments, the provision of grit 2–3 mm in diameter is recommended for the reduction of lead ingestion by waterbirds. Increasing water depth of wetlands during the hunting season to inhibit foraging by waterfowl in lead shot affected areas is proposed. In Spain, it is recommended that hunter-killed large animals that are currently abandoned in the field should be removed or buried, thus removing the main source of metallic lead for raptors and scavengers.

Lead poisoning of swans in England is due mainly to ingestion of discarded lead weights used in angling. The use of lead weights for this purpose has been sharply curtailed through legislation since 1987, and there has been a substantial reduction in lead poisoning in the swans of the Thames River. In order to protect swans, loons, and other large waterfowl in the U.S., replacement of lead fishing weights with nontoxic substitutes is recommended.

The level of human exposure in lead-using industries has been reduced considerably in recent years; associated with this observation is the reduction in lead content of gasoline, the removal of lead-based paints for interior household use, and the reduction in lead content of outside paints. These actions will undoubtedly prove beneficial in reducing the elevated lead concentrations now observed in communities of flora and fauna along heavily traveled roads, and in providing additional protection to captive zoo animals and other animals held in enclosures with lead-painted bars and walls. The decreased use of leaded gasoline has resulted in a significant decline in lead concentrations in streams, and in whole body burdens of lead in starlings collected nationwide, among which the decline was most pronounced in birds from urban areas. Continued nationwide monitoring of lead in fish and wildlife is necessary to determine if this is a continuing downward trend, and also to identify areas of high or potential lead contamination.

Data for lead effects on mammalian wildlife are scarce. Lead residues in soils have been used successfully to predict lead concentrations in kidneys and livers of wood mice and field voles; however, this could not be demonstrated for shrews. In view of the large interspecies differences in lead responses reported for domestic livestock and laboratory populations of small animals, more research is needed to determine if lead criteria for these groups are applicable to sensitive species of mammalian wildlife.

One of the more insidious effects documented for lead in warm-blooded organisms is neurobehavioral deficits (including learning impairments) at dose levels producing no overt signs of toxicity, i.e., apparently normal growth and developmental skills, and sometimes, nonelevated blood lead levels. Behavioral deficits have been reported for young rats when blood lead levels exceeded 0.1 mg/L, and in children with blood lead concentrations of 0.4–0.5 mg/L, and in birds when lead was administered early in development. Behavioral impairment was recorded in 3-year-old monkeys that received 50.0 or 100.0 μg Pb/kg BW from birth to age 200 days. Blood lead levels immediately after exposure, and at time of testing, were 0.15–0.25 mg/L (age 200 days), and 0.11–0.13 mg/L (age 3 years); this is the first report of behavioral impairment in a primate species at blood lead concentrations that are considered to be well within the bounds of safety for children. This subject appears to constitute a high priority research need for wildlife species of concern.

18.8 Summary

Lead (Pb) and its compounds have been known to humans for about 7000 years, and lead poisoning has been recognized for at least 2500 years. All credible evidence indicates that lead is neither essential nor beneficial to living organisms, and that all measured effects are adverse – including those on survival, growth, reproduction, development, behavior, learning, and metabolism. Various living resources are at increased risk from lead: migratory waterfowl that frequent hunted areas and ingest shot; avian predators that eat game wounded by hunters; domestic livestock near smelters, refineries, and lead battery recycling plants; captive zoo animals and domestic livestock held in enclosures coated with lead-based paints; wildlife that forage extensively near heavily traveled roads; aquatic life in proximity to mining activities, areas where lead arsenate pesticides are used, metal finishing industries, organolead industries, and areas of lead aerosol fallout; and crops and invertebrates growing or living in lead-contaminated soils.

Adverse effects on aquatic biota reported at waterborne lead concentrations of 1.0–5.1 μg/L include reduced survival, impaired reproduction, reduced growth, and high bioconcentration from the medium. Among sensitive species of birds, survival was reduced at doses of 50.0–75.0 mg Pb^{2+}/kg body weight (BW) or 28.0 mg organolead/kg BW, reproduction was impaired at dietary levels of 50.0 mg Pb/kg, and signs of poisoning were evident at doses as low as 2.8 mg organolead/kg BW. In general, forms of lead other than shot (or ingestible lead objects), or routes of administration other than ingestion, are unlikely to cause clinical signs of lead poisoning in birds. A notable exception to this generalization is the recent evidence linking waterfowl poisoning with ingestion of lead-contaminated sediments. Data for toxic and sublethal effects of lead on mammalian wildlife are missing. For sensitive species of domestic and laboratory animals, survival was reduced at acute oral lead doses of 5.0 mg/kg BW (rat), at chronic oral doses of 5.0 mg/kg BW (dog), and at dietary levels of 1.7 mg/kg BW (horse). Sublethal effects were documented in monkeys exposed to doses as low as 0.1 mg Pb/kg BW daily (impaired learning at 2 years post-administration) or fed diets containing 0.5 mg Pb/kg (abnormal social behavior). Signs of lead exposure were recorded in rabbits given 0.005 mg Pb/kg BW and in mice given 0.05 mg Pb/kg BW. Tissue lead levels were elevated in mice given doses of 0.03 mg Pb/kg BW, and in sheep given 0.05 mg Pb/kg BW. In general, organolead compounds were more toxic than inorganic lead compounds, food chain biomagnification of lead vas negligible, and younger organisms were most susceptible. More research seems merited on organolead toxicokinetics (including effects on behavior and learning), and on mammalian wildlife sensitivity to lead and its compounds.

Legislation limiting the content of lead in paints, reducing the lead content in gasoline, and eliminating the use of lead shot nationwide (lead shot phaseout program/schedule starting in 1986, and fully implemented by 1991) in waterfowl hunting areas will substantially reduce environmental burdens of lead and may directly benefit sensitive fishery and wildlife resources. Continued nationwide monitoring of lead in living resources is necessary in order to correlate reduced emission sources with reduced tissue lead concentrations.

MERCURY[a]

Chapter 19

19.1 Introduction

The element mercury, also known as quicksilver (symbol Hg for *hydrargyrum*), and its compounds have no known normal metabolic function. Their presence in the cells of living organisms represents contamination from natural and anthropogenic sources; all such contamination must be regarded as undesirable and potentially hazardous. Accumulation of mercury in tissues is reportedly associated with an excess risk of myocardial infarction, increased risk of death from coronary heart disease and cardiovascular disease, and accelerated progression of carotid atherosclerosis.

The most important ore of mercury, cinnabar (mercuric sulfide), has been mined continuously since 415 BCE. In the period before the industrial revolution, mercury was used extensively in gold extraction and in the manufacture of felt hats and mirrors; in the 1800s, it was used in the chloralkali industry, in the manufacture of electrical instruments, and as a medical antiseptic; and since 1900, it has been used in pharmaceuticals, in agricultural fungicides, in the pulp and paper industry as a slimicide, and in the production of plastics. World use of mercury is estimated at 10,000–15,000 metric tons annually, of which the U.S. accounts for about 18%. Today, mercury is a leading public health concern, as judged by the increases in regulations governing mercury emissions, in mercury fish advisories, in clinical studies, and in media attention.

[a]All information in this chapter is referenced in the following source:

Eisler, R. 2006. *Mercury Hazards to Living Organisms*. CRC Press, Boca Raton, Florida, 312 pages.

The first cases of fatal inorganic mercury poisoning in humans were reported for two men in a European chemical laboratory in 1865, and the first documented human poisoning from agricultural exposure to an organomercury compound was in 1940. Human exposure to mercury compounds via dermal, dietary, and respiratory routes have severe consequences. For example, 1600 infants in Argentina showed symptoms of mercury poisoning after a laundry treated their diapers with a mercury disinfectant; numerous poisonings resulted from ingestion of mercury-contaminated fish, pork, seafood, and grains; and from occupational exposure in Nicaraguan mercury fungicide applicators via respiration. Sporadic incidences of human poisoning have occurred in the U.S., the Soviet Union, and Canada; and major epidemics have been reported in Japan, Pakistan, Guatemala, Ghana, Yugoslavia, and Iraq. In Iraq, for example, a major mercury poisoning occurred in the early 1970s, with 6530 hospital admissions and 459 hospital deaths. In 1970, the Iraqi government decided to import seed grain from Mexico owing to a serious wheat shortage throughout Iraq. About 73,200 metric tons of seed wheat was imported and distributed between September and November of 1971. The seed had been treated with a methylmercury fungicide and colored with a red dye; it was delivered in 50 kg sacks with a warning written in Spanish. A skull and crossbones was painted on the outside of each sack to indicate that the seed had been treated with poison. However, both Spanish and the skull and crossbones insignia were totally unfamiliar to the Iraqi farmers and the people decided to make home-made bread instead of using the grain for seed; about half the wheat was consumed for this purpose. The water-soluble red dye was washed off the wheat, with the assumption that the mercury would be equally soluble (it was not). Before the wheat was consumed by humans, the farmers fed the wheat for a few days to chickens and other livestock without apparent effect; it was not realized that a lengthy latency period was involved. The wheat flour contained 4.8–14.6 mg methylmercury/kg, far in excess of the 0.5 mg Hg/kg regulatory limit. Consumption of the bread lasted 1–2 months, and the first cases of poisoning appeared at the end of December 1971. In 1972, there were 6530 reported cases within 18 months, including 459 deaths, among Iraqi farmers who ate bread made from seed wheat treated with a methylmercury fungicide. Babies congenitally affected by methylmercury born in Iraq during this outbreak showed clinical features of mental retardation, motor disorders resembling cerebral palsy, and other physical and mental disturbances. Histopathological examination of the brain of victims in cases of death showed degenerative neuronal damage and abnormal development of the cerebral cortex. There was no effective antidote to counteract the effects of methylmercury on the central nervous system.

At Oak Ridge, Tennessee, liquid metallic mercury was used during the 1950s and 1960s for uranium processing in a nuclear weapons plant. Accidental releases of mercury of about 100 tons – equivalent to that released at Minamata Bay, Japan, in which hundreds died – resulted in heavy accumulations of mercury in the waste pipes and as a "lake of Hg" of about 7000 L under the plant sites. However, unlike Minamata, there have been no documented human health problems at the Oak Ridge site. The mercury at Oak Ridge is released into local waters at about 75 kg annually, at which rate it will take more than 1000 years to disappear, although microbial activities transform the mercury found in local waters within hours.

Mercury contamination is the most serious environmental threat to fishery and wildlife resources in the southeastern U.S., with fish consumption advisories issued in the ten states comprising this region. In March 1989, the Florida Department of Health and Rehabilitative Services issued a health advisory prohibiting consumption of predatory fishes, such as largemouth bass (*Micropterus salmoides*), bowfin (*Amia calva*), and gar (*Lepisosteus* spp.) in southern Florida, and the entire Everglades watershed has been closed to hunting of alligators (*Alligator mississippiensis*) due to excessive mercury in edible tissues of alligators. Methylmercury concentrations of 2.0–3.0 mg/kg fresh weight (FW) muscle were documented in largemouth bass and other fish-eating species, or four to ten times the allowable limit in foods for human consumption. About 0.91 million ha of the South Florida Everglades ecosystem are currently under fish consumption advisories because of mercury contamination.

The number of mercury-contaminated fish and wildlife habitats has, in general, progressively increased worldwide, almost all as a direct result of anthropogenic activities. Long-range atmospheric transport of mercury since 1940 has resulted in elevated mercury loadings in remote Canadian lakes up to 1400 km from the closest industrial centers. Since 1985, annual mercury accumulation rates in flooded Florida Everglades soils averaged 53.0 $\mu g/m^2$, which is about 4.9 times greater than rates observed around the turn of the century; the increase is attributed to increased global and regional deposition and is similar to increases reported for lakes in Sweden and the northern U.S. In 1967, the Swedish medical board banned the sale of fish that contained high concentrations of organomercury salts, originating from about 40 lakes and rivers. More than 10,000 Swedish lakes have been closed to fishing because of excessive mercury loadings. In 1970, after the discovery of high levels of mercury in fish from Lake St. Clair, Canada, restrictions on fishing and the sale of fish were imposed in many areas of the U.S. and Canada. Since 1970, a total of 26 of the 48 States in the conterminous U.S. have reported mercury pollution in their waters as a direct result of human activities. These States have banned sport or commercial fishing in mercury-contaminated waters, or have issued health warnings about the consequences of eating mercury-contaminated fish or seafood from selected watercourses, or have placed restrictions on fish consumption from certain streams, lakes, or rivers polluted with mercury. At present, more than 40 states have issued health advisories restricting fish consumption based on unacceptable mercury contents; similar advisories have been issued in North America, Europe, and Asia. Fish consumption advisories on mercury issued by the U.S. Food and Drug Administration (FDA) are effective in reducing fish consumption; in Massachusetts, for example, there was an average monthly decline in fish consumption of 1.4 servings among pregnant women following an FDA advisory on the health risks of mercury.

Poisoning of game birds and other wildlife in Sweden, apparently by seeds treated with organomercurials, was first noted in 1960. Massive killings of the grey heron (*Ardea cinerea*) in the Netherlands during 1976, were attributed to a combination of low temperatures, undernourishment, and high body burdens of mercury. Mercury contamination has resulted in the closure of pheasant and partridge hunting areas in Alberta, Canada. Declining numbers of wading birds in southern Florida are attributed to mercury contamination of their food supply.

Most authorities on mercury hazards to living organisms now agree on six points. First, mercury and its compounds have no known biological function, and its presence in living organisms is undesirable and potentially hazardous. Second, forms of mercury with relatively low toxicity can be transformed into forms with very high toxicity through biological and other processes. Third, methylmercury can be bioconcentrated in organisms and biomagnified through food chains, returning mercury directly to human and other upper-trophic-level consumers in concentrated form. Fourth, mercury is a mutagen, teratogen, and carcinogen, and causes embryocidal, cytochemical, and histopathological effects. Fifth, high body burdens of mercury normally encountered in some species of fish and wildlife from remote locations emphasize the complexity of natural mercury cycles and human impacts on these cycles. And finally, the anthropogenic use of mercury should be curtailed, because the difference between tolerable natural background levels of mercury and harmful effects in the environment is exceptionally small.

19.2 Mercury Uses and Sources

Mercury is a metallic element easily distinguished from all others by its liquidity at even the lowest temperatures occurring in moderate climates. Mercury was unknown to the ancient Hebrews or the early Greeks, and the first mention of mercury is by Theophrastus, writing around 300 BCE, stating that mercury was extracted from cinnabar by treatment with copper and vinegar.

The most important ore of mercury, cinnabar (mercuric sulfide), has been mined continuously since 415 BCE. Historically, the five primary mining areas for mercury were the Almeden district in Spain, the Idrija district in Slovenia, the Monte Amiata district in Italy, and various locales in Peru, the U.S., especially California and Texas, as well as sites in Russia, Hungary, Mexico, and Austria. The Almeden mines over the past 2500 years were the most important, having produced about 280,000 metric tons of mercury, or about 35%, of the estimated total global production of about 800,000 tons. Major producers of mercury now include the Former Soviet Union, Spain, Yugoslavia, and Italy. In the U.S., mercury consumption rose from 1305 metric tons in 1959 to 2359 tons in 1969. Mining of mercury in the U.S. decreased in recent years owing to decreasing demand and falling prices and the last operating mine in the U.S. that produced mercury as its main product closed in 1990. In the late 1970s, world use of mercury was estimated at 10,000–15,000 metric tons annually, of which the U.S. used about 18%. Accurate data on recent mercury consumption in the U.S. are difficult to obtain. In 1987, the U.S. imported 636 metric tons; this fell to 329 tons in 1988, and 131 tons in 1989. Domestic production of mercury produced as a by-product was 207 metric tons in 1990, 180 tons in 1991, and 160 tons in 1992; during this same period, the U.S. imported 15 tons in 1990, 56 tons in 1991, and 100 tons in 1992.

Atmospheric input of mercury has tripled over the past 150 years. The atmosphere plays an important role in the mobilization of mercury with 25–30% of the total atmospheric mercury burden of anthropogenic origin, although more recent estimates of 67% are significantly higher. As a direct result of human activities, mercury levels in river sediments have increased fourfold since precultural times, and twofold to fivefold in sediment cores from lakes and estuaries. Analysis of sediment cores of North American mid-continental lakes show that mercury deposition rates increased by a factor of 3.7 since 1850, at a rate of about 2% annually. During the past 100 years, more than 500,000 metric tons of mercury entered the atmosphere, hydrosphere, and surface soils, with eventual deposition in subsurface soils and sediments.

19.2.1 Uses

Cinnabar, HgS, the main mercury ore was used as a red pigment long before refining processes for elemental mercury were implemented. In the 16th century, elemental mercury in combination with other compounds was considered a powerful medicinal agent. Mercuro zinc cyanide [Zn_3HgCNO_8], also known as Lister's antiseptic, was used in the early days of antiseptics, circa 1880s, in the form of "cyanide gavage" or "cyanide wool" in general surgery. The stronger mercurial ointments were used to kill cutaneous parasites and to control itching. Until the advent of antibiotics, mercury salts were a key treatment for syphilis and other venereal diseases. Mercuric salts, especially the chloride and iodide, are powerful antiseptics and until the 1940s were routinely used in surgery to kill bacteria; however, their use was contraindicated in patients with actual or potential renal inflammation, patients with scarlet fever (risk to throat), and eclampsia (uterus sensitivity). Mercurochrome, di-sodium hydroxy-mercuro dibromo fluorescein, has been widely used in the U.S. and the U.K. as an antiseptic, and is still sold in drugstores. The volatility of mercury and many of its compounds causes their absorption by the lungs and is an unintended consequence of external application; this could account for the occurrence of chronic mercurial poisoning in certain trades. Mercury is largely used in affectations of the alimentary canal and allegedly has value in many cases of heart disease and arterial degeneration. In cases of intestinal obstruction, elemental mercury has been administered – up to 454.0 g – without ill effects; the weight of the mercury being sufficient to remove the obstruction.

In the period before the industrial revolution, mercury was used extensively in gold extraction and the manufacture of felt hats and mirrors; in the 1800s, it was used in the chloralkali industry, the manufacture of electrical instruments, and as a medical antiseptic; and since 1900, it has been used in pharmaceuticals, agricultural fungicides, the pulp and paper industry as a slimicide, and the production of plastics. In 1892, the process of producing chlorine and caustic soda from brine (sodium chloride) was developed. Electrolysis of brine using a liquid mercury cathode to produce chlorine at the anode and a sodium–mercury amalgam at the cathode is still used worldwide, with significant mercury contamination of the biosphere; however, the process is increasingly under replacement using mercury-free components. Until the late 1940s, mercury was used in the manufacture of fur-felt hats from furs of rabbit, hare, muskrat, nutria, and beaver. Mercuric nitrate was used to remove fur from the hides. The fur fiber is harvested and the hide, now denuded of fur and useless for hat manufacturing purposes, discarded. A "mad hatter" syndrome was sometimes reported among workers in the manufacture of fur-felt hats, with symptoms consistent with that of inorganic mercury poisoning, namely, tremors, excessive salivation, irritability, and excitement.

Mercury catalysts were used in the production of acetaldehyde, acetic acid, and vinyl chloride. The major use of mercury in the decade 1959–1969 has been as a cathode in the electrolytic preparation of chlorine and caustic. In 1968, this use accounted for about 33% of the total U.S. demand for mercury. In that same period, electrical apparatus accounted for about 27% of U.S. mercury consumption; industrial and control instruments, such as switches, thermometers, barometers, and general laboratory appliances, 14%; antifouling and mildew-proofing paints, 12%; mercury formulations to control fungal diseases of seeds, bulbs, and vegetables, 5%; and dental amalgams, pulp and paper manufacturers, pharmaceuticals, metallurgy and mining, and catalysts, 9%. Mercury, however, is no longer registered for use in antifouling paints, or for the control of fungal diseases of bulbs. In India and other countries, however, $HgCl_2$ is commonly used for the preservation of seeds and by farmers in fruit preservatives after harvest and to inhibit growth of microorganisms. In Kenya, certain soaps containing skin-lightening agents also contained high concentrations of inorganic mercury. And in Mexico and other developing countries, merthiolate (an ethylmercury thiosalicylate compound) is currently used as a preservative in medical vaccines and as a skin antiseptic.

Mercury in its various forms is still available widely in thermometers, fungicides, in hearing aid and watch batteries, paints, mercurial drugs, antiquated cathartics, and in ointments. The most recent uses of mercury and its compounds are in the manufacture of lighting fixtures such as fluorescent, metal halide, and mercury vapor lamps; dental amalgams; mining and reprocessing of gold; batteries; and paint manufacture.

19.2.2 Sources

Major inputs of mercury to the environment are mainly from natural sources with significant and increasing amounts contributed from human activities. The atmosphere plays an important role in the mobilization of mercury with an estimated 25–30% of the total atmospheric burden of anthropogenic origin. The global anthropogenic atmospheric emission of mercury is estimated at 900–6200 tons annually, of which the U.S. contributed 300 metric tons in 1990 with 31% of the total from combustion of fossil fuels by power plants. Atmospheric deposition is generally acknowledged as the major source of mercury to watersheds. In northern Minnesota watersheds, for example, atmospheric deposition was the primary source of mercury. Geologic and point source contributions were not significant. Transport from soils and organic materials may also be important, but the mercury from these sources probably originates from precipitation and direct atmospheric sorption by watershed components. In Sweden, increased mercury concentrations in lakes are attributed to increased atmospheric emissions, deposition of mercury, and to acid rain. Airborne particulates may contribute to the high mercury levels found in some marine dolphins and whales. A total of 60–80 metric tons of mercury is deposited from the atmosphere into the Arctic each year; the main sources of mercury to the Arctic are Eurasia and North America from combustion of fossil fuels to produce electricity and heat. However, elevated mercury concentrations in fish muscle (0.5–2.5 mg/kg fresh weight [FW]) from remote Arctic lakes over extended periods (1975–1993) are sometimes due to natural sources of mercury. Atmospheric deposition of mercury into the Great Lakes from sources up to 2500 km distant are documented at annual deposition rates of 15.0 μg Hg/m^2. In South Florida, 80.0–90% of the annual mercury deposition occurs during the summertime–wet season. Dry deposition processes are important for the flux of inorganic mercury and methylmercury to Swedish forested ecosystems; for methylmercury, the most important deposition route is from the air to a relatively stable form in litter.

19.2.2.1 Natural Sources

The total amount of mercury in various global reservoirs is estimated at 334.17 billion metric tons; almost all of this amount is in oceanic sediments (98.75%) and oceanic waters (1.24%), and most of the rest is in soils. Living aquatic organisms are estimated to contain only 7.0 metric tons of mercury.

The largest pool of methylmercury in freshwater biota is found in fish tissues, and fly larvae are alleged to play an important role in mercury cycling from feeding on beached fish carcasses, as judged by observations on blowfly (*Calliphora* sp.) adult egg layers, eggs, larvae, pupae, and emerging adults feeding on carcasses of brook trout (*Salvelinus fontinalis*). Specifically, methylmercury that accumulated in blowfly larvae is retained in pupae but eliminated by adults following emergence.

Mercury from natural sources enters the biosphere directly as a gas, in lava (from terrestrial and oceanic volcanic activity), in solution, or in particulate form; cinnabar (HgS), for example, is a common mineral in hot spring deposits and a major natural source of mercury. The global cycle of mercury involves degassing of the element from the Earth's crust and evaporation from natural bodies of water, atmospheric transport – mainly in the form of mercury vapor – and deposition of mercury back onto land and water. Oceanic effluxes of mercury are tied to equatorial upwelling and phytoplankton activity and may significantly affect the global cycling of this metal. If volatilization of mercury is proportional to primary production in the world's oceans, oceanic phytoplankton activity represents about 36% of the yearly mercury flow to the atmosphere, or about 2400 tons per year. Riverine input – with sediments containing up to 0.5 mg Hg/kg DW – of mercury influences mercury content of outer shelf areas in southeastern Brazil where most of the offshore oil fields are located. Mercury finds its way into sediments, particularly oceanic sediments, where the retention time can be lengthy, and where it may continue to contaminate aquatic organisms. Estimates of the quantities of mercury entering the atmosphere from degassing of the surface of the planet vary widely, but a commonly quoted figure is 30,000 tons annually.

Mercury is emitted from volcanoes into the atmosphere, along with large quantities of lead, cadmium, and bismuth. About 6000 tons of mercury is discharged into the atmosphere every year from all sources, and from all volcanoes about 60 tons or about 1% of the total.

Virtually all mercury in the Florida Everglades from natural sources (39% of the total mercury deposited) is attributed to release from the soil through natural processes, including microbial transformations of inorganic and organic mercury to methylmercury. Major sources of mercury in humans – other than residing in mercury-contaminated environments – include consumption of large predatory fish, such as tunas and swordfish, caught hundreds of kilometers offshore in clean ocean waters. The naturally elevated concentration of mercury in these species is discussed in greater detail later.

Terrestrial vegetation functions as a conduit for the transport of elemental mercury from the geosphere to the atmosphere. Estimated mercury emissions from plants in the Carson River Drainage Basin of Nevada – an area heavily contaminated with mercury from historical gold mining activities – over the growing season (0.5 mg Hg/m^2) add to the soil mercury emissions (8.5 mg Hg/m^2) for a total landscape emission in that area of 9.0 mg Hg/m^2. In one species (tall whitetop, *Lepidium latifolium*), as much as 70% of the mercury, taken up by the roots during the growing season, was emitted into the atmosphere. Factors known to increase the flux of elemental mercury from terrestrial plants growing in soils with high (34.0–54.0 mg Hg/kg soil DW) levels of mercury include increasing air temperature in the range 20.0–40.0°C, increasing irradiance, increasing soil mercury concentrations, and increasing leaf area.

19.2.2.2 Anthropogenic Sources

In 1995, approximately 1900 metric tons of anthropogenic mercury entered the atmosphere (mostly 75%) from the combustion of fossil fuels. About 56% of global mercury atmospheric emissions came from Asian countries, with Europe and North America combined contributing less than 25%; gaseous elemental mercury (Hg) comprised 53% of total atmospheric emissions, gaseous Hg^+ 37%, and particle-associated mercury the remainder.

Large-scale mining of mercury in North America ceased around 1990 because of low prices and stringent environmental regulations. In the U.S., mercury is now produced only as a by-product from presently operating gold mines where environmental regulations require its recovery, and from the reprocessing of precious metal mine tailings and gold-placer sediments. Many of the mercury mines in the California Coast Ranges contain waste rock that contributes mercury-rich sediments to nearby watersheds. At some mines, the release of mercury in acidic drainage is a significant source of mercury to watersheds, where it is taken up by fish and other organisms.

Atmospheric transport of anthropogenic mercury may contaminate remote ecosystems. In one case, a remote lake in northern Wisconsin with no surface inflow and negligible groundwater inflow received about 100.0 mg Hg/ha during 1988–90, an input that could account for the elevated mercury burdens found in water, sediments, and fish.

Several human activities that contribute significantly to the global input of mercury include the combustion of fossil fuels; mining and reprocessing of gold, copper, and lead; operation of chloralkali plants; runoff from abandoned cinnabar mines; wastes from nuclear reactors, pharmaceutical plants, oil refining plants, and military ordnance facilities; incineration of municipal solid wastes and medical wastes; offshore oil exploration and production; disposal of batteries and fluorescent lamps, and the mining, smelting, use and disposal of mercury. In one case, more than 5.5 million kg of elemental mercury was released into the Carson River Drainage Basin in Nevada during historic mining operations – now closed – in which mercury was used to amalgamate gold and silver ore. Mercury concentrations in sample tailings were as high as 1570.0 mg/kg. The air directly over the site contained 1.0–7.1 ng Hg/m^3, and was as high as 240.0 ng/m^3 in October 1993. The estimated range of mercury flux to the atmosphere from the site was 37.0–500.0 ng/m^2 hourly.

Mercury emission from electric utilities is the largest uncontrolled source of mercury release into the atmosphere, and globally it accounts for up to 59% of the total annual atmospheric loading of mercury from both natural and anthropogenic sources. Coal-fired power plants are now considered the greatest source of environmental mercury in the U.S., and the only significant source that continues to be unregulated. In 1994, about 50 metric tons of mercury was emitted into the biosphere from coal-burning power plants in the U.S., with lesser amounts from oil- and gas-combustion units. Available technologies now installed in waste combustion and medical incinerators are recommended for installation in coal-fired plants and may reduce mercury emissions by as much as 90%. In fact, the U.S. Environmental Protection Agency (EPA) made steady progress throughout the 1990s in reducing mercury emissions from power plants, although this effort abated in recent years owing to the high costs of pollution abatement. The economic cost of methylmercury neurotoxicity from these plants was estimated using a hypothetical model. In that scenario, it was stated that methylmercury-induced loss of intelligence would affect between 316,000 and 637,000 children each year in the U.S., as judged by umbilical cord blood concentrations $>5.8\,\mu g$ methylmercury/L, a level associated with IQ loss. This intelligence loss is translated into diminished economic activity equivalent to at least $8.7 billion annually (range 2.2–43.8 billion); this model requires verification.

Logging and forest fires can contribute to the bioavailability of mercury. Watersheds impacted by clear-cut logging, or burnt forest ecosystems, release mercury into the biosphere with significant increases in flesh of predatory fish from impacted drainage lakes when compared to reference watersheds.

Most of the daily intake of mercury compounds is in the form of methylmercury derived from dietary sources, primarily fish, and to a lesser extent elemental mercury from mercury vapor in dental amalgams, and ethylmercury added as an antiseptic to vaccines. Dental amalgams, which may contain up to 50% by weight of metallic mercury, may also constitute a significant source of mercury in some cases. Amalgam mercury is imperfectly stable, slowly leaching from the mercury–silver or mercury–gold amalgam through the action of oral bacteria and exacerbated by chewing. Following placement or removal of fillings, up to 200.0 mg of mercury is eliminated in the feces, with subsequent selection of mercury-resistant bacteria for degradation. Normal mastication may result in body accumulations of 10.0 μg daily through either intestinal uptake or respiratory intake of mercury vapor released during chewing.

Consumption of fish contaminated with mercury was associated with elevated mercury concentrations in blood and hair of about 1200 Amerindians in northwestern Ontario, Canada. The source was inorganic mercury catalysts from a chloralkali plant. In this case, about 9 tons of mercury was discharged into wastewaters between 1962 and 1979, at which time mercury usage for the production of chlorine and caustic was drastically reduced. In aquatic ecosystems, removal of the source of anthropogenic mercury – such as chloralkali plants – results in a slow decrease in the mercury content of sediments and biota. The rate of loss depends, in part, on the initial degree of contamination, the chemical form of mercury, physical and chemical conditions of the system, and the hydraulic turnover time. Between 1986 and 1989, throughout all Canada, a total of five mercury-cell chloralkali plants were still operating. By comparison, during the period 1935 through the mid-1970s, about 15 chloralkali plants were operating; the development of alternative technologies to replace the mercury cell was largely responsible for the plant closures. Total mercury loss to the environment from these five operating plants during the 4-year period 1986–1989 was about 6.5 metric tons.

Mercury contamination of more than 500,000 miners, adjacent Indian populations, and numerous populations of fish and wildlife is one consequence of the gold rush that took place in the early 1980s in the Amazon region of Brazil. Metallic mercury was used to agglutinate the fine gold particles through amalgamation. During this process, large amounts of mercury were lost to the river and soil; additional mercury was lost as vapor to the atmosphere during combustion of the amalgamated gold to release the gold. Elemental mercury used in seals of three trickling filters in municipal wastewater treatment plants – each seal contained several hundred kg of mercury – leaked repeatedly, discharging 157.0 g of mercury and 0.4 g of methylmercury daily; the use of mercury seals for this purpose should be discontinued. Also discontinued, in 1967, by Finland is the use of phenyl mercury compounds as slimicides in the pulp industry. In Canada and the U.S., mercurial compounds were used to control fungi in pulp paper processing plants, with subsequent mercury contamination of edible fish muscle.

Abandoned mercury mines may contribute excess mercury loadings and other contaminants to the environment. For example, mercury mines in western Turkey that were gradually abandoned owing to low demand, low prices, and increasing environmental concern over mercury adversely affected adjacent water resources. One abandoned mine located 5 km west of Beydag, Turkey, that operated from 1958 through 1986 with a total production of 2045 metric tons of mercury during this period released metal-rich, acidic drainage affecting groundwater and adjacent stream water quality through decreasing pH; elevated levels of silicon, aluminum, magnesium, calcium, and potassium; increasing precipitation of iron oxides; and increasing sulfates, manganese, iron, and arsenic. Most of the mine water and groundwater samples exceeded drinking water standards for aluminum, iron, manganese, arsenic, nickel, and cadmium. Mercury concentrations in all samples were below the Turkish drinking water standard of 1.0 μg/L for human health; however, two samples contained 0.3 and 0.5 μg Hg/L and were above the USEPA mercury criterion for aquatic life protection of <0.012 μg/L.

In the Florida Everglades, 61% of the mercury is due to atmospheric deposition from anthropogenic sources, especially municipal solid waste combustion facilities (15%), medical waste incinerators (14%), paint manufacturing and application (11%), electric utility industries (11%) and private residences (2%) through combustion of fossil fuels, and electrical apparatus including fluorescent, metal halide, and mercury vapor lights (6%). All other anthropogenic sources combined – including sugarcane processing, the dental industry, open burning, and sewage sludge disposal – accounted for about 3% of the total mercury emitted to the environment. Recent and more extensive studies on mercury monitoring in the Florida Everglades ecosystem show that more than 95% of the mercury loading is from atmospheric deposition; that 92% of the total deposition came from local sources, such as municipal waste combustion, medical waste incinerators, electric utility boilers (coal, oil, gas), commercial and industrial boilers, and hazardous waste incinerators; and that long-distance transport of mercury wastes from certain sources are becoming increasingly important in Florida, especially mercury from waste incinerators and other atmospheric emitters, increased release of mercury from drainage and soil disturbance, and from hydrologic changes. More data are needed on sources of the mercury deposition in the Florida Everglades, especially atmospheric mercury loadings from non-local sources.

Reduction of mercury emissions and other contaminants from municipal waste incinerators using the best available technology is not always completely successful. For example, the Angers, France, solid waste incinerator plant in operation since 1974 was upgraded in 2000 to comply with the new European standards. Mean mercury emissions were reduced about 13% from 26.8 (18.6–69.4) kg/year to 23.3 (<1.0–72.5) kg/year, and mean ambient air concentrations by about 62% from 0.0004 $\mu g/m^3$ (max. 0.001 $\mu g/m^3$) to 0.00015 $\mu g/m^3$ (max. 0.0028 $\mu g/m^3$); however, the maximum values in both comparisons were larger in 2000–2001 than in 1975–79.

19.3 Properties

Mercury, a silver-white metal that is liquid at room temperature and highly volatile, can exist in three oxidation states: elemental mercury (Hg), mercurous ion (Hg_2^{2+}), and mercuric ion (Hg^{2+}). It can be part of both inorganic and organic compounds. All mercury compounds interfere with thiol metabolism, causing inhibition or inactivation of proteins containing thiol ligands and ultimately to mitotic disturbances. The mercuric species is the most toxic inorganic chemical form, but all three forms of inorganic mercury may have a common molecular mechanism of damage in which Hg^{2+} is the toxic species.

19.3.1 Physical Properties

Pure mercury is a coherent, silvery-white mobile liquid with a metallic luster. In thin layers it transmits a bluish-violet light. It freezes at about minus 39°C with contraction, forming a white, ductile, malleable mass easily cut with a knife, and with cubic crystals. When heated, the metal expands uniformly, boiling at 357.01°C at 760 mm, and vaporizing at about 360.0°C. The vapor is colorless. Mercury forms two well-defined series of salts: the mercurous salts derived from the oxide Hg_2O, and the mercuric salts from the oxide HgO. Mercuric oxide occurs in two forms: a bright red crystalline powder, and as an orange-yellow powder. The yellow form is the most reactive and is transformed into the red when heated at 400.0°C. Heating the red form results in a black compound, which regains its color on cooling; on further heating to 630.0°C, it decomposes into elemental mercury and oxygen.

Mercurous and mercuric chloride, known respectively as calomel and corrosive sublimate, are two of the most important salts of mercury. Other halogenated mercury salts include mercurous bromide, Hg_2Br_2, a yellowish-white powder insoluble in water; mercuric bromide, $HgBr_2$, comprised of white crystals sparingly soluble in cold water but readily soluble in hot water; mercurous iodide, Hg_2I_2, a yellowish-green powder; and mercuric iodide, which exists in two crystalline forms. Other mercuric and mercurous compounds include nitrates, nitrites, sulfides, sulfates, phosphides, phosphates, and ammonium salts.

19.3.2 Chemical Properties

Elemental mercury is relatively inert in dry air, oxygen, nitrous oxide, carbon dioxide, ammonia, and some other gases at room temperatures. In damp air, it slowly becomes coated with a film of mercurous oxide. When heated in air or oxygen, it is transformed into the red mercuric oxide, which decomposes into mercury and oxygen on continued heating at higher temperatures. Mercury dissolves many metals to form compounds called amalgams.

Chemical speciation is probably the most important variable influencing mercury toxicity, but mercury speciation is difficult to quantify, especially in natural environments. Mercury compounds in an aqueous solution are chemically complex. Depending on pH, alkalinity, redox, and other variables, a wide variety of chemical species are liable to be formed, having different electrical charges and solubilities. For example, $HgCl_2$ in solution can speciate into $Hg(OH)_2$, Hg^{2+}, $HgCl^+$, $Hg(OH)^-$, $HgCl_3^-$, and $HgCl_4^{2-}$; anionic forms predominate in saline environments. In the aquatic environment, under naturally occurring conditions of pH and temperature, mercury may also become methylated by biological or chemical processes, or both – although biological methylation is limited. Methylmercury (CH_3Hg^+) is the most hazardous mercury species due to its high stability, its lipid solubility, and its possession of ionic properties that lead to a high ability to penetrate membranes in living organisms. In general, essentially all mercury in freshwater fish tissues is in the form of methylmercury; however, methylmercury accounts for less than 1% of the total mercury pool in a lake.

All mercury discharged into rivers, bays, or estuaries as elemental (metallic) mercury, inorganic divalent mercury, phenylmercury, or alkoxyalkyl mercury can be converted into methylmercury compounds by natural processes. The mercury methylation in ecosystems depends on mercury loadings, microbial activity, nutrient content, pH and redox condition, suspended sediment load, sedimentation rates, and other variables. Net methylmercury production was about ten times higher in reduced sediments than in oxidized sediments. The finding that certain microorganisms are able to convert inorganic and organic forms of mercury into the highly toxic methylmercury or dimethylmercury has made it clear that any form of mercury is highly hazardous to the environment. The synthesis of methylmercury by bacteria from inorganic mercury compounds present in the water or in the sediments is the major source of this molecule in aquatic environments. This process can occur under both aerobic and anaerobic conditions, but seems to favor anaerobic conditions. Transformation of inorganic mercury to an organic form by bacteria alters its biochemical reactivity and hence its fate. Methylmercury is decomposed by bacteria in two phases. First, hydrolytic enzymes cleave the C–Hg bond, releasing the methyl group. Second, a reductase enzyme converts the ionic mercury to the elemental form, which is then free to diffuse from the aquatic environment into the vapor phase. These demethylating microbes appear to be widespread in the environment; they have been isolated from water, sediments, and soil and from the gastrointestinal tract of mammals – including humans. Some strains of microorganisms contain mercuric reductase – which transform inorganic mercury to elemental mercury – and organomercurial lyase – which degrade organomercurials to elemental mercury.

Humic substances can reduce inorganic divalent mercury (Hg^{2+}) to elemental mercury (Hg). In aquatic environments, Hg was highest under anoxic conditions, in the absence of chloride, and at pH 4.5. Under these conditions, about 25% of 400.0 μg Hg^{2+}/L was reduced to Hg in 50 h. Production of Hg was reduced in the presence of europium ions and by methylated carboxyl groups in the humic substances. Mercury is efficiently transferred through wetlands and forests in a more reactive form relative to other land use patterns, resulting in an increased uptake by organisms inhabiting these rivers or downstream impoundments and drainage lakes. The behavior and accumulation of mercury in forest soils of Guyana, South America, is related to the penetration of humic substances and the progressive adsorption onto iron oxy-hydroxides in the mineral horizons; flooding of these soils may lead to a release of 20% of the mercury initially present.

Methylmercury is produced by methylation of inorganic mercury present in both freshwater and saltwater sediments, and accumulates in aquatic food chains in which the top-level predators usually contain the highest concentrations. The percent of total mercury accounted for by methylmercury generally increases with higher trophic levels, confirming that methylmercury is more efficiently transferred to higher trophic levels than inorganic mercury compounds.

Organomercury compounds other than methylmercury decompose rapidly in the environment, and behave much like inorganic mercury compounds. In organisms near the top of the food chain, such as carnivorous fishes, almost all mercury accumulated is in the methylated form, primarily as a result of the consumption of prey containing methylmercury; methylation also occurs at the organism level by way of mucus, intestinal bacteria, and enzymatic processes, but these pathways are not as important as diet. In tissues of marine flounders, inorganic mercury compounds are strongly bound to metallothioneins and high-molecular-weight ligands; however, methylmercury has a low affinity for metallothioneins and is strongly lipophilic.

19.3.3 Biological Properties

The biological cycle of mercury is delicately balanced, and small changes in input rates, geophysical conditions, and the chemical form of mercury may result in increased methylation rates in sensitive systems. For example, the acidification of natural bodies of freshwater is statistically associated with elevated concentrations of methylmercury in the edible tissues of predatory fishes. Acidification has a stronger effect on the supply of methylmercury to the ecosystem than on specific rates of uptake by the biota. In chemically sensitive waterways, such as poorly buffered lakes, the combined effects of acid precipitation and increased emissions of mercury to the atmosphere (with subsequent deposition) pose a serious threat to the biota if optimal biomethylation conditions are met. In remote lakes of the Adirondack mountain region in upstate New York, fish contain elevated mercury concentrations in muscle; mercury loadings in fish were directly associated with decreasing water column pH and increasing concentrations of dissolved organic carbon (DOC), although high DOC concentrations may complex methylmercury, diminishing its bioavailability. At high concentrations of monomeric aluminum, the complexation of methylmercury with DOC decreases, enhancing the bioavailability of methylmercury.

Mercury excretion in mammals is through the urine and feces, and is dependent on the form of mercury, dose, and time post-exposure. With mercury vapor, there is minor initial loss from exhalation and major loss via fecal excretion. With inorganic mercury, fecal loss is predominant after initial exposure and renal excretion increases with time. With methylmercury, about 90% is excreted in feces after acute or chronic exposure and does not change over time. Based on animal studies, all forms of mercury can cross the placenta to the fetus. Fetal uptake of elemental mercury in rats – possibly due to its high solubility in lipids – is 10–40 times higher than uptake after exposure to inorganic mercury salts. Concentrations of mercury in the fetus after exposure to alkylmercuric compounds, when compared to elemental mercury, are twice those found in maternal tissues, and methylmercury levels in fetal red blood cells are 30% higher than in maternal cells. The positive fetal maternal gradient and increased mercury concentration in fetal erythrocytes enhances fetal toxicity to mercury, especially after exposure to alkylmercury. Maternal milk contains about 5% of the mercury concentration of maternal blood; however, neonatal exposure to mercury may be greatly augmented by nursing.

Elemental or metallic mercury is oxidized, probably via catalases, to divalent mercury after absorption into body tissues. Most inhaled mercury vapor absorbed into erythrocytes is transformed into divalent mercury, but some is also transported as metallic mercury to the brain, where biotransformation may occur. Some metallic mercury may cross the placenta into the fetus. Oxidized metallic mercury is then accumulated by brain and fetus. Organomercurials also undergo biotransformation into divalent mercury compounds in tissues by cleavage of the carbon-mercury bond, with no evidence of organomercury formation by mammalian tissues. Phenylmercurials are converted to inorganic mercury more rapidly than the shorter-chain methylmercurials. Phenyl and methoxyethyl mercurials are excreted at about the same rate as inorganic mercury whereas methylmercury excretion is slower. The half-time persistence of methylmercurials in mammalian tissues is about 70 days and seems to follow a linear pattern. For inorganic mercury salts, the biological half-time is about 40 days. And for elemental mercury or mercury vapor the half-time persistence in tissues ranges between 35 and 90 days and also seems to be linear.

19.3.4 Biochemical Properties

Mercury binds strongly with sulfhydryl groups, and has many potential target sites during embryogenesis; phenylmercury and methylmercury compounds are among the strongest known inhibitors of cell division. In mammalian hepatocytes, the L-alanine carrier contains a sulfhydryl group that is essential for its activity and is inhibited by mercurials. In the little skate (*Raja erinacea*), $HgCl_2$ inhibits Na^+-dependent alanine uptake and Na^+/K^+-ATPase activity, and increases K^+ permeability. Inhibition of Na^+-dependent alanine in skate hepatocytes by $HgCl_2$ is attributed to three different concentration-dependent mechanisms: (1) direct interaction with the transporters; (2) dissipation of the Na^+ gradient; and (3) loss of membrane integrity. Organomercury compounds, especially methylmercury, cross placental barriers and can enter mammals by way of the respiratory tract, gastrointestinal tract, skin, or mucus membranes. When compared with inorganic mercury compounds, organomercurials are more completely absorbed, are more soluble in organic solvents and lipids, pass more readily through biological membranes, and are slower to be excreted. Biological membranes, including those at the blood-brain interface and placenta, tend to discriminate against ionic and inorganic mercury, but allow relatively easy passage of methylmercury and dissolved mercury vapor. As judged by membrane model studies, it appears that electrically neutral mercurials are responsible for most of the diffusion transport of mercury, although this movement is modified significantly by pH and mercury speciation. It seems, however, that the liposolubility of methylmercury is not the entire reason for its toxicity and does not play a major role in its transport. This hypothesis needs to be examined further in studies with living membranes.

In liver cells, methylmercury forms soluble complexes with cysteine and glutathione, which are secreted in bile and reabsorbed from the GI tract. In general, however, organomercurials undergo cleavage of the carbon-mercury bond releasing ionic inorganic mercury. Mercuric mercury induces synthesis of metallothioneins, mainly in kidney cells. Mercury within renal cells becomes localized in lysosomes.

The finding of naturally elevated mercury concentrations in marine products of commerce is worrisome to regulatory agencies charged with protection of human health. This topic is discussed later. At this time, however, many authorities argue that mercury–selenium interactions are the key to methylmercury bioavailability in human diets. In the case of marine mammals and seabirds, for example, mercury is accumulated from the diet mainly as methylmercury and then transformed into the less toxic inorganic mercury. Accordingly, most of the tissue mercury found in high concentrations of these two marine groups is inorganic mercury. Many authorities aver that selenium detoxifies inorganic mercury by forming complexes in a 1:1 molar ratio. Field studies on marine mammal livers corroborate the equimolar ratio of mercury and selenium; moreover, mercuric selenide (HgSe) – an inert end product of mercury detoxification in marine mammals – was found in livers of marine mammals and birds. A proposed mercury detoxification model in higher-trophic marine animals involves transformation of dietary methylmercury to inorganic mercury by reactive oxygen species, gut microflora, and selenium. Inorganic mercury binds to metallothioneins or forms an equimolar Hg–Se complex, and subsequently combines with high-molecular-weight compounds in liver. Glutathione molecules solubilize HgSe; the complex then binds to a specific protein. The protein-bound Hg–Se complex is thought to be the precursor of mercuric selenide, which occurs in macrophages of marine mammals.

19.3.5 Mercury Transport and Speciation

The global mercury cycle involves mercury release from geological and industrial processes into water and the atmosphere, followed by sedimentation via rainfall and by microbial metabolism that releases mercury from soil and sediments and transforms mercury from one chemical form to another.

Atmospheric-borne mercury, including anthropogenic mercury is deposited everywhere including remote areas of the globe, hundreds of kilometers from the nearest mercury source, as evidenced by its presence in ancient lake sediments and glacial ice. In Amituk Lake in the Canadian Arctic, recent annual deposition of mercury was estimated at 15.1 kg, about 56% from snowpack, and the rest from precipitation. This represents a dramatic increase from historic annual burdens of 6.0 kg of mercury annually in this remote area; the effects of this increase on Arctic watersheds are unknown.

Many important sources affecting global mercury cycles emit elemental metallic mercury (Hg) in gaseous form, and to a lesser extent gaseous and particulate species of Hg^{2+}. Gaseous and particulate Hg^{2+} are removed from the atmosphere through rainfall and dry deposition, limiting long-range transport. Inorganic Hg^{2+} can be readily reduced to Hg by natural processes in terrestrial and aquatic ecosystems. Elemental mercury can be oxidized in the atmosphere to Hg^{2+}, which is removed through wet and dry deposition. About 67% of the mercury in global fluxes is a result of human activities and the rest from natural emissions. Soils and sediments are the primary sinks for atmospherically derived mercury; however, these enriched pools can be remobilized through volatilization, leaching, and erosion.

Mercury speciation varies in atmospheric, aquatic, and terrestrial environments. In the atmosphere, mercury is in the form of gaseous elemental Hg (95%), Hg^{2+} (called reactive gaseous mercury), and trace amounts of methylmercury. Particulate and reactive mercury in the atmosphere travels short distances, usually less than 50 km, and has a residence time of about 1 year. Reactive gaseous mercury is assumed to be $HgCl_2$, with some $Hg(NO_3)_2 \cdot H_2O$ in the gas phase. Reactive gaseous mercury may comprise a majority of atmospheric gaseous mercury at some locations, for example, springtime in Alaska, and this component is rapidly removed from the atmosphere by wet and dry deposition and available for methylation once deposited. Mercury point sources and rates of particle scavenging are key factors in atmospheric transport rates to sites of methylation and subsequent entry into the marine food chain. Airborne soot particles transport mercury into the marine environment either as nuclei for rain-drop formation or by direct deposition on water. In early 1990, both dimethylmercury and monomethylmercury species were found in the subthermocline waters of the equatorial Pacific Ocean; the formation of these alkylmercury species in the low oxygen zone suggests that Hg^{2+} is the most likely substrate.

In aquatic environments, Hg, and methylmercury species are the most common, with concentrations low, usually in the picogram/L to microgram/L range, except in the vicinity of anthropogenic or natural mercury sources. The speciation of mercury in water is influenced by redox, pH, and ligands. In most aerated surface waters near pH 7.0, ion-pair formation for CH_3^+ and Hg^{2+} is dominated by dissolved organic matter and chloride. Under anoxic conditions, Hg^{2+} and CH_3^+ are present mainly as sulfide and sulfhydryl ion pairs. Complexed Hg^{2+} sulfides are less available for methylation. Concentrations of total mercury in uncontaminated, unfiltered freshwater samples range from 0.3 to 8.0 ng/L, but range from 10.0 to 40.0 ng/L near mercury sources, and up to 1000.0 ng/L in waters contaminated by mercury tailings from gold mines.

Sulfate-reducing bacteria are the most important mercury-methylating agents in aquatic environments, with the most important site of methylation at the oxic–anoxic interface in sediments; a similar pattern is documented for wetlands. In sediments, microbial methylation of mercury is fastest in the upper profiles where rate of sulfate reduction is greatest. Methylcobalamin, produced by bacteria, is the active methyl donor to the Hg^{2+} ion; methylcobalamin reacts with Hg^{2+} to form CH_3Hg^+. Methylation also occurs, but to a lesser degree, in aerobic freshwater and seawater, in aquatic plants, and in mucosal slime and intestines of fish. Abiotic formation of CH_3Hg^+ compounds in sediments is documented; however, amounts formed are small when compared to biotic processes.

Demethylation occurs via abiotic and biotic processes in the near-surface sediments and in the water column. An oxidative demethylation pathway similar to that of the degradation of methanol or monomethylamine by methanogens has been proposed for methylmercury degradation. Photodegradation of methylmercury in surface waters of freshwater lakes is documented at rates up to 18% daily and is quantitatively important in mercury budgets of that ecosystem; the end products of mercury photodemethylation are not known with certainty. Frequently, however, methylmercury concentrations in aerobic lake-water surfaces increase during sunlight hours or remain unchanged. This phenomenon is linked to dissolved organic matter (DOM) and solar radiation. Specifically, to certain fractions of DOM that generates methylmercury when exposed to sunlight, especially in water from lakes with logged watersheds. The mechanism to account for methylmercury production is not clear at present; however, it corrects the conventional wisdom that methylmercury is rapidly photodegraded.

Terrestrial soils are a significant contributor of mercury to surface waters. Moreover, up to 60% of the atmospherically deposited mercury that reaches lakes originates from the associated terrestrial watershed. The main biogeochemical reactions affecting the transport and speciation of mercury in the terrestrial watershed include formation of mercury ligands, mercury adsorption and desorption, and elemental mercury reduction and volatilization. In terrestrial environments, OH^-, Cl^-, and S^- ions have the greatest impact on inorganic mercury-ligand formation. Under oxidized surface soil conditions, $Hg(OH)_2$, $HgCl_2$, $HgOH^+$, HgS (cinnabar), and Hg^0 are the dominant inorganic mercury forms and usually bound to organic and mineral ions. Under reducing conditions, common mercury forms are $HgSH^+$, HgOHSH, and HgClSH, and many are further bound to both inorganic and organic ligands. The following organomercurials predominated in terrestrial soils: $CH_3HgCl > CH_3HgOH >$ free CH_3Hg^+.

In upland soils, mercury is mostly in the form of Hg^{2+} species sorbed to organic matter in the humus layer and to a lesser extent to soil minerals. The overall adsorption of mercury to mineral and organic particles is positively correlated, in order of importance, with surface area, organic content, cation exchange capacity, and grain size. Mercury adsorption and desorption to mineral and organic surfaces is strongly influenced by pH and dissolved ions; for example, increased Cl^- and decreased pH – alone or together – can decrease mercury adsorption, and clays and organic soils have the highest capability of adsorbing mercury. Common forms of methylated mercury in soils depend on pH. At pH 2 to about 4.7, the most common forms were $CH_3HgCl >$ free $CH_3Hg^+ > CH_3HgOH$; at pH 4.7 to about 7.5, the most common forms were $CH_3HgCl > CH_3HgOH >$ free CH_3Hg^+; and at pH 7.5–10, this order was $CH_3HgOH > CH_3HgCl \gg$ free CH_3Hg^+. Reduction of abiotic inorganic mercury is increased with increasing electron donors, low redox potential, and sunlight intensity. Factors that increase mercury volatilization from soils include increased soil permeability, higher temperatures, and increased sunlight intensity; therefore, increased volatilization is expected in tropical climates. A decrease in mercury adsorption and an increase in soil moisture can also increase volatilization.

Additional research is recommended on inorganic mercury-ligand formation in water and runoff and its effects on methylmercury formation in soils, and on quantification of the sources and transport characteristics of methylmercury in terrestrial environments. The mercury-ligand form exiting the terrestrial watershed will strongly influence the mercury/methylmercury bioaccumulation potential in surface waters. Accordingly, more analyses are needed to determine the mercury forms in terrestrial watershed runoff in dissolved and particulate fractions.

19.3.6 Mercury Measurement

Techniques for analysis of different mercury species in biological samples and abiotic materials include atomic absorption, cold vapor atomic fluorescence spectrometry, gas-liquid chromatography with electron capture detection, neutron activation, and inductively coupled plasma mass spectrometry. Methylmercury concentrations in marine biological tissues are detected at concentrations as low as 10.0 μg Hg/kg tissue using graphite furnace sample preparation techniques and atomic absorption spectrometry.

19.4 Mercury Poisoning and Treatment

The toxicity of mercury has been recognized since antiquity. No other metal demonstrates the diversity of effects caused by different biochemical forms than mercury. Toxicologically, there are three forms of mercury: elemental mercury, inorganic mercury compounds, and organomercurials. Among the various forms of mercury and its compounds, elemental mercury in the form of vapor, mercuric mercury, and methylmercury have the greatest toxicological potential. Metallic or elemental mercury volatilizes to mercury vapor at ambient air temperatures, and most human exposure is by inhalation. Mercury vapor is lipid soluble, readily diffuses across the alveolar membranes, and concentrates in erythrocytes and the central nervous system. Inorganic mercury salts may be divalent (mercuric) or monovalent (mercurous). Gastrointestinal absorption of inorganic salts of mercury from food is less than 15% in mice and about 7% in a study of human volunteers; however, methylmercury absorption in the GI tract is 90–95%. Methylmercury, when compared to inorganic mercury compounds, is about five times more soluble in erythrocytes than in plasma, and about 250 times more abundant in hair than in blood.

19.4.1 Poisoning

The primary mode of action of both inorganic and organic mercury compounds is associated with interference of membrane permeability and enzyme reactions through binding of mercuric ion to sulfhydryl groups, although organomercurials may penetrate membranes more readily. Early accounts of acute and chronic mercury poisoning and their treatment follow; however, it is cautioned that treatment should be under the guidance of a physician.

19.4.1.1 Elemental Mercury

Metallic or elemental mercury volatilizes to mercury vapor at ambient air temperatures, and most human exposure is by way of inhalation. The saturated vapor pressure at 20.0°C is 13.2 mg/m^3. This value far exceeds the threshold limited value (TLV) of 0.05 mg/m^3; accordingly, mercury intoxication due to inhalation of the vapor readily occurs in various occupational and environmental situations. Mercury vapor readily diffuses across the alveolar membrane and is lipid soluble so that it has an affinity for the central nervous system and red blood cells. Metallic mercury, unlike mercury vapor, is only slowly absorbed by the GI tract (0.01%) at a rate related to the vaporization of the elemental mercury and is of negligible toxicological significance.

Inhaled mercury vapor (Hg) is readily oxidized via the catalase–H_2O_2 complex and converted to Hg^{2+}, mainly in liver and erythrocytes. Although this reaction is rapid, some Hg crosses the blood-brain barrier and accumulates to a greater extent than does Hg^{2+} after ionic mercury exposure. Because Hg^{2+} is reduced to Hg, there is probably an oxidation-reduction cycle of mercury in the body.

In typical Hg vapor poisonings, excessive bronchitis and bronchiolitis occur in a few hours after heavy exposure, i.e., direct inhalation of mercury vapor generated from heating metallic mercury. This is followed by pneumonitis and respiratory distress, excitability, and tremors. If the amount inhaled is sufficiently large, renal failure will develop.

In one Japanese factory producing sulfuric acid using Hg in the process, a few of the workers died of respiratory distress associated with renal failure. Moderate and repeated exposure results in classical mercury poisoning. Inhalation of mercury vapor, if not fatal, is associated with an acute, corrosive bronchitis, interstitial pneumonitis, tremors, and increased excitability. With chronic exposure to mercury vapor the major effects are on the central nervous system. Early signs are nonspecific and have been termed "micromercurialism" or "asthenic-vegetative syndrome." Micromercurialism is characterized by weakness, fatigue, anorexia, weight loss and GI disturbance. This syndrome is characterized clinically by at least three of the following: tremors, thyroid enlargement, increased radioiodine uptake by thyroid, tachycardia, unstable pulse, dermographism, gingivitis, changes in blood chemistry, and increased excretion of mercury in urine. With increasing exposure the symptoms include tremors of the fingers, eyelids, and lips and may progress to generalized trembling of the entire body and violent chronic spasms of the arms and legs. This is accompanied by changes in personality and behavior, with loss of memory, increased excitability, severe depression, delirium, and hallucination. Another characteristic feature of mercury toxicity is severe salivation. Tremors, increased excitability, and gingivitis have been recognized historically as the major manifestation of mercury poisoning from inhalation of mercury vapor and exposure in the fur, felt, and hat industry to mercuric nitrate.

Effects of mercury vapor exposure lasts long after cessation of exposure, although typical symptoms such as tremors, gingivitis, and salivation usually disappear quickly. Residual effects due to previous exposure have been documented in workers with a peak urinary mercury concentration >0.6 mg/L; neurobehavioral disturbances were observed in these workers 20–35 years post exposure.

19.4.1.2 Inorganic Mercurials

Inorganic mercury salts may be divalent (mercuric) or monovalent (mercurous). Chronic exposure to low levels of inorganic mercury compounds are associated with psychological changes including abnormal irritability (erethismas mercurialis), colored mercury compounds in the anterior lens capsule of the eye (mercurialentis), tremors, and excessive salivation. Inorganic forms of mercury are corrosive and produce symptoms that include metallic taste, burning, irritation, salivation, vomiting, diarrhea, upper GI tract edema, abdominal pain, and hemorrhage. These effects are seen acutely and may subside with subsequent lower GI tract ulceration. Large ingestion of the mercurial salts may produce kidney ingestion, such as nephrosis, oliguria, and anuria.

19.4.1.2.1 Mercuric Mercury

Acute poisoning by mercurials usually occurs in the case of mercuric perchlorides, with intense gastrointestinal inflammation, vomiting, diarrhea, and extreme collapse. Treatment is usually with albumin, which forms an insoluble compound with the perchloride. Chronic poisoning, or mercurialism, is marked by tenderness of the teeth while eating and offensive breath. Later the gums become inflamed, salivary glands are swollen and tender, and saliva pours from the mouth. The teeth may become loose and fall out. The symptoms are aggravated until the tongue and mouth ulcerate, the jawbone necroses, hemorrhages occur in various parts of the body, and death results from anemia, septic inflammation, or exhaustion. Treatment includes administration of potassium iodide in low, repeated doses.

Chronic exposure to low levels of inorganic mercuric compounds produces tremors, excess salivation, and psychological changes characterized by irritability and excitement. Collectively, this is often described as the "mad hatter syndrome." Mercuric mercury (Hg^{2+}) is a potential toxic chemical, although it is poorly absorbed by the GI tract and other body parts. Accidental or suicidal ingestion of mercuric chloride or other mercuric salts produces corrosive ulceration, bleeding, necrosis of the intestinal tract, and are usually accompanied by shock and circulatory collapse. If the patient survives the gastrointestinal damage, renal failure may occur within 24 h, owing to necrosis of the proximal tubular epithelium followed by diminished secretion of urine, and kidney pathology. These may be followed by ultrastructural changes consistent with irreversible cell injury. Regeneration of the tubular lining is possible if the patient can be maintained by dialysis.

The pathogenesis of chronic mercury kidney damage has two phases: an early phase with antibasement membrane glomerulonephritis, followed by a superimposed immune-complex glomerulonephritis. The pathogenesis of the nephropathy in humans appears similar, although early glomerular nephritis may progress to an interstitial immune-complex nephritis. Injection of mercuric chloride produces necrosis of kidney epithelium.

Cellular changes include fragmentation of the plasma membrane, disruption of cytoplasmic membranes, loss of ribosomes, and mitochondrial swelling; however, all of these changes are associated with renal cell necrosis from a variety of insults. High doses of mercuric chloride are directly toxic to renal tubular lining cells and chronic low-level doses to mercuric salts may induce an immunologic glomerular disease that is the most common form of mercury-induced nephropathy. Chronic mercury-induced kidney damage seldom occurs in the absence of detectable damage to the nervous system. Although kidneys contain the highest concentrations of mercury following exposure to inorganic mercury salts and mercury vapor, it is emphasized that organomercurials concentrate in the posterior cortex of the brain.

19.4.1.2.2 Mercurous Mercury

Mercurous (Hg^{+}) mercury compounds are unstable and easily break down to Hg and Hg^{2+}. Mercurous compounds are less corrosive and less toxic than mercuric compounds, probably because they are less soluble. Calomel – a mercurous-chloride-containing powder with a long history of medical use – is known to be responsible for acrodynia or "pink-disease" in children when used as a teething powder. This is probably a hypersensitivity response by skin to mercurous salts producing vasodilation, hyperkeratosis, and excessive sweating. Afflicted children had fever, pink-colored rashes, swollen spleen and lymph nodes, and hyperkeratosis and swelling of fingers. Effects were independent of dose and are thought to be a hypersensitivity reaction.

19.4.1.3 Organomercurials

Among the organomercurials, alkylmercurials – especially methylmercurials (CH_3Hg^{+}) – are the most environmentally and ecologically significant. Methylmercury is naturally produced from inorganic mercury by microbial activity; methylmercurials are lipid soluble and readily cross blood-brain and placental barriers. The sensitivity of the developing brain to methylmercury is due to placental transfer of lipophilic methylmercury to the central nervous system. The blood-brain barrier is incomplete during the first year of life in humans, and methylmercury can cross this barrier during that time.

Phenylmercurials ($C_6H_5Hg^{+}$) and methoxyethyl mercurials ($CH_3OC_2H_4Hg^{+}$) have been used as fungicides and pesticides and readily transform into inorganic mercurials in living organisms with toxic properties similar to those of inorganic mercurials.

Ingestion of organomercurials, such as ethylmercury, may produce symptoms of nausea, vomiting, abdominal pain, and diarrhea, but in most cases the main toxicity is neurological involvement presenting with paresthesias, visual disturbances, mental disturbances, hallucinations, ataxia, hearing defects, stupor, coma, and death. Symptoms may occur for several weeks after exposure. Exposure and poisoning can occur after ingestion of mercury-contaminated seafood, grains, or inhalation of vaporized organomercurials.

The toxic signs of alkylmercury compounds, such as methylmercury, are different from those of inorganic mercurials owing to greater penetration of the organomercurials into the brain. Methylmercury causes necrosis of the granule cell layer of the cerebellum, which is associated with carbohydrate metabolism and kidney disorders. Focal atrophy of the cortex, with sensory disturbances, ataxia, and dysarthria is found after methylmercury intoxication. The emotional changes and autonomous nervous system involvement with inorganic mercury compounds are not seen with organomercurials.

Sensory nerve fibers are selectively damaged. The primary mode of action of both inorganic and organic mercury compounds may be interference with membrane permeability and enzyme actions by binding of mercuric ion to sulfhydryl groups. Small neurons in the CNS are more likely to be damaged than large neurons in the same area by methylmercury. The major clinical features of methylmercury toxicity are neurological, consisting of paresthesia, ataxia, dysarthria, and deafness, appearing in that order. The main pathologic features include degeneration and necrosis of neurons in focal areas of the cerebral cortex and in the granular layer of the cerebellum. Studies of both inorganic- and organic-mercury-related neuropathy show degeneration of primary sensory ganglion cells. Lesion distribution in the CNS suggests that mercury damages small nerve cells in cerebellum and visual cortex. Methylmercuric chloride, as an environmental pollutant, has produced renal damage in humans and animals through inhibition of mitochondrial and other enzyme systems.

19.4.2 Mercury Treatment

Treatment of mercury-poisoned victims is complex, and should be supervised by a physician. Therapy of mercury poisoning is directed to lowering the concentration of mercury at the critical organ or site of injury. For the most severe cases, particularly with acute renal failure, hemodialysis together with infusion of mercury-chelating agents such as cysteine and penicillamine is warranted. For less severe cases of inorganic mercury poisoning, chelation with BAL is recommended. Chelation therapy, however, is not effective for poisoning with organomercurials. In such cases, oral administration of a nonabsorbable thiol resin that binds mercury and enhances intestinal excretion, or surgical establishment of gallbladder drainage have proven satisfactory.

Treatment usually consists of emesis or lavage followed by administration of activated charcoal and a saline cathartic. Cow's milk may be given to help precipitate the mercury compound. Blood and urine levels of mercury may be useful in determining whether chelating agents, such as D-penicillamine or BAL (dimercaprol) should be administered. D-penicillamine is given at 250.0 mg orally four times daily in adults. For children, D-penicillamine is given at 100.0 mg/kg body weight daily to a maximum recommended dose of 1000 mg daily for 3–10 days with continuous monitoring of mercury urinary excretion. In patients unable to tolerate penicillamine, BAL may be administered at a dose of 3.0–5.0 mg/kg body weight (BW) every 4 h by deep intramuscular (im) injection for the first 2 days, then 2.5–3.0 mg/kg BW im every 6 h for 2 days, followed by 2.5–3.0 mg/kg BW im every 12 h for 1 week. Adverse reactions associated with BAL, such as skin eruptions (urticaria), can often be controlled with antihistamines such as diphenylhydramine. The development of renal failure contraindicates use of penicillamine because the kidney is the main excretory route for penicillamine. BAL therapy can be used cautiously in spite of renal failure since BAL is excreted in the bile; however, BAL toxicity, which consists of fever, rash, hypertension, and CNS stimulation must be closely monitored. Dialysis is not recommended because it does not remove chelated or free mercury.

Mercury-antagonistic and mercury-protectant drugs and compounds now include 2,3-dimercaptopropanol, polythiol resins, selenium salts, thiamin, vitamin E, metallothionein-like proteins, and sulfhydryl agents. Thiols (R-SH), which compete with mercury for protein binding sites, are the most important antagonists of inorganic mercury salts, and have been used extensively in attempts to counteract mercury poisoning in humans. Thiamin was the most effective of the Group VIB derivatives (which includes sulfur, selenium, and tellurium) in protecting against organomercury poisoning in higher animals. The protective action of selenium (Se) against adverse or lethal effects induced by inorganic or organic mercury salts is documented for algae, aquatic invertebrates, fish, birds, and mammals. For example, selenium, as sodium selenite, that was introduced into a nonacidified-mercury-contaminated lake in Sweden to concentrations of 3.0–5.0 μg Se/L (from 0.4 μg Se/L) and sustained at this level for 3 years resulted in declines of 50–85% in mercury concentrations in fish muscle. The mercury-protective effect of selenium is attributed to competition by selenium for mercury-binding sites associated with toxicity, formation of a Hg–Se complex that diverts mercury from sensitive targets, and prevention of oxidative damage by increasing the amount of selenium available to the selenium-dependent enzyme glutathione peroxidase. In seabirds, an equivalent molar ratio of 1:1 between total mercury and selenium was found in livers of individual seabirds, which contained more than 100.0 mg Hg/kg DW; this relation was unclear in other individuals which had relatively low mercury levels. The selenium-protective mechanism in birds is explained by a strong binding between mercury and selenium, possibly by the formation of a selenocystamine methylmercury complex ($CH_3HgSeCH_2CH_2NH_3^+$), mercury binding to selenocysteine residues ($CH_3HgSeCH_2CH(NH_3)$ (COO) H_2O), the formation of insoluble mercuric selenide (HgSe), or binding of mercury to SeH residues of selenoproteins, notably metallothioneins with thiols replaced by SeH. However, high selenium concentrations in tissues of marine wading birds do not have their origin in elevated levels of mercury. The Se:Hg ratio in marine wading birds from the Wadden Sea is 32:1 and greatly exceeds the 1:1 ratio found when selenium is accumulated to detoxify mercury. In marine mammals and humans, selenium and mercury concentrations are closely related, almost linearly in a 1:1 molar ratio. The molar ratio between mercury and selenium in marine mammals suggests that the major mechanism of detoxification is through the formation of a complex Hg–Se that leads to mercury demethylation. The site of this process is the liver in which mercury appeared mainly as inorganic, whereas in the muscle the percent of organic to total mercury was much higher. Detoxification is limited in lactating female whales, and sometimes in all the individuals of one school. Selenium does not, however, protect against mercury-induced birth defects, such as cleft palate in mice. It is clear that more research is needed on mercury protectants.

19.5 Mercury Concentrations in Abiotic Materials

Mercury burdens in sediments and other nonbiological materials are estimated to have increased up to five times pre-human levels, primarily as a result of anthropogenic activities. Maximum increases are reported in freshwater and estuarine sediments and in freshwater lakes and rivers, but estimated increases in oceanic waters and terrestrial soils have been negligible. Methylmercury accounts for a comparatively small fraction of the total mercury found in sediments, surface waters, and sediment interstitial waters of Poplar Creek, Tennessee, which was initially contaminated with mercury in the 1950s and 1960s. Mercury measurements in Poplar Creek in 1993–94 showed that methylmercury accounted for 0.01% of the total mercury in sediments, 0.1% in surface waters, and 0.3% in sediment interstitial waters. The residence time of mercury in nonbiological materials is variable, and depends on a number of physicochemical conditions. Estimated half-time residence values for mercury are 11 days in the atmosphere, 1000 years in terrestrial soils, 2100–3200 years in ocean waters, and >250 million years in oceanic sediments; however, this estimate was only 1 month to 5 years for water from the contaminated Saguenay River in Quebec. This section documents mercury concentrations in air, coal, pelagic clays, sediments, sewage sludge, snow, soils, suspended particulate matter, seawater, freshwater, groundwater, and sediment interstitial waters from selected geographic locales.

19.5.1 Air

Most (up to 59.1%) of the mercury contributed to the atmosphere each year is from anthropogenic sources such as combustion of fossil fuels from power plants, with natural sources such as oceans, land runoff, and volcanoes contributing almost all the remainder. Atmospheric concentrations of total mercury in the northern hemisphere are about three times higher than those sampled in the southern hemisphere owing to greater sources from human activities in the comparatively industrialized and populated north. Different mercury species are dominant at different Japanese locations. For example, in 1977–1978, Hg^{2+} was the dominant species in air over hot springs, volcanoes, and urban centers; however, Hg was dominant in air over chloralkali plants and rural areas. Enrichment of toxic metals in respirable particulate matter emissions from a coal-fired power plant in Central India is documented, especially for mercury that was enriched 4.8 times over the coal. In general, mercury concentrations were low in the atmosphere over the open ocean (1.0–3.0 ng/m^3), up to 100.0 ng/m^3 in the air of large cities, and highest (495.0 ng/m^3, 74% Hg^{2+}) in the atmosphere over open volcanoes.

19.5.2 Coal

Mercury concentrations were highest in lignite coal (0.12 mg/kg DW), lowest in subbituminous coal (0.03 mg/kg DW) and intermediate (0.07 mg/kg DW) in bituminous coal samples measured. More recent information indicates that coal contains, on an average 0.2 mg Hg/kg and may contain as much as 1.0 mg/kg. Most of the mercury in coal is associated with arsenic-bearing pyrite; other forms include organically bound mercurials, elemental mercury, and mercuric sulfides and selenides. In coal samples with low pyrite, mercury selenides may be the primary form.

It is noteworthy that installation of available pollution control technology can significantly lower mercury concentrations in surface soils near coal-fired power plants in the United Kingdom. Thus, surface soils near a coal-fired power plant with a flue gas desulfurization (FGD) system contained 0.297 mg total Hg/kg DW vs. 0.495 mg total Hg/kg DW in a coal-fired power plant without FGD, a 40% reduction.

19.5.3 Sediments

Much of the mercury that enters freshwater lakes is deposited in bottom sediments. Sedimentary pools of mercury in these lakes greatly exceed the inventories of mercury in water, seston, and fish, and the release of mercury from the sediments would significantly increase bioavailability and uptake. The dry weight mercury concentrations of sediments seem to underrepresent the significance of the shallow water sediments as a reservoir of potentially available mercury when compared to the mass per volume of wet sediment, which more accurately portrayed the depth distribution of mercury in Wisconsin seepage lakes. The increase in the mercury content of recent lake sediments in Wisconsin is attributed to increased atmospheric deposition of mercury, suggesting that the high mercury burdens measured in gamefish in certain Wisconsin lakes originated from atmospheric sources. Levels of mercury in sediments may be reflected by an increase in mercury content in epibenthic marine fauna. For example, mercury concentrations in sediments near the Hyperion sewer outfall in Los Angeles, which ranged up to 820.0 μg/kg and decreased with increasing distance from the outfall, were reflected in the mercury content of crabs, scallops, and whelks. Concentrations of mercury were highest in organisms collected nearest the discharge, and lowest in those collected tens of kilometers away.

In sediments that were anthropogenically contaminated with mercury, concentrations were significantly elevated (usually >20.0 mg/kg) when compared with uncontaminated sediments (usually <1.0 mg/kg). Significant mercury enrichment in sediments of Newark Bay, New Jersey, may represent a hazard to aquatic life. In Finland, sediments near a pulp and paper mill – where mercury was used as a slimicide – contained up to 746.0 mg Hg/kg dry weight. In Florida, methylmercury in sediments from uncontaminated southern estuaries in 1995 accounted for 0.77% of the total mercury and was not correlated with total mercury or organic content of sediments.

19.5.4 Sewage Sludge

Concentrations of total mercury in sewage sludge from 74 publicly owned treatment works in Missouri ranged from 0.6 to 130.0 mg/kg DW, and strongly indicates that sewage sludge applications to agricultural soils should be carefully monitored.

19.5.5 Snow and Ice

Total mercury concentrations in Siberian snow ranged between 8.0 and 60.0 ng/L; the maximum methylmercury concentration was 0.25 ng/L.

Mercury concentrations in Arctic ice 4000–12,000 years ago during the precultural period were about 20% that of present day concentrations; however, 13,000–30,000 years ago during the last glacial period, they were about five times higher than precultural levels. Ice cores taken in Wyoming representing the 270-year period 1715–1985 demonstrate that annual concentrations between 1715 and 1900 were usually <5.0 ng/L, except for volcanic eruptions in 1815 (Tambora; up to 15.0 ng/L) and 1883 (Krakatoa; up to 25.0 ng/L), and the California gold rush between 1850 and 1884 (up to 18.0 ng/L). Between 1880 and 1985, mercury concentrations ranged up to 10.0 ng/L annually (1880–1950) due to industrialization and World War II manufacturing (1939–1945), up to 30.0 ng/L owing to the eruption of Mount St. Helena in 1980, and around 15.0 ng/L for the remainder due to increased industrialization.

19.5.6 Soils

In general, soil mercury concentrations were higher in the vicinity of acetaldehyde plants, solid waste landfill sites, urban areas, coal-fired power plants, waste incinerators, and crematoriums. The highest mercury concentrations recorded in soils were in soils receiving mercury-containing wastes of a Chinese acetaldehyde plant; these soils contained up to 329.9 mg total mercury (0.045 mg methylmercury)/kg dry weight vs. up to 1.78 mg total mercury/kg dry weight from reference sites.

Jamaican agricultural soils contained up to 0.83 mg total mercury/kg DW, mean 0.221 mg/kg DW; this was far in excess of Danish and Canadian guidelines for mercury in crop soils, i.e, <0.007 mg/kg DW. Jamaican soils also exceeded Danish and Canadian proposed limits for arsenic, cadmium, copper, and chromium in agricultural soils.

Total mercury concentrations in surface soils near combustion plants range from 0.3 to 1.47 mg/kg DW, and in subsurface soils from 0.09 to 2.3 mg/kg DW. Total mercury in both topsoils and subsoils is dominated by elemental mercury and mercuric sulfide, with increasing sulfur content in soil associated with increasing HgS; solubility and pH conditions also influence the occurrence and distribution of mercury in soils near combustion plants. Uptake from the soil is probably a significant route for the entrance of mercury into vegetation in terrestrial ecosystems. In Italy, elevated mercury concentrations in soils near extensive cinnabar deposits and mining activities were reflected in elevated mercury concentrations in plants grown on those soils. Mercury concentrations in tissues of different species of vascular plants growing on flood plain soils at Waynesboro, Virginia, were directly related to soil mercury concentrations that ranged between <0.2 and 31.0 mg Hg/kg DW soil. In a study conducted in Fulton County, Illinois, it was shown that repeated applications of sewage sludge to land would significantly increase the concentration of mercury in surface soil. However, 80–100% of the mercury remained in the top 15 cm and was not bioavailable to terrestrial vegetation. The authors concluded that models developed by the U.S. Environmental Protection Agency overpredict the uptake rates of mercury from sludge-amended soils into grains and animal forage, and need to be modified.

19.5.7 Water

Mercury-sensitive ecosystems are those where comparatively small inputs or inventories of total mercury, i.e., 1.0–10.0 g Hg/ha, result in elevated concentrations of methylmercury in natural resources; these systems are characterized by efficient conversion of inorganic mercuric mercury to methylmercury sufficient to contaminate aquatic and wildlife food webs. Known sensitive ecosystems include surface waters adjoining wetlands, low alkalinity or low pH lakes, wetlands, and flooded terrestrial areas.

In southern Nevada groundwaters, mercury concentrations were usually less 20.0 ng/L in unfiltered samples and less than 10.0 ng/L in filtered samples. Mining activities in southern Nevada have not significantly increased mercury concentrations in groundwater, as was the case in parts of northern Nevada. In seawater, most authorities agree that mercury exists mainly bound to suspended particles; that the surface area of the sediment granules is instrumental in determining the final mercury content; that mercury conversion and transformation occur in the surface layer of the sediment or on suspended particles in the water; and finally, that mercury-containing sediments require many decades to return to background levels under natural conditions. High concentrations of methylmercury in sub-thermocline low-oxygen seawater were significantly and positively correlated with median daytime depth (<200 m to >300 m) of eight species of pelagic fishes; mean total mercury concentrations in whole fishes ranged between 57.0 and 377.0 μg/kg DW. The enhanced mercury accumulation in the marine mesopelagic compartment is attributable to diet and ultimately to water chemistry that controls mercury speciation and uptake at the base of the food chain.

Total mercury concentrations in uncontaminated natural waters (presumably unfiltered) now range from about 1.0 to 50.0 ng/L. Concentrations as high as 1100.0 ng/L are reported in freshwaters near active gold mining facilities in Ecuador, and up to 450,000.0 ng/L in drainage water from abandoned mercury mines in California. Mercury and methylmercury from mercury mine drainage is adsorbed onto iron-rich precipitates and seasonally flushed in streams during periods of high water. Maximum concentrations of 89.7 ng/L in Swedish rain, 78.0 ng total mercury (47.0 ng methylmercury)/L in coastal seawater of New York, and 600.0 ng/L in sediment interstitial waters are documented. Suspended particulate matter in the River Elbe, Germany, contained up to 150.0 mg total mercury (2.7 mg methylmercury)/kg dry weight as a result of mercury-contaminated wastes from a chloralkali plant.

19.6 Mercury Concentrations in Plants and Animals

Information on mercury residues in field collections of living organisms is especially abundant. Elevated concentrations of mercury occur in aquatic biota from areas receiving high atmospheric depositions of mercury, or when mercury concentrations in the diet or water are elevated. Mercury levels are comparatively elevated in fish-eating fishes, birds, and mammals. In general, mercury concentrations in biota were usually less than 1.0 mg/kg FW tissue in organisms collected from locations not directly affected by human use of the element. However, concentrations exceed 1.0 mg/kg – and are sometimes markedly higher – in animals and vegetation from the vicinity of chloralkali plants; agricultural users of mercury; smelters; mining operations; pulp and paper mills; factories producing mercury-containing paints, fertilizers, and insecticides; sewer outfalls; sludge disposal areas; and other anthropogenic point sources of mercury. In some Minnesota lakes, mercury concentrations in fish are sufficiently elevated to be potentially hazardous when ingested by mink, otters, loons, and raptors.

An elevated concentration of mercury (i.e., >1.0 mg/kg FW), usually as methylmercury, in any biological sample is often associated with proximity to human use of mercury. The elimination of mercury point-source discharges has usually been successful in improving environmental quality; however, elevated levels of mercury in biota may persist in contaminated areas long after the source of pollution has been discontinued. For example, mercury remains elevated in resident biota of Lahontan Reservoir, Nevada, which received about 7500 tons of mercury as a result of gold and silver mining operations during the period 1865–1895. It is noteworthy that some groups of organisms with consistently elevated mercury residues may have acquired these concentrations as a result of natural processes rather than from anthropogenic activities. These groups include older specimens of long-lived predatory fishes, marine mammals (especially pinnipeds), and organisms living near natural mercury–ore–cinnabar deposits. In general, concentrations of mercury in feral populations of marine vertebrates – including elasmobranchs, fishes, birds, and mammals – are clearly related to the age of the organism. Regardless of species or tissue, all data for mercury and marine vertebrates show increases with increasing age of the organism. Factors that may account, in part, for this trend include differential uptake at various life stages, reproductive cycle, diet, general health, bioavailability of different chemical species, mercury interactions with other metals, metallothioneins, critical body parts, and anthropogenic influences.

19.6.1 Algae and Macrophytes

Concentrations of total mercury were almost always below 1.0 mg/kg dry weight in aquatic and terrestrial vegetation except for those areas where human activities have contaminated the environment with mercury. In general, mercury concentrations were highest in mosses, fungi, algae, and macrophytes under the following conditions: after treatment with mercury-containing pesticides, near smelter emissions, in sewage lagoons, near chloralkali plants, exposure to mercury-contaminated soils, and proximity to industrialized areas. Samples of the marine flowering plant *Posidonia oceanica* collected near a sewer outfall in Marseilles, France, had elevated concentrations of mercury – in mg/kg dry weight – of 51.5 in leaves, 2.5 in rhizomes, and 0.6 in roots. Also, water hyacinth (*Eichornia crassipes*) from a sewage lagoon in Mississippi contained up to 70.0 mg Hg/kg DW. Both *Posidonia* and *Eichornia* may be useful in phytoremediation of mercury-contaminated aquatic environments.

Highest concentrations of mercury (90.0 mg Hg/kg FW) were found in roots of alfalfa (*Medicago sativa*) growing in soil containing 0.4 mg Hg/kg, in bark of a cherry tree (*Prunus avium*) from a factory area in Slovenia (59.0 mg/kg FW), in leaves of water hyacinth (*Eichornia crassipes*; 70.0 mg/kg DW) from a sewage lagoon, in mosses near a chloralkali plant (16.0 mg/kg FW), in fungi near a smelter (35.0 mg/kg DW), and leaves of *Posidonia oceanica* (51.5 mg/kg DW) near a sewer outfall.

Certain species of macrophytes strongly influence mercury cycling. For example, *Spartina alterniflora* – a dominant salt marsh plant in Georgia estuaries – accounted for almost half the total mercury budget in that ecosystem. Mercury entered the estuary primarily in solution, delivering about 1.5 mg annually to each square meter of salt marsh. Annual uptake of mercury by *Spartina* alone was about 0.7 mg/m^2 salt marsh. Mangrove vegetation plays a similarly important role in mercury cycling in the Florida Everglades. These findings suggest that more research is needed on the role of higher plants in the mercury cycle.

Creation of reservoirs by enlargement of riverine lakes and flooding of adjacent lands has led to a marked rise in rates of methylmercury production by microorganisms in sediments. This process has resulted mainly from increased microbial activity via increased use of organic materials under conditions of reduced oxygen. Increased net methylation in flooded humus and peat soils, especially in anoxic conditions, was determined experimentally and judged to be the main reason for increased methylmercury concentrations in reservoirs.

19.6.2 Invertebrates

In general, all species of invertebrates sampled had elevated concentrations of mercury (up to 10.0 mg/kg FW, 38.7 mg/kg DW) in the vicinity of industrial, municipal, and other known sources of mercury when compared to conspecifics collected from reference locations. The finding of 202.0 mg Hg/kg FW in digestive gland of *Octopus vulgaris* needs verification. Larvae of terrestrial insects (i.e., larvae of blowflies *Calliphora* sp.) play an important role in mercury cycling from feeding on beached fish carcasses.

Comparatively high mercury concentrations of 5.7 mg/kg FW in crayfish abdominal muscle from Lahontan Reservoir, Nevada, an area heavily contaminated with mercury from gold mining operations some decades earlier, and 41.0 mg/kg DW in sea anemones and up to 100.0 mg/kg DW in crustaceans, both from the heavily contaminated Minamata Bay, Japan, are discussed later.

Marine bivalve mollusks can accumulate mercury directly from seawater; uptake was greater in turbulent waters than in clear waters. Mollusks sampled before and immediately after their substrate was extensively dredged had significantly elevated tissue mercury concentrations after dredging, which persisted for at least 18 months. Mercury concentrations in scallops are influenced by reproductive status, sex, and inherent species differences. Mercury–sediment–water interactions influence uptake dynamics by marine benthos. Organisms feeding in direct contact with sediments have higher overall mercury levels than those feeding above the sediment-water interface. Mercury levels in mussels along European coasts tend to reflect mercury levels in water and sediments to a greater degree than does size of mussel, season of collection, or position in the intertidal zone. Reduced mercury inputs to coastal areas as a result of legislation and effective enforcement actions are reflected in mercury levels of common mussels, *Mytilus edulis*, in Bergen Harbor, Norway. In 2002, mussels from Bergen Harbor contained a maximum of 0.04 mg Hg/kg FW soft parts; this was about 60% lower than mercury levels in mussels collected from the same area in 1993. The reduced mercury was attributed to reductions in mercury content to Bergen Harbor of municipal wastewater, urban runoff, and especially of mercury-containing dental wastes. In every case reported wherein mercury concentrations in molluscan soft parts exceed 1.0 mg/kg FW it was associated with mercury pollution from human activities.

In marine crustaceans, total mercury concentrations were always less than 0.5 mg/kg FW edible tissues except in organisms collected from certain areas heavily impacted by mercury-containing industrial wastes, such as Minamata, Japan. Methylmercury concentrations in hepatopancreas of Chinese mitten crabs declined with increasing crab size – possibly through molting – suggesting a mechanism for mercury excretion, with important implications for crab predators that select larger crabs. In echinoderms, mercury concentrations in whole organisms from non-polluted areas are low, never exceeding 0.4 mg Hg/kg FW or 0.92 mg Hg/kg DW.

19.6.3 Elasmobranchs and Bony Fishes

Data on mercury concentrations in field collections of teleosts are especially abundant. Examination of these and other data leads to several conclusions. First, mercury tends to concentrate in the edible flesh of finfish, with older fish containing more mercury per unit weight than younger fish. This is particularly well documented in spiny dogfish (*Squalus acanthias*), squirefish (*Chrysophrys auratus*), European eel (*Anguilla anguilla*), European hake (*Merluccius merluccius*), striped bass (*Morone saxatilis*), and bluefish (*Pomatomus saltatrix*).

Second, most of the mercury in the fish flesh was in the organic form, mainly methylmercury. This is because fish assimilate inorganic mercury less efficiently than methylmercury from the ambient medium and from their diet, and eliminate inorganic mercury more rapidly than methylmercury. Maximum concentrations of total mercury in shark and fish muscle usually did not exceed 2.0 mg Hg/kg FW; however, forms of mercury with very low toxicity can be transformed into forms of very high toxicity, namely, methylmercury, through biological and other processes.

Third, levels of mercury in muscle from adult tunas, billfishes, and other marine carnivorous teleosts were higher than those in younger fishes having a shorter food chain. This indicates associations among predatory behavior, longevity, and mercury accumulation. Oceanic tunas and swordfish caught in the 1970s had mercury levels similar to those of museum conspecifics caught nearly 100 years earlier. It is speculated that mercury levels in fish were much higher 13,000–20,000 years ago during the last period of glaciation, when ocean mercury concentrations were four to five times higher than today.

Fourth, total mercury was uniformly distributed in edible muscle of finfish, demonstrating that a small sample of muscle tissue taken from any region is representative of the whole muscle tissue when used for mercury analysis.

Finally, elevated levels of mercury in wide-ranging oceanic fish were not solely the consequence of human activities but also resulted from natural concentrations. This last point is apparently not consistent with the rationale underlying U.S. Seafood guidelines regulating mercury levels in comestibles and formulated in the 1970s. When the U.S. Food and Drug Administration (USFDA) introduced safety guidelines – which eventually were instrumental in the temporary removal of all swordfish and substantial quantities of canned tuna from market – it acted essentially under the assumption that the fish product was "adulterated" by an "added substance."

It is noteworthy that muscle from two species of recreationally important fish spotted seatrout (*Cynoscion nebulosus*; red drum, *Sciaenops ocellatus*) collected from coastal bays in Texas considered "minimally impacted" by mercury exceeded the current recommended value in the U.S. of 0.3 mg total Hg/kg FW muscle. And walleye (*Stizostedium vitreum vitreum*) collected from Clay Lake, Ontario – a water body heavily contaminated by mercury wastes from a chloralkali plant between 1962 when discharges began and 1970 when the plant closed – contained 2.7 mg total Hg/kg FW muscle in samples collected 28 years after plant closure, a concentration in excess of the Canadian mercury criterion of <0.5 mg total Hg/kg FW edible fish portions.

Of the 159 species of finfish, including sharks and rays, from coastal waters of Alaska, Hawaii, and the conterminous U.S., most muscle samples had mean concentrations less than 0.3 mg total Hg/kg FW. However 31 species contained more than 0.5 mg total Hg/kg FW, the "action level" set by the USFDA. These 31 species represented about 0.65% of the weight of the catch from the 159 species intended for human consumption. Extrapolation of these results indicate that less than 2% of the U.S. catch intended for human consumption may be in excess of the USFDA action level. Of the 31 species containing more than 0.5 mg total Hg/kg FW in muscle, ten were sharks and four were billfishes.

Inshore marine biota often contained higher mercury concentrations than the same or similar species collected offshore. In Sweden, marine fish caught near shore often had elevated methylmercury levels, with many values in the range of 5.0–10.0 mg Hg/kg FW; concentrations above 1.0 mg Hg/kg FW in Swedish fish were usually associated with industrial discharges of mercury compounds. In Mediterranean fishes, the mercury body burden was about twice that of conspecifics of the same size from the Atlantic Ocean. It is speculated that the higher body burdens of mercury in Mediterranean species is due to the elevated natural geochemical levels of mercury in the Mediterranean.

Mercury concentrations in edible muscle of teleosts from German fishing grounds seldom exceeded 0.1 mg total Hg/kg FW. Of the total mercury in German fish muscle, methylmercury comprised 70–98%, which is in general agreement with Japanese, Swedish, and other reports on this subject.

The efficiency of mercury transfer through natural marine food chains among lower levels was comparatively low; however, higher trophic levels including teleosts and fish-eating birds and mammals, show marked mercury amplification. The variability in concentrations is explainable, in part, by collection locale wherein samples were taken from areas receiving anthropogenic mercury wastes resulting in elevated mercury loadings in the aquatic environment and a significant increase in mercury content of endemic fauna. Not all investigators agreed that diet was the most important mercury-concentrating mechanism for marine teleosts. For example, some investigators report that mercury in suspended solids and bottom sediments were not transferred in significant amounts to squirefish (*Chrysophrys major*), that accumulation via the food chain was very low, and that dissolved methylmercury in seawater was the critical pathway for methylmercury accumulation in that species. Others stated that there was a positive correlation between mercury content of edible bottomfish tissues in various U.K. fishing grounds and mean mercury concentrations of water samples from the same localities. A similar case was made for Japanese waters, although this link needs verification. Ionic mercury predominates in water; however, fish tissues contained >80% methylmercury; suggesting that dissolved mercury was removed rapidly from seawater by particulate matter and subsequently to sediments where methylation more readily occurred. It was concluded that mercury variations in fish tissues were attributed to the availability of food and its mercury content; the chemical form and concentration of dissolved mercury; the fish species and trophic level; and the growth rate, sex, and age of the animal. Changes in latitude also seem important for species distributed over a wide latitudinal range, and trends in mercury levels from these species were often opposite to that reported for other fish species from the same geographical area.

Mercury was detectable in the tissues of almost all freshwater fishes examined, with the majority of the mercury (>80–99%) present as methylmercury. Methylmercury is absorbed more efficiently than inorganic mercury from water, and probably from food, and is retained longer regardless of the uptake pathway. Three important factors modifying mercury uptake in aquatic organisms are the age of the organism, water pH, and the dissolved organic carbon content. In fish, for example, mercury tends to accumulate in muscle tissues of numerous species of freshwater and marine fishes and to increase with increasing age, weight, or length of the fish. Mercury concentrations in muscle of freshwater teleosts were significantly higher in acidic lakes than in neutral or alkaline lakes. Highest levels of mercury in fish muscle were from lakes with a pH near 5.0; liming acidic lakes resulted in as much as an 80% decrease in muscle mercury content after 10 years. And mercury concentrations in fish muscle were positively correlated with dissolved organic carbon concentration. Mercury content in edible portions of all species of freshwater fish sampled from River Nitra in the Slovak Republic in 2003 exceeded that country's allowable limit of 0.5 mg total Hg/kg FW muscle by factors of 4–13 times; authors have recommended posting of fish consumption advisories.

In addition to age, water pH, and dissolved organic carbon, other variables known to modify mercury accumulation rates in aquatic organisms include water temperature, sediment mercury concentrations, lake size, season, diet, chemical speciation of mercury, and sex. Elevated water temperatures were associated with elevated accumulations of mercury. Rates of mercury methylation were positively dependent on water temperature, and mercury demethylation rates were inversely related to water temperature. Elevated mercury concentrations in fish muscle were positively correlated with sediment mercury concentrations; a similar case is made for benthic marine invertebrates. Mercury concentrations were inversely related to lake size in planktivorous, omnivorous, and piscivorous fishes from remote lakes in northwestern Ontario; lakes ranged in size from 89 to 35,000 surface ha and were far from anthropogenic influences. Mercury levels in muscle of marine flatfishes were higher during spring than in autumn. In the yellow perch (*Perca flavescens*), seasonal variations in uptake rate of methylmercury and in the proportion of uptake from aqueous and food sources is attributed to seasonal variations in water temperature, body size, diet, and prey availability; methylmercury uptake was primarily from aqueous sources during the spring and fall and was dominated by food sources in the summer. Food chain transfer of mercury from benthic invertebrates to fishes was dependent primarily on the consumption rate of benthivorous fishes, and secondarily to the total invertebrate mercury pools. In the absence of pelagic forage fish, mercury concentrations in muscle of lake trout (*Salvelinus namaycush*), are likely to be depressed. Trophic transfer of methylmercury is much more efficient than that of Hg^{2+}. Sometimes, fish pellets fed to laboratory fish may contain elevated (0.09 mg Hg/kg DW) concentrations of mercury, resulting in elevated blood mercury levels (0.06 mg Hg/L) after 10 weeks, as was the case for the Sacramento blackfish, *Orthodon microlepidotus*. Sexually mature female centrarchids had significantly higher concentrations of mercury in muscle tissue than did sexually mature males, although this has not been reported for other aquatic species. Mercury concentrations in muscle of 14 species of freshwater fishes from Lake Chad, Africa, in December 2000, were highest in fish-eating species, three to four times lower in fish that fed upon insects and other invertebrates, and lowest in herbivores. Mercury concentrations in fish muscle were higher in fish from humic lakes, from lakes of low mineralization, and from lakes with low concentrations of dissolved iron, calcium, alkalinity, chlorophyll *a*, magnesium, phosphorus, and nitrogen. It is noteworthy that low atmospheric depositions of selenium did not affect mercury concentrations in muscle of brown trout (*Salmo trutta*); that mercury and selenium in muscle of marine fishes were not correlated; and that mercury and selenium concentrations in blood of tunas were independent of each other.

Diet, age, logging, and forest fires were all significant factors affecting mercury concentration in fish collected from Canadian drainage lakes in 1996–97 wherein muscle contained between 0.2 and 20.2 mg total Hg/kg DW. Mercury concentrations tended to increase with increasing fish length and were higher in fish-eating fish such as northern pike (*Esox lucius*), and walleye (*Stizostedium vitreum*); concentrations were highest when surrounding forest were clear-cut or fire-impacted and may reflect increased exposure to mercury when compared to conspecifics from lakes with undisturbed watersheds.

In adult fish, females often contain higher mercury concentrations than males, possibly because they consume more food than males in order to support the energy requirements of egg production. The increased feeding rate in females causes greater dietary uptake of methylmercury; however, the transfer to egg mass is a small fraction of the maternal body burden. Mercury concentrations in spiny dogfish (*Squalus acanthias*), were influenced by dogfish sex, length, and area of collection.

Concentrations were higher in males, higher in specimens with body length >65 cm when compared to smaller dogfish, and higher in dogfish from estuarine areas than from offshore locations. Mercury levels in fetuses of the California dogfish (*Squalus suckley*), were 21–42 times lower than maternal tissues, suggesting than mercury is uniquely absent from the fetal environment and may even be selectively excluded.

Nationwide (U.S.) monitoring of whole freshwater fish during the period 1969–1981 demonstrated that the highest mercury concentrations (0.33–1.7 mg/kg FW) were in northern squawfish (*Ptychocheilus oregonensis*) from the Columbia River basin in the Pacific Northwest. Elevated mercury concentrations in this piscivorous species were attributed primarily to the presence of major cinnabar deposits and with mercury use associated with mineral mining in the Columbia River basin. Northern squawfish may have a natural tendency to accumulate high concentrations of mercury in their flesh – as is well known for older specimens of long-lived predatory fishes such as tunas, billfishes, bluefish, striped bass, northern pike, and many species of sharks; however, mercury uptake kinetics in squawfish requires further research.

In the Florida Everglades ecosystem, mercury concentrations in muscle of largemouth bass (*Micropterus salmoides*) are directly correlated with atmospheric deposition of mercury. A model was formulated that showed a positive correlation between total mercury in largemouth bass muscle (in the range 0.3–1.8 mg/kg FW) with atmospheric Hg^{2+} wet plus dry deposition (in the range 5.0–35.0 $\mu g/m^2$ deposition per annum). In the absence of changes to this ecosystem other than mercury cycling, e.g., changes in sulfur cycling, nutrient cycling, and hydrology, a reduction of about 80% of current total annual mercury deposition rates would be needed for the mercury concentration in muscle from a 3-year-old largemouth bass at a heavily contaminated site to be reduced to <0.5 mg/kg FW muscle, which is Florida's present fish consumption advisory action level. Mercury concentrations in muscle from 3-year-old bass – currently averaging 2.5 mg Hg/kg FW – are predicted to achieve 50% of their long-term steady state response following sustained mercury load reduction within 10 years, and 90% within 30 years.

In the Florida recreational fishery for red drum (*Sciaenops ocellatus*), the current maximum size limit of 565 mm standard length or 689 mm total length is an effective filter that limits consumption of large fish containing elevated mercury concentrations. About 94% of all adult red drum from waters adjacent to Tampa Bay, Florida, contain mercury levels in muscle greater than 0.5 mg/kg FW muscle – the Florida Department of Health threshold level – and 64% contained >1.5 mg Hg/kg FW muscle, the Florida "no consumption" level. All fish from this area containing >1.5 mg Hg/kg FW muscle were >689 mm standard length.

Reservoir construction is thought to be a cause of elevated mercury concentrations in fish. Reservoir conditions facilitating the bioavailability of mercury include upstream flooding and leaching of terrestrial sediments, relatively high pH and conductivity of the water, high bacterial counts in the water, complete thermal mixing, low clay content, and low concentrations of sulfur and iron and magnesium oxides in bottom sediments. It is hypothesized that increases in mercury levels observed in fish were due to bacterial methylation of naturally occurring mercury in the flooded soils. Methylation and transfer of methylmercury from flooded soils to suspended particulate matter and zooplankton is rapid and involves the bioaccumulation of methylmercury by phytoplankton and the ingestion of suspended soil-derived organic particles by zooplankton. Suspended particulate matter and zooplankton are disproportionate contributors to methylmercury contamination of aquatic food chains in Quebec reservoirs. In general, mercury levels are higher in fish from younger oligotrophic reservoirs, and lower in fish from older eutrophic reservoirs; in both situations, tissue mercury levels usually decline as the reservoirs age. Mercury concentrations (greater than 0.5 mg/kg FW but less than 1.0 mg/kg) have been reported in trout from several wilderness lakes in northern Maine and from the Adirondacks region of New York; these values are considerably higher than might be expected for fish inhabiting remote lakes. Elevated mercury concentrations in fish tissues were usually associated with lakes of low pH, low calcium, low dissolved organic carbon concentrations, and low water hardness and alkalinity. Enlargement of northern Manitoba lakes to form hydroelectric reservoirs caused a rise in the mercury content of native fishes owing to stimulation of mercury methylating bacteria by submerged terrestrial organic matter. Increased organic substrates beyond a critical amount mitigated this effect via promotion of mercury demethylation and production of mercury-binding agents such as sulfides. Variability in mercury concentrations between fish species was high and was due to differences in habitat preference, metabolic rate, age, growth rate, size, biomass, diet, and excretory pathways. Elevated mercury levels in fish flesh found after impoundment of a reservoir are predicted to decline as the reservoir ages. In Labrador, Canada, mercury concentrations in muscle of omnivorous species of fishes reached background levels in 16–20 years; however, mercury in piscivorous species remained elevated 21 years after impoundment.

19.6.4 Amphibians and Reptiles

Several freshwater marshes in the Florida Everglades are contaminated with mercury and more than 900,000 ha are currently under fish consumption advisories because of high mercury concentrations, namely, >1.5 mg total Hg/kg FW muscle.

Consumption of leg muscle of the pig frog (*Rana grylio*) from certain areas in South Florida under fish consumption advisories may present a risk to human health. Total mercury in frog leg muscle was highest (max. 2.05 mg/kg FW) from areas protected from harvest in the Everglades National Park. Total mercury burdens in frog leg muscle from most harvested areas were <0.3 mg/kg FW, an acceptable level; however, mean concentrations in other areas regularly harvested for human consumption were >0.3 mg total Hg/kg FW muscle.

Elevated concentrations of mercury in amphibian tissues were also found in frogs and toads collected near a mining area in Yugoslavia. Maximum concentrations, in mg/kg fresh weight, were 2.3 in egg, 2.9 in lung, 24.0 in kidney, and 25.5 in liver; conspecifics from a reference site contained <0.08 mg Hg/kg fresh weight in all tissues.

Highest concentrations of mercury in reptiles collected were found in tissues of the American alligator (*Alligator mississippiensis*) from the Florida Everglades. Maximum concentrations, in mg Hg/kg fresh weight, were 6.1 in alligator muscle, 13.1 in spleen, 65.3 in kidney, and 99.5 in liver; other tissues contained 1.3–4.6 mg/kg FW. Mercury concentrations in spleen, kidney, and liver tissues of farm-raised alligators were always <0.2 mg/kg FW. Based on available data, mercury concentrations in all reptiles were highest in liver, followed by kidney, muscle, and egg, in that order; in all tissues sampled, organomercurials comprised 60–90% of the total mercury.

In cottonmouths (*Agkistrodon piscivorus*), from Texas, mercury tissue concentrations were higher in males than females for kidney and liver; one male 96.3 cm in length had 8.6 mg Hg/kg liver FW – the highest mercury concentration ever reported for a serpent.

19.6.5 Birds

It is generally acknowledged that mercury concentrations in avian tissues and feathers are highest in species that eat fish and other birds. Mercury contamination of prey in the diet of nestling wood storks (*Mycteria americana*), an endangered species, may represent a potential concern to the recovery of this species in the southeastern U.S. Increased concentrations of total mercury in livers of diving ducks were associated with lower weights of whole body, liver, and heart, and decreased activities of enzymes related to glutathione metabolism and antioxidant activity. In seabirds, mercury concentrations were highest in tissues and feathers of species that ate fish and benthic invertebrates and lowest in birds that ate mainly pelagic invertebrates. In seabirds, the relation between tissues and total mercury concentrations is frequently 7:3:1 between feather, liver, and muscle; however, there is much variability and these ratios should be treated with caution. Factors known to affect these ratios include the chemical form of mercury present in liver, the sampling date relative to the stage of the molt sequence, and the types of feathers used for analysis. Mercury concentrations in feathers of wading birds collected in Florida between 1987 and 1990 were highest in older birds that consumed large fishes. And wading birds whose prey base consisted of larger fish had four times more mercury in livers than did species which consumed smaller fish or crustaceans. Wading birds with minimal to moderate amounts of body fat had two to three times more mercury in liver than did birds with relatively abundant body fat reserves. Essentially all mercury in body feathers of all seabirds studied was organic mercury; however, more than 90% of the mercury in liver is inorganic. Mercury residues are usually highest in kidney and liver, but total mercury contents are significantly modified by food preference and availability, and by migratory patterns. Also, there is an inverse relation between total mercury and percent methylmercury in tissues of various avian species – a pattern that seems to hold for all vertebrate organisms for which data are available. Diet and migration are the most important mercury modifiers in birds. For example, the higher levels of mercury in juveniles than in adults of wood ducks (*Aix sponsa*) from Tennessee were related to dietary patterns: juveniles preferred insects, whereas adults preferred pondweed tubers; mercury residues were higher in the insects than in the pondweeds.

Factors that modify mercury concentrations in birds include age, tissue, migratory patterns, diet, and season. Adults of the double-crested cormorant (*Phalacrocorax auritus*) contained elevated levels of mercury in liver and whole body when compared to nestlings: 0.3 mg Hg/kg FW in nestling liver vs. 8.0 in adults, and 0.06 in nestling whole body vs. 0.64 mg Hg/kg FW in adults. Among highly migratory birds, dramatic seasonal changes in mercury content are common, and are attributed, in part, to ingestion of mercury-contaminated food. Seasonal variations and diet affect mercury concentrations in avian tissues. Seasonal variations in mercury levels are reported in livers of aquatic birds, being higher in winter when birds were exclusively estuarine and drastically lower in summer when birds migrated to inland and sub-Arctic breeding grounds. It is possible that the wintering populations, for example knots *Calidras* spp., may previously have accumulated mercury while molting in western European estuaries, notably on the Dutch Waddenzee. For example, three species of knots, *Calidras* spp. contained >20.0 mg Hg/kg FW liver during winter while molting in western European estuaries and <1.0 mg/kg during summer when birds migrated to inland Arctic and sub-Arctic breeding grounds. Concentrations of mercury in livers of Antarctic birds reflected mercury body burdens accumulated during migration, while the birds were overwintering near industrialized areas. Concentrations were highest in species that ate higher trophic levels of prey and were especially pronounced for skuas, *Catharacta* spp.; however, significant inherent interspecies differences were evident. Birds that feed on aquatic fauna show elevated mercury concentrations in tissues when compared to terrestrial raptors. Thus, mercury was highest in liver of cormorants *Phalacrocorax* spp. And pelicans *Pelecanus* spp., with concentration factors for mercury of 14 over prey fish in body of cormorants and 6 for pelicans.

The recorded value of 97.7 mg Hg/kg dry weight in liver of dead or dying gannets (*Sula bassana*) requires explanation. It is possible that mercury accumulations of that magnitude were a contributory factor to death in this instance; however, the main cause of death was attributed to poisoning by polychlorinated biphenyls. Further, large variations in mercury content in gannet liver were linked to liver size (positive correlation) and to fat content (inverse relation). The highest values observed of 67.5 and 130.0 mg Hg/kg fresh weight in liver and kidney, respectively, of osprey *Pandion haliaetus* are attributed to a single bird. These levels are clearly excessive, reflect high environmental exposure, and are similar to concentrations found in mercury-poisoned birds. Of the 18 ospreys examined, except for the aberrant observation, highest values were 6.2 mg Hg/kg FW liver and 6.5 in kidney.

Eggs of fish-eating birds, including eggs of herons and grebes, collected near mercury point source discharges contained abnormally high levels of mercury: 29% of eggs contained more than 0.5 mg Hg/kg FW, and 9% contained more than 1.0 mg Hg/kg FW. The main source of mercury in estuaries is probably from the direct discharge of effluent from manufacturing and refining industries into rivers. The high levels of mercury detected in eggs of the gannet *Morus bassanus* are within the range associated with negative influence on hatchability in pheasants and other sublethal effects in mallard ducks; however, the gannets appear to reproduce normally at these levels. Eggs of the common loon (*Gavia immer*) from Wisconsin in 1993–96 had 0.9 mg Hg/kg FW, which is within the range associated with reproductive failure in sensitive avian species.

Most authorities agree that feathers contain most of the total body load of mercury, while constituting usually less than 15% of the weight. Mercury excretion is mainly via the feathers in both sexes, and also in the eggs. Mercury concentrations in feathers of little tern chicks, *Sterna albifrons*, were higher in smaller chicks than larger chicks and higher in early broods (1–3) than later broods (4–7), suggesting depletion of maternal transfer of mercury. In Sweden, fish-eating birds had higher levels of mercury in feathers than did terrestrial raptorial species. Ospreys, which prey almost exclusively on larger fish of about 0.3 kg, show higher mercury levels in feathers than grebes, *Podiceps cristata*, which eat smaller fish and insect larvae. Since larger fish contain more mercury per unit weight than smaller fish, diet must be considered an important factor to account for differences in mercury concentrations of these two fish-eating species.

Based on samples from museum collections, it was demonstrated that mercury content in feathers from fish-eating birds were comparatively low in the years 1815 through 1940. However, since 1940, or the advent of the chloralkali industry (wherein mercury is used as a catalyst in the process to produce sodium hydroxide and chlorine gas from sodium chloride and water, with significant loss of mercury to the biosphere), mercury concentrations in feathers were eight times higher on average. Mercury levels were also elevated in feathers and tissues of aquatic and fish-eating birds from the vicinity of chloralkali plants; these increased levels of mercury were detectable up to 300 km from the chloralkali plant.

Bird feathers have been used for some time as indicators of mercury loadings in terrestrial and marine environments. Feathers represent the major pathway for elimination of mercury in birds and body feathers are useful for assessment of whole bird mercury burdens with almost all mercury present as methylmercury. The keratin in bird feathers is not easily degradable, and mercury is probably associated firmly with the disulfide bonds of keratin. Consequently, it has been possible to compare mercury contents of feathers recently sampled with those from museum birds, thereby establishing a time series. There is considerable variability in mercury content of seabird feathers. Concentrations in adults were higher than those in chicks and independent of adult age or sex, and were lower in spring breeders than in autumn breeders. After the completion of molting, new feathers contained up to 93% of the mercury body burden in gulls. The most probable source of elevated mercury residues in feathers of the Finnish sparrowhawk (*Accipiter nisus*) was from consumption of avian granivores that had become contaminated as a result of eating seeds treated with organomercury compounds; in 1981, 5.6 tons of methoxyethylmercury compounds were used in Finnish agriculture for protection of seeds against fungi. Concentrations of mercury in feathers of herring gulls (*Larus argentatus*) from the German North Sea coast were higher in adults than in juveniles and two times higher after 1940 than in earlier years. A maximum of 12.0 mg/kg FW in feathers during the 1940s was recorded, presumed to be due to high discharges of mercury during the Second World War (1939–45). Concentrations dropped in the 1950s, increased in the 1970s to 10.0 mg Hg/kg FW, before falling in the late 1980s. This pattern correlates well with known discharges of mercury into the Elbe and Rhine. Captive Swedish eagle-owls (*Bubo bubo*), with low mercury content in feathers (<1.0 mg/kg DW), that were introduced into coastal areas quickly reflected the high (6.5 mg/kg) mercury levels in feathers of wild eagle-owls from that region. Captive birds released into inland territories, where mercury levels were near background, did not accumulate mercury in feathers. Mercury levels in feathers of nestling Swedish gyrfalcons (*Falco rusticolus*) showed a better correlation with mercury levels in actual food items than with levels based on adult feathers. Mercury concentrations in feathers were higher in nestlings fed partly with aquatic bird species containing more than 0.07 mg Hg/kg in pectoral muscle than in nestlings fed willow grouse (*Lagopus lagopus*) and ptarmigan (*Lagopus mutus*), both of which contained less than 0.01 mg Hg/kg in pectoral muscle. In some instances, there was a substantial time lag, up to 10 years, between the introduction of a pesticide, such as alkylmercury, its subsequent banning, and measurable declines of mercury in feathers of several species of Swedish raptors; this was the case for various species of *Falco, Haliaeetus, Bubo, Buteo*, and *Accipiter*. Accordingly, a reduction in mercury content in feathers of free-living birds may be sufficient to establish an improved situation.

Methylmercury concentrations in feathers from two species of north Atlantic seabirds increased about 4.8% yearly between 1885 and 1994; increases were attributed to global increases in mercury loadings rather than to local or regional sources. Mercury concentrations in breast feathers of the king penguin (*Aptenodytes patagonicus*) were significantly lower in 2000–01 (1.98 mg Hg/kg DW) than were feathers collected from the same colony in (1966–74) (2.66 mg/kg DW), and suggests that mercury concentrations in southern hemisphere seabirds do not increase – which conflicts with trends observed in the northern hemisphere. Molting is a major excretory pathway for mercury. Down and feathers were effective excretion routes of mercury in contaminated gull and tern chicks. Some seabirds demethylate methylmercury in the liver and other tissues, and store mercury as an immobilizable inorganic form in the liver; species with a high degree of demethylation capacity and slow molting pattern had low mercury burdens in feathers. Egg laying is an important route in reducing the female's mercury burden, especially the first egg because egg mercury levels decline with laying sequence in gulls and terns. In gulls and terns, 90% of the mercury in eggs is in the form of methylmercury. In kittiwakes (*Rissa tridactyla*), mercury concentrations in feathers and tissues of nestlings decreased with increasing age, suggesting that egg contamination was more important in chicks than consumption of mercury-contaminated food items.

Significant downward trends in mercury liver burdens of raptors in England between 1960 and 1990 are useful indicators of the prohibitions placed on mercury discharges in that region. The full significance of mercury residues in birds, however, is not fully understood. For example, all eggs of the bald eagle (*Haliaeetus leucocephalus*) collected nationwide (U.S.) contained detectable levels of mercury, but the mean was 0.15 mg Hg/kg (fresh weight basis) in eggs from unsuccessful nests vs. 0.11 in eggs from successful nests. Many other contaminants – especially organochlorine compounds – were in eagle eggs, and several were present at levels that potentially interfere with eagle reproduction. It is not now possible to implicate mercury as a major cause of unsuccessful eagle reproduction. In the Great Lakes, mercury has no apparent effect on reproduction or nesting success of bald eagles. Livers of 30–80% of some species of wading birds – such as the great blue heron, *Ardea herodias* – contained mercury concentrations greater than 30.0 mg Hg/kg FW; these herons appeared normal although concentrations >30.0 mg Hg/kg FW liver are typically associated with overt neurological signs and reproductive impairment in ducks and pheasants. If reproductive disorders are expected when concentrations in feathers of adult birds approach 9.0 mg total Hg/kg DW, then mercury in southern Florida may be sufficiently high to reduce productivity of wading bird populations, although this needs to be verified.

Many factors are known to modify mercury concentrations in tissues of the common loon (*Gavia immer*). Total mercury concentrations were higher in tissues of emaciated loons when compared with apparently healthy birds, and sometimes exceeded 100.0 mg Hg/kg DW in tissues of loons that were in poor condition. There was a strong positive correlation between total mercury and selenium concentrations in livers and kidneys. As total mercury concentrations increased in liver and kidney of loons, the fraction that was methylmercury decreased. Livers and kidneys with the highest total mercury concentrations had only 5–7% of the total as methylmercury. Concentrations of methylmercury were always less than 10.0 mg/kg DW, regardless of total mercury concentration in liver or kidney. In contrast, methylmercury contributed 80–100% of the total mercury in breast muscle, which ranged between 0.7 and 35.0 mg/kg DW. In general, males had higher mercury concentrations in blood and feathers than did their female mates. The possible transfer of mercury to eggs by females during egg laying may account for some of the sexual discrepancy. Adult loons had higher blood mercury concentrations (up to 13 times higher) than their chicks. Adult and chick blood mercury concentrations were correlated with mercury concentrations in their fish diet. Blood mercury concentrations of loon chicks near Wisconsin lakes increased with decreasing lake pH. Blood and feather mercury concentrations from the same individuals were correlated, especially in loons with the highest blood mercury levels. Common loons had aberrant nesting behavior and low reproductive success when mercury concentrations in prey – small fish and crayfish – exceeded 0.3 mg Hg/kg FW, levels known to occur in fish from many lakes in central Ontario; up to 30% of Ontario lakes exceeded the mercury threshold for loon reproductive impairment. Populations of the common loon are also declining in the northeastern U.S., and this may be due to mercury, in part, as evidenced by the high concentrations in their feathers (9.7–20.2 mg Hg/kg DW), being twice that of other species.

Mercury–selenium interactions are significant in marine mammals and seem to be a factor in loons; however, this is not the case in marine birds. Mercury concentrations in oceanic birds were not correlated with selenium concentrations, as evidenced by values in livers of murres, *Uria* spp. and razorbills *Alca torda*, and in breast muscle of sooty terns *Sterna fuscata*.

19.6.6 Humans

Elevated total mercury concentrations in various human tissues and body fluids are associated with increasing consumption of fish, use of skin-lightening creams containing mercuric ammonium chloride, and recipients or users of ethylmercury compounds as wound disinfectants; moreover, hair concentrations >3.1 mg methylmercury/kg DW were found in Iraqis that died in 1971 from consuming methylmercury-contaminated wheat.

Urine mercury concentrations in women are associated with age, race, smoking, fish consumption, blood mercury concentrations, and dental fillings. Urinary mercury concentrations in this nationwide (U.S.) cohort of women are significantly associated with increasing number of restored mercury dental amalgam surfaces.

Elevated, but not life-threatening, accumulations of mercury in human tissues are common among coastal populations, especially in fishermen who subsist to a large degree on marine fish and shellfish. Some groups, however, routinely ingest diets containing greatly elevated concentrations of total mercury without apparent effect. For example, pregnant Inuit women living in close proximity to the sea and consuming seal meat and blubber on a regular basis displayed greater mercury accumulations in maternal and fetal blood and tissues than did similar populations further from the sea; however, the elevated mercury concentrations reported in Inuit infants from mothers who ate seals or fish every day of their pregnancy were reportedly far below the acknowledged toxic level.

Early estimates of mercury exposure in the general population ranged from 23.0–78.0 μg/day: 1.0 μg/day from air, about 2.0 μg/day from water, and 20.0 to 73.0 μg/day from food, depending on the amount of fish in the diet. Human intake of total mercury from the diet normally ranges between 7.0 and 16.0 μg daily. Fish consumption accounts for much of this exposure in the form of methylmercury: 27% of the intake, and 40% of the absorbed dose. Intake of inorganic mercury arises primarily from foods other than fish, and is estimated at 1.8 μg daily with 0.18 μg absorbed daily. In certain areas of India, blood mercury concentrations of people who ate fish were three to four times higher than non-fish eaters. In some countries, mercury in dental amalgams account for 2.8 μg daily, equivalent to as much as 36% of the total mercury intake and 42% of the absorbed dose.

Total blood mercury is considered the most valid biomarker of recent methylmercury exposure. Some Canadian aboriginal people had grossly elevated blood mercury concentrations of >100.0–660.0 μg Hg/L, although there was no definitive diagnosis of methylmercury poisoning. And, in a study in the U.S. involving older adults aged 50–70 years, with median blood mercury levels of 2.1 μg/L (range 0.0–16.0 μg/L), increasing blood mercury was associated with decreasing performance on a test of visual memory; however, increasing blood mercury level was also associated with increasing manual dexterity. It was concluded that blood mercury concentrations in older adults are not strongly indicative of declining neurobehavioral performance. Among Greenland (Denmark) residents, mercury intake from maritime foods may be associated with cardiovascular disease. Mercury concentration in the blood of Greenlanders was significantly higher in those consuming traditional Greenland foods (seal, whale) when compared to those consuming European diets; moreover, blood pressure increased with increasing blood mercury content.

Hair has been proposed as a diagnostic indicator of mercury exposure because it is a recognized excretory pathway, is formed in a relatively short time, and mercury content is unaffected by ongoing metabolic events. In Agra, India, higher mercury concentrations were found in the hair of vegetarians when compared to non-vegetarians; higher in hair of non-smokers when compared to smokers; and higher in hair of male alcoholics when compared to male non-alcoholics, suggesting that plants may be a primary source of mercury in that population. In Kenya, habitual use of skin-lightening soaps containing elevated concentrations of inorganic mercury was associated with high residues of mercury in hair, tremors, lassitude, neurasthenia, and other symptoms of inorganic mercury poisoning. The U.S. Environmental Protection Agency proposed a Reference Dose of 1.0 mg total Hg/kg DW hair as indicative of mercury exposure, and the concentration at which women of child-bearing age are advised to stop consumption of fish that may have elevated mercury levels. A survey shows that hair levels in excess of 1.0 mg Hg/kg DW are related to increasing fish consumption, to age over 20 years, and to certain geographic areas, especially New York and Florida.

Recent mothers who were employees of the largest steel producing plant in India, located in the Steel City (Bhilai), had elevated concentrations of total mercury in breast milk and blood when compared to recent mothers who did not work at the plant but were residents of the Steel City; both groups had higher levels than recent mothers from a reference site 100 km distant. In Bhilai, mercury concentrations in breast milk and blood increased with increasing age of the mother, and this may be related to increasing respiratory exposure to mercury-contaminated dust.

19.6.7 Other Mammals

Among nonhuman mammals, marine pinnipeds contained the highest reported concentrations of mercury in tissues. Total mercury content in all tissues examined of marine mammals – including muscle, brain, blubber, kidney, and liver – generally increased with increasing age of the animal. This was especially pronounced in livers of northern fur seals (*Callorhinus ursinus*), grey seals (*Halichoerus grypus*), California sea lions (*Zalophus californianus*), and harbor seals (*Phoca groenlandica*) and in various tissues of the Baikal seal (*Phoca sibirica*). Unlike teleosts, a high percentage of the mercury in seal liver occurs in the inorganic form. Moreover, most of the mercury in seal liver is biologically unavailable through complexation with selenium in a 1:1 atomic ratio. Total mercury concentrations were usually highest in livers of marine mammals, intermediate in muscle, and lowest in blubber. With some exceptions, mercury content in meat, blubber, and especially liver of adult and new-born marine mammals exceeded mercury safety guidelines established by regulatory agencies for foodstuffs. The relatively high concentrations appeared to be a result of natural processes rather than of anthropogenic activities, and probably did not represent a significant risk to pinniped health. In one case, liver from an older grey seal contained 387.0 mg Hg/kg FW, a level substantially in excess of levels found toxic to humans. However, in contrast to fish, a high percentage of the total mercury occurs in the inorganic form in seal and whale liver. Inuit sled dogs, subsisting largely on seal meat, contained elevated levels of mercury in liver, up to 11.5 mg/kg FW, without apparent harm.

Mercury in pinniped muscle, unlike liver, was mostly methylmercury in both mothers and pups; pups acquired most of their mercury during gestation. The percentage of methylmercury in any tissue from any marine mammal appears to be inversely correlated with total mercury content. For example, liver of harbor seals from Maine contained a maximum of 7.8 mg total Hg/kg FW vs. 50.9 from those collected from New Brunswick, Canada; methylmercury accounted for 13–37% of total mercury in Maine (U.S.) seal livers but only 2–11% in Canadian seals. Among healthy California sea lions, *Zalophus californianus*, concentrations of total mercury in tissues in mg/kg FW and percent methylmercury were as follows: liver 74.0 and 3.7%; kidney 7.0 and 17.2%; muscle 1.2 and 88.6%; and heart 0.59 and 88.1%.

Many factors are known to modify mercury accumulation and retention in marine pinniped mammals, including diet, age of mammal, sex, general health, proximity to urban areas, selenium residues, and migrations through areas of high tectonic activity. Diet, for example, is an important concentrating mechanism in seals. Grey seals (*Halichoerus grypus*), hood seals (*Cystophora cristata*), and harbor seals (*Phoca vitulina*), which feed on large fish and cephalopods, contained up to ten times more mercury in their tissues than harp seals (*Pagophilus groenlandicus*), which feed on small pelagic fish and crustaceans. Although concentrations of copper, zinc, cadmium, and lead in muscle and liver tissues of prey fish did not differ significantly from corresponding organs of marine mammals, mercury concentrations were considerably higher in liver of whales and seals than in fish.

Mercury and selenium concentrations in livers of marine pinnipeds seem to be positively correlated. It is alleged that selenium protects these species by completely binding to sub-cellular S sites, the presumed location of mercury's toxic action. Both ringed (*Phoca hispida*) and bearded (*Erignathus barbatus*) seals accumulate high levels of naturally occurring mercury in their livers; however, in the large sample of seals that were examined, there was no indication of mercury intoxication. The presence of selenium in a 1:1 atomic ratio with mercury in seal liver indicates a biochemical binding process. The mechanism of detoxification by selenium is not understood with certainty, but it appears that when selenium is ingested along with mercury, some mechanism operates in seals, causing both metals to combine and become immobilized in the liver.

The mechanisms to account for mercury accumulation in pinnipeds are similar to those reported for the striped dolphin (*Stenella coeruleoalba*). Tissue concentrations of mercury in striped dolphins increased with increasing age of the animal, reaching a plateau in 20–25 years; were highest in liver, although muscle accounted for about 90% of the total body mercury burden; were present in the methylated form in fetal and suckling stages, but the proportion of methylmercury decreased over time with no absolute increase after age 10 years; were excreted slowly by all developmental stages, and slowest in older dolphins (resulting in higher accumulations); and were correlated strongly with selenium concentrations in all age groups. It is probable that inorganic mercury and selenium were complexed in a 1:1 molar ratio, in a form biologically unavailable to marine mammals and (probably other mammals), thereby significantly decreasing the risk of mercury toxicosis to individuals with grossly elevated mercury body burdens. The Hg/Se ratio was close to 1.0 in adults of four species of Norwegian seals provided that tissue mercury concentrations were greater than 15.0 mg/kg FW. Total mercury in livers of pinniped mothers – but not pups – was correlated positively with selenium. In grey seals (*Halichoerus grypus*), mercury concentrations were higher in liver, kidney, and muscle of mature males and females when compared to immature individuals; this was especially pronounced in liver, where maximum mercury concentrations of 199.0 mg/kg FW and 238.0 mg/kg FW were recorded in mature males and females, respectively. A strong correlation between cadmium, mercury, and zinc in kidney suggests the presence of a detoxification process involving metallothionein proteins; another strong and positive correlation between mercury and selenium and a molecular Hg:Se ratio close to 1.0 in liver of grey seals suggests a demethylation process leading to the formation of mercuric selenide granules. Large colonies of pinnipeds and, to a lesser extent, marine birds along the western coast of the U.S. may make mercury available to California mussels (*Mytilus californianus*) through fecal elimination of large amounts of mercury, resulting in abnormally high mercury levels in mussels from several west coast sites.

Increasing mercury concentrations in tissues of marine mammals were also associated with poor health due to leptospirosis, with proximity to urbanized areas, and with starvation. Accumulations were usually highest in adult females, than adult males; placental transfer of mercury to developing pups is low. Methylmercury concentrations in seal pups were lower than that of their mothers. Moreover, the seal fetus does not show a preference for mercury over that of the mother's tissues, suggesting that seals may possess enzyme systems capable of demethylating methylmercury.

A source of mercury in livers of pilot whales (*Globicephala macrorhyncha*) may be due to volcanic activity in areas of mercury ore deposits and through which whales migrate; or exposure over time rather than food web. If this is correct, the elevated mercury levels in pilot whales of St. Lucia probably reflect accumulation resulting from existence in a tectonically active region with a higher than average environmental level of mercury, and from exposure to a small fraction of air-transported mercury from outside the region, possibly industrial in origin.

Mercury concentrations in mammals other than pinnipeds are modified by age, sex, sexual condition, diet, season of collection, and other variables. Increasing concentrations of total and organic mercury in muscle and liver were observed with increasing age of fin whales (*Balaenoptera physalus*) and striped dolphins, and in livers of white-tailed deer (*Odocoileus virginianus*), otters, and the endangered Florida panther (*Felis concolor coryi*). However, in harbor porpoises (*Phocoena phocoena*), total mercury – but not methylmercury – increased in tissues with increasing age. Pregnant or lactating sperm whales (*Physeter macrocephalus*) had significantly higher mercury concentrations in muscle than nonbreeding females. In river otters (*Lutra canadensis*), mercury concentrations in liver and kidney were higher in males than in females. No sexual differences in liver mercury concentrations were evident in white-tailed deer or European otters (*Lutra lutra*). Polar bears (*Ursus maritimus*) probably obtain mercury from ringed seals (*Phoca hispida*), which is their main food. Polar bear cubs had lower concentrations of mercury in fur than did yearlings or adults, and the low concentrations in adult fur in summer is attributed to molting. Florida panthers found dead contained as much as 110.0 mg total Hg/kg FW liver, a level found lethal to feral cats (*Felis cattus*) in Minamata, Japan. Florida panthers feed primarily on raccoons that contain as much as 3.0 mg Hg/kg FW, but it is not known if this is the source of the elevated liver mercury in panthers.

Among furbearers in the Wisconsin River drainage system, mercury burdens were higher in fish-eating than in herbivorous species – i.e., river otter > mink (*Mustela vison*) > raccoon (*Procyon lotor*) > red fox (*Vulpes fulva*) > muskrat (*Ondatra zibethicus*) > beaver *Castor* (*canadensis*). In general, fur contained the highest mercury levels, followed by liver, kidney, muscle, and brain, in that order. Mercury levels in fish-eating furbearers collected from the Wisconsin River basin paralleled mercury levels in fish, crayfish, and bottom sediments from that system; levels in all compartments were highest about 30 km downstream from an area that supported 16 pulp and paper mills and a chloralkali plant. Mercury concentrations in tissues of minks and otters trapped from various locations in Ontario between 1983 and 1985 varied by as much as sixfold; mercury levels in fish and crayfish from the study areas followed a similar pattern. Mink and river otter accumulated about ten times more mercury than did predatory fishes from the same drainage areas – suggesting that these furbearers can serve as sensitive indicators of mercury, even at very low levels of mercury contamination. The shorttail shrew (*Blarina brevicauda*) from certain mercury-contaminated sites in Tennessee have extremely high concentrations in kidney (38.8 mg Hg/kg FW) and may be ingesting nephrotoxic levels of mercury through the diet (8.8 mg Hg/kg FW ration).

In the serow (*Capriocornis crispus*), a free-ranging bovine ruminant, about 40% of the total mercury body burden was in the fleece at 0.37 mg Hg/kg FW fleece. Domestic sheep (*Ovis* sp.) allowed to graze for 23 months on grass contaminated with mercury (up to 6.5 mg/kg dry weight) caused by atmospheric emissions of a nearby chloralkali site, retained about 0.1% of the total mercury taken in by ingestion and inhalation, although residues in flesh were negligible. It was concluded that contamination of grass as a result of atmospheric discharges of inorganic mercury from chloralkali sites causes no hazard, either directly to grazing animals or indirectly to humans who might ultimately consume their flesh.

19.6.8 Integrated Collections

Mercury data for abiotic samples as well as living organisms representative of different ecosystems taken from a single collection locale, usually at the same time by the same research group, are particularly valuable. These data may illustrate food web biomagnification and other phenomena more readily than isolated data bits drawn over several years from disparate locales using different collection methods, various sample preparations, and noncomparable chemical methodologies for mercury analysis. One integrated data set demonstrated that mercury from point-source discharges, such as sewer outfalls and chloralkali plants, was taken up by sediments, and the sediment mercury levels were then reflected by an increased mercury content of epibenthic fauna. In another case, analysis of the effluent from the Hyperion sewer outfall in Los Angeles showed a mercury concentration slightly below 0.001 mg/L. Concentrations of mercury in sediment samples near this outfall were as high as 0.82 mg/kg but decreased with increasing distance from the outfall; mercury levels in epibenthic fauna, including crabs, whelks, and scallops, were also highest at stations near the discharge and lowest at stations tens of kilometers distant.

Data sets for mercury are now available for locations in the Adriatic Sea, Alaska, Antarctica, Brazil, Canada, China, Cuba, Florida, Greenland, India, Italy, Korea, Malaysia, New Jersey, Puerto Rico, Spain, Taiwan, Tennessee, Thailand, and Vietnam.

19.7 Lethal Effects of Mercurials

Death is the only biological variable now measured that is considered irreversible by all investigators. Nevertheless, time of death is modified by a host of physical, chemical, biological, metabolic, and behavioral variables and it is unfortunate that some regulatory agencies still set mercury criteria to protect natural resources and human health on the basis of death – usually concentrations producing 50% mortality – and some variable uncertainty factor. Mercury criteria for protection of natural resources and human health, as discussed later, should be based – at a minimum – on the highest dose tested or highest tissue concentration found that does not produce death, impaired reproduction, inhibited growth, or disrupted well-being.

19.7.1 Aquatic Organisms

Lethal concentrations of mercury salts ranged from less than 0.1 μg Hg/L to more than 200.0 μg/L for representative sensitive species of marine and freshwater organisms. The lower concentrations of less than 2.0 μg/L recorded were usually associated with early developmental stages, long exposures, and flowthrough tests. Among teleosts, females and larger fish were more resistant to lethal effects of mercury than were males and smaller fishes. Among metals tested, mercury was the most toxic to aquatic organisms, and organomercury compounds showed the greatest biocidal potential. In general, mercury toxicity was higher at elevated temperatures, at reduced salinities for marine organisms, and in the presence of other metals such as zinc and lead. Salinity stress, for example, especially abnormally low salinities, reduced significantly the survival time of mercury-exposed isopod crustaceans, suggesting that species adapted to a fluctuating marine environment – typical of the intertidal zone – could be more vulnerable to the added stress of mercury than species inhabiting more uniformly stable environments.

19.7.1.1 Invertebrates

The marine ciliate protozoan *Uronema marinum*, with an LC50 (24 h) value of 6.0 μg/L, failed to develop resistance to mercury over an 18-week period. However, another marine ciliate protozoan, *Uronema nigricans*, acquired tolerance to mercury after feeding on mercury-laden bacteria and subsequently exposed to increasing levels of mercury in solution. The phenomenon of acquired mercury tolerance in *U. nigricans* occurred in a single generation. Among coral colonies of *Porites asteroides*, the LC50 (72 h) value was 180.0 μg Hg/L, as $HgCl_2$. Death was preceded by polyp contraction during the first 8 h, color loss within 24 h, and tissue loss within 48 h.

In general, salts of mercury and its organic compounds have been shown in short-term bioassays to be more toxic to marine organisms than salts of other heavy metals. To oyster embryos, for example, mercury salts were more toxic than salts of silver, copper, zinc, nickel, lead, cadmium, arsenic, chromium, manganese, or aluminum; to clam embryos, mercury was the most toxic metal tested, followed, in order, by silver, zinc, nickel, and lead. An LC50 48 h) value of 5.7 μg Hg/L, as inorganic mercury, is reported to embryos of the Pacific oyster (*Crassostrea gigas*); however, embryos were relatively insensitive to mercury 24 h post-fertilization, and survival was enhanced by a variety of factors, including ambient selenium concentrations.

Mercury toxicity to crustaceans was markedly influenced by developmental stage, diet, sex, salinity, tissue sensitivity, and selenium. Larvae and newly molted crustaceans were more sensitive to mercury toxicity than were adults of the same species. Starved larvae of the grass shrimp had lower survival rates than fed larvae when subjected to mercury insult. Also, male adult fiddler crabs (*Uca pugilator*) were more sensitive to mercury salts than females. Lethality of mercury salts to the porcelain crab (*Petrolisthus armatus*) were most pronounced at lower salinities within the range of 7–35‰. A similar pattern was recorded for the fiddler crab, *Uca pugilator*. Adult prawns (*Leander serratus*) held in lethal solutions of mercury (50.0 mg inorganic Hg/L; 1.0 mg organic mercury/L) for 3 h contained at death 320.0–460.0 mg Hg/kg DW in antennary gland. High levels of selenium (>5.0 mg/L) increased mercury toxicity to larvae of dungeness crab, *Cancer magister* to levels below the LC50 (96 h) value of 6.6 μg Hg/L; however, moderate selenium values of 0.01–1.0 mg/L tended to decrease mercury toxicity.

Many acute toxicity bioassays were of 96 h duration, a duration that allows the senior investigator and technicians alike the opportunity to enjoy an uninterrupted weekend. But it is clear that assays of 168-h duration produced LC50 values up to 45 times lower (more toxic) than did the 96-h assays, as was shown for mud snails. It is recommended that acute toxicity bioassays with mercury and other toxicants and estuarine fauna should consist of a minimal 10-day continuous exposure followed by a 10-day observation period.

19.7.1.2 Vertebrates

Signs of acute mercury poisoning in fish included flaring of gill covers, increased frequency of respiratory movements, loss of equilibrium, excessive mucous secretion, darkening of coloration, and sluggishness. Signs of chronic mercury poisoning included emaciation (due to appetite loss), brain lesions, cataracts, diminished response to change in light intensity, inability to capture food, abnormal motor coordination, and various erratic behaviors. Total mercury concentrations in tissues of mercury-poisoned adult freshwater fish that died soon thereafter ranged (in mg/kg fresh weight) from 6.0 to 114.0 in liver, 3.0 to 42.0 in brain, 5.0 to 52.0 in muscle, and 3.0 to 35.0 in whole body. Whole body concentrations up to 100.0 mg/kg FW were reportedly not lethal to rainbow trout (*Oncorhynchus mykiss*), although 20.0–30.0 mg/kg FW in that species were associated with reduced appetite, loss of equilibrium, and hyperplasia of gill epithelium. Brook trout (*Salvelinus fontinalis*), however, showed toxic signs and death at whole body residues of only 5.0–7.0 mg/kg FW.

Some fish populations have developed a resistance to methylmercury, but only in the gametes and embryonic stage. Thus, eggs of the mummichog (*Fundulus heteroclitus*), an estuarine cyprinodontiform fish, from a mercury-contaminated creek, when compared to a reference site, were more than twice as resistant to methylmercury (LC50 values of 1.7 mg Hg/L vs. 0.7 mg Hg/L) when exposed for 20 min prior to combination with untreated sperm. Eggs from the polluted creek that were subjected to 1.0 or 2.5 mg CH_3HgCl/L produced embryos with a 5–7% malformation frequency vs. 32% malformations at 1.0 mg/L and little survival at 2.5 mg/L in the reference group. Genetic polymorphism in mosquitofish (*Gambusia* sp.) at specific enzyme loci are thought to control survival during mercury exposure. In one population of mosquitofish during acute exposure to mercury, genotypes at three loci were significantly related to survival time, as was heterozygosity. But neither genotype nor heterozygosity was related to survival in a different population of mosquitofish during acute mercury exposure.

Embryo-larva tests with amphibians and inorganic mercury showed that 6 of the 21 species tested were more sensitive than rainbow trout embryo-larva tests and 15 were less sensitive; however, all 21 amphibian species were more sensitive than largemouth bass embryos. Amphibian embryos were the most sensitive stage tested to mercury and other chemicals owing to the relatively high permeability of the egg capsule at this time. In general, organomercurials were three to four times more lethal than inorganic mercury compounds to amphibians when the same species and life stage were tested.

Exposure pathways for adult amphibians include soils (dermal contact, liquid water uptake), water (dermal contact with surface water), air (cutaneous and lung absorption), and diet (adults are carnivores). All routes of exposure are affected by various physical, chemical, and other factors. Dietary exposure in adults, for example, is related to season of year, activity rates, food availability, consumption rate, and assimilation rates. Knowledge of these modifiers is necessary for adequate risk assessment of mercury as a possible factor in declining amphibian populations worldwide.

19.7.2 Terrestrial Invertebrates

Methylmercury compounds at concentrations of 25.0 mg Hg/kg in soil were fatal to all tiger worms (*Eisenia foetida*) in 12 weeks; at 5.0 mg/kg, however, only 21% died in a similar period. Inorganic mercury compounds were also toxic to earthworms (*Octochaetus pattoni*); in 60 days, 50% died at soil Hg^{2+} levels of 0.79 mg/kg, and all died at 5.0 mg/kg.

19.7.3 Reptiles

Data on mercury lethality in reptiles are scarce, and those available suggest that sensitivity may be both species- and age-dependent. For example, juveniles of the corn snake (*Elaphe guttata*) fed diets containing 12.0 mg methylmercury/kg FW dietall died within days. However, adults and offspring from treated adults of the garter snake (*Thamnophis sirtalis*), fed diets containing up to 200.0 mg methylmercury/kg FW diet, all survived and showed no sign of toxicity.

19.7.4 Birds

Signs of mercury poisoning in birds include muscular incoordination, falling, slowness, fluffed feathers, calmness, withdrawal, hyporeactivity, hypo-activity, and eyelid drooping. In acute oral exposures, signs appeared as soon as 20 min post-administration in mallards (*Anas platyrhynchos*), and 2.5 h in ring-necked pheasants (*Phasianus colchicus*). Deaths occurred between 4 and 48 h in mallards and 2 and 6 days in pheasants; remission took up to 7 days. In studies with coturnix (*Coturnix* sp.), methylmercury was always more toxic than inorganic mercury, and young birds were usually more sensitive than older birds. Further, some birds poisoned by inorganic mercury recovered after treatment was withdrawn, but chicks that were fed methylmercury and later developed toxic signs usually died, even if treated feed was removed. Coturnix subjected to inorganic mercury, regardless of route of administration, showed a violent neurological dysfunction that ended in death 2–6 h posttreatment. The withdrawal syndrome in coturnix poisoned by Hg^{2+} was usually preceded by intermittent, nearly undetectable tremors, coupled with aggressiveness towards cohorts; time from onset to remission was usually 3–5 days, but sometimes extended to seven days. Coturnix poisoned by methylmercury appeared normal until 2–5 days posttreatment; then, ataxia and low body carriage with outstretched neck were often associated with walking. In advanced stages, coturnix lost locomotor coordination and did not recover; in mild to moderate clinical signs, recovery usually took at least 1 week.

Mercury toxicity to birds varies with the form of the element, dose, route of administration, species, sex, age, and physiological condition. For example, in northern bobwhite chicks fed diets containing methylmercury chloride, mortality was significantly lower when the solvent was acetone than when it was another carrier such as propylene glycol or corn oil. In addition, organomercury compounds interact with elevated temperatures and pesticides, such as DDE and parathion, to produce additive or more-than-additive toxicity, and with selenium to produce less-than-additive toxicity. Acute oral toxicities of various mercury formulations ranged between 2.2 and about 31.0 mg Hg/kg body weight for most avian species tested. Similar data for other routes of administration were 4.0–40.0 mg/kg for diet and 8.0–15.0 mg/kg body weight for intramuscular injection.

Residues of mercury in experimentally poisoned passerine birds usually exceeded 20.0 mg/kg FW, and were similar to concentrations reported in wild birds that died of mercury poisoning. In one study with the zebra finch (*Poephila guttata*), adults were fed methylmercury in the diet for 76 days at dietary levels of <0.02 (controls), 1.0, 2.5, or 5.0 mg Hg/kg DW ration. There were no signs of mercury intoxication in any group except the high-dose group that experienced 25% dead and 40% neurological impairment. Dead birds from the high dose group had 73.0 mg Hg/kg FW in liver, 65.0 in kidney, and 20.0 in brain; survivors without signs had 30.0 in liver, 36.0 in kidney, and 14.0 mg Hg/kg FW in brain; impaired birds had 43.0 mg Hg/kg FW in liver, 55.0 in kidney, and 20.0 in brain.

Mercury levels in tissues of poisoned wild birds were highest (45.0–126.0 mg/kg FW) in red-winged blackbirds (*Agelaius phoeniceus*), intermediate in European starlings (*Sturnus vulgaris*) and cowbirds (*Molothrus ater*), and lowest (21.0–54.0) in common grackles (*Quiscalus quiscula*); in general, mercury residues were highest in the brain, followed by the liver, kidney, muscle, and carcass. Some avian species are more sensitive than passerines. Liver residues (in mg Hg/kg FW) in birds experimentally killed by methylmercury ranged from 17.0 in red-tailed hawks (*Buto jamaicensis*) to 70.0 in jackdaws (*Corvus monedula*); values were intermediate in ring-necked pheasants, kestrels (*Falco tinnunculus*), and black-billed magpies (*Pica pica*). Experimentally poisoned grey herons (*Ardea cinerea*) seemed to be unusually resistant to mercury; lethal doses produced residues of 415.0–752.0 mg Hg/kg dry weight of liver. However, levels of this magnitude were frequently encountered in livers from grey herons collected during a massive die-off in the Netherlands during a cold spell in 1976; the interaction effects of cold stress, mercury loading, and poor physical condition of the herons are unknown.

19.7.5 Mammals

Mercury is easily transformed into stable and highly toxic methylmercury by microorganisms and other vectors. Methylmercury affects the central nervous system in humans – especially the sensory, visual, and auditory areas concerned with coordination; the most severe effects lead to widespread brain damage, resulting in mental derangement, coma, and death. Methylmercury has long residence times in aquatic biota and consumption of methylmercury-contaminated fish is implicated in more than 150 deaths and more than 1000 birth defects in Minamata, Japan, between 1956 and 1960 (see later). By 1987, more than 17,000 people had been affected by methylmercury poisoning in Japan, with 999 deaths. Worldwide, it is estimated that mercury poisoning from ingestion of contaminated food is responsible for more than 1400 human deaths and 200,000 sublethal cases. Excess mercury in human tissues is associated with an increased risk of acute myocardial infarction, and increased death rate from coronary heart disease and carotid atherosclerosis.

In mule deer (*Odocoileus hemionus hemionus*), after acute oral mercury poisoning was induced experimentally, signs included belching, bloody diarrhea, piloerection (hair more erect than usual), and loss of appetite. The kidney is the probable critical organ in adult mammals due to the rapid degradation of phenylmercurials and methoxyethylmercurials to inorganic mercury compounds and subsequent translocation to the kidney, whereas in the fetus the brain is the principal target. Most human poisonings were associated with organomercury compounds used in agriculture as fungicides to protect cereal seed grain; judging from anecdotal evidence, many wildlife species may have been similarly afflicted. Organomercury compounds, especially methylmercury, were the most toxic mercury species tested. Among sensitive species of mammals, death occurred at daily organomercury concentrations of 0.1–0.5 mg/kg body weight, or 1.0–5.0 mg/kg in the diet. Larger animals such as mule deer and harp seals (*Pagophilus groenlandica*) appear to be more resistant to mercury than were smaller mammals such as mink, cats, dogs, pigs, monkeys, and river otters; the reasons for this difference are unknown, but may be related to differences in metabolism and mercury detoxification rates. Tissue residues in fatally poisoned mammals (in mg Hg/kg fresh weight) were approximately 6.0 in brain, 10.0–55.6 in liver, 17.0 in whole body, about 30.0 in blood, and 37.7 in kidney.

The lethal effects of methylmercury in various species of mammals are influenced by ambient temperature, dietary selenium, ethanol, and especially hypertension. Tests with a genetic strain of rat with high blood pressure showed that this strain was more sensitive to methylmercury than were control strains; they died earlier and with higher tissue mercury burdens. Because hypertension and borderline hypertension is common among human inhabitants of mercury-polluted areas, with estimates as high as 56% among individuals 40 years old and older, more research seems warranted on the role of hypertension as a significant health problem in methylmercury-impacted populations.

Mercury interactions with other compounds should be considered. Adverse effects on growth and survival of kits of the mink (*Mustela vison*) are reported for diets containing 1.0 mg Hg/kg ration as methylmercury and Aroclor 1254–a polychlorinated biphenyl – at 1.0 mg/kg ration.

19.8 Sublethal Effects

Significant sublethal effects of mercury include an increased frequency of cancers, birth defects, and chromosomal aberrations in laboratory animals and wildlife. Adverse sublethal effects of mercurials also include growth inhibition, abnormal reproduction, histopathology, high mercury accumulations and persistence, and disrupted biochemistry, metabolism, and behavior. These – and other aspects of exposure to various mercurials by living organisms – are documented and discussed for representative species of bacteria and other microorganisms, aquatic and terrestrial plants and invertebrates, fishes, amphibians, birds, and mammals.

19.8.1 Carcinogenicity, Genotoxicity, and Teratogenicity

The social significance of cancer, inherited effects, and birth defects cannot be underestimated. For example, during the Minamata Bay, Japan, methylmercury poisoning outbreak (to be discussed in detail later) – in which thousands of people were afflicted and many died – dozens of fetuses affected in utero as a result of maternal ingestion of methylmercury-contaminated food were born with mental retardation and motor disorders. There was much concern in the neighboring population that irreversible genetic damage could produce malformations and other adverse effects in subsequent generations; accordingly, some parents prohibited marriages of their sons and daughters with persons from Minamata. Although studies with mice indicated that intra-uterine exposure effects to methylmercury were practically nil in humans, the stigma remains.

19.8.1.1 Carcinogenicity

Mercury has been assigned a weight-of-evidence classification of D, which indicates that it is not classifiable as to human carcinogenicity. Beluga whales (*Delphinapterus leucas*) in the St. Lawrence estuary have a high incidence of cancer. Studies with isolated skin fibroblasts of beluga whales exposed to mercuric chloride or methylmercury induced a significant dose-response increase of micronucleated cells. Concentrations as low as 0.5 mg Hg^{2+}/L and 50.0 μg CH_3Hg^+/L – comparable to concentrations present in certain whales of this population – significantly induced proliferation.

19.8.1.2 Genotoxicity

Chromosomal aberrations and sister chromatid exchanges have been produced by mercury compounds in larvae and embryos of amphibians and fishes; in various species of bacteria, yeasts, and molds; and in cultured somatic cells of fish, insects, echinoderms, and mammals – including somatic cells of rodents, dolphins, cats, and humans. Toxicant stress to individuals may modify genetic characteristics of mosquitofish populations. Allozyme polymorphisms existing in mosquitofish populations with no history of contaminant stress can be subject to selection and provide the basis for adaptation to anthropogenic stress. For example, metabolic differences in glucosephosphatase isomerase genotypes, or closely related loci, may be expressed in degrees of environmental stress or fluctuation. Allozyme genotypes could be responsible for transient genotype effects noted in electrophoretic surveys of mercury-stressed mosquitofish populations.

Methylmercury induces increased mutagenic effects in killifish (*Fundulus heteroclitus*) embryos after exposure to 50.0 μg Hg/L for up to 7 days postfertilization. Chromosomal aberrations were 13% in the treated group vs. 5.1% in controls. Mutagenic effects in killifish embryos occurred at levels below that measured in some marine sediments.

Genetic background may influence the frequency of hydrocephaly (an abnormal increase in cerebrospinal fluid within the cranial cavity accompanied by expansion of cerebral ventricles, enlargement of the forehead, and atrophy of the brain) in mice following prenatal exposure to methylmercury. In pregnant mice of B10.D2 strain – in which hydrocephaly is rare – a single oral dose of 10.0 mg CH_3HgCl/kg BW on one of days 14–17 of pregnancy resulted in 48–75% frequency of gross hydrocephaly in offspring at age 30 days, vs. 5% in sham-treated and 4% in untreated controls. When the study was repeated with C57BL/10 or DBA/2 strains on day 15 of pregnancy, the incidence of hydrocephaly in dosed C57BL/10 mice was 54% vs. 0.8% in controls. Hydrocephaly did not develop in the DBA/2 strain, suggesting that methylmercury-induced hydrocephaly is under genetic control in mice.

There is no evidence that mercury is genotoxic in humans. Dental amalgam has been used for more than 150 years and it is well established that patients with amalgam dental fillings are chronically exposed to mercury and that the number of amalgam fillings correlates positively with mercury levels in blood and plasma. Amalgam removal, however, did not have a measurable effect on proliferation of peripheral blood lymphocytes, suggesting that mercury contributed by amalgams was not mitogenic to lymphocytes. Mercury did not adversely affect the number or structure of chromosomes in human somatic cells in workers occupationally exposed to mercury compounds by inhalation or accidentally exposed through ingestion. In the laboratory, however, mercury compounds often exerted genotoxic effects, especially by binding SH groups and acting as spindle inhibitors, resulting in abnormal chromosome numbers.

19.8.1.3 Teratogenicity

Teratogenicity of methylmercury has been confirmed in mice, hamsters, and rats.

In humans, methylmercury, CH_3HgCl^+, unlike Hg and Hg^{2+}, is efficiently absorbed through the intestinal tract and skin, crosses the placenta resulting in fetal blood levels in excess of the mother, with elevated risks to the fetus. Pathological features of children's brains affected by prenatal methylmercury exposure include microcephaly, dilated lateral ventricles, abnormal neuronal migration causing derangement in the gray matter architecture, and degeneration of formed nerve cells. Liquid metallic mercury, Hg, is poorly absorbed from the intestinal tract of mammals and teratogenic potential of Hg via this route is low; however, inhaled Hg vapor is efficiently absorbed. Inorganic mercury, Hg^{2+}, is also poorly absorbed (about 2%) from the GI tract, and uptake of Hg^{2+} by the fetus is low. Injected inorganic mercury salts, however, are teratogenic in laboratory animals. Inorganic mercury transfers inefficiently to the fetus but accumulates in the placenta. In mice, a significant portion of the Hg^{2+} is locked in the yolk sac; it was concluded that observed fetal defects resulted from inhibition of metabolite transport and from maternal kidney dysfunction and not from direct action of Hg^{2+} on the fetus.

Methylmercury enhanced the teratogenic action of mitomycin in mice (*Mus* sp.), and may act additively to induce cleft palate at doses not known to produce this effect. This conclusion was based on a laboratory study in which pregnant mice were given oral nonteratogenic doses of 0.0, 2.5, 5.0, or 10.0 mg/kg BW of methylmercuric chloride on day 9 of pregnancy and then injected intraperitoneally with a teratogenic dose (4.0 mg/kg BW) of mitomycin-C on day 10. Malformations produced by mitomycin-C alone included cervical rib and vertebral anomalies, polydactyly of the hindlimb, and tail anomalies. Combined treatment significantly increased the incidence of these malformations in dose-dependent pattern. Cleft palate was also evident in combined treatment, but not by either chemical alone.

Studies with mice indicated teratogenic effects in subsequent generations, i.e., surviving males from methylmercury-intoxicated dams mated with untreated females produced progeny with 8.2% malformations, including exencephaly and brain hernia. Subsequent experiments were conducted wherein all progeny survived parturition. Results showed the following: F_1 females retained mercury for 2–3 weeks after birth following maternal exposure to methylmercury during pregnancy; >85% of F_1 males survived to adulthood and all were fertile; and F_1 males mated with untreated females produced some offspring with growth retardation (<4%) but no significant difference in malformation frequency from that of controls. It was concluded that male-mediated fetal effects subsequent to intra-uterine exposure to methylmercury may be practically nil in humans.

In terms of ability to affect normal development of embryos of the horseshoe crab (*Limulus polyphemus*), mercury – as the acetate or chloride – was the most toxic metal tested followed in decreasing order by tributyltin chloride, hexavalent chromium, cadmium, copper, lead, and zinc. Mercury was associated with a high frequency of segment-defective embryos and could be replicated with SH inhibitors, and by compounds inhibiting SH-SS exchange.

19.8.2 Bacteria and Other Microorganisms

Mercury-tolerant strains of bacteria and protozoa have been reported. Mercury-resistant strains of bacteria are common. In Chesapeake Bay, for example, at least 364 strains of bacteria that were isolated were resistant to $HgCl_2$. Most were pseudomonads, and almost all were from seven genera. Other groups of bacteria known to materially influence mercury fluxes in saline waters include strains of *Escherichia coli* that convert Hg^{2+} to Hg, and strains of anaerobic methanogenic bacteria that enhance the transfer of methylcobalamin to Hg^{2+} under mild reducing conditions. Some mercury-resistant strains were reported to degrade petroleum as well. Mercury-resistant strains of bacteria have been recommended as bioindicators of environmental mercury contamination and as markers of methylmercury in biological samples. The mercury-resistant strains of bacteria cultured or discovered may have application in mobilization or fixation of mercury from contaminated aquatic environments to the extent that polluted areas may become innocuous.

Microorganisms may develop resistance to mercury compounds after prolonged exposure and this phenomenon is well documented in procaryotes such as bacteria and blue-green algae. Bacteria resistant to mercury compounds possess two inducible enzymes – mercuric reductase and organomercurial lyase – both of which reduce Hg^{2+} to Hg. In eukaryotic species of fungi, mercury resistance is reported by growing the organism in media supplemented with increasing concentrations of mercury compounds, including strains of *Botrytis*, *Penicillum*, *Sclerotinia*, *Stemphylium*, *Pyrenophora*, *Neurospora*, and *Cryptococcus*. In addition, mercury^{2+}-resistant yeast strains of *Rhodotorula* and *Saccharomyces* were isolated from rotting guava fruit containing 28.0–1067.0 μg Hg^{2+}/kg FW note: >500.0 μg Hg/kg FW is in violation of the food criterion set by the World Health Organization. Mercury was not volatilized by these yeasts, indicating that the pattern of mercury resistance in yeasts is different from that of bacteria; the high resistance of mercury-resistant yeasts towards mercury may be associated with the binding of Hg^{2+} by yeast cell wall or cytoplasmic membrane. The marine protozoan *Uronemia nigricans*, after feeding on mercury-laden bacteria, acquired mercury tolerance within a single generation.

Mercury-resistant and metabolizing strains of bacteria show a wide range of metabolic rates dependent on the chemical species of mercury, the bacterial biomass, and species competition, as was the case for *Escherichia coli* and *Staphylococcus aureus*. Mercury-resistant strains of bacteria which have been developed or discovered could form the basis of a mercury-removal technology in mercury-contaminated estuaries. In one case, a mercury-resistant strain of *Pseudomonas* reduced 0.025 mg of $HgCl_2$ to elemental mercury from solutions containing 6.0 mg $HgCl_2$/L. A mercury-accumulating strain of *Pseudomonas* took up 12.0 mg $HgCl_2$/kg cells every 15 min from a solution containing 6.0 mg Hg/Cl_2.

19.8.3 Terrestrial Plants

Seedlings of rice (*Oryza sativa*) grown on mercury-contaminated waste soil from a chloralkali factory for 75 days showed increasing mercury concentrations over time with increasing soil mercury content. At 2.5% waste soil in gardens, rice seedlings contained 8.0 mg Hg/kg FW; at 10%, 15.2 mg Hg/kg; and at 17.5% waste soil in gardens, seedlings had 19.1 mg Hg/kg FW. Content of nucleic acids and proteins in rice shoots decreased with increasing mercury concentrations in waste soils and with time, but free amino acid content increased. An increase in the RNA/DNA ratio occurred, indicating an enhanced synthesis of RNA per molecule of DNA. Seedlings of spruce (*Picea abies*) exposed to solutions containing up to 0.2 mg Hg/L as inorganic mercury or methylmercury showed a dose-dependent inhibition in root growth, especially for methylmercury; dose-dependent decreases in concentrations of potassium, manganese, and magnesium were evident in roots and root tips, and increases in iron in root tips.

Some species of mushrooms have been recommended as sentinel species of mercury contamination because of their ability to accumulate very high concentrations of mercury from the ambient air. In one study, shiitake mushrooms, *Lentinus edodus*, exposed to Hg vapor at 172.0 μg Hg/m^3 for up to 7 days had grossly elevated concentrations of mercury in caps and stalks. After 3 days caps had 125.0 mg Hg/kg DW and stalks 10.0 mg Hg/kg DW. After 7 days, these values were 310.0–400.0 mg/kg DW in caps and 20.0–30.0 mg/kg DW in stalks.

19.8.4 Terrestrial Invertebrates

Methylmercury compounds have induced abnormal sex chromosomes in the fruit fly (*Drosophila melanogaster*). Earthworms (*Eisenia foetida*) exposed to soil containing methylmercury concentrations of 5.0 mg Hg/kg – typical of soil Hg levels near chloralkali plants – showed a significant reduction in the number of segments regenerated after 12 weeks, and contained 85.0 mg Hg/kg on a whole body fresh weight basis. Regeneration was normal at soil Hg levels of 1.0 mg/kg, although body burdens up to 27.0 mg/kg were recorded. It was concluded that soil contaminated with methylmercury posed a greater hazard to the predators of earthworms than to the earthworms. Studies with a different species of earthworm (*Octochaetus pattoni*) and mercuric chloride, demonstrated a progressive initial increase in reproduction as soil mercury levels increased from 0.0 to the 60-day lethal level of 5.0 mg/kg.

19.8.5 Aquatic Plants

It is emphasized that phytotoxic effects are always more pronounced when mercury is present in the organic form than in the inorganic form. Growth inhibition was recorded in freshwater algae after exposure of 24 h to 10 days to 0.3–0.6 μg organomercury/L, and in the marine alga (*Scripsiella faeroense*) exposed to 1.0 μg Hg^{2+}/L for 24 h. At water concentrations of 1.0 μg Hg^{2+}/L for 24 h there was an increased incidence of frustule abnormalities and burst thecae in two species of marine algae.

Aquatic plants show great variation in sensitivity to mercury insult. Using marine algae as an example, phytotoxic effects – including reduced growth, developmental abnormalities, photosynthesis inhibition, and death – ranged from 1000.0 μg Hg/L and higher for the comparatively resistant *Plumaria elegans* sporelings and *Isochrysis galbana* to 1.0 μg Hg/L for the comparatively sensitive *Scripsiella faeroense* and *Nitzschia* sp. Intermediate in sensitivity were *Macrocystis pyrifera*, with effects evident at 50.0 μg Hg/L, eighteen species of unicellular algae exposed for 18 days to 5.0–25.0 μg Hg/L, and various species of *Laminaria*, *Pelvetia*, *Ascophyllum*, *Dunaliella*, *Chlamydomonas*, *Skeletonema* and *Fucus* at 50.0–100.0 μg Hg/L.

Rapid accumulation of mercury, especially organomercury compounds by various species of algae, has been documented. However, uptake of mercury and its compounds is modified by a number of biological, chemical, and physical factors, such as initial mercury concentration, water pH, exposure time, age of organism, salinity of medium, biological surface area, variability in mercury detoxification mechanisms, cell density, and accumulation after death. Among various species of marine algae, for example, *Croomonas salina* took up 1400.0 mg Hg/kg dry weight after 48 h in solutions containing 164.0 μg Hg/L; *Chaetoceros galvestonensis* and *Phaeodactylum tricornutum* contained 7400.0 and 2400.0 mg Hg/kg, respectively, when cultured in media containing 100.0 μg Hg/L (*Chaetoceros*) or 50.0 μg Hg/L (*Phaeodactylum*), and *Isochrysis galbana* contained up to 1000.0 mg Hg/kg after exposure to 15.0 μg Hg/L. Salt marsh plants show enhanced mercury concentrations in roots at lower salinities and higher pH.

Using artificial ecosystems, reduced initial growth of natural phytoplankton from Saanach Inlet, British Columbia, Canada, was noted after addition of 1.0–5.0 μg Hg/L; however, recovery to above control levels appeared to occur in 21 days.

19.8.6 Aquatic Animals

Sublethal concentrations of mercury are known to adversely affect aquatic fauna through inhibition of reproduction, reduction in growth rate, increased frequency of tissue histopathology, impairment of ability to capture prey, impairment of olfactory receptor function, alterations in blood chemistry and enzyme activities, and disruption of thyroid function, chloride secretion, and other metabolic and biochemical functions.

Aquatic animals take up CH_3Hg^+ from diet, water, and sediment and retain mercury with continued exposure because elimination is slow relative to uptake rate. Methylmercury concentrations in fish may be 6–7 orders of magnitude higher than methylmercury concentrations in ambient surface waters. In aquatic invertebrates, methylmercury is more readily taken up than inorganic mercury with accumulations highest in predators and lowest in herbivores. Accumulation ratios for mercury were >10,000 between seston and water, <10 for predatory fish muscle and their whole prey, and 125–225 for seabird feathers and their diet. In general, the accumulation of mercury by aquatic biota is rapid and depuration is slow. It is emphasized that organomercury compounds, especially methylmercury, were significantly more effective than inorganic mercury compounds in producing adverse effects and accumulations. For example, mercury-accumulation studies with American oysters, *Crassostrea virginica*, show that adults exposed for 74 days to seawater solutions containing 1.0 μg inorganic Hg/L had 2.0 mg Hg/kg FW soft tissues after 20 days, and 10.0 mg/kg after 60 days. When the experiment was repeated using methylmercury or phenylmercury compounds, oysters contained 10.0 mg total Hg/kg FW soft tissues after 20 days and 30.0 mg/kg after 60 days.

19.8.6.1 Invertebrates

Reproduction was inhibited among the most sensitive species of aquatic animals at water concentrations of 0.03–1.6 μg Hg/L. For less sensitive marine invertebrates such as hydroids, protozoans, and mysid shrimp, reproduction was inhibited at concentrations between 1.1 and 2.5 μg Hg^{2+}/L; this range was 5.0–71.0 μg/L for more resistant species of marine invertebrates.

19.8.6.1.1 Planarians

In the planarian *Dugesia dorotocephala*, asexual fission was suppressed at 0.03–0.1 μg organomercury/L; at 80.0–100.0 μg methylmercury/L, behavior was modified and regeneration retarded.

19.8.6.1.2 Coelenterates

Studies with coral colonies of *Porites asteroides* showed that zooxanthellae and skeleton accumulated total mercury, as $HgCl_2$, in a dose-dependent manner. However, polyp tissue accumulated more mercury at 37.0 μg/L than at 180.0 μg/L, suggesting either saturation of mercury in polyps, activation of mechanisms of detoxification, or both. Most of the mercury in colonies exposed to the highest mercury concentration of 180.0 μg/L for 3 days was in zooxanthellae (89%), with the rest in polyps (7%) and skeleton (4%). Increasing concentrations of mercury was associated with decreasing biomass of polyps and zooxanthellae, and decreasing pigment concentration. Authors concluded that the capacity of zooxanthellae and the skeleton to concentrate mercury and the decline of zooxanthellae density support the hypothesis that polyps may divert mercury to these two compartments as a detoxifying mechanism. Studies with coelenterate hydrozoans (*Campanularia flexuosa*), and sea anemones (*Bundosoma cavernata*) show that concentrations as low as 0.2 μg Hg/L had an adverse effect on lysosomal enzymes of *Campanularia*, 1.7 μg Hg/L depressed growth of *Campanularia*, and that exposure of *Bundosoma* to 1.2 μg Hg/L for 7 days increased concentrations of free amino acids.

19.8.6.1.3 Mollusks

In the slipper limpet (*Crepidula fornicata*) – the most sensitive mollusk tested – exposure to 0.25 μg Hg^{2+}/L caused a delay in spawning, a reduction in fecundity, and after 16 weeks a reduction in growth rate of adults. In marine mollusks exposed to water concentrations of 6.0–10.0 μg Hg^{2+}/L for 96 h, the feeding of adults ceased and the swimming rate of larval stages declined. Juveniles of the freshwater rainbow mussel (*Villosa iris*) showed a reduction in growth rate during exposure for 21 days to sublethal concentrations of 8.0–114.0 μg Hg^{2+}/L.

Rapid accumulation of mercury compounds, especially organomercury compounds, by various species of filter-feeding marine bivalve mollusks is welldocumented. Accumulation of organomercury complexes is especially rapid in molluscan tissues with high lipid content. Accumulation is modified by the chemical form of mercury administered; water temperature; salinity of the medium; presence of selenium; sexual condition; tissue specificity; previous acclimatization to mercury salts; season of year; soluble protein content of organism; and presence of mercury-resistant strains of bacteria. Marine mollusks – unlike certain species of oceanic mammals, elasmobranchs, and teleosts – accumulate mercury only when the environment is contaminated with this element as a direct result of human activities.

Adults of the American oyster (*Crassostrea virginica*) accumulate low levels of mercury in the water column to comparatively elevated concentrations in tissues and retain the mercury for extended periods. In one study, oysters exposed to 10.0 μg Hg/L as mercuric acetate for 45 days contained 28.0 mg Hg/kg FW vs. <0.2 mg Hg/kg FW in controls. By day 60 (15 days post-exposure), soft parts of oysters contained 18.0 mg/kg FW, possibly owing to spawning. After an additional 160 days in mercury-free seawater (day 22), levels declined to 15.0 mg/kg FW during the first 18 days but remained unchanged thereafter. Authors concluded that oysters can concentrate 10.0 μg Hg/L by a factor of 2800 and that total self purification was not achieved over a 6-month period in mercury-free seawater. In another study with American oysters, it was found that continuous exposure to 1.0 μg Hg/L as inorganic or organic mercury resulted in accumulations considered hazardous >0.5 mg Hg/kg FW) to human health. And juveniles of bay scallops, *Argopecten irradians*, held for 96 h in solutions containing 40.0 μg Hg/L contained 48.9 mg Hg/kg FW soft parts, a concentration far in excess of the 0.5 mg Hg/kg FW limit in foods considered hazardous to human health.

Persistence of mercury in molluscan tissues is about intermediate between that of fish and crustaceans. Time to eliminate 50% of biologically assimilated mercury and its compounds (Tb1/2) ranged from 20 to 1200 days for various species of teleosts, 297 days for the crayfish (*Astacus fluviatilis*), 435 days for mussels, and 481 days for the clam *Tapes decussatus*. In 24-h radiotracer studies with ^{203}Hg, clams, *Tapes decussata*, contained ten times more radiomercury than did the medium. Radiomercury in clams had a half-time persistence of 10 days if accumulated via the diet, but only 5 days if taken up from the water; however, retention time of radiomercury-203 in most species of marine mollusks were comparatively lengthy, being up to 1000 days for mussels (*Mytilus galloprovincialis*); phylogenetically related species follow a similar pattern of methylmercury excretion, with half-time persistence dependent on water temperature and mode of entry into the organism. Accumulation of mercury by oysters more than doubled in the presence of mercury-resistant strains of *Pseudomonas* spp. Accumulation of inorganic mercury and methylmercury compounds by green mussels, *Perna viridis*, was modified by chemical species, water salinity, dissolved organic carbon (DOC), and colloidal organic carbon (COC). Methylmercury was taken up eight times faster than inorganic mercury at 30‰. At 15‰ salinity, uptake increased 75% for inorganic mercury and 117% for methylmercury. The presence of biogenic DOC derived from diatom decomposition decreased uptake of inorganic mercury, but humic acid increased uptake. Methylmercury uptake, however, was only weakly influenced by differences in DOC quality or quantity. COC increased uptake of inorganic mercury up to sevenfold when compared to low molecular weight-complexed mercury, but decreased the uptake of methylmercury by 42–73%. It was concluded that facilitated transport and direct colloidal ingestion are involved in the uptake of mercury species under different conditions of DOC and COC, and each needs to be considered separately when measuring mercury uptake.

19.8.6.1.4 Crustaceans

In mysid shrimp (*Mysidopsis bahia*), a sensitive species, the abortion rate increased and population size decreased after lifetime i.e., 28 days exposure to 1.6 μg/L of mercury as mercuric chloride.

Among marine crustaceans, variables known to modify mercury accumulation and other effects include tissue specificity, diet, feeding niche, sex, life stage, general health, temperature and salinity of the medium, duration of exposure, chemical form of mercury, and the presence of other compounds. Selected studies illustrating these points follow.

American lobsters (*Homarus americanus*) immersed in seawater containing up to 6.0 μg Hg/L for 30 days had significant accumulations of mercury in various tissues: 15.2 mg Hg/kg FW in digestive gland, up from 0.12 mg Hg/kg FW at start; 85.3 in gills, up from 0.14; and 1.0 mg Hg/kg FW in muscle, up from 0.23. With increasing exposure of 60 days, all tissues showed mercury increases; for example, gill contained 119.5 mg Hg/kg FW. Mercury accumulations from solution were, in general, more rapid and more pronounced than were accumulations from food sources. In an alga (1400.0 mg Hg/kg DW) to copepod food link, for example, copepods showed no impairment of egg laying or egg development, and no retention of mercury in tissues, eggs, or feces.

Mercury effects on fiddler crabs (*Uca pugilator*) immersed in solutions containing high (180.0 μg Hg/L) sublethal concentrations were modified by life stage, sex, and thermosaline regimen. Adults were more resistant than larvae, adult females more resistant than adult males, and resistance for all stages was lowest at low temperatures and salinities. Exposure of fiddler crabs for 3 h to 180.0 μg Hg/L, as $HgCl_2$, demonstrated that gill was the major site of mercury accumulation; lesser amounts accumulated in hepatopancreas and green gland. Survival was lower at 5°C than at 33°C. The inability to transfer mercury from gill to hepatopancreas at low temperatures could be a factor in the toxicity of mercury to fiddler crabs at low temperatures. However, negligible uptake or effects were observed in whole *Uca pugilator* after exposure for 2 weeks in seawater solutions containing 100.0 μg Hg/L. Similarly, pre-exposure of larvae of white shrimp (*Penaeus setiferous*) for 57 days to 1.0 μg Hg/L had no effect on measured parameters during subsequent mercury-stress experiments with these larvae. Although increased toxicity of mercury salts were demonstrable to crustaceans at low salinities within the range of 7–35‰, no increase in mercury uptake was noted in shrimp *Palaemon debilis* at salinities as low as 6‰.

Mercury accumulation in viscera of euphausiids (*Meganyctiphanes norvegica*) was decreased by moderate concentrations of selenium (10.0–1000.0 μg Se/L) in the medium. In studies with larvae of the Dungeness crab (*Cancer magister*), selenium concentrations of 10.0–1000.0 μg/L tended to decrease mercury toxicity; however, high levels of selenium (>5000.0 μg/L) increased mercury toxicity.

Larvae of the barnacle *Elminius modestus* took up 920.0 mg Hg/kg DW during exposure for 3 h in 200.0 μg Hg/L as mercuric chloride. Accumulation of amyl mercuric chloride, n-$C_5H_{11}HgCl$, was 700.0 mg Hg/kg DW during exposure for 3 h in only 1.0 μg Hg/L solutions. Similar trends were observed for brine shrimp (*Artemia salina*). When adults of the spiny spider crab (*Maia squinado*) were immersed in seawater containing added mercuric chloride, mercury concentrated in gills and other sites. Eventually, the concentration of mercury in blood rose above that in the external medium, the concentration in antennary gland above that in blood, and the crabs excreted small, but increasing, amounts of mercury in the urine. Most (95%) of the mercury in blood was attached to protein, with mercury concentrations in blood remaining constant for several weeks following mercury exposure. When *Maia* were poisoned with methylmercury chloride, mercury again concentrated in gills and various internal organs; however, the amount in blood was comparatively low, and none was found in urine. The marine copepod (*Acartia clausi*), subjected to concentrations of 0.05 μg/L of mercury and higher, reached equilibrium with the medium in only 24 h. In that study, bioconcentration factor (BCF) values for whole *Acartia* after 24-h exposures were 14,360 for inorganic mercuric ion at 0.05 μg/L and, for methylmercury, 179,200 at 0.05 μg/L, and 181,000 at 0.1 μg/L.

In the shrimp (*Lysmata seticaudata*), about 45% of all mercury was localized in viscera, 39% in muscle, 15% in exoskeleton, and 1.8% in molts. It was probable that most of the mercury was in methylated form, as was the case in tissues of blue crabs, fiddler crabs, and other benthic crustaceans. It is generally agreed that organomercury compounds are accumulated more rapidly by crustaceans and to a higher degree than inorganic compounds. Some evidence exists showing that iodide salts of mercury have greater biomagnification potential than chloride salts in copepods and brine shrimp, but this requires verification.

Mercury-tolerant strains of crustaceans are documented. The white shrimp (*Penaeus setiferus*), pre-exposed for 57 days to 1.0 μg Hg/L, did not differ from controls during either exposure or subsequent mercury stress experiments; this observation suggested that nonsensitization or adaptation mechanisms are involved. The fiddler crab (*Uca pugilator*) seemed unusually resistant and showed negligible uptake or effects during exposure to 100.0 μg Hg/L for 2 weeks. Daphnids (*Daphnia magna*) developed tolerance to Hg^{2+} after pre-exposure to 0.5 and 5.1 μg Hg^{2+}/L for 4 days followed by 4 days in mercury-free media. These daphnids had elevated body levels of mercury and metallothioneins, and higher tolerance to mercury toxicity than controls. F_1 offspring from parents exposed to 5.1 μg/L also had mercury tolerance and about 25% of mercury concentrations of parents, but not elevated metallothioneins. Mercury tolerance was not evident in the F_2 generation, suggesting that tolerance would disappear quickly on transfer to mercury-free environments, unlike *Tubifex* worms where tolerance lasted for several generations.

19.8.6.1.5 Annelids

Tolerance of freshwater oligochaetes (*Tubifex tubifex*) to mercury was reportedly developed in a single generation, and inherited through several generations without sustained exposure.

19.8.6.1.6 Echinoderms

Sea urchin development was inhibited at 10.0–23.0 μg Hg^{2+}/L, with a high proportion of abnormal plutei at 30.0 μg/L. Inhibition was more pronounced when similar concentrations of organomercury salts were tested. Arrested development of sea urchin larvae was documented at medium concentrations as low as 3.0 μg Hg^{2+}/L and exposure for 40 h. Mercury can also inhibit motility of spermatozoa of the sea urchin, *Arbacia punctata*, although effects were reversed by EDTA.

19.8.6.2 Vertebrates

Growth reduction, impaired reproduction, and enzyme derangement in teleosts, and metamorphosis inhibition in amphibians were observed at medium concentrations less than 1.0 μg Hg/L.

19.8.6.2.1 Fishes

Reduced growth of sensitive species of teleosts is recorded at water concentrations of 0.04–1.0 μg Hg/L. The rainbow trout (*Oncorhynchus mykiss*) was the most sensitive species tested; growth reduction was observed after 64 days in 0.04 μg Hg/L as methylmercury, or 0.11 μg Hg/L as phenylmercury. Growth reduction was documented in brook trout (*Salvelinus fontinalis*) alevins after exposure for 21 days to 0.79 μg organomercury/L.

Methylmercury can impair reproduction in freshwater teleosts through inhibition of gonadal development, spawning success, and survival of embryos and larvae. In the zebrafish (*Brachydanio rerio*), hatching success was reduced at 0.1 μg Hg^{2+}/L and egg deposition was reduced at 0.8 μg/L. Fathead minnows (*Pimephales promelas*) exposed to 0.12 μg methylmercury/L for 3 months failed to reproduce. Impaired egg production and spawning of goldfish (*Carassius auratus*) was associated with a total mercury concentration of 0.76 mg/kg FW gonad; for rainbow trout, 0.49 mg total Hg/kg FW gonad was associated with impaired gamete function and a reduction in survival of early life stages. Impairment of testicular lipid metabolism in catfish (*Clarias batrachus*) at sublethal concentrations of inorganic and organic mercury compounds may account for the mercury-induced inhibition of steroidogenesis and spermatogenesis. Inorganic and organic mercury compounds produce different morphological effects on the micropyle of fish eggs. Reduced insemination success of killifish (*Fundulus heteroclitus*) eggs exposed to methylmercury was due to rupture of cortical vesicles and blockage of the micropyle. However, reduced insemination due to inorganic mercury was due to swelling of the micropylar lip and a decrease in micropyle diameter.

Adverse effects of mercury to fishes, in addition to those listed on reproduction and growth, have been documented at water concentrations of 0.88–5.0 μg/L: enzyme disruption in brook trout (*Salvelinus fontinalis*) embryos immersed for 17 days in solutions containing 0.88 μg/L, as methylmercury; decreased rate of intestinal transport of glucose, fructose, glycine, and tryptophan in the murrel (*Channa punctatus*) at 3.0 μg Hg^{2+}/L for 30 days; altered blood chemistry in striped bass (*Morone saxatilis*) at 5.0 μg Hg^{2+}/L in 60 days; and decreased respiration in striped bass 30 days post-exposure after immersion for 30–120 days in 5.0 μg Hg^{2+}/L. In largemouth bass, elevated liver metallothioneins are indicative of elevated muscle mercury concentrations, suggesting that mercury-induced metallothioneins may be useful biomarkers of mercury exposure.

At 44.0 μg Hg^{2+}/L for 30 days, the freshwater fish *Notopterus notopterus* showed generalized metabolic derangement. And at high sublethal concentrations of methylmercury, rainbow trout were listless and darkly pigmented; appetite was reduced, and digestion was poor. Olfactory receptor function in Atlantic salmon (*Salmo salar*) is highly vulnerable to brief exposures of sublethal concentrations of mercuric chloride or methylmercury chloride. In both cases, mercury was deposited in the olfactory system along its whole length from receptor cell apices to the brain. Inorganic mercury inhibited the olfactory response after exposure to 10.0 μg Hg/L for 10 min, with effects reversible; however, methylmercury-induced inhibition was not reversible.

Rate of accumulation of methylmercury in fish affects toxicity, with slow accumulation rates associated with higher tolerances. Laboratory studies indicate that fish accumulate high concentrations of methylmercury directly across the gills when subjected to elevated concentrations of methylmercury in the medium. After crossing the fish gut, CH_3Hg^+ binds to erythrocytes and is transported to cell organs and tissues via the circulatory system, readily crossing internal membranes. Eventually, most of the methylmercury is translocated to skeletal muscle, where it accumulates bound to sulfhydryl groups in protein. Skeletal muscle of teleosts appears to function as a reservoir for methylmercury. Maximum concentration factors of radiomercury were reached in skeletal muscle, brain, and eye lens after 34, 56, and >90 days, respectively; maximum values in most other tissues were reached in about 7 days. One pathway by which anadromous fishes, such as salmon, accumulate mercury from the medium is through the gills: up to 90% of the mercury taken up on the gills is subsequently bound to erythrocytes within 40 min. It is hypothesized that methylmercury storage in muscle serves as a protective mechanism in fishes because sequestration in muscle reduces central nervous system exposure to methylmercury.

At lower trophic levels, the efficiency of mercury transfer was low through natural aquatic food chains; however, in animals of higher trophic levels, such as predatory teleosts and fish-eating birds and mammals, the transfer was markedly amplified. High uptake and accumulation of mercury from the medium by representative species of marine and freshwater teleosts and invertebrates are extensively documented. Accumulation patterns were enhanced or significantly modified by numerous biological and abiotic factors. In general, the accumulation of mercury was markedly enhanced at elevated water temperatures, reduced water salinity or hardness, reduced water pH, increased age of the organism, and reduced organic matter content of the medium; in the presence of zinc, cadmium, or selenium in solution; after increased duration of exposure; and in the presence of increased nominal concentrations of protein-bound mercury. Uptake patterns were significantly modified by sex, sexual condition, prior history of exposure to mercury salts, the presence of complexing and chelating agents in solution, dietary composition, feeding niche, tissue specificity, and metabolism; however, trends were not consistent between species and it is difficult to generalize. In one example, rainbow trout held in solutions containing 0.1 μg Hg/L, as methylmercury for 30 days, showed bioconcentration factors (BCF) that ranged from 28,300 in brain to 238,000 in spleen; values were intermediate in muscle (30,000), whole fish (36,000), blood (102,000), liver (110,000), kidney (137,000), and gill (163,000). These values may have been higher if exposure had extended beyond 30 days; whole body mercury residues in rainbow trout subjected to mercury insult continued to increase for the first 66 days before stabilizing. When mercury was presented as inorganic mercuric ion at 0.1 μg/L for 30 days, BCF values were usually lower than in trout exposed to methylmercury: 2300 for muscle; 6800 for brain; 7000 for whole trout; 14,300 for blood; 25,000 for liver; 53,000 for kidney; 68,600 for gill; and 521,000 for spleen. The high BCF values recorded for rainbow trout were probably due to efficient uptake from water, coupled with slow depuration.

Mercury accumulation by two species of freshwater teleosts eastern mosquitofish (*Gambusia holbrooki*; lake chubsucker, *Erimyzon sucetta*) in an artificial wetland ecosystem was significantly enhanced by sulfate addition. Authors conclude that sulfate additions result in elevated production of methylmercury in sediment and porewater – possibly due to increased mercury methylation by sulfate-reducing bacteria – which is readily evident in fish and water with subsequent increases through the food web.

Total mercury concentrations, in mg/kg FW, in tissues of adult freshwater fishes with signs of methylmercury intoxication ranged from 3.0 to 42.0 in brain, 6.0 to 114.0 in liver, 5.0 to 52.0 in muscle, and 3.0 to 35.0 in whole body. Whole-body levels up to 100.0 mg Hg/kg were reportedly not lethal to rainbow trout, although 20.0–30.0 mg/kg were associated with reduced appetite, loss of equilibrium, and hyperplasia of gill epithelium. However, brook trout showed toxic signs and death at whole body residues of only 5.0–7.0 mg/kg.

Elimination of accumulated mercury, both organic and inorganic, from teleosts and other aquatic animals is a complex multicompartmental process, but appears to be largely dependent on its rate of biological assimilation. This rate, in turn, varies widely (20–90%) between species, for reasons as yet unexplained. For example, mercury associated with dietary components that are not assimilated is eliminated rapidly with feces. The rest is absorbed across the gut and incorporated into tissues. This assimilated fraction of mercury is depurated much more slowly, at a rate positively correlated with the organism's metabolism. Route of administration is also important. Bioavailability estimates of methylmercury from orally administered doses to channel catfish (*Ictalurus punctatus*) tend to overestimate the true bioavailability, and strongly indicate that data based on this route of administration needs to be re-examined. Time to eliminate 50% of biologically assimilated mercury and its compounds (Tb1/2) is variable. Among various species of freshwater teleosts (Tb1/2 values in days) were 20 for guppies (*Poecilia reticulatus*), 23 for goldfish (*Carassius auratus*), 100 for northern pike, and 1000 each for mosquitofish (*Gambusia affinis*), brook trout, and rainbow trout. For eels (*Anguilla vulgaris*) and flounders (*Pleuronectes flesus*), these values were 1030 days and 1200 days, respectively.

There are many laboratory studies on uptake, retention, and translocation of mercury by marine teleosts. Rapid accumulation of mercury by marine teleosts is documented, especially for organomercury compounds. Eggs of plaice (*Pleuronectes platessa*), for example, concentrated ambient environmental levels of mercury by a factor of 465 in 12 days; for plaice larvae, this factor was 2000 in 8 days; and adult plaice had mercury concentration factors of 600 in 64 days, with most of the mercury sequestered in muscle. Adults of mullet (*Mugil auratus*) held in seawater solutions containing 100.0 μg Hg/L as $HgCl_2$ for 57 days had

2.2 mg Hg/kg wet muscle vs. 0.11 in controls. When mercury was in the form of methylmercury, exposure for 45 days in only 8.0 μg Hg/L produced 2.6 mg Hg/kg FW muscle in *Mugil*.

Accumulation of mercury and its compounds from seawater by fish can be modified by the chemical form of mercury administered; the mode of administration; the presence of chelating or complexing agents in the medium; the initial mercury concentration; tissue specificity; salinity of the medium; and length of exposure. In one study, accumulation of methylmercury by freshwater and seawater-adapted Japanese eels, *Anguilla japonica*, held in 10.0 μg Hg/L as methylmercury chloride for 72 h, was significantly greater in marine-adapted eels than freshwater-adapted eels for liver (2.8 mg/kg FW vs. 2.0 mg/kg FW), spleen (3.4 vs. 2.4), kidney (1.6 vs. 1.4), pancreas (1.0 vs. 0.5), bile (0.4 vs. 0.06), and blood (4.1 mg/kg FW vs. 1.9 mg/kg FW). In freshwater-adapted eels, but not marine-adapted eels, methylmercury induced stimulation of GSH in both liver and kidney and this may interfere with mercury uptake in that group.

In the food chain algae 6 detritus 6 worm 6 prey fish 6 predator fish, mercury had a long biological half-life: >1000 days in predatory fish compared with about 55 days in prey fish. The transfer efficiency of inorganic mercury from prey to predator fish was about 40%, but from worm to prey fish, only 12%; worms assimilated about 60% of the inorganic mercury contained in the algae-detritus compartment. It was concluded that mercury accumulation through the food chain could account for a significant percentage of the mercury body burden in fish. Methylmercury accumulation was studied in the food chain of diatom, *Skeletonema costatum* 6 copepod, *Acartia clausi* 6 fish, *Chrysophrys major*. Diatoms held for 24 h in seawater containing 5.0 μg methylmercury/L contained 3.45 mg Hg/kg. Copepods that fed on these diatoms for 4 days had 3.14 mg Hg/kg, and fish that fed on these copepods for 10 days had 3.1 mg Hg/kg. Additional studies showed that the most effective pathway for methylmercury accumulation is not via the food chain but rather from the medium by way of the gills. In the food chain clam 6 eel, no mercury magnification was recorded. When dogfish meal containing up to 2.3 mg total Hg/kg, of which 1.9 mg/kg was methylmercury, replaced low-mercury fish rations used in salmon culture, the flesh of the salmon contained >0.5 mg Hg/kg FW within 240 days. However, when dogfish meal comprised less than half the diet, mercury levels in salmon flesh were <0.5 mg Hg/kg FW. Similar results were observed in sablefish fed dogfish meal.

The retention time of mercury in marine teleosts depends on several factors. Thus, gill, heart, and swim-bladder tissues of fish lost mercury at a faster rate than did whole animals, but some tissues, including brain, liver, muscle, and kidney showed no significant decrease over time. Retention times of radiomercury-203 in most marine species were comparatively lengthy, ranging up to 267 days for the fish *Serranus scriba*. Phylogenetically related species follow a similar pattern of methylmercury excretion, with biological half-life dependent on water temperature and mode of entry into organisms; it is longer after intramuscular injection than after peroral administration. A study using juvenile sockeye salmon indicated that fish treated repeatedly with mercurials during their freshwater phase (to control parasites), accumulated and retained elevated levels of mercury for several months, but after 4 years at sea, the returning salmon contained normal levels of mercury in all tissues examined. Fish heavily contaminated by methylmercury, both naturally and through laboratory-administered feeding studies, reportedly eliminated the accumulated mercury to safe levels for human consumption in 3 to 7 weeks (shorter depuration periods were unsuccessful) when maintained in a mercury-free media and fed a mercury-free diet. In the case of *Serranus scriba*, mercury-contaminated fish eliminated mercury from gill and digestive tract when placed in clean seawater for 10 days, but mercury content was essentially unchanged in liver, kidney, and muscle tissues.

At sublethal concentrations, methylmercury can impair the ability of fish to avoid predators, and interfere with their ability to locate, capture, and consume prey. Long-term dietary exposure of teleosts to methylmercury is associated with impaired coordination, diminished appetite, inhibited swimming activity, starvation, and sometimes death. Mercury-tolerant strains of fish are reported. Reasons to account for mercury adaptation of the estuarine cyprinodontiform teleost *Fundulus heteroclitus* to both methylmercury and inorganic mercury are unclear.

19.8.6.2.2 Amphibians

Dietary exposure of amphibians in sites receiving mercury mainly via atmospheric deposition – estimated to range from 1.5 to 3.0 mg Hg/kg FW ration – may be sufficient to adversely affect survival, growth, and development. Studies with tadpoles of the southern leopard frog (*Rana sphenocephala*) fed mercury-containing diets for 254 days showed that about 28% died at the highest concentrations fed of 0.5 and 1.0 mg Hg^{2+}/kg FW ration; malformation rates were dose-related, with 5.0% in controls, 5.6% in the 0.1 mg Hg/kg FW diet, 11.1% in the 0.5 mg/kg and 27.8% in the 1.0 mg/kg diet. Malformations included dose-related scoliosis. Arrested growth and development, and tail resorption were also positively dose-related. Total mercury body burdens were also dose-related with about 50.0 μg Hg/kg DW whole body in controls, 100.0 in the low-mercury diet, 250.0 in the medium diet, and about 450.0 μg Hg/kg DW in the 1.0 mg Hg^{2+}/kg FW diet. In all cases, methylmercury accounted for <20.0 μg/kg dry body weight. Dietary accumulation of mercury and methylmercury under these conditions is not governed by simple partitioning processes, and bioaccumulation factors may be limited as predictors of dietary uptake of methylmercury and inorganic mercury.

The leopard frog (*Rana pipiens*) did not metamorphose during exposure to 1.0 μg methylmercury/L for 4 months. In the South African clawed frog (*Xenopus laevis*), a total mercury concentration of 0.49 mg/kg FW gonad was associated with impaired gamete function and reduction in survival of early life stages. At 0.65 mg inorganic mercury/L, there was a significant decrease in ovary weight of the river frog (*Rana heckscheri*).

The significance of mercury concentrations in amphibian tissues is not known with certainty and requires additional research for satisfactory risk assessment. The following areas are recommended for study: acclimatization and adaptation to mercury; mercury remobilization during periods of metamorphosis, hibernation, estivation, and reproduction; critical organ concentrations; and biomarkers of adverse mercury effects. These studies should also consider the influence of exposure duration and dose, mercury speciation, and mercury interaction with other metals.

19.8.7 Birds

Sublethal effects of mercury on birds, administered by a variety of routes, included adverse effects on growth, development, reproduction, blood and tissue chemistry, metabolism, and behavior; histopathology and bioaccumulation were also noted. The dietary route is the most extensively studied pathway of avian mercury intake, especially in the mallard (*Anas platyrhynchos*).

Accumulation and retention of mercury in mallards from mercury-laden diets was investigated. Ducklings of mallards fed diets containing 8.0 mg Hg/kg diet for 2 weeks contained, in mg Hg/kg fresh weight (FW), 16.5 in liver, 17.6 in kidney, 9.1 in carcass, and 4.4 in whole body. After 16 weeks on a mercury-free diet, values in all tissues decreased more or less progressively to 25% of the 2-week values. In another study, adults of the black duck (*Anas rubripes*) were fed diets containing 3.0 mg Hg/kg. Final maximum concentrations, in mg Hg/kg FW, were 23.1 in liver, 65.7 in feathers, 3.8 in brain, 4.5 in muscle, and 16.0 in kidney. Breeding pairs fed this diet, when compared to controls, produced fewer eggs, reduced egg hatch, and lower duckling survival. Brains of dead ducklings from mercury-insulted parents contained 3.2–7.0 mg Hg/kg FW; brain lesions were also evident. Methylmercury in the diet of game-farm and wild strains of mallards at 0.5 mg Hg/kg diet for three generations had little measurable effect – aside from slightly increased tissue levels of mercury – when compared to controls. Mercury-fed ducks weighed about the same as control ducks, but laid fewer eggs and produced fewer ducklings. Eggshell thinning in mallards may be associated with increasing concentrations of dietary methylmercury.

The dietary concentration of 0.5 mg Hg/kg dry weight (equivalent to about 0.1 mg/kg fresh weight) in the form of methylmercury was fed to three generations of mallards. Females laid a greater percentage of their eggs outside nest boxes than did controls, and also laid fewer eggs and produced fewer ducklings. Ducklings from parents fed methylmercury were less responsive than controls to tape-recorded maternal calls, but were hyperresponsive to a fright stimulus in avoidance tests. Mallard hens fed diets containing 3.0 mg Hg/kg ration as methylmercury for two reproductive seasons produced eggs with elevated mercury concentrations (5.5–7.2 mg Hg/kg FW); hatched ducklings from this group had brain lesions of the cerebellum and reduced survival. Adult female mallards fed diets containing 1.0 or 5.0 mg Hg/kg ration, as methylmercury chloride, produced eggs that contained 1.4 mg Hg/kg FW (1.0 mg/kg ration) or 8.7 mg Hg/kg FW (5.0 mg Hg/kg diet); breast muscle had 1.0 or 5.3 mg Hg/kg FW; the addition of 5.0 mg DDE/kg diet did not affect mercury retention in breast muscle or eggs. Lesions in the spinal cord were the primary effect in adult female mallards fed diets containing 1.5 or 2.8 mg Hg/kg DW ration as methylmercury.

The tissues and eggs of ducks and other species of birds collected in the wild have sometimes contained levels of mercury equal to, or far exceeding, those associated with reproductive and behavioral deficiencies in domestic mallards (e.g., 9.0–11.0 mg/kg in feathers and >2.0 mg/kg in other tissues); therefore, it is possible that reproduction and behavior of wild birds have been modified by methylmercury contamination. Tissue mercury residues of wild-strain mallards and game-farm mallards were not significantly different after the birds were fed diets containing 0.5 mg Hg/kg as methylmercury for extended periods – indicating that game-farm mallards are suitable substitutes for wild mallards in toxicological evaluations.

Mercury exposures by immersion and oral administration have caused reproductive and behavioral modifications. Brief immersion of mallard eggs in solutions of methylmercury resulted in a significant incidence of skeletal embryonic aberrations at dosages of 1.0 μg Hg/egg, and higher; no increases in embryonic malformations were noted at 0.3 μg Hg/egg.

Interaction effects of mercury with other metals need to be considered. Pekin duck (*Anas platyrhynchos*) – a color variant of the mallard – age 6 months, fed a diet containing 8.0 mg Hg/kg ration as methylmercury chloride, for 12 weeks had kidney histopathology; damage effects were exacerbated when diets also contained 80.0 mg lead acetate/kg, 80.0 mg cadmium chloride/kg, or both. Studies with mallard adults show significant interaction effects of mercury and selenium. In one 10-week-study, mallard adults were fed diets containing 10.0 mg Hg/kg DW ration as methylmercury chloride, 10.0 mg Se/kg as seleno-DL-methionine, or a mixture containing 10.0 mg Hg/kg plus 10.0 mg Se/kg. One of the 12 adult mallards fed the 10.0 mg Hg/kg diet and 8 others suffered paralysis of the legs; however, none of the 12 males on the mixture diet became sick. Both selenium and mercury diets lowered duckling production through reduced hatching success and survival; the mixture diet was worse than either Hg or Se alone. Controls produced 7.6 ducklings/female, females fed 10.0 mg Se/kg produced 2.8, females fed 10.0 mg Hg/kg produced 1.1 young, and the mixture diet resulted in 0.2 ducklings/females. Deformity frequency was 6.1% in control ducklings, 16.4% for the mercury group, 36.2% for the selenium group, and 73.4% for the group fed the methylmercury and selenomethionine mixture. The presence of mercury in the diet enhanced selenium storage; however, selenium did not enhance Hg storage. In another study with mallard adult males fed various mercury-containing diets for 10 weeks, the 10.0 mg Hg/kg ration group had decreased hematocrit and hemoglobin, and decreased activities of various enzymes involved in glutathione metabolism. Selenium in combination with methylmercury partially or totally alleviated effects of mercury on various glutathione activities. The ability of seleno-DL-methionine to restore the glutathione status involved in antioxidative defense mechanisms may be important in protecting against the toxic effects of methylmercury.

The dietary route of administration is the most extensively studied pathway of avian mercury intake. Domestic chickens fed diets containing as little as 50.0 μg/kg of mercury, as methylmercury, contained elevated total mercury (2.0 mg/kg fresh weight) residues in liver and kidney after 28 weeks; at 150.0 μg/kg diet, residues ranged from 1.3 to 3.7 mg/kg in heart, muscle, brain, kidney, and liver, in that general order; at 450.0 μg/kg in diets, residues in edible chicken tissues (3.3–8.2 mg/kg) were considered hazardous to human consumers, although no overt signs of mercury toxicosis were observed in the chickens. High inorganic mercury levels (500.0 mg/L) in drinking water of chickens decreased growth rate and food and water consumption, and elevated hemoglobin, hematocrit, and erythrocyte content within 3 days.

Dietary concentrations of 1.1 mg total Hg/kg have been associated with kidney lesions in juvenile starlings (*Sturnus vulgaris*) and with elevated residues in the liver (6.5 mg/kg dry weight) and kidney (36.3 mg/kg), after exposure for 8 weeks. In American black ducks (*Anas rubripes*) fed diets containing 3.0 mg Hg/kg as methylmercury for 28 weeks, reproduction was significantly inhibited; tissue residues were elevated in kidney (16.0 mg/kg fresh weight) and liver (23.0 mg/kg); and brain lesions characteristic of mercury poisoning were present. Japanese quail (*Coturnix japonica*) fed diets containing 8.0 mg Hg/kg of inorganic mercury for 3 weeks had depressed gonad weights; those fed 3.0 mg/kg inorganic mercury or 1.0 mg/kg methylmercury for 9 weeks showed alterations in brain and plasma enzyme activities. Grossly elevated tissue residues of 400.0 mg/kg in feathers and 17.0–130.0 mg/kg in other tissues were measured in gray partridge (*Perdix perdix*) after dietary exposure of 20.0–25.0 mg total Hg/kg for 4 weeks. Reduced reproductive ability was noted in gray partridges ingesting 640.0 μg Hg as organomercury)/kg body weight daily for 30 days; similar results were observed in ring-necked pheasants.

Behavioral alterations were noted in rock doves (*Columba livia*) given 3.0 mg inorganic Hg/kg body weight daily for 17 days or 1.0 mg/kg body weight of methylmercury for 5 weeks. Observed behavioral changes in posture and motor coordination of pigeons were permanent after the brain accumulated >12.0 mg Hg/kg fresh weight, and were similar to the "spastic paralysis" observed in wild crows during the Minamata, Japan, outbreak of the 1950s, although both species survived for years with these signs.

Tissue concentrations >15.0 mg Hg/kg FW brain and >30.0–40.0 mg/kg FW liver or kidney are associated with neurological impairment. Mercury residues of 0.79–2.0 mg/kg in egg, and 5.0–40.0 mg/kg in feathers, are linked to impaired reproduction in various bird species. Residues in eggs of 1.3–2.0 mg Hg/kg fresh weight were associated with reduced hatching success in white-tailed sea-eagles (*Haliaeetus albicilla*), the common loon (*Gavia immer*), and in several seed-eating species; this range was 0.90–3.1 mg/kg for ring-necked pheasants, and 0.79–0.86 mg/kg for mallards. Residues of 5.0–11.0 mg Hg/kg in feathers of various species of birds have been associated with reduced hatch of eggs and with sterility. Sterility was observed in the Finnish sparrow hawk (*Accipiter nisus*) at mercury concentrations of 40.0 mg/kg in feathers. Great skuas *Catharacta skua*) fed mercury-contaminated prey excrete mercury via feathers. Chicks of the common tern (*Sterna hirundo*) from a colony in Long Island, New York, with abnormal feather loss, had significantly elevated mercury levels in blood and liver; however, the linkage of feather loss to mercury contamination requires further examination.

Interaction effects of mercury with other contaminants, such as herbicides and pesticides, could intensify hazards to avian populations. For example, a striking parallel exists between levels of mercury and of DDT and its metabolites in tissues of birds of prey, suggesting the existence of common ecotoxicological mechanisms; additional research is clearly needed.

19.8.8 Mammals

Mercury has no known physiological function. In humans and other mammals, it causes teratogenic, mutagenic, and carcinogenic effects; the fetus is the most sensitive life stage. Methylmercury irreversibly destroys the neurons of the central nervous system. Frequently, a substantial latent period intervenes between the cessation of exposure to mercury and the onset of signs and symptoms; this interval is usually measured in weeks or months, but sometimes in years. Alterations in open-field operant and swimming behaviors, learning deficits, and depression in spontaneous locomotor activity have all been demonstrated in mice and rats after in utero exposure to methylmercury. Methylmercury disrupts amino acid metabolism in neural tissues of male mice; alterations were associated with specific neural cell dysfunction and appearance of muscular incoordination. Mercury can cause measurable brain developmental deficits in children exposed to even relatively low levels prior to birth. At high sublethal doses in humans, mercury causes cerebral palsy, gross motor and mental impairment, speech disturbances, blindness, deafness, microcephaly, intestinal disturbances, tremors, and tissue pathology. Pathological and other effects of mercury may vary from organ to organ, depending on factors such as the effective toxic dose in the organ, the compound involved and its metabolism within the organ, the duration of exposure, and the other contaminants to which the animal is concurrently exposed. Many compounds – especially salts of selenium – protect humans and other animals against mercury toxicity, although their mode of action is not clear. Adverse effects of organomercury compounds to selected species of mammals have been recorded at administered doses of 0.25 mg Hg/kg body weight daily, dietary levels of 1.1 mg/kg, and blood Hg levels of 1.2 mg/L.

Retention of mercury by mammalian tissues is longer for organomercury compounds (especially methylmercury) than for inorganic mercury compounds. Excretion of all mercury species follows a complex, variable, multicompartmental pattern; the longer-lived chemical mercury species have a biological half-life that ranges from about 1.7 days in human lung to 1.36 years in whole body of various pinnipeds. In humans, increased urinary excretion and blood levels of mercury were observed in human volunteers who used phenylmercuric borate solutions or lozenges intended for the treatment of throat infections.

Organomercurials were readily transferred across placental membranes of rats, mice, and hamsters whereas inorganic mercury compounds were blocked in the placenta. Because human embryos lack a yolk-sac placenta, results of these studies with inorganic mercurials should be conducted with suitable experimental animals. A review of nephrotoxic effects of mercury in humans and mammals show that all forms of mercury administered by a variety of routes can accumulate in the kidney to produce renal damage and sometimes acute renal failure. The basic biochemical mechanism by which mercury produces renal damage is unclear, but may include inhibition of oxidative pathways and general disruption of the plasma membrane. Mercury vapor has a greater predilection for the CNS than does inorganic mercury salts, but less than organic forms of mercury. Kidneys contain the greatest concentrations of mercury following exposure to inorganic salts of mercury and mercury vapor, whereas organomercurials have greater affinity for the brain, especially the posterior cortex.

Respiratory absorption of elemental mercury in air (1.0–30.0 μg Hg/m^3) by humans ranged from 74 to 100% when inhaled through the nose and exhaled through the mouth; for dogs, this value was 25%. However, respiratory absorption in humans was only about 20% when inspiration and expiration was through mouth only. Two human males exposed to radiolabeled methylmercury as $CH_3^{203}HgCl$ via inhalation had variable half-life retention times (Tb1/2) of mercury. One subject had a Tb1/2 value of 103 days for the early period of observation (days 1–44 post-administration) and 39 days thereafter; the second had a

Tb1/2 value of 107–122 days for the entire post-administration observation period.

Organomercury uptake and retention in rats is modified by metallothioneins, carbon tetrachloride, and various amino acids. Methylmercury induces metallothionein synthesis in rat liver. Subcutaneous doses of 19.0 mg CH_3HgCl/kg BW daily for three days caused a threefold increase of hepatic zinc and about a fourfold increase in cysteine bound to metallothioneins, but no increase in mercury bound to hepatic metallothioneins. Carbon tetrachloride, at 0.5 mL/kg BW delivered intraperitoneally every 48 h for 6 days, enhanced initial uptake of methylmercury (single subcutaneous injection of 10.0 mg/kg BW) in rat brain, liver, kidney, and muscle; results suggest that CCl_4-induced liver damage is responsible for retention time of mercury in rat tissues. Methylmercury uptake in brain of Wistar-strain male rats, *Rattus* sp., was influenced by certain amino acids. It was enhanced threefold by L-cysteine; this effect was depressed by L-phenylalanine and L-isoleucine, which are neutral amino acids, and not by L-lysine (basic) or L-glutamic acid (acidic). Cysteine seems to be an important factor in methylmercury uptake and retention in rat brain but not other tissues, suggesting that the methylmercury–cysteine complex readily penetrates the blood-brain barrier transport system. Cysteine accelerates the intestinal absorption of methylmercury, increases methylmercury uptake in rat brain, in brain microvessels, and in astroglia. Since methylmercury easily reacts with cysteine to form a conjugate similar to methionine, the conjugate is suggested to be taken up by the cells through the L-neutral amino acid carrier system. Studies with cultured rat astroglia suggest that conjugation with glutathione is a major pathway for methylmercury efflux in rat brain cells and that elevation in cellular glutathione levels has application in cases of methylmercury poisoning in promoting accelerated elimination of methylmercury from critical tissues.

Increasing glutathione levels were positively correlated with increasing methylmercury resistance of rat PC12 cells and decreasing methylmercury accumulations. In that study, methylmercury-resistant sublines of rat cells (PC12) were obtained by repeated exposure to stepwise increased concentration of methylmercury; one subline (PC12/TM) was eight to ten times more resistant to methylmercury than were parent PC12 cells. PC12/TM cells accumulated less methylmercury than did parent PC12 cells. Intracellular glutathione level in PC12/TM cells was four times higher than that of PC12 cells. Pretreatment of PC12/TM cells with buthionine sulfoximine reduced the glutathione level to that of parent cells and increased the sensitivity of the cells to methylmercury without affecting methylmercury accumulation.

Mercury granules were detected in brains from rats exposed to $HgCl_2$ but not CH_3HgCl, suggesting that brain mercury granules are inorganic mercury. In that study, Wistar rats were dosed daily with subcutaneous injections of 5.0 mg Hg/kg BW for up to 12 consecutive days with either inorganic or organic mercury. With methylmercury after 8 days, brain contained 12.5 mg total Hg/kg FW of which 0.23 mg/kg was inorganic mercury. After 12 days, these values were 23.6 mg total Hg/kg FW and 0.32 mg/kg inorganic mercury. Mercury granules were detected in brain only after 8 daily doses. Histochemical analysis of rat and of mouse brain after rodents had been subjected to mercury-intoxicated rodents showed that the lowest concentration of inorganic mercury in brain with mercury granules was 0.12 mg/kg FW in $HgCl_2$-treated rats, 0.14 mg/kg in CH_3HgCl-treated rats, and 0.12 mg/kg in CH_3HgCl-treated mice; these concentrations were similar to the levels of inorganic mercury in brain from human autopsy cases from Minamata, Japan.

Mercury can alter the metabolism of essential elements in rats. In one study, rats given a high sublethal dose of methylmercury chloride or mercuric chloride showed rapid changes in certain elements within 24 h postinjection, as measured by inductively coupled plasma emission spectrography. Zinc in brain and kidney was highest in the methylmercury group; copper was highest in liver of the inorganic mercury group. Total mercury in brain of the $HgCl_2$ group was 0.049 mg/kg FW 3 h post-injection; for the methylmercury group it was 0.95 mg/kg FW, and for the controls 0.02 mg/kg FW. For liver, these values were 0.2, 3.4, and 0.013 mg/kg FW, respectively. Total mercury in kidney was 4.7 mg/kg FW in the $HgCl_2$ group, 14.9 for the methylmercury group, and 0.02 for the controls.

There is evidence that marine teleosts and mammals contain high levels of methylmercury and selenium, and that selenium in tuna and marlin can protect against methylmercury neurotoxicity. In one study, weanling rats were given diets for 12 weeks containing 17.5 methylmercury/kg FW diet plus either 0.3 or 0.6 mg selenium/kg FW. The mercury and selenium originated from flesh of a marine sea bass (*Sebastes iracundus*) or sperm whale (*Physeter catodon*), or from sodium selinite. Selenium in blood, brain, and spinal cord of rats was positively correlated with neurological protection (tail rotation, hind limb paralysis), while total mercury and methylmercury in these tissues were negatively correlated. Fish muscle provided greater protection against mercury neurotoxicity than did whale meat.

The developing central nervous system, particularly the cerebellum, of fetuses and neonates is a primary target of mercury toxicity. Mercury neurotoxicity in cerebellar granular cells from transgenic mice has been demonstrated using heat shock protein 70 as a biomarker. Effects were observed at 0.2 mg/L and higher of inorganic mercury and 0.06 mg/L and higher of methylmercury. In the C57B/6N strain of mice, mercury concentrates in fetal tissues, especially brain, when compared to their dams given 2.5–20.0 mg CH_3HgCl/kg BW on day 13 of pregnancy and killed on day 14 or 18. The ratio was especially high at the highest dose tested of 20.0 mg/kg BW as this was a toxic dose and brain weight was reduced. The concentration of methylmercury in the fetal brain was 1.6–4.9 times higher than maternal brain. At the lowest dose tested of 2.5 mg/kg BW, mercury concentrations were higher in male fetuses, but this was not evident at higher doses. Studies with guinea pigs, *Cavia* spp., showed that developmental disturbances of the fetal brain – including abnormal neuronal migration – resulted when dams were exposed to methylmercury in early pregnancy, and focal degeneration of the cerebral cortex when dams were exposed in later pregnancy. In Minamata, where pregnant women were chronically exposed to methylmercury throughout pregnancy, both developmental disturbances and focal degeneration of neurons might be induced in the same fetal brain.

Dose-dependent effects of methylmercury on mouse development in vitro include decreases in heart beat, axial rotation, blood circulation, yolk sac diameter, crown–rump length, and number of somites. Dose-dependent abnormalities of the mouse embryo in vitro include increases in growth retardation, neural tube closure, hind limb hypoplasia, edema, stunted head, and eye hypoplasia. For mercuric chloride, dose-dependent effects and abnormalities in mice were similar to those of methylmercury, but at lower frequencies and at higher doses. Fetal Minamata disease caused by exposure to methylmercury during mid- and late-gestation periods is associated with cerebral palsy, blindness, deafness, microcephaly, and other adverse effects. Because human brain development in late gestation seems comparable to that of postnatal stages in rodents, a study was conducted using artificial rearing systems for infant rats wherein nutrient factors were excluded, methylmercury administered, and brain architecture studied. Artificially reared rats given 40.0 mg methylmercury chloride/kg BW on day 6 after birth were killed 3 days later. These dosed rats, when compared to controls, had decreased brain weight and increased degeneration of nerve cells in the inner granule layer of the cerebellum. This system shows promise for simulating mercury effects in human embryos.

Glutathione (GSH) is a major cytosolic, low-molecular-weight sulfhydryl compound that acts as a protective agent against methylmercury. The high affinity of methylmercury to the sulfhydryl group suggests that the fate of methylmercury is closely related to GSH metabolism. In the case of C57BL/6N female mice challenged by a sublethal oral dose of methylmercury (41.0 mg/kg BW), GSH levels in liver decreased 16%, and in blood 20%, 24 h after dosing. GSH half-time persistence in liver (74 min) was reduced by 17% from controls (89 min) but increased by 28%

in kidney (46 min) from controls (36 min). Methylmercury-induced alterations of GSH metabolism in mouse tissues might reflect a defense mechanism against methylmercury insult.

Low dietary protein levels (7.5%) affected methylmercury and glutathione metabolism in C57BL/6N male mice when compared to a normal protein diet of 24.8%. Mice fed the low protein diet had about four times less mercury in urine than those fed a normal protein diet for 7 days, although fecal mercury levels were the same in both groups; diets of both groups contained 5.14 mg CH_3HgCl/kg. Tissue mercury levels in the low protein group were significantly higher than those in the normal protein group, except for liver. Liver glutathione in the low-protein diet group was lower than the high protein group, but other tissues were the same. Efflux rate of liver glutathione was significantly lower in the low protein diet group, but efflux rates of renal glutathione were the same in both groups. When methylmercury-treated mice were injected with acivicin – a specific inhibitor of gamma-glutamyltranspeptidase – the urinary mercury levels increased by 60-fold in the low-protein diet group and by 36-fold in the high-protein diet group. Authors aver that mice fed diets low in protein would show decreased urinary excretion of mercury via increased retention of mercury metabolites in kidney, and that dietary protein status – which could modulate thiol metabolism – is important in determining the fate of methylmercury. Additional studies by the same group suggest that the mechanism by which dietary protein levels affect the fate of methylmercury in mice is related to sulfur amino acids. In those studies, a low protein diet (LPD) supplemented by methionine and cysteine was fed to methylmercury-dosed mice for 5 days. These mice had increased levels of brain mercury and liver mercury that was further enhanced by the amino acid supplemented diet (ASD). Mercury levels in kidney, blood, and plasma decreased with ASD feeding. The urinary mercury level that decreased with LPD feeding was recovered and far exceeded control levels with ASD feeding. Insufficiency of sulfur amino acids in mice fed LPD could account for observed changes in the fate of methylmercury; changes in neutral amino acid transport caused by LPD feeding is also involved in the fate of methylmercury. Although the increased brain uptake of mercury would enhance the neurotoxic action of methylmercury, the stimulated renal elimination would reduce whole body methylmercury toxicity, including neural tissues. These opposite effects might be related to the alteration in sulfur amino acid metabolism caused by excessive mercury levels.

Methylmercury in combination with ionizing radiation can affect developing mouse cerebellum. In one study, pregnant mice given a low oral dose of methylmercury 9.0 mg Hg/kg BW) on day 17 of gestation delivered pups that contained 4.0–8.0 mg Hg/kg FW cerebellum. Male pups were subsequently exposed to X-radiation of 0.125 or 0.5 gy on the day following birth. Cell deaths in the external granular layer (EGL) of pup cerebellum were similar in groups exposed to 0.125 gy alone or in combination with mercury. In the high radiation dose of 0.5 gy, 14% of these cells were killed by radiation, regardless of methylmercury exposure. Restoration of the EGL was slightly retarded by methylmercury. Authors concluded that methylmercury does not modify radiation-induced cell death in the EGL, but does retard tissue restoration from the damage.

Mercury transfer and biomagnification through mammalian food chains is well documented, but considerable variation exists. Among terrestrial mammals, for example, herbivores such as mule deer, moose (*Alces alces*), caribou (*Rangifer tarandus*), and various species of rabbits usually contained less than 1.0 mg Hg/kg fresh weight in liver and kidney, but carnivores such as the marten (*Martes martes*), polecat (*M. putorius furo*), and red fox (*Vulpes vulpes*) frequently contained more than 30.0 mg/kg. The usually higher mercury concentrations in fish-eating furbearers than in herbivorous species seemed to reflect the amounts of fish and other aquatic organisms in the diet. In river otter (*Lutra canadensis*) and mink (*Mustela vison*) from the Wisconsin River drainage system, mercury levels paralleled those recorded in fish, crayfish, and bottom sediments at that location. Highest mercury levels in all samples were recorded about 30 km downstream from an area that supported 16 pulp and paper mills and a chloralkali plant; residues were highest in the fur, followed by the liver, kidney, muscle, and brain. Mercury exposure has been linked to population declines of mink in Georgia, North Carolina, and South Carolina owing to elevated mercury concentrations in kidney in mink from these areas (25.0 mg/kg FW) when compared to conspecifics from reference areas (4.0 mg/kg FW). The higher mercury levels in mink kidney were toxic to this species in laboratory studies. Dietary methylmercury exposure posed a moderate risk to female mink and otters in Tennessee.

In marine mammals, more than 90% of the mercury content is inorganic; however, enough methylmercury occurs in selected tissues to result in the accumulation of high tissue concentrations of methylmercury in humans and wildlife consuming such meat. The liver of the ringed seal (*Phoca hispida*) normally contains 27,000.0–187,000.0 μg Hg/kg fresh weight, and is a traditional and common food of the coastal Inuit people. Although levels of total mercury in hair (109,000.0 μg/kg) and blood (37.0 μg/L) of Inuits were grossly elevated, no symptoms of mercury poisoning were evident in the coastal Inuits. Similar high concentrations have been reported for Alaskan Eskimo mothers who, during pregnancy, ate seal oil twice a day, and seal meat or fish from the Yukon–Kuskokwim Coast every day. Despite the extremely high total Hg content of seal liver, only a small organomercury component was absorbed and appeared in the tissues. Cats fed a diet of seal liver (26,000.0 μg Hg/kg fresh weight) for 90 days showed no neurological or histopathological signs. It seems that the toxic potential of seal liver in terms of accumulated tissue levels in cats (up to 862.0 μg total Hg/L blood, and 7600.0 μg total Hg/kg hair) is better indicated by the organomercury fraction in seal liver than by the concentration of total mercury.

Under laboratory conditions, gray seals (*Halichoerus grypus*) dosed with methylmercury showed a time-related increase of both mercury and selenium in liver and kidney; however, in other tissues examined (brain, thyroid, blood), only the concentrations of mercury increased. Atomic ratios of mercury and selenium were near 1.0 in wild seals, as expected, but this ratio exceeded 1.0 in seals fed additional methylmercury. Mercury excretion rates in ringed seals, *Pusa hispida*, followed a biphasic or polyphasic pattern. The fastest excreted component took 20 days for complete elimination and the slowest excreted component fraction, which comprised 45% of all mercury, 500 days for 50% elimination. In a 90-day, controlled-feeding study, harp seals were fed 0.25 or 25.0 mg methylmercury/kg body weight (BW) daily. At 0.25 mg/kg BW daily, tissue mercury concentrations at 90 days, in mg Hg/kg FW, ranged between 42.7 and 82.5 in liver, kidney and muscle; between 13.1 and 25.0 in adrenal glands, claws, brain spleen, lung, small intestine, heart, and blood; 1.6 in hair and 0.2 in blubber. At 25.0 mg methylmercury/kg BW daily, one seal died on day 20 and another on day 26. Blood mercury values in these animals shortly before death were 26.8 and 30.3 mg/L. In addition, these seals were diagnosed with toxic hepatitis, uremia, and renal failure. Histopathological damage was also evident. In the low dose group, 5% of the sensory cells in the middle cochlea of the ear were damaged (24% in the 25.0 mg/kg group), as well as sensory hair cells along the full length of the cochlea in both groups. It is noteworthy that dead harbor seals, *Phoca vitulina*, contained 9.9–31.0 mg total Hg/kg FW in brain. These brain mercury concentrations were of the same order as those in brain tissue of animals of various species poisoned experimentally by methylmercury. In view of the numerous reported strandings and beachings of many species of large marine mammals, the significance of these findings should not be discounted.

19.9 Minamata

One of the earliest and most extensively documented cases of mercury poisoning occurred in the 1950s at Minamata Bay, in southwestern Kyushu, Japan – especially among fishermen and their families. Deaths and congenital birth defects in humans were attributed to long-term ingestion of marine fish and shellfish highly contaminated with methylmercury compounds. Abnormal mercury content of >30.0 mg/kg FW was measured in fish and shellfish from the Bay. The source of the mercury was in the waste discharged from an acetaldehyde plant that used inorganic mercury as a catalyst; between 1932 and 1968, Minamata Bay received at least 260 tons of mercury, and perhaps as much as 600 tons. A severe neurological disorder was recognized in late 1953 and had reached epidemic proportions by 1956. At that time, the mercury level in sediments near the plant outfall was about 2010.0 mg/kg FW; this decreased sharply with increasing distance from the plant, and sediments in the Bay contained between 0.4 and 3.4 mg Hg/kg FW. Concentrations of mercury in fish, shellfish, and other organisms consumed by the Japanese decreased with increasing distance from the point of effluence and appeared to reflect sediment mercury levels.

The first recognition of Minamata Disease as a new syndrome occurred in 1956, but it was not until 1959 that mercury poisoning was proposed as the cause of the disease and when fishing within one km of the shore was prohibited. Mercury catalysts were first used beginning in 1932 in a factory producing acetaldehyde, acetic acid, and vinyl chloride; about 82 tons of mercury was discharged into Minamata Bay from this factory alone – one of the several using mercury catalysts – between 1932 and 1971 through the Hyakkon Drainage Outlet. It is noteworthy that production of vinyl chloride using mercury catalysts at this factory continued until 1971.

19.9.1 Minamata Disease

Minamata Disease is defined as neuropathy arising from intake of fish and shellfish containing high concentrations of methylmercury. The outbreak is dependent on factors that include mercury concentrations in water, bioconcentration and biomagnification of mercuric compounds by aquatic plants and animals, and continuous daily intake of mercury-contaminated fish in large quantities. Minamata Disease patients have neurological symptoms that include paresthesia, visual field constriction, impaired handwriting, unsteady gait, tremors, impaired hearing and speech, mental disturbances, and excessive salivation and perspiration. Severe fetal Minamata Disease may result in cerebral palsy, blindness, deafness, microcephaly, and loss of speech and motor coordination. Pathological lesions in Minamata Disease include the visual, auditory and post- and pre-central cortices of the cerebrum, and cerebellar atrophy; in mild cases, the lesions tend to localize in the area striata of the occipital lobe and post central gyrus. These findings suggest that exposure to methylmercury leads to visual, auditory, somatosensory, and autonomic system dysfunction.

19.9.1.1 Human Health

The first recorded case of Minamata Disease was that of a child in 1953. In 1954, a total of 12 victims were documented: 7 adults and 5 children. In 1954, total mercury concentrations in brain from Minamata Disease victims ranged between 0.35 and 5.3 mg/kg FW. In 1955, the total number dead was 15, and in 1956 it had risen to 50: 22 adults, 21 children, and 7 fetal cases. Death rate of victims was 36.9%, being higher in summer and lower in winter, and was correlated with fish landings.

The mortality rate for acutely affected patients of the disease in 1957 was 32.8%. Infants born between 1955 and 1958 and diagnosed with mercury poisoning had a mental retardation rate of 29.1%, excluding congenital cases. By the end of 1960, 111 cases of poisoning were reported; by August 1965, 41 of the 111 had died. All congenital cases showed mental disturbance, lack of coordination, speech difficulty, and impaired chewing and swallowing. Among afflicted children, all had mental disturbance, disturbed coordination, and impaired walking; the most frequent symptoms among adults were visual field constriction (100%), difficulty in chewing and swallowing (94%), speech difficulty (88%), and impaired coordination (85%). A variety of ocular symptoms occur in Minamata Disease, most typically concentric constriction of the visual field and disturbances of eye movements. Based on studies with marmoset monkeys, it was concluded that impaired visual disturbances in patients with methylmercury intoxication is attributed mainly to lesions in the visceral nuclei of the forebrain.

Between 1973 and 1981, total mercury content in kidney, liver, cerebrum, and cerebellum was significantly higher in Minamata victims when compared to similar data from control populations in western Japan, but no statistically significant difference in methylmercury values were found between the two groups. There was some overlap in total mercury concentrations in Minamata victims and other residents in the Minamata Bay area not diagnosed with Minamata Disease. By March 1978, the total number of Minamata Disease patients had risen to 1303 including 155 deaths, being highest among residents nearest Minamata Bay who had consumed fish and shellfish from the Bay more frequently than nonvictims. By 1980, there were 378 deaths among 1422 Minamata disease patients in Kumamoto Prefecture. Of these 378, the first death occurred in 1954 with a peak incidence in 1956. The number of deaths increased rapidly after 1972 with a second peak in 1976. The mean age at death was 67.2 years. Most deaths after 1969 were from a combination of causes including Minamata disease, noninflammatory diseases of the central nervous system, pneumonia, cardiovascular and cerebrovascular diseases, and malignant neoplasms.

Concentrations of mercury in blood, urine, and especially hair are generally recognized as the best indicators of methylmercury exposure. Concentrations of total mercury in the hair of persons with known occupational exposure to mercury and with a low consumption of fish are usually less than 5.0 mg/kg FW. However, persons in Sweden and Finland with high consumption of mercury-contaminated fish and without symptoms of mercury intoxication often contain hair mercury levels >30.0 mg/kg, and in one case 180.0 mg/kg.

At the end of 1981, mortality analysis of 439 victims among 1483 patients with Minamata Disease in Kumamoto Prefecture showed that mortality rate for all causes of death was significantly higher in both sexes when compared to the general population and that older patients had significantly lower survival. Male patients dying of Minamata Disease had significantly higher frequencies of liver and kidney diseases. Obesity and alcohol consumption significantly influenced the frequency of liver dysfunction in Minamata Disease victims, being higher in obese females and alcoholic males; moreover, there was no obvious relation between methylmercury exposure and liver disease.

Analysis of age-specific mortality rates show a significant increase in Minamata Disease patients under age 30 years owing to high mortality from in-utero Minamata Disease or childhood Minamata Disease. Over age 30 years, mortality tended to be slightly higher in Minamata Disease patients than a reference population, though death rates were statistically the same. Death rate from renal and liver diseases, as mentioned earlier, was significantly higher in Minamata Disease patients under age 30 than in controls. Most of the deaths from liver dysfunction were due to cirrhosis and chronic hepatitis, and from renal diseases it was from nephritis and nephrosis. Damage to human kidney and liver function are probably associated with the tendency of these organs to accumulate mercury.

A 71-year-old male severely afflicted with Minamata Disease in 1956 died in 1982. On autopsy, methylmercury concentration in brain was within the normal range; however, the total mercury remained high in the brain and mercury was clearly demonstrated in macrophages over wide areas of the brain and in neurons of specific brain areas. The half-time persistence of methylmercury in brain was estimated at 240–245 days.

By 1982, there were 1800 verified human victims of mercury poisoning in a total regional population of 200,000. Symptoms evidenced by human victims included sensory impairment, constriction of visual fields, hearing loss, ataxia, and speech disturbances. Congenital cases were accompanied by disturbance of physical and mental development; about 6.0% of babies born in Minamata had cerebral palsy (vs. 0.5% elsewhere). Some recovery was evident in 1986 as judged by the finding that mercury concentrations in erythrocytes of Minamata Disease victims were not significantly different from that of nearby inhabitants.

In 1987, afflicted humans displayed symptoms of peripheral neuropathy (70.1%), ataxia 22.9%), constriction of the visual field (17.4%), tremor of the digits (10.2%), and dysarthria (9.2%); another 12.1% had no symptoms. Nearly all patients complained of fatigue, numbness of parts of the body, and muscle cramps. In 1987, a female Minamata disease victim born in 1957, presented with seizures in 1959, and died at age 30 of cerebral palsy. Four of her eight siblings and her parents were diagnosed with Minamata Disease. The total mercury in her mother's head hair was 101.0 mg/kg FW in 1959. The mother died of rectal cancer in 1972 at age 55 years. The victim was presumed to have been exposed to methylmercury in utero. On autopsy, the victim's brain showed marked cerebral atrophy and severe atrophy of nerve cells in cerebellum; mercury granules, mostly inorganic, were present in brain, kidney, and liver, suggesting that biotransformation of methylmercury to inorganic mercury had occurred. The total mercury content in victim's hair was 62.0 mg/kg FW in 1959 at age 2 years and 5.4 mg/kg in 1974 at age 17 years. At death, she weighed 23 kg (50.6 pounds). Total mercury concentrations (methylmercury) measured at death in this case, in mg/kg FW, were 0.3 (0.01) in cerebral cortex, 0.55 (0.007) in thalamus, 0.6 (0.025) in cerebellum, 2.8 (0.025) in kidney, and 0.72 (0.01) in liver. For comparison, healthy adults contained 0.64 mg total Hg/kg FW (0.07 mg methylmercury/kg FW) in liver, 1.0 (0.015) in kidney, and (0.078 0.009) in cerebellum.

Mean mercury concentrations in organs of Minamata Disease victims dying between 1973 and 1985 remained elevated over those of residents not afflicted with Minamata Disease and dying between 1973 and 1991. Minamata Disease victims always had mean total mercury (methylmercury) concentrations, in mg/kg FW, >1.51 (0.05) in kidney, >0.48 (0.035) in liver, >0.10 (0.016) in brain cerebrum, and 0.05 (0.026) in brain cerebellum.

As of 1988, 12,336 persons had applied for compensation as being victims of Minamata Disease. Of these, only 1750 cases were approved, 6653 cases rejected, and 3993 were still under investigation. By 1989, over 20,000 people were thought to have been affected; symptoms were mainly neurological and resulted in death, chronic disability, and congenital abnormalities. By June 1989, 1757 patients were officially diagnosed with Minamata disease, of which 765 had died and their families awarded compensation. Another 7621 people were disapproved for compensation. Another group of 918 (94 dead) were under investigation. A group of 2347 (320 dead) were awaiting official examination, and a final group of 1876 patients received health costs compensation only. It is alleged that a large proportion of the population residing near Minamata Bay, especially the older population, is still incapacitated to varying degrees by the disease and that the more chronic effects of the disease are still becoming apparent.

By 1989, about 2000 individuals in the Minamata area have been officially certified to have Minamata Disease and eligible for financial compensation. Histopathological changes in brain were clearly linked to organomercury insult in Minamata Disease victims and the distribution of lesions in the nervous system was characteristic, especially in the cerebral cortices and the cerebellum. Pathological studies of 112 Minamata Disease victims also showed elevated frequencies of sepsis and malignant neoplasms of the thyroid gland when compared to 112 sex-age matched pair control deaths between 1970 and 1983 from various causes (senile dementia of Alzheimer's type, Parkinson's disease of idiopathic type, amyotrophic lateral sclerosis) in western parts of Japan.

By 1991, after extensive re-examination of Minamata Disease patients in Kumamoto and Kagoshima Prefectures, only 1385 were officially certified as victims and received compensation; however, an additional 6000 applications were still pending.

In 1993, more than 2000 patients have been officially designated with Minamata Disease, and 59 had congenital Minamata Disease in Kumamoto Prefecture. Pathological findings of fetal Minamata Disease indicated that the central nervous system was affected by methylmercury during gestation. Cases diagnosed before 1970 were similar to adult cases except for the higher frequency of speech disturbances, primitive reflex, salivation, and cerebellar abnormalities. Cases designated after 1970 had no specific clinical symptoms; however, symptoms were especially severe for cerebral palsy and mental deficiency, and death rate was higher in this group than in reference populations. In cases of congenital Minamata Disease, the incidence rate of cerebral palsy was comparatively elevated in fishing villages around Minamata Bay: 1.0–2.0 percent vs. 0.06–0.6 percent in the general Japanese population. And hair mercury levels were elevated in children with a maximum recorded of 100.0 mg total mercury/kg FW. Concentrations of methylmercury in dried umbilical cords of infants born with congenital Minamata Disease between 1955 and 1960 were significantly higher than that of the general Japanese population: 3.1 mg methylmercury/kg DW vs. <0.02 mg total mercury/kg DW. Total mercury concentrations in umbilical cords decreased from maxima of 1.1 mg/kg DW in the 1960–1970 decade, to 0.045 in the 1970–1980 decade, to <0.02 from 1981–88, at which point these levels were the same or lower than that of the general Japanese population.

19.9.1.2 Natural Resources

Minamata disease resulted from the discharge of methylmercury from chemical factories into Minamata Bay. It is emphasized that Minamata disease is from direct methylmercury contamination rather than methylation of environmental sources of inorganic mercury. Once diluted and diffused in the water, it was concentrated to a high level in fish and filter-feeding shellfish by several routes, including bioconcentration and food chain biomagnification. When these fish and shellfish were consumed by humans, methylmercury gradually accumulated to exceed a threshold value, causing intoxication. Spontaneously poisoned cats, dogs, rats, waterfowl, and pigs behaved erratically and died; flying crows and grebes suddenly fell into the sea and drowned; and large numbers of dead fish were seen floating on the sea surface. In laboratory studies, cats and rats fed shellfish from the Bay developed the same signs as those seen in animals affected spontaneously. Abnormal mercury content – i.e., more than 30.0 mg/kg fresh weight – was measured in fish, shellfish, and muds from the Bay, and in organs of necropsied humans and cats that had succumbed to the disease. Total mercury concentrations in tissues of fish from Minamata Bay with signs of methylmercury poisoning were about 15.0 mg/kg FW in liver and 8.0–24.0 mg/kg FW in muscle. Mercury contamination of fish and sediments was still evident in 1981, although discharges from the acetaldehyde plant ceased in 1971. Plans are underway to reopen Minamata Bay for fishing in the future; however, certain species of fish are still likely to contain unsafe concentrations of mercury for human consumption for some years.

Most mercury found in fish occurred as methylmercury, even when – as was the case at the Minamata Bay site – the initial release of mercury is inorganic. Inorganic mercury at Minamata was methylated chemically with methylcobalamin as the carbon methyl group donor. The methylcobalamin is synthesized by a range of bacterial species, especially *Desulfovibrio desulfuricans*, which is considered a major methylating source in anaerobic sediments. At low (0.1 mg Hg^{2+}/kg) sediment mercury concentrations, up to 37% of the added mercury was methylated during fermentative growth of *D. desulfuricans*. But under conditions of sulfate reduction, mercury methylation was less efficient. The fermentative mercury methylating cultures were comparatively sensitive to Hg^{2+} whereas the sulfate reducing cultures that did not methylate at a high rate were more resistant to Hg^{2+}. Some strains of *Desulfovibrio desulfuricans* – in addition to producing methylmercury – can also convert methylmercury to methane and Hg, as was true for aerobic species of bacteria isolated from Minamata Bay sediments.

There is a strong relation between the food of birds from Minamata and the mercury content in feathers; the content is highest in fish-eating seabirds and lowest in herbivorous waterfowl. This same relation held in birds collected from China and Korea, although concentrations were significantly lower in those nations. There are close correlations between mercury contents of zooplankton and suspended particulate matter, and of sediments and fish muscle, suggesting a pathway from sediment to fish by way of suspended matter and zooplankton. The conversion from inorganic mercury to methylmercury is believed to have occurred primarily in zooplankton.

19.9.2 Mitigation

In aquatic environments where point sources of industrial contamination have been identified, the elimination of mercury discharges has usually improved environmental quality. Such improvement has been reported for Minamata Bay; for sediments in Saguenay Fjord, Quebec, when chloralkali wastes were limited; for fish residues in Lake St. Clair, Canada, after two chloralkali plants were closed; and in various sections of Europe and North America when industrial discharges were eliminated.

In Minamata Bay, 72 strains of mercury-resistant *Pseudomonas* spp. bacteria were isolated from sediments near the drainage outlet to the Bay. *Pseudomonas* spp. dominated the bacteria with the highest resistance to mercury although *Bacillus* spp. strains were the most numerous among all bacterial species isolated from Minamata Bay sediments. Total bacterial concentrations in Minamata Bay were the same as a nearby reference site, but mercury-resistant bacteria were found in higher numbers in Minamata Bay. In 1984, additional bacteria were isolated. *Bacillus* strains dominated in sediments containing up to 23.0 mg total Hg/kg and *Pseudomonas* strains in sediments with higher (>52.0 mg total Hg/kg) mercury concentrations. The mercury-resistant *Pseudomonas* strains, when compared to *Bacillus* strains, were more resistant to inorganic mercury, methylmercury, and phenylmercury.

The gradual decrease in mercury content of Minamata Bay sediments from 1959 – when organomercury was first suggested as the cause of Minamata Disease – and the late 1980s when dredging removed most remaining mercury, suggests that natural processes, mainly microbial, had removed 75–90% of the mercury in Minamata Bay sediments. However, the mechanisms of action were not known with certainty at that time. Multiple organomercurial-volatilizing bacteria that can volatilize Hg from methylmercury chloride, ethylmercury chloride, phenylmercury acetate, fluorescein mercuric acetate, and *p*-chloromercuric benzoate were found in the sediments of Minamata Bay. The bacteria detoxify mercury compounds by two separate enzymes, organomercurial lyase and mercuric reductase, acting sequentially. Organomercurial lyase cleaves the C–Hg bond of certain mercurials, and mercuric reductase then reduces Hg^{2+} to volatile Hg; bacteria remove mercury from the environment as mercury vapor.

In 1988, a total of 4604 bacterial strains were isolated from Minamata Bay (1428 strains) and from non-contaminated environs (3176 strains) and screened for their ability to volatilize mercuric chloride. Up to 38% of all bacterial strains isolated from Minamata Bay could grow on agar media containing 40.0 mg mercuric chloride/L vs. only 0.8% of bacteria from reference locations. A total of 67 *Pseudomonas* strains and 91 *Bacillus* strains from Minamata Bay were significantly more resistant to mercuric chloride than were 64 strains of *Pseudomonas* and 100 strains of *Bacillus* from reference locations. *Pseudomonas* spp. were more resistant to mercuric chloride than were *Bacillus* spp., but both detoxify the mercury compounds in Minamata Bay by organomercurial lyase and mercuric reductase enzymes. The organomercurials-volatilizing bacterial strains were found only in sediments of Minamata Bay. A total of 78 strains of *Bacillus* that can degrade both inorganic and organic mercurials were collected from Minamata Bay and were identified as *Bacillus subtilis*, *B. firmus*, *B. lentus*, *B. badius*, and two unidentified strains. The ability to detoxify mercury compounds is chromosomally encoded and similar to that of *Bacillus* spp. isolated from Boston Harbor, Massachusetts.

Bacterial detoxification and volatilization of mercury may accelerate natural processes in contaminated environments; however, social concerns were associated with bioremediation of mercury from sewage with subsequent release of mercury into the atmosphere. Mercury releases from these limited sources do not seem to be a public health problem, although major concerns center on systems that retain the detoxified mercury. Retention of the reduced elemental mercury in bioreactors may allay these concerns. Genetically engineered bacteria that degrade organomercurials with subsequent retention on alginate beads are also suggested.

Enzymatic detoxification was determined to be the major resistance mechanism in all species of mercury-resistant bacteria. For example, mercuric reductase was essential for volatilization of Hg from Hg^{2+} and various organomercurial hydrolases were responsible for volatilization of methane (CH_4) from methylmercury, for ethane (C_2H_4) from ethylmercury, and for benzene from phenylmercury. Minamata Bay bacterial isolates can also volatilize Hg from added inorganic and organic mercurials. Genes which govern the chemistry of mercury detoxification were abundant in bacteria found in Minamata Bay and other mercury-polluted sites; these genetic strains of mercury-resistant bacteria show promise for bioremediation of mercury pollution.

In 1984, the clearance rate for mercury in Minamata Bay sediments was estimated at 18.2 years, with 90% clearance via natural processes estimated by the year 2000. This natural cleanup rate was judged unacceptably low and in 1984 dredging was initiated to remove all sediments containing more than 25.0 mg total mercury/kg. By 1987, 1.5 million m^3 of contaminated sediments had been removed from $2.09\,km^2$ of Bay areas and used as landfill at an isolated 58-ha site. This site, with an estimated 7.5 tons of mercury, is now the site of Minamata Disease Park, replete with playing fields and a museum. The landfill was capped with a layer of vinyl plastic sheet, then by volcanic ash, and topped with soil. The mercury at the site will be exposed to physical and microbial activities and subsequently volatilized to the atmosphere or leached into Minamata Bay. Near this site, a hospital and rehabilitation area was established in 1965 for victims of Minamata Disease and in 1978 the National Institute for Minamata Disease was established to research this disease.

By 1989, it was shown that mercury-volatilizing bacterial strains comprised 5.3% of all bacterial strains isolated from Minamata Bay or three times more abundant than control isolates; moreover, the number of bacterial isolates from Minamata Bay able to volatilize Hg from phenylmercury was twenty times greater than reference isolates. The development of mutant bacterial strains with the ability to detoxify inorganic and organic mercurials is continuing.

Mercury bioremediation research is now mainly limited to aqueous sources because mercury bound to soils and sediments is largely unavailable to bacterial cells over short exposures of hours or days. Techniques to measure bioavailable mercury in water using mercury-resistant bacteria show promise for use in decontamination of soils and sediments. Also, more research is needed on isolation of anaerobic strains of bacteria from mercury-contaminated sediments and their ability to methylate and demethylate mercury under field conditions.

19.10 Mercury Hazards from Gold Mining

The use of liquid mercury (Hg) to separate microgold (Au) particles from sediments through formation of amalgam (Au–Hg) with subsequent recovery and reuse of mercury is a technique in force for at least 4700 years; however, this process is usually accompanied by massive mercury contamination of the biosphere. It is estimated that gold mining currently accounts for about 10% of the global mercury emissions from human activities. This chapter documents the history of mercury in gold production and ecotoxicological aspects of the amalgamation process in Brazil and the U.S. Gold mining is covered in greater detail in Chapter 17.

19.10.1 History

The use of mercury in the mining industry to amalgamate and concentrate precious metals dates from about 2700 BCE when the Phoenicians and Carthaginians used it in Spain. The technology became widespread with the Romans in 50 CE and is similar to that employed today. In 177 CE, the Romans banned elemental mercury use for gold recovery in mainland Italy, possibly in response to health problems caused by this activity. Gold extraction using mercury was widespread until the end of the first millennium. In the Americas, mercury was introduced in the 16th century to amalgamate Mexican gold and silver. In 1849, during the California gold rush, mercury was widely used, and mercury poisoning was allegedly common among miners. In the 30-year period 1854–1884, gold mines in California's Sierra Nevada range released between 1400 and 3600 tons of mercury to the environment; dredge tailings from this period still cover more than 73 km^2 in the Folsom-Natomas region of California, and represents a threat to current residents. In South America, mercury was used extensively by the Spanish colonizers to extract gold, releasing nearly 200,000 metric tons of mercury into the environment between 1550 and 1880 as a direct result of this process. At the height of the Brazilian gold rush in the 1880s, more than six million people were prospecting for gold in the Amazon region alone.

It is doubtful whether there would have been gold rushes without mercury. Supplies that entered the early mining camps included hundreds of flasks of mercury weighing 34.5 kg each, consigned to the placer diggings and recovery mills. Mercury amalgamation provided an inexpensive and efficient process for the extraction of gold, and the process can be learned rapidly by itinerant gold diggers. The mercury amalgamation process absolved the miners from any capital investment on equipment, and this was important where riches were obtained instantaneously and ores contained only a few ounces of gold per ton and could not be economically transported elsewhere for processing. Mercury released to the biosphere between 1550 and 1930 due to gold mining activities, mainly in Spanish colonial America, and in Australia, Southeast Asia, and England may have exceeded 260,000 metric tons. Exceptional increases in gold prices in the 1970s concomitant with worsening socioeconomic conditions in developing regions of the world resulted in a new gold rush in the southern hemisphere involving over ten million people in all continents. At present, mercury amalgamation is used as the major technique for gold production in South America, China, Southeast Asia, and some African countries. Most of the mercury released into the biosphere through gold mining may still participate in the global mercury cycle through remobilization from abandoned tailings and other contaminated areas.

From 1860 to 1925, amalgamation was the main technique for gold recovery worldwide, and was common in the U.S. until the early 1940s. The various procedures in current use can be grouped into two categories:

(1) Recovery of gold from soils and rocks containing 4.0–20.0 g of gold per ton. The metal-rich material is passed through grinding mills to produce a metal-rich concentrate. In Colonial America, mules and slaves were used instead of electric mills. This practice is associated with pronounced deforestation, soil erosion, and river siltation. The concentrate is moved to small amalgamation ponds or drums, mixed with liquid mercury, squeezed to remove excess mercury, and taken to a retort for roasting. Any residue in the concentrate is returned to the amalgamation pond and reworked until the gold is extracted.

(2) Gold is extracted from dredged bottom sediments. Stones are removed by iron meshes. The material is then passed through carpeted riffles for 20–30 h, which retains the heavier gold particles. The particles are collected in barrels, amalgamated, and treated (as in 1). However, residues of the procedure are released into the rivers. Vaporization of mercury and losses due to human error also occur.

The organized mining sector abandoned amalgamation because of economic and environmental considerations. But small-scale mine operators in South America, Asia, and Africa, often driven by unemployment, poverty, and landlessness, have resorted to amalgamation because they lack affordable alternative technologies. Typically, these operators pour liquid mercury over crushed ore in a pan or sluice. The amalgam, a mixture of gold and mercury (Au–Hg), is separated by hand, passed through a chamois cloth to expel the excess mercury – which is reused – then heated with a blowtorch to volatilize the mercury. About 70% of the mercury lost to the environment occurs during the blowtorching. Most of these atmospheric emissions quickly return to the river ecosystem in rainfall and concentrate in bottom sediments.

Residues from mercury amalgamation remain at many stream sites around the globe. Amalgamation should not be applied because of health hazards and is, in fact, forbidden almost everywhere; however, it remains in use today, especially in the Amazon section of Brazil. In Latin America, more than one million gold miners collect between 115 and 190 tons of gold annually, emitting more than 200 tons of mercury in the process. The world production of gold is about 225 tons annually with 65 tons of the total produced in Africa. It is alleged that only 20% of the mined gold is recorded officially. About one million people are employed globally on non-mining aspects of artisanal gold, 40% of them female with an average yearly income of $600. The total number of gold miners in the world using mercury amalgamation to produce gold ranges from 3 to 5 million, including 650,000 from Brazil, 250,000 from Tanzania, 250,000 from Indonesia, and 150,000 from Vietnam. To provide a living – marginal at best – for this large number of miners, gold production and mercury use would come to thousands of tons annually; however, official figures account for only 10% of the production level. At least 90% of the gold extracted by individual miners in Brazil is not registered with authorities for a variety of reasons, some financial. Accordingly, official gold production figures reported in Brazil and probably most other areas of the world are grossly under-reported. Cases of human mercury contamination have been reported from various sites around the world ever since mercury was introduced as the major mining technique to produce gold and other precious metals in South America hundreds of years ago. Contamination in humans is reflected by elevated mercury concentrations in air, water, diet, hair, urine, blood, and other tissues. However, only a few studies actually detected symptoms, or clinical evidence of mercury poisoning in gold mining communities.

After the development of the cyanide leaching process for gold extraction, mercury amalgamation disappeared as a significant mining technology. But when the price of gold soared from $58/troy ounce in 1972 to $430 in 1985, a second gold rush was triggered, particularly in Latin America, and later in the Philippines, Thailand, and Tanzania. In modern Brazil, where there has been a gold rush since 1980, at least 2000 tons of mercury were released, with subsequent mercury contamination of sediments, soils, air, fish, and human tissues; a similar situation exists in Columbia, Venezuela, Peru, and Bolivia. Estimates of global anthropogenic total mercury emissions range from 2000 to 4000 metric tons per year, of which 460 tons are from small-scale gold mining. Major contributors of mercury to the environment from recent gold mining activities include Brazil (3000 tons since 1979), China (596 tons since 1938), Venezuela (360 tons since 1989), Bolivia (300 tons since 1979), the Philippines (260 tons since 1986), Columbia (248 tons since 1987), the U.S. (150 tons since 1969), and Indonesia (120 tons since 1988).

The most mercury-contaminated site in North America is the Lahontan Reservoir and environs in Nevada. Millions of kilograms of liquid mercury used to process gold and silver ore mined from Virginia City, Nevada, and vicinity between 1859 and 1890, along with waste rock, were released into the Carson River watershed. The inorganic elemental mercury was readily methylated to water-soluble methylmercury. Over time, much of this mercury was transported downstream into the lower reaches of the Carson River, especially the Lahontan Reservoir and Lahontan wetlands near the terminus of the system, with significant damage to wildlife.

19.10.2 Brazil

High mercury levels found in the Brazilian Amazon environment are attributed mainly to gold mining practices, although elevated mercury concentrations are reported in fish and human tissues in regions far from any anthropogenic mercury source. Since the late 1970s, many rivers and waterways in the Amazon have been exploited for gold using mercury as an amalgamate in the mining process to separate the fine gold particles from other components in the bottom gravel. Between 1979 and 1985, at least 100 tons of mercury were discharged into the Madeira River basin with 45% reaching the river and 55% passing into the atmosphere. As a result of gold mining activities using mercury, elevated concentrations of mercury were measured in bottom sediments from small forest streams (up to 157.0 mg Hg/kg DW), in stream water (up to 10.0 μg/L), in fish (up to 2.7 mg/kg FW muscle), and in human hair (up to 26.7 mg/kg DW). Mercury transport to pristine areas by rainwater, water currents, and other vectors, could be increased with increasing deforestation, degradation of soil cover from gold mining activities, and increased volatilization of mercury (from gold mining practices). Population shifts due to gold mining are common in Brazil. For example, from 1970–85, the population of Rondonia, Brazil, increased from about 111,000 to 904,000, mainly due to gold mining and agriculture. One result was a major increase in deforested areas and in gold production from 4 kg Au/year to 3600 kg/year. Mercury is lost during two distinct phases of the gold mining process. In the first phase, sediments are aspirated from the river bottom and passed through a series of seines. Metallic mercury is added to the seines to separate and amalgamate the gold. Part of this mercury escapes into the river, with risk to fish and livestock that drink river water, and to humans from occupational exposure and from ingestion of mercury-contaminated fish, meat, and water. In the second phase, the gold is purified by heating the amalgam – usually in the open air – with mercury vapor lost to the atmosphere. Few precautions are taken to avoid inhalation of the mercury vapor by the workers.

In Brazil, four stages of mercury poisoning were documented leading to possible occurrence of Minamata Disease. The first route involves inorganic mercury poisoning among miners and gold shop workers directly exposed to elemental mercury used for gold extraction. Inhalation of mercuric vapor via the respiratory tract and absorption through the skin are considered the major pathways. Miners and gold shop workers who have been exposed directly to mercury vapors show clinical symptoms of inorganic mercury poisoning including dizziness, headache, palpitations, tremors, numbness, insomnia, abdominal pain, dyspnea, and memory loss. Serious cases also show hearing difficulty, speech disorders, gingivitis, impotence, impaired eyesight, polyneuropathy, and disturbances in taste and smell. In the second stage, inorganic mercury discharged into the biosphere is converted to organomercurials via bacterial and other processes with resultant contamination of air, soil, and water. In stage three, the organomercurials are bioaccumulated and biomagnified by fish and filter-feeding bivalve mollusks. Finally, humans who consume mercury-contaminated fish and shellfish evidence increased concentrations of mercury in blood, urine, and hair which, if sufficiently high, are associated with the onset of Minamata Disease.

19.10.2.1 Mercury Sources and Release Rates

All mercury used in Brazil is imported, mostly from the Netherlands, Germany, and England, reaching 340 tons in 1989. For amalgamation purposes, mercury in Brazil is sold in small quantities (200.0 g) to a great number (about 600,000) of individual miners. Serious ecotoxicological damage is likely since much – if not most – of the human population in these regions depend on local natural resources for food. In 1972, the amount of gold produced in Brazil was 9.6 tons and in 1988 it was 218.6 tons; an equal amount of mercury is estimated to have been discharged into the environment. In Brazil, industry was responsible for almost 100% of total mercury emissions to the environment until the early 1970s, at which time existing mercury control policies were enforced with subsequent declines in mercury releases. Mercury emissions from gold mining were insignificant up to the late 1970s, but by the mid 1990s it accounted for 80% of the total mercury emissions. About 210 tons of mercury is now released to the biosphere each year in Brazil: 170 tons from gold mining, 17 tons from the chloralkali industry, and the rest from other industrial sources. Emissions into the atmosphere are the major pathway of mercury releases to the environment, with the gold mining industry accounting for 136 tons annually in Brazil. During an 8-year period in the 1980s, about 2000 metric tons of mercury were used to extract gold in Brazil.

About 55% of the mercury used in gold mining operations is lost to the atmosphere during the burning of amalgam. The resulting mercury vapor (Hg) may be transported over considerable distances. Atmospheric transport of mercury from gold mining activities coupled with high natural background concentrations of mercury may produce mercury contamination in pristine areas of the Amazon.

At least 400,000 – and perhaps as many as one million – small-scale gold miners, known as garimpos, are active in the Brazilian Amazon region at more than 2000 sites. It is estimated that each garimpo is indirectly responsible for another 4–5 people, including builders and operators of production equipment, dredges, aircraft (at least 1000), small boats or engine-driven canoes (at least 10,000), and about 1100 pieces of digging and excavation equipment. It is conservatively estimated that this group discharges 100 tons of mercury into the environment each year. The production of gold by garimpos (small-and medium-scale, often clandestine and transitory, mineral extraction operations) is from three sources: extraction of auriferous materials from river sediments; from veins where gold is found in the rocks; and alluvial, where gold is found on the banks of small rivers. The alluvial method is most common and includes installation of equipment and housing, hydraulic pumping (high-pressure water to bring down the pebble embankment), concentration of gold by mercury, and burning the gold to remove the mercury. The latter step is responsible for about 70% of the mercury entering the environment. The gold is sold at specialized stores where it is again fired. Metallic mercury can also undergo methylation in the river sediments and enter the food chain.

Elemental mercury discharged into the Amazon River basin due to gold mining activities is estimated at 130 tons annually. Between 1987 and 1994 alone, more than 3000 metric tons of mercury was released into the biosphere of the Brazilian Amazon region from gold mining activities, especially into the Tapajos River basin. Local ecosystems receive about 100 tons of metallic mercury yearly, of which 45% enters river systems and 55% the atmosphere. Mercury lost to rivers and soils as Hg is comparatively unreactive and contributes little to mercury burdens in fish and other biota. Mercury entering the atmosphere is redeposited with rainfall at 90.0–120.0 $\mu g/m^2$ annually, mostly as Hg^{2+} and particulate mercury; these forms are readily methylated in floodplains, rivers, lakes, and reservoirs. Health hazards to humans include direct inhalation of mercury vapor during the processes of burning the Hg–Au amalgam, and consuming mercury-contaminated fish. Methylmercury, the most toxic form of mercury, is readily formed. High levels of methylmercury in fish collected near gold mining areas and in the hair of humans living in fishing villages downstream of these areas suggest that the reaction that converts discharged Hg to Hg^{2+} is present in nature before Hg^{2+} is methylated to CH_3Hg^+. Oxidation of Hg to Hg^{2+} occurs in the presence of sulfhydryl compounds, including L-cysteine and glutathione. Since sulfhydryl compounds are known to have a high affinity for Hg^{2+}, the conversion of Hg to Hg^{2+} may be due to an equilibrium shift between Hg and Hg^{2+} induced by the added sulfhydryl compounds.

About 130 tons of Hg is released annually by alluvial gold mining to the Amazonian environment, either directly to rivers or into the atmosphere, after reconcentration, amalgamation, and burning. In the early 1980s, the Amazon region in northern Brazil was the scene of the most intense gold rush in the history of Brazil. Metallic mercury was used to amalgamate particulate metallic gold. Refining of gold to remove the mercury is considered to be the source of environmental mercury contamination; however, other sources of mercury emissions in Amazonia include tailings deposits and burning of tropical forests and savannahs. In 1989 alone, gold mining in Brazil contributed 168 metric tons of mercury to the environment.

Natural sources of mercury and natural biogeochemical processes contribute heavily to reported elevated mercury concentrations in fish and water samples collected up to 900 km downstream from local gold mining activities. Based on analysis of water, sediments, and fish samples systematically collected along a 900-km stretch of the Madeira River in 1997, it was concluded that the elevated mercury concentrations in samples were due mainly to natural sources and that the effects of mercury released from gold mining sites were localized. This needs to be verified.

19.10.2.2 Mercury Concentrations in Abiotic Materials and Biota

Since 1980, during the present gold rush in Brazil, at least 2000 tons of mercury was released into the environment. Elevated mercury concentrations are reported in virtually all abiotic materials, plants, and animals collected near mercury-amalgamation gold mining sites. Mercury concentrations in samples show high variability, and this may be related to seasonal differences, geochemical composition of the samples, and species differences. In 1992, more than 200 tons of mercury was used in the gold mining regions of Brazil. One area, near Pocone, has been mined for more than 200 years. In the 1980s, about 5000 miners were working 130 gold mines in this region. Mercury was used to amalgamate the preconcentrated gold particles for the separation of the gold from the slag. Mercury-contaminated wastes from the separation process were combined with the slag from the reconcentration process and collected as tailings. The total mercury content in tailings piles in this geographic locale was estimated at about 1600 kg, or about 12% of all mercury used in the past 10 years. Surface runoff from tropical rains caused extensive erosion of tailings piles – some 4.5 m high – with contaminated material reaching nearby streams and rivers. In the region of Pocone, mercury concentrations in waste tailings material ranged from 2.0 to 495.0 μg/kg, occupied 4.9 km^2, and degraded an estimated 12.3 km^2.

Tropical ecosystems in Brazil are under increasing threat of development and habitat degradation from population growth and urbanization, agricultural expansion, deforestation, and mining. Where mercury has been released into the aquatic system as a result of unregulated gold mining, subsequent contamination of invertebrates, fish, and birds was measured and biomagnification of mercury was documented from gastropod mollusks (*Ampullaria* spp.) to birds snail kite (*Rostrhamus sociabilis*) and from invertebrates and fish to waterbirds and humans. Indigenous people of the Amazon living near gold mining activities have elevated levels of mercury in hair and blood. Other indigenous groups are also at risk from mercury contamination as well as from malaria and tuberculosis. The miners, mostly former farmers, are also victims of hard times and limited opportunities. Small-scale gold mining offers an income, and an opportunity for upward mobility. Throughout the Brazilian Amazon, about 650,000 small-scale miners are responsible for about 90% of Brazil's gold production and for the discharge of 90–120 tons of mercury to the environment every year. About 33% of the miners had elevated concentrations in tissues over the tolerable limit set by the World Health Organization [WHO]. In Brazil, it is alleged that health authorities are unable to detect conclusive evidence of mercury intoxication due to difficult logistics and the poor health conditions of the mining population which may mask evidence of mercury poisoning. There is a strong belief that a silent outbreak of mercury poisoning has the potential for regional disaster.

In the Madeira River Basin, mercury levels in certain sediments were 1500 times higher than similar sediments from non-mining areas, and dissolved mercury concentrations in the water column were 17 times higher than average for rivers throughout the world. High concentrations of mercury were measured in fish and sediments from a tributary of the Madeira River affected by alluvial small-scale mining. The local safety limit of 0.1 mg Hg/kg DW sediment was exceeded by a factor of 25 and the safety level for fish muscle of 0.5 mg Hg/kg FW muscle was exceeded by a factor of 4. Both sediments and fish act as potential sinks for mercury because existing physicochemical conditions in these tropical waters (low pH, high organic load, high microbial activity, elevated temperatures) favor mercury mobilization, methylation, and availability. In Amazonian river sediments, mercury methylation accounts for less than 2.2% of the total mercury in sediments. In soils, mercury mobility is low, in general. There is an association between the distribution of mercury-resistant bacteria in sediments and the presence of mercury compounds. Between 1995 and 1997, mercury concentrations were measured in sediment along the Carmo stream, Minas Gerais, located in gold prospecting areas. Most sediments contained more than the Brazilian allowable limit of 0.1 mg Hg/kg DW. Mercury-resistant bacteria were present in sediments at all sites and ranged from 27 to 77% of all bacterial species, with a greater percentage of species showing resistance at higher mercury concentrations.

The Pantanal is one of the largest wetlands in the world and extends over 300,000 km^2 along the border area of Brazil, Bolivia, Argentina, and Paraguay. Half this surface is flooded annually. Since the 18th century, gold has been extracted from quartz veins in Brazil using amalgamation as a concentration process, resulting in metallic mercury releases into the atmosphere, soil, and sediments. The availability to aquatic biota of Hg released by gold mining activities is limited to its oxidation rate to Hg^{2+} and then by conversion to methylmercury (CH_3^+), which is readily soluble in water. The Pantanal in Brazil, at 140,000 km^2, is an important breeding ground for storks, herons, egrets, and other birds, as well as a refuge for threatened or endangered mammals including jaguars (*Panthera onca*), giant anteaters (*Myrmecophaga tridactyla*), and swamp deer (*Cervus duvauceli*). Gold mining is common in the northern Pantanal. There are approximately 700 operating gold-mining dredges along the Cuiba River. Unregulated gold mines have contaminated the area with mercury and 35–50% of all fishes collected from this area contain more than 0.5 mg Hg/kg FW muscle, the current Brazilian and international (World Health Organization) standard for fish consumed by humans. Gastropod mollusks that are commonly eaten by birds contained 0.02–1.6 mg Hg/kg FW soft parts. Mercury concentrations in various tissues of birds that ate these mollusks were highest in the anhinga (*Anhinga anhinga*) at 0.4–1.4 mg/kg FW and the snail kite at 0.3–0.6 mg/kg FW, and were lower in the great egret (*Casmerodius* [formerly *Ardea*] *albus*) at 0.02–0.04 mg/kg FW and limpkin (*Aramus guarauna*) at 0.1–0.5 mg/kg FW. The high mercury levels detected, mainly in fishes, show that the mercury used in gold mining and released into the environment has reached the Pantanal and spread throughout the ecosystem with potential biomagnification.

Floating plants accumulate small amounts of mercury, but their sheer abundance make them likely candidates for mercury phytoremediation. For example, in the Tucurui Reservoir in the state of Para, it is estimated that 32 tons of mercury are stored in floating plants, mostly *Scurpus cubensis*. Mercury methylation rates in sediments and floating plants were evaluated in Fazenda Ipiranga Lake, 30 km downstream from gold mining fields near Pantanal during the dry season of 1995. Sediments and roots of dominant floating macrophytes (*Eichornia azurea*, *Salvina* sp.) were incubated in situ for 3 days with about 43.0 μg Hg^{2+}/kg DW added as $^{203}HgCl_2$. Net methylation was about 1% in sediments under floating macrophytes, being highest at temperatures in the 33–45°C range and high concentrations of sulfate-reducing bacteria. Methylation was inhibited at >55°C, under saline conditions, and under conditions of low sulfate. Radiomercury-203 was detectable to a depth of 16 cm in the sediments, coinciding with the depth reached by chironomid larvae. Methylation was up to nine times greater in the roots of floating macrophytes than in the underlying surface sediments: an average of 10.4% of added Hg^{2+} was methylated in *Salvina* roots in 3 days and 6.5% in *Eichornia* roots. Using radiomercury-203 tracers, no methylation was observed under anoxic conditions in organic-rich, flocculent surface sediments due to the formation of HgS – a compound that is much less available for methylation than is Hg^{2+}. It was concluded that floating macrophytes should be considered in evaluation of mercury methylation rates in tropical ecosystems.

Clams collected near gold mining operations had elevated concentrations of mercury (up to 0.64 mg Hg/kg FW) in soft tissues. Laboratory studies suggest that mercury adsorbed to suspended materials in the water column is the most likely route for mercury uptake by filter-feeding bivalve mollusks.

Mercury concentrations in fish collected near gold mining activities in Brazil were elevated, and decreased with increasing distance from mining sites. In general, muscle is the major tissue of mercury localization in fishes, and concentrations are higher in older, larger, and predatory species. In the Tapajos River region, which receives between 70 and 130 tons of mercury annually from gold mining activities, mercury concentrations in fish muscle were highest in carnivorous species, lowest in herbivores, and intermediate in omnivores. However, only 2.0% of fish collected in 1988 (vs. 1.0% in 1991) from the Tapajos region exceeded the Brazilian standard of 0.5 mg total mercury/kg FW muscle, and all violations were from a single species of cichlid (tucunare/speckled pavon), *Cichla temensis*. Tucunare from the contaminated Tapajos region, when compared to a reference site, accumulated mercury 3.5–4.0 times more rapidly (0.8–1.4 μg daily vs. 0.2–0.38), and had significantly lower erythrocyte counts, hematocrits, and leukocyte counts. In a 1991 survey of 11 species of fishes collected from a gold mining area in Cachoeira de Teotonio, it was found that almost all predatory species had >0.5 mg Hg/kg FW muscle vs. <0.5 mg/kg FW in conspecifics collected from Guajara, a distant reference site. Limits on human food consumption were set for individual species on the basis of mercury concentrations in muscle – specifically, no restrictions on some species, mostly herbivores and omnivores; some restrictions on some omnivores and small predators; and severe restrictions on larger predators.

Fish that live near gold mining areas have elevated concentrations of mercury in their flesh and are at high risk of reproductive failure. Mercury concentrations of 10.0–20.0 mg/kg FW in fish muscle are considered lethal to the fish, and 1.0–5.0 mg/kg FW sublethal; predatory fishes frequently contain 2.0–6.0 mg Hg/kg FW muscle. Mercury-contaminated fish pose a hazard to humans and other fish consumers, including the endangered giant otter (*Pteronura brasiliensis*), and the jaguar. Giant otters eat mainly fish and are at risk from mercury intoxication: 1.0–2.0 mg Hg/kg FW diet is considered lethal. Jaguars consume fish and giant otters; however, no data are available on the sensitivity of this top predator to mercury. Caiman crocodiles are also threatened by gold mining and related mercury contamination of habitat, increased predation by humans, extensive agriculture, and deforestation. Caiman crocodiles, *Paleosuchus* spp., from the Tucurui Reservoir area had up to 3.6 mg Hg/kg FW muscle and 30.0 mg Hg/kg FW liver. Extensive habitat destruction and mercury pollution attributed to mining activity was observed at 19 localities in Mato Grosso and environs. Crocodiles (*Caiman* spp., *Melanosuchus niger*, *Crocodilus crocodilus*) captured from these areas were emaciated, algae-covered, in poor body condition, and heavily infested with leeches.

High levels of mercury in urine (81.0–102.0 μg/L) and blood (25.0–39.0 μg/L) were positively associated with the amount of amalgam burnt each week; for residents of a fishing village, mercury urine concentrations were highest among residents who refined amalgam by burning and consumed fish frequently with maximum levels recorded of 108.0 μg/L in urine and 254.0 μg/L in blood. The hair mercury concentration in gold miners in the Tapajos River basin averaged 22.2 mg/kg FW, with a maximum of 113.3 mg/kg FW. Also, blood mercury levels were high (max. 291.0 μg/L FW) with organomercurials accounting for about 5.3%; however, the inorganic mercurials are quickly excreted. Gold miners, gold shop workers, and neighbors of gold shops had abnormally high mercury levels in urine (up to 151.0 μg/L) and blood (up to 130.0 μg/L), and some had symptoms indicative of mercury intoxication. However, residents of the same area with no previous contact with metallic mercury and its compounds have shown elevated concentrations of total mercury in hair (up to 151.0 mg/kg FW, about 90% methylmercury). Motor difficulties were seen in some individuals with >50.0 mg total Hg/kg FW hair. In the Tucurui Reservoir area, where the main source of mercury is from gold mining activities upstream, a human male needs 0.3 mg of mercury daily to reach a hair mercury concentration of 50.0 mg/kg DW. To receive this amount of mercury from fish containing 1.0 mg total Hg/kg FW muscle, a daily ingestion of 330.0 g of fish muscle is calculated, although this needs verification.

19.10.2.3 Mitigation

There are two populations at significant risk from mercury intoxication in Brazilian gold mining communities: (1) riverine populations – with high levels of mercury in hair – that routinely eat mercury-contaminated fish; and (2) gold dealers in indoor shops exposed to Hg vapors. These two critical groups should receive special attention regarding exposure risks. Riverine populations, especially children and women of child-bearing age, should avoid consumption of carnivorous fishes. And in gold dealer shops, adequate ventilation and treatment systems for mercury vapor retention need to be installed.

Confounding variables in evaluating mercury risk to these groups include the amount of mercury spilled, transportation and methylation rates, mercury uptake among fish species, date at which mining began in a given area, and patterns of fish consumption in rural Amazonia. Analysis of mercury samples alone is not sufficient in evaluating risk and should be complemented by a questionnaire survey, as well as various clinical, neurological, and psychological tests. Risks are greatest from Hg vapor to gold traders when the Hg–Au amalgam is refined through burning indoors, and from organomercurials to children from riverine populations consuming carnivorous fish contaminated by mercury spillage in mining areas. Hair sampling from pregnant women is the most appropriate initial indicator of organomercury contamination. Children aged between 7 and 12 years old and pregnant women are the most important target groups. Children, rather than infants or toddlers, are recommended because tests of neurological function require children to understand simple instructions and to cooperate with researchers.

Tests measuring memory and coordination are also desirable, provided that they are easy to administer, require only basic equipment, and some training of a nurse or paramedic who will actually administer the test. The full battery of tests need only be applied to children, and hair samples and questionnaires covering dietary habits and clinical history should be sufficient for women at risk. In isolated communities, the research team is also expected by the population to provide general health care and research teams should be accompanied by health professionals working primarily to provide general health care. The minimum team necessary for conducting mercury-related epidemiological work over a 7-to-10-day period, in a community of up to 1500 people should consist of eight individuals: one physician to perform general clinical examinations; one physician to perform neurological tests; one nurse/paramedic to apply memory and physical coordination tests; one nurse to apply clinical history and dietary questionnaire; one physician and two nurses to provide general health care to the community; and one coordinator to control patient access.

An occupational health and safety program was launched to educate Brazilian adolescents about industrial hazards including health risks associated with mercury amalgamation of gold. Adolescents, together with young adults, constitute a large proportion of the garimpos and are in a critical physical and psychological growth phase. The number of adolescents participating in gold extraction increases with decreasing family income and with the structure of the labor market as approved by public policy and the courts. The program was designed to transfer information on risks associated with mining and other dangerous occupations, apprise them of their choices, and to promote training. The method had been successfully tested earlier in selected urban and rural schools of different economic strata. The community selected was Minas Gerais, where gold was discovered more than 300 years ago and is the main source of community income. The municipality does not have a sewer system, and the water supply is not treated. The program was initiated in April 1994 over a 5-week period. The 70 students were from the fifth to eighth grade. Almost all worked, most in mining – where exposure to mercury was illegal. Students successfully improved safety habits and recognition of potential accident sites. A generalized occupational safety program, such as this one, is recommended for other school districts. Continuous and systematic monitoring of mercury levels in fish and fishermen is also recommended.

19.10.3 The United States

Gold mining in the U.S. is ubiquitous; however, persistent mercury hazards to the environment were considered most severe from activities conducted during the latter portion of the 19th century, especially in Nevada. In the 50-year period 1850–1900, gold mining in the U.S. consumed 268–2820 tons of mercury yearly, or about 70,000 tons during that period. Mercury contamination from gold mining in the Sierra Nevada region of California during the late 1800s and early 1900s was extensive in watersheds where placer gold was recovered and processed by amalgamation. Elemental mercury from these operations continues to enter local and downstream water bodies via transport of contaminated sediments by river flooding.

A 1998 study that measured total mercury accumulations in predatory fishes collected nationwide showed that mercury levels in muscle were significantly correlated with methylmercury concentrations in water, pH of the water, percent wetlands in the basin, and the acid volatile content of the sediment. These four variables – especially methylmercury levels in water, accounted for 45% of the variability in mercury concentrations of fish, normalized by total length. A methylmercury water concentration of 0.12 ng/L was, on an average, associated with a fish fillet concentration of 0.3 mg Hg/kg FW for an age 3 (years) fish when all species were considered. Sampling sites with the highest overall mercury concentrations in water, sediment, and fish were highest in the Nevada Basins and environs (from historic gold mining activities), followed, in order, by the South Florida Basin, Sacramento River Basin in California, Santee River Basin and Drainages in South Carolina, and the Long Island and New Jersey Coastal Drainages. Elevated mercury concentrations in fish, except for Nevada, are not necessarily a result of gold mining activities. The mercury criterion for human health protection set by the U.S. Environmental Protection Agency (EPA) in 2001 is now 0.3 mg Hg/kg FW diet, down from 0.5 mg/kg FW previously.

Beginning in 1859, gold from the Comstock Lode near Virginia City, Nevada, was processed at 30 sites using a crude mercury amalgamation process, discharging about 6,750,000 kg of mercury to the environment during the first 30 years of mine operation. Over time, mercury-contaminated sediments were eroded and transported downstream by fluvial processes. The most heavily contaminated wastes – with a total estimated volume of 710,700 m^3 – contained 31,500 kg of mercury, 248 kg of gold, and 37,000 kg of silver. If site remediation is conducted, extraction of the gold and silver – worth about \$12 million – would defray a significant portion of the cleanup costs.

In the Carson River-Lahontan Reservoir (Nevada) watershed, approximately 7100 tons of metallic mercury was released into the watershed between 1859 and 1890 as a by-product of silver and gold ore refining. Between 1859 and 1920, about 5 million troy ounces of gold (180 kg) were produced from this area, as well as 2500 kg of silver, mostly from using amalgamation. In 1901, cyanide leaching was introduced and eventually replaced amalgamation as the dominant gold recovery process. During the past 130 years, mercury has been redistributed throughout 500 km^2 of the basin, and mercury concentrations at this site rank among the highest reported in North America. Mercury contamination was still severe in this region in 1993. Nevada authorities have issued health advisories against fish consumption from the Carson River; an 80-km stretch of the river has been declared a Superfund site. Mercury-contaminated tailings were dispersed throughout the lower Carson River, Lahontan Reservoir, and the Carson Sink by floods that occurred 19 times between 1861 and 1997. Low levels of methylmercury in surface waters of the Carson River-Lahontan Reservoir are attributed to increasing pH and increasing concentrations of anions of selenium [SeO_4^{-2}], molybdenum [MoO_4^{-2}], and tungsten [WO_4^{-2}]), all of which are inhibitory to sulfate-reducing bacteria known to play a key role in methylmercury production in anoxic sediments. Methylmercury concentrations ranged from 0.1 to 0.7 ng/L in this ecosystem and were positively associated with total suspended solids. Removal of mercury from the water column was attributed to binding to particles, sedimentation, and volatilization of dissolved gaseous mercury.

The Lahontan Reservoir supports game and commercial fisheries and the Lahontan Valley wetlands – home to many species of birds – is considered to be the most mercury-contaminated natural system in the U.S. Mercury concentrations in sediments from this site and environs ranged from 10.0 to 30.0 mg/kg, or about 80 times higher than uncontaminated sediments; elevated concentrations of mercury were documented from the site in annelid worms, aquatic invertebrates, fishes, frogs, toads, and birds. In 1997–98, three species of fish-eating birds nesting along the lower River Carson were examined for mercury contamination: double-crested cormorants (*Phalacrocorax auritus*), snowy egrets (*Egretta thula*), and black-crowned night-herons (*Nycticorax nycticorax*). The high concentrations of total mercury observed in livers (means, in mg/kg FW, of 13.5 in herons, 43.7 in egrets, and 69.4 in cormorants) and kidneys (6.1 in herons, 11.1 in egrets, 69.4 in cormorants) of adult birds were possible due to a threshold-dependent demethylation coupled with sequestration of the resultant inorganic mercury. Demethylation and sequestration processes also appeared to have reduced the amount of methylmercury distributed to eggs – although the short time spent by adults in the contaminated area was a factor in the lower-than-expected mercury concentrations in eggs. Most eggs had mercury concentrations, as methylmercury, below 0.8 mg/kg FW, the threshold concentration where reproductive problems may be expected. After hatching, young birds were fed diets by parent birds averaging 0.36–1.18 mg methylmercury/kg FW for 4–6 weeks through fledging; this exceeds recommended avian dietary intakes of <0.1 mg Hg/kg FW ration. Mercury concentrations in organs of the fledglings were much lower than those found in adults, but evidence was detected of toxicity to immune, detoxicating, and nervous systems. Immune deficiencies and neurological impairment of fledglings may affect survival when burdened with stress associated with learning to forage and ability to complete the first migration. Oxidative stress was noted in young cormorants containing the highest concentrations of mercury, as evidenced by increasing thiobarbituric acid-reactive substances, and altered glutathione metabolism. Based on studies with fish-eating birds, the following conclusions were arrived at:

(1) Adults tolerate relatively high levels of mercury in critical tissues through demethylation processes that occur above threshold concentrations.

(2) Adults demethylate methylmercury to inorganic mercury, which is excreted or complexed with selenium and stored in liver and kidney. This change in form and sequestering process reduces the amount of methylmercury circulating in tissues and in the amount available for deposition in eggs.
(3) The low concentrations of methylmercury in eggs is attributed to the short duration of time spent in the area prior to egg laying, and to the demethylation and sequestering processes within the birds.
(4) The young of these migratory fish-eating birds experience neurological and histological damage associated with exposure to dietary mercury.

In the southeastern U.S., gold mining was especially common between 1830 and 1849, and again during the 1880s. Historical gold mining activities contributed significantly to mercury problems in this region, as evidenced by elevated mercury concentrations in surface waters. Mercury in surface waters was positively correlated with total suspended solids and with bioavailable iron. Vegetation in the southeast is comparatively heavy and controlling erosion may reduce the total amount of mercury released from contaminated mining sites to the rivers. Mercury concentrations in surface waters of southeastern states were significantly lower – by orders of magnitude – than those from western states where amalgamation extraction techniques were also practiced, and this may reflect the higher concentrations of mercury used and the sparser vegetation of the western areas.

In northern Georgia, gold was discovered in 1829 and mined until about 1940. Extensive use of mercury probably began in 1838 when stamp mills (ore crushers) were introduced to help recover gold from vein ore, with about 38% of all mercury used in gold mining escaping into nearby streams. Mercury concentrations in historical floodplain sediments near the core of the mining district were as high as 4.0 mg/kg DW, but decreased with increasing distance downstream to <0.1 mg/kg at 10–15 km from the source mines. Near Dahlonega, Georgia, historical sediments contained as much as 12.0 mg Hg/kg, and mean values in stream banks near the mining district of 0.2–0.6 mg Hg/kg DW. Mercury-contaminated floodplain sediments pose a potential hazard to wildlife. Clams and mussels, for example, collected from these areas more than 50 years after mining had ceased contained elevated (0.7 mg/kg DW) mercury concentrations in soft parts. Erosion of channel banks and croplands, with subsequent transport of mercury-contaminated sediments from the mined watersheds, is likely to occur for hundreds or thousands of years. Mercury was the only significant trace metal contaminant resulting from former gold mining activities in Georgia, exceeding the U.S. Environmental Protection Agency's "heavily polluted" guideline for sediments of >1.0 mg Hg/kg. Other metals examined did not exceed the "heavily polluted" sediment guidelines of >50.0 mg/kg for copper, >200.0 mg/kg for lead, and >200.0 mg/kg for zinc.

In South Dakota, most of the fish collected from the Cheyenne River arm of Lake Oahe in 1970 contained elevated (>0.5 mg/kg FW muscle) concentrations of mercury. Elemental mercury was used extensively in this region between 1880 and 1970 to extract gold from ores and is considered to be the source of the contamination. In 1970, when the use of mercury in the gold recovery process was discontinued at this site, liquid wastes containing 5.5–18.0 kg of mercury were being discharged daily into the Cheyenne River arm.

In Nome, Alaska, gold mining is responsible, in part, for the elevated mercury levels (max. 0.45 mg/kg DW) measured in modern beach sediments. However, higher concentrations (max. 0.6 mg/kg) routinely occur in buried Pleistocene sediments immediately offshore and in modern nearby unpolluted beach sediments (1.3 mg Hg/kg). This suggests that the effects of mercury contamination from mining are less than natural concentration processes in the Seward Peninsula region of Alaska.

19.11 Proposed Mercury Criteria for the Protection of Natural Resources and Human Health

Proposed mercury criteria for the protection of representative crops, aquatic organisms, birds, and mammals are shown in Table 19.1 and for human health in Table 19.2. These criteria vary widely between nations and even between localities in the same nation. In almost every instance, these criteria are listed as concentrations of total mercury, with most, if not all, the mercury present as an organomercury species. In some cases, recommended mercury criteria are routinely exceeded, as is the case for brown bears (*Ursus arctos*) in the Slovak Republic, and in Italian seafood products recommended for human consumption.

Table 19.1. Proposed mercury criteria for the protection of selected natural resources.

Resource and Other Variables	Criterion or Effective Mercury Concentration
CROPS	
Irrigation water, Brazil	<0.2 μg/L
Land application of sludge and solid waste; maximum permissible concentration	
Europe; soils with pH 6.0–7.0	<1.0–<1.5 mg/kg sludge
Iowa, Maine, Vermont	10.0 mg/kg waste
California	20.0 mg/kg waste
Vermont, sewage sludge	
Loamy sand and sandy loam soils	<6.0 kg Hg/ha
Fine sandy loam, loam, and silt loam soils	<11.0 kg Hg/ha
Clay loam, clay, silty clay	<22.0 kg hg/ha
Soils, Australia; urban areas	<1.0 mg/kg FW
Soils, Germany	<2.0 mg/kg dry weight (DW)
Soils, Canada, agricultural lands	<0.5 mg/kg DW
Soils, Finland	
Recommended	<0.2 mg/kg DW
Maximum allowable	<5.0 mg/kg DW
Soils, former Soviet Union	<2.1 mg/kg DW
Soils, Japan, contaminated	>3.0 mg/kg DW
Soils, Netherlands	
Target	<0.3 mg/kg DW
Normal	<0.5 mg/kg DW
Moderate contamination: requires additional study	>2.0 mg/kg DW
Contaminated; immediate cleanup required	>10.0 mg/kg DW
Soils, New Jersey	<1.0 mg/kg DW
AQUATIC LIFE	
Diet, piscivorous fishes	<100.0 μg total mercury/kg fresh weight (FW) whole prey fish[e]
Freshwater	
Total mercury	<0.00057 μg/L, 24 h average; not to exceed 0.0017 μg/L at any time
Total mercury	<0.1 μg/L

Table 19.1. cont'd

Resource and Other Variables	Criterion or Effective Mercury Concentration
Total mercury; adverse effects expected with chronic exposure	>0.012 μg/L
Inorganic mercury	Adverse effects at >0.23 μg Hg^{2+}/L
Methylmercury	<0.01 μg/L
Total mercury	<0.012 μg/L, 4-day average (not to be exceeded more than once every 3 years); <2.4 μg/L, 1-h average (not to be exceeded more than once every 3 years)[a]
Inland surface waters, India	<10.0 μg/L from point source discharge
Public water supply, Wisconsin	<0.079 μg/L
Saltwater	Total recoverable mercury <0.025 μg/L, 24-h average; not to exceed 3.7 μg/L at any time[f]
Saltwater	<0.025 μg/L, 4-day average; <2.1 μg/L, 1-h average
Sediments	
California	
Low toxic effect	>0.15 mg/kg DW
Acceptable	<0.51 mg/kg DW
Bivalve mollusks, abnormal larvae	>0.51 mg/kg DW
Hazardous	>1.2–1.3 mg/kg DW
Canada, marine and freshwater	<0.14 mg/kg DW
Safe	
Adverse effects expected	>2.0 mg/kg DW
Ecuador	<0.45 mg/kg DW
Great Lakes	
Nonpolluted	<1.0 mg/kg DW
Heavily polluted	>1.0 mg/kg DW
Wisconsin sediment disposal into Great Lakes	<0.1 mg/kg DW
Ontario guideline for disposal of dredged sediments into lake	<0.3 mg/kg DW
Washington state, safe	<0.41 mg/kg DW
Tissue residues	
Goldfish (*Carassius auratus*); impaired egg production and spawning expected; gonad	>0.76 mg/kg FW
Rainbow trout (Oncorhynchus mykiss)	
Lethal	
Eggs	>70.0 μg/kg FW
Muscle, adults	>10.0 mg/kg FW
Whole body	10.0–20.0 mg/kg FW
Adverse effects probable, whole body	1.0–5.0 mg/kg FW
Impaired spawning and reduced survival of early life stages expected; gonad	0.49 mg/kg FW

Continued

Table 19.1. cont'd

Resource and Other Variables	Criterion or Effective Mercury Concentration
Brook trout (*Salvelinus fontinalis*); whole body, nonlethal	<5.0 mg/kg FW
Various species of freshwater adult fishes; adverse effects expected	
Brain	
Toxic	>3.0 mg/kg FW
Potentially lethal	>7.0 mg/kg FW
Muscle	
Toxic	>5.0-8.0 mg/kg FW
Lethal	10.0–20.0 mg/kg FW
Whole body	
Adverse effects	>1.0 mg/kg FW
No observed effect	<3.0 mg/kg FW
Toxic	5.0–10.0 mg/kg FW
AMPHIBIANS	
South African clawed frog, *Xenopus laevis*; impaired gamete function and reduced early life survival expected; gonad	>0.48 mg/kg FW
BIRDS	
Tissue residues	
Safe	
Brain, muscle	<15.0 mg/kg FW
Feather	<5.0 mg/kg FW
Kidney, seabirds	<30.0 mg/kg FW
Kidney, not seabirds	<20.0 mg/kg FW
Liver	
Normal	1.0–10.0 mg/kg FW
Toxic to sensitive species	>5.0–6.0 mg/kg FW
Hazardous, possibly fatal	>20.0 mg/kg FW
Egg	
Mallard (Anas platyrhynchos); safe	<0.8–<1.0 mg/kg FW
Ring-necked pheasant, Phasianus colchicus; safe	0.5–<0.9 mg/kg FW
Common tern, Sterna hirundo; normal reproduction vs. reduced hatching and fledging success	<1.0 mg/kg FW vs. 2.0–4.7 mg/kg FW
Various species; safe	<0.5 to <2.0 mg/kg FW
Waterbirds; adverse effects	1.0–3.6 mg/kg FW
Feather, acceptable	<9.0 mg/kg FW
Methylmercury-poisoned	
Brain	15.0–20.0 mg/kg FW
Liver	20.0–60.0 mg/kg FW
Kidney	20.0–60.0 mg/kg FW
Muscle	15.0–30.0 mg/kg FW

Table 19.1. cont'd

Resource and Other Variables	Criterion or Effective Mercury Concentration
Diet, fish-eating birds	<20.0 μg Hg/kg FW ration, as methylmercury
Diet, fish-eating birds	<100.0 μg total Hg/kg FW in prey fish[e]
Diet, non fish-eating birds	50.0–<100.0 μg Hg/kg FW ration, as methylmercury
Diet, loon	<300.0 μg/kg FW ration[b]
Diet	<1.0–<3.0 mg/kg DW
Daily intake	<640.0 μg/kg body weight (BW)
Daily intake	<32.0 μg/kg BW[c]
MAMMALS	
Daily intake	<250.0 μg/kg BW
Diet; fish-eating mammals	<100.0 μg Hg/kg FW ration, as methylmercury
Diet; fish-eating mammals	<100.0 μg total Hg/kg FW whole prey fish[e]
Diet	<1.1 mg/kg FW ration
Diet; methylmercury poisoning of minks and otters	>1.0 mg/kg FW ration
Diet; minks and otters; adverse effects	>0.12–1.4 mg total mercury/kg FW ration
Drinking water	
Feral and domestic animal water supply, Wisconsin	<0.002 μg/L
Terrestrial vertebrate wildlife	<0.0013 μg/L[d]
Soils; terrestrial ecosystem protection; agricultural and residential land use vs. commercial and industrial use	<2.0 mg/kg DW vs. <30.0 mg/kg DW
Tissue residues	
Acceptable, most species	
Kidney	<1.1 mg/kg FW
Liver, kidney	<30.0 mg/kg FW
Blood	<1.2 mg/kg FW
Brain	<1.5 mg/kg FW
Hair	<2.0 mg/kg FW
Acceptable; otters, *Lutra* spp.	
Hair	<1.0 to <5.0 mg/kg DW
Liver	<4.0mg/kg FW
Mercury-poisoned minks and otters	
Brain	>10.0 mg/kg FW
Liver	>20.0 to >100.0 mg/kg FW
Florida panther, *Felis concolor coryi*; blood	
Reproduction normal (1.46 kittens per female annually)	<250.0 μg/kg FW
Reproduction inhibited (0.167 kittens per female annually)	>500.0 μg/kg FW

Continued

Table 19.1. cont'd

Resource and Other Variables	Criterion or Effective Mercury Concentration
European otter, *Lutra lutra*; liver	
Normal	<4.0 mg/kg FW
Adverse sublethal effects possible	>10.0 mg/kg FW
Wildlife protection, Slovak Republic	
Fat	<1.0 μg/kg FW
Muscle	<50.0 μg/kg FW
Liver, kidney	<100.0 μg/kg FW

[a] All mercury that passes through a 0.45 micrometer membrane filter after the sample is acidified to pH 1.5–2.0 with nitric acid. Derived from bioconcentration factor of 81,700 for methylmercury and the fathead minnow, *Pimephales promelas*.

[b] Reproduction declined in loons, *Gavia immer*, when mercury in prey exceeded 300.0 μg total mercury/kg FW.

[c] No observed adverse effect level with uncertainty factor of 20.

[d] Based on food chain biomagnification in aquatic webs.

[e] Provisional U.S. Fish and Wildlife Service standard for protection of fish-eating wildlife and animals that feed on them. Also considered safe for propagation and maintenance of healthy well-balanced populations of fish and other wildlife.

[f] Based on bioconcentration factor of 40,000 for methylmercury and the American oyster (*Crassostrea virginica*).

Table 19.2. Proposed mercury criteria for the protection of human health.

Variables	Criterion or Effective Mercury Concentration
AIR; SAFE	
California	<0.00 μg/m^3
North Dakota	<0.0005 μg/m^3 for 8 h
Kansas	0.0024 μg/m^3 annually
Montana	<0.008 μg/m^3 for 24 h
Texas	<0.05 μg/m^3 for 1 year
New York	<0.167 μg/m^3 per year
Connecticut	1.0 μg/m^3 for 1 h; <1.0 μg/m^3 for 8 h
Arizona	<1.5 μg/m^3 for 1 h
Virginia	<1.7 μg/m^3 for 24 h
General population; metallic mercury vapor	<15.0 μg/m^3 for 24 h
Workplace	
Organic mercury	<10.0 μg/m^3
Metallic mercury vapor	<50.0 μg/m^3
AIR; ADVERSE EFFECTS POSSIBLE	
Skin	>30.0 μg/m^3
Emissions from individual industrial sites	>2300.0–3200.0 g of mercury daily
French municipal waste incinerator	>0.3 μg/m^3 for >24 h
CHLORALKALI PLANTS, CANADA	
Wastewater effluents	<2.5 g of mercury daily per ton of chlorine produced
Air emissions	<2.0 μg/m^3 daily

Table 19.2. cont'd

Variables	Criterion or Effective Mercury Concentration
DRINKING WATER	
Brazil	<0.2 μg/L
Turkey	<1.0 μg/L
International	<1.0 μg/L
USA, most states	<2.0 μg/L
Bottled water	<2.0 μg/L
USA EFFLUENT LIMITATIONS FROM WASTEWATER TREATMENT PLANTS (μg/L)	<5.0 μg/L
Illinois, Wisconsin	<0.5 μg/L
New Jersey	<2.0 μg/L
Tennessee	<50.0 μg/L
DIET	
Australia	
General diet	<10.0 to <100.0 μg/kg fresh weight (FW) ration
General diet	<20.0 μg/kg FW
Seafoods	<500.0 μg/kg FW
Benelux countries	<30.0 μg/kg FW ration
Brazil	<50.0 μg/kg FW ration
Canada	<500.0 μg/kg FW ration
Malaysia; vegetables; fruit; vegetable and fruit juices; tomato pulp, paste, and puree; tea; coffee, cocoa	<50.0 μg/kg "as consumed"
Thailand	
General foods	<20.0 μg/kg FW
Seafood	<500.0 μg/kg FW
USA	<1000.0 μg/kg FW ration
Japan	
Total mercury intake	
Adverse effects expected	>250.0 μg daily
Nontoxic	<25.0 μg daily; <0.5 μg/kg BW daily
Symptoms of mercury poisoning	Consumption of >500.0 g fish muscle daily containing 10.0 mg methylmercury chloride/kg FW muscle
Methylmercury	<600.0 μg/kg BW daily[b]
PERMISSIBLE TOLERABLE WEEKLY INTAKE	
Total mercury	<5.0 μg/kg BW
Total mercury	Maximum of 4.28 μg/kg BW
Methylmercury	<3.3 μg/kg BW
Methylmercury	<170.0 μg
Methylmercury; Japan; estimated weekly intake	42.0–52.0 μg
Methylmercury	<200.0 μg/60-kg person

Continued

Table 19.2. cont'd

Variables	Criterion or Effective Mercury Concentration
FISH CONSUMPTION ADVISORY; FLORIDA VS. MOST STATES	>0.5 mg total Hg/kg FW edible aquatic product vs. >1.0 mg total Hg/kg FW edible aquatic product
FISH AND SEAFOOD, EDIBLE PARTS	
USA	<300.0 μg total Hg/kg FW
USA	<300.0 μg methylmercury/kg FW
USA; Food and Drug Administration action level	>1.0 mg Hg/kg FW
Florida	
Safe	<500.0 μg/kg FW[c]
Limited consumption	>500.0 μg/kg FW and <1500.0 μg/kg FW[d]
No consumption advised	>1500.0 μg/kg FW
Japan	
Total mercury	<400.0 μg/kg FW
Methylmercury	<300.0 μg/kg FW
Slovak Republic	<500.0 μg/kg FW
Acceptable intake	
60-kg adult	25.0 μg daily
70-kg adult	200.0 μg weekly
Adult	500.0 μg weekly
PREGNANT WOMEN, DIET	
All	<250.0 μg/kg FW
Japan	<400.0 μg/kg FW
Canada, Germany, USA, Brazil	<500.0 μg/kg FW
Italy	<700.0 μg/kg FW
Finland, Israel, Sweden	<1000.0 μg/kg FW
FLORIDA; CONSUMPTION OF CONTAMINATED FISH CONTAINING 2.0–3.0 mg METHYLMERCURY/kg FW MUSCLE	
Pregnant women	Less than 454.0 g of fish muscle weekly
Women of childbearing age and children less than 15 years of age	Less than 454.0 g of fish muscle monthly
SHARK FLESH	
Containing 0.5–1.5 mg Hg/kg FW	Consumption limited to once weekly by healthy nonpregnant adults
Containing more than 1.5 mg Hg/kg FW	Consumption prohibited
FOODS OF ANIMAL ORIGIN	
Livestock tissues	<500.0 μg/kg FW
Wildlife tissues	<50.0 μg/kg FW
Breast muscle	
Domestic poultry	<500.0 μg/kg FW
Ducks wildlife	<1000.0 μg/kg FW

Table 19.2. cont'd

Variables	Criterion or Effective Mercury Concentration
FOODS OF VEGETABLE ORIGIN	
Mercury-treated grain	<1000.0 μg/kg FW
Vegetables	<50.0 μg/kg DW
ALL FOODS; ADULT WEEKLY INTAKE	
As methylmercury	<100.0 μg
As total mercury	<150.0 μg
As methylmercury	<200.0 μg
As total mercury	<300.0 μg
ADULT DAILY INTAKE; NONPREGNANT VS. PREGNANT	<4.3 μg/kg BW vs. <0.6–1.1 μg/kg BW
ORAL DOSE; MAXIMUM TOLERATED CONCENTRATION; ADMINISTERED AS PHENYLMERCURIC ACETATE	8.4 mg Hg/kg body weight daily[a]
HUMAN TISSUE RESIDUES	
Blood	
Recommended, children	<5.0 μg/L
No adverse neurodevelopmental effects in children exposed during pregnancy	<5.8 μg methylmercury/L umbilical cord blood
Acceptable	<5.8 μg/L
Associated with loss of IQ in children exposed during pregnancy	>5.8 μg methylmercury/L umbilical cord blood
Developmental effects observed	>5.8 μg methylHg/L umbilical cord blood
Safe	<200.0 μg/kg FW
No symptoms observed	<200.0 μg/kg FW
Asymptomatic infants, but serious	120.0–630.0 μg/kg FW
Neurological symptoms	>200.0 μg/kg FW
Critical infants	1050.0–>3000.0 μg/kg FW
Some deaths expected adults	>3100.0 μg/kg FW
Brain, neurological symptoms	>6.0 mg/kg FW
Hair	
Developmental effects observed	>1.0 mg/kg DW
Recommended, children	<2.0 mg/kg DW
Safe, China	<4.0 mg/kg FW
Safe	<6.0 mg/kg FW
USEPA reference dose indicative of human exposure	>1.0 mg/kg DW
No observable symptoms	0.0–300.0 mg/kg FW
Three-point reduction in Wechsler Intelligence Scale IQ of New Zealand children exposed as fetus during pregnancy	>6.0 mg/kg DW
No observable symptoms	5.0–30.0 mg/kg FW

Continued

Table 19.2. cont'd

Variables	Criterion or Effective Mercury Concentration
Safe	<50.0 mg/kg FW
Minamata Disease victims	50.0–200.0 mg/kg FW
Mildly affected numbness of extremities, slight tremors; mild ataxia)	120.0–600.0 mg/kg FW
Hearing difficulties, tunnel vision, partial paralysis	200.0–600.0 mg/kg FW
Severe some combination of complete paralysis, loss of vision, loss of hearing, loss of speech, coma)	400.0–1600.0 mg/kg FW
Urine	
Recommended, children	<3.0 μg/L
Recommended, children	<5.0 μg/L
Neurological disturbances	>0.6 mg total Hg/L
Whole body	
Zero clinical effect	<0.7 mg/kg body weight
Effects observed	>100.0 mg accumulated methylmercury
Fatal	>1000.0 mg accumulated methylmercury
SOILS	
Residential and Parklands	<1.0 mg/kg DW
Commercial and Industrial	<2.0 mg/kg DW
Agricultural; Malaysia	<0.362 mg/kg DW
EFFLUENT LIMITATIONS FROM WASTEWATER TREATMENT PLANTS	
Delaware, Oklahoma, Texas	<5.0 μg/L
Illinois, Wisconsin	<0.5 μg/L
New Jersey	<2.0 μg/L
Tennessee	<50.0 μg/L

[a]Assuming continuous exposure until steady-state balance for methylmercury is achieved. This process will take up to 1 year in most cases.
[b]Based on monkey experiments that showed no symptoms at 30.0 mg methylmercury/kg BW daily for 2 years via diet, and an uncertainty factor of 50.
[c]Women of childbearing age should consume <227.0 g (8 ounces) of freshwater fish for over 7 days; children <10 years old should eat <113.0 g during this same period.
[d]Fish consumption should be limited to once monthly by children or women of childbearing age nor more than once weekly by other adults.

19.11.1 Agricultural Crops

Proposed mercury criteria for crop protection (Table 19.1) include <0.2 μg/L in irrigation water, and <0.2–<0.5 mg/kg DW in soils of several countries, although higher levels are allowable in cropland soils of New Jersey (<1.0 mg/kg DW) and the Former Soviet Union (<2.1 mg/kg DW). Sludge and other wastes applied to European soils should contain <1.5 mg Hg/kg, but higher levels of <10.0–25.0 mg Hg/kg are permissible in solid wastes applied to agricultural soils of Iowa, Maine, Vermont, and California (Table 19.1).

19.11.2 Aquatic Life

In 1980, the U.S. Environmental Protection Agency's proposed mercury criteria for freshwater aquatic life protection were 0.00057 μg/L (24-h average), not to exceed 0.0017 μg/L at any time; these criteria seemed to afford a high degree of protection to freshwater biota, as judged by survival, bioconcentration, and biomagnification. Since mercury concentrations in water of 0.1–2.0 μg/L were fatal to sensitive aquatic species and concentrations of 0.03–0.1 μg/L were associated with significant sublethal effects, the 1980 proposed freshwater criteria provided safety factors for acute toxicities of 175–3508 based on the 24-h average, and 58–1176 based on the maximum permissible concentration (Table 19.1). For protection against sublethal effects, these values were 53–175 based on the 24-h mean, and 18–59 based on the maximum permissible concentration (Table 19.1). However, more recent freshwater criteria of 0.012 μg/L, not to exceed 2.4 μg/L (Table 19.1), dramatically reduces the level of protection afforded to aquatic biota: safety factors for acute toxicities are now 8–167 based on the 96-h average, and only 0.04–0.8 based on the maximum permissible concentration. For protection against sublethal effects, these values were 2–8 based on the 4-day average, and only 0.01–0.04 based on the maximum permissible concentration, or essentially no significant protection. The proposed 1980 saltwater criteria for mercury and marine life were unsatisfactory. Proposed saltwater values of 0.025 μg/L (24-h average), not to exceed 3.7 μg/L at any time (Table 19.1), provided safety factors of 4–8O against acute toxicity (based on 24-h average), but less than 0.5 based on the maximum permissible level. For protection against sublethal damage effects, the safety factors computed were 1.2–4 (based on 24-h average) and less than 0.03 based on maximum allowable concentrations). The more recent saltwater criteria of 0.025 μg/L, not to exceed 2.1 μg/L (Table 19.1), does not appear to offer a substantive increase in protection to marine life, when compared to criteria proposed earlier. It seems that some downward modification is needed in the proposed mercury saltwater criteria if marine and estuarine biotas are to be provided even minimal protection.

The significance of elevated mercury residues in tissues of aquatic organisms is not fully understood. Induction of liver metallothioneins and increased translatability of MRNA are biochemical indicators of the response of fish to mercury exposure, and more research is recommended on this and other indicators of mercury stress. Concentrations exceeding 1.0 mg Hg/kg fresh weight can occur in various tissues of selected species of fish and aquatic mammals eaten by humans. But it would be incorrect to assume that aquatic food chains – especially marine food chains – incorporate mercury exclusively from anthropogenic activities. Some organisms, however, do contain mercury tissue residues associated with known adverse effects to the organism and its predators. Thus, whole body residues of 5.0–7.0 mg Hg/kg fresh weight in brook trout eventually proved fatal to that species. To protect sensitive species of mammals and birds that regularly consume fish and other aquatic organisms, total mercury concentrations in these food items should probably not exceed 100.0 μg/kg for avian protection, or 1100.0 μg/kg for small mammals (Table 19.1).

Since long-lived, slow-growing, high-trophic-position aquatic organisms usually contain the highest tissue mercury residues, some fisheries managers have proposed a legal maximum limit based on fish length or body weight, or alternatively, constraining the mean mercury concentration of the entire catch to a nominated level. In the Australian shark fishery, for example, implementation of a maximum length restriction (to a nominated level of 500.0 μg Hg/kg), would result in retention of less than half the present catch of seven species. Also in Australia, a maximum total length of 92 cm is proposed for the taking of yellowtail kingfish (*Seriola grandis*) and would effectively remove 23% of the total weight of the catch and 9% of the numbers. If the total length of the yellowtail kingfish is reduced to 73 cm, a length that ensures that almost all fish contained <500.0 μg Hg/kg FW muscle, this would preclude 59% by weight and 30% by numbers. Other strategies to control mercury burdens in predatory fishes include control of forage fish, overfishing, and various chemical treatments. In lakes with pelagic forage fish, there is less than a 5% probability of finding elevated mercury levels in muscle of lake trout less than 30 cm in total length vs. 45 cm in lakes where pelagic forage fish were absent. In the case of lake trout lakes with no pelagic forage fish, every effort should be made to avoid their introduction. Overfishing of top-level predators is recommended as a means of lowering methylmercury levels in certain types of lakes, and is attributed to the more rapid growth of the predators and by changes in the dietary intake of methylmercury. Treatment of lakes with selenium compounds is one of the few known methods of lowering the mercury content of fish muscle to <1.0 mg Hg/kg FW. Treatments that have achieved partial success in reducing mercury content in fish tissues include liming of lakes, wetlands, and drainage areas. More research is needed on mercury protectants because several are known to cause substantial reductions in tissue mercury concentrations in fishes and plants. Thiamine and various group VI derivatives, including sulfur, selenium, and tellurium compounds protect against organomercury poisoning by their antagonistic effects; thiamine was the most effective of the derivatives against the widest spectrum of organisms and test systems.

More research is also needed on mercury removal technology. In the Florida Everglades, for example, using prototype wetlands of 1545 ha, removal of agricultural nutrients from stormwater reduced total mercury and methylmercury concentrations in water by as much as 70% in the first 2 years of operation; moreover, total mercury concentrations in largemouth bass were about 0.1 mg Hg/kg FW muscle throughout the project site vs. 0.5 mg Hg/kg FW in adjacent areas.

19.11.3 Birds

Tissue residues of mercury, as methylmercury, considered harmful to adult birds ranged between 8.0 mg/kg FW in brain to 15.0 in muscle to 20.0 mg/kg FW in liver and kidney. Among sensitive avian species, adverse effects, mainly on reproduction, have been reported at total mercury concentrations (in mg/kg fresh weight) of 5.0 in feather, 0.9 in egg, 0.05–0.1 in diet, and daily administered doses of 0.64 mg/kg body weight (Table 19.1). The proposed mercury concentration to protect sensitive species of birds that regularly consume fish and aquatic invertebrates is <0.1 mg/kg FW in these food items (Table 19.1).

Although low mercury concentrations – e.g., 0.05 mg/kg in the diets of domestic chickens – sometimes produced no adverse effects on chickens, the tissue residues of mercury were sufficiently elevated to pose a hazard to human consumers. In contrast, with eggs of the bald eagle (containing 0.15 mg Hg/kg FW and low hatch) it is probable that other contaminants present – especially organochlorine compounds – had a greater effect on hatchability than did mercury.

19.11.4 Mammals

Mammals, such as the domestic cat and the harp seal, showed birth defects, histopathology, and elevated tissue residues at doses of 250.0 μg hg/kg body weight daily (Table 19.1). Tissue residues of mercury, as methylmercury, considered harmful to adult inland mammals ranged between 8.0 mg/kg FW in brain to 15.0 in muscle to 20.0 mg/kg FW in liver and kidney. Mink fed dietary levels of 1.1 mg Hg/kg had signs of mercury poisoning; mercury residues in mink brain at this dietary level ranged from 7.1 to 9.3 mg/kg. Tissue residues in kidney, blood, brain, and hair in excess of 1.1 mg Hg/kg in nonhuman mammals are usually considered presumptive evidence of significant mercury contamination (Table 19.1). In order to protect sensitive species of small mammals that regularly consume fish and other aquatic organisms, total mercury concentrations in these food items should probably not exceed 100.0 μg total mercury/kg FW (Table 19.1).

For most species of mammals, recommended mercury criteria include daily intake of <250.0 μg total mercury/kg body weight, diets that contain <1.1 mg total mercury/kg FW, and for livestock, <0.002 μg total mercury/L in the drinking water supply (Table 19.1). Tissue mercury concentrations in sensitive mammals, in mg total mercury/kg FW, should probably not exceed 10.0 in liver, 2.0 in hair, 1.5 in brain, and 0.5 in blood (Table 19.1).

19.11.5 Human Health

Proposed mercury criteria for the protection of human health are numerous and disparate (Table 19.2). Proposed mercury air criteria, for example, in the workplace range from <10.0 μg/m^3 for organomercury compounds to <50.0 μg/m^3 for elemental mercury vapor; however, much lower criteria are proposed by Texas (<0.05 μg/m^3 for 1 year), New York (<0.167 μg/m^3 per year), and other jurisdictions (Table 19.2). Drinking water criteria for total mercury range between <1.0 and <2.0 μg/L, except for Brazil with <0.2 μg/L. Current fish consumption recommendations are based on risk assessments for children and women of child-bearing age. Dietary criteria for mercury range between 10.0 and 1500.0 μg/kg FW, with lower values associated with seafoods, organomercurials, and pregnancy. Accordingly, proposed mercury levels in fish and seafood should not exceed 250.0 μg/kg FW for expectant mothers, and 400.0–1000.0 μg/kg FW for adults worldwide (Table 19.2). It is again emphasized that total mercury concentrations exceeding 1.0 mg/kg fresh weight naturally occur in edible tissues of some species of fish and aquatic mammals regularly eaten by humans. Proposed tolerable weekly intake ranges between <3.3 and 5.0 μg/kg body weight (Table 19.2). Daily mercury intake for pregnant women should not exceed <0.6–1.1 μg total mercury/kg body weight vs. 4.3 μg/kg BW for others. It has been suggested that humans may safely ingest up to 8.4 mg Hg/kg body weight daily, but this requires verification. The minimum toxic intake for humans is estimated to range between 0.6 and 1.1 μg methylmercury/kg body weight daily. Recommended total mercury concentrations in human tissues include <5.8 μg/L in blood, <0.7 mg/kg whole body, and <6.0–50.0 mg/kg in hair (Table 19.2).

Methylmercury concentration in scalp hair during pregnancy is considered to be the most reliable indicator for predicting the probability of psychomotor retardation in the child. The U.S. Environmental Protection Agency has established a Reference Dose of 1.0 mg total Hg/kg DW in hair as indicative of mercury exposure; at this level, women of child-bearing age are advised to stop consumption of fish that may have elevated mercury levels; however, these results should be considered preliminary pending analysis of additional data and because current mercury criteria for human hair range between <6.0 and 50.0 mg/kg DW (Table 19.2). Other studies suggest that methylmercury concentration in cord blood is a sensitive biomarker of exposure *in utero* and correlates well with neurobehavioral outcomes; moreover, concentrations >5.8 μg methylmercury/L cord blood are associated with loss of IQ in children exposed during pregnancy (Table 19.2).

A major source of mercury vapor in the general population is dental amalgam and several European countries – Austria, Germany, Finland, Norway, United Kingdom, Sweden, etc. – have advised their dentists to specifically curtail installation of mercury-containing amalgam fillings in pregnant women. In a study conducted in Washington State, there was no evidence that mercury-containing dental fillings placed during pregnancy increased the risk of low birth weight. Further evaluations of the safety profiles of dental amalgams, and other types of materials used intraorally, are needed to establish guidelines for use of dental materials in expectant mothers.

In terms of potential health hazards, the dangerous mercury level in human blood is about 2.0 mg/kg; the normal background in about 80% of people without occupational exposure to mercury is <0.005 mg/kg. Since methylmercury has a biological half-life in humans of 70–200 days, it was not unexpected that mutagenic and teratogenic effects of mercury have been reported at levels well below those associated with poisoning. The findings of elevated blood mercury levels, increased chromosome breakage, and easy passage of mercury through placental membranes among women who regularly eat fish containing 1.0–17.0 mg Hg/kg fresh weight (FW) suggest to some health authorities that the previous mercury recommendation level in the U.S. of 0.5 mg Hg/kg FW (recently reduced to 0.3 mg Hg/kg FW) does not have an appreciable level of safety for pregnant women. In Sweden, the official limit for mercury in fish muscle is 1.0 mg/kg FW. Human risk in consumption of mercury-contaminated fish within this guideline is considered negligible in that country; however, some Scandinavian scientists recommend that fish consumption should be limited to one meal a week. It has been advocated that human use of mercury should be curtailed because the difference between tolerable background levels of mercury and harmful levels in the environment is exceptionally small. Specifically, it has been recommended that all usages of methylmercury compounds should be prohibited and that production of other organomercury compounds be drastically reduced.

In Canadian aboriginal people, a 20-year follow-up study on methylmercury levels has been initiated, with emphasis on age, sex, location, relation between maternal and fetal levels, and a reassessment of potential risk in communities where the highest known methylmercury levels have been found. Similar studies are recommended for avian and mammalian wildlife.

At this point, it seems that four courses of action are warranted for protection of human health and sensitive natural resources. First, toxic mercurials in agriculture and industry should be replaced with less toxic substitutes. In Sweden, for example, clinical mercury thermometers have been prohibited since January 1, 1992, for import, manufacture, and sales. Since January 1993, the same prohibition was applied to other measuring instruments and electrical components containing mercury. Since 2000, Sweden has prohibited mercury in all processes and products, including thermometers and sphygmomanometers, and replaced them with available substitutes. In Quebec (Canada) hospitals, medical instruments containing mercury are being replaced with mercury-free instruments because of inadequate maintenance and disposal of existing instruments. Second, controls should be applied at the point of origin to prevent the discharge of potentially harmful mercury wastes. Point sources need to be identified and regulated. In Sweden, discharges from point sources in the 1950s and 1960s averaged 20–30 metric tons annually. Since the end of the 1960s, the annual emission of mercury in Sweden has been reduced to about 3.5 tons through better emission control legislation, improved technology, and reduction of polluting industrial production. Third, continued periodic monitoring of fishery and wildlife resources is important, especially in areas with potential for reservoir development, in light of the hypothesis that increased flooding increases the availability of mercury to biota. The use of museum collections for mercury analysis is strongly recommended for monitoring purposes. For example, the Environmental Specimen Bank at the Swedish Museum of Natural History constitutes a base for ecotoxicological research and for spatial and trend monitoring of mercury and other contaminants in Swedish fauna. And finally, additional research is needed on mercury accumulation and detoxification in comparatively pristine ecosystems. Key uncertainties in understanding the process of mercury uptake in aquatic ecosystems, for example, include relations between water chemistry and respiratory uptake, quantitative estimates of intestinal tract methylation and depuration, and degree of seasonal variability in mercury speciation and methylation-demethylation processes.

In the specialized case of environmental mercury contamination from historic and current gold mining activities, more research seems necessary on physical and biological mercury removal technologies, development of nonmercury technologies to extract gold with minimal environmental damage, measurement of loss rates of mercury through continued periodic monitoring of fishery and wildlife resources in mercury-contaminated areas, and mercury accumulation and detoxification rates in comparatively pristine ecosystems. In view of the demonstrable adverse effects of uncontrolled mercury releases into the biosphere from gold production, it is imperative that all use of liquid mercury in gold amalgamation should cease at the earliest opportunity and that the ban be made permanent.

19.12 Summary

Mercury has been mined continuously for at least 2400 years for use in gold recovery (until the present time), in the manufacture of felt hats and mirrors (1700s), in the chloralkali industry to manufacture chlorine and caustic (since the 1800s), as a fungicide in agriculture and paper production (1900s), and currently in lighting fixtures, batteries, paints, dentistry, and in medicine to kill bacteria and cutaneous parasites. Major anthropogenic sources of mercury to the biosphere include combustion of fossil fuels from power plants, municipal solid waste combustion facilities, medical waste incinerators, paint manufacturing, and gold extraction. Combustion of mercury-containing fossil fuels may account for up to 60% of the global mercury burden from human activities. Use of mercury to amalgamate gold in Amazonia has resulted in mercury contamination of at least 500,000 miners and numerous fish and wildlife populations. And use of inorganic mercury catalysts in chloralkali plants has caused contamination of fish in waterways and elevated blood and mercury levels in Canadian Amerindians who consumed these fish.

World production of mercury in recent years is estimated at 10,000–15,000 metric tons annually; major producers of mercury now include the former Soviet Union, Spain, the former Yugoslavia, and Italy. The total amount of mercury in various global reservoirs is estimated at 334 billion metric tons, mostly in ocean sediments (98.75%) and ocean waters (1.24%). Only 7 metric tons of mercury is believed to be present in living aquatic organisms. During the past 100 years, an estimated 500,000 tons of mercury entered the biosphere, with eventual deposition in subsurface sediments. Mercury inputs to the biosphere are mainly from natural sources, but with significant and increasing amounts contributed from human activities. Soil bacteria and terrestrial plants expedite flux rates of elemental mercury from the geosphere to the atmosphere. The atmosphere plays an important role in the mobilization of mercury, with about half the total atmospheric mercury burden of anthropogenic origin. Atmospheric transport of mercury may contaminate remote ecosystems hundreds of kilometers distant.

Physical, chemical, biological, and biochemical properties of mercury are briefly reviewed; mercury transport and speciation processes are summarized; and analytical techniques listed for mercury measurement. In mammals, all forms of mercury can cross the placenta to the fetus and interfere with thiol metabolism. Chemical speciation is probably the most important variable influencing mercury toxicity; however, speciation is difficult to quantify. In freshwater lakes, for example, mercury speciation depends, in part, on pH, alkalinity, redox, and microbial activity. Most authorities agree that all mercury species discharged into natural bodies of water can be converted into methylmercurials – the most toxic form – at rates influenced, in part, by mercury loadings, nutrient content, sedimentation rates, and suspended sediment loadings. Bioavailability of methylmercury to aquatic biota is highly dependent on lake chemistry, mercury deposition rates, dissolved organic carbon, and other variables.

Methylmercury, in turn, is decomposed abiotically and by bacteria containing mercuric reductase and organomercurial lyase; these demethylating strains of bacteria are common in the environment and have been isolated from water, sediments, soils, and the GI tract of humans and other mammals. Other mechanisms can reduce inorganic Hg^{2+} to Hg in freshwater, with Hg production higher under anoxic conditions, in a chloride-free environment, and at pH 4.5. Methylmercury concentrations in tissues of marine fishes can now be detected at levels $>10.0\,\mu g/kg$ tissue using graphite furnace sample preparation techniques and atomic absorption spectrometry.

Both inorganic and organomercurials interfere with membrane permeability and enzyme reactions through binding of mercuric ion to sulfhydryl groups; organomercurials usually penetrate membranes more readily. Symptoms of acute and chronic mercury poisoning caused by elemental mercury, mercuric mercury, mercurous mercury, and organomercurial compounds are listed, mechanisms of action discussed, and treatment regimes prescribed. For elemental mercury, inhalation of Hg is the primary toxicological route. Inhaled Hg vapor is readily oxidized within the body, mainly in liver and erythrocytes, and converted to Hg^{2+}. Neurobehavioral disturbances were observed in some Hg vapor poisoning cases 20–35 years after exposure. For inorganic mercuric compounds, exposure routes include inhalation and ingestion, with primary damage to the renal system. Mercurous (Hg^{+}) mercury compounds are unstable and degrade to Hg and Hg^{2+}. Mercurous compounds are less corrosive and less toxic than mercuric compounds and this could be associated with their comparatively low solubility. Among the organomercurials, methylmercury compounds (CH_3Hg^{+}) are the most significant toxicologically because they are produced naturally from inorganic mercury by microbial activity and are lipid soluble, thus readily crossing blood-brain and placental barriers. Ingestion is the main route of administration for methylmercurials and the primary target organs are brain and other neurological tissues.

Treatment of mercury-poisoned victims is complex and should be supervised by a physician. Therapy is directed to lowering the mercury concentration at the critical organ or site of injury through emesis, lavage, cathartics, administration of activated charcoal and various mercury chelating agents, and – in the most severe cases – dialysis. The development of mercury-antagonistic and mercury-protectant drugs is proceeding, and some already available have been used to treat cases of inorganic mercury poisoning (thiols) and organomercurial poisoning (thiamin, and selenium-, sulfur-, and tellurium-containing drugs).

Maximum concentrations of mercury recorded were less than 50.0 ng/L in uncontaminated natural waters; 60.0 ng/L in snow; 78.0 ng/L in coastal seawater; 89.7 ng/L in rain; 600.0 ng/L in groundwater; 1100.0 ng/L in freshwaters near active gold mining sites; 450,000 ng/L in drainage water from mercury mines; 495.0 ng/m^3 in air over Japanese volcanoes (74% as Hg^{2+}); 0.12–1.0 mg/kg dry weight in coal; 0.39 mg/kg fresh weight in pelagic clays; 130.0 mg/kg dry weight in Missouri, sewage sludge; 150.0 mg total mercury (2.7 mg methylmercury)/kg dry weight suspended particulate matter in water contaminated by chloralkali wastes in Germany; 329.9 mg total mercury/kg dry weight (0.045 mg methylmercury/kg dry weight) in soils contaminated by mercury-containing wastewater from a Chinese acetaldehyde plant; and 746.0 mg/kg dry weight in sediments near a Finnish pulp and paper mill where mercury was used as a slimicide.

Mercury concentrations in world-wide field collections of plants and animals are documented; factors known to modify mercury accumulations and their significance are discussed.

Aquatic and terrestrial plants usually contain less than 1.0 mg total mercury/kg dry weight (DW), except in the vicinity of naturally occurring cinnabar deposits and various anthropogenic activities such as smelters, sewage lagoons, chloralkali plants, newly formed reservoirs, and applications of mercury-containing agricultural chemicals. Some vegetation from impacted environments had up to 70.0 mg Hg/kg DW and 59.0 mg Hg/kg fresh weight (FW); these species may be useful in mercury phytoremediaton removal from mercury-contaminated sites. Invertebrates collected near industrial, municipal, and other known sources of mercury contained up to 10.0 mg Hg/kg FW and 38.7 mg Hg/kg DW; conspecifics from reference sites usually had <0.5 mg Hg/kg FW. Passage of environmental legislation and effective enforcement were considered instrumental in reducing mercury concentrations in mussels from Bergen Harbor, Norway, by 60% in 10 years.

In bony fishes and elasmobranchs, mercury tends to accumulate in muscle, mainly as methylmercury. Accumulations in muscle and other tissues increase with increasing age of the fish, and is highest in carnivorous species.

Concentrations exceeding 2.0 mg Hg/kg FW in muscle of some wide-ranging oceanic fishes – such as tuna, marlin, and swordfish – were common owing to a combination of human activities and natural processes. For most fish products sold, muscle mercury concentrations were less than 0.3 mg Hg/kg FW; however, about 2% of the total catch landed may contain more than 0.5 mg Hg/kg FW.

Amphibians usually contained less than 0.5 mg Hg/kg FW in various organs; however, frogs collected near a mercury mine in the former Yugoslavia had 24.0 mg Hg/kg FW in kidney and 25.5 mg Hg/kg FW in liver. In reptiles, most of the mercury in tissues was organic mercury; concentrations were highest in liver, kidney, muscle, and egg, in that order. Alligators from the mercury-contaminated Florida Everglades had the highest mercury concentrations recorded in reptiles: 6.1 mg/kg FW in muscle (which is eaten locally), 65.3 in kidney, and 99.5 mg Hg/kg FW in liver. Farm-raised alligators in Florida always had <0.2 mg Hg/kg FW in tissues. In birds, the highest mercury concentrations measured were: 187.0 mg/kg FW in liver of male loons found dead in New England; 175.0 mg/kg FW in liver of great blue herons from the mercury-contaminated Lake. St. Clair; 175.0 mg/kg FW in liver of diving ducks from a mercury-contaminated ecosystem; 130.0 mg/kg FW in kidney of ospreys from the eastern U.S.; 96.0 mg/kg FW in liver of grey herons from England; 306.0 mg/kg DW in liver of seabirds; and from New Zealand 295.0 mg/kg DW in albatross liver and 140.0 mg/kg DW in petrel liver. Elevated concentrations of mercury in avian tissues were associated with low organ and body weight, inhibited reproduction, and decreased activities of enzymes related to glutathione metabolism and antioxidant activities. Feathers usually had the highest mercury concentrations, followed by liver, then muscle, often in the ratio of 7:3:1; however, almost all mercury in feathers is organic and in liver it is inorganic mercury. Many factors modify mercury concentrations in avian tissues, including food preference and availability, migration patterns, age and sex, mercury loadings in the immediate biosphere, season of collection, proximity to industrialized areas, inherent species differences, molting stage, general health, and interactions with selenium.

In humans, increasing mercury concentrations in hair and tissues are primarily a result of increasing consumption of fish and shellfish, and to a lesser degree to dental amalgams.

Pregnant aboriginal women who routinely consume seal meat and blubber with high mercury concentrations throughout pregnancy had elevated mercury concentrations in maternal and fetal blood and other tissues without apparent effect on the fetus or the resultant infant; however, this requires verification. Total mercury concentrations in hair >1.0 mg/kg DW are now considered indicative of mercury exposure and is the recommended concentration at which women of child-bearing age should restrict consumption of fish, according to the U.S. Environmental Protection Agency. However, more than 45% of New Yorkers and 34% of Floridians now exceed 1.0 mg Hg/kg DW hair, suggesting reexamination of this value. Mercury poisoning in humans is associated with total mercury concentrations in hair >249.0 mg/kg FW and methylmercury concentrations in blood >3.1 mg/L. Among nonhuman mammals, mercury concentrations were highest 143.0–765.0 mg total Hg/kg (FW) in tissues of marine pinnipeds, especially in livers of older seals and sea lions; accumulations were not attributed to anthropogenic activities and did not seem to pose a significant threat to pinniped health or to human consumers. Among land mammals, livers of the endangered Florida panther had up to 110.0 mg Hg/kg FW, possibly from consumption of mercury-contaminated raccoons. Mercury data sets on abiotic materials and biological tissues from a single collection area, a short sampling period, and chemical analysis by the same research team are particularly useful in establishing food chain dynamics, identifying areas of probable health concerns, measuring geographic areas of impact, and in predicting future problem locations.

For all organisms tested, early developmental stages were the most sensitive, and organomercury compounds – especially methylmercurials – were more toxic than inorganic forms. Numerous biological and abiotic factors modify the lethality of mercury compounds, sometimes by an order of magnitude or more, but the mechanisms of action are not clear. Lethal concentrations of total mercury to sensitive, representative organisms varied from 0.1 to 2.0 μg/L of medium for aquatic fauna; from 2.2 to 31.0 mg/kg body weight (acute oral) and 4.0 to 40.0 mg/kg (dietary) for birds; and from 0.1 to 0.5 mg/kg body weight (daily dose) and 1.0 to 5.0 mg/kg diet for mammals.

Mercury is a known mutagen, teratogen, and carcinogen. At comparatively low concentrations in vertebrate animals it adversely affects reproduction, growth, development, behavior, blood and serum chemistry, motor coordination, vision, hearing, histology, and metabolism. Mercury has a high potential for bioaccumulation and biomagnification, and is slow to depurate. Organomercury compounds were more effective in producing adverse effects than were inorganic mercury compounds; however, effects were significantly enhanced or ameliorated by numerous biotic and nonbiological modifiers. For sensitive aquatic species, adverse effects on growth and reproduction were observed at water concentrations of 0.03–0.1 μg Hg/L. For sensitive species of birds, adverse effects – mainly on reproduction – were associated with daily intake of 640.0 μg Hg/kg body weight, dietary levels of 50.0–100.0 μg Hg/kg fresh weight diet, and with total mercury concentrations >5.0 mg Hg/kg fresh weight feather, and >0.9 mg/kg FW egg. Sensitive nonhuman mammals showed significant adverse effects of mercury when daily intakes were >250.0 μg/kg body weight, when dietary levels were >1.1 mg/kg fresh weight diet, or when tissue concentrations exceeded 1.1 mg/kg fresh weight.

Up to 600 tons of methylmercury were discharged into Minamata Bay, Japan, between 1932 and 1971 from acetaldehyde manufacturing plants. Human fatalities were documented beginning in 1953 from consumption of methylmercury-contaminated fish and shellfish from the Bay. By 1993, about 2000 victims of Minamata Disease were identified including more than 100 deaths and 59 congenital birth defects from a total regional population of about 200,000; however, at least 10,000 additional cases are pending. Mercury levels in the Minamata Bay ecosystem are now near normal as a result of dredging and natural processes.

Mercury contamination of the environment from historical and ongoing mining practices that rely on mercury amalgamation for gold extraction is widespread. Contamination was particularly severe in the immediate vicinity of gold extraction and refining operations; however, mercury – especially in the form of water-soluble methylmercury – may be transported to pristine areas by rainwater, water currents, deforestation, volatilization, and other vectors. Examples of gold mining-associated mercury pollution are shown for Brazil and the U.S. In parts of Brazil, for example, mercury concentrations in all abiotic materials, plants, and animals – including endangered species of mammals and reptiles – collected near ongoing mercury-amalgamation gold mining sites were far in excess of allowable mercury levels promulgated by regulatory agencies for the protection of human health and natural resources. Although health authorities in Brazil are unable to detect conclusive evidence of human mercury intoxication, the potential exists in the absence of mitigation for epidemic mercury poisoning of the mining population and environs. In the U.S., environmental mercury contamination is mostly from historical gold mining practices; however, portions of Nevada now remain sufficiently mercury-contaminated to pose a hazard to reproduction of carnivorous fishes and fish-eating birds.

In order to protect sensitive agricultural crops, mercury concentrations should not exceed 0.2 μg/L in irrigation water, 0.2–<0.5 mg/kg dry weight (DW) in soils, or 1.5 mg/kg DW in sludges applied to croplands. For protection of aquatic life, different mercury criteria have been recommended. However, many regulatory authorities argue that satisfactory protection is afforded with total mercury concentrations less than 0.012 μg/L in freshwater, <0.025 μg/L in marine environments, <0.15 mg/kg DW in marine sediments, and mercury tissue concentrations less than 70.0 μg/kg fresh weight (FW) in eggs, <480.0 μg/kg FW in gonads, and <1.0 mg/kg

FW in whole organism. Mercury concentrations considered acceptable to protect representative species of birds, in mg total mercury/kg FW, include less than <0.5–<1.0 in eggs, <5.0 in feather, <5.0 in liver, and <20.0 in kidneys. Diets of sensitive fish-eating birds should contain less than 20.0 μg Hg as methylmercury/kg FW or <100.0 μg total mercury/kg FW; daily intake should not exceed 640.0 μg total mercury/kg body weight. For most species of nonhuman mammals, recommended mercury levels include daily intake of <250.0 μg total mercury/kg FW body weight, diets that contain <1.1 mg total mercury/kg FW, and <0.002 μg total mercury in drinking water supply of livestock; however, diets of fish-eating mammals should contain <100.0 μg total mercury/kg FW. Tissue mercury concentrations in sensitive mammals, in mg total mercury/kg FW, should probably not exceed 10.0 in liver, 2.0 in hair, 1.5 in brain, and 0.5 in blood. All these proposed criteria provide, at best, minimal protection.

Proposed mercury criteria for human health protection are numerous and disparate. For example, mercury criteria for air in the workplace range from <10.0 μg/m^3 for organomercurials to <50.0 μg/m^3 for metallic mercury vapor; however, much lower criteria are proposed by Texas (<0.05 μg/m^3 for 1 year), New York (0.0167 μg/m^3 per year), and others. Mercury drinking water criteria range between <1.0 and <2.0 μg total mercury/L, except for Brazil with <0.2 μg/L. Proposed dietary criteria for mercury range between 10.0 and 1500.0 μg mercury/kg FW ration (exception <250.0 μg/kg FW diet for pregnant women), with lower values associated with seafood and organomercurials. Permissible tolerable weekly mercury intakes proposed for the general population range between <3.3 and 5.0 μg/kg body weight; however, daily intake for pregnant women should be restricted to <0.6–1.1 μg total mercury/kg body weight. Proposed tolerable mercury concentrations include <0.2 mg/kg FW in blood, <0.7 mg/kg whole body, and <6.0–50.0 mg/kg in hair; however, adverse neurobehavioral effects have been associated with cord blood concentrations >5.8 μg methylmercury/L and >1.0 mg methylmercury/kg DW hair.

MIREX[a]

Chapter 20

20.1 *Introduction*

Fish and wildlife resources associated with approximately 51 million ha (125 million acres) in the southeastern U.S., and with the Great Lakes, especially Lake Ontario, have been negatively affected by intensive or widespread use of mirex, a chlorinated hydrocarbon compound. Contamination of the Southeast and of Lake Ontario by mirex probably occurred between 1959 and 1978. During that period, mirex was used as a pesticide to control the red imported fire ants (*Solenopsis invicta*) and black imported fire ants (*Solenopsis richteri*), which infested large portions of Alabama, Arkansas, Florida, Georgia, Louisiana, Mississippi, North Carolina, South Carolina, and Texas. Under the trade name of Dechlorane, mirex was used as a fire retardant in electronic components, fabrics, and plastics; effluents from manufacturing processes resulted in the pollution of Lake Ontario. Regulatory agencies, environmentalists, and the general public became concerned as evidence accumulated demonstrating that mirex was associated with high death rates, numerous birth defects, and tumors and that it disrupted metabolism in laboratory mammals, birds, and aquatic biota. Mirex also tends to bioaccumulate and to biomagnify at all trophic levels of food chains. Field studies corroborated the laboratory findings and showed that mirex appeared to be one of the most stable and persistent organochlorine compounds known, being resistant to chemical, photolytic, microbial, metabolic, and thermal degradation processes. Upon degradation, a series of potentially hazardous metabolites are formed, although it is generally acknowledged that the fate and effects of the degradation products are not fully understood. Mirex was also detected in human milk and adipose tissues at low concentrations, the levels related to the degree of environmental contamination. In 1978, the U.S. Environmental Protection Agency (EPA) banned all uses of mirex. It is probable that mirex and its metabolites will continue to remain available to living organisms in this country for at least 12 years, although some estimates range as high as 600 years.

20.2 *Chemical Properties*

Mirex is a white, odorless, free-flowing, crystalline, nonflammable, polycyclic compound composed entirely of carbon and chlorine; the empirical formula is $C_{10}Cl_{12}$, and the molecular weight 545.54. In the U.S., the common chemical name is dodecachloro-octahydro-1,3,4-metheno-2H-cyclobuta (c,d) pentalene; the systematic name is dodecachloropentacyclo $5.3.0.0^{2,6}.0^{3,9}.0^{4,8}$decane (Figure 20.1). Mirex was first prepared in 1946, patented in 1955 by Allied Chemical Company, and introduced in 1959 as GC 1283 for use in pesticidal formulations against hymenopterous insects, especially ants. It was also marketed under the trade name of Dechlorane for use in flame-retardant coatings for various materials. Mirex is also known as ENT 25719, CAS 2385-85-5, Dechlorane 510, and Dechlorane 4070.

[a]All information in this chapter is referenced in the following sources:

Eisler, R. 1985. Mirex hazards to fish, wildlife, and invertebrates: a synoptic review. *U.S. Fish Wildl. Serv. Biol. Rep. 85(1.1)*, 42 pp.

Eisler, R. 2000. Mirex. Pages 1133–1157 in *Handbook of Chemical Risk Assessment: Health Hazards to Humans, Plants, and Animals. Volume 2, Organics*. Lewis Publishers, Boca Raton, Florida.

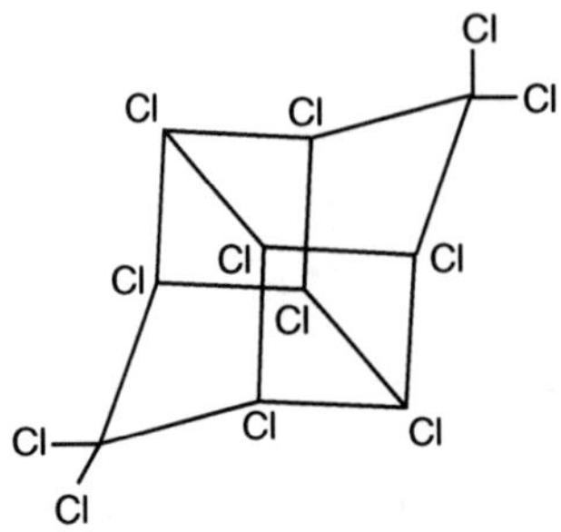

Figure 20.1.
Structural formula of mirex.

Technical grade preparations of mirex consists of 95.19% mirex and less than 2.58×10^{-7}% contaminants, mostly kepone $C_{10}Cl_{10}O$. Mirex is comparatively soluble in various organic solvents, such as benzene, carbon tetrachloride and xylene, with solubilities ranging from about 4000.0 to 303,000.0 mg/L. However, mirex has a very low solubility in water, not exceeding 1.0 µg/L in freshwater or 0.2 µg/L in seawater. In biological systems, mirex lipophilicity would account for the high concentrations observed in fatty tissues and reserves.

Mirex, which is composed of 22% carbon and 78% chlorine, is highly resistant to chemical, thermal, and biochemical degradation. It is reportedly unaffected by strong acids, bases, and oxidizing agents, and is resistant to photolysis in hydrocarbon solvents, but less so in aliphatic amines. Thermal decomposition begins at about 550°C and is rapid at 700°C; degradation products include hexachlorobenzene, hexachlorocyclopentadiene, and kepone. Several additional degradation products of mirex have been isolated, but not all have been identified. At least one photodegradation product, the 8-monohydro analog, sometimes accumulates in sediments and animals, but the fate and effects of these photoproducts is unclear.

Mirex is rapidly adsorbed onto various organic particles in the water column, including algae, and eventually is removed to the sediments. Not surprisingly, mirex has a long half-life in terrestrial and aquatic sediments; large fractional residues were detected at different locations 12 and 5 years after initial application. Some degradation of mirex to the 10-monohydro analog was reported in anaerobic sewage sludge after 2 months in darkness at 30°C. Other studies with mirex-contaminated anaerobic soils, anaerobic lake sediments, and soil microorganisms showed virtually no bacterial degradation over time. In Lake Ontario, mirex from contaminated sediments remained available to lake organisms for many years and, as judged by present sedimentation rates, mirex may continue to be bioavailable for 200–600 years in that system. Disappearance of mirex from baits over a 12-month period was about 41% for those exposed on the ground, 56% from those exposed in soil, and 84% from those exposed in pond water. Mirex disappearance was probably related to uptake by biological organisms, as has been demonstrated in marine ecosystems contaminated with mirex, and not to degradation.

Mirex is a highly stable chlorinated hydrocarbon with lipophilic properties, and its accumulation and persistence in a wide variety of nontarget biological species has been well documented. The biological half-life of mirex reportedly ranges from 30 days in quail to 130 days in fish and to more than 10 months in the fat of female rats; this will be discussed later. At this juncture, it is sufficient to state that most authorities agree on two points: (1) there is little evidence of significant mirex metabolism and (2) mirex ranks among the more biochemically stable organic pesticides known.

20.3 Lethal Effects

Biocidal properties of mirex to aquatic organisms, birds, and mammals are listed below.

20.3.1 Aquatic Organisms

Aquatic organisms are comparatively resistant to mirex in short-term toxicity tests. Among various species of freshwater biota, LC50 (96 h) values were not obtained at the highest nominal concentrations tested of 1000.0 µg/L for insects, daphnids, and amphipods, and 100,000.0 μg/L for five species of fish. Similar results were reported for other species of freshwater invertebrates and fishes, although waterborne mirex at concentrations of 1000.0 μg/L was lethal to post-larval freshwater prawns (*Macrobrachium rosenbergerii*) in 24 h. It is probable that bioavailable concentrations from the water in each test did not exceed 1.0 μg/L. However, delayed mortality frequently occurs for extended periods after exposure, and the potential for adverse effects at the population level remains high. Latent biocidal properties of mirex were documented for fish. Crustaceans were the most sensitive group examined. For example, the crayfish (*Procambarus blandingi*) immersed in nominal concentrations of 0.1–5.0 μg mirex/L for periods of 6–144 h died 5–10 days after initial exposure. Immature crayfish were more sensitive than adults were, and mortality patterns were similar when mirex was administered in the water or in baits.

20.3.2 Birds and Mammals

Acute oral toxicity of mirex to warm blooded organisms was low, except for rats and mice, which died 60–90 days after treatment with 6.0–10.0 mg mirex/kg body weight. Birds were comparatively resistant. The red-winged blackbird (*Agelaius phoeniceus*) was unaffected at 100.0 mg mirex/kg body weight, even though it was considered the most sensitive of the 68 species of birds tested with 998 chemicals for acute oral toxicity, repellency, and hazard potential.

Mortality due to dietary mirex is variable among species, although high death rates were usually associated with high dietary concentrations and long exposure periods. One significant effect of mirex fed to breeding adult chickens, voles, and rats was a decrease in survival of the young. Prairie voles (*Micropterus ochrogaster*) fed diets containing 15.0 mg mirex/kg ration bred normally, but all pups died by day 21. Survival of the pups of prairie voles decreased in the first litter when the diet of the parents contained 10.0 mg mirex/kg ration, in the second litter when it contained 5.0 mg/kg, and in the third litter when it contained 0.1, 0.5, 0.7, or 1.0 mg/kg.

20.4 Sublethal Effects

A variety of adverse sublethal effects of mirex to aquatic organisms, birds, and mammals are documented, including effects on growth, reproduction, embryonic development, behavior, and metabolism.

20.4.1 Aquatic Organisms

The maximum acceptable toxicant concentration (MATC) values calculated for mirex and various freshwater species were <2.4 μg/L for amphipods (*Gammarus* sp.), based on growth inhibition at higher concentrations; 2.0–3.0 μg/L for fathead minnows (*Pimephales promelas*), as judged by disruption of swim bladder hydroxyproline content, vitamin C metabolism, and bone collagen; 34.0 μg/L for fathead minnows, based on impaired reproduction; and >34.0 μg/L for daphnids (*Daphnia* sp.) and midges (*Chaoborus* sp.), predicated on daphnid reproduction and midge emergence. Other mirex-induced sublethal effects included reduced photosynthesis in freshwater algae, gill and kidney histopathology in the goldfish (*Carassius auratus*), reduced growth in the bluegill (*Lepomis macrochirus*), cessation of reproduction in *Hydra* sp., and disrupted behavior in the blue crab (*Callinectes sapidus*) and the marine annelid (*Arenicola cristata*). Channel catfish (*Ictalurus punctatus*) are particularly resistant to high dietary concentrations of mirex; juveniles fed 400.0 mg mirex/kg ration for 4 weeks showed no significant changes in enzyme-specific activities of brain, gill, liver, or muscle. However, yearling coho salmon (*Oncorhyncus kisutch*) fed 50.0 mg mirex/kg ration for 3 months showed significant reduction in liver weight and whole-body lipid content. Additional studies with coho salmon and rainbow trout (*Oncorhynchus mykiss*) fed 50.0 mg mirex/kg ration for 10 weeks demonstrated a significant depression in serum calcium, and significant elevation of skeletal magnesium in salmon; trout showed no measurable changes in calcium and magnesium levels in serum, muscle, or skeleton, although growth was reduced, muscle water content was elevated, and muscle lipid content was reduced. Interaction effects of mirex with other anthropogenic contaminants are not well studied, despite the observation that mixtures of DDT and mirex produced more than additive deleterious effects on fish survival and reproduction.

20.4.2 Birds

Among captive American kestrels (*Falco sparverius*) fed 8.0 mg mirex/kg ration for 69 days, there was a marked decline in sperm concentration and a slight compensatory increase in semen volume, but an overall net decrease of 70% in sperm number. These investigators believed that migratory raptors feeding on mirex-contaminated food organisms could ingest sufficient toxicant to lower semen quality in the breeding season which, coupled with altered courtship, could reduce the fertility of eggs and the reproductive fitness of the individual. Altered courtship in ring-necked doves (*Streptopelia capicola*) fed dietary organochlorine compounds is documented.

Most investigators, however, agree that comparatively high dietary concentrations of mirex had little effect on growth, survival, reproduction, and behavior of nonraptors, including chickens (*Gallus* sp.), mallards, several species of quail, and red-winged blackbirds. For domestic chickens, levels up to 200.0 mg mirex/kg ration were tolerated without adverse effects on various reproductive variables, but 300.0 mg mirex/kg diet for 16 weeks was associated with reduced chick survival, and 600.0 mg/kg for 16 weeks reduced hatching by 83% and chick survival by 75%. Mallard ducklings experienced temporary mild ataxia and regurgitation when given a single dose of 2400.0 mg/kg body weight but not when given 1200.0 mg/kg or less. Mallards fed diets containing as much as 100.0 mg mirex/kg ration for prolonged periods showed no significant differences from controls in egg production, shell thickness, shell weight, embryonation, hatchability, or duckling survival. However, in other studies with mallards fed 100.0 mg mirex/kg diet, eggshells were thinned and duckling survival was reduced, suggesting that 100.0 mg mirex/kg ration may not be innocuous to mallards. No adverse effects on reproduction were noted in the common bobwhite at 40.0 mg mirex/kg diet, or in two species of quail fed 80.0 mg mirex/kg ration for 12 weeks. Red-winged blackbirds were not repelled by foods contaminated with mirex, but consumed normal rations; a similar observation was recorded for bobwhites.

20.4.3 Mammals

Mirex has considerable potential for chronic toxicity since it is only partly metabolized, is eliminated very slowly, and is accumulated in the fat, liver, and brain. The most common effects observed in small laboratory mammals fed mirex included weight loss, enlarged livers, altered liver enzyme metabolism, and reproductive failure. Mirex reportedly crossed placental membranes and accumulated in fetal tissues. Among the progeny of mirex-treated mammals, developmental abnormalities included cataracts, heart defects, scoliosis, and cleft palates.

Mirex has caused liver tumors in mice and rats and must be considered a potential human carcinogen. Long-term feeding of 50.0 and 100.0 mg mirex/kg ration to rats of both sexes was associated with liver lesions that included neoplastic nodules and hepatocellular carcinomas; neither sign was found in controls.

Adults of selected mammalian species showed a variety of damage effects of mirex: enlarged livers in rats at 25.0 mg mirex/kg diet or at a single dose of 100.0 mg/kg body weight; liver hepatomas in mice at 10.0 mg mirex/kg body weight daily; decreased incidence of females showing sperm in vaginal smears, decreased litter size, and thyroid histopathology in rats fed 5.0 mg mirex/kg diet since weaning; elevated blood and serum enzyme levels in rats fed 0.5 mg mirex/kg ration for 28 days; and diarrhea, reduced food and water consumption, body weight loss, decreased blood glucose levels, and disrupted hepatic microsomal mixed function oxidases in mice receiving 10.0 mg/kg body weight daily. In studies of field mice, decreased litter size was observed at 1.8 mg mirex/kg diet, and complete reproductive impairment at 17.6 mg/kg diet after six months. At comparatively high sublethal concentrations of mirex, various deleterious effects were observed: thyroid histopathology and decreased spermatogenesis in rats fed 75.0 mg mirex/kg diet for 28 days; abnormal blood chemistry, enlarged livers, reduced spleen size, and loss in body weight of beagles fed 100.0 mg mirex/kg ration for 13 weeks; and decreased hemoglobin, elevated white blood cell counts, reduced growth, liver histopathology, and enlarged livers in rats fed 320.0 mg/kg ration for 13 weeks.

Cataract formation, resulting in blindness, in fetuses and pups from maternal rats fed comparatively low concentrations of dietary mirex is one of the more insidious effects documented. Mirex fed to maternal rats at 6.0 mg/kg body weight daily on days 8–15 of gestation, or at 10.0 mg/kg body weight daily on days 1–4 postpartum, caused cataracts in 50% of fetuses on day 20 of gestation, and in 58% of pups on day 14 postpartum. Plasma glucose levels were depressed in fetuses with cataracts, and plasma proteins were depressed in neonates; both hypoproteinemia and hypoglycemia are physiological conditions known to be associated with cataracts. Mirex-associated cataractogenicity has been reported in female pups from rats fed 5.0 mg mirex/kg ration since weaning, in rat pups from females consuming 7.0 mg mirex/kg ration on days 7–16 of gestation or 25.0 mg/kg diet for 30 days prior to breeding, and in mice fed 12.0 mg mirex/kg ration. Offspring born to mirex-treated mothers, but nursed by untreated mothers showed fewer cataracts. Other fetotoxic effects in rats associated with dietary mirex included: edema and undescended testes; lowered blood plasma proteins, and heart disorders, including tachycardia and blockages; hydrocephaly; decreases in weight of brain, lung, liver, and kidney; decreases in liver glycogen, kidney proteins and alkaline phosphatase; and disrupted brain DNA and protein metabolism.

In prairie voles exposed continuously to dietary mirex of 0.5, 0.7, 1.0, 5.0, or 10.0 mg/kg ration, the numbers of litters produced decreased. Maximum numbers of litters per year were four at 1.0 mg mirex/kg ration; three at 5.0 mg/kg; and two at 10.0 mg/kg ration. Furthermore, the number of offspring per litter also decreased progressively. Concentrations as low as 0.1 mg mirex/kg ration of adults were associated with delayed maturation of pups and with an increase in number of days required to attain various behavioral plateaus such as bar-holding ability, hind-limb placing, and negative geotaxis. On the basis of residue data from field studies, results strongly suggest that mirex was harmful to the reproductive performance and behavioral development of prairie voles at environmental levels approaching 4.2 g mirex/ha, a level used to control fire ants before mirex was banned.

20.5 Bioaccumulation

Mirex accumulations by aquatic organisms, birds, and mammals via a variety of routes are listed. Variables modifying uptake and retention are discussed.

20.5.1 Aquatic Organisms

All aquatic species tested accumulated mirex from the medium and concentrated it over ambient water levels by factors ranging up to several orders of magnitude; uptake was positively correlated with nominal dose in the water column. Investigators have reported bioconcentration factors from water of 8025 in daphnids, 12,200 in bluegills, 56,000 in fathead minnows, and 126,600 in the digestive gland of crayfish. Rapid uptake of mirex by marine crabs, shrimps, oysters, killifishes, and algae was reported after the application of mirex baits to coastal marshes. Mirex was also accumulated from the diet but not as readily as from the medium. Dietary bioaccumulation studies with guppies and goldfish show that mirex and other persistent hydrophobic chemicals are retained in the organism and biomagnify through food chains because of their hydrophobicity. Mirex may also be accumulated from contaminated sediments by marine teleosts, but such accumulation has not been established conclusively. Although terrestrial plants, such as peas and beans, accumulate mirex at field application levels, mangrove seedlings require environmentally high levels of 11.2 kg mirex/ha before accumulation occurs.

There is a general agreement that aquatic biota subjected to mirex-contaminated environments continue to accumulate mirex, and that equilibrium is rarely attained before death of the organism from mirex poisoning or from other causes. There is also a general agreement that mirex resists metabolic and microbial degradation, exhibits considerable movement through food chains, and is potentially dangerous to consumers at the higher tropic level. Marine algae, for example, showed a significant linear correlation between amounts accumulated and mirex concentrations in the medium. If a similar situation existed in nature, marine unicellular algae would accumulate mirex and, when grazed upon, act as passive transporters to higher trophic food chain compartments. The evidence for elimination rates of mirex from aquatic biota on transfer to mirex-free media is not as clear. Biological half-times of mirex have been reported as 12 h for daphnids, more than 28 days for fathead minnows, about 70 days in Atlantic salmon (*Salmo salar*), 130 days for mosquitofish (*Gambusia affinis*), and 250 days for pinfish. Biological half-times may be much longer if organism growth is incorporated into rate elimination models. For example, brook trout (*Salvelinus fontinalis*) fed 29.0 mg mirex/kg ration for 104 days contained 6.3 mg/kg body weight or a total of 1.1 mg of mirex in whole fish. At day 385 post-exposure, after the trout had tripled in body weight, these values were 2.1 mg/kg body weight, an apparent loss of 67%; however, on a whole fish basis, trout contained 1.2 mg, thus showing essentially no elimination on a total organism basis.

No mirex degradation products were detected in whole fathead minnow or in hydrosoils under aerobic or anaerobic conditions. In contrast, three metabolites were detected in coastal marshes after mirex bait application, one of which, photomirex, was accumulated by fish and oysters. The fate and effects of mirex photoproducts in the environment is unclear and merits additional research.

The significance of mirex residues in various tissues is unresolved, as is the exact mode of action of mirex and its metabolites. Mirex is a neurotoxic agent, with a mode of action similar to that of other chlorinated hydrocarbon insecticides, such as DDT. In studies with crayfish and radiolabeled mirex, mirex toxicosis was associated with neurotoxic effects that included hyperactivity, uncoordinated movements, loss of equilibrium, and paralysis. Before death, the most significant differences in mirex distributions in crayfish were the increases in concentrations in neural tissues, such as brain and nerve cord, by factors up to 14 (or 0.4 mg/kg) in low-dose groups held in solutions containing 7.4 μg mirex/L, and up to 300 (or 6.2 mg/kg) in high-dose groups held in solutions with 74.0 μg/L. With continued exposure, levels in the green gland and neural tissues approached the levels in the hepatopancreas and intestine. Mirex also accumulates in the crustacean hepatopancreas, but other tissues, such as muscle and exoskeleton, have specific binding sites that, once filled, shunt excess mirex to hepatopancreas storage sites.

20.5.2 Birds and Mammals

Like aquatic organisms, birds and mammals accumulated mirex in tissue lipids, and the greater accumulations were associated with the longer exposure intervals and higher dosages. Sexual condition of the organism may modify bioconcentration potential. For example, in adipose fat of the bobwhite, males contained ten times mirex dietary levels and females only five times mirex dietary levels; the difference was attributed to mirex loss through egg laying.

Data on excretion kinetics of mirex are incomplete. Prairie voles fed mirex for 90 days contained detectable whole body levels four months after being placed on a mirex-free diet. Levels of mirex in voles after four months on uncontaminated feed were still far above levels in their mirex diets. Humans living in areas where mirex has been used for ant control contained 0.16–5.94 mg/kg in adipose fat; 60% of the mirex was excreted and most of the rest was stored in body tissues, (especially fat 28%), and in lesser amounts of 0.2–3% in muscle, liver, kidney, and intestines. Almost all excretion of mirex takes place through feces; less than 1% is excreted in urine and milk. The loss rate pattern is biphasic; the fast phase was estimated at 38 h and the slow phase at up to 100 days. Mirex binds firmly to soluble liver proteins and appears to be retained in fatty tissues, a property that may contribute to its long biological half-life. Chickens given single doses of mirex at 30.0 mg/kg intravenously or 300.0 mg/kg orally demonstrated a biphasic decline in blood concentrations. The fast component, constituting about 25% of the total, was lost during the first 24 h; the loss of the slow component was estimated to be at a constant rate of about 0.03% daily, suggesting a half-life of about 3 years. Growing chicks fed 1.0 or 10.0 mg/kg dietary mirex for 1 week lost the compound rather rapidly; disappearance half-times were 25 days for skin and 32 days for fat. It is clear that much additional research is warranted on loss rate kinetics of this persistent compound and its metabolites.

20.6 Mirex in the Southeastern U.S.

Between 1961 and 1975, about 400,000 kg of mirex were used in pesticidal formulations, of which approximately 250,000 kg were sold in the southeastern U.S. for control of native and imported fire ants (*Solenopsis* spp.); most of the rest was exported to Brazil for use in fire ant control in that country. Mirex was also used to control big-headed ant populations in Hawaiian pineapple fields, Australian termites, South American leaf cutter ants, South African harvester termites, and, in the U.S., western harvester ants and yellow jackets. Chemical control measures for imported fire ants began in the southeastern U.S. during the 1950s with the use of heptachlor, chlordane, and dieldrin. The large mounds built by ants in cultivated fields were believed to interfere with mowing and harvesting operations, the "vicious sting" of the insects presented a hazard to workers harvesting the crops, and the species was considered to be a pest in school playgrounds and homes. In 1965, the use of organochlorine insecticides to control fire ants was discontinued, due partly to their high acute toxicity to nontarget biota and their persistence. Previously used compounds were replaced by mirex 4X bait formulations, consisting of 0.3% mirex by weight, dissolved in 14.7% soybean oil, and soaked into corncob grits (85%). Initially, the 4X baits were deployed from low-flying airplanes at a total yearly rate of 1.4 kg bait/ha (1.25 pounds total bait/acre) or 4.2 g mirex/ha. Usually, three applications were made yearly. More than 50 million ha in nine southeastern States were treated over a 10-year period. Later, dosages were modified downward, and mirex was applied to mounds directly. Ecologically sensitive areas, such as estuaries, prime wildlife habitats, heavily forested areas, and State and Federal parks, were avoided. In 1977, for example, the formulation was changed to 0.1% mirex and the application rate lowered to 1.12 g/ha; about 8200 kg of the lower concentration bait were manufactured in 1977. Under ideal aerial application conditions, about 140 particles of mirex-impregnated bait were distributed per square meter. When an infested area is treated, the bait is rapidly scavenged by the oil-loving fire ant workers, placed in the mound, and distributed throughout the colony, including queen and brood, before any toxic effects become evident; death occurs in several days to weeks. The exact mode of action is unknown, but is believed to be similar to that of other neurotoxic agents, such as DDT.

Widespread use of mirex may lead to altered population structure in terrestrial systems, with resurgence or escalation of nontarget pests due to selective mirex-induced mortality of predators. For example, populations of immature horn flies and rove beetles, two species of arthropods normally preyed upon by fire ants, were higher in mirex-treated areas than in control areas. Conversely, other species, such as crickets, ground beetles, and various species of oil-loving ants, were directly affected and populations were still depressed or eliminated 14 months post-treatment, whereas fire ants recovered to higher than pretreatment levels, as judged by mound numbers and mound size.

Field results from aquatic and terrestrial ecosystems receiving mirex bait formulations indicated, with minor exceptions, that mirex accumulates sequentially in food complexes and concentrates in animals at the higher trophic levels. In both ecosystems, omnivores and top carnivores contained the highest residues. In South Carolina, where the 4X formulation was used to control fire ants from 1969–1971, mirex was translocated from treated lands to nearby marshes and estuarine biota, including crustaceans, marsh birds, and raccoons. Juvenile marine crustaceans showed delayed toxic effects after ingesting mirex baits, or after being exposed to low concentrations in seawater. About 18 months post-treatment, mirex residues of 1.3–17.0 mg/kg were detected in shrimp, mammals, and birds; however, 24 months after the last mirex treatment, less than 10% of all samples collected contained detectable residues. A similar study was conducted in pasturelands of Bahia grass (*Paspalium notatum*). Within a month after application, the target fire ant colonies were dead. Of the 4.2 g mirex/ha applied to the 164 ha block, 100% was accounted for on day 1, 63% at 1 month, and 3% at 1 year. Unaccounted mirex residues could include loss through biodegradation; through movement out of the study area by migratory insects, birds, other fauna, and groundwater; and through photodecomposition and volatilization.

Mirex residues in bobwhites from South Carolina game management area were documented after treatment with 4.2 g mirex/ha. Pretreatment residues in bobwhites ranged from undetectable to 0.17 mg mirex/kg breast muscle on a dry weight basis, and 0.25–2.8 mg/kg in adipose tissues on a lipid weight basis. Mirex residues in adipose tissue increased up to five times within one month, posttreatment, and declined thereafter; however, another residue peak was noted in the spring after mirex treatment and corresponded with insect emergence.

Mirex concentrations in muscle and liver of mammalian wildlife in Alabama and Georgia during the period 1973–76 from reference areas were always less than 0.04 mg mirex/kg FW in muscle and less than 0.07 mg/kg FW in liver. In mirex-treated areas, conspecifics were collected up to 2 years post-treatment. Maximum concentrations of mirex in muscle and liver from mirex-treated areas were always less than 1.0 mg/kg FW in raccoons, bobcats (*Lynx rufus*), mink (*Mustela vison*), and foxes (*Urocyon* sp., *Vulpes* sp.). Higher concentrations of 3.7 mg/kg FW in muscle and 1.1 mg/kg FW in liver were measured in the river otter (*Lutra canadensis*) 1 year posttreatment, 3.5 mg/kg FW in muscle of skunks (*Spilogale* sp., *Mephitis* sp.) 6 months posttreatment, and 1.1–1.5 mg/kg FW in embryos and muscle of the opossum (*Didelphius marsupialis*) 6–12 months after treatment.

Heavily treated watershed areas in Mississippi were investigated. After treatment, mirex residues were elevated in crayfish and stream fish; among mammals, residues were highest in carnivores and insectivores, lower in omnivores, and lowest in herbivores. Mirex residues in liver and eggs were substantially higher in the box turtle (*Terrapene carolina*) an omnivorous feeder, than in the herbivorous slider turtle (*Chrysemys scripta*); mirex did not accumulate for protracted periods in tissues of these comparatively long-lived reptiles. Among migratory reptiles, mirex was detected in only 11% of the eggs of the loggerhead turtle (*Caretta caretta*) and not at all in eggs of the green turtle (*Chelonia mydas*) corrected during summer 1976 in Florida. However, DDT or its isomers were present in all eggs of both species, and PCB's were detected in all loggerhead turtle eggs. The low levels of mirex and other organochlorine contaminants suggest that these turtles, when not nesting, live and feed in areas remote from Florida lands treated with mirex and other insecticides.

A 10–5 bait formulation containing 0.1% mirex was designed to make more of the toxicant available to the fire ant and less to non-target biota. In one study, the 10–5 formulation was applied to a previously untreated 8000-ha area near Jacksonville, Florida, infested with fire ants. The bait was applied by airplane at 1.12 kg/ha, or 1.12 g mirex/ha. Insects accumulated mirex to the greatest extent during the first 6 months after application, and most of the mirex was lost by 12 months; other invertebrates accumulated only low levels during the first 9 months, and no residues were detected after 12 months. Fish also showed low concentrations for 9 months and no detectable residues afterward; amphibians contained detectable residues after 12 months, but not at 24; reptiles contained measurable, but low, residues for the entire 24-month study period. Mammals had higher residue levels than reptiles, particularly in fat, whereas birds contained low to moderate residues. After 24 months, mirex was found infrequently and only at low concentrations in birds, mammals, reptiles, and insects. It was concluded that 10–5 mirex formulations were as effective in controlling fire ants as the 4X formulation and that residues in nontarget species were reduced from that following 4X treatment, or were lacking.

Eggs of the American crocodile (*Crocodylus acutus*) from the Florida Everglades contained up to 2.9 mg/kg fresh weight of DDE and 0.86 mg/kg of polychlorinated biphenyls, but less than 0.02 mg mirex/kg. Livers of the deep-sea fish (*Antimora rostrata*), collected in 1971–74 from a depth of 2500 m off the U.S. east coast, contained measurable concentrations of DDT and its degradation products, and dieldrin, but no mirex.

20.7 Mirex in the Great Lakes

Between 1959 and 1975, 1.5 million kg of mirex were sold, of which 74% or more than 1.1 million kg were predominantly Dechlorane, a compound used in flame-resistant polymer formulations of electronic components and fabrics. The total amounts are only approximate because almost half the mirex sold from 1962 to 1973 could not be accounted for. Mirex loadings to Lake Ontario were estimated at 200.0 kg per year in 1960–62, which decreased to 28.0 kg in 1980. Mirex entered Lake Ontario mainly from the Niagara River and Oswego River. About 700.0 kg of mirex were present in the bottom sediments of Lake Ontario in 1968, 1600.0 kg in 1976, and 1784.0 kg in 1981. All fish species in Lake Ontario were contaminated with mirex, and concentrations in half the species exceeded the Food and Drug Administration guideline of 0.1 mg/kg; other aquatic species had mirex residues near this level. Reproduction of the herring gull (*Larus argentatus*) on Lake Ontario was poor; mirex levels were an order of magnitude higher in gull eggs from Lake Ontario than in eggs from other Great Lakes locations. The probable source of contamination was a chemical manufacturer that used mirex (Dechlorane) as a flame retardant, but only Lake Ontario was contaminated. Until 1988, mirex had been reported for only a few locations in the Great Lakes, primarily Lake Ontario and the St. Lawrence River. Since 1988, however, mirex concentrations in water and fish samples have been measured from the other Great Lakes.

Poor reproductive success was reported together with declines in colony size of the herring gull at Lake Ontario at a time when dramatic increases of this species were reported along the Atlantic seaboard. In 1975, herring gull reproduction in Lake Ontario colonies was about one-tenth that of colonies on the other four Great Lakes. In addition, in Lake Ontario colonies, there were reductions in nest site defense, in number of eggs per clutch, in hatchability of eggs, and in chick survival. Hatching success of Lake Ontario gull eggs was 23–26%, compared with 53–79% for eggs from other areas. Analysis of herring gull eggs from all colonies for organochlorine compounds and mercury demonstrated that eggs from Lake Ontario colonies had mean mirex levels of 5.06 mg/kg fresh weight (range, 2.0–18.6), or about ten times more mirex than any other colony. Mean PCB and mercury levels were up to 2.8 and 2.3 times higher, respectively, in gull eggs from Lake Ontario than in those from other colonies, but only mirex levels could account for the colony declines. Short-term deviations from long-term trends in mirex concentrations in eggs of herring gulls from Lake Ontario seem to be correlated with weather patterns, i.e., warm spring weather conducive to phytoplankton growth produces relatively uncontaminated plankton, which results in less contamination for gulls during the critical period of egg yolk formation – and the reverse for cold spring weather. As judged by log-linear regression models, the half-life for mirex in herring gull eggs was 1.9–2.1 years, or essentially none was lost during egg incubation. Reproductive success of the Lake Ontario herring gull colonies improved after the early 1970s, an improvement that was directly paralleled by a decline in mirex, other organochlorine pesticides, and PCBs.

Concentrations of mirex and other contaminants in eggs of the Caspian tern (*Sterna caspia*) from the Great Lakes are declining, and tern populations are increasing. In Lake Huron, mirex concentrations in Caspian tern eggs declined from 0.51 mg/kg FW in 1976 to 0.12 mg/kg FW in 1991, equivalent to a decline of 8.6% annually. In Lake Ontario, mirex concentrations in tern eggs declined from 1.6 mg/kg FW in 1981 to 0.77 mg/kg FW in 1991, a decline of 7.1% annually. Similar trends are reported for eggs of herring gulls from Lakes Michigan, Huron, and Ontario, for whole lake trout (*Salvelinus namaycush*) from Lake Ontario, and whole young-of-the-year spottail shiners (*Notropis hudsonicus*) throughout the Great Lakes.

The fate of mirex in the environment and the associated transfer mechanisms have not been well defined. In one study, levels of mirex and its degradation products in herring gull eggs collected from Lake Ontario in 1977 are documented; authors concluded that photodegradation was the only feasible mechanism for production of the degradation compounds, although mirex and its photoproducts rapidly become sequestered in the ecosystem and protected from further degradation. Mirex degradation products were found in herring gull eggs from all the Great Lakes, suggesting that a high proportion of mirex and related compounds in herring gull eggs from Lakes Erie and Huron originated from Lake Ontario fish, whereas lower levels in eggs from Lakes Superior and Michigan originated from other sources. Mirex in sediments was considered an unlikely source because it was not being recycled into the ecosystem at an appreciable rate. Migrating salmon (*Oncorhynchus* spp.) make a significant contribution to the upstream transport of mirex from Lake Ontario, estimated at 53.0–121.0 g annually. Ingestion of salmon eggs by brown trout, decomposition of salmon carcasses by blowfly larvae, and ingestion of carcasses by aquatic and terrestrial scavengers are all means by which mirex is introduced into upstream environments. A harvest rate of 50% by fishermen represents removal of an additional 61.0 g of mirex annually from Lake Ontario.

Biomagnification of mirex through food chains was investigated. The basic assumption was that both herring gulls and coho salmon ate alewives (*Alosa pseudoharengus*) and rainbow smelt (*Osmerus mordax*). Mirex residues in these organisms, in mg/kg (parts per million) fresh weight, were 4.4 in gull eggs, 0.23 in salmon muscle, 0.10 in salmon liver, and 0.09 in whole alewives and smelt retrieved from stomachs of salmon. Bioconcentration factors (BCF) from prey to predator ranged up to 50, but those from water to gull egg were estimated to be near 25,000,000. Salmon muscle and gull eggs are complementary indicators of organochlorine contamination in the Great Lakes.

Among Great Lakes fishes, the highest mirex value recorded was 1.39 mg/kg FW in whole American eels (*Anguilla rostrata*) collected from Lake Ontario and was substantially in excess of the tolerated limit of 0.3 mg/kg FW for human consumption at that time. In the early 1980's, mirex was detected in 100% of the American eels sampled from Lake Ontario. High mirex values were also reported in chinook salmon (*Oncorhynchus tshawytscha*) and coho salmon (*Oncorhynchus kisutch*) from South Sandy Creek, a tributary of Lake Ontario, during autumn 1976; as a consequence, possession of all fish from that area was prohibited by the State of New York. Mirex concentrations in coho and chinook salmon tissues from Lake Ontario in 1977–78 ranged between 0.07 and 0.24 mg/kg FW tissue and increased with individual fish weight in direct relation to lipid content. The significance of mirex residues in salmonid fishes is unclear. Laboratory studies with brook trout showed that whole-body residues of 6.3 mg/kg fish weight were not associated with adverse effects on growth or survival, suggesting that long-lived species, such as the lake trout, would probably continue to accumulate mirex in Lake Ontario as long as they were exposed and may continue to contain residues for most of their lives even after the source has been eliminated.

There was no widespread mirex contamination of urban environments near Lake Ontario as a result of Dechlorane use, although local contamination of the Lake Ontario area was high when compared with other Great Lake areas. Among humans living in the Great Lakes area, there was great concern that mother's milk might be contaminated, owing to the high lipophilicity of mirex. Mirex was found in mother's milk from residents of New York State at 0.07 μg/L in Albany, 0.12 μg/L in Oswego, and 0.16 μg/L in Rochester, confirming that mirex was present in human milk but that concentrations were sufficiently low to be of little toxicological significance. It is noteworthy that none of the mothers had eaten Lake Ontario fish or any freshwater fish, and only a few had eaten marine fishes. For a 5.0-kg infant consuming 500.0 g of milk daily, this amount would approximate a dietary intake of 0.01 μg mirex/kg body weight daily or about 1/10,000 of the lowest recorded dietary value, causing delayed maturation in prairie voles. It is not known if a safety factor of 10,000 is sufficient to protect human health against delayed toxic effects of mirex, but it now appears reasonable to believe that it is.

20.8 Mirex in Other Geographic Areas

Mirex residues were determined in birds collected nationwide or from large geographic areas of the U.S.; however, aside from the Southeast and the Great Lakes, concentrations were low, considered nonhazardous, and occurred in a relatively small proportion of the samples collected. Among wings of mallards and American black ducks (*Anas rubripes*) collected from the four major flyways during 1976–77, mirex concentrations were highest and percent occurrence greatest in samples from the Atlantic Flyway: mallards, 50% occurrence, 0.14 mg/kg fresh weight; black ducks, 19% and 0.04 mg/kg. Data for mallards collected from other flyways follow: Mississippi, 29% and 0.03 mg/kg; Central, 14% and 0.06 mg/kg; and Pacific 4% and 0.03 mg/kg. Carcasses of several species of herons found dead or moribund nationwide from 1966 to 1980 were analyzed for a variety of common organochlorine pesticides; mirex was detected in less than 15% of the carcasses, a comparatively low frequency, and only in nonhazardous concentrations. However, about 20% of all herons found dead or moribund had lethal or hazardous concentrations of dieldrin or DDT. In bald eagles (*Haliaeetus leucocephalus*) found dead nationwide, elevated mirex levels were recorded in carcass lipids (24.0 mg/kg) and in fresh brain tissues (0.22 mg/kg). Among endangered species such as the bald eagle, it was determined that the most reliable indicator for assessing risk of organochlorine compounds was the ratio of carcass to brain residues on a lipid weight basis. Wings from American woodcocks (*Philohela minor*) collected from 11 states in 1970–71 and 14 states in 1971–72 were analyzed for mirex and other compounds. Mirex residues in the 1971–72 wings showed the same geographical pattern of recovery as those observed in 1970–71: residues were highest in the southern states and New Jersey, and lowest in the northern and Midwestern states. Mirex residues were significantly lower in 1971–72 than in 1970–71. As judged by the analysis of wings of immature woodcocks in Louisiana, mirex residues were significantly lower in immature members than in adults: 2.48 mg/kg lipid weight vs. 6.20 mg/kg.

Mirex concentrations in bald eagle eggs collected nationwide between 1969 and 1979 ranged from 0.03 to 2.0 mg/kg FW, and were highest in Florida and the Chesapeake Bay region. Up to 87% of bald eagle eggs from Florida and the Chesapeake Bay had detectable mirex residues, whereas this value was as low as 17% in Alaska. Eggs from successful bald eagle nests had 0.03 mg mirex/kg FW and lower, but eggs from unsuccessful nests had 0.05 mg/kg FW and higher. Eggs of Cooper's hawk, (*Accipiter cooperi*), collected in 1980 from various locations, all contained more than 0.05 mg mirex/kg FW; concentrations were highest in Pennsylvania with 0.84 mg/kg FW and Wisconsin with 1.6 mg/kg FW. Eggs of the loggerhead shrike

(*Lanius ludovicianus*) from the Shenandoah Valley region of Virginia in 1985–86 contained an average of 0.04 mg mirex/kg FW, with a 63% frequency of occurrence; loggerhead shrike populations in that region are declining but the cause of the decline is not known with certainty. Eggs of the ring-necked grebe (*Podiceps grisigena*) from Manitoba, Canada, in 1980–81, had as much as 28.6 mg mirex/kg lipid weight, and this may account, in part, for the high nesting loss of 79% observed in grebes at that time. Mirex and other organochlorine compounds in eggs of anhingas (*Anhinga anhinga*) and 17 species of waders (including herons, egrets, bitterns, ibises, and storks) were measured in various locations throughout the eastern U.S. during 1972–73. The highest mean concentration of 0.74 mg mirex/kg, range 0.19–2.5, was found in eggs of the green heron (*Butorides striatus*) from the Savannah National Wildlife Refuge in South Carolina; a single egg of the cattle egret (*Bubulucus ibis*) analyzed from there contained 2.9 mg mirex/kg. However, the overall frequency of mirex occurrence was higher in eggs collected from the Great Lakes region (24%) than in those from the South Atlantic Coast (15.6%), inland areas (10.7%), Gulf Coast (4.4%), or North Atlantic region (3.2%). Measurable mirex residues were detected in migratory birds collected from a variety of locations, including areas far from known sources or applications of mirex. For example, 22% of all eggs from 19 species of Alaskan seabirds collected in 1973–76 contained mirex. The highest concentration was 0.044 mg/kg in eggs of a fork-tailed storm petrel (*Oceanodroma furcata*) from the Barren Islands. Mirex residues were low compared with those of other organochlorine compounds. Eggs from the clapper rail (*Rallus longirostris*) collected in New Jersey from 1972–74 contained 0.16 to 0.45 mg mirex/kg. Eggs from the greater black-backed gull (*Larus marinus*) collected from Appledore Island, Maine, in 1977 contained up to 0.26 mg/kg, but no mirex was detected in eggs of common eider (*Somateria mollissima*) or herring gull from the same area. The greater black-backed gull is an active carnivore; 36–52% of its diet consists of small birds and mammals, whereas these items compose less than 1% in eider and herring gull diets. The higher mirex levels in black-backed gulls are attributed to its predatory feeding habits. In New England, eggs of the black-crowned night-heron (*Nycticorax*) contained between 0.28 and 0.66 mg mirex/kg wet weight in 1973; in 1979, this range was 0.11–0.37 mg/kg. Falcon eggs contained detectable mirex; levels were highest in the pigeon hawk (*Falco columbarius*) (0.25 mg/kg) and in the peregrine falcon (*Falco peregrinus*) (0.43 mg/kg), two species that feed on migratory birds or migrate to mirex-impacted areas. Active mirex was also found in eggs of a cormorant (*Phalacrocorax* sp.) from the Bay of Fundy on the Atlantic coast; the suspected source of contamination was the southern wintering range.

Mirex residues in 20 great horned owls (*Bubo virginianus*) found dead or dying in New York State in 1980–82 contained concentrations of mirex and PCBs higher than those reported for great horned owls elsewhere. Owls in "poor flesh" contained higher residues than those in "good flesh"; these values were 6.3 mg/kg FW vs. 0.07 mg/kg FW for brain, and 5.6 mg/kg FW vs. 0.1 mg/kg FW for liver. Waterfowl collected from upstate New York between 1979 and 1982 had about 0.07 mg mirex/kg FW breast muscle and 0.28 mg/kg FW subcutaneous fat.

Mink (*Mustela vison*) collected from the Northwest Territories of Canada between 1991 and 1995 had liver mirex concentrations between 0.08 and 0.39 μg/kg FW. These extremely low mirex concentrations were, nevertheless, higher than liver mirex concentrations in prey species (snowshoe hare, *Lepus americanus*, 0.08–0.13 μg/kg FW; northern red-backed vole. *Clethrionomys rutilus*, 0.32 μg/kg FW), suggesting that mirex biomagnification in mammalian wildlife food chains is possible.

20.9 Recommendations

Mirex is classified as a Group 2B carcinogen, indicating that it is a possible human carcinogen. For the protection of human health, oral intake should not exceed 0.0002 mg/kg

BW daily, equivalent to 0.014 mg daily for a 70-kg person. In 1995, the recommended concentration of mirex in water should not exceed 0.001 μg/L in order to protect human health, freshwater and marine life, irrigated crops, and watered livestock. Fish in the human diet should not contain more than 0.1 mg mirex/kg fresh weight. Average acceptable ambient air concentrations recommended for the protection of human health range between 0.03 μg/m^3 in New York to 0.88 μg/m^3 in Pennsylvania; in Kentucky, air emission levels of mirex products should not exceed 232.0 μg per hour.

Before the banning of mirex for all uses in 1978, the tolerance limits in food for human consumption were 0.1 mg/kg for eggs, milk, and fat of meat from cattle, goats, hogs, horses, poultry, and sheep, and 0.01 mg/kg for all other raw agricultural commodities; higher limits of 0.3 mg mirex/kg in fish and shellfish and 0.4 mg/kg in crabs were tolerated. The maximum recommended allowable concentration of mirex in edible portions of domestic fish for human consumption was 0.1 mg/kg FW at that time. Avoidance of larger and older fish to minimize ingestion of fat-soluble contaminants, including mirex, was recommended. Trimming the fatty tissues from muscle of salmon and trout from Lake Ontario prior to consumption resulted in a mirex reduction of at least 44% in the trimmed fillet – reflecting loss of fat content – and a product considered safe (i.e., <0.1 mg mirex/kg FW) by the U.S. Food and Drug Administration. However, mirex concentrations as low as 0.1 mg/kg in diets of adult prairie voles were associated with delayed maturation of pups, and with significant delays in the attainment of various early development behaviors such as bar-holding ability, hind-limb placing, and negative geotaxis. It is not known whether or not prairie voles can serve as a model for protection of health of humans or various wildlife species. In the absence of supporting data, however, it seems prudent now to establish a dietary threshold of mirex at some level lower than 0.1 mg/kg. A maximum concentration of 0.01 mg/kg total dietary mirex, which is a recommended level for most raw agricultural commodities, appears reasonable and conservative for the protection of fish, wildlife, and human health. This value could be modified as new data become available.

Although mirex is extremely persistent in the environment, research findings suggest that some degradation occurs and that some of the degradation products, such as photomirex, are biologically active. Accordingly, additional research is warranted on the fate and effects of mirex degradation products, with special emphasis on biomagnification through aquatic and terrestrial food chains.

Alternate means of controlling imported fire ants are under consideration. One approach has been to reduce the concentrations of active mirex in bait formulations from 0.3% to some lower, but effective, level. It has been demonstrated that mirex baits containing 0.07% mirex were effective in controlling Australian termites, with a 90% kill in 9 days; baits containing as little as 0.01% mirex were also reported effective, although termite mortality was delayed considerably. Some studies indicated that alternate chemical control agents, such as chlorpyrifos, diazinon, dimethoate, or methyl bromide may be suitable and that nonbiocidal chemicals, such as various pheromones and hormones, which are capable of disrupting reproductive behavior of fire ants, are also under active consideration. Another proposal was to modify mirex chemically to a more water soluble and rapidly degradable product. The formulation Ferriamicide, which consisted of 0.05% mirex, ferrous chloride, and a small amount of long-chain alkyl amines, was formulated in baits during 1978–79 for ant control. Ferriamicide degraded within a few days after initial application; however, approval was revoked in 1980 when it was learned that the toxicity of various degradation products to mammals, especially that of photomirex, exceeded that of 4X bait formulations.

Mirex replacements should not manifest the properties that led to the discontinuance of mirex for all uses; namely, delayed mortality in aquatic and terrestrial fauna; numerous birth defects; tumor formation; histopathology; adverse effects on reproduction, early growth, and development; high biomagnification and persistence; disrupted energy metabolism; degradation into toxic metabolites; population alterations; and movement through aquatic and terrestrial environmental compartments. It is emphasized that mirex replacement compounds must be thoroughly tested before widespread application in the environment; if testing is incomplete, it is almost certain that the nation's fish and wildlife resources will be adversely affected. In 1980, the use of Amdro (tetrahydro-5,5-dimethyl-2 (1H)- pyrimidine) was conditionally approved by the U.S. EPA. Amdro reportedly has good ant control properties, degrades rapidly in sunlight, has a biological half-life of less than 24 h, is nonmutagenic, and is relatively nontoxic to other than targeted species, except fish. Amdro was more acutely toxic than mirex to fish.

20.10 Summary

Mirex (dodecachlorooctahydro-1,3,4-metheno-2H-cyclobuta(c, d)pentalene) has been used extensively in pesticidal formulations to control the red imported fire ant (*Solenopsis invicta*), and as a flame retardant in electronic components, plastics, and fabrics. One environmental consequence of mirex was the severe damage recorded to fish and wildlife in nine southeastern States and the Great Lakes, especially Lake Ontario. In 1978, the U.S. EPA banned all further use of mirex, partly because of the hazards it imposed on nontarget biota. These included delayed mortality and numerous birth defects in aquatic and terrestrial fauna; tumor formation; histopathology; wildlife population alterations; adverse effects on reproduction, early growth, and development; high biomagnification and persistence; degradation into toxic metabolites; movement through aquatic and terrestrial environmental compartments; disrupted mammalian energy metabolism; and detection of residues in human milk and adipose tissues.

Among susceptible species of aquatic organisms, significant damage effects were recorded when concentrations of mirex in water ranged from 2.0 to 3.0 μg/L. The current recommended concentration of 0.001 μg mirex/L affords an unusual degree of protection. Evidence suggests that sensitive species of wildlife are adversely affected at 0.1 mg/kg of dietary mirex. For comparison, current tolerance limits for mirex in food for human consumption range from 0.01 mg/kg for raw agricultural commodities to 0.1 mg/kg for eggs, milk, animal fat, and various seafood products. Additional research is needed on the fate of mirex degradation products and their effects on natural resources. Further, it is strongly recommended that environmental use of all mirex replacement compounds be preceded by intensive ecological and toxicological evaluation.

MOLYBDENUM[a]

Chapter 21

21.1 Introduction

Molybdenum (Mo) is present in all plant, human, and animal tissues, and is considered an essential micronutrient for most life forms. The first indication of an essential role for molybdenum in animal nutrition came in 1953 when it was discovered that a flavoprotein enzyme, xanthine oxidase, was dependent on molybdenum for its activity. It was later determined that molybdenum is essential in the diet of lambs, chicks, and turkey poults. Molybdenum compounds are now routinely added to soils, plants, and waters to achieve various enrichment or balance effects.

There are certain locations where plants will not grow optimally because of a deficiency in molybdenum, and other places where the levels of molybdenum in plants are toxic to livestock that grazing on the plants. Molybdenum poisoning in cattle was first diagnosed in England in 1938; molybdenosis was shown to be associated with consumption of herbage containing large amounts of this element, and to be controllable by treatment with copper sulfate. Molybdenum poisoning of ruminants, especially cattle, has been reported in at least 15 states in the U.S., and in Canada, England, Australia, New Zealand, Ireland, the Netherlands, Japan, and Hungary. Molybdenosis was most pronounced in areas where soils were alkaline, high in molybdenum and low in copper, or near industrial point sources such as coal, aluminum, uranium, or molybdenum mines; steel alloy mills; or oil refineries. All cattle are susceptible to molybdenosis, milking cows and young stock being the most sensitive. Industrial molybdenosis in domestic cattle and sheep, which usually involved a single farm or pasture, has been widely documented: in Colorado in 1958 from contaminated river waters used in irrigation, in Alabama in 1960 from mine spoil erosion, in North Dakota in 1968 from flyash from a lignite burning plant, in Missouri in 1970–1972 from clay pit erosion, in Pennsylvania in 1971 from aerial contamination by a molybdenum smelter, in South Dakota in 1975 from molybdenum-contaminated magnesium oxide, and in Texas in 1965–1972 from uranium mine waste leachate. In humans, a gout-like disease in two villages in Armenia was attributed to the ingestion of local foods high in molybdenum and grown in soils high in molybdenum. Esophageal cancer was prevalent in various parts of southern Africa where food was grown in low molybdenum soils; it was reported in China in a low frequency rate that was significantly correlated with increasing molybdenum concentrations in cereals and drinking water.

21.2 Environmental Chemistry

Molybdenum is a comparatively rare element that is used primarily in the manufacture of steel alloys for the aircraft and weapons industries. Most of the global production of about 100,000 tons annually comes from the U.S. – primarily Colorado. Anthropogenic activities that have contributed to environmental molybdenum contamination include combustion of fossil fuels, and smelting, mining, and milling

[a]All information in this chapter is referenced in the following sources:

Eisler, R. 1989. Molybdenum hazards to fish, wildlife, and invertebrates: a synoptic review. *U.S. Fish Wildl. Serv. Biol. (Rep. 85 1.19)*. 61 pp.

Eisler, R. 2000. Molybdenum. Pages 1613–1647 in *Handbook of Chemical Risk Assessment: Health Hazards to Humans, Plants, and Animals. Volume 3. Metalloids, Radiation, and Cumulative Index to Chemicals and Species*. Lewis Publishers, Boca Raton, Florida.

operations for steel, copper, and uranium, as well as for molybdenum. In general, the chemistry of molybdenum is complex and inadequately known. Its toxicological properties are governed to a remarkable extent by interactions with copper and sulfur, although other metals and compounds may confound this interrelation.

21.2.1 Sources and Uses

Molybdenum is used in the manufacture of high-strength low-alloy steels and other steel alloys in the aircraft and weapons industries, and in the production of spark plugs, X-ray tubes and electrodes, catalysts, pigments, and chemical reagents. The most important industrial compound is the trioxide, MoO_3, which is resistant to most acids, and is oxidized in air at >500°C.

Molybdenum, discovered about 200 years ago, entered the commercial market in the 1920s as a result of extensive metallurgical research into its alloying properties and to the finding at Climax, Colorado, of the largest proven reserves of molybdenum worldwide. Molybdenum does not occur free in nature and is found only in combination with sulfur, oxygen, tungsten, lead, uranium, iron, magnesium, cobalt, vanadium, bismuth, or calcium. The most economically important ores are molybdenite (MoS_2), jordisite amorphous (MoS_2), and ferrimolybdate ($FeMoO_3 \cdot H_2O$); less important are wulfenite ($PbMoO_4$), powellite ($CaMoO_4$), and ilsemannite (Mo_3O_8).

World molybdenum production has increased from about 90 metric tons in 1900 – half from Australia and Norway, half from the U.S. – to 136 tons in 1906, 1364 in 1932 (an increase of several orders of magnitude in 26 years), 10,909 in 1946, and 91,000 tons in 1973. Through the years, molybdenum has been produced in about 30 countries; in 1973, about 60% of the worldwide production was from the U.S., 15% from Canada, 15% from the USSR and China combined, and 10% from other nations – Chile, Japan, Korea, Norway, and Mexico. By 1979, the U.S. produced about 62% of the world production of 103,000 metric tons, and exported about half, chiefly to Western Europe and Japan; other major producers in 1979 were Canada, Chile, and the USSR. In the U.S., only three mines in Colorado account for almost 70% of domestic production. Other active molybdenum mining sites in North America are in Arizona, Nevada, New Mexico, Utah, and California; molybdenum reserves have also been proven in Idaho, Alaska, Pennsylvania, and British Columbia. About 65% of domestic molybdenum is recovered from ores rich in molybdenum; the rest is a by-product from ores of copper, tungsten, and uranium.

As a result of various human activities, molybdenum enters the environment from many sources. Coal combustion is the largest atmospheric source of molybdenum, contributing about 550 metric tons annually, or 61% of all atmospheric molybdenum worldwide that comes from anthropogenic sources. In Sweden alone, about 2.5 tons of molybdenum is emitted into the atmosphere yearly from oil combustion. Molybdenum mining and milling are the source of about 100 metric tons annually to aquatic systems. At the world's largest molybdenum mine in Climax, Colorado, where about 36,000 tons of tailings are generated daily, the operation releases up to 100 tons of molybdenum annually as aqueous effluent. Other sources are molybdenum smelting, uranium mining and milling, steel and copper milling, oil refining, shale oil production, and claypit mining.

21.2.2 Chemical Properties

Molybdenum, which can function both as a metal and metalloid, is an essential component in a large number of biochemical systems – including xanthine oxidase. At least four metalloenzymes are known that are molybdenum-dependent, and all are molybdoflavoproteins. Molybdenum is characterized by the following physical and chemical properties: atomic number 42; atomic weight 95.94; density 10.2; melting point 2617°C; boiling point 4612°C; oxidation states 0, +2, +3, +4, +5, and +6; crystalline forms as gray–black powder, or silver–white metal; mass numbers (percent contribution of naturally occurring molybdenum) of 92 (15.86%), 94 (9.12%), 95 (15.7%), 96 (16.5%), 97 (9.45%), 98 (23.75%), and 100 (9.62%.); and radioactive isotopes of mass number 90, 91, 93, 99 (Tb1/2 of 67 h, frequently used as a tracer), 101, 102, and 105. In water at pH > 7, molybdenum exists primarily as the molybdate ion, MoO_4^{2-}; at pH < 7, various polymeric compounds are formed, including the paramolybdate ion, $Mo_7O_{24}^{6-}$. In soils, molybdate was sorbed most readily to alkaline, high calcium, high chloride soils; retention was least in low pH, low sulfate soils. There is general agreement that molybdenum chemistry is complex and inadequately known.

21.2.3 Mode of Action

Interactions among some trace metals are so pervading and so biologically influential that the results of nutritional and toxicological studies conducted with a single element can be misleading unless the dietary and body tissue levels of interacting elements are clearly defined. For molybdenum, interactions are so dominant – especially in ruminant species – that a particular level of intake in the diet can lead to molybdenum deficiency or to molybdenum toxicity in the animal, depending on the relative intakes of copper and inorganic sulfate.

The first indications of interaction between copper and molybdenum came more than 50 years ago from studies of grazing cattle in certain areas of England. Afflicted animals lost weight, developed severe diarrhea, and (in extreme cases) died. The disease is sometimes called teart (rhymes with heart) or molybdenosis, and is caused by eating herbage rich in molybdenum – i.e., 20.0–100.0 mg/kg dry weight diet compared to <5.0 mg/kg in nearby healthy pastures – and low or deficient in copper and inorganic sulfate. Molybdenosis is a copper-deficiency disease that occurs particularly in cattle and sheep and is usually caused by the depressing effect of molybdenum on the physiological availability of copper. The disease was treated successfully with copper sulfate at 1.0–2.0 g daily in the diet, or 200.0–300.0 mg daily by intravenous injection. When ruminant diets contained copper at 8.0–11.0 mg/kg weight – a normal range – cattle were poisoned at molybdenum levels of 5.0–6.0 mg/kg and sheep at 10.0–12.0 mg/kg. When dietary copper was low (i.e., <8.0 mg/kg) or sulfate ion level was high, molybdenum at 1.0–2.0 mg/kg ration was sometimes toxic to cattle. Increasing the copper in diets to 13.0–16.0 mg/kg protected cattle against concentrations up to 150.0 mg/kg of dietary molybdenum. Studies of molybdenum metabolism are of limited value unless one knows the status in the diet of inorganic sulfate, which alleviates molybdenum toxicity in all known species by increasing urinary molybdenum excretion.

Copper prevents the accumulation of molybdenum in the liver and may antagonize the absorption of molybdenum from food. The antagonism of copper to molybdenum depends on sulfate, which may displace molybdate. In certain sheep pastures, for example, the herbage may contain up to 15.0 mg copper/kg dry weight and <0.2 mg Mo/kg dry weight – conditions favoring the development of a high copper status that may lead to copper poisoning. Treatment consists of providing molybdate salt licks, which are highly effective in reducing copper levels in grazing sheep. A low copper:molybdenum ratio (i.e., <2), rather than the absolute dietary concentration of molybdenum, is the primary determinant of susceptibility to molybdenum poisoning; molybdenosis is not expected when this ratio is near 5. Ratios of copper to molybdenum in sweet clover (*Melilotus* spp., a known molybdenum accumulator plant) growing in coal mine spoils in the Dakotas, Montana, and Wyoming ranged from 0.4 to 5, suggesting that molybdenosis can be expected to occur in cattle and sheep grazing in low Cu:Mo areas. A similar situation existed in British Columbia, where 19% of all fodders and grains had a Cu:Mo ratio <2.

There are several explanations for the high sensitivity of ruminants to increased dietary molybdenum and sulfur, the most plausible being the role of thiomolybdates. Thiomolybdates are compounds formed by the progressive substitution for sulfur and oxygen in the molybdate (MoO_4^{2-}) anion when hydrogen sulfide and MoO_4^{2-} interact in vitro at neutral pH. Di-, tri-, and tetrathiomolybdates are formed, but only the last of these effectively impairs copper absorption. When sufficient tetrathiomolybdate (MoS_4) is formed in the rumen, it combines with copper in the gut and the resultant complex is bound strongly to proteins of high molecular weight. The molybdoproteins so formed are strong chelators of copper, and may be the agents responsible for copper deficiency through formation of biologically unavailable copper complexes in gut, blood, and tissues of animals that consume diets containing high concentrations of molybdenum. To confound matters, the complex molybdenum–copper–sulfur interrelation can be modified, or disrupted entirely, by many compounds or mixtures. These include the salts of tungsten, zinc, lead, manganese, iron, vanadium, chromium, phosphorus, cystine, methionine, fluoride, and proteins.

21.3 Concentrations in Field Collections

Molybdenum levels tend to be elevated in nonbiological materials and in terrestrial flora in the vicinity of molybdenum mining and reclamation activities, fossil-fuel power plants, and disposal areas for molybdenum-contaminated sewage sludge, flyash, and irrigation waters. Concentrations of molybdenum in fish, wildlife, and invertebrates were low when compared to those in terrestrial plants, although certain aquatic invertebrates were capable of high bioconcentration. Concentrations of molybdenum alone, however, were not sufficient to diagnose molybdenum deficiency or toxicosis.

21.3.1 Nonbiological Samples

Elevated levels of molybdenum in nonbiological materials have been reported near certain mines, power plants, and oil shale deposits, as well as in various sewage sludges, fertilizers, and agricultural drainwaters. Molybdenum is concentrated in coal and petroleum, and the burning of these fuels contributes heavily to atmospheric molybdenum. Combustion of fossil fuels contributes about 5000 metric tons of molybdenum annually to the atmosphere; atmospheric particulates contain about 0.001 μg Mo/m^3 air.

Natural molybdenum concentrations in ground and surface waters rarely exceed 20.0 μg/L; significantly higher concentrations are probably due to industrial contamination. Existing wastewater and water treatment facilities remove less than 20% of the molybdenum; accordingly, drinking water concentrations are near those of the untreated source. Molybdenum concentrations in saline waters appear to be directly related to salinity. In the Wadden Sea, for example, molybdenum concentrations were 0.08, 0.4, and 1.0 μg/L at salinities of 0.07, 1.2, and 3.3%, respectively.

The molybdenum content of soil may vary by more than an order of magnitude, causing both deficient and excessive concentrations for plants and ruminants in some parts of the world. Native soils may contain enough molybdenum to cause molybdenosis in range livestock in some areas of the U.S., particularly in Oregon, Nevada, and California. Elevated soil molybdenum levels can result from both natural and industrial sources. Usually when soil molybdenum levels exceed 5.0 mg/kg dry weight, a geological anomaly or industrial contamination is the likely explanation. Molybdenum is more available biologically to herbage plants in alkaline soils than in neutral or acidic soils. Liming of acidic soils or treatment with molybdenum-containing fertilizers can effectively raise the molybdenum content of herbage.

The disposal of sewage sludge, flyash from coal combustion, and molybdenum-contaminated irrigation waters to agricultural fields may result in the production of molybdenum-rich herbage. Sewage sludges rich in molybdenum and applied to agricultural soils resulted in elevated molybdenum content in corn and soybeans in a dose-dependent pattern. Similarly, flyash from coal combustion applied to pasture and croplands at rates sufficient to provide molybdenum at concentrations of 40.0 g/kg and higher resulted in potentially hazardous levels in vegetation to ruminant grazers. Molybdenum in flyash applied to soils remained biologically available for extended periods, especially in calcareous soils. Irrigation has also been proposed as a possible disposal method for large quantities of water having molybdenum concentrations of 5.0–100.0 mg/L that result from mining and reclamation activities. This method of disposal is not recommended unless all animals are kept off irrigated sites and the vegetation can be harvested and destroyed until molybdenum levels in the plants remain below 10.0 mg/kg dry weight.

21.3.2 Biological Samples

All plants contain molybdenum and it is essential for the growth of all terrestrial flora. Molybdenum concentrations were elevated in terrestrial plants collected from soils amended with flyash, liquid sludge, or molybdenum-contaminated irrigation waters, in naturally occurring teart pastures, and in the vicinity of molybdenum mining and ore processing activities, steelworks, and other metal processors; molybdenum concentrations greater than 20.0 mg/kg dry weight were frequently documented in plants from contaminated areas. Legumes, especially trefoil clovers (*Lotus* sp.) selectively accumulated molybdenum; concentrations of 5.0–30.0 mg/kg dry weight were common in molybdenum-contaminated areas. The molybdenum levels were sometimes high and potentially toxic in legumes from poorly drained acidic soils. Some terrestrial grasses displayed copper:molybdenum ratios between 0.5 and 3.7. Since ratios greater than 2 were within the range where molybdenosis is likely, and since most of the molybdenum concentrations were greater than the maximum tolerable level of 6.0 mg/kg dry weight, molybdenosis in cattle was expected. Major sources of molybdenum overload in fodder were in plants grown on high-molybdenum alkaline soils and from industrial contamination by coal and uranium mines and alloy mills. Variations in molybdenum content of pasture species ranged from 0.1 to 200.0 mg/kg dry weight, and most variations were due to soil and species differences. Pasture plants collected from mountainous areas of southern Norway were usually deficient in copper, and low to partly deficient in molybdenum. As a result, the copper–molybdenum ratios were generally high and may explain the occurrence of chronic copper poisoning in grazing sheep in that region.

Except in terrestrial plants, molybdenum concentrations were low in all groups examined; maximum concentrations reported from all sampling locales were about 6.0 mg/kg dry weight in aquatic plants, about 4.0 mg/kg fresh weight in aquatic invertebrates, 2.0 mg/kg fresh weight in fishes (except for rainbow trout liver and kidney – 26.0–43.0 mg/kg fresh weight – from fish collected near a molybdenum tailings outfall), 4.0 mg/kg dry weight in birds, 30.0 mg/kg dry weight in domestic ruminant liver, 85.0 mg/kg dry weight in the horse, and <4.0 mg/kg dry weight in mammalian wildlife and humans.

No food chain biomagnification of molybdenum was found in aquatic organisms from the San Joaquin River, California. Maximum concentrations of molybdenum recorded in the San Joaquin River in 1987 were 10.0 μg/L in water, and – in mg/kg DW – 3.1 in detritus, 1.4 in algae, 0.54 in chironomid larvae, 0.64 in whole crustaceans, and 0.51 in bluegills. Molybdenum did not biomagnify in the fish/crustacean/hump-backed dolphin (*Sousa chinensis*) food chain. Molybdenum concentrations from cetacean liver tissues were 4.1 times lower than that of whole prey organisms, i.e., 0.8 vs. 3.3 mg Mo/kg DW. Hump-backed dolphins from Hong Kong consume about 4.3 mg of molybdenum daily, equivalent to about 0.02 mg Mo/kg BW. Large interspecies differences were evident among aquatic organisms in their ability to accumulate molybdenum from the medium. Marine bivalve mollusks usually contained 30–90 times more molybdenum than the ambient seawater; however, some species from Greek waters had bioconcentration factors up to 1300. Marine plankton accumulated molybdenum from seawater by factors up to 25. But growth in aquatic phytoplankton populations was inhibited under conditions of low or missing molybdenum, nitrogen, and organic matter concentrations; the role of molybdenum in this process requires clarification. In rainbow trout

(*Oncorhynchus mykiss*), residues of molybdenum in tissues were affected only slightly by the concentrations in water; tissue residues ranged from 5.0 to 118.0 μg/kg fresh weight in water containing trace (<6.0 μg/L) concentrations, 10.0–146.0 μg/kg in water containing low (6.0 μg/L) concentrations, and from 13.0 to 322.0 μg/kg in water containing high (300.0 μg/L) concentrations. A similar pattern was reported for kokanee salmon, *Oncorhynchus nerka*. Rainbow trout held for 2 weeks in live traps 1.6 km downstream from a molybdenum mine tailings outfall survived, but liver and kidney had significantly elevated levels of molybdenum, calcium, manganese, iron, zinc, strontium, and zirconium, and 10% less potassium; the observed mineral changes may have been due to outfalls from non-molybdenum mines discharged into the river system.

In moles (*Talpa europaea*), adults had significantly higher concentrations of molybdenum in liver than did juveniles; the significance of this observation is imperfectly understood. Molybdenum mining operations are not detrimental to mammalian wildlife, as judged by normal appearance and low molybdenum levels in liver and kidney of nine species – including deer, squirrel, chipmunk, badger, beaver, marmot, and pika – collected from areas of high environmental molybdenum levels. It is emphasized that molybdenum concentrations in animal tissues give little indication of the dietary molybdenum status, and are of little diagnostic value for this purpose unless the sulfate, protein, and copper status of the diet are also known. This point is discussed in greater detail later.

21.4 Effects

Trace quantities of molybdenum are beneficial and perhaps essential for normal growth and development of plants and animals. In mammals, molybdenum can protect against poisoning by copper, mercury, and probably other metals, and may have anticarcinogenic properties. For all organisms, the interpretation of molybdenum residues depends on knowledge of molybdenum, copper, and inorganic sulfate concentrations in diet and in tissues. Some molybdenum compounds have insecticidal properties at low concentrations and have been proposed as selective termite control agents.

Aquatic flora and fauna seem to be comparatively resistant to molybdenum salts; adverse effects on growth and survival were usually noted only at water concentrations of 50 mg Mo/L, and higher. However, one study with newly fertilized eggs of rainbow trout produced an LC50 (28 day) value of 0.79 mg Mo/L compared to an LC50 (96 h) value of 500.0 mg/L for adults. Also, bioconcentration of molybdenum by selected species of algae and invertebrates (up to 20.0 g/kg dry weight) poses questions on risk to higher trophic level organisms.

In birds, adverse effects of molybdenum have been reported on growth at dietary concentrations of 200.0–300.0 mg/kg, on reproduction at 500.0 mg/kg, and on survival at 6000.0 mg/kg. In mammals, cattle are especially sensitive to molybdenum poisoning, followed by sheep, under conditions of copper and inorganic sulfate deficiency. Cattle were adversely affected when grazing pastures with a copper–molybdenum ratio <3, or when fed low-copper diets containing 2.0–20.0 mg Mo/kg diet, or when total daily intake approaches 141.0 mg molybdenum; cattle usually die at doses of 10.0 mg Mo/kg body weight. Other mammals, including horses, pigs, rodents, and ruminant and nonruminant wildlife are comparatively tolerant to molybdenum. Deer, for example, are at least ten times more resistant than domestic ruminants to molybdenum; no adverse effects in deer were noted at dietary levels of 1000.0 mg/kg after 8 days, slight effects at 2500.0 mg/kg after 25 days, and reduction in food intake and diarrhea at 5000.0 mg/kg diet after 15 days.

21.4.1 Terrestrial Plants

It is generally agreed that molybdenum is essential for plant growth due to its role in the fixation of nitrogen by bacteria using the enzymes nitrogenase and nitrate reductase, and that plants readily accumulate MoO_4^{2-} except under conditions of low pH, high sulfate, and low phosphate, and in some highly organic soils. Molybdenum deficiency is recorded in a variety of crops worldwide, but there is an extremely narrow range between adequacy and deficiency. In lettuce (*Lactuca sativa*), for example, adverse effects were noted at 0.06 mg/kg (dry weight), but sufficiency was attained at 0.08–0.14 mg/kg; a similar case is made for *Brassica* spp., i.e., Brussels sprouts, cabbage, and cauliflower. In certain species, such as beets (*Beta vulgaris*) and corn (*Zea mays*), the ratio between deficiency and sufficiency may differ by more than ten times.

Okra (*Abelmoschus esculentus*), grown in soils supplemented with molybdenum at 1.0, 2.0, or 3.0 mg/kg, as sodium molybdate, showed increasing growth and yields when compared to nonsupplemented soils; fruiting occurred earlier and persisted longer with increasing molybdenum concentration. The cashew (*Anacardium occidentale*) – one of the most valuable plantation crops in India – developed yellow-leaf spots accompanied by low molybdenum levels and excess manganese in low-pH soils; in extreme cases the tree was defoliated. The disorder was corrected by foliar spraying of molybdenum salts or by liming the soil. A similar case was reported for Florida citrus in the 1950s, which was shown to be due to molybdenum deficiency.

Soils amended with sewage sludge containing 12.0–39.0 mg Mo/kg dry weight (soil contained 2.0 mg Mo/kg dry weight at start and 4.8–6.0 mg/kg after treatment) were planted with corn and bromegrass (*Bromus inermis*). A lime-treated sludge increased molybdenum concentrations in plant tissues after several years of sludge application; maximum values recorded were 1.9 mg Mo/kg dry weight in bromegrass and 3.7 in corn. No toxicity of molybdenum has yet been observed in field-grown crops, although forages containing 10.0–20.0 mg/kg dry weight are considered toxic to cattle and sheep.

21.4.2 Terrestrial Invertebrates

Sodium molybdate and other molybdenum compounds in toxic baits have potential for termite control. Baits containing 1000.0 mg Mo/kg were fatal to 99% of the termite *Reticulitermes flavipes* in 48 days. After 8–10 days, termites became steel–gray in color, but appeared otherwise normal; mortality began only after day 16. Termites did not avoid the poisoned bait, even at concentrations of 5000.0 mg Mo/kg. Similar results were reported for other termite species; in one case, sodium molybdate killed 100% of the workers in a colony of *Copotermes formosanus* within 24 hours after they ate filter paper treated with a 5% solution. Some other species of insects – including fire ants (*Solenopsis* sp.) and various species of beetles and cockroaches – were not affected when exposed to baits containing 5000.0 mg Mo/kg for 48 days.

21.4.3 Aquatic Organisms

Aquatic plants are comparatively resistant to molybdenum; in sensitive species adverse effects were evident on growth at 50.0 mg/L, and on development at 108.0 mg/L. Bioconcentration of molybdenum from the medium by certain freshwater algae can result in residues up to 20.0 g/kg dry weight without apparent damage; the implications of this phenomenon to waterfowl and to other species that consume molybdenum-laden algae need to be explored.

Molybdenum is considered essential for aquatic plant growth, but the concentrations required are not known with certainty and are considered lower than those for any other essential element. Molybdenum starvation restricts nitrogen fixation in algae, thereby limiting photosynthetic production during depleted conditions. Blue–green alga (*Anabaena oscillaroides*) cultured in molybdenum-deficient media containing 0.004–0.005 μg Mo/L rapidly depleted molybdenum in the medium; this ability was lost at higher concentrations of added molybdenum, when *Anabaena* began to accumulate the element. The addition of tungstate to molybdenum-deficient media enhances dinitrogenase inactivation, resulting in inhibited algal growth; this process is reversed at molybdenum levels of 0.005–0.04 μg/L. On the other hand, algal growth was significantly enhanced when vanadium (V) was present at 12.5 μg/L, although higher concentrations of V were growth inhibitory in 7 days. Algal uptake of molybdenum is rapid during the first 2 h, and slower thereafter; the sequential biological reduction of hexavalent to pentavalent to trivalent molybdenum occurs intracellularly in green algae. Uptake is greater in freshwater than in seawater, greater at increased doses, and greater at reduced algal densities; it is also greater at elevated temperatures.

Molybdenum occurs naturally in seawater as molybdate ion, MoO_4^{2-}, at about 10.0 μg/L. Despite the high concentrations of dissolved molybdenum in offshore seawater, phytoplanktons from offshore locales contain extremely low molybdenum residues, almost typical of molybdenum-deficient terrestrial plants. This phenomenon is attributed to the high concentrations of sulfate in seawater; sulfate inhibits molybdate assimilation by phytoplankton, making it less available in seawater than in freshwater. As one result, nitrogen fixation and nitrate assimilation – processes that require molybdenum – may require greater energy expenditure in marine than in freshwaters and may explain, in part, why marine ecosystems are usually nitrogen-limited and lakes are not. Experimentally increasing the ratio of sulfate to molybdate inhibits molybdate uptake by marine algae, slows nitrogen fixation rates, and slows the growth of organisms that use nitrate as a nitrogen source.

Limited data suggested that aquatic invertebrates were very resistant to molybdenum; adverse effects were observed on survival at >60.0 mg Mo/L and on growth at >1000.0 mg Mo/L. Bioconcentration factors were low, but depending on initial dose, measured residues (mg/kg fresh weight) were as high as 16.0 in amphipods, and were 3.0 in clams, 18.0 in crayfish muscle, and 32.0 in crayfish carapace. The host organisms seemed unaffected under these molybdenum burdens, but effects on upper trophic level consumers were not clear. Tailings from a pilot molybdenum mine on the North American Pacific coast were acutely lethal at concentrations of >61,000.0 mg tailings solids/L seawater to larvae of the mussel *Mytilus edulis*, and to adults of the amphipod *Rhepoxynius abronius* and the euphausiid *Euphausia pacifica*; acute sublethal effects were observed at >277,000.0 mg/L. All species of invertebrates tested in this preliminary study were more sensitive than juvenile coho salmon, *Oncorhynchus kisutch*. In another study, zooplankton exposed to molybdenum mine tailings <8 um in diameter at high sublethal concentrations ingested and excreted these particles. The lowest tailing concentration tested at which a deleterious effect was observed was 100.0 mg/L for depression of respiration in the copepod *Calanus marshallae*, and 560.0 mg/L for increased mortality in copepods and the euphausiid *Euphausia pacifica*; concentrations of molybdenum mine tailings were always <15.0 mg/L at 0.5 km downstream from a molybdenum tailings outfall.

Freshwater and marine fishes were – with one exception – extremely resistant to molybdenum; LC50 (96 h) values ranged between 70.0 mg/L and <3000.0 mg/L. The exception was newly fertilized eggs of rainbow trout exposed for 28 days through day 4 post-hatch; the LC50 (28 day) value was only 0.79 mg/L, and suggests that additional research is needed on the sensitivity of early life stages to molybdenum. In general, molybdenum was more toxic to teleosts in freshwater than in seawater and more toxic to younger fish than to older fish; in rainbow trout it bioconcentrated up to 16.0 mg/kg fresh weight in liver, 18.0 in spleen, 7.0 in muscle, 6.0 in gill, and 2.0 in gastrointestinal tract. Environmental levels of molybdenum as molybdate measured in the molybdenum mining areas of Colorado were not considered harmful to rainbow trout. Molybdenum enrichment of Castle Lake, California (a high mountain lake in which molybdenum was determined to be the limiting micronutrient), coupled with favorable environmental conditions, led to record high yields of trout. The addition of 16.0 kg of sodium molybdate, or 6.4 kg molybdenum, to Castle Lake in July 1963 was followed by larger standing crops of zooplankton and bottom fauna, which probably promoted survival of the 1965-year class and resulted in record yields to the angler of rainbow trout and brook trout (*Salvelinus fontinalis*) in 1967. Enrichment of molybdenum-deficient waters to improve angler success merits additional research.

21.4.4 Birds

Data are missing on the effects of molybdenum on avian wildlife under controlled conditions. All studies conducted with birds have been restricted to domestic poultry. Signs of molybdenum deficiency in domestic chickens included loss of feathers, lowered tissue molybdenum concentrations, reduced xanthine dehydrogenase activity in various organs, decreased uric acid excretion, disorders in ossification of long bones, and changes in joint cartilage that led to complete immobility; signs were eliminated when diets were supplemented with molybdenum at concentrations of 0.2–2.5 mg/kg. Efforts to produce a molybdenum deficiency syndrome in birds and mammals by feeding diets low in molybdenum have been unsuccessful. Thus, it has been necessary to introduce a compound with a known property of inhibiting molybdenum, namely wolframate (Na_2WO_4), a tungsten compound. Wolframate increases molybdenum excretion, leading to molybdenum deficiency in rats and chickens. With this technique it has been possible to produce an assumed molybdenum deficiency in chicks consisting of reduced weight gain and sometimes death. Dietary requirements to maintain normal growth in rats and chicks were probably less than 1.0 mg Mo/kg food, and thus substantially less than that of any other trace element recognized as essential. In fact, birds may require molybdenum at concentrations up to 6.0 mg/kg in their diets for optimal growth. Dietary molybdenum counteracts adverse effects in chicks on growth and survival induced by hexavalent chromium. Chicks fed 900.0 mg chromium/kg ration for 4 weeks showed significantly depressed growth, 25% mortality, and elevated liver chromium; however, diet supplementation to 150.0 mg Mo/kg resulted in normal growth and liver chromium values, and no deaths.

Early studies with chicks and turkey poults showed that the addition of only 13.0–25.0 μg Mo per kg – as molybdate or molybdic acid – to basal diets containing 1.0–1.5 mg Mo/kg resulted in a growth advantage of 14–19% in 4 weeks over that in unsupplemented groups. Roosters given dietary supplements of 100.0 or 400.0 μg molybdenum per bird daily for 4 weeks to basal diets containing 0.51 mg Mo/kg had reduced serum uric acid values when compared to those of controls; the significance of this finding is not clear. Birds are relatively resistant to molybdenum. For example, day-old chicks fed diets containing 20% molybdenum mine tailings for 23 days were unaffected, and those fed diets containing 40% molybdenum mine tailings showed only a slight reduction in body weight during the same period. Dietary levels of 200.0 mg Mo/kg ration results in minor growth inhibition of chicks; and at 300.0 mg/kg feed, the growth of turkey poults was reduced. Dietary supplements of 500.0 mg Mo/kg ration produced a slight decrease in growth rate of chicks after 4 weeks; hens, however, laid 15% fewer eggs than controls, and all eggs contained embryolethal concentrations of 16.0–20.0 mg Mo/kg. At dietary supplements of 1000.0 mg Mo/kg, egg production was reduced 50% in domestic chickens. Dietary loadings of 2000.0 mg/kg induced severe growth depression and a hundredfold increase in molybdenum content in tibia, and an 80% reduction in egg production. At 4000.0 mg/kg diet, severe anemia was reported in chickens. Mortality of chicks fed 6000.0 mg Mo/kg diet for 4 weeks was 33%; at 8000.0 mg Mo/kg diet for 4 weeks, 61% of the chicks died and survivors weighed only 16% as much as the controls. Chicks, unlike mammals, did not experience molybdenum reduction in tissues after sulfate administration – although sulfate markedly reduced the signs of molybdenum toxicity.

21.4.5 Mammals

Almost all studies conducted to date on molybdenum effects under controlled conditions have been on livestock, especially cattle and sheep. Molybdenum is beneficial and perhaps essential to adequate mammalian nutrition;

moreover, it can protect against poisoning by copper or mercury, and may be useful in controlling cancer. Evidence of functional roles for molybdenum in the enzymes xanthine oxidase, aldehyde oxidase, and sulfite oxidase suggests that molybdenum is an essential trace nutrient for animals. Signs of molybdenum deficiency include decreased intestinal and liver xanthine oxidase activity. Molybdenum prevents damage to the liver in sheep receiving excess copper; accumulations of copper and molybdenum in kidney were present in a biologically unavailable form and of negligible physiological significance. Dietary supplements of 70.0 mg molybdenum per day for a restricted period is recommended for reduction of liver copper in sheep, provided dietary copper levels are simultaneously reduced. Molybdenum, as sodium molybdate, protects against acute inorganic mercury toxicity in rats by altering the metabolism of cysteine-containing proteins in the cytoplasm of liver and kidney, resulting in lowered mercury content in these organs. Anticarcinogenic properties of molybdenum in rats have been reported, although the mechanisms of action are unknown. In one study, 2.0 or 20.0 mg Mo/L in drinking water significantly inhibited cancer of the esophagus and forestomach experimentally induced by *N*-nitrososarcosine ethyl ester. In another study with virgin female rats, 10.0 mg Mo/L in drinking water reduced by half the number of mammary carcinomas experimentally induced by *N*-nitroso-*N*-methylurea. Additional research seems warranted on the role of molybdenum in cancer inhibition.

Molybdenosis has been produced experimentally in many species of mammals, including cattle, sheep, rabbits, and guinea pigs. Signs of molybdenum poisoning vary greatly among species, but generally include the following: copper deficiency, especially in serum; reduced food intake and growth rate; liver and kidney pathology; diarrhea and dark-colored feces; anemia; dull, wiry, and depigmented hair; reproductive impairment, including delayed puberty, female infertility, testicular degeneration, and abnormal or delayed estrus cycle; decreased milk production; joint and connective tissue lesions; bone abnormalities; and loosening and loss of teeth. These authorities also agree on three additional points: first, early signs of molybdenosis are often irreversible, especially in young animals; second, the severity of the signs depends on the level of molybdenum intake relative to that of copper and inorganic sulfate; and third, if afflicted animals are not removed promptly from molybdenum-contaminated diets and given copper sulfate therapy, death may result.

Molybdenum poisoning in ruminants, or teart disease, has been known since the mid 1800s and affects only ruminants of special pastures. Degree of teartness varies from field to field and season to season, and is usually proportional to the molybdenum content in herbage. Molybdenum levels in typical teart pastures range from 10.0 to 100.0 mg/kg dry weight compared to normal levels of 3.0–5.0 mg/kg. If herbage contains more than 12.0 mg Mo/kg dry weight, problems should be expected in cattle, and to a lesser extent in sheep. In situations where cattle are accidentally exposed to high molybdenum levels, the administration of copper sulfate should result in molybdenum excretion, up to 50% in 10 days. Aside from cattle and sheep, all evidence indicates that other mammals are comparatively tolerant of high dietary intakes of molybdenum, including horses, pigs, small laboratory animals, and mammalian wildlife. Cattle excrete molybdenum primarily through feces, but other (more tolerant) species such as pigs, rats, and man, rapidly excrete molybdenum through urine and this may account, in part, for the comparative sensitivity of cattle to molybdenum. Cattle normally excrete about 67% of all administered MoO_3 in feces and urine in 7 days; guinea pigs excreted 100% in urine in 8 days; and swine excreted 75% in urine in 5 days. Cattle are adversely affected when they graze copper-deficient pastures containing 2.0–20.0 mg/kg molybdenum, and the copper to molybdenum ratio is less than 3; or when they are fed low copper diets containing 5.0 mg (or more) Mo/kg dry weight; or when total daily intake approaches 141.0 mg molybdenum; or when body weight residues exceed (a fatal) 10.0 mg Mo/kg. It is clear that both the form of molybdenum administered and the route of exposure affect molybdenum metabolism and survival. By comparison, adverse effects (some deaths) were noted at 250.0 mg Mo/kg body weight (BW) in guinea pigs, at 50.0 mg/kg BW in domestic cats (central nervous system impairment), at 10.0 mg/L drinking water in mice (survival), at 10.0–15.0 mg total daily intake in humans (high incidence of gout-like disease), and at 3.0 mg/m^3 air in humans for 5 years (respiratory difficulties), or 6.0–19.0 mg/m^3 in humans for 4 years.

In newborn lambs from ewes that consumed high-molybdenum diets during pregnancy, demyelinization of the central nervous system was severe, accompanied by low copper contents in the liver. Sheep are more tolerant than cattle to molybdenum poisoning due, in part, to a lower turnover of ceruloplasmin, a copper-transporting enzyme that is inhibited by molybdenum; however, this characteristic makes sheep more sensitive than cattle to copper poisoning. For example, chronic copper poisoning in sheep in several districts in Norway is probably due to molybdenum-deficient forages rather than to excess copper intake. Swayback is a spastic paralysis in lambs born of ewes that were copper deficient during pregnancy. In northern Ireland, where cases have been reported, pastures were not copper deficient and swayback was due to an imbalance of copper, molybdenum, and sulfur. Very severely affected lambs were paralyzed in all limbs and died shortly after birth because they were unable to stand and suckle. Lambs less severely affected developed signs in about 2 weeks, but usually only the hind limbs were affected. Brain and spinal cord lesions were present, resulting in demyelination of the spinal cord and cavitation of brain tissues; lesions were irreversible, but death might have been avoided with adequate copper therapy.

Horses are generally considered to be tolerant of dietary copper deficiencies and of copper and molybdenum excesses that affected ruminants. Yet molybdenum accumulated in equine liver and has been implicated as a possible contributory factor in bone disorders in foals and yearlings grazing pastures containing 5.0–25.0 mg Mo/kg. Cattle and horses are highly susceptible to pyrrolizidine alkaloids, an ingredient in certain poisonous plants such as tansy ragwort (*Senecio jacobaea*). Signs of poisoning included elevated copper levels in liver followed by fatal hemolytic crisis. Sheep are more resistant to alkaloids than equines or bovines, and sheep grazing has been recommended as a means of controlling tansy ragwort. However, dietary supplements of 10.0 mg Mo/kg increased the susceptibility of sheep to tansy ragwort intoxication, despite the observed increase in copper excretion.

In rodents, molybdenum is neither teratogenic nor embryocidal to golden hamsters at doses up to 100.0 mg/kg body weight, and has no measurable effect on fertility or gestation of female rats given similar high doses. Voluntary rejection of high-molybdenum diets by rats results in anorexia. This phenomenon implies sensory, probably olfactory, recognition of molybdate in combination with other dietary constituents to form compounds with a characteristic odor detectable by rats. The ability to reject high-molybdenum diets requires a learning or conditioning period because it is lacking or weak with freshly prepared diets and extends to a discrimination between a toxic (high molybdenum) and nontoxic (high molybdenum plus sulfate) diet. Rats may associate a gastrointestinal disturbance with a sensory attribute of diets containing toxic levels of molybdenum.

Data on molybdenum effects to mammalian wildlife are scarce, although those available strongly suggest that domestic livestock are at far greater risk. Studies with mule deer (*Odocoileus hemionus*) showed that this species was at least an order of magnitude more tolerant to high levels of dietary molybdenum than were domestic ruminants, and at least as resistant as swine, horses, and rabbits. Female mule deer showed no visible effects after 33 days on diets containing up to 200.0 mg Mo/kg feed, or after 8 days at 1000.0 mg/kg. Only slight effects – some reduction in food intake and some animals with diarrhea – were observed at diets of 2500.0 mg/kg for 25 days. At feeding levels of 5000.0 and 7000.0 mg/kg for periods of 3–15 days, signs were more pronounced; however, recovery began almost immediately after transfer to uncontaminated feed. Signs of copper deficiency and of molybdenosis are very similar, and careful diagnosis is necessary to ensure use of the correct remedial action. For example, some populations of Alaskan moose (*Alces alces gigas*) showed faulty hoof keratinization and decreased reproductive rates, but this was attributed to copper-deficient browse growing on low copper soils, and not to increased molybdenum levels in herbage. In another case, a high proportion of white-tailed deer (*Odocoileus virginianus*) feeding near uranium-mine spoil deposits in several Texas counties – areas in which extreme molybdenosis has been documented in grazing cattle – had antlers that were stunted, twisted, and broadened or knobby at the tips. However, the copper levels in liver of these deer were similar to those of deer in a control area – 16.7 mg/kg fresh weight vs. 18.0 – and only 1 of the 19 deer examined from the mining district had a detectable molybdenum concentration in liver (0.7 mg/kg fresh weight) vs. none in any control sample. On the basis of low contents of copper in soils and vegetation, it was concluded that white-tailed deer examined were experiencing copper deficiency (hypocuprosis), with signs similar to molybdenosis.

In humans, molybdenum is low at birth, increases until age 20 years, and declines thereafter. Although conclusive evidence that molybdenum is required by humans is lacking, there is general agreement that it should be considered as one of the essential trace elements. The absence of any documented deficiencies in humans indicates that the required level is much less than the average daily intake of 180.0 μg molybdenum in the U.S. Human discomfort has been reported in workers from copper-molybdenum mines, and in those eating food products containing 10.0–15.0 mg Mo/kg and <10.0 mg copper/kg and grown on soils containing elevated molybdenum of 77.0 mg/kg and 39.0 mg copper/kg. Symptoms include general weakness, fatigue, headache, irritability, lack of appetite, epigastric pain, pain in joints and muscles, weight loss, red and moist skin, tremors of the hands, sweating, dizziness, renal xanthine calculi, uric acid disturbances, and increased serum ceruloplasmin. The typical human adult contains only 9.0 mg of molybdenum, primarily in liver, kidney, adrenal, and omentum. Most of the ingested molybdenum is easily absorbed from the GI tract and excreted within hours or days in urine, mostly as molybdate; excesses may be excreted also by the bile, particularly as hexavalent molybdenum. At high dietary levels molybdenum reportedly prevents dental caries, but this requires verification.

21.5 Recommendations

Although molybdenum is generally recognized as an essential trace metal for plants and animals, and may reduce the incidence and severity of carcinomas in rats and dental caries in humans, there is no direct evidence of molybdenum deficiency being detrimental to animal health. The minimum daily molybdenum requirements in diets are not yet established due to problems in preparing molybdenum-free rations. As a consequence, no regulatory agency recognizes molybdenum as safe and necessary, and molybdenum cannot be legally incorporated into animal feeds.

The richest natural sources of molybdenum (i.e., 1.1–4.7 mg Mo/kg fresh weight) are plants unusually high in purines such as legumes and whole grains, followed by leafy vegetables, liver, and kidney; the poorest sources are fruits, sugars, oils, and fat.

The greatest economic importance of molybdenosis is associated with subclinical manifestations of copper deficiency resulting from forages containing a low copper–molybdenum ratio. Unfortunately, these conditions are often difficult to diagnose accurately, and animal response to copper may be difficult to demonstrate. One recommended treatment for afflicted cattle is 2.0 g daily of copper sulfate to cows and 1.0 g daily to young stock, or intravenous injection of 200.0–300.0 mg of copper sulfate daily for several days.

The animals most sensitive to molybdenum insult are domestic ruminants, especially cattle. Diets containing more than 15.0 mg Mo/kg dry weight and with a low copper to molybdenum ratio, or drinking water levels more than 10.0 mg Mo/L were frequently associated with molybdenosis in cattle. By contrast, adverse effects were documented in birds at dietary levels more than 200.0 mg Mo/kg ration, in ruminant wildlife at dietary levels greater than 2500.0 mg Mo/kg, and in aquatic organisms – with one exception – at more than 50.0 mg Mo/L (Table 21.1). The exception was newly fertilized eggs of rainbow trout, which were about 21 times more sensitive to molybdenum than were zygotes about 1/3 through embryonic development, and about 90 times more sensitive than adult fish.

Proposed criteria for human health protection include drinking water concentrations of less than 50.0 μg Mo/L, and daily dietary intake of less than 7.0 μg Mo/kg food – based on a 70-kg adult (Table 21.1). Molybdenum concentrations in blood of "healthy" people

Table 21.1. Proposed molybdenum criteria for the protection of living resources and human health.

Resource, Criterion, and Other Variables	**Molybdenum Concentration, in mg/kg or mg/L Unless Indicated Otherwise**
TERRESTRIAL PLANTS	
Okra, increased growth	3.0 in soil
Lettuce	
Molybdenum deficiency	about 0.06 dry weight (DW) plant
Molybdenum sufficiency	>0.08 DW plant
Corn, no adverse effects	3.7 DW plant
Agricultural soils, Bangladesh, molybdenum deficiency	<0.1 DW surface soil
TERRESTRIAL INVERTEBRATES	
Toxic baits	
Termites	about 1000.0 DW
Other insect pests	>5000.0 DW
AQUATIC LIFE	
Algae	
Deficiency levels	<0.000005-0.0177 fresh weight (FW)
High bioconcentration	>0.000014 FW
Growth reduction	>50.0 FW
Invertebrates, reduced survival	>60.0 FW
Fish	
Adults	
High bioconcentration	>0.000014 FW
Reduced survival	70.0 FW
Eggs	
Newly fertilized	
Reduced survival	>0.79 FW
No adverse effects	<0.028 FW
Eyed, adverse effects	>17.0 FW
BIRDS	
Molybdenum deficiency	0.013–0.2 FW diet
Normal growth	about 1.0 FW diet
Optimal growth	6.0 FW diet

Continued

Table 21.1. cont'd

Resource, Criterion, and Other Variables	Molybdenum Concentration, in mg/kg or mg/L Unless Indicated Otherwise
Growth reduction	200.0–300.0 FW diet
Reproductive impairment	500.0 FW diet
Reduced survival	6000.0 FW diet
MAMMALS	
Cattle	
Forage	
Healthy pasture	3.0–5.0 DW
Possibility of molybdenosis	10.0–20.0 DW
Probability of molybdenosis	20.0–100.0 DW
Toxic	15.0–30.0 DW
Maximum tolerable level	6.0 DW
Recommended	0.1–0.5 DW
Ratio of copper to molybdenum in diet	
Molybdenosis probable	<0.4
Critical	<2.0 to >20.0
Optimal for growth and reproduction	6.1–10.1
Drinking water	
Safe	<10.0 FW
Minimum toxic concentration for calves	10.0–50.0 FW
Guinea pig; no effect on survival	80.0 body weight FW
Domestic cat; adverse nonlethal effects	25.0–50.0 body weight FW
Mule deer	
No effect	200.0–1000.0 DW diet
Reduction in food intake	2500.0 DW diet
Nonlethal adverse effects	5000.0–7000.0 DW diet
Domestic sheep; recommended in forage	<0.5 DW
Laboratory white rat	
Minimum daily need, all sources	0.0005 mg
Disrupted calcium metabolism and elevated tissue residues	10.0 FW drinking water
Cancer inhibition	2.0–20.0 FW drinking water
Food avoidance	50.0 DW diet
HUMAN HEALTH	
Air; maximum permissible concentration	
Former Soviet Union	6.0 mg/m^3
USA; 8 h daily, 5 days weekly	9.5–10.0 mg/m^3
Blood, normal	0.0147 FW
Total daily intake, 70-kg adult	
Minimal need	0.12 mg
Average	0.1–0.5 mg
Maximum	10.0–15.0 mg
In molybdenum mining areas	>1.0 mg

Table 21.1. cont'd

Resource, Criterion, and Other Variables	Molybdenum Concentration, in mg/kg or mg/L Unless Indicated Otherwise
Intake from diet; average daily intake	
USA	0.335 (0.17–0.46) mg
Former Soviet Union	
Children	0.159 mg
Adults	0.353 mg
United Kingdom	0.128 (0.11–1.0) mg
Intake from drinking water	
Average	<0.005 mg daily
No effect level	<0.5 mg daily
Adverse effects	
Biochemical	0.5–10.0 mg daily
Clinical	10.0–15.0 mg daily
Safe	<0.05 FW
Irrigation water, safe	<0.01 FW

averaged 14.7 μg Mo/L, distributed between the plasma and erythrocytes. Anemic people had significantly lower blood molybdenum levels; in leukemia patients, molybdenum levels increased significantly in whole blood and erythrocytes but not in plasma. Additional work is recommended on the use of blood in fish and wildlife as an indicator of molybdenum stress and metabolism.

Increasing problems associated with marginal mineral deficiencies and unfavorable mineral interaction – as has been the case in the older agricultural areas of northern Europe – can be anticipated as pasture and forage production becomes more intensive. Research has been recommended in areas having high molybdenum content in soils and vegetation, and also in non-contaminated areas where consumption habits favor a high molybdenum intake and an imbalance in relation to other dietary constituents of importance, such as copper. In some parts of the world where molybdenum has been substituted for lime, the soils have become more acidic, thus making them difficult to farm. Liming under these conditions may elevate soil molybdenum from levels previously considered safe to levels potentially hazardous to grazing animals through high molybdenum herbage. The addition of molybdenum fertilizers to sheep pastures resulted in small increments in molybdenum content with negligible risk of induced copper deficiency. But it would be unwise to apply molybdenum fertilizers to temperate grasslands grazed by animals of low initial copper status, as judged by growth retardation of lambs from pastures supplemented with molybdenum.

21.6 Summary

The element molybdenum (Mo) is found in all living organisms and is considered to be an essential or beneficial micronutrient. However, the molybdenum poisoning of ruminants has been reported in at least 15 states in the U.S. and 8 foreign countries. Molybdenum is used primarily in the manufacture of steel alloys. Its residues tend to be elevated in plants and soils near molybdenum mining and reclamation sites, fossil-fuel power plants, and molybdenum disposal areas. Concentrations of molybdenum are usually lower in fish and wildlife than in terrestrial macrophytes.

Aquatic organisms are comparatively resistant to molybdenum salts: adverse effects on growth and survival usually appeared only at water concentrations >50.0 mg Mo/L. But in one study, 50% of newly fertilized eggs of rainbow trout (*Oncorhynchus mykiss*) died in 28 days at only 0.79 mg Mo/L. High bioconcentration of molybdenum by certain species of aquatic algae and invertebrates – up to 20 g of Mo/kg dry weight – has been recorded without apparent harm to the accumulator; however, hazard potential to upper trophic organisms (such as waterfowl) that may feed on bioconcentrators is not clear. Data on molybdenum effects are missing for avian wildlife and are inadequate for mammalian wildlife. In domestic birds, adverse effects on growth have been reported at dietary molybdenum concentrations of 200.0 mg Mo/kg, on reproduction at 500.0 mg/kg, and on survival at 6000.0 mg/kg.

Molybdenum chemistry is complex and inadequately known. Its toxicological properties in mammals are governed to a remarkable extent through interaction with copper and sulfur; residues of molybdenum alone are not sufficient to diagnose molybdenum poisoning. Domestic ruminants, especially cattle, are especially sensitive to molybdenum poisoning when copper and inorganic sulfate are deficient. Cattle are adversely affected – and die if not removed – when grazing on pastures where the ratio of copper to molybdenum is <3, or if they are fed low copper diets containing molybdenum at 2.0–20.0 mg/kg diet; death usually occurs when tissue residues exceed 10.0 mg Mo/kg body weight. The resistance of other species of mammals tested, including domestic livestock, small laboratory animals, and wildlife, was at least tenfold higher than that of cattle. Mule deer (*Odocoileus hemionus*), for example, showed no adverse effects at dietary levels of 1000.0 mg/kg.

Additional research is recommended in several areas: the role of molybdenum on inhibition of carcinomas and dental caries; the establishment of minimum, optimal, and upper daily requirements of molybdenum in aquatic and wildlife species of concern; the improvement in diagnostic abilities to distinguish molybdenum poisoning from copper deficiency; and the determination of sensitivity of early developmental stages of fishes to molybdenum insult.

NICKEL[a]

Chapter 22

22.1 Introduction

In Europe, nickel (Ni) is listed on European Commission List II (Dangerous Substances Directive) and regulated through the Council of European Communities because of its toxicity, persistence, and affinity for bioaccumulation. In Canada, nickel and its compounds are included in the Priority Substances List under the Canadian Environmental Protection Act. The World Health Organization (WHO) classifies nickel compounds in Group 1 (human carcinogens) and metallic nickel in group 2B (possible human carcinogen). The U.S. Environmental Protection Agency (USEPA) classifies nickel refinery dust and nickel subsulfide as Group A human carcinogens and nickel oxides and nickel halides as Class W compounds, that is, compounds having moderate retention in the lungs and a clearance rate from the lungs of several weeks. Nickel and its compounds are regulated by USEPA's Clean Water Effluent Guideline for many industrial point sources, including the processing of iron, steel, nonferrous metals, and batteries; timber products processing; electroplating; metal finishing; ore and mineral mining; paving and roofing; paint and ink formulating; porcelain enameling; and industries that use, process, or manufacture chemicals, gum and wood, or carbon black.

Nickel is ubiquitous in the biosphere. Nickel introduced into the environment from natural or human sources is circulated through the system by chemical and physical processes and through biological transport mechanisms of living organisms. Nickel is essential for the normal growth of many species of microorganisms and plants and several species of vertebrates, including chickens, cows, goats, pigs, rats, and sheep.

Human activities that contribute to nickel loadings in aquatic and terrestrial ecosystems include mining, smelting, refining, alloy processing, scrap metal reprocessing, fossil fuel combustion, and waste incineration. Nickel mining and smelting in the Sudbury, Ontario, region of Canada is associated with denudation of terrestrial vegetation and subsequent soil erosion, and gradual ecological changes, including a decrease in the number and diversity of species and a reduction in community biomass of crustacean zooplankton. At nickel-contaminated sites, plants accumulate nickel, and growth is retarded in some species at high nickel concentrations. But nickel accumulation rates in terrestrial and avian wildlife near nickel refineries are highly variable; nickel variability is also reported for plants, soils, and interstitial sediment waters.

The chemical and physical forms of nickel and its salts strongly influence bioavailability and toxicity. In general, nickel compounds have low hazard when administered orally. In humans and other mammals, however, nickel-inhalable dust, nickel subsulfide, nickel oxide, and especially nickel carbonyl induce acute pneumonitis, central nervous system disorders, skin disorders such as dermatitis, and cancer of the lungs and nasal cavity (Table 22.1). Nickel carbonyl is acutely lethal to humans and animals within 3–13 days of exposure; recovery is prolonged in survivors. An excess

[a]All information in this chapter is referenced in the following sources:

Eisler, R. 1998. Nickel hazards to fish, wildlife, and invertebrates: a synoptic review. *U.S. Geol. Surv. Biol. Sci. Rep. USGS/BRD/BSR–1998-0001*, 1–76.

Eisler, R. 2000. Nickel. Pages 411–497 in *Handbook of Chemical Risk Assessment: Health Hazards to Humans, Plants, and Animals. Volume 1, Metals*. Lewis Publishers, Boca Raton, Florida.

Table 22.1. Nickel chronology.

Date	Event
220 BCE	Nickel alloys made by the Chinese
1500s	Toxicity observed in miners of nickel
1751	Nickel isolated and identified. The name nickel was derived from "Old Nick," a gremlin to whom miners ascribed their problems
Early 1800s	Purified nickel obtained
1826	Nickel toxicity in rabbits and dogs demonstrated experimentally. High doses of nickel sulfate given by stomach gavage caused gastritis, convulsions, and death; sublethal doses produced emaciation and conjunctivitis
1840s	Commercial nickel electroplating initiated
1850s	Commercial exploitation of nickel begins after development of technology to remove copper and other impurities
1850–1900	Nickel used therapeutically in human medicine to relieve rheumatism (nickel sulfate) and epilepsy (nickel bromide)
1880s	Excess nickel found lethal to animals under controlled conditions
1889	Skin dermatitis in humans caused by chemicals used in nickel plating
1890	Extraordinary toxicity of nickel carbonyl ($Ni(CO)_4$) established
1893	Excess nickel found lethal to plants
1912	Nickel dermatitis documented
1915–1960	Nickel applied as fungicide found to enhance plant growth and increase yield
1926	Nickel dust caused skin dermatitis, especially in hot industrial environments
1932	Increased frequency of lung and nasal cancers reported among English nickel refinery workers exposed to high concentrations of nickel carbonyl
1939–1958	Certain forms of nickel found to be carcinogenic to humans
1943	Certain forms of nickel found to be carcinogenic to animals
1965–1967	Nickel found beneficial to plants
1970s	Nickel deficiency leads to adverse effects in microorganisms and plants
1980s	Nickel found to be constituent of various essential plant enzymes

number of deaths from lung cancer and nasal cancer occur in nickel refinery workers, usually from exposure to airborne nickel compounds. At one nickel refinery, workers had a five-fold increase in lung cancer and a 150-fold increase in nasal sinus cancer when compared to the general population. Pregnant female workers at a Russian nickel hydrometallurgy refining plant, when compared to a reference group, showed a marked increase in frequency of spontaneous and threatening abortions and in structural malformations of the heart and musculoskeletal system in live-born infants with nickel-exposed mothers. Nickel is also a common cause of chronic dermatitis in humans as a result of industrial and other exposures, including the use of nickel-containing jewelry, coins, utensils, and various prostheses.

22.2 *Sources and Uses*

About 250,000 people in the United States are exposed annually to inorganic nickel in the workplace. This group includes workers in the mining, refining, smelting, electroplating, and petroleum industries and workers involved in the manufacture of stainless steel, nickel alloys, jewelry, paint, spark plugs, catalysts, ceramics, disinfectants, varnish, magnets, batteries, ink, dyes, and vacuum tubes. Nonoccupational exposure to nickel and its compounds occurs mainly by ingestion of foods and liquids and by contact with nickel-containing products, especially jewelry and coins. Food processing adds to nickel already present in the diet through leaching from nickel-containing alloys in food-processing equipment made from stainless steel, milling of flour, use of nickel catalysts to hydrogenate fats and oils, and use of nickel-containing fungicides in growing crops. Nickel contamination of the environment occurs locally from emissions of metal mining, smelting, and refining operations; from combustion of fossil fuels; from industrial activities, such as nickel plating and alloy manufacturing; from land disposal of sludges, solids, and slags; and from disposal as effluents. In Canada in 1988, the mining industry released a total of 11,664 tons of nickel into the air (9.4%), water (0.5%), and on land as sludges or solids (15.4%) and slags (74.7%). The global nickel cycle is unknown, but recent estimates suggest that 26,300–28,100 tons are introduced each year into the atmosphere from natural sources and 47,200–99,800 tons from human activities; airborne nickel is deposited on land at 50,800 tons and in the ocean at 21,800 tons annually.

22.2.1 Sources

More than 90% of the world's nickel is obtained from pentlandite ($(FeNi)_9S_8$), a nickel-sulfitic mineral, mined underground in Canada and the former Soviet Union. One of the largest sulfitic nickel deposits is in Sudbury, Ontario. Nickeliferous sulfide deposits are also found in Manitoba, South Africa, the former Soviet Union, Finland, western Australia, and Minnesota. Most of the rest of the nickel obtained is from nickel minerals such as laterite, a nickel oxide ore mined by open pit techniques in Australia, Cuba, Indonesia, New Caledonia, and the former Soviet Union. Lateritic ores are less well defined than sulfitic ores, although the nickel content (1–3%) of both ores is similar. Important deposits of laterite are located in New Caledonia, Indonesia, Guatemala, the Dominican Republic, the Philippines, Brazil, and especially Cuba which holds 35% of the known reserves. Nickel-rich nodules are found on the ocean floor, and nickel is also present in fossil fuels.

Total world mine production of nickel increased steadily from 7500 metric tons in 1900 to 2 million tons in 2000. (Table 22.2). In 1980, nickel mine production in the United States was 14,500 tons or about 1.8% of the world total. In 1986, primary nickel production ceased in the United States; secondary nickel production from scrap became a major source of nickel for industrial applications. In 1988, the United States imported 186,000 tons of primary nickel; Canada supplied 58% of the total and Norway 14%. In 1990, Canada produced 196,606 metric tons of nickel. About 63% of the total Canadian production was exported, mostly (56%) to the United States.

Natural sources of airborne nickel include soil dust, sea salt, volcanoes, forest fires, and

Table 22.2. World mine production of nickel.

Year	Metric Tons
1900	7500
1925	42,700
1950	141,000
1970	694,100
1975	753,000[a]
1980	784,100
1985	821,000[b]
2000	>2,000,000

[a]About 32% from Canada, 18% from New Caledonia, 17% from the former Soviet Union, 10% from Australia, 5% from Cuba, 4% from the Dominican Republic, 3% from the Republic of South Africa, 2% each from Greece, Indonesia, and the United States, and 5% from other countries.
[b]Mostly from Canada, the former Soviet Union, Australia, and Cuba in that order. The United States produced 6900 tons in 1985.

vegetation exudates and account for about 16% of the atmospheric nickel burden. Human sources of atmospheric nickel – which account for about 84% of all atmospheric nickel – include emissions from nickel ore mining, smelting, and refining activities; combustion of fossil fuels for heating, power, and motor vehicles; incineration of sewage sludges; nickel chemical manufacturing; electroplating; nickel–cadmium battery manufacturing; asbestos mining and milling; and cement manufacturing. In Canada in 1975, human activities resulted in the release of about 3000 tons of nickel into the atmosphere, mostly from metallurgical operations. Between 1973 and 1981, atmospheric emissions of nickel from stacks of four smelters in the Sudbury Basin, Canada, averaged a total of 495 tons annually. Industrial nickel dust emissions from a single Canadian stack 381 m high averaged 228 tons annually (range 53–342) between 1973 and 1981; this stack accounted for 396 tons annually (range 53–896) between 1982 and 1989. Three other emission stacks of Canadian nickel producers emitted an average of 226, 228, and 396 tons of nickel, respectively, each year between 1973 and 1989. Industrial emissions of nickel to the Canadian atmosphere in 1982 were estimated at 846 tons, mostly from nickel production in Ontario (48% of total) and Quebec (14%) and from industrial fuel combustion (17%). Nickel released into the air in Canada from smelting processes is likely in the form of nickel subsulfide (52%), nickel sulfate (20%), and nickel oxide (6%). Fuel combustion is also a major contributor of airborne nickel in Canada, mostly from combustion of petroleum. In the United States, yearly atmospheric emissions from coal and oil combustion are estimated at 2611 metric tons.

Chemical and physical degradation of rocks and soils, atmospheric deposition of nickel-containing particulates, and discharges of industrial and municipal wastes release nickel into ambient waters. Nickel enters natural waterways from wastewater because it is poorly removed by treatment processes. The main anthropogenic sources of nickel in water are primary nickel production, metallurgical processes, combustion and incineration of fossil fuels, and chemical and catalyst production. The primary human sources of nickel to soils are emissions from smelting and refining operations and disposal of sewage sludge or application of sludge as a fertilizer. Secondary sources include automobile emissions and emissions from electric power utilities. Weathering and erosion of geological materials release nickel into soils, and acid rain may leach nickel from plants into soils as well.

22.2.2 Uses

Most metallic nickel produced is used to manufacture stainless steel and other nickel alloys with high corrosion and temperature resistance. These alloys are used in ship building, jet turbines and heat elements, cryogenic installations, magnets, coins, welding rods, electrodes, kitchenware, electronics, and surgical implants; other nickel compounds are used in electroplating, battery production, inks, varnishes, pigments, catalysts, and ceramics. Some nickel compounds are preferred for use in nickel electroplating (nickel sulfate, nickel ammonium sulfate, nickel chloride, nickel fluoborate, nickel sulfamate), refining (nickel carbonyl), nickel–cadmium batteries (nickel hydroxide, nickel fluoride, nickel nitrate), manufacture of stainless steel and alloy steels (nickel oxide), electronic components (nickel carbonate), mordant in textile industry (nickel acetate),catalysts and laboratory reagents (nickel acetate, nickel hydroxide, nickel nitrate, nickel carbonate, nickel monosulfide, nickelocene), and some – such as nickel subsulfide – are unwanted toxic by-products.

In 1973, global consumption of nickel was 660,000 tons and that of the United States 235,000 tons. End uses of nickel in the United States in 1973 were transportation (21%), chemicals (15%), electrical goods (13%), fabricated metal products (10%), petroleum (9%), construction (9%), machinery (7%), and household appliances (7%); a similar pattern was evident for 1985. In 1988, 40% of all nickel intermediate products consumed was in the production of steel; 21% was in alloys, 17% in electroplating, and 12% in super alloys.

The pattern for 1985 was similar. In Canada, nickel is the fourth most important mineral commodity behind copper, zinc, and gold. In 1990, Canada produced 197,000 tons of nickel worth 2.02 billion dollars and was the second largest global producer of that metal. Most of the nickel used in the United States is imported from Canada and secondarily from Australia and New Caledonia.

Various nickel salts – including the sulfate, chloride, and bromide – were used in human medicine during the mid- to late 1800s to treat headache, diarrhea, and epilepsy and as an antiseptic. Therapeutic use of nickel compounds was abandoned in the early 1900s after animal studies demonstrated acute and chronic toxicity of these salts. Some nickel salts have been incorporated into fungicides to combat plant pathogens, although their use has not been approved by regulatory agencies.

22.3 *Properties*

Nickel normally occurs in the 0 and +2 oxidation states, although other oxidation states are reported. In natural waters, Ni^{2+} is the dominant chemical species in the form of $(Ni(H_2O)_6)^{2+}$. In alkaline soils, the major components of the soil solution are Ni^{2+} and $Ni(OH)^+$; in acidic soils, the main solution species are Ni^{2+}, $NiSO_4$, and $NiHPO_4$. Most atmospheric nickel is suspended onto particulate matter. Nickel interacts with numerous inorganic and organic compounds. Some of these interactions are additive or synergistic in producing adverse effects, and some are antagonistic.

Toxic and carcinogenic effects of nickel compounds are associated with nickel-mediated oxidative damage to DNA and proteins and to inhibition of cellular antioxidant defenses. Most authorities agree that albumin is the main transport protein for nickel in humans and animals and that nickel is also found in nickeloplasmin – a nickel-containing alpha-macroglobulin – and in an ultrafilterable serum fraction similar to a nickel–histidine complex. Normal routes of nickel intake for humans and animals are ingestion, inhalation, and absorption through the skin. Nickel absorption is governed by the quantities inhaled or ingested and by the chemical and physical forms of the nickel. Following oral intake by mammals, nickel was found mainly in the kidneys after short-term or long-term exposure to various soluble nickel compounds; significant levels of nickel were also found in the liver, heart, lung, and fat. Nickel also crosses the placental barrier, as indicated by increases in the levels of nickel in the fetuses of exposed mothers. Inhaled nickel carbonyl results in comparatively elevated nickel concentrations in lung, brain, kidney, liver, and adrenals. Parenteral administration of nickel salts usually results in high levels in kidneys and elevated concentrations in endocrine glands, liver, and lung. Nickel concentrations in whole blood, plasma, serum, and urine provide good indices of nickel exposure.

22.3.1 Physical and Chemical Properties

Nickel was first isolated in 1751, and a relatively pure metal was prepared in 1804. In nature, nickel is found primarily as oxide and sulfide ores. Nickel has high electrical and thermal conductivities and is resistant to corrosion at environmental temperatures between −20 and +30°C. Nickel, also known as carbonyl nickel powder or C.I. No. 77775, has a CAS number of 7440-02-0. Metallic nickel is a hard, lustrous, silvery white metal with a specific gravity of 8.9, a melting point of about 1455°C, and a boiling point at about 2732°C. It is insoluble in water and ammonium hydroxide, soluble in dilute nitric acid or aqua regia, and slightly soluble in hydrochloric and sulfuric acid. Nickel has an atomic weight of 58.71. Nickel is a composite of five stable isotopes: Ni-58 (68.3%), −60 (26.1%), −61 (1.1%), −62 (3.6%), and −64 (0.9%). Seven unstable isotopes have been identified: ^{56}Ni (half-life of 6 days), ^{57}Ni (36 h), ^{59}Ni (80,000 years), ^{63}Ni (92 years), ^{65}Ni (2.5 h), ^{66}Ni (55 h), and ^{67}Ni (50 s). Radionickel-59 (^{59}Ni) and ^{63}Ni are available commercially. In addition to the 0 and +2 oxidation states, nickel can also exist as −1, +1, +3, and +4.

Nickel enters surface waters from three natural sources: as particulate matter in rainwater, through the dissolution of primary bedrock materials, and from secondary soil phases. In aquatic systems, nickel occurs as soluble salts adsorbed onto or associated with clay particles, organic matter, and other substances. The divalent ion is the dominant form in natural waters at pH values between 5 and 9, occurring as the octahedral, hexahydrate ion $(Ni(H_2O)_6)^{2+}$. Nickel chloride hexahydrate and nickel sulfate hexahydrate are extremely soluble in water at 2400.0–2500.0 g/L. Less soluble nickel compounds in water include nickel nitrate (45.0 g/L), nickel hydroxide (0.13 g/L), and nickel carbonate (0.09 g/L). Nickel forms strong, soluble complexes with OH^-, SO_4^{2-}, and HCO_3^-; however, these species are minor compared with hydrated Ni^{2+} in surface water and groundwater. The fate of nickel in freshwater and marine water is affected by the pH, pE, ionic strength, type and concentration of ligands, and the availability of solid surfaces for adsorption. Under anaerobic conditions, typical of deep groundwater, precipitation of nickel sulfide keeps nickel concentrations low.

In alkaline soils, the major components of the soil solution are Ni^{2+} and $Ni(OH)^+$; in acidic soils, the main solution species are Ni^{2+}, $NiSO_4$, and $NiHPO_4$. Atmospheric nickel exists mostly in the form of fine respirable particles less than 2 μm in diameter, usually suspended onto particulate matter.

Nickel carbonyl ($Ni(CO)_4$) is a volatile, colorless liquid readily formed when nickel reacts with carbon monoxide; it boils at 43°C and decomposes at more than 50°C; this compound is unstable in air and is usually not measurable after 30 min. The intact molecule is absorbed by the lung and is insoluble in water but soluble in most organic solvents.

Analytical methods for detection of nickel in biological materials and water include various spectrometric, photometric, chromatographic, polarographic, and voltammetric procedures. Detection limits for the most sensitive procedures – depending on sample pretreatment, and extraction and enrichment procedures – were 0.7–1.0 ng/L in liquids, 0.01–0.2 $\mu g/m^3$ in air, 1.0–100.0 ng/kg in most biological materials, and 12.0 μg/kg in hair.

22.3.2 Metabolism

In mammalian blood, absorbed nickel is present as free hydrated Ni^{2+} ions, as small complexes, as protein complexes, and as nickel bound to blood cells. The partition of nickel among these four components varies according to the metal-binding properties of serum albumin, which is highly variable between species. A proposed transport model involves the removal of nickel from albumin to histidine via a ternary complex composed of albumin, nickel, and L-histidine. The low-molecular-weight L-histidine nickel complex can then cross biological membranes. Once inside the mammalian cell, nickel accumulates in the nucleus and nucleolus, disrupting DNA metabolism and causing cross-links and strand breaks. The observed redox properties of the nickel–histidine complex are crucial for maximizing the toxicity and carcinogenicity of nickel.

The acute toxicity and carcinogenicity of Ni_3S_2 and Ni_3S_2-derived soluble nickel (Ni^{2+}) in mice depend, in part, on the antioxidant capacity of target organs, which varies among different strains. Experimental evidence now supports the conclusion that the nickel-dependent formation of an activated oxygen species – including superoxide ion, hydrogen peroxide, and hydroxy radical – is a primary molecular event in acute nickel toxicity and carcinogenicity. For example, the superoxide radical (O_2^-) is an important intermediate in the toxicity of insoluble nickel compounds such as NiO and NiS. One of the keys to the mechanism of nickel-mediated damage is the enhancement of cellular redox processing by nickel. Accumulated nickel in tissues elicits the production of reactive oxygen species, such as the superoxide radical, as the result of phagocytosis of particulate nickel compounds and through the interaction of nickel ions with protein ligands, which promote the activation of the Ni^{2+}/Ni^{3+} redox couple. Thus, NiS and NiO can elicit the formation of O_2^-.

The most serious type of nickel toxicity is that caused by the inhalation of nickel carbonyl. The half-time persistence of nickel carbonyl in air is about 30 min. Nickel carbonyl can pass across cell membranes without metabolic alteration because of its solubility in lipids, and this ability of nickel carbonyl to penetrate intracellularly may be responsible for its extreme toxicity. In tissues, nickel carbonyl decomposes to liberate carbon monoxide and Ni^0, the latter being oxidized to Ni^{2+} by intracellular oxidation systems. The nickel portion is excreted with urine and the carbon monoxide is bound to hemoglobin and eventually excreted through the lungs. Nickel carbonyl inhibits DNA-dependent RNA synthesis activity, probably by binding to chromatin or DNA and thereby preventing the action of RNA polymerase, causing suppression of messenger-RNA-dependent induction of enzyme synthesis. The lung is the target organ in nickel carbonyl poisoning. Acute human exposures result in pathological pulmonary lesions, hemorrhage, edema, deranged alveolar cells, degeneration of bronchial epithelium, and pulmonary fibrosis. The response of pulmonary tissue to nickel carbonyl is rapid: interstitial edema may develop within 1 h of exposure and cause death within 5 days. Animals surviving acute exposures show lung histopathology.

Gastrointestinal intake of nickel by humans is high compared to some other trace metals because of contributions of nickel from utensils and from food processing machinery; average human dietary values range from 300 to 500 μg daily with absorption from the gastrointestinal tract of 1–10%. In humans, nearly 40 times more nickel was absorbed from the gastrointestinal tract when nickel sulfate was given in the drinking water (27%) than when it was given in the diet (0.7%). Uptake was more rapid in starved individuals. Dogs and rats given nickel, nickel sulfate hexahydrate, or nickel chloride in the diet or by gavage rapidly absorbed 1–10% of the nickel from the gastrointestinal tract, while unabsorbed nickel was excreted in the feces.

During occupational exposure, respiratory absorption of soluble and insoluble nickel compounds is the major route of entry, with gastrointestinal absorption secondary entry. Inhalation exposure studies of nickel in humans and test animals show that nickel localizes in the lungs, with much lower levels in liver and kidneys. About half the inhaled nickel is deposited on bronchial mucosa and swept upward in mucous to be swallowed; about 25% of the inhaled nickel is deposited in the pulmonary parenchyma. The relative amount of inhaled nickel absorbed from the pulmonary tract is dependent on the chemical and physical properties of the nickel compound. Pulmonary absorption into the blood is greatest for nickel carbonyl vapor; about half the inhaled amount is absorbed. Nickel in particulate matter is absorbed from the pulmonary tract to a lesser degree than nickel carbonyl; however, smaller particles are absorbed more readily than larger particles. Large nickel particles (>2.0 μm in diameter) are deposited in the upper respiratory tract; smaller particles tend to enter the lower respiratory tract. In humans, 35% of the inhaled nickel is absorbed into the blood from the respiratory tract; the remainder is either swallowed or expectorated. Soluble nickel compounds were more readily absorbed from the respiratory tract than insoluble compounds. In rodents, the half-time persistence of nickel particles was a function of particle diameter: 7.7 months for particles 0.6 μm in diameter, 11.5 months for particles 1.2 μm in diameter, and 21 months for particles 4.0 μm in diameter. In rodents, a higher percentage of insoluble nickel compounds was retained in the lungs for a longer time than soluble nickel compounds, and the lung burden of nickel decreased with increasing particle size. Nickel retention was 6–10 times greater in rodents exposed to insoluble nickel subsulfide compared to soluble nickel sulfate. Lung burdens of nickel generally increased with increasing duration of exposure and increasing concentrations of various nickel compounds in the air. Animals exposed to nickel carbonyl by inhalation exhale some of the respiratory burden in 2–4 h. The remainder is slowly degraded to divalent nickel, which is oxidized, and carbon monoxide, which initially binds to hemoglobin, with nickel eventually undergoing urinary excretion.

Dermal absorption of nickel occurs in animals and humans and is related to nickel-induced hypersensitivity and skin disorders. Absorption of nickel sulfate from the skin is reported for guinea pigs, rabbits, rats, and humans. Nickel ions in contact with the skin surface diffuse through the epidermis and combine with proteins; the body reacts to this conjugated protein. Nickel penetration of the skin is enhanced by sweat, blood and other body fluids, and detergents. Absorption is related to the solubility of the compound, following the general relation of nickel carbonyl, soluble nickel compounds, and insoluble nickel compounds, in that order; nickel carbonyl is the most rapidly and completely absorbed nickel compound in mammals. Anionic species differ markedly in skin penetration: nickelous ions from a chloride solution pass through skin about 50 times faster than do nickelous ions from a sulfate solution. Radionickel-57 (^{57}Ni) accumulates in keratinous areas and hair sacs of the shaved skin of guinea pigs and rabbits following dermal exposure. After 4 h, ^{57}Ni was found in the stratum corneum and stratum spinosum; after 24 h, ^{57}Ni was detected in blood and kidneys, with minor amounts in liver. As much as 77% of nickel sulfate applied to the occluded skin surface of rabbits and guinea pigs was absorbed within 24 h; sensitivity to nickel did not seem to affect absorption rate. In humans, some protection against nickel may be given by introducing a physical barrier between the skin and the metal, including fingernail polish, a polyurethane coating, dexamethasone, or disodium EDTA.

Nickel retention in the body of mammals is low. The half-time residence of soluble forms of nickel is several days, with little evidence for tissue accumulation except in the lung. Radionickel-63 (^{63}Ni) injected into rats and rabbits cleared rapidly; most (75%) of the injected dose was excreted within 24–72 h. Nickel clears at different rates from various tissues. In mammals, clearance was fastest from serum, followed by kidney, muscle, stomach, and uterus; relatively slow clearance was evident in skin, brain, and especially lung. The half-time persistence in human lung for insoluble forms of nickel is 330 days.

The excretory routes for nickel in mammals depend on the chemical forms of nickel and the mode of nickel intake. Most (>90%) of the nickel that is ingested in food remains unabsorbed within the gastrointestinal tract and is excreted in the feces. Urinary excretion is the primary route of clearance for nickel absorbed through the gastrointestinal tract. In humans, nickel excretion in feces usually ranges between 300.0 and 500.0 μg daily, or about the same as the daily dietary intake; urinary levels are between 2.0 and 4.0 μg/L. Dogs fed nickel sulfate in the diet for as long as 2 years excreted most of the nickel in feces and 1–3% in the urine. Biliary excretion occurs in rats, calves, and rabbits, but the role of bile in human metabolism of nickel is not clear. Absorbed nickel is excreted in the urine regardless of the route of exposure. The excretory route of inhaled nickel depends on the solubility of the nickel compound. Inhalation studies show that rats excrete 70% of the nickel in soluble nickel compounds through the urine within 3 days and 97% in 21 days. Less soluble nickel compounds (nickel oxide, nickel subsulfide) are excreted in urine (50%) and feces (50%); 90% of the initial dose of nickel subsulfide was excreted within 35 days, and 60% of the nickel oxide – which is less soluble and not as rapidly absorbed as nickel subsulfide – was excreted in 90 days. The half-time persistence of inhaled nickel oxide is 3 weeks in hamsters. In addition to feces, urine, and bile, other body secretions including sweat, tears, milk, and mucociliary fluids are potential routes of excretion. Sweat may constitute a major route of nickel excretion in tropical climates. Nickel concentrations in sweat of healthy humans sauna bathing for brief periods were 52.0 μg/L in males and 131.0 μg/L in females. Hair deposition of nickel also appears to be an excretory mechanism (as much as 4.0 mg Ni/kg dry weight (DW) hair in humans), but the relative magnitude of this route, compared to urinary excretion, is unclear. In the case of nickel compounds administered by way of injection, tests with small laboratory animals show that nickel is cleared rapidly from the plasma and excreted mainly in the urine. About 78% of an injected dose of nickel salts was excreted in the urine during the first 3 days after injection in rats and during the first day in rabbits. Exhalation via the lungs is the primary route of excretion during the first hours following injection of nickel carbonyl into rats, and afterwards via the urine.

In microorganisms, nickel binds mainly to the phosphate groups of the cell wall. From this site, an active transport mechanism designed for magnesium transports the nickel. In microorganisms and higher plants, magnesium is the usual competitor for nickel in the biological ion-exchange reactions. In lichens, fungi, algae, and mosses, the active binding sites are the carboxylic and hydroxycarboxylic groups fixed on the cell walls. Nickel in hyperaccumulating genera of terrestrial plants is complexed with polycarboxylic acids and pectins, although phosphate groups may also participate. In terrestrial plants, nickel is absorbed through the roots.

22.3.3 Interactions

In minerals, nickel competes with iron, cobalt, and magnesium because of similarities in their ionic radius and electronegativity. At the cellular level, nickel interferes with enzymatic functions of calcium, iron, magnesium, manganese, and zinc. Binding of nickel to DNA is inhibited by salts of calcium, copper, magnesium, manganese, and zinc. In toads (*Bufo arenarum*), ionic nickel interferes with voltage-sensitive ionic potassium channels in short muscle fibers. Among animals, plants, and microorganisms, nickel interacts with at least 13 essential elements: calcium, chromium, cobalt, copper, iodine, iron, magnesium, manganese, molybdenum, phosphorus, potassium, sodium, and zinc. Nickel interacts noncompetitively with all 13 elements and also interacts competitively with calcium, cobalt, copper, iron, and zinc. Quantification of these relationships would help clarify nickel-essential mineral interactions and the circumstances under which these interactions might lead to states of deficiency or toxicity. Mixtures of metals (arsenic, cadmium, copper, chromium, mercury, lead, zinc) containing nickel salts are more toxic to daphnids and fishes than are predicted on the basis of individual components. Additive joint action of chemicals, including nickel, should be considered in the development of ecotoxicologically relevant water quality criteria.

Nickel may be a factor in asbestos carcinogenicity. The presence of chromium and manganese in asbestos fibers may enhance the carcinogenicity of nickel, but this relation needs to be verified. Barium–nickel mixtures inhibit calcium uptake in rats, resulting in reduced growth. Pretreatment of animals with cadmium enhanced the toxicity of nickel to the kidneys and liver. Simultaneous exposure to nickel and cadmium – an industrial situation common in nickel and cadmium battery production – caused a significant increase in beta-2-macroglobulin excretion. Nickel or cadmium alone did not affect calcium kinetics of smooth muscle from bovine mesenteric arteries. However, mixtures of cadmium and nickel at greater than 100.0 nM inhibited the calcium function and may explain the vascular tension induced by nickel and other cations. Smooth muscle of the ventral aorta of the spiny dogfish (*Squalus acanthias*) contracted significantly on exposure to cadmium or nickel but not to other divalent cations. Cadmium-induced vasoconstriction of shark muscle, but not nickel, was inhibited by atropine. Nickel toxicity in soybeans (*Glycine max*) was inhibited by calcium, which limited the binding of nickel to DNA. Chromium–nickel mixtures were more-than-additive in toxicity to guppies (*Poecilia reticulata*) in 96-h tests. Rabbits (*Oryctolagus* sp.) exposed by inhalation to both nickel and trivalent chromium had more severe respiratory effects than did rabbits exposed to nickel alone. In natural waters, the geochemical behavior of nickel is similar to that of cobalt. It is therefore not surprising that nickel–cobalt mixtures in drinking water of rats were additive in toxicity and that there is a high correlation between nickel and cobalt concentrations in terrestrial plants.

Copper–nickel mixtures have a beneficial effect on growth of terrestrial plants but are more-than-additive in toxic action to aquatic plants. Nickel interacts with iron in rat nutrition and metabolism, but the interaction depends on the form and level of the dietary iron. Weanling rats fed diets containing nickel chloride and ferric sulfate had altered hematocrit, hemoglobin level, and alkaline phosphatase activity which did not occur when a mixture of ferric and ferrous sulfates were fed. In iron-deficient rats, nickel enhanced the absorption of iron administered as ferric sulfate, and nickel acted as a biological cofactor in facilitating gastrointestinal absorption of ferric ion when iron was given as ferric sulfate. Mice given a lead–nickel mixture in drinking water (57.0 mg Ni/L–200.0 mg Pb/L) for 12 days had increased urinary excretion of delta aminolevulinic acid and increased delta aminolevulinic dehydratase activity in erythrocytes when compared to groups given lead alone or nickel alone.

Magnesium competes with nickel in isolated cell studies. Treatment with magnesium reduces nickel toxicity, presumably through inhibition of nickel binding to DNA. Manganese also inhibits the binding of nickel to DNA, and manganese administration reduces the accumulation of nickel in some organs. Manganese dust inhibits nickel subsulfide-induced carcinogenesis in rats following simultaneous intramuscular injection of the two compounds. Also, nickel–manganese mixtures are less-than-additive in producing cytotoxicity of alveolar macrophages in rats. Nickel compounds enhance the cytotoxicity and genotoxicity of ultraviolet radiation, X-rays, and cytostatic agents such as *cis*-platinum, *trans*-platinum, and mitomycin C. Nickel is less-than-additive in toxicity to aquatic algae in combination with zinc. Treatment with zinc lessens nickel toxicity, presumably by competing with nickel in binding to DNA and proteins. Zinc binding sites of DNA-binding proteins, known as "finger loop domains," are likely molecular targets for metal toxicity. Ionic nickel has a similar ionic radius to Zn^{2+} and substitution is possible. Such substitution may disrupt nickel-induced gene expression by interfering with site-specific free radical reactions, which can result in DNA cleavage, formation of DNA protein crosslinks, and disturbance of mitosis.

Nickel also interacts with chelating agents, phosphatases, viruses, vitamins, and polycyclic aromatic hydrocarbons (PAHs). Chelating agents mitigate the toxicity of nickel by stimulating the excretion of nickel. Chelators reduced the toxicity of nickel to aquatic plants, presumably by lowering nickel bioavailability. Lipophilic chelating agents, such as triethylenetetramine and Cyclam (1,4,8,11-tetraazacyclotetradecane) are more effective in reducing toxicity than hydrophilic chelating agents such as EDTA, cyclohexanediamine tetraacetic acid, diethylenetriamine pentaacetic acid, and hydroxyethylenediamine triacetic acid. The greater efficacy of the lipophilic agents may be due to their ability to bind to nickel both intracellularly and extracellularly, while the hydrophilic agents can only bond extracellularly. Nickel irreversibly activates calcineurin, a multifunctional intracellular phosphatase normally activated by calcium and calmodulin. With nickel present, Newcastle Disease virus suppresses mouse L-cell interferon synthesis, suggesting virus–nickel synergism. Nickel interacts with vitamin C and has a synergistic effect on the carcinogenicities of various PAHs. Rats given intratracheal doses of nickel oxide and 20-methylcholanthrene develop squamous cell carcinomas more rapidly than with 20-methylcholanthrene alone. Simultaneous exposure of rats to benzopyrene and nickel subsulfide reduced the latency period of sarcomas by 30% and induced lung histopathology at a higher frequency than either agent alone. Also, tissue retention of PAH carcinogens is prolonged with nickel exposure.

22.4 Carcinogenicity, Mutagenicity, and Teratogenicity

Some forms of nickel are carcinogenic to humans and animals. Carcinogenicity of nickel compounds varies significantly with the chemical form of nickel, route of exposure, animal model used (including intraspecies strain differences), dose, and duration of exposure. In tests with small laboratory mammals, inducement of carcinomas of the types found in humans has only been accomplished following exposures by the respiratory route. Inhalation studies with nickel subsulfide and nickel oxide show evidence of carcinogenicity in mammals and humans; however, the evidence based on oral or cutaneous exposure to these and other nickel compounds is either negative or inconclusive. Nickel carbonyl and metallic nickel are carcinogenic in experimental animals, but data regarding their carcinogenicity in humans are inconclusive.

Certain nickel compounds are weakly mutagenic in a variety of test systems, but much of the evidence is inconclusive or negative. Mutagenicity – as measured by an increased frequency of sister chromatid exchange, chromosome aberrations, cell transformations, spindle disturbances, and dominant lethal effects – is induced by various nickel compounds at high concentrations in isolated cells of selected mammals including humans; however, these effects have not been observed in vivo. Nickel mutagenesis is thought to occur through inhibition of DNA synthesis and excision repair, resulting in an increased frequency of cross-links and strand breaks. DNA strand breaks occur in rat cells exposed to 5.0–40.0 mg Ni/kg medium as nickel carbonate; similar effects occur in hamster cells at 10.0–2000.0 mg Ni/kg medium as nickel chloride and nickel subsulfide, and in human cells with nickel sulfate. The ability of a particular nickel compound to cause mutations is considered proportional to its cellular uptake; however, data on nickel bioavailability to cells is scarce.

No teratogenic effects of nickel compounds occur in mammals by way of inhalation or ingestion except from nickel carbonyl. However, injection of low nickel doses results in consistent fetal malformations, particularly when nickel is administered during the organogenic stage of gestation of mammals or during the early development of domestic chick embryos. Injected doses causing teratogenic effects in rodents were as low as 1.0–1.2 mg Ni/kg body weight (BW), although more malformations resulted at higher dosages (2.3–4.0 mg/kg BW), which also increased fetal mortality and toxicity in the dam. Possible causes of nickel-induced malformations include direct toxicity from high transplacental nickel levels, reduced availability of alpha-fetoprotein to fetuses, or an increase in maternal glucose levels, which induces hyperglycemia in fetuses.

22.4.1 Carcinogenicity

Epidemiological studies conducted some decades ago in England, Canada, Japan, Norway, Germany, Russia, New Caledonia, and West Virginia indicated that humans working in the nickel processing and refining industries – or living within 1 km of processing or refining sites – had a significantly increased risk of developing fatal cancers of the nose, lungs, larynx, and kidneys, and a higher incidence of deaths from nonmalignant respiratory disease. Nasal cancers in nickel refinery workers were similar to those of the general population; however, lung cancers of nickel refinery workers had a higher frequency of squamous cell carcinomas. Smoking of tobacco contributed to the development of lung cancers in the nickel-exposed workers. Smoking about 15 cigarettes daily for one year adds about 1930 μg of nickel, as nickel carbonyl, to the human lung; this amount is equivalent to a carcinogenic dose of nickel for rats. Symptoms of cancer in humans may occur 5–35 years after exposure. The incidence of human lung and nasal cancers in occupationally exposed workers are related to nickel concentration and duration of exposure. Nickel compounds implicated as carcinogens include insoluble dusts of nickel subsulfide (Ni_3S_2) and nickel oxides (NiO, Ni_2O_3), the vapor of nickel carbonyl ($Ni(CO)_4$), and soluble aerosols of nickel sulfate ($NiSO_4$), nickel nitrate ($NiNO_3$), and nickel chloride. Soluble nickel compounds, though toxic, have relatively low carcinogenic activities. In general, carcinogenicity of nickel compounds is inversely related to its solubility in water, the least soluble being the most active carcinogen. The highest risk to humans of lung and nasal cancers comes from exposure to respirable particles of metallic nickel, nickel sulfides, nickel oxide, and the vapors of nickel carbonyl. Cancers were most frequent when workers were exposed to soluble nickel compounds at concentrations greater than 1.0 mg Ni/m^3 air and to exposure to less soluble compounds at greater than 10.0 mg Ni/m^3 air. Nickel subsulfide appears to be the nickel compound most carcinogenic to humans, as judged by animal studies and epidemiological evidence. The death rate of nickel workers from cancer has declined significantly since the mid-1920s because of improved safety and awareness.

The underlying biochemical mechanisms governing the carcinogenicity of various nickel compounds are imperfectly understood. There is general agreement that intracellular nickel accumulates in the nucleus, especially the nucleolar fraction. Intracellular binding of nickel to nuclear proteins and nuclear RNA and DNA may cause strand breakage and other chromosomal aberrations, diminished RNA synthesis and mitotic activity, and gene expression. A key mechanism of the transformation of tumorous cells involves DNA damage resulting from mutation caused by hydroxy radical or other oxidizing species. Alterations in cytokine (also known as tumor necrosis factor) production is associated with fibrotic lung injury in rats. Inhaled nickel oxide is known to increase cytokine production in rats.

Nickel entering the digestive tract of mammals is likely to be noncarcinogenic. Chronic ingestion studies of various nickel compounds that lasted as long as 2 years using several species of mammals show no evidence of carcinogenesis. Inhalation is the dosing route most relevant to human occupational exposure and probably an important route for wildlife exposure in the case of nickel powder, nickel carbonyl, and nickel subsulfide.

Inhalation of airborne nickel powder at 15.0 mg Ni/m^3 air causes an increased frequency of lung anaplastic carcinomas and nasal cancers in rodents and guinea pigs, especially when the particles are less than 4.0 μm in diameter. Rats exposed to airborne dusts of metallic nickel at 70.0 mg Ni/m^3 air for 5 h daily, 5 days weekly over 6 months had a 40% frequency of lung cancers; the latent period for tumor development was 17 months. A similar case is made for nickel sulfide and nickel oxide. In Canada, however, metallic nickel is considered "unclassifiable with respect to carcinogenicity" due to the limitations of identified studies. Inhaled nickel carbonyl is carcinogenic to the lungs of rats, a species generally considered to be peculiarly resistant to pulmonary cancer. Pulmonary cancers developed in rats 24–27 months after initial exposure to nickel carbonyl, and growth and survival of rats during chronic exposure were markedly reduced. Rats exposed to air containing 250.0 μg nickel carbonyl/L for only 30 min had a 4% incidence of lung cancer in 2-year survivors vs. zero percent in controls; rats exposed to 30.0–60.0 μg/L air for 30 min, three times weekly for 1 year had a 21% incidence of lung cancer in 2-year survivors. Inhaled nickel oxides do not seem to be tumorigenic to hamsters at concentrations of 1.2 mg Ni/m^3 air during exposure for 12 months. Hamsters did not develop lung tumors during lifespan inhalation exposure to nickel oxide; however, inhaled nickel oxide enhanced nasal carcinogenesis produced by diethylnitrosamine. Inhalation of nickel subsulfide produced malignant lung tumors and nasal cancers in rats in a dose-dependent manner. Rats develop benign and malignant lung tumors (14% frequency vs. zero percent in controls) after exposure for 78 weeks (6 h daily, 5 days weekly) to air containing 1.0 mg Ni/m^3 (as nickel subsulfide; particles <1.5 μm in diameter) and during a subsequent 30-week observation period.

Local sarcomas may develop in humans and domestic animals at sites of nickel implants and prostheses made of nickel. Latency of the implant sarcomas varies from 1 to 30 years in humans (mean, 10 years) and from 1 to 11 years in dogs (mean, 5 years). No cases of malignant tumors are reported at sites of dental nickel prostheses.

Injection site tumors are induced by many nickel compounds that do not cause cancer in animals by other routes of exposure. In fact, most of the published literature on nickel carcinogenesis concerns injected or implanted metallic nickel or nickel compounds; these data are of limited value in determining carcinogenic exposure levels for avian and terrestrial wildlife. The applicability of these studies to a recommendation for human workplace exposure is also questionable. Nevertheless, injection or implantation site sarcomas have been induced by many nickel compounds after one or repeated injections or implantations in rats, mice, hamsters, guinea pigs, rabbits, and cats. Nickel compounds known to produce sarcomas or malignant tumors by these routes of administration (implantation, intratracheal, intramuscular, intraperitoneal, subcutaneous, intrarenal, intravenous, intratesticular, intraocular, intraosseus, intrapleural, intracerebral, intrahepatic, intraarticular, intrasubmaxillary, intraadipose, intramedullary) include nickel subsulfide, nickel carbonyl, nickel powder or dust, nickel oxide, nickel hydroxide, nickel acetate, nickel fluoride, nickelocene, nickel sulfate, nickel selenide, nickel carbonate, nickel chromate, nickel arsenide, nickel telluride, nickel antimonide, nickel–iron matte, nickel ammonium sulfate and nickel monosulfide.

Some parenteral routes of administration were less effective than others in producing an increase in the frequency of benign or malignant tumors, including intravenous, submaxillary, and intrahepatic injection routes. Some nickel compounds are more effective at inducing tumors than others, for example, nickel sulfate and nickel acetate induce tumors in the peritoneal cavity of rats after repeated intraperitoneal injections but nickel chloride does not. Likewise, some species are more sensitive to tumor induction by injection than others; rats, for example, are more sensitive than hamsters. Most nickel compounds administered by way of injection usually produce responses at the site of injection; however, nickel acetate injected intraperitoneally produced pulmonary carcinomas in mice. Some carcinogenic nickel compounds produce tumors only when a threshold dose is exceeded, and some strains of animals are more sensitive than others. In one study, three strains of male mice (*Mus* sp.) were given a single intramuscular injection of 0.5, 2.5, 5.0, or 10.0 mg nickel subsulfide per mouse – equivalent to 19.0, 95.0, 190.0, or 380.0 mg Ni_3S_2/kg BW – and observed for 78 weeks for tumor development. Nickel subsulfide is a water-insoluble compound suspected to damage cells through oxidative mechanisms. The highest dose injected was lethal (53–93% dead) within 7 days. The final incidence of sarcomas in the 5.0 mg/mouse groups ranged between 40 and 97%, with decreased survival and growth noted in all test groups. In the most sensitive strain tested, there was a dose-dependent increase in tumor frequency, with a significant increase in tumors at the lowest dose tested.

Carcinogenic properties of nickel are modified by interactions with other chemicals. Nickel–cadmium battery workers exposed to high levels of both nickel and cadmium have an increased risk of lung cancer when compared to exposure from cadmium alone. Some nickel compounds interact synergistically with known carcinogens. Nickel chloride enhances the renal carcinogenicity of *N*-ethyl-*N*-hydroxyethyl nitrosamine in rats. Metallic nickel powder enhances lung carcinogenicity of 20-methylcholanthrene when both are administered intratracheally to rats. Nickel subsulfide in combination with benzo[a]pyrene shortens the latency time to local tumor development and produces a disproportionately higher frequency of malignant tumors. Nickel sulfate enhanced dinitrosopiperazine carcinogenicity in rats. And nickel potentiated the specific effects of cobalt in rabbits by enhancing the formation of lung nodules. Some chemicals inhibit nickel-induced carcinogenicity. Carcinogenicity induced by nickel subsulfide is reduced by manganese dust. Manganese protects male guinea pigs against tumorigenesis induced by nickel subsulfide, possibly due to the stimulating effect of manganese on macrophage response and by displacing nickel from the injection site. Sodium diethyldithiocarbamate reduced tumor incidence in rats implanted with nickel subsulfide. And magnesium acetate and calcium acetate inhibit lung adenoma formation in mice treated intraperitoneally with nickel acetate. Nickel interactions with other suspected carcinogens, such as chromium, merit additional research. Nickel and other trace metals in asbestos fibers are responsible, in part, for the pulmonary carcinogenicity found in asbestos workers. Nickel–sulfur mineral complexes may also have carcinogenic potential; a similar case is made for the corresponding arsenides, selenides, and tellurides.

22.4.2 Mutagenicity

Nickel salts gave no evidence of mutagenesis in tests with viruses, and bacterial mutagenesis tests of nickel compounds have consistently yielded negative or inconclusive results. However, nickel chloride and nickel sulfate were judged to be mutagenic or weakly mutagenic in certain bacterial eukaryotic test systems. Nickel subsulfide was positively mutagenic to the protozoan *Paramecium* sp. at 0.5 mg Ni/L. Ionic Ni^{2+} was mutagenic to *Escherichia coli*; mutagenesis was enhanced by the addition of both hydrogen peroxide and tripeptide glycyl-L-histidine, suggesting that short-lived oxygen free radicals are generated. Nickel chloride hexahydrate induced respiratory deficiency in yeast cells, but this may be a cytotoxic effect rather than a gene mutation.

Nickel is weakly mutagenic to plants and insects. Abnormal cell divisions occur in roots of the broad bean (*Vicia faba*) during exposure to various inorganic nickel salts at nickel concentrations of 0.1–1000.0 mg/L. All nickel salts tested produced more abnormal cell divisions than did controls. In beans, nickel nitrate was the most effective inorganic nickel compound tested in producing deformed cells, abnormal arrangement of chromatin, extra micronuclei, and evidence of cell nucleus disturbances; however, nickel salts showed only weak mutagenic action on rootlets of peas (*Pisum* sp.). Nickel sulfate induced chromosomal abnormalities in root tip cells of onions, *Allium* sp. and caused sex-linked recessive mutations in the fruit fly (*Drosophila melanogaster*) at 200.0–400.0 mg Ni/L culture medium.

Human cells exposed to various nickel compounds have an increased frequency of chromosomal aberrations, although sister chromatid exchange frequency is unaffected. Cells from nickel refinery workers exposed to nickel monosulfide (0.2 mg Ni/m^3) or nickel subsulfide (0.5 mg Ni/m^3) showed a significant increase in the incidence of chromosomal aberrations. No correlation was evident between nickel exposure level and the frequency of aberrations.

In Chinese hamster ovary cells, nickel chloride increased the frequency of chromosomal aberrations and sister chromatid exchanges. The cells with aberrations increased from 8% at about 6.0 μg Ni/L to 21% at about 6.0 mg Ni/L in a dose-dependent manner. There is a large difference in the mutagenic potential of soluble and insoluble nickel compounds, which seems to reflect the carcinogenic potential of these forms of nickel. For example, insoluble particles less than 5.0 μm in diameter of crystalline nickel subsulfide – a carcinogen – produced a strong dose-dependent mutagenic response in Chinese hamster ovary cells up to 80 times higher than untreated cells; however, soluble nickel sulfate produced no significant increase in mutational response over background in Chinese hamster ovary cells. A similar response is reported for Syrian hamster embryo cells. Interactions of carcinogens and soluble nickel salts need to be considered. Benzo[a]pyrene, for example, showed a comutagenic effect with nickel sulfate in hamster embryo cells.

In rats, nickel carbonyl is reported to cause dominant lethal mutations, but this needs verification. Nickel sulfate, when given subcutaneously at 2.4 mg Ni/kg BW daily for 120 days causes infertility; testicular tissues are adversely affected after the first injection. Nickel salts given intraperitoneally to rats at 6.0 mg Ni/kg BW daily for 14 days did not produce significant chromosomal changes in bone marrow or spermatogonial cells.

In mice, nickel chloride produces a dose-dependent increase in abnormal lymphoma cells. Mice given high concentrations of nickel in drinking water, equivalent to 23.0 mg Ni/kg BW daily and higher, have an increased incidence of micronuclei in bone marrow. However, mice injected once with 50.0 mg Ni/kg BW as nickel chloride show no evidence of mutagenicity.

22.4.3 Teratogenicity

Nickel carbonyl at high doses is a potent animal teratogen. Inhalation exposure to nickel carbonyl caused fetal death and decreased weight gain in rats and hamsters and eye malformations in rats. Studies on hamsters, rats, mice, birds, frogs, and other species suggest that some individuals are susceptible to reproductive and teratogenic effects when given high doses of nickel by various routes of administration. Intravenous injection of nickel sulfate to hamsters at 2.0–25.0 mg/kg BW on day 8 of gestation produces developmental abnormalities. Teratogenic malformations – including poor bone ossification, hydronephrosis, and hemorrhaging – occur in rats when nickel is administered during organogenesis, and these malformations are maximal at dose levels toxic for the dam. A dose of 4.0 mg/kg BW given intraperitoneally on day 12 or 19 of pregnancy is teratogenic in rats. Rats exposed continuously for three generations to drinking water containing 5.0 mg Ni/L produce smaller litters, higher offspring mortality, and fewer males; an increase in the number of runts suggests that transplacental toxicity occurs.

Divalent nickel is a potent teratogen for the South African clawed frog (*Xenopus laevis*). Frog embryos actively absorb Ni^{2+} from the medium and develop ocular, skeletal, craniofacial, cardiac, and intestinal malformations. A Ni^{2+}-binding serpin, *pNiXa*, is abundant in clawed frog oocytes and embryos; binding of Ni^{2+} to *pNiXa* may cause embryotoxicity by enhancing oxidative reactions that produce tissue injury and genotoxicity. Another Ni^{2+}-binding protein, *pNiXc*, isolated from mature oocytes of the clawed frog, was identified as a monomer of fructose-1,6-bisphosphate aldolase A and raises the possibility that aldolase A is a target enzyme for nickel toxicity.

Nickel is embryolethal and teratogenic to white leghorn strains of the domestic chicken (*Gallus* sp.), possibly due to the mitosis-inhibiting activity of nickel compounds. Fertilized chicken eggs injected with 0.02–0.7 mg Ni/egg as nickel chloride on days 1–4 of incubation show a dose-dependent response. All dose levels of nickel tested were teratogenic to chickens. Malformations include poorly developed or missing brain and eyes, everted viscera, short and twisted neck and limbs, hemorrhaging, and a reduction in body size. Toxicity and teratogenicity are highest in embryos injected on day 2. Mallard (*Anas platyrhynchos*) ducklings from fertile eggs treated at age 72 h with 0.7 μg Ni as nickel mesotetraphenylporphine show a marked decrease in survival. Among survivors, there is a significant increase in the frequency of developmental abnormalities, a reduction in bill size, and a reduction in weight.

Changes in employment practices in North America and Europe have increased the proportion of women among workers in nickel mines and refineries and in nickel-plating industries and has increased the concern regarding possible fetal toxicity associated with exposures of pregnant women to nickel during gestation. One preliminary report strongly suggests that exposure to nickel of Russian female hydrometallurgy workers causes significantly increased risks for abortion, total defects, cardiovascular defects, and defects of the musculoskeletal system. Nickel was observed to cross the human placenta and produce teratogenesis and embryotoxicity, as judged by studies with isolated human placental tissues. Nickel disrupts lipid peroxidative processes in human placental membranes, and this metabolic change may be responsible for the observed decrease in placental viability, altered permeability, and embryotoxicity.

Nonteratogenic reproductive effects of nickel include increased resorption of embryos and fetuses, reduced litter size, testicular damage, altered rates of development and growth, and decreased fertility. Nickel compounds can penetrate the mammalian placental barrier and affect the fetus. Intravenous administration of nickel acetate (0.7–10.0 mg Ni/kg BW) to pregnant hamsters on day 8 of gestation resulted in dose-dependent increases in the number of resorbed embryos. Rats injected intramuscularly with nickel chloride on day 8 of gestation with 12.0 or 16.0 mg Ni/kg BW produced significantly fewer live fetuses than did controls. Three generations of rats given nickel in their diets at 250.0–1000.0 mg Ni/kg ration had increased fetal mortality in the first generation and reduced body weights in all generations at 1000.0 mg/kg. Litter sizes were reduced in pregnant rats fed nickel in various forms at 1000.0 mg Ni/kg ration. Rodents exposed to nickel during gestation show a decline in the frequency of implantation of fertilized eggs, enhanced resorption of fertilized eggs and fetuses, an increased frequency of stillbirths, and growth abnormalities in live-born young. Exposure of eggs and sperm of rainbow trout to 1.0 mg Ni/L as nickel sulfate for 30 min did not affect fertilization or hatchability; however, most exposed zygotes hatched earlier than the controls. Nickel salts produced testicular damage in rats and mice given oral, subcutaneous, or intratesticular doses of 10.0–25.0 mg Ni/kg BW; nickel-treated male rats were unable to impregnate females. Nickel sulfate at 25.0 mg Ni/kg BW daily for 120 days via the esophagus selectively damaged the testes of rats (inhibition of spermatogenesis) and resulted in a reduced procreative capacity; males were permanently infertile after 120 days on this regimen.

22.5 *Concentrations in Field Collections*

Nickel is ubiquitous in the biosphere and is the 24th most abundant element in the earth's crust with a mean concentration of 75.0 mg/kg. Nickel enters the environment from natural and human sources and is distributed throughout all compartments by means of chemical and physical processes and biological transport by living organisms. Nickel is found in air, soil, water, food, and household objects; ingestion or inhalation of nickel is common, as is dermal exposure. In general, nickel concentrations in plants, animals, and abiotic materials are elevated in the vicinity of nickel smelters and refineries, nickel–cadmium battery plants, sewage outfalls, and coal ash disposal basins. A global inventory estimate of nickel shows that living organisms contain about 14 million metric tons of nickel, mostly (98.8%) in terrestrial plants (Table 22.3). But plants and animals account for only 0.00000031% of the total nickel inventory estimate of 4500 trillion metric tons, the vast majority of the nickel being present in the lithosphere and other abiotic materials (Table 22.3).

Table 22.3. Inventory of nickel in various global environmental compartments.

Compartment	Mean Concentration, in mg/kg	Nickel in Compartment, in Metric Tons
Lithosphere, Down to 45 km	75.0	4,300,000,000,000,000
Sedimentary Rocks	48.0	120,000,000,000,000
Soils, to 100 cm	16.0	5,300,000,000,000
Oil Shale Deposits	30.0	1,400,000,000,000
Dissolved Oceanic	0.0006	840,000,000
Nickel Ore Reserves	>2000.0	160,000,000
Coal Deposits	15.0	150,000,000
Terrestrial Litter	15.0	33,000,000
Terrestrial Plants	6.0	14,000,000
Suspended Oceanic Particulates	95.0	6,600,000
Crude Oil	10.0	2,300,000
Terrestrial Animals	2.5	50,000
Swamps and Marshes	7.0	42,000
Lakes and Rivers, Total	0.001	34,000
Consumers/Reducers (Biological)	3.5	11,000
Atmosphere	0.3	1500
Oceanic Plants	2.5	500
Lakes and Rivers, Plankton	4.0	230

22.5.1 Abiotic Materials

Nickel concentrations are elevated in air, water, soil, sediment, and other abiotic materials in the vicinity of nickel mining, smelting, and refining activities; in coal flyash; in sewage sludge; and in waste-water outfalls. Maximum concentrations of nickel found in abiotic materials were 15,300.0 ng/L in air under conditions of extreme occupational exposure, 19.2 μg/L in seawater, 30.0 μg/L in rain, 240.0 μg/L in sewage liquids, 300.0 μg/L in drinking water near a nickel refinery, 500.0 μg/kg in snow, 4430.0 μg/L in groundwater, 27,200.0 μg/L in wastewater from nickel refineries, 1600.0 mg/kg in coal flyash, 183.0 mg/L in freshwater near a nickel refinery, 2000.0 mg/kg in ultramific rocks, 24,000.0 mg/kg in soils near metal refineries, 53,000.0 mg/kg in sewage sludge, more than 100,000.0 mg/kg in lake sediments near a nickel refinery, and 500,000.0 mg/kg in some meteorites.

Nickel in the atmosphere is mainly in the form of particulate aerosols resulting from human activities. Air concentrations of nickel are elevated near urbanized and industrialized sites and near industries that process or use nickel. The greatest contributor to atmospheric nickel loadings is combustion of fossil fuels in which nickel appears mainly as nickel sulfate, nickel oxide, and complex metal oxides containing nickel. Nickel concentrations in the atmosphere of the United States are highest in winter and lowest in summer, demonstrating the significance of oil and coal combustion sources. Nickel in the atmosphere is removed through rainfall and dry deposition, locating into soils and sediments; atmospheric removal usually occurs in several days. When nickel is attached to small particles, however, removal can take more than a month. Cigarette smoke contributes significantly to human intake of nickel by inhalation; heavy smokers can accumulate as much as 15.0 μg of nickel daily from this source.

Most unpolluted Canadian rivers and lakes sampled between 1981 and 1992 contained 0.1–10.0 μg Ni/L; however, natural waters near industrial sites may contain 50.0–2000.0 μg Ni/L. Nickel concentrations in snow from Montreal, Canada, are high compared with ambient air; nickel burdens in Montreal snow are positively correlated with those of vanadium, strongly suggesting that combustion of fuel oil is a major source of nickel. In drinking water, nickel levels may be elevated due to the corrosion of nickel-containing alloys used in the water distribution system and from nickel-plated faucets. Nickel concentrations in uncontaminated surface waters are usually lower with increasing salinity or phosphorus loadings. Nickel tends to accumulate in the oceans and leaves the ocean as sea spray aerosols which release nickel-containing particles into the atmosphere.

Sediment nickel concentrations are grossly elevated near the nickel–copper smelter at Sudbury, Ontario, and downstream from steel manufacturing plants. Sediments from nickel-contaminated sites have between 20.0 and 5000.0 mg Ni/kg DW; these values are at least 100 times lower at comparable reference sites. A decrease in the pH of water caused by acid rain may release some of the nickel in sediments to the water column. Transfer of nickel from water column to sediments is greatest when sediment particle size is comparatively small and sediments contained high concentrations of clays or organics.

In soils, nickel exists in several forms, including inorganic crystalline minerals or precipitates, as free ion or chelated metal complexes in soil solution, and in various formulations with inorganic cationic surfaces. Soil nickel is preferentially adsorbed onto iron and manganese oxides; however, near Sudbury, Ontario, soil nickel is mostly associated with inorganic sulfides. The average residence time of nickel in soils is estimated at 3500 years, as judged by nickel concentrations in soils and estimates of the loss of nickel from continents. Natural levels of soil nickel are augmented by contamination from anthropogenic activities including atmospheric fallout near nickel-emitting industries, automobile traffic, and treatment of agricultural lands with nickel-containing phosphate fertilizers or municipal sewage sludge. Soils with less than 3.0 mg Ni/kg DW are usually too acidic to support normal plant growth.

Nickel availability to plants grown in sludge-amended soils is correlated with soil-solution nickel. Sewage-derived fertilizers from industrial areas may contain 1000.0 mg Ni/kg DW or more. In sewage sludge, a large percentage of the nickel exists in a form that is easily released from the solid matrix. Water solubility of nickel in soils and its bioavailability to plants are affected by soil pH, with decreases in pH below 6.5 generally mobilizing nickel.

22.5.2 Terrestrial Plants and Invertebrates

Nickel is found in all terrestrial plants, usually at concentrations of less than 10.0 mg/kg DW. The majority of terrestrial plants are nickel-intolerant species and are restricted to soils of relatively low nickel content; some plants without specific nickel tolerance can accumulate anomalous levels of nickel, but at a cost of reduced metabolism. Plants grown on nickel-rich soils can accumulate high concentrations of nickel and crops grown in soils amended with sewage sludge may contain as much as 1150.0 mg Ni/kg DW. Vegetation near point sources of nickel, such as nickel refineries, have elevated nickel concentrations that decline with increasing distance from the source. Fruits and vegetables grown near nickel smelters contain 3–10 times more nickel in edible portions than those grown in uncontaminated areas. Trees, ferns, and grasses near nickel smelters had elevated concentrations of nickel: as much as 174.0 mg/kg DW in trees and ferns and 902.0 mg/kg DW in wavy hairgrass (*Deschampsia flexuosa*). Nickel concentrations in lichens and other vegetation were elevated when grown on nickeliferous rocks, serpentine soils, near nickel smelters, near urban and industrial centers, and near roadsides treated with superphosphate fertilizers.

Terrestrial vegetation within 3.5 km of one of the Sudbury, Ontario, smelters had as much as 140.0 mg Ni/kg DW; concentrations decreased with distance from the smelter, reaching a mean concentration of about 12.0 mg Ni/kg DW at a distance of 60 km. Some vegetation near a Sudbury smelter – including lawn grasses, timothy (*Phleum pratense*), and oats (*Avena sativa*) – showed signs of nickel toxicosis; concentrations in these species ranged between 80.0 and 150.0 mg Ni/kg DW. Vegetables – beets (*Beta vulgaris*), radishes (*Raphanus* spp.), cabbages (*Brassica oleracea capitata*), and celery (*Apium graveolans*) – grown in soils about 1 km from a nickel refinery had 40.0–290.0 mg Ni/kg DW in their top portions. All these vegetables had reduced yield, stunted growth, and chlorosis and necrosis, which is attributed to the high levels of nickel in local soils.

Mosses and lichens accumulate nickel readily and at least nine species are used to monitor environmental gradients of nickel. Maximum concentrations of nickel found in whole lichens and mosses from nickel-contaminated areas range between 420.0 and 900.0 mg/kg DW vs. 12.0 mg/kg DW from reference sites. Nickel concentrations in herbarium mosses worldwide have increased dramatically during this century. In one case, nickel concentrations in *Brachythecium salebrosum* from Montreal, Canada, rose from 6.0 mg/kg DW in 1905 to 105.0 mg/kg DW in 1971.

Nickel-tolerant or accumulator species of plants are likely to be found only on nickel-rich soils. Hyperaccumulator species usually grow on relatively infertile nickel-rich serpentine soils and contain more than 10,000.0 mg Ni/kg DW. Leaves from some genera of nickel hyperaccumulator plants, including *Alyssum*, *Homalium*, and *Hybanthus*, growing on soils derived from volcanic rocks, which are rich in nickel, accumulate nickel to concentrations of 120,000.0 mg kg DW. Nickel is bound as a citrate complex in hyperaccumulator plants from New Caledonia; however, nickel accumulator plants from other locations do not contain unusually high levels of citrate, and nickel is not present as a citrate complex but as a carboxylic acid complex.

Terrestrial plants take up nickel from soil primarily via the roots. The nickel uptake rate from soil is dependent on soil type, pH, humidity, organic content, and concentration of extractable nickel. For example, at soil pH less than 6.5 nickel uptake is enhanced due to breakdown of iron and manganese oxides that form stable complexes with nickel. The exact chemical forms of nickel that are most readily accumulated from soil and water are unknown; however, there is growing evidence that complexes of nickel with organic acids are the most favored. In addition to their uptake from the soils, plants consumed by humans may receive several milligrams of nickel per kilogram through leaching of nickel-containing alloys in food-processing equipment, milling of flour, and catalytic hydrogenation of fats and oils by use of nickel catalysts. Nickel reportedly disrupts nitrogen cycling and this could have serious ecological consequences for forests near nickel smelters, although adverse effects of nitrogen disruption by nickel need to be verified.

Data are limited on nickel concentrations in terrestrial invertebrates. Earthworms from uncontaminated soils may contain as much as 38.0 mg Ni/kg DW and workers of certain termite species may normally contain as much as 5000.0 mg Ni/kg DW. Larvae of the gypsy moth (*Porthetria dispar*) near a nickel smelter had 20.4 mg Ni/kg DW; concentrations in pupae and adults were lower because these stages have higher nickel elimination rates than larvae.

22.5.3 Aquatic Organisms

Nickel concentrations are comparatively elevated in aquatic plants and animals in the vicinity of nickel smelters, nickel–cadmium battery plants, electroplating plants, sewage outfalls, coal ash disposal basins, and heavily populated areas. For example, at Sudbury, Ontario, mean nickel concentrations, in mg/kg DW, were 22.0 in larvae of aquatic insects, 25.0 in zooplankton, and 290.0 in aquatic weeds; maximum concentrations reported were 921.0 mg/kg DW in gut of crayfish (*Cambarus bartoni*) and 52.0 mg/kg fresh weight (FW) in various fish tissues. For all aquatic species collected, nickel concentrations were highly variable between and within species; this variability is attributable, in part, to differential tissue uptake and retention of nickel, depth of collection, age of organism, and metal-tolerant strains.

The bioaccumulation of nickel under field conditions varies greatly among groups. Bioconcentration factors (BCFs, which equals the milligrams of nickel per kilogram fresh weight of the sample divided by the milligrams of nickel per liter in the medium) for aquatic macrophytes range from 6 in pristine areas to 690 near a nickel smelter; for crustaceans these values are 10–39; for mollusks, 2–191; and for fishes, 2–52. BCFs of 1700 have been reported for marine plankton, 800 and 40 for soft parts and shell, respectively, of some marine mollusks, and 500 for brown algae, suggesting that some food chain biomagnification may occur.

Concentrations of nickel in roots of *Spartina* sp. from the vicinity of a discharge from a nickel–cadmium battery plant on the Hudson River, New York, ranged between 30.0 and 500.0 mg/kg DW and reflected sediment nickel concentrations in the range of 100.0–7000.0 mg Ni/kg DW. The detritus produced from dead algae and macrophytes is the major food source for fungi and bacteria, and in this way nickel can again enter the food chain. Nickel concentrations in tissues of sharks from British and Atlantic water range between 0.02 and 11.5 mg/kg FW; concentrations were highest in fish-eating, mid-water species such as the blue shark (*Prionace glauca*) and tope shark (*Galeorhinus galeus*). Concentrations of nickel in livers of tautogs (*Tautoga onitis*) from New Jersey significantly decreased with increasing body length in both males and females; however, this trend was not observed in bluefish (*Pomatomus saltatrix*) or tilefish (*Lopholatilus chamaeleonticeps*).

22.5.4 Amphibians

In Maryland, U.S., nickel concentrations in tadpoles of northern cricket frogs (*Acris crepitans*) and gray treefrogs (*Hyla versicolor*) increased with increasing soil nickel concentrations, with maximum nickel concentrations recorded of 7.1 mg/kg DW in gray treefrogs and 10.0 mg/kg DW in northern cricket frogs. In study sites 9–66 km from Sudbury, Ontario, populations of treefrogs (*Hyla crucifer*) and American toads (*Bufo americanus*) declined. Population abundance of adult treefrogs declined with increasing atmospheric deposition of nickel, and abundance of toad tadpoles declined as nickel concentrations in pond water rose from 3.3 μg Ni/L at more distant sites to 19.5 μg Ni/L at sites near Sudbury.

22.5.5 Birds

Nickel concentrations in the organs of most avian wildlife species in unpolluted ecosystems range from about 0.1 to 2.0 mg/kg DW and occasionally reach 5.0 mg/kg DW. In nickel-contaminated areas, nickel concentrations were elevated in feathers, eggs, and internal tissues of birds when compared to conspecifics collected at reference sites. In contaminated ecosystems, mean nickel concentrations between 31.0 and 36.0 mg/kg DW occur in primary feathers of mallards (*Anas platyrhynchos*) collected 20–30 km from a nickel smelter, bone of the common tern (*Sterna hirundo*) from Hamilton Harbor, Ontario, and eggshell of the tree swallow (*Tachycineta bicolor*) from the Hackensack River, New Jersey.

Waterfowl feeding in areas subjected to extensive nickel pollution – such as smelters and nickel–cadmium battery plants – are at special risk because waterfowl food plants in those areas contain 500.0–690.0 mg Ni/kg DW. Dietary items of the ruffed grouse (*Bonasa umbellus*) near Sudbury, Ontario, had 32.0–95.0 mg Ni/kg DW, whereas nickel concentrations in grouse body tissues usually contain less than 10% of the dietary level. Nickel concentrations in aspen (*Populus tremula*) from the crop of ruffed grouse near Sudbury ranged from 62.0 mg/kg DW in May to 136.0 mg/kg DW in September, which shows the role of season in dietary nickel composition.

22.5.6 Mammals

Mammalian wildlife from uncontaminated habitats usually contain less than 0.1 to about 5.0 mg Ni/kg DW in tissues; in nickel-contaminated areas, these same species have 0.5 to about 10.0 mg Ni/kg DW in tissues, with a maximum of 37.0 mg/kg DW in kidneys of the common shrew (*Sorex araneus*). Nickel accumulations in wildlife vary greatly between species. For example, tissues of mice have higher concentrations of nickel than rats and other rodents while beavers and minks have higher nickel concentrations in their liver than birds in similar sites near Sudbury.

The highest concentrations in wildlife tissues from nickel-contaminated locales are associated with tissues exposed to the external environment, such as fur and skin; nickel concentrations in internal organs are usually similar, regardless of degree of contamination. However, nickel concentrations in bone, reproductive organs, and kidneys in certain herbivorous species of wildlife and livestock are elevated when compared to other internal tissues, especially in the vicinity of nickel smelters and other nickel point sources. Trophic position in the food chain, sex, and reproductive state do not seem to significantly influence the nickel body burdens of mammals, but age is an important variable and nickel generally increases in various organs with increasing age of terrestrial and marine mammals. Fetuses of a variety of wildlife and domestic species contain concentrations of nickel significantly lower than those in their mothers or in juveniles, suggesting that placental transfer of nickel is restricted. Nickel concentrations in aquatic macrophytes and lower plants in the vicinity of nickel smelters may approach or exceed dietary levels known to cause adverse effects in young animals. Sensitive species of wildlife ingesting this vegetation for extended periods could experience nickel-related toxicity or risk alterations in community structure as nickel-sensitive taxa are eliminated or their abundance is reduced.

Elevated nickel concentrations in Norwegian wildlife are linked to emissions from Russian nickel smelters. In Norway, nickel concentrations were elevated in livers and kidneys of moose (*Alces alces*) and caribou (*Rangifer tarandus*) because of atmospheric transport of wastes from nickel-processing plants of nearby Russian towns. In Russia between 1974 and 1992, three species of voles (*Clethrionomys glareolus*, *Clethrionomys rutilus*, *Lemmus lemmus*) were eliminated from the immediate vicinity of a copper–nickel smelter that discharged 2700 metric tons of nickel annually to the atmosphere, and these species were scarce at a moderately contaminated area 28 km south of the smelter. Declines were associated with a decrease of important food plants: lichens for *C. glareolus* and *C. rutilus*, mosses for *L. lemmus*, and seed plants for other species of *Clethrionomys*. Close to the smelter, direct toxic effects of accumulated nickel and other metals also may have reduced population densities. Nickel concentrations are also elevated in rodents, shrews, soil, vegetation, and earthworms in the vicinity of roads with high automobile density. In ruminant mammals, tissue nickel concentrations were higher in winter, presumably owing to increased combustion of fossil fuels.

Nickel is normally present in human tissues, and under conditions of high exposure, these levels may increase significantly. Nickel enters the human body through the diet, through inhalation, by absorption through the skin, and in medications. The diet accounts for about 97% of the total intake and drinking water about 2.5%. Foods rich in nickel include tea (7.6 mg/kg DW), cereals (6.5), vegetables (2.6), and fish (1.7 mg/kg DW). The daily dietary intake of nickel by humans in the United States ranges between 0.15 and 0.6 mg, almost all of which is excreted in the feces. Minor amounts are also excreted in sweat, urine, and hair. Residents of the Sudbury, Ontario, area who consume home-grown garden products ingest an average of 1.85 mg of nickel daily, of which 0.6 mg comes from the drinking water. Inhalation intake of nickel for residents of New York City is estimated at 2.4 μg daily; for Chicago, a maximum value of 13.8 μg daily is recorded; and 14.8 μg are inhaled daily by smokers of 40 cigarettes. Canadians in urban areas inhale 0.06–0.6 μg Ni daily; near nickel smelters this may increase to 15.0 μg daily. In Connecticut, serum nickel levels in newborns were normal (3.0 μg/L) and similar to those of their mothers. Nickel concentrations in human serum, however, are modified by disease and stress. Concentrations are usually elevated after strokes, pregnancy, and extensive burns and are depressed in cases of cirrhosis, hypoalbuminemia, extremes of heat, and uremia.

About 727,000 workers were potentially exposed to nickel metal, nickel alloys, or nickel compounds during the period 1980–83. Worker exposure differs from that of the general population in that the major route of exposure for nickel workers is inhalation and for the general population it is dermal contact. Nickel workers with lung cancer had elevated concentrations of 1.97 mg/kg DW in their lungs when compared to the general population (0.03–0.15 mg/kg DW). Plasma concentrations of nickel quickly reflect current exposure history to nickel. Mean nickel concentrations in plasma of humans occupationally exposed to nickel have declined by about 50% since 1976, suggesting decreased exposure due to improved safety.

22.5.7 Integrated Studies

Beaver ponds downstream from an abandoned copper–nickel ore roast yard near Sudbury, Ontario, were devoid of fish and had reduced macroinvertebrate taxon richness and diversity when compared to upstream ponds. Nickel water concentrations, in μg Ni/L, were 57.0 in upstream ponds, 82.0 in downstream ponds, and 1800.0 at the station directly on the roast pit. Beavers (*Castor canadensis*) near nickel smelters had elevated nickel concentrations in livers and kidneys when compared to conspecifics from a reference site; accumulations were attributed to food chain contamination.

Nickel contamination in the Sudbury, Ontario, region is the result of aerial transport and terrestrial drainage from mining and smelting activities. Nickel concentrations in soils were elevated as far as 52 km from the source. Erosion of soils following the death of vegetation was widespread and affected an area of more than 820 km^2. Soils increased in acidity, increasing the solubility of nickel. In aquatic ecosystems, nickel was accumulated from the water column by periphyton, rooted aquatic macrophytes, zooplankton, crayfish, clams, and fishes. However, there was no evidence of food chain biomagnification of nickel in the Sudbury ecosystem. For example, in the nickel-contaminated Wanapitei River,

BCFs during summer 1974 were highest for whole periphyton (19,667), followed by whole pondweeds (11,429), sediments (5333), whole crayfish (929), whole zooplankton (643), muscle of carnivorous fishes (329), soft tissues of clams (262), and muscle of omnivorous fishes (226). Higher BCF values are recorded for acid- and metal-tolerant flora.

There is little convincing evidence for the biomagnification of nickel in the food chain. Most authorities agree that nickel concentrations do not increase with ascending trophic levels of food chains and that predatory animals do not have higher concentrations. The potential for biomagnification exists because algae and macrophytes have comparatively elevated concentrations of nickel; however, animals seem to be able to regulate the nickel content of their tissues by controlled uptake and increased excretion.

22.6 Nickel Deficiency Effects

Nickel is reportedly an essential micronutrient for maintaining health in certain species of plants, invertebrates, birds, and mammals, including humans. However, nickel essentiality for humans has not yet been proven, and the evidence for marine tunicates and land snails is inconclusive. To prevent nickel deficiency in rats and chickens, diets should contain at least 50.0 μg Ni/kg ration; cows and goats require more than 100.0 μg Ni/kg ration, perhaps reflecting the increased use of nickel by rumen bacteria. In humans, nickel deficiency is not a public health concern because daily oral intake normally exceeds 170.0 μg of nickel.

Nickel is considered essential to animals because it is present in the fetus or newborn, is homeostatically regulated, the metabolic pool of nickel is specifically influenced by hormonal substances or pathologic processes, certain metalloproteins contain nickel, and because nickel deficiency has been induced experimentally in certain species of birds and animals. In general, the nickel deficiency syndrome can be cured or prevented by trace amounts of nickel. However, nickel administration may not be successful in reversing all abnormalities produced by nickel deprivation.

Nickel deficiency effects from dietary deprivation of nickel is now documented in at least 17 animal species, including chickens, cows, goats, pigs, rats, and sheep. Some investigators aver that nickel deficiency can be induced only by very low nickel concentrations in the diet – not by its bioavailability. Signs of nickel deficiency include delayed gestation periods and fewer offspring; decreased growth and sometimes dwarfism; anemia; skin eruptions; brittle hair; reduced oxygen consumption; decreased levels of serum proteins; enhanced urinary nitrogen excretion; reduced tissue iron and zinc concentrations; reduced hemoglobin and hematocrit values; abnormal liver morphology and lipid metabolism; reduced liver glucose, lipids, glycogen, and triglycerides; and reduced activity of several enzymes, including dehydrogenases, transaminases, and alpha-amylases.

22.6.1 Bacteria and Plants

Nickel is essential for the active synthesis of urease in plant cells and of various hydrogenases in bacteria. In several species of higher plants, including jack beans (*Canavalia* sp.), soybeans (*Glycine max*), rice (*Oryza sativa*), and tobacco (*Nicotiana tabacum*), nickel is required for effective urea metabolism and urease synthesis. Root growth of onions (*Allium* sp.) is stimulated at 60.0–600.0 μg Ni^{2+}/L culture solution. Some terrestrial plants, such as *Alyssum* spp., accumulate nickel and require it for growth. In bacteria, nickel is required for the growth of *Oscillatoria* sp. and *Alcaligenes* sp., for the synthesis of carbon monoxide dehydrogenase in *Clostridium posterianum*, and as a component of coenzyme F_{430} in *Methanobacterium* spp. Nickel deficiency in bacteria may adversely affect reproductive processes, such as endospore formation, and cause a decrease in nickel-containing intracellular pigments in strains of *Bacillus cereus*; however, both of these observations require verification.

22.6.2 Birds

All studies demonstrating nickel deficiency in birds were conducted on a single species; specifically, chicks of the domestic chicken, *Gallus* sp. The relevance of these results to avian wildlife species is unknown. Chicks grew normally when fed nickel-deficient diets (2.0–15.0 μg Ni/kg ration) for 3–4 weeks. But these chicks had liver histopathology, decreased concentrations of yellow lipochrome pigments in liver, low hematocrit, skin dermatitis, leg thickening, altered lengths of leg bones, and decreased plasma cholesterol. Adverse effects of nickel-deficient diets (<20.0 μg Ni/kg ration) were reversed by the addition of nickel to the diet. Chicks fed diets containing 25.0–2500.0 μg Ni/kg ration for 3–4 weeks grew normally and all organs appeared normal. Nickel-deficient chicks (40.0–80.0 μg Ni/kg ration), when compared to controls (3.0–5.0 mg Ni/kg ration), had swollen hock joints, reduced length-to-width ratios of tibias, scaly dermatitis of the legs, orange–yellow discoloration of the legs, fat-depleted livers, altered liver metabolism, and elevated concentrations of nickel in liver, spleen, and aorta. Chicks fed nickel-deficient diets of 44.0 μg Ni/kg ration for 30 days had markedly lower nickel concentrations in serum and livers than did controls fed diets containing 3.4 mg Ni/kg ration; nickel-deficient chicks had 1.6 μg Ni/L in serum vs. 4.2 μg Ni/L in controls and 64.0 μg Ni/kg DW liver vs. 82.0 μg Ni/kg DW in controls. Livers of nickel-deficient chicks had an altered gross appearance, reduced oxidative ability, and decreased lipid phosphorus concentrations. Nickel deficiency in chicks may be associated with thyroid hormone imbalance, but this needs verification.

22.6.3 Mammals

In humans, there is no evidence of a nickel deficiency syndrome or proof that nickel is essential. Cows (*Bos* sp.) fed nickel-deficient diets containing less than 100.0 μg Ni/kg ration had reduced growth and survival. Nickel-deficiency in cows was exacerbated when diets were also low in protein, but effects were lessened when diets were supplemented with 5.0 mg Ni/kg ration. Lambs from domestic sheep (*Ovis aries*) fed a low nickel diet (30.0 μg Ni/kg ration) for 97 days had lower growth, higher mortality, and altered blood and tissue chemistry when compared to controls fed a diet containing 5.0 mg Ni/kg ration. Lambs given diets containing 65.0 μg Ni/kg DW ration had disrupted metabolism.

Adults and offspring of breeding goats (*Capra hircus*) and swine (*Sus* sp.) fed nickel-deficient diets (<100.0 μg Ni/kg ration) or control diets (10.0 mg Ni/kg ration) for 6 years had normal conception and abortion rates. However, nickel-deficient goats and pigs had delayed pregnancies, reduced litter sizes, lower birth rates, lower weight gains during suckling, and significant increases in mortality during the suckling period; mortality was 41% higher than controls in kids and 51% higher than controls in piglets. Nickel-deficient adult goats had lower nickel concentrations in kidneys, liver, and other tissues than did controls, specifically, 0.2–0.6 mg Ni/kg DW tissue vs. 0.6–1.2 mg Ni/kg DW in controls. Kids of nickel-deficient ewes (100.0 μg Ni/kg DW ration for 6 years vs. control diet of 300.0 μg Ni/kg ration) had inhibited growth starting at age 8 weeks and reduced survival. During lactation, hemoglobin concentrations and hematocrits of nickel-deficient goats were significantly lower than control values. Nickel-deficient pigs had rough coats, decreased growth, and impaired reproduction.

Signs of nickel deficiency in the laboratory white rat (*Rattus* sp.) include retarded growth, anemia, a reduction in hematocrit and hemoglobin values, decreased enzyme activities (malate dehydrogenase, glucose-6-phosphate dehydrogenase, alpha amylase), a reduction in liver total lipids and phospholipids, and altered tissue concentrations of fatty acids, iron, copper, and zinc. Nickel concentrations in fur, kidneys, and muscle of rats fed nickel-deficient diets (15.0 μg Ni/kg DW ration) were about 66% lower than those of controls given 20.0 mg Ni/kg ration. Signs of nickel deficiency in rats were usually reversed by supplementing the diet with nickel at more than 50.0 μg Ni/kg ration. Rats fed nickel-deficient diets (<5.0 μg Ni/kg ration) for three generations produced offspring that were anemic and grew poorly in the first two generations and had impaired reproduction in all generations. In another three-generation study, rats fed nickel-deficient diets containing 2.0–15.0 μg Ni/kg ration had increased perinatal morality, unthrifty appearance of young rats, decreased physical activity, decreased liver cholesterol, and liver histopathology when compared to controls fed diets containing 3.0 mg Ni/kg ration.

22.7 *Lethal and Sublethal Effects*

Nickel toxicity reduces photosynthesis, growth, and nitrogenase activity of algae; fermentative activity of a mixed rumen microbiota; growth rate of marine bacteria; metabolism of soil bacteria; and mycelial growth, spore germination, and sporulation of fungi. Adverse effects of excess nickel have also been observed with yeasts, higher plants, protozoans, mollusks, crustaceans, insects, annelids, echinoderms, fishes, amphibians, birds, and mammals. As discussed later, sensitive species of aquatic organisms are adversely affected at nominal concentrations of 11.0–113.0 μg Ni^{2+}/L.

In birds, mortality occurred in young individuals of sensitive species when rations contained more than 500.0 mg Ni/kg. Nickel accumulated in avian tissues at dietary loadings as low as 0.7–12.5 mg Ni/kg ration; however, nickel intoxication in some species tested was not always reflected by elevated tissue nickel concentrations.

In mammals, the toxicity of nickel is a function of the chemical form of nickel, dose, and route of exposure. Exposure to nickel by inhalation, injection, or cutaneous contact is more significant than oral exposure. Toxic effects of nickel to humans and laboratory mammals are reported for respiratory, cardiovascular, gastrointestinal, hematological, musculoskeletal, hepatic, renal, dermal, ocular, immunological, developmental, neurological, and reproductive systems.

22.7.1 Terrestrial Plants and Invertebrates

In general, the effects of long-term, low-level exposure to nickel are shown in growth inhibition with no other visible signs. However, many species of plants growing on soils contaminated with excess nickel show stunted and discolored roots and tops, wilting, chlorosis, necrosis, twisted stalks, thickening of leaf tissues, and failure of leaves to fold to form compact heads. In solution culture, 1.0 mg of soluble nickel/L is toxic to sensitive plants. Accumulations of 50.0 mg Ni/kg DW plant and higher are toxic to most plants. Depending on soil conditions and chemical form, nickel in soil is toxic when concentrations exceed 500.0 mg Ni/kg DW soil with more than 25.0 mg Ni/L extractable in a 2.5% acetic acid solution. Accumulation and toxic effects occur in vegetables grown on soils treated with sewage sludge and in vegetation close to nickel-emitting sources. Nickel was shown experimentally to decrease growth of soybeans (*Glycine max*) when administered as particulate nickel through the atmosphere or in the rooting medium. Crop plants is the most sensitive group of terrestrial vegetation tested against nickel. Adverse effects on chlorophyll metabolism and growth occur at soil water concentrations as low as 1.0 mg Ni/L. Radishes, beets, cabbages, celery, and lettuce planted in organic soils contaminated by aerial fallout from a nearby nickel smelter and containing between 1570.0 and 6550.0 mg Ni/kg DW soil have decreasing yields with increasing soil nickel concentrations. No radishes or cabbages were suitable for marketing. Celery, lettuce, and beets were reduced from a normal yield on soil with 1300.0 mg Ni/kg to zero on soils with 4800.0 mg/kg. Dried cabbage heads and celery tops had as much as 400.0 mg Ni/kg. Decreased yields of alfalfa (*Medicago sativa*) occur when plant nickel content exceeds 44.0 mg/kg DW. Decreased yield of oats (*Avena sativa*) was associated with nickel concentrations more than 60.0 mg/kg DW grain, more than 28.0 mg/kg DW oat straw, or more than 500.0 mg Ni/kg DW soil. Signs of nickel toxicity in oats decrease in severity with increasing magnesium concentrations in culture solution during exposure for 35 days.

Temperature, pH, chlorophyll, and various metals all modify the toxicity of nickel to fungi. A reduction in the toxicity of nickel to the mycelial growth rates of five species of filamentous fungi occurs when pH increases from acidic to alkaline (*Achyla* sp., *Saprolegnia* sp.); at elevated concentrations of magnesium, zinc, or lead (*Achyla* sp.); at chlorophyll or humic acid contents equivalent to 1% (*Saprolegnia* sp., *Cunninghamella blakesleeana*, *Aspergillus clavatus*); and at increased temperatures of 33°C vs. 23°C (*Aspergillus flavus*). Growth of sensitive species of filamentous fungi is inhibited at 10.0 mg Ni/L and abnormal mycelia occurs at 50.0 mg/L. Histidine may govern nickel accumulation in the approximately 400 known species of nickel-hyperaccumulating plants. Nickel hyperaccumulator plants, including 48 of 170 species of *Alyssum* spp., contain as much as 3% of the dry leaf biomass as nickel. Exposing hyperaccumulator species of *Alyssum* to nickel elicits a large and proportional increase in the levels of free histidine, which is shown to be coordinated with nickel in vitro. Supplying histidine to a nonaccumulating species greatly increases both nickel tolerance and capacity for nickel transport to the shoot, indicating that enhanced production of the amino acid histidine is responsible for the nickel hyperaccumulation phenotype in *Alyssum*.

Data on nickel toxicity to terrestrial invertebrates are scarce. A soil concentration of 757.0 mg/kg DW soil is lethal to 50% of earthworms (*Eisenia foetida*) in 14 days, and higher concentrations of 1200.0–12,000.0 mg/kg DW soil for shorter periods produced reduced growth and survival in the same species. Earthworms are less sensitive to nickel if the medium is rich in microorganisms and organic matter, thus, making the nickel less bioavailable.

22.7.2 Aquatic Organisms

Signs of nickel poisoning in fishes include surfacing, rapid mouth and opercular movements and, prior to death, convulsions and loss of equilibrium. Destruction of the gill lamellae by ionic nickel decreases the ventilation rate and may cause blood hypoxia and death. Other signs of nickel poisoning in fishes include decreased concentrations of glycogen in muscle and liver with simultaneous increases in levels of lactic acid and glucose in blood, depressed hydrogen peroxide production in tissues and a reduction in superoxide dismutase, and contractions of vascular smooth muscle – signs similar to those associated with hypertension in mammals. Ionic nickel is lethal to sensitive species of aquatic organisms at 11.0–113.0 μg/L. Deaths occur among embryos of rainbow trout at 11.0–90.0 μg/L, daphnids at 13.0 μg/L, embryos of channel catfish at more than 38.0 μg/L, embryos of the narrow-mouthed toad at 50.0 μg/L, and embryos of largemouth bass at 113.0 μg/L. Species intermediately resistant to nickel died at 150.0–410.0 μg Ni/L, including mysid shrimp at 150.0 μg/L, freshwater snails at 237.0 μg/L, clam embryos at 310.0 μg/L, and embryos of salamanders at 410.0 μg/L. Aquatic bacteria and yeasts are comparatively tolerant to nickel. Sensitive species of freshwater eubacteria and actinomycetes show reduced growth at 5.0 mg Ni/L; for marine eubacteria, growth inhibition begins at 10.0–20.0 mg/L. Sensitive species of yeasts show growth inhibition at 1.0 mg Ni/L (*Torulopsis glabrata*); resistant species of yeasts (*Rhodotorula* sp., *Cryptococcus terrens*) show a reduction in growth at 5.0–20.0 mg Ni/L.

The biocidal properties of nickel are modified by many variables. For example, nickel is most lethal at pH 8.3 and least lethal to freshwater crustaceans and fishes at pH 6.3; less toxic to algae when copper is reduced or absent and chelating agents, such as EDTA, are present; most lethal to echinoderm embryos prior to gastrulation; and more toxic to estuarine amphipods and clams under conditions of decreased salinity in the 0.5–3.5% range and increased temperature in the 5–15°C range.

Representative nickel-sensitive aquatic species show sublethal effects at 11.7–125.0 μg Ni/L. These effects include altered immunoregulatory mechanisms in tissues of the rainbow trout at 11.7 μg/L, inhibited reproduction of daphnids at 30.0 μg/L, growth inhibition of freshwater and marine algae at 30.0–125.0 μg/L, reduced growth of rainbow trout at 35.0 μg/L, accumulation from the medium by mussels at 56.0 μg/L, and abnormal development of sea urchin embryos at 58.0 μg/L.

BCFs for nickel vary among organisms under laboratory conditions. For freshwater species, typical BCF values for nickel are about 10 for algae, 61 for fathead minnows, and 100 for cladocerans; for marine mussels and oysters, typical BCF values range between 299 and 414. The alga *Thalassiosira rotula* can accumulate as much as 90.0 mg Ni/kg DW. Other species of aquatic plants can extract nickel from water and concentrate it to as much as 10,000.0 mg/kg DW. The alga *Anacystis nidulans* can develop tolerance to nickel and other metals under laboratory conditions, and this may account for high BCF values in this species. Nickel at 50.0 μg/L was accumulated from seawater by softshell clams (*Mya arenaria*) more rapidly during summer at water temperatures of 16–22°C than during winter at 0–10°C; no accumulations occurred at 10.0 μg Ni/L in winter, but clams accumulated twice as much nickel over controls in summer. Embryos of sea urchins actively accumulate nickel from seawater at all dose levels tested. BCFs for rainbow trout after exposure for 6 months to 1.0 mg Ni/L were 0.8 for muscle, 2.9 for liver, and 4.0 for kidneys. Fish can accumulate nickel from food and water. Levels up to 13.0 mg Ni/kg DW occurred in northern pike (*Esox lucius*) and pickerel (*Esox* sp.) from a contaminated river. Common carp (*Cyprinus carpio*) and tilapia (*Tilapia nilotica*) exposed for 16 days to 1.0 mg Ni/L had elevated concentrations in livers of 49.0–77.0 mg Ni/kg DW. Goldfish (*Carassius auratus*) that died during immersion in solutions containing more than 35.0 mg Ni/L showed elevated concentrations in tissues; however, most of the nickel was washed off with water, and it is not clear if accumulation occurred after death. Nickel accumulates in fish tissues and causes alterations in gill structure, including hypertrophy of respiratory and mucus cells, separation of the epithelial layer from the pillar cell system, cauterization and sloughing, and necrosis of the epithelium. Although aquatic organisms can accumulate nickel from their surroundings, there is little evidence of significant biomagnification of nickel levels along food chains.

22.7.3 Birds

In mallards (*Anas platyrhynchos*), nickel accumulates in tissues when diets contain as little as 12.5 mg Ni/kg DW ration. Metabolic upset and altered bone densities occur in mallards fed diets containing 800.0 mg Ni/kg ration for 90 days. Inhibited growth and reduced survival occur in mallards at dietary loadings of 1200.0 mg Ni/kg ration. Dietary nickel concentrations of 0.074 mg Ni/kg ration have no adverse effects on Coturnix quail (*Coturnix risoria*). However, Japanese quail (*Coturnix japonica*) fed diets containing 0.71 mg Ni/kg ration – when compared to controls fed diets containing 0.48 mg Ni/kg – have significantly elevated nickel concentrations in liver. Increased concentrations of nickel in the diets of domestic chickens (*Gallus* sp.) were associated with decreased growth and survival and increased nickel concentrations in bone and kidney. Dietary loadings of 500.0 mg Ni/kg ration and higher were associated with reduced growth and high mortality in some strains of chickens, but not others. No developmental abnormalities occurred in chicks from survivors challenged by nickel during embryogenesis. Chick embryos receiving a single injected dose of 3.6 mg Ni/kg embryo, however, experienced 50% mortality within 18 days. Chicks are more resistant than embryos to injected nickel. Chicks injected with 10.0 mg Ni/kg BW survived but had disrupted glucose metabolism; effects were exacerbated by starvation.

22.7.4 Mammals

Most nickel researchers agree on six points:

(1) Lifetime exposure of resistant species of mammals to diets containing 2500.0 mg Ni/kg DW or to drinking water containing 10,000.0 mg Ni/L are not lethal.

(2) Lethal nickel doses in mammals are usually derived from studies with laboratory animals injected with nickel and its compounds, not from realistic exposure regimens.
(3) Inhaled nickel is at least 100 times more toxic than ingested nickel because it is more readily absorbed from the lungs than from the gastrointestinal tract, and death is more often the result of respiratory failure than of nervous system effects. For example, oral ingestion of 0.05 mg Ni/kg BW and inhalation at 0.005 mg Ni/m^3 are equally effective threshold doses in rats.
(4) Large differences in sensitivity to nickel exist between closely related taxonomic species, such as rats and mice.
(5) Threshold effects on lung function or morphology in several species of laboratory mammals occur at airborne nickel concentrations of 0.1–0.2 mg/m^3, depending on nickel compound and duration of exposure.
(6) Juveniles were usually more sensitive to nickel than were adults.

Nickel salts administered by intravenous or subcutaneous injection are comparatively toxic. For all routes of parenteral administration, the LD50 (lethal dose to 50% of the sample) range for injected nickel salts is 6.0 mg Ni/kg BW for dogs given nickel oxide intravenously to 600.0 mg Ni/kg BW for mice given nickel disodium EDTA intraperitoneally.

Several trends were evident among sensitive species of mammals tested against nickel through administration routes other than injection.

(1) Nickel carbonyl is lethal to mice, rats, and cats at 0.067–0.24 mg Ni/L.
(2) Inhalation of nickel compounds other than nickel carbonyl causes significant effects in humans, rats, mice, rabbits, and dogs, with respiratory effects being most common.
(3) Nickel-contaminated drinking water has adverse effects on rat reproduction and may neurologically affect the eyes of humans, although this needs to be verified.
(4) Diets containing nickel carbonate, nickel chloride, or nickel sulfate cause reduced growth, disruptions of food intake and thyroid function, and emphysema and pneumonia in calves, dogs, mice, or rats.
(5) Dermal application of nickel sulfate hexahydrate causes skin and testicular damage.
(6) Single oral doses of 136.0–410.0 mg Ni/kg BW as nickel acetate are lethal to mice.

Nickel carbonyl ($Ni(CO)_4$) is the only nickel compound known to cause severe acute effects, such as pulmonary damage and death; acute toxic effects of other nickel compounds to mammals are a minor risk. In fatal cases, death occurs 3–13 days after exposure; recovery from nickel carbonyl poisoning usually occurs within 70 days after exposure, but sometimes may take up to 6 months. Nickel carbonyl is a volatile, colorless liquid formed when finely divided nickel or its compounds come into contact with carbon monoxide. It is unstable under atmospheric conditions, and if inhaled, nickel is deposited in highly active form on the respiratory mucosa on contact. Nickel carbonyl is widely used commercially as a catalyst but is one of the most toxic gases encountered in industrial operations. Exposure to air containing more than 50.0 mg $Ni(CO)_4/m^3$ for 0.5–2.0 h may be fatal to humans. Intraperitoneal injection of nickel carbonyl was the most toxic route of administration; for all routes of administration, LD50 values to various tested mammals ranged between 13.0 and 65.0 mg/kg BW. Nickel carbonyl toxicity is due, in part, to its volatility and lipophilicity. Signs of nickel carbonyl poisoning – which strongly resemble those of viral pneumonia – include headache, vertigo, nausea, vomiting, insomnia, and irritability followed by an asymptomatic interval and then the onset of insidious, persistent signs that include chest pains, dry coughing, cyanosis, sweating, visual and gastrointestinal disturbances, severe weakness, paralysis of the hind limbs, and convulsions; the lungs are the primary target organs in all animals tested, although the liver, kidneys, adrenal glands, spleen, and brain are also affected.

Adverse effects in mammals by inhalation of nickel compounds other than nickel carbonyl occur with aerosols of both soluble and insoluble nickel compounds. Inhalation of nickel by humans and other mammals produces respiratory, hepatic, renal, dermal, immune system, and body weight effects. Respiratory effects of nickel include asthma, nasal septal perforations, chronic rhinitis and sinusitis, and increased risk for chronic respiratory tract infections; immunological, genotoxic, and carcinogenic effects were also observed. Lung reactions in the form of asthma were attributed to sensitization by nickel. Insoluble forms of inhaled nickel are more persistent in lungs than are soluble forms, as judged by 90-day studies with nickel chloride (soluble) and nickel oxide (insoluble) given to rodents by intratracheal administration. Severity of respiratory toxicity was higher with increasing solubility of the nickel compound tested and not with increasing burden of nickel on the lung; insoluble nickel oxide had the lowest toxicity but the highest lung burden. Nickel sulfate was more toxic than nickel subsulfide which was more toxic than nickel oxide.

Local effects noted in guinea pigs, rats, mice, and hamsters caused by inhalation of metallic nickel powder (15.0 mg/m^3), nickel subsulfide (0.97 mg/m^3), or nickel oxide (53.0 mg/m^3) include nasal sinus inflammations, ulcers, lung irritation, nickel accumulations in lungs, emphysema, and increased viral respiratory infections. Rats inhaling nickel subsulfide at 2.5 mg/m^3 for 22 days had nasal and lung histopathology within 4 days and disrupted enzyme activities and elevated nickel accumulations within 7 days. Repeated inhalation of nickel subsulfide by rats for 3 months resulted in chronic inflammation in the lung and atrophy of the olfactory epithelium. Rats exposed via inhalation of nickel sulfate hexahydrate of 635.0 μg Ni/m^3 for 6 h daily over 16 days had no outward signs of toxicity; however, internal examination revealed lesions on the olfactory epithelium. Rats and mice died following inhalation exposure for 16 days to equal doses of nickel sulfate or nickel subsulfide, but not nickel oxide. Rats showed epithelial hyperplasia after inhalation exposure to aerosols of nickel chloride or nickel oxide and pulmonary fibrosis after inhalation exposure to nickel subsulfide; a similar syndrome was reported in rabbits after high-level inhalation exposure to nickel–graphite dust. Dogs exposed to nickel powder for 6 months by way of inhalation developed lung pneumosclerosis causing cardiac insufficiency. Rats exposed to airborne nickel dusts (100.0 μg Ni/m^3, 12 h daily for 2 months) had respiratory irritation. Single exposures of mice to 250.0 μg Ni/m^3 for 2 h depressed the humoral immune response. Rats exposed to 1000.0 μg Ni $dust/m^3$ (5 days/week for 3–6 months) had high accumulations of nickel in the lungs and kidneys and interstitial fibrosis.

Nickel and nickel salts are comparatively nontoxic when taken orally because of homeostatic mechanisms that control nickel metabolism and limited intestinal absorption. In cattle, young calves fed nickel carbonate at concentrations as high as 1000.0 mg Ni/kg ration for 8 weeks had nephritic kidneys, with degree of severity increasing with dietary nickel level. However, dietary nickel did not affect growth or food consumption of calves or cause histopathology of the rumen, abomasum, duodenum, liver, or testes. Human and animal data indicate that death is unlikely from oral nickel exposure except when exposed accidentally to high levels. Oral exposure studies for humans were limited to acute intoxication and include death (due to cardiac arrest) and the effects of gastrointestinal (nausea, cramps, diarrhea, vomiting), hematological (increase in reticulocytes), hepatic (increase in serum bilirubin), renal (albuminuria), and neurological damage. A child that accidentally ingested 20.36 g of Ni/kg BW as crystals of nickel sulfate died from heart failure. Oral LD50 doses of nickel chloride to rats produced depression of the nervous system, edema of the mucous membranes of the mouth and nose, diffusions from the oral cavity, lacrimation, bleeding from the nose, and diarrhea. Prior to death, rats were lethargic, ataxic, and with irregular breathing and cool body temperatures.

Nickel is a reproductive toxicant in animals. Specific effects of nickel on reproduction include degenerative changes in the testes, epididymis, and spermatozoa of rats;

adverse effects on embryo viability of rats and hamsters; and delayed embryonic development of rodents. Nickel salts given by injection cause intrauterine mortality and decreased weight gain in rats and mice. Inhibited testosterone and reduced growth occur in male rats given 2.32 mg Ni/kg BW as nickel acetate via intramuscular injection. Females given the same treatment had increased uterine weights. Nickel given in drinking water of rats for three generations at concentrations which do not interfere with growth or survival (i.e., 5.0 mg/L) were intolerable for normal reproduction. All generations of rats given nickel in drinking water had increased proportions of runts and increased neonatal mortality when compared to controls. In the third generation of nickel-treated rats, there were reductions in litter size and a reduction in the proportion of males. Excess nickel inhibits prolactin secretion in rats. Because prolactin influences milk production, the observation that suckling pups from nickel-exposed dams were most severely affected lends support to the concept that nickel plays a role in lactation at the pituitary level.

The most commonly observed toxic reaction to nickel and nickel compounds in the general human population is nickel dermatitis and skin sensitivity arising from dermal contact with metals containing nickel. Studies on occupational dermatitis – which is the most prevalent occupational disease – show that 8% of the cases are due to nickel. Nickel dermatitis in occupational exposure begins as an itching or burning in the web of the fingers, spreading to the fingers, the wrists, and the forearms; the eruption is similar to atopic dermatitis. Once an individual is dermally sensitized to nickel, even minimal contact (i.e., 0.007–0.04 mg Ni/kg BW daily) by any route of exposure may elicit a reaction. Nickel, in fact, is the most common allergin tested in North America; about 1–5% of human males and 7–14% of females are contact sensitized to nickel. Nickel contact hypersensitivity has been documented worldwide with 10% of the female population and 1% of the male population affected. Of these, 40–50% have vesicular hand eczema that, in some cases, can be severe and lead to loss of working ability. Nickel contact dermatitis is decreasing in occupational exposure, but increasing elsewhere due to increasing contact with nickel alloys in jewelry, coins, zippers, tools, pots and pans, stainless steel, detergents, prostheses, and certain hair dressings. Nickel is a major allergen for women, and between 1970 and 1980 there was a two- to threefold increase in the number of cases. In recent years, the incidence of nickel allergy has increased disproportionately in young females due to an increased frequency of ear piercing by this group to accommodate nickel-plated jewelry.

Although contact allergy to nickel is common in humans, experimental sensitization in animals is only successful under special conditions. Dermal studies with nickel salts and small laboratory mammals show that primary nickel sensitization typically takes place beneath nickel-containing metal objects that are in contact with the skin for hours and exposed to friction and sweating; nickel is released from nickel-containing objects by the action of blood, sweat, or saliva; ionic nickel diffuses through the skin at sweat-duct and hair-follicle openings, with a special affinity for keratin; and that nickel subsequently binds to proteins, including amino and carboxyl groups of keratin and serum albumin. Rats, guinea pigs, and rabbits absorbed and subsequently distributed 55–77% of nickel applied dermally. Dermal effects in animals after dermal exposure to nickel include distortion of the dermis and epidermis, hyperkeratinization, atrophy of the dermis, and biochemical changes. For example, in rats treated dermally with more than 40.0 mg Ni/kg BW daily as nickel hexahydrate for 30 days, distortion of the epidermis and dermis occurred by day 15 and hyperkeratinization, vacuolization, hydropic degeneration of the basal layer, and atrophy of the epidermis occurred by day 30. Skin irritation and death from nickel salts is reported in rabbits when nickel was applied dermally to abraded skin; no negative effects occurred in rabbits when the same dose was applied to intact skin. As was the case for humans, allergic reactions occur in laboratory animals after oral nickel challenge in sensitized individuals.

Nickel affects endocrine and enzymatic processes. Nickel-induced endocrine effects include inhibition of insulin production in pancreas, prolactin in hypothalamus, amylase excretion in parotid gland, and iodine uptake in thyroid. Inhibition of enzyme activity by nickel is reported for RNA polymerase, ATPase, dialkyl fluorophosphate, and aspartase. Inhibition of ATPase is associated with neurological abnormalities, such as tremors, convulsions, and coma; altered hormone release or action; and internal rearrangement of calcium ions in muscle that might cause paralysis and abnormal heart rhythm. Nickel increases the duration of the action potential of excitable membranes of nerve and muscle tissues; this effect is competitive with and imitative of those of calcium. Nickel hexahydrate at 14.8 mg Ni/kg BW disrupts hepatic monooxygenases; mice were more sensitive to this disruption than rats or guinea pigs. Nickel is also reported to activate various enzymes, including bovine pancreatic ribonuclease, pancreatic deoxyribonuclease, carboxypeptidase, arginase, phosphoglucomutase, and calcineurin – a calmodulin-dependent phosphoprotein phosphatase. Nickel affects the activity of heme oxygenase, thereby affecting the absorption of hemoglobin iron. Nickel, like many other metals and metalloids, induces heme oxygenase activity in tissues of mice, hamsters, and guinea pigs in a dose-related manner.

Systemic effects of nickel exposure include hyperglycemia, increased levels of plasma glucagon, damage to the pancreatic islet cells, decreased body weight, reduced food and water intake, and hypothermia. Acute administration of nickel salts caused prompt hyperglucagonemia and subsequent hyperinsulinemia in rats, rabbits, and guinea pigs. Nickel chloride given orally to young male rabbits at 500.0 μg daily for 5 months produced a decrease in liver glycogen and an increase in muscle glycogen, with prolonged hyperglycemia. Nickel increased glucose metabolism in rats injected intratracheally with 0.5 mg ionic nickel. This phenomenon probably reflected the influence of nickel on the production or secretion of insulin through decreased production of pituitary hormone secretions – specifically, prolactin – which control insulin concentrations. Nickel significantly affects the activity of hepatic glutathione-*S*-transferases (GSTs); these compounds play important roles in the detoxification of electrophilic xenobiotics, such as nickel, epoxides, and diolepoxides, and readily eliminate the cytotoxic products of lipid peroxidation, particularly the organic peroxides. The influence of nickel chloride on hepatic GST activity levels depends on the animal species tested, being depressed in mice, unchanged in rats, and increased in guinea pigs. In humans, nickel toxicity is not related to GST depletion or increased lipid peroxidases in vitro, whereas in rat kidney, nickel toxicity may be due to GST depletion and stimulation of lipid peroxidases.

Nickel affects the immune, cardiac, and excretory systems. Nickel adversely affects the immune system by reducing host resistance to bacterial and viral infections, suppressing phagocytic activity of macrophages, reducing the number of T-lymphocytes (thereby suppressing the natural killer cell activity) and increasing susceptibility to allergic dermatitis. In mice, nickel chloride suppresses the activity of natural killer cells within 24 h of a single intramuscular injection. Nickel-induced cardiovascular effects include vasoconstriction, inhibition of contraction by myocardial muscle, and a reduction in coronary vascular flow. Nickel salts are demonstrably cardiotoxic in dogs. Cats injected intravenously with $NiCl_2$ had altered heart rhythms, conductivity, and calcium metabolism. Nickel is a nephrotoxin with greatest adverse effect on the glomerular epithelium of the kidney. Kidneys from mammals exposed to nickel showed renal tubular damage, protein loss, and weight changes.

Nickel accumulations in tissues and organs of mammals vary significantly with species, route of administration, sex, and general health. No significant accumulations of nickel were observed in liver or kidney of Holstein calves fed diets containing 1000.0 mg Ni/kg ration for 21 weeks. In lactating dairy cows, no transfer of soluble nickel was observed from diet to tissues. In rats, guinea pigs, rabbits, sheep, dogs, and other species of mammals, nickel tends to accumulate in kidneys and other tissues after nickel exposure. Nickel-poisoned rats had elevated accumulations primarily in myocardium (5.7 vs. 2.2 mg/kg FW in controls) and spleen (2.1 vs. 0.6 mg/kg FW), followed by kidney, bone, and other tissues. In rats, nickel accumulated mainly in lung and secondarily in heart tissues after intratracheal administration of nickel chloride; nickel was retained for at least 40 days after dosing. In rodents, nickel accumulates in endocrine tissues, including the pituitary, adrenals, and pancreas. High nickel concentrations in the pituitary gland of rodents were associated with inhibition of insulin release and decreased prolactin secretion. Rat weanlings fed diets containing 500.0 mg Ni/kg ration as nickel acetate show elevated nickel accumulations in plasma, erythrocytes, heart, liver, testes, and especially kidneys; high accumulations were associated with reductions in growth, hematocrit, hemoglobin, cytochrome oxidase, and alkaline phosphatase. Male guinea pigs accumulated higher concentrations of nickel in hair than did females after exposure for 4 months to drinking water containing 2.5 mg Ni/L. Invading microorganisms can change the distribution of ^{63}Ni in mice infected with coxsackie B3 virus. Infected mice had high accumulations of ^{63}Ni in the pancreas and the wall of the ventricular myocardium. Healthy mice had almost no ^{63}Ni accumulations in these tissues, but residues were elevated in blood, kidney, and lung.

Excretion of ingested nickel by rats, regardless of amount ingested, usually occurs through the feces within 48 h. Most nickel administered to rats through a variety of routes, and irrespective of chemical form, is usually excreted within a few days; however, excretion is slower for nickel powder and from lungs. Nickel caused a twofold increase in urinary corticoid excretion in guinea pigs, increased urinary excretion of protein in rats, and increased urinary excretion of B-2-macroglobulin in nickel refinery workers. Nickelemia was associated with increased urinary B-2-macroglobulin levels, and 5 of 11 workers with urinary nickel concentrations more than 100.0 μg/L had increased urinary B-2-macroglobulin (>240.0 μg/L).

22.8 Proposed Criteria and Recommendations

While nickel may be carcinogenic, perhaps in all forms, there is little or no detectable risk in most sectors of the nickel industry at current exposure levels, including some processes that had previously been associated with very high lung and nasal cancer risks. More research is in progress to clarify the hazards of nickel to humans, including chronic inhalation carcinogenicity studies of nickel subsulfide, nickel oxide, and nickel sulfate hexahydrate in rats and mice. Nevertheless, additional research on nickel-induced cancer has been proposed, including research on (1) route of administration; (2) oxidative state of nickel; (3) effect of nickel on nucleic acid synthesis; (4) interaction effects with asbestos, zinc and magnesium, tobacco smoke, and agents thought to inhibit nickel carcinogenesis, such as manganese, copper, and aluminum; (5) role of diet in nickel carcinogenesis and specificity and mechanism of uptake of nickel ion from the gastrointestinal tract; and (6) nickel immunosuppressive mechanisms, especially effects of nickel on natural killer cell activity and the relation between suppression of these cells and the known carcinogenesis of nickel compounds. Large-scale studies are needed to establish the upper limits of cancer risk from nickel.

Humans have been shown to develop sensitivity to nickel. The use of nickel in products that may release the metal when in contact with the skin should be regulated. Among various subgroups of the U.S. population who may be at special risk for adverse effects of nickel are those who have nickel hypersensitivity and suffer chronic flare-ups of skin disorders with frank exposure. The role of oral nickel exposure in dermatic responses by sensitive individuals suggests that nickel-limited diets resulted in marked improvement of hand eczema and that nickel added to the diets appeared to aggravate the allergic response. More research is needed on the role of nickel in contact dermatitis, including the role of oral nickel exposure, and the pathogenesis and therapy of nickel dermatitis. Additional dermal exposure studies are needed to determine if testicular effects result from both oral and dermal exposure to nickel.

Animal experimental models of nickel-induced skin sensitivity are few and have been conducted only under very specialized conditions. Studies examining the mechanism of nickel contact sensitization and its extent in wildlife are needed. The importance of the surface properties and crystalline structure of nickel compounds in relation to their reactivity and protein-binding activities is well documented. It is therefore necessary to identify clearly the nickel compounds to which exposure occurs. Acute and chronic dermal and inhalation studies using all nickel compounds would determine if certain compounds are more effective in eliciting allergic dermatitis.

To protect terrestrial vegetation against decreased growth and other toxic effects, nickel residues in leaves should contain less than 44.0 to less than 50.0 mg/kg DW, soils should contain less than 50.0 to less than 250.0 mg Ni/kg DW, and sewage sludge applied to agricultural soils should be limited to 30.0–140.0 kg Ni/surface ha at the low end and 50.0–560.0 kg/ha at the high end (Table 22.4). Research is needed on the direct effects on vegetation of nickel from airborne deposition, the effects of soil acidification on mobility and toxicity of nickel in soil, differences in nickel metabolism between tolerant and nickel-sensitive plants, and on the interactions of nickel and organic acids

Table 22.4. Proposed nickel criteria for the protection of natural resources and human health.

Resource, Criterion, and Other Variables	Effective Nickel Concentration
AQUATIC LIFE, FRESHWATER	
Sediments	
Great Lakes	
Safe	Less than 20.0 mg/kg dry weight (DW)
Moderately polluted	20.0–50.0 mg/kg DW
Heavily polluted	More than 50.0 mg/kg DW
Wisconsin; for disposal in water	Less than 100.0 mg/kg DW
Water	
Canada; safe level	Less than 25.0 μg/L
Rainbow trout, *Oncorhynchus mykiss*; safe level vs. toxic effects expected	Less than 29.0 vs. 30.0–50.0 μg/L
Ontario, Canada; from sediment disposal in water; final water concentration	Less than 50.0 μg/L
The Netherlands; safe level	Less than 50.0 μg/L
USA; water hardness of 50 mg $CaCO_3$/L	24-h average not to exceed 56.0 μg total recoverable nickel/L; maximum concentration not to exceed 1100.0 μg/L at any time
Sweden; safe level	Less than 80.0 μg/L
USA; water hardness of 100 mg $CaCO_3$/L	24-h average not to exceed 96.0 μg total recoverable Ni/L; maximum concentration not to exceed 1800.0 μg/L at any time
USA; water hardness of 200 mg $CaCO_3$/L	24-h average not to exceed 160.0 μg total recoverable Ni/L; maximum concentration not to exceed 3100.0 μg/L at any time

Table 22.4. cont'd

Resource, Criterion, and Other Variables	Effective Nickel Concentration
AQUATIC LIFE, MARINE	
Water	24-h average not to exceed 7.1 μg total recoverable Ni/L; maximum concentration not to exceed 140.0 μg/L at any time
BIRDS	
Diet	
Domestic chicken, *Gallus* sp.; to prevent nickel deficiency in chicks	More than 50.0 μg/kg ration
Mallard, *Anas platyrhynchos*	
Ducklings; no adverse effects	Less than 200.0 mg/kg ration
Adults; no adverse effects	Less than 800.0 mg/kg ration
Adults; adverse effects	More than 800.0 mg kg fresh weight (FW) ration
Tissue concentrations	
Adverse effects expected; most species	
Kidney	More than 10.0 mg/kg DW
Liver	More than 3.0 mg/kg DW
Internal organs, most species	
Normal	Less than 3.0 mg/kg DW
Nickel-contaminated environments	As much as 30.0 mg/kg DW
Mallard; liver or kidney; significant exposure to dietary nickel that may be harmful	More than 1.0 mg/kg FW
CROPS AND OTHER TERRESTRIAL VEGETATION	
Plant residues	
Alfalfa, *Medicago sativa*	
Normal	0.3–3.2 mg/kg DW
Decreased growth	44.0 mg/kg DW
Terrestrial vegetation	
Hyperaccumulator plants	More than 1000.0 mg/kg DW
Most species	
Normal	0.05–5.0 mg/kg DW
Toxic	More than 50.0 mg/kg DW
Sewage sludge; maximum addition to agricultural soils	
Europe	30.0–75.0 kg sludge/ha soil
South Africa	200.0 mg/kg DW sludge
USA; soils with low exchange capacity vs. soils with high exchange capacity	
Maryland	140.0 vs. 280.0 kg/ha
Massachusetts	56.0 vs. 112.0 kg/ha
Minnesota and Vermont	56.0 vs. 112.0–224.0 kg/ha
Missouri	140.0 vs. 280.0–560.0 kg/ha

Continued

Table 22.4. cont'd

Resource, Criterion, and Other Variables	Effective Nickel Concentration
New York, all soils	34.0–50.0 kg/ha
Oregon	50.0 vs. 100.0–200.0 kg/ha
Wisconsin	50.0–100.0 vs. 150.0–200.0 kg/ha
Soils; suitability for crop production	
Canada; Alberta; acidic soils; acceptable	Less than 250.0 mg/kg DW
The Netherlands	
Background	50.0 mg/kg DW
Moderate contamination	100.0 mg/kg DW
Unacceptable and requires cleanup	More than 500.0 mg/kg DW
Russia; maximum acceptable concentration; extractable by ammonium acetate buffer at pH 4.6	4.0 mg/kg soil
South Africa, no phytotoxicity or elevated nickel concentrations in crops	More than 38.0 mg/kg DW soil
USA; New Jersey; acceptable	Less than 100.0 mg/kg DW soil
MAMMALS, EXCEPT HUMANS	
Air	
Laboratory white rat, *Rattus* sp.	
Adverse effects; nickel sulfate	More than 0.1 mg/m^3
No adverse effects	
Nickel refinery dust	Equivalent to less than 0.84 mg/kg BW daily
Nickel subsulfide	Equivalent to less than 1.7 mg/kg BW daily
Nickel sulfate	Less than 0.1 mg/m^3
Rodents, *Mus* spp., *Rattus* spp.	
Adverse effects; nickel oxide, nickel sulfate	More than 0.02 mg/m^3
No adverse effects; nickel chloride, nickel subsulfide	Less than 0.1 mg/m^3
Diet	
To prevent deficiency	
Rats, Rattus spp.	More than 50.0 μg/kg ration
Ruminants (Bos spp.*), swine (Sus* spp.*)*	More than 100.0 μg/kg DW ration[a]
No observable adverse effects during chronic exposure	
Cattle, Bos spp.	Less than 0.5 mg/kg DW ration
Dogs (Canis sp.*), rats (Rattus* spp.*), monkeys (Macaca* spp.*)*	Less than 1.0 mg/kg ration
Rat	Equivalent to 16.7 μg/kg BW daily[b]
Various species	Less than 100.0 mg/kg ration, equivalent to 0.8–<40.0 mg/kg BW daily
Adverse effects expected	
Cattle	
Adults	More than 50.0 mg/kg ration
Calves	More than 5.0 mg/kg ration, equivalent to more than 0.16 mg/kg BW daily

Table 22.4. cont'd

Resource, Criterion, and Other Variables	Effective Nickel Concentration
Dogs	Equivalent to more than 1.3 mg/kg BW daily
Mammals, most species	More than 500.0–2500.0 mg/kg diet, equivalent to 10.0–50.0 mg Ni/kg BW daily
Drinking water	
Adverse effects observed	
Rat	5.0 mg/L, equivalent to 0.35 mg/kg BW daily
Most species	200.0–225.0 mg/L
Tissue residues	
Evidence of significant nickel exposure	
Kidney	More than 10.0 mg/kg DW
Liver	More than 3.0 mg/kg DW
HUMAN HEALTH	
Air	
Cancer risk	
Increased risk; soluble nickel compounds	More than 1.0–2.0 mg/m^3
No increased risk; metallic nickel	Less than 0.5 mg/m^3
Industrial plant; USA; nickel carbonyl	
Safe	Daily average less than 1.0 μg/L; single air sample less than 40.0 μg/L
Discontinue operations	More than 1.0–5.0 μg/L daily average; single air sample more than 200.0–2000.0 μg/L
Shut down plant	Daily average more than 5.0 μg/L; single air sample more than 2000.0 μg/L
Outside industrial plant; nickel carbonyl	
Acceptable	Less than 0.3 μg/L monthly average
Shut down plant	More than 1.0 μg/L monthly average
Safe	
Canada	
Soluble nickel compounds	Less than 0.1 mg/m^3
Sparingly soluble nickel compounds	Less than 1.0 mg/m^3
Nickel carbonyl	Less than 0.12 mg/m^3 (equivalent to less than 0.35 mg $Ni(CO)_4/m^3$)
Former Soviet Union	
Nickel metal, nickel monoxide and sulfide dust, soluble nickel compounds	Less than 0.5 mg/m^3
Nickel carbonyl	Less than 0.005 mg/m^3
Germany; nickel carbonyl	Less than 0.7 mg/m^3
Sweden; nickel metal	Less than 0.01 mg/m^3
USA	
Nickel carbonyl	Less than 0.007 mg/m^3
Nickel metal and relatively insoluble nickel compounds; 8-h daily, 40-h weekly	Less than 1.0 mg/m^3

Continued

Table 22.4. cont'd

Resource, Criterion, and Other Variables	Effective Nickel Concentration
Inorganic nickel in workplace (elemental and all nickel compounds except organonickel compounds with a covalent C–Ni bond, such as nickel carbonyl); 10-h work shift, 40-h workweek, over a working lifetime	Less than 0.015 mg/m^3
Water-soluble nickel compounds; 8-h daily, 40-h weekly	Less than 0.1 mg/m^3
Oral, via diet and drinking water	
Safe chronic exposure via diet or drinking water; soluble nickel compounds	Less than 0.002 mg/kg BW daily
Diet; Australia; marine fish muscle; acceptable limit	Less than 1.0 mg/kg FW
Drinking water	
Acceptable daily intake for 70-kg person (with a safety factor of 1000)	0.031 mg daily (equivalent to 0.443 μg/kg BW daily)
Concentrations developed for noncarcinogenic effects	
Daily intake, lifetime exposure, 70-kg adult (safety factor of 100)	Less than 350.0 μg/L
Daily intake, 10-day health advisory for a 10-kg child (with safety factor of 100)	Less than 1.0 mg/L
Daily intake, 10-day health advisory for a 70-kg adult (with safety factor of 100)	Less than 3.5 mg/L
Water containing edible fishery products	
From ingestion through water and nickel-contaminated fishery products	Less than 13.4 μg total recoverable Ni/L
From consumption of fish and shellfish products alone	Less than 101.0 μg/L
Tissue residues	
Plasma; total nickel; nickel workers; considered elevated	More than 11.9 μg/L
Serum; total nickel	
Normal	Less than 2.6 μg/L, excretion of 2.6 μg daily
Elevated (near nickel mine)	More than 4.6 μg/L, excretion of 7.9 μg daily
Urine; nickel carbonyl	
Mild exposure	Less than<0.1 mg/L during the first 8 h after exposure
Significant exposure	More than 0.1 mg/L during the first 8 h after exposure
Urine; total nickel; nickel workers; considered elevated	More than 129.0 μg/L

[a]Elevated requirement may reflect increased use by rumen bacteria.

[b]Based on no observable adverse effects during chronic exposure to diets containing 100.0 mg Ni (as soluble salts) per kg ration (= 5.0 mg Ni/kg BW daily) divided by uncertainty factor of 300.

in nickel-accumulating plants and in the surrounding soils.

To protect freshwater plants and animals against nickel, a proposed range of less than 25.0–96.0 μg total recoverable Ni/L is recommended by various authorities (Table 22.4). This range will protect most species of freshwater biota; however, certain species have reduced survival within this range, including embryos of rainbow trout (*Oncorhynchus mykiss*) at 11.0 μg/L, daphnids (*Ceriodaphnia dubia*) at 13.0 μg/L, and embryos of the narrow-mouthed toad (*Gastrophryne carolinensis*) at 50.0 μg/L. Mixtures of metals are additive or more-than-additive in toxicity and, in some cases, will exceed the recommended water quality criteria based on the individual metals. Such additive effects were demonstrated for daphnids and rainbow trout using water quality criteria developed in the Netherlands for mixtures of nickel salts and those of arsenic, cadmium, chromium, copper, lead, mercury, or zinc. To protect marine life, the 24-h average for total recoverable Ni/L should not exceed 7.1 μg/L, and the maximum concentration should not exceed 140.0 μg/L at any time (Table 22.4). The maximum concentration level for marine life protection needs to be reexamined because 30.0 μg Ni/L adversely affects growth of marine diatoms, 56.0 μg/L results in nickel accumulations in mussels, 58.0 μg/L causes abnormal sea urchin development, and 59.0 μg/L has adverse effects on motility of sperm of sea urchins. In aquatic systems, research is needed to determine the mechanisms of nickel toxicity to biota, the transport of nickel, the interaction of nickel with other inorganic and organic chemicals, and the mobility of nickel in sediments under various environmental conditions.

To protect birds, diets should contain at least 50.0 μg Ni/kg ration to prevent nickel deficiency but less than 200.0 mg Ni/kg ration in the case of young birds and less than 800.0 mg/kg ration in the case of adults to prevent adverse effects on growth and survival (Table 22.4). Nickel residues in avian kidneys in excess of 10.0 mg/kg DW or in liver in excess of 3.0 mg/kg DW are sometimes associated with adverse effects; however, nickel accumulates in kidneys of mallards (*Anas platyrhynchos*) at dietary concentrations as low as 12.5 mg Ni/kg ration. In general, tissue concentrations of nickel were not reliable indicators of potential toxicity in mammals and birds because adverse effects, including death, frequently occurred in the absence of elevated tissue nickel concentrations. For monitoring birds, analysis of kidneys, bones, and feathers is the most likely to reveal elevated exposure to environmental nickel contamination; nickel concentrations in liver and spleen often do not reflect elevated exposure.

To protect humans and other mammals, proposed air quality criteria range from 0.01 to less than 1.0 mg/m^3 for metallic nickel and slightly soluble nickel compounds, 0.015–0.5 mg/m^3 for water-soluble nickel compounds, and 0.005–0.7 mg/m^3 for nickel carbonyl (Table 22.4). Inhalation of nickel subsulfide concentrations (0.11–1.8 mg Ni/m^3) near the current threshold limit value of 1.0 mg Ni/m^3 can produce detrimental changes in the respiratory tract of rats after only a few days of exposure. Additional animal studies are recommended to identify minimally effective inhalation exposure levels for the various nickel compounds. Continued monitoring of nickel refining, nickel–cadmium battery manufacture, and nickel powder metallurgy installations is recommended because ambient air levels of bioavailable nickel at these installations in excess of 1.0 mg/m^3 can sometimes still be found.

Most species of mammals had normal growth and survival during chronic exposure to diets equivalent to 0.8–40.0 mg Ni/kg BW daily. Reduced growth and survival sometimes occurred when sensitive species of wildlife were fed diets containing 500.0–2500.0 mg Ni/kg ration, equivalent to 10.0–50.0 mg Ni/kg BW daily. Proposed criteria for nickel by way of the diet or drinking water range from 2.0 μg total Ni/kg BW daily to 443.0 μg total Ni/kg BW daily for soluble nickel compounds, less than 1.0 mg Ni/kg FW diet, and less than 350.0 μg Ni/L drinking water (Table 22.4). Further research is needed to clarify the role of nickel in mammalian nutrition, including dietary requirements of nickel and identification of the chemical forms of nickel present in foods and their bioavailability. Studies are needed on the absorption and cellular uptake, transport, and metabolism of well-characterized nickel species following different routes and types of administration and on the transfer of dietary nickel to tissues of lactating dams and juveniles. Because young female laboratory mice were more susceptible to dietary nickel than were adults, it is possible that no-observable-adverse-effect-levels (NOAELs) derived from adult animals may be inappropriately high for neonates and juveniles. Studies that compare the toxicokinetics of humans and animals concurrently could be helpful in determining which species of animal is the most appropriate model for assessing the effects of nickel in human health. Animal studies designed to examine neurological effects after inhalation or oral exposure are needed to determine, in part, if human exposure to nickel will cause permanent neurological damage.

Nickel affects reproduction of selected mammals. Drinking water containing 5.0 mg Ni/L – equivalent to 0.2–0.4 mg Ni/kg BW daily – had adverse effects on rat reproduction and iron metabolism. Dogs given the equivalent of 1.3 mg Ni/kg BW daily had decreased litter survival. Nickel is known to cross the placental barrier and reach the fetus in mammals and humans. More information is needed on the effects of in utero nickel exposure in pregnant women. Such information may be obtained using appropriate animal models. Multigenerational inhalation studies are recommended to determine if developmental effects result from both inhalation and oral exposure.

Biomarkers of nickel exposure and effects include nickel concentrations in feces and urine and changes in serum antibodies and serum proteins. Levels of carnosine, a dipeptide, seem to reflect the extent of nickel-induced damage to olfactory mucosa of rats, although the rodent olfactory system is more resilient than is the human. Studies on the availability of trace levels of nickel in food and water and in air would be helpful to relate levels of nickel found in the hair, nails, blood, and urine to levels of nickel in internal organs. Nickel concentrations in human tissues now considered elevated include 4.6 μg/L in serum, 11.9 μg/L in plasma, and 100.0–129.0 μg/L in urine (Table 22.4). Treatment of mammals suffering from nickel poisoning is usually through administration of various classes of chelating agents, including dithiocarb (sodium diethyl-dithiocarbamate – the drug of choice in the management of nickel carbonyl poisoning), EDTA salts, BAL (2,3-dimercaptopropanol), and penicillamine. In all cases, the agents accelerate urinary excretion of absorbed nickel before extensive tissue injury occurs.

The nomenclature of nickel compounds should be further standardized. Analytical methods must be developed and standardized in order to facilitate speciation of nickel compounds in atmospheric emissions, biological materials, and in other environmental samples. Studies are needed to elucidate the biogeochemical nickel cycle on a global scale and determine its potential for long-range transport.

22.9 Summary

Nickel is found in air, soil, water, food, and household objects; ingestion or inhalation of nickel is common, as is dermal exposure. Recent estimates suggest that as much as 28,100 tons of nickel are introduced into the atmosphere each year from natural sources and as much as 99,800 tons from human activities. In the atmosphere, nickel is mostly suspended onto particulate matter. In natural waters, the dominant chemical species is Ni^{2+} in the form of $(Ni(H_2O)_6)^{2+}$. In alkaline soils, the major components of the soil solution are Ni^{2+} and $Ni(OH)^+$; in acidic soils, the main solution species are Ni^{2+}, $NiSO_4$, and $NiHPO_4$.

Nickel is an essential micronutrient for maintaining health in certain species of plants and animals. Nickel deficiency effects from dietary deprivation of nickel has been induced experimentally in many species of birds and mammals. To prevent nickel deficiency in rats and chickens, diets should contain at least 50.0 μg Ni/kg ration, while cows and goats require more than 100.0 μg Ni/kg rations, perhaps reflecting the increased use by rumen bacteria. Nickel deficiency is not a public health concern for humans because daily oral intake is sufficient to prevent deficiency effects.

Nickel contamination from anthropogenic activities occurs locally from emissions of metal mining, smelting, and refining operations; combustion of fossil fuels; nickel plating and alloy manufacturing; land disposal of sludges, solids, and slags; and disposal as effluents. Nickel concentrations in living organisms and abiotic materials tend to be elevated in the vicinity of nickel smelters and refineries, nickel–cadmium battery plants, sewage outfalls, and coal ash disposal basins.

Adverse effects of excess nickel are documented for bacteria, algae, yeasts, higher plants, protozoans, mollusks, crustaceans, insects, annelids, echinoderms, fishes, amphibians, birds, and mammals. To protect terrestrial vegetation against decreased growth and other toxic effects, nickel concentrations in leaves should contain less than 50.0 mg Ni/kg DW (and in some cases less than 44.0 mg Ni/kg DW), growing soils should contain less than 250.0 mg Ni/kg DW (and in some cases <50.0 mg Ni/kg DW), and sewage sludge applied to agricultural soils should be limited to 30.0–140.0 kg Ni/surface ha at the low end and 50.0–560.0 kg/surface ha at the high end. To protect freshwater plants and animals against nickel, a proposed range of less than 25.0–96.0 μg total recoverable Ni/L is recommended by various authorities; however, certain species have reduced survival within this range. To protect marine organisms, the 24-h average for total recoverable nickel per liter should not exceed 7.1 μg/L and the maximum concentration should not exceed 140.0 μg/L at any time; however, certain marine organisms show adverse effects to as little as 30.0 μg Ni/L.

To protect young birds against adverse effects of excess nickel on growth and survival, diets should contain less than 200.0 mg Ni/kg ration; diets of older birds should contain less than 800.0 mg Ni/kg ration. Nickel concentrations in avian tissues in excess of 10.0 mg/kg DW kidney or 3.0 mg/kg DW liver are sometimes associated with adverse effects.

Toxic effects of nickel to humans and laboratory mammals are documented for respiratory, cardiovascular, gastrointestinal, hematological, musculoskeletal, hepatic, renal, dermal, ocular, immunological, developmental, neurological, and reproductive systems. Nickel toxicity in mammals is governed by the chemical form of nickel, dose, and route of exposure. Mammalian exposure to nickel by inhalation or cutaneous contact was more significant than oral exposure. To protect humans and other mammals against respiratory effects, proposed air quality criteria are 0.01 to less than 1.0 mg/m^3 for metallic nickel and sparingly soluble nickel compounds and 0.005–0.7 mg/m^3 for nickel carbonyl. Most species of mammals tested had normal growth and survival during chronic exposure to dietary nickel (equivalent to 0.8–40.0 mg Ni/kg BW daily) and reduced growth and survival when fed diets containing 500.0–2500.0 mg Ni/kg ration (equivalent to 10.0–50.0 mg Ni/kg BW daily). Proposed nickel criteria for sensitive species by way of the diet or drinking water now range from 2.0 to less than 443.0 μg total Ni/kg BW daily for soluble nickel compounds, less than 1.0 mg Ni/kg FW diet, and less than 350.0 μg Ni/L in drinking water. Nickel concentrations in human tissues now considered elevated include 4.6 μg/L serum, 11.9 μg/L plasma, and 100.0–129.0 μg/L urine; comparable data for mammalian wildlife are lacking.

Some forms of nickel are carcinogenic to humans and animals, but only when exposure is by the respiratory route. Toxic and carcinogenic effects of nickel compounds are associated with nickel-mediated oxidative damage to DNA and proteins and to inhibition of cellular antioxidant defenses. Some nickel compounds are weakly mutagenic in a variety of test systems, but much of the evidence is inconclusive or negative. In mammals, no teratogenic effects of nickel compounds occur by way of inhalation or ingestion, except from nickel carbonyl. Inhaled nickel carbonyl results in comparatively elevated nickel concentrations in lung, brain, kidney, liver and adrenals, and is the most hazardous form of nickel.

Overall, nickel is not an immediate threat to the health of plants, animals, and humans at environmentally encountered levels, except in the case of nickel carbonyl, and progress has been made toward minimizing or eliminating occupational nickel exposure.

PARAQUAT[a]

Chapter 23

23.1 Introduction

Paraquat (1,1′-dimethyl-4,4′-bipyridinium) is one of the most widely used herbicidal chemicals in the world and is now available in more than 130 countries. Its chemical structure was first described in 1882, its oxidizing and reducing properties in 1933, and its herbicidal properties in 1955. Paraquat was marketed commercially in the United Kingdom in 1962 and registered for use in the United States in 1964. As the dichloride salt it has found wide use as a nonselective contact herbicide at application rates of 1.12 kg/ha (1.0 pound/acre) and lower. Paraquat kills plants by affecting the green parts, not the woody stems, and is usually completely and rapidly inactivated by contact with clay in the soil. In its bound form, paraquat is biologically inert and innocuous to plants and animals. In 1977, the discovery by narcotics authorities that some marijuana imported from Mexico had been treated with paraquat as a control agent generated much interest in the media. Up to 70% of the paraquat in paraquat-treated marijuana, on smoking, is converted to bipyridine, a respiratory irritant. Frequent consumption of heavily contaminated cigarettes may result in cyanosis and possibly death. The use of paraquat for this purpose has been largely discontinued.

Numerous human injuries and deaths have resulted from intentional ingestion of the concentrated commercial product. For example, in the first 10 years following paraquat's commercial use, 232 human deaths from paraquat poisoning were reported, about half suicidal, and almost all were due to the drinking of concentrated material. Most poisonings resulted from the ingestion of the 21% cation concentrate, which had been decanted and stored in empty beer, soft drink, or lemonade bottles; paraquat is a reddish-brown liquid that resembles root beer or cola drinks. One individual sprinkled paraquat on French fried potatoes, thinking that it was vinegar. He died 25 days later. Another died after applying the concentrated solution to his beard and scalp to treat a lice infestation. In Japan, more than 1000 persons each year are reportedly poisoned by paraquat. Initially, paraquat may produce multiorgan toxicity of kidneys, liver, heart, central nervous system, adrenal glands, skeletal muscle, and spleen, but the ultimate target organ is the lung, in which progressive irreversible pulmonary fibrosis develops. This effect has been described in man, rats, mice, guinea pigs, and dogs. There is no specific antidote for paraquat poisoning. In normal use as a spray, minor reversible injuries are reported to abraded skin, eyes, nose, and fingernails; it is not absorbed through intact skin. Paraquat is fetotoxic as judged by deliberate ingestion of concentrated solutions by nine pregnant Taiwanese women. Paraquat crosses the placenta and concentrates there to levels 4–6 times that of maternal blood. All fetuses died whether or not emergency cesarean operation was performed. Research has focused on the tendency of paraquat to accumulate in neuromelanin of mammals and amphibians and to cause lesions in the pigmented nerve cells, leading to effects very similar to those of Parkinson's disease.

---

[a]All information in this chapter is referenced in the following sources:

Eisler, R. 1990. Paraquat hazards to fish, wildlife, and invertebrates: a synoptic review. *U.S. Fish Wildl. Serv. Biol. Rep. 85(1.22)*, 28 pp.

Eisler, R. 2000. Paraquat. Pages 1159–1192 in *Handbook of Chemical Risk Assessment: Health Hazards to Humans, Plants, and Animals. Volume 2, Organics.* Lewis Publishers, Boca Raton, Florida.

23.2 Uses

Paraquat is a broad-spectrum contact weed killer and herbage desiccant that is used widely in agriculture and horticulture. Paraquat was originally formulated in 1882, but its herbicidal properties were not discovered until 1955. Since its introduction in the early 1960s, paraquat has been used extensively in about 130 countries, including the United Kingdom, Canada, and the United States, on a wide variety of agricultural crops.

Primary uses of paraquat include: weed control in orchards, plantation crops, and forests; weed control before sowing or before crop emergence; pasture renovation; preharvest desiccation; and aquatic weed control, although use as an aquatic herbicide in the United States is not permitted. In New Zealand, use of paraquat for aquatic weed control (2.0 mg/L for 30 min) in 1966–67 on the Waimakariri River severely reduced amphipod populations; paraquat is no longer used for this purpose in that country. Paraquat is registered domestically for preplant or preemergence use for cotton, barley, corn, lettuce, melons, peppers, safflower, soybeans, sorghum, sugar beets, tomatoes, potatoes, and wheat. It is also registered for use on noncrop areas, such as roadsides, highway margins, rights-of-way, around commercial buildings, power plants, storage yards, fence lines, and parkways. In Switzerland, it is used to control voles (*Arvicola terrestris*) in fruit orchards. Paraquat application to corn using a manual knapsack sprayer is considered unsafe to the human operators if lances are less than 1.0 m. Lances >1.0 m in length are recommended, as well as additional protective garments for legs, feet, and hands and switching the paraquat spraying operation to the back of the worker's body.

Paraquat is available as the dichloride or dimethylsulfate salt; both compounds are extremely soluble in water (Figure 23.1). In the United States, paraquat dichloride is available as a 29% liquid concentrate containing 240.0 g/L (2 pounds/gallon) of paraquat cation, or as a 42% liquid concentrate. Elsewhere, it is sold as Gramoxone liquid containing 20–24% of paraquat dichloride. Paraquat dichloride concentrates usually contain various wetting agents (condensation products of ethylene oxide and alkyl phenols), spreaders, humectants to promote moisture retention (calcium chloride, glycerol, polyethylene glycol), plant adhesion materials (carboxymethylcellulose, polymethacrylates), and antifoaming agents.

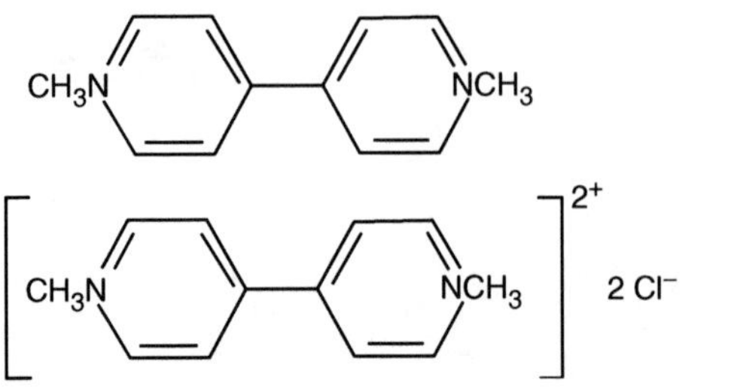

Figure 23.1.
Structural formula of paraquat cation (*upper*) and of paraquat dichloride salt (*lower*).

The recommended field application rates for terrestrial weed control usually range between 0.28 and 1.12 kg paraquat cation/ha (0.25 and 1.0 pounds/acre), or 0.56 and 2.24 kg paraquat dichloride/ha (0.5 and 2.0 pounds/acre) – both applied as an aerosol – and 0.1 and 2.0 mg/L for aquatic weed control, although sensitive aquatic plants may be affected between 0.019 and 0.372 mg/L. Paraquat is frequently used in combination with other herbicides. Water solutions of the dichloride salt, which usually contain 240.0 g/L, have been successfully mixed with 2,4-D, substituted ureas, dalapon, amitrol, and various triazines.

23.3 Concentrations in Field Collections

Data are scarce on ecosystems treated with paraquat. It is clear, however, that both terrestrial and aquatic plants accumulate paraquat, and that the compound disappears rapidly from the water column and tends to concentrate in surface muds.

Water from irrigation channels, rivers, and lagoons from Spanish marshes in 1996 bordering the Mediterranean Sea contained an average of 0.01 μg paraquat/L with a maximum recorded value of 3.95 μg/L. Paraquat values were highest in the summer owing to high application rates, low rainfall, and high evaporation rates. At these comparatively low concentrations, paraquat is not easily degraded chemically or biologically and persists in river waters with more than 80% remaining after 56 days of incubation.

23.4 *Environmental Chemistry*

Paraquat is a nonvolatile, ionic compound that is almost completely soluble in organic solvents, which is typical of the bipyridyl group of chemicals. As discussed later, the biochemical mechanism of paraquat toxicity is due to the cyclic oxidation and reduction that occurs in various tissues, especially lung, leading to production of superoxide anion and other free radicals; these chemical species react with polyunsaturated free radicals, eventually forming the highly destructive hydrogen peroxide. Excretion of paraquat is rapid in living organisms, but delayed toxic effects, including death, are not unusual. No treatment or chemical has proven completely successful in protecting against paraquat-induced lung toxicity.

Paraquat is strongly adsorbed to soils and sediments and is biologically unavailable in that form; however, it is not degraded significantly for many years, except in surface soils. In surface soils, paraquat loss through photodecomposition approaches 50% in 3 weeks. In freshwater ecosystems, loss from water column is rapid: about 50% in 36 h, and 100% in 4 weeks. In marine ecosystems, 50–70% loss of paraquat from seawater was usually recorded within 24 h.

23.4.1 Chemical Properties

Paraquat is a nonvolatile, ionic compound that is almost completely insoluble in fat, and therefore not likely to be accumulated in food chains. The compound belongs to the bipyridyl group of chemicals and is typical of the many hundreds that have been synthesized, variation usually being the result of introducing different quaternizing groups on the nitrogen atoms, which also shift. Paraquat dichloride is produced from pyridine in the presence of sodium in anhydrous ammonia, then quaternizing the 4,4′-dipyridyl with methyl chloride. The common paraquat salts are all fully ionized, and experiments have shown that the anions (e.g., chloride, sulfate, methyl sulfate) do not affect the toxicity of paraquat. Chemical and other properties of paraquat are briefly summarized (Table 23.1).

23.4.2 Mode of Action

Paraquat is absorbed systemically in mammals, following different routes of exposure; absorption is greatest for the pulmonary route, followed by intragastric and dermal routes. Administration of paraquat by every route of entry tested frequently results in irreversible changes in lung. In the intestinal tract, where some microbial degradation occurs, most paraquat (95–100%) is usually excreted unchanged in feces and urine within 2 days. Absorption in the gastrointestinal tract ranges from 0.26% in cow, to 5% in man, 8% in guinea pig, 16% in cat, and up to 20% in rat; the half-time persistence (Tb1/2) of paraquat in certain tissues ranges between 20 and 30 min, but up to 4 days in muscle and 2 days in plasma. Delayed toxic effects of paraquat occurring after the excretion of virtually all of the material have caused it to be classified as a "hit and run" compound, that is, a compound causing immediate damage, the consequences of which are not readily apparent.

Most authorities agree that free radical pathology is the most likely mechanism by which paraquat is cytotoxic. The biochemical mechanism of paraquat toxicity is related to the cyclic oxidation and reduction of paraquat that occurs in lung cells, which leads to continued production of high levels of superoxide anion (O_2^-) and other cytotoxic oxygen free radicals. Superoxide anion and other oxygen free radicals initiate the peroxidation of membrane lipids, causing tissue damage and death. Paraquat oxidation is coupled with the reduction of molecular oxygen, forming superoxide

Table 23.1. Chemical and other properties of paraquat.

Variable	Datum
CHEMICAL NAME	
Paraquat (cation)	1,1′-dimethyl-4,4′-bipyridinium
Paraquat dichloride (salt)	1,1′-dimethyl-4,4′-bipyridinium dichloride
MOLECULAR WEIGHT	186.2 (cation), 257.2 (salt)
CAS NUMBER	4685-14-17 (cation), 1910-42-5 (salt)
EMPIRICAL FORMULA	$C_{12}H_{14}N_2$ (cation), $C_{12}H_{14}Cl_2N_2$ (salt)
ALTERNATE NAMES	Cekuquat, Crisquat, Dextrone, Dextrone X, Dexuron, Dual Paraquat, Esgram, Gramonol, Gramoxone, Gramuron, Herbaxon, Herboxone, Methyl Viologen, Ortho Paraquat, Orvar, Paracol, Paraquat CL, Pathclear, Pillarquat, Pillarxone, Preeglone, PP 148, PP 910, Sweep, Tenaklene, Totacol, Toxer, Total, Weedol
SOLUBILITY AT 20°C	
Water	561.0 g/L
Methanol	144.0 g/L
Ethanol	1.7 g/L
Acetone	200.0 mg/L
Most organic solvents	Insoluble or sparingly soluble
PHYSICAL STATE	White (pure), yellow (technical) solid
MAIN USES	Herbicide, desiccant
SPECIFIC GRAVITY	1.24–1.26
MELTING POINT	175–180°C, decomposes at 345°C
STABILITY	Stable on exposure to hot acids, unstable in alkalis at pH > 10
FLASH POINT	Nonexplosive, nonflammable
VOLATILITY	Nonvolatile

anion, singlet oxygen, and hydroxyl radicals. These molecular species react with polyunsaturated fatty acid free radicals and, on further oxidation, with lipid hydroperoxide radicals. The hydroperoxide radicals then maintain the formation of new fatty acid radicals while being converted to lipid peroxides in a chain reaction. Various enzymes in the cells catabolize the superoxide radical and reduce the lipid hydroperoxides to less-toxic lipid alcohols. The superoxide anions are converted to hydrogen peroxide and oxygen; hydrogen peroxide is further inactivated to water and oxygen by catalases and peroxidases. In the presence of reduced nicotinamide adenine dinucleotide phosphate (NADPH), paraquat is reduced by microsomal NADPH-cytochrome reductase. The reduction of lipid peroxides by glutathione peroxidase requires reduced glutathione. Because the reduction of oxidized glutathione is coupled with NADPH oxidation via glutathione reductase, it seems that the availability of NADPH is essential for paraquat detoxification, and that the critical depletion of NADPH may render the cell more susceptible to lipid peroxidation.

The lung is the organ most severely affected in paraquat poisoning. Pulmonary injury is largely due to the preferential accumulation of paraquat in lung – mediated by an energy-dependent system for uptake of endogenous polyamines – and to the continuous exposure of lung to atmospheric oxygen. Characteristic signs of poisoning include severe anoxia, marked and widespread fibroblastic proliferation in the alveolar walls around the terminal bronchi and blood vessels, and frequently death. The specific toxicity to the lung can be explained by the accumulation of paraquat in the alveolar Type II cells. These cells are responsible for the synthesis of pulmonary surfactant, the surface-active material lining the alveolar epithelium. The pulmonary surfactant is secreted after storage in cytoplasmic organelles known as lamellar bodies. Thus, any damage to the alveolar epithelium could alter synthesis and secretion of the pulmonary surfactant. The pulmonary effects of paraquat are probably related to the conversion of paraquat to a free radical followed by conversion to a long-lived dihydroderivative, which causes transformation of normal alveolar epithelial cells to fibroblasts. The increase in toxicity of paraquat by oxygen supports the hydroperoxide theory, in which the reversible action of the free radical's oxidation–reduction gives rise to hydrogen peroxide. Paraquat also depletes NADPH in the isolated lung to the extent of mixed function oxidation impairment. Depletion of NADPH would impair fatty acid and lipoprotein synthesis and inhibit various detoxification and biosynthetic functions.

Other organs and systems affected by paraquat include the kidney (pathology of proximal tubules), liver (hepatocellular necrosis), spleen and thymus (pathology), circulatory system (irregular and feeble heart beat, myocardial congestion, increase in erythrocytes and leucocytes, external pericarditis, myocardium edema), brain (neuronal depletion, myelin destruction), gastrointestinal tract (esophagitis; ulceration of buccal cavity, pharynx, gastric mucosa; mucosal erosion), skin (erythema, hyperkeratosis), reproductive system (degeneration), nervous system (hyperexcitability, irritability, incoordination, convulsions), various enzyme systems, and the eye.

Several early indicators of paraquat-induced stress have been proposed, including alkaline phosphatase activity, fibronectin levels, and intracellular calcium uptake. Alkaline phosphatase activity is associated with the lamellar body, and changes in this variable are suggested as indicative of toxicity to Type II alveolar epithelium cells. Levels of fibronectin, an extracellular matrix glycoprotein, were elevated in patients with fibrotic lung diseases and in monkeys given multiple injections of paraquat. Lung intracellular calcium uptake was significantly disrupted, even at doses that normally produce significant increases in lung water content. These subjects seem to merit additional research, as does the role of polyamines in mediating fibrotic changes in the lung; paraquat-altered synthesis of proteins, DNA, collagen, and pentose phosphate metabolism; and hyperoxia, that is, increased oxygen free-radical generation.

Certain treatments or chemicals provide varying degrees of protection against paraquat-induced lung toxicity and lethality, although no treatment or chemical has proven completely successful. Present treatment of paraquat-poisoned animals and humans is directed to elimination of the material from the body using repeated doses of adsorbents such as Fuller's earth or bentonite, cathartics to reduce paraquat absorption, and hemodialysis, forced diuresis, and hemofiltration to enhance excretion. The use of 100% oxygen is contraindicated, as mortality is greatly increased. Toxicity mediated by free radicals can be moderated by several cellular defense mechanisms, including superoxide dismutase, catalase, glutathione peroxidase, vitamin E, and reduced glutathione. A low-molecular-weight superoxide dismutase mimic, based on manganese, was found to protect mammalian cells against the cytotoxic effects of the superoxide radical produced by paraquat. Certain chemicals reportedly provide limited protection to small laboratory animals under carefully controlled conditions of administration: nicotinic acid; niacin; cysteine; *N*-acetylcysteine; metallothionein – a metal-binding low-molecular-weight protein rich in cysteine; D-penicillamine; clofibrate; lipid-soluble antioxidants; various amino acids; phenobarbital; methyl prednisolone; and certain anti-inflammatory drugs. In bluegills, 1,10-phenanthroline, a chelator of ionic iron, reduced the toxicity of paraquat through the prevention of hydroxyl radical formation.

In tilapia, the paraquat-induced increase in gill carbonic anhydrase activity was not observed when ionic lead was present at 47.0 mg/L. In plants, the pea (*Pisum sativum*) is protected by cerium chloride, in part, through counteracting peroxide formation.

Paraquat toxicity is increased and its effects otherwise exacerbated in organisms fed diets deficient in selenium or vitamin E, although high levels of these substances in diets did not provide protection; by methyl prostaglandins or diethyl maleate; and by increased iron and copper. Dietary changes that do not result in nutrient deficiency or toxicity may affect the biocidal properties of paraquat and other compounds. In studies with rodents subjected to paraquat insult, survival was higher in those fed cereal-based diets vs. purified diets, and higher in egg-white (protein) purified diet vs. a casein diet, suggesting a need to use strictly defined diets in the study of paraquat toxicity to control for any paraquat–diet interactions.

Paraquat causes tissue damage and increased stress in the common carp (*Cyprinus carpio*), as judged by the increased enzyme activities of lactic dehydrogenase, glutamic oxaloacetic transaminase, and glutamate dehydrogenase, and by the elevated blood sugar levels. Paraquat and copper sulfate when administered together to carp were synergistic in terms of tissue damage and stress effects, especially liver damage.

Paraquat adhering to the plant surface is usually degraded photochemically. Paraquat is phytotoxic through inhibition of processes involving photosynthesis and respiration. Its mode of action in plants is similar to that in animals, that is, lipid peroxidation of membranes due to formation of the superoxide radical and related species. Photosynthetic tissues reduce paraquat to stable free radicals that, upon reoxidation, produce hydrogen peroxide. Unsaturated lipids in the cells are oxidized by the peroxide, and damage is dependent largely on production of hydrogen peroxide. These reactions are dependent on light and oxygen.

In bacteria (*Escherichia coli*), paraquat is concentrated, reduced to the monocation radical, and combines with molecular oxygen to produce the superoxide radical within the cell. Copper and iron are essential mediators in bacteriocidal effects; the cytoplasmic membrane is the target organelle in paraquat toxicity to *E. coli*, and extent of damage correlates positively with levels of these metals.

23.4.3 Fate in Soils and Water

In contact with soil, paraquat is rapidly adsorbed – usually in the clay mineral lattice sheets – and inactivated by base exchange; the process is facilitated by the flat and highly polarizable nature of the paraquat ion. The strong binding of paraquat to soil constituents reduces the mobility of the herbicide due to leaching, although paraquat is displaced from binding sites by low concentrations of ions of ammonium, potassium, sodium, and calcium. Paraquat adsorption is not significantly affected by soil pH, but is modified by soil porosity, moisture content, residence time, and adsorption capacity. Paraquat applied to a sandy loam soil at field application rates between 0.56 and 2.24 kg/ha was adsorbed by organic matter and clays, usually in the top centimeter of soil. Typical soils contain about 300.0 mg paraquat/kg after treatment at recommended applications; however, adsorption capacity varies among soils. Clay minerals, such as kaolinite, can adsorb 2500.0–3500.0 mg/kg, whereas others, such as montmorillonite, adsorb up to 85,000.0 mg/kg after paraquat treatment.

Paraquat is not degraded significantly in soil during incubation periods up to 16 months at 25°C by chemical or microbiological vectors. For example, paraquat dichloride applied once annually at 4.48 kg/ha, or 4 times annually at 1.12 kg/ha, remained essentially undegraded in the soil for 6 years. Massive applications to soils of 3000.0 kg/ha can persist for at least 6 months without significant degradation. Bacterial degradation – which occurs only slowly in soils – consists of demethylation, followed by ring cleavage to eventually form the carboxylated 1-methylpyridinium ion (Figure 23.2). Photochemical decomposition of paraquat is the predominant mechanism of paraquat degradation in soils. In surface

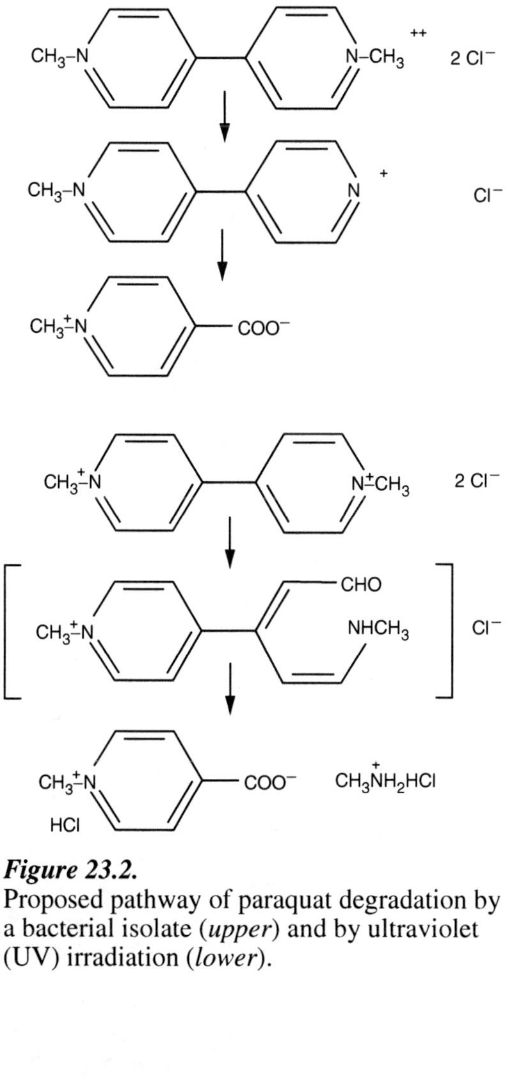

Figure 23.2.
Proposed pathway of paraquat degradation by a bacterial isolate (*upper*) and by ultraviolet (UV) irradiation (*lower*).

soils, paraquat loss through photodecomposition was 20–50% in 3 weeks. Photochemical degradation products of paraquat include 4-carboxy-1-methylpyridium ion and methylamine hydrochloride (Figure 23.2). Laboratory studies have demonstrated that paraquat in soils slated for disposal can be degraded by ultraviolet (UV) irradiation in the presence of oxygen or ozone. Reaction products identified were 4-carboxy-1-methylpyridium ion, 4-picolinic acid, hydroxy-4-picolinic acid, succinic acid, *N*-formylglycine, malic acid and oxalic acid (as trimethylsilicon derivatives), and 4,4′-bipyridyl.

Paraquat is used to control aquatic weeds. It also passes into aquatic environments through rain, where it is rapidly accumulated by aquatic organisms, especially fish. Paraquat applied to control aquatic weeds is accumulated by aquatic macrophytes and algae, and it is adsorbed to sediments and suspended materials. Initial applications of 1.0–5.0 mg/L in the water column are usually not detectable under field conditions after 8–27 days. The half-time persistence of paraquat in water column at normal doses for weed control (i.e., 0.5–1.0 mg/L) was 36 h; less than 0.01 mg/L was detectable in 2 weeks. In solution, paraquat was subject to photodecomposition and microbial metabolism, degrading to methylamine and 4-carboxy-1-methylpyridium ion. In freshwater, without sediment or plants, 100% of the initial concentration of 0.5 mg paraquat/L was degraded in 35 weeks. When sediments were present, 100% loss from the water column occurred in 6–8 weeks, and when both sediment and aquatic plants were present, paraquat was not detectable in the water column in 3–4 weeks. Mud cores taken from a paraquat-treated lake had elevated paraquat residues, but showed no phytotoxic effects on barley seedlings germinated on them.

Paraquat loss from seawater in 24 h was 70% at an initial concentration of 1.0 mg/L, 68% at 5.0 mg/L, and 76% at 10.0 mg/L; most of the loss occurred within the first 60 min.

23.5 Lethal and Sublethal Effects

Adverse effects of paraquat in sensitive species of terrestrial plants and soil microflora have been documented at application rates of 0.28–0.6 kg/ha (death, inhibited germination of seeds, reduced growth), at soil concentrations of 10.0–25.0 mg/kg (growth inhibition), and at soil-water concentrations as low as 1.6 mg/L (reduced growth, inhibited synthesis of protein and RNA). Among terrestrial invertebrates, certain species of mites were sensitive to paraquat at recommended rates of application, and the sensitive honeybee died when its diet contained 100.0 mg/kg.

However, paraquat in soils was not accumulated by earthworms and other species of soil invertebrates after applications up to 112.0 kg/ha. These points are discussed later.

Freshwater algae and macrophytes usually die at paraquat concentrations between 0.25 and 0.5 mg/L; marine algae, however, are relatively resistant and usually require 5.0 mg/L or higher for significant inhibition in growth to occur in 10 days. Aquatic invertebrates, especially crustaceans, seem to be the most sensitive group, with effects most pronounced at elevated temperatures in early developmental stages. Adverse effects were noted in crab larvae at nominal water concentrations between 0.9 and 5.0 μg/L, although 1000.0 μg/L and higher were needed to produce similar effects in other species of aquatic invertebrates. Amphibians and fishes were usually unaffected at concentrations below 3000.0 μg/L, although sensitive species such as frog tadpoles and northern carp (*Cyprinus carpio*) were impacted at 500.0 μg/L. There was little accumulation of paraquat from the medium by aquatic fauna.

Paraquat is embryotoxic to sensitive species of birds. Concentrations equivalent to 0.056 kg/ha applied in oil solution to the surface of eggs of the mallard (*Anas platyrhynchos*) inhibited development; when applied in aqueous solution, paraquat was toxic at a dose equivalent to 0.56 kg/ha. In each case, adverse effects occurred below the recommended field application rate of about 1.0 kg/ha. The lowest doses of paraquat that produced harmful effects in sensitive birds were 10.0 mg/kg body weight (BW) in nestlings of the American kestrel (*Falco sparverius*), 20.0 mg/kg in the diet of the common bobwhite (*Colinus virginianus*), 40.0 mg/L in the drinking water of domestic chickens (*Gallus* sp.), and 199.0 mg/kg BW in mallards (acute oral LD50).

Sensitivity of mammals to paraquat was variable, owing to inherent differences in interspecies resistance. Representative mammals were measurably affected at aerosol concentrations of 0.4–6.0 μg/L, acute oral doses of 22.0–35.0 mg/kg BW, dietary concentrations of 85.0–100.0 mg/kg ration, and drinking water levels of 100.0 mg/L.

23.5.1 Terrestrial Plants and Invertebrates

In terrestrial plants, paraquat's action is at the point of local absorption. Characteristic damage signs to susceptible species include wilting and general collapse in herbaceous plants. Regrowth may occur in some perennial plants, but in resistant species temporary scorch may be the most marked effect. In sugarcane (*Saccharum officinarum*), paraquat application severely desiccated the plant within 72 h, and disrupted activity of leaf amylase and sucrose. Paraquat, once absorbed in plants, is likely to persist. The addition of cationic or nonionic surface-active agents increases the phytocidal effectiveness of paraquat, but in combination with various herbicides paraquat was markedly less phytotoxic to certain cereal grains.

Paraquat adsorbed to soils is usually unavailable to crops. In the case of wheat (*Triticum aestivum*), effects from contaminated soils were negligible until soil residues surpassed 600.0–1000.0 kg/ha, causing growth reduction of 10%, or 1650.0 kg/ha, causing elevated residues in leaves but not in grain.

Three species of grains (barley, *Hordeum vulgare*; wheat; oat, *Avena sativa*) died (>95% kill) following application of 0.28 kg paraquat/ha. At 0.6 kg/ha, paraquat inhibited germination and growth in seeds of six species of grasses (Kentucky bluegrass, *Poa pratensis*; perennial ryegrass, *Lolium perenne*; bentgrass, *Agrostis tenuis*; tall fescue, *Festuca arundinacea*; red fescue, *Festuca rubra*; orchard grass, *Dactylus glomerata*), but two species of legumes (alfalfa, *Medicago sativa*; red clover, *Trifolium pratense*) were comparatively resistant. Paraquat was phytotoxic to several species of terrestrial plants (rape, *Brassica rapa*; ryegrass; white clover, *Trifolium repens*) for several days following application of 1.1–2.2 kg/ha. Transpiration rate of soybean (*Glycine max*) was lowered at 1.0 mg/kg. Paraquat is not considered to be carcinogenic or teratogenic, but is weakly mutagenic to some plants (e.g., 4.1% chromosomal aberrations in seeds of wheat at 9.3 mg/kg). Spray solutions containing 0.6 g paraquat/L applied to crowns of eastern red cedar (*Juniperus virginiana*) killed up to 90% of small trees and up to 30% of large trees; at 0.3 g/L up to 60% of small trees were affected. Seedlings of corn (*Zea mays*) sprayed with 0.2% paraquat ion solution for 6 h had decreased rates of total protein synthesis and some polysome dissociation, suggesting that additional research is needed on mutagenicity of paraquat in plants.

Paraquat resistance has been documented in several genera of weeds. For example, paraquat-resistant strains of barley grass (*Hordeum glaucum*) were first noted in 1982 in Australia; resistant strains (based on chromosome counts and resistance to paraquat) were confined to a small number of lucerne fields where paraquat had been used consistently for at least 10 years. However, the potential exists for this biotype to be transferred and established in other areas by the movement of livestock, machinery, hay, and seeds. Paraquat-resistant strains of weeds have been reported in England, Japan, Egypt, and Australia. Paraquat-resistant strains of bacteria, ferns, and other species of flora have been documented. Paraquat-tolerant ferns (*Ceratopteris richardii*) were 10–20 times more resistant than sensitive wild-type strains. Paraquat-resistant strains of perennial rye were up to 10 times more resistant than normal susceptible strains. In the case of barley grass, survival was reduced by 50% at 0.025 kg/ha in normal susceptible biotypes, but in resistant biotypes 3.2 kg/ha was required. Paraquat-tolerant plants may enjoy certain advantages over nonresistant plants, including resistance to various air pollutants. For example, paraquat-tolerant tobacco plants (*Nicotiana tabacum*), which had higher superoxide dismutase activity than controls, were tolerant to aerosol sprays of 2.0 mg SO_2/L, while controls experienced severe damage.

In every case of resistance, paraquat had been applied 2 or 3 times annually during the preceding 5–11 years; in some cases, a cross-resistance to atrazine was also reported. Paraquat resistance mechanisms in plants include increased epicuticular wax (preventing penetration), binding of paraquat to cell walls, restricted movement into chloroplasts, and altered redox potential. For example, sequestration of paraquat within the apoplast of the leaf seems to be inheritable and controlled by a single nuclear gene with incomplete dominance. Studies with paraquat-tolerant strains of various plants, including perennial rye and tobacco, suggest that tolerance is related to their general ability to rapidly detoxify the generated oxygen species through increased levels of superoxide dismutase, glutathione reductase, and other antioxidants.

At recommended field concentrations, paraquat had negligible effect on soil microflora or soil fertility, although it did cause a temporary suspension of soil nitrification. A concentration as low as 1.0 mg/L completely inhibited ammonium and nitrite oxidation for 40 days in a mixed culture of nitrifying bacteria isolated from soil. Paraquat at 1.6 mg/L adversely affected *Escherichia coli* in 6 h, as judged by diminished growth rate and inhibited synthesis of RNA and protein; at a higher concentration of 18.6 mg/L interference with metabolism of glucose and DNA synthesis was observed. Four species of soil bacteria had 50% growth inhibition at paraquat concentrations between 93.0 and 18,600.0 mg/kg soil; moreover, the mode of action in some species of microorganisms may differ from the generally accepted mechanisms for paraquat toxicity in mammals. Sensitive species of soil fungi experienced marked growth inhibition between 10.0 and 25.0 mg paraquat/kg soil. In various genera of soil fungi (*Rhizopus*, *Ophiobolus*, *Helminthosporium*, *Fusarium*, *Eurotium*), paraquat concentrations up to 100.0 mg/L could be tolerated; at higher concentrations spore germination was suppressed, mycelial growth was inhibited, and spore development was abnormal.

Terrestrial invertebrates show varying degrees of sensitivity to paraquat. In honeybees (*Apis mellifera*), 100.0 mg paraquat/kg syrup (diet) produced toxic signs, 4.4 kg/ha applied as a spray killed 90% in 3 days, and 1000.0 mg/L in drinking water killed most in a few days and 100% within 5 weeks. In soils, adsorbed paraquat may be ingested by soil invertebrates, such as earthworms, but it was not absorbed from the gut into tissues and was rapidly lost when the earthworms were transferred to clean soil. For example, earthworms (*Lumbricus terrestris*) fed soil treated with 112.0 kg paraquat/ha had 111.0 mg paraquat/kg in gut contents, but <0.3 mg/kg in the carcass without gut. Two species of collembolid insects (*Folsomia candida*, *Tullbergia granulata*) fed diets containing 600.0 mg paraquat/kg for 22 weeks survived without measurable adverse effects, but higher dietary levels of 1000.0 and 5000.0 mg/kg were associated with decreased survival, lengthier instar development, decreased egg production, and decreased egg viability. Adults and larvae of the German cockroach (*Blattella germanica*) died after consuming diets containing 1000.0 mg paraquat/kg. Also, paraquat was lethal to two species of mites (*Tetranchus urticae*, *Typhlodromus* sp.) at concentrations below recommended field application rates.

23.5.2 Aquatic Organisms

In general, paraquat is more toxic to aquatic fauna in soft water than in hard water, more toxic to early developmental stages than to juveniles or adults, and more toxic in formulations containing wetting agents than in formulations without these agents. In water, paraquat is taken up rapidly by plants or adsorbed to particulate matter in the water column; however, paraquat is not bioconcentrated by aquatic fauna. Paraquat effects on aquatic biota show several trends. Early developmental stages of certain species of crustaceans are extremely sensitive, and significant adverse effects occur in the range of 0.9–100.0 μg/L, although most species of crustaceans and all other species of invertebrates tested were relatively unaffected at concentrations below 1000.0 μg/L. Freshwater algae and macrophytes are eliminated after treatment with 250.0–500.0 μg/L, but marine algae are relatively resistant and require 5000.0 μg/L or higher to produce significant growth inhibition. Aquatic vertebrates usually are not adversely affected and show little accumulation at 1000.0 μg/L or lower, but at 500.0 μg/L frog tadpoles have low survival and a high frequency of developmental abnormalities, and carp experience biochemical upset.

Paraquat controlled *Typha* and *Phragmites* weeds in Egyptian irrigation canals, drains, and marshes without apparent harm to fishes. Paraquat residues in decomposed plants become available for adsorption to sediments and bottom muds and are not readily available for microbial degradation. Indirect fish kills may occur from anoxia due, in part, to consumption of dissolved oxygen by decaying weeds. Paradoxically, it has been suggested that paraquat may be helpful in improving the oxygen status of aquatic environments at a concentration of 1.0 mg/L by restricting nitrate production due to inhibition of bacterial nitrification. At effective herbicidal concentrations, paraquat was also toxic to eggs, but not adults, of three species of gastropod vectors of bilharzia (*Bulinus truncatas*, *Biomphalaria alexandrina*, *Lymnaea calliaudi*); newly hatched snails were the most sensitive.

Changes in fauna of a reservoir following use of paraquat for weed control are likely to be indirect effects caused by decomposition of angiosperms. Planktonic invertebrates closely associated with aquatic macrophytes were either eliminated by paraquat or survived at lower densities for at least a year posttreatment; analysis of fish stomachs showed dietary changes following weed control and reflected availability of many invertebrate species associated with aquatic plants.

Paraquat can induce activities of antioxidant enzymes such as superoxide dismutase, glutathione peroxidase, and catalase in many species of plants, invertebrates, and vertebrates. Results of studies with ribbed mussels (*Geukensia demissa*) support the hypothesis that these bivalve molluscs can activate redox cycling compounds and demonstrate responses typical of oxidative stress observed in other species. Paraquat also disrupts glucose metabolism and acetylcholinesterase activity and accumulates in melanin. Disrupted glucose metabolism in paraquat-stressed carp was attributed to a high level of circulating epinephrine. Paraquat-induced acetylcholinesterase inhibition in erythrocytes and electric organs of the electric eel (*Electrophorus electricus*) was reversible. Paraquat tended to concentrate in melanin, as judged by accumulation in neuromelanin of frogs (*Rana temporaria*) after intraperitoneal injection, with important implications for research on Parkinson's disease. It seems that paraquat has a structural similarity to a metabolite of 1-methyl-4-phenyl-1,2,3,6-tetrahydropyridine (MPTP), which may induce a Parkinson-like condition. MPTP and its metabolites, like paraquat, have melanin affinity.

23.5.3 Birds

Signs of oral paraquat intoxication in birds include excessive drinking and regurgitation, usually within 10 min of exposure. Other signs appeared after 3 h: diarrhea, ruffled feathers, muscular incoordination, imbalance, wing drop, hyporeactivity, slowness, weakness, running and falling, constriction of the pupil, and terminal convulsions. Additional signs reported after dermal exposure include blistering and cracking of skin, lacrimation, wing spread, and wing shivers. Deaths usually occurred between 3 and 20 h postexposure; remission took up to 12 days. The blood chemistry pattern of paraquat-intoxicated Japanese quail (*Coturnix japonica*) suggested adrenal gland impairment, although recovery from hematologic effects was rapid. Paraquat causes pseudofeminization of male chicken and quail embryos; testes showed intersexual phenomena and Mullerian duct abnormalities; both sexes had a reduction in gonocyte number.

The lowest doses of paraquat causing measurable adverse effects in sensitive species of birds were: 0.2 mg/kg BW administered by single intravenous injection to Japanese quail, causing anemia; 0.25 mg/kg applied in oil solution to the surface of eggs of the mallard, producing reduced survival, reduced growth, and increased frequency of developmental abnormalities; 10.0 mg/kg BW administered orally for 10 days to nestlings of the American kestrel, causing reduced growth; 20.0 mg/kg in diet of the common bobwhite, producing a reduction in egg deposition; 40.0 mg/L in drinking water of the domestic chicken, causing elevated tissue residues and an increase in the number of abnormal eggs produced; and 199.0 mg/kg BW in mallards, producing an acute oral LD50.

Paraquat is highly toxic to avian embryos but is less toxic to adult birds. It is toxic by several routes of administration, including injection and topical application. Paraquat was the most toxic to eggs of the mallard of 42 herbicides and insecticides tested. Paraquat applied to eggshell surfaces in nontoxic oil vehicles was significantly more embryotoxic than were aqueous paraquat solutions, presumably owing to greater penetration of oil past the shell and membranes. The LC50 values for paraquat and mallard eggs were 1.68 kg/ha (1.5 pounds/acre) in aqueous emulsion and 0.11 kg/ha (0.1 pound/acre) in an oil vehicle. The computed LC50 aqueous value was about 1.5 times higher than the recommended field application rate of about 1.0 kg/ha; however, paraquat in aqueous solution caused some deaths at only half the field level of application, and survivors showed impaired growth and some developmental abnormalities.

Nestlings of altricial species, such as the American kestrel, were more sensitive to paraquat exposure than were young or adults of precocial species. There are several food items of kestrels (e.g., grasshoppers, small rodents, passerine birds) that are readily contaminated by paraquat through direct contact during agricultural spraying or by ingestion of contaminated vegetation. From a comparative viewpoint, however, lungs of nestling kestrels were less sensitive to paraquat than were mammalian lungs.

Bobwhite hens immediately exposed to simulated field application rates of paraquat took longer to lay a clutch of eggs once laying had commenced; the completed clutch appeared 10 days later in the season than birds free from paraquat exposure. It is uncertain whether the paraquat-induced delay in sexual maturation produced experimentally will also be reflected in nonlaboratory situations. Turkeys (*Meleagris gallopavo*), for example, held in field plats sprayed 24 h earlier with paraquat at 100 times the recommended agricultural application rate (i.e., up to 200 ounces cation/acre or 14.0 kg/ha) showed no signs of toxicity for 30 days after spraying.

Acute toxicity of paraquat in the domestic chicken was highly responsive to nutritional selenium status and not to vitamin E status; as little as 0.01 mg Se/kg ration protected 8-day-old chicks against acute paraquat poisoning. Paraquat administered to chickens by way of diet was less toxic than the same amount administered in drinking water.

23.5.4 Mammals

Resistance to paraquat among mammals varied substantially owing to inherent differences in sensitivity between species, to route of administration, and to reproductive state. The lowest recorded doses of paraquat causing measurable adverse effects on growth, survival, or reproduction were: aerosol concentrations of 0.4–6.0 μg/L (rat, guinea pig); 0.05 mg administered directly in lung (rat); intravenous injection of 1.0–12.0 mg/kg BW (sheep, dog, rat); subcutaneous injection of 2.4–28.0 mg/kg BW (rat, mouse, monkey); intraperitoneal injection of 3.0–10.0 mg/kg BW (mouse, guinea pig, goat); acute oral dose of 22.0–35.0 mg/kg BW (dog, cat, hare, guinea pig); dermal application of 70.0–90.0 mg/kg BW (rat); dietary levels of 85.0–100.0 mg/kg ration (dog, mouse, rat); and drinking water concentration of 100.0 mg/L (mouse). In microtine rodents, feed aversion and toxicant avoidance were the most significant behaviors elicited by feed tainted with paraquat.

In general, intraperitoneal and intravenous injection were the most sensitive administration routes. LD50 dermal values, however, are often not true percutaneous values because of oral contamination from normal grooming. Aerosol exposure to paraquat produced a concentration-dependent rapid, shallow breathing pattern in guinea pigs (*Cavia* sp.) 18 h after exposure. Aerosol LC50 values in paraquat toxicity tests with mammals were directly related to the duration of exposure, paraquat concentration in spray, and particle size; particles of 3.0 μm diameter seemed most effective.

Paraquat produces rapidly progressive, fatal, interstitial inflammation, and fibrosis of the lung in humans following accidental ingestion, and this has been produced experimentally in several species of laboratory animals. Initial symptoms of paraquat poisoning include burning of the mouth and throat followed by nausea and vomiting. After a latent period of up to several days increasing respiratory distress develops; death is usually the result of a progressive fibrosis and epithelial proliferation that occurs in the lungs. Paraquat poisoning in humans has a mortality of 30–70%, depending largely on the dose ingested. It causes multiorgan failure, including the heart, lungs, kidney, liver, and brain. Although recovery may follow mild involvement of any of these organs, many patients die from progressive untreatable pulmonary fibrosis. This illness usually succeeds renal failure and relates in part to active pulmonary uptake of paraquat. In one case, a 15-year-old boy accidentally ingested a mouthful of paraquat and developed severe respiratory distress, necessitating transplantation of one lung; paraquat-induced rejection of the graft resulted in death 2 weeks after the operation. Paraquat cannot be absorbed significantly through intact human skin, but in the event of broken or abraded skin brief exposure to a concentration of 5.0 g of paraquat per liter may result in death.

Paraquat tends to rapidly localize in selected tissues of injected mice, including melanin, alveolar type cells of the lung, choroid plexus, muscle, proximal tubules of the kidney, liver, gallbladder, and intestinal contents. Half-time persistence of paraquat in rat tissues ranged from 20 to 30 min in plasma to about 5 days in muscle.

Acute effects of paraquat poisoning in livestock and small laboratory animals are similar to those in humans. Signs of acute paraquat toxicosis included hyperexcitability leading to convulsions or incoordination, inflammation of the mouth and throat, vomiting, reluctance to eat or drink, diarrhea, tachycardia, eye irritation of the conjunctiva, corneal lesions, skin reddening, skin ulceration, skin necrosis, histopathology of liver and kidney, and respiratory failure. Paraquat was selectively accumulated in lung of canines, primates, and rodents regardless of route of administration. Lung pathology included congestion, hemorrhage, edema, and collapse; this was associated with degeneration of alveolar and bronchial cells. Death may occur within 10 days of acute exposure.

Chronic administration of small doses or repeated injections usually produce no clinical signs for several weeks. Signs develop suddenly, in general, and include weight loss, anorexia, and death – usually within 10 days of onset of signs. Decreased food consumption and consequent loss of body weight are common in paraquat-poisoned rats and dogs. The area postrema of the hindbrain is an important neural site for detection of blood-borne chemicals and is speculated to control paraquat-induced taste aversion formation and weight loss.

Rabbits are comparatively resistant to paraquat-induced lung damage, regardless of the route of administration. But the closely related hare (*Lepus europaeus*) is comparatively sensitive to paraquat. Hares placed on alfalfa plants within a few hours after the fields were treated with 0.6 kg paraquat/ha experienced 50% mortality in 120 h; survivors that were killed 2 weeks later showed lung damage and ulceration of the lingual mucous membrane. Plant residues were about 30.0 mg/kg fresh weight for alfalfa and 60.0 mg/kg for weeds; residues were negligible in tissues of the hare. In a similar incident in Italy, investigators found that only 1 of 56 hares found dead had lung damage, although all had elevated urine paraquat levels of 0.5 mg/L. It was concluded that paraquat alone was not the causative agent of death, and that paraquat interactions with other chemicals applied at the same time on other crops in the same area may be responsible.

Paraquat applications to spruce plantations for grass control had no effect on the movement or density of field mice (*Microtus arvalis*) and voles (*Microtus agrestis*), but shrews (*Sorex* sp.) migrated from treated areas to untreated ones.

Paraquat is poorly absorbed from the gut and readily excreted. Typical gut absorption rates (%) and peak concentrations in blood (mg/L) were 15–20% and 3.0–4.0 mg/L in rat, 5–10% and 1.0 mg/L in guinea pig, 16% and 13.0 mg/L in cat, 0.26% in cow, and 1–5% in man. Paraquat is actively secreted by a renal mechanism that is vulnerable to paraquat toxicity; poisoning of the secretory component removed a large part of the excretory capacity for paraquat. In rats, a single oral LD50 dose produced a reduction in renal function within 24 h. This effect is probably secondary to a decrease in plasma volume with a consequent reduction in renal blood flow. Paraquat caused mild renal tubular damage in the rat; within 24 h of injection of 20.0 mg/kg BW, there was marked diuresis, sugar and albumin in the urine, and increased plasma urea concentrations. Paraquat-poisoned mice showed a decreased ability to excrete organic acids and bases, probably reflecting interference with proximal tubule function because no change in glomerular filtration rate was observed.

Paraquat was not mutagenic, as judged by noninterference with DNA metabolism, and no reverse mutation-inducing capability. Paraquat had little or no teratogenicity to mammals. No teratogenic effects were observed in rats fed diets containing 400.0 mg paraquat/kg for three generations. The most common malformations in paraquat-stressed rats were those involving costal cartilage. Studies demonstrated transplacental transfer of paraquat in pregnant rats and guinea pigs. High concentrations of radiolabeled paraquat were found in the placenta and throughout the fetuses within 30 min of intravenous administration; concentrations in placenta, maternal blood, and fetal blood were in the ratio of 16:4:1, suggesting that additional research is needed on paraquat embryotoxicity.

23.6 Recommendations

Criteria have not yet been adopted by regulatory agencies for the protection of sensitive species of fish and wildlife against paraquat, although many criteria have been proposed (Table 23.2).

Degradation rate of paraquat in certain soils can be slow, and the compound can persist for years – reportedly in a form that is biologically unavailable. But data are missing or incomplete on flux rates of paraquat from soil into food webs and on interaction dynamics of paraquat with other herbicides frequently

Table 23.2. Proposed paraquat criteria for the protection of natural resources and human health.

Resource, Criterion, and Other Variables	Concentration
AQUATIC ORGANISMS	
Adverse effects level, in mg/L	
Algae and macrophytes	
Freshwater	>0.25
Marine	>5.0
Invertebrates	
Most species	>0.1
Sensitive species	>0.001
Vertebrates	
Most species	>1.0
Sensitive species	>0.5
BIRDS; ADVERSE EFFECTS LEVEL	
Egg surface, in mg/kg egg	
Oil solution	>0.25
Aqueous solution	>2.8
Oral administration, in mg/kg body weight (BW) daily	>10.0
Diet, in mg/kg ration	>20.0
Drinking water, in mg/L	>40.0
Acute oral dose, in mg/kg BW	>199.0
MAMMALS	
No-observable-effect level	
Livestock; forage (alfalfa, clover, pasture, and range grasses), in mg/kg	<5.0
Laboratory rodents	
Diets, in mg/kg ration	
Males	<30.0
Females	<100.0
Diet, in mg/kg BW daily	
Males	<1.1–<6.6
Females	<4.3–<7.1
Air, in μg/L	<0.1
Adverse effects level	
Blood, in mg/L	
Guinea pig	5.0
Rat	22.0
Cat	70.0
Lung, in μg directly into lung	6.0
Lung, in mg/kg BW	0.05
Diet, in mg/kg ration	85.0–100.0
Air, in μg/L, particle size range of 2.5–5 μm	0.4–6

Table 23.2. cont'd

Resource, Criterion, and Other Variables	Concentration
Drinking water, in mg/L	100.0
Acute oral dose, in mg/kg BW, sensitive species	25.0–35.0
TERRESTRIAL PLANTS	
Safe, Argentina, L/ha	<4.0; not to exceed two applications annually
HUMAN HEALTH	
Permissible residues in food items, in mg/kg fresh weight	
Eggs, milk, meat, meat by-products of domestic animals	<0.01
Most fruits and vegetables	<0.05
Fresh hops	<0.1
Passion fruit	<0.2
Almond hulls, cotton seed, beans, hop vines, potatoes, sugar beets, sugarcane	<0.5
Sunflower seeds	<2.0
Aerosol standard, in mg/m^3	0.5
Acute poisoning level; blood, in mg/L	<7.4

applied at the same time. It seems prudent at this time to keep under close surveillance the residues of paraquat in soils in situations where repeated applications have been made over long periods of time.

In Argentina, paraquat concentrations used to control aquatic weeds is set at 0.1–0.2 mg/L, with no more than 4 applications each year. However, some aquatic invertebrates, especially early developmental stages of crustaceans, are unusually sensitive to paraquat. Adverse effects to aquatic invertebrates are documented in the range of 1.0–100.0 μg/L. For this reason, paraquat should be used with caution in estuarine and marshy areas. Fish seem to be "safe" against aquatic weed control concentrations of <1.0 mg paraquat/L. But aquatic plants tend to accumulate paraquat from the medium; accordingly, more research is needed on the effects of ingestion of contaminated plants and plant detritus by amphibians, reptiles, and other aquatic fauna.

Eggs of migratory waterfowl seem to be especially sensitive to paraquat at recommended application rates in an oil vehicle, but were significantly more resistant to the same dose applied in water. Application of paraquat in oil solution appears contraindicated in areas containing nesting waterfowl.

Among mammals, one of the most sensitive species is humans, and permissible dietary residues are low when compared to no-observed-effect levels in other warm-blooded species (Table 23.2). An air concentration of 0.4 mg/m^3 may exceed a safe level for certain mammals, particularly if the size of the aerosol particle is submicroscopic and capable of penetrating the lung. Accordingly, the proposed paraquat aerosol standard – now set at 0.5 mg/m^3 – may have to be modified downwards. Misuse of paraquat has raised the question of cancellation of its registration, cancellation of its use in homes and recreational areas, or changing its packaging to prevent individuals from drinking it. In almost all cases of fatal human poisonings, death was due to the ingestion of a concentrated (20%) solution. More dilute formulations (5% paraquat) are usually not fatal if swallowed accidentally, suggesting that a dilute form of paraquat should be the only formulation permitted commercially.

23.7 Summary

Paraquat (1,1′-dimethyl-4,4′-bipyridinium) and its dichloride salt (1,1′-dimethyl-4,4′-bipyridinium dichloride) are broad-spectrum contact plant killers and herbage desiccants that were introduced commercially during the past 40 years. Today, they rank among the most widely used herbicides globally, and are frequently used in combination with other herbicides. The recommended field application rates for terrestrial weed control usually range between 0.28 and 1.12 kg paraquat/ha (0.25 and 1.0 pounds/acre), and for aquatic weed control it is 0.1–2.0 mg/L. Target plant species are unable to metabolize paraquat and tend to contain elevated residues; paraquat-resistant strains of terrestrial flora, whose numbers are increasing, require greater concentrations for control and may contain proportionately greater residues. Paraquat from decayed flora is usually adsorbed to soils and sediments. Paraquat in surface soils generally photodecomposes in several weeks, but paraquat in subsurface soils and sediments may remain bound, and biologically unavailable, for many years without significant degradation.

Paraquat is not significantly accumulated by earthworms and other species of soil invertebrates, and is usually excreted rapidly by higher animals; however, delayed toxic effects – including death of birds and mammals – are common. At concentrations below the recommended application rate, paraquat is embryotoxic to developing eggs of migratory waterfowl (0.056 kg/ha), and adversely affects sensitive species of freshwater algae and macrophytes (250.0 μg/L), larvae of crustaceans (0.9–5.0 μg/L), and frog tadpoles and carp (500.0 μg/L). Sensitive species of birds are negatively affected at dose rates of 10.0 mg paraquat/kg BW daily, or when fed diets containing 20.0 mg/kg ration, or drinking water containing 40.0 mg/L. Adverse effects in sensitive mammals were observed at dietary levels of 85.0–100.0 mg/kg ration and higher, or 100.0 mg/L in drinking water. Acute oral LD50 values for sensitive species of birds were near 200.0 mg/kg BW, and for mammals 22.0–35.0 mg/kg BW. *Homo sapiens* is one of the more sensitive species, and numerous human poisonings have resulted from accidental or intentional ingestion of a concentrated paraquat formulation.

The biochemical mechanism of paraquat toxicity is due to the cyclic oxidation and reduction in tissues, leading to production of superoxide anion and other free radicals, and eventually the highly destructive hydrogen peroxide. The lung is the organ most severely affected in paraquat poisoning, largely due to the preferential accumulation of paraquat in lung alveolar cells. Although many organs are affected by paraquat, death is usually due to progressive pulmonary fibrosis. There is no completely successful treatment for paraquat-induced lung toxicity.

More information is needed in several areas in order to establish effective criteria for the protection of sensitive species of fish and wildlife against paraquat. These include: flux rates of paraquat from soil into terrestrial food chains; biomagnification potential of paraquat in aquatic food chains, with special reference to plants, plant detritus, amphibians and reptiles; toxicokinetics of mixtures of paraquat and other herbicides applied concomitantly; and the implications of the high sensitivities of crustacean larvae and waterfowl embryos to paraquat.

PENTACHLOROPHENOL[a]

Chapter 24

24.1 Introduction

Pentachlorophenol (PCP) and its water-soluble salt, sodium pentachlorophenate, are commercially produced organochlorine compounds used primarily as preservatives of wood and wood products, and secondarily as herbicides, insecticides, fungicides, molluscicides, and bactericides. Both compounds have been sold for these purposes since 1936 under a variety of trade names. Because of its widespread use, animals and humans are exposed to significant amounts of PCP; detectable PCP levels are found in most people living in industrialized societies, probably as a result of food chain exposure to PCP-treated wood products. In Japan, PCP has been widely used as a herbicide in rice fields, but owing to its high toxicity to fishes, its use was limited (beginning in 1971) to upland fields. The use of PCP in Japan has resulted in the contamination of all surface water in that country to concentrations of 0.01–0.1 μg/L. The chemical and its degradation products bioconcentrate in fish and are among the phenolic compounds known to taint fish flesh. It has been detected in the atmosphere over industrial as well as remote areas, lake sediments, aquatic biota, drinking water, and human blood and urine. All samples of human milk from nursing mothers tested in Bavaria from 1979 to 1981 contained PCP. In the United States between 1976 and 1980, PCP was present in the urine of 71.6% of the general population, suggesting that almost 112 million individuals aged 12–74 years had been exposed to PCP.

In humans, illnesses and deaths have been reported after exposure to PCP through diet or by direct contact with PCP-treated products. For example, 20 of 80 infants who wore, for 8 days, diapers rinsed in an antimicrobial laundry neutralizer containing sodium pentachlorophenate developed enlarged livers and spleens, had high fevers, and sweated profusely; although most recovered spontaneously, 7 died. At least 24 industrial PCP fatalities have been reported. The first deaths occurred at a wood preservative plant in France in 1952. Others were recorded at a chemical factory in Japan in 1953, during herbicide spraying in Australia in 1956, at a sawmill in Indonesia in 1958, in South Africa in 1961, and in Canada and the United States in 1965. The acute toxic action of PCP in humans and experimental animals is caused by the uncoupling of oxidative phosphorylation mechanisms, resulting in marked increases in metabolism.

Data are scarce on PCP effects on wildlife, although it is speculated that no wildlife losses should occur under normal PCP application conditions and that chronic toxicity would not be serious because PCP is rapidly excreted. However, mortality was heavy in two species of bats that came into contact with PCP-treated timbers up to 14 months after treatment. Wood preservatives, including PCP, are implicated in decline of bat population in the United Kingdom. Furthermore, evidence accumulating on the harmful effects of PCP to domestic animals suggests that the chemical may have considerable adverse effects on other

---

[a]All information in this chapter is referenced in the following sources:

Eisler, R. 1989. Pentachlorophenol hazards to fish, wildlife, and invertebrates: a synoptic review. *U.S. Fish Wildl. Serv. Biol. Rep. 85(1.17)*, 72 pp.

Eisler, R. 2000. Pentachlorophenol. Pages 1193–1235 in *Handbook of Chemical Risk Assessment: Health Hazards to Humans, Plants, and Animals. Volume 2, Organics*. Lewis Publishers, Boca Raton, Florida.

species of wildlife. In the poultry industry, for example, PCP has been implicated in the cause of musty taint in chicken meat and eggs and in increased morbidity in chickens housed on PCP-contaminated wood shavings or given PCP-contaminated food. Pentachlorophenol is repellent to animals; diets containing PCP have been rejected by rats, cats, and cattle. In farm animals, PCP intoxication has been reported as a result of confinement in buildings recently treated with a PCP wood preservative, and through dermal contact with PCP-treated fences and feed bunks. Dairy cattle contaminated by PCP produced less milk, grew poorly, and developed skin lesions. The issue is confounded by the presence of various amounts of toxic impurities – primarily dioxins and dibenzofurans – in technical and commercial preparations of PCP; these contaminants are mainly responsible for its observed toxicity in rabbits, rats, pigs, cattle, and chickens. In one example in 1957, millions of chickens died in the southeastern United States after eating poultry feeds containing fat from hides preserved with PCP. Nine dioxins were detected in the toxic animal fat, including the potent 1,2,3,7,8,9-hexachlorodibenzo-*p*-dioxin isomer.

24.2 Environmental Chemistry

Pentachlorophenol and its water-soluble salt, sodium pentachlorophenate, are used extensively in agriculture and industry. Most – about 80% – of the 50 million kg of PCP manufactured each year is used in the protection and preservation of wood products. Commercial samples of technical grade PCP are heavily contaminated with many compounds, including chlorophenols, dioxins, dibenzofurans, hexachlorobenzene, and phenoxyphenols; the relative toxicities and accumulation potentials of some of these contaminants may exceed those of PCP by several orders of magnitude. Pentachlorophenol interferes with the production of high-energy phosphate compounds essential for cell respiration. In general, it readily degrades in the environment by photochemical, chemical, and microbiological processes.

24.2.1 Sources and Uses

In the United States, PCP was one of the most heavily used compounds and was found in all environmental media as a result of its past widespread use. In addition, a number of other chemicals are known to be metabolized to PCP including hexachlorobenzene, pentachlorobenzene, and benzenehexachloride isomers. Although PCP was first synthesized in 1841, it was not produced commercially until 1936. It has since been registered for use as an insecticide, fungicide, herbicide, algicide, and disinfectant, and as an ingredient in antifouling paint; at least 578 products contain PCP. By 1967, PCP and its sodium salt, sodium pentachlorophenate (Na-PCP), were used extensively in industry and agriculture, due in large part to the solubility of PCP in organic solvents and of Na-PCP in water. The major commercial application of technical grade preparations of PCP is in wood preservation formulations, where its fungicidal and bactericidal actions inhibit the growth of wood-destroying organisms.

In the United States, about 80% of the 23 million kg of technical PCP produced annually – or about 46% of worldwide production – is used mainly for wood preservation, especially utility poles. It is the third most heavily used pesticide, preceded only by the herbicides atrazine and alachlor. Pentachlorophenol is a restricted use pesticide and is no longer available for home use. Before it became a restricted use pesticide, annual environmental releases of PCP from production and use were 0.6 million kg to the atmosphere from wood preservation plants and cooling towers, 0.9 million kg to land from wood preservation use, and 17,000 kg to aquatic ecosystems in runoff waters of wood treatment plants. There are about 470 wood preservative facilities in the United States, scattered among 45 states; they are concentrated in the south, southeast, and northwest – presumably due to the availability of preferred timber species in those regions. Livestock facilities are often constructed of wood treated with technical PCP; about 50% of all dairy farms in Michigan used PCP-treated wood in the construction of various components of livestock facilities.

The chemical is usually applied to wood products after dilution to 5% with solvents such as mineral spirits, No. 2 fuel oil, or kerosene. More than 98% of all wood processed is treated with preservative under pressure; about 0.23 kg of PCP is needed to preserve one cubic foot of wood. Lumber treated with PCP retains its natural appearance, has little or no odor, and can be painted as readily as natural wood.

In addition to its extensive use by the construction and lumber industries to control damage by mold, termites, powder post beetles, and wood-boring insects, PCP has been used as a bactericide and fungicide to protect many products, such as adhesives, paper and paperboard, cable coverings, leather, paints, textiles, rope, ink, rubber, and petroleum drilling muds. It has been used to control algae and fungi in cooling towers at electric plants. It has also been added to fabrics for moth proofing, though derivatives such as pentachlorophenol laurate are more widely used for this purpose because their resistance to dry cleaning and washing exceeds that of PCP, and their toxicity to mammals is lower. It has been applied in agriculture and around industrial sites as an herbicide and preharvest desiccant, on pastureland, and in pineapple, rice, and sugarcane fields. A Japanese manufacturer has added PCP to soy sauce – in violation of the law – as a preservative. It has also been used as a bird repellant: PCP discourages woodpeckers when it is mixed as a pellet and plugged into holes drilled by the bird.

In Canada, the main use of PCP is in the protection and preservation of wood, and secondarily as an herbicide and insecticide for agricultural purposes. A total of 50 wood preserving plants – mostly in British Columbia, Alberta, and Ontario – used about 2.7 million kg of PCP in 1978. Treatment with PCP significantly increased the life of timbers, construction lumber, telephone poles, and railway ties; for example, jackpine poles treated with PCP lasted at least 35 years, compared to 7 years for untreated poles.

Sodium pentachlorophenate has been used to control schistosomiasis by eliminating snails that are intermediate hosts of human schistosomes. It is also used as a fungicide, bactericide, and algicide in construction materials, emulsion polymers, paints, textiles, and finished paper products; as a preservative for ammonium alginate; and at concentrations of 15.0–40.0 mg/L, to control microbial growth in secondary oil recovery.

24.2.2 Properties

Pentachlorophenol (Figure 24.1) is readily soluble in most organic solvents, oils, and highly aromatic and olefinic petroleum hydrocarbons (Table 24.1), a property that makes it compatible for inclusion in many pesticide formulations. Purified PCP, however, is practically insoluble in water; therefore, the readily water-soluble sodium pentachlorophenate salt is substituted in many industrial applications (Table 24.1).

The solubility of sodium and potassium penta-chlorophenate in water is pH-dependent; it increases from 79.0 mg/L at pH 5.0 to >4000.0 mg/L at pH 8.0. But differential toxicity of PCP in solution is primarily attributable to variations in uptake as a function of pH, and not to water solubility. At pH 4.0, for example, PCP is fully protonated and therefore highly lipophilic, and has its greatest accumulation potential. Conversely, PCP is completely ionized at pH 9.0; lipophilicity is markedly reduced as is its toxicity to the alga *Selenastrum capricornutum* and the midge, *Chironomus riparius*.

In recent years it has become clear that many commercial samples of technical grade PCP are heavily contaminated with a large number of potentially toxic

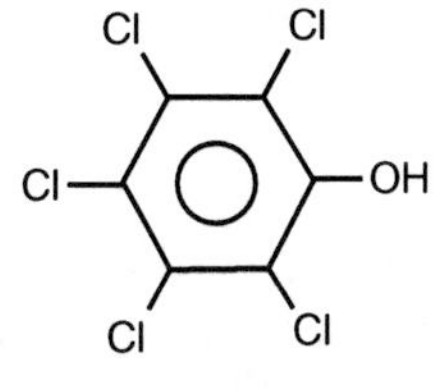

Figure 24.1.
Structural formula of pentachlorophenol (PCP).

Table 24.1. Chemical and other properties of pentachlorophenol (PCP).

Variable	Datum
CHEMICAL NAME	Pentachlorophenol, CAS-87-86-5
ALTERNATE NAMES	Chem-Penta, Chemtrol, Chlorophen, Dow Pentachlorophenol, Dowicide 7, Dowicide EC-7, Dowicide G, DP-2, Durotox, Lauxtol A, Ontrack WE-1, PCP, Penchlorol, Penta, 2,3,4,5,6-pentachlorophenol, Penta General Weed Killer, Pentacon, Penta-kil, Pentanol, Pentasol, Penwar, Permacide, Permaguard, Permasem, Permatox, Priltox, Santobrite, Santophen, Sinituho, Term-l-trol, Weed-Beads, Weedone
PRIMARY USES	Wood preservative, preharvest defoliant, herbicide, molluscicide, insecticide, fungicide
PRODUCERS	Dow Chemical Company, Monsanto Company, Reichold Chemical Company, Vulcan Materials Company
EMPIRICAL FORMULA	C_6Cl_5OH
PHYSICAL STATE	White solid with needle-like crystals. Produced by chlorination of molten phenol. Technical grade material is dark gray to brown
MOLECULAR WEIGHT	266.35
MELTING POINT	190.0–191.0°C
BOILING POINT	309.0–310.0°C (decomposes)
SPECIFIC GRAVITY	1.978 g/mL at 22°C
VAPOR PRESSURE	
25°C	0.0016 mm Hg
100°C	0.02 mm Hg
211°C	40.0 mm Hg
SOLUBILITY	
Water	
0°C	5.0 mg/L
20°C	14.0 mg/L
30°C	20.0 mg/L
50°C	35.0 mg/L
Carbon tetrachloride	20.0–30.0 g/L
Benzene	110.0–140.0 g/L
Ethanol	470.0–520.0 g/L
Methanol	570.0–650.0 g/L
SOLUBILITY OF SODIUM SALT (SODIUM PENTACHLOROPHENATE, CAS 131-52-2) IN WATER AT 25°C	330.0 g/L
LOG K_{ow} (OCTANOL–WATER PARTITION COEFFICIENT)	5.01–5.12

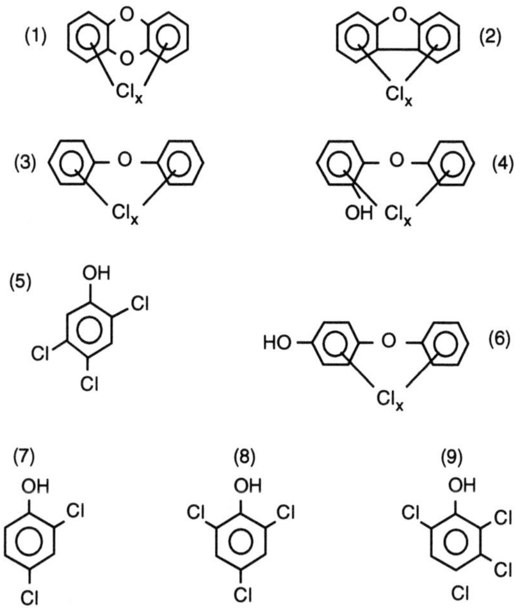

Figure 24.2.
Some impurities found in technical grade pentachlorophenol (PCP). Compounds are: (1) dibenzo-*p*-dioxins, (2) dibenzofurans, (3) chlorinated diphenyl ethers, (4) chlorinated 2-phenoxyphenols, (5) 2,4,5-trichlorophenol, (6) 4-phenoxyphenols, (7) 2,4-dichlorophenol, (8) 2,4,6-trichlorophenol, (9) 2,3,4,6-tetrachlorophenol.

compounds and materials (Figure 24.2). These contaminants include, in part, various isomers of chlorophenols, dibenzofurans, dioxins, hexachlorobenzene, and phenoxyphenols, as well as various chlorinated diphenyl ethers, dihydroxybiphenyls, anisoles, catechols, guaiacols, and other chlorinated dibenzodioxin and dibenzofuran isomers. In relative toxicity and accumulation potential, some contaminants in technical grade PCP may exceed the parent compound by several orders of magnitude. For example, some isomers of hexachlorodibenzodioxin, which are present in technical grade PCP at concentrations of 1000.0–17,300.0 μg/kg, produce LD50 values in guinea pigs of 60.0–100.0 μg/kg body weight (BW) – thus ranking them as extremely toxic chemicals.

24.2.3 Fate

Pentachlorophenol may be absorbed into the body through inhalation, diet, or skin contact. PCP residues in blood serum, adipose tissue, and urine are useful biomarkers of PCP exposure. Acute toxicity of PCP results from its ability to interfere with the production of high-energy phosphate compounds essential for cell respiration. This interference, or uncoupling, causes stimulation of the cell's metabolism to the toxic stage, which is accompanied by fever and other signs of stress. The metabolic consequences resemble those of vigorous exercise in some species. In addition to the proven uncoupling effects on oxidative phosphorylation, the overall inhibitory effects on a variety of enzymes, metabolism of lipids and carbohydrates, ion transport, and protein synthesis may account for the broad-spectrum biocidal effects of PCP and its salts.

Pentachlorophenol is fetotoxic and teratogenic during early gestation; however, evidence of its mutagenic or carcinogenic properties is incomplete. PCP is not mutagenic in bacteria or *Drosophila*; however, positive mutagenicity was reported in yeast and in mice, indicating a need for additional research. Tetrahydroquinone (TCHQ), a major metabolite of PCP, can induce mutations in Chinese hamster cells through formation of reactive oxygen species. PCP was carcinogenic in the B6C3F1 mouse strain. Purified PCP fed to mice in chronic dietary studies was carcinogenic. Human case histories suggest a possible association with occupational exposure to technical PCP and cancer – including Hodgkin's disease, soft tissue sarcoma, and acute leukemia – but no convincing epidemiological evidence exists to indicate that inhalation of PCP in any form produces cancer in humans. Nevertheless, the International Agency for Research on Cancer has assigned PCP to a Group 2B classification, indicating that PCP is possibly carcinogenic in humans.

Pentachlorophenol readily degrades in the environment by chemical, microbiological, and photochemical processes. Its suggested metabolic fates include oxidation and dechlorination to tri- and tetrachloro-*p*-hydroquinones, and glucuronide conjugation to PCP- and tetrachloro-*p*-hydroquinone conjugates. In soils, reductive dehalogenation appears to be the most significant PCP degradation pathway, producing mono-, di-, tri-, and tetrachlorophenols, as well as various tetrachlorocatechols and tetrachlorohydroquinones. Further degradation results in ring cleavage, liberation of chloride, and carbon dioxide evolution; degradation is more rapid in flooded or anaerobic soils than in aerobic moist soils. Irradiation of PCP solutions with sunlight or ultraviolet light produces photodegradation products that include chlorinated phenols, tetrachlorodihyroxyl benzenes, and nonaromatic fragments such as dichloromaleic acid. Subsequent irradiation of the tetrachlorodiols produces hydroxylated trichlorobenzoquinones, trichlorodiols, dichloromaleic acid, and nonaromatic fragments. Prolonged irradiation of PCP or its degradation products yielded colorless solutions containing no ether-extractable volatile materials; evaporation of the aqueous layer left no observable polymeric residue. Photolytic condensation of PCP to form octachlorodioxins was observed on a wood substrate. Octachlorodioxin residues ranged from 4.0 mg/kg for purified PCP, to about 1500.0 mg/kg for technical grade PCP.

Pentachlorophenol can be degraded by microbial flora both aerobically and anaerobically; degradation is more rapid under aerobic conditions but slows significantly at temperatures <19°C. Several strains of aerobic bacteria can metabolize and degrade PCP: *Flavobacterium* sp., a pseudomonad, a coryneform bacterium, and a strain of *Arthrobacter*. Microbial degradation under aerobic or anaerobic conditions was the major process by which PCP was degraded in estuarine sediments; tidal transport and photodegradation played minor roles. The biotic process requires a moderately long adaptive response by the aquatic microflora, but eventually becomes the predominant mechanism of PCP removal. Several significant observations were recorded when the degradation and transformation of PCP were documented in freshwater streams continuously dosed with PCP for 16 weeks: photolysis accounted for a 5–28% decline in initial PCP concentrations and was most rapid at the water surface under conditions of bright sunlight; adsorption to sediments and uptake by biota accounted for less than 5% loss in acclimatized waters and probably less than 15% in unacclimatized waters; and microbial degradation of PCP became significant about 3 weeks after dosing and eventually became the primary mechanism of PCP removal, accounting for up to a 46% decline in initial PCP. A Gram-negative bacterium, *Pseudomonas* sp. strain SR3, can degrade PCP from 39.0 to 40.0 mg PCP/L (a concentration lethal to many species of aquatic organisms) to a nonlethal 0.0006 mg PCP/L within 5 days. However, biodegraded PCP may still be environmentally hazardous. Bioassays with embryonic inland silversides (*Menidia beryllina*) showed that the biodegraded PCP samples were embryotoxic or teratogenic, suggesting that toxic intermediate metabolites were present.

The half-time (Tb1/2) of PCP in water ranged from 0.15 to 1.5 days; degradation was most rapid under conditions of high incident radiation, high dissolved oxygen, and elevated pH. The Tb1/2 in the water column controlled by microbial degradation alone is usually 5–12 h. Pentachlorophenol solutions in water at the appropriate pH and dissolved oxygen content decompose in sunlight, and this makes a strong case for the likelihood of essentially total PCP destruction in aquatic environments. The short residence time of PCP in an aquatic system before degradation further suggests that biological effects would be most pronounced in localized areas that receive PCP continuously from a point source. Technical grade PCP was initially degraded at the same rate as reagent grade PCP by anaerobic microorganisms in municipal sewage sludge, but was later degraded more slowly. Dechlorination and mineralization (to carbon dioxide and methane) of the reagent grade PCP was complete in 7–9 days, but only half the technical grade PCP had been transformed in 6–10 days. At nontoxic concentrations, PCP was readily biodegradable in activated sludge after an adaptation period of 10–20 days. Preexposure of activated sludge to PCP drastically eliminated the acclimatization period and increased the tolerance of the sludge to PCP toxicity.

In soils, PCP persisted for 15 to more than 60 days, depending on soil conditions and application rate. At initial concentrations of 100.0 mg PCP/kg soil, the Tb1/2 was 10–40 days at 30°C under flooded conditions; however, in aerobic soils there was virtually no degradation after 2 months. In rice paddy soils, initial concentrations of 4.0 mg PCP/kg fell to 2.0 mg/kg in 7 days. Pentachlorophenol was still measurable after 12 months in warm, moist soils. In estuarine sediments, degradation was most rapid under conditions of increased oxygen and a pH of 8.0.

A variety of analytical techniques are used to measure PCP, including gas chromatography-mass spectrometry (GC-MS) which has a detection limit of 7.6 μg/kg honey, liquid chromatography with fluorescence detection, and liquid chromatography-electrochemistry (LC-ED) procedures. GC-MS is the most accurate, but LC-ED is used most frequently.

24.3 Concentrations in Field Collections

Measurable PCP concentrations in field collections of living and nonliving materials over widespread geographic areas are almost certainly due to anthropogenic activities, especially to the use of the chemical as a wood preservative.

Pentachlorophenol-contaminated air, rain, snow, surface waters, drinking waters, groundwaters, and aquatic biota are common in the United States. Residues of PCP in food, water, and mammalian tissues may result from the direct use of PCP as a wood preservative and pesticide or as a result of use of other chemicals that form PCP as degradation products – i.e., hexachlorobenzene and lindane. To confound matters, PCP was judged to be the source of dioxin and dibenzofuran contamination in chickens in Canada. More than 50% of all chickens sampled contained hexachlorinated dibenzo-*p*-dioxins (hexa CDDs) at concentrations of 27.0 ng/kg fat and higher; 62% contained hepta CDDs at more than 52.0 ng/kg, and 46% contained octa CDDs at more than 90.0 ng/kg; concentrations of hexa-, and heptachlorinated dibenzofurans were similar.

Pentachlorophenol was found at high concentrations in all samples of sediments, waters, and biota collected near industrial facilities that used PCP as a wood preservative. Fish can bioconcentrate PCP from water up to 10,000 times. However, similar concentrations were measured in the common mussel, *Mytilus edulis*, and the softshell clam, *Mya arenaria*, from the vicinity of PCP-contaminated wastewater discharges as well as from more distant collection sites; thus, PCP bioaccumulation in marine bivalve mollusks does not appear to be dose related.

Pentachlorophenol in terrestrial ecosystems clears rapidly. In one case, a terrestrial ecosystem was given a single surface application of radiolabeled PCP equivalent to 5.0 kg of sodium pentachlorophenate/ha. PCP residues on foliage decreased rapidly, with 50% metabolized within 15 days. After 131 days (autumn), most of the remaining PCP was in the top soil and plant litter. After 222 days (winter), 39% of the radiocarbon remained. There was little bioconcentration in the resident fauna due to rapid metabolism and excretion.

There are differences between the United States, Germany, and the United Kingdom in human intake and absorption of PCP. In the United Kingdom, the typical non-occupationally exposed individual has a total daily intake of 5.7 μg of PCP, of which 4.5 μg is absorbed. Most of the PCP (92%) comes from the diet, 1% via inhalation, and the rest from drinking water. Occupationally exposed individuals in the UK absorb 21.2 μg daily from an intake of 39.0 μg daily; most of the PCP (78.8%) in this group is via inhalation and only 19.6% from the diet. In Germany, the main route of PCP uptake by non-occupationally exposed humans is by ingestion of contaminated food, including drinking water, fish, sugar, pork, chicken, and other foodstuffs. The chickens and pigs were contaminated by PCP because they were reared on PCP-treated wooden floors. The average German diet contains 13.9 (2.7–27.6) μg

PCP/kg ration. The total amount of PCP in an average non-occupationally exposed German citizen (weight 65 kg, age 45 years) is about 532.0 μg: 204.0 μg in body fat, 143.0 μg in liver, 103.0 μg in blood, 63.0 μg in brain, 14.0 μg in kidneys, 4.0 μg in spleen, and about 1.6 μg in urine in the bladder. In the United States, it was estimated that the long-term average daily intake of PCP by humans is 16.0 μg, and this is in good agreement with other values; however, the source of PCP in that study was attributed almost exclusively (99.9%) to fruits, vegetables, grains and other crops grown in PCP-contaminated soils. PCP concentrations in U.S. citizens were usually higher in adipose tissues than in nonfatty tissues, and between the ages of 22 and 75 will increase by a factor of at least 2.

Residues (mg/kg fresh weight) in birds found dead from PCP poisoning were >11.0 in brain, >20.0 in kidney, >46.0 in liver, and 50.0–100.0 in egg.

Data are scarce on the toxicity of PCP to mammalian wildlife, but studies with livestock and small laboratory animals show that the chemical is rapidly excreted. However, there is great variability between species in their ability to depurate PCP, as well as in their overall sensitivity. Acute oral LD50s in laboratory animals ranged from 27.0 to 300.0 mg/kg BW. Tissue residues were elevated at dietary levels as low as 0.05 mg/kg feed and at air levels >0.1 mg/m^3. Histopathology, reproductive impairment, and retarded growth were evident at doses of 0.2–1.25 mg/kg BW, and when the diets fed contained >30.0 mg PCP/kg.

24.4 Effects

The toxicity of commercial or technical grades of PCP significantly exceeds that of analytical or purified PCP. Some of this added toxicity is attributed to impurities such as dioxins, dibenzofurans, chlorophenols, and hexachlorobenzene. Pentachlorophenol is rapidly accumulated and rapidly excreted, and has little tendency to persist in living organisms. It acts by uncoupling oxidative phosphorylation. Terrestrial plants and soil invertebrates were adversely affected at 0.3 mg PCP/L (root growth), and at 1.0–5.0 g PCP/m^2 soil (reduction in soil biota populations).

Pentachlorophenol was most toxic and most rapidly metabolized in aquatic environments at elevated temperatures and reduced pH. Adverse effects on growth, survival, and reproduction of representative sensitive species of aquatic organisms occurred at PCP concentrations of about 8.0–80.0 μg/L for algae and macrophytes, about 3.0–100.0 μg/L for invertebrates (especially mollusks), and <1.0–68.0 μg/L for fishes, especially salmonids.

Fatal PCP doses for birds were 380.0–504.0 mg/kg BW (acute oral), >3850.0 mg/kg in diets, and >285.0 mg/kg in nesting materials. Adverse sublethal effects were noted at dietary levels as low as 1.0 mg/kg ration.

24.4.1 Terrestrial Plants and Invertebrates

Pentachlorophenol is toxic to plant mitochondria; the mode of action is similar to that in other organisms – i.e., uncoupling of oxidative phosphorylation. At 267.0 μg PCP/L, 50% uncoupling was noted in isolated mitochondria of potato, *Solanum tuberosum*, and mung bean, *Phaseolus aureus*. Both PCP and its metabolite tetrachlorohydroquinone adversely affect cell growth and synthesis of RNA and ribosome in yeast, *Saccharomyces* sp., in a dose-related manner. Uptake of PCP by rice (*Oryza sativa*) grown over a two-year period under flooded conditions was studied after a single application of radiolabeled PCP was applied to the soil at 23.0 kg/ha. During the first year, PCP uptake was 12.9% of the application. Roots contained about 5.0 mg PCP/kg, distributed as follows (mg/kg): 3.95 as unextractable residues, 0.48 as polar nonhydrolyzable substances, 0.43 as free and conjugated lower chlorinated phenols, 0.14 as free PCP, 0.07 as anisoles, 0.06 as conjugated PCP, 0.03 as hydroxymonomethoxytetrachlorobenzenes, and 0.01 as dimethoxytetrachlorobenzenes. In the second year, PCP uptake was reduced to 2.5%, and soil residues corresponded to 8.4 kg/ha; the amounts of unextractable residues in plants increased, and lower chlorinated conjugated phenols were identified. Root growth in rice seedlings was inhibited by 50% at 0.3 mg PCP/L.

Laboratory studies with adult earthworms (*Lumbricus terrestris*) exposed for 96 h to filter paper containing 5.0–50.0 μg PCP/cm^2 were conducted. The LC50 (96 h) concentration was 25.0 μg PCP/cm^2 filter paper; the LD50 (96 h) value was 878.0 mg PCP/kg DW whole worm. At 10.0 μg PCP/cm^2, all worms survived for 96 h with an average whole body concentration of 501.0 mg PCP/kg DW. At 20.0 μg PCP/cm^2, 20% died and survivors had 804.0 mg/kg DW. At 35.0 μg PCP/cm^2, none survived 96 h; dead worms had approximately 1060.0 mg PCP/kg DW.

Pentachlorophenol applied to beech forest soils every 2 months for two years at the rate of 1.0 g/m^2 markedly reduced populations of soil organisms; at 5.0 g/m^2, it drastically reduced most of the soil animal species, and also the microflora. Reduction of the soil metabolism by PCP retards decomposition and affects the overall nutrient balance of forest ecosystems. Pentachlorophenol is more toxic to earthworms in soils with comparatively low levels of organic materials. The LC50 (14 day) value for *Lumbricus rubellus* was 1094.0 mg PCP/kg DW soils with 6.1% organic matter and 883 mg/kg DW soils with 3.7% organic matter. The earthworm *Eisenia fetida andrei* is more sensitive than *Lumbricus rubellus*; the LC50 (14 day) values were 143.0 mg/kg DW for soils containing 6.1% organic matter and 94.0 mg/kg DW for soils of 3.7% organic matter. Both species accumulated similar concentrations of PCP. Other studies with earthworms (*Allolobophora caliginosa*, *Lumbricus terrestris*) demonstrated differences between species in sensitivity to PCP, differences in PCP accumulation rates, and rapid metabolism of Na-PCP when compared to PCP.

24.4.2 Aquatic Biota

In general, the most sensitive stages are embryo and larval stages of invertebrates and the late larval premetamorphosis stage of fish; fishes are more sensitive to PCP than invertebrates. Pentachlorophenol affects energy metabolism by partly uncoupling oxidative phosphorylation and increasing oxygen consumption, by altering the activities of several glycolytic enzymes and the citric acid cycle enzymes, and by increasing the consumption rate of stored lipid. Collectively, these events could reduce the availability of energy for maintenance and growth and thereby reduce the survival of larval fish and the ability of prey to escape from a predator. Pentachlorophenol has a dose-dependent inhibitory effect on chemiluminescence activity of isolated phagocyte cells of the Japanese medaka (*Oryzias latipes*) and the mummichog (*Fundulus heteroclitus*). Since chemiluminescence is due to the production of reactive oxygen intermediates (ROIs), such as superoxide anion (O_2^-) and H_2O_2, suppression of ROI may adversely affect the teleost immune system.

The accumulation of PCP in fishes is rapid, and primarily by direct uptake from water rather than through the food chain or diet. Signs of PCP intoxication in fish include rapid swimming at the surface and increased opercular movements, followed by loss of balance, settling to the bottom, and death. The PCP is rapidly excreted by fishes after conjugates of PCP-glucuronide and PCP-sulfate are formed; half-lives in tissues are less than 24 h. Major roles were played by gall bladder and bile in PCP-glucuronide depuration kinetics, and by gill in PCP-sulfate depuration. It has been suggested that the efficient elimination of PCP should allow vertebrates to tolerate periodic low doses of PCP without toxic effects.

Many species of aquatic organisms were found dead in rice fields of Surinam, South America, after they were sprayed with PCP to control populations of snails. Residues of PCP in dead organisms (mg PCP/kg fresh BW) were 8.1 in frogs (*Pseudis paradoxa*); 36.8 in snails (*Pomacea* spp.); and, in three species of fish, 31.2 in krobia (*Cichlasoma bimaculatum*), 41.6 in kwi (*Hoplosternum littorale*), and 59.4 in srieba (*Astyanax bimaculatus*). Pentachlorophenol was also implicated in fish kills in Europe and North America, all of which were associated with the pulpwood industry. In December 1974, near Hattiesburg, Mississippi, water containing PCP in fuel oil that overflowed the banks of a holding pond of a wood-treatment wastewater facility killed all fish in a 24-ha lake. Concentrations of PCP in water and fish returned to background concentrations 10 months after the spill; however, the chemical persisted in leaf litter and sediments for at least 17 months after the spill. In December 1976, another fish kill was observed near the same facility. Residues of PCP in surviving fish – including bluegills (*Lepomis macrochirus*), largemouth bass (*Micropterus salmoides*), and channel catfish (*Ictalurus punctatus*) – were greatly elevated one month later: 8.0–19.0 mg PCP/kg fresh weight in muscle, 42.0–48.0 mg/kg in gill, and 130.0–221.0 mg/kg in liver. Pentachlorophenol persisted in fish for 6–10 months before reaching background concentrations.

Studies with experimental ecosystems have indicated that the effects of PCP on community structure and activity are profound. These included a reduction in the number of individuals and species of estuarine macrobenthos after exposure to 55.0–76.0 μg PCP/L for 5–9 weeks or 15.8 μg/L for 13 weeks; a decrease in periphyton biomass, fish growth, and larval drift, and a suppression of community metabolism at 48.0 μg PCP/L after 3 months exposure; elevated levels of PCP in postlarval shrimp, *Penaeus vannamei*, after chronic exposure to 10.0 μg PCP/L; and bioconcentration factors (BCFs) after exposure to radiolabeled PCP of 5 for an alga (*Oedogonium cardiacum*), 21 for a snail (*Physa* sp.), 26 for a mosquito larva (*Culex pipiens quinquefasciatus*), 132 for mosquitofish (*Gambusia affinis*), and 205 for *Daphnia magna*. In laboratory studies, increased accumulation and adverse effects on growth, survival, and reproduction were seen in sensitive species of aquatic organisms: in algae and macrophytes at water concentrations (μg PCP/L) of 7.5–80.0; in a wide variety of invertebrates, especially mollusks, at 2.5–100.0; and in fishes, especially salmonids, at less than 1.0–68.0.

Biocidal properties of PCP were significantly modified by water pH, dissolved oxygen, salinity, and temperature, and by the purity of PCP compounds tested. In general, PCP was most toxic and was metabolized most rapidly at elevated water temperatures, reduced pH, and decreased dissolved oxygen. Increasing the pH of the water column decreases the hazard of PCP to aquatic biota: at pH values above 4.8, for example, hydroxyl proton is dissociated and penetration in aquatic organisms is reduced. In studies with goldfish (*Carassius auratus*) exposed to 5.0 μg PCP/L for 96 h, a pH increase from 7 to 9 resulted in lesser amounts of PCP in the fish, fewer metabolites formed, and a decreased ability to deplete the PCP from the exposure water; the uptake of PCP at pH 7, 8, and 9 seemed to be controlled by its K_{ow} (ratio of solubility in *n*-octanol–water) at these pH values. Water salinity affects PCP accumulations in euryhaline teleosts. In the Japanese medaka (*Oryzias latipes*), for example, increasing salinity was associated with decreased PCP uptake rates and increased clearance rates; the greater excretion of PCP-glucuronide by seawater-adapted medakas may be responsible for the rapid elimination of PCP. All authorities agree that commercial or technical grades of PCP are significantly more toxic to aquatic organisms than is purified PCP. The sublethal effects of low concentrations of commercial PCP to aquatic biota are primarily due to impurities composed mostly of octa- and nonachlorophenoxyphenols, and also to relatively large quantities of hexachlorobenzene, dioxins, and dibenzofurans.

PCP metabolism by fishes is modified by salicylamide and PCP exposure duration, although diet seems unimportant. Salicylamide is an inhibitor of PCP metabolism in fry of rainbow trout, producing elevated BCF values. Fry exposed to 5.0 μg PCP/L for 96 h in the presence of 25.0 mg salicylamide/L had higher concentrations of PCP than did fry exposed without salicylamide. Bluegills exposed to PCP at continuous low-level sublethal concentrations were at greater risk for decreased growth than were bluegills exposed to a more concentrated short-term pulse of PCP resulting in some deaths. Fat content of diets of juvenile rainbow trout, however, were relatively unimportant in PCP acute toxicity bioassays. Rainbow trout juveniles fed high fat (21%) or low fat (13%) diets for 11 weeks had similar LC50 (144 h) values (109.0

vs. 108.0 μg/L), although the high fat group had 3% more body lipids than the low fat group.

24.4.3 Birds

Signs of PCP intoxication in birds include excessive drinking and regurgitation, rapid breathing, wing shivers or twitching, jerkiness, shakiness, ataxia, tremors, and spasms. Signs sometimes appear within 10 min. Mallards usually die 2–24 h posttreatment, and ring-necked pheasants 3–5 days posttreatment; remission in pheasants requires up to 2 weeks.

Pentachlorophenol killed various species of birds at single oral doses of 380.0–504.0 mg/kg BW, at dietary concentrations of 3850.0 mg/kg ration fed over a 5-day period, and when nesting materials contained >285.0 mg/kg. Residues (mg/kg fresh weight tissue) in birds found dead from PCP poisoning were 11.0 in brain, 20.0 in kidney, 46.0 in liver, and 50.0–100.0 in egg. Sublethal effects, including liver histopathology and diarrhea, were reported in domestic chickens at dietary levels as low as 1.0 mg PCP/kg feed over an 8-week period; significant accumulations in tissues were measured after consumption for 14 days of diets containing 10.0 mg PCP/kg. Residues in chickens fed PCP-containing diets for 8 weeks were dose related and highest in kidney and liver and lower in other tissues; the high residues may reflect the principal routes of metabolism and excretion. The loss of body fat in chickens, accomplished by feeding bile-acid-binding resins, hastens PCP excretion.

Spraying of PCP to control populations of water snails in rice fields of Surinam resulted in the death of fish and birds, including snail kites (*Rostrhamus sociabilis*), certain egrets and herons, and wattled jacanas (*Jacana jacana*). Levels of PCP in these birds and their food items suggested that PCP-contaminated food probably caused the deaths.

Pentachlorophenol is widely used as a wood preservative, which often results in residues in wood shavings used as poultry litter. A moldy smell and taste in chicken tissue has been traced to the presence of chloroanisoles formed from PCP and tetrachlorophenol in the bedding. Several dioxins, diphenyl ethers, dibenzofurans, and 2-phenoxyphenols have also been identified. For example, PCP-contaminated (134.0 mg/kg) commercial wood shavings used as chicken litter contained detectable levels of heptachlorinated diphenyl ethers (18.0 μg/kg), octachlorinated diphenyl ethers (12.0 μg/kg) nonachlorinated diphenyl ethers (6.0 μg/kg), octachlorinated 2-phenoxyphenols (299.0 μg/kg), nonachlorinated 2-phenoxyphenols (50.0 μg/kg), heptachlorinated dibenzodioxins (19.0 μg/kg), and octachlorinated dibenzodioxins (143.0 μg/kg). After 9 weeks, PCP was detectable in liver, fat, and muscle; chlorinated diphenyl ethers were detectable in fat, but not in muscle or liver; octa- and nonachlorinated 2-phenoxyphenols were found in all three tissues; and dioxins only in liver and fat. Exposure of domestic chickens to litter contaminated with PCP enhanced susceptibility to common poultry pathogens, perhaps due to immunosuppression by the chemical.

24.4.4 Mammals

Data are scarce on the biological effects of PCP on mammalian wildlife, although evidence continues to accumulate on this subject for humans, livestock, and small laboratory animals. Data on PCP and domestic mammals are available, but it is not now clear if these findings are applicable to representative species of sensitive mammalian wildlife. Pentachlorophenol tends to accumulate in mammalian tissues unless it is efficiently conjugated into a readily excretable form. The ability to conjugate PCP varies widely among species. For example, both laboratory rats (*Rattus* sp.) and humans eliminate about 75% of all PCP in the urine in an unconjugated form, but rhesus monkeys (*Macaca mulatta*) are unable to excrete PCP efficiently, whereas mice were the most efficient. As one result, Tb1/2 values were low (about 24 h) for mice, high (up to 360 h) for rhesus monkeys, and intermediate for rats and humans. In humans, however, the observed elimination half-life indicates that steady-state body burdens are 10–20 times higher than values extrapolated from animal pharmacokinetic data.

Biotransformation of PCP in mammals occurs via conjugation, reductive dechlorination, hydrolytic dechlorination, and oxidation. In the process, a number of metabolites are formed, some of which are demonstrably toxic. Metabolites of PCP found in rat urine and identified (acute oral LD50 to mice, in mg/kg BW) include: PCP (74.0); 2,3,5,6-tetrachlorophenol (109.0); 2,3,4,6-tetrachlorophenol (131.0); tetrachlorocatechol (325.0–612.0); trichlorohydroquinone and tetrachlorohydroquinone (376.0–500.0); 2,3,4,5-tetrachlorophenol (400.0); tetrachlororesorcinol (752.0); and traces of trichloro-1,4-benzoquinone and tetrachloro-1,4-benzoquinone.

Pentachlorophenol is not a carcinogen, and the evidence for mutagenicity is mixed. No carcinomas were produced in rodents, regardless of the composition of the PCP solution tested or route of exposure. Some studies suggested that PCP may be mutagenic in the bacterium *Bacillus subtilis*, the yeast *Saccharomyces cerevisiae*, and in laboratory mice (*Mus* sp.), but not in two other species of bacteria tested: *Salmonella typhimurium* and *Escherichia coli*. The PCP metabolite tetrachlorohydroquinone – found in urine of occupationally exposed humans and experimentally dosed rodents – is twice as toxic to cultured Chinese hamster ovary cells as PCP (based on growth) and induced significant dose-related increases in micronuclei and DNA single strand breaks. But tetrachlorohydroquinone was 5 times less toxic than PCP to mice in vivo, with no evidence of mutagenicity.

The primary sources of PCP in humans include direct intake by way of diet, air, or water and through contact with PCP-contaminated materials. It is now established that PCP is taken up by female rhesus monkeys via the skin from PCP-contaminated soils; monkeys accumulated up to 24% of the PCP in contaminated California soils over a 24-h period. In humans, the chief routes of exposure in an industrial setting are by way of inhalation and skin contact. Percutaneous absorption is significantly enhanced when PCP is dissolved in organic solvents, such as fuel oil, or when PCP comes in contact with open cuts and scratches. Pentachlorophenol has resulted in death of humans through suicide and occupational and accidental exposures. Cases of PCP intoxication in humans, including fatalities, have been described in farmers, handlers of preserved wood, and industrial workers. Symptoms always included high fever, renal insufficiency, profuse perspiration, rapid heart beat and breathing, abdominal pain, dizziness, nausea, spasms, and sometimes death 3–25 h after onset of symptoms. Postmortem examination showed kidney degeneration, inflamed gastric mucosa, edematous lungs, and centrilobular degeneration of liver. Symptoms of nonfatal PCP intoxication in humans include conjunctivitis, chronic sinusitis, nasal irritation, upper respiratory complaints, sneezing, coughing, recurring headache, neurological complaints, weakness, several types of skin lesions, chloracne, aplastic anemia, leukemia, and other hematologic disorders. All symptoms were related to proximity to PCP-treated wood, and sometimes to elevated PCP residues in serum and urine. Chloracne and hematologic disorders may be associated with various contaminants in the PCP including various chlorophenols, phenoxy acids, dioxins, and dibenzofurans. At the cellular level, PCP – like other halogenated phenols – uncouples oxidative phosphorylation. A possible antidote to PCP poisoning is the administration of cholestyramine, a compound that interferes with the enterohepatic cycle of PCP, and also increases its elimination directly across the intestinal wall. Cholestyramine is known to bind phenols and to enhance fecal elimination of PCP in rats, humans, and rhesus monkeys. Phenobarbital increases the biotransformation of PCP, and its action is enhanced in combination with cholestyramine. Substances to be avoided in suspected cases of PCP poisoning include atropine and salicylates, such as aspirin, because they exacerbate the toxicity of phenolic substances.

The exposure of livestock to PCP can result from ingestion of feeds stored or fed in PCP-treated wooden structures, licking of treated wood, cutaneous absorption by direct contact with treated wood, and inhalation of air containing preservative chemicals – particularly volatile chlorophenols. Acute signs of PCP intoxication in various domestic and laboratory animals include elevated blood sugar, vomiting, elevated blood pressure, increased respiration rate, tachycardia, motor weakness, weakened eye reflex, frequent defecation, high fever, collapse, asphyxial convulsions, and death followed by rapid rigor mortis. In domestic cattle (*Bos* sp.), PCP has also been associated with decreased milk production, skin lesions, increased mastitis, persistent infections, reduced survival, disrupted reproductive and metabolic hormones, and oviductal cysts.

Among sensitive species of mammals tested against PCP, acute oral LD50 values ranged from 27.0 to 300.0 mg/kg BW, but most values were between 55.0 and 150.0 mg/kg BW. Sublethal effects were noted at much lower concentrations than those causing death. They included elevated tissue residues at dietary intake equivalent to 0.05 mg/kg BW, or atmospheric concentrations >0.1 mg/m^3; organ damage at 0.2–2.0 mg/kg BW; reproductive impairment at >1.25 mg/kg BW; and retarded growth and reproduction in animals fed rations containing >30.0 mg/kg. PCP was responsible for the deaths of pipistrelle bats (*Pipistrellus pipestrellus*) roosting on PCP-treated wooden surfaces; PCP concentrations on the roosting structures were similar to those applied in remedial timber treatment.

Many commercial lots of technical PCP are known to contain small – but possibly biologically significant – amounts of highly toxic dioxins, dibenzofurans, and hexachlorobenzene. These contaminants may be responsible for most of the toxicity of technical PCP preparations. However, both technical and analytical grade PCP can induce hepatic mixed function oxidases in intoxicated rats and cattle. In cattle, this effect was observed in both calves and adults, and in hepatic as well as pulmonary microsomes, and seemed to be dose related.

24.5 Recommendations

Commercial PCP preparations often contain variable amounts of chlorophenols, hexachlorobenzene, phenoxyphenols, dioxins, dibenzofurans, chlorinated diphenyl ethers, dihydroxybiphenyls, anisoles, catechols, and other chlorinated dibenzodioxin and dibenzofuran isomers. These contaminants contribute to the toxicity of PCP, sometimes significantly, although the full extent of their interactions with PCP and with each other in PCP formulations are unknown. Unless these contaminants are removed or sharply reduced in existing technical and commercial grade PCP formulations, efforts to establish sound PCP criteria for protection of natural resources may be hindered.

Proposed PCP ambient water quality criteria to protect freshwater and marine life range from 48.0 to 55.0 μg/L for acute effects, 3.2–34.0 μg/L for chronic effects, and daily mean concentrations of 6.2 μg/L, not to exceed 140.0 μg/L (Table 24.2). Available data, however, suggest that significant adverse effects occur at much lower PCP concentrations, that is, between 0.035 and 19.0 μg/L. In rainbow trout (*Oncorhynchus mykiss*), for example, concentrations of 0.035–1.0 μg/L produced elevated tissue residues, 7.4 μg/L caused growth inhibition in 28 days, and 10.0–19.0 μg/L produced adverse effects and some deaths. Other sensitive fish species include sockeye salmon (*Oncorhynchus nerka*), showing growth inhibition after prolonged exposure to 1.8 μg PCP/L; larvae of common carp (*Cyprinus carpio*), having a 96-h LC50 of 9.5 μg/L; and largemouth bass (*Micropterus salmoides*), exhibiting reduced food conversion efficiency at 10.0 μg/L. Among sensitive species of plants and invertebrates, American oysters (*Crassostrea virginica*) have elevated tissue residues after exposure for 28 days to 2.5 μg/L; cladocerans have impaired reproduction at 4.1 μg/L; alga show chlorosis inhibition in 72 h at 7.5 μg/L; and estuarine macrobenthos populations decreased in abundance and species after exposure for 13 weeks to 15.8 μg/L. Also, an air concentration of 0.2 mg/m^3 – a tolerable level to humans (Table 24.2) – interfered with photosynthesis in duckweed, *Lemna minor*. As judged by these studies, it seems appropriate to suggest modification of certain aquatic PCP criteria. A maximum PCP concentration of 3.2 μg/L is indicated, and this level would probably protect most aquatic species, although it

Table 24.2. Proposed pentachlorophenol (PCP) criteria for the protection of natural resources and human health.

Resource and Criterion	Concentration or Dose
AQUATIC BIOTA	
Freshwater life	
Acute (former)	48.0–<55.0 μg/L
Chronic (former)	<3.2 μg/L
24-h mean (former)	<6.2 μg/L
Current	
pH 6.5	4-day mean concentration not to exceed 3.5 μg/L more than once every 3 years and 1-h mean concentration does not exceed 5.5 μg/L[d]
pH 7.8	4-day mean concentration not to exceed 13.0 μg/L more than once every 3 years and 1-h mean concentration does not exceed 20.0 μg/L[d]
pH 9.0	4-day mean concentration not to exceed 43.0 μg/L more than once every 3 years and 1-h mean concentration does not exceed 68.0 μg/L
Maximum concentration	<140.0 μg/L
Fish	
Warmwater species	10.0–<15.0 μg/L
Coldwater species	20.0–<40.0 μg/L
Marine life	
Acute (former)	<53.0 μg/L
Chronic (former)	<34.0 μg/L
Current	4-day mean concentration not to exceed 7.9 μg PCP/L more than once every 3 years, and 1-h concentration does not exceed 13.0 μg/L more than once every 3 years
BIRDS	
Tissue residues	
Contaminated	>2.0 mg/kg fresh weight (FW)
Life threatening	>11.0 mg/kg FW
Diets	
Adverse effects	>1.0 mg/kg diet
Fatal	>3850.0 mg/kg diet
Wood shavings	
Litter	<134.0 mg/kg
Nesting materials	<285.0 mg/kg
LIVESTOCK AND LABORATORY MAMMALS	
No measurable adverse Effects	
Rat	
Females	3.0 mg/kg BW daily for 24 months
Males	10.0 mg/kg BW daily for 22 months
Rabbit	3.0 mg/kg BW daily for 90 days

Table 24.2. cont'd

Resource and Criterion	Concentration or Dose
Adverse effects	
Elevated tissue residues, cattle	
Blood plasma	0.05 mg/kg BW, single dose
Internal organs	0.2 mg/kg BW for 95 days
Histopathology, Cattle	
Internal organs	0.2 mg/kg BW daily for about 80 days, then 2.0 mg/kg BW daily for about 59 days
Internal organs	50.0 mg/kg diet for 12 weeks, equivalent to 3.0 mg/kg BW daily
Reproductive impairment	
Hamster	1.25–20.0 mg/kg BW, single dose
Rat	5.0–5.8 mg/kg BW daily or 50.0 mg/kg diet daily, chronic
Increased tumor frequency, mice	50.0 mg/kg diet, chronic
Death, various species	
Acute oral LD50	55.0–200.0 mg/kg BW
Acute dermal LD50	60.0–200.0 mg/kg BW
Contaminated wood shavings, dermal contact	470.0 mg/kg
HUMAN HEALTH	
Estimated current exposure levels, 70-kg adult	
Food	15.0 μg daily, or 0.21 μg/kg BW daily
Water	0.12 μg daily, or 1.7 μg/kg BW daily
No adverse effect levels	
Food; upper safe limit; 70-kg adult	30.0 μg/kg BW, or 2.1 mg per person
Wood, in contact with food	Up to 50.0 mg/kg
Drinking water	
Recommended	<0.021 mg/L[a]
Global proposed	<0.01 mg/L
Upper safe limit	1.01 mg/L
Arizona	<0.2 mg/L
California	<0.03 mg/L
Kansas, Montana	<0.22 mg/L
Maine	<0.006 mg/L
New York	<0.021 mg/L
Air, 8-h exposure daily, 5 days weekly	<0.5 mg/m^3
Blood	
Allowable	<1.0 mg/L
Normal	<0.1 mg/L
Blood plasma	<0.5 mg/L
Total intake	<3.0 μg/kg BW daily[b]
Total intake, 70-kg adult	<30.0 μg/kg BW daily or 2.1 mg daily[c]

Continued

Table 24.2. cont'd

Resource and Criterion	Concentration or Dose
Expectant mothers	No safe level established to guard against fetal toxicity
Adverse effects expected	
Air	$>1.0\,mg/m^3$
Dermal solutions	4000.0 mg/L
Tissue residues associated with acute toxicosis	
Kidney	>28.0 mg/kg FW
Blood	>40.0–80.0 mg/L
Liver	>62.0 mg/kg FW
Tissue residues associated with death	
Stomach contents	8.0 mg/kg FW
Urine	29.0–75.9 mg/L
Liver	52.0–225.0 mg/kg FW
Lung	116.0 mg/kg FW
Blood	162.0–176.0 mg/L
Kidney	116.0–639.0 mg/kg FW
Bile	1130.0 mg/kg FW
Effluent discharges to water	
New York	<0.021 mg/L
Wisconsin	<0.000001 mg/L

[a] Based on no-observable-adverse-effect level of 3.0 mg/kg BW daily in rat study, uncertainty factor of 1000 and water consumption of 2 L daily.

[b] Based on animal data and uncertainty factor of 1000.

[c] Based on rat chronic oral no-observable-effect level of 3.0 mg/kg BW daily and uncertainty factor of 100.

[d] At pH 6.6, a water concentration of 1.74 μg PCP/L caused a 50% reduction in growth of yearling sockeye salmon in an 8-week test.

would not prevent accumulations and growth inhibition in salmonids or accumulations in oysters. Some downward modifications have been proposed (Table 24.2). However, additional research is needed to establish sound water quality criteria for PCP, and also to interpret the significance of its residues and their metabolites in tissues of representative species.

Microorganisms and other bioremediation processes followed by hyperfiltration may be useful in removing PCP from groundwater and yielding a final effluent acceptable for discharge into the environment or into a municipal sewage treatment facility. At present, diluted PCP-contaminated groundwater may be discharged into the sewerage collection system of Escambia County, Florida, provided that 1% solutions were not teratogenic or embryotoxic to embryos of the inland silverside (*Menidia beryllina*) or lethal to the daphnid (*Ceriodaphnia dubia*). More research is encouraged on the use of sentinel organisms to assess PCP water quality.

Dietary concentrations of 1.0 mg/kg and higher produced diarrhea and liver histopathology in chickens (*Gallus* spp.) after 8 weeks, and deaths occurred at relatively high dietary concentrations, i.e., 3850.0 mg/kg, in Japanese quail (*Coturnix japonica*) after 5 days. Wood shavings contaminated with PCP produced elevated residues when used as litter for domestic chickens, and death in canaries (*Serinus canarius*), when used as nesting materials. Tissue residues >2.0 mg/kg fresh weight are considered to be indicative of significant environmental PCP contamination, and those >11.0 mg/kg fresh weight were associated with birds that died or were recovering from PCP exposure (Table 24.2). No data are now available on avian wildlife and PCP contamination in their diets, residues in their tissues, or frequency of use of PCP-contaminated wood shavings for nesting materials and other purposes.

As judged by studies with domestic and small laboratory mammals, no observable adverse effects have been noted at dietary levels equivalent to 3.0–10.0 mg PCP/kg BW (Table 24.2). Variability is great among species, however, and adverse effects have been documented in some species (Table 24.2) at doses as low as 0.05–0.2 mg/kg BW (elevated tissue residues), 0.2–2.0 mg/kg BW or 50.0 mg/kg diet (histopathology, reproductive impairment, increased tumor frequency), and 55.0–60.0 mg/kg BW (death). Based on guidelines for carcinogen risk assessment and inadequate evidence for animal carcinogenicity or absence of human cancer data, PCP is classified as group D, meaning that it is not classified as a human carcinogen. More research is recommended on the genotoxic and carcinogenic potential of PCP and its metabolites, with special reference to tetrachlorohydroquinone.

Data for humans show that adverse effects occur at concentrations in air >1.0 mg PCP/m^3 and in tissues at more than 8.0 mg/kg fresh weight (Table 24.2). No adverse effects were noted at daily intakes of 2.1 mg per 70-kg adult or 30.0 μg/kg BW, up to 1.01 mg/L in drinking water, <0.5 mg/m^3 in air, <0.5 mg/L in blood plasma, and <1.0 mg/L in blood. It is noteworthy that the recommended PCP air concentration of 0.5 mg/m^3 results in a daily intake of 2.5–3.8 mg (based on 15–23 m^3 of air inhaled daily, 8-h exposure), equivalent to 42.0–63.0 μg/kg BW for a 60-kg female. These levels are higher than the recommended no adverse effect level of 30.0 μg/kg BW daily (Table 24.2), and overlap or exceed the 58.0–74.0 μg/kg BW daily range. Air concentrations >1.0 mg PCP/m^3 can produce respiratory irritation in unacclimatized individuals, but concentrations as high as 2.4 mg/m^3 can be tolerated by conditioned individuals. The recommended biological tolerance value of <1000.0 μg PCP/L in blood is based on occupational air exposure studies: exposure to maximum average air concentrations of 0.18 mg PCP/m^3 for up to 34 years produced blood PCP residues of 23.0–775.0 μg/L, with no measurable adverse effects. The authors concluded that PCP and its impurities in occupationally relevant concentrations below the maximum concentration in the workplace and below the biological tolerance value do not produce genotoxic damage that can be detected on the chromosomal level, either in vivo or in vitro.

The human taste threshold for PCP in drinking water is about 30.0 μg/L, a level far below the upper safe limit of 1.01 mg/L and near the no-observable-effect level of 21.0 μg/L (Table 24.2). Odor detection is not as sensitive as taste: the odor threshold for PCP ranges from about 857.0 μg/L at 30°C, to 1600.0 μg/L at 20–22°C, to 12,000.0 μg/L at 60°C. It is not clear whether the determined organoleptic threshold values made the water undesirable or unfit for consumption. If fish and wildlife species of concern have PCP organoleptic thresholds that are similar to those of humans, or lower, will they too avoid PCP-contaminated habitats or diets?

Data for PCP and terrestrial wildlife are incomplete and – in view of the large interspecies variations in sensitivity – need to be collected. Research is needed on reproductive effects in animals following inhalation exposure to PCP; additional acute and intermediate toxicity testing; chronic duration exposure studies on cancer induction, genotoxicity, and immunotoxicity; and the development of alternate biomarkers of PCP exposure and antidotes. Until the results of these studies become available, it seems reasonable to apply to wildlife the same levels recommended for human health protection.

24.6 Summary

Pentachlorophenol (PCP) is a synthetic organochlorine compound that was first manufactured commercially in 1936 and is now used primarily as a wood preservative and secondarily as a herbicide, insecticide, fungicide, molluscicide, and bactericide. Global production of PCP is estimated at 50.0 million kg annually. Widespread use of PCP has resulted in the detection of residues in air, rain, snow, groundwater, surface water, drinking water, fish, and aquatic invertebrates, as well as in human urine, blood, and milk. Pentachlorophenol may be incorporated into animal tissues through inhalation, diet, or contact; its toxic action results from its ability to interfere with the production of high energy phosphate compounds essential for cell respiration. Pentachlorophenol has caused numerous occupational illnesses and deaths, and has had significant adverse effects on domestic animals. It is fetotoxic and teratogenic, but evidence for mutagenicity and carcinogenicity is incomplete or negative. Commercial PCP preparations often contain variable amounts of toxic impurities – including chlorophenols, hexachlorobenzene, phenoxyphenols, dioxins, and dibenzofurans – that contribute to its toxicity. Pentachlorophenol is rapidly accumulated and rapidly excreted, and has little tendency to persist in living organisms; it is readily degraded in the environment by chemical, microbiological, and photochemical processes.

In sensitive aquatic species, PCP adversely affected growth, survival, and reproduction at media concentrations of 8.0–80.0 μg PCP/L in algae and higher plants, at 3.0–100.0 μg/L in invertebrates, and <1.0–68.0 μg/L in fish. In birds, PCP was fatal at 380.0–580.0 mg/kg BW in oral doses, >3580.0 mg/kg in the diet, and >285.0 mg/kg in contaminated nesting materials (i.e., wood shavings). Residues >11.0 mg PCP/kg fresh weight in bird tissues were associated with acute toxicosis. Adverse sublethal effects in birds were noted at dietary levels as low as 1.0 mg/kg ration. In small laboratory mammals and domestic livestock, acute oral LD50s ranged from 27.0 to 300.0 mg/kg BW. Tissue residues in mammals were elevated at PCP doses as low as 0.05 mg/kg BW, and at air levels >0.1 mg/m^3. Histopathology, reproductive impairment, growth retardation, and other effects were evident in sensitive mammals at PCP concentrations of 0.2–1.25 mg/kg BW, and at dietary levels >30.0 mg/kg ration.

Pentachlorophenol is an undesirable pollutant whose use patterns should be carefully regulated to avoid contamination of soil, water, and food. Recommendations for protection of sensitive fishery and wildlife resources follow; however, it is emphasized that some of these recommendations are markedly lower than those proposed by regulatory agencies. For protection of aquatic life, it is recommended that the PCP water concentration should not exceed 3.2 μg/L; however, even at this level certain species of fishes and oysters accumulate enough of the toxicant to retard their growth. In birds, dietary concentrations greater than 1.0 mg/kg feed and tissue residues greater than 2.0 mg/kg fresh weight should be viewed as presumptive evidence of significant environmental PCP contamination. Data are scarce for PCP and mammalian wildlife; until more data are collected, PCP levels recommended for human health protection (i.e., "no adverse effects" levels) are suggested as reasonable substitutes. In humans, no adverse effects were noted at daily PCP intakes equivalent to 39.0 μg/kg BW in food, or at concentrations of 21.0 μg/L in drinking water, 0.5 mg/m^3 in air, 0.5 mg/L in blood plasma, and 1.0 mg/L in blood.

POLYCHLORINATED BIPHENYLS[a]

Chapter 25

25.1 Introduction

Polychlorinated biphenyls (PCBs), a group of 209 synthetic halogenated aromatic hydrocarbons, are used extensively in the electricity generating industry as insulating or cooling agents in transformers and capacitors. Because of human activities and the chemical characteristics of the products, PCBs are now distributed worldwide, and measurable concentrations occur in aquatic organisms and wildlife from North America, Europe, the United Kingdom, and the Atlantic and Pacific Oceans. PCBs elicit a variety of effects including death, birth defects, reproductive failure, liver damage, tumors, and a wasting syndrome. They bioaccumulate and biomagnify in the food chain. Legislation has prohibited virtually all uses of PCBs and their manufacture in the United States since 1979; the ban has been accompanied by declines in PCB residues in fishes and wildlife. But the current environmental burden of PCBs in water, sediments, disposal sites, deployed transformers, and other PCB containers – now estimated at more than 374 million kg, much of it localized – continues to represent a potential hazard to associated natural resources.

[a]All information in this chapter is referenced in the following sources:

Eisler, R. 1986. Polychlorinated biphenyl hazards to fish, wildlife, and invertebrates: a synoptic review. *U.S. Fish Wildl. Serv. Biol. Rep. 85(1.7)*, 72 pp.

Eisler, R. 2000. Polychlorinated biphenyls. Pages 1237–1341 in *Handbook of Chemical Risk Assessment: Health Hazards to Humans, Plants, and Animals. Volume 2, Organics*. Lewis Publishers, Boca Raton, Florida.

Eisler, R., and A. A. Belisle. 1996. Planar PCB hazards to fish, wildlife and invertebrates: a synoptic review. *U.S. Natl. Biol. Serv. Biol. Rep. 31*, 75 pp.

25.2 Sources and Uses

The main domestic manufacturer of PCBs began production in 1929; commercial mixtures of PCBs – including the Aroclors – were also produced in Western Europe and Japan. PCBs have been used in dielectric fluids; in waxes for metal castings; as heat transfer agents; as plasticizers in paints, coatings, and carbonless copy paper; in cutting oils; in sealants and caulking compounds; and as pesticide extenders. The use of PCBs was curtailed in the United States in 1971, and sales were limited to manufacturers of capacitors and transformers; all new uses were banned in 1976. In 1977, all production of PCBs was halted and no shipments were made after October. Direct and indirect sources of PCB contamination may include aerial transport of combustion products, vaporization from continental and marine areas, current and historic industrial and municipal waste discharges, precipitation, land runoff, concealed dumping, transformer fires, and accidental spills.

The ubiquity of PCBs is indicated by their presence in environmental samples from the polar regions of air, snow, ice, water, and in living organisms. The presence of PCBs in such remote areas suggests the importance of atmospheric transport. The Committee on the Assessment of Polychlorinated Biphenyls in the Environment estimated that 50–80% of the PCBs derived from the United States were now in sediments and waters of the North Atlantic Ocean. Of the estimated total world PCB production of 1.2 million tons to date, about 374,000 tons are now in various portions of the terrestrial, coastal, and open ocean ecospheres (Table 25.1). Another 783,000 tons are still in use in electrical equipment and other products or deposited in landfills and dumps and represent a potential source of

Table 25.1. Estimated PCB loads in the global environment.

Ecosystem	PCB Loads, in Metric Tons
TERRESTRIAL AND COASTAL	
Air	500
River and Lakewater	3500
Seawater	2400
Soil	2400
Sediment	130,000
Biota	4300
OPEN OCEAN	
Air	790
Seawater	230,000
Sediment	110
Biota	270
TOTAL	374,000

environmental contamination. An additional 43,000 tons have been degraded or incinerated. Long-range atmospheric and oceanic transport seem to be the primary mechanism of global PCB dispersal.

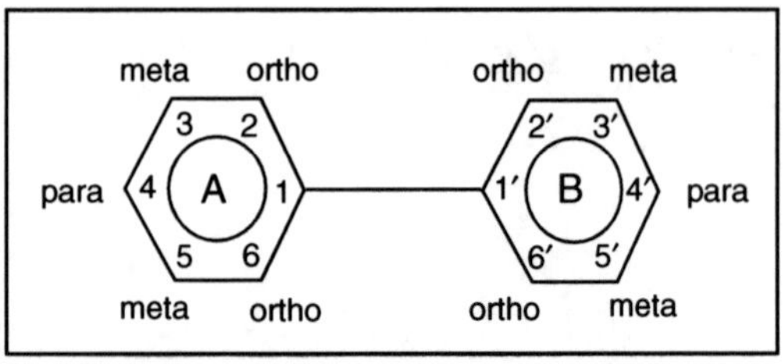

Figure 25.1.
Structure of biphenyl. Polychlorinated biphenyls (PCBs) are commercially produced by chlorination of a biphenyl with anhydrous chlorine in the presence of iron filings or ferric chloride as the catalyst. Depending on the conditions under which chlorination occurred, the purified product is a complex mixture of chlorobiphenyls containing 18–79% chlorine. Ten possible degrees of chlorination of the biphenyl group produce 10 PCB congener groups: mono-, di-, tri-, tetra-, penta-, hexa-, hepta-, octa-, nona-, and decachlorobiphenyl. Within any congener group, a number of positional isomers are possible. For example, the tetrachlorobiphenyl group consists of 30 possible isomers and the pentachlorobiphenyl congener group contains 46 possible isomers. Not all of the 209 possible isomers are likely to be formed during the manufacturing process. In general, the most common PCB isomers formed have either an equal number of chlorine atoms on both rings, or a difference of one chlorine atom between rings. Chlorine substitution is favored at the *ortho* and *para* positions; however, commercial products are complex mixtures of isomers and congeners with no apparent positional preference for halogen substitution.

25.3 Chemical and Biochemical Properties

Polychlorinated biphenyls, a highly lipophilic group of global pollutants, consist of 209 congeners (Figure 25.1) with widely different toxicity and other biological effects. Each of the congeners has been assigned to an isomeric group and a specific PCB number based on the number and placement of chlorine atoms (Table 25.2). All PCBs listed are 5–6 orders of magnitude more soluble in octanol than in water, as judged by K_{ow} values (Table 25.2); this is discussed later.

In vertebrates, toxicological effects of PCBs have been related to their ability to induce the cytochrome P450-dependent monooxygenase system (P450). This varies with the degree of chlorination and the arrangement of chlorine atoms on the biphenyl molecule. Transformation of PCBs to hydroxylated metabolites by the cytochrome P450 system is the major pathway of PCB metabolism and occurs mainly in the liver. The rate of cytochrome P450-catalyzed hydroxylation of PCBs decreases as the number of chlorines increases and as the number of unsubstituted adjacent carbon atoms decreases. Some animals metabolize PCBs at different rates, and this is related in

Table 25.2. Polychlorinated biphenyls (PCBs): isomeric group, PCB number, structure, and octanol–water partition coefficients (log K_{ow}).

Isomeric Group and PCB Number	Structure (Chlorine-Filled)	Log K_{ow} [a]	Isomeric Group and PCB Number	Structure (Chlorine-Filled)	Log K_{ow} [a]
MONOCHLOROBIPHENYLS			**38**	3,4,5	5.767
1	2	4.601	**39**	3,4′,5	5.897
2	3	4.421	TETRACHLOROBIPHENYLS		
3	4	4.401	**40**	2,2′,3,3′	5.561
DICHLOROBIPHENYLS			**41**	2,2′,3,4	6.111
4	2,2′	5.023	**42**	2,2′,3,4′	5.767
5	2,3	–[b]	**43**	2,2′,3,5	5.757
6	2,3′	5.021	**44**	2,2′,3,5′	5.811
7	2,4	5.15	**45**	2,2′,3,6	5.537
8	2,4′	5.301	**46**	2,2′,3,6′	5.537
9	2,5	5.18	**47**	2,2′,4,4′	6.291
10	2,6	5.311	**48**	2,2′4,5	5.787
11	3,3′	5.343	**49**	2,2′,4,5′	6.221
12	3,4	5.295	**50**	2,2′,4,6	5.637
13	3,4′	–[b]	**51**	2,2′,4,6′	5.637
14	3,5	5.404	**52**	2,2′,5,5′	6.091
15	4,4′	5.335	**53**	2,2′,5,6′	5.627
TRICHLOROBIPHENYLS			**54**	2,2′,6,6′	5.904
16	2,2′,3	5.311	**55**	2,3,3′,4	6.117
17	2,2′,4	5.761	**56**	2,3,3′,4′	6.117
18	2,2′,5	5.551	**57**	2,3,3′,5	6.177
19	2,2′,6	5.481	**58**	2,3,3′,5′	6.177
20	2,3,3′	5.577	**59**	2,3,3′,6	5.957
21	2,3,4	5.517	**60**	2,3,4,4′	5.452
22	2,3,4′	5.421	**61**	2,3,4,5	5.943
23	2,3,5	5.577	**62**	2,3,4,6	5.897
24	2,3,6	5.671	**63**	2,3,4′,5	6.177
25	2,3′,4	5.677	**64**	2,3,4′,6	5.957
26	2,3′,5	5.667	**65**	2,3,5,6	5.867
27	2,3′,6	5.447	**66**	2,3′,4,4′	5.452
28	2,4,4′	5.691	**67**	2,3′,4,5	6.207
29	2,4,5	5.743	**68**	2,3′,4,5′	6.267
30	2,4,6	5.504	**69**	2,3′,4,6	6.047
31	2,4′,5	5.677	**70**	2,3′,4′,5	6.231
32	2,4′,6	5.751	**71**	2,3′,4′,6	5.987
33	2′,3,4	5.572	**72**	2,3′,5,5′	6.267
34	2′,3,5	5.667	**73**	2,3′,5′,6	6.047
35	3,3′,4	5.827	**74**	2,4,4′,5	6.671
36	3,3′,5	4.151	**75**	2,4,4′,6	6.057
37	3,4,4′	4.941	**76**	2′,3,4,5	6.137

Continued

Table 25.2. cont'd

Isomeric Group and PCB Number	Structure (Chlorine-Filled)	Log K_{ow}[a]	Isomeric Group and PCB Number	Structure (Chlorine-Filled)	Log K_{ow}[a]
77	3,3′,4,4′	6.523	**119**	2,3′,4,4′,6	6.587
78	3,3′,4,5	6.357	**120**	2,3′,4,5,5′	6.797
79	3,3′,4,5′	6.427	**121**	2,3′,4,5′,6	6.647
80	3,3′,5,5′	6.583	**122**	2′,3,3′,4,5	6.647
81	3,4,4′,5	6.367	**123**	2′,3,4,4′,5	6.747
PENTACHLOROBIPHENYLS			**124**	2′,3,4,5,5′	6.737
82	2,2′3,3′,4	6.142	**125**	2′,3,4,5,6′	6.517
83	2,2′,3,3′,5	6.267	**126**	3,3′,4,4′,5	6.897
84	2,2′,3,3′,6	6.041	**127**	3,3′,4,5,5′	6.957
85	2,2′,3,4,4′	6.611	**HEXACHLOROBIPHENYLS**		
86	2,2′,3,4,5	6.204	**128**	2,2′,3,3′,4,4′	6.961
87	2,2′,3,4,5′	6.371	**129**	2,2′,3,3′,4,5	7.321
88	2,2′,3,4,6	7.516	**130**	2,2′,3,3′,4,5′	7.391
89	2,2′,3,4,6′	6.077	**131**	2,2′,3,3′,4,6	6.587
90	2,2′,3,4′,5	6.367	**132**	2,2′,3,3′,4,6′	6.587
91	2,2′,3,4′,6	6.137	**133**	2,2′,3,3′,5,5′	6.867
92	2,2′,3,5,5′	6.357	**134**	2,2′,3,3′,5,6	7.304
93	2,2′,3,5,6	6.047	**135**	2,2′,3,3′,5,6′	7.151
94	2,2′,3,5,6′	6.137	**136**	2,2′,3,3′,6,6′	6.511
95	2,2′,3,5′6	6.137	**137**	2,2′,3,4,4′,5	>7.711
96	2,2′,3,6,6	5.717	**138**	2,2′,3,4,4′,5′	7.441
97	2,2′,3′,4,5	6.671	**139**	2,2′,3,4,4′,6	6.677
98	2,2′,3′,4,6	6.137	**140**	2,2′,3,4,4′,6′	6.677
99	2,2′,4,4′,5	7.211	**141**	2,2′,3,4,5,5′	7.592
100	2,2′,4,4′,6	6.237	**142**	2,2′,3,4,5,6	6.517
101	2,2′,4,5,5′	7.071	**143**	2,2′,3,4,5,6′	6.607
102	2,2′,4,5,6′	6.167	**144**	2,2′,3,4,5′,6	6.677
103	2,2′,4,5′6	6.227	**145**	2,2′,3,4,6,6′	6.257
104	2,2′,4,6,6	5.817	**146**	2,2′,3,4′,5,5′	6.897
105	2,3,3′,4,4′	6.657	**147**	2,2′,3,4′,5,6	6.647
106	2,3,3′,4,5	6.647	**148**	2,2′,3,4′,5,6′	6.737
107	2,3,3′,4′,5	6.717	**149**	2,2′,3,4′,5′,6	7.281
108	2,3,3′,4,5′	6.717	**150**	2,2′,3,4′,6,6′	6.327
109	2,3,3′,4,6	6.487	**151**	2,2′,3,5,5′,6	6.647
110	2,3,3′,4′,6	6.532	**152**	2,2′,3,5,6,6′	6.227
111	2,3,3′,5,5′	6.767	**153**	2,2′,4,4′,5,5′	7.751
112	2,3,3′,5,6	6.457	**154**	2,2′,4,4′,5,6′	6.767
113	2,3,3′,5′,6	6.547	**155**	2,2′,4,4′,6,6′	7.123
114	2,3,4,4′,5	6.657	**156**	2,3,3′,4,4′,5	7.187
115	2,3,4,4′,6	6.497	**157**	2,3,3′,4,4′,5′	7.187
116	2,3,4,5,6	6.304	**158**	2,3,3′,4,4′,6	7.027
117	2,3,4′,5,6	6.467	**159**	2,3,3′,4,5,5′	7.247
118	2,3′,4,4′,5	7.121	**160**	2,3,3′,4,5,6	6.937

Table 25.2. cont'd

Isomeric Group and PCB Number	Structure (Chlorine-Filled)	Log K_{ow}[a]	Isomeric Group and PCB Number	Structure (Chlorine-Filled)	Log K_{ow}[a]
161	2,3,3′,4,5′,6	7.087	**187**	2,2′,3,4′,5,5′,6	7.177
162	2,3,3′,4′,5,5′	7.247	**188**	2,2′,3,4′,5,6,6′	6.827
163	2,3,3′,4′,5,6	6.997	**189**	2,3,3′,4,4′,5,5′	7.717
164	2,3,3′,4′,5′,6	7.027	**190**	2,3,3′,4,4′,5,6	7.467
165	2,3,3′,5,5′,6	7.057	**191**	2,3,3′,4,4′,5′,6	7.557
166	2,3,4,4′,5,6	6.937	**192**	2,3,3′,4,5,5′,6	7.527
167	2,3′,4,4′,5,5′	7.277	**193**	2,3,3′,4′,5,5′,6	7.527
168	2,3′,4,4′,5′,6	7.117	**OCTACHLOROBIPHENYLS**		
169	3,3′,4,4′,5,5′	7.427	**194**	2,2′,3,3′,4,4′,5,5′	8.683
HEPTACHLOROBIPHENYLS			**195**	2,2′,3,3′,4,4′,5,6	7.567
170	2,2′,3,3′,4,4′,5	7.277	**196**	2,2′,3,3′,4,4′,5′,6	7.657
171	2,2′,3,3′,4,4′,6	6.704	**197**	2,2′,3,3′,4,4′,6,6′	7.307
172	2,2′,3,3′,4,5,5′	7.337	**198**	2,2′,3,3′,4,5,5′,6	7.627
173	2,2′,3,3′,4,5,6	7.027	**199**	2,2′,3,3′,4,5,6,6′	7.207
174	2,2′,3,3′,4,5,6′	7.117	**200**	2,2′,3,3′,4,5′,6,6′	7.277
175	2,2′,3,3′,4,5′,6	7.177	**201**	2,2′,3,3′,4′,5,5′,6	7.627
176	2,2′,3,3′,4,6,6′	6.767	**202**	2,2′,3,3′,5,5′,6,6′	8.423
177	2,2′,3,3′,4′,5,6	7.087	**203**	2,2′,3,4,4′,5,5′,6	7.657
178	2,2′,3,3′,5,5′,6	7.147	**204**	2,2′,3,4,4′,5,6,6′	7.307
179	2,2′,3,3′,5,6,6′	6.737	**205**	2,3,3′,4,4′,5,5′,6	8.007
180	2,2′,3,4,4′,5,5′	7.367	**NONACHLOROBIPHENYLS**		
181	2,2′,3,4,4′,5,6	7.117	**206**	2,2′,3,3′,4,4′,5,5′,6	9.143
182	2,2′,3,4,4′,5,6′	7.207	**207**	2,2′,3,3′,4,4′,5,6,6′	7.747
183	2,2′,3,4,4′,5′,6	7.207	**208**	2,2′,3,3′,4,5,5′,6,6′	8.164
184	2,2′,3,4,4′,6,6′	6.857	**DECACHLOROBIPHENYL**		
185	2,2′,3,4,5,5′,6	7.933	**209**	2,2′,3,3′,4,4′,5,5′,6,6′	9.603
186	2,2′,3,4,5,6,6′	6.697			

[a] K_{OW} (octanol–water partition coefficient) = C_{OW}/C_{WO}, where C_{OW} is the concentration of the solute in octanol saturated with water, and C_{WO} is the concentration of the solute in water saturated with octanol.
[b] No data.

part to differences in the basal level of particular isozymes of cytochrome P450 present in liver. For example, the dog eliminates PCBs more rapidly than other species because it has higher levels of a constitutive isozyme of cytochrome P450 with activity toward the slowly metabolized, bioaccumulated 2,2′,4,4′,5,5′-hexachlorobiphenyl. The reactive arene oxides formed during biotransformation can bind covalently to tissue macromolecules or conjugate with glutathione. Derivatives of glutathione conjugates and glucuronides of the hydroxylated products are major PCB metabolites. Biotransformation of xenobiotics by cytochrome P450 is not always beneficial to the organism because metabolites can be more toxic or biologically active than the parent compound. The carcinogenic effect of certain xenobiotics depends on the conversion by cytochrome P450 to a reactive carcinogenic metabolite. In general, PCB metabolites are less toxic than their parent compounds, although hydroxylated metabolites of 3,3′,4,4′-tetrachlorobiphenyl can disrupt thyroxine levels and serum transport of vitamin A in rodents. Hydroxylated metabolites and methylsulfonyl metabolites of PCBs were found in plasma and liver of the Laysan albatross (*Diomedea immutabilis*) and the black-footed albatross (*Diomedea nigripes*). Total concentrations of hydroxylated metabolites accounted for 20–100% of the total PCBs, but methylsulfonyl metabolites accounted for only 0.004% of the total PCBs. Metabolites and possible degradation pathways of selected PCBs in mammals are known in detail.

Sedimentation and volatilization are the dominant processes that determine the fate of PCBs in lakes. Both processes remove PCBs from the water, but the relative importance of the transferred amount is influenced by particulate dissolved-phase partitioning that determines the relative size of the particulate pool for sedimentation and the soluble pool for volatilization of PCBs. PCB-contaminated sediments from the St. Lawrence River (300.0 mg total PCBs/kg DW sediment) dechlorinated rapidly (36% loss during the first 4 months) under laboratory incubations over a 39-month period, and more slowly afterwards. High productivity of algae increased the proportion of added PCBs that is absorbed to particulate matter and sedimented. In general, PCB volatilization losses increase under conditions of high mixing and low productivity. Volatilization of mono-, di-, and trichlorobiphenyls from Aroclor 1248-contaminated sediments occurs at ambient environmental conditions with volatilization enhanced by microbial reductive dechlorination. (Note: Aroclors are manufactured PCB-containing compounds, with Aroclor 1248 containing about 48% total PCBs, Aroclor 1242 with about 42% total PCBs, and so on in this series.) Aroclor 1242 and 1254 were anaerobically dechlorinated by microorganisms eluted from PCB-contaminated sediments. Dechlorination occurred mainly from the *meta* position and suggest that dioxin-like toxicities of PCB mixtures are markedly reduced by microbial reductive dechlorination.

25.3.1 Physical Properties

The retention times and response factors relative to a reference standard (octachloronaphthalene) of all 209 congeners was determined using temperature-programmed, high-resolution gas chromatography, and electron-capture detection methods (HRGC/ECD). The relative retention times (the ratio of congener retention time: reference standard retention time) of 187 of the 209 congeners differed and permitted HRGC column identification contingent on full or partial separation. Of the 209 congeners, 20 can assume a planar configuration because of the absence of chlorine substitution in the *ortho* positions (Figure 25.1). Approximately 1% of the non-*ortho*-substituted biphenyl molecules adopt the planar configuration. Among the isomers of each homolog, the planar PCBs had the longest retention times. The presence of *ortho*-chloro substituents reduces planarity of the rings. However, congeners with 1 or 2 *ortho* chlorines can also assume a planar ring position.

The transport and fate of PCBs in aquatic systems and their partitioning between sediment, water, and organisms depend largely on sorption reactions. In soils, the sorption and retention of PCB congeners are influenced by the number of chlorine atoms in the molecule, and the more highly chlorinated PCBs tend to be more strongly bound. Relative sorption capacity and other properties of congeners also depend on PCB configuration. The soil sorption capacities of PCB congeners and their bioconcentration factors (BCFs) were related to octanol–water partition coefficients as follows:

$$K_{\mathrm{ow}} = \frac{C_{\mathrm{ow}}}{C_{\mathrm{wo}}},$$

where K_{ow} is the octanol–water partition coefficient, C_{ow} the concentration of the solute in octanol saturated with water, and C_{wo} the concentration of the solute in water saturated with octanol.

The K_{ow} values are used to estimate BCFs (bioaccumulation after uptake from water), soil and sediment organic carbon–water partition coefficients, toxicities, and aqueous solubilities. Techniques for measuring the K_{OW} values include the shake-flask method, the reverse-phase thin-layer chromatography, reverse-phase high-performance liquid chromatography, and the generator column technique, or it may be calculated by an estimation technique based on correlation with properties of compounds with known K_{OW} values. From a strong linear relation calculated between known log K_{OW} values and calculated total surface areas (TSAs; correlation coefficient of 0.951 for 30 congeners), estimated K_{OW} values of individual congeners were determined from the calculated TSA. Estimated K_{OW} values were previously unavailable for many PCB congeners, including toxic non-*ortho*-substituted congeners found only in relatively small amounts in commercial Aroclors. Reported K_{OW} values of some congeners vary in the literature and may not be comparable when different measuring techniques are used. Despite many uncertainties, the K_{OW} is routinely used in hazard evaluation and risk assessment of most organic chemicals.

25.3.2 Toxic Equivalency Factors

A significant part of the toxicity associated with commercial PCB mixtures is related to the presence of the small number of planar congeners. These compounds induce several similar toxic effects in mammals and birds, such as hepatotoxicity, immunotoxicity, and reproductive toxicity. Planar halogenated aromatic compounds act, in part, by a common mechanism initiated by binding to a cytosolic aryl hydrocarbon receptor. The relative toxicities of planar halogenated hydrocarbons are calculated by expressing their toxicity in relation to 2,3,7,8-TCDD, the most potent compound in this class of chemicals. Toxic equivalency factors (TEFs) are fractional potencies that relate a compound's potency to that of 2,3,7,8-TCDD. The 2,3,7,8-TCDD TEF has been used to estimate the relative toxic potencies of individual planar halogenated hydrocarbons. According to most authorities, TEF values should be derived from data on the following effects in descending order of priority: (1) long-term carcinogenicity studies; (2) reproductive studies; (3) subchronic studies that measure Ah receptor-mediated responses, such as thymic atrophy, loss in body weight, and immunotoxicity; (4) acute toxicity studies; and (5) in vivo or in vitro biochemical responses such as enzyme induction and receptor binding. Relative toxic potencies are modified by many variables including age, sex, species, and strain of the animal; the efficiency of the chemical to induce cytochrome P450 and associated monooxygenase enzyme activities, glucuronosyl and glutathione transferases, and other drug-metabolizing enzymes; and the efficiency of the organism to modulate steroid-metabolizing enzymes, induce delta aminolevulinic acid synthetase, inhibit porphyrinogen decarboxylase, decrease Ah receptor binding activity, and alter vitamin A and thyroid hormone levels.

The in vitro induction of the cytochrome P450c-dependent monooxygenases, AHH (aryl hydrocarbon hydroxylase), or EROD (ethoxyresorufin *O*-deethylase) by 2,3,7,8-TCDD and related halogenated aryl hydrocarbons in rat liver cells was developed as a short-term quantitative bioassay for these chemicals; aromatics that do not fit this correlation are considered as congeners that are readily metabolized in vivo. Induction of either AHH or EROD activity in the H4IIE rat hepatoma cell line by PCB, PCDF, and PCDD congeners, either singly or in combination, correlates well with the in vivo toxicity of these compounds to rats. The proposed mean TEF values range from 0.01 to 0.1 in non-*ortho*-substituted planar PCBs to 0.001 in mono-*ortho*-substituted planar PCBs to 0.00002 in di-*ortho*-substituted planar PCBs (Table 25.3). Although the concentration of non and mono-*ortho*-substituted PCBs in animal tissues ranges from about 0.01 μg/kg to several micrograms per kg (about 1000–100,000 times lower than the sum of total PCBs), it is significantly higher than the concentration of the highly toxic 2,3,7,8-TCDD and 2,3,4,7,8-pentachlorodibenzofuran. Accordingly, the non- and mono-*ortho* PCBs–despite their lower toxic potency–often contribute as much or more to the 2,3,7,8-TCDD-like activity than either dioxins or furans. However, the overall importance of mono-*ortho* PCBs to the TEF is questioned by some authorities, and this

Table 25.3. Proposed toxicity equivalency values (TEFs) relative to 2,3,7,8-TCDD of non-*ortho*, mono-*ortho*, and di-*ortho* planar PCBs.

PCB Congener	TEF
NON-*ORTHO* PLANAR PCBs	
PCB 77	0.0005
PCB 126	0.1
PCB 169	0.01
MONO-*ORTHO* PLANAR PCBs	
PCB 105	0.0001
PCB 114	0.0005
PCB 118	0.0001
PCB 123	0.0001
PCB 156	0.0005
PCB 157	0.0005
PCB 167	0.00001
PCB 189	0.0001
DI-*ORTHO* PLANAR PCBs	
PCB 170	0.0001
PCB 180	0.00001

must be considered in future risk assessment evaluations.

Mammal-derived TEFs underestimate the potency of planar PCB mixtures in fish. TEF values of non-*ortho* PCB congeners based on mortality of rainbow trout in early life stages are as much as 1000 times lower than TEFs proposed for human risk assessment. TEFs for PCBs were compared using cytochrome P4501A1 induction measured as EROD activity in a rainbow trout liver cell line and in a rat hepatoma cell line. EROD activity in trout liver cells was induced only by PCBs 77, 81, 118, and 126 under normal growth conditions, and in some cultures with PCBs 105, 156, and 169. The trout TEFs were 0.023 for PCB 126, 0.064 for PCB 81, 0.0034 for PCB 77, 0.00016 for PCB 169, and 0.000017 for PCB 118; TEFs for PCBs 105 and 156 were <0.00003. In rat cells, however, all seven PCBs clearly induced EROD activity. It was concluded that the toxic impact of PCBs on rainbow trout is overestimated by risk assessment TEFs based on rat cells and assessed more appropriately with TEFs derived from rainbow trout cells.

25.3.3 Structure–Function Relations

Three classes of PCB congeners are described on the basis of their ability to induce benzo[a]pyrene hydroxylase (also known as AHH) and EROD activities: (1) planar PCBs; (2) mono-*ortho* analogs of the planar PCBs; and (3) di-*ortho* analogs of the planar PCBs. Among the 20 possible planar PCB congeners and their analogs (Figure 25.2), the most toxic in rats were PCBs 77, 126, and 169; these three congeners are approximate isostereomers of 2,3,7,8-tetrachlorodibenzo-*p*-dioxin (2,3,7,8-TCDD) and have similar toxic effects, including induction of 3-methylcholanthrene-type drug-metabolizing enzymes, body weight loss, thymic atrophy, dermal disorders, hepatic damage, high binding affinity to hepatic cytosolic receptor proteins, immunotoxicity, reproductive impairment, and teratogenicity. PCBs 77, 126, and 169 have been detected in eggs of terns from Lake Michigan, in marine mammals and humans, and in fish from the Hudson and Ohio Rivers. However, they are difficult to detect without proper methods.

The structural characteristics of individual PCB congeners influence their induction of various P450 activities. In mammals, PCB congeners have been characterized as 3-methylcholanthrene-type inducers, phenobarbital-type inducers, or mixed-type inducers of both. AHH and EROD activities (which are preferentially catalyzed by the P450IA gene subfamily) have been induced by planar PCBs in fish and mammals and by some mono- and di-*ortho* analogs of planar PCBs in mammals. The mechanism of toxic action of planar and mono-*ortho* planar PCBs is linked to an interaction with the 2,3,7,8-TCDD (or Ah) receptor protein. But this mechanism does not account for all observed PCB toxicities. Toxic responses unrelated to Ah receptor effects have been reported of PCBs 4, 28, 31, 49, 52, 84, 95, 110, 136, and 153. For example, PCB 153 is less cytotoxic than PCB 169 but is a more effective inhibitor of intercellular communication. PCB 52 caused moderate chick embryotoxicity; however, PCBs 18 and 153 were inactive, and PCBs 84 and 118 were severely toxic but by different mechanisms. The induction of cytochrome P4501A activity in the presence of both an inducer (PCB 126) and low concentrations of an inhibitor (tributyltin (TBT)) indicates that TBT does not interfere with the Ah receptor binding. Potentiation of EROD activity and cytotoxicity as a result of coexposure to PCB 126 and TBT is significant because they both accumulate in a variety of marine organisms.

Group I planar PCBs are 10 times more toxic and 100 times more effective as inducers of cytochrome P450c-dependent monooxygenase and 70 times more effective in competitively displacing 2,3,7,8-TCDD from a rat cytosol receptor protein than Group II planar PCBs (Table 25.3; Figure 25.2). Mono-*ortho* analogs of the planar PCBs have one substituent in the *ortho* (2 or 2′) position; these compounds possess diminished yet significant EROD- or AHH-inducing capacity and also induce P450 forms that are induced by the phenobarbital class of compounds. Mono-*ortho* derivatives (PCBs 105, 118, 156, and 189) may be more important in terms of 2,3,7,8-TCDD-like activity and in occurrence. Di-*ortho* analogs of the planar PCBs, that is, those with *ortho-*, *meta-*, and *para*-substituents, possess still weaker but significant AHH-inducing activity. Certain di-*ortho* derivatives of the 3,3′,4,4′ pattern (PCBs 128, 138, 153, 170, 180) are significant components of PCB residues; however, PCBs 128, 138, and 170 have reduced 2,3,7,8-TCDD-like effects. In rats, several PCBs (105, 114, 118, 123, 126, 156, 157, 169) produced a linear correlation between the EC50 response (in vitro) of AHH induction against the ED50 (in vivo) of body weight loss, thymic atrophy, hepatic AHH, and EROD induction. The planar mono- and di-*ortho* derivatives (PCBs 105, 118, 156, 189, 128, 138, 153, 170, 180) are referred to as mixed inducers because they elicit effects similar to coadministration of phenobarbital plus methylcholanthrene.

PCB 156, a mixed inducer of microsomal enzymes, significantly increases the incidences of cleft palates by 2,3,7,8-TCDD in rodents. Interactions among polychlorinated congeners may range from antagonism to additivity to synergism, and the toxicity of individual PCBs can be raised by interaction with other PCBs.

25.3.4 Quantitation

PCBs are chemically inert, nonpolar compounds and relatively stable during collection and storage; however, PCB concentrations in environmental samples vary with different measurement techniques, with types of Aroclor standards used for calibration, and with oven drying techniques. Reports of interlaboratory comparison studies for PCB analysis show wide variations and strongly indicate a need for more rigorous quality control and assurance. In one multilaboratory study – wherein PCBs were determined in water, soil, and sediments – percent recoveries from analysis of fortified waters averaged 60% in the high-concentration sample containing 148.0 μg/L total PCBs and 55% in the low-concentration sample of 37.0 μg/L. No single combination of extraction and cleanup was best for all solid samples. An analytical intercomparison exercise was conducted on solutions containing 10 PCB congeners. Of the 61 evaluated laboratories, a group of 47 laboratories produced between-group standard deviations of 1.10–1.13 by all PCB congeners except PCB 52; a group of 11 laboratories were identified as an outlier group. Only three laboratories were able to quantify PCBs 110 and 77, which coelute in most GC columns. Peak height measurements gave better reproducibility than peak area methods. In another intercomparison study, the analysis by 11 laboratories of a solution containing 12 pure congeners resulted in a variation of as much as 20% for analysis of a single congener. For the analysis of the same congeners in a fish oil, the coefficient of variation ranged from 17.4 to 132% (66.2% without outliers) and had a median of 47.1%.

PCBs in biological samples are usually extracted by a Soxhlet column and with a non-polar solvent such as hexane; the sample

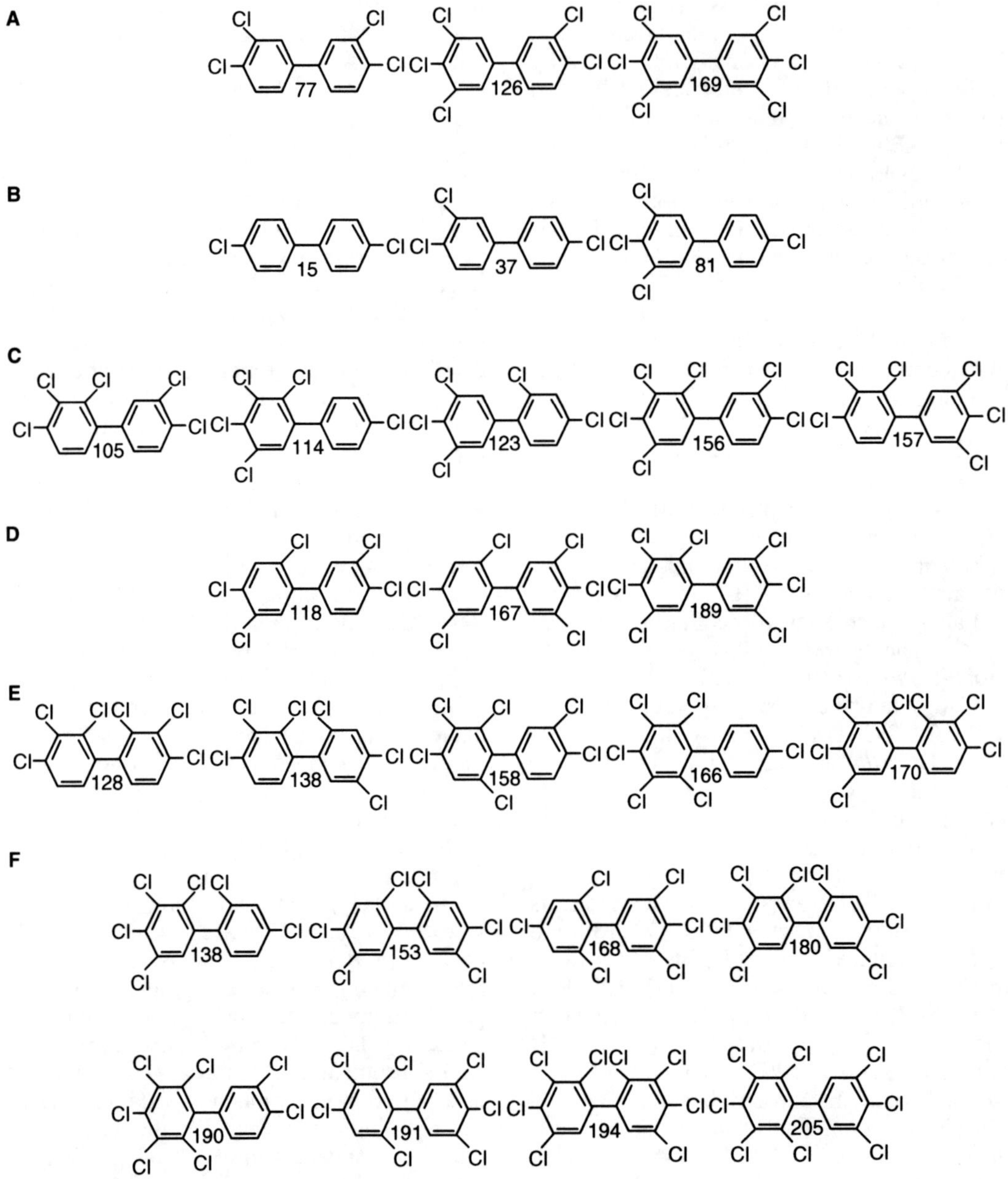
A
77
126
169
B
15
37
81
C
105
114
123
156
157
D
118
167
189
E
128
138
158
166
170
F
138
153
168
180
190
191
194
205

Figure 25.2.

is first mixed with sodium sulfate to remove moisture. The extraction of PCBs from sediments was tested with sonication, with two sonications interspersed at a 24-h quiescent interval, with steam distillation, or with Soxhlet extraction. Comparison of the recoveries of various PCB mixtures from dry and wet sediments by the four techniques and the extraction efficiency of four solvents showed that the best overall recoveries were obtained by Soxhlet extraction and the two sonication procedures. In comparisons of solvent systems of acetone, acetonitrile, acetone–hexane (1 + 1), and water–acetone–isooctane (5 + 1.5 + 1), recoveries of lower chlorinated congeners (dichloro- to tetrachloro-) were usually higher with acetonitrile and recoveries of higher chlorinated congeners (tetrachloro- to heptachloro-) extracted with acetone were superior. The completeness of extraction from a sample matrix does not seem to discriminate against specific isomers; however, discrimination in the cleanup and fractionation process may occur and must be tested.

With most analytical techniques for the quantification of PCB residue levels, chromatographic separations were used, most frequently electron-capture gas chromatography (GC/ECD). With early methods, selected peaks were used to estimate total PCBs. Low-resolution separations were satisfactory when packed columns that produced a pattern of peaks with measured areas were used. The patterns were compared with known amounts of Aroclor mixtures. If the Aroclor peaks in a sample closely resembled a particular Aroclor reference mixture of known weight, the total area or peak height of the sample PCBs was compared to those of the reference mixture and the weight of calculated sample PCBs. Other investigators used selected peaks to report Aroclor equivalents, but these methods are not useful when samples and Aroclor standards are dissimilar. For another procedure, response factors were used for individual Aroclor peaks as determined by GC and GC-MS procedures. The sample peaks were compared with peaks from common Aroclors obtained on packed

Figure 25.2.
Planar polychlorinated biphenyls (PCBs) and their derivatives. The general order of biological activity is: Group I non-*ortho* planar PCBs > Group II non-*ortho* planar PCBs > mono-*ortho* planar PCBs > di-*ortho* planar PCBs. **A.** Group I. Potent non-*ortho* planar PCBs. This group contains no *ortho*, 2 *para* and at least 2 *meta*-chlorines (PCBs 77, 126, and 169), and are approximate stereoisomers of 2,3,7,8-tetrachlorodibenzo-*p*-dioxin (2,3,7,8-TCDD). They are less potent than 2,3,7,8-TCDD but elicit similar biological responses including induction of cytochrome P450 and aryl hydrocarbon hydroxylase (AHH). This group is the most biologically active of all planar PCBs and their derivatives. **B.** Group II. Less-potent non-*ortho* planar PCBs (PCBs 15, 37, and 81). Group II non-*ortho* planar PCBs, when compared to Group I, were only 0.1 times as toxic, 0.01 times as effective in inducing cytochrome P455c-dependent monooxygenase, and about 0.015 times as effective in competitively displacing 2,3,7,8-TCDD from a cytosol-receptor protein in rat liver. **C.** Potent mono-*ortho* planar PCBs. The addition of a single *ortho*-chlorine substituent to non-*ortho* planar PCBs 77, 81, 126, and 169 yields 8 derivatives, of which 5 (PCBs 105, 114, 123, 156, 157) were more potent when the *ortho* chlorine was adjacent to a *meta* hydrogen. **D.** Less-potent mono-*ortho* planar PCBs. The toxicity of non-*ortho* planar PCBs is reduced by the introduction of an *ortho*-chloro substituent, especially when it is adjacent to a *meta*-chlorine (PCBs 118, 167, 189). **E.** Potent di-*ortho* planar PCBs. The di-*ortho* planar PCBs are less toxic than the mono-*ortho* planar PCBs. At least 5 di-*ortho* derivatives (PCBs 128, 138, 158, 166, 170) compete with 2,3,7,8-TCDD for receptor-binding sites in rat liver cytosol to induce cytochrome P450c. **F.** Less-potent di-*ortho* planar PCBs. These 8 di-*ortho* derivatives (PCBs 137, 153, 168, 180, 190, 191, 194, 205) are less potent than those figured in **E.**

columns; retention time windows for the early, middle, and late-eluting peaks were assigned to 3 Aroclors (1242, 1254, 1260) based on the presence or absence of specific peaks. The weight of PCB in the sample peak was calculated by multiplying its peak area by the appropriate response factor; all peaks were added to obtain a total weight of PCB. However, packed columns failed to separate and to identify many congeners because several congeners usually eluted under a single peak. These methods do not account for changes in composition from interfering compounds; to congener changes from hydrolysis, photodegradation, and biodegradation; to selective evaporation and adsorption of certain isomers; or to solubility differences resulting in different partitioning ratios among the various environmental compartments. Some researchers, who used capillary columns, estimated PCB residues on the basis of a relatively simple cleanup and analysis by high-resolution GC-ECD or by HRGC/MS; others emphasized total PCBs, homolog subgroups, or individual congeners present in substantial amounts in the sample or in commercial mixtures. And still others used total PCBs derived from the sum of homolog subgroup concentrations, concentrations of individual subgroups, or the contribution of all congeners in each subgroup.

Gas chromatography/mass spectrometry in the electron ionization mode (GC/MS-EI) has been used to a more limited extent for routine analysis of PCBs. Some interference with quantification ions can occur with EI when compounds such as PCBs 77 and 110 coelute. GC/MS has also been operated in the negative chemical ionization (NCI) mode for PCB determinations. Increased selectivity is observed in the NCI mode, although interference from other compounds that readily form stable negative ions may be observed. Improvements have also been made in capillary column separations. However, the application of single long, narrow bore capillary columns does not enable investigators to attain full separation of PCB congeners. Multidimensional GC (MDGC) analysis of PCBs drastically improved congener separations and the analysis of all sample constituents. Using MDGC techniques, chemists analyzed PCBs 28, 52, 101, 138, 153, and 180 and positively identified PCBs 52, 101, and 180 as well as coeluting peaks of PCBs 24, 26, 29, 44, 49, 81, 84, 114, 128, 151, 169, 177, 183, 187, 189 and 194. Others used the same MDGC technique to fully resolve all congeners in commercial PCB mixtures of Clophen A30, A40, A50, A60 and in Aroclors 1221, 1016, 1242, 1254, and 1260. A column of SE-54 with ECD was used to monitor the eluate, and a second, more polar column (usually OV-210) with ECD was used to obtain fully resolved single peaks. The procedure required exact timing of cuts based on retention time but fully separated the PCB congeners by GC for the first time. A total of 132 congeners at concentrations above 0.05% (w/w) were eluted as fully resolved single peaks and measured.

Pattern recognition techniques that incorporate statistical methods have been used to determine spatial and temporal patterns in PCB residue data. One researcher, for example, characterized sample PCB profiles from 105 individual congeners, measured parameters of similarity between sample and standard mixtures, and determined whether the sample residue pattern corresponded to an Aroclor mixture. Modeling environmental samples with the individual congener concentrations provided more accurate estimations of Aroclor profiles than homolog concentrations. Intact commercial Aroclors have been characterized by automated mass spectrometric determination of weight percent distribution by homolog groups. One isomer from each level of chlorination was normally used to calibrate the MS response to all measurable isomers in that group. In another example, researchers applied pattern recognition techniques to assess biomagnification and to characterize source patterns of multicomponent pollutants such as PCBs, PCDDs, and PCDFs in eggs of the herring gull (*Larus argentatus*) from the Great Lakes between 1983 and 1990. A linear regression was fitted to PCB concentrations in herring gull eggs from the Great Lakes during 1970–85 and from more recent measurements. Aroclor equivalents in eggs were determined with

PCB 138 as a single peak estimate for the older data and more recent PCB data as the sum of 41 congeners. The earlier choice of PCB 138 for quantitation seemed fortunate; uptake of heptachloro isomers maintained a stable percentage of total PCB egg residues over time, whereas less-chlorinated congeners generally declined and more-chlorinated congeners increased over time.

Methods were developed for routine analysis of AHH-inducing and other PCB congeners in fish by using a comparatively simple gel permeation chromatography (GPC) cleanup and GC/MS-NCI. Methane was used as the reagent gas for NCI, source temperature and pressure were optimized (1100°C and 0.9 Torr), and fragmentation of the major ions was reduced as much as 10%. A peak area correction factor for interfering ion fragmentation was obtained by instrument calibration with standards and manually applied to coeluting ions as needed. Additional methods were developed for the analysis of planar congeners. Using a carbon foam adsorbent, one research group devised procedures for the analysis of toxic non-*ortho*-chloro-substituted PCB congeners and trace planar impurities in Aroclors. With carbon-foam chromatography (carbon particles suspended on a polyurethane substrate), concentrations of planar PCB congeners in Aroclors 1016, 1242, 1248, 1254, and 1260 were detected by high-resolution electron capture gas chromatography (HRGC/ECD). With current chemical methods for analysis of non-*ortho*-chloro-substituted planar PCBs, extraction and a preliminary cleanup, carbon chromatography, and HRGC/ECD or HRGC/MS are generally used. The brands of carbon available for isolating planar PCBs include Amoco AX-21 (PX-21), Alltech SK-4, Serva SP-1, and Wako active carbon. All methods shared at least three characteristics: (1) they require some form of adsorption chromatography (usually with carbon) to isolate planar compounds; (2) they include several cleanup steps; and (3) they are complicated.

A semipermeable membrane device (SPMD) with a nonpolar, low-density polymeric film is used to separate PCBs from large amounts (20.0–>50.0 g) of lipids by dialysis in an organic solvent prior to chemical analysis. Liquid–liquid phase partitioning of extracts in organic solvent with sulfuric acid has been used to convert fats and pigments into water-soluble compounds that can be separated and removed from the target analytes with water rinses. Fats may be removed by refluxing or partitioning extracts with alcoholic potassium hydroxide to form water-soluble hydrolysis products. Some chemists combined alkali digestion, active carbon column chromatography, fuming sulfuric acid cleanup, HRGC/ECD, and HRGC/MS confirmation for the analysis of PCBs 77, 126, and 169 in porpoise blubber. Measurement of these three highly toxic PCBs in Aroclor and Kanechlor mixtures is reported. Others separated 5 non-*ortho*-substituted PCBs from a synthetic mixture containing Aroclors, organochlorine and organophosphorus pesticides, dioxins, and dibenzofurans on an HPLC column packing of porous graphite carbon (PGC). Soil samples required prior cleanup to remove coextracted organics, and elution from a multilayered column containing acid, base, and silica was followed by elution from Florisil. Procedures are available for separating mono- and non-*ortho* chloro PCBs. Sulfuric acid cleanup, low-pressure liquid chromatography with activated carbon/silica gel and HRGC/ECD were used to analyze 4 non-*ortho* and 8 mono-*ortho*-substituted PCBs. A non-carbon HPLC column of 2-(1-pyrenyl)ethyldimethylsylated silica (PYE) column was used to isolate mono- and non-*ortho* chloro PCBs from tissue samples. Isolation with this column required almost complete prior removal of lipids by sulfuric acid partitioning and subsequent GPC with Bio-Beads S-X3 to remove remaining lipids. An automated dioxin analysis system was adapted with programmable pump and valve setup for the sequential processing of non-*ortho*-substituted PCBs in 5 blubber samples. The procedure included ball/mill tissue extraction, preliminary GPC separation and cleanup, silica gel chromatography, and automated separation of non-*ortho*-substituted PCBs on AX21 activated carbon/glass fiber with 3 solvent systems and was followed by GC-ECD and GC/MS-EI analysis with selected ion monitoring (SIM). The application of high-resolution capillary column techniques combined with the development of carbon adsorbents for the separation of planar aromatics has become a powerful tool for the identification and measurement of single congeners in complex PCB mixtures and have facilitated the resolution and more accurate quantification of individual congeners and the correction of the presence of non-PCB interferences. But the introduction of such methods in long-term monitoring programs, where one, several, or all peaks in Aroclor standard mixtures were used to estimate total PCBs, raises the problem of comparing results among the data sets. For such purposes a value for total Aroclor can also be reported.

25.4 Concentrations in Field Collections

An increasing number of reports indicate the widespread presence of toxic planar PCB congeners such as the non-*ortho*-substituted planar PCBs 77, 126, and 169. These PCB congeners were detectable in all Finnish food commodities of animal origin sold in Helsinki; in all samples of salmon muscle, cod liver, and seal blubber from the Baltic Sea and environs in the 1980s; and at concentrations between 0.03 and 30.0 μg/kg fat fresh weight (FW) in a wide variety of vertebrates, including fish, marine mammals, dogs, cats, and humans. These and other planar congeners contribute the majority of the toxic potency to PCB mixtures as judged by their ability to induce AHH and EROD. Detection of these toxic residues in field collections from remote areas suggests that planar PCBs are now as widely distributed as other PCB isomers. The clear positive correlation between concentrations of total PCBs and PCBs 77, 126, and 169 in all analyzed mammals suggests that the sources of planar PCB contamination to the environment are mainly commercial PCB formulations.

25.4.1 Nonbiological Materials

Relatively little contamination from PCBs was found in sediments from riverine and pothole wetlands at national wildlife refuges and waterfowl production areas (WPAs) in the north central United States during 1980–82. PCBs were above detection levels (20.0 μg/kg) in less than 4% of the sediments; a similar case was recorded in fish from WPAs. Maximum total PCB concentrations in field collections of nonbiological materials were 0.000028 μg/kg in ice, 0.000125 μg/kg in snow, 12.3 μg/m^3 in air, 233.0 μg/L in seawater, 3860.0 μg/L in sediment interstitial waters, and 1800.0 mg/kg in sediments. Concentrations were comparatively elevated in urban areas, near anthropogenic activities, and at known sites of PCB contamination.

Atmospheric transport is a major route in PCB distribution. Deposition and evaporation studies of 17 PCB congeners in Siskiwit Lake on Isle Royale, Michigan, showed that PCB input fluxes to the lake from rain, snow, and aerosol equalled output fluxes from sedimentation and evaporation. The magnitude of the net vapor flux, calculated by difference and with the assumption that inputs equal outputs, was large and positive for almost all 17 congeners. Low but measurable PCB concentrations (measured as Aroclor 1254) were found in air, snow, ice, seawater, and sediment in the Arctic Ocean north of Axel Heiberg Island off Canada's northern outer coast. PCB residues were relatively high in melted snow (8.0–125.0 pg/L), indicating efficient scavenging from air. The ice island area from which surface seawater samples were collected was well removed from any direct influence of river drainage, and PCB concentrations probably reflect those over much of the Arctic Ocean. PCBs were also monitored in the atmosphere of Ross Island, Antarctica, from March 1988 to January 1990. The geometric mean of total PCBs in air during this period was 15.2 pg/m^3 and the maximum concentration was 12,300.0 pg/m^3. The geometric mean during the Antarctic summers of 1988–89 and 1989–90 was 21.0 pg/m^3. During one sampling period (December 16–28, 1988), PCB levels were about 100 times higher than during any other period, suggesting either irregular, long-range transport of atmospheric pollutants or volatilization of PCBs from a local dumpsite at McMurdo Base on Ross Island. PCB levels did not correlate with seasonal temperature changes, although changes in atmospheric levels were recorded. The absence of seasonal differences may be due to the cold climate and the low vapor pressure of PCBs.

PCB concentrations in sediment cores from Lake Ontario were similar to production and sales data of PCBs in the United States. Annual PCB accumulation rates in the sediments rose from about 2.0 ng/cm^2 in 1950 to about 40.0 ng/cm^2 in the 1966–69 peak years and declined in 1980 to 10.0–20.0 ng/cm^2; about 50% of the 1966–69 load was attributed to the upward mixing by oligochaete worms. Each of several PCB congeners in Lake Ontario sediments contributed 5.7–7.9% of the total PCBs (PCBs 55, 60, 66, and 110) and others (including PCBs 44, 52, 70, 76, 101, and 153) contributed 4.0–4.7%. Surficial sediments of Lake Ontario during 1981–86 contained elevated concentrations of organochlorine compounds, including mirex, chlorobenzenes, octachlorostyrenes, DDT, 2,3,7,8-TCDD, fluorinated aromatic compounds, and PCBs. Among the identified congeners, PCBs 60 (5.7%) and 118 (2.6%) are AHH active. The frequency of isomers of low chlorination decreased with core depth in sediments, remained fairly stable of hexa isomers, and increased with core depth of the more highly chlorinated congeners. This pattern with depth reflects the change in use pattern over time to less highly chlorinated PCBs such as Aroclor 1016. The more highly chlorinated congeners that are disproportionately present in deeper sediments may be due to stronger partitioning to sediment and higher hydrophobicity than those of less-chlorinated congeners and to the positive correlation with their high octanol–water partition coefficients. Anaerobic dechlorination was not evident in these sediments. This contradicts the findings of others who maintain that relatively high levels of less-chlorinated PCBs in the bottom (anaerobic) core sections of Hudson River sediments were due to anaerobic dechlorination. It was concluded that the total mass of PCBs in Lake Ontario sediments was about 50 tons and sufficient to impact Lake Ontario for many years. Lake Michigan sediments, analyzed for 18 planar AHH-active PCBs, total PCBs, 2,3,7,8-TCDD, and 2,3,7,8-tetrachloro-*p*-dibenzofuran (2,3,7,8-TCDF), had elevated levels of PCBs 77, 126, and 169. Results suggest that the contribution of toxic equivalents – that is, the sums of congeners after the raw congener concentrations were normalized by the TEF – was greater from PCBs 77, 126, and 169 than from 2,3,7,8-TCDD and 2,3,7,8-TCDF, even in environments with significant concentrations of dioxins and dibenzofurans. Sediments from the Waukegan Harbor, Illinois, in 1978 contained weathered mixtures of Aroclors 1242, 1248, and 1254. Total PCB concentrations were variable between stations and ranged from 10.6 to 13,360.0 mg/kg DW; 3,3′,4,4′-TCB residues ranged from 5.0 to 27,500.0 μg/kg DW, or about 0.16% of the total PCBs; concentrations were higher of 2,3,3′,4,4′-pentaCB (102.0–131,000.0 μg/kg DW, or 0.66%).

PCB homologs and total PCBs in water, sediments, and biota were measured in Hamilton Harbor on Lake Ontario and Wheatley Harbor on Lake Erie. Hamilton Harbor receives inputs from steel mills and an incinerator plant; Wheatley Harbor receives fish-processing plant wastes. Total PCBs in sediments ranged from 608.0 to 14,185.0 μg/kg DW in Hamilton Harbor and from 166.0 to 1177.0 μg/kg DW in Wheatley Harbor. The high PCB value in Hamilton Harbor is similar to values reported earlier in Lake Erie sediments. Concentrations of lower chlorinated homologs were greater in water than in sediment in both harbors. Homolog patterns in biota (oligochaetes, snails, isopods, and fish) reflected patterns in sediments but not in water. Concentrations of penta- and hexachlorobiphenyls were dominant in the oligochaetes and sediment samples from Wheatley Harbor; concentrations of these homologs also predominated in Hamilton Harbor, but the differences were less pronounced. Particle size distribution of 82–98% silt and clay (63 μm) was similar in both harbors. The concentrations of several metals (Zn, Pb, Cu, and Cr) were much higher in Hamilton Harbor, but their effect on PCB bioaccumulation is unknown.

Contaminated sediments are a major source of PCBs in aquatic environments. PCB discharges prior to 1977 from capacitor-manufacturing plants on the Hudson River at Ft. Edward and Harbor Falls, New York (14.0 kg PCBs daily for 30 years, mostly Aroclors 1242 and 1016), contaminated 306 km of river bed between Hudson Falls and New York Harbor. By 1978, an estimated 63,500.0 kg of PCBs were in the river bank deposits, 134,000.0 kg in the upper river, and 91,000.0 kg in the lower river. PCB concentrations in biota and the water column are largely controlled by PCB concentrations in surficial sediments. Declines in PCB levels in Hudson River sediments corresponded to decreased use of PCBs at local capacitor-manufacturing plants. The stabilization of highly contaminated upper stream banks and reduced PCB releases from bed sediments contributed to lower concentrations in the water column. A model of the fate and accumulation of 7 PCB homologs in the Hudson River estuary showed that 66% of the 270,454 kg of total PCBs discharged into the estuary between 1947 and 1987 had volatilized, 6% was stored in sediments, and the rest was either dredged or lost by boundary transport to the New York Bight and Long Island Sound. Total PCBs peaked in the mid-1970s and upstream loading now has a relatively small effect. The upstream PCB load above Troy, New York, accounts for about 20% of the PCB concentrations in striped bass. A steady decline of Aroclors 1016 and 1254 in upstream contribution was projected to the year 2005 and suggests that 95% of 3- to 6-year-old striped bass in the lower Hudson would have residues below the USFDA action level of 2.0 mg/kg FW within 10 years. PCB congener patterns in Hudson River sediments differed from the Aroclors discharged into the river. Sediments had comparatively enriched concentrations of certain lower chlorinated congeners, especially *ortho*-substituted congeners, and some congeners not usually detected in commercial Aroclor mixtures. Changes in PCB composition of sediments were greater in lower portions of sediment cores than in surficial sediments. Two dechlorination mechanisms seemed to be operating: *meta*, *para*-dechlorinations with stepwise, selective dechlorination at the *meta* and *para* positions of certain *ortho*-substituted di-through tetra-*ortho* chlorinated biphenyls; and *ortho*, *meta*, *para*-dechlorinations, in which the dechlorinations occur at *ortho*, *meta*, or *para* positions and reactivity is favored by increasing electron affinity and relatively positive reduction potential. Both mechanisms preferentially removed toxic congeners. Residues of individual PCB congeners in the upstream water were similar to residues in caddisfly larvae from that system, although residues in the larvae were enriched with lower chlorinated congeners. Because of their relatively high water solubility and low affinity for sediment, higher concentrations of lower chlorine homologs were expected to preferentially dissolve in water. But PCB profiles of dissolved residues transported downstream in the water samples during low flow season were dominated by only three low chlorinated PCBs; more than half of the residues consisted of 2-chlorobiphenyl and 2,2′- and 2,6′-dichlorobiphenyl congeners.

Anoxic Hudson River sediments contaminated with PCBs and dredged sediments in clay encapsulation were treated to induce or raise anaerobic biodegradation of PCBs. Variations in sediment type, bacterial flora, and treatment influenced dechlorination. The addition of biphenyl under anoxic nitrogen was the most effective treatment for raising biodegradation. No evidence was found on accumulation of less-chlorinated PCBs from biodegradation of more highly chlorinated congeners, although a faster degradation rate of less-chlorinated congeners than the more highly substituted PCBs would have allowed this phenomenon to occur unobserved. Hudson River sediments incubated at room temperature (25°C) under a nitrogen atmosphere or incubated with biphenyl enrichment under nitrogen for 7 months decreased significantly in chlorobiphenyl compounds of 33–65%. In general, mono- and dichlorobiphenyl congeners from Hudson River sediments were not significantly degraded by the biphenyl, but biphenyl addition significantly decreased the higher chlorinated congeners. In earlier studies with the same mixed bacterial culture without biphenyl addition, significant reductions were evident in Hudson River sediments of PCB congeners with as many as three chlorines.

Sediments in oceans, estuaries, rivers, and lakes concentrate PCBs. Organisms accumulate PCBs by way of the water column, from interstitial sediment waters, and from consumption of contaminated prey by predator species. Coastal sediments with the highest total PCB concentrations between 1984 and 1989 are usually within 20 km of population centers of more than 100,000 people and directly correlate with sediment content of total organic carbon. Variability in PCB content of sediments is great, and large differences between adjacent stations are not uncommon. The approximate percent PCB distribution in sediments by level of chlorination was di- 3.9%, tri- 11%, tetra- 20%, penta- 24%, hexa- 21%, hepta- 12%, octa- 3.9%, and nona- 1.3%. The mean PCB concentration in sediments at 233 sites was 39.0 μg/kg, and concentrations greater than 200,000.0 μg/kg were considered elevated. In most regions of New Jersey, for example, PCB contamination of sediments and water column was negligible. But PCBs in estuarine and coastal locations of New Jersey showed elevated sediment contamination in the northeastern region – an area that included the Hudson–Raritan estuary and adjacent ocean waters of the New York Bight – where direct discharges from capacitor-manufacturing plants on the upper Hudson River, dredged material, and sewage-sludge dumping were the principal sources of PCB contamination. PCB levels in teleosts from the northeastern region in 1986–87 exceeded the USFDA action level of 2000.0 μg/kg and were consistent with data from previous years. Action levels were also exceeded in fishes from Camden and from the northern coast regions. PCB sediment residues in Raritan Bay, New Jersey, in the 1970s averaged 100.0 μg/kg DW in the range of 3.4–2035.0 μg/kg. Water column concentrations were usually not detectable but approached 0.04 μg/L in parts of the New York Bight after sewage sludge and dredged material were deposited in the 1970s and early 1980s. Sediments near dumpsites in the New York Bight contained up to 15,000.0 μg/kg PCBs, and bottom sediments in the upper Hudson River had localized contaminated areas containing more than 50,000.0 μg/kg. Nationwide, PCB loadings in surficial sediments and mussels (*Mytilus edulis*) in the Long Island Sound were greater than in other sites.

Surficial stream sediments in the San Francisco estuarine system during 1972 had high residues (350.0–1400.0 μg/kg DW) at several locations; however, residues were highly variable between sites. In 1984, low to intermediate concentrations of PCBs (9.0–60.0 μg/kg DW) were measured at four locations in the San Francisco Bay. Overall, congener profiles resembled Aroclor 1254 and were dominated by pentachlorobiphenyl isomers. The patterns of the 11 reported congeners were variable between areas, indicating many sources of different PCB mixtures.

Lake Hartwell in South Carolina received Aroclor 1016 and 1254 discharges for 21 years from a capacitor-manufacturing plant. Analysis of sediment cores collected between 1984 and 1987 showed that samples nearest to the source were relatively high in lower chlorinated PCB congeners, and downstream samples were enriched with the higher congeners; in all samples, total PCB concentrations decreased with increasing distance from the point source. The estimated total remaining PCBs was 41,000.0 kg. The highest PCB concentrations were in subsurface sediments, although PCB levels in surficial sediments were also elevated. A survey of total PCBs in sediments and invertebrates in major estuaries of South Carolina showed no significant contamination. PCBs were found only in sediments (622.0 μg/kg) and in oysters (87.0 μg/kg) from the Wando River near a large ship repair and retrofitting facility; PCB residues in crabs, when present, ranged from 95.0 to 375.0 μg/kg.

25.4.2 Marine Mammals

Marine mammals are the most vulnerable and most probable target organisms to PCBs. The metabolic potential to degrade organochlorine contaminants and therefore accumulate relatively high concentrations of persistent PCBs is lower in marine mammals, particularly in cetaceans, than in terrestrial mammals. Dead harbor porpoises (*Phocoena phocoena*) along the Dutch coast contained 51,350.0–139,790.0 μg total PCBs/kg blubber; 35–50% of the total PCBs were PCBs 138 and 153; a similar pattern was noted in harbor porpoises from Scandinavia during 1987–91, and harbor seals from the northern hemisphere during 1988–95. The intrinsic toxicity of PCBs mainly resulted from the planar PCB congeners that imposed a greater toxic threat to marine mammals than chlorinated dioxins and furans. The toxic threat of planar PCBs to higher aquatic predators was primarily assessed by 2,3,7,8-tetrachlorodibenzo-*p*-dioxin toxic equivalent analysis, which is based on the induction of AHH and EROD. The concentrations of planar PCBs in marine mammals were higher (in the order of di-*ortho*>mono-*ortho*>non-*ortho* congeners) and significantly higher than the levels of toxic dioxins and furans. In particular, the accumulation of PCBs 105 and 126 in carnivorous aquatic mammals is a cause for considerable concern. TCDD-like toxicity is relatively serious in Baikal seals (*Phoca sibirica*) because of the enrichment of toxic PCB congeners in tissues.

Declines in total PCB levels in blubber of marine mammals between the late 1960s and 1990–92 have been noted worldwide, possibly because of the PCB ban in the mid-1970s. PCB levels in animal tissues will probably not decline in the near future because of the greater quantities of PCBs still in use than the quantity that already escaped into the open environment. Temporal changes of PCBs in remote marine waters are slow and could be attributable to the large PCB load in the marine environment. The geographical distribution of planar PCBs with reference to total PCBs did not vary in terrestrial, coastal, and open ocean mammals, whereas those of dioxins and furans decreased from land to ocean. PCB concentrations – including planar PCBs – in blubber of striped dolphins (*Stenella coeruleoalba*) from the Mediterranean Sea during 1990–92 are among the highest reported in the literature. Striped dolphins affected by the western Mediterranean morbillivirus epizootic also contained extremely high concentrations of PCBs (including non- and mono-*ortho* planar congeners) and low immune suppression, suggesting that PCBs were a major factor in this epizootic; however, this needs verification. Planar PCBs in blubber of striped dolphins accounted for about 53% of total PCBs and for virtually all of the potential toxicity. Di-*ortho* planar PCBs accounted for 93.7% of all planar PCB residues (especially PCB 170) in striped dolphins and for 21% of the potential toxicity, mono-*ortho* planars for 6.3% of the residues and 70.8% of the potential toxicity, and non-*ortho* planars (especially PCB 126) for 0.03% of the residues and 8.2% of the potential toxicity. Concentrations of PCBs in tissues of adult male striped dolphins of the same age from the Pacific coast of Japan did not change between 1979 and 1986. However, PCB concentrations in blubber of male striped dolphins were about twofold higher than in females and this is attributed to PCB transfer to offspring by females.

Beluga whales (*Delphinapterus leucas*) of the St. Lawrence River estuary were severely contaminated by DDT metabolites and PCBs; the highest residues were in blubber. High concentrations of PCBs and other organochlorines may impair the immune function in these animals. Several PCB congeners known to be AHH inducers – including PCBs 138 and 153 – were among the major PCB congeners detected in tissues of beluga whales. Because these compounds are not metabolized and persist indefinitely in tissues of the beluga whale, the integrated total exposure to these AAH inducers may be significant. Considerable variations in PCB concentrations by sex, age, and lipid content of the tissues were observed in beluga whales. Qualitative and quantitative differences in the PCB profiles of beluga whales were not only solely related to the respective lipid content of the tissues but also to the specific nature of the lipids, which varied from one tissue to the other. The major PCB components in various tissues of beluga whales were PCBs 52, 99, 129, 137, 141, 153, 165, 180, and 185. Organ-specific retention of some PCB congeners occurs in beluga whales. PCBs 165 and 179, which were minor in blubber, were more abundant in kidney, liver, and lung. PCB 129 was more abundant in kidney and liver than in lung. Atlantic cod (*Gadus morhua*) composed a minor part of the beluga whale diet; however, PCB patterns in cod positively correlated with those in beluga whale tissues, suggesting a common source of intake. PCB congeners 52, 91, 99, 108, 118, 128, 138, 144, 149, 153, 163, and 180 were among the most abundant chlorinated biphenyls in cod liver oil and in the beluga whale blubber, liver, lung, kidney, and milk. Beluga whales from the St. Lawrence River estuary had elevated levels of PCBs in blubber during 1987–90 due to elevated levels in diet, as judged by bioaccumulation ratios. PCB 126 was the most prominent planar PCB in blubber; other biologically active congeners found were PCBs 105 and 118. Significant declines in concentrations of Aroclor PCBs were measured in blubber of male – but not female – beluga whales from the St. Lawrence estuary between the 1982–85 and 1993–94 collection periods; these declines were also evident in eels, harp seals, and seabirds.

When compared to other cetaceans, PCB residues in blubber of harbor porpoises (*Phocoena phocoena*) from the Black Sea showed a measurable sexual difference. PCB concentrations were lower in older female porpoises possibly due to lactational transfer to their calves, while in males the PCB concentrations were positively correlated with increasing age. In stellar sea lions (*Eumetopias jubatus*), the transfer rate of PCBs through lactation was estimated at 80% of the total body PCB burden of adult females. PCB concentrations in liver of the stellar sea lion from the Bering Sea increased with increasing age and correlated positively with those of blubber. In ringed seals (*Phoca hispida*) from the Russian Arctic, lactational transfer of PCBs was estimated at 25% of whole body burden in the mature female, or about one-third that of stellar sea lions. In grey seals (*Halichoerus grypus*), PCBs were transferred from blubber to milk by way of the circulatory system. Of the total concentration of PCBs in various tissues of maternal grey seals, PCB 153 accounted for about 30% in blubber, 20% in serum, and 30% in milk; PCB 180 accounted for 20% in blubber, 20% in serum, and 10% in milk; PCB 77 accounted for 20% in serum. Most of the toxicity in grey seal milk was attributed to PCB 126; the mono-*ortho* chlorinated PCBs 105, 118, 156 and the *ortho*-chlorinated PCBs 170 and 180 – especially PCB 170 – were responsible for 47% of the toxicity. In whales (*Ziphius* sp.), the dominant PCB congeners in blubber had 5–8 chlorines; PCBs 138, 153, 149, 180, and 201 accounted for about 70% of the total PCBs. In the Canadian Arctic food chain of Arctic cod (*Boreogadus saida*) to ringed seal (*Phoca hispida*) to polar bear (*Ursus maritimus*), the total PCB concentrations in μg/kg FW ranged from 4.0 in cod muscle to 680.0 in seal blubber to 4500.0 in bear fat. The hexachlorobiphenyl PCB 153 accounted for 42% of the total PCBs in bear fat and for only 20% in seal blubber and 7% in cod muscle. Tri- and tetrachlorobiphenyl homologs predominated in cod, penta- and hexachlorobiphenyl congeners predominated in seal blubber, and hexa- and heptachlorobiphenyl congeners in fat of polar bears.

25.4.3 Other Aquatic Organisms

In comparison to conspecifics from noncontaminated sites, aquatic invertebrates from PCB-contaminated sites contained elevated concentrations of PCBs in tissues. Adult aquatic insects are one of the groups considered useful and reliable indicators of PCB contamination. Mayflies (*Hexagenia* spp.) from Lake St. Clair had PCB concentrations that reflected sediment PCB concentrations. Observed mayfly sediment concentration ratios from PCBs linearly correlated with K_{OW} when expressed on a logarithmic basis. PCB congener distributions in lake biota showed that no particular trophic level consistently accumulated the highest PCB concentrations and suggest that accumulations were associated with the organism's lipid concentrations. A relation was consistent between the concentration of dissolved PCBs and tissue concentrations in mussels from PCB-contaminated sites, such as New Bedford Harbor, Massachusetts. Uptake of PCB congeners in common mussels (*Mytilus edulis*) from the dissolved phase of seawater was predictable from the log of the BCF and the log K_{OW} of the congeners. American oysters (*Crassostrea virginica*) from Galveston Bay, Texas, contained as much as 1100.0 μg/kg total PCBs DW soft parts, whereas conspecifics from Tampa Bay, Florida contained only 580.0 μg/kg DW soft parts; most (54–94%) of the relative toxicity in both groups was due to PCBs 77, 126, and 169.

The partitioning of individual, highly chlorinated PCB congeners with small differences in K_{ow} values may not adequately explain the accumulations in aquatic organisms. Hydrophobic chemicals, such as PCBs, are accumulated as a consequence of chemical partitioning between the water column, the organic phase of sediment, and biotic lipids or from biomagnification, a process reflecting the ratio between uptake rate from food and elimination rate from the organism. Accumulations of six PCB congeners (PCBs 28, 52, 101, 138, 153, 180) in surficial sediments (0–20 cm) and in an aquatic food chain in Lake Nieuwe Meer – a freshwater lake near Amsterdam containing contaminated dredged materials discharged over 30 years – were low in sediments; elevated in carnivores, plankton, mollusks, crustaceans, and eels; and independent of fat content. With concentrations in organisms expressed on the basis of lipid content (C_{org}) and concentrations in sediments expressed on the basis of organic carbon (C_{sed}) the median C_{org}/C_{sed} PCB accumulation patterns in aquatic organisms showed significant differences and indicated that mechanisms other than partitioning were operating. In plankton and mollusks, C_{org}/C_{sed} ratios seemed to be independent of hydrophobicity of PCB congeners. But with ascending trophic level from plankton to mollusks to crustaceans to eels, the median C_{org}/C_{sed} ratios of higher chlorinated congeners (PCBs 138, 153, 180) increased. Differences in accumulations of individual congeners were attributed to (1) increased biomagnification of higher chlorinated congeners because increasing hydrophobicity decreased elimination rates but not uptake efficiency; (2) greater mobility of eels and different feeding habits of eels and crustaceans that impact accumulation patterns because of biomagnification and partitioning; (3) variability in the time period (limited by lifespan) available in a particular trophic level for equilibration between uptake and clearance; and (4) the tendency of equilibrium to be established at faster rates in less-chlorinated congeners. In crustaceans, C_{org}/C_{sed} ratios decreased with decreasing hydrophobicity; the opposite occurred in eels and is attributed to differences in uptake efficiencies and low elimination rates of lower chlorinated congeners in crustaceans.

A correlation exists between the concentration of lipophilic, hydrophobic, chlorinated hydrocarbons in benthic fishes and the concentration of these compounds in sediments. The correlation is affected by the solubility of the contaminants, as reflected by the octanol–water partition coefficient K_{ow} and the carbon content of the sediment. One study concluded that surface sediments, which change more slowly than the water column, are useful for averaging spatial and temporal contaminant inputs; however, correlations between PCB concentrations in sediment and those in nonbenthic carnivores with limited home ranges are extremely variable. PCBs in a tidal creek in Georgia were traced to Aroclor 1268 used at a former chloralkali plant near the creek. Sediment-ingesting forage fishes, such as the striped mullet (*Mugil cephalus*), efficiently accumulate PCBs and are an important link in the food web transfer of sediment-associated contaminants.

Concentrations of total PCBs – measured as Aroclors 1248, 1254, and 1260 – in adult freshwater fishes in the United States from noncontaminated sites declined between 1976 and 1984, and more than 90% of all analyzed samples contained measurable quantities of PCBs during this period. Total PCB concentrations in domestic freshwater fishes in 1986–87 from contaminated sites were as high as 124,000.0 μg/kg FW. In general, total PCB concentrations in domestic freshwater fishes sampled between 1976 and 1984 were highest in the industrialized regions of the northeast, the Great Lakes, the upper Mississippi River, and the Ohio River. A review was conducted of PCB concentrations (as judged by residues of Aroclors 1016, 1221, 1232, 1242, 1248, 1254, and 1260) in sediments and fish tissues in the United States during 1978–87. During 1978–81, total PCB concentration rankings in sediments were highest in the lower Mississippi, Tennessee, South Atlantic–Gulf of Mexico, and lower Colorado regions and lowest in the Great Lakes, Arkansas, mid-Atlantic,

Pacific Northwest, and Rio Grande regions. During this same period, PCB concentrations in fish tissues were highest in the Missouri, upper Colorado, California, and the Great Lakes regions and lowest in the upper Mississippi, New England, Ohio, Pacific Northwest, Tennessee, lower Mississippi, and Rio Grande regions. Sediment PCB rankings during 1982–87 were highest in the Arkansas, California, Ohio, and Missouri regions and lowest in the South Atlantic–Gulf of Mexico, and Colorado regions. Total PCBs were highest in fish tissues in the upper Mississippi Region during 1982–87. The fish tissue rankings in descending order assigned to 12 regions with sufficient total PCB data were the upper Mississippi and Missouri regions, Ohio, South Atlantic–Gulf of Mexico, Arkansas, Great Basin, lower Colorado, California, Pacific Northwest, Ohio, upper Colorado, and lower Mississippi. PCB rankings between fish tissues and sediments were not necessarily comparable because high levels in sediment do not necessarily result in high levels in fishes if bioconversion was significant. Total PCB concentrations in coastal sediments and fish liver in the United States were highest in the Boston Harbor (17.1 mg/kg DW sediment, 10.5 mg/kg in liver of winter flounder, *Pleuronectes americanus*), San Diego Harbor (0.42 mg/kg DW sediment, 19.7 mg/kg in barred sand bass, *Paralabrax nebulifer*), and Elliot Bay, Washington (0.33 mg/kg sediment, 14.7 mg/kg in flathead sole, *Hippoglossoides elassodon*). Trichloro PCBs were in sediments at many sites but did not accumulate in fish livers except in the Boston Harbor. Sediments in the Boston Harbor, western Long Island Sound, and Raritan Bay were contaminated with PCB mixtures that were relatively high in tri- and tetrachlorobiphenyl isomers, although penta- and particularly hexachlorobiphenyls were the dominant isomers at most sediment sites. As expected, levels of hexachlorobiphenyls in fish livers were dominant because of the more persistent and lipophilic characteristics of increasingly chlorinated PCBs.

Variations in PCB concentrations in sediment and water in the Great Lakes can largely account for the variability in fish PCB residues between different bodies of water. Other variables include fish lipid content, position of the fish species in the food web, and trophic structure of the food chain. Collectively, these variables explain 72% of the variation in PCB concentrations of 25 species of Great Lakes fishes. There is a strong linear correlation between total PCB concentration and percent lipid in five species of Lake Michigan salmonids between 1984 and 1994; however, there is considerable variability among individuals, especially among nonspawning individuals. Tissue concentrations of PCBs in benthic and lower trophic organisms in lakes can be estimated by assuming equal lipid-normalized concentrations in biota and sediment; however, food chain transport had a greater effect on PCB concentrations in higher trophic levels. Total PCB concentrations in whole body of lake fishes were higher among older piscivores and higher with increasing lipid concentration and seemed to reflect exposure conditions at the capture site.

Eggs of chinook salmon (*Oncorhynchus tshawytscha*) from Lake Michigan in 1986 contained AHH-active PCB congeners – including PCBs 77, 105, 118, and 126 – at concentrations from 0.9 to 262.0 μg/kg FW; concentrations of these congeners did not correlate with survival. Mortality of chinook salmon eggs was not related to total PCB concentrations as high as 7020.0 μg/kg FW. In Lake Ontario, the overall trend in total PCB concentrations in whole lake trout (*Salvelinus namaycush*) between 1977 and 1988 was a gradual decline with a half-time persistence of about 10 years. In Lake Superior, the PCB congener fingerprint in eggs of the lake trout differed from that of lake trout eggs of other Great Lakes. A difference between residue patterns was also identified between eggs and the parent fish, suggesting preferential deposition of congeners other than AHH-active congeners. Concentrations of individual congeners in lake trout have been declining at similar rates in the Great Lakes during a 10-year period. In Lake Michigan, total PCB concentrations declined 64% in bloaters (*Coregonus hoyi*) from 5700.0 mg/kg FW in 1972 to 1600.0 μg/kg FW in 1986; however, PCB concentration trends may have been influenced by sampling methodology.

During this period, the bloater population increased 40-fold resulting in a diet shift and a density-dependent decline in growth. The lower growth rates during the 1980s placed older, more contaminated bloaters in the size range most vulnerable to predators and also in the size range sampled by PCB monitoring programs. Future sampling should include a representative sample of bloaters of known age from the population.

Distribution patterns of PCB congeners in water, sediment, and four groups of biota from two lakes in Ontario contaminated by known point sources of PCBs (Lake Clear, Rice Lake) were compared with the congener distribution in Lake Scugog, a relatively clean control lake exposed only to atmospheric inputs of PCBs. Samples were analyzed for 19 PCBs. Those from Lake Clear had a distribution pattern similar to Aroclor 1254 and dominant concentrations of congeners 87, 101, and 118; this lake was contaminated with a PCB mixture similar to Aroclor 1254 in the mid- to late 1970s. The sources to Rice Lake were less clear. Lake Scugog contained a higher proportion of less-chlorinated PCBs, in agreement with another study of atmospheric deposition to isolated lakes. Because the sediments contained elevated levels of organic carbon, the sediments were expected to also hold relatively large concentrations of the higher, more hydrophobic PCBs, in accord with previous reports. But this was not the case; subsequent deposition of total and higher chlorinated congeners into the bottom sediments (organic carbon basis) was unexpectedly low. The proportion of higher chlorinated congeners in sediments were also lower than in biota (lipid weight basis) in all three lakes. Because dissolved organic carbon (DOC) increases the solubility of PCBs in water, the high DOC levels may have caused partitioning of more PCBs into the water and less sorbed onto sediments. The sediments were not efficient at accumulating PCBs, although bottom sediment concentrations were higher in contaminated lakes. Adsorption of PCBs on suspended particles occurred, as anticipated; PCBs on total suspended solids were higher in contaminated lakes (978.0 μg/kg) than in the control lake (49.0 μg/kg) and reflected lake concentrations. In a related study, the concentration and distribution of 19 PCB congeners was determined in biota, sediments, water, and suspended solids of isolated oligotropic lakes in central Ontario that were contaminated by atmospheric deposition. The range of the total congener concentrations was 1.0–2.0 ng/L dissolved in water, 10.0–50.0 μg/kg DW in sediment, 5.0–10.0 μg/kg FW in lower trophic levels, and 10.0–30.0 μg/kg in fishes from upper trophic levels. The high proportion of trichlorobiphenyls previously reported in vapor and hexachlorobiphenyl congeners 153 and 138 in particulate-bound PCBs were reflected in the four study lakes. PCB concentrations (lipid basis) were higher in teleosts than in invertebrate prey organisms.

Winter flounder from the PCB-contaminated harbor in New Bedford, Massachusetts, had grossly elevated concentrations of PCBs in their livers (as high as 333,000.0 μg/kg DW); concentrations were about 5 times higher than in any other fish sample collected worldwide. PCB patterns in the New Bedford Harbor showed high agreement between the exposure environment (water and sediments) and ribbed mussels (*Geukensia demissa*) and mummichogs (*Fundulus heteroclitus*); however, agreements with American eels (*Anguilla rostrata*) or grass shrimp (*Palaemonetes pugio*) were poor because, in part, of differential metabolism of PCBs by these species. PCB concentrations in four species of catfishes from the Mississippi River and its tributaries in summer 1987 were highest from the Illinois River; the Ohio River at Olmsted; and the Mississippi River at Helena, Arkansas, and Arkansas City, Arkansas. These sites seem to be point sources of PCB pollution because PCB residues in catfishes above and below these sites were lower. Although PCBs were banned in 1978, the elevated levels in catfishes suggest PCB leakage from hazardous waste sites with transformer and hydraulic fluids and flame-resistant plasticizers. Findings of high (greater than 4000.0 μg/kg FW) total PCB levels in mature roe samples of the paddlefish (*Polyodon spathula*) from the Ohio River warranted warnings of the general public about consuming this domestic caviar.

The upper Hudson River was massively contaminated with PCBs from an industrial plant for several decades prior to 1975. All fishing in this section in 1976 was banned because of PCB contamination. The prohibition is still in effect because, in part, of measurable PCB residues in caged fishes from this area. Striped bass (*Morone saxatilis*) collected near Troy and Albany, New York, contained higher concentrations in muscle of PCB 77 (37.0 μg/kg FW) and PCB 126 (8.0 μg/kg FW) than conspecifics from other locations in New York. Almost all (99%) the PCB toxicity in muscle of striped bass was attributed to PCBs 77, 105 (62.0 μg/kg FW), and 126.

The most prominent PCB congeners in muscle from 14 species of Wisconsin fishes in 1986–87 were PCBs 28, 31, 66, 70, 76, 95, 101, 105, 110, 118, 138, 146, 149 and 180. Congeners 105 and 118 were found in the greatest amount in fishes at 1–5% of the total PCB concentration of each. Congeners with responses similar to 2,3,7,8-TCDD, that is, the planar PCBs, were seldom present above detection levels. The sum of the individual congeners measured in Wisconsin fish muscle was similar to total recorded PCB values. Increased fish consumption by Wisconsin anglers in 1985 positively correlated with increased human serum PCB concentrations. Human consumers of Wisconsin game fishes (chinook salmon, *Oncorhynchus tshawytscha*; yellow perch, *Perca flavescens*; walleye, *Stizostedion vitreum*) in 1986 contained various PCB congeners in their sera. PCB 153 (78% frequency of occurrence) was present at 1.46 (0.6–7.3) μg/L human serum, PCB 138 (56%) at 1.32 (0.6–6.0) μg/L, PCB 180 (42%) at 1.06 (0.6–3.5) μg/L, PCB 118 (34%) at 1.12 (0.6–5.7) μg/L, PCB 187 (11%) at 0.98 (0.6–2.2) μg/L, PCB 170 (5.8%) at 0.86 (0.6–1.4) μg/L, PCB 28 (1.2%) at 0.8 μg/L, PCB 101 (0.58%) at 0.8 μg/L, PCB 70 (0.58%) at 0.7 μg/L, and a single planar PCB – PCB 77 – (0.58% frequency of occurrence) at 1.3 μg/L. PCBs 118, 138, and 180 are potentially most toxic to human consumers, as judged by the concentrations of these congeners in human sera.

Concentrations of PCBs in female northern pike (*Esox lucius*) from a Scandinavian lake decreased with increasing age, weight, or body length. Seasonal elimination of the lipophilic contaminants in roe – which contained as much as 10 times more fat than muscle and more than 10 times the amount of pollutants than muscle – is the major route of PCB loss. Male northern pike contained higher concentrations of PCBs than females because of the lower elimination by way of gonadal products; males showed no significant relation between age and PCB burdens in tissues. Total PCB levels of 7700.0–34,000.0 μg/kg LW in eggs of the Arctic char (*Salvelinus alpinus*) from Lake Geneva, Switzerland, correlated with a mortality rate of 29–100%.

On the Pacific coast and in adjacent areas of Mexico, data from more than 150 survey and monitoring programs were summarized on contamination of sediments, invertebrates, and fishes. PCBs in sediments seem to be reflected in mussels, and PCB residues from mussels collected at harbor entrances remained unchanged or were increasing. The harbors in Los Angeles-Long Beach and San Diego remained contaminated with PCBs, and PCB concentrations in sediments were reflected in fish livers. Waste management seems to have been effective in the Palos Verdes outfall area. Sediments and mussel samples in Palos Verdes from 1974 to 88 showed decreasing PCB levels that reflected a 100-fold reduction in PCB wastewater emissions during that period. Contamination of the coastal zone declined to levels found 30 and 40 years ago. PCB levels had declined at least one order of magnitude in teleosts and shellfish at offshore sites since the 1970s. Bays and harbors were more contaminated than the open coastal zone and must be monitored more closely; lower detection levels (0.001–0.01 mg/kg FW vs. the current analytical limits of 0.02 mg/kg FW) were proposed to monitor the effectiveness of current source control programs.

PCBs in Puget Sound, Washington, were measured in sediments, fish livers, and benthic invertebrates. Maximum total PCB concentrations were 2100.0 μg/kg DW in sediments near Tacoma and 32,000.0 μg/kg DW in crab hepatopancreas and 35,000.0 μg/kg DW in fish liver near Seattle. PCB concentrations in sediments of the Puget Sound in May 1988 positively correlated with PCB concentrations in livers of several species of flatfishes in these sediments. Increased sediment PCB concentrations also correlated well with increased hepatic AHH and EROD activities and with increases in total hepatic GSH, all of which are acknowledged early indicators of chemical contamination by PCBs and other organic contaminants.

PCB residues in liver of the European flounder (*Platichthys flesus*) were extremely variable, but residues of individual congeners were usually higher in fall, higher in females, and higher in flounders captured inland near a PCB point source; a similar pattern was documented in the Atlantic cod (*Gadus morhua*). In the dab (*Limanda limanda*), a marine flatfish, the accumulation of PCBs 128, 138, and 163 differs significantly by sex. Depletion of lipids from the liver of female dabs during ovary maturation is an important excretory pathway for PCBs during spawning. PCB levels in liver of dabs were higher in spring than in winter; livers and ovaries were dominated by penta- and hexachlorobiphenyls, but the dominant PCBs in testes were tri- and tetrachlorobiphenyls.

The most prominent PCB congeners at 280.0–323.0 μg/kg DW in the tilefish (*Lopholatilus chamaeleonticeps*) from Georges Bank during 1981–92 were PCBs 138 and 153 in gonad and liver; at 69.0–82.0 μg/kg DW the most prominent PCB congeners in the tilefish from New Jersey during this same period were PCBs 138 and 153 in liver. Total PCB concentrations in marine coastal fishes were dominated by the hexachlorobiphenyls, but trout from isolated mountain lakes had tri-, tetra-, and pentachlorobiphenyls as the major components of total PCBs.

25.4.4 Reptiles

PCBs accumulate in the fat, testes, and brain of snapping turtles (*Chelydra serpentina*), and concentrations seem to reflect the lipoprotein solubility of individual congeners. With increasing hydrophobicity (increasing K_{OW}) of PCB congeners, accumulations increased in livers of snapping turtles; total liver PCB concentrations in adults increased with increasing age, length, and weight. PCB loadings in snapping turtle eggs were not related to the body size of females or to the number of eggs in the clutch. However, a positive relation between PCB loadings in liver of adult female snapping turtles and their eggs was significant.

PCB 105 may be an important contributor to the toxic burden of snapping turtle populations. Eggs of snapping turtles from the Great Lakes had a lower hatch rate and a significantly increased frequency of deformed hatchlings than eggs from a control site, and this seemed to be strongly associated with total PCB concentrations and PCB 105. Of the 5 toxic PCB congeners measured in the yolks, egg whites, and shells of snapping turtle eggs, PCBs 105 and 167 accounted for more than 99% of the total toxicity – as measured by 2,3,7,8-TCDD TEF equivalents – and 95% of the total toxicity resided in the yolk. Large reserves of fat in eggs of the snapping turtle do not seem to protect against toxic PCB congeners from being dispersed into egg components low in fat.

In loggerhead turtles (*Caretta caretta*), PCB concentrations in the chorioallantoic membrane correlate closely with whole egg PCB concentrations. The authors recommend the use of chorioallantoic membrane tissues as a nonlethal methodology for predicting PCB concentrations in sea turtle hatchlings.

25.4.5 Birds

Embryos of the double-crested cormorant (*Phalacrocorax auritus*) exposed in ovo to elevated mixtures of PCBs in the environment were 25 times more likely to hatch with asymmetric brains than did those from reference sites. A high frequency of dead and deformed embryos of double-crested cormorants and Caspian terns are documented in the upper Great Lakes during 1986–91. Half the embryos found dead in eggs were deformed. Only 1 of 10 cross-billed cormorant embryos survived to hatch and no bill-deformed terns hatched, although tern embryos had a higher frequency of crossed bills than did cormorants. Planar PCB congeners were present at sufficient concentrations to cause the observed effects.

A severely deformed bill of the type associated with high environmental levels of PCBs was observed in a newly hatched chick of the shag (*Phalacrocorax aristotelis*); the two dominant congeners in the chick, accounting for about 57% of the total PCB body burden, were PCBs 153 (35%) and 138 (22%), both of which are known to show selective biomagnification.

In general, total PCB concentrations in birds were usually higher in males and in eggs than in livers, in adipose tissues, in fish-eating species, and at PCB-contaminated sites; PCBs 138 and 153 tended to predominate in all samples. The change in PCB content in livers of Norwegian raptors between 1965 and 1983 was not significant despite a marked reduction in the use of these compounds. When total PCB concentrations declined, for example, in eggs of red-breasted mergansers (*Mergus serrator*) between 1977 and 1990, the relative potency of the mixture of PCBs – as measured by 2,3,7,8-TCDD equivalents – was unchanged.

Commercial PCB mixtures frequently contain impurities that may contribute to the 2,3,7,8-TCDD toxic equivalency factor (TEF); these impurities may include other PCBs, dioxins, dibenzofurans, naphthalenes, diphenyl ethers and toluenes, phenoxy and biphenyl anisoles, xanthenes, xanthones, anthracenes, and fluorenes. PCB concentrations in avian tissues sometimes positively correlate with DDE concentrations. Eggs of peregrine falcons (*Falco peregrinus*) from California, for example, contained measurable quantities of various organochlorine compounds, including dioxins, dibenzofurans, mirex, hexachlorobenzene, and *p,p′* DDE at 7.1–26.0 mg/kg FW; PCB 126 accounted for 83% of the 2,3,7,8-TCDD equivalents, but its interactions with other detectable organochlorine compounds is largely unknown.

There is a relation between PCB uptake and the position of the species in the food chain. In a 3-step central-European oak-forest food chain involving the great tit (*Parus major*), caterpillars (*Tortrix viridana*, *Operophtera brumata*, *Erannis defoliara*), and leaves of the red oak (*Quercus* sp.), mean concentrations of PCB 153 – the most abundant measured congener – rose from about 1.0 μg/kg DW in leaves to 10.0 μg/kg in caterpillars to 170.0 μg/kg in bird eggs. Older juvenile tits contained 307.0 μg of PCB 153/kg whole body DW; these birds received PCBs from the mother during egg transfer and from the caterpillar food source during the nesting period. PCBs 101, 138, and 180 were also present in most samples but at lower concentrations than PCB 153. Populations of *Parus major* in this area declined in recent years, and the influence of anthropogenic contaminants may be a factor. Fish-eating waterfowl and seabirds had comparatively high total PCB and high planar PCB concentrations in eggs and tissues; waterfowl and seabirds that feed mainly on invertebrates had lower PCB concentrations. PCB concentrations were higher in adipose tissues of the Arctic tern (*Sterna paradisaea*) than in those of their fish and invertebrate food items. PCB concentrations in adipose tissues of cormorants, when compared to their diet of fishes, were 10–100 times higher than marine fishes and 100–1000 times higher than freshwater fishes. Double-crested cormorants (*Phalacrocorax auritus*) biomagnify total PCBs from their fish diet to their eggs – based on 2,3,7,8-TCDD equivalents – by a factor of 31.3. Higher chlorinated PCBs accumulated in tissues of the herring gull (*Larus argentatus*) to a greater extent than were present in the alewife (*Alosa pseudoharengus*), a primary food item; lower chlorinated biphenyls, including the tetra- and penta-CBs, did not biomagnify.

PCBs can move from local sediments into the avian food web, as judged by PCB accumulation rates of tree swallows (*Tachycineta bicolor*) from contaminated and reference sites. Patterns of relative concentrations of PCB congeners change from sediment to invertebrates, and from tree swallow eggs to nestlings. Dioxin-like activity (TEF) measured in tree swallow tissues could predict TEF in sediments and the reverse. Models of dioxin-like activity in the sediments of Saginaw Bay, Michigan, predicted that sediments were not harmful to tree swallows from that area.

Declining populations of Caspian terns (*Sterna caspia*) – especially populations nesting in Green Bay and Saginaw Bay between 1986 and 1990 – were associated with elevated PCB concentrations in blood; the frequency of developmental abnormalities and deformities in Caspian tern populations at Saginaw Bay was almost 100 times above that recorded in the same area between 1962 and 1972. High concentrations of PCB 126 found in eggs of the bald eagle (*Haliaeetus leucocephalus*) are nearly 20-fold higher than the lowest toxic concentration tested in American kestrels (*Falco sparverius*) and may be a factor in the decline of some eagle populations. High PCB concentrations in tissues of white-tailed eagles (*Haliaeetus albicilla*) are directly connected to high concentrations in eggs and associated with eggshell thinning and low reproductive success. A total lack of reproduction among white-tailed sea eagles in the coastal area of the southwestern Baltic Sea in the 1960s and 1970s may be related, in part, to high concentrations of PCBs 105, 118, 126, and 156 in tissues of adult eagles. It is noteworthy that concentrations of planar PCBs in adult white-tailed sea eagles were among the highest reported in wildlife and that total PCB concentrations in this species were similar to those reported in dead eagles from Sweden and Finland in the 1960s and 1970s.

PCB 153 is the most widespread PCB in the environment because it is easily stored and retained in adipose tissue; PCB 153 was the main PCB congener in eggs of eight examined species of Italian waterfowl and accounted for 11.4–21.2% of the total PCB concentration. Infertile eggs of the endangered imperial eagle (*Aquila heliaca adalberti*) contained as much as 28.9 mg total PCBs/kg FW; PCB 153 constituted 13.5% of the total PCB loading, PCB 180 13%, PCB 138 3.2%, PCB 101 3.2%, and PCB 118 0.7%. In the endangered Audouin's gull (*Larus audouinii*), most (62%) of the total PCB burden consisted of PCBs 138, 153, 170, and 180; other important congeners were PCBs 118, 194, and 203, and each contributed about 5%. PCBs 138, 153, and 180 comprised more than 50% of the total PCB burden in eggs of the yellow-legged herring gull (*Larus cachinnans*); a similar case is made for eggs of other species of marine birds. PCBs 138, 153, and 180 were also dominant in tissues of most birds collected in Great Britain between 1988 and 1990, although total PCB concentrations ranged from 0.02 to 105.0 mg/kg FW and also differed considerably in different tissues from individual birds. PCBs 138 and 153 were the most prominent congeners in eggs of 3 species of gulls collected in Spain during 1988, accounting for 10.5 and 8.7%, respectively, of the total PCB burden; other important congeners were PCBs 180 (7.5%), 170 (3.2%), 101 (1.9%), 151 (1.1%) and 194 (0.9%).

PCB signatures in bird eggs are not constant. Eggs of the dipper (*Cinclus cinclus*) from Wales and Ireland were dominated by PCB 118 in 1990, PCB 170 in 1991, and PCB 153 in 1992; 6 congeners accounted for 26–35% of the total PCBs in Welsh eggs and for 10–26% of the total in eggs from Ireland. In tissues of birds in Great Britain, the mono-*ortho* congeners – PCBs 105 and 118 – made a high contribution (70%) to the TEF, whereas the non-*ortho* congeners (PCBs 77, 126, 169) contributed 20% and the di-*ortho* congeners (PCBs 138, 153, 180) contributed 10%. Young of all avian species sampled in Wisconsin accumulated PCBs 77, 105, 126, and 169. Chicks of Forster's terns (*Sterna forsteri*) had daily uptakes of 15.0 μg total PCBs, 0.07 μg PCB 77, 0.2 μg PCB 105, 0.006 μg for PCB 126, and 0.00014 μg PCB 169.

Concentrations of mono-*ortho* PCBs in yolk-sac of cormorants ranged from 10.0 to 250.0 mg/kg lipid weight (LW); high PCB residues in yolk were associated with increased cytochrome P450 and EROD activities and decreased thyroid hormone activity. Embryos of the black-crowned night heron (*Nycticorax nycticorax*) with the greatest burdens of total PCBs had increased cytochrome P450-associated monooxygenase activities and cytochrome P450 proteins, which suggest that cytochrome P450 may be a useful biomarker of exposure to some PCB mixtures. The absence of established thresholds for P450 induction indicates that more research is needed to make this a useful technique for evaluating PCB exposure.

25.4.6 Terrestrial Mammals

The highest total PCB concentrations recorded in terrestrial mammalian wildlife occurred in fat and liver tissues of species collected near urban areas; di-*ortho* congeners were the major contributors to PCB tissue burdens. Atmospheric transport of PCBs governed uptake in terrestrial mammalian herbivores and predators; for example, PCB residues in tissues of voles and shrews in the Scandinavian Peninsula directly correlated with fallout loadings. An increase in atmospheric deposition of PCBs increased PCB burdens in plants, herbivores, and predators of the herbivores. But herbivores and predators differentially metabolized PCBs, raising concentrations of highly chlorinated congeners in predators and concentrations of the more easily metabolized low-chlorinated PCBs in herbivores.

Populations of mink (*Mustela vison*) declined in many areas of the world, and the declines were linked to exposures to synthetic halogenated hydrocarbons. In the Great Lakes region, mink density is lower along the shores of the Great Lakes and their tributaries where minks have access to fishes from the Great Lakes. Tissue PCB concentrations and their dioxin TEFs were considered critical in the hazard assessment of PCBs. Mink that consumed fishes below dams in Michigan were 10–20 times more likely to suffer PCB damage than mink consuming fishes from above the dams, as judged by the elevated concentrations of total PCBs and dioxin TEFs in fishes from below the dams. European polecats (*M. putorius furo*) collected in the Netherlands between 1985 and 1990 had PCB patterns that were independent of diet and seemed to be controlled by anal gland secretions containing elevated PCB residues. Juvenile polecats contained higher PCB concentrations than adult males and females, and this is attributed to an increased elimination of PCBs by adults through anal gland secretions. In all examined polecat tissues, PCB 126 accounted for 63–98% of the 2,3,7,8-TCDD toxic equivalents. Polecats – unlike weasels (*Mustela nivalis*), stoats (*Mustela erminea*), and otters (*Lutra lutra*) – can readily metabolize PCBs 126 and 169. As a consequence, PCBs 126 and 169 are selectively retained in the livers of weasels, stoats, and otters when compared to polecats. Polecats are less sensitive to PCBs than other mustelids when dietary PCB intakes are similar; however, otters are exposed to high dietary concentrations of PCBs when compared to other species of mustelids and are considered the most vulnerable of all mustelids to PCBs.

The use of PCBs in Germany was prohibited in 1983. From 1983 to 1991, the body fat of red foxes (*Vulpes vulpes*) in Germany showed a reduction in the mean concentration of highly chlorinated PCBs (PCBs 138, 153, and 180) but an increase in the lower chlorinated congeners (PCBs 24, 49, and 52). These findings suggest a trend toward a reduction of environmental contamination with highly chlorinated biphenyls since 1983, perhaps as a consequence of metabolic degradation, whereas contamination with lower chlorinated biphenyls from diverse sources is increasing. Low-chlorinated congeners that are metabolized via reactive intermediates must be evaluated because they show weak tumor-initiating properties. At present, PCB congeners that are considered as indicators of contamination in Germany include PCBs 28, 31, 52, 101, 138, 153, and 180.

Populations of bats in Europe have been declining, and PCBs together with pesticides and wood preservatives are the suspected main causes of the decline. Three species of bats collected in Spain during 1988–90 contained only a few dominant PCB congeners; PCBs 138, 153, and 180 accounted for about 80% of the total PCB burden in whole bats. But the most abundant PCB congeners in brain and liver of European otters (*Lutra lutra*) were in the descending order of PCBs 163, 153, 138, and 170, each constituting at least 10% of the total PCB burden.

Total PCBs in adipose tissue of polar bears (*Ursus maritimus*) throughout their known range, and collected between 1989 and 1993, are as high as 24.3 mg/kg on a lipid weight basis and are dominated by PCBs 153 (46% of total PCBs), 180 (18.5%), 170 and 190 (8.6%), and 99 (8.3%). Total PCBs are higher in adult males than females, and there was no significant trend with age. PCB content was higher in bears from Svalbard, East Greenland, and the Arctic Ocean near Prince Patrick Island than most other areas. Atmospheric, oceanic, and ice transport may contribute to the high concentrations of total PCBs. The PCB composition in tissues of polar bears suggest that polar bears – unlike other mammals – can readily metabolize PCB congeners with unsubstituted *para* positions and unsubstituted adjacent *ortho–meta* positions. Six PCB congeners (PCBs 99, 138, 153, 170, 180, 194) – all with a minimum 2,2′,4,4′-chlorine substitution – accounted for about 99% of the total PCB content in liver and 87% in fat; PCB 153 accounted for 37% of the total PCB loading in liver. The PCB congener pattern in polar bear liver and adipose tissue is similar and seems to be independent of sex, age, nutritional status, collection locale, and PCB body burden.

In humans, total PCB concentrations in maternal milk were elevated (>1.1 mg/kg milk fat) in 4 of 122 cases in the New Bedford Harbor vicinity, Massachusetts. At least one female was occupationally exposed, as judged by the congener profile and history. PCB exposures from fish consumption were likely, but not from residence adjacent to a PCB-contaminated site. In all 4 cases, their newborns were full-term and healthy.

25.5 Lethal and Sublethal Effects

In all tested organisms, PCBs – especially PCBs with 2,3,7,8-TCDD-like activity – adversely affected patterns of survival, reproduction, growth, metabolism, and accumulation. Common manifestations of PCB exposure in animals include hepatotoxicity (hepatomegaly, necrosis), immunotoxicity (atrophy of lymphoid tissues, suppressed antibody responses), neurotoxicity (impaired behavior and development, catecholamine alterations), increased abortion, low birth weight, embryolethality, teratogenicity, gastrointestinal ulceration and necrosis, bronchitis, dermal toxicity (chloracne, edema, hyperplasia), weak mutagenicity at high doses, and preneoplastic changes at low doses. PCBs can mimic female hormones in human cell lines and other mammalian systems and may be associated with learning deficiencies and lowered IQs in children. At concentrations above a threshold, PCBs are potent promoters of hepatic carcinogenesis in laboratory rodents; however, there is no clear evidence of carcinogenicity of PCBs to human and animal populations from natural exposure. Induction of hepatic microsomal enzymes is one of the earliest and most sensitive responses to PCBs.

PCB-induced toxicity patterns are highly variable. Variability, as discussed later, is attributed in part to differences between species and strains in the ability to metabolize PCBs and in primary sites of action; in the age, growth rate, biomass, and lipid content of the species; in dose rate, duration of exposure, route of administration, and tested congeners; in physicochemical characteristics of the habitat during exposure; and in PCB interactions with other PCBs, other organochlorine compounds, and heavy metals. Chinook salmon (*Oncorhynchus tshawytscha*) had decreased hatch when eggs contained the equivalent of 0.1 μg 2,3,7,8-TCDD/kg fresh weight (FW); domestic chickens (*Gallus* sp.) had decreased survival and increased developmental abnormalities when embryos had 20.0 μg PCB 77/kg FW; mink (*Mustela vison*) had reduced growth when fed 100.0 μg Aroclor 1254/kg BW daily and reduced survival at 50.0 μg PCB 169/kg diet; rhesus macaques (*Macaca mulatta*) had reproductive impairment when fed more than 8.0 μg Aroclor 1016/kg BW daily; and rats (*Rattus* sp.) had reduced litter sizes and survival when given 10.0 μg PCB 126/kg BW daily during gestation.

25.5.1 Aquatic Organisms

PCBs influence patterns of survival, reproduction, growth, enzyme activities, and accumulation in representative aquatic organisms. Some PCB congeners at laboratory concentrations that were several orders of magnitude higher than those encountered under field conditions killed 47–83% of tested freshwater fishes and invertebrates in 24–48 h; however, most PCB congeners tested produced negligible mortality under these conditions. Mortality increased when PCB 133 or PCB 177 concentrations in whole guppies (*Poecilia reticulata*) exceeded 1.0 μM/g, equivalent to more than 200.0 mg PCB/kg whole body FW or about 4000.0 mg PCB/kg on a lipid weight basis. PCBs – especially those with TCDD-type activity – adversely affect reproductive success of spawning female chinook salmon. Chinook salmon eggs that contained total PCB concentrations equivalent to 0.1 μg 2,3,7,8-TCDD equivalents/kg eggs and higher had a dose-dependent decrease in hatching success. PCBs also impair the reproductive capacities of marine mammals.

The relation between PCB accumulation by the freshwater alga *Scenedesmus* sp. and the compound's octanol–water partition coefficient (K_{ow}) was measured with 40 PCB compounds in a log K_{ow} range of 4.46–8.18, PCB concentrations between 0.03 and 1.1 μg/L, and exposures between 20 and 30 days. The accumulation process was consistent with partitioning from water into cell lipids but was slower than the growth of *Scenedesmus* (i.e., no significant uptake of PCBs from congeners with log $K_{ow}>5.0$ under conditions of rapid growth or >log 7.0 under conditions of slow growth). Thus, under nonwinter field conditions, many PCB congeners never reached equilibrium concentrations. Similar results are reported for other species of freshwater algae, including *Selenastrum capricornutum*, *Anabaena* sp., and *Synedra* sp. High algal densities alter BCFs of PCBs. Bioconcentration factors for PCBs 15, 52, and 153 by the alga *Chlorella pyrenoidosa* were higher at comparatively low algal densities, suggesting that BCFs determined at high algal densities should be applied with caution to field situations.

Zebra mussels (*Dreissena polymorpha*) accumulated PCB 77 from their diet and from the surrounding lake sediments. An uptake rate of PCB 77 by zebra mussels followed the descending order of sediment, food, and water. Tissue concentrations in mussels peaked after 10–14 days at 3.4–3.7 mg PCB 77/kg FW soft parts; equilibrium levels of PCB 77 were near 1.0 mg/kg FW. Zebra mussels are more efficient accumulators of PCBs than other bivalve molluscs to which they are attached; accordingly, high densities of zebra mussels probably influence contaminant dynamics. A freshwater crustacean (*Mysis relicta*) plays an important role in the transfer of PCBs from sediments into the Lake Champlain food web, and freshwater grazing and shredding benthic invertebrates promote downstream transport of PCB 153. Marine invertebrates accumulated PCBs 52, 101, 128, 138, 151, 153, 180, 194, 206, and 209 from PCB-contaminated sediments. Clams (*Macoma nasuta*) reached a steady-state equilibrium in 10 days, but sandworms (*Nereis virens*) took 70–120 days. Clams showed preferential accumulation of lower molecular weight PCB congeners, and this may be due to the comparatively low lipid content in this species. Sandworms and grass shrimp (*Palaemonetes pugio*) metabolized PCBs 52, 101, and 151.

Golden shiners (*Notemigonus crysoleucas*) rapidly accumulated radiolabeled PCBs from water during a 96-h exposure. The uptake rate of PCBs from water was controlled by gill blood-flow rate. About 50% of the accumulated PCBs in shiners was eliminated in 4.9 days, and this is similar to PCB elimination rates in striped bass (*Morone saxatilis*) and channel catfish (*Ictalurus punctatus*). In guppies, residual PCB concentrations increased with increasing duration of exposure; however, steady-state concentrations did not occur during dietary exposure for 65 days. PCB congeners with *ortho*-chlorine substitutions (PCBs 77, 105, 118, 128, 138) were effective inducers of EROD (7-ethoxyresorufin *O*-deethylase) and AHH activities in marine mammals and freshwater fishes but were ineffective at doses as high as 10.0 mg/kg BW in scup (*Stenotomus chrysops*), a marine teleost. Industrial mixtures containing both planar and nonplanar PCBs induced AHH in fishes. Mixtures of planar PCBs and dioxins, however, produced synergism of AHH activity in fish liver at low doses and antagonism at high doses; the possible antagonistic effects of nonplanar halogenated compounds may further complicate these interactions. Liver is the primary target organ for the induction of cytochrome P450-dependent monooxygenases by PCBs in fishes, and the most frequently examined organ; however, in salmonids, muscle tissue is also suitable for evaluation of hepatic monooxygenase induction, as judged by PCB concentrations in muscle.

25.5.2 Birds

PCB 126 is among the most toxic of all PCB congeners to birds and the domestic chicken is the most sensitive tested species. However, adverse effects of PCBs in birds vary markedly between species and tissues. And birds react differently to different PCB congeners and to PCB–metal mixtures. American kestrels (*Falco sparverius*) differ from Japanese quail (*Coturnix japonica*) after exposure to PCBs 105, 126, and 153; quail accumulated porphyrins in liver but kestrels did not. Japanese quail dosed with PCB 47 or 77 showed marked differences between the abilities of the small intestines and liver to metabolize porphyrin and the induction of cytochrome P450 isozymes and associated monooxygenases. Japanese quail fed different dietary concentrations of Aroclor 1260 for 25 days had a significant dose-dependent correlation between liver monooxygenase activity and PCB levels in blood. EROD and porphyrin-induction responses of primary hepatocytes to various PCBs in newly hatched chickens, gulls, terns, and cormorants followed a concentration-dependent increase with significant variability in response between species. Pure hexachlorobiphenyls (HCBs) caused uroporphyrin accumulations and increased delta-aminolevulinic acid synthetase activity in chicken livers, and some HCBs significantly increased cytochrome P450 and *p*-nitrophenol glucuronyl transferase when given in the feed for 3 weeks at 400.0 mg/kg ration; however, PCBs 128 and 169 caused greater accumulations of hepatic porphyrins than PCBs 138, 155, and 156. PCBs 77, 136, 153, and 159 produced acute histopathological changes in chick embryo livers and selectively induced cytochrome P448-mediated mixed function oxidases; the degree of histopathologic change produced by each of the tested PCBs positively correlated with the degree of P448 inhibition. Interactions of metals with PCBs are not well documented, although some studies with Japanese quail showed that cadmium raises PCB uptake from the diet. In one study, cadmium fed at high dietary levels of 100.0 mg/kg ration interfered with high levels of dietary PCBs (100.0 mg/kg ration). In that study, muscle of treated quails had increased loadings of congeners chlorinated in the 2,4,5 position (such as PCBs 138, 153, 170, and 180), and these are comparatively toxic and resistant to metabolic degradation. More research is needed on variables known to modify PCB uptake, retention, translocation, and toxicity.

Embryos of the domestic chicken (*Gallus* sp.) are unusually sensitive to PCB 77. Mortality and a high incidence of developmental abnormalities – including microphthalmia, beak deformities, edema, and retarded growth – were recorded in chicks at 0.02 mg PCB 77/kg egg FW, but no deaths or abnormalities were recorded in embryos of ducks, pheasants, and gulls at 0.1–1.0 mg PCB 77/kg egg FW. Chicken embryos were 20–100 times more sensitive to PCB 77 than embryos of the wild turkey (*Meleagris gallopavo*), and this sensitivity emphasizes the uncertainties of applying toxicity data from one species of bird to predict toxic effects in other avian species. Differences in sensitivity of birds to PCB 77 may be related to differences in metabolism and in the formation of toxic metabolites and to the increased availability of Ah receptors in chicks. For example, Ah receptors were detected in livers of 7-day-old chicken embryos but not in livers of 9-day-old turkey embryos. Chicken embryos were extremely sensitive to PCB 126 when compared to embryos of the American kestrel and common tern; LD50 values, in μg/kg FW egg, were 0.4 in chickens, 65.0 in kestrels, and 104.0 in terns; EROD induction in chicken embryo liver by PCB 126 was about 800 times more responsive than in tern and at least 1000 times more responsive than in kestrels.

Birds and mammals exposed to PCB mixtures frequently react differently. PCBs generally elicit large-colloid goiters in birds; these goiters are inherently different from hyperplastic goiters produced in mammals exposed to PCBs. Fish-eating seabirds, such as the razorbill (*Alca torda*), can rapidly metabolize PCB congeners that have at least one pair of adjacent unsubstituted *meta–para* combinations in the biphenyl moiety. Razorbills metabolized 4-chlorobiphenyl to 4-chloro-4′ hydroxybiphenyl; however, mice (*Mus* sp.) metabolized the same compound 15 times faster. PCB-exposed razorbills and rock doves (*Columba livia*) had similar concentrations of cytochrome P450 and glutathione-*S*-transferase enzymes, but concentrations were significantly higher in rats (*Rattus* sp.) than either avian species. The relative potencies of tested PCBs – as measured by EROD activity of microsomal liver enzymes – in fertile eggs of the domestic chicken were 0.02 for PCB 77, and less than 0.001 for PCB 169; these values differ somewhat from those proposed for mammals of 0.0005 for PCB 77, and 0.01 for PCB 169. Lymphoid development in the bursa of Fabricius of the avian embryo is inhibited by TCDD-like congeners. PCB 77, for example, affects the immune system by interacting with the Ah receptor, causing inhibition of lymphoid development in the mammalian thymus and in the avian bursa of Fabricius. More research seems needed on the relation of avian Ah receptors to natural physiological processes.

25.5.3 Mammals

Deleterious effects were significant on growth, survival, reproduction, or metabolism from chronic exposures of sensitive species of tested rodents, primates, and mustelids to daily concentrations as low as 0.008 mg/kg BW of Aroclor 1016, 0.01–0.02 mg/kg BW of PCB 126, 0.01–0.05 mg/kg diet or 0.1 mg/kg BW of PCB 169, 0.09 mg/kg BW of Aroclor 1246, 0.1 mg/kg BW of Aroclor 1254, and 0.3–1.0 mg/kg diet or 0.6 mg/kg BW of PCB 77. Several PCBs had negligible adverse effects on tested mammals during chronic daily exposures to doses of at least 5.0 mg/kg ration or 5.0 mg/kg BW, specifically, PCBs 4, 15, 47, 52, 80, 136, 153, 155, and 167. Although subhuman primates seem to be more sensitive to reproductive and other adverse effects of PCBs than humans, no clear and convincing evidence associated PCB exposures with human cancers and reproductive problems. At present, no meaningful reproductive problems have been identified in female capacitor workers and no carcinogenicity was evident in humans having more than 1.0 mg total PCBs/L serum or more than 400.0 mg total PCBs/kg adipose fat.

The mink (*Mustela vison*) is among the most sensitive mammals to PCB toxicity. Reproductive failure, especially fetal death and resorption, of PCB-fed mink is well documented. The mechanisms of intrauterine death of mink fetuses after PCB exposure are not fully understood, although planar PCB congeners seem to be implicated. Studies showed that EROD and AHH activities were maximally induced in adult mink by PCB fractions containing non-*ortho* or mono-*ortho* chlorobiphenyls and that mink kits are about 10 times more sensitive to P450 inducers than adults. PCB-dosed mink also show altered blood chemistry, abnormal liver metabolism and histology, raised cortisol excretion, enhanced EROD activity, and altered metabolism of vitamin A. Selective retention of certain PCBs and their hydroxylated metabolites occurred in mink muscle after 3 months of a diet containing a PCB mixture. Retention of PCBs 99, 105, 118, 138, 153, 156, and 180 was high in muscle of mink on the diet. Not retained and presumably metabolized were PCBs 44, 49, 52, 91, 92, 95, 97, 107, 132, 149, and 174.

Estrogen receptors bind to many compounds other than natural estrogen. Several organochlorine compounds, including certain PCBs, reportedly act as estrogen mimics. Many estrogen mimics are persistent, lipid-soluble compounds that are defined by their ability to stimulate the proliferation of cells in the uterus of the mouse (*Mus* spp.); in males, these mimics may also inhibit sperm production and testes growth. PCBs and their metabolites may be estrogenic to wildlife, although the evidence is not conclusive. PCB mixtures and congeners with a significant degree of *ortho*-substitution have elevated estrogenic activity; Aroclor 1221, for example, rich in *ortho*-chlorobiphenyl, has estrogenic activity in female rats. Hydroxylated metabolites of PCBs also show estrogenic hormonal activity. Hydroxylation of PCBs occurs in amphibians and teleosts but at a much slower rate than that of mammals. Hydroxylated metabolites of PCB 3 include 4′-chloro-4-biphenylol, 4′-chloro-3,4-biphenyldiol, and 4′-chloro-3-methoxy-4-biphenylol; PCB 15 gave

4,4′-dichloro-3-biphenylol; and Aroclor 1254 yielded mono-, di-, and tri-chlorophenylols. PCB 77 is detoxified by metabolic hydroxylation to hydroxy biphenyl metabolites. Hydroxylated PCB metabolites may be more toxic than the parent product because of their (1) estrogenic properties; (2) tendency to accumulate in the fetus, and (3) interference with thyroxine metabolism. The ability of hydroxylated PCB metabolites to bind to the uterine estrogen receptor in rats was in increasing order of effectiveness 4-hydroxy 3,5,4′-trichlorobiphenyl, 4,4′-dihydroxy 3,5,3′,5′-tetrachlorobiphenyl, 4-hydroxy 2-chlorobiphenyl, 4-hydroxy 4′-chlorobiphenyl, 4,4′-dihydroxy 2′,3′,5′,6′-tetrachlorobiphenyl, 4,4′-dihydroxybiphenyl, and 4-hydroxybiphenyl; PCB compounds that demonstrated appreciable receptor-binding activity were also active in stimulating uterine weight increases. More research seems needed on the estrogenic properties of hydroxylated PCB metabolites.

Long-term neurobehavioral changes were reported in children, monkeys, and rodents exposed to commercial PCB mixtures during fetal and neonatal development. Perinatal exposure of rats (*Rattus* sp.) to *ortho*-substituted PCBs (PCBs 28, 118, 153) can cause long-lasting deficits in learning in females; males were not affected. The PCB-induced deficit in spatial learning did not appear to be mediated by decreased thyroid hormone levels; however, neonatal hyperthyroid rats exposed to low concentrations of PCBs 77 and 126 in utero and during lactation had enhanced spatial learning.

In comparison to primates and mustelids, rodents are only moderately sensitive to intoxication by most PCBs. In rodents, PCBs 77 and 169 cause effects that are characteristic of the dioxins and furans, including P450 induction, porphyrin accumulations, and atrophy of the lymphatic organs, as well as teratogenicity in mice. Exposure of rats to PCB 126 in utero and through lactation produced fetotoxic effects, delayed physical maturation, and induced liver xenobiotic metabolizing enzymes without causing neurobehavioral effects. In mice, PCBs 77, 126, and 169 were teratogenic in a high percentage of the fetuses from treated dams but without apparent effect on the dams. PCB 77, for example, was toxic to the conceptus at dose levels below those toxic to the dam when administered to pregnant CD-1 mice on days 6–15 of gestation. The predominant PCB-induced fetal malformations in mice were cleft palate and hydronephrosis. PCB 77 also interferes with retinyl ester hydrolase (REH), the enzyme responsible for the hydrolysis of vitamin A into free retinol, lowering levels of vitamin A in liver and decreasing serum concentrations of REH and retinol.

Toxicokinetics of PCBs in rodents were altered when administered in mixtures. PCBs 153, 156, and 169 produced biphasic elimination patterns in mice when administered in combinations but single-phase elimination when administered alone; elimination of all PCBs was more rapid after coadministration. Mixtures of PCBs 153 and 156 raised EROD activity and lengthened retention of each congener in liver; however, a mixture of PCBs 153 and 169 lowered EROD activity. Selected PCBs of low acute toxicity may increase the toxicity of compounds such as 2,3,7,8-TCDD. Thus, PCBs 153 or 157 at sublethal dosages (20.0–80.0 mg/kg BW) did not produce cleft palate deformities in mouse embryos. But a mixture of PCB 157 and 2,3,7,8-TCDD produced a 10-fold increase in the incidence of palate deformities that were expected of 2,3,7,8-TCDD alone; palate deformities did not increase with a mixture of PCB 153 and 2,3,7,8-TCDD. The widespread environmental occurrence of PCB-PCDD and PCB-PCDF combinations suggests a need for further evaluation of the mechanism of this interaction.

Intraspecific variability to PCBs was high in fish-eating mammals, and genetically inbred AHH-responsive and AHH-nonresponsive strains of mice. In fish-eating mammals, the ability to metabolize PCB congeners with H-atoms only in the *ortho-* and *meta*-positions and with one *ortho*-chlorine substituent were in the order phocid seals > cetaceans > otters. But the metabolism of PCB congeners with H-atoms in the *meta-* and *para*-positions and with two *ortho*-chlorines were in the order otter > seals > cetaceans. Both categories of congeners were probably metabolized by different families of cytochrome P4501A and P4502B, of which levels were different between the cetaceans, pinnipeds and otters. Within-species PCB patterns differed in a concentration-dependent manner; the induction of cytochrome P450 enzymes is the most likely explanation for this phenomenon, although starvation could have a similar effect. In mice, PCB 77 was metabolized and excreted more rapidly than 2,3,7,8-TCDD; in rats, PCB 77 was excreted more rapidly than PCB 47. PCBs that lack adjacent hydrogen atoms in at least one of the rings are enriched in rat tissues, indicating that accumulation exceeds elimination by metabolism and excretion. PCBs with a tendency to accumulate were non-*ortho*- and mono-*ortho*-substituted congeners; however, PCBs with *meta–para* unsubstituted carbon atoms in at least one ring were not enriched in tissues. Uptake and retention of individual PCB congeners in rats are related to properties associated with K_{ow} and high chlorination, especially in the tetra- and penta-chlorobiphenyls. The highly chlorinated hexa- and octa-chlorobiphenyls produced morphological changes in rats comparable to those produced by DDT, Aroclor 1254, and Aroclor 1260.

25.6 Recommendations

Proposed PCB criteria for the protection of natural resources are predicated on total PCBs and selected Aroclors (Table 25.4) and, unfortunately, offer minimal insight

Table 25.4. Proposed PCB criteria for the protection of natural resources and human health.

Resource, Criterion, and Other Variables	Effective Concentration
AQUATIC LIFE PROTECTION	
Fish, total PCBs	
Diet	<500.0 μg/kg fresh weight (FW)
Eggs	<300.0 μg/kg FW
Whole body	<400.0 μg/kg FW
Adverse effects, total PCBs	
Invertebrates, whole	>25.0 mg/kg FW
Fish, whole	>50.0 mg/kg FW
Marine mammals	
Blubber	
Total PCBs	<70.0 mg/kg FW
PCB 101	<0.6 mg/kg FW
PCB 138	<7.0 mg/kg FW
PCB 153	<10.0 mg/kg FW
PCB 180	<3.0 mg/kg FW
Blood lipid, total PCBs	<25.0 mg/kg FW
Medium, total PCBs	
Freshwater	
Acute	<2.0 μg/L
Chronic	<0.014 μg/L
Saltwater	
Acute, most species	<10.0 μg/L
Chronic, sensitive species	<0.03 μg/L
Filter-feeding shellfish	<0.006 μg/L

Continued

Table 25.4. cont'd

Resource, Criterion, and Other Variables	Effective Concentration
Birds, total PCBs	
Brain	<300.0 mg/kg FW
Mammals	
Aroclor 1016	
Rhesus macaque	<0.008–<0.028 mg/kg body weight (BW) daily
Aroclor 1248	
Rhesus macaque	<0.009 mg/kg BW daily
Aroclor 1254	
Mink	<0.1– <0.115 mg/kg BW daily
Mouse	<1.3 mg/kg BW daily
Rat	<0.25 mg/kg BW daily
Total PCBs	
Fat	<10.0 mg/kg FW
Liver	<4.0 mg/kg FW
HUMAN HEALTH PROTECTION, TOTAL PCBs (UNLESS INDICATED OTHERWISE)	
Air	
Aroclor 1242, acceptable vs. hazardous	<0.001 vs. 10.0 mg/m^3
Aroclor 1254, acceptable vs. hazardous	<0.001 vs. 5.0 mg/m^3
Arizona; total PCBs, acceptable	<0.3 $\mu g/m^3$ for 1 h; <0.079 $\mu g/m^3$ or 24 h; 0.0061 $\mu g/m^3$ annually
Florida; acceptable	
Aroclor 1254	5.0 $\mu g/m^3$ for 8 h; 1.2 $\mu g/m^3$ for 24 h; 0.00083 $\mu g/m^3$ annually
Aroclor 1242	10.0 $\mu g/m^3$ for 8 h; 2.4 $\mu g/m^3$ for 24 h
South Carolina; total PCBs	<2.5 $\mu g/m^3$ for 24 h
Virginia, 24-h exposure; acceptable	
Total PCBs	<8.0 $\mu g/m^3$
Aroclor 1242	<17.0 $\mu g/m^3$
Aroclor 1254	<8.3 $\mu g/m^3$
Diet	
Acceptable daily intake, total PCBs, Scandinavia	<0.2 mg, equivalent to about 2.8 μg/kg BW daily for a 70-kg person
Drinking water	
Acceptable, AZ, KS, ME, NH, VT	0.008–0.05 μg/L
Acceptable, CT, MA, NJ, NY, RI	0.5–1.0 μg/L
Connecticut	
Action level	>1.0 μg/L
Hazardous	>2.0 μg/L
USA	
Child	<1.0 μg/L
Adult	<4.0 μg/L

Table 25.4. cont'd

Resource, Criterion, and Other Variables	Effective Concentration
Public water supplies, Wisconsin	
Warm water sport fish communities	0.49 ng/L
Cold water communities	0.15 ng/L
Great Lakes communities	0.15 ng/L
Land disposal restriction, USA; wastewater vs. nonwastewater	
Aroclors 1016, 1232, or 1248	0.013 vs. 0.92 mg/kg
Aroclor 1221	0.014 vs. 0.92 mg/kg
Aroclor 1242	0.017 vs. 0.92 mg/kg
Aroclor 1254	0.014 vs. 1.8 mg/kg
Reduced risk from cancer, total PCBs	<7.7 mg/kg BW daily
Commercial milk	<1.5 mg/kg, fat basis
Edible poultry, total PCBs	
Canada	<0.5 mg/kg LW
USA	<3.0 mg/kg LW
Fish, edible portion	
Canada	<2.0 mg/kg FW
Great Lakes, gamefish	<1.9 mg/kg FW
USA	<2.0 mg/kg FW
Great Lakes	<0.5 mg/kg FW
New York State	<2.0 μg/kg FW

into PCB toxicokinetics. Most authorities now agree that future PCB risk assessments require (1) analysis of non-*ortho* PCBs and selected mono-*ortho* PCBs; (2) exposure studies of individual species to specific congeners alone or in combination with other compounds, including other PCB congeners, dioxins, and dibenzofurans; (3) clarification of existing structure–induction relations, and (4) more refined analytical techniques. Because component-based analysis of PCBs has limited the usefulness of the historical database for current environmental research and in formulation of regulatory criteria, procedures were developed to accurately detect, speciate, and quantify mixtures of multiple Aroclors. Aroclor conversion factors have been calculated for 14 PCB congeners (PCBs 18, 28, 31, 99, 118, 128, 138, 149, 153, 180, 194, 195, 201, and 203), allowing quantification of PCB profiles as Aroclors 1248, 1254, and 1260 using measurements of these congeners.

Marked differences between species in their abilities to metabolize specific PCB congeners must be considered in toxicity testing. Foxes and dogs, for example, in contrast to monkeys and rats, can degrade the otherwise highly persistent PCB 153 because they possess an unusual cytochrome P450 isozyme that metabolizes PCB 153. Also, low-chlorinated congeners that are metabolized via reactive intermediates must be critically evaluated because they show weak tumor-initiating properties. Intraspecies differences are also documented. Three populations of the harbor porpoise (*Phocoena phocoena*) from Newfoundland, St. Lawrence River, and the Gulf of Maine are distinguishable based on PCB levels in blubber.

To complement PCB chemical residue analyses, the rat hepatoma cell bioassay was useful for assessing the toxic potency of PCBs in extracts from environmental samples. This in vitro bioassay of cytochrome P450IA1 catalytic activity in the H4IIE cells in response to planar halogenated hydrocarbons was considered accurate and precise. Comparison of the responses of the H4IIE cells was calibrated against their responses to 2,3,7,8-TCDD. EROD and porphyria induction measurements also potentially complement PCB chemical residue analyses and have been used to determine the toxic potencies of complex mixtures of PCBs and other halogenated aromatic hydrocarbons extracted from wildlife tissues. In one case, extracts of PCB-contaminated eggs of herring gulls (*Larus argentatus*) from the Great Lakes and of great blue herons (*Ardea herodias*) from British Columbia induced EROD and porphyria in primary cultures of chicken embryo hepatocytes. Porphyrins in feces as a nondestructive biomarker is recommended as an alternative to the traditional liver porphyrins in the hazard assessment of birds contaminated with PCBs. Hepatic cytochrome P450-associated monooxygenases and cytochrome P450 proteins in embryos of the black-crowned night-heron (*Nycticorax nycticorax*) were associated with concentrations of total PCBs and 11 PCB congeners that express toxicity through the Ah receptor, and also should be considered as biomarkers for assessing PCB contamination of wetlands. Hepatic cytochromes P450, P420 (degraded P450) and mixed function oxidases are recommended as biomarkers of PCB exposure in harbor seals (*Phoca vitulina*); total PCB burdens in both blubber and liver had positive correlations with P450, P420, and MFO activity levels.

The interpretation of PCB residue data is challenging from several perspectives, as judged by analysis of eggs of Forster's terns (*Sterna forsteri*) from Wisconsin: (1) data from a single analysis frequently contained measurable concentrations of 100–150 PCB congeners; (2) a single sample was not sufficient to understand the environmental distribution of PCBs; (3) source profiles of PCB inputs into the environment were poorly characterized; (4) PCB congeners in the original polluting material often merged with congeners from other sources; and (5) the contaminant mixture may have been altered by metabolism and subsequent partition into multiple environmental compartments that may be further changed by weathering or degradation. To understand these processes and to correlate residue profiles with specific toxic responses, congener-specific methods of analysis and complex statistical techniques (principal component analysis) are required. Using these techniques, it was established that eggs of Forster's terns of two colonies differed significantly in PCB composition. Similar techniques were used to identify various PCB-contaminated populations of harbor seals (*Phoca vitulina*) in Denmark.

Selected congeners should be quantified in human foodstuffs and tissues, as determined from a survey of PCB congener frequency in commercial formulations, environmental and biological samples and human tissues, and a consideration of the relative toxicity and persistence of the congeners. PCBs 28, 74, 77, 99, 105, 118, 126, 128, 138, 153, 156, 169, 170, 179, and 180 reportedly account for more than 70% of the total PCB burden in any sample and should be quantified. Additionally, PCBs 8, 37, 44, 49, 52, 60, 66, 70, 82, 87, 101, 114, 158, 166, 183, 187, and 189 should be considered for quantification because of their reported occurrence or toxicity. Some PCBs are particularly prevalent in aquatic animals, especially PCBs 95, 101, 110, 118, 138, 149, 153, 180, and 187; also detected in aquatic biota and reported as important components were PCBs 26, 52, 66, 70, 99, 105, 132, 151, 170, 177, 201, and 206. In the Netherlands, maximum PCB limits in fishes as dietary items for human health protection are now derived from the sum of PCBs 28, 95, 101, 138, 149, 153, and 180. From a toxicological viewpoint, other congeners may be more important. These have been identified (on the basis of ability to induce AHH) as the most toxic planars (PCBs 15, 37, 77, 81, 126, and 169), the mono-*ortho* analogs of the planar PCBs (PCBs 105, 114, 123, 156, and 189), and the di-*ortho* analogs (PCBs 128, 138, 158, 166, and 170). Of these compounds, PCBs 37, 105, 114, 128, 138, 156, and 158 occur in human tissues and PCBs 15, 77, 81, 123, 126, 166, and 169 do not occur. In Germany, PCB congeners 49, 118, 156, and 170 are considered carcinogen initiators; these four congeners and PCBs 28, 31, 52, 101, 138, 153 and 180 are proposed as indicators of PCB contamination in that country. In vitro studies with PCBs 138 and 169 indicate a relation between PCB structure, bioavailability, and the capacity to stimulate oncogene expression, strongly indicating the need for more research in this subject area.

Ecological considerations in setting limits in foods for human consumption are complex. In Lake Ontario, for example, stocking rates of the alewife (*Alosa pseudoharengus*) – the main food of all Great Lakes sportfishes – necessary to achieve PCB consumption advisories of <0.5 mg total PCBs/kg FW fish muscle, carry about a 90% probability of an alewife population crash. It is postulated that increases of 25% in current stocking rates of chinook salmon would decrease PCB concentrations of chinook salmon without a large increase in the probability that the alewife population would crash. These scenarios are applicable to species of salmonids in the Great Lakes because they, too, exhibit size-selective predation and their recruitment is largely determined by stocking.

25.7 Summary

All production of PCBs in the United States ceased in 1977. Of the 1.2 million tons of PCBs manufactured to date, about 65% are still in use in electrical equipment and 31% in various environmental compartments, and 4% were degraded or incinerated. The 209 PCB congeners and their metabolites show wide differences in biological effects. A significant part of the toxicity associated with commercial PCB mixtures is related to the presence of about 20 planar congeners, i.e., congeners without chlorine substitution in the *ortho* positions. Toxic planar congeners, like other PCB congeners, have been detected in virtually all analyzed samples, regardless of collection locale. Planar PCB concentrations were usually highest in samples near urban areas and in fat and liver tissues, filter-feeding bivalve molluscs, fish-eating birds, and carnivorous marine mammals. Adverse effects of planar PCBs on growth, survival, and reproduction are highly variable because of numerous biotic and abiotic modifiers, including interaction with other chemicals. In general, embryos and juveniles were the most sensitive stages tested to planar PCBs, and the chinook salmon (*Oncorhynchus tshawytscha*), domestic chicken (*Gallus* sp.), mink (*Mustela vison*), rhesus macaque (*Macaca mulatta*), and laboratory white rat (*Rattus* sp.) were among the most sensitive species. For protection of natural resources, most authorities now recommend (1) analyzation of environmental samples for planar and other potentially hazardous congeners; (2) exposure studies with representative species and specific congeners, alone and in combination with other environmental contaminants; (3) clarification of existing structure–induction–metabolism relations; and (4) more research on physiological and biochemical indicators of PCB stress.

POLYCYCLIC AROMATIC HYDROCARBONS[a]

Chapter 26

26.1 Introduction

Several polycyclic aromatic hydrocarbons (PAHs) are among the most potent carcinogens known to exist, producing tumors in some organisms through single exposures to microgram quantities. PAHs act at both the site of application and at organs distant to the site of absorption; their effects have been demonstrated in nearly every tissue and species tested, regardless of the route of administration. The evidence implicating PAHs as an inducer of cancerous and precancerous lesions is overwhelming, and this class of substances is probably a major contributor to the increase in cancer rates reported for industrialized nations. PAHs were the first compounds known to be associated with carcinogenesis. Occupational skin cancer was first documented in London chimney sweeps in 1775 and in German coal tar workers in the late 1800s. By the early 1900s, soot, coal tar, and pitch were all found to be carcinogenic to humans. By 1918, it was shown that topical applications of coal tar produced skin tumors in mice and rabbits; benzo[a]pyrene, a PAH, was identified as one of the most carcinogenic compounds in coal tar. The carcinogenic activity to humans of soots, tars, and oils is beyond dispute. In addition to the skin cancers noted initially, higher incidences of respiratory tract and upper gastrointestinal tract tumors were associated with occupational exposures to these carcinogens. PAH-induced cancers in laboratory animals is well documented. Benzo[a]pyrene, for example, has produced tumors in mice, rats, hamsters, guinea pigs, rabbits, ducks, and monkeys following administration by oral, dermal, and intraperitoneal routes. Teratogenic or carcinogenic responses have been induced in sponges, planarians, echinoderm larvae, teleosts, amphibians, and plants by exposure to carcinogenic PAHs. An unusually high prevalence of oral, dermal, and hepatic neoplasms have been observed in bottom-dwelling fish from polluted sediments containing grossly elevated PAH levels. Hepatic disorders, including adenomas and carcinomas, were found in common carp (*Cyprinus carpio*) from West Point Lake in Georgia in 1991 and are linked to elevated concentrations in lake sediments. PAH compounds have damaged chromosomes in cytogenetic tests, have produced mutations in mammalian cell culture systems, and have induced DNA repair synthesis in human fibroblast cultures. While some PAHs are potent mutagens and carcinogens, others are less active or suspected carcinogens. Some, especially those of biological origin, are probably not carcinogens. Certain lower molecular weight, noncarcinogenic PAHs, at environmentally realistic levels, were acutely toxic to aquatic organisms, or produced deleterious sublethal responses. However, few generalizations can be made about the class of PAH compounds because of the extreme variability in toxicity and physicochemical properties of

---

[a]All information in this chapter is referenced in the following sources:

Eisler, R. 1987. Polycyclic aromatic hydrocarbon hazards to fish,wildlife, and invertebrates: a synoptic review. *U.S. Fish Wildl. Serv. Biol. Rep. 85(1.11)*, 81 pp.

Eisler, R. 2000. Polycyclic aromatic hydrocarbons. Pages 1343–1411 in *Handbook of Chemical Risk Assessment: Health Hazards to Humans, Plants, and Animals. Volume 2, Organics*. Lewis Publishers, Boca Raton, Florida.

PAHs and their various effects on individual species.

PAHs are widely distributed in the environment, and have been detected in animal and plant tissues, sediments, soils, air, surface water, drinking water, industrial effluents, ambient river water, well water, and groundwater. Humans have probably always been exposed to PAHs from the natural background level in soils and plants; avoiding exposure to nanogram quantities of these substances on a daily basis is now considered essentially impossible for all living resources. Ever since benzo[a]pyrene was recognized as a carcinogen in the early 1900s, the presence of it and of other PAHs in the environment has received continuous attention. As one consequence, many reviews have been published on ecological and toxicological aspects of PAHs in the environment, with special reference to their carcinogenic properties.

26.2 *Environmental Chemistry, Sources, and Fate*

Physical, chemical, and biological properties of selected PAHs are summarized; major sources of PAHs in the environment are listed; and metabolism and degradation of selected PAHs discussed.

26.2.1 Properties

Polycyclic aromatic hydrocarbons (PAHs), also known as polynuclear aromatic hydrocarbons (PNAs) and polycyclic organic matter (POM), are composed of hydrogen and carbon arranged in the form of two or more fused benzene rings in linear, angular, or cluster arrangements, which may or may not have substituted groups attached to one or more rings. In some cases, the newly defined substituted PAH has strikingly greater toxicological effects than does the parent compound. The nomenclature of PAH compounds has been ambiguous in the past due to different peripheral numbering systems. The currently accepted nomenclature is shown in Figure 26.1.

Of major environmental concern are mobile PAHs that vary in molecular weight from 128.16 (naphthalene, $C_{10}H_8$) to 300.36 (coronene, $C_{24}H_{12}$). Higher molecular weight PAHs are relatively immobile because of their large molecular volumes and their extremely

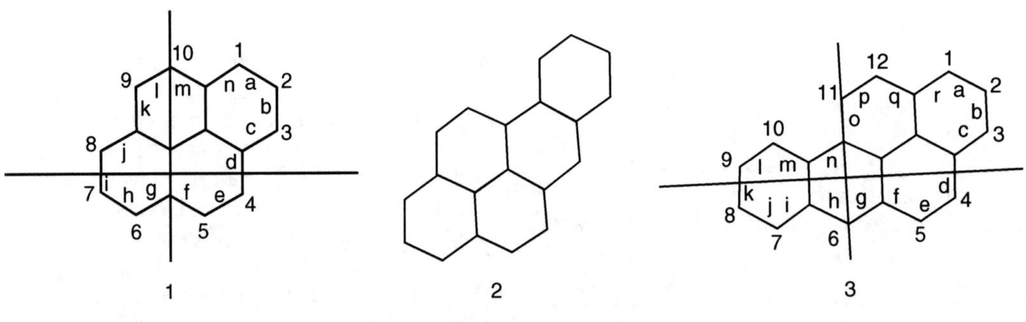

Figure 26.1.
Nomenclature of PAHs. The PAH formula is oriented so that the greatest number of rings are in a horizontal row and a maximum number of rings are above and to the right of the horizontal row. The first carbon atom that belongs to the uppermost ring and is not engaged in ring fusion with another ring is given the number C-1; numbering continues in a clockwise direction omitting those carbon atoms which do not carry a hydrogen atom. The bond between C-1 and C-2 is designated as side "a"; other peripheral sides continue in clockwise direction in alphabetical order. Examples are: (1) pyrene (correctly oriented, numbered, and lettered), (2) benzo[a]pyrene (not oriented correctly), and (3) benzo[a]pyrene (correctly oriented, numbered, and lettered).

low volatility and solubility. Among the mobile forms are thousands of compounds that differ in the number and position of aromatic rings, and in the position of substituents on the basic ring system. The lower molecular weight unsubstituted PAH compounds, containing 2–3 rings, such as naphthalenes, fluorenes, phenanthrenes, and anthracenes (Figure 26.2), have significant acute toxicity to some organisms, whereas the higher molecular weight 4- to 7-ring aromatics do not. However, all known PAH carcinogens, co-carcinogens, and tumor producers are in the high-molecular-weight PAH group (Figure 26.3).

Physical and chemical characteristics of PAHs generally vary with molecular weight. With increasing molecular weight, aqueous solubility decreases, and melting point, boiling point, and log K_{ow} (octanol–water partition coefficient) increases (Table 26.1), suggesting increased solubility in fats, a decrease in resistance to oxidation and reduction, and a decrease in vapor pressure. Accordingly, PAHs of different molecular weight vary substantially in their behavior and distribution in the environment and in their biological effects. Measurements of various PAH metabolites, especially naphthalene and benzo[a]pyrene-type metabolites, in bile and other tissues are now conducted routinely using high performance liquid chromatography with fixed wavelength fluorescence.

26.2.2 Sources

About 43,000 metric tons of PAHs are discharged into the atmosphere each year, and another 230,000 tons enter aquatic environments (Table 26.2). PAHs are ubiquitous in nature as a consequence of synthesis in terrestrial vegetation, microbial synthesis, and volcanic activity, but quantities formed by these natural processes are small in comparison with those produced from forest and prairie fires and anthropogenic sources. Anthropogenic activities associated with significant production of PAHs include: coke production in the iron and steel industry; catalytic cracking in the petroleum industry; the manufacture of carbon black, coal tar pitch, and asphalt; heating and power generation; controlled refuse incineration; open burning; and emissions from internal combustion engines used in transportation. Thus, the formation of PAHs in the environment is due to an endogenous synthesis by microorganisms, algae, and macrophytes which provide natural background, and to a second process which is connected to human-controlled high-temperature (> 700°C) pyrolysis of organic materials, to open burning, and to natural volcanic activities. The discovery in fossil fuels of complex mixtures of PAHs spanning a wide range of molecular weights has led to the conclusion that, given sufficient time (i.e., millions of years), pyrolysis of organic materials at temperatures as low as 100–150°C can also lead to production of PAHs.

Forest and prairie fires release much greater amounts of PAHs to the atmosphere than does fossil fuel burning. Nearly all of the airborne PAHs produced by flame pyrolysis are associated with the particulate fraction produced during combustion, and these are significantly modified by the chemical composition of the fuel, the pyrolysis temperature, the duration of exposure to elevated temperature, and to other factors. In one study, a PAH profile was established for a series of laboratory fires simulating the prescribed burning of pine needle litter. Heading fires (moving with wind) produced more total particulate matter than backing fires (moving against wind), but backing fires produced significantly higher amounts of PAHs, with the actual amounts formed dependent on fuel loading and the residence time of combustible gases in the burning zone. Emission factors for benzo[a]pyrene varied from 238.0 to 3454.0 μg/kg in backing fires and 38.0–97.0 μg/kg in heading fires.

PAHs present in the atmosphere enter rain as a result of in-cloud and below-cloud scavenging. Total PAHs deposited on land and water are almost equivalent to PAH content in rainfall; significant quantities of PAHs are found in presumed pollution-free areas, indicating the importance of rain in transport and distribution of PAHs.

PAHs may reach aquatic environments in domestic and industrial sewage effluents, in surface runoff from land, from deposition

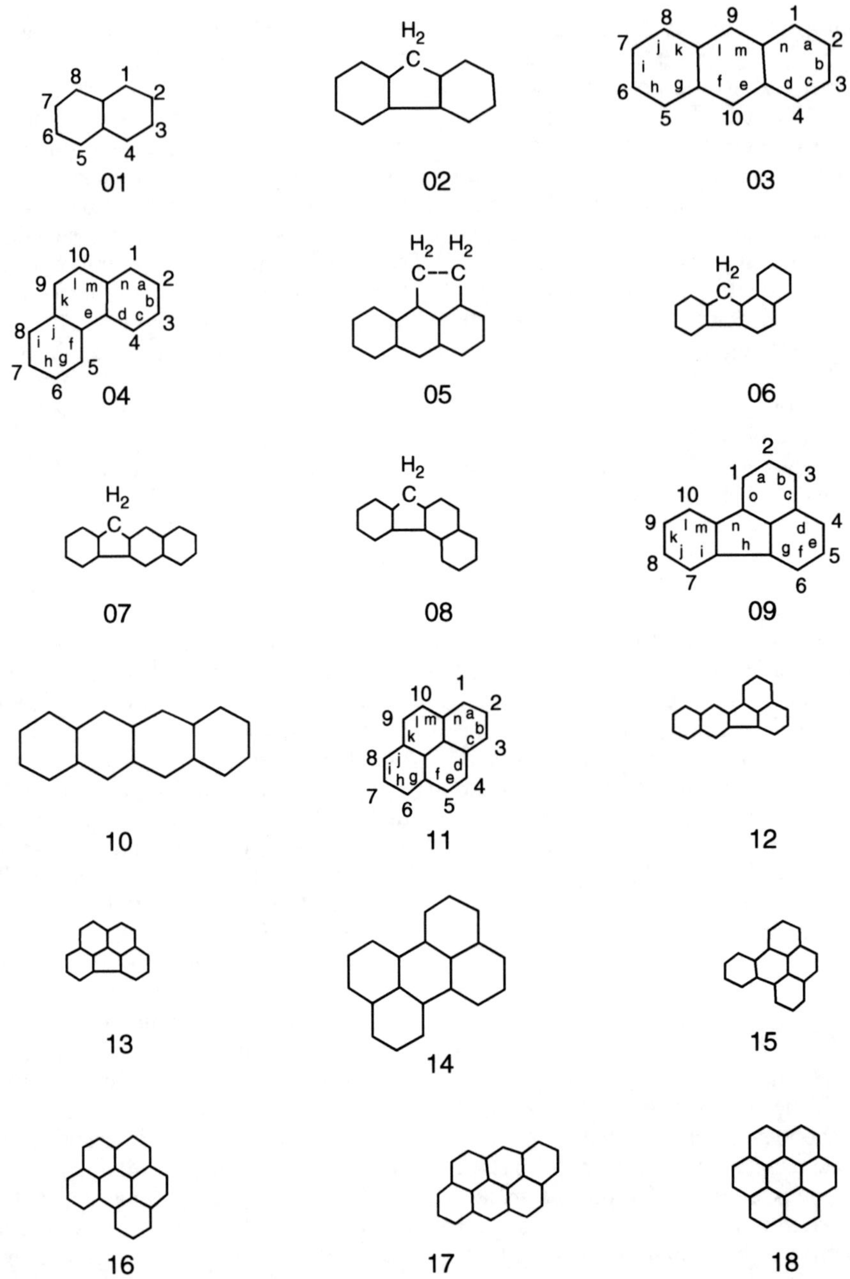

Figure 26.2.
Ring structures of representative noncarcinogenic PAHs. The numbering and lettering system for several PAHs is also given. Compounds are: (1) naphthalene, (2) fluorene, (3) anthracene, (4) phenanthrene, (5) aceanthrylene, (6) benzo[a]fluorene, (7) benzo[b]fluorene, (8) benzo[c]fluorene, (9) fluoranthene, (10) naphthacene, (11) pyrene, (12) benzo[a]fluoranthene, (13) benzo[g,h,i]fluoranthene, (14) perylene, (15) benzo[e]pyrene, (16) benzo[g,h,i]perylene, (17) anthanthrene, and (18) coronene.

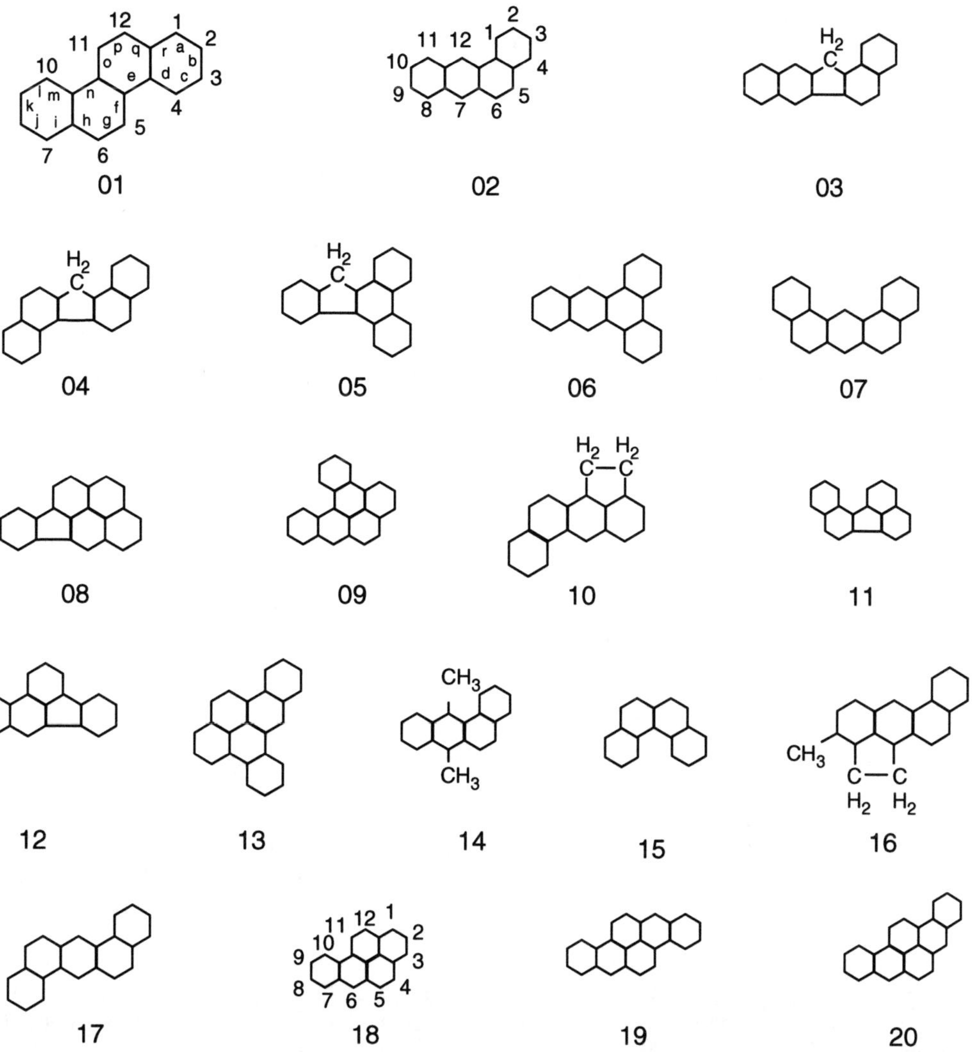

Figure 26.3.
Ring structures of representative tumorigenic, co-carcinogenic, and carcinogenic PAHs. The numbering and lettering system for several PAHs is also given. Compounds are: (1) chrysene, (2) benz[a]anthracene, (3) dibenzo[a,h]fluorene, (4) dibenzo[a,g]fluorene, (5) dibenzo[a,c]fluorene, (6) dibenz[a,c]anthracene, (7) dibenz[a,j]anthracene, (8) indeno[1,2,3-cd] pyrene, (9) dibenzo[a,1]pyrene, (10) cholanthrene, (11) benzo[j]fluoranthene, (12) benzo[b]fluoranthene, (13) dibenzo[a,e]pyrene, (14) dimethylbenz[a]anthracene, (15) benzo[c]phenanthrene, (16) 3-methylcholanthrene, (17) dibenz[a,h]anthracene, (18) benzo[a]pyrene, (19) dibenzo[a,h]pyrene, (20) dibenzo[a,i]pyrene. Compounds 1–9 are weakly carcinogenic, co-carcinogenic, or tumorigenic; compounds 10–13 are carcinogenic; compounds 14–20 are strongly carcinogenic.

Table 26.1. Some physical and chemical properties of selected PAHs.

Compound	Number of Rings	Approximate Molecular Weight	Melting Point (°C)	Solubility in Water (mg/L)	Log K_{ow}
Naphthalene	2	128	80	30.0	3.37
Anthracene	3	178	216	0.07	4.45
Benz[a]anthracene	4	228	158	0.014	5.61
Benzo[a]pyrene	5	252	179	0.0038	6.04
Benzo[g,h,i]perylene	6	276	222	0.00026	7.23

of airborne particulates, and especially from spillage of petroleum and petroleum products into water bodies. The majority of PAHs entering aquatic environments remains close to sites of deposition, suggesting that lakes, rivers, estuaries, and coastal marine environments near centers of human populations are the primary repositories of aquatic PAHs. Large variations in aquatic PAH contents were evident due to localized source inputs and physicochemical conditions. For example, urban runoff from stormwater and highways to Narragansett Bay, Rhode Island, accounted for 71% of the total inputs for higher molecular weight PAHs, and 36% of the total PAHs. More than 30% of all combustion-derived PAHs in coastal sediments of Washington State is supplied by riverine transport of suspended particulate materials, while direct atmospheric input accounts for a maximum of 10%. In contrast, concentrations of PAHs in sediments from the vicinity of Georges Bank, off the U.S. northeastern coast, varied from 1.0 to 100.0 μg/kg dry weight (DW), and were directly related to total organic carbon, silt, and clay contents in sediments; combustion-derived PAHs dominated at the higher concentrations, while lower levels were often associated with a fossil fuel origin.

Human activities have resulted in exposure of Antarctic fishes to petroleum-derived PAHs. Fish captured near Palmer station on the Antarctic peninsula had induced EROD activities and elevated concentrations of biliary PAH metabolites of phenanthrene and naphthalene when compared to conspecifics from reference sites. Artificial reefs consisting of oil and coal flyash stabilized with cement and lime in Florida waters near Vero Beach contained elevated PAH levels ranging from as high as 1.2 mg fluoranthene/kg and 0.25 mg naphthalene/kg. But there is negligible leaching because seawater is not an effective medium for removing PAHs from reef bricks or the ash.

Discharge water from hydrostatic testing of natural gas pipelines is a significant source of PAH loading into aquatic environments, contributing as much as 32,000.0 μg PAHs/L of discharge water, mostly as naphthalenes. More than 25 PAHs, primarily anthracenes and pyrenes, were detected in pipeline residues on inner walls of natural gas pipelines at concentrations up to 2400.0 $\mu g/m^2$ of inner surface; the same compounds may be reasonably expected in aqueous waste from pipeline maintenance. Release of these, or similar, discharge waters directly into aquatic environments will result in contamination similar to that caused by oil spills; however, these sites for pollution may occur in locations far distant from oil production and refinery activities. PAHs are also present in tap water at concentrations of 0.1–1.0 ng/L, primarily as mono- and di-chlorinated derivatives of naphthalene, phenanthrene, fluorene, and fluoranthene. The presence of PAHs and chlorinated PAHs in tap water indicates the reaction of PAHs with chlorine; however, their significance to human health and to aquatic biota is unknown.

Creosote (about 85% PAHs) has been used extensively as a long-term wood preservative for marine and freshwater support structures

Table 26.2. Major sources of PAHs in atmospheric and aquatic environments.

Ecosystem and Sources	Annual Input, in Metric Tons
ATMOSPHERE	
Total PAHs	
Forest and prairie fires	19,513
Agricultural burning	13,009
Refuse burning	4769
Enclosed incineration	3902
Heating and power	2168
Benzo[a]pyrene	
Heating and power	
Worldwide	2604
USA only	475
Industrial processes (mostly coke production)	
Worldwide	1045
USA only	198
Refuse and open burning	
Worldwide	1350
USA only	588
Motor vehicles	
Worldwide	45
USA only	22
AQUATIC ENVIRONMENTS	
Total PAHs	
Petroleum spillage	170,000
Atmospheric deposition	50,000
Wastewaters	4400
Surface land runoff	2940
Biosynthesis	2700
Total benzo[a]pyrene	700

such as pilings, railway ties, and utility poles. About 75% of the creosote applied to marine pilings will remain in the wood after 40 years of service. Most of the PAHs leached into water from creosote-treated pilings are rapidly lost owing to volatility, photodegradation, and microbial degradation pathways.

26.2.3 Fate

Concern about PAHs in the environment is due to their persistence and to the fact that some are known to be potent mammalian carcinogens, although environmental effects of most noncarcinogenic PAHs are poorly understood. Prior to 1900, a natural balance existed between the production and the degradation of PAHs. Synthesis of PAHs by microorganisms and volcanic activity and production by man-made high-temperature pyrolytic reactions and open burning seemed to be balanced by PAH destruction via photodegradation and microbial transformation. With increased industrial development and increased emphasis of fossil fuels as energy sources, the balance has been disturbed to the extent that PAH production and introduction into the environment greatly exceeds known PAH removal processes. Mass balance models that quantitatively estimate PAH cycling in various aquatic systems seem promising. In one case, a steady-state model for naphthalene, phenanthrene and benzo[a]pyrene in Saguenay Fjord in Quebec, has been constructed incorporating atmospheric and industrial sources, and transport and transformation rates. More research on mass balance models is needed.

When released into the atmosphere, PAHs become associated with particulate materials. Their residence time in the atmosphere and transport to different geographic locations are governed by particle size, meteorological conditions, and atmospheric physics. The highly reactive PAHs photodecompose readily in the atmosphere by reaction with ozone and various oxidants; degradation times range from several days to six weeks for PAHs adsorbed onto particulates $<1.0\,\mu m$ in diameter (in the absence of rainfall) to <1 day to several days for those adsorbed to larger particles. Smaller atmospheric particulates containing PAHs are easily inhaled, and may pose special problems, as yet unevaluated, for airborne organisms such as birds, insects, and bats. Photooxidation, one of the most important processes in the removal of PAHs from the atmosphere, can also produce reaction products that are carcinogenic or mutagenic, although little is known of their persistence. One of the more common photooxidation reactions of PAHs is the formation of endoperoxides that ultimately undergo a series of reactions to form quinones. Various parameters may modify chemical and photochemical transformation of PAHs in the atmosphere, including light intensity, concentration of gaseous pollutants (O_3, NO_x, SO_x), and chemicophysical characteristics of particulates or substrates into which the PAHs are adsorbed; depending on these variables, the half-life of benzo[a]pyrene in the atmosphere varies from 10 min to 72 days. Atmospheric PAHs are transported over relatively long distances from industrial areas and from natural forest and prairie fires; however, sites nearer urban centers have much higher PAH deposition rates than more rural areas.

Much of the PAHs released into the atmosphere eventually reach the soil by direct deposition or by deposition on vegetation. In soils, adsorption of naphthalene mainly occurs on the organic matter and is not related to the size of the soil particles.The PAHs may be adsorbed or assimilated by plant leaves before entering the animal food chain, although some adsorbed PAHs may be washed off by rain, chemically oxidized to other products, or returned to the soil as the plants decay. PAHs assimilated by vegetation may be translocated, metabolized, and possibly photodegraded within the plant. In some plants growing in highly contaminated areas, assimilation may exceed metabolism and degradation, resulting in an accumulation in plant tissues.

In water, PAHs may either evaporate, disperse into the water column, become incorporated into bottom sediments, concentrate in aquatic biota, or experience chemical oxidation and biodegradation. The most important degradation processes for PAHs in aquatic systems are photooxidation, chemical oxidation, and biological transformation by bacteria and animals. Most PAHs in aquatic environments are associated with particulate materials; only about 33% are present in dissolved form. PAHs dissolved in the water column will probably degrade rapidly through photooxidation and degrade most rapidly at higher concentrations, at elevated temperatures, at elevated oxygen levels, and at higher incidences of solar radiation. Microcosm studies with creosote-associated PAHs showed that initial concentrations of 7.0 μg total PAHs/L degraded to 0.8 μg/L in 84 days, and 5803.0 to 14.0 μg/L in the same time frame. Low- and high-molecular weight PAHs were lost first from the water column followed by PAHs of intermediate molecular weight, that is, those with 4–5 aromatic rings. The half-time persistence in water of most PAHs in that study was about 1 week; in sediments, PAHs peaked at about 8 weeks with little degradation during the next 6 weeks.

The ultimate fate of those PAHs that accumulate in sediments is believed to be biotransformation and biodegradation by benthic organisms. PAHs in aquatic sediments, however, degrade very slowly in the absence of penetrating radiation and oxygen, and may persist indefinitely in oxygen-poor basins or in anoxic sediments. Persistence of phenanthrene and naphthalene is high in Cook Inlet, Alaska, where the sediments are organic rich and contain low numbers of microbial degraders. Some scientists aver that the presence of microorganisms accelerates degradation of phenanthrene and other PAHs and is a better predictor of maximum biodegradation rate than media without microorganisms. PAH degradation in aquatic environments occurs at a slower rate than that in the atmosphere, and the cycling of PAHs in aquatic environments, as is true for other ecological systems, is poorly understood. Degradation rates of fluoranthene and other PAHs are dependent on the presence of sediments and other PAHs. In sediment-free systems, fluoranthene biodegradation did not occur when it was present alone or in combination with acenaphthene, but was degraded when combined with naphthalene; naphthalene and acenaphthene degradation were not influenced by fluoranthene. In sediment-containing systems, fluoranthene degradation occurred only in the presence of naphthalene. After complete degradation of naphthalene, fluoranthene degradation ceased.

Animals and microorganisms can metabolize PAHs to products that may ultimately experience complete degradation. The degradation of most PAHs is not completely understood. Those in the soil may be assimilated by plants, degraded by soil microorganisms, or accumulated to relatively high levels in the soil. High PAH concentrations in soil can lead to increased populations of microorganisms capable of degrading the compounds. Of equal importance to PAH cycling dynamics is the physical state of the PAH, i.e., whether in vapor phase or associated with particles such as flyash. Particles may increase or decrease the susceptibility of PAHs to degradation, depending on the PAH and particles involved.

PAHs can be taken into the mammalian body by inhalation, skin contact, or ingestion, although they are poorly absorbed from the gastrointestinal tract. The main routes of elimination of PAHs and their metabolites include the hepatobiliary system and the gastrointestinal tract. In mammals, an enzyme system variously known as the cytochrome P450-dependent mixed-function oxidase, mixed-function oxidase, mixed-function oxygenase, aryl hydrocarbon hydroxylase, or drug metabolizing system, is responsible for initiating the metabolism of various lipophilic organic compounds, including PAHs. The primary function of this system is to render poorly water-soluble lipophilic materials more water soluble, and therefore more available for excretion. Some PAHs are transformed to intermediates, which are highly toxic, mutagenic, or carcinogenic to the host. Oxidative metabolism of PAHs in this system proceeds via high electrophilic intermediate arene oxides, some of which bind covalently to cellular macromolecules such as DNA, RNA, and protein. Most authorities agree that metabolic activation by the mixed-function oxidase system is a necessary prerequisite for PAH-induced carcinogenesis and mutagenesis. This enzyme system is known to be present in rodent tissues, and human liver, skin, placenta, fetal liver, macrophages, lymphocytes, and monocytes. Studies with rodents have shown that the mixed-function oxidase system can convert PAHs to various hydroxylated derivatives including phenols, quinones, and epoxides, and can also activate PAHs to produce carcinogenic metabolites. Oxygenated groups of xenobiotics are then conjugated with glutathione-*S*-transferase and other transferase enzymes. The resulting polar and water-soluble end product is excreted via the bile or urine. Fish and most crustaceans tested to date possess the enzymes necessary for activation, but some mollusks and other invertebrates are unable to efficiently metabolize PAHs. Although many aquatic organisms possess the requisite enzyme systems for metabolic activation of PAHs, it is not certain in many cases whether these enzymes produce the same metabolites as those produced by mammalian enzymes.

Virtually all organisms possess biotransformation or detoxification enzymes which convert lipophilic xenobiotics to water-soluble and excretable metabolites. In the metabolic process, PAHs are altered by Phase I metabolism into various products such as epoxides, phenols, quinones, dihydrodiols, dihydrodiol epoxides, tetrahydrotriols, and tetrahydrotetrols. The intermediate metabolites have been identified as the mutagenic, carcinogenic, and teratogenic agents. The activation mechanisms occur by hydroxylation or production of unstable epoxides of PAHs which damage DNA, initiating the carcinogenic process. Metabolic formation of bay region diol epoxides represents an important pathway by which PAHs are activated to carcinogens (Figure 26.4). Such metabolic activation proceeds via initial formation of the dihydrodiol with the bay region double bond, followed by subsequent oxidation of the dihydrodiol to the bay region diol epoxide. Active epoxides may be converted to less toxic products by various enzymatic and other reactions. In the case of benzo[a]pyrene, the "ultimate carcinogen" (7 beta, 8 alpha-dihydroxy-7,8,9,10 tetrahydrobenzo[a]pyrene-9 alpha, 10 alpha-epoxide) reacts with the guanine of RNA and DNA, the linkage taking place between the C-10 atom of benzo[a]pyrene and the C-2 amino group of guanine. In Phase II metabolism, these products are converted into highly water-soluble conjugates with a large water-soluble moiety,

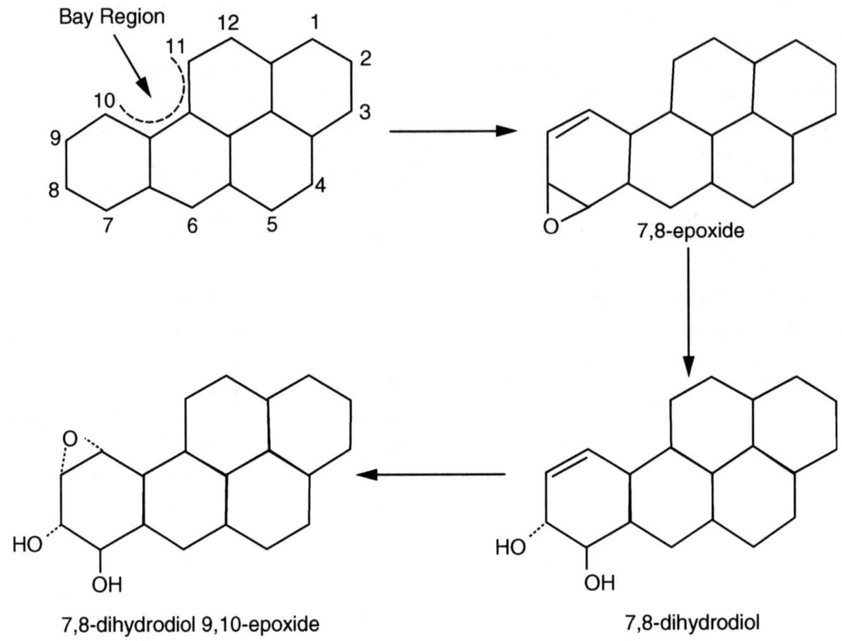

Figure 26.4.
The bay region dihydrodiol epoxide route of benzo[a]pyrene.

such as the tripeptide glutathione or sugar derivative glucuronic acid.

There is considerable variability between species in their ability to metabolize benzo[a]pyrene. For example, two bottom-dwelling species of fish (common carp *Cyprinus carpio*, brown bullhead *Ictalurus nebulosus*) differ by a factor of about 12 in the rate at which liver microsomes degrade benzo[a]pyrene to metabolites, with carp degrading BaP more rapidly. Major BaP metabolites from both species included BaP-7,8-diol, BaP-8,9-diol, 3-hydroxy-BaP, 9-hydroxy-BaP, BaP-1,6-quinone, BaP-3,6-quinone, and BaP-6,12-quinone. Carp liver microsomes converted a much greater proportion of BaP to benzo-ring dihydrodiols than did bullhead liver microsome; bullhead liver microsomes formed a significantly higher percentage of BaP-quinones. In other species, glutathione conjugation represented the major hepatic detoxification pathway of benzo[a]pyrene, as was the case for white suckers (*Catostomus commersonii*).

26.3 Concentrations in Field Collections

PAHs are ubiquitous in the environment. In nonbiological materials, concentrations are elevated in the vicinity of urban industrialized locales, and from areas of significant wood burning activities such as forest fires and residential home heating. Terrestrial vegetation and aquatic invertebrates can accumulate significant concentrations of PAHs, possibly due to inefficient or missing mixed-function oxidase systems. Fish do not appear to contain grossly elevated PAH residues; this may be related to their efficient degradation system. At present, data are scarce on PAH background concentrations in natural populations of birds and other wildlife – although it seems unlikely that significant accumulations will occur. Some investigators have shown that aquatic invertebrates, fish, and amphibians collected from areas of high sediment PAH content show elevated frequencies of hyperplasia and neoplasia, and that hepatic carcinoma has been induced in rainbow trout (*Oncorhynchus mykiss*) by benzo[a]pyrene through dietary and intraperitoneal injection routes.

26.3.1 Nonbiological Samples

Total PAH levels in air are usually much higher in winter than in summer, higher in urban communities than in rural areas, and appear to be related primarily to the weight of total suspended particulates in the atmosphere. PAH levels in precipitation are signi-ficantly higher in winter than in summer, primarily due to emissions from household heating. Among industrial sources, the production of metallurgical coke is the single most significant source of atmospheric PAHs in Ontario, Canada. Coke production in 1977 represented about 52% of all PAH emissions from Ontario sources vs. about 46% formed as a result of forest fires. Beyond 2 km distance from the coke point source, PAH concentrations in air were typical of those measured in major urban nonindustrialized areas. A variety of PAHs have been detected in ambient air in the U.S. and elsewhere. Benzo[a]pyrene, because of its carcinogenic properties, has been monitored extensively, and has frequently been used as an indicator of PAHs. In general, total PAHs in air are about 10 times higher than benzo[a]pyrene levels, although this relation is extremely variable. Benzo[a]pyrene levels, like total PAHs, were higher in winter than summer, probably due to residential and industrial heating; air levels in urban areas with coke ovens were 40–70% higher than in cities without coke ovens, but this may be related to higher industrial emissions in those cities. In one case, benzo[a]pyrene levels in air from the center of a remote mountain community in Colorado were several times higher than what is usually found in U.S. metropolitan areas, and was attributed to extensive residential wood burning. Average concentrations of benzo[a]pyrene in urban air nationwide declined from 3.2 ng/m^3 in 1966 to 0.5 ng/m^3 in 1978, an 80% decrease. These decreases are believed to be primarily due to decreases in coal consumption for commercial and residential heating, improved disposal of solid wastes, and restrictions on open burning.

A major source of PAHs in soils and soil litter is from emissions and deposition from forest fires. In a controlled burn study, it was shown that lower molecular weight PAHs, such as phenanthrene and fluorene, which had been deposited in soil litter, degraded to non-detectable levels within 2 years after burning. Higher molecular weight PAHs such as benzo[k]fluorene, benzo[a]pyrene, benzo[g,h,i]perylene, perylene, and indeno[1,2,3-cd]pyrene, were more persistent in litter, decreasing after 5 years to about 20% of initial deposition. Although movement into the top 2 cm of the soil profile was initially more pronounced for lower molecular weight PAHs, all compounds appeared to reach equilibrium between litter and soil on the basis of organic content within one year postburn. Differential persistence and fate of PAHs on slash burn sites is explained by solubility, K_{ow}, and other physicochemical properties. PAHs from vehicle emissions constitute a minor, but measurable, source of soil PAHs. The majority of highway-derived PAHs appears to be deposited within 3.8 m of the road, but the influence of the highway may extend to nearly 70 m. The use of composted municipal wastes for conditioning of agricultural soils is not recommended, as these contain at least nine identified carcinogenic PAHs.

PAH-contaminated sediments are involved in epizootics of neoplasms in native fishes at contaminated sites, and other adverse biological effects. In the Detroit River, Michigan, brown bullheads (*Ameiurus nebulosus*) from sediments with the highest PAH concentration of 346.0 mg/kg DW had the highest prevalence of external abnormalities (lip and skin lesions, truncated barbels) and liver lesions and these were positively correlated with concentrations of total PAHs in the sediments. In the Black River, Ohio, between 1980 and 1982, PAH concentrations declined by 65% and in tissues of brown bullheads by 93%. By 1987, sediment PAHs declined an additional 99% coincident with closure of a coking facility in 1983. Liver cancer frequency in 3- to 4-year-old bullheads declined to 10% frequency in 1987 vs. 39% frequency in 1982; livers without lesions increased from 20% frequency in 1982 to 42% in 1987. Sediments and sediment extracts from the Buffalo River, New York, contained elevated levels of carcinogenic PAHs (1000.0–16,000.0 μg/kg). Brown bullheads, in response to repeated applications of Buffalo River sediment extracts, showed epidermal hyperplasia and neoplasia when compared to controls. Extraction of PAHs from industrially contaminated sediments of Lake Erie and its tributaries (27,000.0–363,000.0 mg/kg DW total PAHs) established the presence of chemical mutagens that could be correlated with neoplasms in fish from many of the sites. In the Netherlands, high molecular weight PAHs dominated the sediments, fluoranthene and pyrene in freshwater isopods, and naphthalene in water; lipid-based bioconcentration factors (BCFs) increased with increasing hydrophobicity, that is, with increasing K_{OW}. Elevated PAH concentrations in sediments from the North Sea were positively correlated with increasing DNA breaks in the pyloric caeca of resident starfish (*Asterias rubens*). PAH concentrations in sediments from the Great Barrier Reef, Australia, were always <0.8 μg/kg DW, except in small areas close to sites frequently visited by power boats; in those instances, total PAH levels exceeded 13,400.0 μg/kg. Highest PAH levels measured in sediments of Cayuga Lake, New York, were found in marinas or areas of the lake receiving urban runoff, and were apparently not related to stack emissions from a nearby coal-fired power plant; it is probable that stack emissions were either masked by other sources or were atmospherically transported and deposited elsewhere. Coastal and offshore sediments are subject to highly elevated PAH levels from a variety of sources, mostly unknown, relative to preindustrial times. For example, PAH levels in sediments of Penobscot Bay, Maine, fell within the range found in sediments near industrialized regions, and were significantly higher than expected for an area previously considered to be uncontaminated.

Sewage effluents usually contained measurable levels of PAHs, although extreme variability between and among sites is common. During a heavy storm, individual PAH levels in a sewage works may increase more than 100 times over a dry weather period. Conventional sewage treatment plant processes remove up to 90% of carcinogenic PAHs, and this may be increased to 99% using percolating filters and activated sludge processes. Tiger salamanders (*Ambystoma tigrinum*), collected in 1975 from a 13-ha sewage effluent lagoon at Reese Air Force Base, Texas, showed a remarkably high incidence (53%) of neoplastic and other lesions. Analysis of sludge composites showed elevated PAH levels, especially perylene; levels of organochlorine and organophosphorus pesticides, nitrosamines, and heavy metals were judged to be nonelevated. Careful disposal of used motor oils is warranted, as these contain high quantities of mutagenic and carcinogenic PAHs.

All but the most heavily contaminated fresh and marine waters contain total PAH concentrations in the parts-per-trillion or low parts-per-billion range. A large proportion of the PAH content in water is probably adsorbed onto suspended solids. In Lake Michigan, concentrations of total PAHs in the surface microlayer varied from 0.15 to 0.45 μg/L, representing on a relative scale 106 times the concentration in air, suggesting that aerosols are a major source of these compounds and that the microlayer is a repository until the PAHs are removed by adsorption and sedimentation.

26.3.2 Biological Samples

Carcinogenic PAHs have been extracted from a large variety of fresh plants, including root and leaf vegetables, fruits, grains, and edible mushrooms, as well as from various marine bacteria and phytoplankton under circumstances suggesting that PAHs were present due to local biosynthesis. PAH concentrations in terrestrial vegetation were lower at higher altitudes and higher near highways when compared to conspecifics from lower altitudes and further from the highway. Vegetation and soil near known PAH sources are more highly contaminated with PAHs than those collected at greater distances. PAH levels in lettuce (*Lactuca sativa*) grown in Sweden seemed to be directly related to its proximity to local recognized point sources of PAH emitters. Washing lettuce with water had little effect on phenanthrene levels, but significantly reduced other PAHs, such as benzo[a]pyrene, benz[a]anthracene, and benzo[g,h,i]perylene by 68–87%. Fruits and vegetables grown in polluted atmospheres may contain up to 100-fold higher levels of total PAHs than those grown in unpolluted environments. PAH concentrations for plants are generally greater on plant surfaces than internal tissues, greater in above-ground plant parts than those below ground, and greater in plants with broad leaves (greater surface area) than those with narrow leaves. Plants can become contaminated with PAHs through environmental pollution, particularly through deposition from the atmosphere, and also through food processing. For example, the bran portion of milled wheat, as well as finished bran cereal, had a considerably higher PAH content than other fractions or finished products. Enrichment of PAHs in plants is associated with deposition of atmospheric particulate matter with relatively small particle sizes; thus, PAH content is usually in the order of humus > mosses > lichens. Mosses appear to be good indicators of regional PAH air pollution and have been recommended for this purpose. Concentrations of total PAHs in soils, usually the sum of 5–20 PAHs, typically exceeded benzo[a]pyrene levels by at least one order of magnitude; however, concentrations of benzo[a]pyrene in vegetation were generally less than those in soil where plants were growing.

PAH accumulations in marine mollusks have been reported; however, some of these data may be misleadingly low. For example, lengthy cold storage of 10 months can result in loss of volatile PAHs, such as anthracene, in tissues of mussels; accordingly, background concentrations in these organisms may be underreported. Bivalve mollusks tend to accumulate high PAH levels due to their inability to metabolize and excrete them, presumably due to inefficient or missing mixed-function oxidase systems. Cellular proliferative disorders, resembling neoplastic conditions in vertebrates, were found in mussels with the greatest PAH concentrations: 9.5% vs. 0.7% in control sites. Baseline levels of PAHs in indigenous bivalve mollusks reflected the degree of human onshore activity at the various sample sites, and presumably the level of water contamination; however, little relation was evident between accumulated levels of individual PAHs and total PAHs. Elevated PAH concentrations, especially benz[a]anthracene, chrysene, fluorene, phenanthrene, and pyrene in oyster tissues and sediments were measured in samples from the vicinity of marinas, and were higher in oysters in cooler months, when lipids and glycogen were being stored in the preparatory stage to spawning. In general, PAH concentrations in marine clams and mussels were highest in areas adjacent to industrialized bayfronts, and lowest in more remote areas; higher in larger animals and in those with higher lipid concentrations; and lowest in autumn–winter, and highest during spring–summer. A similar pattern was observed in mussels, *Mytilus edulis*, with the more water-soluble, lower molecular weight, PAHs bioconcentrated 10–100 times above that of the higher molecular weight, less water-soluble PAHs; PAH levels in mussels seemed to be independent of water salinity. Clams contaminated with PAHs and removed to clean seawater for 24 h showed significant depuration of unsubstituted 3- and 4-ring PAHs; in contrast, concentrations of all 5-, 6-, and 7-ring compounds, which includes most of the carcinogenic PAHs, were not significantly depurated. A positive relation exists between PAH isomers in sediments, soft tissues of the mussel *Mytilus edulis*, and a seaweed (*Fucus* sp.) collected at Vancouver, British Columbia. For mussels, the general trend towards lower levels of higher molecular weight PAHs relative to levels in sediments suggests an uptake mechanism which involves the solution of PAHs in water; superimposed on this pattern is the more rapid turnover and shorter half-life of lower molecular weight PAHs in mussels.

PAH residues were higher than expected in American lobsters (*Homarus americanus*) collected offshore (mean weight 3.6 kg) when compared to smaller (0.6 kg) lobsters collected inshore, suggesting that age or body size are important modifiers in PAH accumulation dynamics. PAH concentrations in sediments collected near a coking facility in Nova Scotia in 1980 contained up to 2830.0 mg/kg DW, or more than 20 times the levels recorded in Boston (Massachusetts) Harbor; concentrations in excess of 100.0 mg/kg DW sediment were recorded for phenanthrene, fluorene, pyrene, benz[a]anthracene, chrysene, benzo[e]pyrene, benzo[b]fluoranthene, and benzo[a]pyrene, and these seemed to reflect the elevated tissue levels in American lobsters collected from that locale. PAH residues in digestive glands of American lobsters collected in 1979 in Nova Scotia from the vicinity of a major oil spill were higher than those from coastal control sites; however, PAH contents of edible muscle from control and oiled lobsters were similar.

PAH levels in fish are usually low because this group rapidly metabolizes PAHs; furthermore, higher molecular weight PAHs, which include the largest class of chemical carcinogens, do not seem to accumulate in fish. Tissue lipid concentration was the primary factor in determining PAH concentrations in fishes. Raw fish from unpolluted waters usually do not contain detectable amounts of PAHs, but smoked fish contain elevated concentrations of PAHs. The concentration of benzo[a]pyrene in skin of cooked fish was much higher than in other tissues, suggesting that skin may serve as a barrier to the migration of PAHs in body tissues.

In many cases, aquatic organisms from PAH-contaminated environments have a higher incidence of tumors and hyperplastic diseases than those from nonpolluted environments. Carcinogenic PAHs have not been unequivocally identified as the causative agent for an increased incidence of cancer in any natural population of aquatic organisms, according to several authorities. However, a growing body of evidence links PAHs to cancer in feral fish populations especially bottom-dwelling fish from areas with sediments heavily contaminated with PAHs.

Sediments and biota collected from the Hersey River, Michigan, in 1978, were heavily contaminated with phenanthrene, benz[a]anthracene, and benzo[a]pyrene when compared to a control site. Elevated PAH concentrations were recorded in sediments, whole insect larvae, crayfish muscle, and flesh of lampreys (family Petromyzontidae), brown trout (*Salmo trutta*), and white suckers (*Catostomus commersoni*), in that general order. The polluted collection locale was the former site of a creosote wood preservation facility between 1902 and 1949, and, at the time of the study, received Reed City wastewater treatment plant effluent, described as an oily material with a naphthalene-like odor. In San Francisco Bay, elevated PAH concentrations in fish livers reflected elevated sediment PAH concentrations. In Chesapeake Bay, spot (*Leiostomus xanthurus*) collected from a PAH-contaminated tributary (up to 96.0 mg PAHs/kg DW sediment) had elevated cytochrome P450 and EROD activity in liver and intestine microsomes. Intestinal P450 activity was 80–100 times higher in fish from highly contaminated sites than in conspecifics from reference sites; intestinal EROD activity had a similar trend. Liver P450 and EROD activity was about 8 times higher in spot from the contaminated sites when compared to the reference sites. Liver P450 activity correlated positively with sediment PAH but intestinal P450 activity seemed to reflect dietary exposure. The poor correlation between hepatic concentrations of PAHs and P4501A is attributed to the rapid metabolism of these compounds.

Embryos of the common tern (*Sterna hirundo*) from eight colonies in the Netherlands and Belgium were analyzed for PAHs in May–July 1991. In general, eggs containing embryos with elevated PAH concentrations in the yolk sac and with elevated hepatic EROD activity levels were laid later, had a more prolonged incubation period, were of smaller volume, and produced smaller chicks.

Norway rats (*Rattus norvegicus*) from a PAH-contaminated site near Lyon, France, when compared to conspecifics from a reference site, had altered liver and lung monooxygenases and altered antioxidant enzyme activities in liver, lungs, and erythrocytes. PAH concentrations in muscle tissues of ten species of marine mammals from Newfoundland–Labrador waters in 1988–89 were highest in harbor porpoise (*Phocoena phocoena*) and lowest in large whales and hooded seals

(*Cystophora cristata*). Benzo[a]pyrene was specifically related to the presence of tumors in beluga whales (*Delphinapterus leucas*) from the Gulf of St. Lawrence in Canada.

26.4 Lethal and Sublethal Effects

A wide variety of PAH-caused adverse biological effects have been reported in numerous species of organisms under laboratory conditions, including effects on survival, growth, reproduction, metabolism, and especially tumor formation. Inter- and intraspecies responses to carcinogenic PAHs were quite variable, and were significantly modified by many chemicals including other PAHs that are weakly carcinogenic or noncarcinogenic. Until these interaction effects are clarified, the results of single substance laboratory tests may be extremely difficult to apply to field situations of suspected PAH contamination.

26.4.1 Fungi

Fungal degradation of PAHs may be important in the detoxification and elimination of PAHs in the environment. The fungus *Cunninghamella elegans*, for example, inhibited the mutagenic activity of benzo[a]pyrene, 3 ethyl cholanthrene, benz[a]anthracene, and 7,12-dimethylbenz[a]anthracene, as judged by results of the Ames test using *Salmonella typhimurium*. The rate of decrease in mutagenic activity in bacterial cultures incubated with PAHs was coincident with the rate of increase in fungal metabolism. *Cunninghamella elegans* metabolized PAHs to dihydrodiols, phenols, quinones, and dihydrodiol epoxides, and to sulfate, glucuronide, and glucoside conjugates of these primary metabolites in a manner similar to that reported for mammalian enzyme systems, suggesting that this organism (and perhaps other fungi) is important in PAH metabolism and inactivation.

26.4.2 Terrestrial Plants

Biological effects of PAHs on terrestrial vegetation have been extensively reviewed. In general, these authorities agreed on several points. First, plants and vegetables can absorb PAHs from soils through their roots, and translocate them to other plant parts such as developing shoots. Uptake rates were governed in part, by PAH concentration, PAH water solubility, soil type, and PAH physicochemical state (vapor or particulate). Lower molecular weight PAHs were absorbed by plants more readily than higher molecular weight PAHs. Under laboratory conditions, some plants concentrated selected PAHs above that of their immediate geophysical surroundings, but this has not been conclusively demonstrated in field-grown cultivated crops or other vegetation. Second, above-ground parts of vegetables, especially the outer shell or skin, contained more PAHs than underground parts, and this was attributed to airborne deposition and subsequent adsorption. Externally deposited PAHs in vegetables were difficult to remove with coldwater washings; not more than 25% were removed from lettuce, kale, spinach, leeks, and tomatoes using these procedures. Third, PAH-induced phytotoxic effects were rare; however, the database on this subject is small. Fourth, most higher plants can catabolize benzo[a]pyrene, and possibly other PAHs, but metabolic pathways have not been clearly defined. Finally, the biomagnification potential of vegetation in terrestrial and aquatic food chains needs to be measured; this work should be conducted with a variety of PAHs in both field and laboratory experiments.

Some plants contain chemicals known to protect against PAH effects. Certain green plants contain ellagic acid, a substance that can destroy the diol epoxide form of benzo[a]pyrene, inactivating its carcinogenic and mutagenic potential. PAHs synthesized by plants may act as plant growth hormones. Some vegetables, such as cabbage, brussels sprouts, and cauliflower, contain naturally occurring antineoplastic compounds including benzyl isothiocyanate and phenethyl isothiocyanate; these compounds are known to inhibit mammary cancers, stomach tumors, and pulmonary edemas induced in rats by benzo[a]pyrene and 7,12-dimethylbenz[a]anthracene. Decreased activation of carcinogens has also been demonstrated in animals fed diets that were high in protein, low in carbohydrate, and containing adequate choline; the reverse was observed in diets high in carbohydrate, low in protein, or containing certain organophosphorus insecticides, piperonyl butoxide, carbon tetrachloride, nickel carbonyl, or tin. In cases where dietary constituents can alter the metabolism of foreign agents, such as PAHs, the anticarcinogenic effect may result from an alteration of steady-state levels of activated vs. detoxified metabolites. The implications of these observations to herbivorous wildlife are unknown at present.

26.4.3 Aquatic Biota

PAHs vary substantially in their toxicity to aquatic organisms. In general, toxicity increases as molecular weight increases (although high molecular weight PAHs have low acute toxicity, perhaps due to their low solubility in water) and with increasing alkyl substitution on the aromatic ring. Toxicity is most pronounced among crustaceans and least among teleosts. In all but a few cases, PAH concentrations that are acutely toxic to aquatic organisms are several orders of magnitude higher than concentrations found in even the most heavily polluted waters. Sediments from polluted regions, however, may contain PAH concentrations similar to those which are acutely toxic, but their limited bioavailability would probably render them substantially less toxic than PAHs in solution.

Several PAHs accumulated by aquatic organisms during exposure are severely toxic when the contaminated organisms were exposed to sunlight or ultraviolet (UV) radiation. PAHs that have exhibited photoinduced toxicity in fishes, daphnids, frogs, or algae and macrophytes at PAH concentrations well below their aqueous solubility due to simultaneous UV radiation exposure include fluoranthene, pyrene, benzo[a]pyrene, benz[a]anthracene, acridine, and benz[a]fluorene. Anthracene, for example, is not acutely toxic to fish and algae within its aqueous solubility range (about 35.0 μg/L at 25°C) unless the exposure is performed in the presence of UV radiation from simulated or natural sunlight. Anthracene is phototoxic in this concentration range to mosquito larvae, bluegills, and leopard frog embryos. Anthracene is phototoxic to alga (*Selenastrum capricornutum*), with growth inhibition of 50% at 16.1 μg/L in 28 h and 10% inhibition at 8.3 μg/L. PAH phototoxicity is due to PAH concentrations in tissues, length of exposure to radiation and absorption, efficiency of the photoconversion process, and the probability of the excited intermediary reacting with a target molecule. Studies with larvae of the bullfrog (*Rana catesbeiana*) and phototoxic effects of fluoranthene suggest that behavioral and histopathological endpoints – especially skin histology – are more sensitive than survival. Sediments contaminated with complex mixtures of PAHs also show enhanced toxicity to aquatic species under conditions of UV light representative of sunlight. In one case, interstitial sediment pore water from sediments near an oil refinery discharge was toxic to a freshwater oligochaete worm (*Lumbriculus variegatus*) following exposure to UV light, but worms exposed to the same pore water without UV treatment were unaffected. Use of electron density shape features has been used to model photoinduced toxicity of 16 PAHs to duckweed (*Lemna gibba*), with good correlation between detailed molecular shape features and toxicity.

A growing literature exists on uptake, retention, and translocation of PAHs by aquatic plants and animals. Authorities generally agree that most species of aquatic organisms studied to date rapidly accumulate (i.e., bioconcentrate) PAHs from low concentrations in the ambient medium. Uptake of PAHs is highly species specific, being higher in algae, mollusks, and other species which are incapable of metabolizing PAHs. Bioconcentration factors (BCFs) tend to increase with increasing molecular weight of the PAH, with increasing octanol–water partition coefficient values, with time until approaching an apparent equilibrium level (sometimes within 24 h), and with increases in dissolved organic matter in the medium, lipid content of organism, and a variety of endogenous and exogenous factors.

BCF values have been determined for selected PAHs and aquatic organisms; additional BCF data for aquatic biota are available for plants, crustaceans, tunicates, mollusks, annelids, and fishes. Algal accumulation of benzo[a]pyrene increased linearly in a 24-h exposure period, and correlated positively with surface area, suggesting adsorption rather than absorption. Algae readily transform benzo[a]pyrene to oxides, peroxides, and dihydrodiols. Photosynthetic rates of algae, and presumably PAH accumulations, were significantly modified by light regimens. For reasons still unexplained, algae grown in "white" light (major energy in blue–green portion of the spectrum) were more sensitive to benzo[a]pyrene than were cultures grown in "gold" light.

Accumulation by American oysters (*Crassostrea virginica*) and clams (*Rangia cuneata*) of naphthalene, phenanthrene, fluorene, and their methylated derivatives increased with increasing methylation and PAH molecular weight; uptake was more rapid under conditions of continuous flow than in static tests. When returned to PAH-free seawater, mollusks released PAHs to non-detectable levels in about 60 days, with high molecular weight PAHs depurated more slowly than low molecular weight compounds; brown shrimp (*Penaeus aztecus*) and longnose killifish (*Fundulus similis*), which can metabolize PAHs, lost PAHs more quickly than did clams and oysters, which apparently lack the detoxifying enzymes. Pink shrimp (*Penaeus duorarum*) exposed to 1.0 μg chrysene/L for 2 days and then transferred to unpolluted seawater for an additional 28 days contained concentrations of chrysene (91.0 μg/kg fresh weight (FW) in abdomen, 48.0 μg/kg in cephalothorax) that were considered potentially hazardous to human consumers over extended periods. Eggs of the sand sole (*Psettichthys melanostictus*) exposed to 0.1 μg benzo[a]pyrene/L for 5 days showed reduced and delayed hatch and, when compared to controls, produced larvae with high accumulations (2.1 mg/kg FW) and gross abnormalities, such as twinning and tissue overgrowths, in 50% of the test larvae. Naphthalene and benzo[a]pyrene were rapidly accumulated from the medium by three species of California marine teleosts; loss was rapid, being >90% for naphthalene in 24 h, and 20% (muscle) to 90% (gill) for benzo[a]pyrene in a similar period. Phenanthrene is metabolized by many species of aquatic organisms, including fish. A marine flounder *Platichthys flesus*, given a single oral dose of 0.7 mg phenanthrene/kg body weight (BW), contained elevated phenanthrene concentrations in lipids, melanin-rich tissues (such as skin), and the eye lens; most was eliminated within 2 weeks. Studies on uptake rates of radiolabeled phenanthrene by marine polychaetes (*Aberenicola pacifica* from PAH-contaminated sediments of high or low organic content showed that rates increased when worms were acclimatized for 48 h to sediments of low organic content. Different rates of accumulation and depuration of benzo[a]pyrene and naphthalene in bluegill (*Lepomis macrochirus*) and *Daphnia magna* have been documented. Benzo[a]pyrene accumulations in bluegill, for example, were 10 times greater than naphthalene, but benzo[a]pyrene is extensively metabolized, whereas naphthalene is not. Consequently, postexposure accumulations of naphthalene greatly exceeded that of the parent benzo[a]pyrene. Because the more hydrophobic PAHs, such as benzo[a]pyrene, show a high affinity for binding to dissolved humic materials and have comparatively rapid biotransformation rates, these interactions may lessen or negate bioaccumulation and food chain transfer of hydrophobic PAHs.

Time to depurate or biotransform 50% of accumulated PAHs (Tb1/2) varied widely. The Tb1/2 values for *Daphnia pulex* and all PAH compounds studied ranged between 0.4 and 0.5 h. In rainbow trout given an intraarterial injection of 10.0 mg/kg BW of 2-methylnaphthalene, fluorene, or pyrene, the Tb1/2 values ranged between 9.6 and 12.8 h; when route of exposure was intragastric and doses were 50.0 mg/kg BW, there was negligible uptake. For marine copepods and naphthalene, a Tb1/2 of about 36 h was recorded. For most marine bivalve mollusks, Tb1/2 values ranged from 2 to 16 days. Some species, such as the hardshell clam (*Mercenaria mercenaria*), showed little or no depuration, while others, such as oysters, eliminated up to 90%

of accumulated PAHs in 2 weeks – although the remaining 10% was released slowly, and traces may remain indefinitely. Percent loss of various PAHs in American oysters (*Crassostrea virginica*), 7 days postexposure, ranged from no loss for benzo[a]pyrene to 98% for methylnaphthalene; intermediate were benz[a]anthracene (32%), fluoranthene (66%), anthracene (79%), dimethylnaphthalene (90%), and naphthalene (97%). Teleosts and arthropods usually had low Tb1/2 values. In bluegill, 89% loss of benzo[a]pyrene was recorded 4 h postexposure; for midge larvae it was 72% in 8 h, and for daphnids it was 21% in 18 h.

The role of sediments in PAH uptake kinetics should not be discounted. Yellow perch (*Perca flavescens*) and northern pike (*Esox lucius*) exposed to sediments contaminated with up to 3.7 mg PAHs/kg had pituitary damage and were unable to increase serum cortisol in response to the acute stress of capture when compared to conspecifics from uncontaminated sites, suggesting that life-long exposure to PAHs will destabilize the cortisol-producing endocrine system. Male winter flounders (*Pleuronectes americanus*) exposed to PAH-contaminated sediments at concentrations as low as 1.0 mg/kg for 4 months had altered mixed-function oxygenase levels and fat content of liver. Sediment-associated anthracene contributed about 77% of the steady-state body burden of this compound in the amphipod *Hyalella azteca*. For benzo[a]pyrene and the amphipod *Pontoporeia hoyi*, the sediment source (including interstitial water) accounted for 53% in amphipods collected at 60 m, but only 9% at 23–45 m. Benthos from the Great Lakes, such as oligochaete worms (*Limnodrilus* sp., *Stylodrilus* sp.) and amphipods (*Pontoporeia hoyi*), obtain a substantial fraction of their PAH body content from the water when sediment PAH concentrations are low. However, when sediment PAH concentrations are elevated, benthos obtain a majority of their PAHs from that source through their ability to mobilize PAHs from the sediment–pore water matrix; the high concentrations of phenanthrene, fluorene, benzo[a]pyrene, and other PAHs measured in these organisms could provide a significant source of PAHs to predator fish. Great Lakes benthos appear to contain as much PAHs as the fine grain fraction of the sediment which serves as their food, although overlying water or pore water appears to contribute a larger proportion of PAHs to the organism's body burden than does sediments. Marine mussels (*Mytilus edulis*) and polychaete annelid worms (*Nereis virens*) exposed for 28 days to sediments heavily contaminated with various PAH compounds accumulated significant concentrations (up to 1000 times control levels) during the first 14 days of exposure, and little thereafter; during a 5-week postexposure period, depuration was rapid, with the more water-soluble PAHs excreted most rapidly; PAH levels usually remained above control values to the end of the postexposure period. English sole (*Parophrys vetulus*), during exposure for 11–51 days to PAH-contaminated sediments, showed significant accumulations of naphthalenes in liver (up to 3.1 mg/kg DW) after 11 days, with concentrations declining markedly thereafter; uptake of phenanthrene, chrysene, and benzo[a]pyrene was negligible during the first 7 days.

Selected studies on effects of benzo[a]pyrene to aquatic vertebrates demonstrate dose-dependent increases in death, enzyme disruption, liver and intestinal histopathology, carcinogenic and mutagenic effects, and behavioral deficits. In general, effects of benzo[a]pyrene were exacerbated by poor health, and by interactions with radiation or cadmium; effects were modified by route of exposure, diet, and age of organism. Exposure of fishes to BaP at early life stages has potential life-long significance. Embryonic sublethal doses of BaP are capable of inducing sublethal changes in behavior, weeks or even months following hatching of salmon and trout.

Fluorene effects in freshwater pond ecosystems have been evaluated. In ponds exposed to initial fluorene concentrations of 0.12–2.0 mg/L, Tb1/2 values in water ranged from 6 to 11 days. Ten weeks after fluorene introduction, little degradation had occurred in the organic bottom sediments; fluorene residues were present in fish, invertebrates, and rooted submerged macrophytes. Studies with fingerling bluegills showed that 0.062 mg fluorene/L adversely affected their ability to capture chironomid prey, 0.12 mg/L reduced growth, and 1.0 mg fluorene/L increased their vulnerability to predation by largemouth bass (*Micropterus salmoides*). The authors concluded that fluorene, at concentrations well below its solubility and at levels that could realistically occur in the environment, represents a potential hazard to aquatic organisms.

Large interspecies differences in ability to absorb and assimilate PAHs from food have been reported. For example, crustaceans and fish readily assimilated PAHs from contaminated food, whereas mollusks and polychaete annelids were limited. In all cases where assimilation of ingested PAHs was demonstrated, metabolism and excretion of PAHs were rapid. Thus, little potential exists for food chain biomagnification of PAHs. In laboratory aquatic ecosystem studies, benzo[a]pyrene can be accumulated to high, and potentially hazardous, levels in fish and invertebrates. In the case of mosquitofish (*Gambusia affinis*), almost all the accumulated benzo[a]pyrene was from its diet, with negligible accumulations from the medium. However, mosquitofish degraded benzo[a]pyrene about as rapidly as it was absorbed, in contrast to organisms such as snails (*Physa* sp.) which retained most (88%) of the accumulated benzo[a]pyrene for at least 3 days postexposure, presumably due to deficiencies in their mixed-function oxidase detoxication system. Benzo[a]pyrene, when administered to northern pike (*Esox lucius*) through the diet or the medium, followed similar pathways: entry via the gills or gastrointestinal system, metabolism in the liver, and excretion in the urine and bile. Benthic marine fishes exposed to naphthalene or benzo[a]pyrene, either in diet or through contaminated sediments, accumulated substantial concentrations in tissues and body fluids. The tendency of fish to metabolize PAHs extensively and rapidly may explain why benzo[a]pyrene, for example, is frequently undetected, or only detected in low concentrations in livers of fish from environments heavily contaminated with PAHs. Extensive metabolism of benzo[a]pyrene plus the presence of large proportions of polyhydroxy metabolites in liver of English sole indicates the formation of reactive intermediates such as diol epoxides and phenol epoxides of benzo[a]pyrene, both of which are implicated in mammalian mutagenesis and carcinogenesis.

Cytotoxic, mutagenic, and carcinogenic effects of many PAHs are generally believed to be mediated through active epoxides formed by interaction with microsomal monooxygenases. These highly active arene oxides can interact with macromolecular tissue components and can further be metabolized or rearranged to phenols or various conjugates. They can also be affected by epoxide hydrolase to form dihydrodiols, which are precursors of biologically active diol epoxides – a group that has been implicated as ultimate carcinogens. Investigators generally agree that marine and freshwater fishes are as well equipped as mammals with liver PAH-metabolizing enzymes; rapidly metabolize PAHs by liver mixed-function oxidases, with little evidence of accumulation; translocate conjugated PAH metabolites to the gall bladder prior to excretion in feces and urine; and have mixed-function oxidase degradation rates that are significantly modified by sex, age, diet, water temperature, dose–time relationships, and other variables. In addition, many species of fishes can convert PAHs, benzo[a]pyrene for example, to potent mutagenic metabolites, but because detection of the 7,8-dihydrodiol, 9,10-epoxide by analytical methods is difficult most investigators must use biological assays, such as the Ames test, to detect mutagenic agents. The interaction effects of PAHs with inorganic and other organic compounds are poorly understood. Specific examples of the above listed phenomena for PAH compounds and teleosts are documented for benzo[a]pyrene, benz[a]anthracene, chrysene, pyrene, fluoranthene, and 7,12-dimethylbenz[a]anthracene.

Increasing frequencies of liver tumors in wild populations of fish is documented during the past decades, especially in brown bullhead (*Ictalurus nebulosus*) from the Fox River, Illinois (12% tumor frequency), in Atlantic hagfish (*Myxine glutinosa*) from Swedish estuaries (6%), in English sole from the Duwamish estuary, Washington (32%), and in tomcod (*Microgadus tomcod*) from the Hudson River, New York (25%). In all these instances, significant levels of contaminants were present in the sediments, including PAHs. PAHs have been identified as genotoxic pollutants in sediments from the Black River, Ohio, where a high incidence of hepatoma and other tumors has been observed in ictalurid fishes. Reports of tumors in Great Lakes fish populations have been increasing. Tumors of thyroid, gonad, skin, and liver are reported, with tumor frequency greatest near areas contaminated by industrial effluents such as PAHs; liver tumors were common among brown bullhead populations at sites with large amounts of PAHs in sediments. A positive relationship was finally established between sediment PAH levels and prevalence of liver lesions in English sole in Puget Sound, Washington, and sediment levels and liver tumor frequency in brown bullheads from the Black River, Ohio. Sediment PAH levels in the Black River, Ohio, from the vicinity of a coke plant outfall, were up to 10,000 times greater than those from a control location: concentrations were greater than 100.0 mg/kg for pyrene, fluoranthene, and phenanthrene; between 50.0 and 100.0 mg/kg for benz[a]anthracene, chrysene, and benzofluoranthenes; and between 10.0 and 50.0 mg/kg for individual naphthalenes, benzo[e]pyrene, benzo[a]pyrene, perylene, indeno[1,2,3-cd]pyrene, benzo[g,h,i] perylene, and anthanthrene. Brown bullheads from this location contained >1.0 mg/kg of acenaphthalene (2.4), phenanthrene (5.7), fluoranthene (1.9), and pyrene (1.1), and lower concentrations of heavier molecular weight PAHs; bullheads also exhibited a high (33%) liver tumor frequency, which seemed to correspond to their PAH body burdens. Investigators concluded that the elevated frequency of liver neoplasia in Black River bullheads was chemically induced, and was the result of exposure to PAHs.

Neoplasms in several species of fishes have been produced experimentally with 3-methylcholanthrene, acetylaminofluorene, benzo[a]pyrene, and 7,12-dimethyl-benz[a]-anthracene, with tumors evident 3–12 months postexposure. Under laboratory conditions, liver neoplasms were induced in two species of minnows (*Poeciliopsis* spp.) by repeated short-term exposures (6 h once a week, for 5 weeks) to an aqueous suspension of 5.0 mg/L of 7,12-dimethylbenz[a]anthracene. About 44% of the fish surviving this treatment developed hepatocellular neoplasms 6–9 months postexposure. Guppies developed hepatic and extrahepatic neoplasms within 6 months following brief 6 h exposure, once weekly for 4 weeks) waterborne exposures to very low (20.0–35.0 μg/L) concentrations of 7,12-dimethylbenz[a]anthracene. Eastern mudminnows (*Umbra pygmaea*) kept in water containing up to 700.0 μg PAHs/L for 11 days showed increased frequencies of chromosomal aberrations in gills: 30% vs. 8% in controls. High dietary benzo[a]pyrene levels of 500.0 mg/kg produced significant elevations in hepatic mixed-function oxidase levels in rainbow trout after 9 weeks. Rainbow trout fed diets containing 1006.0 mg benzo[a]pyrene/kg for 12 months developed liver tumors. About 25% of rainbow trout kept on diets containing 1000.0 mg benzo[a]pyrene/kg for 18 months had histologically confirmed liver neoplasms as compared to 15% after 12 months, with no evidence of neoplasia in controls. Young English sole may activate and degrade carcinogenic PAHs, such as benzo[a]pyrene, to a greater extent than adults, but additional research is needed to determine if younger fish are at greater risk than older sole to PAH-induced toxicity. In English sole, a high significant positive correlation between PAH metabolites (1- and 3-hydroxy benzo[a]pyrene, hydroxy and dehydrodiol metabolites of pyrene and fluoranthene) in bile, and idiopathic liver lesions, prevalence of neoplasms, megalocytic hepatosis, and total number of hepatic lesions suggests that selected PAH metabolites and key organs or tissues may be the most effective monitors of PAH contamination in aquatic organisms.

In addition to those effects of PAHs emphasizing survival, uptake, depuration, and carcinogenesis previously listed, a wide variety of additional effects have been documented for aquatic organisms. These include: inhibited reproduction of daphnids and delayed emergence of larval midges by fluorene;

reproductive impairment in fish by anthracene, and elevated concentrations of total PAHs in sediments; generalized disruption of cell membrane function in fishes by anthracene; decreased respiration and heart rate in mussels (*Mytilus californianus*) by benzo[a]pyrene; increased weight of liver, kidney, gall bladder, and spleen of sea catfish (*Arius felis*) by 3-methylcholanthrene, which was dose related; photosynthetic inhibition of algae and macrophytes by anthracene, naphthalene, phenanthrene, pyrene, and fluorene; immobilization of the protozoan, *Paramecium caudatum*, by anthracene, with an EC50 (60 min) of 0.1 μg/L; perylene accumulation by algae; accumulation without activation of benzo[a]pyrene and benzo[a]anthracene by a marine protozoan (*Parauronema acutum*), and biotransformation of various fluorenes by *P. acutum* to mutagenic metabolites; interference by toluene and anthracene with benzo[a]pyrene uptake by freshwater amphipods; abnormal blood chemistry in oysters (*Crassostrea virginica*) exposed for one year to 5 μg 3-methylcholanthrene/L; and enlarged livers in brown bullheads from a PAH-contaminated river.

26.4.4 Amphibians and Reptiles

In vitro studies with abdominal skin of *Rana pipiens* demonstrated that naphthalene inhibited sodium transport after exposure to 4.4 mg/L for 30 min. Some data were available on biological effects of benzo[a]pyrene, 3-methylcholanthrene, and perylene to reptiles and amphibians. Implantation of 1.5 mg of benzo[a]pyrene crystals into the abdominal cavity of adult South African clawed toads (*Xenopus laevis*) produced lymphosarcomas in 11 of the 13 toads (85%) after 86–288 days. Immature toads were more resistant with only 45% bearing lymphoid tumors of liver, kidney, spleen, or abdominal muscle 272–310 days after implantation of 1.5 mg of benzo[a]pyrene crystals in the dorsal lymph sac or abdominal cavity. Implantation of 3-methylcholanthrene crystals into *X. laevis* provokes development of lymphoid tumors similar to those occurring naturally in this species; moreover, these tumors are readily transplantable into other *Xenopus* or into the urodele species *Triturus cristatus*. Intraperitoneal injection of perylene into tiger salamanders can result in hepatic tumors.

A critical point of interaction between PAHs and reptiles/amphibians involves the transformation of these compounds by cytochrome P450-dependent monooxygenase systems; in general, reaction rates in this group are considerably slower than those observed in hepatic microsomes from mammals. Mixed-function oxidation systems can be induced in liver and skin of tiger salamanders by perylene and 3-methylcholanthrene, and in liver of the leopard frog (*Rana pipiens*) and garter snake (*Thamnophis* sp.) by benzo[a]pyrene and 3-methylcholanthrene. A single dose of 40.0 mg/kg BW of 3-methylcholanthrene was sufficient to induce mixed-function oxidase activity for several weeks in the leopard frog. Amphibians, including tiger salamanders, are quite resistant to PAH carcinogenesis when compared to mammals. This conclusion was based on studies with *Ambystoma* hepatic microsomes and their inability to produce mutagenic metabolites of benzo[a]pyrene and perylene (as measured by bacterial *Salmonella typhimurium* strains used in the Ames test); however, rat liver preparations did produce mutagenic metabolites under these procedures.

26.4.5 Birds

In birds, PAHs were associated with impaired reproduction, growth retardation, morphological abnormalities, behavioral changes, and alterations in vitamin A and thyroid hormone metabolism. In one study, fertilized eggs of the chicken (*Gallus domesticus*) and the common eider duck (*Somateria mollissima*) were injected with a mixture of 16 PAHs at 0.2 mg PAH/kg egg FW on day 4 of incubation. In chicken eggs, 94% of the administered dose was metabolized within the egg by day 18. In embryos of both chickens and eiders, PAH concentrations were highest in gallbladder, followed by liver, kidney, and adipose tissues. Eiders had 40% of the total PAH content in these organs vs. 16% for chickens. Chick embryos, eider embryos, and juvenile eiders had similar PAH concentrations and PAH profiles. The largest studied PAH molecule, coronene, was not taken up from the yolk by the embryo as efficiently as other PAHs, but once taken up it was metabolized as readily as the other PAHs; AHH activities of chick and eider embryos were of similar magnitude. Studies with European starlings (*Sturnus vulgaris*) and 7,12-dimethylbenz[a]anthracene show that effects on the immune function and hepatic mixed-function oxygenase activity were more pronounced when administered by subcutaneous injection when compared to an oral route of exposure, and more pronounced in nestlings than in adults. Serious adverse effects were noted in adults at total administered doses of 125.0 mg/kg BW via injection or orally, and 100.0 mg/kg BW in nestlings.

Mallards (*Anas platyrhynchos*) fed diets that contained 4000.0 mg PAHs/kg (mostly as naphthalenes, naphthenes, and phenanthrene) for a period of 7 months had normal survival and no visible signs of toxicity during exposure; however, liver weight increased 25% and blood flow to liver increased 30%, when compared to controls. In another study with mallards, embryotoxicity was measured of various PAHs applied externally, in a comparatively innocuous synthetic petroleum mixture, to the surface of mallard eggs. The most embryotoxic PAH tested was 7,12-dimethylbenz[a]anthracene: approximately 0.002 μg/egg (equivalent to about 0.036 μg/kg FW, based on an average weight of 55 g per egg) caused 26% mortality in 18 days, and, among survivors, produced significant reduction in embryonic growth and a significant increase in the percent of anomalies, e.g., incomplete skeletal ossification, defects in eye, brain, liver, feathers, and bill. At 0.01 μg 7,12-dimethylbenz[a]anthracene/egg, only 10% survived to day 18. Similar results were obtained with 0.015 μg (and higher) chrysene/egg. For benzo[a]pyrene, 0.002 μg/egg did not affect mallard survival, but did cause embryonic growth reduction and an increased incidence of abnormal survivors. At 0.01 μg benzo[a]pyrene/egg, 60% died in 18 days; at 0.05 μg/egg, 75% were dead within 3 days of treatment. Embryos may contain microsomal enzymes that can metabolize PAHs to more highly toxic intermediates than can adults, and avian embryos may have a greater capacity to metabolize PAHs in this manner than do mammalian embryos and fetuses; this observation warrants additional research. Several investigators have suggested that the presence of PAHs in petroleum, including benzo[a]pyrene, chrysene, and 7,12-dimethylbenz[a]anthracene, significantly enhances the overall embryotoxicity in avian species, and that the relatively small percent of the aromatic hydrocarbons contributed by PAHs in petroleum may confer much of the adverse biological effects reported after eggs have been exposed to microliter quantities of polluting oils.

26.4.6 Mammals

Numerous PAH compounds are distinct in their ability to produce tumors in skin and in most epithelial tissues of practically all animal species tested; malignancies were often induced by acute exposures to microgram quantities. In some cases, the latency period can be as short as 4–8 weeks, with the tumors resembling human carcinomas. Carcinogenic potential values of individual PAHs via dermal exposure, when compared to benzo[a]pyrene with an assigned value of 1.0, were 5.0 for dibenz[a,h]anthracene; 0.1 for benzo[a]anthracene, benzo[b,k]fluoranthene, and indeno[1,2,3-cd]perylene; 0.01 for benzo[g,h,i] perylene, and chrysene; and 0.001 for fluoranthene, fluorene, phenanthrene, and pyrene. Certain carcinogenic PAHs are capable of passage across skin, lungs, and intestine, and can enter the rat fetus, for example, following intragastric or intravenous administration to pregnant dams. In most cases, the process of carcinogenesis occurs over a period of many months in experimental animals, and many years in humans. The tissue affected is determined by the route of administration and species under investigation. Thus, 7,12-dimethylbenz[a]anthracene is a potent carcinogen for the mammary gland of young female rats after oral or intravenous administration; dietary benzo[a]pyrene leads to leukemia, lung adenoma, and stomach tumors in mice and both PAH compounds can induce hepatomas in skin of male mice when injected shortly after birth. Mammary tumors were observed in female rats 6 weeks after they were given the first of 4 weekly intragastric doses of 5.0 mg 7,12 dimethylbenz[a]anthracene/rat; exposure to magnetic fields of low flux density may promote the growth and development of mammary tumors. Acute and chronic exposure to various carcinogenic PAHs have resulted in destruction of hematopoietic and lymphoid tissues, ovotoxicity, antispermatogonic effects, adrenal necrosis, changes in the intestinal and respiratory epithelia, and other affects. For the most part, however, tissue damage occurs at dose levels that would also be expected to induce carcinomas, and thus the threat of malignancy predominates in evaluating PAH toxicity. There is a scarcity of data available on the toxicological properties of PAHs which are not demonstrably carcinogenic to mammals.

Target organs for PAH toxic action are diverse, partly due to extensive distribution in the body and also to selective attack by these chemicals on proliferating cells. Damage to the hematopoietic and lymphoid system in experimental animals is a particularly common observation. In rats, the target organs for 7,12-dimethylbenz[a]anthracene are skin, small intestine, kidney, and mammary gland, whereas in fish the primary target organ is liver. Application of carcinogenic PAHs to mouse skin leads to destruction of sebaceous glands and to hyperplasia, hyperkeratosis, and ulceration. Tumors are induced in mouse skin by the repeated application of small doses of PAHs, by a single application of a large dose, or by the single application of a subcarcinogenic dose (initiation) followed by repeated application of certain noncarcinogenic agents (promotion). Newborn mice were highly susceptible to 3-methylcholanthrene, with many mice dying from acute or chronic wasting disease following treatment; some strains of mice eventually developed thymomas, but other strains showed no evidence despite serious damage to the thymus.

In general, PAH carcinogens transform cells through genetic injury involving metabolism of the parent compound to a reactive diol epoxide. This, in turn, can then form adducts with cellular molecules, such as DNA, RNA, and proteins, resulting in cell transformation. In the case of benzo[a]pyrene, one isomer of the 7,8-diol, 9,10-epoxide is an exceptionally potent carcinogen to newborn mice and is believed to be the ultimate carcinogenic metabolite of this PAH. One of the most toxicologically significant processes involved in the response to PAH absorption is the interaction with drug metabolizing enzyme systems. Increased production of mixed-function oxidase enzymes in various small mammals has been induced by halogenated naphthalenes, 3-methylcholanthrene, and numerous other PAHs. PAH metabolites produced by microsomal enzymes in mammals can be arbitrarily divided into water-soluble groups, and organosoluble groups such as phenols, dihydrodiols, hydroxymethyl derivatives, quinones, and epoxides. In the case of benzo[a]pyrene, the diol epoxides are usually considered as the ultimate carcinogens. Other microsomal enzymes convert epoxide metabolites to easily excretable water-soluble compounds, with excretion primarily through feces and the hepatobiliary system. Interspecies differences in sensitivity to PAH-induced carcinogenesis are largely due to differences in levels of mixed-function oxidase activities, and these will directly affect rates at which active metabolites are converted to less active products.

Investigators agree that unsubstituted aromatic PAHs with less than 4 condensed rings have not shown tumorigenic activity; that many, but not all, 4-, 5-, and 6-ring PAH compounds are carcinogenic; and that only a few unsubstituted hydrocarbons with 7 rings or greater are tumorigenic or carcinogenic. Many PAH compounds containing 4 and 5 rings and some containing 6 or more rings, provoke local tumors after repeated application to the dorsal skin of mice; the tumor incidence exhibited a significant dose–response relationship. Among unsubstituted PAHs containing a nonaromatic ring, e.g., cholanthrene and acenaphthanthracene, all active carcinogens retained an intact phenanthrene segment. The addition of alkyl substituents in certain positions in the ring system of a fully aromatic PAH will often confer carcinogenic activity or dramatically enhance existing carcinogenic potency. For example, monomethyl substitution of benz[a]anthracene can lead to strong carcinogenicity in mice, with potency depending on the position of substitution in the order $7 > 6 > 8 = 12 > 9$; a further enhancement of carcinogenic activity is produced by appropriate dimethyl substitution, with 7,12-dimethylbenz[a]anthracene among the most potent PAH carcinogens known. Alkyl substitution of partially aromatic condensed ring systems may also add considerable carcinogenic activity as is the case with 3-methylcholanthrene. With alkyl substitutes longer than methyl, carcinogenicity levels decrease, possibly due to a decrease in transport through cell membranes.

A good correlation exists between skin tumor initiating activities of various benzo[a]pyrene metabolites and their mutagenic activity in mammalian cell mutagenesis systems, although variations in chromosome number and structure may accompany tumors induced by various carcinogenic PAHs in rats, mice, and hamsters. Active PAH metabolites, e.g., dihydrodiols or diol epoxides, can produce sister chromatid exchanges in Chinese hamster ovary cells. When exchanges were induced by the diol epoxide, a close relationship exists between the frequency of sister chromatid exchanges and the levels of deoxyribonucleoside-diol-epoxide adduct formation. In general, noncarcinogenic PAHs were not mutagenic. Laboratory studies with mice have shown that many carcinogenic PAHs adversely affect the immune system, thus directly impacting an organism's general health, although noncarcinogenic analogs had no immunosuppressive effect; further, the more carcinogenic the PAH, the greater the immunosuppression. Destruction of oocytes and follicles in mice ovary is documented following intraperitoneal injection of benzo[a]pyrene; 3-methylcholanthrene, and 7,12-dimethylbenz[a]anthracene; the rate of destruction was proportional to the activity of the ovarian cytochrome P450-dependent monooxygenase, as well as the carcinogenicity of the PAH. However, no information is presently available to indicate whether PAHs present a hazard to reproductive success. In those cases where teratogenic effects are clearly evident, e.g., 7,12-dimethylbenz[a]anthracene, the required doses were far in excess of realistic environmental exposures. Numerous studies show that unsubstituted PAHs do not accumulate in mammalian adipose tissues despite their high lipid solubility, probably because they tend to be rapidly and extensively metabolized.

Biological half-life (Tb1/2) of PAHs is limited, as judged by rodent studies. In the case of oral doses of benzo[a]pyrene and rat blood and liver, Tb1/2 values of 5–10 min were recorded; the initial rapid elimination phase was followed by a slower disappearance phase lasting 6 h or more. The Tb1/2 value of benzo[a] pyrene from blood of rats given 15.0 mg/kg BW by intravenous injection was 400 min. In that study, adipose and lung tissues had comparatively high concentrations of benzo[a]pyrene 32 h after dosing, and fecal excretion was the dominant route of BaP loss being 4–10 times higher than urinary excretion. Tb1/2 values from the site of subcutaneous injection in mice were 1.75 weeks for benzo[a]pyrene, 3.5 weeks for 3-methylcholanthrene, and 12 weeks for dibenz[a,h]anthracene; the relative carcinogenicity of each compound was directly proportional to the time of retention at the injection site.

Many chemicals are known to modify the action of carcinogenic PAHs in experimental animals, including other PAHs that are weakly carcinogenic or noncarcinogenic. The effects of these modifiers on PAH metabolism appear to fall into three major categories: those which alter the metabolism of the carcinogen, causing decreased activation or increased detoxification; those which scavenge active molecular species of carcinogens to prevent their reaching critical target sites in the cell; and those which exhibit competitive antagonism. For example, pyrene given to rats by intravenous injection or orally at 20.0 mg/kg BW alone or in combination with other PAHs produce the following observations: bioavailability of pyrene is increased in combination with fluoranthene or benz[a]anthracene; pyrene enhances the carcinogenic effect of benzo[a]pyrene and carcinogenicity of benzo[a]pyrene can be inhibited by pyrene, phenanthrene, or anthracene; and some PAHs present in the environment accelerate the clearance of pyrene but reduce the level of 1-hydroxypyrene in urine. Benz[a]anthracene, a weak carcinogen when applied simultaneously with dibenz[a,h]anthracene, inhibits the carcinogenic action of the latter in mouse skin; a similar case is made for benzo[e]pyrene or dibenz[a,c]anthracene applied to mouse skin shortly prior to initiation with 7,12-dimethylbenz[a]anthracene, or 3-methylcholanthrene. Benzo[a]pyrene, a known carcinogen, interacts synergistically with cyclopenta [cd]pyrene, a moderately strong carcinogen found in automobile exhausts, according to results of mouse skin carcinogenicity studies. Other PAH combinations were co-carcinogenic, such as benzo[e]pyrene, pyrene, and fluoranthene applied repeatedly with benzo[a]pyrene to the skins of mice. Effective inhibitors of PAH-induced tumor development include selenium, vitamin E, ascorbic acid, butylated hydroxytoluene, and hydroxyanisole. In addition, protective effects against PAH-induced tumor formation have been reported for various naturally occurring compounds such as flavones, retenoids, and vitamin A. Until these interaction effects are clarified, the results of single substance laboratory studies may be extremely difficult to apply to field situations of suspected PAH contamination. Additional work is also needed for PAH dose–response relationships, testing relevant environmental PAHs for carcinogenicity, and elucidating effects of PAH mixtures on tumor formation.

26.5 Recommendations

No standards have been promulgated for PAHs by any regulatory agency for the protection of sensitive species of aquatic organisms or wildlife, although some criteria have been proposed for aquatic life protection (Table 26.3). This observation is not unexpected in view of several factors: (1) the paucity of data on PAH

Table 26.3. Proposed PAH criteria for the protection of human health and aquatic life.

Criterion, PAH Group, and Units	Concentration
HUMAN HEALTH	
Air	
Total PAHs (μg/m^3)	<0.0109
Total PAHs, daily intake (μg)[f]	0.164–<0.251
Cyclohexane extractable fractions; coke oven emissions, coal tar products (μg/m^3), 8–10-h weighted average	100.0–<150.0
Benzene-soluble fractions; coal tar pitch volatiles (μg/m^3), 8-h average	<200.0
Benzo[a]pyrene	
Daily intake (μg/m^3)	<0.0005
Daily intake (μg)[f]	0.005–<0.0115
Acceptable (μg/m^3)	
Arizona	<0.79 for 1 h; <0.21 for 24 h; <0.00057 for 1 year
Connecticut, Indiana	<0.1 for 8 h
Florida, Kansas, Michigan	<0.0003 for 1 year

Continued

Table 26.3. cont'd

Criterion, PAH Group, and Units	Concentration
Maine	<0.0006 for 1 year
North Carolina	<0.03 for 1 year
Pennsylvania	<0.0007 for 1 year
Texas	<0.03 for 30 min; 0.003 for 1 year
Carcinogenic PAHs	
Daily intake ($\mu g/m^3$)	<0.002
Daily intake (μg)[f]	0.03–<0.046
Drinking Water	
Total PAHs	
Daily intake ($\mu g/L$)[a]	0.0135–<0.2
Daily intake (μg)[a]	0.027–<0.4
Yearly intake (μg)	<4.0
Europe ($\mu g/L$)[h]	<0.2
Benz[a]anthracene ($\mu g/L$)	<0.1
Benzo[a]pyrene	
USA ($\mu g/L$)	<0.2
Most states ($\mu g/L$)	0.03–<0.003
Daily intake (μg)	<0.0011
Dibenzanthracene ($\mu g/L$)	<0.3
Carcinogenic PAHs	
Daily intake ($\mu g/L$)[b]	<0.0021
Daily intake (μg)[b]	<0.0042
Daily intake ($\mu g/L$)[c]	
Cancer risk 10^{-5}	<0.028
Cancer risk 10^{-6}	<0.0028
Cancer risk 10^{-7}	<0.00028
Daily intake (μg)[c]	
Cancer risk 10^{-5}	<0.056
Cancer risk 10^{-6}	<0.0056
Cancer risk 10^{-7}	<0.00056
Food	
Total PAHs	
Daily intake (μg)[d]	1.6–<16.0
Yearly intake (μg)	<4,150.0
Benzo[a]pyrene; daily intake (μg)[e]	0.16–<1.6
Low-risk range; oral intake; mg/kg BW daily for up to 364 days	
Acenaphthene	<0.06
Anthracene	<0.03
Fluoranthene	<0.04
Fluorene	<0.04
Pyrene	<0.03

Table 26.3. cont'd

Criterion, PAH Group, and Units	Concentration
All sources	
Total PAHs; daily (μg)	1.79–<16.6
Benzo[a]pyrene	
Daily intake (μg)	0.166–<1.61
Daily allowable limit (μg)	<0.048
Carcinogenic PAHs; daily intake (except diet) (μg)[b]	<0.086–0.102
Land disposal restrictions (Proposed)	
Maximum allowed; nonwastewaters (mg/kg FW)	
Benz[a]anthracene, benzo[a]pyrene, chrysene, or indeno[1,2,3-cd]pyrene (=I(1,2,3-cd)P	3.4
Benzo[b]fluoranthene, benzo[k]fluoranthene	6.8
Dibenz[a,h]anthracene	8.2
Maximum allowed; wastewaters (mg/L)	
Benzo[a]anthracene, chrysene	0.059
Benzo[a]pyrene	0.061
Benzo[b]fluroanthene, benzo[k]fluoranthene	0.11
I(1,2,3-cd)P	0.0055
Dibenz[a,h]anthracene	0.055
AQUATIC LIFE	
Sediment criteria to protect freshwater organisms (in mg/kg DW)[g]	
Naphthalene	<0.01
Phenanthrene	<0.04
Benzo[a]pyrene	<0.06
Acenaphthene	<0.15
Fluorene	<0.2
Benz[a]anthracene	<0.2
Anthracene	<0.6
Acridine	<1.0
Fluoranthene	<2.0
Sediment criteria to protect marine life, in (mg/kg DW)[g]	
Naphthalene	<0.01
Benzo[a]pyrene	<0.06
Acenaphthene	<0.15
Fluorene	<0.2
Chrysene	<0.2
Water criteria (μg/L), to protect freshwater biota under conditions of chronic exposure	
Benzo[a]pyrene	<0.01
Acridine	<0.05 (phototoxic effects); <3.0 (no phototoxic effects)

Continued

Table 26.3. cont'd

Criterion, PAH Group, and Units	Concentration
Benz[a]anthracene	<0.1
Anthracene	<0.1 (phototoxic effects); <4.0 (no phototoxic effects)
Fluoranthene	<0.2 (phototoxic effects); <4.0 (no phototoxic effects)
Naphthalene	<1.0
Acenaphthene	<6.0
Fluorene	<12.0
Total PAHs ($n = 13$); Ohio	<310.0
Marine water criteria (in μg/L) to protect biota under conditions of chronic exposure	
Benzo[a]pyrene	<0.01
Chrysene	<0.1
Naphthalene and methylated naphthalenes	<1.0
Acenaphthene	<6.0
Fluorene	<12.0
Anthracene	
Fathead minnow (*Pimephales promelas*)	
Reproduction impaired	>15.0 mg/kg BW
Reproduction normal	<15.0 mg/kg BW; <6.0 μg/L medium
Bluegill, (*Lepomis macrochirus*); no adverse effect	<13.5 mg/kg BW

[a]Total of 6 PAHs: fluoranthene, benzo[a]pyrene, benzo[g,h,i]perylene, benzo[b]fluoranthene, benzo[k] fluoranthene, and indeno[1,2,3-cd]pyrene.

[b]Total of 3 PAHs: benzo[a]pyrene, benzo[j]fluoranthene, and indeno[1,2,3-cd]pyrene.

[c]Based on all carcinogenic PAHs.

[d]Assuming 1600 g food daily, 70-kg adult, 1.0–10.0 mg total PAHs/diet.

[e]As above, except 0.1–1.0 mg benzo[a]pyrene/diet.

[f]Assuming average of 15–23 m^3 of air inhaled daily.

[g]To derive PAH criteria, a safety factor of 0.1 was used with most chronic data but ranged between 0.01 and 0.1 of the lowest observed effective concentration in freshwater and 0.1 and 0.5 in marine waters.

[h]Total of 6 PAHs: benzo[a]pyrene, benzo[b]fluoranthene, benzo[a]fluoranthene, benzo[k]fluoranthene, fluoranthene, and indeno[1,2,3-cd]pyrene.

background concentrations in wildlife and other natural resources; (2) the absence of information on results of chronic oral feeding studies of PAH mixtures and the lack of a representative PAH mixture for test purposes; and (3) the demonstrable – and as yet, poorly understood – effects of biological modifiers, such as sex, age, and diet, and interaction effects of PAHs with inorganic and other organic compounds, including other PAHs. Nevertheless, the growing database for aquatic life indicates a number of generalizations: (1) many PAHs are acutely toxic at concentrations between 50.0 and 1000.0 μg/L; (2) deleterious sublethal responses are sometimes observed at concentrations in the range of 0.1–5.0 μg/L; (3) uptake can be substantial, but depuration is usually rapid except in some species of invertebrates; and (4) whole body burdens in excess of 300.0 μg benzo[a]pyrene/kg (and presumably other PAHs) in certain teleosts would be accompanied by a rise in the activity of detoxifying enzymes.

Aquatic research has focused on PAHs because of their known relationship with carcinogenesis and mutagenesis. Many reports exist of high incidences of cancer-like growths and developmental anomalies in natural populations of aquatic animals and plants, but none conclusively demonstrate the induction of cancer by exposure of aquatic animals to environmentally realistic levels of carcinogenic PAHs in the water column, diet, or sediments. However, studies by several research groups have now established that sediments heavily contaminated with PAHs from industrial sources were the direct cause of elevated PAH body burdens and elevated frequencies of liver neoplasia in fishes from these locales. Only a few sites containing high PAH concentrations in sediments have been identified, suggesting a need to identify and to evaluate other PAH-contaminated aquatic sites. Most fishery products consumed by upper trophic levels, including humans, contain PAH concentrations similar to those in green vegetables and smoked and charcoal-broiled meats, and would probably represent a minor source of PAH toxicity; however, consumption of aquatic organisms, especially filter-feeding bivalve mollusks, from regions severely contaminated with petroleum or PAH-containing industrial wastes, should be avoided. Repeated consumption of PAH-contaminated shellfish may pose a cancer risk to humans, and consumption of PAH-contaminated fish from the Arabian Gulf poses a risk to human health in that region. In order to determine if similar risks exist to wildlife, studies should be initiated with seabirds, pinnipeds, and other wildlife groups which feed extensively on mollusks and teleosts that are capable of accumulating high burdens of carcinogenic PAHs.

For avian wildlife, data are incomplete on PAH background concentrations and on acute and chronic toxicity. Studies with mallard embryos and PAHs applied to the egg surface showed toxic and adverse sublethal effects at concentrations between 0.036 and 0.18 μg PAH/kg whole egg. Additional research is needed on petroleum-derived PAHs and their effects on developing embryos of seabirds and other waterfowl. There is an urgent need for specific avian biomarkers of PAH exposure.

PAH criteria for human health protection (Table 26.3) were derived from tests with small laboratory mammals, primarily rodents. Accordingly, these proposed criteria should become interim guidelines for protection of nonhuman mammalian resources pending acquisition of more definitive data. The proposed PAH criteria are controversial because there is no way at present to quantify the potential human health risks incurred by the interaction of any PAH with other PAHs or with other agents in the environment, including tumor initiators, promoters, and inhibitors. The problem arises primarily from the diversity of test systems and bioassay conditions used for determining carcinogenic potential of individual PAHs in experimental animals, and is confounded by the lack of a representative PAH mixture for test purposes, the absence of data for animal and human chronic oral exposures to PAH mixtures, and the reliance on data derived from studies with benzo[a]pyrene to produce generalizations concerning environmental effects of PAHs – generalizations which may not be scientifically sound. Further, only a small percentage of PAH compounds are known to be carcinogenic and measurements of total PAHs (i.e., the sum of all multiple fused-ring hydrocarbons having no heteroatoms) cannot be equated with carcinogenic potential. The term "total PAHs" is unclear; the compounds being considered should be specified for each case. An analysis of dose–response relationships for PAH-induced tumors in animals shows, in some cases, deviation from linearity in dose–response curves, especially at low doses, suggesting a two-stage model consistent with a linear nonthreshold pattern. Because overt tumor induction follows a dose–response relationship consistent with a multihit promotion process, the multihit component of carcinogenesis may be supplied by environmental stimuli not necessarily linked or related to PAH exposure.

The well-documented existence of carcinogenic and anticarcinogenic agents strongly suggests that a time assessment of carcinogenic risk for a particular PAH can be evaluated only through a multifactorial analysis. One of the most toxicologically significant processes involved in the response to PAH absorption is the interaction with drug metabolizing enzyme systems. The induction of this enzyme activity in various body tissues by PAHs and other xenobiotics is probably critical to the generation of reactive PAH metabolites at the target site for tumor induction. At present, wide variations occur in human and animal carcinogen-metabolizing capacity. Moreover, it has not yet been possible to definitely correlate enzyme activity with susceptibility to carcinogenesis. The obligatory coupling of metabolic activation with PAH-induced neoplasia in animals indicates that the modulation of drug metabolizing enzymes is central to carcinogenesis.

PAHs from drinking water contribute only a small proportion of the average total human intake. The drinking water quality criterion for carcinogenic PAH compounds is based on the assumption that each compound is as potent as benzo[a]pyrene, and that the carcinogenic effect of the compounds is proportional to the sum of their concentrations. Based on an oral feeding study of benzo[a]pyrene in mice, the concentration of this compound estimated to result in additional risk of one additional case for every 100,000 individuals exposed (i.e., 10^{-5}) is 0.028 μg/L. Therefore, with this assumption, the sum of the concentrations of all carcinogenic PAH compounds should be less than 0.028 μg/L in order to keep the lifetime cancer risk below 10^{-4}. The corresponding recommended criteria which may result in an incremental cancer risk of 10^{-6} and 10^{-7} over the lifetime are 0.0028 and 0.00028 μg/L, respectively (Table 26.3). If the above estimates are made for consumption of aquatic organisms only, the levels are 0.311 (10^{-5}), 0.031 (10^{-6}), and 0.003(10^{-7}) μg/kg, respectively. The use of contaminated water for irrigation can also spread PAHs into vegetable foodstuffs. When vegetables grown in a PAH-polluted area are thoroughly washed and peeled, their contribution to total PAH intake in humans is not significant. Herbivorous wildlife, however, may ingest significant quantities of various PAHs from contaminated vegetables – but no data were available on this subject.

PAHs are widely distributed in the environment as evidenced by their detection in sediments, soils, air, surface waters, and plant and animal tissues. However, the ecological impact of PAHs is uncertain. PAHs show little tendency for bioconcentration despite their high lipid solubility, probably because most PAHs are rapidly metabolized. Additional research on PAHs in soil–plant systems is recommended. Specifically, research is needed to establish: the rates of PAH decomposition in soils; the soil PAH levels above which PAH constituents adversely affect the food chain; and enhancement factors that increase degradation rates of PAHs, especially PAHs with more than 3 rings. Once these factors have been determined, PAH disposal into soils may become feasible at environmentally nonhazardous levels.

Diet is the major source of PAHs to humans. Authorities agree that most foods contain 1.0–10.0 μg total PAHs/kg FW, that smoking or barbecuing fish and meats increases total PAH content up to 100-fold, that contaminated mollusks and crustaceans may contribute significantly to PAH intake, and that PAH carcinogenic risk to health has existed at least since humans began to cook food over fire. A total of 22 PAHs has been identified in foods, of which 11 have been found to be carcinogenic in experimental animals. Of these, only 5 (benzo[a]pyrene, benz[a]anthracene, 3-methylcholanthrene, dibenz[a,h]anthracene, and 7,12-dimethylbenz[a]anthracene) have been demonstrated to induce tumors following oral administration to rats and mice, and only 3 of the 11 exhibited positive dose–response relationships in chronic studies with mice. There is no credible evidence that any of the 11 known carcinogenic PAHs or their combinations can cause cancer in human beings via the oral route, especially in quantities likely to be present in foods. There is, however, a need for a complete risk assessment to human health from consumption of PAH-contaminated foods, especially in areas of the Middle East where PAH concentrations for anthracene and pyrene in food items greatly exceed the carcinogenic and mutagenic risk based on benzo[a]pyrene equivalents.

In view of the carcinogenic characteristics of many PAH compounds, their increasing concentrations in the environment should be considered alarming, and efforts should be made to reduce or eliminate them wherever possible. Current research, not unexpectedly, has focused on PAH removal from contaminated environments and on bioindicators of PAH exposure. Irradiation is recommended with UV or daylight of certain PAHs or hydrocarbon-rich mixtures prior to disposal into the environment to help reduce their harmful effects. Flushing with anionic surfactants to remove anthracene from contaminated soils shows promise, and surfactant soil flushing to remove PAHs is recommended as an alternative new technology to groundwater flushing with water. Biomarkers proposed as indicators of PAH exposure include elevated excretion rate of 1-OH pyrene in bile of eels from contaminated sites, intestinal P450 activity as an indicator of dietary PAH exposure and liver P450 activity as an indicator of sediment PAH concentrations in marine teleosts, loss rate of 5-methyl deoxycytidine from DNA, and elevated hepatic EROD activity and concentrations of biliary PAH metabolites. However, levels of bile conjugates as a biomarker of PAH exposure should be interpreted with caution, particularly if some fish are known to have chronic liver disease. Synchronous fluorescent spectroscopy (SFS) has been used successfully to identify pyrene- and benzo[a]pyrene-type metabolites in bile of brown bullheads from Lake Erie tributaries in 1990–91; SFS is recommended for screening large numbers of fish samples for evidence of PAH exposure. Increased benzo[a]pyrene monooxygenase activity was evident in several species of freshwater fishes exposed to benzo[a]pyrene, as reflected by measurement of bile fluorescence. Finally, there is a need for additional PAH toxicokinetic models that emphasize uptake, retention, translocation, and biotransformation rates.

26.6 Summary

Polycyclic aromatic hydrocarbons (PAHs) consist of hydrogen and carbon arranged in the form of two or more fused benzene rings. There are thousands of PAH compounds, each differing in the number and position of aromatic rings, and in the position of substituents on the basic ring system. Environmental concern has focused on PAHs that range in molecular weight from 128.16 (naphthalene, 2-ring structure) to 300.36 (coronene, 7-ring structure). Unsubstituted lower molecular weight PAH compounds, containing 2–3 rings, exhibit significant acute toxicity and other adverse effects to some organisms, but are noncarcinogenic; the higher molecular weight PAHs, containing 4–7 rings, are significantly less toxic, but many of the 4- to 7-ring compounds are demonstrably carcinogenic, mutagenic, or teratogenic to a wide variety of organisms, including fish and other aquatic life, amphibians, birds, and mammals. In general, PAHs show little tendency to biomagnify in food chains, despite their high lipid solubility, probably because most PAHs are rapidly metabolized. Inter- and intraspecies responses to individual PAHs are quite variable, and are significantly modified by many inorganic and organic compounds, including other PAHs. Until these interaction effects are clarified, the results of single substance laboratory tests may be extremely difficult to apply to field situations of suspected PAH contamination.

PAHs are ubiquitous in nature – as evidenced by their detection in sediments, soils, air, surface waters, and plant and animal tissues – primarily as a result of natural processes such as forest fires, microbial synthesis, and volcanic activities. Anthropogenic activities associated with significant production of PAHs – leading, in some cases, to localized areas of high contamination – include high-temperature (>700°C) pyrolysis of organic materials typical of some processes used in the iron and steel industry, heating and power generation, and petroleum refining. Aquatic environments may receive PAHs from accidental releases of petroleum and its products, from sewage effluents, and from other sources. Sediments heavily contaminated with industrial PAH wastes have directly caused elevated PAH body burdens and increased frequency of liver neoplasia in fishes.

At present, no criteria or standards have been promulgated for PAHs by any regulatory agency for the protection of sensitive species of aquatic organisms or wildlife. This observation was not unexpected in view of the paucity of data on PAH background concentrations in wildlife and other natural resources, the absence of information on results of chronic oral feeding studies of PAH mixtures, the lack of a representative PAH mixture for test purposes, and the demonstrable – and, as yet, poorly understood – effects of biological and nonbiological modifiers on PAH toxicity and metabolism. By contrast, criteria for human health protection and total PAHs, carcinogenic PAHs, and benzo[a]pyrene have been proposed for drinking water and air, and for total PAHs and benzo[a]pyrene in food: drinking water, 0.01–<0.2 μg total PAHs/L, <0.002 μg carcinogenic PAHs/L, and <0.0006 μg benzo[a]pyrene/L; air, <0.01 μg total PAHs/m^3, <0.002 μg carcinogenic PAHs/m^3, and <0.0005 μg benzo[a]pyrene/m^3; food, 1.6–<16.0 μg total PAHs daily, and 0.16–<1.6 μg benzo[a]pyrene daily. In view of the carcinogenic characteristics of many PAH compounds and their increasing concentrations in the environment, it now seems prudent to reduce or eliminate them wherever possible, pending acquisition of more definitive ecotoxicological data.

RADIATION[a]

Chapter 27

27.1 Introduction

Life on earth has evolved under the ubiquitous presence of environmental solar, X-ray, gamma, and charged-particle radiation. On a global basis, radiation from natural sources is a far more important contributor to radiation dose to living organisms than radiation from anthropogenic sources. However, ionizing radiation can harm biological systems, and this harm can be expressed (1) in a range of syndromes from prompt lethality to reduced vigor, shortened life span, and diminished reproductive rate by the irradiated organism and (2) by the genetic transmission of radiation-altered genes that are most commonly recessive and almost always disadvantageous to their carriers. Direct effects of radiation were documented in lampreys in 1896 – soon after H. Becquerel discovered radioactivity – and in brine shrimp (*Artemia* sp.) in 1923. Genetic effects of ionizing radiation and thus X-rays as a mutagenic agent were first documented in 1927 in fruit flies, *Drosophila melanogaster*. The discovery of radioactivity of nuclear particles and the discovery of uranium fission resulted in a great upsurge of nuclear research. During and shortly after World War II, nuclear reactors, testing of nuclear weapons, and use of radionuclides as tracers in almost all scientific and technical fields were developed rapidly. Environmental radiation from anthropogenic sources caused serious concerns beginning with the early 1940s when fission of uranium and transuranic nuclei became possible in reactors and in explosions of nuclear weapons. The first nuclear explosion resulted from a 19-kiloton (TNT-equivalent) source in New Mexico in July 1945. On August 6, 1945, about 75,000 people were killed when the United States Army Air Corps dropped a uranium nuclear bomb on Hiroshima, Japan; on August 9, 1945, about 78,000 Japanese were killed and more than 100,000 injured when a plutonium nuclear bomb was detonated at Nagasaki. The Former Soviet Union detonated its first nuclear device in August 1949, and in 1952 the United Kingdom exploded a device in Australia. Since 1960, nuclear devices have also been detonated by France, India, Pakistan, The People's Republic of China, and possibly others. Nuclear devices have been developed that can release energy in the megaton range. The first such device was detonated by the United States in 1954 at Bikini Atoll and accidentally contaminated Japanese fishermen and Marshall Island natives. Between 1945 and 1973, an estimated 963 nuclear tests were conducted by The People's Republic of China, France, the former Soviet Union, the United Kingdom, and the United States; 47% of them were atmospheric, and 53% subsurface. Today, the most important environmentally

---

[a]All information in this chapter is referenced in the following sources:

Eisler, R. 1994. Radiation hazards to fish, wildlife, and invertebrates: a synoptic review. *U.S. Natl. Biol. Serv. Biol. Rep. 26*, 124 pp.

Eisler, R. 1995. Ecological and toxicological aspects of the partial meltdown of the Chernobyl nuclear plant reactor. Pages 549–564 in D.J. Hoffman, B.A. Rattner, G.A. Burton, Jr., and A.J. Cairns, Jr., eds. *Handbook of Ecotoxicology*. Lewis Publishers, Boca Raton, Florida.

Eisler, R. 2000. Radiation. Pages 1707–1828 in *Handbook of Chemical Risk Assessment. Health Hazards to Humans, Plants, and Animals. Volume 3, Metalloids, Radiation, Cumulative Index to Chemicals and Species*. Lewis Publishers, Boca Raton, Florida.

Eisler, R. 2003. The Chernobyl nuclear power plant reactor accident: ecotoxicological update. Pages 703–736 in D.J. Hoffman, B.A. Rattner, C.A. Burton, Jr., and J. Cairns, Jr., eds. *Handbook of Ecotoxicology, Second Edition*. Lewis Publishers, Boca Raton, Florida.

damaging anthropogenic radiation comes from atmospheric testing of nuclear weapons conducted 30–40 years ago, authorized discharges to the sea from nuclear reprocessing plants, and from the Chernobyl accident in 1986. In the year 2000, the United States had an estimated 40,000 tons of spent nuclear fuel stored at some 70 sites and awaiting disposal; by 2035, after all existing nuclear plants have completed 40 years of operation, an estimated 85,000 metric tons will be awaiting disposal.

27.2 Physical Properties of Radiation

Radiation is usually defined as the emission and propagation of energy through space in the form of waves and subatomic particles. For regulatory purposes in the United States, radiation is narrowly defined as α, β, γ, or X-rays; neutrons; and high-energy electrons, protons, or other atomic particles; but not radio waves nor visible, infrared, or ultraviolet light. Readers may wish to consult the glossary at this time.

In current atomic theory, all elementary forms of matter consist of small units called atoms. All atoms of the same element have the same size and weight; atoms of different elements differ in size and weight. Atoms of the same or different elements may unite to form compound substances called molecules. Each atom consists of a central nucleus and several negatively charged electrons in a cloud around the nucleus. The nucleus is composed of positively charged particles called protons, and particles without charge called neutrons. Electrons are arranged in successive energy levels around the nucleus, and the extranuclear electronic structure of the atom is characteristic of the element. Electrons in the inner shells are tightly bound to the nucleus but can be altered by high-energy waves and particles. Atoms are classified chemically into 92 naturally occurring elements and another dozen or so artificial elements based on the number of protons in their nucleus (=the atomic number). Atoms of the same element may occur as isotopes that differ in the number of neutrons accompanying the protons in the nucleus. The sum of the number of protons and neutrons in the nucleus is called the mass number (see Glossary), and is indicated by a superscript that precedes the chemical symbol of the element. For example, three isotopes of hydrogen (one proton) are denoted as ^{1}H (no neutrons), ^{2}H (1 neutron, also known as deuterium), and ^{3}H (2 neutrons, also known as tritium). A nuclide is an elemental form distinguished from others by its atomic and mass numbers. Some nuclides, such as ^{238}U and ^{137}Cs, are radioactive and spontaneously decay to a different nuclide with the emission of characteristic energy particles or electromagnetic waves; isomers of a given nuclide that differ in energy content are metastable (i.e., ^{115m}Cd) and characterized in part, by the half-life of the isomer.

Chemical forms with at least one radioactive atomic nucleus are radioactive substances. The capability of atomic nuclei to undergo spontaneous nuclear transformation is called radioactivity. Nuclear transformations are accompanied by emission of nuclear radiation. The average number of nuclei that disintegrates per unit time (=activity) is directly proportional to the total number of radioactive nuclei; the time for 50% of the original nuclei to disintegrate (=half-life or Tb1/2) is equal to ln 2/(decay constant) for that element. Radiations that have sufficient energy to interact with matter to produce charged particles are called ionizing radiations. Radiation injury is related to the production of ions inside the cell. Ionizing radiations include electromagnetic radiation such as gamma (γ) and X-rays and particulate or corpuscular radiation such as alpha (α) particles, beta (β) particles, electrons, positrons, and neutrons. Ionizing radiation may be produced from manufactured devices such as X-ray tubes or from the disintegration of radioactive nuclides. Some nuclides occur naturally, but others may be produced artificially, for example, in nuclear reactors. The basic reaction of ionizing radiation with molecules is either ionization or excitation. In ionization, an orbital electron is ejected from the molecule and forms an ion pair. Directly ionizing particles are charged and possess the energy to produce ionizations along their path from impulses imparted to orbital electrons via electrical forces between the charged particles and electrons. During excitation, an electron is raised to a higher energy level. Indirectly ionizing radiations are not charged and penetrate a medium until they collide with elements of the atom and liberate energetically charged ionizing particles.

27.2.1 Electromagnetic Spectrum

The electromagnetic spectrum is defined as the ordered array of known electromagnetic radiations including cosmic rays; gamma rays; X-rays; ultraviolet, visible, and infrared radiations; and radio waves. The energy transfer by electromagnetic waves can be described by discrete processes with elementary units called photons. Their energy, E, is given by $E = h\nu$, where h is Planck's constant and ν is the frequency. Because velocity C, wavelength λ, and frequency υ are related ($C = \lambda\nu$), $E = hc/\lambda$. The relations between E, ν, and λ for parts of the total spectrum of the electromagnetic waves are shown in Figure 27.1. The high-energy radiation that enters the earth's atmosphere from outer space is known as primary cosmic rays. On interaction with the nuclei of atoms in the air, secondary cosmic rays and a variety of reaction products (cosmogenic nuclides) such as ^{3}H, ^{7}Be, ^{10}Be, ^{14}C, ^{22}Na, and ^{24}Na are produced.

27.2.2 Radionuclides

Radioactive nuclides contain atoms that disintegrate by emission of subatomic particles and gamma or X-ray photons. In alpha decay, a helium nucleus of 2 protons and 2 neutrons is emitted and reduces the mass number by 4 and the atomic number by 2. In beta decay, an electron – produced by the disintegration of a neutron into a proton, an electron, and an antineutrino – is emitted from the nucleus and increases the atomic number by 1 without changing the mass number. Sometimes a positron together with a neutrino is emitted. And sometimes an electron may be captured from the K (outermost) shell of the atom; the resultant electron hole in the K shell is filled by electrons from outer orbits and causes the emission of X-rays. Alpha and beta decay generally leave the resultant daughter nuclei in an

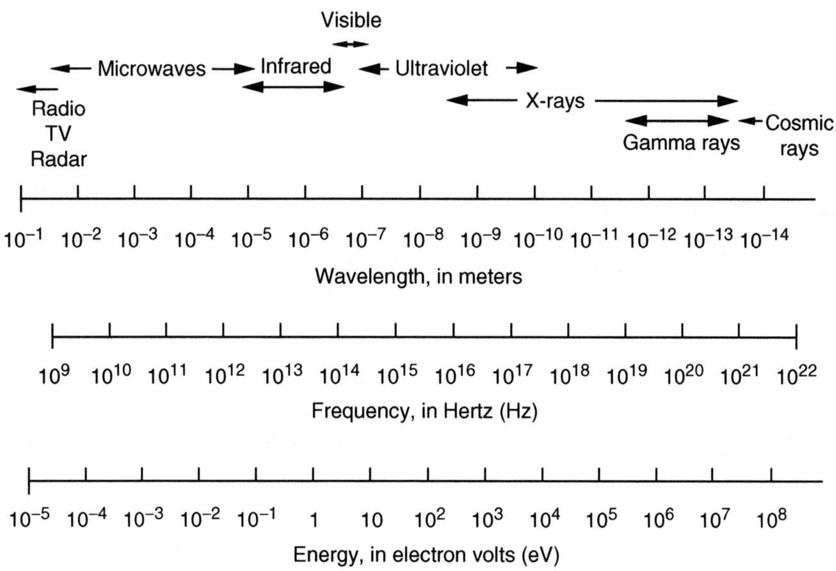

Figure 27.1.
The spectrum of electromagnetic waves, showing relation between wavelength, frequency, and energy.

excited state that is deactivated by emission of γ photons. Although γ emission accompanies most decays, it is not always detected, especially not with light β emitters such as ^{3}H, ^{14}C, ^{32}P, and ^{35}S. The half-life of individual radionuclides can be measured (i.e., the time during which half the atoms of the radionuclide spontaneously decay to a daughter nuclide). Another form of nuclear breakdown is fission, in which the nucleus breaks into two nuclides of approximately half the parent's size. Each radionuclide is characterized by mass number, atomic number, half-life, and decay mode.

Four general groups of radionuclides are distinguished: (1) a long half-life group (i.e., Tb1/2 > 10^9 years) of elements including ^{238}U, ^{235}U, ^{232}Th, ^{40}K, ^{87}Rb, and ^{143}Sm formed about 4.5 billion years ago; (2) shorter lived daughters of U and Th such as Ra and Rn that form as a result of the decay of their long-lived parents; (3) nuclides (i.e., ^{14}C and ^{3}H) formed by continuing natural nuclear transformations driven by cosmic rays, natural sources of neutrons, or energetic particles that are formed in the upper atmosphere by cosmic rays; and (4) nuclides formed as a result of nuclear weapons tests, nuclear reactor operations, and other human activities. Important members of this group include ^{90}Sr, ^{137}Cs, ^{14}C, and ^{3}H; note that many members of the third group – such as ^{14}C and ^{3}H – are also formed in this fourth fashion.

Radioactive decay usually does not immediately lead to a stable end product but to other unstable nuclei that form a decay series. The most important examples of unstable nuclei are started by very heavy, naturally occurring nuclei. Because the mass number changes only with α-decay, all members of a series may be classified according to their mass numbers (see the uranium-238 decay series in Figure 27.2). A total of three natural decay series – formed at the birth of our planet – are named after their parent isotope: ^{232}Th, ^{235}U, and ^{238}U (Figure 27.3). Several shorter decay series also exist. For example, ^{90}Sr decays with a Tb1/2 of 28 years by β emission to ^{90}Y, which in turn disintegrates (β emission) with a Tb1/2 of 64 h to the stable ^{90}Zr. Other examples of known radionuclides since the earth's origin include ^{40}K and ^{87}Rb. In hazard assessments, all members of a decay series have to be considered.

27.2.3 Linear Energy Transfer

The deposition of energy in an exposed body is mediated almost exclusively by charged particles. These particles cause ionizations but lose energy with each ionization until they reach the end of their range. Depending on the type of particle, the ionizations are more or less closely spaced and described by the energy loss of a traversing particle. The linear energy transfer (LET) is defined as the amount of locally absorbed energy per unit length, that is, only the energy fraction that leads to ionizations or excitations in the considered site is counted. Because radiation effects are dependent on the nature of the radiation, a weighting factor is used to modify the absorbed dose and to define the dose equivalent; this factor – now called the Radiation Weighting Factor – is a function of LET. Approximate weighting values range from 1 (X-rays, electrons, gamma rays) to 10 (neutrons, protons, singly charged particles of rest mass greater than one atomic mass of unknown energy), and to 20 (alpha particles and multi-charged) particles of unknown energy.

27.2.4 New Units of Measurement

A variety of units have been used for the assessment of exposures to ionizing radiation. The current international standard terminology is shown in Table 27.1. This chapter uses the new terminology exclusively; this frequently necessitated data transformation of units from early published accounts into the currently accepted international terminology.

27.3 Sources and Uses

Most external exposure of living organisms to radiation is from naturally occurring electromagnetic waves, and most internal exposure

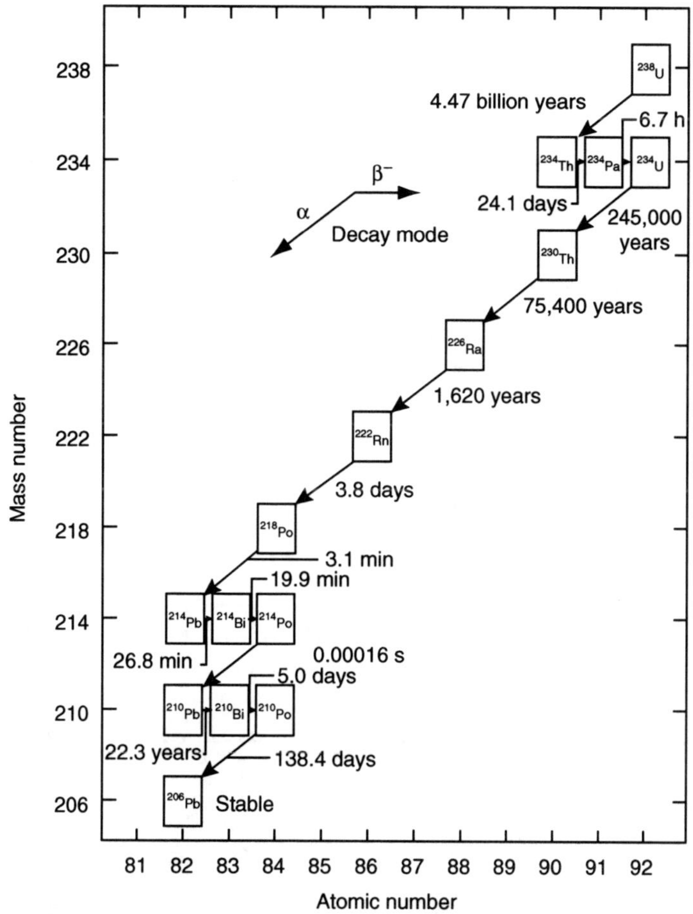

Figure 27.2.
The principal uranium-238 decay series, indicating major decay mode and physical half-time of persistence.

from naturally occurring radionuclides, such as potassium-40. Natural radiation doses vary significantly with altitude, radionuclide concentrations in the biogeophysical environment, and uptake kinetics. The major source of global anthropogenic radioactivity is fallout from military atmospheric weapons testing; locally, radiation levels tend to be elevated near nuclear power production facilities, nuclear fuel reprocessing plants, and nuclear waste disposal sites. Dispersion of radioactive materials is governed by a variety of physical, chemical, and biological vectors, including winds, water currents, plankton, and avian and terrestrial wildlife.

27.3.1 Natural Radioactivity

Exposure to natural sources of radiation is unavoidable. Externally, individuals receive cosmic rays, terrestrial X-rays, and gamma radiation. Internally, naturally occurring radionuclides of Pb, Po, Bi, Ra, Rn, K, C, H, U, and Th contribute to the natural radiation dose from inhalation and ingestion. Potassium-40 is

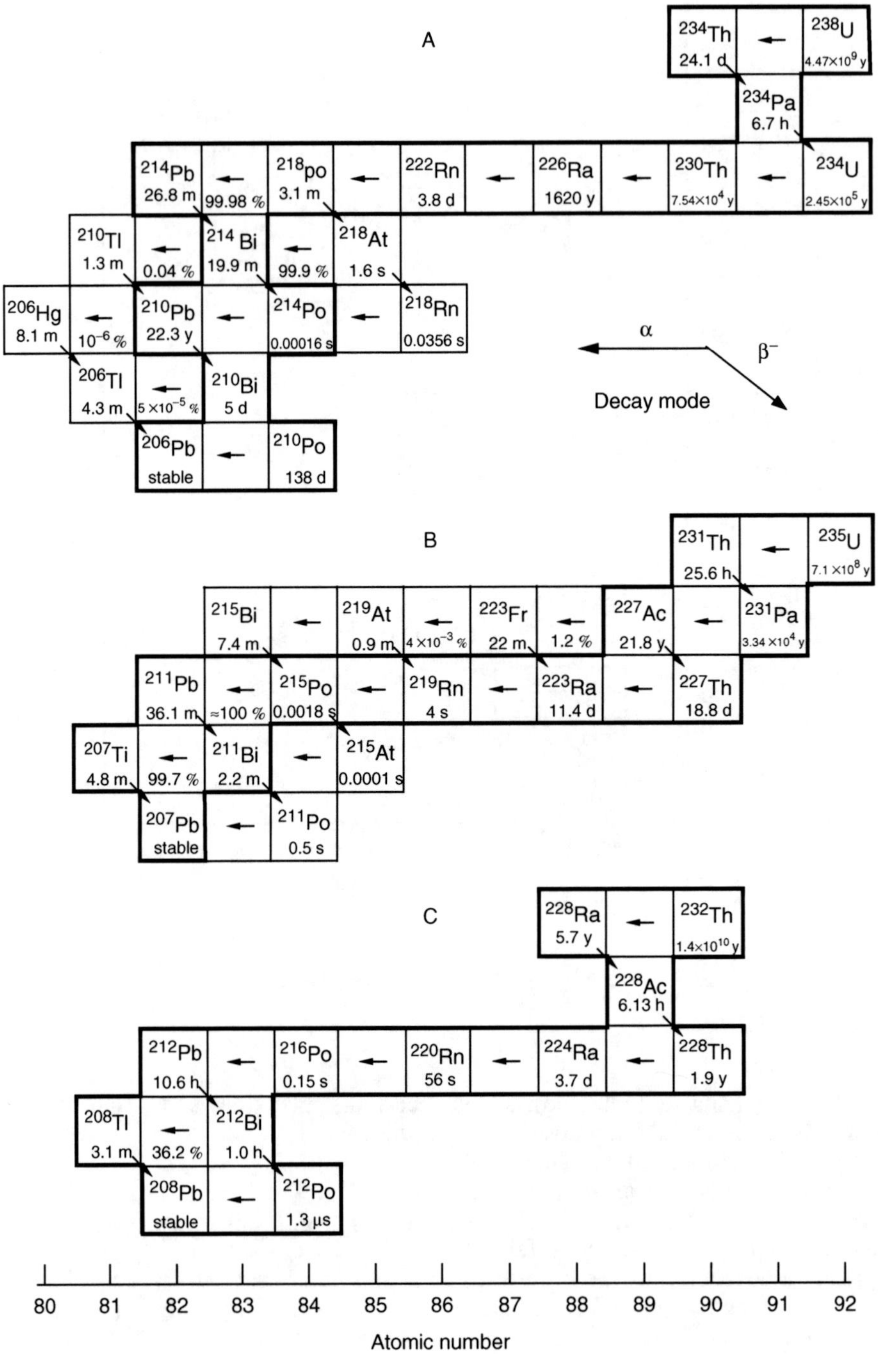

Figure 27.3.
The three still existing natural decay series. A. Uranium-238; B. Uranium-235; and C. Thorium-232. Principal decay products occur within the heavy borders outlined.

Table 27.1. New units[a] for use with radiation and radioactivity.

Variable	Old Unit	New Unit	Old Unit in Terms of New Unit
Activity	Curie (Ci) = 3.7×10^{10} disintegrations per s (dps)	Becquerel (Bq) = 1 dps	1 Ci = 3.7×10^{10} Bq
Exposure	Roentgen (R) = 2.58×10^{-4} C/kg	Coulomb/kg (C/kg)	1 R = 2.58×10^{-4} C/kg
Absorbed dose	Rad = 100 erg/g	Gray (Gy) = 1 J/kg	1 Rad = 0.01 Gy
Dose equivalent	Rem = damage effects of 1 R	Sievert (Sv) = 1 J/kg	1 Rem = 0.01 Sv

[a] See Glossary.

the most abundant radionuclide in foods and in all tissues. The mean effective human dose equivalent from natural radiations is 2.4 milliSieverts (mSv); this value includes the lung dose from radon daughter products and is about 20% higher than a 1982 estimate that did not take lung dose into account.

The dose of natural radiation that an organism receives depends on height above sea level, amount and type of radionuclides in the soil of its neighborhood, and the amount taken up from air, water, and food. Natural radiations in various ecosystems result in radiation dose equivalents that usually range between <0.005 and 2.07 mSv annually (Figure 27.4). Radiation doses are substantially higher at atypically elevated local sites, such as Denver, and sometimes exceed 17 mSv annually in mountainous regions of Brazil and the former Soviet Union.

27.3.2 Anthropogenic Radioactivity

Nuclear explosions and nuclear power production are the major sources of anthropogenic activity in the environment. But radionuclide use in medicine, industry, agriculture, education, and production and transport, use, and disposal from these activities present opportunities for wastes to enter the environment. Radiation was used as early as 1902 in the treatment of diseases, including enlarged thymus, tinea capitis, acne, and cancers of childhood and adolescence. The use of X-rays by physicians and dentists represents the largest source of annual dose equivalent of the U.S. population to artificial radiation: 0.78–1.01 mSv to bone marrow and 0.016 mSv to the upper GI tract; radiopharmaceuticals contribute an additional 0.14 mSv or a yearly total mean dose of 0.94–1.17 mSv to bone marrow.

Atmospheric testing of nuclear weapons is an important human source of environmental radiation. The first test explosion of a nuclear weapon took place in 1945. Atmospheric tests by the United States, the former Soviet Union, and the United Kingdom continued until they were banned in 1963. France and the People's Republic of China continued to conduct limited atmospheric tests, although no atmospheric nuclear explosions have taken place since 1980. Large nuclear explosions in the atmosphere carry most of the radioactive material into the stratosphere where it remains for 1–5 years, depending on the altitude and latitude; fallout can occur years after an explosion injected material into the atmosphere. Smaller explosions carry the radioactive material only into the troposphere, and fallout occurs within days or weeks. Fallout was highest in the temperate regions and in the Northern hemisphere where most of the testing was done. The most abundant radionuclides from atmospheric tests to date are $^{14}C > ^{137}Cs > ^{95}Zr > ^{90}Sr > ^{106}Ru > ^{144}Ce > ^{3}H$. Of the many radionuclides produced in nuclear and thermonuclear explosions, the primary contributors to human radiation exposure include ^{14}C, ^{89}Sr, ^{90}Sr, ^{95}Zr, ^{106}Ru, ^{131}I, ^{137}Cs, ^{141}Ce, and

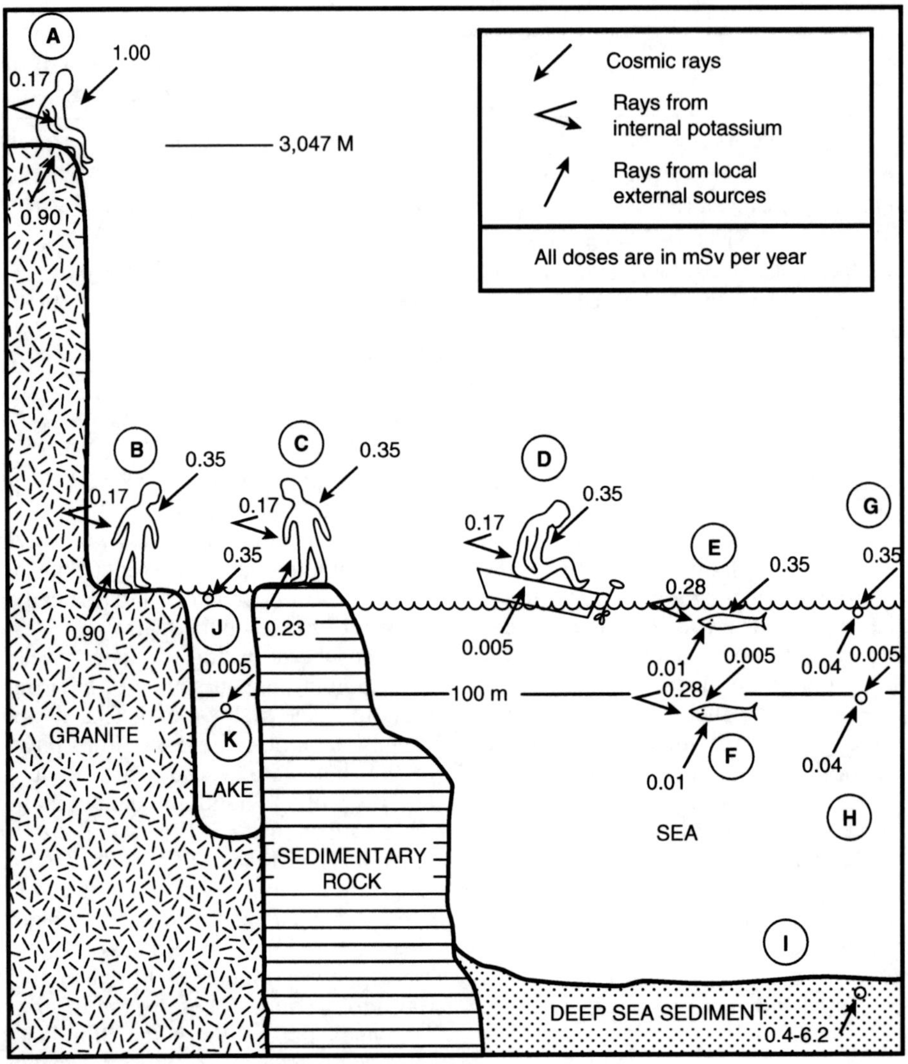

Figure 27.4.
Natural radiations in selected radiological domains. **A.** Human over granite at 3047 m (10,000 feet) elevation above sea level; total annual dose equivalent of 2.07 mSv (cosmic rays 1.00, granite 0.90, internal emitters 0.17); **B.** Human over granite at sea surface; total annual dose of 1.42 mSv; **C.** Human over sedimentary rock at sea level; total annual dose of 0.75 mSv; **D.** Human over sea; total annual dose of 0.525 mSv; **E.** Large fish in sea near surface; total annual dose of 0.64 mSv; **F.** Large fish in sea at depth of 100 m; total annual dose of 0.295 mSv; **G.** Microorganism in water near sea surface; total annual dose of 0.39 mSv; **H.** Microorganism in water >100 m deep in sea; total annual dose of 0.045 mSv; **I.** Microorganism buried in deep sea sediments; total annual dose between 0.4 and 6.2 mSv; **J.** Microorganism near freshwater surface; total annual dose of 0.35 mSv; **K.** Microorganism 100 m deep in a freshwater lake; total annual dose of 0.005 mSv.

^{144}Ce; isotopes of plutonium and americium – although present in quantity – are not significant contributors because of their low solubility. The primary dose from fallout radiation is through external gamma radiation, assimilation through the food chain, or beta radiation of the skin.

Radioisotope thermoelectric generators (RTGs) are sometimes used as power sources for space systems. In April 1964, a United States RTG navigational satellite, SNAP 9A, reentered the atmosphere and burned up at high altitude over the Mozambique Channel, releasing 629 trillion becquerels (TBq), equivalent to 17,000 Ci, of ^{238}Pu and 0.48 TBq of ^{239}Pu. In January 1978, a Soviet RTG satellite, Kosmos 954, reentered the atmosphere over Canada and spread radiouranium across parts of that country. The amount of radioactive materials in space applications is expected to increase.

Significant amounts of radioactivity are present in the Great Lakes basin, which has numerous nuclear reactors and uranium-mine waste areas. The prevailing low levels of artificially produced radionuclides, arising largely from previous fallout, provide small doses of radiation to residents who consume lake water. Radionuclides enter the Great Lakes ecosystem from natural and anthropogenic processes. The main natural processes that introduce radioactivity are the weathering of rocks, which contain uranium- and thorium-series radionuclides, and fallout of cosmic-ray-produced radionuclides such as ^{3}H, ^{7}Be, and ^{14}C. Anthropogenic radioactivity is created, for example, by uranium mining, milling, and fuel fabrication; releases of artificially produced radionuclides through nuclear power reactors and nuclear fuel processing plants; medical uses of radioisotopes; and coal-fired electrical generating plants.

Production of power from nuclear reactors involves uranium mining, fuel fabrication, the reactor operations, and storage of wastes. All of these processes may expose humans and the environment to radiation. Uranium production in the United States was 12,300 tons of U_3O_8 in 1977, primarily from western states, Texas, and Florida. Mining from deep shafts or open pits is the preferred method of uranium extraction, although in Florida it is produced as a by-product of phosphate mining. Mines disperse radionuclides of uranium, thorium, and radium, which are associated with dust particles, and radon, which emanates from ore as a gas and decays to create a series of radioactive daughters. Groundwater also contains radionuclides of the uranium series. As many as 18 uranium mills were in operation, located close to major mining centers in the western states. Collectively, these mills process or processed about 30,000 tons of ore daily and used acid or alkali leach methods to extract 90–95% of the uranium from ore. Uranium is barreled at the mill for shipment as uranium oxide or as salt concentrates (yellowcake) that contain 70–90% U_3O_8 by weight. Residues of the uranium extraction process are usually pumped as a slurry to liquid-retention impoundments; about 0.55 TBq of ^{230}Th and ^{226}Ra enter tailings each day from milling operations. Radium-226 produces gaseous ^{222}Rn; daughters of ^{222}Rn, such as ^{210}Pb, expose the surrounding biota to measurable radiation. Purification of yellowcake to UF_6 (uranium hexafluoride) and its enrichment to ^{235}U causes a loss of about 0.55 TBq annually. Nuclear reactor fuel contains about 3% ^{235}U. A nuclear explosion in a nuclear reactor is highly unlikely because the nuclear fuel suitable for weapons must contain >90% ^{235}U. Following enrichment, UF_6 is hydrolyzed to uranyl fluoride, converted to ammonium diuranate, and calcined to the dioxide UO_2. Uranium dioxide pellets at one time were prepared by as many as 10 commercial fuel fabrication plants and subsequently transported to nuclear reactors. In the current light-water cooled reactors, the most abundant radionuclides in the reactor effluents under normal conditions are ^{3}H, ^{58}Co, ^{60}Co, ^{85}Kr, ^{85}Sr, ^{90}Sr, ^{130}I, ^{131}I, ^{131}Xe, ^{133}Xe, ^{134}Cs, ^{137}Cs, and ^{140}Ba. Gaseous and volatile radionuclides, such as ^{85}Kr, ^{131}Xe, and ^{133}Xe, contribute to the external gamma dose whereas the others contribute to the dose externally by surface deposition and internally by way of the food chain. The mean dose from environmental releases of all radionuclides from nuclear reactors in the United States is <0.01 mSv/year. Nuclear fission follows the capture of a neutron by an atom of fissionable material, such as ^{235}U or ^{239}Pu.

The fission releases 1–3 neutrons and, if additional fissionable material is present in sufficient quantity and in the right configuration, a chain reaction occurs. Radionuclides formed per megaton of fission include fission products (^{89}Sr, ^{90}Sr, ^{95}Zr, ^{103}Ru, ^{106}Ru, ^{131}I, ^{137}Cs, ^{144}Ce), and activation products in air (^{3}H, ^{14}C, ^{39}Ar) and soil (^{24}Na, ^{32}P, ^{42}K, ^{45}Ca, ^{55}Fe, ^{59}Fe). Fission-product radionuclides of potential biological importance include ^{90}Sr, ^{137}Cs, ^{131}I, ^{129}I, ^{144}Ce, ^{103}Ru, ^{106}Ru, ^{95}Zr, ^{140}Ba, ^{91}Y, ^{143}Ce, ^{147}Nd, and others.

Most of the world's supply of uranium consists of about 0.7% ^{235}U and 99% ^{238}U. In theory, about 2.27 kg of ^{235}U can release energy equivalent to 20,000 tons of TNT. Uranium-238 and ^{232}Th can be converted into fissionable material following neutron capture. Radionuclides of biological significance produced by neutron activation in nuclear reactors include ^{3}H, ^{14}C, ^{24}Na, ^{32}P, ^{35}S, ^{45}Ca, ^{54}Mn, ^{55}Fe, ^{57}Co, ^{58}Co, ^{60}Co, ^{65}Zn, ^{239}Pu, ^{239}Np, ^{241}Am, and ^{242}Cm. Nuclear energy can also be released by fusion of smaller nuclei into larger nuclei that is accompanied by a decrease in mass. Fusion reactors – which do not yet exist – require very high temperatures of several million degrees; no fission products are produced in the fusion process.

Radioactive wastes are usually stored in underground tanks or in temporary storage at reactor sites for recycling or disposal. For low level wastes, containment and isolation are the preferred disposal options, including burial, hydraulic injection into deep geological strata, and ocean disposal. Options for the disposal of high-level wastes include retrievable surface storage and entombment in deep geological strata; many risks are associated with these options, and more suitable alternative disposals are needed. Spent nuclear fuel elements are usually stored for about 3 months to allow the decay of shorter lived radionuclides before reprocessing or disposal. Reprocessing involves extractions to separate uranium and plutonium from the fission products into UF_6 and plutonium dioxide. Longer lived fission products, such as ^{90}Sr and ^{137}Cs, are sometimes chemically separated and encapsulated for storage or disposal. Fuel reprocessing tends to release measurable quantities of various radionuclides that are detected in fish, wildlife, and food for humans. Liquid discharges from the Sellafield reprocessing plant have been reduced by a factor of more than 100 since the mid-1970s. Human populations that consume higher than average quantities of marine fish and shellfish from the Sellafield area theoretically receive about 3.5 mSv annually from radioactivity associated with nuclear power production. Human populations in the vicinity of nuclear power production discharging directly into the marine environment – except for Sellafield – generally receive <0.05 mSv annually from this source.

Radioactive transuranic elements with atomic numbers that are greater than 92 have been introduced into the environment since the 1940s from atmospheric testing of nuclear weapons, discharges of nuclear wastes, and nuclear fuel reprocessing. Transuranic isotopes with half-lives of more than 10,000 years (i.e., ^{247}Cm, ^{248}Cm, ^{239}Pu, ^{242}Pu, ^{244}Pu, ^{237}Np) will persist over geologically significant time periods. Transuranics at detectable but considered nonhazardous levels to biota are now widely dispersed throughout the environment in most waters, soils, sediments, and living organisms including humans. Of current primary concern are ^{244}Cm, ^{241}Am, ^{238}Pu, ^{239}Pu, ^{240}Pu, ^{241}Pu, and ^{237}Np – especially americium-241, which is increasing globally as a result of ^{241}Pu decay. However, the estimated peak dose received from Pu and Am radioisotopes seems to be decreasing in the vicinity of the Sellafield nuclear fuel reprocessor. Miscellaneous exposures include radiations from television sets, luminous dial watches, smoke detectors, electron microscopes, building materials, and air travel. Most of the exposure in building materials is due to naturally occurring radionuclides; similarly, air travel increases radiation exposure of travelers from increased exposure to cosmic radiations. Cigarette smokers may receive dose equivalent rates up to 3 times higher than nonsmokers because of inhalation of ^{210}Po and ^{210}Pb from the cigarette. Some of the lung dose is also received from radionuclides released during combustion of fossil fuels, which contain small quantities of naturally occurring radionuclides.

27.3.3 Dispersion

Radioactive materials are cycled throughout the environment by a variety of physical, chemical, and biological vectors. Dispersion through the atmosphere is governed by the magnitude, frequency, and direction of the wind; in the hydrosphere, transport is modified by water depth, motion, temperature, winds, tides, and groundwater. Deposition from the atmosphere is a function of particle size, precipitation, and dry deposition. Small radioactive particles may be elevated into the airstream from the ground surface; resuspension is a function of disturbances by wind at the soil surface, atmospheric variables (i.e., velocity, turbulence, density, viscosity), and soil-ground variables, such as texture, cohesiveness, moisture content, density, vegetation cover, ground surface roughness, and topography. Only 1.0 kg of the original 15.0 kg of Pu was fissioned from the dropping of the plutonium nuclear bomb on Nagasaki, Japan, on August 9, 1945. The remaining 14.0 kg of Pu escaped into the environment. Local fallout accounted for about 37.0 g or 0.26% of the total global fallout; the highest ^{239}Pu and ^{240}Pu concentration measured was 64.0 Bq/kg soil about 2.8 km from ground zero.

Biological agents can also transport radioactive wastes. Birds, especially waterfowl, disperse accumulated radiocesium and other radionuclides along their migratory flyways. Native mammalian herbivores and their predators that have come in contact with radioactivity in food or soils disperse the material in their feces, urine, or regurgitated pellets. For example, the black-tailed jackrabbit (*Lepus californicus*) in the vicinity of radioactive waste-disposal trenches dispersed radioactive fecal pellets over an area of 15 km^2; elevated radioactivity readings were recorded in jackrabbits and in their predators, including feces of coyotes (*Canis latrans*) and bones of hawks.

Biological transport of trace elements and radionuclides in the sea is provided mainly through phytoplankton and zooplankton because of their (1) ability to accumulate these elements to high levels, (2) diurnal vertical migration, and (3) production of detritus in the form of fecal pellets, molts, and carcasses. Considerations related to biomass, feeding rates, conversion efficiencies, migratory habits of zooplankton, and the chemical properties of trace elements suggest that the major downward transport of these elements and radionuclides is through gravitational action on fecal pellets, molts, and carcasses; direct biological transport accounts for $<10\%$ of the total downward movement. In estuarine and nearshore regions, the bottom sediments and their associated epiphyton often significantly influence the distribution of added radionuclides. Large populations of sessile filter feeders may drastically increase the rate of sedimentation of added trace elements and radionuclides.

In some coastal areas, some of the radionuclides discharged into coastal waters from industrial establishments are recycled via the air–sea interface back onto land. At the sea surface, aerosol is generated by bubble bursting and wave shearing. The aerosol is advected to land by onshore winds and deposited in coastal regions. Sea-to-land transfer has been documented from the vicinity of nuclear fuel reprocessing facilities in England, Scotland, and France; however, the sea-to-land transfer pathway was only about 8% of that from the seafood pathway. The solubility of different radionuclides at the sediment–seawater interface is variable. Plutonium solubility, for example, depends on pH, Eh, ionic strength, complexing ions, organic chelators, living accumulator organisms, and oxidation state. The oceanic distributions of many nuclides are strongly controlled by interactions with particulate matter. Thorium is an extreme case; the high reactivity of this element accounts for its residence of only a few decades in the ocean from where it is removed largely by vertical transport in association with settling particulate matter. Lead-210 and ^{231}Pa are also particle-reactive but to a lesser extent than Th. Their oceanic mean residence time is about 100 years. The mean oceanic residence time of ^{227}Ac and Ra isotopes is about 1000 years because of particulate scavenging; these nuclides are supplied by insoluble parents in underlying sediments and are released to overlying waters by porewater diffusion. Radium-228 can serve as a novel tracer in ocean circulation for about 30 years; ^{227}Ac can be used for about 100 years. The distribution of ^{226}Ra is largely governed by biogeochemical cycling, much like dissolved silica.

27.4 Radionuclide Concentrations in Field Collections

The wide dispersion of anthropogenic radiocontaminants has significantly altered natural background levels of radioactivity in many parts of the globe. Radionuclide concentrations in selected abiotic materials and living organisms were usually elevated in samples from the vicinity of human nuclear activities, especially atmospheric military tests. Radionuclide concentrations in organisms were significantly modified by its age, sex, diet, metabolism, trophic level, proximity to point source, and many other biological, chemical, and physical variables, as discussed later.

27.4.1 Abiotic Materials

Most authorities agree that radionuclide concentrations in nonliving materials are elevated in samples from the site of repeated nuclear detonations, near nuclear fuel reprocessing and waste facilities, and from locations receiving radioactive fallout from atmospheric military tests. Rocks, especially granite, have high levels of naturally occurring radionuclides such as ^{40}K. Concentrations are usually low or negligible in drinking water and cow's milk for human consumption. Nuclear-weapons testing has resulted in large environmental releases of radionuclides. Between 1961 and 1966, for example, the Republic of Korea received fallout from nuclear tests by the former Soviet Union in 1961 and by the United States in 1962 and from 3 explosions by the People's Republic of China. The highest levels of total combined β and γ activity in various Korean samples during 1962–64, in Bq/L or Bq/kg, were: 0.0002 in air, 133 in water, 1572 in milk, 2023 in rain, 16,428 in plants, and 99,345 in soils.

Water in the Great Lakes in 1981 contained measurable concentrations of ^{137}Cs, ^{3}H, and ^{90}Sr and detectable – but extremely low – concentrations of ^{241}Am, ^{113m}Cd, ^{144}Ce, ^{210}Pb, ^{239}Pu, ^{240}Pu, ^{226}Ra, ^{125}Sb, and ^{228}Th. Radiocesium-137 in water from the Hudson River estuary, New York, decreased tenfold between 1964 and 1970, but the ^{137}Cs content in fish and in sediments remained relatively constant. The effluent from the United Kingdom's Atomic Energy Agency Sellafield facility on the Cumberland Coast of the Irish Sea contained ^{90}Sr and ^{137}Cs, which are soluble in seawater and tend to remain in solution, and ^{106}Ru, ^{144}Ce, and $^{95}Zr/^{95}Nb$, which are relatively insoluble in seawater and coprecipitate or adsorb on free inorganic and organic surfaces.

Soils in the vicinity of an English nuclear fuel reprocessing facility in the period 1979–85 contained as much as 42 times more ^{241}Am, 12 times more ^{137}Cs, 13 times more ^{90}Sr, and 87 times more ^{239}Pu and ^{240}Pu than soils from a reference site. In the United States, radiological trends in abiotic materials were difficult to interpret. For example, one nationwide monitoring program for radionuclide concentrations in air, drinking water, milk, groundwater, and precipitation was not consistent in the selection of measured radionuclides, frequency of sampling, and types of samples analyzed.

27.4.2 Aquatic Ecosystems

Field studies indicate that effects of radiation on marine ecosystems cannot be demonstrated at prevailing dose rates. Two major periods of worldwide fallout occurred in Arctic ecosystems. The first and most sustained occurred during 1953–59 and the second during 1961–64, reflecting the atmospheric nuclear-weapons test regimes of Great Britain, the former Soviet Union, and the United States. Military accidents created localized radiocontamination of the Arctic environment. In one case, a B-52 aircraft from the U.S. Air Force crashed on the ice in northwestern Greenland in January 1968. Plutonium from the nuclear weapons on board contaminated the benthos (Figure 27.5). The ^{239}Pu and ^{240}Pu concentrations in various environmental samples declined at a much faster rate than the physical half-life of ^{239}Pu (24,000 years), suggesting

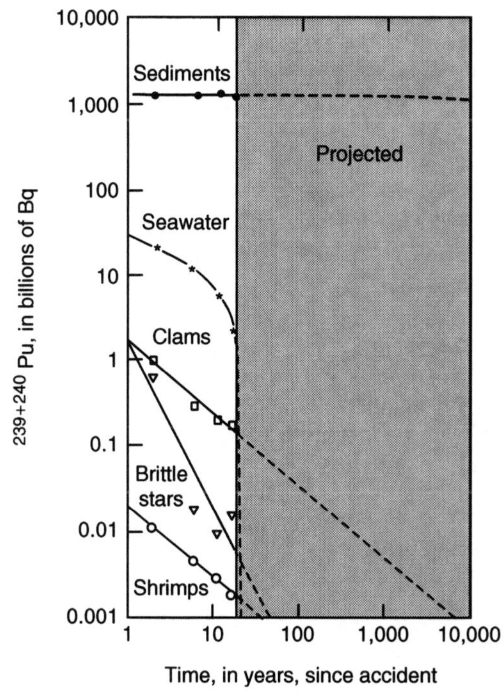

Figure 27.5.
Plutonium-239, -240 in environmental samples at Thule, Greenland, between 1970 and 1984, after a military accident in 1968. Within the contaminated area of 3.2×10^9 m^2, the FW biomass of shrimps was 0.11×10^9 kg, of brittle star echinoderms 0.06×10^9 kg, and of clam (*Macoma balthica*) soft parts 0.32×10^9 kg. The seawater mass was 3×10^{14} kg, and the DW of the upper 15 cm sediment layer was 3×10^{11} kg.

that Pu becomes increasingly unavailable to the benthos over time as a result of dispersion from the epicenter and a dilution effect.

In marine environments, the major portion of the background dose rate in plankton and fish arises from the incorporated activity of natural alpha emitters, such as ^{210}Po, and from ^{40}K; in mollusks, crustaceans, and benthos, the gamma radiation from the seabed provides the major background dose. The situation is similar in freshwater environments, although water containing appreciable levels of ^{222}Rn and its daughter radionuclides may exert an additional burden, especially to phytoplankton. Artificial radionuclides that contribute significantly to background concentrations of marine organisms include ^{239}Pu and ^{90}Sr; of freshwater organisms, ^{137}Cs and ^{90}Sr. The total natural radiation received by a marine flounder (*Pleuronectes platessa*) in the Irish Sea consisted of 63% from radiations from seabed sediments, 16% from ^{40}K in seawater, 15% from internal ^{40}K, and 6% from cosmic radiation. The estimated dose rates in aquatic environments from natural background are as high as 3.5 mGy annually and of the same order as those in most terrestrial environments. By 1976, the estimated dose rates from global fallout had declined to the same range as natural dose rates, although environments receiving radioactive wastes had variable responses.

Muscle of largemouth bass (*Micropterus salmoides*) collected from lakes in South Carolina in May and June of 1993 contained 109–4607 Bq ^{137}Cs/kg DW; increasing concentrations of ^{137}Cs were correlated with increasing DNA damage. Increasing levels of ^{137}Cs in the muscle of fish in Minnesota between 1954 and 1966 reflect fallout from atmospheric nuclear testing. The effective half-time for ^{137}Cs in these lakes, as judged from small fish, is about 30 months. In game fish from Colorado, ^{137}Cs in muscle was up to 7 times higher in 1968 than in 1965; higher in fish from mountain lakes than in fish from reservoirs in the plains, foothills, lakes, and rivers; and highest in trout from alpine lakes and reservoirs. In 1966, ^{137}Cs levels in trout from Colorado alpine lakes were 8–18 times higher than mean levels in muscle of deer from Colorado during the same period, and 20–300 times higher than domestic meat products. Radionuclides in livers of tunas from southern California during the period 1964–70 originated mainly from weapons tests in 1961–62, although ^{65}Zn may have reached southern California waters from nuclear reactors in Hanford (Washington) and from French or Chinese nuclear tests.

Many variables are known to modify radionuclide concentrations in biota. In general, lower trophic levels of aquatic organisms are likely to have greater concentrations of radionuclides than higher trophic levels. However, radionuclide concentrations in biota are modified significantly by the organism's age, size, sex, tissue, season of collection, and other variables – and these have to be acknowledged when integrating radiological analyses. For example, older *Fucus vesiculosus* had higher radioactivity concentrations than younger algae; concentrations of ^{60}Co and ^{54}Mn were highest in older parts of plants during spring and summer; and ^{137}Cs and ^{40}K were highest in receptacles and new vegetative fronds. Changes in concentrations of ^{60}Co and ^{137}Cs in freshwater plankton from the discharge canal of an Italian nuclear power station seem to reflect changes in water concentrations of these isotopes; changes were lowest in winter and highest in summer. Marine bivalve mollusks and algae from Connecticut in 1960 and 1961 had the highest levels of gross beta radioactivity in spring and summer and the lowest in winter; natural ^{40}K probably accounted for most of the beta radioactivity. Similar seasonal variations in gross beta radioactivity in other species of marine algae and mollusks were documented, suggesting a correspondence with periods of dormancy and activity. Although fat in the liver of crabs accounted for 47% of the fresh weight (FW) (74% on a dry weight (DW) basis), the gross beta activity of the fat fraction amounted to <0.5% of the total radioactivity, suggesting that radiological liver analyses be compared on the basis of nonfat solids. In mosquitofish (*Gambusia holbrooki*) from some locations in a ^{137}Cs-contaminated reservoir, males contained higher ^{137}Cs concentrations than females, and smaller females contained more ^{137}Cs than larger females. Strontium-90 concentrations in the carapace bone of turtles from 5 southwestern states in 1970 were used as indicators of ^{90}Sr fallout. However, older turtles tended to have lower concentrations of ^{90}Sr and concentrations differed geographically; concentrations were highest in Georgia and increasingly lower in Tennessee, Mississippi, Arkansas, and Florida.

Consumption of shellfish represents a negligible radiological risk to humans, although bivalve mollusks seem to be effective accumulators of radioisotopes. After the Chinese nuclear tests in May and December 1966, concentrations of ^{144}Ce, ^{103}Ru, ^{95}Zr, ^{95}Nb, ^{140}Ba, and ^{140}La in 3 species of bivalves in the Neuse River, North Carolina increased suddenly. In 1973, Pacific oysters (*Crassostrea gigas*) from the discharge canal of a nuclear power plant in Humboldt Bay, California, rapidly accumulated ^{54}Mn, ^{60}Co, ^{65}Zn, and ^{137}Cs within 30 min of release. Isotope uptake correlated positively with particulates in the water, including living microorganisms, organic detritus, inorganic materials, and especially resuspended bottom sediments. Although concentrations of cesium and plutonium in mussels (*Mytilus edulis*) from most Irish estuaries are essentially the same as global fallout levels, concentrations were elevated in mussels from the northeast coast.

27.4.3 Birds

Television and newspaper reporters attributed radionuclides to a decline in bird numbers at the Ravenglass estuary, England, particularly of the black-headed gull (*Larus ridibundus*), although the concentrations of radionuclides in the avian diet, body tissues, and general environment were at least 1000 times too low to have had any effect. Although oystercatchers (*Haematopus ostralegus*) and shelducks (*Tadorna*) had the highest concentrations of ^{137}Cs in their tissues, the breeding success and population size of these birds were not affected. Black-headed gulls had less radiocontamination than other birds at Ravenglass, but their population continued to decline. The most likely cause was a combination of an uncontrolled fox population, a severe outbreak of myxomatosis in rabbits (normal fox prey), and a drought – all in the same year. Nesting success of birds was unaffected in the vicinity of nuclear power plants. For example, nesting barn swallows (*Hirundo rustica*) near radioactive leaching ponds had normal nesting success despite their consumption of arthropods from the pond and use of contaminated mud for nest construction. Adult swallows received a total internal dose rate of 219 μGy/day, mostly

(72%) from ^{24}Na; daily dose rates for eggs and nestlings during the nesting season were 840 μGy and 2200 μGy, respectively. The total dose to eggs and nestlings (54 mGy) and adults (450 mGy) had no measurable effect on survival and was below accumulated doses reported to cause death of passerines.

Strontium-90 behaves much like calcium in the biological environment. In birds, ^{90}Sr is expected to occur in bone and in the calcium-rich eggshell. In one case, a positive relation was demonstrated between reactor releases of ^{90}Sr to the Columbia River and ^{90}Sr concentrations in reed canary grass (*Phalaris arundinacea*) and eggshells of the Canada goose (*Branta canadensis moffitti*).

No human health problem is anticipated from consumption of ruffed grouse (*Bonasa umbellus*) contaminated with ^{226}Ra in Canada or American coot (*Fulica americana*) contaminated with ^{137}Cs in Washington state. Tissues of ruffed grouse collected near discharged uranium tailings in Canada in 1987–88 did not contain grossly elevated levels of ^{226}Ra over controls; consumption of grouse by humans did not present a radiological health problem. Based on ^{137}Cs alone, humans who consumed a single contaminated American coot captured at Hanford, Washington, would receive about 1.1% of the annual radiation protection dose of 1.70 mSv for individuals and populations in uncontrolled areas.

27.4.4 Mammals

Diets in Denmark contained elevated loadings of ^{137}Cs in 1964 because of the intensive atmospheric nuclear test series by the United States and the former Soviet Union in 1961 and 1962. Total ^{137}Cs intake declined in the Danish population from 72 Bq/kg BW in 1964 to <2 in 1985 but rose to about 13 in 1986 from the effects of debris from Chernobyl on dietary ^{137}Cs during the first year after the accident. The estimated dose equivalent from ^{137}Cs to human consumers of fish from the Great Lakes is about 0.01 μSv/kg FW muscle from fish in Lakes Erie and Ontario, and 0.06–0.07 μSv/kg from fish in Lakes Superior and Huron. The guide for the protection of the general public from radiation is <5 mSv annually, and consumption of fish containing a dose equivalent greater than 0.02 μSv/kg fish flesh is not recommended. Some Scandinavians now receive a dose equivalent of about 5 mSv/year from intake of radiocesium in the diet. In Finland, uptake of radionuclides by humans in Finnish Lapland and in other areas with an arctic climate is attributed to ecological factors and to a high amount of local fallout. For example, reindeer-herding Finnish Lapps contained about 50 times more ^{137}Cs and 10 times more ^{55}Fe in tissues than other Finns during 1961–67. For ^{137}Cs, this disparity is attributed mainly to the reliance by Finns on reindeer meat – which contains high levels of ^{137}Cs as a result of reindeer feeding on lichens – and secondarily, on freshwater fish and cow's milk.

In the United States, the estimated annual whole body human radiation dose equivalent is 1.61 mSv, mostly from natural sources (0.85 mSv) and medical sources (0.70 mSv) and also from fallout (0.03 mSv), miscellaneous sources (0.02 mSv), occupational hazards (0.008 mSv), and nuclear power (0.0001 mSv). Radiation doses to people living near the Hanford nuclear industrial and research site in the state of Washington are well below existing regulatory standards. Only trace amounts of radionuclides from Hanford have been detected in the offsite environment. In December 1984, radon levels up to 130 times greater than considered safe under the current guideline for underground uranium miners were discovered in human residences in eastern Pennsylvania, New Jersey, and New York. About 25% of all residences in 10 states exceeded the action level for radon of 0.185 Bq/L air. The significance of this observation to avian and terrestrial wildlife merits investigation.

As a result of nuclear weapons testing, mandibles of Columbian black-tailed deer (*Odocoileus hemionus columbianus*) from California increased from <9 Bq ^{90}Sr/kg ash weight (AW) to >204 Bq/kg AW between 1952 and 1960. Age and season affected strontium kinetics in male mule deer (*Odocoileus hemionus*) during the period of antler growth; these variables did not affect strontium kinetics in females. The concentrations of ^{90}Sr in forage of mule deer were higher in summer than in winter and the differences were of sufficient magnitude to account for the ^{90}Sr variations in mule deer antlers; ^{137}Cs concentrations were similar in the forage and flesh of the white-tailed deer (*Odocoileus virginianus*). Levels of iodine-129 in thyroids of mule deer and pronghorns (*Antilocapra americana*) increased with proximity to nuclear fuel reprocessing plants in Colorado, Idaho, New Mexico, and Wyoming during 1972–76, although levels were considered of no consequence to the health of the animals.

Radium-226, a bone-seeking α-emitter with a half-life of 1600 years may cause tissue damage and possibly subsequent osteosarcoma. Elevated ^{226}Ra concentrations have been reported in tissues of the common beaver (*Castor canadensis*) from the Serpent River watershed, Canada, the recipient of uranium tailings during 1984–87. Measurable levels of ^{226}Ra were also found in feces of snowshoe hares (*Lepus americanus*) from this area and in black cutworms (*Agrotis ipsilon*) eaten by herring gulls (*Larus argentatus*) on the tailings. Maximum levels in tissues of beavers from this watershed were <5 Bq ^{232}Th/kg DW in all tissues, 15 Bq ^{228}Th/kg DW bone, <5 Bq ^{228}Th/kg DW muscle and liver, 70–160 Bq ^{210}Po/kg DW bone, 11–75 Bq ^{210}Po/kg DW muscle, and 35–65 Bq ^{210}Po/kg DW liver. Consumption of these beavers would not be hazardous to human health. In the worst case, humans who consume substantial (71 kg) amounts of flesh of beavers from the Serpent River drainage system would receive <10% of the annual limits set by Canadian regulatory authorities.

Cesium-137 levels in gray seals (*Halichoerus grypus*) in 1987 seem to reflect ^{137}Cs levels in their fish diet, but there is no biomagnification of ^{137}Cs and other radionuclides. An estimated 29% of the ^{137}Cs in the diets of gray seals is from the Chernobyl accident and 71% from the nuclear facility at Sellafield, United Kingdom. The dose to gray seals from their diet is about 36 mSv annually and higher than the permissible dose limit of 5 mSv/year allowed the general public but below the current limit for radiation workers of 50 mSv/year.

The weekly dose rates from internal radionuclides were markedly different in muskrats (*Ondatra zibethicus*) and cotton rats (*Sigmodon hispidus*) collected at Oak Ridge, Tennessee, in August 1960 (20-1,112 mSv for muskrats vs. 3 mSv for cotton rats); the difference is probably due to differences in diets and habitats. Foxes and wildcats contain 2–16 times more ^{137}Cs than their prey organisms, such as rats and rabbits, suggesting food-chain magnification. The biological half-time of ^{137}Cs is about 30 days in foxes, dogs, and pigs but about 60 days in humans. Jackrabbits (*Lepus californicus*) in 1958, one year after contamination at the Nevada test site, averaged 1908 Bq ^{90}Sr/kg AW bone within a 160-km radius from ground zero; in 1961, the average for the same population was only 984 Bq ^{90}Sr/kg AW bone, and the few higher values were restricted to older animals. It was concluded that ^{90}Sr from fallout in jackrabbits is at its maximum at an early time after contamination and that biological availability is later reduced by natural (unspecified) mechanisms. Jackrabbits at the Nevada test site also contained certain neutron activation products, including isotopes of Co, Mn, and W.

Radionuclide concentrations in sheep and cattle grazing near a nuclear fuel reprocessing facility amounted to a small fraction of the recommended limits. Americium-241, ^{137}Cs, ^{239}Pu, and ^{240}Pu in the bone, liver, lung, and muscle of beef cattle were quite low from the vicinity of a nuclear fuel reprocessing facility in England between September and December 1986 and practically indistinguishable from control samples. Maximum concentrations, in Bq/kg FW, were 0.0015 ^{239}Pu and ^{240}Pu in lung, 0.019 ^{241}Am in liver, and 3.1 ^{137}Cs in muscle. Levels of ^{129}I were elevated in thyroids of cows near Mol, Belgium, in 1978 in the vicinity of a nuclear reprocessing plant closed in 1974.

27.5 Case Histories

Military weapons tests conducted at the Pacific Proving Grounds in the 1940s and 1950s resulted in greatly elevated local concentrations of radionuclides, and an accident at the Chernobyl nuclear power plant in the former Soviet Union in 1986 resulted in comparatively low concentrations of radionuclides dispersed over a wide geographical area. Both cases are briefly reviewed.

27.5.1 Pacific Proving Grounds

The first artificial, large-scale introduction of radionuclides into a marine environment was at Bikini Atoll in 1946. In succeeding years through 1958, Bikini and Eniwetok became the Pacific Proving Grounds where 59 nuclear and thermonuclear devices were detonated. Gross radiation injury to marine organisms has not been documented, possibly because seriously injured individuals do not survive and the more subtle injuries are difficult to detect. On land, the roof rat (*Rattus rattus*) survived heavy initial radiation by remaining in deep burrows. Terrestrial vegetation was heavily damaged by heat and blast, although regrowth occurred in 6 months. The land-dwelling hermit crab (*Coenobita* sp.) and coconut crab (*Birgus latro*) were subjected to higher levels of chronic radiation from internally deposited radionuclides than any other studied Atoll organism; levels remained constant in *Coenobita* at 166,000 Bq of ^{90}Sr/kg skeleton and 16,835 Bq ^{137}Cs/kg muscle over 2 years; *Birgus* contained 25,900 Bq ^{90}Sr/kg skeleton and 3700 Bq ^{137}Cs/kg muscle over 10 years. A survey in August 1964 at Eniwetok and Bikini Atolls (Table 27.2) showed that general levels of radioactivity were comparatively

Table 27.2. Radionuclide concentrations in selected samples from the Pacific Proving Ground. Concentrations are in becquerels/kg fresh weight (FW) or dry weight (DW).

Location, Sample, Radionuclide, and Other Variables	Concentration, in Bq/kg or Bq/L
BIKINI ATOLL	
Samples with highest concentrations, August 1964	
^{207}Bi, sediments	Max. 6660 DW
^{144}Ce, marine algae	Max. 1739 DW
^{137}Cs, land invertebrates	Max. 14,060 DW
^{57}Co, sediments	Max. 3400 DW
^{60}Co, marine invertebrates	Max. 35,150 DW
^{54}Mn, sediments	Max. 962 DW
^{106}Ru, sediments	Max. 10,360 DW
^{125}Sb, groundwater	Max. 12,950 DW
Seawater, 1972, ^{55}Fe	Max. 0.025 FW
Sediments	
1958 vs. 1972, ^{55}Fe	Max. 777,000 DW vs. 11,100 DW
August 1964, ground zero	
^{207}*Bi*	6660 DW
^{57}*Co*	3404 DW
^{60}*Co*	9620 DW
^{54}*Mn*	962 DW
^{106}*Ru*	10,360 DW
^{125}*Sb*	3663 DW

Continued

Table 27.2. cont'd

Location, Sample, Radionuclide, and Other Variables	Concentration, in Bq/kg or Bq/L
ENIWETOK ATOLL, AUGUST 1964	
Whole marine algae vs. whole marine fishes	
^{207}Bi	181 DW vs. 74 DW
^{144}Ce	814 DW vs. non-detectable (ND)
^{137}Cs	52 DW vs. 21 DW
^{60}Co	355 DW vs. 888 DW
^{54}Mn	48 DW vs. 70 DW
^{106}Ru	96 DW vs. ND
^{125}Sb	34 DW vs. ND
Terrestrial invertebrates vs. terrestrial vegetation	
^{207}Bi	6 DW vs. 10 DW
^{144}Ce	5 DW vs. 888 DW
^{137}Cs	No data vs. 12,580 DW
^{60}Co	888 DW vs. 141 DW
^{54}Mn	281 DW vs. 296 DW
^{106}Ru	15 DW vs. 19 DW
^{125}Sb	ND vs. 8 DW
Seabirds (whole) vs. shorebirds (whole)	
^{207}Bi	ND vs. ND
^{57}Co	12 DW vs. ND
^{60}Co	340 DW vs. 4810 DW
^{137}Cs	ND vs. 4440 DW
^{54}Mn	81 DW vs. ND
^{106}Ru, ^{125}Sb	ND vs. ND
Roof rat, *Rattus rattus*; whole	
^{207}Bi	5 DW
^{144}Ce	362 DW
^{60}Co	888 DW
^{137}Cs	19,980 DW
^{54}Mn	1 DW
^{106}Ru, ^{125}Sb	ND
Samples with highest concentrations	
^{207}Bi, marine plankton	Max. 333 DW
^{144}Ce, soils	Max. 2109 DW
^{137}Cs, rats	Max. 19,980 DW
^{57}Co, sediments	Max. 740 DW
^{60}Co, marine invertebrates	Max. 6290 DW
^{54}Mn, land plants	Max. 296 DW
^{106}Ru, soils	Max. 4440 DW
^{125}Sb, soils	Max. 703 DW

Table 27.2. cont'd

Location, Sample, Radionuclide, and Other Variables	Concentration, in Bq/kg or Bq/L
Soils vs. sediments	
^{207}Bi	20 DW vs. 218 DW
^{144}Ce	2109 DW vs. no data
^{137}Cs	2072 DW vs. 814 DW
^{57}Co	No data vs. 740 DW
^{60}Co	2849 DW vs. 1073 DW
^{54}Mn	44 DW vs. 148 DW
^{106}Ru	4440 DW vs. 3700 DW
^{125}Sb	703 DW vs. 407 DW
ENIWETOK ATOLL, RUNIT ISLAND (8 nuclear detonations between 1948 and 1958)	
Roof rat, whole	
Immediate vicinity of detonations; 1967 vs. 1973; ^{137}Cs	
Bone	21,978 DW vs. 81,363 DW
Intestine	137,344 DW vs. no data
Kidney	189,958 DW vs. 126,799 DW
Liver	83,657 DW vs. 83,583 DW
Muscle	137,122 DW vs. 156,880 DW
Skin	13,209 DW vs. 77,256 DW
^{60}Co	
200 m vs. 2,460 m; 1967	
Bone	185 DW vs. ND
Intestine	8251 DW vs. no data
Kidney	110,223 DW vs. 333 DW
Muscle	499 DW vs. 266 DW
Skin	259 DW vs. ND
Soils	
^{137}Cs, 1967	
Ground zero vs. 200 m	1258 DW vs. 399 DW
1030 m vs. 2460 m	88 DW vs. 18 DW
^{137}Cs, 1971	
Ground zero vs. 200 m	4736 DW vs. 403 DW
1030 m	44 DW
^{60}Co, 1967	
Ground zero vs. 200 m	1221 DW vs. 66 DW
1030 m	25 DW
^{60}Co, 1971	
Ground zero vs. 200 m	1110 DW vs. 133 DW
1030 m vs. 2460 m	40 DW vs. 4 DW

Continued

Table 27.2. cont'd

Location, Sample, Radionuclide, and Other Variables	Concentration, in Bq/kg or Bq/L
1973, 2460 m	
^{137}Cs	11 DW
^{60}Co	52 DW
Terrestrial vegetation	
Ground zero, 1967 vs. 1971	
^{137}Cs	16,199–93,380 DW vs. 34,780–94,239 DW
^{60}Co	Max. 1221 DW vs. Max. 2775 DW
1030 m, 1967 vs. 1971	
^{137}Cs	296–2035 DW vs. 333–1961 DW
^{60}Co	Max. 14 DW vs. Max. 48 DW

elevated and highest in soils and increasingly lower in aquatic invertebrates, groundwater, shorebirds, plants, rats, zooplankton, algae, fish, sediments, seawater, and seabirds. Cobalt-60 was found in all samples of animals, plants, water, sediments, and soils, and was the major radionuclide in the marine environment; on land cesium-137 and ^{90}Sr predominated. All samples contained traces of ^{54}Mn; ^{106}Ru and ^{125}Sb were detected in groundwater and soil and trace concentrations in animals and plants. Trace amounts of ^{207}Bi and ^{144}Ce were usually detected in algae, soils, and land plants. Iron-55 was comparatively high in vertebrates, and ^{239}Pu was found in the soil and in the skin of rats and birds.

27.5.2 Chernobyl

The partial meltdown of the 1000 MW reactor at Chernobyl, Ukraine, on April 26, 1986 released large amounts of radiocesium and other radionuclides into the environment (Table 27.3), causing widespread radioactive contamination of Europe and the former Soviet Union. Among the reactors operating in the former Soviet Union are 13 identical to the one in Chernobyl, Ukraine, including units in Chernobyl, Leningrad, Kursk, and Smolensk. Of the three remaining reactors in Chernobyl, one was closed in 1991 after a fire swept through the turbine hall causing extensive damage, though not to the reactor. A second unit was closed in 1994 by the Russian president after pledges were received from the Group of Seven to help finance an overall plan to decommission the Chernobyl complex. And the last unit was closed in late 2000.

At least 3,000,000 TBq were released from the fuel during the April 1986 accident, dwarfing – by orders of magnitude – radiation releases from other highly publicized reactor accidents at Windscale (UK) and Three-Mile Island (USA). (Note: 1 Bq = 1 disintegration/second; about 0.037 Bq = 1 picoCurie.) The Chernobyl accident happened while a test was being conducted during a normal scheduled shutdown and is attributed mainly to human error. About 25% of the released radioactive materials escaped during the first day of the accident; the rest over a nine-day period (Figure 27.6). The initial explosions and heat from the fire carried some of the radioactive materials to an altitude of 1500 m where they were transported by prevailing winds. Long-range atmospheric transport spread the radioactivity through the Northern hemisphere where it was initially detected in Japan on May 2, in China on May 4, in India on May 5, and in Canada and the United States

Table 27.3. Selected fission products in the Chernobyl reactor core, and their estimated escape into the environment.

	Trillions of becquerels (TBq)	
Radionuclide	**In Core**	**Escaped**[a]
Krypton-85 (^{85}Kr)	33,000	33,000
Xenon-133 (^{133}Xe)	1,700,000	1,700,000
Iodine-131 (^{131}I)	1,300,000	260,000
Tellurium-129 (^{132}Te)	320,000	48,000
Cesium-134 (^{134}Cs)	190,000	19,000–50,000
Cesium-137 (^{137}Cs)	290,000	37,700–100,000
Molybdenum-99 (^{99}Mo)	4,800,000	110,400
Zirconium-95 (^{95}Zr)	4,400,000	140,800
Ruthenium-103 (^{103}Ru)	4,100,000	118,900
Ruthenium-106 (^{106}Ru)	2,000,000	35,000–58,000
Barium-140 (^{140}Ba)	2,900,000	162,400
Cerium-141 (^{141}Ce)	4,400,000	101,200
Cerium-144 (^{144}Ce)	3,200,000	89,600
Strontium-89 (^{89}Sr)	2,000,000	80,000
Strontium-90 (^{90}Sr)	200,000	8000
Neptunium-239 (^{239}Np)	140,000	4200
Plutonium-238 (^{238}Pu)	1000	30
Plutonium-239 (^{239}Pu)	850	25
Plutonium-240 (^{240}Pu)	1200	36
Plutonium-241 (^{241}Pu)	170,000	5100
Curium-242 (^{242}Cm)	26,000	780

[a]Other radionuclides in the Chernobyl escapement include 1500 TBq of Silver-110 (^{110}Ag), 3000 TBq of Antimony-125 (^{125}Sb), 6 TBq of Americium-241 (^{241}Am), and 6 TBq of Curium-243, -244 (^{243}Cm and ^{244}Cm).

on May 5–6, 1986. Airborne activity was also detected in Turkey, Kuwait, and Israel in early May. No airborne activity from Chernobyl has been reported south of the equator.

Effective dose equivalents from the Chernobyl accident in various regions of the world were highest in southeastern Europe (1.2 mSv), Northern Europe (0.97 mSv), and Central Europe (0.93 mSv). (Note: 1 mSv = 0.1 rem.) In the first year after the accident, whole body effective dose equivalents were highest in Bulgaria, Austria, Greece, and Romania (0.5–0.8 mSv); Finland, Yugoslavia, Czechoslovakia, Italy (0.3–0.5 mSv): Switzerland, Poland, U.S.S.R, Hungary, Norway, Germany, and Turkey (0.2–0.3 mSv); and elsewhere (<0.2 mSv). This value was 0.81 mSv in the former Soviet Union, <0.2 mSv in southwest Asia and Western Europe, and <0.1 mSv elsewhere. By comparison, the recommended whole-body annual effective dose equivalent for the general public is <5 mSv. Thyroid dose equivalents were significantly higher than whole-body effective dose equivalents because of significant amounts of iodine-131 in the released materials.

Thyroid dose equivalents were as high as 25 mSv to infants in Bulgaria, 20 mSv in Greece, and 20 mSv in Romania; the adult

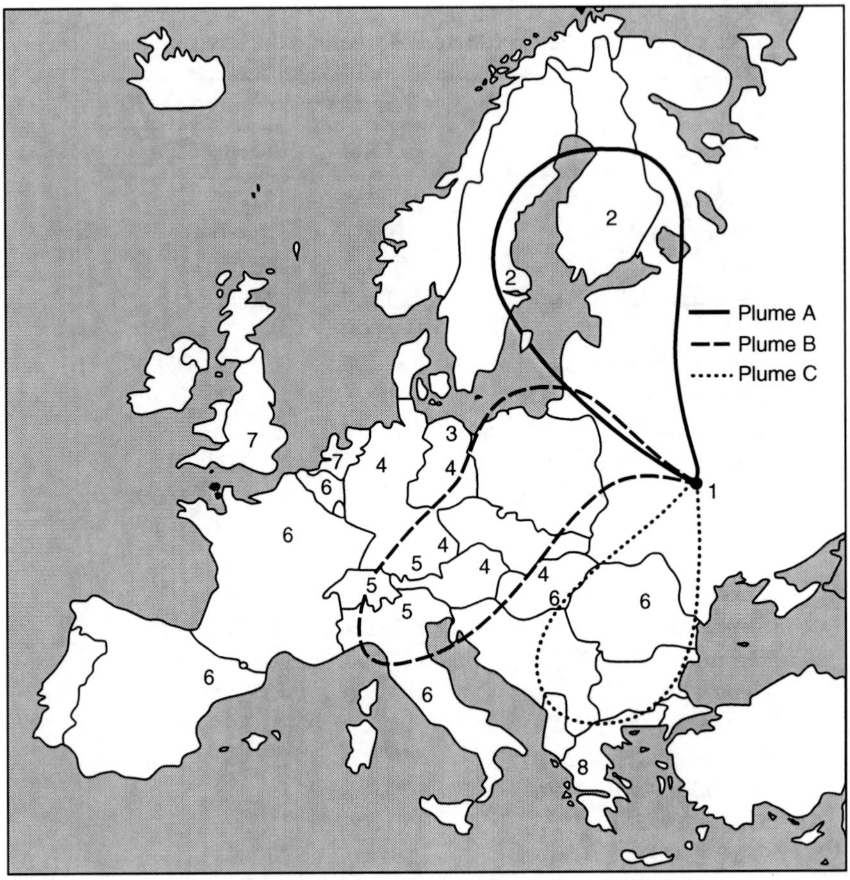

Figure 27.6.
Chernobyl air plume behavior and reported initial arrival times of detectable radioactivity. Plume A originated from Chernobyl on April 26, 1986, Plume B on April 27–28, and Plume C on April 29–30. The numbers indicate initial arrival times: April 1, 26; April 2, 27; April 3, 28; April 4, 29; April 5, 30; May 6, 1; May 7, 2; and May 8, 3.

thyroid dose equivalents were usually 80% lower than the infant dose equivalents.

27.5.2.1 Local Effects

The initial contamination rate in the 30-km exclusion zone surrounding the site of the nuclear accident was estimated at 37,000,000 Bq/km^2 (1000 Ci/km^2); isotopes included iodine-131, tellurium-127, -132, barium-140, lanthanum-140, cerium-141, -144, zirconium-95, niobium-95, ruthenium-103, -106, praseodymium-144, cesium-134, -137, molybdenum-99, strontium-89, -90, plutonium-238, -239, -240, -241, silver-110, antimony-125, and others. About 28,000 km^2 of land and 2225 settlements in Belarus, Russia, and the Ukraine were officially declared contaminated with radiocesium, i.e., levels were >185,000 Bq ^{134}Cs/m^2 and ^{137}Cs/m^2 (>5 Ci/km^2). Approximately 850,000 people are still living in these contaminated areas. About 105,000 km^2 were contaminated with 37,000 Bq/m^2 (1 Ci/km^2) or more. In the first years of the catastrophe, 144,000 ha of agricultural land and 492,000 ha of forest were withdrawn from use. More than 4 million people in these three countries were affected by the accident. Current estimates of the eventual toll from cancer deaths as a direct result of Chernobyl range from a minimum of 14,000 to a maximum of 475,000.

27.5.2.1.1 Acute Effects

At Chernobyl, at least 115 humans received acute bone marrow doses >1 Gray (Gy), as judged by lymphocyte aberrations. (Note: 1 Gy = 100 rad.) The death toll within 3 months from the accident was at least 30 individuals, usually from groups receiving >4 Gy, especially the reactor's operating staff and the fire-fighting crew. Humans from highly-contaminated areas within 30 km of Chernobyl received mean thyroid doses from radioiodines of 1.6 Gy for adults, 2.1 Gy for those aged 7–17 years, and 4.7 Gy for infants. Residents were evacuated from a 30-km exclusion zone surrounding the reactor because of increasing radiation levels; more than 115,000 people, including 27,000 children, were evacuated from the Kiev region, Byelorussia, and the Ukraine. Children evacuated from the contaminated areas showed – in July 1986 – significantly elevated levels of dicentric chromosomal aberrations when compared to resident children of noncontaminated areas. Vitamin E and A deficiencies observed in some children were correlated with high Chernobyl radiation loads of their mothers. Tens of thousands of cattle were also removed from the contaminated area, and consumption of locally produced milk and other foods was banned. Agricultural activities were halted and large scale decontamination was undertaken. Humans residing in areas where topsoil contamination exceeded 555,000 Bq/m^2 were cautioned against inclusion of forest products in their diets and to avoid cattle grazing on wet floodplain meadows without remediation. The forced evacuation and health concerns contributed to severe sociopsychological impacts in all age groups of the displaced population.

The most sensitive ecosystems affected at Chernobyl were the soil fauna and pine forest communities; the majority of the terrestrial vertebrate communities were not adversely affected by released ionizing radiation. Pine forests seemed to be the most sensitive ecosystem. One 400 ha stand of *Pinus silvestris* died and probably received a dose of 80–100 Gy. Other stands experienced heavy mortality of 10–12-year-old trees and as much as 95% necrotization of young shoots; these pines received an estimated dose of 8–10 Gy. Abnormal top shoots developed in some *Pinus*, and these probably received 3–4 Gy. In contrast, leafed trees in the Chernobyl Atomic Power Station zone, such as birch, oak, and aspen, survived undamaged, probably because they were about 10 times more radioresistant than pines. Extremely high radioresistance was documented in genetically adapted strains of the filamentous fungus *Alternaria alternata* isolated from the reactor of the Chernobyl power plant; other strains of this species are supersensitive to radiation. There was no increase in mutation rate of spiderwort (*Arabidopsis thaliana*), a radiosensitive plant, suggesting that the dose rate was <0.05 Gy/h in the Chernobyl locale.

Populations of soil mites were reduced in the Chernobyl area, but no population showed a catastrophic drop in numbers. By 1987, soil microfauna – even in the most heavily contaminated plots – was comparable to controls. Flies (*Drosophila* spp.) collected at various distances from the accident site and bred in the laboratory had highest incidences of dominant lethal mutations (14.7%, estimated dose of 0.8 mGy/h) at sites nearest to the accident and higher incidences than controls (4.2%).

The most contaminated water body in the Chernobyl emergency zone was the Chernobyl cooling pond ecosystem, an area of about 30 km^2. On May 30, 1986, the total amount of radioactivity in the water of this system was estimated at 806 TBq and in sediments 5657 TBq. In water, ^{131}I contributed about 31% of the total radioactivity, ^{140}Ba–^{140}La 25%, ^{95}Zr–^{95}Nb 15%, ^{134}Cs and ^{137}Cs 11%, ^{141}Ce and ^{144}Ce 10%, ^{103}Ru and ^{106}Ru 7%, and ^{90}Sr <1%. The distribution pattern in sediments was significantly different: about 41%

of the total radioactivity was contributed by ^{95}Zr–^{95}Nb, 27% by ^{141}Ce and ^{144}Ce, 16% by ^{103}Ru and ^{106}Ru, 12% by ^{140}Ba–^{140}La, 3% by ^{134}Cs and ^{137}Cs, 1% by ^{90}Sr, and 0.5% by ^{131}I.

Fish populations seemed unaffected in July–August 1987, and no grossly deformed individuals were found; however, ^{134}Cs and ^{137}Cs levels were elevated in young fish. The most heavily contaminated teleost in May 1987 was the carp (*Carassius carassius*). But carp showed no evidence of mutagenesis, as judged by incidence of chromosomal aberrations in cells from the corneal epithelium of carp as far as 60 km from Chernobyl.

In 1986, total radioactivity in muscle of birds near Chernobyl after the accident exceeded the temporarily permitted limits for human food consumption (598 Bq/kg FW) by about 100 times. In 1987, radionuclide concentrations in bird muscle had decreased by a factor of about 7 due, in part, to physical decay of short-lived isotopes.

Several rodent species composed the most widely distributed and numerous mammals in the Chernobyl vicinity. It was estimated that about 90% of rodents died in an area receiving 60 Gy and 50% in areas receiving 6–60 Gy. Rodent populations seemed normal in spring 1987, and this was attributed to migration from adjacent nonpolluted areas. The most sensitive small mammal was the bank vole (*Clethrionomys glareolus*), which experienced embryonic mortality of 34%. The house mouse (*Mus musculus*) was one of the more radioresistant species. *Mus* from plots receiving 0.6–1.0 mGy/h did not show signs of radiation sickness, were fertile with normal sperm, bred actively, and produced normal young. Some chromosomal aberrations were evident, namely, an increased frequency of reciprocal translocations. New data on the house mouse suggests that fertility was dramatically reduced in the 30-km zone around the Chernobyl nuclear power plant station in 1986–87 and that survivors had high frequencies of abnormal spermatozoa heads and dominant lethal mutations. Elevated incidences of germ line mutations were observed in four species of *Apodemus* mice from the vicinity of Chernobyl when compared to control areas; however, variability between species was large.

During the early period after the accident, there was no evidence of increasing mortality, decline in fecundity, or migration of vertebrates as a result of the direct action of ionizing radiation. The numbers and distribution of wildlife species were somewhat affected by the death of the pine stand, the evacuation of people, termination of cultivation of soils (the crop of 1986 remained standing), and the forced transfer of domestic livestock. There were no recorded changes in survival or species composition of game animals and birds. In fact, because humans had evacuated and hunting pressure was negligible, many game species – including fox, hare, deer, moose, wolf, and waterfowl – moved into the zone in autumn 1986–winter 1987 from the adjacent areas in a 50–60 km radius.

27.5.2.1.2 Latent Effects: Humans

The frequency of childhood thyroid cancer in areas of Belarus and Ukraine most affected by the Chernobyl accident is significantly higher than any other region of the world. Thyroid cancers in children were induced by ^{131}I and ^{132}I, although the mechanisms of action are imperfectly understood. During the period 1985–95, childhood thyroid cancer rates in Belarus were higher after a minimum latent period of three years, higher in children under age 10 years than those age 10–15 years, and higher in females than males. Between 1990 and 1997, the group most affected were those younger than 5 years old in 1986; the largest number of cases occurred in patients living in areas of thyroid radiation doses >0.5 Gy. The post-Chernobyl thyroid carcinomas were usually papillary, more aggressive at presentation, and frequently associated with thyroid autoimmunity. Gene mutations involving the receptor tyrosine kinase (RET) proto-oncogene were the causative agents specific for papillary cancer.

Cutaneous radiation syndrome was the primary cause of death in most of the 32 adult patients who died shortly after the accident.

Excessive cutaneous fibrosis in 8 survivors who participated in the Chernobyl cleanup was treated successfully with interferon over a period of 36 months. There is general agreement that cleanup workers and populations residing in heavily contaminated areas (>555,000 Bq $^{137}Cs/m^2$) had an increased frequency of thyroid cancers in the period 1986–93, but current epidemiological evidence does not conclusively support an increased incidence of other types of cancers.

Increased frequencies of chromosome breakage were evident in plants and small mammals from the Chernobyl-contaminated zone in the period 1986–91. For humans, chromosome and chromatid aberration frequencies in lymphocytes of Chernobyl evacuees (receiving 330–420 mGy) and cleanup personnel (receiving a maximum of 940 mGy) one year after the accident were 3–4 times higher than in controls. By 1994–95, aberration frequency in evacuees receiving <250 mGy was the same as the controls; however, this value was 1.7 times higher than controls in cleanup personnel, 1.95 times higher in evacuees receiving 250 mGy, and 2.37 times higher in evacuees receiving >250 mGy in 1986. In some patients, the increased level of chromosome breakage was associated with high cancer susceptibility.

Russian adult males who participated in Chernobyl cleanup activities ($n = 126$) received an estimated 0.14–0.15 Gy (maximum of 0.56–0.95 Gy, depending on statistical model); however, dose estimates could be overestimated if age and smoking status were ignored, particularly for older subjects who smoke. This group had a significantly increased frequency of chromosomal translocations when compared to Russian controls ($n = 53$). Cancer incidence between 1986 and 1996 for cleanup workers ($n = 114,504$) with no known oncology before arrival in 1986 to the 30-km zone increased significantly for solid tumors and malignant neoplasms of the digestive tract (but not of the respiratory system) when compared to controls. By 1996, Chernobyl cleanup workers experienced a variety of eye pathologies that seemed to increase in frequency with increasing time post-accident and with high initial doses of absorbed radiation. Eye pathology patterns are still developing because the latency period for irradiation cataract can exceed 10 years. The incidence of urinary bladder cancer in the Ukraine increased from 26.2 per 100,000 population in 1986 to 36.1 in 1997. Individuals from the most severely radiocontaminated areas showed the greatest incidence of early malignant transformation of bladder epithelium. In 1995, the endocrine status and spermatogenesis of Chernobyl cleanup workers was normal except for lower cortisol and higher testosterone. A woman resident of Kiev during the period 1986–92 who subsequently emigrated to the United States was diagnosed in 1996 with radiation-induced cancer of the ovaries, kidneys, and bile duct.

In certain rural portions of Russia receiving Chernobyl contamination on April 28–29, 1986, ^{137}Cs concentrations in the human body in 1986–87 were positively correlated with consumption of meat and dairy products. Domestic livestock were fed clean feed as much as possible for milk production, and were fed clean feed for 40–120 days before slaughter. Beginning in 1993, however, and persisting to at least 1996, the content of ^{137}Cs in whole humans correlated positively with the levels of consumption of naturally occurring foodstuffs, such as mushrooms, wild berries, fish, and game.

27.5.2.1.3 Latent Effects: Plants and Animals

Total radiocesium-137 deposited in soils at Chernobyl sites 2–15 km from the reactor was estimated at 1,660,000 Bq/m^2, mainly as insoluble fuel particles. The half-time persistence of ^{137}Cs in surface soils 0–2 cm in depth decreased from 9 years in 1987 to 3 years in 1994; but the residence time of this isotope increased with increasing depth over time. This increase in deeper layers is attributed to the progressive fixation of radiocesium by clay minerals of the soil.

In 1992, bacteria were isolated from soils within 30 km of the power plant. Spore-forming bacilli collected nearest to the power plant were more resistant to X-radiation, ultraviolet radiation, and 4-nitroquinoline 1-oxide than were isolates of the same species from control sites. In 1993–95, populations of cellulolytic, nitrifying, and sulfate-reducing bacteria collected within the 10-km zone from contaminated soils (11,000–629,000 Bq/kg DW soil) were 10–100 times lower in abundance and diversity than were populations from control soils (<222 Bq/kg DW soil).

On heavily contaminated agricultural lands – about 2.4 million ha – about 388,000 ha were withdrawn from use, and the remainder planted with crops of low coefficients of radionuclide transfer from soil, such as grains, potatoes, and corn. In addition, the upper polluted layer of soil (5–6 cm) was plowed to a depth of 40–50 cm to inhibit transfer to root systems, especially those of grasses. Application of potassium and phosphorus fertilizers in combination with liming further reduced uptake of ^{90}Sr and ^{137}Cs by plants. From autumn 1986–91, ^{137}Cs in vegetable and animal agricultural products from various contaminated areas of Russia decreased at an observed half-time persistence of 0.7–1.5 years. Beginning in 1991 and extending through 1995, the average values of transfer factors for ^{137}Cs from all sources from soil to milk and potatoes were similar to those of the pre-Chernobyl period. Dose rates from soil to the house mouse between 1986 and 1993 ranged from 0.0002 to 2 mGy/h, and these were positively correlated with the frequency of reciprocal translocations in mouse spermatocytes. The frequency of mice heterozygous for recessive lethal mutations decreased over time after the accident.

In 1991, pine forests (*Pinus silvestris*) within 10 km of Chernobyl exposed initially to doses of 10–60 Gy had low survival and no regeneration since 1987; mortality was exacerbated by pathogenic insect invaders. Samples of wood and bark from these trees had significant histological changes in resin ducts and radial rays, and this was correlated with radionuclide content of bark. Stands exposed initially to 0.1–1.0 Gy had reduced growth; trees receiving initial doses of less than 0.1 Gy seemed outwardly normal. Between 1986 and 1996, radiocesium-137 concentrated in the bark of pine trees grown in the exclusion zone from about 10,000 Bq/kg DW in 1986 to 37,000 Bq/kg DW in 1996; the high ^{137}Cs concentrations were associated with growth suppression. In the period 1986–94, ^{137}Cs dynamics in forests within the 30-km zone around the Chernobyl reactor were influenced mainly by the size of radioactive particles in the fallout, humidity, soil type, and tree age. In June 1996, abnormal development in three species of plants (black locust tree, *Robina pseudoacacia*; rowan, *Sorbus aucuparia*; camomile, *Matricaria perforata*) was observed between Chernobyl and an uncontaminated area 225 km to the southeast. Abnormalities, including leaf or flower asymmetry, were three to four times more frequent near Chernobyl than more distant sites and were directly related to ^{137}Cs concentrations in soils. In 1993–94, about 55% of all young (2- to 9-year old) plants of pine and spruce growing within 10 km of Chernobyl had abnormal needles due, in part, to radiation-induced alterations in protein composition. In 1993–95, timber products within 30 km of Chernobyl were sufficiently contaminated with radiocesium and other isotopes as to preclude human use, as was the case for berries and mushrooms. Economic damage to forest products within the 30-km exclusion zone is estimated at $278 million (U.S.) annually with a total estimated loss between 1986 and 2015 of $8.4 billion (U.S.).

Concentrations of radioactivity in water, sediments, and biota of the Chernobyl cooling pond ecosystem declined between 1986 and 1990, as judged by ^{137}Cs concentrations. Carp and other fish species held in Chernobyl cooling pond waters had reproductive system anomalies that were more pronounced in males than females. Silver carp (*Hypophthalmichthys molitrix*) survivors from the cooling pond of the Chernobyl nuclear power station received 7–11 Gy between 1989 and 1992 and displayed reproductive system disorders that included sterility, changes in gonadal morphology, and degeneration of reproductive cells. Silver carp born in 1989 from parents reared in Chernobyl cooling pond waters had a marked increase of 17–26% above controls in reproductive system anomalies in 1989–92. Anomalies included degenerative changes in oocytes, spermatogonia, and spermatocytes, and the appearance of bisexual and sterile fish.

The gonadal abnormalities are attributed to the high radiation dose of 7–10 Gy received by the parent fish during gonad formation and the continuing exposure of 0.2 Gy annually to this generation. Mature silver carp dwarf individuals age 4+, descendants of fish irradiated in 1986, were observed in 1991. In 1992, second generation fish hatched from 3-year-old fish of the first generation. Female reproductive systems were not significantly different from controls; however, males had decreased ejaculate volume and concentration and destructive changes in testes caused by irradiation. Silver carp from this ecosystem also had a dose-dependent decrease in hormonal control over the Na^+, K^+-pump in erythrocytes, with increased passive permeability of the erythrocyte membrane to radioactive analogues of sodium and potassium.

Radionuclide levels in soil invertebrates, soil, litter, and terrestrial insects within the 30-km exclusion zone declined sharply between 1986 and 1987, usually by a factor of 10 or more. Radionuclide concentrations in muscle of fishes from the Kiev Reservoir decreased significantly during the period 1986–97. Female northern pike, *Esox lucius*, of the 1986 year class from this locale had gonadal abnormalities during the period 1987–97. Abnormality frequency rate was about 34%, and included asymmetry, histopathology, and roe resorption. The internal radiation dose from ^{134}Cs and ^{137}Cs decreased from about 0.1–0.2 Gy in 1986 to 0.001–0.002 Gy during 1993–97. Amphibians and reptiles within the 30-km exclusion zone, in the 5-year period following the accident had gamma radiation levels that remained stable and sometimes increased due to accumulation of radionuclides. One year after the accident, terrestrial birds within the 30-km exclusion zone had gamma radiation levels 5–7 times higher than did conspecifics from uncontaminated areas. Waterfowl, however, during the summer of 1987, had radiation levels that were 2–5 times higher than a year previously, with highest levels in young and females.

In June 1991, male barn swallows (*Hirundo rustica*) collected within 50 km of Chernobyl – when compared to control areas 100 km distant and with museum samples from both areas – had significant differences in length of tail feathers between the right and left side (fluctuating asymmetry), and in morphology of feathers. The degree of fluctuating asymmetry in male tail length and the frequency of deviant morphology in the tails of male barn swallows were associated with a delay in the start of the breeding season. In 1996, barn swallows near Chernobyl, when compared to conspecifics from distant sites, had decreased lymphocyte and immunoglobulin concentrations, reduced spleen size, and reduced intensity of carotenoid-based coloration. Length of outermost tail feathers of males – important secondary sexual characteristics – was positively related to coloration in controls but not in the Chernobyl population, and this may affect breeding success of Chernobyl swallow populations. In 1994, mallard tissues collected within 10 km of Chernobyl had 8000–30,000 Bq/kg FW. Conspecifics collected 45 km from the reactor in 1994 had tissue concentrations that were 40–100 times lower. Population numbers of mallards and teals decreased in the 30-km zone in 1994, perhaps due to decreased food sources of farmlands, and invasion of nest areas by plant communities not used in agriculture. In 1994, there were no cases of radiation-induced pathology in any bird examined. Also in 1994, there were increased sightings of rarely seen species of egrets, cranes, and eagles.

In 1991, five years after the accident, a female root vole (*Microtus oeconomus*) with an abnormal karyotype (reciprocal translocation) was found within the 30-km radius of the Chernobyl nuclear power plant. These chromosomal aberrations were probably inherited and did not affect the viability of vole populations. Population density of Chernobyl rodents in 1988–89 was about twice that predicted from previous cycles, and was attributed, in part, to increasing radioresistance and abundance of their food supplies. In 1994–95, the diversity and abundance of the small mammal population (12 species of rodents) at the most radioactive sites at Chernobyl were the same as reference sites. Rodents from the most radioactive areas did not show gross morphological features other than enlargement of the spleen. There were no gross chromosomal arrangements, as judged by examination of the karyotypes. Also observed within the most heavily contaminated site were red fox (*Vulpes vulpes*), gray wolf (*Canis lupus*), moose (*Alces alces*), river otter (*Lutra canadensis*), roe deer (*Capreolus capreolus*), Russian wild boar (*Sus scrofa*), brown hare (*Lepus europaeus*), and feral dogs. The rich biodiversity and abundance of individuals in the 10-km exclusion zone is attributed, in part, to the absence of human habitation. In 1994–96, small mammals collected within 8 km of the Chernobyl reactor were experiencing substantial radiation dose rates from ^{134}Cs and ^{137}Cs in muscle and ^{90}Sr in bone. Radiocesium concentrations averaged 3,200,000 Bq/kg muscle DW and for ^{90}Sr in bone 297,000 Bq/kg AW. Dose rates from radiocesium averaged 2.4 mGy daily (max. 60 mGy daily) and from radiostrontium 1.0 mGy daily. Conspecifics captured 30 km southeast of the reactor averaged only 2000 Bq/kg muscle DW and were receiving 0.014 mGy daily from internal radiocesium. Estimated dose rates in certain areas of the Chernobyl region now exceed those reported to interfere with mammalian reproduction.

27.5.2.2 Nonlocal Effects

Radioactive material from the Chernobyl accident became widely dispersed throughout Europe and the Northern hemisphere. In Europe alone, about 80,000 TBq of ^{137}Cs was deposited as follows: Belarus 33.5%, Russia 24%, Ukraine 20%, Sweden 4.4%, Finland 4.3%, Bulgaria 2.8%, Austria 2.7%, Norway 2.3%, Romania 2%, and Germany 1.1%. It is probable that the full impact of the Chernobyl reactor accident on natural resources will not be known for several decades, primarily because of data gaps on long-term genetic and reproductive effects and on radiocesium cycling and toxicokinetics.

27.5.2.2.1 Soil and Vegetation

The radiocesium fallout in Sweden was among the highest in Western Europe – exceeding 60,000 Bq/m^2 on Sweden's Baltic coast – and involved mainly upland pastures and forests. In Norway, radiocesium deposition from the Chernobyl accident ranged from <5000 to $>200,000$ Bq/m^2, and greatly exceeded deposition from prior nuclear weapons tests. In Italy, heavy rainfall coincided with the passage of the Chernobyl radioactive cloud and caused high local deposition of radionuclides in soil, grass, and plants. The Chernobyl plume reached Greece on May 1, 1986. A total of 14 gamma emitters were identified in Greek soil and vegetation in May 1986, and three (^{134}Cs, ^{137}Cs, ^{131}I) were also detected in milk of free-grazing animals in the area. Radiocesium-134 and ^{137}Cs intake by humans in Germany during the period 1986–87 was mainly from rye, wheat, milk, and beef. In the United Kingdom, elevated concentrations of radionuclides of iodine, cesium, ruthenium, and other nuclides were measured in air and rainwater during May 2–5, 1986. The background activity concentrations were about 3 times normal levels in early May, and those of ^{131}I approached the derived emergency reference level (DERL) of drinking water of 5 mSv ^{131}I (equivalent to a thyroid dose of 50 mSv); however, ^{131}I levels were not elevated in foodstuffs or cow's milk. Syria – 1800 km from Chernobyl – had measurable atmospheric concentrations of ^{137}Cs and ^{131}I and near detection limit concentrations of ^{144}Ce, ^{134}Cs, ^{140}La, and ^{106}Ru. The maximum ^{131}I thyroid dose equivalent received by Syrians was 116 μSv in adults and 210 μSv in children. One year later, these values were 25 μSv in adults and 70 μSv in a 10-year old.

The amount of fallout radioactivity deposited on plant surfaces depends on the exposed surface area, the developmental season of the plants, and the external morphology. Mosses, which have a comparatively large surface area, showed highest concentrations of radiocesium. In northern Sweden, most of the radiocesium fallout was deposited on plant surfaces in the forest ecosystem and was readily incorporated into living systems because of browsing by herbivores and cesium's chemical similarity to potassium. In August 1992, the distribution of ^{137}Cs fallout from Chernobyl in a Swedish forest was 87% in soils, 6% in the bryophyte layer, and 7% in standing biomass of trees;

the mean deposition of ^{137}Cs was estimated at 54,000 Bq/m^2. The mean ^{137}Cs concentration in muscle samples from moose shot within a 10,000 ha area around the forest in October 1992 was 810 (51–2133) Bq/kg FW; for roe deer muscle from this area in autumn 1992, it was 4200 (500–12,000) Bq/kg FW.

Forest plants seemed to show less decrease than agricultural crops in ^{137}Cs activity over time. For example, the effective retention halftime of ^{137}Cs from Chernobyl was 10–20 days in herbaceous plants and 180 days in chestnut trees, *Castanea* spp. In Finland, during summer 1986, pine tree needles lost about 50% of their accumulated ^{137}Cs activity in 72 days; washout by throughfall accounted for 79% of the decrease and shedding of older needles most of the remainder. The radioactive fallout from the Chernobyl accident also resulted in high ^{137}Cs levels in Swedish pasture grass and other forage, although levels in grain were relatively low. Radiocesium isotopes were still easily measurable in grass silage harvested in June 1986 and used as fodder for dairy cows in 1988. The rejection of the first harvests of radiocesium-contaminated perennial pasture and in particular of rye grass (*Lolium perenne*) does not constitute a safe practice because later harvests – even 1 year after the contamination of the field – may contain very high values, as in Greece.

27.5.2.2.2 Aquatic Life

After Chernobyl, the consumption of freshwater fishes by Europeans declined, fishing license sales dropped by 25%, and the sale of fish from radiocesium-contaminated lakes was prohibited. Many remedial measures have been attempted to reduce radiocesium loadings in fish, but none has proven effective.

Bioconcentration factors (BCFs) of ^{137}Cs (Bq/kg FW organism/Bq/L seawater) measured in plankton and soft bottom benthos during 1989–90 in the Adriatic Sea ranged from 29 to 152 for plankton and 100 to 229 for benthos. Radiocesium BCFs in muscle of fishes from the southern Baltic Sea increased 3–4 times after Chernobyl, and ^{134}Cs, ^{137}Cs, and ^{106}Ru in fishes in the Danube River increased by a factor of 5. However, these levels posed negligible risk to human consumers. BCFs of ^{137}Cs in fishes from Lake Paijanne, Finland – a comparatively contaminated area – ranged between 1250 and 3800; the highest BCF values were measured in the predatory northern pike (*Esox lucius*) a full 3 years after Chernobyl; consumption of these fishes was prohibited. Chernobyl radioactivity, in particular ^{141}Ce and ^{144}Ce, entering the Mediterranean as a single pulse, was rapidly removed from surface waters and transported to 200 m in a few days, primarily in fecal pellets of grazing zooplankton.

After the Chernobyl accident, radiocesium isotopes were also elevated in trees and lichens bordering an alpine lake in Scandinavia and in lake sediments, invertebrates, and fishes. Radiocesium levels in muscle of resident brown trout (*Salmo trutta*) remained elevated for at least 2 years. People consuming food near this alpine lake derived about 90% of their effective dose equivalent from the consumption of freshwater fish, reindeer meat, and milk. The average effective dose equivalent of this group during the next 50 years is estimated at 6–9 mSv with a changed diet and 8–12 mSv without any dietary changes.

Between 1988 and 1992, ^{137}Cs levels declined in sediments from 52 lakes in southern Finland by 27% and in fish tissues by 26–39%. Radiocesium-137 concentrations in whole freshwater fishes from Norwegian lakes contaminated by Chernobyl fallout were quite variable. Major sources of ^{137}Cs variations included the fish weight and growth rate, and these were related to fish age and diet. Change over time in dissolved phase ^{137}Cs concentrations of lake water is significantly related to water residence time and mean lake depth; these variables have been incorporated into models to predict estimates of ^{137}Cs decline in freshwater systems. Radiocesium-137 activity in muscle of brown trout (*Salmo trutta*) and Arctic char (*Salvelinus alpinus*) from six lakes in Cumbria, England, between June 1986 and October 1988 were highest between December 1986 and March 1987, with maximum values of about 1200 Bq/kg FW in trout and 350 in char. Maximum ^{137}Cs values were related to the initial concentration of ^{137}Cs in both water and sediments and the maximum monthly geometric mean ^{137}Cs values obtained from the routine water samples; no similar relationships were found with mean calcium or potassium levels in lake water. Ecological half-life for ^{137}Cs in fish flesh from Cumbrian lakes was 132 days in char and 180–240 days in trout.

27.5.2.2.3 Wildlife

Reindeer (*Rangifer tarandus*) – also known as caribou in North America – are recognized as a key species in the transfer of radioactivity from the environment to humans because (1) the transfer factor of radioactivity from reindeer feed to reindeer muscle is high, (2) lichens – which constitute a substantial portion of the reindeer diet – are efficient accumulators of strontium, cesium, and actinide radioisotopes, and (3) reindeer feed is not significantly supplemented with grain or other feeds low in contamination. During 1986–87, about 75% of all reindeer meat produced in Sweden was unfit for human consumption because ^{137}Cs exceeded 300 Bq/kg FW. In May 1987, the maximum permissible level of ^{137}Cs in Swedish reindeer and other game was raised to 1500 Bq/kg FW; however, about 25% of slaughtered reindeer in 1987–89 still exceeded this limit. Concentrations in excess of 100,000 Bq ^{134}Cs and ^{137}Cs/kg FW lichens have been recorded in the most contaminated areas and in the 1986–87 season was reflected in reindeer muscle concentrations >50,000 Bq/kg FW from the most contaminated areas of central Norway. Norwegian reindeer containing 60,000–70,000 Bq ^{137}Cs/kg FW in muscle receive an estimated yearly dose of 500 mSv. The maximum radiation dose to reindeer in Sweden after Chernobyl was about 200 mSv/year with a daily dose rate of about 1 mSv during the winter period of maximum tissue concentrations. In general, reindeer calves had higher ^{137}Cs levels in muscle than adult females (4700 vs. 2700 Bq/kg FW) during September 1988, suggesting translocation to the fetus. Two reindeer herds in Norway that were heavily contaminated with radiocesium had a 25% decline in survival of calves; survival was normal in a herd with low exposure. Several compounds inhibit uptake and reduce retention of ^{137}Cs in reindeer muscle from contaminated diets, but the mechanisms of the action are largely unknown. These compounds include zeolite – a group of tectosilicate minerals – when fed at 25–50 g daily; ammonium hexacyanoferrate – also known as Prussian Blue or Giese salt – at 0.3–1.5 g daily; bentonite – a montmorillonite clay – when fed at 2% of diet; and high intakes of potassium. In Sweden, radiocesium concentrations in domestic reindeer meat were reduced by early slaughter (to avoid grazing on pastures abundant in fungi and lichens), feeding uncontaminated diets for 8–12 weeks prior to slaughter, and adding cesium binders to the feed. Much additional work seems needed on chemical and other processes that will hasten excretion and prevent uptake and accumulation of radionuclides in livestock and wildlife. Reindeer herding is, at present, the most important occupation in Finnish Lapland and portions of Sweden. Swedish Lapland reindeer herders have experienced a variety of sociocultural problems as a result of the Chernobyl accident. The variability of contamination has been compounded by the variability of expert statements about risk, the change in national limits of becquerel concentrations set for marketability of meat, and the variability of the compensation policy for slaughtered reindeer. These concerns may result in fewer Lapps becoming herders and in a general decline in reindeer husbandry.

Caribou in northern Quebec contained up to 1129 Bq ^{137}Cs/kg muscle FW in 1986–87, but only 10–15% of this amount originated from Chernobyl; the remainder is attributed to fallout from earlier atmospheric nuclear tests. The maximum concentration of ^{137}Cs in meat of caribou (*Rangifer tarandus granti*) from the Alaskan Porcupine herd after Chernobyl did not exceed 232 Bq/kg FW, and this is substantially below the recommended level of 2260 Bq ^{137}Cs/kg FW. Radiocesium transfer in an Alaskan lichen–reindeer–wolf (*Canis lupus*) food chain has been estimated. If reindeer forage contained 100 Bq/kg DW in lichens and 5 Bq/kg DW in vascular plants, the maximum winter concentrations – at an effective half-life of 8.2 years in lichens and 2.0 years in vascular plants – were estimated at 20 Bq/kg FW in reindeer–caribou skeletal muscle and 24 Bq/kg FW in wolf muscle. The radioactive body burden in exposed reindeer and the character of chromosomal aberrations – which was different in exposed and nonexposed reindeer – indicate a genetic effect of radiation from the Chernobyl accident. Chromosomal aberrations in Norwegian female reindeer positively correlated with increasing radiocesium concentrations in flesh. The frequency of chromosomal aberrations in reindeer calves from central Norway were greatest in those born in 1987 when tissue loadings were equivalent to fetal doses of 70–80 mSv and lower in 1988 (50–60 mSv) and 1989 (40–50 mSv), strongly suggesting a dose-dependent induction.

For many households in Sweden, moose (*Alces*) are an important source of meat. Radiocesium concentrations in moose foreleg muscle in Sweden during 1987–88 were highest in autumn when the daily dietary intake of the animals was about 25,000 Bq ^{137}Cs and lowest during the rest of the year when mean daily intake was about 800 Bq. Cesium-137 levels in moose flesh did not decrease significantly for about 2 years after the Chernobyl accident. Nine years after the accident many moose in Sweden from areas receiving high deposition still showed ^{137}Cs activity concentrations that exceeded the limit for human consumption. The selection of food by moose is paramount to the uptake of environmental contaminants and the changes in tissue levels over time. Increased foraging on highly contaminated plant species, such as bilberry (*Vaccinium myrtillus*), aquatic plants, and mushrooms, might account for the increased ^{137}Cs radioactivity in moose. Habitat is a useful indicator of ^{137}Cs radioactivity in moose muscle; radioactivity was highest in moose captured in swamp and marsh habitats and lowest in farmlands. For reasons that are not yet clear, transfer coefficients of ^{137}Cs from diet to muscle were about the same in moose (0.03) and beef cattle (0.02), but were significantly higher in sheep (0.24).

Consumption of game or wildlife in Great Britain after the Chernobyl accident probably did not exceed the annual limits of intake (ALI) based on ^{134}Cs and ^{137}Cs concentrations in game and the numbers of animals that can be eaten in 1 year before the ALI is exceeded. For example, a person who eats hares containing 3114 Bq ^{134}Cs and ^{137}Cs/kg FW in muscle would have to consume 99 hares before exceeding the ALI. For the consumption of red grouse (3022 Bq/kg), this number is 441 grouse; and for the consumption of woodcock (55 Bq/kg), it is 45,455 woodcocks. Rabbits (*Oryctolagus* sp.) from northeastern Italy that were fed Chernobyl-contaminated alfalfa meal (1215 Bq ^{134}Cs/kg and ^{137}Cs/kg diet) had a maximum of 156 Bq/kg muscle FW of ^{134}Cs and ^{137}Cs, a value much lower than the current Italian guideline of 370 Bq/kg FW for milk and children's food and 600 Bq/kg FW for other food. More than 85% of the ingested radiocesium was excreted by rabbits in their feces and urine; about 3% was retained.

Radiocesium-137 contamination of muscle of roe deer in 1986–87 in Germany was lowest in March and highest in October and is attributed to increased food intake and the high concentrations in food plants such as ferns and mushrooms consumed at this time of the year. Cesium radioactivity in tissues and organs of the wolverine (*Gulo gulo*), lynx (*Felis lynx*), and Arctic fox (*Alopex lagopus*) in central Norway after the Chernobyl accident was highly variable. In general, cesium-137 levels were substantially lower in these carnivores than in lower trophic levels, suggesting little or no food-chain biomagnification, and at variance with results of studies of the omnivore and herbivore food chain.

Mutagenicity tests were used successfully with feral rodents to evaluate the biological effects of the radiation exposure from the Chernobyl accident. Increased mutagenicity in mice (*Mus musculus domesticus*) was evident, as judged by the bone marrow micronucleus test, at 6 months and 1 year after the accident. Rodents with increased chromosomal aberrations also had ^{137}Cs burdens that were 70% higher 6 months after the accident and 55% higher after 1 year, but elevated radiocesium body burdens alone were not sufficient to account for the increase in mutagenicity. In bank voles, however, mutagenicity (micronucleated polychromatic erythrocytes) correlated well with the ^{137}Cs content in muscle and in the soil of the collection locale. The estimated daily absorbed doses of 4.2–39.4 μGy were far lower than those required to produce the same effect in the laboratory.

Migratory waterfowl could serve as potential vectors of contamination to the human food chain. For example, cooked meat from migratory waterfowl harvested on their wintering grounds as much as 3000 km from the Chernobyl site may exceed the maximum level of radiocesium generally permitted for human consumption (600 Bq/kg FW) depending on deposition rate. Migratory species of game birds in Northern Ireland during the period 1986–88 were contaminated by radiocesium of Chernobyl origin. The songthrush (*Turdus philomelos*) collected in Spain in November 1986 had elevated concentrations of ^{134}Cs, ^{137}Cs, and ^{90}Sr. The contamination probably occurred in Central and Northern Europe before the birds' migration to Spain. In 1994, concentrations in songthrush tissues remained elevated. However, Spaniards who ate songthrushes contaminated with radiocesium isotopes usually received about 58 μSv yearly, which is well below current international guidelines. In 1987–88, ptarmigans (*Lagopus* spp.) from alpine ecosystems in central Norway receiving 20,000–60,000 Bq $^{137}Cs/m^2$ in 1986 usually had <350 Bq ^{137}Cs/kg FW muscle. Concentrations of ^{137}Cs in ptarmigan muscle were higher in juveniles than in adults, higher in summer than in winter, and were correlated with radiocesium concentrations in food plants.

27.5.2.2.4 *Domestic Animals*

Radiocesium isotopes from the Chernobyl accident transferred easily to grazing farm animals. Both ^{134}Cs and ^{137}Cs were rapidly distributed throughout the soft tissues of animals after dietary ingestion and were most highly concentrated in muscle. Radiocesium activity in milk and flesh of Norwegian sheep and goats, for example, increased three- to fivefold 2 years after the accident and coincided with an abundant growth and availability of fungal fruit bodies with ^{134}Cs and ^{137}Cs levels as much as 100 times greater than green vegetation. Norway – almost 2000 km from Chernobyl – condemned $70.9 million (USA) of domestic animal products in the period 1986–95 as a result of the accident. In 1986 alone, large amounts of mutton, reindeer meat, and goat's cheese exceeded the radiocesium content set by Norwegian authorities and was destroyed at an estimated loss of $24 million. Economic loss in 1995 was about $2.1 million. In 1996, after 10 years of expensive countermeasures (changes of pasturing areas, special feeding, cesium binders, change in slaughtering time), there was some decline in the size of the areas and the numbers of animals involved; however, countermeasures are still required.

In cattle, coefficients of radiocesium transfer from diet to muscle were about 2.5% in adults and 16% in calves. The higher value in calves was probably due to a high availability of cesium from the gastrointestinal tract and to daily uptake of potassium in growing animal muscle. There was no correlation between retention of ^{137}Cs and pregnancy stage in cattle. Radiocesium concentrations in pork in Czechoslovakia did not decline between 1986 and 1987 because the feed of pigs during this period contained milk by-products contaminated with ^{134}Cs and ^{137}Cs.

Sheep farming is the main form of husbandry in the uplands of West Cumbria and North Wales, a region that received high levels of radiocesium fallout during the Chernobyl accident. Afterwards, typical vegetation activity concentrations were −6000 Bq/kg (down to −1000 Bq/kg in January 1989). But sheep muscle concentrations exceeded 1000 Bq ^{137}Cs/kg FW, which is the United Kingdom's dietary limit for human health protection. Contaminated lambs – which usually had higher concentrations of ^{137}Cs than ewes – that were removed to lowland pastures (<50 Bq/kg vegetation) rapidly excreted radiocesium in feces and urine, and cesium body burdens had an effective half-life of 11 days. This practice should not significantly increase radiocesium levels in soil and vegetation of lowland pastures. The absorption and retention of radiocesium by suckling lambs is highly efficient, about 66%. Fecal excretion was an important pathway after termination of ^{137}Cs ingestion. In weaned animals, the absorption of added ionic cesium was about twice that of cesium fallout after the accident at Chernobyl. Silver-110m was also detected in the brains and livers of ewes and lambs in the United Kingdom. The transfer of ^{110m}Ag was associated with perennial rye grass harvested soon after deposition in 1986. Silver-110m was taken up to a greater extent than ^{137}Cs in liver; but unlike ^{137}Cs, the ^{110m}Ag was not readily translocated to other tissues. Other than cesium isotopes and ^{131}I, ^{110m}Ag was the only detected nuclide in sheep tissues.

Atmospheric deposition of ^{137}Cs from Chernobyl to vegetation and eventually to the milk of sheep, cows, and goats on contaminated silage was reported in Italy, the Netherlands, Japan, and the United Kingdom. The effective half-life of ^{137}Cs was 6.7 days in silage and 13.6 days in milk. The average daily transfer coefficient of ^{134}Cs and ^{137}Cs from Chernobyl from a 70% grass silage diet to milk of Dutch dairy cows was about 0.25%/L. In goats (*Capra* sp.), about 12% of orally administered ^{137}Cs was collected in milk within 7 days after dosing.

Iodine-131 was one of the most hazardous radionuclides released in the Chernobyl accident because it is easily transferred through the pasture–animal–milk pathway and rapidly concentrated in the thyroid gland to an extent unparalleled by any other organ. Because of its high specific activity, ^{131}I can transmit a high dose of radiation to the thyroid. Iodine-131 levels of 618,000 Bq/kg FW in sheep thyroids from northwestern Greece on July 3, 1986 are similar to maximal ^{131}I concentrations in sheep thyroids in Tennessee in 1957 after global atmospheric fallout from military weapons tests and in London after the Windscale accident. Iodine-131 has an effective whole-body half-life of about 24 h and is rapidly excreted from sheep and cows. The effective half-life of ^{131}I in pasture grass was 3.9 days and 5 days in cow's milk. The daily transfer coefficient of ^{131}I from vegetation to cow's milk was 0.007%/L milk. This value was 57 times higher (0.4) in sheep, but the mechanism that accounts for this large interspecies difference is not clear.

27.6 Effects: Nonionizing Radiations

Living organisms are constantly exposed to nonionizing electromagnetic radiations, including ultraviolet, visible, infrared, radio, and other low energy radiations that form an integral part of the biosphere. Emissions from anthropogenic sources such as radios, microwave ovens, television communications, and radar significantly altered the character of our natural electromagnetic field. Although the primary focus of this review is on ionizing radiations, it would be misleading to assume that low energy electromagnetic waves cannot elicit significant biological responses. For example, behavioral and biochemical changes are reported in rats, monkeys, rabbits, and other laboratory animals after exposure to nonionizing electromagnetic radiations; the severity of the effect is associated with the type and duration of the radiation and various physicochemical variables.

Ultraviolet radiation, for example, should be considered as a plausible factor contributing to amphibian malformations in field settings. Ultraviolet radiation is linked to teratogenesis, growth inhibition, and DNA photodamage in larvae of the South African clawed frog, *Xenopus laevis*. About 50% of newly fertilized eggs of the northern leopard frog (*Rana pipiens*) exposed to UV radiation for 24 h developed hindlimb malformations, such as missing limb segments, missing or reduced digits, and missing or malformed femurs. The higher energy portion of the ultraviolet spectrum (UV-B) was lethal to embryos of Canadian frogs, with death occurring in as short as 30 min; the lower energy portion (UV-A) was not harmful to eggs or larvae at exposures twice that of ambient levels. Exposure to solar radiation – even at low elevations – is lethal to amphibian eggs of species with relatively poor capacity to repair UV-damaged DNA. Increased UV-B radiation in the environment due to decreasing ozone levels has been suggested as a factor in the worldwide decline of sensitive species of amphibians;

however, UV-B radiation was not implicated in the decline of the endangered green and golden bell frog (*Litoria aurea*) in Australia.

Ultraviolet radiation in mammals causes the aging of skin, making it wrinkled and leathery. Dermatologists of the late 19th century described the devastating effects of sunlight on the skin of farmers and sailors when compared with indoor workers. Photoaged skin has a variety of neoplasms, deep furrows, extensive sagging, and profound structural alterations that are quite different from those in protected, intrinsically aged skin. Similar results were documented of skin of guinea pigs and rodents after exposure to ultraviolet radiation. Ultraviolet radiation causes eye cancer in cattle, interferes with wound healing in guinea pig skin, is a potent damaging agent of DNA and a known inducer of skin cancer in experimental animals, and interferes with an immune defense mechanism that normally protects against skin cancer. Aquatic organisms exposed to ultraviolet radiation show disrupted orientation, decreased motility, and reduced pigmentation in *Peridinium gatunense*, a freshwater alga; effects were similar in several species of marine algae. Increased lipid peroxidation rates and a shortening of the life-span after ultraviolet exposure were reported in the rotifer *Asplanchna brightwelli*. Cells of the goldfish (*Carassius auratus*) were damaged, presumably by DNA impairment, from UV exposure.

Visible radiation adversely affected survival and growth of embryos of the chinook salmon (*Oncorhynchus tshawytscha*), chloroplast structure in the symbiotic marine dinoflagellate *Symiodinium* sp., and in vitro growth of cultured mammalian cells. Infrared radiation contributes significantly to skin photoaging, producing severe elastosis; the epidermis and the dermis were capable of self-restoration when the exogenous injury ceased.

Investigations of the cellular effects of radiofrequency radiation provide evidence of damage to various types of avian and mammalian cells. These effects involve radiofrequency interactions with cell membranes, especially the plasma membrane. Effects include alterations in membrane cation transport, Na^+/K^+-ATPase activity, protein kinase activity, neutrophil precursor membrane receptors, firing rates and resting potentials of neurons, brain cell metabolism, DNA and RNA synthesis in glioma cells, and mitogenic effects on human lymphocytes.

Microwaves inhibit thymidine incorporation by DNA blockage in cultured cells of the Chinese hamster; irradiated cells had a higher frequency of chromosome lesions. Microwaves induce teratogenic effects in mice when the intensity of exposure places a thermal burden on the dams and fetuses, resulting in a reduction in fetal body mass and an increased number of resorptions.

Extremely low frequency (ELF) electromagnetic fields – similar to fields that emanate from electrical appliances and the electrical power distribution network, usually <300 Hz – are used therapeutically in the healing of human nonunion bone fractures, in the promotion of nerve regeneration, and in acceleration of wound healing. ELF electric and magnetic fields produce biological effects, usually subtle, and of low hazard in short-term exposure. These effects include altered neuronal excitability, neurochemical changes, altered hormone levels, and changes in behavioral responses. For example, electric field perception has been reported in humans, mice, pigs, monkeys, pigeons, chickens, and insects; altered cardiovascular responses in dogs and chickens; and altered growth rate of chicks. No deleterious effects of ELF fields on mammalian reproduction and development or on carcinogenesis and mutagenesis have been documented. In the vicinity of a powerful radar station, some birds avoided nesting (flycatchers, *Myiarchus* spp.), and the percent of nesting boxes occupied by other species (tits, *Parus* spp.) increased significantly with increasing distance from the radar station. ELF fields had no effect on the growth of bone in chicks. However, adult newts (*Notophthalmus viridescens*), regenerating amputated forelimbs, had grossly abnormal forelimbs 12% of the time when exposed for 30 days to ELF fields of the type reported to facilitate healing of human bone fractures. Additional studies are recommended on the biological effects of nonionizing radiations on fishery and wildlife resources, especially on ELF radiations.

27.7 Effects: Ionizing Radiations

High acute doses of ionizing radiation produce adverse biological effects at every organizational level: molecular, cellular, tissue–organ, whole animal, population, community, and ecosystem. Typical adverse effects of ionizing radiation include cell death, decreased life expectancy, increased frequency of malignant tumors, inhibited reproduction, increased frequency of gene mutations, leukemia, altered blood–brain barrier function, and reduced growth and altered behavior. Species within kingdoms have a wide variation in sensitivity and sometimes at low radiation exposures the response is considered beneficial. Overall, the lowest dose rate at which harmful effects of chronic irradiation have been reliably observed in sensitive species is about 1 Gy/year; this value for acute radiation exposures is about 0.01 Gy.

In general, the primitive organisms are the most radioresistant taxonomic groups, and the more advanced complex organisms – such as mammals – are the most radiosensitive (Figure 27.7). The early effects of exposure to ionizing radiation result primarily from cell death; cells that frequently undergo mitosis are the most radiosensitive, and cells that do not divide are the most radioresistant. Thus, embryos and fetuses are particularly susceptible to ionizing radiation, and very young animals are consistently more radiosensitive than adults. In addition to the evolutionary position and cell mitotic index, many extrinsic and intrinsic factors modify the response of a living organism to a given dose of radiation. Abiotic variables include the type and energy of radiation, exposure rate, length of exposure, total exposure and absorbed dose, dose rate, spatial distribution of dose, season, temperature, day length, and environmental chemicals; biotic variables include the species, type of cell or tissue, metabolism, sex, nutritional status, sensitizing or protective substances, competition, parasitism, and predation.

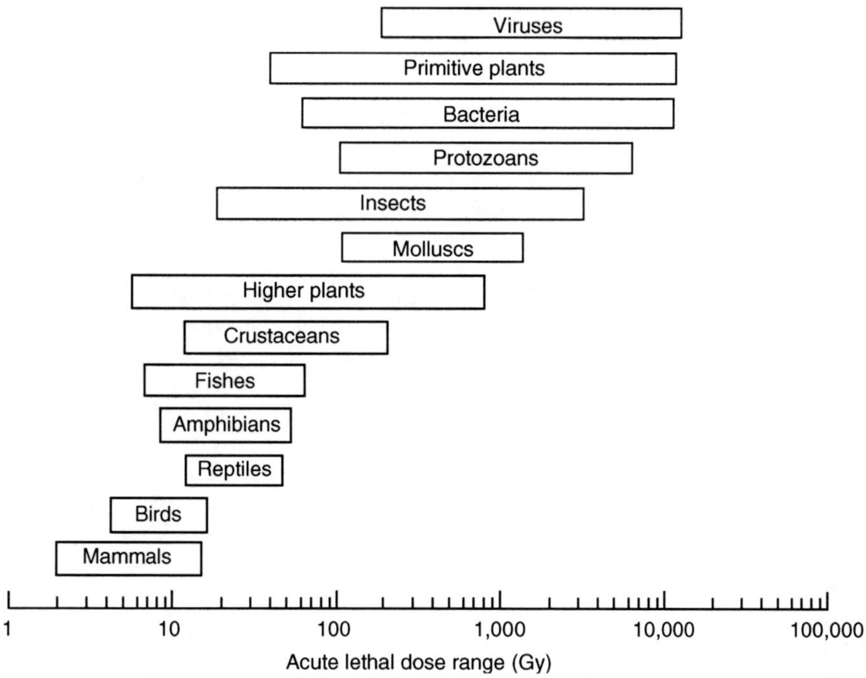

Figure 27.7.
Acute radiation dose range fatal to 50% (30 days postexposure) of various taxonomic groups.

Radiosensitivity of cells is related directly to their reproductive capacity and indirectly to their degree of differentiation. Early adverse effects of exposure to ionizing radiation are due mainly to the killing of cells. Cell death may result from the loss of reproductive integrity, that is, when after irradiation a cell fails to pass through more than one or two mitoses. Reproductive death is important in rapidly dividing tissues such as bone marrow, skin, gut lining, and germinal epithelium. When the whole animal is exposed to a large dose of ionizing radiation, some tissues are more prone to damage than others. Death rates of mammalian reproductive cells from ionizing radiations is modified by variations in the linear energy transfer of the radiation, the stage in the cell cycle, cell culture conditions, and sensitizing and protecting compounds. The chemical form of the main stage of the acute radiation syndrome depends on the size and distribution of the absorbed dose. It is determined mainly by damage to blood platelets and other blood-forming organs at 4–5 Gy, to epithelial cells lining the small intestine at 5–30 Gy, and to brain damage at >30 Gy; death usually occurs within 48 h at >30 Gy.

Cellular DNA is extremely sensitive to ionizing radiation although other cell constituents may approach DNA in sensitivity. Radiation-induced mutations are explainable on the basis of chromatin and DNA organization in cells and the biophysical properties of ionizing radiation. Based on studies of spontaneous and radiation-induced mutations in the mouse, more than 67% of the ionizing radiation-induced mutations are lethal and almost all mutations, including enzyme activity variants, dominant visibles, and dominant skeletal mutations, are lethal. These findings are consistent with the view that most radiation-induced mutations in germ cells of mice are due to DNA deletions.

Experimental animal data clearly demonstrate that ionizing radiation at relatively high doses and delivered at high dose rates is mutagenic. However, radiation-induced genetic damage in the offspring of exposed parents has not been credibly established in any study with humans. In one human population – the ethnically isolated Swedish reindeer-breeding Lapps – elevated concentrations of fallout products have been ingested via the lichen–reindeer–human food chain since the 1950s. However, during 1961–84 no increased incidence of genetic damage was evident in Lapps.

Radiation is carcinogenic. The frequency of death from cancer of the thyroid, breast, lung, esophagus, stomach, and bladder was higher in Japanese survivors of the atomic bomb than in nonexposed individuals, and carcinogenesis seems to be the primary latent effect of ionizing radiation. The minimal latent period of most cancers was <15 years and depended on an individual's age at exposure and site of cancer. The relation of radiation-induced cancers to low doses and the shape of the dose–response curve (linear or nonlinear), the existence of a threshold, and the influence of dose rate and exposure period have to be determined.

Radioactive materials that gain entry to the body, typically through ingestion or inhalation, exert effects that are governed by their physical and chemical characteristics which, in turn, influence their distribution and retention inside the body. The effective half-life includes both physical and biologic half-times. In addition, the type of radiation (i.e., α, β, γ) and its retention and distribution kinetics govern the radiation dose pattern. In general, the radiation dose from internal emitters is a function of the effective half-time, energy released in the tissue, initial amount of introduced radioactivity, and mass of the organ. Retention of radionuclides by living organisms is quite variable and modified by numerous biologic and abiotic variables. For example, ^{137}Cs retention in selected animals varies significantly with the body weight, diet, and metabolism of an organism (Figure 27.8). The time for 50% persistence of ^{137}Cs ranges between 30 and 430 days in ectotherms and it is longer at lower temperatures and shortest in summer and under conditions of inadequate nutrition. In mammals, the ^{137}Cs biological half-life was between 6 and 43 days in rodents, dogs, mule deer, reindeer, and monkeys; in humans, this value ranged from 60 to 160 days. Biological half-life of ^{90}Sr ranges from 122 to 6000 days in ectotherms and is longer at colder temperatures and under laboratory conditions. In mammals and under conditions of

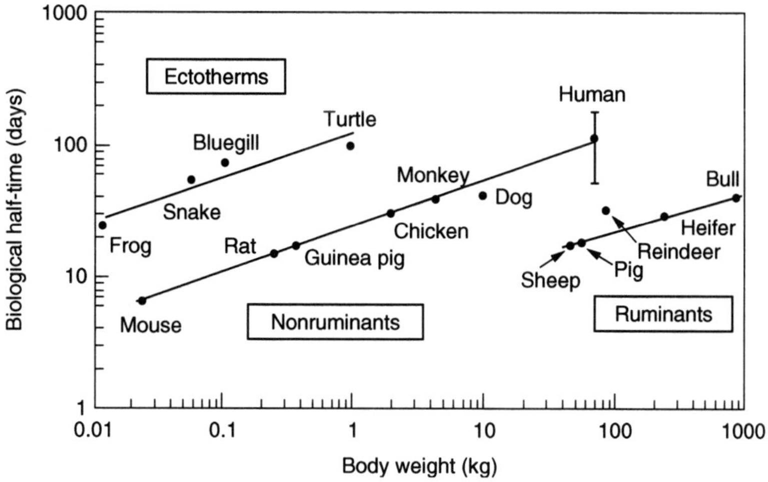

Figure 27.8.
Relation between diet, metabolism, and body weight with half-time retention of longest-lived component of cesium-137. Data are shown for selected ruminant and nonruminant mammals.

chronic intake, the ^{90}Sr biological half-life was 533 days in rat, 750 days in humans, and at least 848 days in beagle dogs.

27.7.1 Terrestrial Plants and Invertebrates

Radiosensitive terrestrial plants exposed to single doses of ionizing radiation had reduced growth at 0.5–1.0 Gy and reduced survival at 3.0–4.1 Gy; chronic exposures of 0.2–0.65 Gy/day adversely affected sensitive forest ecosystems. Chronic gamma irradiation of 131 Gy/year and higher of mixed forest ecosystems caused the disappearance of trees and shrubs and subsequent erosion of the soil. The radiation sensitivity of 5 plant communities suggested that pine forests were the most sensitive and that deciduous evergreen forests, tropical rain forests, herbaceous rock outcrop communities, and abandoned cropland were increasingly less sensitive. Neutrons were 3–4 times more effective than gamma rays in root growth inhibition. Altitude affects the response of vegetation to ionizing radiation. Peas (*Pisum sativum*) in gardens 2225–3750 m above mean sea level and exposed to 0, 5, 10, or 50 Gy had reduced growth from all treatments at increasing altitudes; however, a dose–response growth curve was evident only at <3049 m altitude. Seeds of tobacco (*Nicotiana tabacum*) exposed to cosmic rays aboard a spacecraft had a higher mutation rate than controls; effects occurred at total doses as low as 0.1–0.2 Gy, but this needs verification.

Sometimes, irradiation prevents the usual colonizing vegetation from becoming established. Germination and survival of shrub seedlings have been much slower at nuclear test sites than at non-disturbed sites. The return to its original state of the perennial shrub vegetation takes decades on a radiation-disturbed site, although native annual species and grasses have grown abundantly within 12 months. Transplanting of shrubs into radiation-disturbed areas has been largely unsuccessful because of intense browsing by rabbits and other small mammals. A nuclear detonation damages terrestrial vegetation by heat, blast, or radiation. Plant injury from thermal or ionizing radiation at an above-ground detonation site varied with stem rigidity and stability of the substratum, although radiation effects are ordinarily masked by damage from blasts. A typical nuclear detonation at the

Nevada test site – an airburst of a 20- to 40-kiloton yield – denuded a zone of desert within 0.8-km radius of shrub vegetation. Recovery at the Nevada site seemed complete within 4 years, suggesting little relation between fatal injury, morphological aberration in vegetation, and ionizing radiation from nuclear detonations. A northern Wisconsin forest experimentally subjected to a ^{137}Cs radiation source for 5-months showed several trends: (1) herbaceous and shrub species with a spreading form of growth are more radioresistant than upright forms; (2) larger pine and oak trees are more radioresistant than smaller trees; (3) perennial plants with shielded buds and vigorous asexual reproduction are relatively radioresistant; (4) plants adapted to extreme habitats, such as old fields and granite rock outcrops, and plants typical of early successional stages are relatively radioresistant; (5) all plants are more radiosensitive during the growing season than the dormant season; and (6) reproductive stages are always more radiosensitive than vegetative stages. The recovery of vegetation in a tropical rain forest in Puerto Rico – after plants were deliberately subjected to lethal doses of gamma radiation – closely resembled secondary succession after other types of disturbances, such as mechanical stripping and treatment with the Picloram herbicide.

Hormesis – the beneficial physiological stimulation by low doses of a potentially harmful agent – is documented for ionizing radiation and many species of terrestrial plants and invertebrates. Radiation hormesis in plants includes increased germination, growth, survival, and yield. Some species of terrestrial invertebrates had increased fecundity, growth, survival, disease resistance, and longevity after exposure to low sublethal doses of ionizing radiation. The growth and development of some terrestrial invertebrates are stimulated at comparatively high sublethal acute doses (i.e., 2 Gy in silkworm, *Bombyx mori*), but survival is reduced at 10 Gy; in all cases, younger stages were the most sensitive. Cockroaches (*Blaberus giganteus*) adapted to the dark reportedly can visually detect radiation sources as low as 0.001 mGy, however, the mechanisms are not understood.

Following the successful application of radiation to sterilize male screw-worm flies (*Cochliomyia hominivorax*), various insect pests became the target of similar techniques throughout the world. The technique has suppressed populations of the Mediterranean fruit fly (*Ceratitis capitata*), a major pest of fruits, although results have not been as spectacular as with the screw-worm fly. The pestiferous Caribbean fruit fly (*Anastrepha suspensa*), heavily parasitized by a beetle, became sterile after acute exposures to ionizing radiation, although beetles remained fecund. Mass rearing and inundative release of the radioresistant beetle parasite is now considered an option for control of the Caribbean fruit fly.

27.7.2 Aquatic Organisms

Among aquatic organisms, it is generally acknowledged that primitive forms are more radioresistant than complex vertebrates and that older organisms are more resistant than the young. Developing eggs and young of some species of freshwater fish are among the most sensitive-tested aquatic organisms; death was observed at acute doses of 0.3–0.6 Gy, and minor effects on physiology or metabolism were observed at chronic daily dose rates of 0.01 Gy. Radiosensitivity correlated positively with the metabolic rate of the dividing cell, which accounts for the radioresistance of dormant eggs of aquatic invertebrates and the general sensitivity of early embryonic stages of all aquatic species.

Adverse effects on the fecundity of sensitive aquatic vertebrates were detected at dose rates as low as 0.4 mGy/h; adverse effects on fecundity of resistant species were measured only at dose rates greater than 1.0 mGy/h. Thus, deleterious effects in populations of aquatic vertebrates are probably not detected until the 0.4–1.0 mGy/h dose rate is exceeded. Organisms, such as estuarine organisms, that are exposed to variable physicochemical conditions are more radioresistant than those in buffered environments, and this may be due to a higher degree of genetic polymorphism in species of fluctuating environment. Dose–effect estimates for the induction of chromosomal aberrations in polychaete annelid worms were dependent on cell stage at time of irradiation. For reproduction, dose–effect estimates were dependent on potential for regeneration of gonadal tissue.

Radiation causes dominant lethal mutations in the medaka (*Oryzias latipes*). Mosquitofish (*Gambusia* spp.) from radionuclide-contaminated ponds in South Carolina differed from conspecifics in reference ponds as judged by the frequency of DNA markers and this is consistent with the hypothesis that these DNA markers may originate from genetic elements that provide a selective advantage in contaminated habitats. Ionizing radiation at low-level chronic exposure reportedly has no deleterious genetic effects on aquatic populations because exposure is compensated by density-dependent responses in fecundity; however, this needs verification.

Accumulation of radionuclides from water by aquatic organisms varies substantially with ecosystem, radionuclide, and trophic level; with numerous biological, chemical, and physical variables; and with proximity to sources of radiation. Accumulated radionuclides within embryos of the scorpionfish (*Scorpaena porcus*) and turbot (*Scopthalmus maeoticus*) increased the frequency of nuclear disruptions in these species; ^{90}Sr–^{90}Y and ^{91}Y had greater cytogenetic effects than other radionuclides tested. In the absence of site-specific data, the U.S. Nuclear Regulatory Commission recommends the use of listed concentration ratios – the concentration of the element in the organism (in mg/kg FW) divided by the concentration in the medium (in mg/L) – for various elements in marine and freshwater fishes and invertebrates. However, the Commission clearly indicates that these values are only approximations.

After more than 400 atmospheric nuclear test explosions and the fallout from Chernobyl, ^{137}Cs became the most frequently released nuclear fission product throughout Central Europe. Cesium behaves like potassium; it has a ubiquitous distribution inside the body, especially in soft tissues. In the gastropod *Helix pomatia* the biological half-time after a single 24-h dietary dose was 2.5 days for the short-lived component and 28.5 days for the long-lived component. Concentration factors (CF) of ^{137}Cs in muscle (ratio of Bq/kg FW muscle:Bq/L filtered seawater) of marine fishes from the North Sea between 1978 and 1985 ranged from a low of 39 in the plaice (*Pleuronectes platessa*) to a high of 150 in the whiting (*Merlangius merlangius*); CF values were intermediate in the haddock (*Melanogrammus aeglifinus*; CF of 58) and Atlantic cod (*Gadus morhua*; CF of 92). These data seem to support the use of a CF of 100 for ^{137}Cs in muscle of marine fishes in generalized assessments, although some adjustment is necessary when particular species, such as whiting, form the bulk of a consumer's diet. In the Great Lakes, the maximum CF values of ^{137}Cs range from 1000 to 10,000 in algae, amphipods and fishes, and from 100 to 1000 in zooplankton. Maximum concentration factors of ^{137}Cs in a contaminated creek in South Carolina were 4243 in suspended particulates, 938 in detritus, 4496 in algae and macrophytes, 997 in omnivores, 1292 in primary carnivores, and 1334–2595 in top carnivores, such as redbreast sunfish (*Lepomis auritis*), largemouth bass (*Micropterus salmoides*), and water snakes (*Natrix* spp.). Cesium uptake by oligochaete worms (*Limnodrilus hoffmeisteri*) is inhibited by low temperatures, potassium concentrations >1.0 mg/L, and the presence of bacteria (*Escherichia coli*) that compete with the worms for ^{137}Cs.

Atmospheric fallout from nuclear testing is the main pathway by which transuranic nuclides, such as neptunium (Np), plutonium (Pu), curium (Cm), and americium (Am), enter the aquatic environment. In general, transuranics are strongly partitioned into particulates. Living organisms are less enriched than particulate matter by as much as 1000 times, and concentration factors by marine biota are similar for transuranics beyond neptunium. The uptake of ^{241}Am and ^{244}Cm from contaminated sediments by a freshwater amphipod (*Hyalella* sp.) and oligochaete (*Tubifex* sp.) is reported, presumably by way of adsorption, and this is considered the principal uptake pathway by benthic organisms in freshwater and marine ecosystems. Transuranics ingested with food by various crabs were initially excreted with feces; the remaining transuranics entered a soluble radionuclide pool within the animal that was slowly excreted. Decapod crustaceans assimilate and retain 10–40% of the transuranic nuclides in their diets. Initially, absorbed radionuclides accumulate in the hepatopancr eas but are then translocated to other tissues, particularly to tissues of the exoskeleton; accordingly, molting strongly influences elimination in crustaceans. Neptunium isotopes have a higher potential for environmental transport in aquatic systems and groundwater than other actinides tested. Laboratory studies with ^{235}Np and ^{237}Np, for example, show concentration factors between 275 and 973 in a green alga (*Selenastrum capricornutum*), 32 and 72 in a daphnid (*Daphnia magna*), 2 in an amphipod (*Gammarus* sp.) and in juvenile rainbow trout, 8.7 in carcass, 1.1 in skin, and 0.3 in muscle over a 96-h period. When the much higher biological effectiveness of alpha vs. beta or gamma radiation is considered, plutonium isotopes may contribute more artificial radiation dose equivalent to marine invertebrates than either ^{90}Sr or ^{137}Cs. Concentration factors of ^{239}Pu and marine organisms ranged from 300–100,000 in seaweeds, 250–690 in mollusks, 760–1020 in echinoderms, 2100 in sponges, and as much as 4100 in worms. Concentration factors of ^{239}Pu and ^{240}Pu in Lake Michigan ranged between 1 and 10 in predatory salmonids, 10 and 300 in nonpredatory fish, 900 and 1200 in amphipods and shrimp, about 200 in zooplankton, and about 6000 in algae.

Iodine-131 (half-life of 8 days) may cause deleterious effects in marine teleosts – although ^{131}I concentrations in tissues were not detectable. In one case, coral reef fishes from Eniwetok Atoll collected as long as 8 months after a nuclear explosion had thyroid necroalteration, suggesting a thyrotoxic level of ^{131}I in the environment. Laboratory studies with teleosts injected with ^{131}I showed similar signs of histopathology. Herbivorous fishes and species that habitually consumed bivalve mollusks were the most severely affected.

Strontium-90 is an anthropogenic radionuclide in liquid effluents from some European nuclear power plants. Algae and sediments are the most important accumulators of ^{90}Sr, although levels in gastropods and fish bone and scales are also elevated, suggesting piscine uptake through gills and skin. Fish tend to accumulate calcium more than strontium, even when calcium levels in food and water are low. Gill tissue is the most and gut the least discriminatory against strontium. Strontium assimilation was linked to the Sr:Ca ratio in food and water, amounts of Ca derived from each source, and biological discrimination against Sr relative to Ca. The ability of organisms to discriminate between strontium radioisotopes is also documented. In one case, ^{85}Sr was taken up rapidly in bluegill (*Lepomis macrochirus*) muscle and blood and quickly exchanged with stable strontium; however, ^{90}Sr was retained longer than 35 days in these tissues.

Ruthenium-106 appeared in clams from North Carolina within 2 weeks after the third and fifth Chinese nuclear tests in 1965–67. Its retention was resolved into two rate functions with apparent effective half-lives of 40 days and 7 days. Iron-55 is a neutron-activation product produced in large quantities from ferrous materials in the immediate vicinity of a nuclear detonation. Concentration factors of ^{55}Fe and plankton in Bikini Atoll ranged from 15,000 to 25,000. Silver-110m has been detected in marine organisms after atmospheric weapon tests in the Pacific, in fishes from the Rhone River after the Chernobyl accident, and in fishes near reactor-waste outfalls. Silver-110m is depurated rapidly by brown trout (*Salmo trutta*) after high intake exposures via the water or diet. Radiotungsten is produced in quantity by certain types of nuclear devices. In one case tungsten was the most abundant radionuclide in the environment, accounting for about 90% of the total fallout activity 167 days after the detonation. Tungsten-181 tended to concentrate in the hepatopancreas and gut of the crayfish (*Cambarus longulus longirostris*); whole body elimination consisted of two components: a rapid 1-day component and a second slower component with a biological half-time of 12.2 days. Benthic organisms take up limited amounts of heavy metals and radionuclides associated with bottom sediments and recycle them to benthic and pelagic food webs. For example, polychaete worms (*Nereis diversicolor*) in contact with ^{65}Zn-contaminated sediments for 5 days lost 50% of accumulated ^{65}Zn in about 19 days on transfer to uncontaminated sediments.

27.7.3 Amphibians and Reptiles

Radiation adversely affects limb regeneration of amphibians, alters DNA metabolism, and increases the frequency of chromosomal aberrations and liver lesions. In some species of amphibians and reptiles, as in many mammals, mortality rates after acute exposure to radiation do not stabilize within 30 days – effectively invalidating the conventional LD50 (30 day postexposure) value. In the rough-skinned newt (*Taricha granulosa*), for example, the minimal LD50 dose at 200 days after irradiation was 2.5 Gy compared with 350 Gy at 30 days. Low temperatures seem to prolong the survival of amphibians exposed to ionizing radiation. The survival was greater of leopard frogs (*Rana pipiens*) held at low temperatures (5–6°C) after total body exposure to lethal doses of X-rays than of frogs held at higher temperatures. Prolonged survival at low temperatures was due to a prolongation of the latent period rather than to appreciable recovery.

The South African clawed frog (*Xenopus laevis*) is a useful bioindicator of radioactive contamination because of the greater radiosensitivity of amphibians than fishes, the ease of maintaining *Xenopus* in the laboratory, and the sensitivity of the *Xenopus* liver to radioactive contamination – including ^{45}Ca, which does not accumulate in the liver. *Xenopus* oocytes exposed to X-rays showed single and double strand breaks in DNA and oxidative-type base lesions at a frequency between 85% and 95%. *Xenopus* oocytes repaired X-ray induced damage in plasmid DNA; however, some X-ray lesions can stimulate homologous recombination in these cells. Slider turtles (*Trachemys scripta*) in a radioactive reservoir show evidence of genetic damage, and this was attributed to long-term exposure of low concentrations of long-lived radionuclides, including ^{137}Cs and ^{90}Sr. Natural populations of toads (*Bufo valliceps*) reportedly can survive genetically damaging doses of ionizing radiation without impairment of population integrity. Toads and many other species share a high attrition on the large numbers of young produced each generation, and this provides an agency for intensive selection. Also under this regime, recessive mutants are eliminated as they are exposed through inbreeding in future generations. Sterility in field collections of the leopard lizard (*Crotaphytus wislizenii*) and the whip-tail lizard (*Cnemidophorus tigris*) was reported after long term exposure of 3–5 years to various doses of gamma radiation, i.e., 4–5 Sv annually in *Crotaphytus* and 2.0–2.5 Sv annually in *Cnemidophorus*. However, a third species of lizard in the study area (side-blotched uta, *Uta stansbunana*) reproduced normally.

The retention of selected isotopes by amphibians and reptiles is quite variable. For example, whole body retention of ^{131}I after intraperitoneal injection in the rough-skinned newt showed 2 distinct loss components with biological half-lives of 2 and 210 days; the slower component accounted for 26% of the administered activity; thyroid contained 78% of the total ^{131}I and clearly accounted for the long-term component. However, similar studies with ^{131}I and the leopard frog showed 3 distinct loss components (0.1 day; 1.4–2.9 days; 44.3–69.4 days); loss of each component was greater at 25°C than at 10°C; also, the fast component probably represented plasma clearance through urinary excretion.

27.7.4 Birds

Among birds, as in most other tested species, there is a direct relation between dose and mortality at single high doses of ionizing radiations. For any given total dose, the survival of a bird is higher if the dose is delivered at a lower rate or over a longer period of time and suggests that biological repair processes compensate for radiation-induced cellular and tissue damage over a prolonged period or at a comparatively low dose rate. Nestling bluebirds (*Sialia sialis*) were more resistant to gamma radiation than young domestic chickens (*Gallus* sp.), and nestling great crested flycatchers (*Myiarchus crinitus*) were more sensitive than bluebirds. Passerine nestlings are more resistant to radiation stress than adults of larger-bodied precocial species. But the comparatively resistant passerine nestlings frequently show a disproportional disturbance in radiation-induced growth, resulting in a reduction of overall survival. For example, if feather growth is stunted, death results from the inability to escape predators owing to impaired flight.

Free-living, resident bird populations in the vicinity of sites contaminated with low levels of ionizing radiations generally have negligible genotoxic effects. However, 14% of mallards (*Anas platyrhynchos*) from an abandoned South Carolina reactor cooling reservoir heavily contaminated with ^{137}Cs (mallards contained an average of 2520 Bq ^{137}Cs/kg whole body FW) had abnormal chromosome numbers and unusual variability in the concentration of erythrocyte DNA. Contaminated waterfowl rapidly eliminate accumulated radionuclides, suggesting inconsequential long-term damage to the birds and little hazard to human consumers of waterfowl flesh. This conclusion was from a study wherein mallards were held for 68–145 days on liquid radioactive waste ponds in southeastern Idaho then transferred to an uncontaminated environment for 51 days. The biological half-life in mallards under these conditions was 10 days for ^{131}I and ^{134}Cs, 11 days for ^{137}Cs, 22 days for ^{140}Ba, 26 days for ^{75}Se, 32 days for ^{58}Co, 67 days for ^{60}Co and ^{65}Zn, and 86 days for ^{51}Cr. At the time of removal from the waste ponds, radionuclide concentrations were highest in gut, then feather, liver, and muscle, in that order. After 51 days in a radionuclide-free environment, decreasing order of radionuclide concentrations were feather, liver, muscle, and gut.

Zinc-65 in trace amounts is accumulated by migratory waterfowl in the Pacific flyway of North America from ^{65}Zn discharged into the Columbia River from water-cooled reactors at Hanford, Washington. The retention of ^{65}Zn in mallards was affected by sex and season but not by the age of the duck. Biological retention of ^{65}Zn was greater in males (Tb1/2 of 34.7 days) than in females (29.8 days) and greater in October (38 days) than in the spring (32 days). Egg production accounted for the elimination of 25% of the ^{65}Zn and feather molt of 2–8%. Retention of ^{60}Co and ^{137}Cs – but not ^{109}Cd – in the common bobwhite (*Colinus virginianus*) after either acute or chronic exposure to contaminated food is similar. The biological half-life in bobwhites during exposure for 21 days was 8 days of ^{109}Cd, 11 days of ^{137}Cs, and 13 days of ^{60}Co. When radioisotopes were administered during a single 4-h feeding, Tb1/2 values were 3 days of ^{109}Cd, 10 days of ^{137}Cs, and 15 days of ^{60}Co. The biological half-life of ^{137}Cs in avian tissues is about 6.0 days in domestic chickens, 6.7 days in the bluejay (*Cyanocitta cristata*), 5.6 days in the American wood duck (*Aix sponsa*), and 11.7 days in mallards. Domestic poultry, when compared with mammals, seem to accumulate a higher fraction of the daily ingested ^{137}Cs/kg muscle, but levels were effectively reduced by feeding an uncontaminated ration for at least 10 days prior to slaughter.

27.7.5 Mammals

Data on mammalian sensitivity to acute and chronic exposures of ionizing radiation, ability to retain selected radionuclides, and effect of biological and abiotic variables on these parameters clearly indicate a dose-dependent effect of radiation on growth, survival, organ development, mutagenicity, fatal neoplasms, kidney failure, skeletal development, behavior, and all other investigated parameters. In general, fetuses and embryos were most sensitive to ionizing radiation, and acute or chronic exposures between 0.011 and 0.022 Gy were demonstrably harmful to mice, rats, and guinea pigs.

27.7.5.1 Survival

Survival time is inversely related to dose in whole body, acute exposures to ionizing radiation (Figure 27.9). In general, hematopoietic organs are most sensitive, and gastrointestinal tract and central nervous system are next

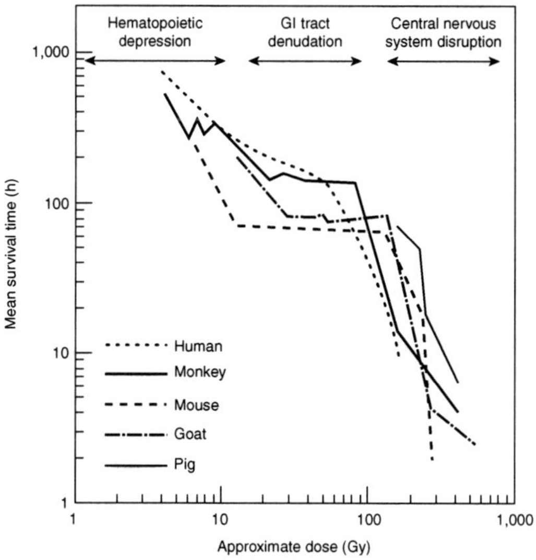

Figure 27.9.
Survival time and associated mode of death of selected mammals after whole body doses of gamma radiation.

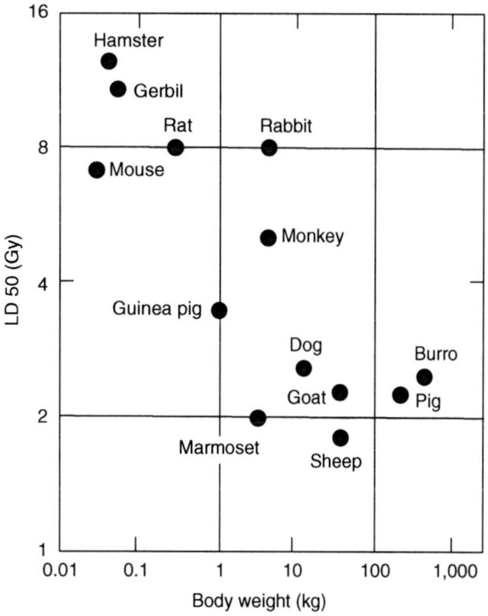

Figure 27.10.
Relation between body weight and radiation-induced LD50 (30 days postexposure) for selected mammals.

most sensitive. Body weight is an important modifier, and heavier mammals are usually most sensitive to radiation (Figure 27.10). Feral rodent populations are at risk from ionizing radiation through the reduction in numbers from direct kill and indirectly from the radiation-caused diminution of reproduction.

Low doses of ionizing radiation are beneficial to many species of mammals; effects of radiation hormesis include increased survival and longevity, lowered sterility, increased fecundity, and accelerated wound healing. Low doses of gamma irradiation cause irreversible injury to the eastern chipmunk (*Tamias striatus*), although the life-span is significantly longer. Acquired radioresistance after exposure to a low dose of ionizing radiation has been described in rats, mice, and yeast. In mice, for example, low doses of X-irradiation (not higher than 0.15 Gy) enhanced 30-day survival if given 2 months prior to a dose of 7.5 Gy. The low-dose exposure seems to stimulate the recovery of blood-forming stem cells after the second irradiation and favors a decrease in the incidence of bone-marrow death. The exact mechanisms of radiation hormesis are unknown because effects are not related to and not predictable from the high-dose exposure.

Irradiated small mammals released into the environment had a lower survival rate than laboratory populations, suggesting that the extrapolation from laboratory results may overestimate the radioresistance of free-ranging voles and other small animals because of the general level of stress in the population. The opposite was observed in eastern chipmunks given high sublethal doses of X-rays. Chipmunks had an overall reduction in mobility when they were released into the environment and a higher survival rate than controls, possibly owing to increased predation on the more mobile controls.

27.7.5.2 Carcinogenicity

The risk of the induction of cancer is a recognized somatic effect of low doses of ionizing radiation, as judged by epidemiological studies of Japanese survivors of the U.S. nuclear bombs and of Marshall Islanders, underground miners, and radium watch-dial workers. However, in a study on the occurrence of malignant tumors in Japanese children <10 years old and born between 1946 and 1982 to survivors of the atomic bombings in 1945, no statistically significant increase in malignant tumors was found in the children of parents exposed to >0.01 Sv whole-body radiation (mean gonadal exposure of 0.43 Sv) at the time of the atomic bombings when compared to a suitable control group. Nutritional status is important when treating malignant tumors. Unlike tumors of nonanemic individuals, tumors in anemic mice and humans frequently do not respond satisfactorily to radiotherapy.

Ionizing radiation induces basal cell carcinomas in skin and is active in the initiation of malignant tumors and in the progression of benign to malignant tumors. Skin has been widely used in studies of carcinogens because of its accessibility and the visibility of its tumors. All data on experimental radiogenic skin cancer in mice are on a relatively narrow and well-defined response curve; however, mouse skin is about 100 times more sensitive than human skin, strongly suggesting that appropriate animal models are necessary in the extrapolation of results to another species.

Thyroidal cancer in dogs and sheep has been induced with repeated administrations of ^{131}I, although single injections of ^{131}I failed to induce thyroid cancer in adult animals except in some strains of laboratory rodents. Humans with a prior history of ^{131}I and other radiation exposure in childhood are at a significantly higher risk of thyroid carcinogenesis, and females are at higher risk than males. The minimum latent period in humans is about 4 years, and neoplastic lesions may develop as late as 40 years after irradiation. Historically, the human thyroid received radiation from irradiation of the scalp for epilation (up to 0.5 Gy), thymus (up to 5 Gy), tonsils and adenoids (8 Gy), and facial acne (15 Gy). Higher doses of external irradiation (up to 50 Gy) were used during 1920–40 for the treatment of hyperthyroidism in adults and are still used for the treatment of cervical malignancies in people of all ages.

Radium-induced bone malignancies after exposure to ^{226}Ra are similar in beagles and humans, and the tibia in dogs is especially sensitive. "Radium jaw" has been described in humans as a late effect of accidental ingestion or therapeutic administration of long-lived radium isotopes, such as ^{226}Ra and ^{228}Ra, and is characterized by bone tumors, spontaneous fractures, and osteosclerosis. However, the short-lived ^{224}Ra (Tb1/2 of 3.6 days) produces similar effects in mice, suggesting that the events that trigger radium-induced bone disorders occur within days of incorporation, even though the consequence is a late effect.

Aerosol exposures of mice, rats, dogs, and hamsters to radon and its decay products resulted in lifetime shortening, pulmonary emphysema, pulmonary fibrosis, and respiratory tract carcinoma; damage to the skin and kidney were also reported, but the lung seems to be the primarily affected organ. Small mammals and birds that live in burrows containing radon-rich soils (9900 Bq/m^3 of ^{222}Rn) are expected to have additional 17 lung cancers per 1000 animals than animals not similarly exposed; however, tumors have not been widely reported in these species. Radon and ^{222}Rn daughters have caused problems to miners who work underground in uranium mines. These miners had an excessive incidence of disease of the respiratory system, including lung cancer. The problem is related to the emanation of radon into the mines and the decay of the radon, the short-lived radioactive daughters (^{216}Po, ^{214}Pb, ^{214}Bi, ^{214}Po) which attach to dust particles, eventually resulting in alpha-radiation exposure of the respiratory airways. A similar pattern was evident in rats exposed to ^{239}Pu. Rats exposed to $^{239}PuO_2$ aerosol of about 3700 Bq/lungs and examined 8–18 months after exposure had a very high frequency (as much as 80%) of malignant pulmonary neoplasms; genetic mutations were evident in 46% of the radiation-induced tumors.

The incidence of ovarian tumors in mice, guinea pigs, and rabbits increased after 3 years of chronic irradiation at doses as low as 1.1 mGy daily. Unlike other tumors, the induction of ovarian tumors depended on a minimum total dose and seemed to be independent of a daily dose. Radiation-induced neoplastic transformation of hamster cells may be associated initially with changes in expression of the genes modifying cytoskeletal elements.

27.7.5.3 Mutagenicity

In general, ionizing radiation has produced mutations in every plant and animal species studied. Some genetic risks are associated with exposures, but the risk of inducing a dominant genetic disease is quite small because radiation-induced mutations are primarily recessive and usually lethal. Residents living near the site of extensive mining and milling of uranium operations in Texas have an increased frequency of chromosomal aberrations and a reduced DNA repair capacity; uranium-238 was much higher in these areas when compared to reference sites, possibly as a result of leaching into the groundwater. The genetic doubling dose of radiation is the amount of acute or chronic radiation that doubles the naturally occurring spontaneous mutation rate each generation. For mice, the estimated genetic doubling dose equivalent is 1.35 Sv from acute exposures and 4.0 Sv from chronic exposures to radiation. For protection from radiation, the estimated genetic risks to humans have largely been based on data from mice. Studies of children of Japanese survivors of nuclear bomb explosions showed that the genetic doubling dose equivalent of acute gonadal radiation is about 2.0 Sv (1.69–2.23); from chronic radiation, this value is about 4.0 Sv. Based on results of the study of Japanese survivors of the nuclear explosions, it was concluded that there was no increase in the spontaneous mutation rate after parents were exposed. The high doubling dose of about 4 Sv estimated from these data is another way of stating that, relative to the assumed spontaneous rates, the rate of induction of mutations leading to the measured effects is too small. The transmission of radiation-induced genetic effects to offspring has not yet been demonstrated in any human population.

Specific point mutations were identified in ^{239}Pu-induced preneoplastic lesions and malignant neoplasms in the lungs of rats. Mice exposed to a single whole-body dose of 3 Gy produced a radiation-induced mutation that simultaneously generated distinct alleles of the limb deformity and agouti (grizzled fur color) loci, 2 developmentally important – but not adjoining – regions on a single chromosome. This phenomenon was probably associated with DNA breaks caused by inversion of a segment in another chromosome. The plasma membrane in immature oocytes of mice is the hypersensitive lethal target in producing radiation-induced genetic damage.

27.7.5.4 Organ and Tissue Damage

In the abdomen, the kidneys are one of the most sensitive organs to serious or fatal radiation-induced damage. The relatively high incidence of kidney disease among mature beagles injected with ^{226}Ra and its accompanying ^{210}Bi and ^{210}Po resulted from alpha irradiation of the kidneys by the substantial amount of ^{210}Po that was in the injected solution. Hepatic injury induced by ionizing radiation can be a life-threatening complication. The main responses of the liver to acute radiation exposure include enlargement, dilation of blood vessels, fluid accumulation, and histopathology. Damaging effects of ionizing radiation on the fetal cerebral cortex has been recognized for many years. The deleterious effects of ionizing radiation on the developing brain are prolonged and progressive. Doses <2 Gy of gamma radiation are harmful to the developing brain, and in humans mental retardation may occur from doses as low as 0.2 Gy between week 8 and 15 of gestation.

Irradiated white-footed mice (*Peromyscus leucopus*) frequently had atrophied gonads, degenerating fetuses in the uterus, and graying hair. High sublethal doses (7 Gy) of radiation to the pine vole (*Microtus pinetorium*) caused pelage graying wherein unpigmented hair from damaged follicles replaces molted pigmented hair. Pelage graying may decrease survival from increased predation, although this needs verification.

Human sperm chromosomes retain a high fertilizing ability after a high dose of X-irradiation, although mammalian spermatozoa have little capacity to repair DNA damage induced by radiation. Radiation-induced death of lymphoid cells in rats is associated with damage to the cell itself but may also be due to secretions from irradiation-activated natural killer cells which induce pycnosis and interphase death in lymphoid cells.

27.7.5.5 Behavior

Numerous behavioral measures have been evaluated for their usefulness in providing a sensitive index of exposure to ionizing radiation. Radiation-related mental retardation is the most likely type of behavioral abnormality in humans; sensitivity peaked between 8 and 15 weeks of conception and doses >0.4 Gy. No specific mechanism for the production of mental retardation has been established, although proposed mechanisms include the loss of cells, migration of neurons, and failure of synaptogenesis. In studies with rats, operant responses decreased (maintained by positive reinforcement such as food or water) at sublethal radiation doses (3.0–6.75 Gy) under various schedules of reinforcement. Disrupted operant responses under shock avoidance at >LD100 levels are reported in pigs and rhesus monkeys.

27.7.5.6 Absorption and Assimilation

The absorption, bioavailability, and retention of radionuclides in mammals are modified by the age, sex, species, and diet of the organism; season of collection; the chemical form of the radionuclide in tissue and blood; residence time in the digestive tract; preferential accumulation by selected organs and tissues; and many other variables.

Many radionuclides preferentially accumulate in certain organs or tissues, but the critical organ is different for different radionuclides: liver for ^{54}Mn, erythrocytes and spleen for ^{55}Fe, liver and kidney for cobalt nuclides, liver and prostate for ^{65}Zn, skeletal muscle for ^{137}Cs, and GI tract for ^{95}Zr. The persistence of radionuclides in mammals varies with the chemical form, kinetics, species, and other variables. Thus, the time for 50% persistence of selected radionuclides in whole-animal studies ranges from 19 h to 14 days for ^{134}Cs; 4 to 35 days for ^{137}Cs; 5 to 12 h for the short-lived component of ^{60}Co, 5 to 21 days for the long-lived component; 25 to 593 days for ^{90}Sr; and 4 to 26 days for ^{131}I.

Some radionuclides act antagonistically when administered together. The combined incorporation of ^{227}Ac and ^{227}Th at levels tested in mice shows a lower biological effect than the sums of the effects of the components administered singly. The less-than-additive effect is in good agreement with experiments with the incorporation of a mixture of β emitters, in which the effects are also less-than-additive. Uptake and retention characteristics of essential biological nutrients (i.e., H, C, P, I, K, Ca, Mn, Fe, Co, Zn) are largely controlled by biological processes. For example, ^{131}I regardless of route of administration is rapidly absorbed into the bloodstream and concentrated in the thyroid. Ionizing radiation associated with high levels of ^{131}I destroy the thyroid and thyroid hormone production.

Alkali metals (K, Rb, Cs) behave similarly and sometimes one is accumulated preferentially when another is deficient; a similar case is made for Sr and Ca. The most important alkali metal isotope is ^{137}Cs because of its long physical half-life (30 years) and its abundance as a fission product in fallout from nuclear weapons and in the inventory of a nuclear reactor or a fuel-reprocessing plant. Cesium behaves much like potassium. It is rapidly absorbed into the bloodstream and distributes throughout the active tissues of the body, especially muscle. The β and γ radiation from the decay of ^{137}Cs and its daughter, ^{137}Ba, result in essentially whole-body irradiation that harms bone marrow.

Beacause ^{226}Ra and ^{90}Sr are metabolic analogs of calcium, they are deposited in the skeleton; both isotopes are associated with bone cancers. In pregnant rats, the total amount of ^{226}Ra transferred from the dam to the

8–10 fetuses in a litter was low after a single injection and did not exceed 0.3% of the maternal content; the retained whole-body burden in dams was 53% at the first, 48% at the second, and 44% at the third pregnancy, mostly in the skeletal system. The rare earths (i.e., ^{144}Ce, ^{152}Eu, ^{140}La, ^{147}Pr, ^{151}Sm) are usually not effectively absorbed from the GI tract, and elimination is rapid. Cerium-144 is one of the more biologically hazardous radionuclides in this group because of its half-life (285 days) and the energetic β emissions from it and its daughter, ^{144}Pr.

The greatest uncertainty in dose estimates from the ingestion of long-lived alpha emitters is the values used for their fractional absorption from the GI tract. For transuranic elements, the fraction of the ingested material that was assimilated by the whole organism was always <0.01% and usually nearer 0.003%. The major hazard of plutonium nuclides to terrestrial organisms comes from inhalation; uptake by plants is low, and further uptake by humans through the gut is low. Americium-241 is an artificial, toxic bone-seeking radionuclide produced through beta decay of ^{241}Pu. Because of its long half-life, its high-energy alpha irradiation and its accumulation in the liver and skeleton, consideration should be given to ^{241}Am in risk estimates of latent effects, such as induction of liver cancers, bone cancers, and leukemias. In comparison with ^{226}Ra, ^{241}Am is 20 times more effective in reducing the lifespan in mice and 13 times more effective in the rate of death from bone cancer.

Although radon has long been known as a health hazard to miners in the uranium industry, it was only in the 1980s that radon contamination of buildings was recognized as widely distributed over the earth; however, years of exposure are required before a health problem develops. Exposure to radon-decay products can be expressed in two different ways: the amount of inhaled decay products (taking into account their potential to emit radiation energy) or the product of the time during which the decay products were inhaled and their concentration in the inhaled air. The potential alpha energy of the inhaled decay products may be expressed in joules (J). The potential alpha energy concentrations in air is expressed in joules per cubic meter; for radon in equilibrium with its decay product, this corresponds to 3700 Bq/m^3.

Rodents dosed with tungsten-185 excreted 80% in 24 h; bone was the major retention site; the half-time persistence ranged from 5.7 days in femurs of mice to 86 days in femurs of rats; some components in the bone of rats persisted with a half-time >3 years. Niobium-95 is produced directly by nuclear fission and indirectly by decay of ^{95}Zr. Routine discharges of ^{95}Nb from a nuclear fuel reprocessing plant in the United Kingdom in 1970 contributed about 5% of the bone-marrow dose to a 10-year-old child living in the vicinity. Gastrointestinal absorption of ^{95}Nb by adult guinea pigs was about 1.1%, and supports the values of 1% absorption in adults and 2% in infants now used to calculate percentage absorption of niobium isotopes by humans.

27.8 Proposed Criteria and Recommendations

For the protection from radiation, effects of radiation have been characterized as stochastic or nonstochastic. The probability of a stochastic effect – and not its severity – varies as a function of dose in the absence of a threshold, i.e., hereditary effects or carcinogenesis. The probability and severity of nonstochastic effects vary with dose, and a threshold for the dose exists, i.e., cataract of lens, nonmalignant damage to the skin, cell depletion in the bone marrow causing hematological deficiencies, gonadal cell damage leading to impairment of fertility, or pneumotis and pulmonary fibrosis following lung irradiation. The prevention of nonstochastic effects is achieved by setting dose-equivalent limits at sufficiently low levels so that no threshold dose is reached, not even after exposure for the whole of a lifetime or for the total period of a working life. Guides for the protection from radiation are also predicated on the effective half-life of each isotope, the critical organ, the fraction that reaches the critical organ by ingestion and inhalation, and the maximum tolerable whole body burdens, as judged by radionuclide concentrations in air, water, and diet.

At present, no radiological criteria or standards have been recommended or established for the protection of fish, wildlife, or other natural resources. All radiological criteria now promulgated or proposed are directed towards the protection of human health. It is generally assumed that humans are comparatively radiosensitive and that guides will probably also protect sensitive natural resources, although this needs verification. Numerous radiological criteria now exist for the protection of human health (Table 27.4). Most authorities agree that some adverse effects to humans are likely under the following conditions: >5 mSv whole-body exposure of women during the first 2 months of pregnancy; >50 mSv whole-body exposure in any single year or >2000 mSv in a lifetime; an annual inhalation intake by a 60-kg individual – in Bq/kg BW – that exceeds 0.67 ^{232}Th, 3.3 ^{241}Am, 3.3 ^{239}Pu, 16 ^{252}Cf, 33 ^{235}U, 1666 ^{90}Sr, 16,666 ^{60}Co, or 166,666 ^{32}P; an annual ingestion intake by a 60-kg individual – in Bq/kg BW – that exceeds 3333 ^{129}I, 16,666 ^{125}I, 16,666 ^{131}I, or 66,666 ^{137}Cs; or a total annual intake from all sources – in Bq/kg BW by a 60-kg person – that exceeds 66 ^{210}Pb, 166 ^{210}Po, 333 ^{226}Ra, 666 ^{230}Th, 833 ^{228}Th, or 1333 ^{238}U (Table 27.4).

Table 27.4. Recommended radiological criteria for the protection of human health.

Criterion and Other Variables	**Concentration or Dose**
AIR	
USA; radon-222	
Average	<0.0555 Bq (<1.5 pCi)/L or <55 Bq/m^3
Acceptable	<0.148 Bq (<4.0 pCi)/L
Allowable emission discharge	<0.74 Bq (<20 pCi)/m^2 per sec; should not increase the radon-222 concentration in air at or above any location outside the disposal site by >0.0185 Bq (>0.5 pCi)/L or >18 Bq (>500 pCi)/m^3
Unacceptable	>0.185 Bq (>5 pCi)/L
ASTRONAUTS	
Age 25–55 years; expected whole body career dose; females vs. males	1.0–3.0 Sv (100–300 rem) vs. 1.5–4.0 Sv (150–400 rem)
Adverse effects expected; lifetime exposure	> 2.0 Sv (>200 rem)
CANCER RISK AND BIRTH DEFECTS	
Projected 0.04% increase in cancers; 0.01% increase in birth defects	0.11 mSv (0.011 rem) whole body maximum per year; 0.69 mSv (0.069 rem) whole body over 30 years; or 1.00 mSv (0.1 rem) bone marrow over 30 years
Projected 0.18% increase in cancers; 0.07% increase in birth defects	0.51 mSv (0.05 rem) whole body maximum per year; 3.30 mSv (0.33 rem) whole body over 30 years; or 4.60 mSv (0.46 rem) bone marrow over 30 years

Table 27.4. cont'd

Criterion and Other Variables	Concentration or Dose
Projected 0.92% increase in cancers; 0.38% increase in birth defects	3.03 mSv (0.3 rem) whole body maximum per year; 19.0 mSv (1.9 rem) whole body over 30 years; or 23.0 mSv (2.3 rem) bone marrow over 30 years
Projected 4.4% increase in cancers; 1.8% increase in birth defects	20.1 mSv (2.0 rem) whole body maximum per year; 91.0 mSv (9.1 rem) whole body over 30 years; or 110.0 mSv (11 rem) bone marrow over 30 years
DIET	
All foods; maximum recommended values	
Adults, Italy	600 Bq (16,200 pCi) cesium-134, -137/kg FW
Children, Italy	370 Bq (10,000 pCi) cesium-134, -137/kg FW
Sweden, pre-Chernobyl	300 Bq (8100 pCi) cesium-134, -137/kg FW
Caribou; muscle; North America	<2260 Bq (<61,000 pCi) cesium-137/kg FW
Fish; Great Lakes; muscle	Dose of <0.02 μSv (0.00002 rem)/kg FW fish flesh equivalent to consumers
Fish; Sweden	<1500 Bq (<40,500 pCi) cesium-137/kg FW
Fraction of ingested dose absorbed; recommended maximum; selected isotopes	
Americium, curium, neptunium, plutonium, thorium	<0.05%
Americium, plutonium	<0.1%
Californium, and higher mass radionuclides	<0.1%
Uranium	<5.0%
Milk; maximum values	
Italy	370 Bq (10,000 pCi) cesium-134, -137/L
Japan	370 Bq (10,000 pCi) cesium-137/L
Sweden	300 Bq (8100 pCi) cesium-137/L
Meat and fish; Sweden; maximum values	1500 Bq (40,500 pCi) cesium-137/kg FW
Reindeer meat, game, animal meat, fish, berries, mushrooms; Sweden; post-Chernobyl; maximum values	1500 Bq (40,500 pCi) cesium-137/kg FW
Sheep, muscle	<1000 Bq (<27,000 pCi) cesium- 134, -137/kg FW
Sheep, muscle	<1000 Bq (<27,000 pCi) cesium-137/kg FW

Continued

Table 27.4. cont'd

Criterion and Other Variables	Concentration or Dose
DRINKING WATER	
Natural radioactivity; maximum allowed	
Radium-226, -228	0.185 Bq (5 pCi)/L
Gross alpha	0.555 Bq (15 pCi)/L
Artificial radioactivity; maximum allowed	
Gross beta	1.85 Bq (50 pCi)/L
Tritium (Hydrogen- 3)	740 Bq (20,000 pCi)/L
Strontium-90	0.296 Bq (8 pCi)/L
Great Lakes; maximum dose to consumers	10 μSv (0.001 rem)/year
GENERAL PUBLIC	
Annual effective dose[a]	<1 mSv (<0.1 rem)
Cesium-137; total intake	
Sweden	<50, 000 Bq (<1,350,000 pCi)/year, equivalent to <1 mSv (<0.1 rem)
North America	<300,666 Bq (<810,000 pCi)/year
United Kingdom	<400,000 Bq (<10,800,000 pCi)/year, equivalent to 5 mSv (0.5 rem)
Maximum permissible dose	
Eye lens	<15 mSv (<1.5 rem)/year
Skin	<50 mSv (<5 rem)/year
Whole body	
Individual, except students and pregnant women	<5 mSv (<0.5 rem)/year
Students	<1 mSv (<0.1 rem)/year
Pregnant women	<5 mSv (<0.5 rem) during the first 2 months of pregnancy
Population dose limits, genetic or somatic	<1.7 mSv (<0.17 rem) yearly average
GROUNDWATER, MAXIMUM ALLOWED	
Radium-226, -228	0.185 Bq (5 pCi)/L
Alpha-emitting radionuclides – including radium- 226, -228, but excluding radon isotopes	0.555 Bq (15 pCi)/L
Total beta and gamma radiation	Total annual whole body dose equivalent, or dose to any internal organ, <0.04 mSv (<0.004 rem), based on individual consumption of 2 L daily of drinking water from a groundwater source

Table 27.4. cont'd

Criterion and Other Variables	Concentration or Dose
RADIOACTIVE WASTES	
Dose limits from spent nuclear fuel or transuranic radioactive wastes	
Whole body	<0.25 mSv (<0.025 rem)/year
Thyroid	<0.75 mSv (<0.075 rem)/year
Any other critical organ	<0.25 mSv (<0.025 rem)/year
Stored for 10,000 years; maximum cumulative release allowed to the accessible environment per 1000 metric tons of heavy metal during storage	
Americium-241	3.7 trillion (T) Bq (100 TpCi)
Americium-243	3.7 TBq (100 TpCi)
Any alpha emitter with physical half-life >20 years	3.7 TBq (100 TpCi)
Any non-alpha emitter radionuclide with physical half-life >20 years	37.0 TBq (1000 TpCi)
Carbon-14	3.7 TBq (100 TpCi)
Cesium-135	37.0 TBq (1000 TpCi)
Cesium-137	37.0 TBq (1000 TpCi)
Iodine-129	3.7 TBq (100 TpCi)
Neptunium-237	3.7 TBq (100 TpCi)
Plutonium-238	3.7 TBq (100 TpCi)
Plutonium-239	3.7 TBq (100 TpCi)
Plutonium-240	3.7 TBq (100 TpCi)
Plutonium-242	3.7 TBq (100 TpCi)
Radium-226	3.7 TBq (100 TpCi)
Strontium-90	37.0 TBq (1000 TpCi)
Thorium-230	0.37 TBq (10 TpCi)
Thorium-232	0.37 TBq (10 TpCi)
Tin-126	37.0 TBq (1000 TpCi)
Uranium-233	3.7 TBq (100 TpCi)
Uranium-234	3.7 TBq (100 TpCi)
Uranium-235	3.7 TBq (100 TpCi)
Uranium-236	3.7 TBq (100 TpCi)
Uranium-238	3.7 TBq (100 TpCi)
Uranium by-product materials; maximum discharge rates allowed into water	

Continued

Table 27.4. cont'd

Criterion and Other Variables	Concentration or Dose
Radium-226, -228	0.185 Bq (5 pCi)/L
Gross alpha-particle activity, excluding radon and uranium isotopes	0.555 Bq (15 pCi)/L
Wastes from uranium fuel cycle entering the environment per billion watts/year of electrical energy produced by the fuel cycle; maximum allowed	
Krypton-85	1.85 TBq (50 TpCi)
Iodine-129	185 million Bq (5 billion pCi)
Plutonium-239 and other alpha emitting transuranics with Tb1/2> 1 year	2.69 million Bq (72.6 million pCi)
OCCUPATIONAL WORKERS	
Annual limit of intake[a]	
Inhalation vs. oral	
Americium-241	200 Bq (5400 pCi) vs. 50,000 Bq
Californium-252	1000 Bq (27,000 pCi) vs. 200,000 Bq
Cesium-137	6 million Bq (162 million pCi) vs. 4 million Bq
Cobalt-60	1 million Bq (27 million pCi) vs. 7 million Bq
Hydrogen-3	3 billion Bq (81 billion pCi) vs. 3 billion Bq
Iodine-125	2 million Bq (54 million pCi) vs. 1 million Bq
Iodine-129	300,000 Bq (8,100,000 pCi) vs. 200,000 Bq
Iodine-131	2 million Bq (54 million pCi) vs. 1 million Bq
Phosphorus-32	10 million Bq (270 million pCi) vs. 20 million Bq
Plutonium-239	200 Bq (5400 pCi) vs. 200,000 Bq
Polonium-210	20,000 Bq (540,000 pCi) vs. 100,000 Bq
Radium-226	20,000 Bq (540,000 pCi) vs. 70,000 Bq
Strontium-90	0.1 million Bq (2.7 million pCi) vs. 1.0 million Bq
Thorium-232	40 Bq (1000 pCi) vs. 30,000 Bq
Uranium-235	2000 Bq (54,000 pCi) vs. 500,000 Bq
Total intake from all sources:	
Canada	
Lead-210	<4000 Bq (<108,000 pCi)
Polonium-210	<10,000 Bq (<270,000 pCi)
Radium-226	<20,000 Bq (<540,000 pCi)
Thorium-228	<50,000 Bq (<1.35 million pCi)
Thorium-230	<40,000 Bq (<1.08 million pCi)

Table 27.4. cont'd

Criterion and Other Variables	Concentration or Dose
Thorium-232	<7000 Bq (<189,000 pCi)
Uranium-238	<80,000 Bq (<2.1 million pCi)
Effective dose[a]	
Average annual	20 mSv (2 rem), not to exceed 50 mSv (5 rem)
Five-year maximum	<100 mSv (<10 rem), not to exceed 50 mSv (5 rem) in any year
Maximum permissible dose	
Whole body	50 mSv (5 rem) in any one year
Long-term accumulation to age *N* years	$(N - 18) \times 50$ mSv (5 rem)
Skin	150 mSv (15 rem) in any one year
Hands	750 mSv (75 rem) in any one year; not to exceed 250 mSv (25 rem) in 3 months
Forearms	300 mSv (30 rem) in any one year; not to exceed 100 mSv (10 rem) in 3 months
Skin and hands	500 mSv (50 rem) annually
Other organs	150 mSv (15 rem) in any one year; not to exceed 50 mSv (5 rem) in 3 months
Pregnant women	5 mSv (0.5 rem) in gestation period
Eye lens	150 mSv (15 rem) annually
SOIL	
Radium-226; maximum allowed	<185 Bq (<5000 pCi)/kg over background in top 15 cm; <555 Bq (<15,000 pCi)/kg in soils at depth >15 cm
Total gamma; maximum allowed	<0.2 μSv (<0.00002 rem)/h over background

[a] The Annual Limit of Intake (ALI) for any radionuclide is obtained by dividing the annual average effective dose limit (20 mSv) by the committed effective dose (E) resulting from the intake of 1 Bq of that radionuclide.

Astronauts between 25 and 55 years of age usually receive an average career dose of 2.0 Sv (1.0–3.0 Sv in females; 1.5–4.0 Sv in males) and, theoretically, this may cause a life shortening of 2000–3000 days. Other environmental variables also result in life shortening and include cigarette smoking (2250 days), coal mining (1100 days), and being 30% overweight (1300 days); thus, models that assess the harm of a single variable – such as radiation – on life expectancy have to incorporate all known data and their interacting effects.

Environmental dose–response models and animal epidemiological data are most frequently used to assess the risk from ionizing radiation. In its ideal form, a risk assessment should clearly present the rationale for an estimate of risk and should include the recognition of the roles of assumptions, approximations, data, theories, models, and deductions in arriving at an inference and a discussion of the involved uncertainties. It now seems clear that current risk assessments of ionizing radiation hazards to all living organisms – not just humans – require additional data and reinterpretation of existing data. Specifically, more effort is needed in the following areas: (1) measurement of concentrations of naturally occurring radionuclides and natural background doses in the environment as a baseline for studies on radiation effects; (2) refinement of models of radionuclide transfer in food chains to aid in the assessment of radioactive releases from nuclear reactors and other point sources – including possible biomagnification by trophic components and turnover rates by receptor organisms; (3) continuance of protracted exposure studies to measure carcinogenesis in animal and human cell lines and the role of secondary factors – especially chemical agents – in radiation carcinogenesis; (4) research on radiation-induced recessive lethal mutations – the predominant type of radiation-induced mutation – and dominant mutation systems; (5) initiation of long-term studies to establish sensitive indicators of radiation stress on individuals and communities, including effects on growth and reproduction; (6) clarification of the role of enzymes and proteins in repair of radiation-damaged cellular DNA and of mechanisms of enzymatic reactions leading to altered nucleotide sequences; (7) reinterpretation of low-level chronic irradiation effects on developing embryos under rigorously controlled conditions; and (8) resolution of mathematical shape(s) of radiation dose–response curve(s).

27.9 Summary

Nuclear explosions and nuclear power production are the major sources of human radioactivity in the environment. Other sources include radionuclide use in medicine, industry, agriculture, education, and production; transport and disposal from these activities present opportunities for wastes to enter the environment. Dispersion of radioactive materials is governed by a variety of biogeochemical factors including winds, water currents, and biological vectors. Living organisms normally receive most of their external exposure to radiation from naturally occurring electromagnetic waves and their internal exposure from naturally occurring radionuclides such as potassium-40. Radiation exposure doses from natural sources of radiation are significantly modified by altitude, amount and type of radionuclides in the immediate vicinity, and route of exposure.

Radionuclide concentrations in representative field collections of biota tend to be elevated in the vicinity of nuclear fuel reprocessing, nuclear power production, and nuclear waste facilities; in locations receiving radioactive fallout from nuclear accidents and atmospheric nuclear tests; and near sites of repeated nuclear detonations. Radionuclide concentrations in field collections of living organisms vary significantly with organism age, size, sex, tissue, diet, and metabolism; season of collection; proximity to point source; and other biological, chemical, and physical variables. To date, no extinction of any animal population has been linked to high background concentrations of radioactivity.

The accident at the Chernobyl, Ukraine, nuclear reactor on April 26, 1986 released large amounts of radiocesium and other radionuclides into the environment, contaminating much of the Northern hemisphere, especially Europe. In the vicinity of Chernobyl, at least 30 people died, more than 115,000 others were evacuated, and consumption of milk and other foods was banned because of radiocontamination. At least 14,000 human cancer deaths are expected in Russia, Belarus, and the Ukraine as a direct result of Chernobyl. The most sensitive local ecosystems, as judged by survival, were the soil fauna, pine forest communities, and certain populations of rodents. Elsewhere, fallout from Chernobyl significantly contaminated freshwater and terrestrial ecosystems and flesh and milk of domestic livestock; in many cases, radionuclide concentrations in biological samples exceeded current radiation protection guidelines. Reindeer (*Rangifer tarandus*) in Scandinavia were among the most seriously afflicted by Chernobyl fallout, probably because their main food during winter (lichens) is an efficient absorber of airborne particles containing radiocesium. Some reindeer calves contaminated with ^{137}Cs from Chernobyl showed ^{137}Cs-dependent decreases in survival and increases in frequency of chromosomal aberrations. Although radiation levels in the biosphere are declining with time, latent effects of initial exposure – including an increased frequency of thyroid and other cancers – are now measurable. The full effect of the Chernobyl nuclear reactor accident on natural resources will probably not be known for at least several decades because of gaps in data on long-term genetic and reproductive effects and on radiocesium cycling and toxicokinetics.

A dose- and dose-rate dependent radiation effect on growth, survival, organ development, mutagenicity, fatal neoplasms, and other parameters exists for almost all organisms tested under laboratory conditions. Some discoveries suggest that low acute exposures of ionizing radiation may extend the life-span of certain species, although adverse genetic effects may occur under these conditions. In living organisms, the sensitivity to radiation is governed by ontogeny and phylogeny. Thus, rapidly dividing cells, characteristic of embryos and fetuses, are most radiosensitive and evolutionarily advanced organisms such as mammals are more radiosensitive than primitive organisms. Between species within each taxonomic grouping are large variations in sensitivity to acute and chronic exposures of ionizing radiation and in ability to retain selected radionuclides; these processes are modified by numerous biological and abiotic variables.

Radiosensitive terrestrial plants are adversely affected at single exposures of 0.5–1 Gy and at chronic daily exposures of 0.2–0.65 Gy. Terrestrial insects are comparatively resistant to ionizing radiation; some species show growth stimulation and development at acute doses of 2 Gy – a demonstrably harmful dose for many species of vertebrates. Among aquatic organisms, the developing eggs and young of freshwater fish are among the most sensitive tested organisms; death was observed at acute doses of 0.3–0.6 Gy and adverse effects on physiology and metabolism at chronic daily exposure rates of 0.01 Gy. The ability of aquatic organisms to concentrate radionuclides from the medium varies substantially with ecosystem, trophic level, radionuclide, proximity to radiation point source, and many other biological, chemical, and physical modifiers. In amphibians, radiation adversely affects limb regeneration, alters DNA metabolism, causes sterility, and increases the frequency of chromosomal aberrations. Mortality patterns in some species of amphibians begin to stabilize about 200 days after exposure to a single acute dose of ionizing radiation and cannot be evaluated satisfactorily in the typical 30-day postexposure period. In birds, adverse effects on growth were noted at chronic daily exposures as low as 0.9–1.0 Gy and on survival and metabolism at single exposures to 2.1 Gy. Genotoxic effects were associated with whole body loadings of 2520 Bq of cesium-137/kg in mallards (*Anas platyrhynchos*). The radionuclide retention in birds was modified by sex, season, and reproductive state. In mammals, embryos and fetuses of sensitive species were adversely affected at acute doses of 0.011–0.022 Gy. Humans exposed as fetuses to 0.18–0.55 Gy scored significantly lower on tests of intelligence.

No radiological criteria now exist for the protection of fish, wildlife, or other sensitive natural resources. All current guides for protection from radiation target human health and are predicated on the assumption that protection of comparatively radiosensitive humans confers a high degree of protection to other life forms. Most authorities agree that significant harmful effects to humans occur under the following conditions: exposure of the whole body of women during the first 2 months of pregnancy to >5 mSv; exposure of the whole body to >50 mSv in any single year or to >2000 mSv in a lifetime; annual inhalation intake by a 60-kg individual, in Bq/kg BW, of more than 0.7 of thorium-232, 3.3 of americium-241, 3.3 of plutonium-239, 16 of californium-252, 33 of uranium-235, 1670 of strontium-90, 16,670 of cobalt-60, or 166,670 of phosphorus-32; annual ingestion intake by a 60-kg individual, in Bq/kg BW, of more than 3330 of iodine-129, 16,670 of iodine-125, 16,670 of iodine-131, or 66,670 of cesium-137; or total annual intake, in Bq/kg BW, from all sources by a 60-kg person exceeds 66 of lead-210, 166 of polonium-210, 333 of radium-226, 670 of thorium-230, 830 of thorium-228, or 1330 of uranium-238.

Current risk assessments of ionizing radiation hazards to living organisms requires additional data and reinterpretation of existing data. Specifically, more effort seems needed in eight areas: (1) establishing a baseline for studies on radiation through measurement of naturally occurring radionuclides and natural background radiation doses; (2) refining radionuclide food chain transfer models; (3) measuring the role of chemical agents in radiation-induced carcinogenesis; (4) accelerating research on radiation-induced lethal mutations; (5) initiating long term studies to establish sensitive indicators of radiation stress on individuals and ecosystems; (6) clarifying the role of enzymes and proteins in repair of radiation-damaged cellular DNA; (7) reinterpreting embryotoxic effects of low level chronic irradiation; and (8) resolving the mathematical shapes of radiation dose–response curves.

Glossary

Actinides. Elements of atomic numbers 89–103 (Ac, Th, Pa, U, Np, Pu, Am, Cm, Bk, Cf, Es, Fm, Md, No, Lw).

Activity. The activity of a radioactive material is the number of nuclear disintegrations per unit time. Up to 1977, the accepted unit of activity was the curie (Ci), equivalent to 37 billion disintegrations/s – a number that approximated the activity of 1 g of radium-226. The present unit of activity is the becquerel (Bq), equivalent to 1 disintegration/s.

Alpha (α) particles. An α particle is composed of 2 protons and 2 neutrons, with a charge of +2; essentially, it is a helium nucleus without orbital electrons. Alpha particles usually originate from the nuclear decay of radionuclides of atomic number >82, and are detected in samples containing U, Th, or Ra. Alpha particles react strongly with matter and consequently produce large numbers of ions per unit length of their paths; as a result, they are not very penetrating and will traverse only a few centimeters of air. Alpha particles are unable to penetrate clothing or the outer layer of skin; however, when internally deposited, α particles are often more damaging than most other types of radiations because comparatively large amounts of energy are transferred within a very small volume of tissue. Alpha particle absorption involves ionization and orbital electron excitation. Ionization occurs whenever the α particle is sufficiently near an electron to pull it from its orbit. The α also loses kinetic energy by exciting orbital electrons with interactions that are insufficient to cause ionization.

Atom. The smallest part of an element that has all the properties of that element. An atom consists of one or more protons and neutrons (in the nucleus) and of one or more electrons.

Atomic number. The number of electrons outside the nucleus of a neutral (nonionized) atom and the number of protons in the nucleus.

Becquerel (Bq). The present accepted unit of activity is the becquerel, equivalent to 1 disintegration/s. About 0.037 Bq = 1 picocurie.

Beta (β) particles. Beta particles are electrons that are spontaneously ejected from the nuclei of radioactive atoms during the decay process. They may either be positively or negatively charged. A positively charged beta (β^+), called a *positron*, is less frequently encountered than its negative counterpart, the *negatron* (β^-). The *neutrino*, a small particle, accompanies beta emission. The neutrino has very little mass and is electrically neutral. But neutrinos conduct a variable part of the energy of transformation and accounts for the variability in kinetic energies of beta particles emitted from a given radionuclide. Positrons (β^+) are emitted by many of the naturally and artificially produced radionuclides; they are considerably more penetrating than α particles, but less penetrating than X-rays and γ rays. Beta particles interact with other electrons as well as nuclei in the travel medium. The ultimate fate of a beta particle depends upon its charge. Negatrons, after their kinetic energy is spent, combine with a positively charged ion or become free electrons. Positrons also dissipate kinetic energy through ionization and excitation; the collision of positrons and electrons causes annihilation and release of energy equal to the sums of their particle masses.

Breeder reactor. A nuclear chain reactor in which transmutation produces a greater number of fissionable atoms than the number of consumed parent atoms.

Cosmic rays. Highly penetrating radiations that originate in outer space.

Curie (Ci). The Ci is equal to that quantity of radioactive material producing 37 billion nuclear transformations/s. One millicurie (mCi) = 0.001 Ci; 1 microcurie (μCi) = 1 millionth of a Ci; 1 picocurie (pCi) = 1 millionth of a millionth Ci = 0.037 disintegrations/s. About 27 pCi = 1 becquerel (Bq).

Decay. Diminution of a radioactive substance because of nuclear emission of α or β particles or of γ rays.

Decay product. A nuclide resulting from the radioactive disintegration of a radionuclide and found as the result of successive transformations in a radioactive series. A decay product may be either radioactive or stable.

Effective dose equivalent. The weighted sum, in Sieverts, of the radiation dose equivalents in the most radiosensitive organs and tissues, including gonads, active bone marrow, bone surface cells, and the lung.

Electron. An electron is a negatively charged particle with a diameter of 10^{-12} cm. Every atom consists of one nucleus and one or more electrons. Cathode rays and negatrons are electrons.

Electron-volt (eV). Energy acquired by any charged particle that carries unit electronic charge when it falls through a potential difference of 1 volt. One eV = 1.602×10^{-19} J.

Fission. The splitting of an atomic nucleus into two fragments that usually releases neutrons and γ rays. Fission may occur spontaneously or may be induced by capture of bombarding particles. Primary fission products usually decay by β particle emission to radioactive daughter products. The chain reaction that may result in controlled burning of nuclear fuel or in an uncontrolled nuclear weapons explosion results from the release of 2 or 3 neutrons/fission. Neutrons cause additional fissile nuclei in the vicinity to fission, producing still more neutrons, in turn producing still more fissions. The speed of the chain reaction is governed by the density and geometry of fissile nuclei and of materials that slow or capture the neutrons. In nuclear reactors, neutron-absorbing rods are inserted to various depths into the reactor core. A nuclear explosion is not physically possible in a reactor because of fuel density, geometry, and other factors.

Fusion. A nuclear reaction in which smaller atomic nuclei or particles combine to form larger ones with the release of energy from mass transformation.

Gamma (γ) rays. Gamma rays have electromagnetic wave energy that is similar to but higher than the energy of X-rays. Gamma rays are highly penetrating and able to traverse several centimeters of lead. See **Photons.**

Genetically significant dose (GSD). A radiation dose that, if received by every member of the population, would produce the same total genetic injury to the population as the actual doses that are received by the various individuals.

Gray (Gy). 1 Gy = 1 J/kg = 100 rad.

Half-life. The average time in which half the atoms in a sample of radioactive element decay.

Hertz (Hz). A measure of frequency equal to 1 cycle/s.

Indirectly ionizing particles. Uncharged particles such as neutrons or photons that directly liberate ionizing particles or initiate nuclear transformations.

Ion. An atomic particle, atom, or chemical radical with an either negative or positive electric charge.

Ionization. The process by which neutral atoms become either positively or negatively charged by the loss or by the gain of electrons.

Isomer. One of two or more radionuclides having the same mass number and the same atomic number but with different energies and radioactive properties for measurable durations.

Isotope. One of several radionuclides of the same element (i.e., with the same number of protons in their nuclei) with different numbers of neutrons and different energy contents. A single element may have many isotopes. Uranium, for example, may appear naturally as ^{234}U (142 neutrons), ^{235}U (143 neutrons), or ^{238}U (146 neutrons); however, each uranium isotope has 92 protons.

Joule (J). $1 J = 10^7$ ergs.

Latent period. Period of seeming inactivity between time of exposure of tissue to an acute radiation dose and the onset of the final stage of radiation sickness.

Linear energy transfer (LET). A function of the capacity of the radiation to produce ionization. LET is the rate at which charged particles transfer their energies to the atoms in a medium and a function of the energy and velocity of the charged particle. See **Radiation dose**.

Linear hypothesis. The assumption that any radiation causes biological damage in the direct proportion of dose to effect.

Mass number. The total number of neutrons and protons in the nucleus of the element, which is equal to the sum of the atomic number and the number of neutrons.

Meson. Particles of mass that are intermediate between the masses of the electron and proton.

Neutrinos. Neutrinos and antineutrinos are formed whenever a positron particle is created in a radioactive decay; they are highly penetrating.

Neutrons. Neutrons are electrically neutral particles that consist of an electron and a proton and are not affected by the electrostatic forces of the atom's nucleus or orbital electrons. Because they have no charge, neutrons readily penetrate the atom and may cause a nuclear transformation. Neutrons are produced in the atmosphere by cosmic ray interactions and combine with nitrogen and other gases to form carbon-14, tritium, and other radionuclides. A free neutron has a lifetime of about 19 min, after which it spontaneously decays to a proton, a β particle, and a neutrino. A high-energy neutron that encounters biological material is apt to collide with a proton with sufficient force to dislodge the proton from the molecule. The recoil proton may then have sufficient energy to cause secondary damage through ionization and excitation of atoms and molecules along its path.

Nucleus. The dense central core of the atom in which most of the mass and all of the positive charge is concentrated. The charge on the nucleus distinguishes one element from another.

Photons. X-rays and gamma (γ) rays, collectively termed photons, are electromagnetic waves with shorter wavelengths than other members of the electromagnetic spectrum such as visible radiation, infrared radiation, and radiowaves. X- and γ photons have identical properties, behavior, and effects. Gamma rays originate from atomic nuclei, but X-rays arise from the electron shells. All photons travel at the speed of light, but energy is inversely proportional to wavelength. The energy of a photon directly influences its ability to penetrate matter. Many types of nuclear transformations are accompanied by γ-ray emission. For example, α and β decay of many radionuclides is frequently accompanied by γ photons. When a parent radionuclide decays to a daughter nuclide, the nucleus of the daughter frequently contains excess energy and is unstable; stability is usually achieved through release of one or more γ photons, a process called isometric transition. The daughter nucleus decays from one energy state to another without a change in atomic number or weight. The most probable fate of a photon with an energy higher than the binding energy of an encountered electron is photoelectric absorption, in which the photon transfers its energy to the electron and photon existence ends. As with ionization from any process, secondary radiations that are initiated by the photoelectron produce additional excitation of orbital electrons.

Planck's constant (h). A universal constant of nature that relates the energy of a photon of radiation to the frequency of the emitting oscillator. Its numerical value is about 6.626×10^{-27} ergs/s.

Positron. A positively charged particle of equal mass to an electron. Positrons are created either by the radioactive decay of unstable nuclei or by collision with photons.

Proton. A positively charged subatomic particle with a mass of 1.67252×10^{-24} g that is slightly less than the mass of a neutron but about 1836 times greater than the mass of an electron. Protons are identical to hydrogen nuclei; their charge and mass make them potent ionizers.

Radiation. The emission and propagation of energy through space or through a material medium in the form of waves. The term also includes subatomic particles, such as α, β, and cosmic rays and electromagnetic radiation.

Radiation absorbed dose (rad). Radiation-induced damage to biological tissue results from the absorption of energy in or around the tissue. The amount of energy absorbed in a given volume of tissue is related to the types and numbers of radiations and the interactions between radiations and tissue atoms and molecules. The fundamental unit of the radiation absorbed dose is the rad; 1 rad = 100 erg (absorbed)/g material. In the latest nomenclature, 100 rad = 1 Gray (Gy).

Radiation dose. The term radiation dose can mean several things, including absorbed dose, dose equivalent, or effective dose equivalent. The absorbed dose of radiation is the imparted energy per unit mass of the irradiated material. Until 1977, the *rad* was the unit of absorbed dose, wherein 1 rad = 0.01 J/kg. The present unit of absorbed dose is the Gray (Gy), equivalent to 1 J/kg. Thus, 1 rad = 0.01 J/kg = 0.01 Gy. Different types of radiation have different Relative biological effectiveness (RBE). The RBE of one type of radiation in relation to a reference type of radiation (usually X or γ) is the inverse ratio of the absorbed doses of the two radiations needed to cause the same degree of the biological effect for which the RBE is given. Regulatory agencies have recommended certain values of RBE for radiation protection and absorbed doses of various radiations are multiplied by these values to arrive at radioprotective doses. The unit of this weighted absorbed dose is the roentgen equivalent man (rem). The dose equivalent is the product of the absorbed dose and a quality factor (Q), and its unit is the rem. The quality factor is a function of the capacity to produce ionization, expressed as the linear energy transfer (LET). A Q value is assigned to each type of radiation: 1 to X-rays, γ rays, and β particles; 10 to fast neutrons; and 20 to α particles and heavy particles. The new unit of the effective dose equivalent is the Sievert (Sv), replacing rem, where 1 Sv = 100 rem. In addition to absorbed dose and dose equivalent, there is also the exposure. Exposure is the total electrical charge of ions of one sign produced in air by electrons liberated by X or gamma rays per unit mass of irradiated air. The unit of exposure is C/kg, but the old unit, the roentgen (R) is still in use. One roentgen = 2.58×10^{-4} C/kg.

Radioactivity. The process of spontaneous disintegration by a parent radionuclide, which releases one or more radiations and forms a daughter nuclide. When half the radioactivity remains, that time interval is designated the half-life (Tb1/2). The Tb1/2 value gives some insight into the behavior of a radionuclide and into its potential hazards.

Radionuclide. An atom that is distinguished by its nucleus composition (number of protons, number of neutrons, energy content), atomic number, mass number, and atomic mass.

Relative biological effectiveness (RBE). The biological effectiveness of any type of ionizing radiation in producing a specific damage (i.e., leukemia, anemia, carcinogenicity). See **Radiation dose.**

Roentgen (R). 1 R = 2.58×10^{-4} C/kg air = production by X or γ rays of one electrostatic unit of charge per cm^3 of dry air at 0°C and 760 mmHg = 0.87 rad in air.

Roentgen equivalent man (rem). The amount of ionizing radiation of any type that produces the same damage to humans as 1 roentgen of radiation. One rem = 1 roentgen equivalent physical (rep)/relative biological effectiveness (RBE). In the latest nomenclature, 100 rem = 1 Sievert (Sv).

Roentgen equivalent physical (rep). One rep is equivalent to the amount of ionizing radiation of any type that results in the absorption of energy of 93 ergs/g, and is approximately equal to 1 roentgen of X-radiation in soft tissue.

Shell. Extranuclear electrons are arranged in orbits at various distances from the nucleus in a series of concentric spheres called shells. In order of increasing distance from the nucleus the shells are designated the K, L, M, N, O, P, and Q shells; the number of electrons that each shell can contain is limited.

Sievert (Sv). New unit of dose equivalent. 1 Sv = 100 rem = 1 J/kg. See **Radiation dose**.

Specific activity. The ratio between activity (in number of disintegrations/min) and the mass (in grams) of material giving rise to the activity. Biological hazards of radionuclides are directly related to their specific activity and are expressed in Bq/kg mass.

Threshold hypothesis. A radiation-dose- consequence hypothesis that holds that biological radiation effects will occur only above some minimum dose.

Transmutation. A nuclear change that produces a new element from an old one.

Transuranic elements. Elements of atomic number >92. All are radioactive and produced artificially; all are members of the actinide group.

X-rays. See **Photons**.

SELENIUM[a]

Chapter 28

28.1 Introduction

Selenium poisoning is an ancient and well-documented disease. Signs of it were reported among domestic livestock by Marco Polo in western China near the borders of Turkestan and Tibet in about the year 1295; among livestock, chickens, and children in Columbia, South America, by Father Pedro Simon in 1560; among human adults in Irapuato, Mexico, in about 1764; and among horses of the U.S. Cavalry in South Dakota in 1857 and again in 1893. In 1907–08, more than 15,000 sheep died in a region north of Medicine Bow, Wyoming, after grazing on seleniferous plants. The incidents have continued, and the recent technical literature abounds with isolated examples of selenosis among domestic animals and wildlife.

Selenium (Se) was first identified as an element in 1817 by the Swedish chemist Berzelius. It is now firmly established that selenium is beneficial or essential in amounts from trace to micrograms per kilo (ppb) concentrations for humans and some plants and animals, but toxic at some concentrations present in the environment. Selenium deficiency was reported among cattle grazing in the Florida Everglades, which showed evidence of anemia, slow growth, and reduced fertility. Selenium deficiency has been demonstrated in Atlantic salmon, *Salmo salar*, in various species of deer in Florida and Washington, and in free-ranging ungulates in Washington State including moose (*Alces alces*) and bighorn sheep (*Ovis canadensis*). Conversely, calves of Indian buffaloes (*Bubalus* sp.) died of selenium poisoning after eating rice husks grown in naturally seleniferous soils. Adverse effects of excess selenium are reported in reproduction of cattle, monkeys, sheep, swine, rats, and hamsters, including fetal and maternal death, and a dramatic increase in developmental abnormalities. Severe reproductive and developmental abnormalities were observed in aquatic birds nesting at selenium-contaminated irrigation drainwater ponds in the San Joaquin Valley, California. Accumulation of more than 8.0 mg Se/kg dry weight in fish gonads is the probable cause of reduced reproduction and subsequent species disappearances in Belews Lake, North Carolina, and the endangered razorback sucker (*Xyrauchen texanus*) from the Green River, Utah, in 1991.

Most authorities agree that: (1) selenium is widely distributed in nature, being especially abundant with sulfide minerals of various metals, such as iron, lead, and copper; (2) the major source of environmental selenium is the weathering of natural rock; (3) the amount of selenium entering the atmosphere as a result of anthropogenic activities is estimated to be 3500 metric tons annually, of which most is attributed to combustion of coal and the irrigation of high-selenium soils for crop production. However, aside from highly localized contamination, the contribution of selenium by human activities is small in comparison with that attributable to natural sources; (4) all authorities agree that selenium may favorably or adversely affect growth, survival,

[a] All information in this chapter is referenced in the following sources:

Eisler, R. 1985. Selenium hazards to fish, wildlife, and invertebrates: a synoptic review. *U.S. Fish Wildl. Serv. Biol. Rep. 85(1.5)*, 1–57.

Eisler, R. 2000. Selenium. Pages 1649–1705 in *Handbook of Chemical Risk Assessment: Health Hazards to Humans, Plants, and Animals. Volume 3, Metalloids, Radiation, Cumulative Index to Chemicals and Species.* Lewis Publishers, Boca Raton, Florida.

and reproduction of algae and higher plants, bacteria and yeasts, crustaceans, mollusks, insects, fish, birds, and mammals (including humans); (5) sensitivity to selenium and its compounds is extremely variable in all classes of organisms and, except for some instances of selenium deficiency or of selenosis, metabolic pathways and modes of action are imperfectly understood. For example, selenium indicator plants can accumulate selenium to concentrations of thousands of parts per million (mg/kg) without ill effects; in these plants selenium promotes growth, whereas in crop plants accumulations as low as 25.0–50.0 mg/kg may be toxic. Thus, plants and waters high in selenium are considered potentially hazardous to livestock and to aquatic life and other natural resources in seleniferous zones.

28.2 Environmental Chemistry

Selenium is characterized by an atomic weight of 78.96, an atomic number of 34, a melting point of 271°C, a boiling point of 685°C, and a density of 4.26–4.79. Chemical properties, uses, and environmental persistence of selenium were documented by a number of researchers with general agreement on four points. First, that selenium chemistry is complex, and that additional research is warranted on chemical and biochemical transformations among valence states, allotropic forms, and isomers of selenium. Second, that selenium metabolism and degradation is significantly modified by interaction with heavy metals, agricultural chemicals, microorganisms, and a variety of physicochemical factors. Third, that anthropogenic activities (including fossil fuel combustion and metal smelting) and naturally seleniferous areas pose the greatest hazards to fish and wildlife. And fourth, that selenium deficiency is not as well documented as selenium poisoning, but may be equally significant.

Selenium chemistry is complex. In nature, selenium exists: as six stable isotopes (Se-74, -76, -77, -78, -80, and -82), of which Se-80 and -78 are the most common, accounting for 50.0 and 23.5%, respectively; in three allotropic forms; and in five valence states. Changes in the valence state of selenium from −2 (hydrogen selenide) through 0 (elemental selenium), +2 (selenium dioxide), +4 (selenite), and +6 selenate) are associated with its geologic distribution, redistribution, and use. Soluble selenates occur in alkaline soils, are slowly reduced to selenites, and are then readily taken up by plants and converted into organoselenium compounds including selenomethionine, selenocysteine, dimethyl selenide, and dimethyl diselenide. In drinking water, selenates represent the dominant chemical species. Selenites are less soluble than the corresponding selenates and are easily reduced to elemental selenium. In seawater, selenites are the dominant chemical species under some conditions. Selenium dioxide is formed by combustion of elemental selenium present in fossil fuels or rubbish. Selenium is the most strongly enriched element in coal, being present as an organoselenium compound, a chelated species, or as an adsorbed element. On combustion of fossil fuels, the sulfur dioxide formed reduces the selenium to elemental selenium. Elemental selenium is insoluble and largely unavailable to the biosphere, although it is still capable of satisfying metabolic nutritional requirements. Hydrogen selenide is highly toxic (at 1.0–4.0 μg/L in air), unstable, acidic, and irritative. Selenides of mercury, silver, copper, and cadmium are very insoluble, although their insolubility may be the basis for the reported detoxification of methylmercury by dietary selenite, and for the decreased heavy metal toxicity associated with selenite. Metallic selenides are thus biologically important in sequestering both Se and heavy metals in a largely unavailable form.

In areas of acid or neutral soils, the amount of biologically available selenium should steadily decline. The decline may be accelerated by active agricultural or industrial practices. In dry areas, with alkaline soils and oxidizing conditions, elemental selenium and selenides in rocks and volcanic soils may oxidize sufficiently to maintain the availability of biologically active selenium. Concentrations of selenium in water are a function of selenium levels in the drainage system and of water pH. In Colorado, for example, streams with pH 6.1–6.9 usually contain <1.0 μg Se/L, but those with pH 7.8–8.2 may contain 270.0–400.0 μg/L. Selenium volatilizes from soils at rates that are modified by temperature, moisture, time, season of year, concentration of water-soluble selenium, and microbiological activity. Conversion of inorganic and organic selenium compounds to volatile selenium compounds (such as dimethyl selenide, dimethyl diselenide, and an unidentified compound) by microorganisms has been observed in lake sediments of the Sudbury area of Ontario. This conversion may have been effected by pure cultures of *Aeromonas*, *Flavobacterium*, *Pseudomonas*, or an unidentified fungus, all of which are found in methylated lake sediments. Production of volatile selenium is temperature dependent. Compared with the amount of $(CH_3)_2$ Se produced at an incubation temperature of 20°C, 25% less was produced at 10°C and 90% less at 4°C. Details of selenium reduction and oxidation by microorganisms are not clear. One suggested mechanism for selenite reduction in certain microorganisms involves attachment to a carrier protein and transformation from selenite to elemental selenium, which in turn may be oxidized to selenite by the action of *Bacillus* spp., as one example. It is apparent that much additional research on this problem is warranted. It now appears that selenates and selenites are absorbed by plants, reduced, and then incorporated in amino acid synthesis. The biological availability of selenium is higher in plant foods than in foods of animal origin. The net effect of soil, plant, and animal metabolism is to convert selenium to inert and insoluble forms such as elemental selenium, metallic selenides, and complexes of selenite with ferric oxides.

Selenium was used in the early 1900s as a pesticide to control plant pests, and is still used sparingly to control pests of greenhouse chrysanthemums and carnations. It has been used to control cotton pests (in Trinidad), mites and spiders that attack citrus, and mites that damage apples. Although no insect-resistant strains have developed, the use of selenium pesticides has been discontinued, owing to their stability in soils and resultant contamination of food crops, their high price, and their proven toxicity to mammals and birds. In Canada and France, sodium selenite applied to the soil to discourage deer from browsing conifer seedlings when deer numbers were high was unsuccessful, and should be avoided. Selenium shampoos, which contain about 1% selenium sulfide, are still used to control dandruff in humans and dermatitis and mange in dogs. Selenium is used extensively in the manufacture and production of glass, pigments, rubber, metal alloys, textiles, petroleum products, medical therapeutic agents, and photographic emulsions.

Domestic consumption of selenium in 1981 exceeded 453,000 kg. About 50% was used in electronic and copier components, 22% in glass manufacturing, 20% in chemicals and pigments, and 8% miscellaneous. In 1987, world production of selenium was about 1.4 million kg. In 1986, 46% of the global selenium produced was used in the semiconductor and photoelectric industries; 27% in the glass industry to counter coloration impurities from iron; 14% in pigments; and 13% in medicine, in antidandruff shampoos, as catalysts in pharmaceutical preparations, in nutritional feed additives for poultry and livestock, and in pesticide formulations.

Air and surface waters generally contain nonhazardous concentrations of selenium. Significant increases of selenium in specific areas are attributed exclusively to industrial sources, and to leaching of groundwater from seleniferous soils. In the United States, about 4.6 million kg of selenium are released annually into the environment: 33% from combustion of fossil fuels, 59% from industrial losses, and 8% from municipal wastes. Of the total, about 25% is in the form of atmospheric emissions, and the rest in ash. Mining and smelting of copper–nickel ores at Sudbury, Ontario, Canada, alone releases about 2.0 metric tons of selenium to the environment daily, and probably represents the greatest point source of selenium release in the world. In 1977, 680,000 kg of selenium was produced at Sudbury, but only about 10% was recovered, suggesting that about 90% was lost to the environment. Of the amount lost, perhaps 50 metric tons was dispensed into the atmosphere, probably as selenium dioxide (airborne Se levels 1–3 km from Sudbury were as high as 6.0 $\mu g/m^3$). The rest was probably associated with mine tailings, wastewater, and scoria, and is a local source of selenium contamination, most notably in lakes. The present annual rates of selenium accumulation in lake sediments in the Sudbury area range from 0.3 to 12.0 mg/m^2; these deposition rates exceed those of pre-colonial times by factors of 3–18, and are among the highest recorded in North America.

Selenium is a serious hazard to livestock and probably to people in a wide semiarid belt that extends from inside Canada southward across the United States into Mexico. Selenium tends to be present in large amounts in areas where the soils have been derived from Cretaceous rocks. Total selenium in such soils averages about 5.0 mg/kg, but is sometimes as high as 80.0 mg/kg. Lack of rainfall has prevented the solution of the selenium minerals and the removal of their salts in drainage waters. In some areas, modern fertilization practices and the buildup of sulfates in the soil due to acid precipitation partly lessen the availability of selenium to plants and forage crops. In the United States, highly seleniferous natural areas (200.0–300.0 μg/kg in forage) are most abundant in the Rocky Mountain and High Central Plains areas; areas with lower concentrations (20.0–30.0 μg/kg) in forage are typical in the Pacific Northwest and the Southeast. However, huge variations are not uncommon from one specific location to another. Among plants, primary and secondary selenium accumulators are almost always implicated in cases of acute or chronic selenium poisoning of livestock. Primary selenium accumulator plants, such as various species of *Astragalus*, *Oonopsis*, *Stanelya*, *Zylorhiza*, and *Machaeranthera*, may require 1.0–50.0 mg Se/kg in either soil or water for growth, and may contain 100.0–10,000.0 mg Se/kg as a glutamyl dipeptide or selenocystanthionine. Secondary accumulator plants (representative genera: *Aster*, *Gutierrezia*, *Atriplex*, *Grindelia*, *Castillaja*, and *Comandra*) grow in either seleniferous or nonseleniferous soils and may contain 25.0–100.0 mg Se/kg. Nonaccumulator plants growing on seleniferous soils contain 1.0–25.0 mg Se/kg fresh weight. Meat and eggs of domestic animals may contain 8.0–9.0 mg Se/kg in seleniferous areas, compared with 0.01–1.0 mg/kg in nonseleniferous areas. Tissues from animals maintained on high-selenium feeds generally contain 3.0–5.0 mg Se/kg fresh weight vs. up to 20.0 mg/kg in animals dying of selenium poisoning.

Selenium is nutritionally important as an essential trace element, but is harmful at slightly higher concentrations. Although normal selenium dietary levels required to ensure human health range from 0.04 to 0.1 mg/kg, toxicity may occur if food contains as little as 4.0 mg/kg. Minimum selenium concentrations required are usually higher in livestock than in humans. In areas with highly seleniferous soils, excess selenium is adsorbed onto a variety of plants and grains and can be fatal to grazing livestock. There is general agreement, however, that selenium inadequacy can be of greater concern to health than is selenium toxicity. Selenium has a comparatively short effectual biological life in various species of organisms for which data are available. Studies with radioselenium-75 indicated that its biological half-life is 10–64 days: 10 in pheasants, 13 in guppies and voles, 15 in ants, 27 in eels, 28 in leeches, and 64 in earthworms. Many investigators concluded that the greatest current and direct use of selenium is in the transportation of grains grown in seleniferous areas to selenium-deficient areas as animal and human food.

28.3 Concentrations in Field Collections

Selenium concentrations in nonbiological materials extend over several orders of magnitude. In terrestrial materials, concentrations in excess of 5.0 mg Se/kg are routinely recorded in meteorites, copper-nickel ores, coal and other fossil fuels, lake sediments in the vicinity of a nickel-copper smeltery, and in sediments of flyash settling ponds. Water concentrations exceeding 50.0 μg Se/L have been documented in groundwater, especially in areas with seleniferous soils, in sewage wastes, in irrigation drainwater, and in water of flyash settling ponds. Selenium concentrations in air samples were >0.5 μg/m^3 in the vicinity of selenium production plants, and these were at least 500 times higher than in a control area.

As a result of natural and anthropogenic processes, comparatively high concentrations of selenium in nonbiological materials may offer protection or pose significant risks to fish and wildlife. In Finland, for example, agricultural fertilizers were supplemented with 6.0–16.0 mg Se/kg beginning in 1985. In Finnish lakes, selenium concentrations have increased in sediments and fish muscle from this activity and from atmospheric fallout, but no adverse biological effects were observed. In Sweden, selenium treatment raised the lake water concentrations from 3.0 μg Se/L at the start to 5.0 μg/L. This treatment lowered mercury concentrations in mercury-contaminated northern pike (*Esox lucius*) and yellow perch (*Perca flavescens*) in treated Swedish lakes by 60–85%. Selenium is normally present in surface waters at about 0.1–0.3 μg/L; however, at 1.0–5.0 μg/L it can biomagnify in aquatic food chains and pose a concentrated dietary source of selenium that is toxic to fish and wildlife.

It is emphasized that selenium concentrations in all organisms tended to be significantly higher when collected from locales having certain characteristics: highly seleniferous soils or sediments; high human population densities; heavy accumulations of selenium-laden wastes, such as effluents from systems used to collect flyash scrubber sludge or bottom ash; and selenium-contaminated subsurface irrigation drainwater. Accumulation, transfer, and release of selenium by aquatic biota may affect the speciation and toxicity of dissolved selenium in aquatic environments. Depletion of dissolved selenite and increased concentrations of organoselenium compounds occur during seasonal peaks in phytoplankton abundance in freshwater and marine systems. For example, green algae (*Chlamydomonas reinhardtii*) previously exposed to inorganic radioselenium-75 produced increased concentrations of organoselenium species during population blooms and crashes.

Among terrestrial plants, selenium accumulations in species of *Aster*, *Astragalus*, and several other genera are sometimes spectacularly high. *Astragalus* is the most widely distributed. About 24 of its more than 200 species are selenium accumulators that require selenium to grow well. The highest reported concentration in plants was 15,000.0 mg Se/kg DW, in loco weed, *Astragalus racemosus*. Consumption of these and other selenium-accumulating forage plants by livestock has induced illness and death from selenium poisoning. Even at much lower concentrations, selenium may harm animals that eat considerable amounts of the forage. Plants that accumulate selenium tend to be deeper rooted than the grasses and survive more severe aridity, thus remaining as the principal forage for grazing in time of drought. There is little danger to human health of selenium toxicity from consuming game that foraged in high selenium environments.

Selenium levels in freshwater biota are relatively low, compared with those in their marine counterparts. In freshwater organisms, about 36% of the total selenium was present as selenate, and the rest as selenite and selenide. In marine samples, only 24% of the total selenium was present as selenate. The implications of this difference are not understood now, but have relevance in the ability of selenium to complex and detoxify various potentially toxic heavy metals, such as mercury and cadmium. In a U.S. Nationwide monitoring survey of selenium and other contaminants in freshwater fishes, selenium ranged from 0.05 to 2.9 mg/kg FW whole fish and averaged about 0.6 mg/kg; stations where concentrations in fish exceeded 0.82 mg/kg (>85th percentile) were in three areas: Atlantic coastal streams; Mississippi River system; and California. Among fish from Atlantic coastal streams, those from the Delaware River near Camden, New Jersey, had elevated whole body concentrations (>1.0 and <3.0 mg Se/kg FW), which were attributed to the industrialized character of the river. In the Big Horn and Yellowstone Rivers, high selenium concentrations in fish may result from geologic sources of the element, including coal, phosphate, and sedimentary rock. Fish from the South Platte River near Denver, Colorado, may receive selenium from industrial effluents, or from natural and anthropogenic activities associated with the removal of deposits of coal, barite, and sulfur. These same trends persisted in more recent Nationwide monitoring of freshwater fishes, with selenium concentrations usually highest in whole fish from stations in Utah, Nevada, Texas, California, Hawaii, and in arid locations of the western United States. In California, where selenium was elevated in fish from the San Joaquin River, it was speculated that Selocide, a selenium-containing pesticide registered for use on citrus fruits in the 1960s, may have been a source, although contaminated irrigation drainwater was considered a more likely possibility. Of the seven species of fishes analyzed from the San Joaquin Valley, California, in 1986–87, mosquitofish (*Gambusia affinis*) had the highest concentrations (11.1 mg Se/kg DW whole body); these fish were collected from canals and sloughs in the Grasslands Water District that received large inflows of subsurface agricultural drainage water. Selenium persisted in the biota of the Grasslands drainage regions for at least one year after the switch to uncontaminated drainage water.

Selenium bioconcentrates and biomagnifies in aquatic food chains from invertebrates to birds. Maximum selenium concentrations reported in Cibola Lake in the lower Colorado River Valley in 1989–90 were 5.0 μg/L in water, and – in mg Se/kg DW – 3.3 in sediments, 1.2 in aquatic plants (*Myriophyllum, Ceratophyllum*), 4.6 in crayfish (*Procambarus clarki*), and 9.2 in bluegills (*Lepomis macrochirus*). Diet is the primary source of selenium to fish, as judged by radioselenium-75 uptake studies in Canadian oligotrophic lakes. Hatchery-reared smolts and adults of silver salmon (*Oncorhynchus kisutch*) had less selenium in livers than did wild fish, and this might account for the higher survival and better health of wild fish.

Belews Lake in North Carolina was contaminated with selenium during the 1970s from coal-fired power plant wastewater, causing mortality and reproductive failure in the fish population. Selenium concentrations in fish tissues were as high as 125.0 mg Se/kg DW and were as much as 100 times higher than those from nearby reference sites. There was a positive relation between tissue selenium concentrations and frequency of developmental malformations for largemouth bass and bluegill over the range 1.0–80.0 mg Se/kg DW tissue and 0–70% deformities. In 1992, selenium residues had declined to less than 20.0 mg/kg DW, but were still 5–18 times higher than those in reference lakes, and deformity frequency was 7 times higher. Alterations in zooplankton species densities and dominance – but not number of species – were observed in Belews Lake between 1970 (uncontaminated), 1976–77 (selenium contamination), and 1984–86. Observed changes are attributed to the dominance of planktivorous fishes.

All reported selenium levels in tissues of marine invertebrates and plants were less than 2.0 mg Se/kg on a fresh weight (FW) basis, or 12.0 mg/kg dry weight (DW). In marine algae, most of the selenium accumulated was associated with proteins and may represent a form of storage prior to detoxification. Higher levels are routinely recorded in liver and kidney tissues of marine and coastal vertebrates, including teleosts, birds, and mammals. Livers from adult seals were comparatively rich in selenium; however, high concentrations in liver of maternal California sea lions were not reflected in liver of newborn pups. In marine mammals, selenium concentrations are positively correlated with increasing age, and with increasing mercury residues in piscivorous mammals. The mercury/selenium ratio was close to 1.0 in tissues of marine mammals at mercury concentrations >15.0 mg Hg/kg FW. Increasing mercury concentrations in tissues of marine teleosts are also positively correlated with selenium, although the evidence is conflicting. Selenium varies seasonally in crustaceans. In general, concentrations of selenium in various tissues are usually higher in older than in younger organisms; among marine vertebrates, selenium increases were especially pronounced among the older specimens of predatory, long-lived species.

Selenium concentrations in avian tissues are modified by the age of the organism, condition, diet, presence of other metals, and other variables. Fish-eating birds had the highest selenium concentrations in livers and herbivorous species the lowest; omnivores were intermediate. Selenium concentrations were elevated in livers of molting birds compared to nonmolting conspecifics, elevated in feathers of older terns and egrets when compared to younger stages, and elevated in tissues of marine birds that consume invertebrate prey animals with elevated selenium burdens. Selenium concentrations in tissues of shorebirds were positively correlated with concentrations of copper, zinc, and iron. Feathers have been proposed as indicators of selenium exposure; however, variability in selenium concentrations in whole feathers is considerable. In shorebirds, for example, the highest selenium concentrations are found in wing feathers, specifically in the outer primaries, notably primary 8. Moreover, within the vane of a single feather, the highest selenium concentrations are in the tip and the lowest at the basis. All of these differences need to be considered before feathers are routinely used as indicators of selenium exposure.

Subsurface agricultural drainage waters from the western San Joaquin Valley, California had elevated selenium concentrations, as selenate. In 1978, these drainage waters were diverted to Kesterson Reservoir, a pond system within the Kesterson National Wildlife Refuge (KNWR), with diversion complete by 1982. In 1983, aquatic birds at KNWR had unusual rates of death and developmental abnormalities attributed to selenium. In 1984–85, selenium-induced recruitment failure was observed at KNWR of American avocets (*Recurvirostra americana*) and black-necked stilts (*Himantopus mexicanus*); unlike a nearby reference area, chicks at KNWR of either species did not survive to fledging. Selenium concentrations in livers of diving ducks from San Francisco Bay in 1982 were similar to those of dabbling ducks in the nearby San Joaquin Valley where reproduction was impaired severely. Mean concentrations of selenium in kidneys of seven species of coastal birds collected from the highly industrialized Corpus Christi, Texas, area usually varied between 1.7 and 5.6 mg Se/kg FW, but were 10.2 mg/kg FW in one bird; selenium concentrations of this magnitude may be sufficient to impair reproduction in shorebirds. Barn swallows (*Hirundo rustica*) nesting at a selenium-contaminated lake in Texas had elevated concentrations of selenium in eggs and tissues when compared to conspecifics at a reference site. However, nest success of barn swallows was significantly higher at the contaminated site; development was normal at both sites.

28.4 Deficiency and Protective Effects

Selenium is an essential nutrient for most plants and animals. It constitutes an integral part of the enzyme glutathione peroxidase and may have a role in other biologically active compounds, especially vitamin E and the enzyme formic dehydrogenase. Some animals require selenium-containing amino acids (viz. selenocysteine, selenocystine, selenomethionine, selenocystathionine, selenium-methylselenocysteine, and selenium-methylselenocysteine, and selenium-methyl-selenomethionine), but reportedly Selenium also forms part of certain proteins, including cytochrome c, hemoglobin, myoglobin, myosin, and various ribonucleoproteins.

The availability of selenium to plants may be lessened by modern agricultural practices, eventually contributing to selenium deficiency in animal consumers. For example, fertilizers containing nitrogen, sulfur, and phosphorus all influence selenium uptake by plants through different modes of action, the net effect being a reduction in selenium uptake. The buildup of sulfur (as sulfates) in the soil, due to acid rain, fertilizers, and other sources, interferes with selenium accumulation by crops. In addition, high dietary levels of various heavy metals (including copper, zinc, silver, and mercury) contribute to selenium deficiency in animals, presumably as a result of selenium binding with the metal into biologically unavailable forms.

Clinical selenium deficiency in ruminants is expressed as white muscle disease, lethargy, impaired reproduction, weight loss and reduced growth, shedding, decreased immune response, decreased erythrocyte glutathione peroxidase, and sudden death. Selenium deficiency – as judged by blood concentrations <0.1 mg Se/L – has been documented in California among domestic cattle, mule deer (*Odocoileus hemionus*), pronghorn antelope (*Antilocapra americana*), elk (*Cervus elaphus*), and bighorn sheep (*Ovis canadensis*). More than 95% of black-tailed deer (*Odocoileus hemionus columbianus*) in northern California had inadequate blood selenium levels (<37.0 μg/kg whole blood FW), as judged by recommended levels for cattle (>40.0 μg/kg) and sheep (>50.0 μg/kg). Selenium deficiency in red deer was reversed with subcutaneous injection of 50.0 mg Se/mL as barium selenate at 2.0 mL per 50.0 kg body weight. Selenium boluses calibrated to release 1.0 mg selenium daily given orally to selenium-deficient adult female black-tailed deer, effectively raised whole blood levels to 121.0 μg Se/kg FW. These selenium-supplemented females produced fawns with increased survival (0.83 fawns/female) when compared to untreated does (0.32 fawns/female). Supplementation of selenium-deficient mule deer does with intrarumenal selenium pellets can triple fawn survivability. Some feral animals selectively prefer plants with comparatively elevated selenium content. The black rhinoceros (*Diceros bicornis*) in Kenya, for example, prefers 10 of 103 plants ingested; preferred vegetation contained 3.0–6.3 μg Se/kg FW vs. 1.8–2.7 μg Se/kg FW in nonpreferred plants.

There is a general consensus that selenium deficiency in livestock is increasing in many countries, resulting in a need for added selenium in the food. Selenium deficiency is considered by some researchers to constitute a greater threat to health than selenium poisoning. Studies with animals and humans have suggested that selenium deficiency, in part, underlies susceptibility to cancer, arthritis, hypertension, heart disease, and possibly periodontal disease and cataracts. These linkages have not yet been demonstrated conclusively; for example, eye lens cataract was induced in 10-day-old male rats by 10-day-old male rats by selenate, selenite, selenomethionine, and selenocystine, with glutathione metabolism. On the other hand, adverse effects of selenium inadequacy have been clearly documented for a wide variety of organisms, including bacteria, protozoans, Atlantic salmon, rainbow trout, Japanese quail, ducks, poultry, rats, dogs, horses, domestic sheep, bighorn sheep, swine, cattle, antelopes, gazelles, deer, monkeys, and humans. Selenium deficiency, whether induced experimentally by use of low selenium feeds supplemented with alpha tocopherol or by chronic ingestion of low selenium diets, has caused a number of maladies: high embryonic mortality in cattle and sheep; anemia in cattle; poor growth and reproduction in sheep and rats; reduced viability of newly hatched quail; nutritional myopathy (white muscle disease) in sheep, swine, and cattle; hepatic necrosis and lameness in dogs, horses, and breeding bulls; hair loss and sterility in rat offspring; and spermatozoan abnormalities in rats. Deficiencies were usually prevented or reversed by supplements with sodium selenate or selenite at 100.0 μg Se/kg ration, or 20.0 μg Se/kg body weight administered parenterally.

The protective action of selenium against the adverse or lethal effects induced by mercury, cadmium, arsenic, thallium, copper, zinc, silver, and various pesticides is well documented for a wide variety of plant and animal species. Among marine organisms, for example, selenium protects against toxic levels of mercury in algae, shrimp, crabs and oysters, fish, and mammals. Similar observations have been recorded for copper and marine algae; cadmium and freshwater snails, marine crabs, earthworms, and rats; mercury or methylmercury and rats, eggs of lake trout, freshwater teleosts and (temporarily) Japanese quail; and arsenic and freshwater and marine teleosts. Not all tests were conclusive. Studies with some species of freshwater teleosts demonstrated negligible antagonism of selenium against mercury or cadmium. Selenium reportedly protects mammals and poikilotherms against poisoning by thallium, the herbicide paraquat, cadmium, mercury, lead, arsenic, and copper. Selenium also protects against fatal biological agents. Juvenile chinook salmon (*Oncorhynchus tshawytscha*) naturally infected with *Renibacterium salmoninarum*, the causative agent of bacterial kidney disease, was protected when diets were supplemented with 2.5 mg Se/kg DW ration and vitamin E.

Reasons to account for the antagonism of selenium and heavy metals (here mercury is used as an example) include dietary source and chemical form of selenium, influence of sulfur, biological translocation of selenium or mercury to less critical body parts, and chemical linkage of selenium to mercury on a linear basis. The exact mode of interaction is probably complex and has not yet been resolved. With regard to diet, selenium of animal origin and in the form of selenate is less effective than selenium from plant and inorganic sources in preventing methylmercury neurotoxicity in experimental animals. Disruption of sulfur metabolism by selenium, the sulfur being replaced by seleno-amino acids and other cell constituents containing selenium in living organisms, is one probable cause of selenosis. It is conceivable that Se–Hg compounds formed within the organism would be sufficiently nonreactive biologically to interfere with sulfur kinetics, presumably –SH groups. Differential redistribution of selenium or mercury to less critical body parts may partly account for observed antagonisms. Pretreatment of marine minnows with selenium protects against mercury poisoning and causes a marked redistribution of mercury among organs, presumably to noncritical body parts, and this transfer may partly account for the observed Se–Hg antagonisms in that species. Some investigators have reported that selenium results in increased mercury accumulations. Increased retention of mercury and other metals may lead to a higher level of biomagnification in the food chain and higher body burden in the individual, which might counteract the positive effect of decreased intoxication. Extensive research is under way on the chemical linkage of selenium and mercury. In marine mammals and humans, selenium and mercury concentrations are closely related, almost linearly in a 1:1 molar ratio, but this relation blurs in teleosts in which selenium is in abundance, and fails in birds.

28.5 Lethal Effects

Lethality of various selenium compounds to selected species of aquatic organisms, birds, and mammals under controlled conditions is briefly documented.

28.5.1 Aquatic Organisms

Among representative species of aquatic organisms, death was observed at water concentrations between 60.0 and 600.0 μg Se/L; early life history stages that were subjected to comparatively lengthy exposures accounted for most of these data. Sensitive species of fishes had reduced survival after extended exposure to 10.0–47.0 μg Se/L (Table 28.1). Adult bluegills (*Lepomis macrochirus*), for example, had reduced survival after exposure to 10.0 μg Se/L for 1 year; those exposed to 30.0 μg Se/L all died. Adult bluegills exposed for 60 days to 10.0 μg Se/L as selenite, plus 33.3 mg Se/kg ration as seleno-L-methionine, were normal but produced fry with reduced survival. Exposure for 1 year to 25.0 μg/L caused reduced survival and reproduction of perch and grass carp. Mortality of 35% occurred at 47.0 μg Se/L, as selenite, in chinook salmon exposed for 90 days at the yolk-sac stage; 70% died at 100.0 μg/L. Survival was reduced in chinook salmon fingerlings when their diets contained >9.6 mg Se/kg ration.

Latent mortality after exposure to comparatively high selenium concentrations has been documented, but not extensively. For example, all embryos of the zebrafish (*Brachydanio rerio*) survived exposure to 3000.0 μg Se/L during development, but more than 90% of the resultant larvae died soon after hatching; at 1000.0 μg/L, survival was similar to that in controls. It has been suggested that selenite is more toxic than selenate and is preferentially concentrated over selenate by mussels, *Mytilus galloprovincialis*. Selenite is generally more toxic to early life history stages and effects are most pronounced at elevated temperatures. Also, selenium salts may be converted to methylated forms by microorganisms, and these are readily accumulated by aquatic vertebrates. Among freshwater algae species, it has been demonstrated that selenite, selenate, selenomethionine, and selenopurine are all toxic, but that sulfur, as sulfate, has

Table 28.1. Toxicity of selenium salts to selected aquatic species. Values shown are in μg Se/L (ppb) in medium fatal to 50% of the organisms during exposure for various intervals.

Medium, Taxonomic Group, Species, and Other Variables	Exposure Interval in Hours (h) or Life Cycle (LC)	LC50 (μg/L)
FRESHWATER		
Coelenterates		
Hydra sp.		
Selenite	96 h	1700.0
Selenate	96 h	7300.0
Crustaceans		
Cladoceran, *Daphnia magna*		
Selenite	48 h	700.0
Selenate	48 h	2560.0
Selenite	MATC[a]	110.0–237.0
Selenate	MATC[a]	1730.0–2310.0
Fishes		
Green River, Utah; 3 endangered species (Colorado squawfish, *Ptychocheilus lucius*; razorback sucker, *Xyrauchen texanus*; bonytail, *Gila elegans*); fry and juveniles		
Selenite	96 h	18,000.0 (14,000.0–22,000.0)
Selenate	96 h	97,000.0 (75,000.0–129,000.0)
Bluegill, *Lepomis macrochirus*		
Seleno[DL]*methionine*	96 h	13.0
Seleno[L]*methionine*	96 h	13.0
Selenite	96 h	7800.0–13,000.0
Selenate	96 h	90,000.0
Striped bass, *Moronesaxatilis*, fingerlings		
Seleno[L]*methionine*	96 h	4.0
Selenite	96 h	1000.0
Selenate	96 h	39,000.0
Fathead minnow, *Pimephales promelas*		
Fry	96 h	2100.0
Juvenile	96 h	5200.0
Adult	96 h	620.0–12,500.0
Adult		
Selenite	MATC[a]	83.0–153.0
Selenate	MATC[a]	390.0–820.0
Rainbow trout, *Oncorhynchus mykiss*		
Selenite	MATC[a]	60.0–130.0
Selenate	MATC[a]	2200.0–3800.0

Table 28.1. cont'd

Medium, Taxonomic Group, Species, and Other Variables	Exposure Interval in Hours (h) or Life Cycle (LC)	LC50 (μg/L)
Chinook salmon, *Oncorhynchus tshawytscha*, fry		
Selenite	96 h	13,800.0
Selenate	96 h	115,000.0
Colorado squawfish, *Ptychocheilus lucius*		
Larvae, selenite	96 h	12,800.0
Larvae, selenate	96 h	24,600.0
Juveniles, selenite	96 h	27,900.0
Juveniles, selenate	96 h	77,500.0
MARINE		
Fishes		
Sheepshead minnow, *Cyprinodon variegatus*		
Adult	96 h	7400.0–67,100.0
Egg through juvenile	MATC[a]	470.0–970.0
Striped bass, *Morone saxatilis*		
Selenite	96 h	1600.0
Selenate	96 h	9800.0

[a]MATC = maximum acceptable toxicant concentration. Lower value in each MATC pair indicates highest concentration tested producing no measurable effect on growth, survival, reproduction, and metabolism during chronic exposure; higher value indicates lowest concentration tested producing a measurable effect.

a significant protective role against selenium toxicity. Numerous additional chemical compounds and mixtures probably protect against selenium toxicity, much as selenium protects against toxic effects of mercury salts and other chemicals, but data are sparse on selenium protective agents.

Signs of selenium poisoning in teleosts include loss of equilibrium, lethargy, contraction of dermal chromatophores, loss of coordination, muscle spasms, protruding eyes, swollen abdomen, liver degeneration, reduction in blood hemoglobin and erythrocyte number, and increase in white blood cells. These signs were observed in selenium-poisoned green sunfish (*Lepomis cyanellus*) together with elevated liver selenium concentrations, reduced blood hematocrit, enlarged liver, histopathology of kidney and heart, swollen gill lamellae with extensive cellular vacuolization, and necrotic and degenerating ovarian follicles. Other signs of selenosis in freshwater fishes include loss of osmotic control and liver histopathology.

28.5.2 Mammals and Birds

The element selenium can be traced in an orderly sequence from its origin in the earth's crust to specific geological formation, to distribution of specific genera and groups of plants which require the element for their growth, to the accumulation in vegetation, and to its subsequent toxicity to birds or mammals that consume the seleniferous foods. Selenosis in warm blooded organisms is modified by numerous factors, including method of administration, chemical form of selenium, dietary composition, and age and needs of the organism. Concurrent ingestion of minerals and rough or high protein feeds reduces selenium toxicity, and exposure by diet is less toxic than exposure parenterally or by inhalation. Many compounds are known to prevent or reduce toxic effects of subacute and chronic selenosis in pigs, beef cattle, and other warm-blooded organisms. A partial list includes arsenic, strychnine sulfate, tungsten, germanium, antimony, beet pectin, high-fat diets, ACTH injections, sulfate, increased dietary proteins, lactalbumin, ovalbumin, wheat protein, dried brewer's yeast, desiccated liver, linseed oil meal, glucosamine, hemocysteine, creatine, methionine, and choline. Not all these compounds afforded equal protection against various selenium formulations; the reasons for the difference are not clear, but it appears that the subject of selenoprotective agents warrants additional research effort.

In livestock, there are three basic types of selenium poisoning: acute, resulting from consumption (usually in a single feeding) of a sufficient quantity of highly seleniferous weeds; blind staggers, from consumption of moderately toxic amounts of seleniferous weeds over an extended period of time; and alkali disease, caused by the consumption of moderately seleniferous grains and forage grasses over a period of several weeks to months. Acute poisoning is associated with plant materials containing 400.0–800.0 mg Se/kg: sheep died when fed amounts of plant material ranging from 8.0 to 16.0 g Se/kg body weight, or about 3.2–12.8 mg Se/kg body weight. The minimum lethal dose of Se administered orally as selenite (milligram Se per kilogram body weight) ranged from 3.3 for horses and mules to 11.0 for cattle and 15.0 for swine. Other modes of administration were more toxic, e.g., 2.0 and 1.2 mg Se/kg body weight given subcutaneously killed swine in 4 h and 5 days, respectively; and 1.5–6.0 mg Se/kg body weight given intravenously or intraperitoneally to rats and rabbits were fatal. Accidental toxicosis of sheep and cattle from overtreatment with commercial mixtures of Se salts and vitamin E are also documented for Australia and New Zealand. Acute Se poisoning in domestic livestock is characterized by abnormal movements, lowered head, drooped ears, diarrhea, elevated temperature, rapid pulse, labored breathing, bloating with abdominal pain, increased urination, and dilated pupils. Before death, which is due to respiratory failure, there is complete prostration and lethargy. Duration of illness extends from a few hours to several days, depending on the toxicity of plant material ingested. In these cases, selenium is distributed by the circulatory system to all body organs, the concentrations being highest in liver, blood, kidney, spleen and brain, and lowest in muscle, skin, hair, and bone. Elimination is primarily in the urine; smaller quantities are excreted with the feces, breath, perspiration, and bile. Postmortem examinations indicate many pathological changes in the heart, lungs, rumen, liver, kidney, and other organs. No effective treatment is known for counteracting toxic effects of large amounts of ingested selenium.

Chronic selenosis in mammals may be induced by dietary exposure to natural selenite, selenate, or seleniferous feedstuffs at dietary concentrations between 1.0 mg/kg (rat) and.0 44 mg/kg (horse), or from water containing 0.5–2.0 mg Se/L. Cattle fed 0.5 mg Se/kg body weight 3 times weekly lost their appetite; sheep fed up to 75.0 mg selenite daily developed myocardial degeneration and fibrosis, pulmonary congestion, and edema. The minimum toxic concentration of selenium in lifetime exposure of rats (a comparatively sensitive species) fed Se-deficient diets fortified with selenium was 0.35 mg Se/kg diet, as judged by changes in liver chemistry; and 0.75 mg Se/kg diet, as judged by longevity, and histological changes in heart, kidney, and spleen. These concentrations are 10 times the nutritional threshold for selenium, and about 25% of the minimum lifetime exposure to selenium in natural feedstuffs that produces similar effects under the same experimental conditions. Signs of chronic selenosis include skin lesions, lymph channel inflammation, loss of hair and nails, anemia, enlarged organs (spleen, pancreas, liver), fatigue, lassitude, and dizziness. Blind staggers is characterized by anorexia, emaciation, and sudden collapse, followed by death. Typically, the upper intestinal tract is ulcerated. In alkali disease in cattle, hogs, and horses that had eaten seleniferous grains, the signs were deformation and sloughing of the hooves, hair loss, lassitude, erosion of the articular cartilages, reduced conception, increased reabsorption of fetuses, and degeneration of heart, kidney, and liver. It is likely that selenium displaces sulfur in keratin, resulting in structural changes in hair, nails, and hooves.

Elevated selenium concentrations were measured in tissues and diet of two captive California sea lions (*Zalophus californianus*) that died shortly after performing at a show in 1988. Selenium concentrations, in mg/kg FW, were 49.0 and 88.0 in liver, 42.0 and 47.0 in kidney, and 5.1 and 5.2 in blood. Selenium concentrations in their fish diet was 2.5 mg/kg FW and in thawed fish fluids 45.0 mg/kg FW.

Fatal chronic selenosis in aquatic birds is characterized by low body weight or emaciation, liver necrosis, enlarged kidneys (up to 40% heavier than normal), and more than 66.0 mg Se/kg DW liver. Selenomethionine was the most toxic form or selenium tested against mallards. Mallard ducklings fed 8.0 mg Se/kg ration as selenomethionine for 120 days had hepatotoxicity as adults; 10.0 mg/kg ration for 120 days inhibited reproduction; 15.0 mg/kg ration for 28 days inhibited growth; and 60.0 mg/kg ration for 60 days was fatal to all ducklings. All fatal cases of selenomethionine-induced poisoning in mallards were characterized by histologic lesions of the liver, pancreas, spleen, and lymph nodes, and severe atrophy and degeneration of fat.

28.6 Sublethal and Latent Effects

Results of laboratory studies and field investigations with fish, mammals, and birds have led to general agreement that elevated concentrations of selenium in diet or water were associated with reproductive abnormalities including congenital malformations, selective bioaccumulation by the organism, and growth retardation. Not as extensively documented, but nevertheless important, are reports of selenium-induced chromosomal aberrations, intestinal lesions, shifts in species composition of freshwater algal communities, swimming impairment of protozoans, and behavioral modifications.

28.6.1 Aquatic Organisms

Adverse effects on reproduction is one of the most insidious effects of selenium in freshwater ecosystems. Adult bluegills exposed to 2.5 μg Se/L as sodium selenite for 319 days in an experimental stream ecosystem produced fry with a high incidence of edema, lordosis, and hemorrhaging. Fathead minnows (*Pimephales promelas*) held in ecosystems containing 10.0 or 30.0 μg Se/L for 1 year produced a high incidence of malformed progeny with humped backs, missing scales, and malformations of the jaw, head, operculum, snout, and mandible. The frequency of malformations in the controls was 0.3%; in the 10.0 and 30.0 μg/L groups, these frequencies were 8.0 and 29.8%, respectively. Selenium concentrations in whole fathead minnows from the 10.0 μg Se/L group was 3.9 mg/kg FW vs. 0.3 mg/kg FW in the controls. Mosquitofish reproduction was inhibited when whole body selenium concentrations were >100.0 mg Se/kg DW vs. normal reproduction in a reference area at 1.5 mg Se/kg DW whole body.

In green sunfish from a lake in North Carolina receiving selenium (as flyash wastes from a coal-fired power station), reproduction failed and the population declined markedly. In these fish, selenium levels were elevated in liver (up to 21.4 mg/kg fresh weight) and other tissues; kidney, heart, liver, and gill showed histopathology; and blood chemistry was altered. Ovaries of fish had numerous necrotic and ruptured egg follicles that may have contributed to the population extinction. It is probable that selenium uptake by plankton (containing 41.0–97.0 mg/kg dry weight) from lake water (9.0–12.0 μg/L) introduced selenium to the food chain where it ultimately reached elevated levels in fish through biomagnification. In laboratory tests, however, eggs of common carp hatched normally when incubated in media containing 5000.0 μg Se/L, as did eggs of lake trout (*Salvelinus namaycush*) at 10,000.0 μg Se/L. In frogs (*Xenopus laevis*), cranial and vertebral deformities and lowered survival were documented during development in water with concentrations of 2000.0 μg Se/L or higher.

Reduced growth of freshwater fishes was associated with tissue concentrations of 3.6–6.7 mg Se/kg DW, and dietary concentrations between 5.3 and 25.0 mg Se/kg DW ration. Toxic sublethal effects of selenium in aquatic systems were more pronounced for organoselenium compounds than inorganic compounds and more pronounced for inorganic selenite than inorganic selenate compounds. In salmon, younger stages were more sensitive than older stages. Adverse effects of selenium stress to freshwater fishes were reduced with increasing water hardness; fishes were more sensitive to selenium stress under conditions of reduced temperature and photoperiod.

At water concentrations of 47.0–53.0 μg/L, selenium was associated with anemia and reduced hatch of rainbow trout, growth retardation of freshwater green algae, and shifts in species composition of freshwater algal communities. At 250.0 μg Se/L, growth was reduced in rainbow trout fry after exposure for 21 days, and goldfish demonstrated an avoidance response after 48 h. At water concentrations of 7930.0–11,000.0 μg Se/L, growth was inhibited in freshwater and marine algae, and swimming rate was reduced in the protozoan *Tetrahymena pyriformis*. Eggs of channel catfish exposed to certain metals (including cadmium, mercury, and copper) produced an increased percentage of albino fry; however, eggs exposed to 250.0 μg Se/L produced fry with normal pigmentation.

A significant number of chromosomal aberrations were induced in the edible goby, *Boleophthalmus dussumieri*, by selenium after intramuscular and water exposures. Intramuscular injections as low as 0.1 mg Se/kg body weight, or 3200.0 μg Se/L in the water column, were associated with a marked enhancement of polyploid cells 76–96 h postadministration; some deaths were recorded at higher test concentrations. Selenite was more effective than selenate in inducing chromosomal aberrations. It was concluded that a relatively narrow range of selenium concentrations lead to a mutagenic rather than a lethal effect.

Accumulation of selenium by aquatic organisms is highly variable. In short-term (48 h) laboratory tests at water concentrations of 0.015–3.3 μg Se/L, biological concentration factors (BCF) ranged from 460 in mosquitofish to 32,000 in a freshwater gastropod; values were intermediate for daphnids (2100), plankton (2600), and *Fundulus kansae* (3300), the freshwater killifish. High BCFs (>680) were recorded for freshwater diatoms subjected to maximum concentrations of 40.0 μg Se/L. Livers from rainbow trout and brown trout may contain from 50.0 to 70.0 mg Se/kg fresh weight during lifetime exposure in seleniferous (12.3–13.3 μg/L) water, and have BCF values of 3759–5691; BCF values were 361–390 for skin, and about 180 for muscle. In short-term 180 for muscle. In short-term exposures, most of the selenium was probably adsorbed to the body surface and rapidly lost on transfer to selenium-free media. In longer exposures, the BCF values in aquatic organisms were lower after immersion in high ambient selenium concentrations over extended periods. Thus, marine crabs exposed to a water concentration of 250.0 μg Se/L for 29 days accumulated selenium over water concentration level by a factor of 25 for carapace, and 3.8 for gill; accumulations in muscle and hepatopancreas were negligible. Cadmium in solution enhanced selenium uptake. Exposure of common carp to 1000.0 μg Se/L for 85 days resulted in a whole body BCF of 6; additional studies of 7 weeks exposure plus 7 weeks postexposure at concentrations between 500.0 and 5000.0 μg Se/L yielded a BCF range of 0.6 (5000 μg/L) to 1.8 (500 μg/L). Highest BCFs in carp were 50 for kidney and 80 for liver after exposure of the fish to 100.0 μg Se/L for 7 weeks plus 7 weeks in selenium-free media. For carp, weeks in selenium-free media. For carp, selenium tended to accumulate in kidney, muscle, in that general order. Studies with freshwater organisms collected from a farm pond contaminated by flyash with high selenium levels, and with marine bivalves and nereid worms held for 4 months in seawater flowing through coal flyash containing 6200.0 μg Se/L, showed that accumulation was slight. However, in contrast, mosquitofish from a drainage system that received high coal flyash concentrations at one end and thermal discharges at the other had up to 9.0 mg Se/kg whole body fresh weight. Of the 40 elements examined, only selenium, zinc, and calcium were accumulated in excess of the levels measured in the water. A concentration of 172.0 mg Se/kg dry weight (range 110.0–280.0 mg/kg) was measured in whole mosquitofish from irrigation drainwater ponds contaminated by about 300.0 μg Se/L; based on a wet/dry factor of 4, the BCF for whole mosquitofish was >91.

Selenium accumulation is modified by water temperature, age of the organism, organ or tissue specificity, mode of administration, chemical species, and other factors. In freshwater fishes, selenium concentrations in tissues increased with increasing exposure, with increasing dose, and decreased with increasing water hardness. Food chain accumulation of selenium can severely affect reproductive success of bluegills. Bluegills tend to accumulate inorganic selenium compounds through the diet and organoselenium compounds through the medium as well as the diet. Dietary selenite accumulates in liver and gonads, and selenomethionine in muscle and whole fish.

Diet is the major route of selenium intake by mussels, *Mytilus edulis*. Studies with radioselenium-75 demonstrated that selenium efficiency was 28–34% from the diet and 0.03% from the medium. It was concluded that 96% of the selenium in mussels is obtained from ingested food under conditions typical of coastal waters. In the marine mussel, *Mytilus galloprovincialis*, an increase in water temperature from 13 to 29°C doubled the bioconcentration factor (BCF) in 13 days. Mussels preferentially accumulated selenite over selenate; however, mussels did not reach a steady state in 63 days, indicating that selenium kinetics in some species are difficult to elucidate in short-term studies. Accumulation rates were higher in small than in large mussels, as they were in freshwater teleosts. However, the reverse was documented for marine mammals and teleosts. When selenium was available from both the diet and the medium, concentrations were highest in liver, kidney, and gills of teleosts, exoskeleton of crustaceans, and visceral mass and gills of mollusks. When selenium was administered in food to marine shrimps, concentrations were highest in viscera and exoskeleton, suggesting that ingested selenium is readily translocated from internal to external tissues. Concentrations of selenium in crustaceans usually were higher in fecal pellets than in the diet; fecal pellets may represent a possible biological mechanism for downward vertical transport of selenium in the sea, as well as in freshwater environments.

The time for 50% excretion of accumulated selenium has ranged from 13 to 181 days in various species of marine and freshwater fauna. Biological half-life of selenium accumulated from the medium has been estimated at 13 days for guppies, 27 days for eels, 28 days for leeches, 28 days for carp, 37 days for the marine euphausid crustacean *Meganctiphanes norvegica*, 58–60 days for the marine shrimp *Lysmata caudata*, and 63–81 days for the marine mussel *Mytilus galloprovincialis*. Studies with bluegills and largemouth bass showed elevated tissue levels after exposure to 10.0 μg Se/L for 120 days. Time for 50% excretion in 30-day elimination trials was about 30-day elimination trials was about 15 days from gill and erythrocytes; however, spleen, liver, kidney, or muscle. It appears that research is needed on preferential tissue retention of selenium and its implications for biochemical and metabolic transport mechanisms. Urine is a major excretory route for selenium in marine mammals. Urine of minke whales (*Balaenoptera acutorostrata*) contains 1.5 mg Se/L, or about 30 times more selenium than human urine. There are at least 5 selenium components in urine of minke whales, including trimethylselonium ion; the significance of this observation is imperfectly understood.

28.6.2 Terrestrial Invertebrates

Concentrations of selenium decreased in whole earthworms from 22.4 to 15.0 mg/kg (dry weight) as the rate of sludge application increased from 15 to 60 metric tons/ha. Concentrations of selenium in soil and sludge were 0.3 and 0.5 mg/kg dry weight, respectively. Other studies indicated that some metals, notably cadmium, decreased in worms living in soils amended with sewage sludge but that selenium concentrations were not affected. The biological half life of selenium in earthworms is estimated to be 64 days, a period consistent with values of 10–81 days documented for ants, birds, mammals, and aquatic biota.

28.6.3 Birds

Embryos of the domestic chicken (*Gallus domesticus*) are extremely sensitive to selenium. The hatchability of eggs is reduced by concentrations of selenium in feeds (6.0–9.0 mg/kg) that were too low to produce poisoning in other avian species. Dietary selenium excess was associated with decreased egg weight, decreased egg production and hatchability, anemia, elevated kidney selenium residues in chicks, and a high incidence of grossly deformed embryos with missing or distorted eyes, beaks, wings and feet. Similar results were observed in Japanese quail at 6.0 and 12.0 mg/kg dietary selenite. Severe reproductive effects in ducks (*Anas* spp.), American coot (*Fulica americana*), and other species of aquatic birds nesting were documented at irrigation drainwater ponds in the San Joaquin Valley, California. Water in these ponds contained abnormally high concentrations of about 300.0 μg Se/L, but low or undetectable levels of silver, chromium, arsenic, cadmium, mercury, lead, and zinc. Of 347 nests examined from this site, about 40% had at least one dead embryo and about 20% had at least one embryo or chick with obvious external anomalies, including missing or abnormal beaks, eyes, wings, legs, or feet. In addition, brain, heart, liver, and skeletal anomalies were recorded. Concentrations of selenium (mg/kg, dry weight) were 2.0–110.0 in eggs and 19.0–130.0 in livers of birds, 12.0–79.0 in plants, 23.0–200.0 in invertebrates, and 110.0–280.0 in fish from the ponds, or 7–130 times those found at a nearby control area. It was concluded that selenium was the probable cause of poor reproduction and developmental abnormalities in the aquatic nesting birds, due to interference with their reproductive processes. The concentrations of selenium in breast muscle of coots were sufficiently high (up to 11.0 mg Se/kg FW) to induce State agencies to post the area, advising against the consumption of more than one meal per week of this species, or of any coots by children or pregnant women.

Selenomethionine is the predominant form of selenium in commercial grains, usually as seleno-L-methionine. Selenomethionine is more teratogenic and embryotoxic to avian waterfowl than inorganic forms of selenium tested, possibly owing to higher uptake of organoselenium. Blood selenium concentrations in American kestrels (*Falco sparverius*) seemed to reflect dietary concentrations of seleno-L-methionine. Embryotoxic and teratogenic effects of selenomethionine were observed in mallards at dietary concentrations exceeding 4.0 mg Se/kg FW ration in the laboratory, causing effects similar to those found in field studies. Excess dietary selenium, as seleno-DL-methionine has a more pronounced effect on hepatic glutathione metabolism and lipid peroxidation than does selenite, and may enhance selenium accumulation. In aquatic birds, developmental malformations were associated with lipid peroxidation in livers. Selenomethionine causes lipid peroxidation in livers of aquatic birds, and this is consistent with the observation that selenomethionine is the primary causative agent of selenium-induced embryonic mortality and overt teratogenesis in waterfowl at Kesterson Reservoir. Dietary selenomethionine effects were modified by salts of boron, arsenic, and protein composition, with significant interactions between mixtures.

28.6.4 Mammals

Pregnant long-tailed macaques (*Macaca fascicularis*) given L-selenomethionine for 30 days at doses of 25.0, 150.0, or 300.0 μg Se/kg body weight daily showed dose-dependent increases in erythrocyte and plasma selenium, glutathione peroxidase activities, hair and fecal selenium, and urinary selenium excretion. Adverse effects, including body weight loss, were associated with daily doses of 150.0 and 300.0 μg Se/kg BW and with concentrations of erythrocyte selenium >2.3 mg/L, plasma selenium >2.8 mg/L, and hair selenium >27.0 mg/kg FW.

Young adult female mice (*Mus* sp.) given intraperitoneal injections of sodium selenate or selenomethionine (in each case, three injections of 2.0 mg Se/kg BW at 2-day intervals) had altered blood composition 24 days after the last injection. Both forms of selenium induced a transient, marked decrease in the number of circulating leukocytes (a condition known as leukopenia) following serial injections. Leukopenia was more extensive and of greater duration for selenomethionine-treated mice.

Nonlethal effects of selenium on mammals include reproductive anomalies. Selenosis caused congenital malformations in rats, mice, swine, and cattle. In general, young born to females with selenosis were emaciated and unable to nurse. Mice given selenium in drinking water reproduced normally for three generations, but litters were fewer and smaller when compared to controls, pups were runts with high mortality before weaning, and most survivors were infertile.

In rats, selenium did not induce cirrhosis or neoplasia; however, intestinal lesions were observed among those fed diets containing 0.8–1.0 mg Se/kg ration during lifetime exposure. The threshold requirement for optimal rat nutrition under similar conditions is about 0.08 mg Se/kg ration, again demonstrating the relatively narrow range separating selenium deficiency from selenium poisoning. Absorption of oral radioselenite by rats was as high as 95–100%. A single dose of radioselenite concentrated, in descending order of accumulation, in pancreas, intestine, erythrocytes, liver, kidney, and testes; tissue distributions from chronic exposure were similar. As expected, levels of selenium in poisoned rats were highest in liver and kidney. Rats, and probably other mammals, can regulate dietary selenium accumulations. Dietary concentrations in excess of 54.0–84.0 μg Se/kg ration were usually excreted in urine. Urine is the major excretory route for selenium. In urine of selenium-challenged rats, the trimethylselonium ion – a metabolic product of selenite or selenoamino acid, such as seleomethionine – is dominant. However, when selenium intake exceeded 1000.0 μg/kg ration, pulmonary excretion was active. Excretion of selenium in feces, bile, saliva, and hair appears to be relatively constant regardless of the amount of exposure. Selenotoxic effects in mice, including abortion and maternal death, were prevented by prior treatment with vitamin E, but exacerbated by reduced glutathione; the mechanisms for these interactions are unknown, and merit additional research.

28.7 Recommendations

All investigators appear to agree on four points. First, that insufficient selenium in the diet may have harmful and sometimes fatal consequences. Second, that exposure to grossly elevated levels of selenium in the diet or water is inevitably fatal over time to terrestrial and aquatic organisms. Third, that there is a comparatively narrow concentration range separating effects of selenium deficiency from those of selenosis. And fourth, that additional fundamental and basic research is required on selenium metabolism, physiology, recycling, interactions with other compounds or formulations, and chemical speciation in order to elucidate its nutritive role as well as its toxic effects. Accordingly, the proposed selenium criteria shown in Table 28.2 for prevention of selenium deficiency and for protection of aquatic life, livestock, crops, and human health, should be viewed as guidelines, pending acquisition of additional, more definitive data.

Regarding selenium deficiency, it appears that diets containing 50.0–100.0 μg Se/kg ration provide adequate protection to humans and to various species of fish, small laboratory mammals, and livestock (Table 28.2). Factors contributing to selenium deficiency in crops include increasing use of agricultural fertilizers and increasing atmospheric fallout of sulfur; further, foliar applications of selenate, although efficient in raising selenium levels in plants, have only short-term value. There is a general consensus that selenium deficiency in livestock in many countries is increasing, resulting in a need for added selenium in the food chain.

Table 28.2. Proposed criteria for prevention of selenium deficiency and for protection against selenosis.

Criterion	Selenium Concentration
PREVENTION OF SELENIUM DEFICIENCY	
Rainbow trout (water levels of 0.4 μg Se/L); diet	>70.0 μg/kg FW; 150.0–380.0 μg/kg DW
Poultry	
Diet	>30.0–50.0 μg/kg FW
Dietary supplement allowed	
Chickens, ducks	100.0 μg/kg ration as sodium selenate or selenite
Turkeys	200.0 μg/kg ration as sodium selenate or selenite
Rat; diet	>54.0–84.0 μg/kg FW
Livestock	
Diet	>20.0 μg/kg FW
Dietary supplement allowed	
Cattle, sheep, adult swine	100.0 μg/kg ration as sodium selenate or selenite
Weanling swine	300.0 μg/kg ration as sodium selenate or selenite
Forage, grazing sheep and cattle	>100.0 μg/kg DW
Blood	>40.0–80.0 μg/L
White-tailed deer	
Heart	>150.0 μg/kg DW
Kidney	>3000.0 μg/kg DW
Liver	>250.0 μg/kg DW
Serum	>30.0 μg/L FW
Humans	
Diet	40.0–200.0 μg/kg FW
Daily intake recommended	
Females	55.0 μg (= 1.0 μg/kg body weight daily)
Males	70.0 μg (= 1.0 μg/kg body weight daily)
Children	0.004 μg/kg body weight
Maximum	500.0 μg
Drinking water	
Bottled water	<10.0 μg/L FW
Assuming water is sole selenium source, 2 L/daily	20.0 μg/L FW
PROTECTION AGAINST SELENOSIS	
Crop protection; irrigation water	<50.0 μg/L
Aquatic life protection: freshwater	
Total dissolved	Average 4-day concentration not to exceed 5.0 μg acid-soluble Se/L more than once every 3 years; average 1-h concentration not to exceed 20.0 μg/L more than once every 3 years

Table 28.2. cont'd

Criterion	Selenium Concentration
Total recoverable[a]	After filtration through 0.45 μm filter, <2.0 μg total Se/L, sometimes <1.0 μg/L for organoselenium compounds
Waterborne selenium	<2.0 μg/L
Adverse effects on fish reproduction possible	>2.0 μg inorganic Se/L; >1 μg organic Se/L; sometimes, <1.0 μg organic Se/L
Fish diet, freshwater fishes	
Acceptable	<3000.0 μg/kg DW ration
Lethal	>6500.0–54,000.0 μg/kg DW feed
Growth reduction	>5000.0–20,000.0 μg/kg DW feed
Kidney damage	>11,000.0 μg/kg DW feed
Juvenile chinook salmon	
Adverse effects	>3000.0–5000.0 μg total Se/kg DW diet
Safe	<3000.0 μg total Se/kg DW diet
Rainbow trout, safe	<3000.0 μg/kg DW diet
Tissue residues, acceptable	
Whole body	<4000–<12,000.0 μg/kg DW
Gonads, eggs	<8000.0–<10,000.0 μg/kg DW
Bluegill, carcass or ovaries	<6000.0 μg/kg FW
Muscle	<8000.0 μg/kg DW
Liver	<12,000.0 μg/kg DW
Great Lakes, water	<10.0 μg/L
Aquatic life protection: marine	
Water, total dissolved	Average 4-day concentration not to exceed 71.0 μg acid-soluble Se/L more than once every 3 years and 1-h concentration not to exceed 300.0 μg/L more than once every 3 years
Birds	
Diet, mallard	
Maximum tolerated	<10,000.0 μg/kg DW
Reproduction inhibited	7000.0–11,000.0 μg/kg DW, as selenomethionine
Malformed embryos	16,000.0 μg/kg DW, as selenomethionine
Fatal	10,000.0–20,000.0 μg/kg, as selenomethionine
Drinking water	
Sensitive fish-eating species	<0.8–<1.9 μg dissolved Se/L
Mallard	<2.1 μg dissolved Se/L
Aquatic birds, most species	
Minimal hazard	<2.3 μg dissolved Se/L
Hazardous	3.0–20.0 μg dissolved Se/L
Maximum allowable	<10.0 μg dissolved Se/L

Continued

Table 28.2. cont'd

Criterion	Selenium Concentration
Egg concentrations; reproductive impairment:	
Unlikely	<2000.0 μg/kg DW; <3000.0–<3300.0 μg/kg FW
Possible	>1000.0 μg/kg FW
Probable	>5000.0 μg/kg FW; >15000.0 μg/kg DW
Liver concentrations	
Acceptable	<3000.0 μg/kg FW; <5200.0–<10,000.0 μg/kg DW
Adverse effects	>10,000.0 μg/kg FW
Poisoned	>66,000.0 μg/kg DW
Lethal	>20,000.0 μg/kg FW
Mammals (nonhumans)	
Livestock protection	
Drinking water	<50.0 μg/L FW
Diet (total)	<2000.0 μg/kg DW
Diet (natural)	<4000.0 μg/kg DW
Feeds (natural)	<2000.0 μg/kg DW
Forage (natural)	<5000.0 μg/kg DW
Tissues	
Blood; adequate vs. toxic	80.0 μg/L FW vs. >3000.0 μg/L FW
Kidney, toxic	>3000.0–6000.0 μg/kg FW
Liver, toxic	>12,000.0–15,000.0 μg/kg FW
Monkeys; tissues; adverse effects	
Erythrocytes	>2300.0 μg/kg FW
Hair	>27,000.0 μg/kg FW
Plasma	>2800.0 μg/L FW
Drinking water	
River otter	<0.7 μg dissolved Se/L
Bats and shrews	<0.9 μg dissolved Se/L
Mink	<1.1 μg dissolved Se/L
Humans	
Seafood	Not to exceed 2000.0 μg/kg FW
Drinking water	
Most states	<10.0 μg/L FW
Minnesota	<20.0 μg/L FW
International	<10.0 μg/L FW
Maximum permissible[b]	<50.0 μg/L FW
Health advisory, chronic	
Child	<31.0 μg/L FW
Adult	<107.0 μg/L FW
Food (natural)	<5000.0 μg/kg FW; <850.0 μg daily
Milk	<500.0 μg/L FW

Table 28.2. cont'd

Criterion	Selenium Concentration
Daily intake (all sources)	
Adults	
Safe	<200.0 μg
Safe, chronic[c]	<5.0 μg/kg body weight (=<350.0 μg for a 70-kg adult)
Normal	60.0–250.0 μg
Maximum tolerable level	<500.0 μg
Infants	4.0–<35.0 μg
Children	
Age 1–3	20.0–<80.0 μg
Age 4–6	30.0–<120.0 μg
Age 7–11+	50.0–<200.0 μg
Air	
Japan	<100.0 μg/m^3
Former Soviet Union	<100.0 μg/m^3
USA	Usually <200.0 μg/m^3; some states 2.0–<5.0 μg/m^3

[a]High potential for biomagnification in aquatic food chains, dietary toxicity, and reproductive toxicity.
[b]Based on a NOAEL (no observed adverse effect level) of 400 μg Se daily, equivalent to 5.7 μg/kg BW daily for a 70-kg person.
[c]Based on a NOAEL of 15.0 μg /kg BW daily for dermal effects and an uncertainty factor of 3 to account for sensitive individuals.

Recommendations for protection of freshwater aquatic life include acid-soluble total selenium concentrations in the water of less than 5.0 μg/L on a daily average, or 20.0 μg/L at any time (Table 28.2). These values are higher for saltwater life: 71.0 μg/L daily average, 300.0 μg/L at any time. The concentration range of 5.0–20.0 μg/L recommended for protection of freshwater aquatic life is below the range of 60.0–600.0 μg/L that is fatal to various sensitive species of marine and freshwater fauna, and in this respect affords an adequate measure of protection. It is also below the range (47.0–53.0 μg/L) that has been associated with growth inhibition of freshwater algae, anemia and reduced hatch in rainbow trout, and shifts in species composition of freshwater algal communities. However, water concentrations of 9.0–12.0 μg Se/L were associated with reduced reproduction of freshwater fishes, and their results strongly indicated that some downward modification of the selenium freshwater aquatic life protection criterion may be appropriate. Furthermore, high bioconcentration and accumulation of selenium from the water column by numerous species of algae, fish, and invertebrates is well documented at levels between 0.015 and 3.3 μg Se/L, which is substantially below the recommended range of 5.0–20.0 μg/L in freshwater. The significance of selenium residues in aquatic biota in terms of bioavailability and selenium receptor sites is imperfectly understood, and it appears that much additional research is warranted on formulating suitable models of selenium biogeochemistry and pharmacokinetics in aquatic environments.

Resource managers and aquatic biologists need data on selenium concentrations in water, food-chain organisms, and fish and wildlife tissues in order to adequately assess the overall selenium status and health of aquatic ecosystems. Because selenium is depurated rapidly in aquatic birds, resource managers should be concerned primarily about current exposure in nature and not previous exposures. The potential for food-chain biomagnification and reproductive impairment in fish and birds are the most sensitive biological responses for estimating ecosystem-level impacts of selenium contamination, and these are best reflected in selenium concentrations in gravid ovaries and eggs of adult fish and aquatic bird populations. More research is recommended on biomarkers of selenium exposure and effect.

Field studies demonstrated that migratory waterfowl were heavily and adversely affected while nesting at selenium-contaminated irrigation drainwater ponds in California, where food chain organisms contained between 12,000.0 and 280,000.0 μg Se/kg. The source and fate of selenium in irrigation drainwater ponds are largely unknown; they must be determined so that alternate technologies for selenium control can be implemented to protect waterfowl in that geographical region.

Livestock appear to be protected against selenosis provided that their diets contain less than 4000.0 μg Se/kg of natural i.e., nonsupplemented selenium (Table 28.2). This concentration is somewhat higher than levels reported for rats. Minimum toxic concentrations of selenium in lifetime exposure of rats given diets containing natural selenium were 1400.0 μg/kg ration, as judged by evidence of liver changes, and 3000.0 μg/kg ration as estimated from longevity and histological changes in heart, kidney, and spleen. These values were only 350.0 and 750.0 μg/kg ration, respectively, when rat diets contained purified, rather than natural, selenium. This relation emphasizes that accidental poisoning of livestock, and presumably fish and wildlife, may occur when soils are deliberately supplemented with purified selenium, or when soils or aquifers are contaminated as a result of faulty waste disposal practices. Although the concentration of <50.0 μg Se/L for livestock drinking water and irrigation water for crop protection is inconsistent with that of <35.0 μg Se/L for aquatic life protection, neither livestock nor crops appear threatened at the higher level. Since many waterways that abut agricultural lands or areas of high anthropogenic loadings of selenium contain valuable and desirable aquatic species, it would appear that some downward modification of the current livestock drinking water concentration of selenium is necessary. Selenium is a proven teratogen in fishes and birds, but this has not been established in mammals; more research is needed on possible teratogenic effects of organoselenium compounds in mammals.

Acute lethal doses for livestock species ranged from 3300.0 μg Se/kg body weight for horses and mules to 15,000.0 μg/kg for swine; appetite loss in cattle was noted at 500.0 μg/kg. For humans, the maximum tolerance level is usually set at 500.0 μg of selenium daily, and the "safe" level at 200.0 μg/day. Selenium dietary levels for humans should not exceed 5000.0 μg/kg ration; however, recommended maximum dietary levels in other mammals ranged from 1000.0 μg/kg for rats to 4000.0 μg/kg for horses. For all species, including humans, there is a tendency to list selenium dosage levels in terms of "natural" and "supplemented" levels, with the tacit understanding that natural levels are about one-fourth as toxic as supplemented values. Given the complexities of selenium metabolism and speciation, it appears that greater precision and clarity are necessary in formulating selenium criteria if these criteria are to become administratively enforced standards through passage of appropriate legislation.

Aerosol concentrations in excess of 4.0 μg Se/m^3 are potentially harmful to human health although concentrations in excess of this value (6.0 μg Se/m^3) were regularly encountered in the vicinity of the smeltery at Sudbury, Canada. It is not now known whether respiration rates of wildlife, particularly birds, are comparable to those of humans, or whether selenium absorption energetics are similar, or whether wildlife species that frequent point sources of air contaminated by high selenium levels for protracted periods are at greater risk than humans. Until additional and more conclusive data become available, aerosol concentrations of less than 4.0 μg Se/m^3 are recommended for the protection of sensitive species of wildlife.

28.8 Summary

Most authorities agree on five points. First, selenium deficiency is not as well documented as selenosis, but may be equally significant. Second, selenium released as a result of anthropogenic activities (including fossil fuel combustion and metal smelting), as well as that in naturally seleniferous areas, pose the greatest threat of poisoning to fish and wildlife. Third, additional research is required on chemical and biological transformations among valence states, allotropic forms, and isomers of selenium. Fourth, metabolism and degradation of selenium are both significantly modified by interaction with various heavy metals, agricultural chemicals, microorganisms, and numerous physicochemical factors, and until these interactions are resolved it will be difficult to meaningfully interpret selenium residues in various tissues. And fifth, documented biological responses to selenium deficiency or to selenosis vary widely, even among closely related taxonomic groups.

It is generally agreed that selenium deficiency may be prevented in fish, small laboratory mammals, and livestock by feeding diets containing 50.0–100.0 μg Se/kg. The concentration range of total acid soluble selenium currently recommended for aquatic life protection – 5.0 μg/L in freshwater to 71.0 μg/L in marine waters – is below the range of 60.0–600.0 μg/L that is fatal to sensitive aquatic species. In freshwater, it is also below the range of 47.0–53.0 μg/L associated with growth inhibition of freshwater algae, anemia and reduced hatching in trout, and shifts in species composition of freshwater algae communities. Accordingly, current recommendations for selenium with respect to aquatic life appear to afford an adequate measure of protection. However, some studies have shown that water concentrations of 9.0–12.0 μg Se/L are associated with inhibited reproduction of certain freshwater teleosts, suggesting that selenium criteria for protection of freshwater life should be revised downward. Also, high bioconcentration and accumulation of selenium from water by numerous species of algae, fish, and invertebrates is well documented at levels of 0.015–3.3 μg Se/L, which are substantially below the recommended range of 5.0–71.0 μg Se/L. The significance of selenium residues in aquatic biota is still unclear, and more research appears to be needed on selenium pharmacokinetics in aquatic environments.

Aerosol concentrations exceeding 4.0 μg Se/m^3 are considered potentially harmful to human health; however, no comparable database for birds and other wildlife species is available at this time. Selenium poisoning in livestock is prevented if diets do not exceed 5.0 mg Se/kg natural forage, or 2.0 mg Se/kg in feeds supplemented with purified selenium. Minimum toxic concentrations of selenium in the rat (a sensitive species) fed diets containing natural selenium were 1400.0 μg Se/kg as judged by evidence of liver changes, and 3000.0 μg Se/kg as estimated from longevity and histopathology; these values were only 350.0 and 750.0 μg/kg, respectively, when diets low in natural selenium were fortified with purified selenium. The evidence is incomplete for migratory waterfowl and other birds, but diets containing more than 3.0 mg Se/kg are demonstrably harmful, as are total selenium concentrations in excess of 5.0 mg/kg FW in eggs and 10.0 mg/kg DW in livers.

SILVER[a]

Chapter 29

29.1 Introduction

Silver (Ag) found in the body of mammals – including humans – has no known biological purpose and is suspected of being a contaminant. Silver, as ionic Ag^+, is one of the most toxic metals known to aquatic organisms in laboratory testing, although large industrial losses to the aquatic environment are probably infrequent because of its economic value as a recoverable resource. Silver, however, is of concern in various aquatic ecosystems because of the severity of silver contamination in the water column, sediments, and biota. San Francisco Bay, for example, is impacted from discharges of silver in wastewater outfalls and from the diagenic remobilization of silver from contaminated sediments in the estuary.

The principal industrial use of silver is as silver halide in the manufacture of photographic imaging materials; other products include jewelry, coins, indelible inks, and eating utensils. In medicine, silver salts are used as caustics, germicides, antiseptics, and astringents; the use of silver nitrate for prophylaxis of ophthalmia neonatorum in the eyes of newborn infants is a legal requirement in some states. Long-term industrial or medical exposure to silver and its compounds may increase blood concentrations of silver to levels which can have toxic effects, such as induction of sarcomas, anemia, and enlargement of the heart. Repeated occupational handling of silver objects, especially after repeated minor injuries, may result in localized argyria – a bluish-gray discoloration of the skin at the exposed site. In humans, the most common noticeable effects of chronic exposure to silver and its compounds are generalized argyria, localized argyria, and argyrosis (argyria of the eye, usually). Generalized argyria consists of a slate-gray pigmentation of the skin and hair caused by deposition of silver in the tissues, a silver coloration of the hair and fingernails, and a blue halo around the cornea and in the conjunctiva. Acute toxic effects in humans have resulted only from accidental or suicidal overdoses of medical forms of silver. Ecological and toxicological aspects of silver are considerable and discussed later.

29.2 Sources and Uses

About 2.47 million kg of silver are lost each year to the domestic biosphere, mostly (82%) as a result of human activities. As discussed later, the photography industry accounts for about 47% of all silver discharged into the environment from anthropogenic sources. In 1990, about 50% of the refined silver consumed domestically was used to manufacture photographic products; 25% in electrical and electronic products; 10% in electroplated ware, sterlingware, and jewelry; 5% in brazing alloys; and 10% in other products and processes.

29.2.1 Sources

Silver is a rare but naturally occurring metal, often found deposited as a mineral ore in association with other elements. Argentite is the

[a]All information in this chapter is referenced in the following sources:

Eisler, R. 1996. Silver hazards to fish, wildlife, and invertebrates: a synoptic review. *U.S. Natl. Biol. Serv. Biol. Rep.* 32, 44 pp.

Eisler, R. 2000, Silver. Pages 499–550 in *Handbook of Chemical Risk Assessment. Health Hazards to Humans, Plants, and Animals. Volume 1, Metals*. Lewis Publishers, Boca Raton, Florida.

main ore from which silver is extracted by cyanide, zinc reduction, or electrolytic processes. Silver is frequently recovered as a by-product from smelting of nickel ores in Canada, from lead–zinc and porphyry copper ores in the United States, and from platinum and gold deposits in South Africa. About 12–14% of the domestic silver output is recovered from lead ores and about 4% from zinc ores. Secondary sources of silver comprise new scrap generated in the manufacture of silver-containing products; coin and bullion; and old scrap from electrical products, old film and photoprocessing wastes, batteries, jewelry, silverware, and bearings.

Silver is produced in 68 countries, but most (75%) of the world's silver (excluding the former Soviet bloc) is mined in the United States, Mexico, Canada, Australia, and Japan; the United States produces about 50% of the world's supply of refined silver. The primary silver mines of the United States are in the Coeur d'Alene mining district in the northern Idaho panhandle. Between 1949 and 1970, the United States consistently produced less than 15% of the global silver production and consumed 34–64%. Since 1951, silver consumption in the United States has exceeded its extraction from ore. In 1979, about 95% of the silver in domestic production was from Idaho, Nevada, Arizona, Colorado, Utah, Montana, and Missouri. End use categories of silver consumed domestically in 1979 included photography (39%), electrical and electronic components (25%), sterlingware and electroplated materials (15%), and brazing alloys and solders (8%). In 1979, the photographic industry was located mainly in New York, and most other end use manufacturers were in Connecticut, New York, Rhode Island, and New Jersey.

World production of silver increased from 7.4 million kg in 1964 to 9.06 million kg in 1972 and to 9.67 million kg in 1982. In 1986, 13.06 million kg of silver was produced globally; the United States produced 1.06 million kg in 1986 but consumed 3.94 million kg. In 1990, the estimated world mine production of silver was 14.6 million kg; major producers were Mexico with 17% of the total, the United States with 14%, Peru with 12%, the former Soviet Union with 10%, and Canada with 9%. In the United States during 1990, about 160 mines produced silver worth an estimated value of $320 million; most (71%) of the 1990 mine production was in Nevada (32%), Idaho (21%), Montana (11%), and Arizona (7%). In 1990, 22 major refiners of commercial grade silver and more than 5000 silver fabricating and manufacturing firms were located primarily in the northeastern states. Of the silver imported into the United States in 1990, 44% came from Mexico, 34% from Canada, 5% from Peru, and 4% from Chile. Most was exported after transformation to sterlingware, coinage, and other finished products. Melting and refining old scrap silver in 1990 accounted for 500,000 kg of silver.

Emissions from smelting operations, manufacture and disposal of certain photographic and electrical supplies, coal combustion, and cloud seeding are some of the anthropogenic sources of silver in the biosphere. Fallout from cloud seeding with silver iodide is not always confined to local precipitation; silver residuals have been detected several hundred kilometers downwind of seeding events. In 1978, the estimated loss of silver to the environment in the United States was 2.47 million kg, mostly to terrestrial and aquatic ecosystems; the photography industry alone accounted for about 47% of all silver discharged into the environment from anthropogenic sources. In California, anthropogenic sources contributed 50% more silver to sediments of coastal basins than natural sources, as judged by sedimentary basin fluxes of 0.09 $\mu g/cm^2$ in anthropogenic sources of silver and 0.06 $\mu g/cm^2$ in natural sources. Sometimes, liquid effluents from the nuclear industry contained significant quantities of radiosilver-110m. In Lake Michigan, storms contribute a large fraction of the annual load of tributary-derived silver; concentrations of particle-bound silver in many rivers during storms were more than 0.1 $\mu g/L$.

Most of the silver lost to the environment each year enters terrestrial ecosystems where it is immobilized in the form of minerals, metal, or alloys; agricultural lands may receive as much as 80,000 kg of silver from photoprocessing wastes in sewage sludge. An estimated 150,000 kg of silver enter the aquatic environment every year from the photography industry, mine tailings, and electroplaters. During processing of photographic paper and film, silver is generally solubilized as the tightly bound thiosulfate complex. Silver thiosulfate in secondary biological waste treatment plants is converted to insoluble silver sulfide, which is removed in the sludge; only trace amounts of complexed and adsorbed silver are discharged into the aquatic environment. The silver incorporated into the sludge is immobile and should not restrict the use of sludge for the enrichment of soils. The atmosphere receives 300,000 kg of silver each year from a variety of sources, but atmospheric concentrations are not known to exceed the occupational threshold limit value of 10.0 μg total Ag/m^3.

Daily intake of total silver from all sources by humans in the United States ranged from 70.0 to 88.0 μg; diet accounted for 35.0–40.0 μg daily. Sources of elevated dietary silver include seafood from areas near sewage outfalls or industrial sources and crops grown in areas with high ambient levels of silver in the air or soil. Most occupational exposures to silver occur through inhalation of silver-containing dusts or dermal exposure to photographic compounds. Dermal routes of human exposure to silver include handling of silver-containing processing solutions used in radiographic and photographic materials, dental amalgams, and silver sulfadiazine cream and solutions for treating burns.

29.2.2 Uses

Silver was used for ornaments and utensils for almost 5000 years, and as a precious metal, a monetary medium, and a basis of wealth for more than 2000 years. Until the late 1960s, it was used extensively for coinage. Since 1970, U.S. coinage has not contained silver, although minting of as many as 45 million silver-clad subsidiary coins has been authorized. Industrial consumption of silver in the United States between 1966 and 1972 totaled 4.67 million kg, primarily in the manufacture of photographic materials, electrical contacts and conductors, and sterlingware. In 1973, silver was used mainly in photographic materials (29%), electrical and electronic components (22%), sterlingware (20%), electroplated ware (10%), brazing wares (20%), dental and medical products, catalysts, bearings, and jewelry (9%). In 1986, photographic materials accounted for 45% of the silver consumption in the United States; electrical and electronic components, 25%; jewelry, sterlingware, and electroplated ware, 11%; alloys and solders, 5%; and mirrors, dental amalgam, medical supplies, chemicals, water purification, and cloud seeding, 14%. Silver, as silver iodide, is used in the United States for weather modification including rain and snow making and hail suppression; as much as 3110 kg of silver is used for this purpose annually. Silver nitrate in hair dyes has been in use regularly for almost 200 years, although its use may lead to argyria. In 1990, about 50% of the refined silver consumed domestically was used to manufacture photographic and X-ray products; 25% in electrical and electronic products; 10% in electroplated ware, sterlingware, and jewelry; 5% in brazing alloys; and 10% in other uses.

Because of its bacteriostatic properties, silver compounds are used in filters and other equipment to purify water of swimming pools and drinking water and in the processing of foods, drugs, and beverages. Activated charcoal filters coated with metallic silver to yield water concentrations of 20.0–40.0 μg Ag/L are used in filtering systems of swimming pools to control bacteria. Silver may also function as an algicide in swimming pools if chlorine, bromine, and iodine are absent; it prevents growth of blue-green algae at 80.0–140.0 μg Ag/L. Aboard orbiting Russian space stations and spaceships, potable water is routinely treated with 100.0–200.0 μg Ag/L to eliminate microorganisms; sterilization is usually complete in 20 min. Silver-containing ceramic water filters are used to purify potable water in Swiss ski resorts, German breweries, British ships, oil tankers, drilling rigs, U.S. home consumption, and more than half the world's airlines. Monovalent and metallic silver compounds are considered excellent disinfectants; however, Ag^{2+} and Ag^{3+} are about 50–200 times more effective than Ag^{+} or Ag^{0}, possibly because of their higher oxidation states.

Silver nitrate was used for many years as eye drops in newborns to prevent blindness caused by gonorrhea. Laws in many states still require that a few drops of a 1–2% silver nitrate solution be applied to the conjunctiva of the eyes of newborn infants to prevent ophthalmia neonatorum by transmittal of gonorrhea from the mother. This treatment is still required in Denmark but not in Japan or Australia. Silver nitrate is not used in many U.S. hospitals because of the dangers of chemical conjunctivitis and has been replaced by antibiotics. In the United States, several silver-containing pharmaceuticals were used topically on skin or mucous membranes to assist in healing burn patients and to combat skin ulcers. Oral medicines containing silver include silver acetate-containing anti-smoking lozenges; breath mints coated with silver; and silver nitrate solutions for treating gum disease. The widespread medical use of silver compounds for topical application to mucous membranes and for internal use became nearly obsolete in the past 50 years because of the fear of argyria and the development of sulfonamide and antibiotic microbials.

29.3 Properties

Silver occurs naturally in several oxidation states, the most common being elemental silver (Ag^0) and the monovalent ion (Ag^+). Soluble silver salts are in general more toxic than insoluble salts; in natural waters, the soluble monovalent species is the form of environmental concern. Sorption is the dominant process that controls silver partitioning in water and its movements in soils and sediments. As discussed later, silver enters the animal body through inhalation, ingestion, mucous membranes, and broken skin. The interspecies differences in the ability of animals to accumulate, retain, and eliminate silver are large. Almost all the total silver intake is usually excreted rapidly in feces; less than 1% of the total silver intake is absorbed and retained in tissues, primarily liver, through precipitation of insoluble silver salts. In mammals, silver usually interacts antagonistically with selenium, copper, and vitamin E; in aquatic environments, ionic or free silver interferes with calcium metabolism in frogs and marine annelids and with sodium and chloride uptake in gills of fishes.

29.3.1 Physical and Chemical Properties

Silver is a white, ductile metal occurring naturally in the pure form and in ores. Silver has the highest electrical and thermal conductivity of all metals. Some silver compounds are extremely photosensitive and are stable in air and water except for tarnishing readily when exposed to sulfur compounds. Metallic silver is insoluble in water but many silver salts, such as silver nitrate, are soluble in water to more than 1220.0 g/L. In natural environments, silver occurs primarily in the form of sulfide or is intimately associated with other metal sulfides, especially those of lead, copper, iron, and gold, which are all essentially insoluble. Silver readily forms compounds with antimony, arsenic, selenium, and tellurium. Silver has two stable isotopes (^{107}Ag and ^{109}Ag) and 20 radioisotopes; none of the radioisotopes of silver occurs naturally, and the radioisotope with the longest physical half-life (253 days) is ^{110m}Ag. Several compounds of silver are potential explosion hazards: silver oxalate decomposes explosively when heated; silver acetylide (Ag_2C_2) is sensitive to detonation on contact; and silver azide (AgN_3) detonates spontaneously under certain conditions.

Silver occurs naturally in several oxidation states, usually as Ag^0 and Ag^+; other possible oxidation states of silver are Ag^{2+} and Ag^{3+}. In surface freshwater, silver may be found as the monovalent ion; in combination with sulfide, bicarbonate, or sulfate; as part of more complex ions with chlorides and sulfates; and adsorbed onto particulate matter. In the aqueous phase, silver at the lowest concentrations exists as either a simple AgSH or as a simple polymer HS–Ag–S–Ag–SH. At higher concentrations, colloidal Ag_2S or polysulfide complexes are formed. Ag^+ binds strongly with S^{2-} in inorganic and organic species, resulting in ng/L aqueous dissolved concentrations. Trace levels of dissolved silver in the presence of ferric sulfide are rapidly adsorbed and silver remaining in solution remains as acanthite (Ag_2S); however, silver thiolate complexes can be the dominant dissolved species in highly contaminated waters near urban centers or in waters with high levels of natural organic matter. The most important and crucial aspect of silver thiolate chemistry is the rapid exchange of Ag^+ among thiolates whereby Ag^+ can transfer onto, or off, particulate materials or the cells of an organism. Silver thiolates also react rapidly with H_2S or HS^- as ligands to form Ag_2S, although the reverse process is slow. Soluble silver salts are more toxic than insoluble salts, and soluble silver ion (Ag^+) is the most toxic chemical species. In natural waters, the soluble monovalent species is the form of environmental concern. The argentous ion (Ag^+) does not hydrolyze appreciably in solution and is considered to be a mild oxidizing agent. Hypervalent silver species, such as Ag^{2+} and Ag^{3+}, are significantly more effective as oxidizing agents than Ag^0 and Ag^+ but are unstable in aqueous environments, especially at water temperatures near 100°C. In natural waters, silver may exist as metalloorganic complexes or adsorbed to organic materials. In freshwater and soils, the primary silver compounds under oxidizing conditions are bromides, chlorides, and iodides; under reducing conditions the free metal and silver sulfide predominates. In river water, one study showed silver present as the monovalent ion (Ag^+) at 53–71% of the total silver, as silver chloride (AgCl) at 28–45%, and as silver chloride ion ($AgCl_2^-$) at 0.6–2.0%. Increasing salinity of brackish and marine waters increased concentrations of silver chloro complexes ($AgCl^0$, $AgCl_2^-$, $AgCl_3^{2-}$, $AgCl_4^{3-}$); these chloro complexes retain some silver in dissolved form, and relatively small anthropogenic quantities can substantially enrich the environment. In the open ocean, the principal dissolved form of silver is $AgCl_2^-$, but the most bioavailable form may be the neutral monochloro complex AgCl.

Sorption is the dominant process that controls silver partitioning in water and its movement in soils and sediments. Silver may leach from soils into groundwater; the leaching rate increases with decreasing pH and increasing drainage. Silver adsorbs to manganese dioxide, ferric compounds, and clay minerals, and these compounds are involved in silver deposition into sediments; sorption by manganese dioxide and precipitation with halides reduce the concentration of dissolved silver, resulting in higher concentrations in sediments than in the water column. Under reducing conditions, adsorbed silver in sediments may be released and subsequently reduced to metallic silver or combine with reduced sulfur to form the insoluble silver sulfide. Sediments may be a significant source of silver to the water column. In one study, anoxic sediments containing 1.0–27.0 g of silver/kg DW and 10.0 mmol of acid volatile sulfide/kg DW were resuspended in oxygenated seawater for several hours to days. The seawater in contact with sediment containing 10.8 g/kg had 20.0 μg Ag/L; seawater in contact with sediments containing 27.0 g Ag/kg had about 2000.0 μg Ag/L, which seems to be the solubility of silver in seawater.

The global biogeochemical movements of silver are characterized by releases to the atmosphere, water, and land by natural and anthropogenic sources, long-range transport of fine particles in the atmosphere, wet and dry deposition, and sorption to soils and sediments. The chief source of silver contamination of water is silver thiosulfate complexes in photographic developing solutions that photofinishers discard directly to sewers. Secondary waste treatment converts most of the silver thiosulfate complex to insoluble silver sulfide and forms some metallic silver. About 95% of the total silver is removed in publicly owned treatment works from inputs containing municipal sewage and commercial photoprocessing effluents; effluents usually contained less than 0.07 μg ionic silver/L and concentrations were independent of the influent silver concentration. Silver in sewage treatment plant effluents may be associated with suspended particles or be present as thiosulfate complex, colloidal silver complex, colloidal silver chloride, silver sulfide, or soluble organic complexes. Silver on suspended matter and in colloidal forms and insoluble salts ultimately settles out in the sediments. At the water treatment plant, most of the silver is precipitated after treatment with lime or adsorbed after treatment with alum-flocculent. Chlorination converts some silver to silver chloride or to a soluble silver chloride complex. Aerobic biodegradation of a photoprocessing wastewater containing 1.85 mg total Ag/L did not adversely affect the activated sludge process. Practically all silver became associated with the sludge solids at 1840.0 mg Ag/kg mixed liquor suspended solids. When fresh sludge and aerobically digested sludge solids were subjected to leaching procedures, the resulting silver concentration was at least 40 times lower than the regulatory limit of 5.0 mg/L.

Forms of silver in atmospheric emissions are probably silver sulfide, silver sulfate, silver carbonate, silver halides, and metallic silver. About 50% of the silver released into the atmosphere from industrial operations is transported more than 100 km and is eventually deposited in precipitation. Minute amounts of ^{110m}Ag have been detected in natural waters and is attributed to atmospheric fallout from nuclear explosions.

A variety of spectrographic, colorimetric, polarographic, and other analytical techniques are used for routine measurement of silver in biological and abiotic samples. The detection limit of silver in biological tissues with scanning electron microscopy and X-ray energy spectrometry is 0.02 μg/kg and sometimes as low as 0.005 μg/kg. In air, water, and soil samples, the preferred analytical procedures include flame and furnace atomic absorption spectrometry, plasma emission spectroscopy, and neutron activation. Sensitive anodic stripping voltammetry techniques have recently been developed to measure free silver ion in surface waters at concentrations as low as 0.1 μg/L.

29.3.2 Metabolism

The acute toxicity of silver to aquatic species varies drastically by the chemical form and correlates with the availability of free ionic silver. In natural aquatic systems, ionic silver is rapidly complexed and sorbed by dissolved and suspended materials that are usually present. Complexed and sorbed silver species in natural waters are at least one order of magnitude less toxic to aquatic organisms than the free silver ion. Thus, silver nitrate – which is strongly dissociated – is extremely toxic to rainbow trout (*Oncorhynchus mykiss*); the 7-day LC50 value is 9.1 μg/L. Silver thiosulfate, silver chloride, and silver sulfide were relatively benign (7-day LC50 values >100,000.0 μg/L), presumably due to the abilities of the anions to remove ionic silver from solution.

For freshwater fish, the acute toxicity of silver is caused solely by Ag^+, interacting at the gills, inhibiting basolateral Na^+, K^+-ATPase activity. Disruption of this enzyme inhibits active Na^+ and Cl^- uptake and therefore osmoregulation by the fish. The primary toxic mechanism of silver in rainbow trout is the interruption of ionic regulation at the gills, stopping active Na^+ and Cl^- uptake without increasing passive efflux, thereby causing net ion loss. However concentrations of silver in the gills of rainbow trout did not correlate to Ag^+ concentrations in the medium and no correlation was found between gill silver levels and either Na^+ influx rates or gill Na^+, K^+-ATPase activity. The sites of action of silver toxicity in rainbow trout may be inside the cells of the gill epithelium rather than at the external surface and linked to carbonic anhydrase – a gill enzyme involved in Na^+ and Cl^- transport. Silver concentrations and metallothionein levels in gills and livers of rainbow trout increased with increasing exposure to silver; internal toxicity associated with increased silver accumulations may be lessened by the formation of silver-induced metallothioneins. A key toxic effect of Ag^+ in freshwater is the inhibition of branchial Na^+, K^+-ATPase activity which leads to blockade of active Na^+ and Cl^- across the gills; increased metabolic ammonia production and internal build up occur as part of this acute stress syndrome. The probable cause of hyperventilation in rainbow trout exposed to silver nitrate was a severe metabolic acidosis manifested in decreased arterial plasma pH and HCO_3^- levels. Lethality of ionic silver to trout is probably due to surface effects at the gills – disrupting Na^+, Cl^-, and H^+ – causing secondary fluid volume disturbance, hemoconcentration, and eventual cardiovascular collapse. Acidosis in rainbow trout – due to a net uptake of acidic equivalents from the water – in the intracellular compartment accounts for the continual loss of K^+ to the water in the absence of any change in plasma K^+.

In seawater, silver nitrate is less toxic than in freshwater. This difference is probably due to the low concentration of free Ag^+ (the toxic moiety in freshwater) in seawater, the high levels of chloride and the predominance of negatively charged silver-chloro complexes. However, high levels of silver nitrate are toxic to marine invertebrates despite the absence of Ag^+, and this is attributed to the bioavailability of stable silver-chloro complexes. In seawater, in contrast to freshwater, plasma Na^+ and Cl^- rise rather than fall, and death may result from the elevated Na^+ and Cl^- concentrations combined with dehydration. Osmoregulatory failure occurs in marine teleosts exposed to high concentrations of Ag^+ and the intestine is the main toxic site of action.

Ionic silver interferes with calcium metabolism of frogs and marine polychaete worms. Silver ions cause muscle fibers of frogs (*Rana* spp.) to deteriorate by allowing excess calcium to enter the cell. Studies with frog skeletal muscle fibers exposed to 1.08 mg/L showed that silver activated the calcium ion channel by acting on sulfhydryl groups in a calcium ion channel protein. In marine polychaetes contaminated with silver, the calcium content of nephridial cells was reduced, although silver was not detected in the calcium vesicles. Silver binds with protein sulfhydryl groups and this process protects the worm against silver poisoning. In marine mollusks, however, sulfide anion was the ligand of silver. In marine gastropods (*Littorina littorea*), silver was stored in the basement membranes of the digestive system; in clams (*Scrobicularia plana*), it was stored in the basement membrane of the outer fold of the mantle edge and in the amoebocytes. The availability of free silver in marine environments was strongly controlled by salinity because of the affinity of silver for the chloride ion. Silver sorbs readily to phytoplankton and to suspended sediments. As salinity increases, the degree of sorption decreases. Nearly 80% of silver sorbed to suspended sediments at low salinities desorb at higher salinities, but desorption does not occur when silver is associated with phytoplankton. Thus, silver incorporation in or on cellular material increases the retention of silver in the estuary, reducing the rate of transport.

Silver may enter the body of mammals through inhalation, ingestion, mucous membranes, or broken skin. In most cases of occupational argyrosis, absorption occurs via the respiratory tract or at the eyes. Silver is retained by all body tissues; tissue concentrations are related to the dose, form of administered silver, and route of exposure. Silver also accumulates in mammalian tissues with increasing age of the individual, even if none is administered intentionally. Inside the body, silver is transported mainly in the protein fractions of plasma, presumably as silver albuminate or silver chloride. In mammals, the highest concentrations of silver are usually found in the liver and spleen and to some extent in the muscles, skin, and brain. The primary sites of silver deposition in the human body are the liver, skin, adrenals, lungs, muscle, pancreas, kidney, heart, and spleen; silver is also deposited in blood vessel walls, the trachea, and bronchi. Dogs exposed to silver by inhalation accumulated most of the administered dose in the liver; concentrations in the lung, brain, skin, and muscle were lower. Intravenous injection of silver produces accumulations in the spleen, liver, bone marrow, lungs, muscle, and skin. Intestinal absorption of silver by rodents, canids, and primates has been recorded at 10% or less after ingestion of radioactive silver; a value of 18% was estimated in a single human given radiosilver acetate and about 3–10% of the absorbed silver is retained in the tissues. In a human given radioactive silver, more than 50% of the whole body burden of silver was found in the liver after 16 days.

Deposition of silver in tissues of warm-blooded animals results from precipitation of relatively insoluble silver salts, such as silver chloride and silver phosphate. These insoluble salts may be transformed into soluble silver sulfide albuminates that bind or complex with RNA, DNA, and proteins, or may be reduced to metallic silver by ascorbic acid or catecholamines. In humans with argyria, the blue or gray skin discoloration is caused by the photoreduction of silver chloride to metallic silver during exposure to ultraviolet light. Metallic silver, in turn, is oxidized and bound as black silver sulfide. Silver sulfide (Ag_2S) is localized in extracellular structures such as basement membranes and in macrophageous cells. Before storage as a stable mineral combination, silver binds to proteins that contain a large proportion of sulfhydryl groups such as metallothioneins. The last stage in the catabolic pathway of these proteins leads to storage of silver after reaction with a sulfur ligand. These mechanisms explain why liver, the most important organ for protein synthesis, shows the highest capacity for silver accumulation. High concentrations of silver in the digestive tract are linked to the numerous basement membranes contained in its tissues. Interspecies differences in the ability to accumulate, retain, and eliminate silver are large.

The enzyme-inhibiting action of silver ions may be due to the binding of sulfhydryl groups of some enzymes. Binding, in certain enzymes, is probably at a histidine imidazole group; in the case of glucose oxidase, silver ions compete with molecular oxygen as a hydrogen acceptor. About 60% of the silver in liver and kidneys of silver-injected rats were in the cytosol fractions bound to the high-molecular-weight proteins and metallothionein fractions; however, in spleen and brain only 30% of the total tissue silver was found in the cytosol fractions. At moderate doses (0.4 mg Ag/kg BW) in rats, the liver handles most of the absorbed silver from the body in the bile; at higher doses, silver deposits are markedly increased in the skin. In house sparrows (*Passer domesticus*), a silver binding protein was identified in liver after radiosilver-110m injection; the protein was heat-stable, resistant to low pH, and of low molecular weight. The properties of the hepatic silver-binding protein in birds were similar to other studied metallothioneins, but more research is needed to distinguish differences from mammalian metallothioneins.

Most absorbed silver is excreted into the intestines by way of the liver into the bile and subsequently excreted in feces; urinary excretion of silver is generally very low. Rodents, monkeys, and dogs given radioactive silver salts by oral and other routes excreted more than 90% of the absorbed dose in the feces. Rats injected intravenously with radioactive silver nitrate excreted silver in bile mainly bound to a low-molecular-weight complex similar to glutathione. Excretion was faster and percentages excreted by mice, rats, monkeys, and dogs were larger when silver was administered orally than by intravenous or intraperitoneal injection.

Among mammals, low doses of ingested silver were eliminated from the body within one week. In rats, mice, and rabbits, about 99% of a single oral dose of silver was eliminated within 30 days. Time for 50% clearance of silver in rats, mice, monkeys, and dogs after oral ingestion was about one day; this short half-time is due, in part, to fecal elimination of unabsorbed silver; the half-times were longer (1.8–2.4 days) after intravenous injection. Rodents dosed with silver accumulated high initial concentrations in the liver, which greatly decreased within 10 days; however, silver concentrations in spleen and brain were retained for longer periods. The biological half-time of radiosilver in rats given a single intraperitoneal injection was 40 h in whole blood, plasma, kidney, and liver; 70 h in spleen; and 84 h in brain. After exposure by inhalation, dogs cleared 59% of an administered dose of radiosilver-110m from the lungs in 1.7 days and from the liver in 9 days. The mean daily intake of silver in humans is about 88.0 μg; about 60.0 μg is excreted daily in the feces. In humans, the whole body effective half-time of persistence was 43 days. The biological half-time of silver in the lungs of an exposed person was about 1 day; in liver it was 52 days. In humans, 80% of the retained silver in lung was cleared in about 1 day; 50% of the remainder was usually cleared in 3 days. In persons who had accidentally inhaled radiosilver-110m, most of the inhaled silver had a half-time persistence of about 1 day, probably because of rapid mucociliary clearance, swallowing, and fecal excretion;

most of the absorbed radiosilver translocated to the liver.

Silver interacts competitively with selenium, vitamin E, and copper and induces signs of deficiency in animals fed adequate diets or aggravates signs of deficiency when diets were lacking one or more of these nutrients; antagonistic effects of silver have been described in dogs, pigs, rats, sheep, chicks, turkey poults, and ducklings. Conversely, the addition of selenium, copper, or vitamin E to diets of turkey poults decreased the toxicity of diets containing 900.0 mg Ag/kg. Dietary administration of silver acetate antagonized selenium toxicity; silver prevented growth depression and death in chicks fed diets containing excess selenium. The addition of selenium to the diets of rats exposed to silver in drinking water prevented growth retardation but increased the concentration of silver in liver and kidneys. Silver deposits in rat liver, kidneys, and other internal organs were in the form of sulfides; under high selenium exposure, the sulfur can be replaced with selenium and formation of silver selenide deposits in the liver may be considered a silver detoxification process.

29.4 Concentrations in Field Collections

Silver is comparatively rare in the earth's crust – sixty seventh in order of natural abundance of elements; the crustal abundance is an estimated 0.07 mg/kg and predominantly concentrated in basalt (0.1 mg/kg) and igneous rocks (0.07 mg/kg). Silver concentrations in nonbiological materials tend to be naturally elevated in crude oil and in water from hot springs and steamwells. Anthropogenic sources associated with the elevated concentrations of silver in nonliving materials include smelting, hazardous waste sites, cloud seeding with silver iodide, metals mining, sewage outfalls, and especially the photoprocessing industry. Silver concentrations in biota were greater in organisms near sewage outfalls, electroplating plants, mine wastes, and silver-iodide seeded areas than in conspecifics from more distant sites.

29.4.1 Abiotic Materials

Maximum concentrations of total silver recorded in selected nonbiological materials were 36.5 ng/m^3 in air near a smelter in Idaho; 2.0 μg/m^3 in atmospheric dust; 0.1 μg/L in oil well brines; 6.0 μg/L in groundwater near a hazardous waste site; 8.9 μg/L in seawater from Galveston Bay, Texas; 260.0 μg/L in the Genesee River, New York – the recipient of photoprocessing wastes; 300.0 μg/L in steam wells; 300.0 μg/L in treated photoprocessing wastewaters; 4500.0 μg/L in precipitation from clouds seeded with silver iodide; 31.0 mg/kg in some Idaho soils; 43.0 mg/L in water from certain hot springs; 50.0 mg/kg in granite; as much as 100.0 mg/kg in crude oils; 150.0 mg/kg in some Genesee River sediments; and 27,000.0 mg/kg in some solid wastes from photoprocessing effluents. It is emphasized that only a small portion of the total silver in each of these compartments is biologically available. For example, typical publicly owned treatment works receiving photoprocessing effluents show silver removal efficiencies greater than 90%; the mean concentration of free silver ion present in the effluents from these plants ranged from 0.001 to 0.07 μg/L.

Silver is usually found in extremely low concentrations in natural waters because of its low crustal abundance and low mobility in water. One of the highest silver concentrations recorded in freshwater of 38.0 μg/L occurred in the Colorado River at Loma, Colorado, downstream of an abandoned gold–copper–silver mine, an oil shale extraction plant, a gasoline and coke refinery, and a uranium processing facility. The maximum recorded value of silver in tapwater in the United States was 26.0 μg/L – significantly higher than finished water from the treatment plant (maximum of 5.0 μg/L) – because of the use of tin–silver solders for joining copper pipes in the home, office, or factory.

In general, silver concentrations in surface waters of the United States decreased between

1970–74 and 1975–79, although concentrations increased in the north Atlantic, Southeast, and lower Mississippi basins. About 30–70% of the silver in surface waters may be ascribed to suspended particles, depending on water hardness or salinity. For example, sediments added to solutions containing 2.0 μg Ag/L had 74.9 mg Ag/kg DW sediment after 24 h in freshwater, 14.2 mg/kg DW at 1.5% salinity and 6.9 mg/kg DW at 2.3% salinity. Riverine transport of silver to the ocean is considerable: suspended materials in the Susquehanna River, Pennsylvania – that contained as much as 25.0 mg silver/kg – resulted in an estimated transport of 4.5 metric tons of silver to the ocean each year. Measurements of silver in rivers, lakes and estuaries using clean techniques show levels of about 0.01 μg/L for pristine, nonpolluted areas and 0.01–0.1 μg/L in urban and industrialized areas.

Emissions of silver from coal-fired power plants may lead to accumulations in nearby soils. Silver in soils is largely immobilized by precipitation to insoluble salts and by complexation or adsorption by organic matter, clays, and manganese and iron oxides.

Silver can remain attached to oceanic sediments for about 100 years under conditions of high pH, high salinity, and high sediment concentrations of iron, manganese oxide, and organic. Estuarine sediments that receive metals, mining wastes, or sewage usually have higher silver concentrations (>0.1 mg/kg DW) than noncontaminated sediments. Silver is tightly bound by sewage sludge, and elevated silver concentrations in sediments are often characteristic of areas near sewage outfalls. In the absence of sewage, silver in oxidized sediments is associated with oxides of iron and with humic substances. Sediments in the Puget Sound, Washington, were significantly enriched in silver, in part, from human activities; concentrations were higher in fine-grained particles. Marine annelids and clams accumulate dissolved and sediment-bound forms of silver. Uptake of silver from sediments by marine polychaete annelids decreased in sediments with high concentrations of humic substances or copper but increased in sediments with elevated concentrations of manganese or iron.

29.4.2 Plants and Animals

Maximum concentrations of total silver recorded in field collections of living organisms, in mg Ag/kg DW, were 1.5 in liver of marine mammals, 2.0 in liver and 6.0 in bone of trout from ecosystems receiving precipitation from silver-iodide seeded clouds, 7.0 in kidneys and 44.0 in liver of birds from a metal-contaminated area, 14.0 in marine algae and macrophytes, 30.0 in whole annelid worms from San Francisco Bay, 72.0 in skin of humans afflicted with argyria, 110.0 in whole mushrooms, 133.0–185.0 in soft parts of clams and mussels near sewage and mining waste outfalls, and 320.0 in whole gastropods from South San Francisco Bay. Silver concentrations in conspecifics from areas remote from anthropogenic contamination were usually lower by one or more orders of magnitude. The strong bioaccumulation of silver in marine benthic organisms from sediments is a major cause for concern, and with potential for adverse effects on reproduction. High accumulation of silver from marine sediments is attributed, in part, to the formation of stable chlorocomplexes of silver with chlorine which, in turn, favor the distribution and accumulation of silver.

Silver is a normal trace constituent of many organisms. In terrestrial plants, silver concentrations are usually less than 1.0 mg/kg ash weight (equivalent to less than 0.1 mg/kg DW) and are higher in trees, shrubs, and other plants near regions of silver mining; seeds, nuts, and fruits usually contain higher silver concentrations than other plant parts. Silver accumulations in marine algae (max. 14.1 mg/kg DW) are due mainly to adsorption rather than uptake; silver bioconcentration factors of 13,000–66,000 in marine algae are not uncommon.

Silver concentrations in mollusks vary widely between closely related species and among conspecifics from different areas. The inherent differences in ability to accumulate silver among bivalve mollusks are well documented (oysters ≫ scallops ≫ mussels). The highest silver concentrations in all examined species of mollusks were in the internal organs, especially in the digestive gland and kidneys.

Elevated concentrations of silver (5.3 mg/kg DW) in shells of limpets from uncontaminated sites suggest that silver may actively participate in carbonate mineral formation, but this needs verification. In general, silver concentrations were elevated in mollusks collected near port cities and in the vicinities of river discharges, electroplating plant outfalls, ocean dumpsites, and urban point sources including sewage outfalls and from calcareous sediments rather than detrital organic or iron oxide sediments. Season of collection and latitude also influenced silver accumulations. Seasonal variations in silver concentrations of Baltic clams (*Macoma balthica*) were associated with seasonal variations in soft tissue weight and frequently reflected the silver content in the sediments. Oysters from the Gulf of Mexico vary considerably in whole body concentrations of silver and other trace metals. Variables that modify silver concentrations in oyster tissues include the age, size, sex, reproductive stage, general health, and metabolism of the animal; water temperature, salinity, dissolved oxygen, and turbidity; natural and anthropogenic inputs to the biosphere; and chemical species and interactions with other compounds. Silver concentrations in whole American oysters (*Crassostrea virginica*) from the Chesapeake Bay were reduced in summer; reduced at increasing water salinities, and elevated near sites of human activity; chemical forms of silver taken up by oysters included the free ion (Ag^+) and the uncharged $AgCl^0$. Declines in tissue silver concentrations of the California mussel (*Mytilus californianus*) were significant between 1977 and 1990; body burdens decreased from 10.0 to 70.0 mg/kg DW to less than 2.0 mg/kg DW and seem to be related to the termination of metal plating facilities in 1974 and decreased mass emission rates by wastewater treatment facilities.

Among arthropods, pyrophosphate granules isolated from barnacles have the capability to bind and effectively detoxify silver and other metals under natural conditions. In a Colorado alpine lake, silver concentrations in caddisflies and chironomid larvae usually reflected silver concentrations in sediments; seston, however, showed a high correlation with lake water silver concentrations from 20 days earlier.

Silver concentrations in fish muscle rarely exceeded 0.2 mg/kg DW and usually were less than 0.1 mg/kg FW; livers contained as much as 0.8 mg/kg FW, although values greater than 0.3 mg/kg FW were unusual; and whole fish contained as much as 0.225 mg/kg FW. Livers of Atlantic cod (*Gadus morhua*) contained significantly more silver than muscles or ovaries; a similar pattern was evident in other species of marine teleosts. Accumulations of silver in offshore populations of teleosts is unusual, even among fishes collected near dumpsites impacted by substantial quantities of silver and other metals. For example, of the 7 species of marine fishes from a disposal site in the New York Bight and examined for silver content, concentrations were highest (0.15 mg/kg FW) in muscle of blue hake (*Antimora rostrata*). Similarly, the elevated silver concentration of 0.8 mg/kg FW in liver of winter flounder (*Pleuronectes americanus*) was from a specimen from the same general area.

Silver concentrations in muscle of Antarctic birds were low (0.01 mg/kg DW) when compared to livers (0.02–0.46 mg/kg DW) or feces (0.18 mg/kg DW). Silver concentrations in avian tissues, especially in livers, were elevated in the vicinity of metal-contaminated areas and in diving ducks from the San Francisco Bay. Birds with elevated concentrations of silver in tissues – as much as 44.0 mg/kg DW in liver in the common eider (*Somateria mollissima*) – seemed outwardly unaffected.

Silver in mammalian tissues is usually present at low or undetectable concentrations. The concentration of silver in tissues of 3 species of seals collected in the Antarctic during 1989 was highest in liver (1.55 mg/kg DW) and lowest in muscle (0.01 mg/kg DW); intermediate in value were kidney (0.29 mg/kg DW) and stomach contents (0.24 mg/kg DW). The mean concentration of silver in livers from normal female California sea lions (*Zalophus californianus*), having normal pups, was 0.5 mg/kg DW. Mothers giving birth to premature pups had only 0.4 mg Ag/kg DW liver. In general, *Zalophus* mothers delivering premature pups had lower concentrations in liver of silver, cadmium, copper, manganese, mercury, and zinc than mothers delivering normal pups. Silver concentrations in tissues of Antarctic seals were related to, and possibly governed by, concentrations of other metals. In muscle, silver inversely correlated with zinc; in liver, silver positively correlated with nickel, copper, and zinc; and in kidney, correlations between silver and zinc and between silver and cadmium were negative. In humans, silver is present in placentas and fetal livers; silver concentrations in tissues increase with age, and variations in tissue concentrations of silver are wide. The average daily intake of silver from all sources by humans is 88.0 μg, but very little of the silver ingested from nontherapeutic sources is retained.

29.5 *Lethal and Sublethal Effects*

As discussed later, free silver ion is lethal to representative species of sensitive aquatic plants, invertebrates, and teleosts at water concentrations of 1.2–4.9 μg/L. Adverse effects occur on development of trout at concentrations as low as 0.17 μg/L and on phytoplankton species composition and succession at 0.3–0.6 μg/L. Aquatic organisms accumulate silver from environmental sources. No data were found on effects of silver on avian or mammalian wildlife and all studied effects were on poultry and small laboratory mammals. Silver was not mutagenic, carcinogenic, or teratogenic to tested animals by normal routes of exposure. Adverse effects of silver on poultry occur at 1.8 mg/kg FW whole egg by way of injection (reduced survival), 10.0 mg/kg in diets (reduced hemoglobin with copper-deficient diet), 200.0 mg/kg in diets (growth suppression with copper-adequate diet), or when given drinking water containing 100.0 mg Ag/L (liver necrosis). Effects of silver on sensitive species of mammals include death at 13.9–20.0 mg/kg BW by intraperitoneal injection, histopathology of kidney and brain at 250.0–450.0 μg Ag/L drinking water, tissue accumulations at 6.0 mg/kg diet, and liver necrosis when fed diets containing more than 130.0 mg/kg. In humans, generalized argyria seems to be declining, which may be due to improved work conditions.

29.5.1 Terrestrial Vegetation

In general, accumulation of silver by terrestrial plants from soils is low, even if the soil is amended with silver-containing sewage sludge or the plants are grown on tailings from silver mines – where silver accumulated mainly in the root systems. Germination was the most sensitive stage to plants grown in solutions containing various concentrations of silver nitrate. Adverse effects on germination were expected at concentrations greater than 0.75 mg Ag/L for lettuce, and 7.5 mg/L for ryegrass (*Lolium perenne*) and other plants tested. Sprays containing 9.8 mg dissolved Ag/L kill corn (*Zea mays*), and sprays containing 100.0–1000.0 mg dissolved Ag/L kill young tomato (*Lycopersicon esculentum*) and bean (*Phaseolus* spp.) plants. Seeds of corn, lettuce (*Lactuca sativa*), oat (*Avena sativa*), turnip (*Brassica rapa*), soybean (*Glycine max*), spinach (*Spinacia oleracea*), and Chinese cabbage (*Brassica campestris*) were planted in soils amended with silver sulfide and sewage sludge to contain as much as 106.0 mg Ag/kg DW soil. All plants germinated and most grew normally at the highest soil concentration of silver tested; yields of lettuce, oat, turnip and soybean were higher on soils amended with silver-laden, waste-activated sludge than control soils. But growth of Chinese cabbage and lettuce was adversely affected at 14.0 mg Ag/kg DW soil and higher. Silver concentrations in edible portions from all plants at all soil levels of silver tested, except lettuce, were less than 80.0 μg/kg DW, suggesting that availability of sludge-borne silver sulfide to most agricultural crops is negligible. Lettuce grown in soil containing 5.0 mg Ag/kg DW had about 0.5 mg Ag/kg DW leaves, and in 120.0 mg/kg DW soil as much as 2.7 mg Ag/kg DW leaves vs. 0.03 mg/kg DW in controls.

29.5.2 Aquatic Organisms

In fish and amphibian toxicity tests with 22 metals and metalloids, silver was the most toxic tested element as judged by acute LC50 values. In solution, ionic silver is extremely toxic to aquatic plants and animals, and water concentrations of 1.2–4.9 μg/L killed sensitive species of aquatic organisms, including representative species of insects, daphnids, amphipods, trout, flounders, sticklebacks, guppies, and dace. At nominal water concentrations of 0.5–4.5 μg/L, accumulations in most species of exposed organisms were high and had adverse effects on growth in algae, clams, oysters, snails, daphnids, amphipods, and trout; molting in mayflies; and histology in mussels.

In general, silver ion was less toxic to freshwater aquatic organisms under conditions of low dissolved Ag^+ concentration, increasing water pH, hardness, sulfides, and dissolved and particulate organic loadings; under static test conditions when compared to flowthrough regimens; and when animals were adequately nourished when compared to starvation. Most authorities agree that increasing concentrations of dissolved organic carbon affords the highest protective effects against ionic silver. Among all tested species, the most sensitive individuals to silver were the young and those exposed to low water hardness or salinity. In the case of seawater-acclimatized rainbow trout, silver-induced mortality was greater at higher salinities. But the increased toxicity with salinity was linked to an incomplete hypoosmoregulatory ability and not to an increase in a more toxic $AgCl_n$ species. Sediment chemistry can affect toxicity of silver to marine amphipods (*Ampelisca abdita*) exposed for 10 days to sediments supplemented with various concentrations of silver. In general, sediments with an excess of acid volatile sulfide (AVS) relative to simultaneously extracted metal (SEM) were generally not toxic to marine amphipods. Sediments with an excess of SEM relative to AVS, and no measurable AVS, were generally toxic. Sediments with measurable AVS were not toxic. It is emphasized that silver-induced stress syndromes vary widely among animal classes. Among marine organisms, for example, silver ion was associated with respiratory depression in marine gastropods and cunners (*Tautagolabrus adspersus*), a teleost; however, silver ion increased oxygen consumption in six species of bivalve mollusks.

Sensitive aquatic plants accumulated silver from water containing as little as 2.0 μg Ag/L to whole cell burdens as high as 58.0 mg Ag/kg DW; grew poorly at 3.3–8.2 μg Ag/L during exposure for 5 days; and died at concentrations greater than 130.0 μg Ag/L. Some metals seem to protect aquatic plants against adverse effects of silver. Algae in small lakes that contained elevated concentrations of metals, especially copper and nickel, had higher tolerances to silver than did conspecifics reared in the laboratory under conditions of depressed concentrations of heavy metals. Species composition and species succession in Chesapeake Bay phytoplankton communities were significantly altered in experimental ecosystems continuously stressed by low concentrations (0.3–0.6 μg/L) of silver. At higher concentrations of 2.0–7.0 μg/L for 3–4 weeks, silver inputs caused disappearance of *Anacystis marina*, a mat-forming blue-green alga; increased dominance by *Skeletonema costatum*, a chain-forming centric diatom; and cell burdens of 8.6–43.7 mg Ag/kg DW. Dissolved silver speciation and bioavailability were important in determining silver uptake and retention by aquatic plants. Silver availability was controlled by the concentration of free silver ion (Ag^+) and the concentrations of other silver complexes, such as AgCl. Silver uptake by phytoplankton was rapid, in proportion to silver concentration, and inversely proportional to water salinity. Silver incorporated by phytoplankton was not lost as the salinity increased, and silver associated with cellular material was largely retained in the estuary. Diatoms (*Thalassiosira* sp.), for example, readily accumulated silver from the medium. Once incorporated, silver was tightly bound to the cell membrane, even after the cells were mechanically disrupted.

The ability to accumulate dissolved silver from the medium ranges widely between species. Some reported bioconcentration factors (mg Ag per kg FW organism/mg Ag per liter of medium) are 210 in diatoms, 240 in brown algae, 330 in mussels, 2300 in scallops, and 18,700 in oysters. Silver is the most strongly accumulated of all trace metals by marine bivalve mollusks. Studies with radiosilver-110m suggest that the half-time persistence of silver is 27 days in mussels, 44–80 days in clams, and more than 180 days in oysters. In oysters and other bivalve mollusks, the major pathway of silver accumulation was from dissolved silver; uptake was negligible from silver adsorbed onto suspended sediments or algal cells, and oysters eliminated adsorbed silver in the feces. Sometimes, benthic bivalve mollusks accumulated silver from certain sediments. Sediment-bound silver was taken up by the Baltic clam (*Macoma balthica*) at 3.6–6.1 times the concentration in calcite sediments but less than 0.85 times from manganous, ferrous, and biogenic $CaCO_3$ sediments. In oysters, silver associated with food was unavailable for incorporation which may be due to the ability of silver to adsorb rapidly to cell surfaces and to remain tightly bound despite changes in pH or enzymatic activity. Silver concentrations in American oysters (*Crassostrea virginica*) held in seawater solutions containing 1.0 mg Ag/L for 96 h rose from 6.1 mg/kg FW soft parts to 14.9 mg/kg FW; in gills, these values were 5.9 and 33.9 mg/kg FW. A similar pattern was evident in common mussels (*Mytilus edulis*) and quahaug clams (*Mercenaria mercenaria*). Adults of surf clams (*Spisula solidissima*) immersed for 96 h in seawater containing 10.0 μg Ag/L had 1.0 mg Ag/kg FW soft tissues vs. 0.08 mg/kg in controls. Oysters accumulated radiosilver-110m from the medium by factors of 500–32,000; uptake of dissolved silver by oysters was higher at elevated temperatures in the range of 15–25°C. American oysters maintained near a nuclear power plant in Maryland that discharged radionuclides on a daily basis into the Chesapeake Bay accumulated radiosilver-110m; accumulations were higher in summer and fall than in winter and spring.

Marine gastropods exposed to concentrations as low as 1.0 μg Ag/L for as long as 24 months showed histopathology and accumulations as high as 34.0 mg Ag/kg FW soft parts; higher exposure concentrations of 5.0 and 10.0 μg Ag/L were associated with inhibited reproduction and whole body burdens as high as 87.0 mg Ag/kg FW. Histopathological findings in silver-exposed mussels (*Mytilus edulis*) were typical of argyria in humans and other mammals that have absorbed organic or inorganic silver compounds. Juvenile Pacific oysters (*Crassostrea gigas*) exposed for 2 weeks to solutions containing 20.0 μg Ag/L had high silver accumulations in tissues and a reduced capacity to store glycogen; however, after 30 days of depuration, glycogen storage capacity was restored and 80% of the soluble silver and 27% of the insoluble forms were eliminated, suggesting recovery to a normal physiological state. About 70% of the insoluble silver in Pacific oysters was sequestered as Ag_2S, a stable mineral form that is not degradable, thereby limiting the risk of silver transfer through the food chain. Most (69–89%) of the silver accumulated from the medium in soft tissues of oysters and clams was sequestered in amoebocytes and basement membranes; in scallops and mussels, silver was stored in basement membranes and pericardial gland. In all species of bivalve mollusks, sequestered silver was in the form of silver sulfide. American oysters excreted about 60% of their accumulated silver in soft tissues within 30 days of transfer to silver-free seawater; soluble forms were preferentially eliminated and insoluble forms retained. Interspecies differences in ability to retain silver among bivalve mollusks are large, even among closely related species of crassostreid oysters. For example, the half-time persistence of silver was about 149 days in American oysters but only 26 days in Pacific oysters.

Among arthropods, grass shrimp (*Palaemonetes pugio*) rapidly incorporate silver dissolved in brackish water in proportion to its concentration but not from planktonic or detrital food sources containing elevated silver burdens. Variations in ability of decapod crustaceans to accumulate radiosilver-110m from seawater are large, as judged by concentration factors that ranged from 70 to 4,000. The reasons for this variability are unknown but may be associated with hepatopancreas morphology. It is generally acknowledged that hepatopancreas or digestive gland is the major repository of silver in decapods. Aquatic insects concentrate silver in relative proportion to environmental levels, and more efficiently than most fish species. Whole body bioconcentration factors (BCFs) of silver in three species of aquatic insects ranged from

21 to 240 in water containing 30.0–65.0 mg $CaCO_3$/L during exposure of 3–15 days; in bluegill sunfish (*Lepomis macrochirus*), this value was less than 1 after exposure for 28 days. Molt frequency of the stonefly (*Isonychia bicolor*) was a sensitive indicator of silver stress over time, and 1.6 μg total Ag/L over a 20-day period inhibited molting. In freshwater sediments supplemented with silver nitrate, a high proportion of the dissolved silver fraction was not readily bioavailable to cause lethality to dipteran insect larvae. Porewater concentrations of dissolved silver that killed 50% of the dipteran larvae were up to 275 times greater than the 10-day water with LC50 value of 57.0 μg/L only, indicating that most of the dissolved fraction was not readily bioavailable to cause death. Concentrations of silver in these sediments that caused significant adverse effects – 200.0–500.0 mg Ag/kg DW – in dipteran larvae were markedly above silver concentrations usually reported in the environment.

Silver ion (Ag^+) was the most toxic chemical species of silver to fishes. Silver ion was 300 times more toxic than silver chloride to fathead minnows (*Pimephales promelas*), 15,000 times more toxic than silver sulfide, and more than 17,500 times more toxic than silver thiosulfate complex; in all cases, toxicity reflected the free silver ion content of tested compounds; a similar pattern was noted in rainbow trout. Silver was less toxic to fathead minnows under conditions of increasing water hardness between 50 and 250 mg $CaCO_3$/L, increasing pH between 7.2 and 8.6, and increasing concentrations of humic acid and copper; starved minnows were more sensitive to ionic silver than minnows fed regularly. Eggs of rainbow trout (*Oncorhynchus mykiss*) exposed continuously to silver concentrations as low as 0.17 μg/L had increased embryotoxicity and hatched prematurely; resultant fry had a reduced growth rate. Removal of the egg capsule of eyed embryos of steelhead trout (*O. mykiss*) significantly lowered the resistance of the embryos to salts of silver, copper, and mercury but not zinc and lead. Silver accumulation in gills of juvenile rainbow trout exposed to 11.0 μg Ag/L for 2–3 h was significantly inhibited by various cations (Ca^{2+}, Na^+, H^+) and complexing agents (dissolved organic carbon, thiosulfate, chloride); these variables must be considered when constructing predictive models of silver binding to gills.

Largemouth bass (*Micropterus salmoides*) and bluegills accumulated silver from the medium; accumulations increased with increasing concentrations of ionic silver and increasing duration of exposure. Dietary silver sulfide exposure at or below 3000.0 mg Ag/kg ration is physiologically benign to juvenile rainbow trout over a 58-day period, although other forms of silver at dietary concentrations as low as 3.0 mg/kg DW ration can accumulate in liver of juvenile trout at concentrations 12-fold greater than controls after 3 months. Bioconcentration factors of radiosilver-110m and various species of teleosts were as high as 40 after 98 days. However, flounders (*Pleuronectes platessa*) and rays (*Raja clavata*) fed nereid polychaete worms labeled with radiosilver-110m retained about 4.2% of the ingested dose after 3 days, which suggests that high silver concentration factors reported may have been due to loosely bound adsorbed silver. Flounders (*Pleuronectes* sp.) held in seawater solutions containing 40.0 μg Ag/L for 2 months had elevated silver concentrations in the gut (0.49 mg Ag/kg FW) but less than 0.05 mg/kg in all other examined tissues. Similarly exposed rays (*Raja* sp.) contained 1.5 mg Ag/kg FW in liver, 0.6 in gut, 0.2 in heart, and 0.05–0.18 mg/kg FW in spleen, kidney, and gill filament; liver is usually considered the major repository of silver in teleosts. In tidepool sculpins (*Oligocottus maculatus*), ionic silver was more toxic at lower salinities, longer exposure durations, and increasing medium ammonia concentrations; however, there was no correlation between whole body silver burden and toxicity at 2.5% salinity, and no uptake at 3.2% salinity.

At concentrations normally encountered in the environment, food chain biomagnification of silver in aquatic systems is unlikely, although regular ingestion of fish from contaminated waters may significantly affect dietary silver intake. Silver – as thiosulfate-complexed silver at nominal concentrations of 500.0 or 5000.0 μg Ag/L – was concentrated and magnified over a 10-week period in freshwater food chains of algae, daphnids, mussels, and fathead minnows, although the mechanisms of accumulation in this study were imperfectly understood. Sediments contaminated with silver sulfide (Ag_2S), however, do not seem to pose a major route of entry into the aquatic food web. Juvenile amphipods (*Hyalella azteca*) held on sediments containing as much as 753.0 mg Ag/kg DW, as silver sulfide, for 10 days had normal growth and survival. And aquatic oligochaetes (*Lumbriculus variegatus*) held on sediments containing 444.0 mg Ag/kg DW, as silver sulfide, for 28 days had a low bioconcentration factor of 0.18. Trophic transfer of silver in marine herbivores, especially mussels, is dependent on the silver assimilation efficiency from ingested food particles, feeding rate, and silver efflux rate. Silver assimilation efficiency is usually less than 30% and lower for sediment than for phytoplankton. Silver assimilation efficiency and distribution from ingested phytoplankton particles is modified by gut passage time, extracellular and intracellular digestion rates, and metal desorption at lowered pH. A kinetic model for mussels predicts that either the solute or particulate pathway can dominate and is dependent on silver partition coefficients for suspended particles, and silver assimilation efficiency.

29.5.3 Birds and Mammals

No data were found on the effects of silver compounds on avian or mammalian wildlife. All controlled studies with silver were with domestic poultry, livestock, or small laboratory mammals. Signs of chronic silver ion intoxication in tested birds and mammals included cardiac enlargement, vascular hypertension, hepatic necrosis, anemia, lowered immunological activity, altered membrane permeability, kidney pathology, enzyme inhibition, growth retardation, and a shortened life span.

Silver affects turkeys (*Meleagris gallopavo*) and domestic chickens (*Gallus* spp.). Turkey poults on diets containing 900.0 mg Ag/kg feed for 4 weeks had reduced growth, hemoglobin, and hematocrit and an enlarged heart. Chicken eggs injected with silver nitrate at 0.1 mg Ag/egg (equivalent to about 1.8 mg Ag/kg egg FW) had a 50% reduction in survival but no developmental abnormalities. Adverse effects of silver were reported in normal chicks fed diets containing 200.0 mg Ag/kg ration (growth suppression) or given drinking water containing 100.0 mg Ag/L (liver necrosis). Chicks on copper-deficient diets had adverse effects at 10.0 mg Ag/kg ration (reduced hemoglobin; reversible when fed copper-adequate diet) and at 50.0–100.0 mg Ag/kg ration (growth suppression and increased mortality). Chicks that were deficient in vitamin E experienced reduced growth when given drinking water containing 1500.0 mg Ag/L. Chickens infected with *Salmonella pullorum-gallinarum* and *Escherichia coli* were cured with aerosol treatments containing 10.0 μg Ag/L air.

Studies with small laboratory mammals – which require verification – show that long-term exposure to high levels of silver nitrate in drinking water may result in sluggishness and enlarged hearts; however, these effects have not been observed in silver-exposed humans. Concentrations as high as 200.0 μg Ag/L in drinking water of test animals for 5 months had no significant effect on animal health or metabolism. But 400.0 μg Ag/L for 5 months caused kidney damage, and 500.0 μg/L for 11 months was associated with impaired conditioned-reflex activities, immunological resistance, and altered brain nucleic acid content. Diets deficient in vitamin E or selenium caused rapidly fatal hepatocellular necrosis and muscular dystrophy to rats if they contained the dietary-intake equivalent of 130.0 mg Ag/kg BW daily, a comparatively high silver ion intake.

The extent of absorption of an administered dose of silver depends on silver speciation, the presence and extent of silver-binding proteins, and other variables. But absorption is dependent mainly on the transit time through the gastrointestinal tract; the faster the transit time, the less silver is absorbed. Transit times ranged from about 8 h in mice and rats to about 24 h in monkeys, dogs, and humans. Route of administration affected the excretion rate of silver. Clearance of silver from mammals 2 days after silver was administered intravenously ranged from 15% in dogs to 82% in mice; clearance rates were intermediate in monkeys and rats. When silver was administered orally, clearance was more rapid, and extended from 90.4% in dogs to 99.6% in mice. The half-time persistence of silver administered orally to mice was 0.1 day for the short-lived component and 1.6 days for the long-lived component. Other species of tested laboratory animals had biphasic or triphasic whole-body silver-excretion profiles that differed significantly from mice. Monkeys, for example, had a biphasic excretion profile with peaks at 0.3 and 3.0 days; rats had a triphasic profile with peaks at 0.1, 0.7, and 5.9 days; and dogs had half-time persistence peaks at 0.1, 7.6, and 33.8 days.

Ionic silver is lethal to mice (*Mus* spp.) at 13.9 mg/kg BW by intraperitoneal injection, to rabbits (*Oryctolagus* spp.) at 20.0 mg/kg BW intraperitoneally, to dogs (*Canis familiaris*) at 50.0 mg/kg BW by intravenous injection, to humans at greater than 166.0 mg/kg BW in a single dose, and to rats (*Rattus* spp.) at 1586.0 mg/L drinking water for 37 weeks. Sublethal effects are reported in rabbits given 250.0 μg Ag/L drinking water (brain histopathology), in rats given 400.0 μg Ag/L drinking water for 100 days (kidney damage), in mice given 95.0 mg Ag/L drinking water for 125 days (sluggishness), in guinea pigs (*Cavia* sp.) given 81.0 mg Ag/cm^2 skin applied daily for 8 weeks (reduced growth), and in rats given diets containing 6.0 mg Ag/kg for 3 months (high accumulations in kidneys and liver) or 130.0–1110.0 mg/kg diet (liver necrosis).

The connections between human cancers and silver as a causal agent are tenuous. All available evidence is negative or inconclusive on silver's ability to induce cancer, mutagenicity, or birth defects in animals by normal routes of exposure. Silver pellets, however, implanted under the skin of rodents, have caused sarcomas, malignant fibrosarcomas, fibromas, fibroadenomas, and invasions of muscle with connective tissue; in these cases, silver seems to act as a nonspecific irritant rather than as a specific carcinogen. Intratumoral injections of colloidal silver promotes cancer growth in rats, possibly by producing an area of lowered tissue resistance that allows resistant cancer cells to grow freely; however, silver nitrate seems to be a tumor inhibitor in mice.

In humans, acute toxic effects of silver have resulted only from accidental or suicidal overdoses of medical forms of silver. Symptoms of acute silver poisoning in patients dying after intravenous administration of Collargo (silver plus silver oxide) included gastrointestinal disturbances, pulmonary edema, tissue necrosis, and hemorrhages in bone marrow, liver, and kidney. High sublethal doses of silver nitrate taken orally cause some patients to experience violent abdominal pain, abdominal rigidity, vomiting, and severe shock; systemic effects among recovering patients are unlikely, although degenerative liver changes may occur. In humans, skin contact with silver compounds may cause mild allergic reactions such as rash, swelling, and inflammation; industrial and medicinal exposures to silver may cause lesions of the kidneys and lungs, and arteriosclerosis; colloidal silver compounds may interfere with nasal ciliary activity; and exposure to dust containing high levels of silver compounds, such as silver nitrate or silver oxide, may cause breathing problems, lung and throat irritation, and stomach pain.

Chronic exposure of humans to silver or silver compounds frequently resulted in generalized argyria (slate-gray pigmentation of the skin and hair caused by deposition of silver), localized argyria (limited areas of pigmentation usually associated with medicinal silver applications), or argyrosis (argyria of the eye). Every silver compound in common chemical use has caused generalized argyria, usually from medical and occupational exposures. In generalized argyria, skin pigmentation was highest in light-exposed areas, although silver concentrations in light-exposed and dark-exposed skin were the same. In severe cases of argyria, the skin may become black with a metallic luster, the eyes affected to the point that vision is disturbed, and the respiratory tract impaired. Individual variability in susceptibility to argyria is great and this is probably explained by the variability in absorption and retention of silver. Generalized argyria as an occupational disease is unusual but has been reported in workers that make silver nitrate or are involved in mirror plating, glass bead silvering, silver Christmas cracker manufacturing, photographic plate manufacturing, and silver mining. Generalized argyria was also associated with chronic inhalation or ingestion of silver fulminate, silver nitrate, silver albuminate, and silver cyanide. Improved workplace ventilation and sanitation among silver nitrate workers effected a decline in general argyria.

Localized argyria is rare and usually occurs when silver compounds contact broken skin or mucous membranes. Localized argyria has been reported in workers who handle metallic silver in filing, drilling, polishing, turning, engraving, forging, soldering, or smelting operations. Silver polishers exposed for 25 years or more (range 2–38 years) sometimes exhibit increased densities in their lung X-rays due to silver impregnation of the elastic membranes of the pulmonary vessels. In one case, an Italian physician who dyed his facial hair with a silver dye for 25 years developed argyria in the conjunctiva of both eyes.

29.6 Recommendations

Most measurements of silver concentrations in natural waters prior to the use of clean techniques are considered inaccurate. Until analytical capabilities that exceed the dissolved-particulate classification are developed, it will be necessary to rely on laboratory and theoretical modeling studies to fully understand chemical speciation of silver in natural waters.

Factors governing the environmental fate of silver are not well characterized, including silver transformations in water and soil and the role of microorganisms. The toxic potential of silver chloride complexes in seawater and the role of sediments as sources of silver contamination for the food web needs more research. Food chain transfer of silver requires more current information on sources and forms of silver and data on concentrations in field collections of flora and fauna, especially near hazardous waste sites. Although silver in sewage sludge is mostly immobilized, data are limited on silver concentrations in flesh and milk of livestock pastured or fed grains raised on soils amended with sewage sludge. Data are needed on partition coefficients and vapor pressures of silver compounds and on silver concentrations in emissions from cement producers and smelters and refineries of copper, lead, zinc, silver, iron, and steel. Also, technology to recapture silver from waste media before it reaches the environment must be improved.

Silver criteria in aquatic ecosystems are under constant revision by regulatory agencies. For example, total recoverable silver is no longer recommended by the U.S. Environmental Protection Agency in silver criteria formulation and should be replaced by dissolved silver. Dissolved silver more closely approximates the bioavailable fraction of silver in the water column than does total recoverable silver. Dissolved silver criteria recommended are about 0.85 times those of total recoverable silver under certain conditions but may vary considerably depending on other compounds present in solution.

The proposed human drinking water criteria of 50.0 to <200.0 μg total Ag/L do not seem to represent a hazard to human health, although much lower concentrations adversely affect freshwater and marine organisms (Table 29.1). Proposed silver criteria for the protection of freshwater aquatic life during acute exposure range from 1.2 to 13.0 μg total recoverable silver per liter (Table 29.1). If all total recoverable silver were in the ionic form, these proposed criteria would overlap the 1.2–4.9 μg/L range found lethal to sensitive species of aquatic plants and animals and indicates that the proposed freshwater acute silver criteria need to be reexamined. For freshwater aquatic life protection during chronic exposure, the proposed criterion of less than 0.13 μg total recoverable silver per liter (Table 29.1) is probably protective. But the proposed silver criterion of 2.3 μg total silver/L to protect marine life needs to be reconsidered because phytoplankton species composition and succession are significantly altered at 0.3–0.6 μg total silver/L and because some species of marine algae and mollusks show extensive accumulations at 1.0–2.0 μg total silver/L. Limited but insufficient data were available on correlations between tissue residues of silver with health of aquatic organisms; additional research seems

Table 29.1. Proposed silver criteria for the protection of natural resources and human health.

Resource, Criterion, and Other Variables	Effective Silver Concentration
AGRICULTURAL CROPS	
Soils	<100.0 mg total silver/kg dry weight soil for most species; <10.0 mg/kg for sensitive species
Groundwater	<50.0 μg total silver/L
FRESHWATER AQUATIC LIFE	
Acute exposure	
Total recoverable silver	<1.32 μg/L
Acid-soluble silver[a]	Four-day average shall not exceed 0.12 μg/L more than once every three years; 1-h average not to exceed 0.92 μg/L more than once every three years
Acute exposure	
Total recoverable silver, in μg/L, should not exceed $e^{(1.72(\ln(\text{hardness}))-6.52)}$ at any time. Examples follow	
50 mg $CaCO_3$/L	<1.2 μg/L
100 mg $CaCO_3$/L	<4.1 μg/L
200 mg $CaCO_3$/L	<13.0 μg/L
Chronic exposure	<0.12–<0.13 μg total recoverable silver/L
Tissue residues	
Adverse effects on growth of the Asiatic clam, *Corbicula fluminea*	>1.65 mg total silver/kg soft tissues, fresh weight basis
MARINE LIFE	
Acute exposure	
Total recoverable silver	<2.3 μg/L at any time
Acid-soluble silver[a]	Four-day average concentration not to exceed 0.92 μg/L more than once every three years on average and the 1-h concentration not to exceed 7.2 μg/L more than once every three years
Tissue residues	
Marine clams, soft parts	
Normal	<1.0 mg total silver/kg dry weight
Stressful or fatal	>100.0 mg total silver/kg dry weight
HUMAN HEALTH	
Air, USA	
Current level of exposure, nationwide	100.0 μg total silver daily per person
Short-term exposure limit (15 min; up to 4 times daily with 60 min intervals at <0.01 mg Ag/m^3 air)	<0.03 mg total silver/m^3

Continued

Table 29.1. cont'd

Resource, Criterion, and Other Variables	Effective Silver Concentration
Threshold limit value (8 h daily, 5 days weekly)	
Aerosol silver compounds	<0.01 mg total silver/m^3
Metallic silver dust	<0.1 mg total silver/m^3
Diet, USA	
Current level of exposure	35.0–40.0 μg daily per person
Drinking water	
USA	
Long-term exposure (>10 days)	<50.0 μg total silver/L
Proposed long-term exposure	<90.0 μg total silver/L
Short-term exposure (1–10 days)	<1142.0 μg total silver/L
California	<10.0 μg/L
Germany	<100.0 μg/L
Space vehicles	
Former Soviet Union	Maximum 200.0 μg total silver/L
USA	100.0 to maximum 200.0 μg total silver/L
Switzerland	<200.0 μg total silver/L

[a] Silver that passes through a 0.45 μm membrane after the sample has been acidified to a pH between 1.5 and 2.0 with nitric acid.

needed on the significance of silver residues in tissues.

In aquatic environments, more research is needed on the chemical speciation of silver to evaluate risk to the organism and its consumers. Most silver criteria formulated for the protection of aquatic life are now expressed as total recoverable silver per liter. But total silver measurements do not provide an accurate assessment of potential hazard. Silver ion (Ag^+), for example, is probably the most toxic of all silver chemical species and must be accurately measured in the assessment of silver risks in aquatic environments, perhaps as acid-soluble silver. Little is known of the biocidal properties of Ag^{2+} and Ag^{3+} that are the active ingredients in disinfectants and used increasingly in water purification systems of drinking water and swimming pools. The effects of these silver species on organism health clearly must be researched. Silver interactions with other metals and compounds in solution are not well defined. For example, mixtures of salts of silver and copper markedly increased the survival of oyster embryos, but only when copper concentrations were less than 6.0 μg/L and total silver less than 11.0 μg/L.

The proposed freshwater acute ambient water quality criterion for silver in the United States recognizes that water calcium levels may modify toxicity; however, studies with rainbow trout suggest that calcium and sodium were not as important as chloride or dissolved organic carbon in ameliorating adverse effects of silver. Dissolved organic carbon was considered more important than hardness for predicting the toxicity of ionic silver in natural waters to daphnids and fishes. Incorporating an organic carbon coefficient into the silver criterion equation will enhance the criterion values for site specificity. The proposed silver criteria for aquatic life protection could be further improved by taking into account the important geochemical modifiers of silver. Thus, complexation of Ag^+ by chloride, dissolved organic carbon and sulfide are important in reducing silver toxicity, and $AgNO_3$ as Ag^+ is less toxic in seawater (LC50 [96 h] range of 320.0–2700.0 μg/L) than in freshwater (5.0–70.0 μg/L). More research is needed on the buildup of ammonia in Ag^+-poisoned teleosts and its effects on swimming ability. In any event, it is debatable whether silver discharges into the freshwater environment ever result in high enough Ag^+ levels to cause acute toxicity. Laboratory tests with $AgNO_3$ reportedly overestimate acute silver toxicity because of the abundance in the field of natural ligands which bind Ag^+ and reduce its toxicity. Also, there is little credible evidence that internal silver accumulations in freshwater biota play any role in acute toxicity or that dietary silver causes toxicity. In marine environments, current acute silver criteria seem overly stringent for aquatic life protection, with more research needed on chronic toxicity to euryhaline fishes – including the toxic potential of accumulated silver.

No studies have been conducted with silver and avian or mammalian wildlife, and it is unreasonable to extrapolate the results of limited testing with domestic poultry and livestock to wildlife to establish criteria or administratively enforce standards. Research on silver and avian and terrestrial wildlife merits the highest priority in this subject area. No silver criteria are available for the protection of avian and mammalian health, and all criteria now proposed are predicated on human health. As judged by the results of controlled studies with poultry and small laboratory mammals, safe concentrations of silver ion were less than 250.0 μg/L in drinking water of mammals, less than 100.0 mg/L in drinking water of poultry, less than 6.0 mg/kg in diets of mammals, less than 10.0 mg/kg in copper-deficient diets of poultry, less than 200.0 mg/kg in copper-adequate diets of poultry, and less than 1.8 mg/kg in chicken eggs. The proposed short-term (10-day) allowable limit of 1142.0 μg Ag/L in drinking water for human health protection (Table 29.1) should be reconsidered because it is 4.6 times higher than the value that produced adverse effects in sensitive laboratory mammals. Additional animal studies are needed to elucidate the effects of silver and silver compounds on reproduction, development, immunotoxicity, neurotoxicity, absorption, distribution, metabolism, and excretion; and on oral, dermal, and inhalation routes of exposure. In animals, there is also the need to establish a target organ for intermediate exposures to silver; to establish suitable biomarkers of silver exposures and effects; and to measure effects of chronic silver exposures on carcinogenicity. These studies should be implemented with suitable sentinel organisms including waterfowl, aquatic mammals, and other species of wildlife.

It is emphasized that silver and its compounds do not pose serious environmental health problems to humans from 50.0 μg/L in drinking water and 10.0 μg/m^3 in air. The only proven effect of chronic exposure to silver is argyria from occupational or therapeutic exposure to much larger amounts of silver (minimum necessary absorption of 910.0 μg, equivalent to about 15.0 μg/kg BW) than can feasibly be ingested or inhaled from environmental sources. Regular ingestion of fish, meat, and plants from silver-contaminated areas probably does not cause argyria. Humans at special risk to argyria include those treated with silver-containing medicinals and people marginally deficient or deficient in copper, selenium, or vitamin E. There is no recognized effective treatment for argyria, although the condition seems to be relatively stationary when exposure to silver is discontinued. Absorption and retention of silver from food and medicinals is imperfectly understood, suggesting the need for additional animal studies.

Finally, alternatives exist to the use of silver in various materials and processes. These include substitution of aluminum and rhodium for silver in mirrors and other reflecting surfaces; tantalum replacement of silver in surgical plates, pins, and sutures; stainless steel as an alternative material to silver in the manufacture of table flatware; and, in photography, film with reduced silver content.

29.7 Summary

Elevated silver concentrations in biota occur in the vicinities of sewage outfalls, electroplating plants, mine waste sites, and silver-iodide seeded areas; in the United States, the photography industry is the major source of anthropogenic silver discharges into the biosphere. Maximum concentrations recorded in field collections, in mg total Ag/kg dry weight (tissue), were 1.5 in mammals (liver), 6.0 in fish (bone), 14.0 in plants (whole), 30.0 in annelid worms (whole), 44.0 in birds (liver), 110.0 in mushrooms (whole), 185.0 in bivalve mollusks (soft parts), and 320.0 in gastropods (whole); humans afflicted with silver poisoning (argyria) contained 72.0 mg total Ag/kg dry weight skin and 1300.0 mg total Ag/kg fresh weight whole body. Silver and its compounds are not known to be mutagenic, teratogenic, or carcinogenic. Under normal routes of exposure, silver does not pose serious environmental health problems to humans at less than 50.0 μg total Ag/L drinking water or 10.0 μg total Ag/m^3 air. Free silver ion, however, was lethal to representative species of sensitive aquatic plants, invertebrates, and teleosts at nominal water concentrations of 1.2–4.9 μg/L; at sublethal concentrations, adverse effects were significant between 0.17 and 0.6 μg/L. No data were found on effects of silver on avian or mammalian wildlife; all studied effects were on poultry, small laboratory animals, and livestock. Silver was harmful to poultry at concentrations as low as 1.8 mg total Ag/kg whole egg fresh weight by way of injection, 100.0 mg total Ag/L in drinking water, or 200.0 mg total Ag/kg in diets; sensitive mammals were adversely affected at total silver concentrations as low as 250.0 μg/L in drinking water, 6.0 mg/kg in diets, or 13.9 mg/kg whole body. Proposed criteria for the protection of living organisms from silver are listed and discussed.

SODIUM MONOFLUOROACETATE[a]

Chapter 30

30.1 Introduction

Sodium monofluoroacetate ($CH_2FCOONa$), also known as 1080 or Compound 1080, belongs to a class of chemicals known as the fluoroacetates. It is a tasteless and odorless water-soluble poison of extraordinary potency that has been used widely against rodents and other mammalian pests. The widespread use of 1080 in pest control has resulted in accidental deaths of livestock, wildlife, pets (cats and dogs), and humans, and several suicides in Asia from drinking 1080 rat poison solutions. There is no effective antidote to 1080. When consumed, fluoroacetate is converted to fluorocitrate, inhibiting the enzymes aconitase and succinate dehydrogenase; the accumulated citrate interferes with energy production and cellular function.

Monofluoroacetic acid (CH_2FCOOH) was first synthesized in Belgium in 1896 but attracted little attention from chemists and pharmacologists at that time. In 1927, sodium monofluoroacetate was patented as a preservative against moths. The toxic nature of monofluoroacetate compounds was first noted in Germany in 1934. In the late 1930s and early 1940s, Polish scientists conducted additional research on the toxic properties of fluoroacetate compounds, especially on the methyl ester of fluoroacetic acid which they had synthesized. In 1942, British scientists further refined this compound to the sodium salt which became known as 1080. In 1944, potassium monofluoroacetate (CH_2FCOOK) was isolated from *Dichapetalum cymosium*, a South African plant, and was the first known example of a naturally occurring organic fluoride; the plant, known locally as Gifblaar, caused many livestock deaths, and was recognized by Europeans as poisonous as early as 1890. Fluoroacetate compounds have since been isolated from poisonous plants in Australia (*Acacia georginae*, *Gastrolobium* spp.), Brazil (rat weed, *Palicourea margravii*), and Africa (*Dichapetalum* spp.). Ratsbane (*Dichapetalum toxicarium*), a West African plant, was known to contain a poison – subsequently identified as a fluoroacetate – that was lethal to rats, livestock, and humans and reportedly used by African natives during the 1800s to poison the wells and water supplies of hostile tribes.

During World War II (1939–45), as a result of acute domestic shortages of common rodenticides, such as thallium, strychnine, and red squill, a testing program was initiated for alternative chemicals. In June 1944, the U.S. Office of Scientific Research and Development supplied the Patuxent Wildlife Research Center (PWRC) – then a U.S. Fish and Wildlife Service laboratory – with sodium monofluoroacetate and other chemicals for testing as rodenticides. Sodium monofluoroacetate received the PWRC acquisition number 1080, which subsequently was adopted as its name by the chemical's manufacturer. Samples of 1080 were also shipped to the Denver Wildlife Research Center, another former U.S. Fish and Wildlife Service laboratory, for testing on additional species. Results of these tests gave evidence of the value of 1080 as an effective

---

[a]All information in this chapter is referenced in the following sources:

Eisler, R. 1995. Sodium monofluoroacetate (1080) hazards to fish, wildlife, and invertebrates: a synoptic review. *U.S. Natl. Biol. Serv. Biol. Rep. 27*, 47 pp.

Eisler, R. 2000. Sodium monofluoroacetate (compound 1080). Pages 1413–1457 in *Handbook of Chemical Risk Assessment: Health Hazards to Humans, Plants and Animals. Volume 2, Organics*. Lewis Publishers, Boca Raton, Florida.

method of controlling animal predators of livestock and other animal pests. During World War II, 1080 protected Allied troops in the Pacific Theater against scrub typhus, also known as "tsut sugamushi," a louse-borne rickettsial disease with rodents as vectors. In the United States, 1080 was first used in 1945 to control rodents, and later coyotes (*Canis latrans*), rabbits, prairie dogs, and gophers. Between 1946 and 1949, at least 12 humans died accidentally in the United States from 1080 poisoning when used as a rodenticide; a child became ill but recovered after eating the cooked flesh of a 1080-poisoned squirrel. Since 1955, 1080 has been used extensively in a variety of baits – especially in Australia and New Zealand – to control European rabbits (*Oryctolagus cuniculus*), dingoes (*Canis familiaris dingo*), feral pigs (*Sus scrofa*), brush-tailed possums (*Trichosurus vulpecula*), and various species of wallabies. In Australia, vegetable baits are sometimes eaten by nontarget herbivores, such as sheep (*Ovis aries*), cattle (*Bos taurus*), and various species of wildlife, causing both primary and secondary poisoning of nontarget animals. In the United States, most uses of 1080 were canceled in 1972 due, in part, to deaths of nontarget animals. At present, the use of 1080 in the United States is restricted to livestock protection collars on sheep and goats (*Capra hircus*) against predation by coyotes.

30.2 Uses

The use of 1080 in the United States is now restricted to livestock collars on sheep and goats for protection against predation by coyotes. Other countries, most notably Australia and New Zealand, use 1080 extensively in a variety of baits to control many species of vertebrate pests.

30.2.1 Domestic Use

Compound 1080 is highly poisonous to all tested mammals as well as humans. There is no known antidote to 1080 and it has been impossible to resuscitate any animal or human poisoned with 1080 once final stages of poisoning have appeared. In 25 years of use in the United States, there have been 4 suicides and at least 12 accidental human deaths; between 1959 and 1969, 37 known incidents of domestic animal poisoning have resulted from federal use of 1080. Compound 1080 is not recommended for use in residential areas or for distribution in places where the public might be exposed; only licensed pest control operators can use 1080. Tull Chemical in Oxford, Alabama, is the sole domestic producer of 1080; none is imported. When handling 1080, human operators should wear protective clothing, including gloves and a respirator; extreme caution is recommended at all times. Each applicator must carry syrup of ipecac to induce vomiting in case of accidental 1080 poisoning when attaching, removing, or disposing of livestock protection collars.

Compound 1080 was first used in the United States in the late 1940s to control gophers, ground squirrels, prairie dogs, field mice, commensal rodents, and coyotes. Coyote damage to livestock in California alone is estimated at $75 million annually. Yearly amounts of 1080 used in the United States for predator control were 23 kg in the early 1960s, 7727 kg in the late 1960s, and only 8 kg in 1971. Total production of 1080 in the United States between 1968 and 1970 averaged about 1182 kg annually. In 1977, 277,545 kg of 1080-containing baits (272 kg of 1080) were used to control ground squirrels (76%), prairie dogs (7%), and mice, rats, chipmunks, and other rodents (17%); California used 83% of all 1080 baits, Colorado 12%, and Nevada and Oregon 5%. About 0.3 kg of 1080 per year is used in the livestock protection collar but only about 35.0 g per year is released into the environment. In March 1972, the use of 1080 for predator control was prohibited on federal lands. Later that year, all uses of 1080 for predator control were banned in the United States because of adverse effects on nontarget organisms including endangered species. In the period since 1080 was banned, the number of grazing livestock reported lost to predation on western national forests has increased. Between 1960 and 1971, 1.42% (range 1.0–1.9%) of all sheep and goats grazed were lost to predators vs. 2.17% (1.7–2.5%) in 1970–78. Until it was banned in 1972, the use of 1080 as a predator control agent in the United States was strictly controlled. The chemical was registered under the Federal Insecticide, Fungicide and Rodenticide Act (61 Stat 163; 7 U.S.C. 135-135K) for use only by governmental agencies and experienced pest control operators. The use of 1080 as a rodenticide was disallowed in 1985 for three reasons; (1) lack of emergency treatment, namely a viable medical antidote; (2) high acute toxicity to nontarget mammals and birds; and (3) a significant reduction in populations of nontarget organisms, and fatalities to endangered species. In 1985, 1080 use was conditionally permitted in livestock protection collars and in single lethal dose baits; a registration for the livestock protection collar was issued to the U.S. Department of the Interior on July 18, 1985. On February 21, 1989, the registration for 1080 was canceled, effectively prohibiting all uses. In June 1989, however, technical 1080 was conditionally approved for use only in the 1080 livestock protection collar. The 30 mL collar is registered for use by the U.S. Department of Agriculture; by the states of Montana, Wyoming, South Dakota, and New Mexico; and by Rancher's Supply, Alpine, Texas.

Compound 1080 was highly effective against all species of rats, prairie dogs, and ground squirrels, and satisfactory for the control of mice. The chemical was formulated in grain baits or chopped greens for crop and range rodents, and in water bait stations to control rats. The concentration of 1080 in baits was lowered to 0.02% both in the range of the California condor (*Gymnogyps californianus*) and for prairie dog control because of possible impacts on the endangered black-footed ferret (*Mustela nigripes*). Commercial 1080 was commonly colored with 0.5% nigrosine and sold as a compound containing >90% sodium monofluoroacetate, to be mixed with foods at 2226.0 mg/kg in preparing baits, or dissolved in water at 3756.0 mg/L for poisoning drinking water in indoor control of rodents. Bait acceptance by rats was not significantly reduced by the dye. Compound 1080 was adequately accepted by rats and mice when present in water; solid food baits poisoned with 1080 were not always accepted as readily and sometimes required special preparation to insure the ingestion of lethal amounts. A water solution of 1080 was the most effective rodenticide tested for rat control in southern states and 1080-grain baits were the most effective field rodenticides against ground squirrels, prairie dogs, and mice in California, South Dakota, and Colorado. Seeds and cereal grains were the most effective baits for small rodents: 1.0 kg of 1080 was sufficient to kill 3.96 million squirrels. Grain baits were colored brilliant yellow or green to heighten repellency to birds; coloring baits did not affect their acceptance by rodents. Rats did not develop any significant tolerance to 1080 from ingestion of sublethal doses; although rats that survived a poisoning incident may develop an aversion to 1080.

To kill coyotes and wolves (*Canis lupus*) in the United States and Canada, meat baits containing 35.0 mg 1080/kg were recommended, usually by injecting a water solution of 1080 into horse meat baits; only 28–56 g of a poisoned bait was sufficient to kill. Meat baits were usually placed during the autumn in areas with maximum coyote use and minimum use by most nontarget carnivores. The most widely publicized technique for poisoning predators was the 1080 large bait station: a 22–45 kg livestock meat bait injected with 35.0 mg 1080/kg bait. The use of 1080 stations peaked in the early 1960s, at which time 15,000–16,000 stations were placed each winter in the western United States. After 1964, the number of stations declined annually to 7289 stations in 1971. Against canine predators of livestock, 1080 was more selective and less hazardous to nontarget species than strychnine or traps. Meat baits used to control coyotes were seldom fatal to hawks, owls, and eagles, even when these raptors gorged themselves on the 1080-poisoned baits. In addition to the large bait stations, an unknown number of U.S. government hunters used 1080 in smaller baits at various stations.

The introduction of 1080-livestock protection collars to protect goats and sheep against coyote depredation was initiated in 1985; its use was limited to certified applicators. The 1080-filled rubber collars are attached to the throats of sheep and goats; 1080 is released when coyotes attack collared livestock with characteristic bites to the throat. The livestock protection collars contain 30 mL of a 1% 1080 solution and tartrazine as a marker. The livestock protection collar may not be used in areas known to be frequented by endangered species of wildlife and this includes various geographic areas in California, Michigan, Minnesota, Montana, Washington, Wisconsin, and Wyoming. Compound 1080 is reportedly more effective and safer in livestock protection collars than sodium cyanide, diphacinone, or methomyl. Pen tests with compound 1080 in livestock protection collars began late in 1976, and field tests in 1978. Under field conditions, 1080 livestock protection collars on sheep seem to protect selectively against predation by coyotes; no adverse effects on humans, domestic animals and nontarget wildlife were recorded. The decision to permit limited use of 1080 in livestock protection collars is now being contested by at least 14 conservation groups because of its alleged hazard to nontarget organisms (bears, badgers, dogs, eagles) and to human health, and to the availability of alternate and more successful methods of coyote control. In Texas, for example, annual predation losses of sheep and goats to coyotes are estimated at $5 million. But very few Texas ranchers have taken advantage of the opportunity to use livestock protection collars and only 23 coyotes were killed in 1989 by the collars vs. 473 by cyanide, snares, aerial gunning, and other control measures. Toxic livestock protection collars in full operation would probably kill <1000 coyotes annually vs. 1 million coyotes killed annually in sport hunting and other control measures.

Compound 1080 was also effective against jackrabbits, foxes, and moles. Baits containing 0.05–0.1% of 1080 on vegetables were used in California to kill jackrabbits (*Lepus* spp.) and various rodents. The Arctic fox (*Alopex lagopus*), intentionally introduced onto the Aleutian Islands in 1835, almost eliminated the Aleutian Canada goose (*Branta canadensis leucoparlia*) by 1967; 1080-tallow baits were successfully used to control fox populations. Earthworm baits are used to kill moles. The earthworms are soaked for 45 min in a 2.5% solution of 1080 and placed in mole burrows. The solution can be used several times for additional lots of worms; however, the use of the manure worm (*Eisenia foetida*) should be avoided because it is seldom eaten by moles.

Secondary poisoning of domestic cats and dogs from consumption of 1080-poisoned rodents was frequently noted. Cats and dogs are highly susceptible to 1080 and may die after eating freshly poisoned rodents, dried carcasses, or 1080-baits, or after drinking 1080-poisoned water. All pets should be confined or removed from the area to be poisoned and released after the entire program has been completed. Pigs and carnivorous wildlife are also at risk from consumption of 1080-poisoned rodents. Secondary poisoning of kit foxes (*Vulpes* spp.) is theoretically possible after eating a single kangaroo rat (*Dipodomys* spp.) that had swallowed or stuffed its cheeks with 1.0 g of a 0.1% vegetable/cereal bait and contained a total whole body burden of about 1.0 mg of 1080 per rat. To prevent secondary poisoning, all uneaten baits and carcasses of poisoned rodents should be recovered and incinerated and no 1080-contaminated animal should be eaten by humans or fed to animals.

30.2.2 Nondomestic Use

Compound 1080 has had limited use as a vertebrate pesticide in Canada, India, Mexico, and South Africa, and extensive use in Australia and New Zealand. In Canada, 1080 was first used in 1950–51 in British Columbia to control wolves and coyotes preying on livestock. Poisoned 1080 baits were used in India to control (67–100% effective) populations of the Indian crested porcupine (*Hystrix indica*) throughout its range because of porcupine-caused damage and losses to agriculture crops; however, 1080-baits were not as effective as fumigants in controlling this species. In Mexico, 1080 was used against rabid coyotes, although many domestic dogs were also killed. In South Africa, beginning in 1961, 1080 was used to control the black-backed jackal (*Canis mesomelas*) preying on livestock, and baboons (*Papio anubis*) and moles that consumed agricultural crops. Livestock protection collars containing 30 mL of a 1% solution of 1080 are now used in South Africa to combat predation by the Asiatic jackal (*Canis aureus*).

Compound 1080 was first used in Australia in the 1950s to kill the introduced European rabbit (*Oryctolagus caniculus*). Principal target species in Australia now include other introductions such as dingoes, foxes (*Vulpes vulpes*), feral pigs (*Sus scrofa*), feral cats (*Felis cattus*), as well as native brush-tailed possums (*Trichosurus vulpecula*), red-necked wallabies (*Macropus rufogriseus*), and pademelons (*Thylogale billardierii*). In Australia, different baits contained different concentrations of 1080; meat baits contained 144.0 mg/kg, grain baits 288.0–300.0 mg/kg, fruits and vegetables 330.0 mg/kg, and pellets 500.0 mg/kg.

One method of killing rabbits in many areas of Australia is to apply 1080-poisoned bait (carrots, oat grains, pellets of bran or pollard) to furrows made in the earth or broadcast across the area from the air or ground. Aerial dropping of diced carrots treated with 1080 was found to be almost 100% effective for rabbits. In Victoria, more than 6.5 million ha were treated with 1080-poisoned carrots. To attract rabbits to the kill area, nonpoisoned carrots were applied to rabbit trails at more than 8.3 kg/km; nonpoisoned baits were offered twice, 3 days apart, followed by 1080-poisoned carrots one week later. Bait avoidance is reported in some populations of European rabbits exposed repeatedly to 1080 baits through sustained control programs. Behavioral resistance may reduce the effectiveness of sustained control and should be considered in pest management plans. Individuals – but not populations – of some native species of Australian animals and birds face a greater risk of being poisoned by 1080 during rabbit-poisoning campaigns than rabbits, particularly herbivorous macropodids, rodents, and birds with no prior exposure history to naturally occurring fluoroacetates. Foxes, dingoes, dogs, and cats seem to be at greater risk of secondary poisoning than native birds and mammals, particularly from eating muscle from poisoned rabbits that contained as much as 5.0 mg of 1080 per rabbit.

The injection method of fresh meat baits for use in control of dingoes produced baits more uniform with respect to the amount of 1080 in the bait when compared with mixed baits prepared by tumbling in 1080 solutions. Both techniques, however, produced baits containing variable quantities of 1080. Use of 1080-poisoned baits to control wild dogs (*Canis familiaris familiaris*) and dingoes were not as successful as traps: 22% control for 1080 vs. 56% control for traps. Factors which reduced the success of poisoned baits included rapid loss in toxicity of the baits after their distribution; the rapid rate at which they were removed by other animals, particularly foxes and birds; and the dogs' apparent preference for natural prey.

Feral pigs in Australia damage crops, degrade pasture, kill and eat lambs, and are potential vectors and reservoirs of exotic pathogens. Control of feral pigs with poisoned baits, including 1080 bait, is difficult because most pigs regurgitate these baits shortly after ingestion. The vomitus may cause secondary poisoning of nontarget species, and pigs surviving sublethal exposure to 1080 as a result of vomiting may develop an aversion to 1080 thus decreasing their susceptibility to subsequent poisoning programs. The incorporation of antiemetics into 1080 baits should reduce or prevent vomiting, but those tested were not completely successful. Feral cats have altered ecosystems and depleted populations of indigenous lizards and birds in Australia, New Zealand, and numerous island habitats throughout the world. Fresh fish baits injected with 2.0 mg of 1080 per bait are used as a humane and lethal poison for feral cats.

The use of 1080 in New Zealand is restricted to licensed operators employed by pest destruction boards and government departments. In Australia, and other locations, the addition of dye to identify toxic baits is standard practice. The main purpose of such addition is to reduce the unintentional poisoning of birds; birds eat significantly less blue- or green-dyed feed than undyed feed. Although birds prefer undyed baits to those dyed green, Canada geese (*Branta canadensis*) when feeding at night are unable to distinguish between dyed and undyed baits and consume both with equal frequency. Carrots used as wallaby baits in New Zealand are dyed with special green or blue pigments: however, the red-necked wallaby (*Macropus rufogriseus*) accepted both dyed and undyed carrots equally. Mice (*Mus* spp.) readily consumed dyed wheat. Compound 1080 is used in jam-type baits to control brush-tailed possums. These baits contained 1080 at concentrations as high as 1500.0 mg 1080/kg FW bait and were dyed green to protect birds. Cinnamon was added to mask the flavor of the 1080 poison, and 800.0 mg potassium sorbate/kg was added as an antifungal bait preservative.

The Norway rat (*Rattus norvegicus*) had a severe effect on island populations of New Zealand birds, reptiles, and invertebrates. In one case, rats on Big South Island exterminated five species of native forest birds within 3 years, including the last known population of the bush wren (*Xenicus longipes*). A paste containing petroleum jelly, soya oil, sugar, green dye, and 800.0 mg 1080/kg remained toxic for 6–9 months to rats preying on gray-faced petrels (*Pterodroma macroptera*) and other birds. Because 1080 produces a poison-shyness in any Norway rat which eats a sublethal dose, complete eradication of this species by 1080 is improbable. The use of anticoagulants – such as warfarin (multiple doses needed), brodifacoum (single dose) or coumatetralyl – seem more promising than 1080 in rat control programs although secondary poisoning of owls and hawks may occur.

In New Zealand, compound 1080 in a gel carrier is sometimes applied to the leaves of broadleaf (*Griselinia littoralis*) to poison red deer (*Cervus elephus*), feral goats, white-tailed deer (*Odocoileus virginianus*) and red-necked wallabies. Use of 1080-gel baits reduced feral goat populations by 90%. Wallaby populations were reduced 87–91% using a 1080 gel applied to the foliage of palatable plants, and this compares favorably to reductions achieved using aerially sown baits. The gel carrier was an effective phytotoxin, causing withering, death, or loss of chlorophyll from leaves within 10 days, and sometimes within 24 h.

Feral pigs are sometimes poisoned by inserting as many as 10 gelatin capsules (each containing 100.0 mg of 1080) into carcass or offal baits. Poisoned carcasses may remain edible for more than 2 months during autumn and winter when poisoning campaigns are conducted; the 1080 is leached out when the carcass has disintegrated. Other techniques to control feral pigs include injection of 1080 gel into beef lung baits or insertion of capsules containing 1080 into apple, potato, or other fruit and vegetable baits. However, these techniques are potentially the most dangerous to applicators because 1080 powder, rather than a diluted solution, is used; also, the baits are lethal to nontarget scavengers.

30.3 Environmental Chemistry

Sodium monofluoroacetate is a whitish powder soluble in water to at least 263.0 mg/L but relatively insoluble in organic solvents. Some aqueous solutions of 1080 retain their rodenticidal properties for at least 12 months, but others lose as much as 54% of their toxicity after 24 days. Compound 1080 is unstable at >110°C and decomposes at >200°C, although 1080 in baits or poisoned carcasses is comparatively stable. Losses of 1080 from meat baits are primarily due to microbial defluorination, and also to leaching from rainfall and consumption by maggots. Leachates from 1080 baits are not likely to be transported long distances by groundwater because they tend to be held in the upper soil layers. Compound 1080 can be measured in water at concentrations as low as 0.6 μg/L and in biological samples at 10.0–15.0 μg/kg. As discussed later, 1080 is readily absorbed through the gastrointestinal tract, mucous membranes, and pulmonary epithelia; once absorbed it is uniformly distributed in the tissues. Metabolic conversion of high concentrations of fluoroacetate to fluorocitrate results in large accumulations of citrate in the tissues and eventual death from ventricular fibrillation or respiratory failure. Regardless of dose and in all tested species, no signs or symptoms of 1080 poisoning were evident during a latent period of 30 min to 2 h; however, death usually occurred within 24 h of exposure. Repeated sublethal doses of 1080 have increased the tolerance of some species of tested birds and mammals to lethal 1080 doses. Reptiles are more resistant to

1080 than mammals because of their low facility to convert fluoroacetate to fluorocitrate and their high defluorination capability. No effective antidote is now available to treat advanced cases of fluoroacetate poisoning; accidental poisoning of livestock and dogs by 1080 is likely to be fatal. Partial protection against 1080 poisoning in mammals has been demonstrated with glycerol monoacetate, a sodium acetate–ethanol mixture, and a calcium glutonate–sodium succinate mixture.

30.3.1 Chemical Properties

Some chemical and other properties of 1080 are summarized in Table 30.1. In water, trace amounts (0.6 μg/L) of 1080 were detected using gas chromatography (GC) with electron capture detection; recoveries from environmental water spiked at 5.0–10.0 μg/L ranged from 93 to 97%. Recent advances make it possible to measure 1080 in solutions at concentrations as low as 0.2 μg/L. In biological tissues, various methods have been used to determine fluoroacetic acid, including colorimetry, fluoride-ion electrodes, gas–liquid chromatography, and high-pressure chromatography; however, these methods involve lengthy extraction procedures, have low recoveries, or show lack of selectivity. A sensitive gas chromatographic technique was developed and used successfully to determine fluoroacetate levels in organs from a magpie (*Gymnorphina tibicen*) that had ingested a bait containing 1080 poison. The procedure involved extraction of 1080 with acetone:water (8:1) followed by derivatization with pentafluorobenzyl bromide. Bait samples were initially screened by thin-layer chromatography and identification of derivatized extracts was confirmed

Table 30.1. Some properties of sodium monofluoroacetate.

Variable	Datum
ALTERNATE NAMES	1080; Compound 1080; fratol; monosodium fluoroacetate; sodium fluoacetate; sodium fluoroacetate; ten-eighty
CHEMICAL FORMULA	$CH_2FCOONa$
MOLECULAR WEIGHT	100.03
PHYSICAL STATE	White, odorless, almost tasteless, hygroscopic powdery salt, resembling powdered sugar, or baking powder
PRIMARY USE	Rodenticide; mammal control agent
PURITY	96.0–98.6%
SOLUBILITY	
Water	263.0 mg/L
Acetone, alcohol, animal and vegetable fats, kerosene, oils	Relatively insoluble
STABILITY	Unstable at >110°C and decomposes at >200°C. Hydrogen fluoride (20% by weight) is a decomposition product which readily reacts with metals or metal compounds to form stable inorganic fluoride compounds

by gas chromatography/mass spectrometry= GC/MS. A new method for fluoroacetate determination in biological samples involves isolation of fluoroacetate as its potassium salt by ion-exchange chromatography and conversion to its dodecyl ester. The ester is quantified by capillary GC with a flame ionization detector for the range 1.0–10.0 mg/kg and by selected ion monitoring using GC/MS for the range 0.01–1.00 mg/kg. The detection limit for 1080 in tissues and baits is 15.0 μg/kg using a reaction-capillary GC procedure with photo-ionization detection; the detection limit is 100.0 μg/kg using flame-ionization procedures. The detection limit using these procedures is less sensitive than GC/MS; however, GC/MS is not normally available in veterinary diagnostic laboratories.

30.3.2 Persistence

Significant water contamination is unlikely after aerial distribution of 1080 baits. In one New Zealand field trial in which >20 metric tons of 1080 baits were aerially sown over a 2300 ha island to control brushtail possums (*Trichosurus vulpecula*) and rock wallabies (*Petrogale penicillata*), no 1080 was detected in surface or groundwater of the island for at least 6 months after baits were dropped. A similar case was made for streams and rivers after 100 metric tons of 1080 baits were sown by airplane over 17000 ha of forest. Laboratory studies on 1080 persistence in solutions suggest that degradation to nontoxic metabolites is most rapid at elevated temperatures and in biologically conditioned media but is highly variable. In general, aqueous solutions of the salt or esters decrease in toxicity over time through spontaneous decarboxylation to sodium bicarbonate and to the highly volatile, relatively nontoxic, methyl fluoride. Solutions refrigerated at 5°C lost about 54% of their initial toxicity to laboratory rats after 24 days and about 40% after 7 days at room temperature, but 1080 solutions remained toxic to yeast for at least 1 month after storage at 3–5°C. In an aquarium containing plants and invertebrates and 0.1 mg 1080/L, water concentration of 1080 declined 70% in 24 h and was not detectable after 100 h; residues in plants were not detectable after 330 h. In a distilled water aquarium without biota, 1080 residues declined only 16% in 170 h. In another study, 1080 solutions prepared in distilled water and stored at room temperature for 10 years showed no significant breakdown; moreover, solutions of 1080 prepared in stagnant algal-laden water did not lose biocidal properties over a 12-month period. More research seems needed on 1080 persistence in aquatic environments.

In soils, 1080 is degraded to nontoxic metabolites by soil bacteria and fungi, usually through cleavage of the carbon–fluoride bond. Soil microorganisms capable of defluorinating 1080 include *Aspergillus fumigatus*, *Fusarium oxysporum*, at least 3 species of *Pseudomonas*, *Nocardia* spp., and 2 species of *Penicillium*. These microorganisms can defluorinate 1080 when grown in solution with 1080 as the sole carbon source, and also in autoclaved soil; the amount of defluorination ranged from 2 to 89% in soils and 2 to 85% in 1080 solutions. Some indigenous soil microflora were able to defluorinate 50–87% of the 1080 within 5–9 days in soil at 10% moisture at 15–28°C. The most effective defluorinaters in solution and in soils were certain strains of *Pseudomonas*, *Fusarium*, and *Penicillium*. *Pseudomonas cepacia*, for example, isolated from the seeds of various fluoroacetate-accumulating plants can grow and degrade fluoroacetate in fluoroacetate concentrations as high as 10,000.0 mg/kg. Biodefluorination of 1080 by soil bacteria was maximal under conditions of neutral to alkaline pH, fluctuating temperatures between 11 and 24°C, and at soil moisture contents of 8–15%; biodefluorination of 1080 by soil fungi was maximal at pH 5.

Losses of 1080 from meat baits were most likely due to consumption of the bait by blowfly maggots, leaching by rainfall, defluorination by microorganisms, and leakage from baits during their decomposition. The 1080 in baits will persist under hot and dry conditions where leaching from rain is unlikely. Baits remained toxic to dogs for over 32 days during winter when maggots were absent and 6–31 days during summer when maggots were present. Baits contained an average LD50 dose to tiger quolls (*Dasyurus maculatus*) – a raccoon-like marsupial – for 4–15 days in winter and 2–4 days in summer. Meat baits that initially contained 4.6 mg of 1080 retained 62% after 3 days, 29% after 6 days, and 28% after 8 days. The persistence of 1080 in fatty meat baits for control of wild dogs in Australia was measured over a period of 226 days. Baits that initially contained 5.4 mg of 1080 retained 73% at day 7, 64% at day 20, 25% at day 48, and 15% at day 226. These baits retained LD50 kill values after 52 days to wild dogs, 93 days to cattle dogs, and 171 days to sheep dogs. In that study, loss of 1080 from the baits was not correlated with rainfall, temperature, or humidity. Losses were attributed to metabolism of 1080 bound to the fatty meat bait, leaching, consumption by maggots, and bacterial defluorination. When it is desirable for baits to remain toxic for long periods, the defluorination activity and microbial growth can be reduced significantly by incorporating bacteriostats and fungistats; conversely, baits may be inoculated with suitable defluorinating microbes that rapidly detoxify 1080-poisoned baits.

Compound 1080 was found to be highly persistent in diets formulated for mink (*Mustela vison*). Mink diets analyzed 30 months after formulation lost 19–29% of the 1080 when the initial concentration ranged between 0.9 and 5.25 mg 1080/kg; loss was negligible at 0.5 mg 1080/kg ration. A paste containing 0.08% 1080 plus petroleum jelly, soya oil, sugar, and green dye retained its rodenticidal properties for 6–9 months. But a rolled oats-cat food 1080 bait, because of its moistness, became fly-infested in warm weather, tended to rot, and lost its rodenticidal properties in a few days. Gel baits set to kill deer were sampled after 45 days of weathering; only 10% of the 1080-treated leaves retained toxic gel after 45 days. About 1.4% of 1080 was lost from the leaves per mm of rainfall; about 90% was lost in 2 trials in which 81 and 207 mm of rainfall were recorded. Compound 1080 decreased from 604.0 mg/bait at the start to 76.0 mg/bait after 30 days and to 5.0 mg/bait after 45 days. Significant losses of compound 1080 also resulted from biodegradation in storage. *Penicillium* spp. from broadleaf samples degraded 1080 at pH 5.4 and 23°C and grew vigorously on 1080-poisoned gels; other species of microorganisms can also degrade 1080.

Leachates from 1080-poisoned baits are not likely to be transported long distances by the leaching water because they are held in the upper soil layers. This statement is predicated on the facts that: (1) salts of monofluoroacetic acid rapidly adsorb to plant tissues and other cellulosic materials; (2) some plants can decompose 29% of the adsorbed 1080 in 48 h; and (3) 1080 in soils is decomposed by soil microorganisms. The percent of 1080 defluorinated from various bait materials after 30 days as a result of microbial action ranged between 0.0 and 7.2% for cereals, eggs, horse meat, and beef; 14% for kangaroo meat; and 71% for oats. The defluorinating ability of fungi and bacteria was low when 1080 was the sole carbon source and high when alternative carbon sources such as peptone-meat extracts were present. The extent of defluorination varied among the different types of organisms associated with the baits. Microorganisms isolated from oats and kangaroo meat had the highest defluorinating activity and those from cereals and eggs the lowest.

30.3.3 Metabolism

Sodium monofluoroacetate is absorbed through the gastrointestinal tract, open wounds, mucous membranes, and the pulmonary epithelium; it is not readily absorbed through intact skin. Once absorbed, it seems to be uniformly distributed in the tissues including the brain, heart, liver, and kidney. All tested routes of 1080 administration are equally toxic: there is no noteworthy difference in the acute toxicity of 1080 when administered orally, subcutaneously, intramuscularly, intraperitoneally, or intravenously. Moreover, the oral toxicity of 1080 is independent of the carrier, including water, meat, grain, oil, gum acacia suspension, or gelatin capsule carriers.

All students of the action of fluoroacetate have been impressed with the unusually long and variable latent period between administration and response. This latent period occurred in all species studied regardless of route of administration. With few exceptions the latent period ranges between 30 min and 2 h and massive doses – such as 50 times an LD95 dose – do not elicit immediate responses. The time between 1080 treatment and death was relatively constant in all tested species, and usually ranged between 1 h and 1 day. The latent period associated with 1080 may result from three major factors: (1) the time required for hydrolysis of monofluoroacetate to monofluoroacetic acid, and its subsequent translocation and cell penetration; (2) the time required for biochemical synthesis of a lethal quantity of fluorocitrate; and (3) the time required for the fluorocitrate to disrupt intracellular functions on a large enough scale to induce gross signs of poisoning.

Many authorities agree that the toxicity of 1080 to mammals is due to its conversion to fluorocitrate, a fluorotricarboxylic acid. These authorities concur that enzymatic conversion of fluoroacetate via fluoroacetyl coenzyme A plus oxalacetate in mitochondria is the metabolic pathway that converts the nontoxic fluoroacetate to fluorocitrate. Fluorocitrate blocks the Krebs cycle, also known as the tricarboxylic acid cycle, which is the major mechanism for realizing energy from food. Fluorocitrate inhibits the enzyme aconitase and thereby inhibits the conversion of citrate to isocitrate. Fluorocitrate also inhibits succinate dehydrogenase, which plays a key role in succinate metabolism. The inhibition of these two enzymes results in large accumulations of citrate in the tissues, blocking glucose metabolism through phosphofructokinase inhibition, and eventually destroying cellular permeability, cell function, and finally the cell itself. The classical explanation of fluorocitrate toxicity through aconitase inhibition has been questioned. A more recent explanation is that fluorocitrate binds with mitochondrial protein, thereby preventing citrate transport and its utilization by cells for energy production, although the underlying biochemical mechanisms are not completely understood. Based on calculated metabolic rates of fluorocarboxylic acids, secondary poisoning of animals that have consumed 1080-poisoned prey is probably due to unmetabolized fluoroacetate rather than to fluorocitric acid.

Dogs, rats, and rabbits metabolize fluoroacetate compounds to nontoxic metabolites, and excrete fluoroacetate and fluorocitrate compounds; peak rate of excretion occurs during the first day after dosing and drops shortly thereafter. Rats dosed with radiolabeled 1080 at 5.0 mg/kg BW had 7 different radioactive compounds in their urine. Monofluoroacetate comprised only 13% of the urinary radioactive material, fluorocitrate only 11%, and an unidentified toxic metabolite 3%; 2 nontoxic metabolites accounted for almost 73% of the urinary radioactivity. Animal muscle usually contained nondetectable residues of any 1080 component within 1–5 days of treatment. Defluorination occurred in the liver by way of an enzymic glutathione-dependent mechanism which in the brush-tailed opossum resulted in the formation of *S*-carboxymethylcysteine and free fluoride ion. A rapid rate of defluorination together with a low reliance on aerobic respiration favored detoxification of fluoroacetate rather than its conversion into fluorocitrate, and may account for the resistance of reptiles to 1080 when compared to mammals.

Sublethal doses of 1080 have led to a tolerance to subsequent challenging doses in certain animals; in other species, however, repeated sublethal doses have resulted in accumulation of a lethal concentration. Repeated sublethal doses of 1080 have increased the tolerance of some eagles, rats, mice, and monkeys, but not dogs. Conversely, repeated sublethal doses of 1080 have accumulated to lethal levels in dogs, guinea pigs, rabbits, and mallards. Continued sublethal doses of 1080 to rats caused regressive changes in the germinal epithelium of the seminiferous tubules. Altered behavior in mice following high sublethal doses of 1080 probably resulted from neuronal damage caused by concurrent energy deficiency further accentuated by the CNS stimulant action of fluoroacetate/fluorocitrate and the brain anoxia that occurred during 1080-induced intermittent convulsions; a similar pattern has been reported in 2 human patients. Anuria in some 1080-dosed mice probably resulted from renal shutdown caused by hypocalcemic tension. Tolerance to 1080 is a time-related phenomenon. Laboratory rats given 0.5 mg 1080/kg BW were more resistant to 5.0 mg/kg BW given >4 and <24 h later than nontested rats. Accumulation of 1080 is also a time-related phenomenon. Domestic dogs given 25.0 μg 1080/kg BW daily were unaffected until the fifth dose, when convulsions and death occurred. Also, larger sublethal doses could be administered to dogs on alternate days without adverse effects.

Fish, amphibians, and reptiles are usually less sensitive to 1080 than warm-blooded animals. Reptiles, for example, are more resistant to 1080 than mammals. The relatively small elevation of plasma citrate levels in skinks (*Tiliqua rugosa*) given 100.0 mg 1080/kg BW reflects the exceptional tolerance of this lizard species. The minimal effect of fluoroacetate on aerobic respiration in *T. rugosa* could be explained by a low conversion of fluoroacetate into fluorocitrate or by a low susceptibility of aconitase to the fluorocitrate formed. Though defluorination in skinks helped to minimize the immediate effects of fluoroacetate in aerobic respiration, it resulted in rapid depletion of liver glutathione levels.

The breakdown in intracellular processes caused by fluorocitrate or decreased energy production may result in death from gradual cardiac failure or ventricular fibrillation, death from progressive depression of the CNS with either cardiac or respiratory failure, or death from respiratory arrest following severe convulsions; signs of 1080 intoxication included labored breathing, vomiting, lethargy, muscular incoordination, weakness, and tremors. Among herbivores, 1080-induced deaths were primarily due to cardiac disorders, among carnivores deaths were from CNS disorders, and among omnivores deaths were from both cardiac and CNS disorders. Other signs of 1080 intoxication included kidney and testicular damage and altered blood chemistry, specifically, elevated concentrations of citrate, glucose, lactic acid, pyruvic acid, acetate, inorganic phosphate, potassium, and fluorine. Some mammals additionally displayed parasympathetic nervous system effects including increased salivation, urination, and defecation, with eventual cardiac failure.

Vomiting probably evolved among carrion eaters as a natural protective mechanism, but it does not necessarily ensure survival from 1080 poisoning. For example, even though 90% of eastern native quolls (*Dasyurus viverrinus*) and 95% of Tasmanian devils (*Sarcophilus harrisii*) vomited within 26–55 min after ingesting 1080, this was still sufficient time for many to absorb a lethal dose. Loud sounds, sudden movements of an observer, or convulsions by another animal nearby sometimes stimulated convulsions; however, variability was great between species and among conspecifics. Signs preceding convulsions usually included restlessness; hyperexcitability or increased response to stimuli; trembling; rapid, shallow breathing; incontinence or diarrhea; excessive salivation; twitching of facial muscles; abnormal eye movements; incoordination; vocalization; and sudden bursts of violent activity. All affected animals subsequently fall to the ground in a tetanic seizure, with hind limbs or all four limbs and sometimes the tail extended rigidly from their arched bodies. This tonic phase is followed by a clonic phase in which the animals kick with the front legs, and eventually begin to relax. After this phase animals either recover gradually, die shortly afterwards, experience additional seizures and then die or recover, or remain comatose until death up to 6 days later.

30.3.4 Antidotes

No highly effective treatment of well-established fluoroacetate poisoning is available, and accidental poisoning of livestock and domestic dogs is likely to be fatal. The following compounds were tested and had no effect on ameliorating 1080 intoxication: salts of fatty acids, anticonvulsants, vitamins, and metabolic intermediates; and nonphysiological sulfhydryl compounds, such as *N*-acetylcysteine and cysteamine. As discussed later, sodium–acetate/ethanol mixtures, barbiturates, glycerol monoacetate, calcium glutonate/sodium succinate mixtures, and 4-methylpyrazole offer partial protection to 1080-poisoned mammals, possibly because they compete with fluoroacetate in the Krebs cycle.

Sodium acetate partially protects mice against 1080, as does ethanol. Ethanol and sodium acetate administered together are twice as effective as either alone, suggesting a synergistic effect. Mixtures of acetate and ethanol reduced mortality of 1080-poisoned mice (given 2 times an LD50 dose) from 80 to 30%. Mice given 170.0 mg 1080/kg BW (about 10 times an LD50 dose) plus an intraperitoneal injection of sodium acetate (2.0–3.0 g/kg BW) dissolved in ethanol (1.6 g/kg BW) reduced mortality by 90%. But the beneficial effect of the acetate–ethanol treatment to mice decreased rapidly with increasing time after the administration of 1080. Ethanol–acetate mixtures had some antidotal effect on 1080-poisoned dogs provided that treatment was administered within 30 min of poisoning. A mixture of 2.0 g sodium acetate/kg BW plus 2.0 g ethanol/kg BW is recommended for treatment of 1080-poisoned monkeys.

Barbiturates were marginally effective in protecting domestic dogs against fluoroacetate poisoning, but not laboratory mice. Barbiturates administered to dogs within 30 min of 1080 poisoning (4 times an LD50 dose) resulted in 80% survival; when therapy was given 3 h after poisoning, survival was 17%. At higher 1080 doses (i.e., 6 times the LD50 value), barbiturates were ineffective. Repeated intravenous injections of 20.0 mg pentobarbital/kg BW to a 1080-poisoned dog (0.3 mg 1080/kg BW) prevented death when administered within 8.5 h of poisoning.

Glycerol monoacetate at 2.0–4.0 g/kg BW partially protects 1080-poisoned rats, rabbits, dogs, and rhesus monkeys. But its effectiveness is apparent only when administered intramuscularly in large amounts immediately after 1080-ingestion. A single dose of magnesium sulfate at 800.0 mg/kg BW given intramuscularly as a 50% solution shortly after 1080 exposure prevented death of rats dosed with marginally lethal amounts of 1080.

A reduced level of blood calcium is one explanation for the toxic effects of fluoroacetate, and may account for the gap between chemical manifestations and the biochemistry of 1080 poisoning. Cats poisoned with 1080 showed a 27% drop in blood calcium levels within 40 min; intravenous administration of calcium chloride prolonged the life of treated cats from 94 min to 167 min. In a search for effective antidotes to fluoroacetate poisoning, calcium gluconate was chosen to antagonize hypocalcemia, and sodium alpha ketoglutarate and sodium succinate were selected to revive the TCA cycle. Effectiveness of each of these antidotes individually and in certain combinations was tested in laboratory mice exposed to lethal doses (15.0 mg/kg BW, intraperitoneal injection) of 1080. Antidotal treatments were administered from 15 min to 36 h after dosing. All three antidotes alone, and a combination of calcium glutonate with sodium alpha ketoglutarate were ineffective in reducing mortality in treated mice. However, a combination of calcium glutonate (130.0 mg/kg BW) and sodium succinate (240.0 mg/kg BW) was effective if the 2 solutions were either injected at separate sites or mixed in the same syringe just prior to injection. Increasing the dose of sodium succinate to 360.0 or 480.0 mg/kg BW with calcium glutonate (130.0 mg/kg BW) was unrewarding. Additional studies are needed to confirm the efficacy and mechanisms of action of this combination.

Intraperitoneal injection of 4-methylpyrazole to rats at 90.0 mg/kg BW, given 2 h prior to 1080 administration, offered partial protection against accumulations of citrate or fluorocitrate in the kidney. The antidotal effects of 4-methylpyrazole are attributed to its inhibition of NAD^+-dependent alcohol dehydrogenase that converts 1,3-difluoro-2-propanol to difluoroacetone, an intermediate in the pathway of erythrofluorocitrate metabolism. A disadvantage of 4-methylpyrazole is that it needs to be administered before significant exposure to fluoroacetate.

First aid treatment for humans accidentally poisoned with 1080 includes immediate emesis and gastric lavage followed by an oral dose of magnesium sulfate or sodium sulfate to remove the poison from the alimentary tract before absorption of lethal quantities can occur. When the stomach is emptied, oral administration of ethanol may be beneficial. The patient should be put at complete rest and given barbiturates having moderate duration of action, such as sodium amytol, to control convulsions. Intramuscular injections of undiluted glycerol monoacetate at 0.5 mg/kg BW are recommended every 30 min for several hours and then at a reduced level for at least 12 h. If intramuscular administration is not feasible, a mixture of 100.0 mL of undiluted glycerol monoacetate in 500.0 mL of water can be given orally and repeated in an hour. If glycerol monoacetate is not available, acetamide or a combination of sodium acetate and ethanol may be given in the same dose. If ventricular fibrillation occurs, the heroic treatment of 5.0 mL of a 1% procaine hydrochloride via intracardiac puncture is justified. Intravenous administration of procainamide is also effective in restoration of normal rhythm in ventricular fibrillations. Symptoms of 1080 poisoning usually subside in 12–24 h, but the patient should be kept in bed for at least 3 days.

30.4 Lethal and Sublethal Effects

Mammals were the least resistant group tested against 1080; individuals of sensitive species died after receiving a single dose of 0.05–0.2 mg/kg BW. As discussed later, adverse sublethal effects included testicular damage in rats (*Rattus* sp.) after drinking water containing 2.2–20.0 mg 1080/L for 7 days (0.07–0.71 mg/kg BW daily), impaired reproduction in mink fed diets containing 0.8 mg 1080/kg ration for 60 days, and altered blood chemistry in European ferrets given diets containing 1.1 mg 1080/kg feed for 28 days. Elevated fluoroacetate residues were measured in some 1080-poisoned mammals, notably European rabbits, of 34.0 mg/kg DW muscle and 423.0 mg/kg DW liver. Birds belonging to sensitive species died after a single 1080 dose of 0.6–2.5 mg/kg BW, daily doses of 0.5 mg/kg BW for 30 days, 47.0 mg/kg diet for 5 days, or 18.0 mg/L drinking water for 5 days. Accumulation and adverse sublethal effects in birds occurred at dietary loadings of 10.0–13.0 mg 1080/kg ration. The risk to human consumers of cooked meat from 1080-poisoned waterfowl seems negligible. Amphibians and reptiles were more resistant to 1080 than mammals and birds because of their greater ability to detoxify fluoroacetate by defluorination, a reduced ability to convert fluoroacetate to fluorocitrate, and an aconitase hydratase enzyme that is comparatively insensitive to fluorocitrate inhibition. LD50 values for amphibians were >44.0 mg 1080/kg BW; for reptiles, this value was >54.0 mg 1080/kg BW. Other studies with 1080 and sensitive species showed death of mosquito larvae at water concentrations of 0.025–0.05 mg/L, death of terrestrial beetle and lepidopteran larvae at 1.1–3.9 mg/kg BW, no phytotoxicity to terrestrial flora at water concentrations of 10.0 mg/L, and – based on limited data – no adverse effects on freshwater fish at 370.0 mg/L.

30.4.1 Terrestrial Plants and Invertebrates

Fluoroacetate was first isolated in South Africa in 1944 from the gifblaar plant (*Dichapetalum cymosum*). Seeds of the South African *Dichapetalum braunii* may contain as much as 8000.0 mg fluoroacetate/kg DW. Several other species of *Dichapetalum* produce fluoroacetate, as well as *Palicourea marcgravii*, a South American species known to be poisonous. In Australia, fluoroacetate occurs naturally in the leaves, flowers, and seeds of more than 35 species of leguminous plants of the genera *Gastrolobium* and *Acacia*. All but two of these species are confined to the southwest corner of Western Australia; the other two species are found in northern and central Australia. Fluoroacetate concentrations varied regionally, seasonally, among species, and among parts of the plants. Fluoroacetate content of these plants is usually greatest in flowers, seeds, and young leaves, and this is consistent with chemically mediated defense strategies in which plants use poisonous compounds to protect those parts most essential to them. In Australia, the highest fluoroacetate concentrations measured were in air-dried leaves and seeds of two species from Western Australia: concentrations reached 2650.0 mg/kg in leaves and 6500.0 mg/kg in seeds of *Gastrolobium* spp. Air-dried samples of the two species from northern and central Australia, *Acacia georginae* and *Gastrolobium grandiflorum*, contained as much as 25.0 mg fluoroacetate/kg leaf and 185.0 mg/kg seed.

Significant economic losses of domestic livestock have occurred in Africa and Australia after ingestion of fluoroacetate-bearing vegetation. Herbivores which have had evolutionary exposure to this vegetation are much less susceptible to fluoroacetate intoxication than geographically separate, unchallenged species. The development of tolerance to fluoroacetate by insects, reptiles, birds, and mammals has evolved on at least three continents where indigenous plants produce fluoroacetate which protects them against herbivores. In Australia, for example, animal populations that have coexisted with fluoroacetate-bearing vegetation for at least several thousands of years have developed varying degrees of tolerance to this potent toxin. Tolerance depends upon their diet and habitat, size of their home range, mobility, and length of evolutionary exposure to fluoroacetate-bearing vegetation. Once developed, this tolerance is retained by animal populations even after isolation from the toxic vegetation for 70–100 centuries. Biochemical mechanisms responsible for the large toxicity differential between conspecifics with and without exposure to fluoroacetate-bearing vegetation are poorly understood.

Fluoroacetate and fluorocitrate have also been isolated from forage crops grown in an environment rich in atmospheric or inorganic fluoride. For example, soybeans (*Glycine max*) can synthesize fluoroacetic acid when grown in an atmosphere containing elevated levels of hydrogen fluoride or in media containing high levels of sodium fluoride. Forage crops, including alfalfa (*Medicago sativa*) and crested wheat grass (*Agropyron cristatum*) found growing near a phosphate plant that discharged inorganic fluoride contained as much as 179.0 mg fluoroacetate/kg DW, 896.0 mg fluorocitrate/kg DW, and 1000.0 mg total fluoride/kg. The plants were not adversely affected, but horses (*Equus caballus*) grazing these crops showed signs of fluoride poisoning, suggesting that the toxic effect of inorganic fluoride adsorbed or absorbed by plants and not incorporated into monofluoroacetic acid was greater than the toxic effect of monofluoroacetic acid synthesized by the plants. Lettuce (*Lactuca sativa*) can absorb radiolabeled 1080 through its roots or leaves, resulting in elevated citrate concentrations and active retention of radioactivity when compared to controls. Plants can degrade 1080 by cleaving the carbon–fluorine bond, as judged by studies with germinating seeds of the peanut, *Arachis hypogea*.

Compound 1080 mixed with gel, paste, or grease carriers smeared on leaves of palatable plants have been used to control ungulate and marsupial pests in New Zealand, including feral goats (*Capra* sp.), red deer (*Cervus elephus*), and white-tailed deer (*Odocoileus virginianus*). The effectiveness of 1080 in carbopol gel or petrolatum grease on leaves of the mahoe (*Melicytus ramiflorus*) was significantly modified by the phytotoxicity of these carriers. Both carriers caused baited leaves to abscise, and the rate of abscission increased when 1080 was included. Petrolatum was one-third as phytotoxic as carbopol and retained 1080 for longer periods – at least 1 year. Carbopol lost about 95% of its 1080 after 64 days of exposure and 100 mm of rain vs. 22% loss in petrolatum under similar conditions. Carbopol with 1080 is recommended for use where its distribution is sufficient to place goats and other target species at immediate risk; petrolatum can be used in areas where a long-lasting bait is needed.

Compound 1080 has systemic insecticidal properties against insects feeding on treated plants. Cabbage (*Brassica oleracea capitata*) that had accumulated 1080 through its roots from solution or soil cultures, or following leaf application, was toxic by contact to eggs and larvae of the large white butterfly (*Pieris brassicae*), and various species of aphids. Compound 1080 was not phytotoxic at 10.0 mg/L or several times the concentration necessary for insecticidal action, but its use as an insecticide is not recommended because of its high mammalian toxicity.

At least nine groups of terrestrial invertebrates are adversely affected by eating 1080-poisoned baits, living in habitats contaminated by residues leaching from 1080 baits, or consuming animal by-products and carcasses contaminated with 1080. Lethal effects are reported in houseflies, moths, aphids, ants, bees, and mites that ate 1080-poisoned baits and in fleas that ate 1080-poisoned rats. Cockroaches, collembolids, and slugs that ate poisoned baits experienced adverse effects. Egg production in wasps was disrupted after a single sublethal dose of 1080, and butterfly eggs treated with 1080 had 98% mortality of resultant larvae. Harvester ants (*Pogono myrmex*) and darkling ground beetles (Tentyridae) removed and consumed 1080 bait, leaving bait and dead ants concentrated on the ground near the nest. In a wasp-control program, German wasps (*Vespula germanica*) and common wasps (*Vespula vulgaris*) fed 1080-poisoned canned sardines in aspic jelly were not affected at concentrations <100.0 mg 1080/kg bait. At 1000.0 mg/kg, however, wasp traffic at nest entrances was reduced 17%; at 5000.0–10,000.0 mg/kg, traffic was reduced 78–89%, and almost all wasps died within 100 m of bait stations after 6 h. Honeybees (*Apis mellifera*) feed readily on 1080-jam baits used to control opossums (*Trichosurus vulpecula*) in New Zealand. Bee kills have been documented in the vicinity of jam baits and dead bees contained 3.1–10.0 mg 1080/kg whole bee. The oral LD50 for the honey bee is 0.8 μg/bee. Because no deaths occur within 2 h after feeding, poisoned bees may make several foraging trips before dying. Molasses or oxalic acid is now added to 1080-jam baits to repel bees. Poisoned insects may cause secondary poisoning of insectivores. Accordingly, 1080 should not be used in the vicinity of susceptible nontarget species of invertebrates or endangered insectivores.

Tested insect larvae showed great variability in sensitivity to 1080 after abdominal injection. The LD50 value, in mg 1080/kg BW – administered by way of fluoroacetate-bearing vegetation – was 1.05 for *Perga dorsalis* (Hymenoptera); for Lepidoptera, these values were 3.9 for *Mnesampla privata*, 42.7 for *Spilosoma* sp., and about 130.0 for *Ochrogaster lunifer*. For all species tested, death occurred within 2–48 h after injection, and total body citrate concentrations were significantly higher than that of unpoisoned conspecifics. Enhanced tolerance to 1080 was shown in larvae of Western Australian insects feeding on fluoroacetate-bearing vegetation.

Populations of terrestrial invertebrates were not adversely affected by 1080 poisoning operations to control brushtail possums in New Zealand, including populations of amphipods, ants, beetles, collembolids, millipedes, mites, weevils, slugs, spiders, and snails. Residues of 1080 in nontarget terrestrial invertebrates were low or negligible after an aerial poisoning campaign. Residues of 1080 were measured in various species of terrestrial invertebrates in New Zealand before and after aerial application of possum baits containing 800.0 mg 1080/kg and sown at 5.0 kg/ha. No residues of 1080 were found in spiders, beetles, millipedes, centipedes, or earthworms at any stage. Residues of 1080 were detectable in some orthopteran insects (2.0 mg/kg FW) and cockroaches (4.0 mg/kg FW). Laboratory studies indicated that 90% of all 1080 was eliminated from insects within 4–6 days after dosing, suggesting low risk to insectivorous birds.

30.4.2 Aquatic Organisms

Despite an intensive literature search, very little data were found on the toxicity of 1080 to aquatic life. One group reported that fingerling bream and bass (species unidentified) tolerated 370.0 mg of 1080/L for an indefinite period with no apparent discomfort. Another group aver that fourth-instar larvae of the mosquito *Anopheles quadrimaculatus* were comparatively sensitive to 1080, and that 1080 was among the most toxic 3% of 6000 organic compounds screened against this life stage. In 48 h, concentrations of 0.025, 0.05, and 0.1 mg 1080/L were fatal to 15, 40, and 65% of these larvae, respectively. The common duckweed (*Spirodela oligorrhiza*) seems to be unusually sensitive to 1080. Growth inhibition of duckweed was recorded at 0.5 mg/L, but this needs verification.

Unpublished data on the acute toxicity of 1080 to rainbow trout (*Oncorhynchus mykiss*), bluegill (*Lepomis macrochirus*), and daphnid (*Daphnia magna*) suggest that these organisms are comparatively tolerant to 1080. For example, bluegills exposed to 970.0 mg 1080/L for 96 h showed no observable adverse effects;

for rainbow trout the no-observable-effect concentration during 96-h exposure was 13.0 mg 1080/L and the LC50 (96 h) value was 54.0 mg/L with a 95% confidence interval of 39.0–74.0 mg/L; for *Daphnia*, no adverse effects were noted at 130.0 mg 1080/L during exposure for 48 h although 50% were immobilized at 350.0 mg/L in 48 h. No data were available on effects of 1080 to aquatic biota during life cycle or long term exposures. Studies need to be initiated on effects of chronic exposure of 1080 to nontarget species of aquatic arthropods and macrophytes.

30.4.3 Amphibians and Reptiles

In general, the onset of action and time to death or recovery was slowest in amphibians and reptiles and they were among the most resistant species to 1080 of all vertebrate animals tested. LD50 values for representative species of amphibians ranged from 54.0 to 2000.0 mg 1080/kg BW and for reptiles 44.0–800.0 mg/kg BW. Frogs and lizards given a lethal oral dose of 1080 did not show signs of poisoning for 22–56 h and survived for 78–131 h. Frogs seem to be more sensitive to 1080 in summer than in winter. Amphibians and reptiles possess an innate tolerance to 1080 when compared to mammals because of their greater ability to detoxify fluoroacetate by defluorination, a reduced ability to convert fluoroacetate to fluorocitrate, and an aconitase hydratase enzyme system which is less sensitive to inhibition by fluorocitrate.

One of the most tolerant reptiles tested against 1080 was the shingle-back lizard (*Tiliqua rugosa*), but populations of *T. rugosa* from Western Australia that coexist with fluoroacetate-bearing vegetation were much less sensitive to 1080 intoxication than conspecifics from South Australia not exposed to the toxic plants. The shingle-back lizard is an omnivore which feeds on flowers, leaves, and seeds, and probably evolved an increased tolerance to fluoroacetate through feeding on toxic plants such as *Gastrolobium* and *Oxylobium* which are abundant in southwestern Australia.

Reptiles are unlikely to be affected by either primary or secondary poisoning during 1080-poisoning campaigns. In Australia, 1080-poisoned baits contained 330.0 mg 1080/kg in carrot baits for rabbits and oat baits for pigs, 400.0 mg 1080/kg in oat baits for rabbits, 500.0 mg 1080/kg in pellet baits for rabbits and pigs, 14.0 mg 1080/kg in meat baits for dingos, and 144.0 mg/kg in meat baits for pigs. These data indicate that most species of reptiles tested would need to ingest unrealistic quantities of bait to be adversely affected by 1080. Most lizards, for example, would need to eat 43–172% of their body weight of poisoned rabbit baits, and 143–393% of their body weight of meat baits intended for pigs. However, Goulds monitor (*Varanus gouldi*) may ingest lethal amounts of meat baits intended for pigs after eating 31% of its body weight of poisoned baits. By comparison, a large pig (130 kg) needs to eat about 2.0 kg of meat baits (1.6% of its body weight) for an LD99 dose.

30.4.4 Birds

Laboratory studies with birds indicated several trends: (1) death occurred in orally dosed sensitive species after a single dose of 0.6–2.5 mg 1080/kg BW, daily doses of 0.5 mg 1080/kg BW for 30 days, 47.0 mg/kg diet for 5 days, or 18.0 mg/L drinking water for 5 days; (2) single doses >10.0 mg/kg BW were usually fatal; (3) 1080 toxicity was enhanced at lower temperatures; (4) younger birds were more sensitive than older birds; (5) birds tended to avoid diets and drinking water containing high sublethal concentrations of 1080; (6) accumulations and adverse effects were noted at dietary concentrations of 10.0–13.0 mg 1080/kg feed; and (7) birds with prior or continuing exposure to naturally occurring fluoroacetates were more resistant to 1080 than conspecifics lacking such exposure. Drinking water LC50 values were about 10 times higher (i.e., 10 times less toxic) than dietary LC50s for mallards (*Anas platyrhynchos*) and common bobwhites (*Colinus virginianus*); however, both species of birds consumed 5–10 times more water than food on a daily mg/kg BW basis. The minimum repeated daily oral dosage that was lethal to mallards in 30-day tests was 0.5 mg/kg BW, suggesting a high degree of cumulative action for this species. But European starlings (*Sturnus vulgaris*) tolerated 13.5 mg 1080/kg diet for extended periods without significant adverse effects. Studies with the galah (*Cacatua roseicapilla*) showed that 1080 lethality was not affected by the age or sex of the bird or route of administration. But breeding adult female Pacific black ducks (*Anas superciliosa*) were more sensitive to 1080 than either males or nonbreeding females.

The most common external signs of avian 1080 poisoning included depression, fluffed feathers, a reluctance to move, and convulsions. Signs of 1080 poisoning first appeared 1–60 h after dosing, and deaths occurred 1 h to almost 11 days after dosing. Death of 1080-poisoned California quail (*Callipepla californica*) usually occurred within 3 h, although birds were inactive within 2 h of dosing and comatose until death. The most common internal sign of 1080 poisoning was a dose-related increase in plasma citrate concentration, and this was a useful indicator of fluoroacetate sensitivity among birds of similar metabolic rates and phylogenetic affinities. Some birds poisoned with 1080 either vomited (little crow, *Corvus bennetti*; emu, *Dromaius novaehollandiae*; wedge-tailed eagle, *Aquila audax*; sulfur-crested cockatoo, *Cacatua galerita*) or had saliva or fluid dripping from their beaks (Pacific black duck, *Anas superciliosa*). Early signs of poisoning, such as vomiting, were seen at oral doses of 10.0 mg/kg BW in various raptors including the rough-legged hawk (*Buteo lagopus*), the ferruginous rough-legged hawk (*Buteo regalis*), the northern harrier (*Circus cyaneaus*), and the great horned owl (*Bubo virginianus*). The onset of convulsions was preceded by rapid panting, squawking, shrieking or other vocalizations and then a brief period (5–120 s) of violent wing flapping, loss of balance, or paddling or running motions with the feet. Birds then fell to the ground while undergoing tetanic seizures, breathing slowly and laboriously, with wings and tail outstretched. Turkey vultures (*Cathartes aura*) fatally poisoned by 1080 died 4–32 h after dosing; prior to death, birds displayed tremors, ataxia, lethargy, wing drooping, and emesis. Turkey vultures were more sensitive to 1080 at colder temperatures of 8–9°C than at 23–28°C; this may be due to inhibition by 1080 of mitochondrial oxidative phosphorylation at colder temperatures, making animals more sensitive at times of increased metabolic demand.

Some bird species probably developed a tolerance to 1080 from eating plants that contain fluoroacetate, or insects and other organisms that have fed on such plants. Birds indigenous to geographic areas of Australia where fluoroacetate-bearing vegetation is abundant were more tolerant to 1080 than birds distributed outside the range of the toxic plants. Fluoroacetate tolerance in birds is postulated to increase with increasing evolutionary exposure to the toxic plants and decreasing mobility. In the low-nutrient environment of Western Australia, fluoroacetate-tolerant herbivores clearly have a potential advantage over nontolerant herbivores in their broadened choice of fluoroacetate-bearing vegetation in the diet. The most sensitive Australian bird tested was the red-browed firetail (*Emblema temporalis*) with an LD50 of 0.63 mg 1080/kg BW (0.007 mg/whole bird); the most resistant bird tested was the emu with an LD50 of about 250.0 mg 1080/kg BW or about 8000.0 mg/whole bird. Emus in the southwest portion of Western Australia with evolutionary exposure to fluoroacetate-bearing vegetation have unusually high tolerance to 1080. Emu tolerance was attributed to: (1) their ability to detoxify fluoroacetate by defluorination; (2) a limited ability to convert fluoroacetate into fluorocitrate; and (3) possession of an aconitase hydratase enzyme which is relatively insensitive to fluorocitrate.

Deaths of nontarget species of birds after eating 1080-poisoned baits have been reported, although population effects have not yet been demonstrated. Birds of several species were found dead after 1080 baits were applied to kill California ground squirrels (*Spermophilus beecheyi*), but only Brewer's blackbird (Euphagus cyanocephalus) contained measurable 1080 residues. Nontarget seed-eating birds that died after eating 1080-poisoned baits included sparrows, blackbirds, towhees (*Pipilo* spp.), horned larks (*Eremophila lapestris*), McCown's longspurs (*Calcarius mccownii*), chestnut-collared longspurs (*Calcarius ornatus*), and western meadowlarks (*Sturnella neglecta*). Individuals of at least 20 species of Australian birds are at risk from dingo and pig poisoning campaigns that use meat baits containing 14.0–140.0 mg 1080/kg bait, and 39 species are at risk from rabbit and pig poisoning campaigns using vegetable baits that contain 330.0–500.0 mg 1080/kg bait. The extent of bird mortality and possible population effects depend on several factors: bait palatability to each species; availability of other foods; the amount of 1080 ingested; the number of birds in each population that consume baits before the target species or other nontarget groups; and the rate of 1080 leaching from baits by dew or rainfall. Birds seen feeding on 1080-poisoned baits for control of wild dogs included the pied currawong (*Strepera graculina*), the Australian raven (*Corvus coronoides*), the Australian magpie, (*Gymnorhina tibicen*), and the wedge-tailed eagle (*Aquila audax*). Avian scavengers such as vultures, condors, hawks, and ravens are likely to find poisoned food items as they search for carcasses.

Secondary 1080 poisoning of birds is documented. Australian birds found dead after eating 1080-poisoned carcasses of pigs (*Sus* sp.) included kites (whistling kite, *Haliastur sphenurus*; black kite, *Milvus migrans*), eagles (Australian little eagle, *Hieraaetus morphnoides*; wedge-tailed eagle), brown falcon (*Falco bevigora*), Australian kestrel (*Falco cenchroides*), brown goshawk (*Accipiter fasciatus*), Australian magpie-lark (*Grallina cyanoleuca*), Australian raven, and crows (Australian crow, *Corvus orru*; little crow, *Corvus bennetti*). Insectivorous birds that may have died after eating 1080-poisoned ants (*Veromessor andrei*, *Liometopum occidentale*) in the United States include acorn woodpeckers (*Melanerpes formicivorus*), the white-breasted nuthatch (*Sitta carolinensis*), and the ash-throated flycatcher (*Myiarchus cinerascens*).

Little or no secondary hazards to raptors were evident – as judged by the absence of carcasses – from 1080 ground squirrel baiting operations among hawks, harriers, eagles, ravens, vultures, and condors. However, some species of owls were comparatively susceptible to 1080, including burrowing owls (*Athene cunicularia*) and barn owls (*Tyto alba*). Raptors are less susceptible to secondary poisoning from 1080 than mammalian predators because birds have higher LD50 values, refuse to eat large amounts of 1080-poisoned meats, and sometimes regurgitate poisoned baits. The reduced hazard of acute 1080 poisoning via secondary sources for raptors is illustrated for the golden eagle (*Aquila chrysaetos*), a bird that normally consumes the internal organs of its prey before consuming other portions of the carcass. Golden eagles fed diets containing 7.7 mg 1080/kg diet – about 3 times the highest concentration of 1080 detected in carcasses of coyotes killed by 1080 livestock protection collars – all survived, although some eagles showed signs of 1080 intoxication including loss of strength and coordination, lethargy, and tremors. For a 3.2-kg golden eagle to obtain an LD50 dose (1.25–5.0 mg 1080/kg BW) it would have to consume the internal organs of 7–30 coyotes killed by 1080, assuming that each coyote ingested 0.1 mg 1080/kg BW and did not excrete, detoxify, or regurgitate any of the toxicant and that, as in rats, about 40% of the dose is present in the internal organs at death. Since the internal organs of a coyote account for 20–25% of its live weight or 2.7–3.2 kg/coyote, and a golden eagle's daily consumption of food is about 30% of its live weight or 0.9 kg, it seems unlikely for raptors to be at great risk from consuming coyotes killed by 1080 livestock protection collars.

Human consumers of meat from 1080-killed ducks would probably not be adversely affected after eating an average cooked portion. Moreover, oven-baking or grilling at temperatures >200°C will cause breakdown of 1080. For example, if a mallard received a triple lethal dose of 1080, then a 1-kg mallard would contain an estimated 14.4 mg of 1080. A 70-kg human would have to consume 25.4 kg of poisoned duck flesh to receive a lethal dose, as judged by LD50 values of 4.8 mg/kg BW for mallards and 5.0 mg/kg BW for humans; theoretically, consumption of only two whole ducks poisoned by 1080 may cause transient toxicity.

Avian populations that were reduced in numbers during 1080 poisoning for possum control usually recovered quickly if they had high potential for reproduction and dispersal. Birds from Australia or New Zealand with poor reproductive potential and poor dispersal had a high risk of nonrecovery; this group includes the 3 species of kiwi (*Apteryx* spp.), the takake (*Notornis mantelli*), kakapo (*Strigops habroptilus*), laughing owl (*Sceloglaux albifacies*), bush wren (*Xenicus longipes*), rock wren (*Xenicus gilviventris*), fernbird (*Bowdleria punctata*), yellowhead (*Mohoua ochrocephala*), stitchbird (*Notiomystis cincta*), saddleback (*Philesturnus carunculatus*), kokako (*Callaeas cinera*), and New Zealand thrush (*Turnagra capensis*). Poison control programs against wild dogs, dingoes, and their hybrids using 1080 meat baits did not significantly affect nontarget populations of birds in the treated areas. Baiting with 1080 to control rabbits and foxes in Australia usually had no significant permanent adverse effects on nontarget birds, although 15 of the 30 bird species in the treated areas during the poisoning campaign showed a temporary negative trend in abundance, especially welcome swallows (*Hirundo neoxena*), tree martins (*Hirundo nigricans*), and crimson rosellas (*Platycercus elegans*). Aerial drops of 1080-laced pellets (11.8 kg/ha) to control brushtail possums and rock wallabies (*Petrogale penicillata*) on Rangitoto Island, New Zealand, had no observed effect on island bird populations over the next 12 months. No species of bird showed a population decline and several showed significant increases in numbers, including greenfinch (*Carduelis chloris*), Australian harrier hawk (*Circus approximans*), and tui (*Prosthemadera novae-seelandiae*). Increases were attributed to the reduction in numbers of mammalian browsers which led to increased vegetation and improved habitat for nontarget bird species.

Mortality of nontarget birds in 1080 poisonings may be underreported because many die in their nests or roosts and are never found. Raptors of several species were found dead shortly after application of 1080 baits; however, no 1080 residues were detected in any of these birds and the cause of death was not established. Application of 1080 baits to control California ground squirrels was associated with deaths of yellow-billed magpies (*Pica nuttalli*) which contained about 1.02 mg 1080/kg FW of internal organs vs. 0.6–0.7 mg 1080/kg FW in stomachs of black-billed magpies (*Pica pica*) treated with lethal doses of 1.6–3.2 mg 1080/kg BW. It is not known if *P. nutalli* ingested the 1080 bait directly, ate other poisoned animals, or both. Risks of 1080 poisoning to birds can be reduced by (1) setting meat baits out just before sunset and removing them early next morning; (2) burying baits for pigs below ground; (3) using baits that only the target animals prefer; (4) reducing the number of available small bait fragments; and (5) masking the appearance of baits and enhancing their specificity by the use of dyes – although some birds in Australia seem to prefer green-dyed meat baits.

30.4.5 Mammals

Studies with mammals showed several trends: (1) individuals of sensitive species died after receiving a single dose between 0.05 and 0.2 mg/kg BW, including species of livestock, marsupials, canids, felids, rodents, and foxes; (2) most individuals of tested species died after a single dose between 1.0 and 3.0 mg/kg BW; (3) a latent period was evident between exposure and signs of intoxication; (4) mortality patterns usually stabilized within 24 h after exposure; (5) species from fluoroacetate-bearing vegetation areas were more resistant than conspecifics from non-fluoroacetate vegetation areas; (6) route of administration had little effect on survival patterns; (7) younger animals were more sensitive than adults; (8) high residues were detected in some 1080-poisoned animals, notably rabbits with 34.0 mg/kg DW muscle and 423.0 mg/kg DW liver; (9) secondary poisoning was evident among carnivores after eating 1080-poisoned mammals; and (10) sublethal effects included testicular damage in rats after drinking water containing 2.2–20.0 mg 1080/L for 7 days (0.07–0.71 mg/kg BW daily), impaired reproduction in mink fed diets containing 0.8 mg 1080/kg ration for 60 days, and altered blood chemistry in ferrets given diets containing 1.1 mg 1080/kg ration for 28 days.

The most sensitive mammal tested was the Texas pocket gopher (*Geomys personatus*) with an LD50 of <0.05 mg 1080/kg BW. In general, carnivorous eutherian mammals were most sensitive to 1080 and amphibians most resistant; intermediate in sensitivity were herbivorous eutherian mammals and marsupials, carnivorous marsupials, herbivorous–granivorous rodents, omnivorous mammals, and birds, in that order. Very young mammals seemed more sensitive to 1080 than other members of their populations; no other differences in sensitivity to 1080 were found that could be related to sex, age, or nutritional status. Route of administration had little effect on 1080 toxicity. Oral dosages were as toxic as subcutaneous, intramuscular, intravenous, and intraperitoneal dosages. There are species differences, as yet unexplained, in fatal 1080 poisonings: dogs died of convulsions or respiratory paralysis, but monkeys, horses, rabbits, and humans died of ventricular fibrillations. Individuals of most species dosed with 1080 died within 7 days but feral pigs and wedge-tailed eagles took longer. Ambient air temperatures in the range 4–33°C modified the sensitivity of small mammals to 1080. In mice (*Mus* spp.) and guinea pigs (*Cavia* spp.), sensitivity was greatest at the extremes of the thermal regimes than at intermediate temperatures. Raccoons (*Procyon lotor*) and feral pigs were more sensitive at elevated ambient temperatures, but opossums and domestic sheep were more sensitive at low temperatures. At elevated temperatures 1080 was more toxic to feral pigs when administered via drinking water vs. oat baits, and in wheat baits vs. pellet baits.

Warm-blooded species varied considerably in response to sodium fluoroacetate, with primates more resistant and rodents and carnivores more susceptible. Based on fatal or near fatal cases of human poisonings, the dangerous dose for humans is 0.5–2.0 mg/kg BW. Among the 171 species of mammals tested, for which there are data, there was considerable variability in the time until signs of poisoning became apparent (0.1 h to >7 days), the time to death (0.1 h to >21 days), and the time until animals began to show signs of recovery (2 h to 18 days). Signs of poisoning among herbivorous species of marsupials first appeared 1–39 h after dosing; death followed 3–156 h after dosing. Australian carnivores did not show signs of 1080 poisoning for 0.6–4.8 h; first deaths occurred between 1.6 and 21 h and recovery in 0.4 to 26 h. Marsupial carnivores generally showed signs of 1080 poisoning earlier and died or recovered quicker than did marsupial herbivores and placental mammals. After the latent period, common signs of 1080 poisoning in caged mammals included hyperexcitation, rapid breathing, and trembling. Some animals then recovered while others began to vomit, convulse, or both. The most common signs of 1080 poisoning in 14 species of Australian rodents were depression, hypersensitivity to stimuli, respiratory distress, and convulsions; signs usually appeared 0.4–38.1 h after dosing and deaths occurred 0.7–206 h after dosing. A few species were more tolerant, perhaps because of exposure to indigenous plants that contained fluoroacetate. Rabbits (*Oryctolagus* sp.) poisoned by 1080 showed increased sensitivity to noise or disturbance; those surviving high sublethal doses began recovering 5–23 h after dosing. Cows (*Bos* spp.) showed no signs of fatal 1080 poisoning until shortly before death; signs appeared in the following sequence: urination, staggering, falling down, slight spasms, and death 1.5–29 h after treatment. Prairie dogs (*Cynomus* sp.) showed no signs of 1080 poisoning for several hours after consuming a fatal dose; death occurred 8–13 h after dosing and was preceded by a rapid respiratory rate, hyperactivity, and convulsions. In feral pigs, signs of poisoning such as vomiting, increasing lethargy, and labored breathing appeared about 6.2 h after dosing (range 1.9–47.3 h), and death after 16.1 h (range 2.8–80 h) after dosing. Vomiting occurred in 98% of poisoned pigs, but was unrelated to dose or bait type. With some animals, particularly the eastern native cat (*Dasyurus viverrinus*), the tiger cat (*Dasyurus maculatus*), and the Tasmanian devil, the first sign of 1080 poisoning is the sudden onset of vomiting. Vomiting was independent of dose ingested or mode of administration. Thereafter, animals may either recover or experience hyperexcitation, convulsions, and death.

Many 1080 control programs report high effectiveness without significant effect on nontarget species. Australian baits used to control various mammal pests usually contain 15.0–110.0 mg 1080/kg bait, although concentrations as high as 1200.0 mg/kg bait are documented. Baiting with 1080 to control European rabbits and red foxes (*Vulpes vulpes*) in New South Wales, Australia, caused a 90% reduction in numbers of rabbits and 75% of foxes; populations of both species began to recover soon after the campaign ended, indicating the need for continued control measures. Populations of nontarget birds and mammals did not appear to be affected and no dead birds or nontarget mammals were found. A similar case is reported for 1080 control programs in Australia against wild dogs, dingoes, and their hybrids. In Tasmania, deliberate poisoning of forest-browsing pests with carrot baits containing 0.014% of 1080 – the same concentration used elsewhere in Tasmania for rabbit control – resulted in 94% mortality of brushtail possum populations, 96% mortality of red-bellied pademelons, and 86% mortality of Bennett's wallabies. The use of 1080 to protect island-dwelling rare or endangered species of herbivorous marsupials – a comparatively tolerant group – to kill more sensitive introduced competitors or predators such as rabbits, foxes, and feral cats was suggested as an interesting possibility.

Compound 1080 is highly toxic to some species of nontarget mammals, including domestic cats and dogs. Hazards to wildlife associated with 1080 baiting for California ground squirrels that reduced squirrel populations by 85% included some deaths of Heermann's kangaroo rats (*Dipodomys heermanni*), the little pocket mouse (*Perognathus longimembris*), the desert woodrat (*Neotoma lepida*), deer mice (*Peromyscus* spp.), and the western harvest mouse (*Reithrodontomys megalotis*); poisoned rodents contained between 5.2 and 23.1 mg 1080/kg BW and 1080-poisoned desert cottontails (*Sylvilagus audubonii*) contained 8.2 mg 1080/kg stomach contents. Nontarget animals found dead in New South Wales State forest areas after 22 rabbit poisoning operations between 1971 and 1975 included, in decreasing order of frequency, foxes, wallabies, possums, gray kangaroos, wombats, rats, hares, birds, cats, sheep, and dogs. This pattern may reflect the relative abundance of each species in the areas involved, their access to and acceptance of baits, and their ease of detection after death by forestry personnel. In Australia, the animals alleged to be most at risk during rabbit or pig-poisoning campaigns using pellet, grain, or carrot baits are the kangaroos, wallabies, and wombats. For example, common wombats (*Vombatus ursinus*) and hairy-nosed wombats (*Lasiorhinus latifrons*) need to consume only 10.0–16.0 g of pellet, grain, or carrot baits containing 0.33–0.5 mg of 1080 to receive an LD50. Hairy-nosed wombats eat 120.0–570.0 g of food daily and common wombats can eat over 500.0 g of unpoisoned carrots daily, indicating that both species could easily consume lethal quantities of bait. Livestock were next at theoretical risk, followed by brushtail possums, pigs, and various rodents and birds. More data are needed on bait consumption rates of nontarget mammals if risk from 1080 poisoning campaigns is to be satisfactorily assessed.

Laboratory studies may overestimate the risk to nontarget species from 1080 baiting. The northern quoll (*Dasyurus hallucatus*), for example, was found to be at highest theoretical risk from aerial baiting programs as judged by LD50 laboratory studies with 15 species of rodents and dasyurids. But no quolls were found dead during aerial baiting to control dingoes, and all seemed to have normal movements as judged by radiotelemetry. Alternatives to LD50 testing now include tissue culture techniques, monitoring of metabolite levels in blood or tissues, and estimating the lowest dose likely to cause death. Monitoring the level of citrate in blood plasma of animals that received a sublethal dose of 1080 has been used successfully with species large enough to provide adequate samples of blood plasma in several bleeds over a 24-h period, but these other alternatives have not been attempted on Australian fauna.

Because 1080 acts as an emetic, especially on coyotes and feral pigs, there is a risk of primary poisoning to nontarget animals from eating the vomitus. Wild pigs poisoned by carrot baits placed for European rabbits were observed to leave trails of vomitus containing carrot and other ingested foods. The antiemetic compound metoclopramide (Maxolon®) prevents vomiting in pigs by blocking dopamine receptors in the chemoreceptor trigger zones. The addition of metoclopramide to 1080 poison baits for wild pigs reduces vomiting and thereby reduces the poisoning risk to nontarget species. The addition of metoclopramide improves the efficiency and percentage of the kill of wild pigs because they will not develop taste aversion to the baits. A similar case is made for dogs. Baits containing this antiemetic at an effective concentration of 1 mg/kg BW shortened the median time for death for dogs from 151 min postdose for 1080 baits without metoclopramide to 132 min. At tested doses (1.0–16.0 mg/kg BW), metoclopramide did not decrease the frequency of vomiting by dogs but decreased the amount of vomitus.

Secondary poisoning is likely among carrion eaters feeding on rabbits and other herbivores poisoned with 1080-treated carrots, especially foxes and dingoes (secondary target species), and dogs and cats. Secondary poisoning was reported for dogs feeding on 1080-treated rodents and prairie dogs, and for cats feeding on treated rats and mice. Some domestic dogs and cats were found dead within 450 m of a 1080-treatment area; signs of 1080 poisoning were evident but no 1080 residues were detected by chemical analyses. Ground squirrel control with 1080 baits caused secondary poisoning of dogs, cats, coyotes, bobcats (*Lynx rufus*), skunks, and kit foxes. The high susceptibility of threatened and endangered species of kit foxes to 1080 rodenticides, as judged by studies with nonthreatened species of kit foxes, suggests that 1080 could be a factor in their population decline. Sodium monofluoroacetate has a high degree of secondary toxicity in mammals, as evidenced by deaths of domestic ferrets that ate 1080-poisoned white-footed mice (*Peromyscus leucopus*). Similarly, coyotes died after ingestion of 1080-poisoned ground squirrels that contained 3.0–6.0 mg of 1080 equivalent to 0.24–0.63 mg/kg BW coyote. Coyotes that ate a single 1080-poisoned squirrel daily for 5 days, for an estimated total dose of 0.12–0.27 mg/kg BW, usually survived, suggesting that there is little secondary hazard from multiple doses when they are small. Carcasses and viscera from coyotes that died after ingesting 5–15 mg of 1080 were fed for 14–35 days to other coyotes, domestic dogs, striped skunks (*Mephitis mephitis*), and black-billed magpies; no evidence of secondary poisoning was seen in any species tested. Maximum residues of 1080 in dead coyote tissues, in mg/kg FW, were 0.66 in muscle, 0.79 in small intestine, and 0.76 in stomach tissue. Tissues of 1080-poisoned coyotes did not produce secondary poisoning in opossums (*Didelphis virginiana*), striped skunks, raccoons, or badgers (*Taxidea taxus*). The hazard of secondary poisoning to predators is minimal after consuming tissues of 1080-killed black-tailed prairie dogs (*Cynomus ludovicianus*), as their tissues contained <0.1 mg fluoroacetate/kg FW. No mink died when fed 1080-poisoned rabbits at 40% of the total diet provided that the rabbit gastrointestinal tract had been removed from the carcass. This suggests that secondary toxicity from 1080 is due primarily to consumption of the unmetabolized compound from the gut of prey species. The risk to different individuals or populations depends on the species' sensitivity to 1080, the number of poisoned animals consumed, and the amounts of different tissues or organs consumed.

Animals in Australia vary greatly in their sensitivity to 1080 poison, with known LD50 values ranging from 0.11 to >800.0 mg/kg BW. Many native species, particularly in Western Australia have evolved tolerances to 1080 through ingestion of native plants that contain fluoroacetate or prey that consume these plants. The degree to which this tolerance is developed depends on the extent of the toxic plants in the microhabitat, the need of each species to include those food species that contain fluoroacetate in its diet, and the length of evolutionary exposure to the toxic plants. This naturally occurring resistance to the toxins allows control programs that use 1080 to be more specific for introduced test species.

Tolerance to fluoroacetate is present in insects, reptiles, mammals, and birds and is in the order of herbivores > omnivores > carnivores. Mammals with lower metabolic rates – such as marsupial carnivores – seem to be more tolerant to a metabolically interfering poison such as 1080 than mammals with a higher metabolism such as eutherian carnivores. Tolerance to gradually increasing doses of fluoroacetate can be induced in the mouse, rat, and rhesus monkey, but not in dog or rabbit; however, the protective effect of prior exposure to 1080 seldom persisted for more than 48 h. Laboratory white rats may acquire tolerance to 1080 by the ingestion of sublethal doses over a period of 5–14 days; cessation of dosing for 7 days caused a loss of tolerance. Some species acquired tolerance to 1080 after repeated sublethal doses and others accumulated the chemical until a lethal threshold was reached. Both phenomena were unpredictable if 1080 residues in the tissues remained between doses. Time required for complete elimination of 1080 from tissues varied among species: dogs required 2–3 days, rats 36 h, and sheep as long as 1 month.

Sublethal concentrations of 1080 may adversely affect reproduction, growth, and behavior. In rats (*Rattus* sp.), the organ most vulnerable to 1080 poisoning is the testes and this is consistent with 1080-impaired energy production via blockage of the Krebs cycle and subsequent impairment of carbohydrate metabolism. Subacute dietary exposure to 1080 caused dose-dependent decreases in body weights and feed consumption in mink and European ferrets. Toxic 1080 meat baits were usually avoided by the majority of tested nontarget dasyurids and rodents when alternative foods were available; 12 of the 24 groups tested did not sample meat baits under these conditions. Adult wild pigs given a sublethal dose of 1080 (0.5 mg/kg BW) in apple baits vomited within 30 min after eating the treated bait and avoided apple baits in future tests. Caged wild Norway rats (*Rattus norvegicus*) and black rats (*Rattus rattus*) developed a gradually increasing aversion to drinking water solutions of 1080, although this aversion was not sufficient to disrupt growth and reproduction.

30.5 Recommendations

It is emphasized that 1080 is a restricted pesticide that can only be used by certified applicators who have received special training, and that carcasses of all organisms found dead from 1080 poisoning must be buried or incinerated. Some authorities aver that continued use of 1080 is justified and desirable, and that risk is minimal to nontarget organisms. As discussed earlier, 1080 is a natural plant product, is generally highly toxic to most pests at low concentrations, is readily lost from baits following heavy dews or rainfall, is biodegraded by fungi and bacteria, and does not persist in soil or water. In New Zealand, 1080 has been used since 1954 and is still considered an essential pesticide for limiting forest and crop damage and for containing the spread of tuberculosis to livestock by brushtailed possums. It has been used to control isolated island populations of mammals that prey upon endangered or threatened species of birds, as was the case for Arctic foxes preying on Aleutian Canada geese in the Aleutian Islands. In Australia and New Zealand, results of field studies suggest that 1080 poisoning campaigns had no significant effect on almost all populations of common nontarget species, although more studies are recommended on vulnerable, rare, endangered, or uncommon species.

There is, however, a growing body of information on 1080 that questions its usefulness in the United States. This database includes adverse effects on some nontarget organisms and endangered species; the confounding effects of the latent period, behavioral alterations, and application routes; and the development of suitable alternative chemicals. On the basis of acute oral toxicity tests, it is likely that sensitive nontarget mammals and birds will consume lethal quantities of 1080 from poisoned baits or from consumption of organisms fatally poisoned with 1080. Field studies record deaths among sensitive nontarget species that ate 1080 baits, including bees, insectivorous birds, rabbits, rodents, cats, dogs, and livestock. Secondary poisoning is reported for carrion eaters and mammalian predators – especially canids and felines – after feeding on 1080-poisoned prey. Sublethal effects of 1080 on growth of ferrets and reproduction of mink are reported. Some endangered species are at risk from direct consumption of the 1080 baits or from secondary poisoning. In general, the use of 1080 within the geographic range of any endangered species is discouraged, or disallowed outright in the case of the California condor, the San Joaquin kit fox (*Vulpes macrotis mutica*), the Aleutian Canada goose; the Morrow Bay kangaroo rat (*Dipodomys heermanni morroensis*), and the salt marsh harvest mouse (*Reithrodontomys raviventris*). When exceptions are made, or when 1080 use is permitted in an area known to be frequented by an endangered species, restrictions are placed on the maximum concentration of 1080 in the baits.

It is unlikely that human consumers of meat from 1080-killed ducks would be adversely affected after eating an average cooked portion. The risk to humans is minimal to low from eating meat of domestic animals accidentally poisoned with high sublethal concentrations of 1080 because it is cleared rapidly from domestic animals, usually within a few days. In the absence of additional data it seems prudent to postpone for at least 3 weeks the slaughter or marketing of livestock that survived 1080 exposure. No livestock in the United States contaminated with 1080 are marketed.

No effective antidote to 1080 is currently available and accidental poisoning of livestock and dogs is likely to be fatal. The lack of emergency human treatment in cases of 1080 poisoning coupled with the observation that monoacetin – potentially the most effective medication for compound 1080 poisoning – is not available in a pharmaceutical grade, strongly indicates the need for a viable 1080 antidote. The search for an effective 1080 antidote is ongoing, and some candidate compounds offer partial protection including mixtures of sodium acetate and ethanol, barbiturates, glycerol monoacetate, a mixture of calcium glutonate and sodium succinate, and 4-methylpyrazole. The development and availability of an effective 1080 antidote should constitute a high research priority. Until such time when this antidote is distributed, it seems reasonable to use 1080 in the United States only after other alternatives have been considered.

The interval between 1080 dosage and signs of intoxication is at least 30 min, regardless of dose or species tested, and needs to be considered when evaluating the efficacy of 1080. Coyotes, for example, may continue to kill livestock after receiving a lethal dose. And coyotes may travel some distance from their prey prior to incapacitation, making carcass recovery and program evaluation difficult, as was the case for 1080-poisoned quolls in Australia. Similarly, many 1080-poisoned nontarget animals may have left the treated area before succumbing, thus leading to underestimation of mortality among this group. Tolerance to fluoroacetates and avoidance of 1080 baits should also be considered in future 1080 poisoning campaigns by wildlife managers and animal damage control operators. Avoidance of 1080 toxic baits by target mammals is documented when alternative foods are available, and among pigs and rats surviving sublethal exposures. Indigenous populations of mammals, birds, and reptiles which coexist with fluoroacetate-bearing vegetation are much less sensitive to 1080 poisoning, perhaps by as much as 2 orders of magnitude, than conspecifics lacking such exposure.

The timing of application of 1080 baits is critical. In one mishap, baits were dropped aerially while many ground squirrels – the targeted species – were still in hibernation underground for the winter and had not emerged. Aerial application of 1080 baits in a ground squirrel control program in California, although effective in controlling the squirrels, resulted in great overuse of the baits. As many as 70–77% of the poisoned baits were not eaten by the squirrels and were not recovered. Also, the yellow dye used to color the baits – as a deterrent to birds – faded rapidly. To protect migratory waterfowl, 1080 baits should not be applied immediately preceding or during the main waterfowl hunting season or whenever birds are abundant. To protect honeybees, 1080-poisoned jam baits should be deposited >400 m from apiary sites. If 1080 baits are dispersed <400 m from apiary sites, then beekeepers should remove their hives to a more distant site. The 1080 toxicity database for aquatic organisms is insufficient for practicable formulation of criteria to protect this ecosystem. This seems to be a high priority research need in geographic areas of intensive 1080 application.

Potential replacement chemicals for 1080 include PAPP (*para*-aminopropiophenone), DFP (1,3-difluoro-2-propanol), and various anticoagulant and nonanticoagulant toxins. PAPP is highly toxic to coyotes and domestic cats (each with LD50s of 5.6 mg/kg BW) and lethal to rats and mice (LD50s of 177.0 and 233.0 mg/kg BW, respectively); intermediate in sensitivity were bobcats (10.0), and kit foxes (14.1 mg/kg BW). DFP is under investigation in Australia as an alternative to 1080 in faunal management programs because it has a mode of action similar to that of 1080 and has an antidote in pyrazole. DFP is the major ingredient of the pesticide gliftor used in Russia to control rodents, particularly voles of the genus *Microtus*. Also deserving of evaluation are 4-methylpyrazole and related compounds to function as antidotes to DFP intoxication. In New Zealand, alternatives to 1080 under evaluation include several nonanticoagulant toxins (gliftor, cholecalciferol, calciferol, alpha-chloralase, nicotine, malathion) and anticoagulants including brodifacoum and pindone.

30.6 Summary

Sodium monofluoroacetate ($CH_2FCOONa$), also known as 1080, was first used in the United States to control gophers, squirrels, prairie dogs, rodents, and coyotes (*Canis latrans*); 1080 domestic use is currently restricted to livestock protection collars on sheep and goats to selectively kill depredating coyotes. However, Australia, New Zealand, and some other nations continue to use 1080 to control rabbits, possums, deer, foxes, feral pigs and cats, wild dogs, wallabies, rodents, and other mammals. The chemical is readily absorbed by ingestion or inhalation. At lethal doses, metabolic conversion of fluoroacetate to fluorocitrate results in the accumulation of citrate in the tissues and death within 24 h from ventricular fibrillation or respiratory failure; no antidote is available. At sublethal doses, the toxic effects of 1080 are reversible. Primary and secondary poisoning of nontarget vertebrates may accompany use of 1080. Sensitive mammals died after receiving a single dose of 0.05–0.2 mg 1080/kg body weight (BW), including representative species of livestock, marsupials, canids, felids, rodents, and foxes; most tested species died after a single dose of 1.0–3.0 mg/kg BW. High residues were measured in some 1080-poisoned target mammals and this contributed to secondary poisoning of carnivores ingesting 1080-poisoned prey organisms. Sublethal effects occurred in sensitive mammals at >2.2 mg 1080/L drinking water or 0.8–1.1 mg 1080/kg diet. Sensitive species of birds died after a single 1080 dose of 0.6–2.5 mg/kg BW, or daily doses of 0.5 mg/kg BW for 30 days, or 47.0 mg/kg in diets for 5 days, or 18.0 mg/L in drinking water for 5 days. Adverse effects occurred in birds at dietary loadings as low as 10.0–13.0 mg 1080/kg ration. Amphibians and reptiles were more resistant to 1080 than birds and mammals. LD50 values were >44.0 mg/kg BW for tested amphibians and >54.0 mg/kg BW for tested reptiles; resistance to 1080 was attributed to their reduced ability to convert fluoroacetate to fluorocitrate and their increased ability to detoxify fluoroacetate by defluorination. Mosquito larvae reportedly died at 0.025–0.05 mg 1080/L but fish seemed unaffected at 13.0 mg/L; however, data on 1080 in aquatic ecosystems are incomplete. Acute LD50 values for terrestrial insects ranged from 1.1–3.9 mg/kg BW to 130.0 mg/kg BW for larvae feeding on fluoroacetate-bearing vegetation. Residues of 1080 in exposed insects were usually low (<4.0 mg 1080/kg fresh weight) or negligible and were usually eliminated completely within 6 days, suggesting low risk to insectivorous birds. Loss of 1080 from baits occurs primarily due to microbial defluorination and secondarily to leaching by rainfall and consumption by insect larvae; leachates from 1080 baits are likely to be held in the upper soil layers. The use of 1080 seems warranted in the absence of suitable alternative control methods.

TIN[a]

Chapter 31

31.1 Introduction

Interest in the toxicity of tin compounds dates to the early 1800s when investigators demonstrated that inorganic tin compounds produced muscular weakness, loss of pain sensation, and immobility in dogs. In humans, organotins can be assimilated by inhalation, absorption through the skin, and from food and drinking water. The first documented case of organotin poisoning of humans was in 1880 when workers complained of headaches, general weakness, nausea, and diarrhea after exposure to triethyltin acetate vapors. Renewed interest in the toxicity of organotin compounds resulted from a medical tragedy in France in 1954. "Stalinon," a proprietary compound of diethyltin diiodide plus linoic acid used to treat furuncles and other skin infections, caused 217 poisonings and 111 deaths. The identified toxic components in Stalinon were triethyltin contaminants; victims received a total dose of 3.0 g over a 6- to 8-week period. Symptoms included constant severe headache, rapid weight loss, vomiting, urine retention, vertigo, hypothermia, abdominal pain, and visual and psychic disturbances. Some of the more severely affected patients had convulsions. Death usually occurred in coma or from respiratory or cardiac failure. In survivors, headaches and diminished visual acuity remained for at least 4 years.

World production of organotin compounds is about 30,000 tons, although relatively few organotin compounds, perhaps only 25, are produced and used to any great extent. Diorganotins are used in the manufacture of antioxidants, whereas triorganotins are used as general biocides against microbial and invertebrate pests and in marine antifouling paints. The first antifouling paints incorporating an organotin compound as a biocide were developed in 1961. Because of their effectiveness and availability in a variety of colors, tributyltin antifouling paints are now the most commonly used type, replacing copper-, mercury-, and lead-based paints. Worldwide synthesis of tributyltin compounds is about 900 metric tons annually for all applications. Tributyltins are highly toxic to aquatic plants and animals, readily accumulate in fish and mollusks from contaminated localities, and are present in some harbors where their release from antifouling paints – found usually on small boats and recreational craft – is the putative source. Tributyltin is a contributory factor and probably a major cause for the reproductive failure of the European flat oyster (*Ostrea edulis*) in some locations. In fact, tributyltins are capable of causing adverse biological effects at levels far below that of any previously reported marine pollutant.

The widespread agricultural applications of trialkyltin biocidal agents have greatly increased the relative exposure risks to workers handling these materials. Internationally, tin was recognized as a potential environmental contaminant at the Paris and Helsinki conventions in 1974; in later conventions, organotin compounds were moved to the "black list." Due to the increasing use of

---

[a] All information in this chapter is referenced in the following sources:

Eisler, R. 1989. Tin hazards to fish, wildlife, and invertebrates: a synoptic review. *U.S. Fish Wildl. Serv. Biol Rep. 85(1.15)*, 1–83.

Eisler, R. 2000. Tin. Pages 551–603 in *Handbook of Chemical Risk Assessment: Health Hazards to Humans, Plants, and Animals. Volume 1, Metals*. Lewis Publishers, Boca Raton, Florida.

organotin compounds as a class, the Canadian government, in 1979, placed organotins on Canada's Category III Contaminant List. Category III indicates that additional data are needed on the occurrence, persistence, and toxicity of organotins for preparation of informed environmental and human health risk assessments. In 1982, the use of tributyltin paints was curtailed in France and in 1985 the United Kingdom introduced regulations to limit sales of tributyltin paints that released the biocide at high rates.

Most authorities agree that inorganic tin compounds are comparatively harmless and that many organotin compounds are potentially very hazardous to natural resources – especially tributyltin compounds to aquatic biota. One rare exception to this generalization involved 113 cases of acute gastrointestinal illness in Washington and Oregon in 1969 associated with ingestion of canned tomato juice contaminated by inorganic tin; detinning in many cans resulted in tin levels as high as 477.0 mg inorganic tin per liter of juice. It seems that excessive use of nitrate fertilizer on one tomato crop was the ultimate cause of the detinning.

31.2 Chemical and Biochemical Properties

The chemical, physical, and biochemical properties of inorganic tin compounds differ dramatically from those of representative organotin compounds. There is general agreement that inorganic tins are not highly toxic due to their poor absorption and rapid turnover rate in tissues and to their being essential for growth in at least one species (rat). Of the 260 known organotin compounds, all but a few are manufactured, and 36 are listed as toxic. Most authorities agree on several points regarding organotin compounds: information concerning the mechanism of toxic action is incomplete; there is no evidence of carcinogenicity; trialkyltin compounds are the most toxic; and there are large differences in resistance between and within species.

31.2.1 Inorganic Tin

Elemental tin has an atomic number of 50, an atomic mass of 118.69, and exists in three allotropic forms: white tin at room temperature, nonmetallic gray tin at $<13.3°C$, and brittle tin at $>161°C$. White tin is a stable silver-white, lustrous, soft metal with a density of 7.27, a melting point of 231.9°C, and a boiling point of 2507°C. Tin has 10 stable isotopes (^{112}Sn, ^{114}Sn, ^{115}Sn, ^{116}Sn, ^{117}Sn, ^{118}Sn, ^{119}Sn, ^{120}Sn, ^{122}Sn, and ^{124}Sn), the most for any element. Inorganic tin compounds exist in the +2 (stannous) and +4 (stannic) oxidation states. Stannous compounds are generally more polar than stannic compounds, are unstable in dilute aqueous solutions, are easily oxidized, and normally contain some Sn^{+4}. Stannic oxide occurs naturally as the mineral cassiterite, has a melting point of 1127°C, and has wide application in industry.

Signs of inorganic tin poisoning in mammals include local effects such as vomiting, diarrhea, and eye and nose irritation; however, these vary considerably among species. The major systemic effects include ataxia, twitching of limbs, weakness of limbs, paralysis, growth retardation, decreased hemoglobin levels, and – at extremely high doses – testicular degeneration, pancreatic atrophy, formation of spongy brain white matter, and kidney necrosis. In humans, symptoms of inorganic tin intoxication include nausea, vomiting, diarrhea, stomach ache, fatigue, and headache. The lowest concentration producing outbreaks was about 250.0 mg Sn per liter in canned orange and apple juice. Ingestion of 50.0 mg of tin through eating canned peaches that contained Sn concentrations of about 450.0 mg/kg caused acute symptoms in 2 of 7 human volunteers. Inhalation of SnO_2 dust is a hazard in the deep-mining of tin; deposits in lungs are easily detectable as "stannosis."

Inorganic tin and its salts are not highly toxic due to their poor absorption, relative insolubility of their oxides, and rapid tissue turnover. The absorption of ingested inorganic tin is usually less than 5%, although up to 20% has been reported. Stannous compounds are more readily absorbed from the gastrointestinal tract than stannic compounds, but absorbed tin leaves the vascular system rapidly. Bone is the main site of tin deposition, followed by lung, liver, and kidney. Penetration of the blood–brain and placental barriers by inorganic tin seems to be very slight. Except for lung, inorganic tin does not accumulate in organs with increasing age. Absorbed inorganic tin is excreted mainly in the urine, although excretion through the bile may account for up to 15% of the total. Tin and its inorganic compounds do not produce significant dermatitis or allergic reactions to skin epithelium, and results of all long-term studies of carcinogenicity, teratogenicity, and mutagenicity have been negative to date.

The half-time (Tb1/2) persistence of inorganic tins is complex. Studies with Sn^{+2} in mouse, rat, monkey, and dog show that in all species, elimination is a four-compartment process regardless of the route of administration (i.e., intraperitoneal or intravenous). The Tb1/2 for the longest-lived tin component was >3 months. In studies with rats, for example, radiotin-113 in skeleton following intramuscular administration had a Tb1/2 of 3–4 months, but for oral administration of Sn^{+2} and Sn^{+4} it was only 28–40 days in bone and 10–20 days in liver and kidney.

Inorganic tin can be biomethylated by microorganisms in the aquatic environment and subsequently mobilized in the ecosystem. The process is slow and usually does not proceed beyond the monomethyltin stage, although dimethyltin formation by *Pseudomonas* bacteria may occur.

Tin is an essential nutrient for growth in the rat, and a tin-deficient diet leads to reduced growth. The mechanism of action is unclear, but involves increasing metabolic activity of liver lysosomes and liver hydrolytic enzymes during regeneration.

31.2.2 Organotins

Organotins are compounds with at least one tin–carbon bond. In most organotin compounds, tin is in the tetravalent oxidation state. Four series of organotin compounds are known: R_4Sn, R_3SnX, R_2SnX_2, and $RSnX_3$ wherein R is usually a butyl, octyl, or phenyl group, and X is commonly chloride, fluoride, oxide, hydroxide, carboxylate, or thiolate. The possible molecular composition and structure of the R groups are virtually unlimited. At least 260 organotin compounds are known, of which 36 are listed as toxic chemicals. Except for some methyltin compounds, all organotins are manufactured. Most commercially used organotins are characterized by low mobility in the environment because of low aqueous solubility, low vapor pressure, and high affinity for soils and organic sediments. Solubility data for organotin compounds are incomplete. In general, their solubility in water is limited to about 5.0–50.0 mg/L, but they are very soluble in many common organic solvents. The presence of chloride in seawater reduces the solubility of tributyltin and triphenyltin compounds, probably by association with the hydrated cation to form the covalent organotin chloride. Organotin compounds are analyzed in aqueous media by spectrophotometric, fluorometric, and electrochemical techniques. However, if picomole per liter concentrations are required, additional techniques must be used. More work needs to be done on analytical detection methods of organotins in sediments and biota.

Methylation of inorganic and methyltin compounds has been reported with the formation of mono-, di-, tri-, and tetramethyltin compounds. In addition, tributylmethyltin and dibutylmethyltin species have been found in harbor sediments, which suggests that some butyltin compounds may be methylated in aquatic systems. Methyltin formation in the environment is due mainly to methyl donation from methylcobalamin and methyl iodide. Photochemical reaction and transalkylation of inorganic tins produce methyltins; methylation of tin increases the toxicity of their original metal form due, in part, to their higher volatility and lipophilicity. Methyltins are ubiquitous in the environment and have been measured in seawater, freshwater, rain, wastewaters, sediments, fish, invertebrates, birds, and humans.

Abiotic and biological degradation of organotins generally occurs through sequential dealkylation or dearylation. Organotin compounds undergo successive cleavage of tin–carbon bonds to ultimately produce inorganic tin as follows: R_4Sn (via k_4) to R_3SnX (k_3) to R_2SnX_2 (k_2) to $RSnX_3$ (k_1) to SnX_4.

The reaction rate, k, usually proceeds as $k_4 > k_3 > k_2 = k_1$. The breaking of a Sn–C bond can occur by a number of different processes, including ultraviolet (UV) irradiation, biological cleavage, chemical cleavage, gamma irradiation, and thermal cleavage. In general, UV and biological cleavage are the most important processes. The main abiotic factors that seem to limit organotin persistence in the environment are elevated temperatures, increased intensity of sunlight, and aerobic conditions.

The tendency of an organotin compound to be concentrated by an organism depends on its partition behavior between lipid and aqueous phases. In general, compounds highly soluble in octanol and only slightly soluble in water have high K_{ow} values. K_{ow} values of organotins increase with number and molecular weight of organic groups attached to the tin atom, with significant bioaccumulation potential for organotins with R groups of butyl and larger. K_{ow} values for tributyltins in seawater vary from 5500 to 7000, but can be significantly modified by salinity and speciation products. Thus, organotins would be expected to accumulate in lipid-rich surface microlayers of natural waters and in biota (as discussed later). However, the ability of microorganisms, algae, and higher organisms to reduce various organotins into less toxic metabolites that can be rapidly excreted seems to preclude food chain biomagnification and to lessen the potential hazards to natural resources from consumption of organisms with elevated organotin residues.

Most authorities now agree on five points: (1) information concerning the mechanism of the toxic action of organotin compounds is inadequate; (2) results of all studies with various organotins for possible carcinogenicity are negative; (3) triorganotin compounds are the most toxic group of organotins; (4) large inter- and intraspecies differences exist in resistance to organotin compounds; and (5) organotins can alter enzyme activity levels in many organs and tissues including brain, liver, and kidney.

In aquatic systems, triorganotins were the most toxic group of organotins tested, followed in decreasing order of toxicity by diorganotins, tetraorganotins, and monoorganotins. Within each series, butyltin and phenyltin compounds seemed most toxic. Organotin toxicity increases with an increase in the length of the linear carbon chain; degradation processes, which involve the gradual breaking on Sn–C bonds, lead to the formation of compounds of low toxicity. Different biochemical mechanisms can account for the toxic effects of each structural class of organotins. For triorganotins, the derangement of energy-coupled processes, which occur at the membrane level seems to be the most probable mode of action in aquatic organisms.

The monoorganotin compounds, $RSnX_3$ have a generally low toxicity and do not seem to have any important biological action in mammals. Dialkylorganotins, R_2SnX_2, are associated with hepatotoxicity (ethyl, propyl, butyl, and pentyltins), immunotoxic effects to T-cells (butyl and octyltins), and skin and eye irritation (methyl, ethyl, propyl, butyl, and octyltins). The diorganotins combine with coenzymes or enzymes possessing dithiol groups and exert their toxic action by inhibiting alpha-keto acid oxidation and blocking mitochondrial respiration. Resistance to diorganotin toxicity varies widely among species. For example, dibutyltins and dioctyltins – unlike other organotins tested – were toxic to rat thymocytes, but did not induce similar effects on lymphoid atrophy in mice, guinea pigs, or Japanese quail. Selected dibutyltins are effective as antihelminthics and are used to kill parasitic worms in chickens and turkeys without harm to host birds.

In any member of the organotin series R_nSnX_{4-n}, progressive substitution of organic groups at tin produces a maximum biological activity for the triorganotin derivatives, R_3SnX. Among triorganotin compounds, trimethyltins are highly toxic to insects, birds, and mammals; triethyltins to mammals; tripropyltins to gram-negative bacteria; tributyltins to fish, mollusks, fungi, and gram-positive bacteria; triphenyltins to fish, fungi, and mollusks; and tricyclohexyltins to mites. In mammals, the lower triorganotin homologues (trimethyltins, triethyltins) are essentially neurotoxic, the intermediate trialkyltins and triphenyltins are primarily immunotoxic, and the higher homologues are only slightly toxic or not toxic. The toxicity of triorganotin compounds is probably due to their ability to bind to proteins and to inhibit mitochondrial oxidative phosphorylation. Triorganotins also interfere with phagocytosis and exocytosis and other pathways where sulfhydryl groups play a pivotal role, and inhibit uptake of gamma-aminobutyric acid and Na^+-K^+-ATPase in brain. Impairment of phagocytosis and related activities of polymorphonuclear leukocytes may enhance susceptibility for infection.

Trimethyltins are the most toxic trialkyltins to mammals, regardless of the nature of the substituent (X group). They induce pathological lesions in brain and overt neurological and behavioral changes in rodents. Trimethyltins are neurotoxins that damage the limbic system, cerebral cortex, and brain stem and can traverse the placenta and accumulate in the fetus. Trimethyltins (but not inorganic, monomethyl, or dimethyltins) inhibit brain protein synthesis by 47% and can cause a decrease of 4.2°C in body temperature of mice within 1 h postadministration of 3.0 mg/kg body weight (BW). Raising the ambient temperature to 35°C prevented hypothermia in treated mice and resulted in only a 20% inhibition in protein synthesis. More research is needed on the role of protein synthesis in organotin-induced neurotoxicity.

Triethyltins modify phosphorylation processes in subcellular fractions of rat brain proteins. Signs of triethyltin poisoning in rodents include weakness of hindlimbs, dyspnea, and peripheral vasodilation. Internally, acute triethyltin intoxication is characterized by a transient edema of the central and peripheral nervous systems manifested by extensive intramyelinic vacuolation due to splitting of myelin lamellae; changes are reversible. Neuronal death is reported following triethyltin intoxication during the neonatal period, possibly as a result of elevated intracranial pressure. In rabbit brain, triethyltins alter activity of pyruvate dehydrogenase.

Studies on tributyltin uptake and depuration from food or water by rats, crabs, oysters, and fish showed that in all species it was accumulated and metabolized, at least partly, within 48 h to dibutyltins, monobutyltins, and more polar metabolites; however, oysters (*Crassostrea virginica*) metabolized significantly less tributlytin than did other species tested. The accumulation of tributyltin compounds in different tissues correlated well with lipid content and supports a partitioning mode of uptake. The mixed function oxygenase system from hepatic tissues was able to metabolize tributyltins by forming hydroxylated metabolites. Tributyltins are also potent cytotoxicants in rabbit erythrocyte and skin cultures.

The potential of tricyclohexyltins to modify the inducibility of cytochrome P450 by various substances, such as 3-methylcholanthrene, is of considerable toxicological importance. Significant metabolic interactions can result from a combination of environmental chemicals and drugs that produce alterations in heme and mixed function oxygenase activity, suggesting that more research is needed on interaction effects of organotins with other environmental substances or contaminants.

The biological effects of the tetraorganotin compounds, R_4Sn, seem to be caused entirely by the R_3SnX derivative that is produced by their rapid in vivo dealkylation. Increasing toxicity of tetra- and triorganotins in mammals has been shown to be associated with decreasing length of their ligands, as reflected by solubility in biological fluids. It is not known if damage is produced by the metal or by its alkyl derivative, but the presence of trialkyl groups seems to enhance the toxicity of tin – probably by increasing its partition into lipids, thus aiding the absorption of the metal and speeding its distribution to the site of action.

31.3 Sources and Uses

Metallic tin is derived mainly from the mineral cassiterite (SnO_2) and to a lesser extent from the sulfide ore stannite Cu_2S–FeS–SnS_2, although it can be derived from rarer minerals such as malayaite, $CaSnSiO_5$. Tin is one of the earliest metals known and has influenced our lifestyle through the ages. Tin alloy artifacts dating from about 5000 years ago have been unearthed at Ur, the site of ancient Babylonia. Today, we are exposed to tin on a daily basis through the use of tinplated food cans; of alloys such as pewter, bronze, brass, and solder; and from toothpaste containing stannous fluoride. Inorganic tin compounds are also used in a variety of industrial processes such as the strengthening of glass, as a base for colors, as catalysts in various chemical reactions, as stabilizers in perfumes and soaps, and as dental anticariogenic agents. Organotin use is extensive in antifouling marine paints, in molluscicides, and in agriculture, which sometimes causes serious adverse effects on nontarget biota. In Italy, however, use of organotin products in agriculture at recommended application levels had no adverse effects – and did not accumulate in tissues – in European hares, *Lepus europaeus*, sampled over a 12-month period. Some treated plants of the family Chenopodiaceae had 3.4 mg Sn/kg fresh weight (FW) in August (range 0.9–13.4) when compared to lower mean values in other families of plants sampled in August (0.3–1.2 mg Sn/kg FW).

In 1975, the total world tin production was 236,000 tons, of which 72% was produced by China (10%), Indonesia (8%), Malaysia (35%), Thailand (7%), and 6% each by the UK and the Former Soviet Union. Annual mine production of tin in the United States is a comparatively low 3300 metric tons. The world production of recycled tin was about 20,000 tons, of which France produced about half. About 25% of the tin used in the United States is recovered from scrap materials containing tin. This secondary production occurs in the United States at 7 detinning plants and 162 processing plants. The production and consumption of tin chemicals, especially organotins, has increased markedly in the past several decades.

The United States is the major consumer of tin and organotin compounds, followed by Japan, the UK, Germany, and France. In 1976, for example, the United States consumed 11,000 tons of organotins, or about 39% of the world organotin production. The total demand for primary tin in the year 2000 is about 7.5 million tons. Total reserves are about 6.5 million tons; however, new discoveries and increases in known reserves resulted in sufficient new tin to meet the demand.

The uses of inorganic and organotin compounds are numerous and increasing. Industrial consumption of organotins, for example, rose from about 5000 tons in 1965 to about 35,000 tons in 1985. The uses of nontoxic organotin compounds (R_2SnX_2 and $RSnX_3$ types) account for about 67% of the total world production, although use of R_3SnX types as selective biocides has increased disproportionately. Tin now has more of its organometallic derivatives in commercial use than any other element.

Biocidal applications of organotins to control marine fouling communities, agricultural pests, and as selective molluscicides merit additional comment. The use of antifoulants on ships is necessitated by the damage some organisms can cause to wooden structures and by the reduced fuel efficiency and speed due to drag when vessels become heavily fouled. Until recently, the most widely used antifouling paint contained a copper base that is biocidally active when copper leaches as an ion from the paint. However, short effective lifetimes and high costs have limited the usefulness of copper-based paints. Organocompounds of arsenic, mercury, or lead have also been used in antifouling paints, but these paints have been removed from the commercial market due to the toxicological risks during preparation and application and to their hazards to the environment. Organotin coatings are promoted because of their excellent antifouling action, long lifetime (up to 4 years), and lack of corrosion. Organotin coatings, especially tributyltins, present potential environmental problems to nontarget aquatic biota due to their extreme toxicity.

Use of organotin antifouling paints on recreational and commercial water craft has increased markedly in recent years. In Maryland, for example, 50–75% of the recreational boats used in Chesapeake Bay are covered with organotin paints. The organotin biocide released by hydrolysis from the surface of the paint film into seawater provides the antifoulant action. In consequence, the depleted outer layer of paint film, containing hydrophilic carboxylate groups, is easily eroded by moving seawater exposing a fresh surface layer of organotin acrylate polymer. In continuing tests by the U.S. Navy, ablative organotin fouling coatings have demonstrated more than 48 months of protection. As discussed later, the use of organotin compounds in antifouling paints has been severely curtailed. Antifouling paints containing triphenyltin (TPT) will not be a suitable substitute for tributyltin (TBT) in paints designed to inhibit microbial biofilms. Up to 80% of bacteria resistant to six organotins, including TBT, isolated from estuarine sediments of Boston Harbor in Massachusetts were also resistant to TPT. All bacteria were resistant to at least six of eight metals tested, suggesting that resistance to metals – including nickel, cadmium, lead, copper, zinc, and mercury – may be associated with resistance to organotins.

Several organotins have been used extensively as agricultural pesticides, especially tricyclohexyltin and triphenyltin compounds. In general, these compounds showed low phytotoxicity, low toxicity to nontarget organisms, no evidence of development of resistant insect strains, and degradation to form harmless tin residues. It is probable that agricultural uses of organotins will increase. The toxicity of triorganotin compounds to aquatic invertebrates, especially slow release formulations of tributyltins, is usually high, and this property has been used advantageously to eradicate certain species of freshwater snails that are intermediate vectors of schistosomiasis, i.e., *Biomphalaria* spp., *Bulinus* spp. Unfortunately, nontarget biota, including some sensitive species of fishes, are killed at recommended application levels.

Organotins enter air, soil, and water primarily as a result of routine agricultural, industrial, municipal, and biocidal operations. Deposition rates of organotins from air into soils and water are unknown at present, but may be significant around urban and industrialized areas. Total tin concentrations, primarily inorganic tin, in the atmosphere of the northern hemisphere are significantly higher than those in the southern hemisphere and are dominated by anthropogenic sources (Table 31.1). The most important of these sources seems to be the incineration of municipal wastes, which accounts for most of the tin flux to the atmosphere. Riverine fluxes of tin to the oceans vary between 36 and 71 million kg annually, almost all of it in particulate fractions (Table 31.1).

Table 31.1. Total tin flux to the atmosphere and hydrosphere.

Environmental Compartment and Other Variables	Annual Flux, in Millions of Kilograms
ATMOSPHERE	
Northern hemisphere	
Anthropogenic	16.6
Natural	1.2
Total	17.8
Southern hemisphere	
Anthropogenic	1.6
Natural	0.7
Total	2.3
HYDROSPHERE	
Riverine flux to oceans	
Dissolved fraction	0.09
Particulate fraction	35.6–71.2

31.4 Concentrations in Field Collections

In aquatic environments, organotin concentrations were elevated in sediments, biota, and surface water microlayers collected near marinas, aquaculture rearing pens, and other facilities where organotin-based antifouling paints were used. In some cases, organotin concentrations in the water column were sufficiently high to pose a substantial risk to sensitive species. Data are limited on concentrations of organotins in environmental samples, especially in samples from terrestrial ecosystems, and this may be attributed, in part, to limitations in routine chemical analytical capabilities.

31.4.1 Abiotic Materials

Tin concentrations in water, air, soils, sediments, and other nonbiological materials are documented but information is scarce except for aquatic systems. In aquatic systems, several trends were evident. First, tin and organotin compounds tend to concentrate in surface microlayers by factors up to 10,000 relative to subsurface water; in the case of organotins, this may be due to partitioning into the film of petroleum hydrocarbons commonly present on water surfaces. Second, organotin concentrations, especially tributyltins, were highest in the vicinity of marinas and harbors, and this is consistent with its use as an antifouling agent in some paints for boats, ships, and docks. Peak tributyltin concentrations occurred in late spring and early summer in association with postwinter launching of freshly painted boats. Third, organotin levels throughout the water column of marinas in numerous freshwater and marine locations were sufficiently elevated to cause chronic toxic effects in sensitive organisms including algae, copepods, oysters, mussel larvae, and fish. Fourth, methyltin species were infrequently detected. Their occurrence was positively correlated with the presence of relatively high concentrations of inorganic tin and was due primarily to biotic and abiotic methylation of both organotin and inorganic tin compounds. Finally, butyltin species were detected in harbor sediments at concentrations that were toxicologically hazardous to benthic fauna. Tributyltin species can be accumulated from the sediments by oligochaetes (*Tubifex tubifex, Limnodrilus hoffmeisterii*), thus making it potentially available to bottom-feeding fish; oligochaetes can also degrade tributyltins by a sequential debutylation, with Tb1/2 estimates of 5 months in water and 4 months in water–sediment mixtures.

Tributyltin antifouling paints were banned for use in tin-based paints in the UK in 1987. Field work on the River Crouch Estuary showed that sediments from areas most contaminated with TBT in 1987 contained 0.16 mg TBT/kg dry weight (DW); however, by 1992 this had declined to 0.02 mg TBT/kg DW. TBT declines were accompanied by increases in abundance and diversity of benthic fauna, especially bivalve mollusks and amphipod crustaceans.

31.4.2 Biological Samples

Information on background concentrations of total tin in tissues of field populations of animals and plants was comparatively abundant when compared to organotin species. Tin concentrations in marine algae and macrophytes varied between 0.5 and 101.0 mg total Sn/kg DW and clearly demonstrated that most species of aquatic flora bioconcentrate tin from seawater. Marine plants are also important in the cycling of tin. Living algae are effective in immobilizing tin from seawater and regulating the formation and degradation of toxic methyltin compounds. Dead and decaying algae accumulate inorganic and organotin compounds, release them, and ultimately remove tin from the estuary to the atmosphere by formation of tetramethyltins.

Organotin content in fish tissues is quite variable, ranging from a low of 3–6% of the total tin body burden to 18% for goatfish (*Upeneus moluccensis*) to 5% for *Mullus barbatus*, another species of goatfish. By contrast, the limpet (*Patella caerulea*) contains 35–75% of its total tin body burden as organotin. In January 1982, France banned organotin compounds for use in antifouling paints. By 1985, tin and organotin concentrations in seawater and Pacific oysters (*Crassostrea gigas*) were 5–10 times lower than those found in 1982. In Arcachon Bay, France, a decrease in the incidence and extent of anomalies in oyster calcification mechanisms was noted that seemed to be correlated with decreases in tin contamination. Crassostreid oysters can accumulate radiotin (Sn-113) to a higher degree than other species of bivalve mollusks, a characteristic that may be useful as a bioindicator in the event of contamination due to this isotope. Tributyltin and its degradation products continue to be detected in tissues of American oysters (*Crassostrea virginica*) from the Gulf of Mexico ten years after the use of TBT-based paints was regulated; the likely sources for the butyltin compounds include sediments, TBT-based paints on vessels greater than 25 m in length, and shipyard wastes. Fish, shellfish, and sediment samples from southwestern British Columbia in 1992–93 contained tributyltin and its metabolites dibutyltin and monobutyltin, strongly suggesting that tributyltin is a widespread contaminant in this geographic area and a continuing cause for concern despite restrictions on the use of organotin-based marine antifouling paints imposed in 1989.

Antifouling paints containing tributyltin compounds are used widely on netting panels of sea cages at fish and shellfish aquaculture units to minimize the obstruction of water exchange through the cages. Under these conditions, tributyltin paints were detrimental to the growth and survival of juvenile scallops and to calcium metabolism and growth of adult oysters and resulted in elevated concentrations of tributyltin in salmon tissues. Scallops (*Pecten maximus*) reared in sea pens for 31 weeks on nets coated with tributyltin oxide contained 2.5 mg total Sn/kg FW soft parts (1.9 mg tributyltin/kg), but lost up to 40% during a 10-week depuration period. Scallop adductor muscle contained 0.53 mg tributyltin/kg, suggesting that this tissue – the one consumed by humans – is a probable tin storage site. Pacific oysters reared for 31 weeks on tributyltin-exposed nets contained a maximum of 1.4 mg tributyltin/kg FW at week 16 (controls 0.12 mg/kg), but lost 90% during a 10-week depuration period. Atlantic salmon (*Salmo salar*) held for 3 months during summer in cages with tributyltin-treated net panels contained 0.75–1.5 mg tributyltin/kg FW muscle vs. 0.28 mg/kg at start. Based on laboratory studies, it is probable that Atlantic salmon were exposed to approximately 1.0 μg tributyltin/L during this interval. Chinook salmon (*Oncorhynchus tshawytscha*), reared in sea pens treated with tributyltin paints, contained <0.013 mg tributyltin/kg muscle FW when introduced into the pens. Concentrations were 0.3 mg/kg after 3 months, 0.8 mg/kg at 13 months, and 0.9 mg/kg at 19 months. Cooking did not destroy or remove organotins from salmon muscle tissues. Pink salmon (*Oncorhynchus gorbuscha*) and chum salmon (*Oncorhynchus keta*) fry cultured in TBT-treated marine net pens for 20–68 days prior to ocean release contained mean concentrations of 1.5 mg Sn/kg FW fry (chum) and 2.7 mg/kg FW (pinks) vs. <0.1 in controls; however, growth and survival were normal and returning adults 1–3 years later had no detectable TBT.

Diet and proximity to tributyltin affect butyltin concentrations in waterfowl. Seaducks that fed mainly on mollusks had higher concentrations of butyltins than did predatory birds feeding on fish, other birds, and small mammals. Continued exposure of birds to butyltin compounds occurs in harbors and marinas where tributyltin is used on vessels >25 m in length.

31.5 Effects

Inorganic tin compounds are of low toxicologic risk due largely to their low solubility, poor absorption, low accumulations in tissues, and rapid excretion. By contrast, some organotin compounds – especially trialkyltins – produce a variety of harmful effects resulting in impaired behavior and lowered growth, survival, and reproduction. Among aquatic organisms, tributyltin compounds were especially potent. Adverse effects were noted in mollusks at water concentrations of 0.001–0.06 μg/L and in algae, fish, and other species of invertebrates at 0.1–1.0 μg/L. Bioconcentration of organotins was high, but degradation was sufficiently rapid to preclude food chain biomagnification. Birds seem to be relatively resistant to organotins, and data suggest that diets containing 50.0 mg of tin as trimethyltin chloride/kg are fatal to ducklings in 75 days; however, no deaths occurred in 75 days at 50.0 mg/kg of eleven other mono-, di-, tri-, and tetraalkyltin compounds. Trimethyltin was lethal to other species of birds tested at doses of 1.0–3.0 mg/kg BW. Trimethyltins and triethyltins were the most toxic organotin compounds tested on small laboratory mammals. Neurotoxicological effects of trimethyltins were usually not reversible, while those caused by triethyltins were reversible after exposure. Adverse effects of trimethyltins were produced at concentrations as low as 0.15 mg/L in drinking water (learning deficits), 0.625 mg/kg BW (diet aversion), and 1.25 mg/kg BW (death).

31.5.1 Aquatic Organisms

Studies on lethal and sublethal effects of tin compounds to representative species of aquatic organisms demonstrate that organotin compounds are more toxic than inorganic tin compounds; triorganotin compounds are more toxic than mono-, di-, or tetraorgano forms; and tributyltin compounds are the most toxic triorganotin compounds tested. Adverse effects of tributyltins were noted at water concentrations of 0.001–0.06 μg/L in marine gastropod and bivalve mollusks, and at 0.1–1.0 μg/L in algae, echinoderms, fish, crustaceans, and coelenterates. In order of toxicity, tributyltins were followed by tripropyltins (harmful effects recorded at 0.001–10.0 μg/L to gastropods, fish, and algae), triphenyltins (0.6–1.0 μg/L to diatoms and annelids), triethyltins (3.8–10.0 μg/L to fish and algae), trimethyltins (20.0 μg/L to algae and crustaceans), and tripentyltins (50.0–100.0 μg/L to gastropods). Because many organotin compounds are slow-acting poisons, short-term toxicity tests seriously underestimate the toxicity of these compounds. Tributyltin chloride was more toxic to larvae of the horseshoe crab (*Limulus polyphemus*) than were salts of other metals tested – mercury, cadmium, chromium, zinc, copper, and lead – as judged by effects on survival, molting, and limb regeneration.

Results of acute toxicity tests with several organotin compounds and *Daphnia magna* indicated several distinct trends: toxicity increased with the length of alkyl group from methyl to butyl; the anion substituents are relatively unimportant; and bioavailability is correlated with increasing solubility in lipids, which is a direct function of K_{ow}, the *n*-octanol–water partition coefficient. Structure–activity relations seem to have high predictive capacity in hazard assessment, and those for organotins seem particularly promising. For example, studies on the biocidal properties of structurally distinct diorganotins (R_2SnX_2) and triorganotins (R_3SnX) to zoeae of a marine crab show, within a homologous series, that diorganotins are less toxic than the corresponding triorganotins (Table 31.2). It was concluded that the toxicity of organotins to crab zoeae seems to be a function of the hydrophobic characteristics conferred by the number and structure of the organic ligands. Studies with yolk-sac fry of the rainbow trout also demonstrate that triorganotins are

Table 31.2. Toxicity of selected diorganotin and triorganotin compounds to zoeae of the marine mud crab (*Rithropanopeus harrisii*) exposed from hatching to age 14 days. Values shown are in milligrams total product and tin only per liter for lowest concentration tested producing at least 50% mortality.

Compound Tested	Total Product	Tin Only
DIORGANOTINS		
Dimethyltin dichloride, $(CH_3)_2SnCl_2$	20.0	10.8
Diethyltin dichloride, $(C_2H_5)_2SnCl_2$	5.0	2.4
Dipropyltin dichloride, $(C_3H_7)_2SnCl_2$	5.0	2.2
Dibutyltin dichloride, $(C_4H_9)_2SnCl_2$	2.0	0.78
Diphenyltin dichloride, $(C_5H_{11})_2SnCl_2$	0.75	0.27
Dicyclohexyltin dichloride, $(C_6H_{13})_2SnCl_2$	0.25	0.082
TRIORGANOTINS		
Trimethyltin hydroxide, $(CH_3)_3SnOH$	0.1	0.067
Triethyltin hydroxide, $(C_2H_5)_3SnOH$	0.1	0.053
Tripropyltin oxide, $(C_3H_7)_3Sn_2O$	0.05	0.03
Tributyltin oxide, $(C_4H_9)_3Sn_2O$	0.02	0.011
Triphenyltin hydroxide, $(C_5H_{11})_3SnOH$	0.02	0.007
Tricyclohexyltin bromide, $(C_6H_{13})_3SnBr$	0.009	0.0023

at least two orders of magnitude more toxic than diorganotins. The no-observable-effect-concentration (NOEC) for rainbow trout fry during exposure for 110 days was 40.0 μg/L for dibutyltin chloride, 60.0 μg/L for diphenyltin chloride, and 0.04–0.05 μg/L for various tributyltin compounds. Alterations of steroid mechanisms, such as ability to metabolize testosterone, by TBT may be a more sensitive indicator of sublethal exposure in *Daphnia magna* than reproductive endpoints.

Signs of tributyltin poisoning in rainbow trout and other freshwater fishes include sluggishness, loss of appetite, altered body pigmentation, air gulping, loss of positive rheotaxis, increased rate of opercular movements, increases in blood hemoglobin and hematocrit, damaged gills and cornea, increases in erythrocyte number, damage to epithelial cells of the bile duct, decreased resistance to bacterial pathogens, and peripheral blood neutrophilia and impaired antibody secreting mechanisms. These changes were consistent with the known inhibitory effects on mitochondrial and oxidative phosphorylation of triorganotin compounds. Exposure to tributyltin may alter both cytochrome P450-dependent metabolism, and induction response to other environmental pollutants. Studies with scup (*Stenotomus chrysops*), a marine fish, show that a single intraperitoneal injection of 3.3, 8.1, or 16.3 mg tributyltin/kg BW results in a dose-dependent decrease in hepatic microsome ethoxyresorufin *O*-deethylase activity, cytochrome P450 degradation to P420, and increase in liver concentrations from 8.0 to 202.0 mg Sn/kg FW. Organotins also act on the microsomal electron transport system, the hemoprotein P450, and the flavoproteins of freshwater fishes. Cytopathological changes in trout brain induced by tributyltin or triphenyltin resemble the reactions observed in brains of mammals after triethyltin insult and similar to symptoms in the brains of patients affected by a disease known as central pontine myelinosis (tin detected in intramyelin vacuoles) or by multiple sclerosis. In rainbow trout erythrocytes, tributyltin inhibits the uptake of Na^+ and Cl^- ions and interacts with the activity of the Na^+/H^+ exchanger. Tributyltin compounds modify calcium flux across the plasma membrane in a dose-dependent manner of oyster toadfish (*Opsanus tau*), a marine species. At 50.0 μg/L, tributyltin facilitated an inward flux of calcium. At 500.0 μg/L, calcium mobilization was inhibited resulting in impaired macrophage function. In trout hepatocytes, tributyltin mobilizes Ca^{+2} from intracellular stores. The cytoplasmic acidification following tributyltin exposure seems to be caused by the combination of intracellular Ca^{+2} and by direct action of tributyltin. Suppression of regeneration in echinoderms, and presumably other aquatic groups of organotins, may be due primarily to neurotoxicological action of organotins, or secondarily by direct action on tissue at the breakage point. In freshwater minnows (*Phoxinus phoxinus*), the toxicity of tributyltin compounds to early life stages is based on its eye- and skin-irritative activity, and on its activity against muscular, renal, and neuronal tissues.

Imposex – the superimposition of male characteristics onto a functionally normal female reproductive anatomy – is a phenomenon documented in populations of marine gastropod mollusks in the vicinity of yacht basins and marinas and is a sensitive indictor of tributyltin contamination. A female with imposex displays one or more male characteristics such as a penis, a vas deferens, or convolution of the normally straight gonadal oviduct. It is measured by frequency of occurrence in the adult females and by the intensity of expression of all male characteristics in bearer females. Imposex is prevalent in mud snails (*Nassarius obsoletus*) near estuarine marinas and has been induced experimentally in that species by exposure for 60 days to three tributyltin compounds at concentrations of 4.5–5.5 μg/L. Imposex has been documented extensively in declining populations of the common dogwhelk (*Nucella lapillus*), especially in southwestern England. There is general agreement on six points: (1) dogwhelk populations near centers of boating and shipping activity show the highest degrees of imposex, coinciding with the introduction and increasing use of antifouling paints containing tributyltin compounds; (2) imposex is not correlated with tissue burdens of arsenic, cadmium, copper, lead, silver, or zinc, but is correlated with increasing concentrations of tributyltin and dibutyltin fractions; (3) transplantation of dogwhelks from a locality with little boating activity to a site near a heavily used marina causes a marked increase in the degree of imposex and in tissue accumulations of tributyltins; (4) imposex can be induced in female dogwhelks by exposure to 0.02 μg Sn/L leached from a tributyltin antifouling paint; after exposure for 120 days, 41% of the females had male characters and whole body residues of 1.65 mg Sn/kg DW soft parts (vs. 0.1 in controls), of which almost all was tributyltin (1.64 vs. 0.08 mg/kg in controls). Concentrations as low as 0.0015 μg tributyltin/L can initiate imposex in immature females; (5) declining dogwhelk populations studied were characterized by a moderate to high degree of imposex, relatively fewer functional females, few juveniles, and a general scarcity of laid egg capsules. Many females in late imposex contained aborted capsules as a result of oviduct blockage, resulting in sterility and premature death; and (6) there was no evidence that loss of tin leads to any remission of imposex; in fact, all evidence indicates that gross morphological changes that occur in late imposex are irreversible. Signs of imposex were severe in some populations of whelks (*Thais* spp., *Vasum turbinellus*) sampled in eastern Indonesia in 1993 and is attributed to unrestricted use of paints containing tributyltin. Studies indicate that imposex in gastropod mollusks is caused by elevated testosterone titers that masculinize TBT-exposed females as a result of competitive inhibition of cytochrome P450-mediated aromatase. TBT-induced masculinization in gastropods, imposex and intersex, is a convincing example of endocrine disruption described in invertebrates that is unequivocally linked to an environmental contaminant. It is clear that additional research is needed on the imposex phenomenon in mollusks and on its implications to vertebrates and other taxonomic groups.

Biological factors known to modify lethal and sublethal effects of organotins include age of the organism, inherent interspecies resistance, and tissue specificity. Abiotic modifiers include exposure route, and physicochemical regimen. Early developmental stages were more sensitive to organotins than later developmental stages in marine annelids, mysid shrimp, and rainbow trout. Mortality of zoeae of fiddler crabs (*Uca pugilator*) to trimethyltins was greatest at elevated temperatures and low salinities. Mussels exposed through a diet of algae showed slow accumulation of organotins when compared to exposure from the medium; the reverse was observed for crabs. A marine diatom (*Thalassiosira pseudonana*) showed no adaptation or resistance to triphenyltins or tributyltins, but another diatom (*Amphora coffeaeformis*) was extremely resistant. Benthic fauna are probably capable of transferring organotins from sediments to bottom-feeding teleosts. For example, sediments spiked with 0.98 mg Sn/kg DW, as tributyltin, resulted in concentrations of 4.41 mg/kg whole body DW in oligochaete annelid worms after 22 weeks, up from 0.38 at the start. Mortality was substantially higher when organisms were exposed simultaneously to organotins through water and sediments; in the case of grass shrimp (*Palaemonetes pugio*), the addition of contaminated sediments increased mortality by up to 1000 times. However, the addition of uncontaminated sediments to assay containers reduced the bioavailability of tributyltin to freshwater fishes by as much as 84%.

Aside from direct toxic effects that antifouling paint residues may have on marine life, there is no evidence of any risk from cytogenetic damage. Tributyltins, for example, were not genotoxic to larvae of the mussel (*Mytilus edulis*) based on results of sister chromatid exchange and analysis of chromosomal aberrations. Teratogenic effects, however, were detected in larvae of the lugworm (*Arenicola cristata*) at sublethal concentrations of tributyltins, and algae (*Nitzschia liebethrutti*) exposed to 15.0 mg inorganic Sn/L for 14 days had frustule abnormalities.

Bioconcentration of inorganic and organic tin compounds from the medium is considerable. Bioconcentration factors (BCFs) for inorganic tin and marine algae were about 1900; moreover, tin-resistant bacteria contained a remarkable 3.7–7.7 g Sn/kg DW. Partitioning or binding may control bioaccumulation of tributyltin by marine phytoplankton. A linear relation is documented for external concentrations of tributyltin compounds and cell burdens in the marine microalgae *Nannochloris* sp., *Chaetoceros gracilis*, and the cyanobacterium *Synechococcus* sp.; however, the relation was not linear for the alga *Isochrysis galbana*. BCFs for organotin compounds varied from about 400–30,000 among various species of mollusks, algae, and crustaceans and were highest when ambient tin concentrations were <1.0 μg/L, when exposure times were comparatively lengthy, and when organism lipid content was elevated. Studies with freshwater minnows (*Phoxinus phoxinus*) demonstrate a high potential for tributyltin bioconcentration in early life stages, and slow metabolism in embryo and yolk-sac larvae. Sheepshead minnows (*Cyprinodon variegatus*) were unable to reach equilibrium with a medium containing 1.61 μg tributyltin/L after 58 days of exposure and maximum BCF values recorded were 2600 in whole fish, 1810 in muscle, 4580 in viscera, and 2120 in the remainder of the carcass; however, whole body loss was 52% after depuration for 7 days and 74% after 28 days. Sheepshead minnows were able to metabolize tributyltins into lower alkyl moieties, which were less toxic. Thus, even though significant bioconcentration occurred, the chronic toxicity of tributyltins to sheepshead minnow was not significantly greater than its acute toxicity.

Uptake of organotins through the medium is a more effective route than the diet, and triphenyltin compounds are more readily accumulated through dietary uptake than are tributyltin compounds. Tributyltin and triphenyltin compounds show BCF values of 1384–1974 in whole goldfish (*Carassius auratus*) after exposure for 21–28 days in media containing 0.13–0.14 μg Sn/L. But goldfish fed diets containing these same compounds at 1.7–1.9 μg Sn/kg ration for 28–35 days had BCF values of 0.04–0.1. A nearly identical pattern was reported for willow shiner (*Gnathopodon caerulescens*), another freshwater teleost. Red sea bream (*Pagrus major*) fed tributyltin or triphenyltin compounds in the diet (8.0–1000.0 μg Sn/kg ration) for 8 weeks and simultaneously challenged with seawater containing 0.067–0.083 μg/L of triorganotin compounds accumulated 25% of their whole body tin burden from the diet, regardless of tin concentration or chemical species in the ration. Tributyltin assimilation in red sea bream was 10% and retention 24%; for triphenyltin, assimilation was 13% and retention 60%. American plaice (*Hippoglossoides platessoides*) given a single oral dose of ^{113}Sn-tributyltin showed ^{113}Sn distribution over the entire body – especially liver and gallbladder – with steady state achieved in 5–10 days; average retention efficiency of TBT over a 6-week period was 44%, with half-time persistence of 15–77 days.

31.5.2 Birds

Information is scarce on the effects of tin and organotin compounds on birds. Limited data suggest that triorganotin compounds, especially trimethyltins – and to a lesser extent triethyltins – are the most toxic. In a 75-day feeding study with mallard ducklings (*Anas platyrhynchos*) and 12 organotin compounds it was concluded that (1) trimethyltin was the most toxic compound tested, (2) a dietary level of 50.0 mg of tin as trimethyltin chloride/kg food was fatal to all ducklings, (3) 5.0 mg trimethyltin chloride/kg ration killed 40% but all survived at 0.5 mg/kg diet, (4) death was preceded by mild to severe tremors, progressing to ataxia and lethargy, (5) trimethyltin-stressed ducklings exhibited degeneration of the large neurons of the pons, medulla oblongata, gray matter of the spinal cord, and pyramidal cells of the cerebral cortex, (6) all ducklings survived exposure to 50.0 mg/kg ration of tetraethyltin, tetrabutyltin, tetraphenyltin, triethyltin chloride, tripropyltin chloride, tributyltin chloride, tributyltin oxide, triphenyltin chloride, tricyclohexyltin chloride, dimethyltin chloride, and dibutyltin chloride. Sublethal effects were recorded at 50.0 mg triethyltin chloride/kg ration (low BW, vacuolization of spinal cord and brain white matter), at 50.0 mg tributyltin chloride/kg (enlarged liver), and at 50.0 mg tetrabutyltin/kg (elevated kidney weight).

Dietary studies with Japanese quail (*Coturnix japonica*) showed that tributyltin oxide affected reproduction at a dose where no overt toxicity was observed. Dietary levels as low as 60.0 mg tributyltin oxide/kg ration for 6 weeks was associated with decreased hatchability and decreased survival of chicks although adults fed and behaved normally.

31.5.3 Mammals

Inorganic tin compounds and some heterocyclic organic tin compounds are of low toxicologic risk to mammals, due largely to their low solubility, poor absorption, low tissue accumulations, and rapid tissue excretion. Inorganic tin compounds accumulate mostly in liver and kidney, rarely in brain, in proportion to dose and regardless of the exposure route. Noncyclic organotin compounds, by contrast, have produced adverse effects on the skin, eyes, gastrointestinal tract, liver, bile duct, kidney, hematopoietic system, central nervous system, reproduction, growth, and chromosomes of small laboratory animals. Effects of diorganotin compounds can be distinguished from those of tri- and tetraorganotin compounds. The chief toxicological difference is that some trialkyltins have a specific effect on the central nervous system resulting in cerebral edema, whereas diorganotins do not produce this effect but are potent irritants that induce inflammatory reactions. The tetraorganotins resemble triorganotins, which are usually more toxic than either mono- and diorganotins.

Diorganotin compounds cause cerebral edema and inhibit mitochondrial respiration by preventing the oxidation of keto acids, presumably through inhibition of alpha-keto oxidase activity. Large interspecies variability exists in the capacity of diorganotins to induce lymphoid atrophy. For example, dioctyltins and dibutyltins were selectively cytotoxic to rat thymocytes after dietary exposures of 50.0–150.0 mg/kg diet for 2 weeks; in contrast, no lymphoid atrophy occurred in mice, guinea pig, or Japanese quail given similar dosages and exposures. Route of exposure can also modify effects of diorganotins. Oral exposure to dibutyltin compounds, for example, produces inflammatory changes in bile duct of rat and necrotic changes in liver of mice and rats; dermal exposure causes bile duct injury in rats and rabbits; and intravenous administration produces pulmonary edema in rats. Intratesticular administration of high doses of some dibutyltins produced marked degeneration in rat testes within 7 days, including atrophy of seminiferous tubules and complete arrest of spermatogenesis; however, similar results have been reported for cadmium, zinc, and copper salts.

Trimethyltin, triethyltin, and tributyltin compounds are highly toxic to animals and humans. Trimethyltin and triethyltin compounds are more toxic to mammals than are the higher triorganotin homologues, probably because of poorer absorption of higher trialkyltin compounds from the gastrointestinal tract. Trimethyltin and triethyltin compounds are potent inhibitors of oxidative phosphorylation in the mitochondria for which these compounds have a high binding affinity. Different triorganotin compounds cause different neuronal patterns of toxicity in adult animals. Trimethyltins, for example, produce largely irreversible behavioral impairments, such as hyperactivity and impaired learning and performance, and these are consistent with reported neuronal cell death in limbic system structures. Triethyltins, with their direct effect on muscle – consistent with reports of myelin vacuolation and cerebral edema – produce largely reversible effects. Differences in chronic toxicity between triethyltins and trimethyltins have resulted in different strategies in assessment of hazard. Evaluations of triethyltin have focused on repeated testing throughout dosing, followed by a recovery period. But evaluations of trimethyltin-induced behavioral impairments have generally focused on testing weeks to months after exposure.

Symptoms of trimethyltin intoxication in humans include irritability, headache, depression, aggressiveness, disorientation, appetite loss, memory deficits, and decreased libido; changes were largely reversible following cessation of exposure. At high doses, trimethyltins cause death in primates and humans, preceded by seizures, anorexia, and emotional lability. Trimethyltin produced adverse effects in laboratory animals over an unusually narrow dose range, with differences of 10-fold or less between doses producing no observable effects and those producing 100% mortality in all species tested. Trimethyltin effects in small laboratory animals are usually not reversible. Signs of trimethyltin poisoning include tremors, hyperexcitability, aggressive behavior, weight loss, neuronal destruction in hippocampus and other portions of the brain, seizures, learning and memory impairment, self-mutilation, altered sensitivity to stimuli, and disrupted patterns of drinking and eating.

Trimethyltin-induced behavioral disruptions usually peak 3–5 days after exposure, but effects persist for extended periods and seem to be irreversible. Rats sometimes survive the trimethyltin behavior syndrome and appear outwardly normal, although later neuropathological examination shows extensive bilateral damage, including hippocampus shrinkage and cell loss.

Triethyltins were the most potent organotins tested on mammals, although other organotins produced similar signs of poisoning. Mammals poisoned by triethyltin compounds showed muscle weakness within hours of dosing; after a short period of recovery, tremors developed, leading to convulsions and death 2–5 days after dosing. Although toxicity produced by triethyltins becomes more pronounced with continued exposure, reversal of behavioral deficits occurs within weeks after dosing is terminated. Initial reaction to triethyltin exposure in rats was fluid accumulation in white matter of the central nervous system, which persisted for as long as the compound was administered; after administration the effects reversed. There is general agreement that triethyltin-induced behavioral changes are accompanied by cerebral edema, neurodegenerative disorders, interference with oxidative phosphorylation, and disrupted metabolism of glucose and enzyme activity.

Triphenyltins are skin and eye irritants to rats and rabbits. They do not accumulate in rats, dogs, and guinea pigs although some triphenyltin acetate was partly absorbed by cattle and sheep – with most excreted in 6–8 weeks. Thymus atrophy was associated with a lymphocyte depletion in the thymic cortex and is the predominant effect of the intermediate trialkyltins. Intermediate trialkyltin homologues caused a dose-related reduction of thymus weight in male rats after 2 weeks on diets containing 150.0 mg organotin/kg; decreases were 19% for triphenyltin, 47% for tripropyltin, and 61% for tributyltin. Phenyltin compounds significantly inhibit natural killer cell function, and possible natural killer cell mediated immunotoxic potential of these compounds in humans. The toxic potential of phenyltins followed the order of triphenyltin > diphenyltin > monophenyltin; however, phenyltins were less toxic than butyltins to human natural killer cells.

Tributyltins and other organotins induce chromosomal aberrations in mammals, although this was not observed in tests with aquatic invertebrates. Studies with isolated rat hepatoma cells, TBT, and PCB 126, show that TBT inhibits cytochrome P4501A activity and PCB 126 induces EROD activity; however, PCB-induced EROD activity was potentiated by coexposure to low noncytotoxic concentrations of TBT. It was concluded that TBT does not interfere with Ah receptor binding, and that potentiation of EROD activity and cytotoxicity as a result of coexposure to PCB 126 and TBT is significant because they coaccumulated in a variety of marine organisms.

Tetraorganotin compounds produce muscular weakness, paralysis, respiratory failure, tremors, and hyperexcitability as acute effects in mice and dogs; latent effects are similar to those seen in triorganotin poisoning. Tetra methyltin, for example, produces the same toxic syndrome as trimethyltin in rats because it is rapidly dealkylated in vitro to the latter compound. Signs of triorganotin poisoning in rabbits were evident shortly after administration of tetraorganotin compounds, suggesting that triorganotins were soon distributed to the site of action in amounts sufficient to produce signs of poisoning. The dealkylation and distribution of tetraorganotins are related to alkyl chain length and to their accumulations in tissues, including brain. In 3-h studies with rabbits, at intravenous dosage rates of 2.0–3.0 mg/kg BW, tetraethyltin was quickly distributed to liver, but tetrapropyltin and tetrabutyltin were slowly distributed. Tetraethyltin was more readily converted into the corresponding trialkyltin than was tetrapropyltin. About 20% of the tetraethyltin, 4% of the tetrapropyltin, and 1% of the tetrabutyltin were converted to their corresponding trialkyltins.

Thus, the extent of formation of triorganotins decreased as the size and stability of the ligand increased. There was poor distribution of tetraorganotins to brain, but the amounts of triorganotin metabolites found in brain increased over time. Particularly, the transfer of triethyltin to the brain was significant and compatible with the appearance of signs of toxicity. It was concluded that the extent of the dealkylation and the toxicity of organotin compounds depends on the length of their alkyl group, which was associated with their rate of absorption and ultimate distribution.

Organotin compounds are not mutagenic, teratogenic, or carcinogenic, as judged by largely negative but incomplete evidence. It has been suggested that some organotins retard the onset and growth of cancer in laboratory animals and that the anticarcinogenic action is mediated through the thymus gland. The absence of tin in tissues may also be associated with tumor development. In one study, mice with cancer-prone mammary glands and transplanted mammary tumors had significantly reduced tumor growth rates after oral dosing with tributyltin fluoride. In another study, tumor growth rates were significantly reduced in mice continuously exposed to various diorganotin compounds in drinking water at 1.0 and 10.0 mg/L. It is hypothesized that the unknown thymic organotins are antagonistic to cancers in mice and possibly humans. Additional research on potential anticarcinogenic properties of organotins is clearly indicated.

31.5.4 Terrestrial Invertebrates

Resistance to organotin acaricides has been reported in several populations of spider mites. After cyhexatin and fenbutatin oxide were used for 10–17 years on pears and apples to control mites, populations of McDaniel spider mite (*Tetranychus mcdanieli*), two-spotted spider mite (*T. urticae*), and European red mite (*Panonychus ulmi*) slowly began to develop strains that were resistant to these chemicals.

31.6 Recommendations

Proposed organotin criteria for the protection of aquatic life, domestic animals, and human health, vary substantially (Table 31.3). The most stringent criteria now proposed are for triorganotins and aquatic life; these vary from 0.002 to 0.008 μg/L (Table 31.3). But even these comparatively low concentrations will not protect certain species of gastropod mollusks or larvae of the sheep sturgeon (*Accipenser nudiventris*) from tributyltin impacts, as discussed earlier. No criteria are currently proposed for protection of mammals against trimethyltins and triethyltins, the most toxic organotins tested in this group. Trimethyltins, for example, produce nonreversible neurotoxicological effects to certain species of small laboratory animals at concentrations as low as 0.15 mg/L drinking water or 0.625 mg/kg BW and are fatal at 1.25 mg/kg BW.

Hazard evaluation posed by organotin compounds to natural resources is predicated partly on their chemical composition, partly on their concentration and persistence in abiotic materials and diet items, and partly on their availability from these materials to organisms. In each of these areas, key data are missing for promulgation of effective regulations. It seems that additional research is needed in eight areas to acquire these data: (1) the development of sensitive and rapid analytical schemes for the extraction and separation of inorganic tin and organic tin compounds and their chemical speciation products from water, sediments, and biological materials; (2) elucidation of mechanisms and modes of toxicity for organotin compounds, especially those involving sublethal chronic exposures and cellular and subcellular impacts; (3) acquisition of data on organotin toxicokinetics, including data on routes of exposure, uptake, retention, and translocation. Studies should emphasize whole organisms, to determine if food chain biomagnification

Table 31.3. Proposed organotin criteria for the protection of natural resources and human health.

Resource, Organotin Compound	Criterion or Effective Tin Concentration
AQUATIC LIFE: FRESHWATER	
Water	
Triorganotins, North Carolina	<0.008 μg/L
Triethyltins, maximum permissible concentration	<100.0 μg/L
Tributyltins	
Acute value	<0.97 μg/L
Chronic value	<0.30 μg/L
Safe level	<0.026–<0.27 μg/L
In schistosomiasis control	<0.1 μg/L
Sediments	
Tributyltins	
4-day average	<30.0 μg/kg
1-h average	<48.0 μg/kg
AQUATIC LIFE: MARINE	
Water	
Dibutyltin dichloride, Former Soviet Union (FSU)	<2000.0 μg/L
Dibutyltin disulfide, FSU	<20,000.0 μg/L
Triorganotins	
North Carolina	<0.002 μg/L
USA, safe level	<0.05 μg/L
Tributyltins	
USA	
4-day average	<0.017 μg/L (not to be exceeded more than once in 3 years)
Chronic value	<0.064 μg/L
Acute value	<0.22 μg/L
1-h average	<0.43 μg/L (not to be exceeded more than once in 3 years)
Safe level	
UK	<0.02 μg/L
USA	<0.002 μg/L to <0.02 μg/L
Tributyltins; USA, Canada	<0.2 μg/L
Tetraethyltin, FSU	<200.0 μg/L
Sediments	
Tributyltins	
4-day average	<7.0 μg/kg
1-h average	<1.0 μg/kg
No effect level on annelids, mysids, clams	<610.0 μg/kg
Antifouling paints	
Organotins	<4.0 g/L
Tributyltin	<4.0 μg/cm^2 daily
DOMESTIC AND LABORATORY ANIMALS	
Diet	
Dibutyltins	
Rat, age 3 months	<40.0 mg/kg diet
Rat, age 6 months	<20.0 mg/kg diet
Triphenyltins	
Guinea pig, daily intake	<0.1 mg/kg body weight (BW)
Rat, daily	<3.0 mg/kg BW
Drinking water	
Rat	<0.007μg/L
HUMAN HEALTH	
Air	
Total tin	
Connecticut	<40.0 μg/m^3, 8-h daily
Virginia	<1.6 μg/m^3, 24-h daily

Continued

Table 31.3. cont'd

Resource, Organotin Compound	Criterion or Effective Tin Concentration	Resource, Organotin Compound	Criterion or Effective Tin Concentration
Inorganic tin compounds	<2000.0 μg/m^3	*Daily intake*	
		From fresh foods	1.0–4.0 mg
Organotins	<100.0 μg/m^3; <200.0 μg/kg BW daily	From water	<0.03 mg
		Daily intake; adult, low tin diet	0.003–<0.13 mg/kg BW
Triethyltins, occupational exposure	<100.0 μg/m^3	*Daily intake; adult, high tin diet*	0.2–<17.0 mg/kg BW
		Composition of diet	
Tricyclohexyltins	<1200.0 μg/m^3	*Inorganic tin*	About 1.0 mg/kg
Diet		*Organic tin*	<2.0 mg/kg
Total tin		*Tricyclohexyltins*	
Daily intake	0.2–<8.8 mg	Peaches	<4.0 mg/kg
Daily intake		Apples, pears	<2.0 mg/kg
Total	4.003 mg	Meat	<0.2 mg/kg
Diet	4.000 mg	Milk	<0.05 mg/L
Air	0.003 mg	Total daily intake	<0.0075 mg/kg BW

is a potential problem; reproductive organs, in which organotin burdens may affect proliferation; and edible tissue, especially muscle and liver, which are selectively consumed by humans and various animal species; (4) determination of the persistence and mobility of organotin compounds – especially in aquatic abiotic materials, such as sediments, sediment interstitial waters, suspended particulates, and the water column – and on the partitioning of these compounds between the surface microlayer and subsurface waters; (5) determination of the extent of tin methylation and the biotransformation and pharmacodynamics of organotins; (6) measurement of biological interaction effects of organotins with other toxic chemicals under stressful environmental conditions of temperature, oxygen, and other variables; (7) development of quantitative structure–activity relations for use in evaluating toxicity of organotin compounds; and (8) initiation of long-term environmental monitoring studies in terrestrial and aquatic ecosystems to establish appropriate baseline concentrations and to separate these from contaminant effects.

Antifouling paints containing organotin compounds have been associated with a number of adverse effects to marine biota, including reductions in natural bacterial assemblages, contamination of salmon farmed in sea cages with treated net panels, and reduced growth of oysters. The U.S. Navy, however, has implemented fleetwide use of organotin antifouling paints that contain tributyltin as a biocide. This procedure will result in a 15% fuel consumption reduction, increase the interval between cleaning ship hulls from 2 years (with cuprous-oxide-based antifouling paints) to about 7 years, and increase ship speed up to 40% as a direct result of reduced drag. Since naval vessels rarely remain moored for extended periods in coastal areas, hazard effects to the environment are minimal – despite the size of the vessel – when compared to boating practices at local marinas. Accordingly, civilian use of marine paints containing organotin compounds has been severely restricted in recent years. France banned tributyltin compounds in antifouling paints in January 1982 for use on vessels under 25 m (82 feet) length. The State of Virginia enacted legislation that prohibits the use of tributyltin paints on nonaluminum vessels under 25 m. Also in Virginia, tributyltin paints applied on large ships and aluminum craft should not exceed a daily leach rate of 4.0 $\mu g/cm^2$. Similar legislation is proposed in at least 12 coastal and Great Lakes States. In April 1987, England banned the retail sale of antifouling paints containing organotin compounds at concentrations greater than 4.0 g of total tin per liter. Because of their hazards, use of the more toxic triorganotin biocides should be curtailed to prevent their entry into the environment. Continued monitoring of tributyltin levels is recommended, especially in areas of extensive boating activity.

31.7 Summary

Tin (Sn) has influenced our lifestyle for the past 5000 years. Today we are exposed to tin on a daily basis; including tinplated baby food cans; alloys such as pewter, bronze, brass, and solder; and toothpaste containing stannous fluoride. These inorganic tin compounds are not highly toxic due to their low solubility, poor absorption, low accumulation, and rapid excretion. Synthetic organotin compounds, however, first manufactured commercially in the 1960s, may present a variety of problems to animals, including impaired behavior and reduced growth, survival, and reproduction. Some triorganotins – for example, in antifouling marine paints, in molluscicides, and in agricultural pesticides – can be harmful to sensitive species of nontarget biota at recommended application protocols. Background concentrations of organotin compounds are frequently elevated, occasionally to dangerous levels, in aquatic organisms collected near marinas and other locales where organotin-based antifouling paints are extensively used. But more information is needed on background concentrations of organotins, especially those from terrestrial ecosystems.

Tributyltin compounds are especially toxic to aquatic organisms. Adverse effects were noted at concentrations of 0.001–0.06 μg/L on mollusks and at 0.1–1.0 μg/L on algae, fish, and crustaceans. In general, bioconcentration of organotins from seawater was high, especially by algae, but degradation was sufficiently rapid to preclude food chain biomagnification. In contrast, current environmental concentrations of some organotins are not likely to be directly toxic to birds and mammals. Birds seem to be relatively resistant to organotins, although data are scarce. Preliminary studies of 75 days duration suggest that diets containing 50.0 mg tin as trimethyltin chloride/kg were fatal to ducklings; 5.0 mg/kg killed 40%, and 0.5 mg/kg was not lethal. Trimethyltin compounds were lethal to other species of birds tested at doses of 1.0–3.0 mg/kg BW. Other tests with ducklings and eleven other mono-, di-, tri-, and tetraalkyltin compounds at dietary levels equivalent to about 50.0 mg Sn/kg showed no adverse effects on survival. Small laboratory mammals were adversely affected by trimethyltin compounds at doses as low as 0.15 mg/L in drinking water (learning deficits), 0.63 mg/kg BW (diet aversion), and 1.25 mg/kg BW (death); neurotoxicological effects of trimethyltins were usually not reversible. Triethyltins were also toxic to small mammals, but effects – which were similar to those of trimethyltins – were usually reversible after cessation of exposure. All evidence to date indicates that organotin compounds are not carcinogenic.

Methodologies and data necessary for the promulgation of effective criteria and standards to protect natural resources seem to be deficient in eight key areas: (1) routine analytical chemical methodologies for extraction, separation, and identification of inorganic and organic tin compounds and their chemical speciation products in biological and other samples; (2) mechanisms of toxicity for organotin compounds;

(3) rates of uptake, retention, and translocation of organotins in biota; (4) persistence and mobility rates of organotins in nonbiological materials; (5) rates of tin methylation and biotransformation in biological and abiotic samples; (6) organotin interactions with other toxic chemicals; (7) quantitative structure–activity relation for use in evaluating organotin toxicity; and (8) long-term environmental monitoring studies in terrestrial and aquatic ecosystems for establishment of baseline concentrations.

TOXAPHENE[a]

Chapter 32

32.1 Introduction

Toxaphene is a complex mixture of chlorinated bornanes and bornenes containing as many as 670 individual compounds. Heavy past usage, environmental persistence, and long-range atmospheric transport caused ecosystem-wide toxaphene contamination of biota from regions with little or no historical toxaphene application – including Arctic invertebrates, fishes, and mammals, Baltic seals and whales, and the atmosphere of the Great Lakes. Environmental hazards and increasing public concerns associated with toxaphene (chlorinated camphene, 67–69% chlorine) are extensively documented.

Toxaphene was introduced in the mid-1940s as a new insecticide, but only a few years elapsed before it was being used commercially on a large scale to effectively control a variety of pests – especially pests of cotton, corn, fruits, vegetables, grains, soybeans, and ectoparasites of livestock. Toxaphene solutions were often mixed with other pesticides, including methyl and ethyl parathion, DDT and lindane. In the mid-1950s, toxaphene was first used in ponds, lakes, and streams as a piscicide. By 1966, toxaphene was the chemical of choice in fish eradication programs in Canada and second in the United States after rotenone. Its use for this purpose was discontinued in the 1960s due to its lengthy persistence in water, high acute toxicity to aquatic biota, and significant bioaccumulation and biomagnification in various environmental compartments. By 1974, cumulative world use of toxaphene, mainly against insect pests of cotton, was estimated at 450,000 metric tons. Production of toxaphene declined from 1973 to 1980; however, annual consumption in 1980 was estimated at 105,000 tons, thus qualifying toxaphene as one of the most heavily utilized agricultural chemicals worldwide at that time. In the early 1980s, toxaphene was extensively applied in California to control fruitworms on tomatoes, bollworm on cotton, and a wide range of pestiferous insects that infested alfalfa, broccoli, celery, beans, clover, lettuce, cauliflower, and pears. In time, toxaphene-resistant strains of cotton pests, including bollworm and lygus bug, appeared in California, Texas, Egypt, and India.

In November 1982, most registered uses of toxaphene were canceled by the U.S. Environmental Protection Agency (EPA), although existing available stocks could be used through 1986. Prior to the EPA action, similar actions that banned or restricted toxaphene use had been implemented in a number of countries, including Canada, England, Sweden, Finland, Denmark, France, Switzerland, Hungary, Italy, Egypt, and Algeria. In 1990, the U.S. EPA banned all uses of toxaphene in the United States or any of its territories because of adverse effects on human and animal health. In 1993, the EPA banned the importation of food containing toxaphene residues in the United States or any of its territories. And on September 1, 1993 all tolerances, interim tolerances, and food additive regulations for toxaphene on all agricultural commodities were revoked. However, toxaphene-like

[a]All information in this chapter is referenced in the following sources:

Eisler, R. 2000. Toxaphene. Pages 1459–1481 in *Handbook of Chemical Risk Assessment: Health Hazards to Humans, Plants, and Animals. Volume 2, Organics*. Lewis Publishers, Boca Raton, Florida.

Eisler, R. and J. Jacknow. 1985. Toxaphene hazards to fish, wildlife, and invertebrates: a synoptic review. *U.S. Fish Wildl. Serv. Biol. Rep. 85(1.4)*, 26 pp.

pesticides are currently produced and used in many countries, including India, many South and Central American countries, Eastern Block countries in the former Soviet Union, and many African countries.

32.2 Environmental Chemistry

The commercial production of toxaphene involves the reaction of camphene, chlorine activated by ultraviolet radiation, and certain catalysts to yield chlorinated camphene with a chlorine content of 67–69% by weight. This product is a relatively stable material composed of at least 177 – as many as 670 are predicted – chlorinated bornanes and bornenes. Of these components, 26 have been isolated and 10 identified; these 26 components comprise 40% of the toxaphene. Information on chemical properties and the fate and effects of the remaining components is missing or incomplete. Several components that have been tested are more toxic to houseflies than the technical mixture, especially di-, tri-, and tetrachlorobornane compounds. Technical toxaphene is a yellow, waxy solid of empirical formula $C_{10}H_{10}Cl_8$ and an average molecular weight of 414. It melts at 65–99°C. Toxaphene is soluble in water to 3.0 mg/L and is readily soluble in fats and organic solvents, based on its high partition coefficient of $1 \times 10^{3.3-6.4}$. Toxaphene has a tendency to adsorb on sediments and to bioaccumulate in aquatic organisms.

Because toxaphene consists of numerous compounds, it seems inappropriate and misleading to continue using the name toxaphene to describe this insecticide. It is known that chemical properties, such as solubility, toxicity, volatility, and other properties, are the sum of the individual contributions of many different compounds in differing relative amounts. A 50-fold difference between toxicities of toxaphene components can occur, and, with a wide range in the polarity of different fractions, there probably are also significant solubility differences. In addition, the composition of toxaphene changes with time, and residues in fat are not of the same composition as parent toxaphene. The metabolism of toxaphene has been an area of limited research activity, owing to the analytical difficulties involved in detecting a multicomponent substance. However, toxaphene has been reported more often in biological samples in recent years. This increased recognition is probably due to better analytical methods for toxaphene analysis, especially electron capture, negative ion mass spectrometry, greater awareness by analysts, and the continuing use of toxaphene while use of potentially interfering organochlorine insecticides has slowly decreased.

Toxaphene was available as an emulsifiable concentrate, wettable power, or dust. The commercial product is relatively stable but may decay upon prolonged exposure to sunlight, alkalis, or temperatures above 120°C. Toxaphene is also known as Agracide Maggot Killer, CAS 8001-35-2, Camphofene, Huilex, Motox, Toxafeen, chlorinated camphene, Synthetic 3956, Octachlorocamphene, Alltox, Geniphene, Toxakil, polychlorocamphene, camphechlor, Clor Chem T-590, Cristoxo, Moto, Phenacide, Phenatox, Strobane-T, Toxon 63, and Vapotone. Chemically, it is known as a mixture of various chlorinated camphenes. The actual global production of toxaphene between 1950 and 1993 is unknown, but estimates range as high as 1.3 million metric tons.

Toxaphene partitions to the atmosphere, surface and groundwater, soil and sediment particulates, and adipose tissue. As a result of its volatility and resistance to photolytic transformation, toxaphene has been transported over long distances in the atmosphere. Toxaphene residues have been detected in various environmental compartments hundreds of kilometers distant from known applications of this insecticide. Prevailing winds, rainfall, and sediment runoff probably account for substantial portions of this transport. Rainfall, for example, has been implicated as a significant toxaphene vector in South Carolina estuaries. During and immediately after the summer use season, toxaphene levels in rain exceeded, by several times, the concentrations reported to produce bone damage in fish under controlled laboratory conditions. Toxaphene becomes sorbed to soils when it is used in agriculture; therefore, a major mode of toxaphene transport in areas planted continuously in cotton is through sediment loss in runoff. In soils, toxaphene is relatively immobile. Under anaerobic conditions toxaphene is rapidly biotransformed in soils and sediments with a half-time persistence of 2–4 months; however, under aerobic conditions toxaphene may persist for years – Tb1/2 of 11 years – in aerated surface soils and sediments. Measurements indicated a linear relation between toxaphene yield and sediment yield in runoff water. Atmospheric transport of toxaphene is well documented. Air samples from the western North Atlantic contained measurable levels of toxaphene at distances up to 1200 km from the nearest point source of application on land. Similarly, Nationwide monitoring of toxaphene in fish showed increases during 1970–74, especially in areas where the insecticide was not used, suggesting that atmospheric transport is essential to widespread distribution. Airborne toxaphene is resistant to photodecomposition; however, selective volatilization of toxaphene components is a major cause of degradation resulting in an estimated half-time of 15 days while in the atmosphere.

Toxaphene degrades more rapidly in most environmental compartments than other chlorinated pesticides, such as DDT and dieldrin. The major metabolic degradation mechanisms for toxaphene in all organisms from bacteria to primates are now believed to be the reductive dechlorination, reductive dehydrochlorination, and in some cases oxidative dechlorination to produce hydroxyl derivatives, acids, or ketones. Toxaphene persistence and degradation in soil, water, and biota is modified by numerous and disparate biological and abiotic factors. Under laboratory conditions, toxaphene in water has a half-time persistence of more than 10 years under conditions of high oxygen content, 25°C, and pH 5–8. In lakes, toxaphene persistence was significantly related to lake depth, stratification, and turnover, but not related to surface area, pH, temperature, sunlight, and oxygen. Data from studies where toxaphene was used to control nongame fish in lakes suggest that it may persist in water from several months to more than 9 years. For example, two mountain lakes in Oregon that were treated with toxaphene in fish eradication programs remained toxic for 1–6 years. Davis Lake, a shallow lake rich in aquatic life, which was treated with 88.0 μg/L toxaphene, could be restocked with rainbow trout (*Oncorhynchus mykiss*) within 1 year when water toxaphene levels were 0.63 μg/L. Trout grew rapidly, although whole body burdens up to 24.0 mg/kg were recorded. Miller Lake, a deep biologically sparse lake, was treated with 40.0 μg/L toxaphene; trout could not be restocked for 6 years until water levels had dropped to 0.8 μg/L toxaphene. Toxaphene at 50.0 μg/L was used to eradicate fish from Clayton Lake, New Mexico. Water concentrations of 1.0 μg/L were measured 250 days posttreatment, but the lake remained toxic for 9 months, with restocking possible only after 12 months. Residues in fish surviving treatment were 3.5 mg/kg whole body wet weight shortly after exposure and 0.3 mg/kg about 5 months postapplication. Some lakes treated with toxaphene to kill fish have remained toxic for 3–4 years. In another study, lake water that contained 1.0 μg/L toxaphene (9 years after toxaphene treatment) supported healthy fish populations. In this lake, particulate matter contained 70.0 μg of toxaphene/kg, and plankton contained 15,000.0 μg/kg. However, there were changes in gas chromatographic profiles of toxaphene residues taken from the lakes, suggesting that the parent toxaphene had been altered or degraded into compounds with lower environmental hazards to biota. Clearly, this subject area merits additional researc effort.

In soils, toxaphene can persist for lengthy periods, with microbial degradation occurring under aerobic and anaerobic conditions. Toxaphene applied at 140.0 mg/kg of soil persisted for more than 6 years; when applied at 50.0 mg/kg, half the toxaphene was measurable after 11 years. Further, in sandy loam soils, 45% of the toxaphene remained 14 years after initial application of 100.0 mg/kg. Some investigators suggest that toxaphene degradation is more rapid under anaerobic conditions. Thus, toxaphene in anaerobic salt marsh sediments generally degraded within a few days to shorter-lived components. Toxaphene accumulated only slightly in anaerobic marsh soils not flooded daily by tides, and the highest pesticide concentrations were associated with roots of dead plants.

Degradation of toxaphene in plant, air, and soil samples was evident following toxaphene application of 9.0 kg/ha to a San Joaquin Valley, California, cotton field. Cotton leaves contained 661.0 mg toxaphene/kg immediately after application and 135.0 mg toxaphene/kg after 58 days, with the greatest loss attributed to components of highest volatility. Air samples were essentially the same at 2 and 14 days postapplication (1.8–1.9 $\mu g/m^3$); this was attributed to a corresponding enrichment of volatile components. Top soil samples immediately after application and 58 days later contained 13.1 and 6.4 mg/kg, respectively; loss was primarily via vaporization, but at least one component was significantly degraded. One year later, soil cores and irrigation ditch samples showed extensive toxaphene degradation resulting in a selective decline of some components; anaerobic reduction occurred in these environmental compartments.

Toxaphene elimination rates vary between species. In rats, the half-time persistence of toxaphene (time to 50% excretion = Tb1/2) was 1–3 days. If the trend persisted, virtually all toxaphene would be eliminated in about 15 days. Elevated blood toxaphene levels in a human subject who had eaten catfish fillets containing 52.0 mg of toxaphene/kg dropped 67% in 11 days. By 14 days after the initial measurement, toxaphene blood levels were below analytical detection limits. Persistence seems to be longer in some fishes. Lake trout (*Salvelinus namaycush*) given a single intraperitoneal injection of 7.0 mg toxaphene/kg body weight (BW) had a Tb1/2 of 322 days; for white suckers (*Catostomus commersoni*) this value was 524 days.

32.3 Concentrations in Field Populations

Long-range atmospheric transport of toxaphene has resulted in measurable toxaphene residues in tissues of fishes from remote lakes in northern Canada, fishes from pristine areas of the North Atlantic, North Pacific, and Antarctic Oceans, and fishes, birds, and seals from the western North Atlantic Ocean, Arctic Ocean, Greenland, Sweden, Canada, and the drainage basin of Chesapeake Bay. In Lake Baikal, Russia, the toxaphene congener pattern in water was similar to that in the air, indicating the importance of atmospheric deposition processes to this water body. Atmospheric toxaphene concentrations vary markedly with season, being lower in winter and higher in summer.

National contaminant monitoring surveys, conducted in the period 1974–76, show that toxaphene was detected in about 6% of all fish sampled; this is a higher percentage than recorded in fruits, vegetables, poultry, and meat. Fish collected nationwide at 109 stations between 1976 and 1979 had measurable toxaphene residues at about 60% of all stations sampled; concentrations in fish from the Great Lakes stations exceeded those in fish from most of the rest of the United States, including locations within the cotton-growing areas. Lake trout (*Salvelinus namaycush*) from Lake Michigan typically contained 5.0–10.0 mg of toxaphene/kg whole body on a wet weight basis; lake trout from Lake Huron contained 9.0 mg/kg. These residues are considered harmful to various sensitive species of freshwater teleosts. Since relatively little toxaphene has been used in the Great Lakes region when compared to cotton-growing areas in the mid-South, Northeast, and Southeast, it is postulated that atmospheric transport from areas to the south and southwest are the sources of toxaphene contamination in the Great Lakes. Between 1976 and 1984, mean toxaphene concentrations in whole fish collected nationwide have declined significantly from 0.34 mg/kg FW, maximum 12.7 mg/kg FW in 1976–77 to 0.14 mg/kg FW, maximum 8.2 mg/kg FW in 1984. Declines in toxaphene concentrations were also recorded in eggs of the guillemot (*Cepphus* sp.) population from the Baltic Sea between 1974 and 1989.

Freshwater fishes of the Arroyo Colorado, a major waterway traversing the lower Rio Grande Valley in Southern Texas, were highly contaminated with toxaphene and DDE residues when compared to fish collected elsewhere in the Valley; toxaphene concentrations ranged up to 31.5 mg/kg wet weight in whole fish composite samples. These values were within or above the range producing adverse effects in sensitive species of fish. In addition, toxaphene residues in carcasses of fish-eating birds contained up to 3.0 mg/kg toxaphene. Unlike fishes, avian species readily metabolize and excrete toxaphene, so that little accumulation occurs in tissues; in any event, these levels of toxaphene in carcasses of piscivorous birds are probably biologically insignificant. In the Arroyo Colorado area, toxaphene was being used, to some extent, on crops such as cotton, not only as an insecticide, but as a carrier for more effective chemicals. Another possible source of contamination is a former pesticide plant at Mission, Texas, near the headwaters of the Arroyo Colorado. Soil at this site contained high concentrations of various pesticides, including toxaphene. Contaminant-laden runoff from this site could eventually reach the Arroyo from storm sewers and other water diversion facilities. The contaminated Arroyo Colorado, in turn, empties into the Laguna Madre, one of the more important breeding and nursery grounds for fish and wildlife in the United States. The Texas Department of Health, in an advisory to consumers, has stated that consumption of fishes from the Arroyo Colorado, especially blue catfish (*Ictalurus furcatus*) and gizzard shad (*Dorosoma cepedianum*), is not advised.

Birds, unlike fish, generally contained low or undetectable levels of toxaphene, and the frequency of occurrence was relatively low when compared with that of other organochlorine pesticides. This generalization held for eggs of the osprey (*Pandion haliaetus*) collected in southern New Jersey in 1974; carcasses of 103 skinned shorebirds from Corpus Christi, Texas, during winter 1976–77; eggs of the brown pelican (*Pelecanus occidentalis*) from 1971 to 1976 in Louisiana; eggs of clapper rails (*Rallus longirostris*), purple gallinules (*Porphyrula martinica*), and limpkins (*Aramus guarauna*) from the Southeast in 1972–74; and all avian tissues in Texas between 1965 and 1988. Among 105 herons found dead nationwide since 1976, only nine contained measurable quantities of toxaphene; for DDE, PCBs, dieldrin, and DDD, these frequencies were 96, 90, 37, and 35, respectively. Levels of toxaphene and other organochlorines in canvasbacks (*Aythya valisineria*) from Chesapeake Bay, Maryland, during 1973–76 were below the levels known to cause problems in other species. However, adipose tissues from 55 male wild turkeys (*Meleagris gallopavo*) killed during the 1974 hunting season in southern Illinois contained 0.2–0.9 mg/kg of toxaphene, suggesting that certain species of birds may selectively accumulate low concentrations of toxaphene.

Two bird kills reported in California have been attributed to toxaphene poisoning. In one case, the apparent route of exposure was from contaminated fish, with bird poisoning the result of toxaphene biomagnification in the food chain. In that case, algae contained 0.1–0.3 mg toxaphene/kg wet weight, snails and daphnids 0.2, fish 3.0–8.0, and fish-eating birds 39.0 mg/kg. The latter value is substantially in excess of 3.0 mg/kg, a concentration considered biologically insignificant to fish-eating birds. The second incident involved some birds that were apparently killed by toxaphene when it was used to control grasshoppers on a shortgrass range. At 2–3 weeks post-spray, bird carcasses contained 0.1–9.6 mg toxaphene/kg.

Toxaphene accumulates in tissues of many aquatic species, especially in lipid-rich tissues of polar fish and marine mammals. Toxaphene concentrations in livers of Arctic fishes of 2.9 mg/kg FW and in Canadian cod liver oil of 28.0 mg/kg FW are recorded. Toxaphene concentrations – up to 39.7 mg/kg FW blubber – in marine mammals from Newfoundland were higher than other groups of organochlorines measured, and were probably due to elevated toxaphene concentrations in their diet of Atlantic cod (*Gadus morhua*) and herring (*Clupea harengus harengus*). Only a few of the many polychlorobornane congeners accumulate in liver and blubber of marine mammals, perhaps because some of the higher chlorinated compounds with more than 6 chlorine atoms remain unmetabolized or are highly accumulated. The two main toxaphene congeners identified in fish liver, dolphin blubber, and human milk were 2-exo, 3-endo, 5-exo, 6-endo, 8,8,9,10,10-nonachlorobornane, and

2-exo, 3-endo, 5-exo, 6-endo, 8,8,10,10-octachlorobornane. The three main congeners found in blubber (two octa- and one nonachloro compounds) accounted for about 30–40% of the total toxaphene burden. The major peak in the Baikal seal chromatogram was an 8-chlorinated bornane similar to that of ringed seals from the Canadian Arctic. But cod liver oils contain up to 10 major congeners and exceed the maximum acceptable concentration for food in Germany of 0.1–0.4 mg total toxaphenes/kg lipid weight. The octochlorobornane and nonaclorobornane congeners which dominated in Arctic amphipods, fishes, and marine mammals (narwhals and belugas) comprised only 8–11% in the ringed seal–polar bear (*Ursus maritimus*) food chain.

Toxaphene is biomagnified through the food web by marine mammals, with factors ranging from about 40–100 for dolphins and porpoises. In Lake Michigan, toxaphene was biomagnified in aquatic food webs by about 5 from plankton to fish (sculpin, *Myoxocephalus* sp.) and 14 from mysids to fish. Biomagnification of toxaphene through food webs was clearly demonstrated in 16 species of organisms collected from oxbow lakes in northeastern Louisiana during 1980. Without exception, residues were highest (3.6 mg/kg whole body wet weight, range 1.7–5.5) in tertiary consumers, such as green-backed heron (*Butorides striatus*), various species of snakes, spotted gar (*Lepisosteus oculatus*), and largemouth bass (*Micropterus salmoides*). Secondary consumers, such as bluegill (*Lepomis macrochirus*), blacktail shiner (*Notropis venustus*), and yellow-crowned night-heron (*Nycticorax violaceus*), contained lower residues (0.9 mg/kg wet weight, range 0.1–1.2). Primary consumers, including crayfish (*Procambarus* spp.) and threadfin shad (*Dorosoma petenense*), contained the lowest levels (0.8 mg/kg wet weight, range 0.6–1.0) of all consumer groups. Toxaphene levels were not detectable in water and sediments from these oxbow lakes. Variability in tissue toxaphene concentrations of fishes from Canadian Yukon lakes are related, in part, to differences in the food chains of the lakes, with emphasis on prey lipid content.

32.4 Lethal Effects

Toxaphene is extremely toxic to freshwater and marine biota. In laboratory tests of 96 h duration, 50% mortality was recorded for the most sensitive species of freshwater and marine teleosts, marine crustaceans, and freshwater insects at nominal water concentrations of less than 10.0 μg/L of toxaphene, and, in several cases, less than 1.0 μg/L. Bioassays of longer duration, based on exposure of aquatic organisms for the entire or most of the life cycle, produced significant adverse effects on growth, survival, and reproduction at toxaphene concentrations between 0.025 and 1.0 μg/L. Toxaphene was most toxic to freshwater fishes in soft water at elevated temperatures. Based on its high toxicity and extensive use, it is not surprising that toxaphene was considered a major cause of nationwide fish kills in 1977.

Warm-blooded organisms are relatively resistant to toxaphene, as determined from results of short-term tests involving oral, dermal, and dietary routes of administration. In acute oral toxicity tests with birds and mammals, LD50 values ranged between 10.0 and 160.0 mg/kg BW. The dietary LC50 values for four species of birds ranged between 538.0 and 828.0 mg toxaphene/kg ration. The acute oral toxicities of toxaphene to rats, mice, dogs, guinea pigs, cats, rabbits, cattle, goats, and sheep extended from 25.0 to 270.0 mg/kg BW; these values are in good agreement with other mammalian data. For reasons unknown, pregnant rodents were 5–10 times more resistant to oral toxaphene than were nonpregnant rodents. Dermal toxicities of toxaphene ranged from 250.0 mg/kg BW for rabbits and 930.0 mg/kg for rats to 25,000.0 mg/kg for cattle. As was true for acute oral and dermal toxicity data, comparatively high levels of dietary toxaphene were required, i.e., 538.0–828.0 mg/kg diet, to produce significant death rates in various species of birds. In a study on four species of gamebirds, each aged 2 weeks, birds were fed diets containing graded concentrations of toxaphene for 5 days, followed by 3 days of untreated food. LD50 values at the end of day 8 were 828.0 mg toxaphene/kg diet for northern bobwhite, 686.0 for Japanese quail

(*Coturnix coturnix japonica*), 542.0 for ring-necked pheasant (*Phasianus colchicus*), and 538.0 for mallard (*Anas platyrhynchos*). It appears that toxaphene is not a major hazard to bird survival at previously recommended field application rates. However, at toxaphene levels not considered life-threatening to birds and mammals, fetotoxic effects have been recorded. For example, ring-necked pheasants fed 100.0 mg/kg dietary toxaphene produced eggs with significantly reduced hatch over controls; similarly, toxaphene administered orally to pregnant rats and mice during organogenesis caused fetal toxicity at 15.0 mg/kg BW.

Some human deaths, especially those of children, have been reported following the ingestion of toxaphene-contaminated foods. Known toxaphene residues in food items of victims ranged from 9.7 to 47.0 mg/kg; a total dose of 2.0–7.0 g of toxaphene is considered acutely toxic to a 70-kg adult. For comparison purposes, a 4.5-kg bird would probably die after consumption of 45.0–450.0 mg of toxaphene.

32.5 Sublethal Effects

Among sensitive species of marine and freshwater fish and invertebrates, water concentrations of 0.054–0.299 μg/L of toxaphene were associated with growth inhibition, reduced reproduction, backbone abnormalities, or histopathology. Aquatic biota are capable of spectacular accumulations of toxaphene from the medium; factors ranged between 1270 and 52,000 those of water under laboratory conditions. A similar pattern was observed in Big Bear Lake, California, where toxaphene was applied at 200.0 μg/L to eradicate goldfish. Biomagnification factors of 365 were calculated for plankton, 1000 for goldfish, and 8500 in pelican fat, representing residues of 73.0 mg/kg toxaphene in phytoplankton, 200.0 in goldfish, and 1700.0 in pelican fat. Accumulation of toxaphene by various species of fish food organisms is dependent on exposure time and concentration. For example, insect nymphs subjected to 20.0 μg/L of toxaphene for less than 24 h did not accumulate doses lethal to fish; however, algae, diatoms, and protozoan ciliates held for 24 h in 20.0 μg/L toxaphene solutions, and *Daphnia magna* held 120 h in 10.0 μg/L, were lethal when fed to fish.

Fish accumulated part-per-million (mg/kg) toxaphene concentrations in various tissues within a few days when placed in toxaphene-treated lakes that contained less than 1.0 μg/L. Freshwater teleosts experienced acute and chronic effects when whole body levels were in excess of 0.4 mg/kg but less than 5.0 mg/kg (this latter value being the Food and Drug Administration "action level" for human consumption). Thus, groups of brook trout (*Salvelinus fontinalis*) eggs containing 900.0 μg toxaphene/kg had drastically reduced survival when compared to controls, and brook trout tissue residues exceeding 400.0 μg toxaphene/kg were associated with reductions in growth, abnormal bone development, and reduced fecundity. Fathead minnows (*Pimephales promelas*) containing more than 400.0 μg toxaphene/kg grew more slowly than controls; similar results were reported in channel catfish (*Ictalurus punctatus*) fry containing 600.0–3400.0 μg toxaphene/kg. Toxaphene retention by aquatic organisms is relatively lengthy when compared to mammals. In one case, American oysters (*Crassostrea virginica*) held for 24 weeks in 10.0 μg/L toxaphene solutions contained 32.4 mg/kg in soft tissues; after 16 weeks in contaminant-free seawater, oysters still contained 3.0 mg/kg.

Sublethal effects of toxaphene observed in mammals, small laboratory animals, and birds were similar to those recorded for aquatic organisms; however, there was general agreement that effects were induced at much higher concentrations. In domestic white leghorn chickens, for example, toxaphene at 100.0 mg/kg in the diet for 30 weeks did not significantly alter egg production, hatchability, or fertility, although some bone deformation and kidney lesions were recorded. The highest dietary dose of toxaphene fed to chickens in life-time exposure studies, which produced no effect on any parameter measured, ranged between 3.8 and 5.0 mg/kg. Several studies with the American black duck (*Anas rubripes*) produced effects similar to those recorded in chickens. In one study, ducklings that were fed diets containing 10.0 or 50.0 mg of toxaphene/kg for 90 days had reduced growth and impaired backbone development. Collagen, the organic matrix of bone, was significantly decreased in cervical vertebrae of ducklings fed a diet containing 50.0 mg toxaphene/kg ration. Calcium concentrations increased in vertebrae of ducklings fed either 10.0 or 50.0 mg/kg dietary toxaphene; effects were observed only in female ducklings. In a long-term feeding study lasting 19 months, which included two breeding seasons, American black ducks, fed 10.0 or 50.0 mg toxaphene/kg in a dry mash diet, showed no significant differences when compared to control birds in survival, egg production, fertility, hatchability, eggshell thickness, or growth and survival of young. The only negative effects recorded included weight loss among treated males during summer and a slight delay in the number of days required to complete a clutch. Carcass toxaphene residues, which seldom exceeded 0.5 mg/kg, were found in only one duck fed with the 50.0 mg/kg diet, suggesting low body accumulations in American black duck. However, toxaphene residues were present in the liver of all birds fed toxaphene. At dietary concentrations of 10.0 or 50.0 mg/kg, there was no change in avoidance behavior of young American black ducks, which, if interrupted, is considered life-threatening.

Ring-necked pheasants (*Phasianus colchicus*) fed diets containing 300.0 mg toxaphene/kg showed decreases in egg deposition, egg hatch, food intake, and weight gain; at 100.0 mg/kg, all of these parameters, except reduced hatch, were the same as controls. In a field study, aerial applications of a DDT–toxaphene mixture in southwestern Idaho during 1970, at recommended concentrations to control pests, had no major impact on penned or feral ring-necked pheasants, suggesting that conformance with recommended application rates should be endorsed whenever possible. However, it is emphasized that recommended toxaphene application rates, varied widely and depended, in part, on the pest species to be controlled, the number and type of other pesticides applied jointly, and climatic conditions. Laboratory studies with mallard eggs suggest that recommended toxaphene application rates in excess of 1.12 kg/ha, which is generally exceeded in most cases, may produce severe embyrotoxic effects, including dislocated joints and poor growth.

Common bobwhite fed 5.0 mg/kg dietary toxaphene for 4 months showed thyroid hypertrophy and interference with the ability of bobwhites to discriminate patterns. In the latter investigation, 3-day-old bobwhites fed diets containing 10.0 or 50.0 mg/kg toxaphene for 20 weeks made 50% more errors than controls on initial testing. These effects appeared as early as 30 days after toxaphene exposure. In a second test, there was no difference between experimentals and controls, indicating that the ability to learn these tasks was not permanently impaired.

Rats, mice, dogs, deer, sheep, and cattle are all relatively resistant to toxaphene. No-effect levels of 20.0–25.0 mg/kg dietary toxaphene were documented during multigeneration exposure of rats and during 2-year feeding studies with mice and dogs. No effects were observed in monkeys over a 2-year period during which they were fed diets containing 0.7 mg toxaphene/kg. Toxaphene is carcinogenic in rodents, inducing malignant neoplasms of the liver and other sites. Cancers were induced in mice and rats by toxaphene when residues in the diet exceeded 50.0 mg/kg during lifetime exposure. Toxaphene increases the frequency of sister chromatid exchanges of chromosomes in a human lymphoid cell culture line. These results, together with the positive mutagenic response (to *Salmonella* bacteria) constitute substantial evidence that toxaphene is likely to be a human carcinogen. Penned and wild deer fed toxaphene at 1000.0 mg/kg appeared normal but showed a decreased digestion rate, which was attributed to a decrease in rumen bacteria. Steers fed alfalfa hay containing 306.0 mg toxaphene/kg for 19 weeks stored 772.0 mg/kg in abdominal fat and 27.0 mg/kg in lean meat without apparent ill effects, demonstrating the lipophilicity of toxaphene and the relatively low accumulation rates. For sheep under an identical regimen, these values were 317.0 mg/kg in fat and 36.0 mg/kg in meat. Milk from dairy cows had measurable concentrations of toxaphene after the cows had been sprayed with emulsions, suspensions, or oil solutions of toxaphene at entomologically effective doses; toxaphene residues in milk were highest on days 1 and 2 postexposure (0.7–0.8 mg/L) and declined to control values of 0.06–0.08 mg/L by days 14–21.

32.6 Recommendations

In November 1982, the U.S. EPA canceled the registration of toxaphene for most uses and, thus joined a growing number of Nations in Western Europe, Scandinavia, North America, and North Africa that previously initiated similar actions. With some restrictions, toxaphene was used domestically through the mid-1980s for treatment of scabies in cattle and sheep; controlling sporadic infestations of armyworms, cutworms, and grasshoppers on cotton, corn, and small grains; and, in Puerto Rico and the Virgin Islands, to control mealy bugs, pineapple gummosis moths, and banana weevils. Existing stocks of toxaphene were used through 1986 for control of sicklepod in soybeans and peanuts, for insects in corn cultivated without tillage, and for pests of dry and southern peas. In 1993, all uses and manufacture of toxaphene were banned in the United States. To understand the potential global budget of toxaphene it is now necessary to obtain information on the present and historical uses in Third World and former East Block countries.

Although toxaphene is not markedly hazardous to most wildlife species for which data were available, the decision to withdraw or curtail agricultural uses of toxaphene was popular with most natural resource managers. Their concerns, apparently shared by others, were based, in part, on the following observations. First, toxaphene causes death and deleterious effects to nontarget aquatic biota at extremely low concentrations, i.e., $<1.0\,\mu g/L$. Second, toxaphene is persistent in soils, water, and other environmental compartments, with residence times measured in years. Third, toxaphene accumulates in aquatic organisms and biomagnifies through food chains. Fourth, toxaphene is widely distributed, even when the initial application point is hundreds of kilometers distant; transport is presumably by atmospheric and other vectors. Fifth, technical difficulties continue to exist in the chemical analysis of toxaphene, a compound with at least 177 isomers. Sixth, there is an imperfect understanding of the fate and effects of individual toxaphene components. Seventh, there is inadequate knowledge of interaction effects of toxaphene with other agricultural chemicals (especially when mixtures are applied simultaneously) and with other persistent compounds in aquatic ecosystems, such as PCBs, DDT and its isomers, and petroleum. Finally, there is the perception that suitable alternative pesticidal chemicals are available, including some carbamates, organophosphorus compounds, and synthetic pyrethroids.

At present, large but unknown quantities of toxaphene that were discharged into the environment over the past several decades remain undegraded and potentially bioavailable. Also, knowledge of toxaphene ecotoxicology is incomplete or inadequate. Limits for toxaphene residues in air, water, biota, and other environmental compartments for the protection of fish, livestock, and human health vary considerably (Table 32.1). The concentration of toxaphene in seawater considered safe for marine life protection is $0.07\,\mu g/L$; for sensitive freshwater species it lies between 0.008 and $0.013\,\mu g/L$. This contrasts sharply with some recommended drinking water criteria for human health protection of $5.0–8.8\,\mu g/L$. Other existing criteria for human health protection, which range in various foods from 0.1 to 7.0 mg/kg, appear adequate at this time to protect sensitive species of wildlife. It is emphasized that these values, and others shown in Table 32.1, are considered criteria and not administratively legislated standards.

No standard reference materials have been certified for analysis of chlorobornanes, although several may be suitable including fish oils and whale blubber. Analytical standards for several chlorobornanes are now available, but there is a need to identify other remaining environmentally significant congeners. More research is now recommended in the following areas: partitioning between air–water (including snow), air–plant, and air–soil interfaces; processes that control degradation in soils and sediments and transformations during

Table 32.1. Proposed toxaphene criteria for the protection of natural resources and human health.

Compartment	Allowable Concentration
AQUATIC LIFE PROTECTION	
Freshwater[a]	0.013 μg/L (24-h average); 1.6 μg/L maximum at any time
Freshwater	
Most states	
Acute	0.073 μg/L
Chronic	0.002 μg/L
Wisconsin	
Acute, Great Lakes	0.61 μg/L
Acute, cold water	0.81 μg/L
Acute, warm water	0.6l μg/L
Chronic; Great Lakes, both warm and cold water	0.01 μg/L
Marine	0.07 μg/L maximum at any time
Marine; medium; most states	
Acute	0.21 μg/L
Chronic	0.0002 μg/L
Residues in fish tissues	Maximum proposed values range from a high of 5.0 mg/kg fresh weight (FW) to a low of 0.4–0.6 μg/kg FW
LABORATORY WHITE RAT; DIET	
No adverse effect level	4.0 mg/kg ration, equivalent to 0.29–0.38 toxaphene/kg (BW) daily
Lowest adverse effect level (histological damage)	20.0 mg/kg diet
HUMAN HEALTH	
Minimum risk level, oral	5.0 μg/kg BW daily for acute duration exposure of 14 days or less[b]; 1.0 μg/kg BW daily for intermediate duration exposure of 15–364 days
Safe daily dose	3.4 μg/kg BW
Acceptable daily intake	1.25 μg/kg BW
Air	
Arizona	
1-h average	3.7 μg/m^3
24-h average	1.5 μg/m^3
New York, 1-year average	1.67 μg/m^3
Oklahoma, 24-h average	5.0 μg/m^3
Texas	
30-min average	0.5 μg/m^3
Immediately hazardous	200.0 mg/m^3

Table 32.1. cont'd

Compartment	Allowable Concentration
Total daily intake	0.00018 μg/kg BW
Drinking water	5.0 μg/L; maximum of 500.0 μg/L for 1 day or 40.0 μg/L for 10 days
California	0.21 μg/L
Maryland	5.0 μg/L
Missouri	0.000071 μg/L
Diet	
Fish consumption	
Germany	0.4 mg/kg lipid weight
USA	5.0 mg/kg FW
Fat of meat from livestock	7.0 mg/kg
Milk and milk products, fat weight basis	0.5 mg/kg
Soybean oil	6.0 mg/L
Sunflower seeds	0.1 mg/kg FW
Citrus fruits	5.0–7.0 mg/kg FW basis in Canada; 0.4 mg/kg FW basis in Germany and the Netherlands
Groundwater	5.0 μg/L
WILDLIFE PROPAGATION	
Nevada, water	
Irrigation	5.0 μg/L
Watering of livestock	0.01 μg/L

[a]The International Joint Commission of the United States and Canada recommend a water quality standard of 0.008 μg/L for the protection of freshwater aquatic life. This standard is based on the finding that toxaphene at 0.039 μg/L in water caused a significant increase in mortality and a significant decrease in growth of surviving brook trout fry over a 90-day period. The standard of 0.008 μg/L is obtained by applying an uncertainty value of 5.

[b]Based on a lowest observed adverse effect level (LOAEL) of 5.0 mg/kg BW daily for decreased hepatic biliary function in rats treated with toxaphene in the diet for 8 days and an uncertainty factor of 1000.

atmospheric transport; effects of inhalation and dermal exposure to toxaphene, especially studies of an intermediate or chronic exposure; the toxicities of specific toxaphene congeners and their chemical analysis; bioconcentration and biomagnification of individual congeners in terrestrial food chains; and the acute management of toxaphene-induced seizures in humans with anticonvulsants, especially diazepam, phenobarbital, and phenytoin.

32.7 Summary

Toxaphene (chlorinated camphene, 67–69% chlorine) is a broad-spectrum insecticide which was formerly one of the most heavily used agricultural chemicals on a global scale, especially against pests of cotton. In 1982, the U.S. EPA canceled the registrations of toxaphene for most uses; stocks of toxaphene could be used, with some restrictions, through 1986. Considerable, but unknown quantities of toxaphene previously discharged into the environment over the past several decades may remain undegraded and potentially available to living resources. Since 1993, all domestic uses and manufacture of toxaphene were prohibited; however, toxaphene use in other countries is continuing.

Toxaphene is extremely persistent in soil and water, with documented half-times of

9–11 years; however, in air and in warm-blooded organisms, toxaphene degradation is rapid with half-times of 15 and 3 days, respectively. Toxaphene is especially hazardous to nontarget marine and freshwater organisms, with death recorded at ambient water concentrations substantially below 10.0 μg/L, and adverse effects observed on growth, reproduction, and metabolism at water concentrations between 0.05 and 0.3 μg/L. Aquatic organisms readily accumulate toxaphene from the ambient medium and diet, sometimes spectacularly, retain it for lengthy periods, and biomagnify the chemical through food chains. These phenomena could account for the numerous fish kills recorded after toxaphene application, as well as the high residues measured in fish from the Rio Grande Valley in southern Texas and other locations of former high agricultural use of toxaphene. Atmospheric vectors, including prevailing winds and rainfall, may transport toxaphene hundreds of kilometers from known point sources of application. This, in part, would explain the levels of 5.0–10.0 mg/kg whole body wet weight recorded in various species of fish from the Great Lakes.

Based on estimated environmental exposure levels, toxaphene does not appear to constitute a major threat to warm-blooded animals, including migratory birds and other wildlife, domestic poultry and livestock, small laboratory mammals, and humans. Wildlife typically contains low or non-detectable levels of toxaphene, except for some species of fish-eating raptors, and the frequency of occurrence is low when compared with that of other organochlorine agricultural compounds. However, toxaphene has been implicated as a human carcinogen and mutagen at relatively high test dosages and was associated with some bird kills following aerial applications.

In water, the concentration of toxaphene considered safe for protection of freshwater life is conservatively estimated to lie between 0.008 and 0.013 μg/L; for marine life, it is <0.07 μg/L. This is in sharp contrast to the current recommended drinking water criterion for human health protection of 5.0–8.8 μg/L. Similarly, residues in fish tissue in excess of 0.4–0.6 mg/kg wet weight may be hazardous to fish health and should be considered as presumptive evidence of significant environmental contamination, although fish may contain up to 5.0 mg/kg before they are considered hazardous to human consumers. Toxaphene criteria for human health protection, which range in various foods from 0.1 mg/kg for sunflower seeds to 7.0 mg/kg in meat, fats, and citrus fruits, also appear adequate to safeguard sensitive species of wildlife.

ZINC[a]

Chapter 33

33.1 Introduction

Zinc is an essential trace element for all living organisms. As a constituent of more than 200 metalloenzymes and other metabolic compounds, zinc assures stability of biological molecules, such as DNA, and of biological structures, such as membranes and ribosomes. Plants do not grow well in zinc-depleted soils, and deficiency has resulted in large losses of citrus in California and pecans in Texas. Clinical manifestations of zinc deficiency in animals include growth retardation, testicular atrophy, skin changes, and poor appetite. The ubiquity of zinc in the environment would seem to make human deficiencies unlikely; however, reports of zinc-associated dwarfism and hypogonadism in adolescent males are now confirmed, and reflect the fact that much of their dietary zinc is not bioavailable. Zinc deficiency was a major factor in the syndrome of nutritional dwarfism in adolescent males from rural areas of Iran and Egypt in 1961; about 3% of the population in these areas were affected – and a similar syndrome was found in Turkey, Tunisia, Morocco, Portugal, and Panama. The use of unleavened bread as a major staple food contributed to severe zinc deficiency in the Middle East. Unleavened bread may contain adequate amounts of zinc for nutrition and intakes may exceed recommended allowances by a wide margin; however, zinc is largely unavailable for absorption due to the high levels of fiber and phytic acid esters in unleavened bread. Marginal deficiency of zinc in man is probably widespread and common throughout the world, including the United States. Dietary zinc replacement usually will reverse the pathologic events of zinc depletion in humans and animals. But zinc repletion seems to be of little value in rat offspring with congenital malfunctions or behavioral abnormalities associated with zinc depletion.

Zinc poisoning has been documented in dogs, cats, ferrets, birds, cattle, sheep, and horses, usually as a result of ingesting galvanized metal objects, certain paints and fertilizers, zinc-containing coins, and skin and sunblock preparations containing zinc oxide. Signs of acute poisoning include anorexia, depression, enteritis, diarrhea, decreased milk yield, excessive eating and drinking and, in severe cases, convulsions and death. Emissions from zinc smelters at Palmerton, Pennsylvania, have destroyed wildlife habitat, reduced prey abundance, poisoned deer, songbirds, and shrews, and eliminated terrestrial amphibians from the mountainside at Lehigh Gap. Aquatic populations are frequently decimated in zinc-polluted waters. Zinc in the aquatic environment is of particular

---

[a]All information in this chapter is referenced in the following sources:

Eisler, R. 1980. Accumulation of zinc by marine biota. Pages 259–351 in J.O. Nriagu, ed. *Zinc in the Environment. Part II. Health Effects*. John Wiley, New York.

Eisler, R. 1981. *Trace Metal Concentrations in Marine Organisms*. Pergamon Press, Elmsford, New York, 697 pp.

Eisler, R. 1993. Zinc hazards to fish, wildlife, and invertebrates: a synoptic review. *U.S. Fish Wildl. Serv. Biol. Rep. 10*, 106 pp.

Eisler, R. 1997. Zinc hazards to plants and animals with emphasis on fishery and wildlife resources. Pages 443–537 in P.N. Cheremisinoff, ed. *Ecological Issues and Environmental Impact Assessment. Advances in environmental control technology series*. Gulf Publ. Co., Houston, Texas.

Eisler, R. 2000. Zinc. Pages 605–714 in *Handbook of Chemical Risk Assessment: Health Hazards to Humans, Plants, and Animals. Volume 1, Metals*. Lewis Publishers, Boca Raton, Florida.

importance because the gills of fish are physically damaged by high concentrations of zinc.

Zinc toxicosis in humans is not a common medical problem, although it may appear in some metal workers and others under special conditions. Industrial processes, such as welding, smelting, or fabrication of molten metals can produce ultrafine metal oxides at harmful concentrations. Inhalation of these metal oxides, including oxides of zinc, causes the industrial malady known as *metal fume fever*. Symptoms occur several hours after exposure and include fever, chills, perspiration, tachycardia, dyspnea, and chest pains. Recovery is normally complete within 24 h, but susceptible workers can have persistent pulmonary impairment for several days after exposure. Most reports of human zinc intoxication have been in response to food poisoning incidents arising from lengthy storage of acidic foods or beverages in galvanized containers.

Historically, zinc has been used by humans for industrial, ornamental, or utilitarian purposes for nearly 2000 years, and may have been used as an ointment to treat skin lesions by the ancient Egyptians and other Mediterranean people. In biblical times, the Romans were known to have produced brass by mixing copper with a zinc ore. In its isolated form, zinc was not recognized until the 15th century when smelting occurred accidentally. The Chinese probably were the first to extract zinc metal, although its first description in 1597 by an occidental traveler, Liborius, related that the process was observed in India. Commercial smelting began in the 18th century, when it was realized that zinc could be obtained from the calamine ore used to make brass; no reports of zinc toxicosis in any form were recorded from these early accounts. The first documented use of zinc administered orally was in 1826 to treat discharges from various body orifices. Zinc was recognized as an essential nutrient for plants and animals in 1869. Its occurrence in biological matter, e.g., human liver, was first described in 1877. In 1934, zinc was conclusively demonstrated to be essential to normal growth and development in animals.

Zinc composes 0.004% of the earth's crust and is twenty-fifth in order of abundance of the elements. Uses of zinc include production of noncorrosive alloys, galvanizing steel and iron products, and in the therapeutic treatment of zinc deficiency. Zinc is found in coal and many manufactured products such as motor oils, lubricants, tires, and fuel oil. Ecological and toxicological aspects of zinc in the environment have been reviewed by many authorities, and is the main thrust of this chapter.

33.2 Sources and Uses

World production of zinc increased from 0.5 million metric tons in 1900 to 6.1 million tons in 1978 to 7.1 million tons in 1987. The principal ores of zinc are sulfides, such as sphalerite and wurtzite. The major world producers include Canada, the former Soviet Union, and Japan – which collectively account for about half the production – and, secondly, the United States, Australasia, Mexico, and Peru. Zinc is now available as ingots, lumps, sheets, wire, shot, strips, granules, and powder. The United States produced 240,000 tons of zinc in 1987 – mostly from Tennessee, Mississippi, and New York, but also 16 other states – and imported an additional 774,000 metric tons, thus consuming 14% of world zinc production while producing 3.4%.

Zinc is mainly used in the production of noncorrosive alloys, brass, and in galvanizing steel and iron products. Zinc undergoes oxidation on the surface, thus protecting the underlying metal from degradation. Galvanized products are widely used in construction materials, automobile parts, and household appliances. Zinc oxide is used to form white pigments, in rubber processing, and to coat photocopy paper. Zinc sulfate is used as a cooperative agent in fungicides and as a protective agent against zinc deficiency in soils. When incorporated with copper compounds or arsenic–lead wettable powders and applied by spraying, it can minimize the toxic effects of these metals on fruits, such as plum, apple, and peach; in Japan alone, about 250 tons of zinc sulfate is sprayed in fields each year. Zinc is used therapeutically in human medicine in the treatment of zinc deficiency, various skin diseases, wound healing, and to reduce pain in sickle cell anemia patients.

Zinc is discharged into the global environment at a yearly rate estimated at 8.8 million tons; 96% of the total is a result of human activities. Major sources of anthropogenic zinc discharges to the environment include electroplaters, smelting and ore processors, drainage from active and inactive mining operations, domestic and industrial sewage, combustion of fossil fuels and solid wastes, road surface runoff, corrosion of zinc alloys and galvanized surfaces, and erosion of agricultural soils. During smelting, large amounts of zinc are emitted into the atmosphere. In the United States alone during 1969, about 50,000 metric tons of zinc were discharged into the atmosphere during smelting operations. Another 20,000 metric tons are discharged annually into U.S. estuaries. Zinc is also dispersed from corroded galvanized electrical transmission towers for at least 10 km by runoff, and by wind-driven spray and water droplets from the towers. Discharges from placer mining activities usually contain 75.0–165.0 μg Zn/L, sometimes up to 882.0 μg Zn/L in active mines, and these concentrations may represent acute hazards to salmonids in areas downstream of placer mine effluents. In Maine, galvanized culverts significantly increased zinc concentrations in stream waters, particularly in newer culverts. Highest zinc concentrations in culverts were found during conditions of elevated temperatures and low flow; levels of zinc sometimes exceeded the avoidance threshold (0.05 mg/L) of Atlantic salmon (*Salmo salar*); invertebrates seemed unaffected, except for a freshwater sponge, *Spongilla* sp. Zinc sources implicated in livestock poisonings include galvanized iron wire and troughs, and zinc-containing fertilizers and fungicides. Zinc toxicosis in humans has been reported from consumption of milk stored in galvanized vessels, and from food contaminated with particles of zinc from a zinc pigment plant. Zinc toxicity is discussed in greater detail later.

33.3 Chemical and Biochemical Properties

Most of the zinc introduced into aquatic environments eventually is partitioned into the sediments. Zinc release from sediments is enhanced under conditions of high dissolved oxygen, low salinity, and low pH. Dissolved zinc usually consists of the toxic aquo ion ($(Zn(H_2O)_6)^{2+}$) and various organic and inorganic complexes. Aquo ions and other toxic species have their greatest biological impacts under conditions of comparatively low pH, low alkalinity, low dissolved oxygen, and elevated temperatures. Zinc has its primary metabolic effect on zinc-dependent enzymes that regulate RNA and DNA. Low molecular weight proteins, metallothioneins, play an important role in zinc homeostasis and in protection against zinc poisoning in animals; zinc is a potent inducer of metallothioneins. The pancreas seems to be a primary target organ of zinc intoxication in birds and mammals, followed by bone; in fish, gill epithelium is the primary target site. Biological effects of excess zinc are modified by numerous variables, especially by interactions with other chemicals. Interactions frequently produce radically altered patterns of accumulation, metabolism, and toxicity, some of which are beneficial to the organism while others are harmful.

33.3.1 Chemical Properties

Zinc is a bluish-white metal which dissolves readily in strong acids. In nature it occurs as a sulfide, oxide, or carbonate. In solution, it is divalent and can form hydrated Zn^{2+} cations in acids, and zincated anions – probably $Zn(OH)_4^{-2}$ – in strong bases. Zinc dust and powder are sold commercially under a variety of trade names: Asarco, Blue powder, CI 77949, CI pigment metal 6, Emanay zinc dust, granular zinc, JASAD Merrillite, L15, and PASCO. Selected physical and chemical properties of zinc, zinc chloride, and zinc sulfate are listed in Table 33.1.

Zinc ligands are soluble in neutral and acidic solutions, so that zinc is readily transported

Table 33.1. Some properties of zinc, zinc chloride, and zinc sulfate.

Property	Zinc	Zinc Chloride	Zinc Sulfate
FORMULA	Zn	$ZnCl_2$	$ZnSO_4$
CAS NUMBER	7440-66-6	7646-85-7	7733-02-0
MOLECULAR WEIGHT	65.38	136.29	161.44
MELTING POINT, °C	419.5	290	Decomposes at 600
BOILING POINT, °C	908	732	–
DENSITY	7.14	2.907	3.54
PHYSICAL STATE	Bluish-white lustrous solid	Solid white granules	Colorless solid
SOLUBILITY	Insoluble in water, soluble in acetic acid and alkali	61.4 g/L water, 769.0 g/L alcohol, 500.0 g/L glycerol	Soluble in water, slightly soluble in alcohol

in most natural waters. But zinc oxide, the compound most commonly used in industry, has a low solubility in most solvents. Zinc mobility in aquatic ecosystems is a function of the composition of suspended and bed sediments, dissolved and particulate iron and manganese concentrations, pH, salinity, concentrations of complexing ligands, and the concentration of zinc. In freshwater, zinc is most soluble at low pH and low alkalinity: 10.0 mg Zn/L of solution at pH 6 that declines to 6.5 at pH 7, 0.65 at pH 8, and 0.01 mg/L at pH 9. Dissolved zinc rarely exceeds 40.0 μg/L in Canadian rivers and lakes; higher concentrations are usually associated with zinc-enriched ore deposits and anthropogenic activities. Marine waters usually contain <10.0 μg Zn/L, most adhering to suspended solids; however, saturated seawater may contain 1.2–2.5 mg Zn/L.

In water, the free zinc ion is thought to coordinate with six water molecules to form the octahedral aquo ion $(Zn(H_2O)_6)^{2+}$ in the absence of other complexing or adsorbing agents. In freshwater, zinc exists almost exclusively as the aquo ion at pH >4 and <7. In freshwater at pH 6, the dominant forms of dissolved zinc are the free ion (98%) and zinc sulfate (2%); at pH 9 the dominant forms are the monohydroxide ion (78%), zinc carbonate (16%), and the free ion (6%). In typical river waters, 90% of the zinc is present as aquo ion, and the remainder consists of $ZnHCO_3^+$, $ZnCO_3$, and $ZnSO_4$.

Zinc bioavailability and toxicity to aquatic organisms are highest under conditions of low pH, low alkalinity, low dissolved oxygen, and elevated temperatures. Soluble chemical species of zinc are the most bioavailable and most toxic. The aquo ion predominates over other dissolved species and is suspected of being most toxic; however, aquo ion concentrations decrease under conditions of high alkalinity, at pH >7.5, and increasing salinity. Under conditions of high alkalinity and pH 6.5, the most abundant species are $ZnHCO_3^+$, Zn^{2+}, and $ZnCO_3$; at low alkalinity and an elevated pH of 8.0, the order of abundance is $Zn^{2+} > ZnCO_3 >$ zinc humic acid $> ZnOH^+ > ZnHCO_3^+$. Water hardness is the principal modifier of acute zinc toxicity. Increased alkalinity or water hardness results in decreased toxicity to freshwater organisms when entire zinc is dissolved; this effect is associated with decreased concentration of aquo ions and is heightened by increased pH. Increased water hardness at pH <8.5 when zinc is in suspension results in increased toxicity associated with increased suspended $ZnCO_3$. Increased water hardness at pH >8.5 when zinc is in suspension produces decreased toxicity and increased suspended $Zn(OH)_2$. Suspended zinc carbonate may also be toxic, although its toxicity decreases under conditions suitable to zinc hydroxide formation; suspended $Zn(OH)_2$ is relatively nontoxic. Thus, $ZnCO_3$ composes <1% of the dissolved zinc at low pH and low alkalinity but is the predominant chemical species at high pH and high alkalinity. Organozinc complexes are not stable, and under reducing conditions may dissociate, liberating Zn^{2+}.

In seawater, zinc exists in a dissolved state, as a solid precipitate, or adsorbed to particle surfaces. Soluble zinc in seawater exists as uncomplexed free (hydrated) ions, as inorganic complexes (the primary form in the open sea), or as organic complexes. In seawater at pH 8.1, the dominant species of soluble zinc are zinc hydroxide (62%), the free ion (17%), the monochloride ion (6.4%), and zinc carbonate (5.8%). At pH 7, the percentage of dissolved zinc present as the free ion increases to 50%. In the presence of dissolved organic materials, most of the dissolved zinc is present as organozinc-complexes. In estuaries and other marine environments, the relative abundance of zinc species changes with increasing salinity. At low salinities, $ZnSO_4$ and $ZnCl^+$ predominate; at higher salinities, the aquo ion predominates. But as salinity decreases, the concentration of free zinc ion increases and the concentration of zinc–chloro complexes decreases, resulting in increased bioavailability of the free metal ion and increased bioconcentration by resident organisms.

In solution, zinc is adsorbed by organic agents such as humic materials and biogenic structures (i.e., cell walls of plankton), and by inorganic adsorbing agents, such as mineral particles, clays, and hydrous oxides of manganese, iron, and silicon. Particulate materials in the medium may contain as little as 2% and as much as 100% of the total zinc. Formation of zinc–ligand complexes increases the solubility of zinc and probably increases the tendency for zinc to be adsorbed. Sorption to particulates was lower at higher salinities due to displacement of sorbed zinc ions by alkali and alkaline earth cations. Increased pH increases zinc sorption to particulates, and seems to be independent of water salinity or hardness.

Most of the zinc introduced into aquatic environments is sorbed onto hydrous iron and manganese oxides, clay minerals, and organic materials, and eventually is partitioned into the sediments. Zinc is present in sediments as precipitated zinc hydroxide, ferric and manganic oxyhydroxide precipitates, insoluble organic complexes, insoluble sulfides, and other forms. As the sediments change from a reduced to an oxidized state, soluble zinc is mobilized and released; however, the bioavailability of different forms of sediment zinc varies substantially and the mechanisms of transfer are poorly understood. Sorption to sediments was complete at pH >7, but was negligible at pH <6. Zinc is dissolved from sediments at low salinities due to displacement of adsorbed zinc ions by alkali and alkaline earth cations which are abundant in brackish waters. Sulfide precipitation in sediments is an important control on zinc mobility in reducing environments, and precipitation of the hydroxide, carbonate, or sulfate may occur when zinc is present in high concentrations.

Extractable concentrations of sediment-bound zinc were positively correlated with zinc concentrations in deposit-feeding clams. Availability of sediment zinc to bivalve mollusks was higher at increased sediment concentrations of amorphous inorganic oxides or humic substances, and lower at increased concentrations of organic carbon and ammonium-acetate-soluble manganese. Zinc uptake by euryhaline organisms was enhanced at low water salinity.

33.3.2 Metabolism

Zinc is ubiquitous in the tissues of plants and animals, and is essential for normal growth, reproduction, and wound healing. More than 200 different enzymes require zinc for maximum catalytic activity, including carbonic anhydrase, alkaline phosphatase, alcohol dehydrogenase, acid phosphatase, lactic dehydrogenase, carboxypeptidase, and superoxide dismutase. Zinc has its primary effect on zinc-dependent enzymes that regulate the biosynthesis and catabolic rate of RNA and DNA. Zinc exerts a protective effect on liver by inhibiting lipid peroxidation and stabilizing lysosomal membranes; aids neurotransmission in brain of fish, birds, reptiles, and mammals;

prolongs muscular contractions, increases oxygen affinity of myoglobin, and is necessary for the growth and differentiation of muscle fiber types; increases numbers and birthweights of lambs of zinc-supplemented ewes; is essential for wound healing in most organisms studied; and is used therapeutically in treating patients with skin diseases, zinc deficiency, and other symptoms.

Zinc enters the gastrointestinal tract as a component of low molecular weight proteins secreted by the salivary glands, intestinal mucosa, pancreas, and liver. Usually, only dissolved zinc is sorbed or bound. But zinc dissolution probably occurs in the alimentary tract of animals after ingestion of particulates containing undissolved zinc. After ingestion, zinc is absorbed across several physiologically active membranes: gut mucosa, alveocapillary membranes, and tissue and organ membranes. The exact transport mechanisms are unknown but may be associated with formation of a tetrahedral quadredentate ligand with a small organic molecule. Some of the zinc taken up by the intestinal epithelial cells is rapidly transferred to the portal plasma where it associates with albumin, α_2 macroglobulin, and amino acids; about 67% of the zinc in plasma is bound to albumin, and about 3% is stored in liver. Soluble organozinc complexes are passively absorbed across the plasma membrane of the mucosa of the intestinal villi; the soluble, nondiffusable complexes are transported in the intestinal products and excreted in feces. Zinc loss from urine and sweat is usually small. In a normal human adult, about 2.0 g of zinc is filtered by the kidneys daily and about 0.3–0.6 mg is actually excreted each day. Zinc homeostasis in rats, unlike most mammals, is maintained by zinc secretion from the intestines rather than by regulation of zinc absorption. Initial uptake of zinc from the rat gastrointestinal tract involves binding to albumin, and transport of the zinc–albumin complex from intestine to liver.

Foods rich in zinc include red meat, milk, gelatin, egg yolks, shellfish, liver, whole grain cereals, lentils, peas, beans, and rice. About 20–30% of zinc in the diet is absorbed, but this is highly variable and ranges from <10% to >90%. Increased zinc absorption, for example, was associated with low body weight (BW), poor zinc status, and various prostaglandins; decreased absorption was caused by excess dietary calcium or phytate, and by a deficiency of pyridoxine or tryptophan. The half-time persistence of zinc in most mammalian tissues ranges between 100 and 500 days; it is longer in bone and muscle, and shorter in liver.

Metallothioneins play an important role in metal homeostasis and in protection against heavy metal toxicity in vertebrates and invertebrates. Metallothioneins are cysteine-rich (>20%), low (about 6000) molecular weight proteins, with high affinity for copper, silver, gold, zinc, copper, and mercury. These heat-stable, metal-binding proteins are found in all vertebrate tissues and are readily inducible by a variety of agents, to which they bind through thiolate linkages. Zinc is a potent inducer of metallothioneins, and a redistribution of zinc from enzymes to metallothioneins is one way to maintain low intracellular zinc concentrations. Metallothioneins also serve as temporary storage proteins for zinc and other metals during early development and may function by maintaining the pool of available zinc at an appropriate concentration. Metallothioneins are quite similar among organisms, that is, all metallothioneins are small proteins of molecular weight 6000–10,000, rich in sulfur and cysteine, and lacking aromatic amino acids. Metallothioneins isolated from cattle, sheep, horses, pigs, and other livestock, contain 61 amino acids; thioneine, the metal-free protein is a single chain polypeptide with a molecular weight of about 6000. Chicken thioneine consists of 63 amino acids, including histidine, an amino acid not present in mammalian metallothioneins. The unusually high cysteine content enables metallothioneins to selectively bind up to 7 zinc and 12 copper atoms per mole of protein.

Metallothioneins are involved in zinc homeostasis in chick, rat, and calf. When zinc is present at high dietary concentrations, a temporary zinc storage protein aids in counteracting zinc toxicity. Zinc absorption in mice is directly proportional to intestinal metallothionein levels and implies a significant role for metallothionein in zinc absorption. Chick embryo hepatic metallothionein is highly responsive to exogenous zinc introduced into the yolk, and increases in a dose-dependent manner; a similar pattern is evident in turkey development. Zinc protects against subsequent exposure to zinc insult, and protection is believed to be mediated by metallothioneins. For example, preexposure of South African clawed frog (*Xenopus laevis*) tadpoles to 5.0 mg $ZnSO_4(7H_2O)$/L for 96 h resulted in no deaths during subsequent exposure to 15.0 mg Zn/L for 90 h vs. 45% dead in the non-treated group; at 20.0 mg Zn/L, 15% died in the pretreated group vs. 50% in the non-treated group. Metallothioneins are an important factor in zinc regulation during the period of exogenous vitellogenesis in rainbow trout (*Oncorhynchus mykiss*). In female rainbow trout, for example, metallothioneins maintain homeostasis of hepatic zinc during egg formation. In plaice (*Pleuronectes platessa*), a marine fish, intraperitoneal injection of zinc raised hepatic metallothionein-like species by a factor of 15; metallothionein levels remained elevated for the next four weeks.

In marine mollusks and crustaceans, excess zinc is usually sequestered by metal-binding proteins and subsequently transported to storage or detoxification sites; soluble proteins and amino acids may contain 20–70% zinc. Metallothioneins are actively involved in zinc regulation during normal growth processes in blue crab (*Callinectes sapidus*), as judged by a decrease in zinc content in hemolymph and digestive gland during molting.

Elevated metallothionein levels are not necessarily indicative of heavy metal insult. Thus, liver metallothionein levels in mice are elevated following acute stress or starvation; this effect is blocked by actinomycin D, a protein synthesis inhibitor. It is further emphasized that not all zinc-binding proteins are metallothioneins. Low molecular weight metal-binding proteins – not metallothioneins – were induced in snails and polychaete annelids in metals-contaminated environments. A high molecular weight protein fraction was detected in the plasma of laying turkey (*Meleagris gallopavo*) hens which bound significant amounts of zinc and which coeluted with vitellogenin; vitellogenin, a metalloprotein, from laying hens contained 0.54 mg Zn/kg protein. In rock oysters (*Saccostrea cuccullata*) collected near an iron-ore shipping terminal, some of the tissue zinc was bound to a high molecular weight (around 550,000) iron-binding protein called ferritin. Ferritin accounts for about 40% of the protein-bound zinc in rock oysters, and most probably in other bivalves containing elevated tissue levels of zinc, although this requires verification. In four species of sediment-feeding marine polychaete annelids, zinc was mainly associated with high molecular weight proteins, suggesting that metallothionein-like proteins may not be satisfactory for monitoring purposes and that other cytosolic components should be studied.

High zinc levels induce copper deficiency in rats and interferes with metabolism of calcium and iron. Excess zinc interferes with normal metabolism of pancreas, bone, gall bladder, and kidney in mammals, and gill in fishes. The pancreas is a target organ for zinc toxicity in birds and mammals. Pancreatic alterations are documented from experimentally produced zinc toxicosis in cats, sheep, dogs, calves, chickens, and ducklings, and naturally in sheep and calves. Pancreatic changes were limited to acinar cells, specifically cytoplasmic vacuolation, cellular atrophy, and eventually cell death. Zinc excess may cause stimulation of bone resorption and inhibition of bone formation in chicks, dogs, monkeys, and rats. By preferentially accumulating in bone, zinc induces osteomalacia – a softening of the bone caused by deficiency of calcium, phosphorus, and other minerals. Zinc plays a role in bone metabolism of ageing rats. Normally, the femoral zinc diaphysis content in rats increases from 50.0 to 150.0 mg/kg FW during the first three weeks of life and remains constant thereafter. Oral administration of zinc (5.0–20.0 mg/kg BW daily for 3 days to 28-week-old rats) increased alkaline phosphatase activity and calcium content in femur, and delayed bone deterioration in ageing rats. Its high affinity for electrons causes zinc to bind covalently to proteins, mostly at imidazole and cysteine residues. In mud puppy (*Necturus maculosus*), zinc blocks apical membrane anion exchange in gallbladder epithelium and blocks chloride channels in nerve and muscle cells. The slow onset and reversal of the effects suggest a covalent modification of the exchanger, or an effect requiring Zn^{2+} transport to the cell interior.

Zinc toxicity to aquatic organisms is dependent on the physical and chemical forms of zinc, the toxicity of each form, and the degree of interconversion among the various forms. Aquatic plants and fish are relatively unaffected by suspended zinc, but many aquatic invertebrates and some fishes may be adversely affected after ingesting enough zinc-containing particulates. Zinc toxicosis affects freshwater fishes by destruction of gill epithelium and consequent tissue hypoxia. Signs of acute zinc toxicosis in freshwater fishes include osmoregulatory failure, acidosis and low oxygen tensions in arterial blood, and disrupted gas exchange at the gill surface and at internal tissue sites. Zinc exerts a critical influence on mammalian and piscine immune systems. Lymphocytes from the pronephros of common carp (*Cyprinus carpio*) were transformed by various mitogenetic agents; zinc added to lymphocyte cultures enhanced thymidine incorporation, and inhibited the response of the mitogenetic agents – although Zn^{2+} was in itself toxic at these concentrations (650.0 μg Zn^{2+}/L).

33.3.3 Interactions

Zinc interacts with numerous chemicals, sometimes producing greatly different patterns of accumulation, metabolism, and toxicity when compared to zinc alone. Recognition of these interactions is essential to the understanding of zinc kinetics in the environment.

33.3.3.1 Cadmium

Cadmium–zinc interactions are typical in that sometimes they act to the organism's advantage and sometimes not, depending on the organism, its nutritional status, and other variables. Dietary cadmium accentuates signs of zinc deficiency in turkeys, chicks, rodents, and pigs. Chicks on a zinc-deficient diet showed an increased frequency of muscle and feather abnormalities when 40.0 mg of cadmium/kg of diet was added; however, supplementation of the diet with 200.0 mg Zn/kg for 14–15 days lessened or reversed the adverse effects of cadmium. But cadmium promotes the growth of zinc-limited phytoplankton. Substitution of trace metals or metalloenzymes could be a common strategy for phytoplankton in trace-metal impoverished environments, such as the ocean, and could result in an effective colimitation of phytoplankton growth by several bioactive elements. Zinc-deficient marine diatoms (*Thalassiosira weissflogi*), for example, can grow at 90% of their maximum rate when supplied with cadmium (which substitutes for zinc in certain macromolecules); cobalt can also substitute for zinc, although less efficiently than cadmium.

Zinc diminishes or negates the toxic effects of cadmium. Specifically, zinc protected embryos of the toad (*Bufo arenarum*) and other amphibian embryos against cadmium-induced developmental malformations. Zinc counteracted adverse effects of cadmium on limb regeneration and growth of the fiddler crab (*Uca pugilator*). Preexposure of a freshwater amphipod (*Gammarus pulex*) to 10.0 μg Zn/L for 2 weeks increased whole body zinc content from 74.0 to 142.0 mg/kg DW and protected against the toxic effects of subsequent cadmium exposure of 500.0 μg Cd/L for 96 h. In crickets (*Acheta domesticus*), excess zinc in diets of larvae protected against cadmium toxicity. Zinc protected rats (*Rattus* sp.) against the toxic effects of cadmium such as testicular lesions, reduced sperm counts, hepatotoxicity and lung damage. Zinc protected mouse (*Mus* sp.) embryos against cadmium toxicity. An effective protection ratio was 1 Cd:1 Zn for mouse embryos, but for free living embryos of the toads (*B. arenarum*), this ratio was 1 Cd:8 Zn. Zinc reversed the toxic action of cadmium on natural killer cells of mice: 500.0 mg Zn/L drinking water negated the toxic action of 50.0 mg Cd/L. The mechanisms of zinc protection against cadmium were variously attributed to metallothionein induction, enhanced detoxification rates of cadmium, and competition with cadmium for the same metalloenzyme sites.

Waterborne solutions of zinc–cadmium mixtures were usually additive in toxicity to aquatic organisms, including freshwater fishes and amphipods, and to marine fishes, copepods, and amphipods. However, mixtures of zinc and cadmium were less toxic than expected to *Daphnia magna*, as judged by acute lethality studies. Zinc exerted antagonistic effects on uptake of cadmium by gills of the freshwater clam (*Anodonta cygea*), but accelerated cadmium transport from gills towards internal organs. Cadmium uptake in tissues of *Anodonta* was reduced by about 50% during exposure for 16 weeks to water containing 25.0 μg Cd/L and 2.5 mg Zn/L. In a marine prawn (*Pandalus montagui*), cadmium exposure had no effect on tissue zinc levels, but zinc enhanced cadmium uptake in hepatopancreas at the expense of carcass. In marine fishes, cadmium was taken up more rapidly at elevated seawater zinc levels; however, zinc concentrations in fish tissues decreased with increasing tissue cadmium burdens, suggesting competition between these two metals for the same physiologically active site. Zinc concentrations in larval shrimp (*Palaemon serratus*), within its threshold regulation range of 75.0–525.0 μg Zn/L, were not affected by the addition of 100.0 μg of cadmium/L. In zebrafish (*Brachydanio rerio*), zinc did not affect cadmium uptake by whole body or gills, but did inhibit intestinal uptake and tended to increase gill cadmium elimination rates. Among marine vertebrates, cadmium is selectively accumulated over zinc.

In ducks, zinc selectively competes with cadmium on high and low molecular weight protein pools in kidney and liver. Once the high molecular weight protein pool is zinc-saturated, excess zinc is stored on metal binding proteins, with serious implications for waterfowl stressed simultaneously with cadmium and zinc. On the other hand, a cadmium-induced disease in bone collagen of chicks was prevented by zinc because of preferential accumulation of zinc.

33.3.3.2 Copper

Mixtures of zinc and copper are generally acknowledged to be more-than-additive in toxicity to a wide variety of aquatic organisms, including oyster larvae, marine fishes, freshwater fishes and amphipods, and marine copepods. But Zn–Cu mixtures were less-than-additive in toxicity to marine amphipods (*Allorchestes compressa*). Zinc added to the ambient water depressed copper accumulations in tissues of juvenile catfish (*Clarias lazera*), but copper added to the medium depressed zinc uptake. A similar situation was reported in barnacles (*Elminus modestas*); however, simultaneous exposure to copper and zinc resulted in enhanced uptake of both metals.

In higher organisms, zinc is a copper antagonist and potentiates the effects of nutritional copper deficiency in rats and chicks. This effect only occurs at extremely high Zn:Cu dietary ratios. The addition of copper to the diet of chicks or rats in physiological amounts counteracted all observed signs of zinc intoxication. No antagonism was evident between dietary copper and zinc fed to channel catfish (*Ictalurus punctatus*) fingerlings; therefore, the high levels of supplemental zinc required in practical feeds should not impair copper status, provided that normal dietary copper levels are present.

High levels of administered zinc limits copper uptake in humans and certain animals, and provides protection against toxicosis produced by copper in pigs and sheep. Excessive zinc in humans interferes with copper absorption from the intestine, resulting in copper deficiency, and eventually to cardiovascular diseases; high zinc intakes also decrease iron bioavailability, leading to a reduction of erythrocyte life span by 67%. Copper deficiency induced by excess dietary zinc is associated with lameness in horses, donkeys, and mules.

33.3.3.3 Lead

Lead–zinc mixtures were more-than-additive in toxicity to marine copepods and significantly delayed development of mud crab (*Rithropanopeus harrissii*) larvae. Lead is accumulated up to 10 times more rapidly by marine fishes at elevated zinc concentrations in seawater.

Among terrestrial animals, zinc protects against lead toxicosis. Dietary zinc reduced the toxic effects of dietary lead to larvae of the house cricket (*Acheta domesticus*) by inhibiting assimilation and stimulating fecal excretion. Zinc at 100.0–200.0 μg/egg (1.0 mg Zn/kg egg) significantly protected developing white leghorn chicks against lead (50.0 μg/egg)-induced deformities and death when injected into the yolk sac on the 7th day of incubation. Zinc also protects against lead toxicity in horses, and against testicular injury induced by lead in rats.

33.3.3.4 Nickel

Nickel–zinc mixtures were additive in toxicity to marine copepods and to the three-spined stickleback (*Gasterosteus aculeatus*). Oral nickel toxicity in chicks was prevented by increased dietary zinc. Nickel is a leading cause of allergic contact dermatitis (ACD) in many industrial nations; about 6% of the general public is sensitive to nickel and about 11% of dermatology clinic patients. Zinc prevents nickel-sulfate-induced ACD in guinea pigs (*Cavia* spp.) through addition of 100.0–200.0 mg Zn/L drinking water for 4 weeks prior to nickel insult. Nickel and other metals that cause ACD penetrate the skin, complex with selected ligands, and stimulate a delayed hypersensitivity. Zinc is thought to block the sites where nickel complexes to the protein.

33.3.3.5 Others

Zinc interacts with a wide variety of inorganic, organic, and biological agents, but in most cases the available information is fragmentary, and the mechanisms of action are unknown.

Mice pretreated with zinc at 6.5 mg Zn/kg BW for 9 days showed increased resistance to arsenic toxicosis over a 30-day observation period. Oral zinc therapy was effective in treating biological agents, such as infectious pododermatis in cattle, ovine foot rot in sheep, sporidesmin in sheep, cattle, and rodents, and the toxins of the fungus *Phomopsis leptostromiformis* in sheep. However, juvenile channel catfish (*Ictalurus punctatus*) fed zinc in the diet for 16 weeks at concentrations as high as 60.0 mg Zn/kg ration as zinc methionine or zinc sulfate were not protected against bacterial infections of *Edwardsiella ictaluri*. Elevated testicular zinc concentrations in bank voles (*Clethrionomys glareolus*) protect the testes from fluoride-induced histopathology caused by 200.0 mg F/L drinking water for 4 months. Calcium modifies zinc toxicity to freshwater aquatic organisms, with increased calcium associated with decreased acute toxicity. Zinc absorption in rat gut is decreased after ingestion of phosphorus as polyphosphate or orthophosphate plus high levels of calcium. Zinc cytotoxicity is blocked by increased calcium or iron, but not magnesium. Zinc reportedly protects rats against carbon tetrachloride poisoning.

Various chelating agents protect mice against zinc acetate poisoning, including disodium ethylene diamine tetraacetic acid (EDTA), disodium calcium cyclohexanediamine tetraacetate, D-penicillamine, 2,3-dimercapto-1-propane sulfonic acid, and 2,3-dimercaptosuccinic acid. Zinc protects toad embryos against agents known to produce malformations, including excess vitamin A, acetazolamide, calcium-EDTA, and acetaminophen. Venom of the jararaca (*Bothrops jararaca*), a venomous Brazilian serpent, contains a zinc metalloprotease called J protease; the proteolytic activity of J protease is inactivated by EDTA and other sequestering agents.

Chromium–zinc mixtures were more-than-additive in toxicity to *Tisbe holothuriae*, a marine copepod. Zinc in combination with chromium was more toxic to copepods than were mixtures of zinc with copper, lead, nickel, or cadmium. Renal tubular absorption of zinc in mice was impaired by certain diuretics and was further influenced by dietary proteins. Zinc absorption in rats was depressed after consumption of high levels of inorganic iron; absorption was normal with organoirons. Mercury–zinc mixtures were more-than-additive in toxicity to oyster larvae. Preexposure of mussels (*Mytilus edulis*) to 50.0 μg Zn/L for 28 days conferred increased tolerance to 75.0 μg inorganic mercury/L. Zinc inhibited the accumulation of mercury in marine snails and crustaceans.

Zinc deficiency places an increased demand on selenium (Se) pools in daphnids. As little as 5.0 μg Se/L in zinc-free water eliminated overt cuticle damage and substantially increased reproduction, but did not alter the shortened life span. Cladocerans at the threshold of selenium deficiency will become overtly selenium-deficient when zinc supplies are lacking. Insufficient copper introduces cuticle problems in daphnids similar to those introduced by insufficient zinc or selenium, increasing the likelihood of a proposed relation between glutathione peroxidase (which contains selenium), and copper–zinc superoxide dismutase. High levels of dietary tin increased zinc loss from rats. Zinc prevented toxic effects of vanadium (10.0 mg/kg BW) on bone metabolism of weanling rats.

33.4 *Carcinogenicity, Mutagenicity, and Teratogenicity*

Zinc can induce testicular sarcomas in birds and rats when injected directly into the testes, but zinc has not been shown to be tumorigenic by any other route. Zinc promotes tumor growth after conditions of zinc deficiency, but excess zinc may suppress or inhibit tumor proliferation, although the mechanisms of action are imperfectly understood. Chromosomal aberrations were observed under conditions of zinc deficiency, but excess zinc was not mutagenic in most tests. Organozinc compounds are effective mutagens when presented to susceptible cell populations in an appropriate form, but the evidence for inorganic zinc is incomplete. Zinc is teratogenic to frog and fish embryos, but conclusive evidence of teratogenicity in mammals is lacking. Zinc may protect against the effects of some mammalian teratogens. Under conditions of mild zinc deficiency, however, diabetes and effects of various teratogens are exacerbated.

33.4.1 Carcinogenicity

Carbamate esters of zinc, zineb, and ziram are carcinogenic and teratogenic in animals, but this is attributed to the action of the carbamate esters and not to zinc. Results of studies with small mammals showed zinc to be co-carcinogenic with 4-nitroquinoline-*N*-oxide on oral cancer, and with *N*-ethyl-*N*-nitrosourea on brain cancer. There is conclusive evidence that repeated intratesticular injections of zinc salts can induce testicular sarcomas in birds and rats. Testicular teratomas in roosters were first produced experimentally in 1926 when zinc salts were injected into the testes as a method of practical castration; tumors could be induced only by intratesticular injection during the spring period of gonadal growth. Teratomas of the testes were observed in fowl given testicular injections of 2 mL of 10% $ZnSO_4$ solution. Teratomas were induced in Japanese quail (*Coturnix japonica*) by intratesticular injections of 3% zinc chloride solutions during a period of testicular growth stimulated by increased photoperiod; tumors were similar to those of domestic fowl and have histological features in common with spontaneous testicular teratomas in humans. Testicular tumors in rats were produced by direct intratesticular injection of zinc; no other carcinogenic effects were produced by any other route, regardless of dose. It is emphasized that zinc and zinc compounds are not conclusively carcinogenic except when injected directly into the testes; no field or experimental evidence exists showing zinc to be tumorigenic through any other route.

Zinc is essential for the growth of rapidly proliferating cells, such as tumors. The high zinc requirements of these cells in tumor disease can result in latent zinc deficiency. Accordingly, growth of animal tumors is stimulated by zinc, and retarded by zinc deficiency. In mouse fibrosarcoma cells, zinc inhibits endonucleases, subsequently blocking DNA fragmentation and tumor cell lysis, allowing tumors to grow. There is no evidence that zinc deficiency causes cancer, although deficiency was associated with decreasing tumor growth. Malignant human tissues, for example, frequently contained less zinc than normal tissue, i.e., 78.0 mg/kg FW normal liver vs. 18.0 in cancerous liver.

Zinc can also inhibit tumor growth, although the mechanisms of zinc suppression of carcinomas are imperfectly understood. Zinc inhibits the growth of mouse melanoma cells at concentrations between 8.2 and 9.9 mg Zn/L culture medium. The addition of 100.0 mg $ZnSO_4$/L to drinking water of hamsters inhibited formation of dimethylbenzanthracene-induced carcinomas. Rats fed high zinc diets of 500.0 mg/kg ration had reduced growth of a chemically induced hepatoma. Intramuscular injections of zinc oxide or zinc acetate administered together with nickel sulfide – a potent muscle carcinogen – delayed, but did not prevent, 100% tumor incidence in rats over a 66-week observation period. Administration of zinc slows the carcinogenic process induced by nickel due to the production of water soluble and water insoluble zinc compounds, despite markedly different retention times in muscle of zinc compounds (Tb1/2 of ZnO = 24 days; Zn acetate = 2.5 days; Ni_3S_2 = 21 days). Zinc in either form exerted no measurable influence on nickel retention at the injection site or early local cellular reactions to nickel. Testicular tumors in rats caused by injection of cadmium were suppressed by zinc injection, provided that the Zn:Cd molar ratio was about 100:1. Zinc inhibition of cadmium carcinogenesis is a complex phenomenon, depending on dose, route, and target site. For example, the number of cadmium-induced testicular tumors in rats was reduced by 50% over a 2-year period after three subcutaneous injections of 65.4 mg Zn/kg BW given within 18 h of initial cadmium insult, although this group had a marked elevation in prostatic tumors when compared to controls; tumor number was reduced by 92% when rats were given 100.0 mg Zn/L in drinking water.

33.4.2 Mutagenicity

Results of mutagenicity studies with whole organisms were usually negative because homeostatic controls of absorption and protein binding preclude the likelihood of zinc being genotoxic under standard feeding conditions. However, zinc is an effective mutagen and clastogen when presented to a susceptible cell population in an appropriate form. Zinc acetate produced dose-related positive responses in the mouse lymphoma assay, and also in a cytogenetic assay with Chinese hamster ovary cells; however, results of mutagenicity assays with inorganic zinc were negative in the *Salmonella* mutation assay, and in unscheduled DNA synthesis on primary cultures of rat hepatocytes. Organozinc compounds have mutagenic potential, as judged by the positive responses with zinc 2,4-pentanedione and *Salmonella.*

Structural chromosome aberrations, particularly chromatid gaps and increased frequency of fragment exchange, were observed in rat bone marrow cells after 14 days of exposure to 240.0 mg Zn/L drinking water. Chromosomal aberrations were observed in bone marrow cells of mice fed diets equivalent to 650.0 mg Zn/kg BW daily, in mice exposed to zinc oxide by inhalation, and in mice maintained on a low calcium diet. Aberrations in bone marrow of mice given 5000.0 mg Zn/kg diet may be associated with calcium deficiency. Calcium is displaced by zinc in calcium-depleted conditions, leading to chromosomal breaks and interference in the repair process.

Zinc chloride induces chromosomal aberrations in human lymphocytes in vitro. A higher incidence of chromosome anomalies in leukocytes occurs among workers exposed to zinc, but these aberrations are likely due to other (unspecified) mutagenic factors in the work environment. Zinc inhibits the mutagenic action of some carcinogens because it is a constituent of mutagen detoxifying enzymes, or because it acts directly on the microsomal monooxygenases forming the ultimate carcinogen. Zinc significantly reduced a genotoxic effect of lead in rat bone marrow cells (500.0 mg Pb/L drinking water followed by 240.0 mg Zn/L for 2 weeks), and also protects against lead accumulations in erythrocytes and Pb-induced inhibition of delta-amino levulinic acid dehydratase. Zinc deficiency can lead to chromosomal aberrations, but excess zinc was not mutagenic in the majority of tests for DNA damage – except for zinc-containing fungicides wherein the organic dithiocarbamate constituents were the mutagenic agents, and for zinc chromate wherein the chromate ion was the active agent. Frequencies of sister chromatid exchanges (SCE) in calves with hereditary zinc deficiency, also known as lethal trait A46, are lower than in healthy normal cows, suggesting a fundamental association between disturbed zinc metabolism and the low incidence of SCE in A46 cattle.

33.4.3 Teratogenicity

Excess zinc is teratogenic to frog and fish embryos, possibly by inhibition of DNA synthesis. Zinc at 150.0 mg/kg in rat diets was associated with inhibited fetal implantation, but this needs confirmation. No conclusive evidence now exists demonstrating that excessive zinc produces any teratogenic effect in mammals. Excess zinc may protect against some teratogens, such as calcium EDTA. Also, teratogenic effects of cadmium salts in golden hamsters were reduced by simultaneous administration of zinc salts.

Zinc deficiency is clearly teratogenic in mammals. Severe maternal zinc deficiency is known to be teratogenic in rats. Fetal malformations – especially calcification defects – due to maternal zinc deficiency affect almost every tissue. Skeletal malformations are most common, possibly due to a reduction in cellular proliferation and in activity of bone alkaline phosphatase. Human zinc deficiency may act teratogenically, either directly or indirectly via other toxic agents. Zinc deficiency may exacerbate effects of several teratogenic agents, such as thalidomide; there is also the possibility that zinc deficiency may increase the incidence of spina bifida and anencephaly, but this needs verification. Diabetes during pregnancy can amplify the effects of a mild maternal zinc deficiency. In one study, diabetic and nondiabetic rat strains were fed a low zinc diet (4.5 mg Zn/kg diet), an adequate zinc diet (24.5 mg/kg) or a high zinc diet (500.0 mg/kg) throughout gestation. Fetuses from diabetic dams were smaller, weighed less, and had less calcified skeletons and more malformations than did fetuses from control dams. In controls, maternal dietary zinc had a minor effect on fetal malformation frequency. In diabetic strains, however, the low zinc diet had a strong teratogenic effect.

33.5 *Concentrations in Field Collections*

Total zinc concentrations in nonbiological samples seldom exceed 40.0 μg/L in water, 200.0 mg/kg in soils and sediments, or 0.5 $\mu g/m^3$ in air. Environments heavily contaminated by anthropogenic activities may contain up to 99.0 mg Zn/L in water, 118.0 g/kg in sediments, 5.0 g/kg in soil, and 0.84 $\mu g/m^3$ in the atmosphere. Zinc measurements in field collections of plants and animals show several trends: (1) zinc is present in all tissues of all organisms measured; (2) concentrations are elevated in organisms near anthropogenic point sources of zinc contamination; (3) concentrations are normally grossly elevated (>4.0 g/kg FW soft parts) in some species of bivalve mollusks and barnacles; (4) zinc-specific sites of accumulation include the frond in algae; kidney in mollusks; hepatopancreas in crustaceans; jaws in polychaete annelids; viscera, gonad, and brain in fishes; liver, kidney, and bone in birds; and serum, pancreas, feces, liver, kidney, and bone in mammals; (5) interspecies variations in zinc content are considerable, even among species closely related taxonomically; (6) intraspecies differences in zinc content vary with age, size, sex, season, and other modifiers; and (7) many species regulate zinc within a threshold range of concentrations.

33.5.1 Abiotic Materials

Zinc concentrations in freshwater, seawater, groundwater, sewage sludge, sediments, and soils are considered reliable but may be excessively high in some cases. Newer clean laboratory techniques suggest that dissolved zinc concentrations in nonpolluted rivers may be 10–100 times lower than previously reported.

Zinc concentrations in water seldom exceed 40.0 μg/L except near mining, electroplating, and similar activities – where concentrations between 260.0 and 954.0 μg/L have frequently been recorded. Drinking water usually contains <10.0 μg Zn/L, although concentrations >2.0 mg/L may occur after passage through galvanized pipes. Zinc-contaminated streams within the Platt River Basin sometimes contain up to 99.0 mg Zn/L, and in Arkansas up to 79.0 mg/L. Zinc concentrations in water downstream of placer mining activities in Alaska sometimes exceed the concentrations found to be toxic to Arctic grayling, *Thymallus arcticus*. The disappearance of stone loach (*Noemacheilus barbatulus*) in the UK from streams receiving industrial wastes was attributed directly to zinc concentrations in the stream rising from 1.0 mg/L to a lethal 5.0 mg/L.

Concentrations of zinc in sediments and soils usually do not exceed 200.0 mg/kg, but can range between 3.0 and 118.0 g/kg as a result of human activities. Atmospheric zinc levels were almost always <1.0 μg/m^3, although they tended to be higher over industrialized areas. Average zinc concentrations, in μg/m^3 atmosphere, were <0.001 at the South Pole, 0.01–0.02 in rural areas of the United States, <0.01–0.84 in U.S. cities, and 0.06–0.35 at various locations in the United Kingdom.

33.5.2 Terrestrial Plants and Invertebrates

Zinc concentrations in forest plants vary considerably. In oaks (*Quercus* spp.), for example, some species are accumulators whereas others may be termed discriminators. For individual species, zinc concentrations tend to follow the pattern of roots > foliage > branch > trunk. Small lateral roots accumulate zinc to much greater levels than other vegetation components and are probably most sensitive to changes in zinc inputs. Half-time persistence of zinc in forest ecosystems varies from about 3 years in organic matter components to >200 years for large soil pools.

Terrestrial plants growing beneath corroded galvanized fencing have been poisoned by zinc. Vegetables are relatively low in zinc, but growing plants can accumulate zinc applied to soils. High soil level of zinc is the primary cause of vegetation damage near zinc smelters. Elevated zinc concentrations in soils near zinc smelters inhibit seedling root elongation and probably prevent establishment of invader species in denuded areas. Lichen species richness and abundance were reduced by about 90% in lichen communities near a Pennsylvania zinc smelter; elevated zinc concentrations were the probable cause of the impoverished lichen flora. Soils and vegetation surrounding zinc smelters in Palmerton, Pennsylvania, were grossly contaminated with zinc, cadmium, and lead. Zinc was primarily responsible for the destruction of trees and subsequent erosion of the soil, reductions in moss and lichen flora, reductions in litter arthropod populations, and reductions in species diversity of soil fungi and bacteria; zinc residues were elevated in slugs and millipedes. Soil litter invertebrates were rare or absent 2 km downwind of the smelter; invertebrates collected up to 10 km upwind of the smelters had significantly elevated zinc concentrations when compared to soil litter invertebrates from more distant sites.

The maximum zinc concentration in earthworms collected from a contaminated site was 1600.0 mg/kg DW whole animal; for uncontaminated sites it was 650.0 mg/kg. Whole body zinc concentrations in earthworms (*Dendrodrilus rubidus*, *Lumbricus rubellus*) tended to reflect zinc concentrations in soil, although zinc accumulations in both species seem to be physiologically regulated when soil zinc values exceeded 1000.0 mg/kg DW. Whole body zinc content of terrestrial isopods seems to reflect soil zinc levels and may be a useful indicator of soil contamination. *Porcellio scaber*, a terrestrial isopod known as a woodlouse, is recommended as a biological indicator of zinc contamination because of the positive correlation between zinc content in soil or leaf litter and woodlouse hepatopancreas. Zinc content in *Porcellio*, litter, and soil near a zinc smelter, in mg/kg DW, was >1000.0

in whole isopod, >9000.0 in hepatopancreas, >10,000.0 in litter, and >50,000.0 in soil.

Interspecies variability in zinc content of terrestrial invertebrates is large, and governed by numerous modifiers. For example, whole body zinc content in closely related species of terrestrial gastropods collected from a single contaminated site ranged between 600.0 and 1200.0 mg/kg DW. Zinc was highest in gray field slugs (*Deroceras reticulatum*) in late spring, lowest in summer, and positively correlated with tissue cadmium concentrations; starvation for 16 days had no effect on body zinc concentrations. Zinc tends to concentrate in mechanical structures of various invertebrates, such as mandibular teeth. High concentrations of zinc are reported in jaws of polychaete worms, cutting edge of the mandibles of herbivorous insects, mandibles of various species of beetles, copepod mandibles, chaetognath teeth and spines, mandibular teeth of ants, and fangs of spiders. Honeybees (*Apis mellifera*) collected near a lead smelting complex at East Helena, Montana, had depressed whole body zinc concentrations despite increased ambient air zinc values; however, whole body burdens of arsenic, cadmium, copper, and lead were significantly elevated and this may have influenced zinc kinetics. Also, pollen was usually the most indicative source of zinc and other heavy metals in bees.

33.5.3 Aquatic Organisms

Concentrations of zinc in tissues of aquatic organisms are usually far in excess of that required for normal metabolism. Much of the excess zinc is bound to macromolecules or present as insoluble metal inclusions in tissues. Diet is the most significant source of zinc to aquatic organisms, and is substantially more important than uptake from seawater. In general, zinc concentrations in sediments and tissues of aquatic organisms are elevated in the vicinity of smelters and other point sources of zinc, and decrease with increasing distance.

Freshwater algae in Canadian mine tailing environments heavily concentrate zinc and other metals, and may retard metal dispersion through the water column. Zinc levels in field collections of marine algae and macrophytes are usually at least several orders of magnitude higher than zinc concentrations in the surrounding seawater. In general, concentrations in marine aquatic flora were high when seawater zinc concentrations were elevated, although the relation was not linear. Marine flora, especially red and brown algae, are among the most effective marine zinc accumulators. Increasing accumulations of zinc in marine algae were associated with decreasing light intensity, decreasing pH, increasing temperature, decreasing levels of DDT, and increasing oxygen. Ionic zinc was accumulated more rapidly than other forms of zinc. Many species of marine algae had zinc concentrations >1.0 g/kg DW. These grossly elevated levels were usually associated with nearby industrial or domestic outfalls containing substantial amounts of zinc. In eelgrass, *Zostera marina*, zinc concentrations increased with age of leaf.

In the Fal estuary, England, long-term metal pollution over the past 120 years has resulted in zinc sediment levels between 679.0 and 1780.0 mg/kg DW, producing benthic communities that favor zinc-tolerant organisms, such as oysters and nereid polychaetes, and a general impoverishment of mussels, cockles, other polychaetes, and gastropods.

Zinc in mollusks is usually associated with high molecular weight proteins, with diet (as opposed to ambient water zinc concentrations), from collection locales with elevated sediment zinc burdens, and with particulate matter resulting from dredging and storm perturbations. Zinc levels in mollusks were highest in animals collected near anthropogenic point sources of zinc. Excess zinc accumulations do not seem to affect normal molluscan life processes, and zinc is frequently accumulated far in excess of the organism's immediate needs. American oysters (*Crassostrea virginica*), for example, may naturally contain up to 4.0 g Zn/kg FW soft parts; this is comparable to accumulations observed in oysters exposed to 0.2 mg Zn/L for 20 weeks. In general, zinc concentrations in American oysters were highest in summer and lowest in winter and spring. Zinc tends to accumulate in molluscan digestive gland and stomach as excretory granules, and in kidney as concretions. Kidney is the preferred storage site in mussels and scallops, and digestive gland in oysters. In oysters, granules may contain up to 60% of the total body zinc, explaining, in part, how some shellfish can exist with such high body burdens. Zinc in molluscan tissues is usually elevated under conditions of increasing water temperature and pH, and decreasing salinity; however, zinc accumulation kinetics in mollusks vary considerably among species. Variations in zinc content of clam tissues were associated with seasonal changes in tissue weights. Gastropods nearest a ferro-nickel smelter had elevated zinc concentrations in hepatopancreas when compared to those collected at more distant sites; however, there were no consistent seasonal variations. Fluctuations in zinc content of mussels (*Mytilus edulis*) related to size or season of collection were sufficient to conceal low chronic or short-term pollution. Diet – which is the primary route of zinc accumulation in most mollusks – had no significant effect on whole body zinc content of certain predatory marine gastropods. Whole body zinc concentration of gastropod oyster drills (*Ocenebra erinacea*) ranged between 1451.0 and 2169.0 mg/kg DW, and remained unchanged after feeding for 6 weeks on Pacific oysters (*Crassostrea gigas*) containing 1577.0 mg Zn/kg DW or mussels (*Mytilus edulis*) containing 63.0 mg Zn/kg DW.

High zinc concentrations in crustaceans are usually associated with industrial contamination. In the case of barnacles (*Balanus* spp.), high (>3.3 g/kg DW soft parts) levels are attributed to inorganic granules that contain up to 38% zinc and which accumulate in tissue surrounding the midgut. The granules consist of phosphorus, zinc, potassium, sulfur, and chlorine, in that order. These insoluble membrane-limited spheres form in response to high zinc levels in the ambient seawater within 12 days of exposure, and concentrate in specialized cells around the gut: the stratum perintestinale. Zinc granules in barnacles represent a detoxification mechanism for surplus zinc. Older barnacles have greater whole body zinc accumulations than younger stages, and accumulations change seasonally. Zinc concentrations in marine crustacean tissues are usually <75.0 mg/kg FW or <100.0 mg/kg DW; exceptions include hepatopancreas, molts, eggs, fecal pellets, and barnacles. In crustaceans, zinc is slightly elevated in hepatopancreas, but most tissues are only 2–3 times higher than muscle. For marine crustaceans, the highest concentration recorded in muscle was 57.0 mg Zn/kg FW in king crab, *Paralithodes camtschaticus*, and was associated with two metal binding proteins of molecular weight 11,500 and 27,000. In crustacean tissues, zinc levels were higher in summer, at lower salinities, and in young animals, although young amphipods had higher zinc residues than older stages. Seasonal accumulations of whole body zinc in the shrimp (*Palaemon serratus*) during spring and summer and loss in winter seem to reflect water zinc concentrations in the range 0.0–9.0 μg/L. Zinc is present in crustacean serum at concentrations >1000 times higher than ambient seawater; in serum, it serves primarily as a cofactor of carbonic anhydrase – the principal enzyme involved in calcification. Serum zinc concentrations in crustaceans seem to be independent of season, and water temperature or salinity. Molting results in a 33–50% loss of total zinc in marine crustaceans; molts, together with fecal pellets, constitute an important vehicle of zinc transfer in marine ecosystems. The freshwater opossum shrimp (*Mysis relicta*) can transport zinc from sediments into the water column, and the reverse, during their migratory cycle. *Mysis relicta* and other benthic invertebrates play an important role in determining the concentration of zinc and other metals in lake sediments. Unlike decapod crustaceans, marine amphipods do not regulate body zinc concentrations; amphipod body burdens of zinc may reflect sediment total zinc levels and suggest that certain groups may be suitable bioindicators. Molting had no effect on body zinc concentration in four species of adult marine amphipods, and this forces a reexamination of the role of cast exuviae in zinc transport.

In annelids, zinc content was highest in nonselective deposit feeders, omnivores, and carnivores, and from animals collected from sediments with elevated zinc levels. Freshwater tubificid worms have the potential to increase zinc concentrations in the water column, particularly during short episodes of high burrowing activity. A high zinc content appears to be a structural characteristic of jaws of marine nereid worms. In the marine polychaete worm *Nereis diversicolor*, zinc is localized in the gut wall, epidermis, nephridia, and blood vessels; most of the body zinc is present in wandering amoebocytic cells of excretory organs. Zinc in *Nereis* may be present as insoluble granules in membrane-bound vesicles; excretion is via exocytosis with the aid of amoebocytes. Unlike the insoluble zinc phosphate granules of mollusks and crustaceans, zinc granules in *Nereis* are very soluble and retained only by sulfide precipitation.

Marine vertebrates, including fishes and elasmobranchs, have low zinc concentrations in tissues (i.e., 6.0–400.0 mg/kg DW) when compared to marine plants and invertebrates. Highest concentrations in muscle of marine fishes (20.1–25.0 mg/kg FW) were recorded in northern anchovy (*Engraulis mordax*) and Atlantic menhaden (*Brevoortia tyrannus*). The highest zinc concentrations measured in whole freshwater fishes in the conterminous United States in 1978–79 were in common carp (*Cyprinus carpio*) from Utah; concentrations in carp from Utah ranged between 70.0 and 168.0 mg Zn/kg FW vs. an average of 63.0 mg Zn/kg FW for this species collected elsewhere. Zinc concentrations in fishes tend to be higher near urban areas; highest in eggs, viscera, and liver; lowest in muscle; positively correlated with metallothionein concentrations; lower in all tissues with increasing age and growth; and relatively unaffected by water salinity, temperature, or copper concentrations. Zinc residue data from marine fishes that were dead on collection are of limited worth because dead fish accumulate zinc from seawater at a substantially higher rate than living teleosts.

Zinc concentrations in fishes and other aquatic vertebrates are modified by diet, age of the organism, reproductive state, and other variables. In fish, diet is the major route of zinc uptake and juveniles accumulate zinc from the medium more rapidly than embryos or larvae. Because the diet of many teleost carnivores changes dramatically with age, and because upper trophic level vertebrates are frequently used as indicators of water quality, it seems that more research is needed on zinc burdens in prey organisms. A reduction in serum zinc during egg formation in a flatfish (*Pleuronectes platessa*) may represent a transfer of zinc to eggs. High (>35.0 mg/kg FW) zinc concentrations in eggs of Atlantic salmon (*Salmo salar*) are sometimes associated with increasing mortality, although low (14.0 mg/kg FW) concentrations seem to have no adverse effect on survival. Zinc concentration in Atlantic salmon milt ranged from 0.5 to 5.5 mg Zn/kg, and was linearly proportional to spermatozoan abundance. In lakes containing 1150.0 mg Zn/kg sediment and 209.0–253.0 μg Zn/L water column, white sucker (*Catostomus commersoni*) females did not grow after sexual maturity and had increased incidences of spawning failure. Alterations in growth and reproduction were related, in part, to nutritional deficiencies as a result of chronic effects of elevated sediment zinc on the food base of the sucker, that is, invertebrate fauna were absent in the uppermost 7 m. Eggs of the white sucker incubated at a metals-contaminated site (400.0 μg Zn/L) produced larvae with a decreased tolerance to copper and with elevated zinc body burdens when compared to a noncontaminated (2.7 μg Zn/L) site; larval size and fertilization rate were the same at both sites.

33.5.4 Birds

Zinc residues were elevated in birds collected near zinc smelters. In general, the highest concentrations of zinc in birds are in liver and kidney, and the lowest in muscle. In giant Canada goose (*Branta canadensis maxima*), red muscle contains more zinc than does white muscle, and slow contracting muscle more than fast muscle. In nestling kittiwakes (*Rissa tridactyla*), zinc concentrations increased in liver and feathers throughout chick growth. Zinc concentrations in marine birds normally range between 12.0 mg/kg FW in eggs to 88.0 mg/kg FW in liver. The highest concentration of zinc recorded in a marine bird was 541.0 mg/kg DW in the liver of a booby (*Sula* sp.) that died from polychlorinated biphenyl poisoning. Elevated zinc levels in these birds may have been a manifestation of toxicant-induced stress (i.e., breakdown in osmoregulatory processes), as is the case for other taxonomic groups. Seabirds with high zinc concentrations in liver and kidney tend to have high cadmium levels in these tissues. In flamingos, zinc in liver was positively correlated with copper levels in liver and kidney, and with metallothionein levels in kidney. In egrets, zinc was positively correlated with metallothionein protein levels in liver. In blue-winged teal (*Anas discors*), zinc concentrations were higher in liver than muscle, higher in males than in females, and higher in autumn than in the spring. Zinc concentrations in liver of black-crowned night-heron (*Nycticorax nycticorax*) were usually higher in younger birds, although weight and sex had no direct effect on zinc content. Zinc concentrations in tissues and feathers of California condors (*Gymnogyps californianus*) found dead on collection from a variety of causes were similar to those found in turkey vultures (*Cathartes aura*), common ravens (*Corvus corax*), and ospreys (*Pandion haliaetus*) and are considered normal. The highest concentration recorded in condor liver of 250.0 mg/kg FW approaches those in livers of mallards (*Anas platyrhynchos*) that died from high dietary loadings of zinc. Zinc concentrations in liver of osprey were similar between age groups and sexes. With the onset of egg production in turkeys (*Meleagris gallopavo*), serum zinc in hens increased from 1.6 to 6.9 mg/L and remained significantly elevated throughout egg laying; during this same period, liver zinc concentration declined from 75.0 to 39.0 mg/kg DW, although total liver zinc increased because of an increase in liver weight.

Zinc concentrations in the sediments of the Rhine River have increased about sixfold between 1900 and 1950 – and have remained stable since then. But migratory waterfowl from this collection locale do not have elevated zinc concentrations in their primary feathers. Zinc content in feathers of the hoopoe (*Upupa epops*) increased from 200.0 mg/kg DW at age 7 days to 1000.0 mg/kg DW at age 35 days. Hoopoe populations are declining in India and this decline is said to be associated with increasing zinc concentrations in feathers. Feathers of the greater flamingo (*Phoenicopterus ruber*) have been proposed as indicators of atmospheric zinc contamination: outer barbs of the black primary feathers – exposed to air pollution – contained 53% more zinc, on average, than did inner barbs. More research seems needed on the use of feathers as indicators of zinc contamination.

Zinc concentrations in seminal plasma of domestic chickens (*Gallus* sp.) are about 100 times lower than those for humans and most other mammals, except sheep. Concentrations of zinc in fowl seminal plasma after in vitro storage of spermatozoa for 24 h at 4°C were near the threshold values toxic to spermatozoa, suggesting that poultry spermatozoa normally function near their lower lethal zinc threshold.

33.5.5 Mammals

White-tailed deer (*Odocoileus virginianus*) collected near a zinc smelter had elevated tissue zinc concentrations when compared to those from more distant sites. Deer with zinc concentrations of 150.0 mg/kg FW (750.0 mg/kg DW) in the renal cortex portion of the kidney had swollen joints, lameness, and joint lesions similar to those seen in zinc-poisoned horses from the same area. Zinc was elevated in kidney cortex of red deer (*Cervus elephus*), and older deer tended to have higher concentrations (up to 184.0 mg/kg DW) than did younger deer (as low as 20.0 mg/kg FW); in the case of older deer, zinc was associated with the metallothionein fraction. Zinc residues were usually elevated in rodents near smelters. Rodents from metals-contaminated forests had zinc loadings in tissues similar to those from control locations, although lead and cadmium were significantly elevated in the contaminated zone. Elevated zinc concentrations in mine tailings reportedly do not represent a notable contamination hazard to the invading mammalian fauna, although zinc concentrations in invertebrates, especially earthworms, and vegetation were elevated.

Otters (*Lutra lutra*) were found only on a single unpolluted tributary of a river system contaminated by zinc mine drainage waste, suggesting that a contaminated food supply may be responsible for avoiding otherwise suitable habitat. Marine mammals collected near heavily urbanized or industrialized areas or near zinc pollution point sources usually had elevated zinc concentrations when compared to individuals of the same species and of similar age from relatively pristine environments. Zinc concentrations in tissues of the ringed seal (*Phoca hispida*) were essentially the same in animals near a lead–zinc mine and in those of a distant reference site, although lead and selenium burdens were elevated in the vicinity of the mine site. Concentrations of zinc in tissues of the steller sea lion (*Eumetopias jubata*) were highest in liver and pancreas, followed by kidney, muscle, heart, spleen, and lung; this rank order is comparable to that in human tissues. There is considerable variation among species in tissue zinc concentrations; 3-fold differences are not uncommon for the same tissue in different species of marine mammals. In bottlenose dolphins (*Tursiops truncatus*), zinc concentrations were higher in juveniles than in adults. Marine mammals contained the lowest zinc concentrations (2.0–505.0 mg/kg DW, elevated in liver) of all groups of marine organisms examined. Because zinc is usually available in sufficient quantity in the marine environment and is usually accumulated in excess of the organism's immediate needs, it remains unclear why zinc is comparatively depressed in tissues of marine mammals.

Zinc toxicosis in horses near a zinc smelter was characterized by lameness, swollen joints, and unkempt appearance, particularly in foals. Zinc concentrations in afflicted foals were elevated in pancreas, liver, kidney, and serum when compared to foals at more distant sites. Foals born near the smelter had joint swellings that were attributable to generalized osteochondrosis; lesions were similar to those induced experimentally in animals fed high-zinc diets, and may have been the result of a zinc-induced abnormal copper metabolism. Concentrations of zinc in tissues of horses from farms near the Palmerton smelter were extremely high, in some cases approaching lethal thresholds; zinc poisoning was a cause of debility and death of foals. Grazing mares managed with standard husbandry had significant monthly variations in plasma zinc due, in part, to dietary factors such as nutritional supplementation and to seasonal variations in the quality of grazing pasture. Peak plasma zinc levels in horses are positively related to age (in weanlings aged 22–52 weeks), and to summer diets.

Dairy cattle near a lead and zinc ore processing facility did not have elevated blood or hair zinc levels, although daily zinc intake was 5.6 mg/kg BW vs. 1.2 mg/kg BW daily for a control area. In cattle, proximity to zinc refineries did not result in significant elevation of liver or kidney zinc concentrations. However, cows living within 6 km of a power plant in Czechoslovakia had elevated zinc loadings in hair, and poor reproduction when compared to a herd 26 km distant. In adult bovines, zinc reserves are usually small and located primarily in the skeleton and muscle, although appreciable hepatic accumulations can occur in the fetus. At 270 days of gestation, for example, 30% of fetal zinc in cattle is present in liver; fetal zinc concentration is about 4 times higher than that in maternal liver. Liver concentrations >120.0 mg Zn/kg DW in cattle are frequently associated with elevated dietary zinc loadings. Concentrations of zinc in milk of cows and goats varied significantly between breeds, with zinc level in diet, and declined markedly after parturition.

A normal 70-kg human male contains 1.5–2.0 g of zinc, or about 21.0–29.0 mg Zn/kg BW; normal zinc uptake is 12.0–15.0 mg daily, equivalent to 0.17–0.21 mg/kg BW. Foods rich in zinc are seafoods, meats, grains, dairy products, nuts, and legumes. About 90% of the total body zinc is in the musculoskeletal system. Highest zinc concentrations of 100.0–200.0 mg/kg occur in prostate, eye, brain, hair, bone, and reproductive organs; intermediate concentrations of 40.0–50.0 mg/kg occur in liver, kidney, and muscle. In blood, about 80% of the total zinc is in red cells, where it is associated with carbonic anhydrase. The mean plasma zinc level is about 0.9 mg/L; about half is in a freely exchangeable form loosely bound to albumin; most of the remainder is tightly bound to macroglobulins and amino acids, especially histidine and cysteine. The greatest zinc concentration in the human body is in prostate, and may be related to the elevated levels of acid phosphatase, a zinc-containing enzyme, in that organ. The prostate gland contributes zinc to spermatozoa in dogs – a necessary process for canine fertility and fecundity; in rats, however, the prostate does not contribute to zinc in spermatozoa, and its function is not essential for rat reproduction.

33.6 Zinc Deficiency Effects

Zinc is important in the metabolism of proteins and nucleic acids and is essential for the synthesis of DNA and RNA. Zinc deficiency has been reported in humans and a wide variety of plants and animals – with severe effects on all stages of reproduction, growth, and tissue proliferation in the young. In early gestation, zinc deficiency may cause severe congenital abnormalities. Later in gestation, deficiency can cause growth inhibition and brain growth impairment, leading to altered behavioral development after birth. Feeding a low zinc diet to lactating dams produces signs of zinc deficiency in suckling pups. In humans, zinc deficiency is associated with delayed sexual maturation in adolescent males; poor growth in young children; impaired growth of hair, skin, and bone; disrupted vitamin A metabolism; and abnormal taste acuity, hormone metabolism, and immune function.

33.6.1 Terrestrial Plants

Zinc deficiencies in citrus groves in California, pecan trees in Texas, and various crops in Australia, have resulted in large crop losses. Applications of zinc salts were effective under acidic soil conditions. But neutral or alkaline soils rendered zinc salts insoluble and zinc therapy ineffective. Sprays of zinc salts applied to leaves, or injections into tree trunks, overcame the problems of soil solubility and have generally been successful. Zinc is usually bound strongly in plants, particularly in grains, markedly decreasing its availability to animal consumers. Binding is attributed mainly to high content of phytate, and also to high levels of fiber, hemicelluloses, and amino acid–carbohydrate complexes. Whole-grain cereals and legumes are considered rich sources of zinc.

33.6.2 Aquatic Organisms

Nutritional zinc deficiency is rare in aquatic organisms, although reports are available of experimentally induced zinc deficiency in algae, sponges, daphnids, echinoderms, fish, and amphibians. Experimental zinc deficiency in euglenoids (*Euglena gracilis*) was associated with arrested growth and abnormal cell differentiation and development, leading to extensive teratological abnormalities. Zinc-deprived *Euglena* survived for extended periods through decreased metabolism. Marine algae stopped growing when ambient zinc concentrations fell below 0.7 μg/L; also, zinc-deficient cultures of freshwater algae were unable to metabolize silicon. A freshwater sponge (*Ephydatia fluviatilis*) grew normally at a concentration of 0.65 μg Zn/L, but growth was reduced at lower concentrations tested. Daphnids (*Daphnia pulex*, *Daphnia magna*) reared for six brood cycles in zinc-free water showed reduced survival, inhibited reproduction, and cuticle damage. Zinc is important in pH regulation of sperm of marine invertebrates. Zinc reduction in semen to <6.5 μg/L adversely affected sperm pH and motility in sea urchins (*Strongylocentrotus purpuratus*, *Lytechinus pictus*), horseshoe crab (*Limulus polyphemus*), and starfish.

Rainbow trout (*Oncorhynchus mykiss*) fry fed diets containing 1.0–4.0 mg Zn/kg ration have poor growth, increased mortality, cataracts, and fin erosion; supplementing the diet to 15.0–30.0 mg Zn/kg alleviate these signs. Rainbow trout fry diets containing 1.0, 90.0, or 590.0 mg Zn/kg ration and simultaneously exposed to a range of waterborne zinc concentrations of 7.0, 39.0, 148.0, or 529.0 μg Zn/L were observed for 16 weeks. At that time, the 7.0 μg Zn/L plus 1.0 mg/kg diet group showed clear signs of deficiency, including a significantly reduced plasma zinc concentration (which was evident as early as the first week of exposure), reduced growth (with no growth after week 12), decreased hematocrit, and reduced plasma protein and whole body zinc concentration. Elevating waterborne zinc to 39.0 or 148.0 μg Zn/L partially corrected the deficiency, but did not restore plasma or whole body zinc to levels seen initially or in fish raised for 16 weeks on a zinc-adequate diet of 90.0 mg Zn/kg ration. There were no toxic effects at any other dietary-waterborne zinc mixture. It was concluded that zinc uptake from water was independent of uptake from diet, since at any dietary zinc level an increase in the waterborne zinc resulted in an increase in whole body zinc. In freshwater, where waterborne concentrations of <10.0 μg Zn/L are most commonly encountered, waterborne zinc contributions to whole body zinc loadings are likely to be insignificant. In cases where dietary zinc was adequate (i.e., 90.0 mg Zn/kg ration), the contribution of waterborne zinc was significant in the case of rainbow trout. In marine teleosts, diet is the major zinc source when seawater contained <15.0 μg Zn/L; at higher ambient concentrations of 600.0 μg Zn/L, waterborne zinc contributed up to 50% of the total body zinc burden.

Experimentally produced zinc deficiency in toad (*Bufo arenarum*) embryos resulted in adults with abnormal ovarian development, altered meiotic and ovulation processes, and embryos with a high incidence of congenital malformations.

33.6.3 Birds

Zinc-deficiency in chickens, turkeys, and Japanese quail is characterized by low survival, reduced growth rate and food intake, poor feathering, shortening and thickening of long bones of legs and wings, reduced egg production and hatchability, skeletal deformities in embryos, an uncoordinated gait, reduced bone alkaline phosphatase activity, and increased susceptibility to infection.

Laying hens (*Gallus* sp.) had low egg hatchability on diets that contained 6.0 mg Zn/kg, and produced chicks that were weak and poorly feathered; these chicks usually died within a few days on 8.0–9.0 mg Zn/kg diets. Zinc-deficient chicks (13.0–16.0 mg Zn/kg DW diet for 4 weeks) had pathological defects in epiphyseal cartilage; no interference with calcification was noted in controls fed diets containing 93.0–96.0 mg Zn/kg feed. Pullets fed diets containing 28.0 mg Zn/kg for 4 months, then 4.0 mg Zn/kg ration for 4.5 months produced few hatchable eggs after 4 months; prevalent malformations included faulty trunk and limb development, missing vertebrae, missing limbs and toes, abnormal brain morphology, small eyes, and skeletal malformations. Most zinc deficiency effects were reversed by increasing dietary zinc concentrations to 96.0–120.0 mg/kg.

Japanese quail fed an excess of zinc (25.0–30.0 mg Zn/kg diet) during their first week of life were protected during a subsequent period of zinc deprivation (1.0 mg Zn/kg diet for 1 week). Birds that received an initial intake of zinc in excess of requirements grew significantly better than those receiving a minimal amount of zinc. Japanese quail may store excess zinc in bones; this zinc store may become available during a subsequent period of zinc deprivation, especially during a period of rapid bone growth, but this requires verification. Egg production constitutes a major loss of zinc and other trace metals by the laying hen. Vitellogenin mediates the transfer of zinc from liver to the maturing oocyte, ultimately resulting in deposition into yolk of the newly formed egg. More research seems needed on the role of zinc in avian reproduction.

33.6.4 Mammals

Compared with zinc toxicity, zinc deficiency is a much more frequent risk to mammals. Zinc is required in all stages of the cell cycle and deficiency adversely affects metabolism of DNA, RNA, proteins, and activity of carbonic anhydrase, lactic dehydrogenase, mannosidase, and other enzymes. In zinc deficiency, the activity of various zinc-dependent enzymes are reduced in testes, bone, esophagus, and kidney of rats, and alkaline phosphatase activity is reduced in bone and plasma from zinc-deficient rats, pigs, and cows. Deficiency leads to loss of appetite and taste, skin disturbances, slow wound healing, impaired brain development, deficient immune system, and disrupted water metabolism. Zinc deficiency adversely affects testicular function in humans and animals and seems to be essential for spermatogenesis and testosterone metabolism. Zinc deficiency in young men with very low zinc intakes resulted in testicular lesions and reduced accessory gland weights, due primarily to reduced food intake and growth. Zinc deficiency during pregnancy has produced low birth weight, malformations, and poor survival in rat, lamb, and pig; the role of zinc in human reproductive problems is still unclear.

Zinc-deficient diets for ruminants and small laboratory animals usually contain <1.0 mg Zn/kg ration, although rats show deficiency at <12.0 mg Zn/kg ration. Zinc deficiency has been documented in humans, small laboratory animals, domestic livestock, mink, and monkeys; signs of severe zinc deficiency in mammals include decreased food intake, growth cessation, fetal malformations, testicular atrophy, swelling of feet, excessive salivation, dermal lesions, parakeratosis of the esophagus, impaired reproduction, hair loss, unkempt appearance, stiffness, abnormal gait, skin and organ histopathology, and hypersensitivity to touch. Selected examples of zinc deficiency in various species follow.

Zinc deficiency in humans is rare, and is usually associated with severe malabsorption, parenteral alimentation lacking zinc, or geophagia. Symptoms of zinc deficiency depend, in part, on age, acuteness of onset, duration and severity of the zinc depletion, and the circumstances in which deficiency occurs. Many of the features of zinc deficiency observed in humans are similar to those seen in zinc-deficient animals. Simple nutritional deficiency due to marginal zinc intake may be common, even in the United States. Factors contributing to zinc deficiency include inadequate dietary intake (protein-calorie malnutrition), decreased availability (high fiber/phytate diets), decreased absorption, excessive losses (increased sweating, burns), increased requirements (rapid growth, pregnancy, lactation), as well as old age, alcoholism, and possible genetic defects. Zinc deficiency may also occur as a result of liver or kidney disease, gastrointestinal disorders, skin disorders, parasitic infections, diabetes, and genetic disorders such as sickle cell disease. Clinical disorders aggravated by zinc deficiency include ulcerative colitis, chronic renal disease, and hemolytic anemia. In the 60 years since human zinc deficiency was demonstrated, it has been observed in a wide variety of geographic areas and economic circumstances. Severe zinc deficiency occurs in some areas of the Middle East and North Africa, and is frequently associated with the consumption of unrefined cereals as a major part of the diet. Chronic zinc deficiency in humans is associated with dwarfism, infantile testes, delayed sexual maturity, birth defects, poor appetite, mental lethargy, immunodeficiency, skin disorders, night blindness, impotence, spleen and liver enlargement, defective mobilization of vitamin A, delayed wound healing, impaired taste acuity, abnormal glucose tolerance, impaired secretion of luteinizing hormone, and iron and folate deficiency. A deficiency of zinc in the growing age period results in growth retardation; a severe zinc deficiency may be fatal if untreated. Zinc-deficient humans excrete $<100.0\,\mu g$ Zn daily in urine compared to a normal $>300.0\,\mu g$ Zn daily. Zinc deficiency may exacerbate impaired copper nutrition; interactions with cadmium and lead may modify the toxicity of these metals. Acrodermatis enterohepatica is a disease characterized by skin eruptions, gastrointestinal disorders, and low serum zinc levels. One causative factor is poor intestinal absorption of zinc; a complete cure was accomplished by oral administration of 135.0 mg Zn daily as 600.0 mg of zinc sulfate. Using radiozinc-65, it was shown that afflicted individuals had a greater turnover of plasma zinc, a smaller pool of exchangeable zinc, and a reduced excretion of zinc in stool and urine. Zinc deficiency in humans is usually treated by oral administration of 1.0 mg Zn/kg BW daily. However, zinc-deficient humans given daily intravenous injections of 23.0 mg of zinc experienced profuse sweating, blurred vision, and hypothermia. An endemic zinc deficiency syndrome among young men has been reported from Iran and Egypt, and is characterized by retarded growth, infantile testes, delayed sexual maturation, mental lethargy, anemia, reduced concentration of zinc in plasma and red cells, enlarged liver and spleen, and hyperpigmentation; oral supplementation of 30.0 mg Zn daily had a prompt beneficial effect. A zinc deficiency syndrome during human pregnancy includes increased maternal morbidity, abnormal taste sensations, prolonged gestation, inefficient labor, atonic bleeding, and increased risks to the fetus. Pregnant women with initially low and subsequently decreasing serum zinc levels had a high frequency of complications at delivery, including congenital malformations in infants. Multiple severe skeletal abnormalities and organ malformations in human fetuses have been attributed to zinc deficiency. In newborns, zinc deficiency is manifested by growth retardation, dermatitis, hair loss, impaired healing, susceptibility to infections, and neuropsychologic abnormalities.

Hereditary zinc deficiency occurs in certain strains of cattle (*Bos* spp.) and affects the skin and mucous membranes of the gastrointestinal tract. The disease – also known as Lethal Trait A46 – is caused by failure of a single autosomal recessive gene regulating zinc absorption from the intestine. Affected animals will die within a few months from secondary bacterial infections unless treated daily with high oral doses of zinc compounds. Certain imported breeds of cattle in the western Sudan with low zinc serum levels (i.e., <0.6 mg/L) showed signs of zinc deficiency, including stunted growth, weakness, skin lesions, and loss of hair pigment. Cows fed a low (25.0 mg/kg ration), but adequate, zinc diet had liver zinc concentrations below the expected 125 mg Zn/kg DW; increasing the total zinc dietary loading to 45.0 or 50.0 mg/kg DW is recommended for counteracting reduced zinc absorption in diets with soybean products. Cows and calves fed low zinc diets of 25.0 mg Zn/kg ration showed a decrease in plasma zinc from 1.02 mg/L at start to 0.66 mg/L at day 90; cows fed 65.0 mg Zn/kg diet had a significantly elevated (1.5 mg Zn/L) plasma zinc level, and increased blood urea and plasma proteins. Biomarkers used to identify zinc deficiency in bovines include zinc concentrations in plasma, unsaturated zinc-binding capacity, ratio of copper to zinc in plasma, and zinc concentrations in other blood factors; indirect biomarkers include enzyme activities, red cell uptake, and metallothionein content in plasma and liver.

Domestic goats (*Capra* sp.) fed a zinc-deficient diet (15.0 mg Zn/kg) developed skin histopathology and alopecia (hair loss) after 177 days; zinc-deficient diets lacking vitamin A hastened the process, with signs evident between 46 and 68 days. No signs were evident in goats fed vitamin A-adequate diets containing 80.0 mg Zn/kg ration.

Guinea pigs (*Cavia* spp.) fed a zinc-deficient diet (1.25 mg Zn/kg FW) for 60 days, had significant reductions in zinc concentration of serum (0.5 mg/L), kidney (10.0 mg/kg FW), testes (9.5 mg/kg FW), and liver (9.4 mg/kg FW). Guinea pigs fed 1.25 mg Zn/kg FW diet for 45 days followed by a zinc-replete diet of 100.0 mg/kg FW for 15 days had normal concentrations of zinc (mg/kg FW) in serum (1.6–2.0), kidney (18.0–20.0), testes (19.0–27.0), and liver (15.0–17.0). Zinc-deficient guinea pigs (<3.0 mg Zn/kg diet, 1.0 mg Zn/L drinking water) exposed from day 30 of gestation to term on day 68, when compared to zinc-adequate animals (<3.0 mg Zn/kg diet, 15.0 mg Zn/L) produced young of low birth weight, with severe skin lesions, sensitive to handling, slow in recovering balance when turned on side, and a peculiar stance; fetal Zn concentrations were depressed 15–33% in liver and placenta. Disrupted immunocompetence responses and disordered protein metabolism were found in guinea pigs fed a zinc-deficient diet of 1.25 mg/kg FW ration for 45 days; marked, though incomplete, restoration, occurred when this group was switched to 100.0 mg Zn/kg ration for 15 days. Neuromuscular pathology was evident in weanling guinea pigs fed a zinc-deficient diet (<1.0 mg Zn/kg) for 4 weeks, as judged by abnormal posture, skin lesions, and disrupted vocalizations; signs became severe after 5–6 weeks, but a single intraperitoneal injection of 1.3 mg Zn/kg BW (as $ZnSO_4$) caused remission within 7 days. Acute experimental allergic encephalomyelitis (EAE) was induced in guinea pigs maintained on low (6.0 mg/kg), normal (20.0 mg/kg), and high (200.0 mg/kg) levels of zinc in the diet. Acute

EAE is usually a fatal disease of the central nervous system (CNS) induced by inoculation with protein found in myelin of the CNS. Guinea pigs on the zinc-deficient diet exhibited the expected signs of zinc deficiency, but unlike other groups did not develop neurological signs of acute EAE. EAE suppression observed in the Zn-deficient guinea pigs is ascribed to the influence of zinc deficiency of the T-cell function. A model of autoimmune CNS disease, such as EAE, that requires a prominent T-lymphocyte sensitization, can be altered or suppressed when the immunoregulatory mechanisms are impaired by Zn deficiency.

Rhesus monkeys (*Macaca mulatta*) fed a marginally deficient zinc diet (4.0 mg Zn/kg diet) between age 5.5 and 30 months had lower plasma Zn levels, delayed onset of accelerated weight gain and linear growth, and no loss of subcutaneous fat – typical of early adolescence – when compared to those fed diets containing 100.0 mg Zn/kg. Marginal dietary zinc deprivation also depressed immune function in rhesus monkeys by about 30%, and impaired both learning and reversal of a visual discrimination task by 33–66%. When pregnant rhesus monkeys are fed a diet marginally deficient in zinc (4.0 mg/kg), perturbations in the mother's immune system can occur. Their infants have reduced immune responsiveness despite the absence of marked differences in plasma or soft tissue zinc concentrations from control (100.0 mg Zn/kg diet) infants. Infant rhesus monkeys from zinc-deprived (4.0 mg Zn/kg ration) pregnant dams and subsequently fed the same low zinc diet showed delayed skeletal maturation during their first year. The condition was most severe at age 6 months but began to return to normal despite continuation of the marginally deficient zinc diet.

Mice (*Mus* sp.) fed a zinc-deficient diet of 0.7 mg Zn/kg ration for 40 days, when compared to mice fed a zinc-adequate diet of 36.5 mg Zn/kg, had a reduced growth rate, impaired phagocytic function, increased susceptibility to lead poisoning, and reduced zinc content in blood (0.7 mg/L vs. 1.0–1.1) and liver (12.0 mg Zn/kg FW vs. 17.0–19.0). Zinc deficiency during early development affects neural tube development through arrested cell growth. Zinc deficiency in mice may disrupt olfactory function through interference with zinc-containing neurons in higher olfactory centers. Adult mice fed a zinc-deficient diet of 5.0 mg Zn/kg ration for 42 days, when compared to mice given 100.0 mg Zn/kg diet, could not distinguish odors, although olfactory epithelia seemed normal.

Mink (*Mustela vison*) kits fed a zinc-deficient diet of 4.1 mg Zn/kg FW ration for 4 days retained 0.49 mg Zn/kit and lost weight. Kits fed a zinc-adequate diet (35.0–45.0 mg Zn/kg FW, 100.0–150.0 mg/kg DW) retained 2.5 mg Zn/kit, and those fed 83.0 mg Zn/kg FW diet retained 7.8 mg Zn/kit. Low-dose kits ate less than other groups. Urine was the most important excretory route in the zinc-deficient group, compared to feces in higher dose groups.

Domestic sheep (*Ovis aries*) fed a low zinc diet (2.2 mg Zn/kg DW diet) for 50 days, when compared to those fed a zinc-adequate diet (33.0 mg Zn/kg DW diet), excreted less zinc (<4.0 mg daily vs. 23.0–25.0 mg), consumed less food (409 g daily vs. 898 g), and had lower plasma zinc concentrations (0.18 mg/L vs. 0.53–0.58 mg/L); a reduction in plasma alkaline phosphatase activity and an increase in plasma zinc binding capacity were also noted. Sensitive indicators of zinc deficiency in lambs include significant reductions in plasma alkaline phosphatase activity and plasma zinc concentrations; signs were clearly evident in lambs fed 10.8 mg Zn/kg DW diet for 50 to 180 days. A normal diet for lambs contains 124.0–130.0 mg Zn/kg DW ration vs. 33.0 for adults. One recommended treatment for zinc-deficient sheep is ruminal insertion of zinc-containing boluses every 40 days; bolus zinc release is about 107.0 mg daily.

Zinc-deficient pregnant laboratory white rats (*Rattus* sp.) have reduced litter size, a high frequency of fetal deformities, low birth weight, and a prolonged parturition; dams are inactive and behavior towards young seems indifferent. Fetal skeletal defects are prominent in rats fed zinc-deficient diets of 10.0 mg/kg ration during a 21-day gestation period. About 91% of zinc-deficient fetuses had multiple skeletal malformations vs. none in controls fed 76.0 mg Zn/kg diet. Zinc-deficient

(1.5 mg Zn/kg diet) pregnant rats also had increased iron levels in liver, kidney, and spleen, depleted liver glycogen, and reduced levels of zinc in pancreas and duodenum. Zinc deficiency causes testicular atrophy and hypogonadism in rats; the effects include spermatic arrest, histopathology of seminiferous tubules and interstitial cells, reduced serum testicular testosterone levels, and reduced testicular zinc concentrations. Zinc is required in Leydig cells for normal testosterone activity. Calcitonin inhibits transmembrane influx of zinc in the isolated rat Leydig cell, but these effects usually take >2 days and are critical only in states of borderline zinc deficiency. Zinc deficiency during pubertal development of rats depresses the activity of dipeptidyl carboxypeptidase in the testes and epididymis; this enzyme is required for maturation and development of sperm cells and reduced activity may cause suppression of sexual maturity. Laboratory white rats fed zinc-deficient diets for 20 days show an aversion to the zinc-deficient diet. They readily consumed a familiar zinc-adequate diet for 15 days, but the previously deficient animals continued to avoid zinc-deficient diets when given a choice. Zinc deficiency in rats (<1.0 mg Zn/kg diet for 26 days) significantly reduced blood pressure and this correlated positively with serum angiotensin converting enzyme activity; increasing the dietary intake of calcium had no effect on these responses. During zinc deficiency, zinc is mobilized from bone in young immature animals and may be available for metabolic processes including growth. Diabetic rats are at risk of developing zinc deficiency, owing to zinc's role in modulating immune system dysfunction in diabetes mellitus. Cadmium toxicity is related to the zinc status of the body. Zinc-deficient rats (<1.0 mg Zn/kg diet) and zinc-adequate rats (40.0 mg/kg) were both challenged with cadmium. The zinc-deficient group had accelerated zinc loss from kidney; enlarged liver, kidney, spleen and lungs; and increased distribution of cadmium in tissues. Other signs noted in zinc-deficient laboratory white rats include the following: decreased food intake and loss of body weight, reduction in serum zinc, altered cholesterol metabolism, increased serum magnesium, low bone (femur) zinc concentrations, degeneration of olfactory epithelium, reduction in serum total proteins, decreases in activity of glutamate, glycine, methionine, arginine, lysine, and proline, and increased dental caries.

Zinc deficiency in domestic pig (*Sus* spp.) is associated with a condition known as porcine parakeratosis, characterized by dermatitis, diarrhea, vomiting, anorexia, severe weight loss, and eventually death; the condition is exacerbated by high calcium levels.

33.7 Lethal and Sublethal Effects

Significant adverse effects on growth, reproduction, and survival are documented for sensitive marine and freshwater species of aquatic plants, invertebrates, and vertebrates at nominal water concentrations between 10.0 and 25.0 μg Zn/L. Sensitive terrestrial plants died when soil zinc concentrations were >100.0 mg/kg, and showed decreased photosynthesis when total plant contained >178.0 mg Zn/kg DW. Representative soil invertebrates showed reduced growth at 300.0–1000.0 mg Zn/kg diet, and reduced survival at 470.0–6400.0 mg Zn/kg soil. Domestic poultry and avian wildlife had reduced growth at >2000.0mg Zn/kg diet, and reduced survival at >3000.0 mg Zn/kg diet or at a single oral dose >742.0 mg Zn/kg BW; younger stages (i.e., chicks, ducklings) were least resistant. Sensitive species of livestock and small laboratory animals were adversely affected at >0.8 mg Zn/m^3 air, 90.0–300.0 mg Zn/kg diet, >90.0 mg Zn/kg BW daily, >300.0 mg Zn/L drinking water, and >350.0 mg Zn/kg BW single oral dose.

33.7.1 Terrestrial Plants and Invertebrates

Sensitive terrestrial plants die when soil zinc levels exceed 100.0 mg/kg or when plant zinc content exceeds 178.0 mg/kg DW. The phytotoxic zinc level for barley (*Hordeum vulgare*) is not known, but zinc content of barley leaf rarely exceeds 100.0 mg/kg DW. Uptake of zinc from soils by plants is dependent on soil type; for example, uptake is lower in coarse loamy soils than in fine loamy soils. Zinc uptake by barley leaf is greater with increasing rate of sludge application, but the relation is not proportional.

Among terrestrial invertebrates, adverse effects on earthworm survival were documented at 470.0–662.0 mg/kg soil, slugs had reduced food consumption at 300.0 mg Zn/kg diet and reduced growth at 1000.0 mg Zn/kg diet; the woodlouse had impaired reproduction at 1600.0 mg Zn/kg soil, and reduced survival at 5000.0 mg Zn/kg diet or 6400.0 mg Zn/kg soil. Woodlouse (*Porcellio scaber*) fed diets containing sublethal concentrations of zinc for 22 weeks at air temperatures of 12, 16, 20, or 24°C had decreased growth and increased accumulations at high dietary loadings and low temperatures; however, the interaction between temperature and zinc toxicity is not attributed to increased accumulations, but to a physiological interaction with energy metabolism.

High zinc concentrations in soils are responsible for reductions in populations of soil invertebrates near brass mills and zinc smelters. Soils in the vicinity of zinc smelters contained up to 35.0 g Zn/kg and had decreased populations of arthropods; experimentally, 20.0 g of total zinc per kg of soil could account for the decreased survival. Zinc concentrations exceeding 1600.0 mg/kg soil litter are associated with reduced natural populations of decomposer organisms in contaminated forest soil litter, and this has been verified experimentally. Poisoning of decomposer organisms, such as the woodlouse, may disrupt nutrient cycling and reduce the number of invertebrates available as wildlife food. The woodlouse contains higher concentrations of zinc than other terrestrial invertebrates: up to 152.0 mg Zn/kg DW whole organism. It is speculated that the large zinc stores in *P. scaber* repels predators which find zinc distasteful.

Slugs (*Arion ater*) are resistant to high dietary zinc intakes (1000.0 mg/kg feed) for 30 days, although zinc accumulations occur in excretory and calcium cells of the digestive gland. Histochemical detection of zinc in digestive gland of *Arion* is an indication of high levels of zinc in the environment. Zinc elimination in *Arion* occurs directly from lipofuscin material of excretory cells and from spherules of calcium cells; excretion of lipofuscin material via feces is the major excretory route.

Zinc normally aids wound healing in terrestrial invertebrates. Wounding of the optic tentacle, foot tissue, and partial shell removal in *Helix aspersa*, a terrestrial gastropod, resulted in deposition of zinc in the wound area after 2–5 days. Increased zinc in *Helix* wound areas may be necessary to promote protein synthesis, collagen formation, and mitotic cell division.

33.7.2 Aquatic Organisms

Significant adverse effects of zinc on growth, survival, and reproduction occur in representative sensitive species of aquatic plants, protozoans, sponges, mollusks, crustaceans, echinoderms, fishes, and amphibians at nominal water concentrations between 10.0 and 25.0 μg Zn/L. Latent effects of zinc intoxication after brief exposures are poorly documented. One study showed that sensitive species of freshwater crustaceans exposed to zinc concentrations as low as 150.0 μg/L for as little as 30 min had delayed effects that included increasing immobility for up to 172 h after exposure.

Acute LC50 (96 h) values for freshwater invertebrates ranged between 32.0 and 40,930.0 μg Zn/L; for fish, this range was 66.0 to 40,900.0 μg/L. For marine invertebrates the LC50 (96 h) range was 195.0 μg/L for embryos of the hard-shelled clam (*Mercenaria mercenaria*) to >320.0 mg/L for adults of the Baltic clam (*Macoma balthica*). For marine teleosts LC50 (96 h) values ranged between 191.0 μg/L for larvae of the cabezon (*Scorpaenichthys marmoratus*) to 38.0 mg/L for juvenile spot, *Leiostomus xanthurus*. Many factors are known to modify the biocidal properties of zinc in aquatic environments. In general, zinc was more toxic to embryos and juveniles than to adults, to starved animals, at elevated temperatures, in the presence of cadmium and mercury, in the absence of chelating agents, at reduced salinities, under conditions of marked oscillations in ambient zinc concentrations, at decreased water hardness and alkalinity, and at low dissolved oxygen concentrations.

Bioconcentration factors (BCF) for zinc accumulation from the medium varied widely between and within species of aquatic organisms. For representative freshwater organisms BCF values ranged from 107 to 1130 for insects, and 51 to 432 for fish. In marine environments the most effective zinc accumulators included red and brown algae, ostreid and crassostreid oysters, and scallops. The ranges of BCF values for representative marine groups were 370–64,000 for algae, 85–1,500,000 for crustaceans, 15–500 for echinoderms, up to 4,000,000 in scallop kidney, and 1900–6900 for fishes. Significant zinc accumulations were reported after death in algae and fish, suggesting that residue data from these and other organisms found dead on collection are of limited worth. Maximum net daily accumulation rates recorded for various whole marine organisms, in mg Zn/kg FW, were 1.3 for the alga *Ascophyllum nodosum*, 7.7 for mussel *Mytilus edulis*, 19.8 for oyster *Crassostrea virginica*, 32.0 for the killifish *Fundulus heteroclitus*, 32.0 for softshell clam *Mya arenaria*, and 223.0 for sandworm *Nereis diversicolor*; in general, accumulation rates and total accumulations were higher at elevated water temperatures, and at higher ambient zinc water concentrations.

33.7.2.1 Algae and Macrophytes

Blue-green algae are among the most zinc-resistant aquatic plants. Algae are classified as very resistant (can tolerate >10.0 mg Zn/L), resistant (2.0–0.0 mg/L), moderately resistant (0.5–2.0 mg/L), low resistance (0.1–0.5 mg/L; *Navicula*, *Synedra*), and very low resistance (<0.1 mg Zn/L; *Diatoma*, *Tabellaria*, *Microspora*, *Ulothrix*).

The most sensitive aquatic plant was *Schroederella schroederi*, a diatom; 19.0 μg Zn/L was sufficient to inhibit growth 50% in 48 h. Freshwater aquatic plants are usually absent from areas containing >2.0 mg Zn/L; in hard waters of artificial streams containing 170 mg $CaCO_3$/L, a water concentration of 1.1 mg Zn/L caused a 50% decrease in the number of algal species. Most freshwater diatom populations decreased in the range of 175.0–380.0 μg Zn/L; this sensitivity may be useful as an indicator of zinc contamination. Zinc and cadmium are strongly synergistic in their toxic action to plants. Any level of cadmium >10.0 μg/L should be suspected of producing a significant increase in the toxicity of available zinc to freshwater plants.

In heavily contaminated zinc environments (i.e., 130.0–6500.0 μg Zn/L), zinc-tolerant species are dominant. Highly tolerant strains of algae require 1.5–1.65 mg Zn/L for normal growth; at least three species of some tolerant strains can live in water containing 3.0 g Zn/L. Highly tolerant mutant strains of *Anacystis nidulans* required 1.5–16.5 mg Zn/L. In France, at least 17 species of freshwater algae seemed to be flourishing at 42.5 mg Zn/L and pH 4.2. Zinc-tolerant strains of aquatic algae tolerate high zinc concentrations with little bioconcentration. A zinc-tolerant strain of *Euglena gracilis*, for example, tolerates >700.0 mg Zn/L, but contains <500.0 mg Zn/kg DW whole organism vs. 50.0 mg Zn/L and 5000.0 mg/kg DW for nontolerant strains. Another zinc-tolerant strain of *Euglena* had normal growth at 300.0 mg Zn/L and residues of about 7000.0 mg Zn/kg DW vs. population decline of nontolerant strains at 300.0 mg Zn/L.

Algae are effective accumulators of zinc. Three species of marine algae had a mean bioconcentration factor (BCF) of 1530 in 12 days, 4680 in 34 days, and 16,600 in 140 days. Bioconcentration factors for zinc and various species of algae are quite variable and usually range between 76 and 163,750. Many species of aquatic plants contain 150.0 mg Zn/kg DW, and higher. In one case, algae (*Mougeotia* spp.) from northern England in Zn-contaminated waters contained a spectacular 219.0 g Zn/kg DW; it is likely that most of the zinc in *Mougeotia* was not biologically incorporated. Algal accumulations of zinc are modified significantly by physicochemical variables. Zinc concentrations in algae were higher under conditions of decreasing light intensity, water pH, DDT levels, copper, cadmium, phosphate, suspended sediments, organic chelators and other complexing agents, calcium, and magnesium, and under conditions of increasing water temperature, dissolved oxygen, duration of exposure, and ambient zinc concentrations.

Unlike algae, submerged aquatic macrophytes play a minor role in cycling of zinc. Rooted aquatic macrophytes may participate in heavy metal cycling in the aquatic environment either as a source or as a sink. But studies with eelgrass (*Zostera marina*) show that zinc exchange between the sediment and the water is insignificant.

33.7.2.2 Mollusks

Zinc was most toxic to representative mollusks at elevated temperatures, in comparatively soft water or in the case of marine mollusks at low salinity, at earlier developmental stages, at low dissolved oxygen concentrations, and with increasing exposure to high zinc concentrations.

High zinc accumulations in mollusks are usually linked to high levels of calcium in tissues, low ambient concentrations of iron or cobalt, exposure to organochlorine or organophosphorus insecticides, low salinity, elevated temperatures, increased particulate loadings in medium, increasing length of exposure to higher doses of zinc, increasing age of the organism, and especially proximity to heavily carbonized and industrialized areas. Radiozinc-65 was rapidly accumulated in southern quahog (*Mercenaria campechiensis*) over a 10-day period; accumulation in kidney was linear over time, and was enhanced at elevated phosphate loadings in the medium. Large variations in daily zinc accumulation rates by marine bivalve mollusks are typical. For example, softshell clams (*Mya arenaria*) immersed in 500.0 μg Zn/L at 16–22°C had daily accumulation rates, in mg/kg FW soft parts, of 2.0 on the first day of exposure, 7.7 between days 1 and 7, and 3.3 between days 7 and 14. At a lower temperature regimen of 0–10°C, immersion in 500.0 μg/L produced daily accumulation rates of 9.9 mg/kg FW soft parts for the first 42 days, but clams lost zinc at a rate of 0.24 mg/kg daily between days 42 and 112. At 2500.0 μg/L and 16–22°C, daily accumulation rates in surviving *Mya* were 32.0 mg Zn/kg FW soft parts on the first day of exposure, and 11.7 between days 1 and 7. Changes in accumulation rates of zinc by *Mya* reflect, at least partially, complex interactions between water temperature, ambient zinc concentrations, duration and season of exposure, and physiological saturation and detoxification mechanisms.

The half-time persistence (Tb1/2) of zinc in whole mollusks is extremely variable, and reported to range from 4 days in the common mussel (*Mytilus edulis*) to 650 days in the duck mussel (*Anodonta nutalliana*); intermediate values were 23–40 days in limpet (*Littorina irroratea*), 76 days in the California mussel (*Mytilus californianus*), and 300 days in the Pacific oyster. Zinc persistence in selected organs also shows considerable variability and may be significantly different from Tb1/2 values seen in whole animal. For example, the Tb1/2 of zinc in *Mytilus edulis* kidney was estimated at 2–3 months vs. 4 days for whole animal.

Mytilus edulis has been used extensively as a model for molluscan zinc kinetics. In mussels, zinc is taken up by the digestive gland, gills, and mantle and rapidly transported via hemolymph to kidney where it is stored as insoluble granules. There is a high degree of variability in soft tissues of *M. edulis* entirely due to an unusually high degree of variability in kidney zinc of 97.0–7864.0 mg/kg DW. This variability in kidney zinc content is largely due to a low molecular weight zinc complex (MW 700–1300), that showed a high degree of variability and a positive correlation with kidney zinc concentration. But at low ambient concentrations of 50.0 μg Zn/L, the most sensitive bioindicators of zinc exposure were gills and labial palps. Food composition had little effect on tissue distribution of radiozinc-65 in mussels as judged by 5-day feeding studies of radiolabeled diatoms (*Thalassiosira pseudonana*), green alga (*Dunaliella tertiolecta*), glass beads, and egg albumin particles. Soft part BCF values ranged between 12 and 35 times, and was probably due to a rapid desorption of radiozinc from the food particles into the acidic gut, followed by binding to specific ligands or molecules. The Tb1/2 in mussel soft parts ranged from 42 to 80 days for all food items – including glass beads – and about 20 days in shell. Zinc concentrations in mussels are proportionately related to zinc loadings in the water column and the assimilation efficiency from ingested particles, which in *Mytilus edulis* ranges from 32 to 41%. Elevated temperatures in the range 10–25°C were associated with increased uptake rates of zinc from seawater by mussels. If the temperature is oscillated through this range over a 6-h period, there is a further enhancement of zinc uptake. This effect parallels decreases in zinc content of cytosol fractions and increases in granular fractions. Mussels were more sensitive to zinc than were other bivalve mollusks tested. The pumping rate of mussels completely stopped for up to 7 h on exposure to 470.0–860.0 μg Zn/L; however, other bivalves tested showed only a 50% reduction in filtration rates in the range 750.0–2000.0 μg Zn/L. *Mytilus edulis* accumulates zinc under natural conditions, but under some conditions does not depurate. This conclusion was based on results of a study wherein mussels were transferred from a pristine environment in the Netherlands to a polluted estuary for 70 days, then back again for 77 days. At the start, zinc concentration was 106.0 mg/kg DW soft parts. By day 70, it had risen to 265.0 mg/kg DW, at a linear daily uptake of 0.47 mg/kg. But mussels contained 248.0 mg/kg DW on day 147, indicating that elimination was negligible. In another study, zinc depressed sperm motility through respiratory inhibition at 6.5 mg/L, a concentration much higher than those normally found environmentally. In mussel spermatozoa, zinc caused reductions of bound calcium and phosphorus in both acrosomes and mitochondria, suggesting increased permeability of organelle membranes to both elements.

33.7.2.3 Arthropods

Arthropods were the most zinc-sensitive group of invertebrates tested. Toxicity was usually greatest to marine crustaceans, to larvae, at elevated temperatures, during extended exposures, in soft water, under conditions of starvation, at salinity extremes above and below the isosmotic point, in summer, at low concentrations of humic acid, in proximity to anthropogenic discharges, and at low sediment particulate loadings. Acquired zinc tolerance is reported in amphipods collected from zinc-contaminated sewage wastes and in fiddler crabs (*Uca* spp.) from a metals-contaminated area. *Uca* from zinc-contaminated areas were more resistant to zinc than were crabs from pristine areas, as judged by increased survival and lower tissue zinc concentrations. More research seems warranted on acquired zinc tolerance.

Adverse effects of zinc insult to crustaceans include gill histopathology in prawns, *Macrobrachium hendersodyanum*; increased tissue total proteins, decreased glycogen, and decreased acid phosphatase activity in crabs, *Portunus pelagicus*; retardation of limb regeneration of fiddler crab, *Uca pugilator*; and elevated tissue residues in American lobster, *Homarus americanus*. For example, tissue zinc residues in *Homarus americanus* exposed for 4 days to 25.0 mg Zn/L were especially high in gill (2570.0 mg Zn/kg DW vs. 126.0 at start), hepatopancreas (734.0 vs. 135.0), and green gland (1032.0 vs. 148.0). After 7 days in uncontaminated media, tissue zinc residues remained elevated in gill (675.0 mg Zn/kg DW), hepatopancreas (603.0), green gland (286.0), and other tissues. Zinc concentrations in crustacean soft tissues usually range between 50.0 and 208.0 mg/kg DW, and exceed soft tissue zinc enzymatic requirements by factors of 1.4–6.0.

Half-time persistence of zinc in the prawn (*Palaemon elegans*) is about 17 days, and between 30 and 270 days for five other crustacean species. Differences in half-time persistence are linked to differences in excretion rates of ionic zinc and complexed zinc. In general, ionic zinc in crustaceans is excreted first, then complexed zinc; surface-adsorbed zinc is turned over faster than internally adsorbed zinc; and molting accounts for 33–50% loss of the total body burden in crabs.

Crustaceans can accumulate zinc from both water and food. In uncontaminated waters, the diet is probably the major source of zinc.

Absorption from the stomach is efficient and occurs, in part, via the hepatopancreas. When a large pulse of zinc reaches the blood from the stomach, some is excreted, but much is resorbed and stored in the hepatopancreas in a relatively nonlabile form. Ultimately, stored zinc is also excreted, although removal via the gut is unimportant. Zinc absorption occurs initially at the gill surface, followed by transport on a saturable carrier in the cell wall, and is most efficient at low dissolved ambient zinc concentrations. Urinary excretion is an important body removal pathway, especially at high dissolved ambient concentrations when it can account for 70–80% of total zinc excretion.

Barnacles (*Elminius modestus*) usually accumulate zinc to high body concentrations with no significant excretion. Barnacle detoxification mechanisms of the stored zinc includes production of metabolically inert zinc phosphate granules. However, *Elminius modestus* transplanted from an area of high ambient zinc (101.0 μg/L) to a low ambient zinc (4.0 μg/L) environment lost zinc slowly (0.3% body burden daily) over an 11-week period. Whole body zinc burdens declined from 1554.0 mg/kg DW to 125.0 mg/kg DW, or about 4.1 mg/kg DW daily. In the case of *Balanus balanoides*, another barnacle, high BCF values were attributed to inorganic granules that contained up to 38% zinc and accumulated in tissues surrounding the midgut.

Crustaceans – and other groups – can regulate body concentration of zinc against fluctuations in intake, although the ways in which regulation is achieved vary among species. Regulation of whole body zinc to a constant level is reported for many crustaceans, including intertidal prawns (*Palaemon* spp.), sublittoral prawns (*Pandalus montagui*), green crab (*Carcinus maenus*), lobster (*Homarus gammarus*), amphipods (*Gammarus duebeni*), isopods (*Asellus communis*), and crayfish (*Austropotamobius pallipes*). The body zinc concentration at which zinc is regulated in crustaceans usually increases with temperature, salinity, molting frequency, bioavailability of the uncomplexed free metal ions, and chelators in the medium. Lobsters (*Homarus gammarus*) are able to equilibrate over a 30-day period with seawater containing between 2.0 and 505.0 μg/L. In response to a 100-fold rise in seawater concentrations (from 5.0 to 500.0 μg/L), zinc levels in whole body, blood, hepatopancreas, excretory organs, and gills approximately doubled but changed little in muscle. Shell zinc concentrations increased about 12 times, largely through adsorption. Regulation of zinc in lobster blood is achieved by balancing uptake through the gills against urinary excretion and loss over the body surface including the gills. The sublittoral prawn, *Pandalus montagui*, can regulate total body zinc concentration to a constant level (75.0 mg/kg DW) in dissolved zinc concentrations up to 22.0 μg/L, beyond which there is net accumulation of body zinc. This threshold of zinc regulation breakdown is lower than that in *Palaemon elegans* (93.0 μg Zn/L) and *Palaemonetes varians* (190.0 μg Zn/L) under the same physicochemical conditions. The authors conclude that regulation of body zinc concentration is most efficient in decapods adapted to the fluctuating environments of littoral habitats, possibly as a result of changes in permeability of uptake surfaces in combination with improved zinc excretion systems.

Freshwater crayfish (*Orconectes virilis*) are among the more resistant crustaceans (LC50 value of 84.0 mg Zn/L in two weeks), and can easily tolerate the recommended water quality criteria of 50.0–180.0 μg/L; nevertheless, some streams in Arkansas and Colorado contain 79.0–99.0 mg Zn/L. *Orconectes virilis* exposed to very high sublethal ambient zinc concentrations of 63.0 mg/L for 2 weeks show whole body BCF values of only 2; a similar pattern was observed at other concentrations tested. In all cases, zinc tended to concentrate in gill and hepatopancreas at the expense of muscle, carapace, and intestine. In freshwater crayfish (*Procambarus acutus*), the major uptake route was the ambient medium when compared to diet, although retention time was greater for dietary zinc. When dietary zinc was the only zinc source, crayfish rapidly reached a steady state; when water was the only zinc source crayfish did not reach a steady state. Freshwater mysidaceans and their particulate wastes may play an important role in zinc cycling. The freshwater opossum shrimp (*Mysis relicta*) feeding on sediments ingested

2–4 times more zinc than did mysids feeding on zooplankton. However, sediment-feeding mysids excreted 3–5 times more zinc than did zooplankton consumers; zinc concentrations in fecal pellets of sediment feeders were up to 24 times higher than in food. In the freshwater crayfish *Austropotamobius pallipes*, fecal excretion is a major zinc removal pathway; a similar case is made for green crab (*Carcinus maenus*). Marine copepods (*Anomalocera, Acartia, Temora*) excreted 52% of the ingested zinc in fecal pellets that subsequently leached all zinc to seawater within 24 h.

Freshwater insects, including many species of mayflies, damselflies, stoneflies, and caddisflies, are relatively tolerant to zinc, with LC50 values usually >1.33 mg/L – although some species were adversely affected at concentrations between 30.0 and 37.0 μg Zn/L. Mayfly (*Epeorus latifolium*) larvae were adversely affected at ambient water concentrations of 30.0 μg Zn/L, but could tolerate dietary loadings of 600.0 mg Zn/kg DW ration without measurable effects on growth or emergence. Chironomid insect populations were reduced or missing immediately downstream from coal mine drainage containing 5.0–10.0 mg Zn/L; populations further downstream recovered numerically, but diversity was reduced when compared to upstream communities.

33.7.2.4 Annelids

Populations of freshwater oligochaetes and leeches were reduced in numbers of individuals and numbers of taxa in mine tailing effluents containing 146.0–213.0 μg Zn/L or sediments containing >20.0 g Zn/kg DW. Leeches (*Erpobdella octoculata*) experienced a reduction in density and reproductive capacity in streams containing 25.0–310.0 μg Zn/L from mine wastes, and did not avoid these harmful concentrations.

The highest rate of net zinc absorption reported for any group of invertebrates was 2230.0 mg Zn/kg BW daily in sandworms (*Nereis diversicolor*) from sediments with low zinc levels during exposure for 34 days in 250.0 mg Zn/L. At 10.0 mg Zn/L, the rate decreased to 55.0 mg Zn/kg BW daily. Zinc uptake in *Nereis* increased with increasing sediment zinc levels, at lower salinities, and elevated temperatures. Zinc had no significant effect on burrowing behavior of *Nereis*, even at acutely lethal concentrations. Sandworms from zinc-contaminated sediments were more resistant to waterborne zinc insult by 10–100 times than those from clean sediments. Tolerance to zinc in sandworms may be a result of acclimatization or genetic adaptation. In either event, the degree of metal tolerance decreases rapidly as the level of zinc contamination declines, suggesting that some zinc-tolerant worms may be competitively inferior to normal individuals in clean environments. More research on zinc-tolerant populations seems merited.

Unlike other major groups of marine benthic organisms, the polychaete *Neanthes arenaceodentata* has a limited capacity to regulate zinc. Uptake in *Neanthes* occurs from the free ionic pool of zinc whereas EDTA- and EDTA-zinc complexes are largely excluded. Zinc accumulates linearly over time (350 h) and the rate decreases with increasing temperature in the range 4–21°C. It is possible that uptake and accumulation of zinc is passive in *Neanthes* and does not require metabolic energy. Zinc transfer across the plasma membrane is by way of diffusion. Within the cell, zinc binds to a variety of existing ligands which maintain an inwardly directed diffusion gradient, preventing zinc efflux. Accumulation rate is determined by the number and binding characteristics of the available ligands and their accessibility to zinc. After 50 h of exposure, worms selectively accumulate zinc over cadmium from the medium by a process requiring metabolic energy, and this is attributed to a change in the turnover rate and to the size and nature of the pool of zinc-binding ligands.

33.7.2.5 Echinoderms

In echinoderms, zinc concentrations are usually higher in detrital feeders than in carnivores, higher in surface feeders than in sediment feeders, and higher in specimens collected inshore than those collected offshore in deeper waters. Sea cucumbers, *Stichopus tremulus*, accumulate radiozinc-65 from seawater by a factor of 1400; however, radiozinc accumulation data should be viewed with caution as addition of stable zinc can reduce zinc-65 accumulations in echinoderm viscera up to 10-fold. Zinc inhibits the formation of the fertilization membrane in sea urchin eggs, possibly by interfering with cortical granule-derived proteases and proteins.

33.7.2.6 Fishes

Most authorities agree that: (1) freshwater fishes are more sensitive to zinc than marine species; (2) embryos and larvae are the most sensitive developmental stages; (3) lethal and sublethal effects occur in the range 50.0–235.0 μg Zn/L for most species, and 4.9–9.8 μg Zn/L for the sensitive brown trout (*Salmo trutta*); and (4) behavioral modifications, such as avoidance, occur at concentrations as low as 5.6 μg Zn/L.

Signs of zinc poisoning in fish include hyperactivity followed by sluggishness; prior to death, fish swam at the surface, were lethargic and uncoordinated, showed hemorrhaging at gills and base of fins, shed scales, and had extensive body and gill mucus. Zinc is most toxic to yearlings of brown trout in soft water at pH 4–6 and pH 8–9; toxicity at alkaline pH is attributed to the formation of $ZnOH^+$, $Zn(OH)_2$, and $ZnCO_3$ in both hard and soft water – suggesting increased entrapment of metal precipitates within mucus and epithelial layers of the gill. Acute zinc poisoning in fish is generally attributed to blockade of gas exchange across the gills, causing hypoxia at the tissue level. Tissue hypoxia in fish is a major physiological change before death once the gas exchange process at the gills is no longer sufficient to meet its oxygen requirements. Cardiorespiratory responses to zinc in the spangled perch (*Leiopotherapon unicolor*) are similar to those induced by hypoxia; zinc-poisoned perch had damaged gill epithelia, resulting in impaired gas exchange and lowered oxygen tension in arterial blood. Acute exposures to high lethal concentrations of zinc also caused histopathology of epithelia lining the oral cavity.

Many factors modify the lethal properties of zinc to fish. Zinc is more toxic under conditions of comparatively low dissolved oxygen concentrations, high sodium concentrations, decreased loadings of organic complexing agents, and low pH. In guppies (*Poecilia reticulata*), females were more resistant than males to acute zinc insult; adults of both sexes were more resistant than were 5-day-old fry. Dominant bluegills (*Lepomis macrochirus*) survived exposure to 32.0 mg Zn/L longer than did submissive fish. Water temperature is also an important modifier and it is generally agreed that zinc is more toxic at elevated temperatures, provided that acclimatization temperature is considered. For example, cold-acclimatized (3°C) Atlantic salmon (*Salmo salar*) survived longer than warm-acclimatized (19°C) salmon when exposed to lethal concentrations of zinc at their respective acclimatization temperatures. However, at test temperatures lower than their prior acclimatization temperatures, salmon were less tolerant of zinc. Fish surviving high sublethal concentrations of zinc had significant alterations in blood and serum chemistry, liver enzyme activity, and muscle glycogen, total lipids, phospholipids, cholesterol, RNA, and proteins.

Reproductive impairment seems to be one of the more sensitive indicators of zinc stress in freshwater teleosts, with effects evident in the range 50.0–340.0 μg Zn/L. In some cases, reproduction was almost totally inhibited at zinc concentrations that had no effect on survival, growth, or maturation of these same fish. Zinc-induced developmental abnormalities were documented in marine teleosts, but concentrations tested were grossly elevated. Eggs of the Baltic herring (*Clupea harengus*), for example, exposed to >6.0 mg Zn/L have an altered rate of development, and produce deformed larvae with cellular disruptions in the brain, muscle, and epidermis.

Avoidance tests with fathead minnow (*Pimephales promelas*) show that almost all fatheads except males with established territories will avoid 284.0 μg Zn/L when given a choice; however, avoidance thresholds were 6.4 times higher for established males.

Limited tolerance to zinc was observed in freshwater fishes preexposed to sublethal levels of zinc. In one case, rainbow trout acclimatized to 50.0 μg Zn/L for 21 days were up to 5 times more tolerant to subsequent zinc exposures than were nonacclimatized trout; this was not evident at 100.0 μg Zn/L; also, acclimatization to zinc produced tolerances to copper and cadmium in trout. The mechanisms to account for this phenomenon are unknown, but several theories are proposed: increased metallothionein synthesis; high mortality during preexposure may have caused the selection of more zinc-tolerant individuals; and tolerance may be limited to strains capable of increased zinc excretion, although no evidence now exists linking genetic mechanisms to zinc resistance.

The half-time persistence (Tb1/2) of zinc in whole mosquitofish (*Gambusia affinis*) was estimated at 215 days. The half-time persistence of zinc in whole marine fishes ranged from 35 to 75 days in mummichog (*Fundulus heteroclitus*) to 295–313 days in a flatfish (*Pleuronectes platessa*); Tb1/2 in mummichog was shortest at 30°C, longest at 10°C, and intermediate at 20°C.

Fish can accumulate zinc from both the surrounding medium and from their diet. The freshwater zebra danio (*Brachydanio rerio*) accumulated zinc from the medium, but there was no additional zinc enrichment from a *Daphnia* diet. In marine fishes, however, diet was considered to be the major route of zinc intake, and significantly more important than water zinc levels. In freshwater fishes, bioconcentration factors for whole individuals range between 51 and 500 times, but are strongly influenced by dose, duration of exposure, water chemistry, and other variables. In mosquitofish, uptake rate from water and zinc elimination rate decreased with increasing age of the fish. In three-spined stickleback (*Gasterosteus aculeatus*), uptake was greater in hard water than in soft water and higher in larger fish, suggesting a surface adsorption mechanism. In brown trout, however, uptake was lower and excretion greater in hardwater of 220 mg $CaCO_3$/L than in soft water of 9 mg $CaCO_3$/L, thereby reducing tissue burdens. Starved rainbow trout accumulated zinc more rapidly than did fed fish, due to an increased contribution of waterborne zinc to total body zinc levels. Rapidly growing chinook salmon (*Oncorhynchus tshawytscha*) fingerlings removed radiozinc-65 from the medium and retained nearly all of it for 63 days after transfer to uncontaminated media. Most of the zinc-65 was translocated to vertebral column, head, and visceral mass. The outer surface of the bone seems to be an ion-exchange medium capable of taking up large quantities of metal ions whether natural or foreign to the system. Metals thus exchanged from serum proteins may be prevented from undergoing further exchange by the overlayering action of growing bone. Channel catfish (*Ictalurus punctatus*) fingerlings fed diets containing up to 200.0 mg Zn/kg FW ration for 12 weeks had elevated bone zinc levels (359.0 mg/kg DW vs. 254.0 mg/kg DW in controls) and reduced hematocrit, but survival and feed conversion efficiency was the same as controls. Plasma zinc levels in four species of freshwater fishes consuming diets containing 100.0–200.0 mg Zn/kg ration ranged between 9.3 and 15.1 mg Zn/L FW; in rainbow trout, zinc tended to concentrate in the erythrocyte membrane.

In marine fishes, zinc residues were usually higher in dead fish than in live or moribund animals, higher in smaller fish, higher in liver and viscera, and higher with decreasing water cadmium levels. Uptake from the medium by adult mummichogs was inversely related to zinc concentration in the water. In mummichogs, zinc accumulates in scales during exposure to 10.0 mg Zn/L, significantly elevating the zinc:calcium ratio; ratios remained elevated for at least 4 months after transfer to low zinc media, and this phenomenon may have application for environmental monitoring. Scale osteoblasts of zinc-exposed mummichogs showed an increase in the number of lysosome-like structures contained by cytoplasm, and suggests that osteoblast lysosomes are involved in zinc accumulation in fish scales via enzymatic degradation of metallothioneins or other metal-binding proteins. Dietary zinc is not well assimilated in marine flatfish. Turbot (*Scopthalmus maximus*) fed diets containing

100.0 (control) or 1000.0 mg Zn/kg DW for 200 days were not different in renal and hepatic metallothionein levels, or in zinc concentrations in liver, kidney, muscle, skin, or bone; a similar case is made for other marine flatfish. However, turbot injected intraperitoneally (2.0 mg Zn/kg BW) had an 18-fold increase in liver metallothionein content, and a 3-fold increase in liver zinc, confirming the ability of this species to synthesize metallothionein rapidly to a high concentration.

33.7.2.7 Amphibians

Amphibian embryos are more sensitive to zinc than older stages; developmental abnormalities were evident in most species at concentrations >1.5 mg Zn/L. Embryos of the narrow-mouthed toad (*Gastrophryne carolinensis*) seem to be especially sensitive, with adverse effects reported at 10.0 μg Zn/L, but this requires verification. Amphibians, along with other taxonomic groups, were rare or absent in the vicinity of zinc smelters when compared to more distant sites.

In tests with isolated skin of frogs (*Rana* spp.), Zn^{2+} stimulates sodium transport and inhibits chloride-related tissue conductance; however, skin of toads (*Bufo* spp.) is relatively insensitive to zinc. In early stages of embryonic development, Zn^{2+} stimulates multiplication of germ cells, but long term treatment with $ZnSO_4$ has a toxic effect upon the larval gonad and especially on the germ cells of the ovarian structure that is developed in frog larvae.

33.7.3 Birds

Ducks (*Anas* spp.) had reduced survival when fed diets containing 2500.0–3000.0 mg Zn/kg ration or when force-fed zinc metal shot equivalent to 742.0 mg Zn/kg BW. Domestic chickens (*Gallus* sp.) were more resistant: 8000.0 mg Zn/kg ration was fatal to chicks, although higher doses were routinely fed to laying hens to induce molting; 2000.0–3000.0 mg Zn/kg ration inhibited chick growth; 178.0 mg Zn/kg feed caused immunosuppression in chicks; and dietary concentrations as low as 100.0 mg Zn/kg caused pancreas histopathology in chicks under conditions of selenium deficiency. Excessive zinc (2000.0 mg/kg diet for 21 days) fed to chicks (*Gallus* sp.) caused zinc accumulations in tissues, reduced tissue turnover of zinc, reduced liver turnover of iron, and reductions in copper content of liver and pancreas, and in iron of tibia. However, hens were less sensitive and when fed diets containing 2000.0 mg Zn/kg for 44 weeks produced chicks that had no apparent alteration in tissue zinc, copper, or iron metabolism.

Zinc-poisoned mallards (Anas platyrhynchos) force fed zinc shot pellets developed ataxia, paresis, and total loss of muscular control of legs, including ability to swim. The muscular weakness associated with zinc intoxication would probably make ducks highly susceptible to predation, and argues against the use of zinc shot as a substitute for lead shot. Mallards fed 3000.0 mg Zn/kg DW ration for 60 days had diarrhea after 15 days, leg paralysis in 20 days, high mortality after 30 days, and zinc residues at day 60 that were 14 times higher in pancreas than in controls, 7 times higher in liver, 15 times higher in kidney, and 2–4 times higher in adrenals, muscle, testes, and ovary.

In Australia, almost all aviary birds are held in cages made of galvanized wire mesh, resulting in sporadic cases of “new wire disease” caused by the ingestion of galvanized metal used in cage construction. In one case, peach-faced lovebirds (*Agapornis roseicollis*) died within 5 weeks of placement in a newly erected wire cage; dead birds had elevated liver zinc concentrations of 75.0–156.0 mg/kg DW vs. normal values of 21.0–33.0 mg/kg DW. Zinc poisoning in a captive Nicobar pigeon (*Caloenas nicobarica*) was attributed to plated zinc metal fragments found in the gizzard – presumably ingested from the galvanized cage bars. In addition to elevated tissue zinc concentrations, this pigeon had a swollen liver and kidneys, and extensive kidney histopathology. A zinc-poisoned blue and gold macaw (*Ara araruana*) showed weakness, ataxia, extreme thirst, diarrhea, cyanosis, and a plasma zinc concentration of 15.5 mg/L after ingesting galvanized hardware cloth that was 24% zinc by weight and 0.2% lead. The bird was treated with 35.0 mg/kg BW calcium versanate intramuscularly (im) and 30 mg thiamine hydrochloride kg BW (im); recovery following chelation therapy took 2 months, at which time plasma zinc was 0.6–0.8 mg/L vs. 1.3–2.0 mg/L for normal birds. New galvanized wire used in aviary construction should weather for 1–2 months then scrubbed with a mild acidic solution, such as vinegar, and rinsed; flakes of galvanized metal – which contain up to 2.4 g Zn/kg – should be removed before birds are put in cages. Zinc toxicosis was diagnosed in a gray-headed chachalaca (*Ortalis cinereiceps*) after it ingested a copper-plated zinc penny; necropsy showed pancreas histopathology and severe gizzard erosion; liver contained 1910.0 mg Zn/kg FW.

Zinc phosphide – a rodenticide – is relatively toxic when compared to elemental zinc or zinc oxide; most of the biocidal action is attributed to the phosphide fraction. Acute oral LD50s for zinc phosphide ranged between 16.0 and 47.0 mg/kg BW for ring-necked pheasant (*Phasianus colchicus*), golden eagle (*Aquila chrysaetos*), mallard, and horned lark (*Eremophila alpestris*). Signs of zinc phosphide poisoning include excessive drinking, regurgitation, muscular incoordination, appetite loss, sluggishness, rapid breathing, and eyelid droop. Signs appeared as soon as 15 min after dosing, and death usually occurred between 2 and 21 h; remission took up to 1 month.

Large amounts of zinc are crucial for new feather growth. Zinc deficiency during this period results in stunted, frayed, easily broken feathers. Studies with the giant Canada goose (*Branta canadensis maxima*) show that zinc was released from the pectoralis muscle during molt-induced atrophy and used for growth of feathers and leg muscles during this period. High dietary levels of zinc are frequently fed to poultry to force molting and reduce egg deposition. Extremely high dietary levels of 20.0 g Zn/kg ration have been used as a commercial management technique to force the molting of laying hens and the subsequent improvement of long-term egg production that molting produces. Laying hens (*Gallus* sp.) given high zinc diets increased their zinc uptakes 5–40-fold in a dose-dependent pattern despite the decreased food intake associated with high zinc dietary levels. Zinc preferentially accumulated in chicken kidney, liver, pancreas, and gizzard; significant increases in egg zinc occurred at dietary levels of 10.0 and 20.0 g Zn/kg. Unlike adults, high dietary levels of zinc adversely affected pancreatic exocrine function in the chick; effects were exacerbated under conditions of selenium deficiency and feeding of purified diets. Impaired enteric absorption and transport of vitamin E as a consequence of zinc-induced pancreatic insufficiency is a major cause of reduced tissue concentrations of alpha-tocopherol produced in chicks by excess dietary zinc; these effects were magnified by diets low in corn, soybean meals, and other materials known to chelate zinc, and thus reduce its biological availability. Excess dietary zinc causes pancreatic damage in the chick, including reduced activities of major digestive enzymes, elevated plasma amylase activities, reduced digestibility of starch, and reduced vitamin A activity; these changes were associated directly with elevated tissue zinc concentrations, especially in the pancreas.

Zinc-sulfate-treated homing pigeons had impaired navigational ability when released approximately 60 km from their roosts. Homing pigeons made anosmic by treating their olfactory mucosa with zinc sulfate solution had significantly poorer homing behavior than did unmanipulated or treated control pigeons, as judged by homeward initial orientation and homing performance.

33.7.4 Mammals

Livestock and small laboratory animals are comparatively resistant to zinc, as judged by their tolerance for extended periods to dietary loadings >100 times the minimum recommended daily zinc requirement. Nevertheless, excessive zinc intake through inhalation or oral exposure can have dramatic effects on survival, metabolism, and well being. Sensitive species of mammals were affected at 90.0–300.0 mg Zn/kg diet, >300.0 mg Zn/L drinking water,

>90.0 mg/kg BW daily, >350.0 mg Zn/kg BW as a single oral dose, and >0.8 mg Zn/m^3 air.

Zinc is relatively nontoxic in mammals. A wide margin of safety exists between normal intakes and those producing deleterious effects. In most cases, dietary levels up to 100 times the daily requirement for extended periods show no discernable effects. The possibility of oral zinc intoxication in adult humans is unusually low, as judged by the low (40%) bioavailability of zinc from the gastrointestinal tract and the high tolerances to zinc reported in domestic livestock and small laboratory animals. Humans ingesting up to 12 g of elemental zinc over a 2-day period, equivalent to 33.0 mg/kg BW for a 60-kg adult, show no evidence of hematologic, hepatic, or renal toxicity.

Excessive zinc intake adversely affects survival of all mammals tested – including humans – and produces a wide variety of neurological, hematological, immunological, hepatic, renal, cardiovascular, developmental, and genotoxic effects. The most sensitive species of mammals tested showed adverse effects at dietary levels of 80.0–90.0 mg Zn/kg in humans, 300.0 mg Zn/kg ration in domestic cats, and 500.0 mg Zn/kg feed in rats; drinking water concentrations of 300.0 mg/L in domestic mice, and 800.0 mg Zn/L in laboratory white rats; daily whole body intakes >90.0 mg Zn/kg in horses; acute oral LD50 doses of 350.0–800.0 mg Zn/kg BW in rats; intraperitoneal injections of 13.0 mg Zn/kg BW in mice; and 0.8 mg Zn/m^3 air in guinea pigs.

Metal fume fever is commonly encountered among industrial workers exposed to zinc fumes, and is characterized by pulmonary irritation, fever, chills, and gastroenteritis. Attacks begin 4–8 h after exposure, and recovery in 24–48 h. The pathogenesis of metal fume fever is unknown, but may be associated with endogenous pyrogens released by cell lysis. Rabbits, rats, and cats exposed to zinc oxide fumes for 3.5 h at concentrations of 110.0–600.0 mg/m^3 reacted with a transient fall in body temperature followed by leucocytosis; heavily exposed animals had signs of bronchopneumonia. The atmospheric threshold limit value (TLV) for zinc is 5.0 mg/m^3; however, results of studies with guinea pigs suggest that the TLV value for zinc oxide should be lowered.

Excessive zinc uptake is associated with lameness, unthrifty appearance, and osteochondrosis in foals and pigs, nephrosis in ferrets, and pancreatic fibrosis in sheep. Zinc-poisoned mammals are usually characterized by a decreased growth rate, subcutaneous hematomas, ulcerative gastritis, hemorrhagic enteritis, lesions of major limb joints, renal lesions, elevated serum and tissue zinc concentrations, acute diarrhea, copper deficiency, impaired reproduction, and decreased activity of cardiac and hepatic cytochrome oxidase. In severe cases histopathological changes in liver and especially pancreas are evident, as are degenerative changes in kidney and gastrointestinal tract, followed by life-threatening hemolytic anemia. The pancreas is key in the diagnosis of zinc toxicity and in estimating the period of exposure; in sheep, it takes about four weeks of continued ingestion of toxic amounts of zinc before the pancreas is affected. More research is needed on the role of the pancreas in zinc toxicokinetics.

Zinc is important to the normal functioning of the central nervous system (CNS). At low concentrations, zinc protects mammalian brain neurons by blocking *N*-methyl-D-aspartate receptor-mediated toxicity. At high concentrations, zinc is a potent, rapidly acting neurotoxicant in the mammalian brain, as judged by zinc-induced neuronal injury of in vitro mature cortical cell cultures. Increased brain levels of zinc are associated with Pick's disease in certain strains of rodents with inherited epileptic seizures. Intravenous injection of zinc in rats with genetically inherited epilepsy produces seizures; a similar response occurs with intracranial injection of zinc in rabbits with inherited audiogenic seizures.

Zinc fed to adult male rats at 500.0 mg/kg diet for 3 weeks or longer negatively impacts the testes and other male accessory organs; effects are a direct result of zinc cytotoxicity from transfer across the blood–testes barrier. Elevated dietary zinc also depresses bone calcium levels and increases fecal calcium loss in rats. Increases in serum zinc levels of rats after acute zinc overload is mainly due to increases in the zinc bound to the albumin fraction, and secondarily to that bound to the globulin fraction. Albumin may play a new physiological role by fitting its binding capacity to serum zinc levels, essentially binding all excess zinc that arrives in the blood.

Zinc toxicosis has been observed in humans and livestock after ingestion of acidic foods or drink prepared and stored in galvanized containers. Symptoms appear within 24 h and include nausea, vomiting, diarrhea and abdominal cramps. The emetic dose for zinc in humans was estimated at 225.0–450.0 mg (3.2–6.4 mg Zn/kg BW), equivalent to 1.0–2.0 g of zinc sulfate. Zinc poisoning in dogs is well documented as a result of ingestion of galvanized metal objects, calamine lotion, skin and sunblock preparations containing zinc oxide, staples, nails, fertilizers, some paints, products containing zinc undecylenate, metallic hardware items with a high zinc content, nuts found on certain types of animal transport cages, and pennies. The propensity of some individuals to throw pennies (U.S. coinage) into animal cages while visiting zoos and animal parks should be considered a potential source of zinc poisoning in captive animals. Pennies minted prior to 1982 contain 95% copper and 5% zinc; however, copper-clad pennies minted after 1981 contain 97.6% zinc and 2.4% copper.

Humans given zinc supplements should be aware of possible complications attendant to their use. Low intakes of 100.0–300.0 mg of zinc daily in excess of the recommended dietary allowance of 15.0 mg Zn daily may produce induced copper deficiency, impaired immune function, and disrupted blood lipid profiles. Patients treated with zinc supplements (150.0 mg daily) to control sickle cell anemia and nonresponsive celiac disease developed a severe copper deficiency in 13–23 months; normal copper status was restored by cessation of zinc supplements and increased dietary copper.

Due to false positives, zinc may confound interpretation of the paralytic shellfish poisoning (PSP) mouse bioassay, one of the routine tests used to measure shellfish safety for human consumption. For example, mice injected intraperitoneally with extracts of healthy oyster tissues showed extreme weakness, a drop in body temperature, cyanosis, and some deaths. The threshold for a toxic PSP response corresponds to a drained tissue zinc level >900.0 mg/kg FW, and this overlaps the zinc concentration range of 230.0–1650.0 mg/kg FW (1900.0–9400.0 mg/kg DW) recorded in healthy oyster soft tissues.

33.8 Recommendations

For growing agricultural crops: (1) sewage sludge may be applied to soils provided that total zinc content does not exceed 150.0–560.0 kg per surface hectare; (2) a maximum permissible extractable soil zinc concentration of 23.0 mg/kg DW is recommended, according to Soviet agronomists; and (3) seedlings of oak (*Quercus* spp.) and red maple (*Acer rubrum*) will eventually die in culture medium containing >100.0 mg Zn/kg, although total zinc concentrations for global crop production routinely exceed 100.0 mg/kg DW soil. Research is needed in standardized methodology for measurement of bioavailable (i.e., extractable) soil zinc and on its relation to other soil measurements such as total zinc, and depth of cultivation in the case of surface application. Data are limited on zinc hazards to terrestrial invertebrates; however, sensitive species are adversely affected at dietary concentrations >300.0 mg Zn/kg, or at soil concentrations >400.0 mg/kg.

Water quality criteria for protection of aquatic life should include both total recoverable zinc and acid-soluble zinc. For example, if total recoverable zinc is substantially above the proposed criteria and acid-soluble zinc is below the limit, there is cause for concern. To protect approximately 95% of freshwater animal genera, the U.S. Environmental Protection Agency recommends water concentrations that average <47.0 μg total recoverable zinc per liter, not to exceed 180.0 μg/L at any time in soft water (i.e., <50 mg $CaCO_3$/L), or a mean concentration of 59.0 μg acid-soluble zinc per liter, not to exceed 65.0 μg/L at any time in soft water (Table 33.2). These criteria are unsatisfactory because lower ambient zinc concentrations between 5.0 and 51.0 μg/L clearly have significant negative effects on

Table 33.2. Proposed zinc criteria for the protection of natural resources and human health.

Resource, Criterion, and Other Variables	Effective Zinc Concentration
CROP PLANTS	
Sewage sludge applied to agricultural soils	
Europe, acceptable	150.0–<300.0 kg/ha at pH 6.0–7.0
Florida	
Maximum permissible	205.0 kg/ha
Unacceptable	>10,000.0 mg/kg DW
Oregon[a], Wisconsin[a], acceptable	250.0–<1000.0 kg/ha
Vermont[a], acceptable	280.0–<1120.0 kg/ha
Maryland[a], Massachusetts[a], acceptable	280.0–<560.0 kg/ha
Minnesota[a], Missouri[a], acceptable	280.0–<1120.0 kg/ha
Illinois, maximum	560.0 kg/ha
Soils	
Soviet Union, maximum permissible	23.0 mg/kg DW, extractable by ammonium acetate buffer at pH 4.8
Alberta, Canada: for growing livestock forage	<100.0 mg/kg DW
Quebec, Canada	
Background	200.0 mg/kg DW
Marginal	500.0 mg/kg DW
Unacceptable	>3000.0 mg/kg DW
Netherlands	
Background	200.0 mg/kg DW
Marginal	500.0 mg/kg DW
Unacceptable	>3000.0 mg/kg DW
Ontario, Canada, acceptable	<220.0 mg/kg DW
Germany, acceptable	<300.0 mg/kg DW
New Jersey, goal	<350.0 mg/kg DW
New York, acceptable	
Agricultural soils	168.0–<250.0 kg/ha DW
Forest soils	<560.0 kg/ha DW
TERRESTRIAL INVERTEBRATES	
Earthworms; soil	
High accumulations, but otherwise safe	97.0 mg/kg DW
Adverse effects	>400.0 mg/kg DW
Slugs; diet, adverse effects	>300.0 mg/kg DW
FRESHWATER AQUATIC LIFE	
Water	
Total recoverable zinc	
50 mg $CaCO_3/L$	47.0 μg/L, 24 h average; not to exceed 180.0 μg/L at any time
100 mg $CaCO_3/L$	47.0 μg/L, 24 h average; not to exceed 320.0 μg/L at any time
200 mg $CaCO_3/L$	47.0 μg/L, 24 h average; not to exceed 570.0 μg/L at any time

Table 33.2. cont'd

Resource, Criterion, and Other Variables	Effective Zinc Concentration
Acid-soluble zinc[b]	4-day average concentration not to exceed the numerical value $e^{((0.8473\,[\ln]\,\text{hardness})+0.7614)}$ more than once every 3 years on average; 1-h concentration not to exceed $e^{((0.8473\,[\ln]\,\text{hardness})+0.8604)}$ more than once every 3 years on average. See below for examples.
50 mg $CaCO_3$/L	4-day average not to exceed 59.0 μg/L; 1-h average not to exceed 65.0 μg/L
100 mg $CaCO_3$/L	4-day average not to exceed 110.0 μg/L; 1-h average not to exceed 120.0 μg/L
200 mg $CaCO_3$/L	4-day average not to exceed 190.0 μg/L; 1-h average not to exceed 210.0 μg/L
Adverse effects, most sensitive species	
Brown trout, Salmo trutta, embryos and fry	4.9–19.6 μg/L
Daphnid, Daphnia magna	5.0–14.0 μg/L
Rainbow trout, Oncorhynchus mykiss	5.6–10.0 μg/L
Narrow-mouthed toad, Gastrophryne carolinensis, embryos	10.0 μg/L
Daphnid, Daphnia galeata mendotae	15.0–30.0 μg/L
Freshwater sponge, Ephydatia fluviatilis	26.0 μg/L
Mayfly, Epeorus latifolium	30.0 μg/L
Midge, Tanytarsus dissimilis	37.0 μg/L
Atlantic salmon, Salmo salar	50.0 μg/L
Cladoceran, Ceriodaphnia reticulata	51.0 μg/L
Flagfish, Jordanella floridae	51.0 μg/L
Diet	
Channel catfish, *Ictalurus punctatus*	
Minimum	20.0 mg/kg DW
Recommended[c]	150.0–200.0 mg/kg DW
Rainbow trout, *Oncorhynchus mykiss*	
Minimum	10.0–30.0 mg/kg DW; 15.0–30.0 mg/kg FW
Adequate	90.0 mg/kg FW
Sediments	
Great Lakes	
Safe	<90.0 mg/kg DW
Marginal	90.0–200.0 mg/kg DW
Unacceptable	>200.0 mg/kg DW
Wisconsin and Ontario, for Great Lakes sediments dredged from harbors and for disposal in water	<100.0 mg/kg DW

Continued

Table 33.2. cont'd

Resource, Criterion, and Other Variables	Effective Zinc Concentration
MARINE AQUATIC LIFE	
Seawater	
Total recoverable zinc	58.0 μg/L, 24-h average; not to exceed 170.0 μg/L at any time
Acid-soluble zinc[b]	4-day average concentration does not exceed 86.0 μg/L more than once every 3 years on average; 1-h average concentration does not exceed 95.0 μg/L more than once every 3 years on average
No adverse effects, most species	
Algae	<1400.0 μg/L
Mollusks	<54.0 μg/L
Crustaceans	<230.0 μg/L
Adverse effects, most sensitive species	
Brown algae, Fucus serratus	8.8–9.5 μg/L
Copepod, Tisbe holothuriae	10.0 μg/L
Pacific oyster, Crassostrea gigas, larvae	10.0–20.0 μg/L
Alga, Rhizosolenia spp.	15.0–25.0 μg/L
Diatom, Schroederella schroederi	19.0 μg/L
Diatom, Skeletonema costatum	19.6 μg/L
Dinoflagellate, Glenodinium halli	20.0 μg/L
Purple sea urchin, Strongylocentrotus purpuratus, embryos	23.0 μg/L
Sand dollar, Dendraster excentricus	28.0 μg/L
Atlantic herring, Clupea harengus harengus, embryos	50.0 μg/L
Mud crab, Rithropanopeus harrissii, larvae	50.0 μg/L
Diet; fish, adequate	90.0 mg/kg FW
Tissue residues; minimum theoretical requirement for whole mollusks and crustaceans	34.5 mg/kg DW
BIRDS	
Mallard, *Anas platyrhynchos*; zinc-poisoned	
Diet	2500.0–3000.0 mg/kg DW ration
Single oral dose	0.64 mg; 517.0–742.0 mg/kg BW
Selected species; tissue concentrations	
Normal	
Liver	21.0–33.0 mg/kg DW
Plasma	1.3–2.0 μg/L
Zinc-poisoned	
Liver	75.0–156.0 mg/kg DW
Plasma	15.5 mg/L

Table 33.2. cont'd

Resource, Criterion, and Other Variables	Effective Zinc Concentration
Japanese quail, *Coturnix japonica*; safe level; diet	25.0–30.0 mg/kg DW diet
Chicken, *Gallus* sp.	
Recommended daily intake	>31.0 mg
Diet	
Adverse effects, zinc deficiency	<38.0 mg/kg DW ration
Adequate	93.0–120.0 mg/kg DW ration
Excessive	>178.0 mg/kg DW ration
Toxic	>2000.0 mg/kg DW ration
MAMMALS	
Cattle, *Bos* spp.	
Diet	
Soluble zinc, recommended level	
Calves	>8.0 mg/kg DW
Adults	
Beef cattle	10.0–30.0 mg/kg DW
Dairy cattle	40.0 mg/kg DW
Total zinc	
Marginal	25.0 mg/kg DW
Recommended	45.0–60.0 mg/kg DW
Maximum tolerated	
Calves	500.0 mg/kg DW
Adults	1000.0 mg/kg DW
Toxic	>900.0–2000.0 mg/kg DW
Tissue residues	
Liver	
Zinc-deficient	<10.0 mg/kg DW
Suboptimal	10.0–30.0 mg/kg DW
Optimal	30.0–120.0 mg/kg DW
Excessive	>120.0 mg/kg DW
Lethal	>500.0 mg/kg DW
Plasma	
Zinc-deficient	<0.66 mg/L
Normal	1.02 mg/L
Elevated	1.5 mg/L
Serum; zinc-deficient	<0.6 mg/L
Recommended daily intake	
Calves	
5-month old	3.0 g (25.0–35.0 mg/kg BW)
14–8-month old	16.0 g (50.0–80.0 mg/kg BW)
Cows	55.0 g (110.0–140.0 mg/kg BW)
Dog, *Canis familiaris;* tissue concentrations, normal vs. zinc-poisoned	
Serum	0.7–1.1 vs. 33.0 mg/L

Continued

Table 33.2. cont'd

Resource, Criterion, and Other Variables	Effective Zinc Concentration
Plasma	0.6–1.0 vs. 16.0–32.0 mg/L
Urine	1.3–2.0 vs. 20.0–25.0 mg/L
Liver	17.0–32.0 vs. 369.0 mg/kg FW
Kidney	9.0–23.0 vs. 295.0 mg/kg FW
Guinea pig, *Cavia* spp.	
Air; adverse effects	0.8–4.0 Zn/m^3
Diet	
Deficient	3.0 mg/kg DW plus 1.0 mg/L drinking water
Adequate	3.0 mg/kg DW plus 15.0 mg/L drinking water
Normal	20.0 mg/kg DW
Adequate	100.0 mg/kg FW
High	200.0 mg/kg DW
Tissue concentrations; zinc deficient vs. normal	
Serum	0.5 vs. 1.6–2.0 mg/L
Liver	9.4 vs. 15–17 mg/kg FW
Testes	9.5 vs. 19–27 mg/kg FW
Kidney	10.0 vs. 18–20 mg/kg FW
Domestic goat, *Capra* sp.	
Diet	
Soluble zinc, recommended	
Adults	>4.0 mg/kg DW
Kids	>7.0 mg/kg DW
Total zinc	
Deficient	<15.0 mg/kg DW
Recommended	80.0 mg/kg DW
Bank vole, *Clethrionomys glareoulus;* diet; recommended	30.0 mg/kg DW
Horse, *Equus caballus*	
Diet	
No adverse effects	250.0 mg/kg DW
Adverse effects	1000.0 mg/ kg DW
Daily intake; adverse effects	>90.0 mg/kg BW
Domestic cat, *Felis domesticus;* diet; adverse effects	300.0 mg/kg DW
Humans, *Homo sapiens*	
Air	
Safe levels	
Zinc chloride, fumes	<1.0 mg/m^3
Zinc oxide, fumes	<5.0 mg/m^3
Zinc and zinc oxides	5.0–10.0 mg/m^3
Zinc oxide, total dust	10.0 mg/m^3
Zinc oxide, fume and dust, ceiling limit	15.0 mg/m^3

Table 33.2. cont'd

Resource, Criterion, and Other Variables	Effective Zinc Concentration
Adverse effects; zinc oxides	600.0 mg/m^3 for 10 min
Daily intake	
Recommended dietary intake, assuming availability of 20%	
Children	
To age 1 year	3.0–6.0 mg
1–10 years	8.0–10.0 mg
No age specified	10.0 mg
Males	
Age 11–17 years	14.0–15.0 mg
Age 18+	11.0–15.0 mg
No age specified	15.0 mg
Females	
Age 10–13 years	13.0–15.0 mg
Age 14+	11.0–15.0 mg
No age specified	12.0 mg
Pregnant	15.0–20.0 mg
Lactating	25.0–27.0 mg
Maximum safe total, adults	
Not zinc deficient	0.3–1.0 mg/kg BW
Zinc deficient	1.0 mg Zn/kg BW, oral administration
Adverse effects level	>160.0 mg (>2.3 mg/kg BW)
Diet	
Safe level; seafoods; Australia	<40.0 mg/kg FW
Adverse effects	
Gastrointestinal disorders	>80.0 mg/kg DW diet for 6 weeks
Severe copper deficiency	150.0 mg Zn daily for 13–23 months
Vomiting	Single dose of 225.0–450.0 mg Zn or 1.0–2.0 g of $ZnSO_4$
Drinking water	
Safe level	5.0 mg/L
Adverse effect (acute GI distress)	>280.0 mg/L
Intravenous injection; adverse effects	23.0 mg/kg BW daily
Soils	
Canada, nonhazardous to human health	
Ontario; residential, parkland, commercial, industrial	< 800.0 mg/kg DW
Alberta; noncrop uses	< 700.0 mg/kg DW
Tissue residues	
Serum	
Normal	0.5–1.29 mg/L
No toxic effects	1.92 mg/L

Continued

Table 33.2. cont'd

Resource, Criterion, and Other Variables	Effective Zinc Concentration
Plasma	
Zinc-deficient	0.4–0.6 mg/L
Normal	0.7–1.1 mg/L
GI disturbances	1.51 mg/L
Rhesus monkey, *Macaca mulatta*; diet	
Deficient	4.0 mg/kg DW
Adequate	100.0 mg/kg DW
Mouse, *Mus* spp.	
Diet	
Zinc-deficient	<5.0 mg/kg DW
Zinc-adequate	36.5 mg/kg DW
Tolerated	100.0 mg/kg DW
Tolerated	682.0 mg/kg DW for 13 weeks (107.0 mg/kg BW)
Harmful	500.0 mg/kg DW for 3 months
Harmful	6820.0 mg/kg DW
Fatal	30,000.0 mg/kg DW for 13 weeks
Drinking water; adverse effects	300.0 mg/L
Tissue residues	
Blood	
Deficient	0.7 mg/L
Normal	1.0–1.1 mg/L
Liver	
Deficient	12.0 mg/kg FW
Normal	17.0–19.0 mg/kg FW
European ferret, *Mustela putorius furo*; diet	
Tolerated	500.0 mg/kg DW
Fatal	1500.0 mg/kg DW
Mink, *Mustela vison*; diet	
Zinc-deficient	4.1 mg/kg FW
Adequate	35.0–45.0 mg/kg FW; 100.0–150.0 mg/kg DW
Domestic sheep, *Ovis aries*	
Diet	
Soluble zinc, adequate	
Adults	>4.0 mg/kg DW
Lambs	>7.0 mg/kg DW
Total zinc	
Adults, adequate	33.0 mg/kg DW
Lambs adequate	124.0–130.0 mg/kg DW
Harmful	>1000.0 mg/kg DW
Recommended daily intake	>18.0 mg

Table 33.2. cont'd

Resource, Criterion, and Other Variables	Effective Zinc Concentration
Tissue residues	
Feces	
Normal	158.0 mg/kg DW
Zinc-poisoned	4900.0 mg/kg DW
Kidney	
Normal	84.0–150.0 mg/kg DW
Elevated	>180.0 mg/kg DW
Zinc-poisoned	274.0–760.0 mg/kg DW
Liver	
Normal	144.0–165.0 mg/kg DW
Elevated	>250.0 mg/kg DW
Zinc-poisoned	463.0–650.0 mg/kg DW
Pancreas	
Normal	88.0 mg/kg DW
Zinc-poisoned	752.0 mg/kg DW
Harbor porpoise, *Phocoena phocoena*; liver	
Normal homeostatic range	20.0–100.0 mg/kg FW
Impaired regulating mechanism	<20.0 mg/kg FW or >100 mg/kg FW
Laboratory white rat, *Rattus* sp.	
Diet	
Soluble zinc, recommended	15.0 mg/kg DW
Total zinc	
Zinc-deficient	<12.0 mg/kg DW
Adequate	76.0 mg/kg DW
Adverse effects	>500.0 mg/kg DW
Fetotoxic	>4000.0 mg/kg DW
Daily intake	
Tolerated	320.0 mg/kg BW
Harmful	640.0 mg/kg BW
Single oral dose, harmful	>350.0 mg/kg BW
Domestic pig, *Sus* sp.	
Diet	
Soluble zinc, safe levels	
Normal diet	14.0–20.0 mg/kg DW
Cassava-rice-bran diet	>40.0 mg/kg DW
Soy base diet	50.0 mg/kg DW
Total zinc, harmful	1000.0 mg/kg DW
Recommended daily intake	>20.0 mg

[a] Higher values permissible for soils with higher cation exchange capacity.

[b] Zinc that passes through a 0.45 μm membrane filter after acidification to pH 1.5–2.0 with nitric acid.

[c] Higher concentrations recommended to compensate for reduced bioavailability caused by excess calcium and phytate in diet.

growth, survival, and reproduction of important species of freshwater fishes, amphibians, insects, sponges, and crustaceans (Table 33.2). Some downward modification seems necessary in the current proposed zinc criteria for freshwater aquatic life protection. To protect important species of marine animals, the U.S. Environmental Protection Agency recommends that total recoverable zinc in seawater should average <58.0 μg/L and never exceed 170.0 μg/L; for acid-soluble zinc, these values are <86.0 and 95.0 μg/L (Table 33.2). As was the case for freshwater biota, there is a growing body of evidence demonstrating that many species of marine plants, crustaceans, mollusks, echinoderms, and fish are adversely affected at ambient zinc concentrations between 9.0 and 50.0 μg/L, or significantly below the current proposed criteria for marine life protection.

Zinc deficiency effects have been produced experimentally in freshwater sponges at <0.65 μg Zn/L, in rainbow trout fed diets containing <15.0 mg Zn/kg FW, in certain species of marine algae at <0.7 μg Zn/L, and in certain species of marine invertebrates at <6.5 μg Zn/L or <34.0 mg Zn/kg DW whole organism. Zinc deficiency in natural aquatic ecosystems has not yet been credibly documented and merits additional research.

In aquatic environments, three research needs are recommended: (1) development of analytical procedures for measurement of individual dissolved zinc species, notably the aquo ion and zinc chloride, and for nondissolved species that occur in natural waters; (2) separation of natural from anthropogenic influences of sediment–water interactions on flux rates, with emphasis on anoxic conditions, the role of microorganisms, and the stability of organozinc complexes; and (3) establishment of toxicity thresholds for aquatic organisms based on bioaccumulation and survival to determine the critical dose and the critical dose rate, with emphasis on aquatic communities inhabiting locales where zinc is deposited in sediments.

Bird diets, in mg Zn/kg DW feed, should contain 25.0–38.0 to prevent zinc deficiency effects, 93.0–120.0 for adequate to optimal growth, <178.0 to prevent marginal sublethal effects, and <2000.0 to prevent the death of chicks and ducklings (Table 33.2). Extremely high dietary levels of 20.0 g Zn/kg ration are fed routinely to laying hens by poultry managers to force molting and to improve long-term egg production; in these cases zinc preferentially accumulates in kidney, liver, pancreas, and eggs. Much additional work now seems warranted on the role of zinc in avian nutrition, and on the significance of tissue concentrations as an indicator of zinc stress.

The normal daily intake for all human age groups ranges between 8.0 and 14.0 mg, but pregnant women require an additional 350.0–375.0 mg of zinc during the course of their pregnancy; however, zinc used therapeutically in humans at >160.0 mg daily may have deleterious effects on copper status. Lower levels – close to the recommended daily allowance of 15.0 mg – are reported to interfere with iron metabolism and with high density lipoprotein cholesterol concentrations, but this requires verification. The proposed air quality criterion for human health protection is 5.0 mg Zn/m^3, although this is demonstrably harmful to guinea pigs (Table 33.2). It is not yet known if guinea pigs are more sensitive than humans to atmospheric zinc, or if some downward modification is needed in the current zinc air quality criterion for protection of human health, and presumably wildlife. Eating seafoods that contain high concentrations of zinc does not seem to present a threat to human health. However, oysters from Tasmania allegedly caused nausea and vomiting in some people who ate them; these oysters contained about 20.0 g of zinc per kg soft parts FW, or about 500 times more than the Australian food regulation of 40.0 mg/kg FW.

Single oral doses >350.0 mg Zn/kg BW were fatal to rats, although doses of 320.0 mg/kg BW were tolerated (Table 33.2), suggesting a rapid breakdown in ability to regulate zinc in a relatively narrow critical threshold range. More research seems needed on zinc regulation of massive doses. Data are scarce in mammals that link zinc concentrations in tissues with environmental zinc perturbations. In harbor porpoises, impaired homeostasis reportedly occurs when zinc exceeds 100.0 mg/kg FW liver; however, livers of many species of marine mammals routinely exceed this value. Elevated zinc concentrations, in mg Zn/kg DW tissue, were >120.0 in cattle liver, >180.0 in sheep kidney, and >250.0 in sheep liver, but their significance is unclear. No international regulations or guidelines applicable to zinc are available. No U.S. Food and Drug Administration "action level" or other maximum acceptable concentration exists for zinc, and therefore no Final Residue Value can be calculated. This seems to be a high priority research need.

In mammals, large differences are evident between and within species in resistance to zinc poisoning and in sensitivity to zinc nutritional needs (Table 33.2). Adverse effects of excess dietary zinc occurred in sensitive species at 80.0 (human) and 300.0 (cat) mg Zn/kg DW; other species tested were significantly more resistant. Daily intake rates considered harmful over long periods ranged from about 2.3 mg/kg BW in humans to >90.0 mg/kg BW in horses. Dietary loadings that optimally prevented zinc deficiency, in mg Zn/kg DW diet, were 30.0 for bank voles, 33.0 for adult sheep (124.0–130.0 for lambs), 37.0 for mice, 45.0–60.0 for cattle, 76.0 for rat, 80.0 for goat, 100.0 for monkey, and 150.0 for mink; recommended daily intake rates in mg/kg BW, ranged from about 0.2 in humans to 110.0–140.0 in cattle. More research is needed on the interaction effects of zinc with proteins, calcium, chloride, and other trace elements, and on the long term consequences of nutrient interactions using animals of various age and nutrient status.

33.9 Summary

World production of zinc is estimated at 7.1 million tons; the United States produces about 4% of the total and consumes 14%. Zinc is used primarily in the production of brass, noncorrosive alloys, and white pigments; in galvanizing iron and steel products; in agriculture as a fungicide, and as a protective agent against soil zinc deficiency; and therapeutically in human medicine. Major sources of anthropogenic zinc in the environment include electroplaters, smelting and ore processors, mine drainage, domestic and industrial sewage, combustion of solid wastes and fossil fuels, road surface runoff, corrosion of zinc alloys and galvanized surfaces, and erosion of agricultural soils.

Zinc has its primary effect on zinc-dependent enzymes that regulate RNA and DNA. The pancreas is a primary target organ in birds and mammals, followed by bone; in fish, gill epithelium is a primary target site. Dietary zinc absorption is highly variable in animals; in general, it increases with low body weight and low zinc status, and decreases with excess calcium or phytate and by deficiency of pyridoxine or tryptophan. Low molecular weight proteins called metallothioneins play an important role in zinc homeostasis and in protection against zinc poisoning; zinc is a potent inducer of metalliothioneins. Zinc interacts with many chemicals to produce altered patterns of accumulation, metabolism, and toxicity; some interactions are beneficial to the organism while others are not, depending on the organism, its nutritional status, and other variables. Knowledge of these interactions is essential to the understanding of zinc toxicokinetics.

In natural waters, dissolved zinc speciates into the toxic aquo ion $(Zn(H_2O)_6)^{2+}$, other dissolved chemical species, and various inorganic and organic complexes; zinc complexes are readily transported. Aquo ions and other toxic species are most harmful to aquatic life under conditions of low pH, low alkalinity, low dissolved oxygen, and elevated temperatures. Most of the zinc introduced into aquatic environments is eventually partitioned into the sediments. Zinc bioavailability from sediments is enhanced under conditions of high dissolved oxygen, low salinity, low pH, and high levels of inorganic oxides and humic substances.

Zinc and its compounds induce testicular sarcomas in birds and rodents when injected directly into the testes; however, zinc is not carcinogenic by any other route. Growth of animal tumors is stimulated by zinc, and retarded by zinc deficiency. Under some conditions excess zinc can suppress carcinoma growth, although the mechanisms are imperfectly understood. Organozinc compounds are effective mutagens when presented to susceptible cell populations in an appropriate form; the evidence for mutagenic potential of inorganic zinc compounds is incomplete. Zinc deficiency can lead to chromosomal aberrations, but excess zinc was not mutagenic in the majority of tests. Excess zinc is teratogenic to frog and fish embryos, but conclusive evidence of teratogenicity in higher vertebrates is lacking. In mammals, excess zinc may protect against some teratogens. Zinc deficiency may exacerbate the teratogenic effects of known teratogens, especially in diabetic animals.

Background concentrations of zinc seldom exceed 40.0 μg/L in water, 200.0 mg/kg in soils and sediments, or 0.5 μg/m^3 in air. Environments heavily contaminated by anthropogenic activities may contain up to 99.0 mg Zn/L in water, 118.0 g/kg in sediments, 5.0 g/kg in soil, and 0.84 μg/m^3 in air. Zinc concentrations in field collections of plants and animals are extremely variable and difficult to interpret. Most authorities agree on six points: (1) elevated concentrations (i.e., >2.0 g Zn/kg fresh weight FW) are normally encountered in some species of oysters, scallops, barnacles, red and brown algae, and terrestrial arthropods; (2) concentrations, in mg Zn/kg dry weight (DW) tissue, are usually <700.0 in fish, <210.0 in birds, and <210.0 in mammals; (3) concentrations are higher in animals and plants collected near zinc-contaminated sites than in the same species collected from more distant sites; (4) zinc content in tissue is not proportionate to that of the organism's immediate surroundings; (5) for individual species, zinc concentration varies with age, sex, season, tissue or organ, and other variables; and (6) many species contain zinc loadings far in excess of immediate needs, suggesting active zinc regulation.

The balance between excess and insufficient zinc is important. Zinc deficiency occurs in many species of plants and animals, with severe adverse effects in all stages of growth, development, reproduction, and survival. In humans, zinc deficiency is associated with delayed sexual maturation in adolescent males; poor growth in children; impaired growth of hair, skin, and bones; disrupted vitamin A metabolism; and abnormal taste acuity, hormone metabolism, and immune function. Severe zinc deficiency effects in mammals are usually prevented by diets containing >30.0 mg Zn/kg DW ration. Zinc deficiency effects are reported in aquatic organisms at nominal concentrations between 0.65 and 6.5 μg Zn/L of medium, and in piscine diets at <15.0 mg Zn/kg FW ration. Avian diets should contain >25.0 mg Zn/kg DW ration for prevention of zinc deficiency effects, and <178.0 mg Zn/kg DW for prevention of marginal sublethal effects.

Sensitive terrestrial plants die when soil zinc levels exceed 100.0 mg/kg (oak and maple seedlings), and photosynthesis is inhibited in lichens at >178.0 mg Zn/kg DW whole plant. Sensitive terrestrial invertebrates have reduced survival when soil levels exceed 470.0 mg Zn/kg (earthworms), reduced growth at >300.0 mg Zn/kg diet (slugs), and inhibited reproduction at >1600.0 mg Zn/kg soil (woodlouse). The most sensitive aquatic species were adversely affected at nominal water concentrations between 10.0 and 25.0 μg Zn/L, including representative species of plants, protozoans, sponges, mollusks, crustaceans, echinoderms, fishes, and amphibians. Acute LC50 (96 h) values ranged between 32.0 and 40,930.0 μg/L for freshwater invertebrates, 66.0 and 40,900.0 μg/L for freshwater teleosts, 195.0 and >320,000.0 μg/L for marine invertebrates, and 191.0 and 38,000.0 μg/L for marine teleosts. Acute toxicity values were markedly affected by the age and nutrient status of the organism, by changes in the physicochemical regimen, and by interactions with other chemicals, especially copper salts. Pancreatic degeneration occurred in ducks fed diets containing 2500.0 mg Zn/kg ration. Ducks died when fed diets containing 3000.0 mg Zn/kg feed, or when given single oral doses >742.0 mg Zn/kg BW. Domestic poultry are routinely fed extremely high dietary levels of 20.0 g Zn/kg ration as a commercial management technique to force the molting of laying hens and the subsequent improvement of long term egg production that molting produces. However, poultry chicks died at 8.0 g Zn/kg diet, had reduced growth at 2.0–3.0 g Zn/kg diet, and experienced pancreas histopathology when fed selenium-deficient but zinc-adequate (100.0 mg Zn/kg) diets. Mammals are comparatively resistant to zinc, as judged by their tolerance for extended periods to diets containing >100 times the minimum daily zinc requirement. But excessive zinc through inhalation or orally will negatively impact mammalian survival, metabolism, and well being. The most sensitive species of mammals were adversely affected at dietary concentrations of 90.0–300.0 mg Zn/kg, drinking water concentrations >300.0 mg Zn/L, daily intakes >90.0 mg Zn/kg BW, single oral doses >350.0 mg Zn/kg BW, and air concentrations >0.8 mg Zn/m^3. Humans are comparatively sensitive to excess zinc. Adverse effects occur in humans at >80.0 mg Zn/kg diet, or at daily intakes >2.3 mg/kg BW.

Proposed criteria for protection of aquatic life include mean zinc concentrations of <47.0–<59.0 μg/L in freshwater, and <58.0–<86.0 μg Zn/L in seawater. However, significant adverse effects of zinc are reported for a growing number of freshwater organisms in the range of 5.0–51.0 μg Zn/L, and to saltwater biota between 9.0 and 50.0 μg Zn/L, suggesting that some downward modification in the proposed criteria is necessary.

Although tissue residues are not yet reliable indicators of zinc contamination, zinc poisoning usually occurs in birds when liver or kidney contains >2.1 g Zn/kg DW, and in mammals when concentrations, in mg Zn/kg DW, exceed 274.0 in kidney, 465.0 in liver, or 752.0 in pancreas. The proposed air quality criterion for human health protection is <5.0 mg Zn/m^3, but guinea pigs were more sensitive with adverse effects evident at >0.8–4.0 mg/m^3.

Current research needs include the development of protocols to: (1) separate, quantitate, and verify the different chemical species of zinc; (2) identify natural from anthropogenic sources of zinc; (3) establish toxicity thresholds based on accumulation; (4) evaluate the significance of tissue concentrations in target organs as indicators of zinc stress; and (5) measure the long-term consequences of zinc interactions with other nutrients using animals of various age and nutrient status.

General Index

B

C

D

E

F

G

H

I

M

N

O

P

Q

R

T

U

V

Species Index

ALGAE AND OTHER PLANTS

AMPHIBIANS

ANNELIDS (Aquatic)

ANNELIDS (Terrestrial)

ARTHROPODS (Aquatic)

ARTHROPODS (Terrestrial)

BACTERIA

BIRDS

COELENTERATES

ECHINODERMS

FISHES AND ELASMOBRANCHS

FUNGI

MAMMALS

MOLLUSCS (Aquatic)

MOLLUSCS (Terrestrial)

NEMATODES

PLANARIANS

PORIFERA

PROTOZOANS

REPTILES

ROTIFERS

VIRUSES

YEASTS

CPSIA information can be obtained at www.ICGtesting.com
Printed in the USA
LVOW09*0335090715

445474LV00009B/198/P

9 780444 531056